HARMONIC ANALYSIS

FOR
ENGINEERS
AND
APPLIED SCIENTISTS

Updated & Expanded Edition

Gregory S. Chirikjian
Alexander B. Kyatkin

Dover Publications, Inc.
Mineola, New York

About the Authors

Gregory S. Chirikjian received his PhD from the California Institute of Technology in 1992. Since then he has been a professor in the Whiting School of Engineering at Johns Hopkins University. His primary faculty appointment is in Mechanical Engineering, and he holds secondary appointments in Computer Science and in the Department of Applied Mathematics and Statistics.

Alexander B. Kyatkin received his PhD in physics from Johns Hopkins University in 1996. He has conducted research in harmonic analysis on the motion group, computer vision, numerical algorithm development, and high energy particle physics. He currently works as a software developer.

Bibliographical Note

Harmonic Analysis for Engineers and Applied Scientists: Updated and Expanded Edition, first published by Dover Publications, Inc., in 2016, is a new, revised edition of *Engineering Applications of Noncommutative Harmonic Analysis: With Emphasis on Rotation and Motion Groups,* originally published in 2001 by CRC Press, LLC.

International Standard Book Number

ISBN-13: 978-0-486-79564-5
ISBN-10: 0-486-79564-0

Manufactured in the United States by LSC Communications
79564002   2019
www.doverpublications.com

To our families,
and with special acknowledgment of
Jack G. Chirikjian

Preface

Preface to the Dover Edition

Preface to the Dover Edition

Advanced mathematical methods, which were known only to mathematicians and theoretical physicists in previous decades, are becoming more widely used by engineering researchers and applied scientists. This general trend is reflected in this book, which covers the basics of group representation theory and noncommutative harmonic analysis (i.e., Fourier analysis on groups), and then demonstrates the utility of these methods by presenting a broad survey of concrete applications from the literature. These applications include areas of engineering (e.g., robotics, image processing, mechanics, etc.) as well as applied science (e.g., polymer theory, rotational Brownian motion, protein-protein docking, etc.). Noticeably absent are the applications to which these methods have been applied extensively in the physics literature over the past century (crystallography, quantum mechanics, and particle physics) since many excellent books already cover those topics. The spirit of this presentation is that mathematical facts are introduced on a "need-to-know" basis, with an emphasis on how to use the underlying mathematics as a modeling tool in new application areas, rather than demonstrating the art of proof.

This book is an updated version of our 2001 book *Engineering Applications of Noncommutative Harmonic Analysis*. That book sought to harness developments in noncommutative harmonic analysis and illustrate how those methods could be applied as a descriptive tool, with an emphasis on computational aspects. Our updated version keeps that emphasis while expanding to include additional material. In order to make it accessible to engineers and applied scientists, a large amount of requisite mathematics above and beyond that usually learned in the first two years in the undergraduate curricula in these fields has been included. Therefore the first third of this book and the appendix cover very classical material. Each topic in these chapters is presented only in as much depth as is required for later use, and references to more detailed treatments of each topic are provided. These introductory chapters each cover multiple topics on which whole monographs are available (e.g., Fourier series, Fourier transforms, orthogonal polynomials, Bessel functions, wavelets, vector calculus, rigid-body kinematics, group theory, etc.). We do not take time out to prove every result but provide references where proofs can be found. It is not until the middle third of the book that the more exotic and modern "noncommutative" part is addressed, and in this part we provide detailed proofs and derivations. The final third is where most of the applications appear, including recent results and our own original work.

Throughout the book we have updated the material and references substantially to reflect developments over the past 15 years, and made corrections that have come to our attention during this time. In addition, we added a new chapter on "protein kinematics," which is an area that the first author has been working in for more than a decade.

The title of this book has been modified to reflect the fact that it is not only about applications, but rather can be used by students and researchers as a general reference. Moreover, in addition to the engineering audience for whom the previous version of the book was written, the material presented here may be of interest to physicists, chemists, structural biologists, and computer scientists. We are absolutely thrilled to be republishing this modified version of our book with Dover, and we hope that readers find the subject matter as useful and enjoyable as we do.

Many people have suggested changes to the original version of this book, and we have updated the book accordingly. Some of the updates draw on recent articles co-authored with others, and we would like to acknowledge them as well. In particular, we thank (in alphabetical order): M. Kendal Ackerman, Bijan Afsari, Hui Dong, Jin Seob Kim, Moonki Kim, Michael Kutzer, Wooram Park, Bernard Shiffman, Dmitri Toptygin, Yunfeng Wang, Kevin C. Wolfe, Yan Yan, and Yu Zhou for their scientific input. We would also like to thank Drs. Hendrik and Nils Napp and Dr. Silvère Bonnabel, respectively, for correcting some of our German and French references. Last but not least, we would like to thank our families for their patience while we have made these revisions. In particular, GC would like to thank Atsuko, John, Margot, and Brian.

G. Chirikjian
A. Kyatkin
Baltimore, May 2016.

Preface to Original Edition

This book is intended for engineers, applied mathematicians, and scientists who are interested in learning about the theory and applications of a generalization of Fourier analysis called *noncommutative harmonic analysis*. Existing books on this topic are written primarily for specialists in modern mathematics or theoretical physics. In contrast, we have made every attempt to make this book self-contained and readable for those with a standard background in engineering mathematics. This means that we present concepts in a heavily coordinate-dependent way. We believe we have done this without sacrificing the rigorous nature of the subject.

Fourier analysis is an invaluable tool to engineers and scientists. There are two reasons for this: (1) A wide range of physical phenomena have been modeled using the classical theory of Fourier series and transforms; (2) The computational benefits that follow from fast Fourier transform (FFT) techniques have had tremendous impact in many fields including communications, and signal and image processing.

While classical Fourier analysis is such a ubiquitous tool, it is relatively rare to find engineers who are familiar with the mathematical concept of a group. Even among the minority of engineers familiar with the group concept, it is rare to find engineers familiar with generalizations of Fourier analysis and orthogonal expansion of functions on groups. We believe that there are two reasons for this: (1) This generalization of Fourier analysis (noncommutative harmonic analysis) has historically been a subject developed by and for pure mathematicians and theoretical physicists. Hence, few with knowledge in this area have had the access to the engineering problems that can be addressed using this area of mathematics; (2) Without knowing of the existence of this powerful and beautiful area of mathematics, it has been impossible for engineers to formulate their problems within this framework, and thus impossible to apply it. Thus, this book can be viewed as a bridge of sorts. On the one hand, mathematicians who are familiar with abstract formulations can look to this book and see group representation theory and harmonic analysis being applied to a number of problems of interest in engineering. Likewise, engineers completely unfamiliar with these concepts can look to this book to gain the vocabulary needed to express their problems in a new and potentially more efficient way.

Chapter 1 is an overview of several engineering problems to which this theory is applicable. In Chapter 2 we review commutative harmonic analysis (classical Fourier analysis) and associated computational issues that have become so important in all areas of modern engineering. Chapter 3 is a review of the orthogonal expansions of classical mathematical physics combined with more modern techniques such as Walsh functions, discrete polynomial transforms, and wavelets. Advanced engineering mathematics with emphasis on curvilinear coordinate systems and basic coordinate-dependent differential geometry of hyper-surfaces in $\mathbb{R}^N$ is the subject of Chapter 4. Chapters 5 and 6 are each devoted to the properties and parameterizations of the two groups of most importance in the application areas covered here: the rotation group and the group of rigid-body motions. For the most part, we do not use group theoretical notation in Chapters 5 and 6. In Chapter 7 we present ideas from group theory that are the most critical for the formulation in the remainder of the book. This includes many geometrical examples of groups and cosets. The distinction between discrete and continuous groups is made, and the classification of continuous groups using matrix techniques is examined. In Chapter 8 we discuss how to integrate functions on continuous (Lie) groups and present the general ideas of representation theory and harmonic analysis on finite and compact Lie groups. Chapter 9 is devoted to orthonormal expansions on the rotation group and related mathematical objects such as the unit sphere in four dimensions. In Chapter 10 we present the theory behind Fourier analysis on the Euclidean motion groups (groups of rigid-body motions in the plane and space). These are groups that we have found to be of great importance in engineering applications (though they are rarely discussed in any detail in mathematics textbooks and research monographs). Chapter 11 develops a fast Fourier transform for the motion groups. Chapters 12 – 18 are devoted exclusively to concrete applications of noncommutative harmonic analysis in engineering and applied science. A chapter is devoted to each application area: robotics; image analysis and tomography; pose determination and sensor calibration; estimation and control with emphasis on spacecraft attitude; rotational Brownian motion and diffusion (including the analysis of liquid crystals); conformational statistics of macromolecules; orientational averaging in mechanics.

This book can serve as a supplement for engineering, applied science, and mathematics students at the beginning of their graduate careers, as well as a reference book for researchers in any field of engineering, applied science, or mathematics. Clearly there is too much material for this book to be used as a one-year course. However, there are many ways to use parts of this book as reference material for semester-long (or quarter-long) courses. We list several options below:

- Graduate Engineering Mathematics: Chapters 2, 3, 4, Appendix
- Robotics/Theoretical Kinematics: Chapters 5, 6, 9, 10, 12
- Mathematical Image Analysis: Chapters 2, 10, 11, 13, 14
- Introduction to Group Theory: Chapters 5, 6, 7, 8
- Computational Harmonic Analysis on Groups: Chapters 7, 8, 9, 10, 11
- Engineering Applications of Harmonic Analysis: Chapters 12, 13, 15, 18
- Applications of Harmonic Analysis in Statistical and Chemical Physics 14, 16, 17.

Writing a book such as this requires the efforts and inputs of many people other than the authors themselves. We would like to thank a number of people for their helpful comments and discussions which have been incorporated into this text in one way or another.

First, we would like to thank (in alphabetical order) those who suffered through the "Orientational Phenomena" course given by the first author based on an early version of this book: Richard Altendorfer, Commander Steve Chism, Lara Diamond, Vijay Kumar, Sang Yoon Lee, Subramanian Ramakrishnan, Ed Scheinerman, Arun Sripati, David Stein, Jackrit Suthakorn, Mag Tan, Yunfeng Wang, Yu Zhou.

We would also like to thank (in alphabetical order) Jonas August, Troels Roussau Johansen, David Joyner, Omar Knio, David Maslen, Bud Mishra, Keizo Miyahara, Joshua Neuheisel, Frank Chongwoo Park, Bahram Ravani, Dan Rockmore, Wilson J. Rugh, Ramin Takloo-Bighash, Panagiotis Tsiotras, Louis Whitcomb, and Hemin Yang for proofreading and/or helpful suggestions.

Special thanks go to Imme Ebert-Uphoff, whose initial work (under the supervision of the first author) led us to study noncommutative harmonic analysis, and to David Stein for preparing many of the figures in this book. We also acknowledge the support of the NSF, without which this work would not have been pursued.

G. Chirikjian
A. Kyatkin
Baltimore, March 2000.

Contents

List of Figures

1

Introduction and Overview of Applications

In this chapter we provide a layman's introduction to the meaning of *noncommutative* operations. We also illustrate a number of applications in which the noncommutative convolution product arises. In subsequent chapters we provide the detail required to solve problems in each application area. Chapters 12 – 18 are each devoted to one of these applications.

1.1 Noncommutative Operations

The word *noncommutative* is used in mathematical contexts to indicate that the order in which some kind of operation is performed matters. For instance, the result of the cross product of two vectors depends on the order in which it is performed because

$$\mathbf{a} \times \mathbf{b} = -\mathbf{b} \times \mathbf{a}.$$

Likewise, the product of two square matrices generally depends on the order in which the matrices are multiplied. For instance if

$$A = \begin{pmatrix} 1 & 3 \\ 2 & 4 \end{pmatrix} \quad \text{and} \quad B = \begin{pmatrix} 3 & 5 \\ 1 & 6 \end{pmatrix}$$

then

$$AB \neq BA.$$

Noncommutative operations are in contrast to commutative ones such as the addition of vectors or multiplication of real numbers. Both commutative and noncommutative operations arise outside the realm of pure mathematics. The following subsections review some common examples.

1.1.1 Assembly Planning

A common problem experienced by engineers is how to design a functional device so that it can be assembled either manually or by an automated factory. In order to make the assembly process as efficient as possible, it is necessary to know which parts need to be assembled in specific orders and for which assembly operations the order does not matter. Once this information is known, the best schemes for assembling a given device can be devised. The determination of such schemes is called *assembly planning* [5, 6].

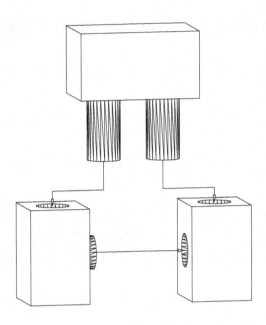

Fig. 1.1. An Assembly Schematic

An assembly plan for relatively simple devices can be expressed in an "exploded" schematic such as in Figure 1.1.

Clearly the assembly operations in this case can be performed only in one order. That is, the bottom blocks must be connected before the top can be inserted.

The importance of the order in which assembly operations are performed is common to our everyday experience outside of the technical world. For instance, when getting dressed in the morning, the order in which socks, shirt and pants are put on does not matter (these operations, which we will denote as SX, ST and PA, commute with each other). However, the operation of putting on shoes (which we will denote as SH) never commutes with SX, and rarely with PA (depending on the kind of pants). If we use the symbol ∘ to mean "followed by" then, for instance,

$$SX \circ ST = ST \circ SX$$

is a statement that the operations commute, and hence result in the same state of being dressed. Using this notation, $SX \circ SH$ works but $SH \circ SX$ does not.

Any number of operations that we perform on a daily basis can be categorized according to whether they are commutative or not. In the following subsections we review some more familiar examples. These provide an intuitive introduction to the mathematics of noncommutative operations.

1.1.2 Rubik's Cube

An easy way to demonstrate noncommutativity is to play with a Rubik's cube. This toy is a cube with each of its six faces painted a different color. Each face consists of nine squares. There are three kinds of squares: corner, edge and center. The corner squares are each rigidly attached to the two other corner squares of adjacent faces. The edge squares are each attached to one edge square of an adjacent face. The center squares are fixed to the center of the cube, and can rotate about their face. A whole face can be rotated about its center by increments of ninety degrees, resulting in a whole row in each of four faces changing color. Multiple rotations of this kind about each of the faces results in a scrambling of the squares.

In order to illustrate the noncommutativity of these operations, we use the following notation [8]. Let U, D, R, L, F, B denote up, down, right, left, front, and back of the cube, respectively. An operation such as U^i for $i = 1, 2, 3, 4$ means rotation of the face U by an angle of $i \times \pi/2$ radians counterclockwise. Figure 1.2 illustrates U^1 and R^1. Since rotation by 2π brings a face back to its starting point, $U^4 = D^4 = R^4 = L^4 = F^4 = B^4 = I$, the identity (or "do nothing") operation.

The composition of two cube operations is simply one followed by another. These operations are noncommutative because, for instance,

$$U^1 \circ R^1 \neq R^1 \circ U^1.$$

This is not to say that all the cube operations are noncommutative. For instance

$$U^1 \circ U^2 = U^3 = U^2 \circ U^1.$$

However, since the whole set of cube operations contains at least some noncommutative elements, the whole set together with composition is called noncommutative.

The study of sets of transformations (or operations) that are either commutative or noncommutative and satisfy some additional properties is called *group theory*. This is the subject of Chapter 7. For now it suffices to say that group theory is a powerful tool in the analysis of a set of operations. It can answer questions such as how many distinct states Rubik's cube can attain (which happens to be approximately 4.3×10^{19}), and what strategies are efficient for restoring the cube to its original state without a priori knowledge of the moves that scrambled the cube. Using the notation $(\cdot)^n$ to mean that the contents of the parentheses is composed with itself n times, we see that

$$(U^1)^4 = U^1 \circ U^1 \circ U^1 \circ U^1 = U^4 = I.$$

Group theory seeks relationships such as

$$(U^2 \circ R^2)^6 = I, \quad \text{and} \quad (U^1 \circ R^1)^{105} = I, \quad \text{and} \quad (U^1 \circ R^3)^{63} = I$$

which clarify the structure of a set of transformations.

Whereas Rubik's cube is a toy that demonstrates the concept of a group, in real engineering applications, a set of transformations of central importance is rigid-body motions. We review such motions in the following subsection.

1.1.3 Rigid-Body Kinematics

Kinematics is the field of study concerned with the geometry of motion without regard to the forces that cause the motion [1]. For instance, when considering the motion of a rigid body, Newton's equation

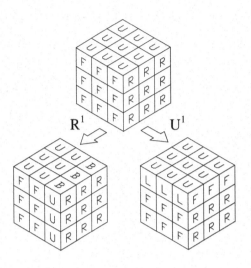

Fig. 1.2. Rubik's Cube

$$m\frac{d^2\mathbf{x}}{dt^2} = \mathbf{F}$$

and Euler's Equation

$$I\frac{d\boldsymbol{\omega}}{dt} + \boldsymbol{\omega} \times (I\boldsymbol{\omega}) = \mathbf{N}$$

both consist of a kinematic part (the left side), and a part describing an external force $\mathbf{F}$ or moment of force $\mathbf{N}$. Here m is mass and I is moment of inertia (see Chapter 5 for definitions).

This separation of kinematic and non-kinematic terms in classical mechanics is possible for equations governing the motion of systems composed of rigid linkages, deformable objects or fluid systems.

The above Newton-Euler equations are differential equations where the variables are the vectors $\mathbf{x}(t)$ and $\boldsymbol{\omega}(t)$ and the non-kinematic terms are either strict functions of time or of time and the kinematic variables. A common problem in some areas of engineering is to determine the value of the kinematic variables as a function of time. The inverse problem of determining forces and moments from observed motions is also of importance.

In the Newton-Euler equations, the variables are vector quantities (and hence are commutative when using the operation of vector addition). The only hint of the non-commutativity of rigid-body motion is given by the cross product in Euler's equations of motion.

In order to illustrate the noncommutative nature of rigid-body motion (without resorting to detailed equations), observe Figure 1.3. This illustrates the composition of two rigid-body motions (or transformations) in two different orders. In Figure 1.3(left) the frame of reference is translated and rotated from its initial pose to a new one by the motion g_1. Then g_2 is the relative motion of the frame from its intermediate pose to

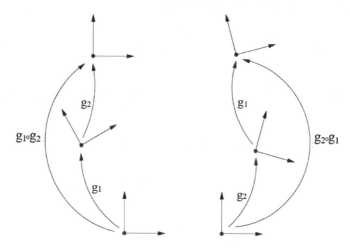

Fig. 1.3. Composition of Motions: (left) $g_1 \circ g_2$; and (right) $g_2 \circ g_1$.

the final one. This composition of motions is denoted as $g_1 \circ g_2$. In contrast, in Figure 1.3(right) g_2 is the motion from the initial pose to the intermediate one, and g_1 is the motion from intermediate to final. This results in the composite motion $g_2 \circ g_1$. The motions g_1 and g_2 in both figures are the same in the sense that if we were to cut out Figure 1.3(right) it would be possible to align Figure 1.3(right) so that the top two frames would coincide with the bottom two in Figure 1.3(left), and vice versa. Needless to say, we do not recommend this experimental verification! The importance of Figure 1.3 is that it illustrates $g_1 \circ g_2 \neq g_2 \circ g_1$.

1.2 Harmonic Analysis: Commutative and Noncommutative

Generally speaking, a *function*, f, is an assignment of values in a given range, R, to a given domain, D. This is denoted as $f : D \to R$. The arrow points in this direction because in the process of assigning values, every point in D is "mapped to" a point in R. For example, if we assign real values to the closed interval $[a, b]$, then we write $f : [a, b] \to \mathbb{R}$. The value of $f(x)$ for any point $x \in [a, b]$ is then a real number, whereas the set $\{f(x) \,|\, a \leq x \leq b\}$ defines the whole function. We will blur the distinction between the value of the function at an arbitrary abstract point and the value at all points in the domain, and refer to both $f(x)$ and $f : [a, b] \to \mathbb{R}$ as a "function on $[a, b]$."

Classical (pre-twentieth-century) harmonic analysis is concerned with the expansion of functions into fundamental (usually orthogonal) components. Of particular interest in classical harmonic analysis is the expansion of functions on intervals of the form $a \leq x \leq b$ where x, a and b are real numbers. Since the addition of real numbers is a commutative operation ($x + y = y + x$ for $x, y \in \mathbb{R}$), we will refer to the terms "classical" and "commutative" interchangeably when referring to harmonic analysis.

Essentially, the goal of commutative harmonic analysis is to expand a function $f(x)$ (that either assigns real or complex values to each value of x) in a series:

$$f(x) \approx \sum_{i=1}^{N} a_i \phi_i(x). \tag{1.1}$$

The choice of the set of functions $\{\phi_i(x)\}$ as well as the meaning of approximate equality ($\approx$) depends on "what kind" of function $f(x)$ is.

For instance, if

$$\int_a^b |f(x)|^p \, w(x) \, dx < \infty$$

for some $w(x) > 0$ for all $a \leq x \leq b$, we say $f \in \mathcal{L}^p([a,b], w(x) \, dx)$. In later chapters the generalization of this concept to functions on multi-dimensional domains, $f : D \to \mathbb{R}$, will be used. When it is clear from the context what the weighting function is for integration on D, we will use the abbreviation $\mathcal{L}^p(D)$.

In the special case when $p = 1$, $f(x)$ is called "absolutely integrable," and when $p = 2$, $f(x)$ is called "square integrable." But these are not the only distinctions that can be made. For instance, a function that is continuous on the interval $[a, b]$ is called $C^0([a,b])$, and one for which the first N derivatives exist for all $x \in [a,b]$ (and the Nth derivative produces a continuous function) is called $C^N([a,b])$. A function $f \in C^\infty([a,b])$ is called *analytic* for $x \in [a,b]$ if it has a convergent Taylor series

$$f(x) = \sum_{n=0}^{\infty} \frac{f^{(n)}(c)}{n!}(x - c)^n$$

for all $c \in (a,b)$. Here we use the notation $f^{(n)} = d^n f/dx^n$.

These definitions also hold for functions on the infinite interval $-\infty < x < \infty$. This is the real line, $\mathbb{R}$, and functions on the real line can be in $\mathcal{L}^p(\mathbb{R}, w(x) \, dx)$, $C^N(\mathbb{R})$, etc. Another class of functions on the line are the so-called "rapidly-decreasing" functions. These have the property

$$\int_{-\infty}^{\infty} f(x) \, |x|^N dx < \infty$$

for any finite integer value of $N > 0$. Essentially this means that the function dies off fast enough as $|x| \to \infty$ to overcome the rate at which a polynomial increases such that the integral of the product converges. We will denote the set of rapidly decreasing functions on the line as $\mathcal{S}(\mathbb{R})$. As with all the classes of functions discussed above, the concept of rapidly decreasing functions extends easily to multiple dimensions.

Throughout the text we refer in many contexts to "nice" functions.[1] These are functions that are analytic (infinitely differentiable at all points, and have convergent Taylor series), absolutely and square integrable, and rapidly decreasing. For functions defined on certain domains it is redundant to specify all of these conditions. For instance, a bounded function with compact support $[a, b] \subset \mathbb{R}$ will necessarily be rapidly decreasing and be contained in $\mathcal{L}^p([a,b], dx)$ for all finite positive integer values of p. The set of nice functions on an arbitrary domain, D, will be denoted as $\mathcal{N}(D)$.

A typical set of nice functions on the real line is

$$f_m(x) = x^m e^{-x^2}$$

for $m = 0, 1, \ldots$.

Whereas in pure mathematics the main goal is to find the broadest possible mathematical truths (and hence "weird" functions serve as the cornerstones of powerful theorems), in engineering and the macroscopic physical world the vast majority of functions vary smoothly and decay as desired for large values of their argument. For example, to

---

[1] Nice functions are also called "well-behaved."

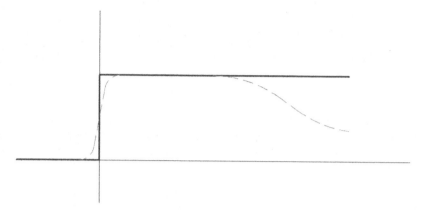

Fig. 1.4. A Step Function and Its Nice Approximation

an engineer working on a particular application, there may be no difference between the unit step function shown in Figure 1.4 with solid lines (which is not continuous, rapidly decreasing, or even in $\mathcal{L}^p(\mathbb{R}, dx)$), and its approximation in Figure 1.4 with dashed lines (which is a nice function).

Hence, while many of the mathematical statements we will make for nice functions also hold for functions that are not so nice, we will not be concerned with the broadest possible mathematical statements. Rather we will focus on twentieth-century generalizations of classical harmonic analysis in the context of twenty-first-century applications with a noncommutative component.

Having said this, harmonic analysis is a useful tool because given a nice function, we can approximate it using a series of the form of (1.1). For a nice function $f(x)$ and a set of functions $\{\phi_i(x)\}$ satisfying certain properties, the values $\{a_i\}$ can be found as the solution to the minimization problem:

$$\min_{\{a_i\}} \int_a^b |f(x) - \sum_{i=1}^N a_i \phi_i(x)|^2.$$

Not any set of nice functions $\{\phi_i\}$ will do if we want the series approximation to accurately capture the behavior of the function. Chapters 2 and 3 provide the requirements for such a set of functions to be acceptable. But before getting into the details of classical harmonic analysis, we review in each of the following sections (at a very coarse and nonmathematical level) the kinds of problems to which noncommutative harmonic analysis is applied in the later chapters of the book.

1.3 Commutative and Noncommutative Convolution

The concept of convolution of functions arises in an amazingly varied set of engineering scenarios. Convolution of functions which take their argument from the real line is an essential concept in signal/image processing and probability theory, just to name a few areas.

The classical convolution integral can be visualized geometrically as follows. One function, $f_2(x)$, is translated past another, $f_1(x)$. A translated version of $f_2(x)$ by amount ξ in the positive x direction takes the form $f_2(x - \xi)$. Then, the convolution is formed as the weighted sum of each of these translated copies, where the weight is $f_1(\xi)$ (value of $f_1(x)$ at the amount of shift $x = \xi$) and the sum, which is over all translations on the real line, becomes an integral. The result of this procedure is a function

$$(f_1 * f_2)(x) = \int_{-\infty}^{\infty} f_1(\xi)\, f_2(x - \xi)\, d\xi. \tag{1.2}$$

As we review in detail in Chapter 2, the classical Fourier transform is an extremely useful tool in simplifying the computation of convolutions, as well as in solving partial differential equations. But in the remainder of this chapter, we will provide an overview of generalizations of the convolution product which arise in engineering applications.

To begin this generalization, instead of describing points and translations on the real line as scalar variables x and ξ, consider them as matrices of the form

$$g_1 = g(x) = \begin{pmatrix} 1 & x \\ 0 & 1 \end{pmatrix} \qquad g_2 = g(\xi) = \begin{pmatrix} 1 & \xi \\ 0 & 1 \end{pmatrix}.$$

Then it is easy to verify that $g(x) \circ g(\xi) = g(x + \xi)$ and $[g(x)]^{-1} = g(-x)$ where g^{-1} is the inverse of the matrix g and $\circ$ stands for matrix multiplication.[2]

Then the convolution integral is generalized to the form

$$(f_1 * f_2)(g_1) = \int_G f_1(g_2)\, f_2(g_2^{-1} \circ g_1)\, d\mu(g_2) \tag{1.3}$$

where G is the set of all matrices of the form $g(x)$ for all real numbers x, and $d\mu(g(\xi)) = d\xi$ is an appropriate integration measure for the set G. Henceforth, since we have fixed which measure is being used, the shorthand dg will be used in place of the rather cumbersome $d\mu(g)$.

A natural question to ask at this point is why we have written $f_2(g_2^{-1} \circ g_1)$ instead of $f_2(g_1 \circ g_2^{-1})$. The answer is that in this particular example both are acceptable since $g(x) \circ g(y) = g(y) \circ g(x)$, but in the generalizations to follow, this equality does not hold, and the definition in (1.3) remains a useful definition in applications.

Denoting the real numbers as $\mathbb{R}$, it is clear that G in this example is in some sense equivalent to $\mathbb{R}$, and we write $G \cong R$. The precise meaning of this equivalence will be given in Chapter 7.

We now examine a generalization of this concept which will be extremely useful later in the book. Consider, instead of points on the line and translations long the line, points in the plane and rigid motions in the plane. Any point in the plane can be described with matrices of the form

$$g_1 = \begin{pmatrix} 1 & 0 & x_1 \\ 0 & 1 & x_2 \\ 0 & 0 & 1 \end{pmatrix},$$

and any rigid body motion in the plane (translation and rotation) can be described with matrices of the form

$$g_2 = g(a_1, a_2, \alpha) = \begin{pmatrix} \cos\alpha & -\sin\alpha & a_1 \\ \sin\alpha & \cos\alpha & a_2 \\ 0 & 0 & 1 \end{pmatrix}$$

---

[2]It is standard to simply omit the $\circ$, in which case juxtaposition of two matrices stands for multiplication, but we will keep the $\circ$ for clarity.

and

$$g_2^{-1} = \begin{pmatrix} \cos\alpha & \sin\alpha & -a_1\cos\alpha - a_2\sin\alpha \\ -\sin\alpha & \cos\alpha & a_1\sin\alpha - a_2\cos\alpha \\ 0 & 0 & 1 \end{pmatrix}.$$

By defining $dg_2 = da_1 da_2 d\alpha$, (1.3) now takes on a new meaning. In this context $f_1(g(a_1,a_2,\alpha)) = \tilde{f}_1(a_1,a_2,\alpha)$ is a function that has motions as its argument, and $f_2(g(x_1,x_2,0)) = \tilde{f}_2(x_1,x_2)$ is a function of position. Both functions return real numbers. Explicitly we write $(f_1 * f_2)(g(x_1,x_2,0))$ as

$$(\tilde{f}_1 * \tilde{f}_2)(x_1,x_2) = \int_{\alpha=-\pi}^{\pi} \int_{a_2=-\infty}^{\infty} \int_{a_1=-\infty}^{\infty} \tilde{f}_1(a_1,a_2,\alpha) \cdot$$

$$\tilde{f}_2((x_1 - a_1)\cos\alpha + (x_2 - a_2)\sin\alpha, -(x_1 - a_1)\sin\alpha + (x_2 - a_2)\cos\alpha)\, da_1 da_2 d\alpha,$$

which is interpreted as the function f_2 being swept by f_1. We can think of this as copies of f_2 which are translated and rotated and then deposited on the plane with an intensity governed by f_1. With a slight modification, so that $g_1 = g(x_1,x_2,\theta)$ and $f_2(g(x_1,x_2,\theta)) = \tilde{f}_2(x_1,x_2,\theta)$, the result is precisely a convolution of functions on the group of rigid body motions of the plane. It is easy to check that in this case $g_1 \circ g_2 \neq g_2 \circ g_1$ in general.

The middle third of this book is devoted to developing tools for the fast numerical computation of such generalized convolutions and the inversion of integral equations of the form $(f_1 * f_2)(g) = h(g)$ where $h(g)$ and $f_1(g)$ are known and $f_2(g)$ is sought.

The remaining sections of this chapter are each devoted to an overview of applications examined in detail in the last third of the book.

1.4 Robot Workspaces, Configuration-Space Obstacles, and Localization

A robotic manipulator is a particular kind of device used to pick up objects from one place and put them in another place. Often the manipulator resembles a mechanical arm and has a gripper or hand-like mechanism at the distal end (called the end effector). The proximal end of the manipulator is usually affixed to a torso, gantry or moving platform. The *workspace* of a manipulator is the set of all positions and orientations that a frame of reference attached to the distal end can attain relative to a frame fixed to the proximal end. Figure 1.5 shows an arm that is highly articulated (has many joints). For such an arm we can ask how the workspace for the whole arm can be found if we know the workspaces of two halves. Superimposed on the arm in this figure are the positional boundaries of the workspaces of the proximal and distal halves (small ellipses), together with the workspace for the whole arm (large ellipse). A sweeping procedure involving the two halves generates the whole as described in [3]. In Chapter 12 we show that this sweeping is in fact a generalized convolution, which can be performed efficiently using the computational and analytical techniques discussed in Chapters 2–11.

Manipulator arms are one of several classes of robotic devices. Another important class consists of mobile robots that can be viewed as single rigid bodies that can move in any direction. A central task in robotics is to navigate such objects through known environments. A popular technique to achieve this goal is to expand the obstacles by the girth of the robot and shrink the robot to a point [10]. Then the motion planning problem becomes one of finding the paths that a point robot is able to traverse in an

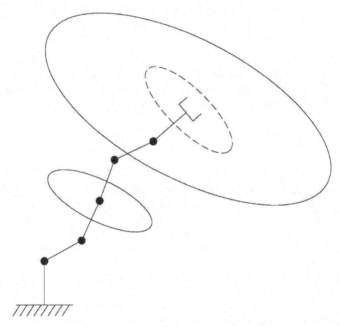

Fig. 1.5. Workspace Generation by Sweeping

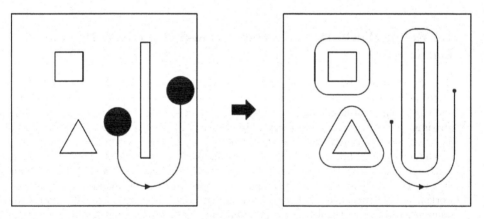

Fig. 1.6. Configuration-Space Obstacles of Mobile Robots

environment of enlarged obstacles. The procedure of generating the enlarged obstacles is again a generalized convolution [9], as shown in Figure 1.6. Similar problems arise in computer aided design [11].

As a practical matter, in motion planning the mobile robot does not always know exactly where it is. Using sensors, such as sonars, that bounce signals off walls to determine the robots distance from the nearest walls, an internal representation of the local environment can be constructed. With this information, the estimated shape of

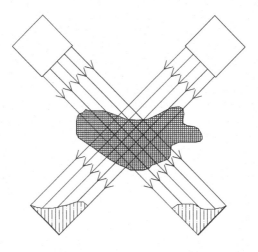

Fig. 1.7. The Forward Tomography Problem

the local free space (places where the robot can move) is calculated. Given a global map of the environment, a generalized convolution of the local free space estimate with the global map will generate a number of likely locations of the robot.

1.5 Template Matching and Tomography

Template matching (also called pattern matching or image registration) is an important problem in many fields. For instance, in automated military surveillance, a problem of interest is to match a building or piece of military hardware with representative depictions in a database. Likewise, in medicine it is desirable to map anatomical features from a database onto images observed through medical imaging. In the case when the best rigid-body fit between the feature in the database and the observed data is desired, the generalized convolution technique is a natural tool.

Two problems in tomography can also be posed as generalized convolutions. The *forward* tomography problem is the way in which data about the three-dimensional structure of an object (typically a person) is reconstructed from multiple planar projections, as in Figure 1.7. See [4] for a detailed description and analysis of this problem. The *inverse* tomography problem (also called the *tomotherapy* problem) is concerned with how much radiation a patient is given by way of x-ray (or other high energy) beams so as to maximize the damage to a tumor, while keeping exposure to surrounding tissues within acceptable limits. Both the forward and inverse tomography problems can be posed as integral equations that are of the form of generalized convolutions.

1.6 Attitude Estimation and Rotational Brownian Motion

A variety of problems in engineering, physics and chemistry are modeled well as rigid bodies that are subjected to random disturbances [12]. This includes everything from satellites in space to the dynamics of liquid crystals composed of many molecules modeled as identical rigid bodies. Regardless of the particular physical situation, they have the common feature that they are described by linear partial differential equations describing processes that evolve on the group of rotations in three dimensional space.

In this context, noncommutative harmonic analysis provides a tool for converting partial differential equations into systems of ordinary differential equations in a different domain. In this other domain, the equations can be solved, and the solution can be converted back to the original domain. This is in exact analogy with the way the classical Fourier transform has been used in engineering science for more than a century. The only difference is that in the classical engineering context, the equations evolve on a commutative object (i.e., the real line with the operation of addition).

1.7 Conformational Statistics of Macromolecules

A macromolecule is a concatenation of simpler molecular units. These units often are assembled so that the macromolecule is a chain with a backbone of serially connected carbon atoms (e.g., many of the man-made polymers). Alternatively, a macromolecule can be a molecule such as DNA that is not strictly serial in nature (it has a structure more like a ladder), but "looks serial" when viewed from far enough away. We shall call both kinds of macromolecules "macroscopically serial."

A problem that has received considerable interest is that of determining the probability density function of the end-to-end distance of a macroscopically serial chain [7]. This information is important in a number of contexts. In the study of man-made polymers, this probability density function is related to the bulk mechanical properties of polymeric materials. For example, the extent to which rubber will stretch can be explained with knowledge of the end-to-end probability density function. In the context of DNA, a vast body of literature exists that addresses the probability that a section of DNA will bend and close to form a ring.

For reasons stated above, a fundamental problem in the statistical mechanics of macromolecules is the determination of the probability density function describing the end-to-end distance. This function may be found easily if we know the probability density of relative positions and orientations of one end of the macromolecule relative to the other. It simply involves integrating over relative orientations and the directional information of one end relative to the other. What remains is the end-to-end distance density.

Much like the workspace generation problem for robotic manipulator arms, the probability density function of relative position and orientation of one end of a macromolecule can be generated by dividing it up into small units and performing multiple generalized convolutions.

In the case of stiff macromolecules (such as DNA) that are only able to bend, rather than kink, it is also possible to write diffusion-type partial differential equations that evolve on the group of rigid-body motions. Noncommutative harmonic analysis serves as a tool for converting such equations into a domain where they can be solved efficiently and transformed back.

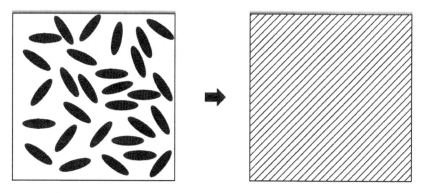

Fig. 1.8. Averaging Strength Properties and Continuum Approximation

1.8 Mechanics and Texture Analysis of Materials

Mechanics is the study of the behavior of solids and fluids under static and dynamic conditions. A useful approximation in mechanics is the *continuum* model, in which a solid or fluid is approximated as a continuous domain (rather than modeling the motion of each atom). Our discussion of applications of noncommutative harmonic analysis in mechanics will be restricted to solids. Applications in fluid mechanics are in some ways similar to rotational Brownian motion and diffusion.

The continuum model has its limitations. For instance a real solid material may be composed of many tiny crystals, it may have many small cracks, or it may be a composite of two very different kinds of materials. A goal in the mechanics of solids is to find an equivalent continuum model that reflects the small scale structure without explicitly modeling the motion of every small component. This goal is depicted in Figure 1.8.

To this end, if the orientation distribution of constituents (e.g. crystals) is known throughout the material, a goal is to somehow average the strength properties of each crystal over all orientations in order to determine the strength of the bulk material. The fields in which such problems have been studied are called *mathematical texture analysis* and *orientational averaging* in the mechanics of solids and materials [2, 14].

1.9 Protein Kinematics

Chapter 19 covers the topic of "protein kinematics." This includes specialized descriptions of rotations and translations that are suited for protein-protein docking and for the analysis of preferred orientations of substructures within individual proteins. Other topics include the molecular theory of liquids, the method of molecular replacement in x-ray crystallography, and modeling the conformational mobility of biomolecular loops.

In Chapters 2–11, we provide the mathematical background required to pose all of the applications discussed in this introductory chapter in the language of noncommutative harmonic analysis (see, e.g., [13]). In some instances harmonic analysis will serve the role of a language in which to describe the problems systematically. In other contexts, it will be clear that harmonic analysis is the only tool available to solve problems efficiently. The various applications are addressed in Chapters 12–18, together with numerous pointers to the literature.

References

1. Bottema, O., Roth, B., *Theoretical Kinematics*, Dover Publications, Inc., New York, reprinted 1990.
2. Bunge, H.-J., *Texture Analysis in Materials Science*, translated by P.R. Morris, Butterworth and Co., Boston, 1982. (also see *Mathematische Methoden der Texturanalyse*, Akademie-Verlag, Berlin, 1969.)
3. Chirikjian, G.S., Ebert-Uphoff, I., "Numerical Convolution on the Euclidean Group with Applications to Workspace Generation," *IEEE Transactions on Robotics and Automation*, 14(1): 123–136, Feb. 1998.
4. Deans, S.R., *The Radon Transform and Some of Its Applications*, John Wiley, New York, 1983. (Dover, 2007).
5. DeFazio, T.L., Whitney, D.E., "Simplified Generation of All Mechanical Assembly Sequences," *IEEE J. Robotics and Automation*, 3(6): 640–658, Dec. 1987.
6. DeMello, L.S.H., Sanderson, A.C., "A Correct and Complete Algorithm for the Generation of Mechanical Assembly Sequences," *IEEE Transactions on Robotics and Automation*, 7(2): 228–240, April 1991.
7. Flory, P.J., *Statistical Mechanics of Chain Molecules*, Wiley-Interscience, New York, 1969
8. Frey, A.H. Jr., Singmaster, D., *Handbook of Cubik Math*, Enslow Publishers, Inc., 1982.
9. Kavraki, L.,"Computation of Configuration Space Obstacles Using the Fast Fourier Transform," *IEEE Transactions on Robotics and Automation*, 11(3): 408–413, 1995.
10. LaValle, S.M., *Planning Algorithms*, Cambridge University Press, 2006.
11. Lysenko, M., Shapiro, V., Nelaturi, S., "Non-commutative Morphology: Shapes, Filters, and Convolutions," *Computer Aided Geometric Design*, 28(8): 497–522, 2011.
12. McConnell, J., *Rotational Brownian Motion and Dielectric Theory*, Academic Press, New York, 1980.
13. Miller, W., Jr., *Lie Theory and Special Functions*, Academic Press, New York, 1968.
14. Wenk, H.-R., ed., *Preferred Orientation in Deformed Metals and Rocks: An Introduction to Modern Texture Analysis*, Academic Press, New York, 1985.

Classical Fourier Analysis

In this chapter we review classical Fourier analysis, which is a tool for generating continuous approximations of a wide variety of functions on the circle and the real line. We also illustrate how the Fourier transform can be calculated in an analog fashion using properties of optical systems, and how the discrete Fourier transform (DFT) lends itself to efficient numerical implementation by way of fast Fourier transform (FFT) techniques. The material in this chapter is known in varying levels of detail by many engineers and scientists. Good understanding of this material is critical for the developments in Chapters $8-11$, which form the foundation on which the applications in Chapters $12-18$ are built.

2.1 Fourier Series: Decomposition of Functions on the Circle

A Fourier series is a weighted sum of trigonometric basis functions used to approximate *periodic* functions.[1]

A periodic function of real-valued argument and period $L > 0$ is one that satisfies the constraint

$$f(x) = f(x + L). \tag{2.1}$$

We use the notation $\mathbb{R}$ for the real numbers, and say $x, L \in \mathbb{R}$ to denote the fact that x and L are real. The set of positive real numbers is denoted $\mathbb{R}_{>0}$. When $L \in \mathbb{R}_{>0}$ is the smallest positive number for which this equation holds for all $x \in \mathbb{R}$, then L is called the *fundamental period* of $f(x)$, and $f(x)$ is called *L-periodic*.

By definition, this means that $f(x + L) = f((x + L) + L) = f(x + 2L)$. Similarly, by the change of variables $y = x + L$, we observe that $f(y - L) = f(y)$ for all $y \in \mathbb{R}$. Thus it follows immediately from (2.1) that $f(x) = f(x + nL)$ for any $n \in \mathbb{Z}$ (the integers). That is, a periodic function with fundamental period L also has period nL.

Instead of viewing such functions as periodic on the line, it is often more convenient to view them as functions on a circle of circumference L. The mapping from line to circle can be viewed as "cutting" an $L-$length segment from the line and "pasting" the ends together. The mapping from circle to line can be viewed as the circle rolling on the line. As explained in Chapter 7, the group theoretical way to say this is that the

---

[1] Jean Baptiste Joseph Fourier (1768-1830) initiated the use of such series in the study of heat conduction. The work was originally rejected by the Academy of Sciences in Paris when presented in 1807, and was finally published in 1822 [14].

1-torus, T, is congruent to a quotient: $T \cong \mathbb{R}/L\mathbb{Z}$. Roughly speaking, this says that the circle with circular addition behaves like a representative piece of the line cut out with the ends glued.

In applications it is common to encounter L-periodic functions which satisfy the inequality[2]

$$\|f\|_2^2 \doteq \int_0^L |f(x)|^2 \, dx < \infty, \tag{2.2}$$

where for a number $c \in \mathbb{C}$ (the complex numbers), the squared modulus is $|c|^2 = c\bar{c}$, where $\bar{c}$ is the complex conjugate of c. Thus, if $c = a + ib$ where $i = \sqrt{-1}$ and $a, b \in \mathbb{R}$, then $\bar{c} = a - ib$ and $|c|^2 = a^2 + b^2$. The set of all square integrable complex-valued functions with respect to integration measure dx and with argument in the interval $[0, L]$ is denoted as $\mathcal{L}^2([0, L], \mathbb{C}, dx)$ or simply $\mathcal{L}^2([0, L])$. Hence we should say that if (2.2) holds, then $f \in \mathcal{L}^2([0, L])$. It is also common to encounter functions whose modulus is integrable:

$$\|f\|_1 \doteq \int_0^L |f(x)| \, dx < \infty.$$

Such functions are said to be in $\mathcal{L}^1([0, L])$. When L is finite, it can be shown that a function $f \in \mathcal{L}^2([0, L])$ is automatically in $\mathcal{L}^1([0, L])$ as well. This follows from the Cauchy-Schwarz inequality given in Appendix C. We will not distinguish between functions on $[0, L]$, functions on the circle of circumference L, and L-periodic functions on the line. In the discussion below, it is also assumed that the functions under consideration have at most a finite number of jump discontinuities.

Observe that for L-periodic functions, the following is true for any $c \in \mathbb{R}$:

$$\int_0^L f(x) \, dx = \int_c^{c+L} f(x) \, dx.$$

That is, integration is independent of shifting the limits. Likewise, if we shift or reflect a periodic function, the value of its integral over a period is unchanged:

$$\int_0^L f(x) \, dx = \int_0^L f(x + c) \, dx = \int_0^L f(-x) \, dx.$$

The *Fourier series* expansion of a complex-valued function with fundamental period L is defined as:

$$f(x) = \sum_{k=-\infty}^{+\infty} \hat{f}(k) \, e^{2\pi i k x/L}. \tag{2.3}$$

$\hat{f}(k)$ is called the kth *Fourier coefficient*. For the moment we will consider only functions of the form in (2.3). We shall see later in this section the conditions under which this expansion holds.

A function $f \in \mathcal{L}^2([0, L])$ for which $\hat{f}(k) = 0$ for all $|k| > B$ for some fixed natural number (nonnegative integer) $B \in \mathbb{N}$ is called a *band-limited* function. If f is not band-limited, it can be approximated with the *partial sum*[3]

$$f(x) \approx f_B(x) \doteq \sum_{k=-B}^{B} \hat{f}(k) \, e^{2\pi i k x/L}.$$

---

[2]Here "$\doteq$" denotes a defining equality.
[3]Here "$\approx$" denotes an approximate equality.

Recall that the exponential functions given previously have the properties $d/d\theta(e^{i\theta}) = ie^{i\theta}$, $e^{i\theta} = \cos\theta + i\sin\theta$, $|e^{i\theta}| = 1$, and $e^{i\theta_1}e^{i\theta_2} = e^{i(\theta_1+\theta_2)}$ for any $\theta, \theta_1, \theta_2 \in \mathbb{R}$. Furthermore, for integers k and n,

$$\int_0^L e^{2\pi ikx/L} \cdot e^{-2\pi inx/L}\, dx = \int_0^L e^{2\pi i(k-n)x/L}\, dx = L\delta_{k,n}$$

where

$$\delta_{k,n} = \begin{cases} 1 \text{ for } k = n \\ 0 \text{ for } k \neq n \end{cases}$$

is the Kronecker delta function.

This means that the Fourier coefficients, $\hat{f}(n)$, can be calculated by multiplying both sides of (2.3) by $e^{-2\pi inx/L}$ and integrating x over the interval $[0, L]$. The result is

$$\hat{f}(n) \doteq \frac{1}{L}\int_0^L f(x)\,e^{-2\pi inx/L}\, dx. \tag{2.4}$$

Sometimes it is useful to denote $\hat{f}(n)$ alternatively as $\mathcal{F}_n(f)$ when we want to emphasize the operation of converting a function to its Fourier coefficients.

Written as one expression, the original function is then

$$f(x) = \sum_{k=-\infty}^{+\infty} \left(\frac{1}{L}\int_0^L f(y)\,e^{-2\pi iky/L}\, dy\right) e^{2\pi ikx/L}. \tag{2.5}$$

It becomes obvious when written this way that there is freedom to insert an arbitrary constant C as follows:

$$f(x) = \frac{1}{C}\sum_{k=-\infty}^{+\infty} \left(\frac{C}{L}\int_0^L f(y)\,e^{-2\pi iky/L}\, dy\right) e^{2\pi ikx/L}.$$

In other words, we could have defined the Fourier series as

$$f(x) = \frac{1}{C}\sum_{k=-\infty}^{+\infty} \tilde{f}(k)\,e^{2\pi ikx/L}$$

from the beginning (with $\tilde{f}(k) = C\hat{f}(k)$), and calculated the corresponding coefficients

$$\tilde{f}(n) = \frac{C}{L}\int_0^L f(x)\,e^{-2\pi inx/L}\, dx.$$

Some authors prefer the choice of $C = L$ to make the Fourier coefficients look neater. Others prefer to choose $C = \sqrt{L}$ to make the Fourier series and coefficients look more symmetrical. For simplicity, we will keep $C = 1$.

Fourier series are useful in applications because they possess certain interesting properties. For instance, operations such as addition and scaling of functions result in addition and scaling of Fourier coefficients. If $f(x)$ is an even/odd function, $f(x) = \pm f(-x)$, then the Fourier coefficients will have the symmetry/antisymmetry $\hat{f}(n) = \pm\hat{f}(-n)$ (which reflects the fact that even functions can be expanded in a cosine series, while odd ones can be expanded in sine series). Similarly, if we define the function $f^*(x) \doteq \overline{f(-x)}$, it is easy to see that the Fourier coefficients of $f^*(x)$ are $\hat{f}^*(k) = \overline{\hat{f}(k)}$. In the case when $f(x)$ is real valued, $f(x) = \overline{f(x)} = f^*(-x)$, and this is reflected in the Fourier coefficients as $\hat{f}(k) = \overline{\hat{f}(-k)}$.

One of the most important properties of Fourier series (and Fourier analysis in general) is the *convolution property*. The (periodic) convolution of two L−periodic functions is defined as

$$(f_1 * f_2)(x) \doteq \frac{1}{L} \int_0^L f_1(\xi)\, f_2(x-\xi)\, d\xi.$$

Expanding both functions in their Fourier series, we observe that

$$(f_1 * f_2)(x) = \frac{1}{L} \int_0^L \left(\sum_{k=-\infty}^{+\infty} \hat{f}_1(k)\, e^{2\pi i k\xi/L} \right) \left(\sum_{n=-\infty}^{+\infty} \hat{f}_2(n)\, e^{2\pi i n(x-\xi)/L} \right) d\xi$$

$$= \frac{1}{L} \sum_{k,n=-\infty}^{+\infty} \hat{f}_1(k)\hat{f}_2(n)\, e^{2\pi i n x/L} \int_0^L e^{2\pi i (k-n)\xi/L}\, d\xi. \tag{2.6}$$

The integral in the above expression has value $L\delta_{k,n}$, and so the following simplification results:

$$(f_1 * f_2)(x) = \sum_{k=-\infty}^{+\infty} \hat{f}_1(k)\hat{f}_2(k)\, e^{2\pi i k x/L}. \tag{2.7}$$

In other words, the kth Fourier coefficient of the convolution of two functions is simply the product of the kth Fourier coefficient of each of the two functions. Note that because the order of multiplication of two complex numbers does not matter, it follows from switching the order of $\hat{f}_1$ and $\hat{f}_2$ in (2.7) that $(f_1 * f_2)(x) = (f_2 * f_1)(x)$.

The *Dirac delta function* (see Appendix E) can be defined as the function for which

$$(\delta * f)(x) = f(x)$$

for any well-behaved (i.e., "nice") $f(x)$. In other words, all of the Fourier coefficients in the expansion of $\delta(x)$ must be $\hat{\delta}(k) = 1$. The obvious problem with this is that the Fourier coefficients of $\delta(x)$ do not diminish as k increases, and so this Fourier series does not converge. This reflects the fact that $\delta(x)$ is not a "nice" function (it is not even in $\mathcal{L}^2([0,L])$). It is nevertheless convenient to view $\delta(x)$ as the limit of the following band-limited function as $B \to \infty$:

$$\delta_B(x) \doteq \sum_{k=-B}^{B} e^{2\pi i k x/L} = \chi_B(2\pi x/L)$$

where

$$\chi_B(\theta) \doteq 1 + 2\sum_{n=1}^{B} \cos(n\theta) = \frac{\sin((B+1/2)\theta)}{\sin(\theta/2)} \tag{2.8}$$

is called the *Dirichlet kernel*.[4]

With the aid of $\delta(x)$, we can show that if $f(x)$ is L-periodic and the sequence $\{\hat{f}(k)\}$ for $k = 0, 1, \ldots$ converges, then $f(x)$ can be reconstructed from its Fourier coefficients. All

---

[4]Named after Peter Gustav Lejeune Dirichlet (1805-1859).

that is required is to rearrange terms in (2.5), and to substitute the series representation of the Dirac delta function:

$$f(x) = \frac{1}{L} \int_0^L f(y) \sum_{k=-\infty}^{+\infty} e^{2\pi i k(x-y)/L} \, dy = (f * \delta)(x). \qquad (2.9)$$

Whereas the limits in the summation are to infinity, the change in order of summation and integration that was performed is really only valid if $\delta_B(x)$ is used in place of $\delta(x)$. Since B can be chosen to be an arbitrarily large finite number, the equality above can be made to hold to within any desired precision.

If we define $f_1(x) = f(x)$ and $f_2(x) = f^*(x) = \overline{f(-x)}$ where $f(x)$ is a "nice" function, then evaluating the convolution at $x = 0$ yields

$$(f_1 * f_2)(0) = \frac{1}{L} \int_0^L |f(\xi)|^2 d\xi.$$

The corresponding k^{th} Fourier coefficient of this convolution is $\hat{f}(k)\overline{\hat{f}(k)} = |\hat{f}(k)|^2$. Evaluating the Fourier series of the convolution at $x = 0$ and equating, we get

$$\frac{1}{L} \int_0^L |f(\xi)|^2 d\xi = \sum_{k=-\infty}^{+\infty} |\hat{f}(k)|^2.$$

This is *Parseval's equality*. It is important because it expresses the "power" of a signal (function) in the spatial and Fourier domains.

Evaluating the convolution of $f_1(x)$ and $f_2^*(x)$ at $x = 0$, we observe the generalized Parseval relationship

$$\frac{1}{L} \int_0^L f_1(\xi)\overline{f_2(\xi)} \, d\xi = \sum_{k=-\infty}^{+\infty} \hat{f}_1(k)\overline{\hat{f}_2(k)}.$$

Finally, we note that certain "operational" properties associated with the Fourier series exist. For instance, the Fourier coefficients of the shifted function $(S_{x_0} f)(x) \doteq f(x-x_0)$ can be calculated from those of $f(x)$ by making a change of variables $y = x-x_0$ and using the fact that integration over a period is invariant under shifts of the limits of integration:[5]

$$\begin{aligned} \mathcal{F}_k(S_{x_0}f) &= \frac{1}{L} \int_0^L f(x - x_0) \, e^{-2\pi i k x/L} \, dx \\ &= \frac{1}{L} \int_0^L f(y) \, e^{-2\pi i k(y+x_0)/L} \, dy \\ &= e^{-2\pi i k x_0/L} \, \hat{f}(k). \end{aligned}$$

Similarly, we can use integration by parts to calculate the Fourier coefficients of df/dx, assuming f is at least a once differentiable function. Or, simply taking the derivative of the Fourier series,

$$\frac{df}{dx} = \sum_{k=-\infty}^{+\infty} \hat{f}(k)\frac{d}{dx}\left(e^{2\pi i k x/L}\right)$$

---

[5]This is an example of when the notation $\mathcal{F}_k(f)$ is a convenient substitute for $\hat{f}_k$.

we find that

$$\frac{df}{dx} = \sum_{k=-\infty}^{+\infty} \left(\hat{f}(k) \cdot 2\pi ik/L \right) e^{2\pi ikx/L},$$

and so the Fourier coefficients of the derivative of $f(x)$ are

$$\mathcal{F}_k \left(\frac{df}{dx} \right) = \hat{f}(k) \cdot 2\pi ik/L.$$

If $f(x)$ can be differentiated multiple times, the corresponding Fourier coefficients are multiplied by one copy of $2\pi ik/L$ for each derivative. Note that since each differentiation results in an additional power of k in the Fourier coefficients, the tendency is for these Fourier coefficients to not converge as quickly as those for the original function, and they may not even converge at all.

2.2 The Fourier Transform: Decomposition of Functions on the Line

A function on the line, $f(x)$, is said to be in $\mathcal{L}^p(\mathbb{R})$ if, for some positive integer p, it holds that

$$(\|f\|_p)^p \doteq \int_{-\infty}^{\infty} |f(x)|^p dx < \infty.$$

In analogy with the way that a periodic function (function on the circle) is expressed as a Fourier series, we can express an arbitrary "nice" function on the line (i.e., one in $\mathcal{L}^2(\mathbb{R}) \cap \mathcal{L}^1(\mathbb{R})$ and which in addition is analytic.) as[6]

$$f(x) = \frac{1}{2\pi} \int_{-\infty}^{\infty} \hat{f}(\omega) e^{i\omega x} \, d\omega \qquad (2.10)$$

where the Fourier coefficients $\hat{f}(\omega)$ (now called the *Fourier transform*) depend on the continuous parameter $\omega \in \mathbb{R}$. The operation of reconstructing a function from its Fourier transform in (2.10) is called the *Fourier inversion* formula, or inverse Fourier transform. The Fourier transform of any nice function, $f(x)$, is defined as[7]

$$\hat{f}(\omega) \doteq \int_{-\infty}^{\infty} f(x) e^{-i\omega x} \, dx. \qquad (2.11)$$

Sometimes it is convenient to write (2.10) and (2.11) as

$$f = \mathcal{F}^{-1}(\hat{f}) \text{ and } \hat{f} = \mathcal{F}(f).$$

The fact that a function is recovered from its Fourier transform is found by first observing that it is true for the special case of $g(x) = e^{-ax^2}$ for $a > 0$. One way to calculate

---

[6]In practice, even functions that are not analytic and have a finite number of finite jump discontinuities and oscillations in any finite interval of $\mathbb{R}$ can be handled in Fourier analysis.

[7]In analogy with the Fourier series in which there is some freedom in where to put L and C, different scaling conventions can be used in the Fourier transform. For example, if ω is replaced with $2\pi\omega$ in the exponent, then the 2π in (2.10) disappears.

$$\hat{g}(\omega) = \int_{-\infty}^{\infty} e^{-ax^2} e^{-i\omega x}\, dx$$

is to differentiate both sides with respect to ω to yield

$$\frac{d\hat{g}(\omega)}{d\omega} = -i \int_{-\infty}^{\infty} x e^{-ax^2} e^{-i\omega x}\, dx = \frac{i}{2a} \int_{-\infty}^{\infty} \frac{dg}{dx} e^{-i\omega x}\, dx.$$

Integrating by parts, and observing that $e^{-i\omega x} g(x)$ vanishes at the limits of integration yields

$$\frac{d}{d\omega}(\hat{g}) = -\frac{\omega}{2a}\hat{g}(\omega).$$

The solution of this first order ordinary differential equation is of the form

$$\hat{g}(\omega) = \hat{g}(0)\, e^{-\frac{\omega^2}{4a}}$$

where

$$\hat{g}(0) = \int_{-\infty}^{\infty} e^{-ax^2}\, dx = \sqrt{\frac{\pi}{a}}.$$

The last equality is well known, and can be calculated by observing that

$$\left(\int_{-\infty}^{\infty} e^{-ax^2}\, dx \right)^2 = \int_{-\infty}^{\infty} \int_{-\infty}^{\infty} e^{-a(x^2+y^2)}\, dx dy = \int_{0}^{2\pi} \int_{0}^{\infty} e^{-ar^2} r\, dr d\theta.$$

The integral in polar coordinates (r, θ) is calculated easily in closed form as π/a.

Having found the form of $\hat{g}(\omega)$, it is easy to see that $g(x)$ is reconstructed from $\hat{g}(\omega)$ using the inversion formula (2.10) (the calculation is essentially the same as for the forward Fourier transform). Likewise, the Gaussian distribution

$$h_\sigma(x) = \frac{1}{\sqrt{2\pi}\sigma} e^{-\frac{x^2}{2\sigma^2}}$$

has Fourier transform

$$\hat{h}_\sigma(\omega) = e^{-\frac{\sigma^2}{2}\omega^2},$$

and the reconstruction formula holds. The Gaussian distribution is a probability density function, i.e., it is non-negative everywhere and

$$\int_{-\infty}^{\infty} h_\sigma(x)\, dx = 1.$$

As σ becomes small, $h_\sigma(x)$ becomes like $\delta(x)$ (See Appendix E).[8] To emphasize this fact, we write $\delta_\epsilon(x) = h_\epsilon(x)$ for $0 < \epsilon \ll 1$. Since it is always possible to choose ϵ small enough that

$$\|f - \delta_\epsilon * f\|_2 < \nu$$

for any given $\nu \in \mathbb{R}_{>0}$, in many applications the approximation $f(x) \approx (\delta_\epsilon * f)(x)$ is good enough, and $\delta_\epsilon(x)$ is effectively indistinguishable from $\delta(x)$.

As with the case of Fourier series, the Dirac δ-function is the one whose Fourier transform is defined as

---

[8] Other "special" functions such as the Heaviside (unit) step function [23] defined as $u(x) \doteq \int_{-\infty}^{x} \delta(y)\, dy$ play an important role in many fields including mechanics and systems theory.

$$\hat{\delta}(\omega) = 1,$$

and has the property that

$$\int_{-\infty}^{\infty} f(\xi)\, \delta(x - \xi)\, d\xi = f(x).$$

Since the Fourier inversion formula works for a Gaussian for all values of $\sigma > 0$, it also works for $\delta_\epsilon(x)$. We therefore write

$$\delta(x) = \frac{1}{2\pi} \int_{-\infty}^{\infty} e^{i\omega x}\, d\omega \quad \text{or} \quad \delta(x - \xi) = \frac{1}{2\pi} \int_{-\infty}^{\infty} e^{i\omega(x-\xi)}\, d\omega. \qquad (2.12)$$

This formula is critical in proving the inversion formula in general, and it is often called a *completeness relation*, because it proves that the functions $e^{i\omega x}$ for all $\omega \in \mathbb{R}$ are complete in $\mathcal{L}^2(\mathbb{R})$ (i.e., any function in $\mathcal{L}^2(\mathbb{R})$ can be expanded in terms of these basic functions). Using the completeness relation, we see that

$$\begin{aligned}
(\mathcal{F}^{-1}(\hat{f}))(x) &= \frac{1}{2\pi} \int_{-\infty}^{\infty} \left[\int_{-\infty}^{\infty} f(\xi)\, e^{-i\omega\xi}\, d\xi \right] e^{i\omega x}\, d\omega \\
&= \int_{-\infty}^{\infty} f(\xi) \left[\frac{1}{2\pi} \int_{-\infty}^{\infty} e^{i\omega(x-\xi)}\, d\omega \right] d\xi \\
&= \int_{-\infty}^{\infty} f(\xi)\, \delta(x - \xi)\, d\xi \\
&= f(x).
\end{aligned} \qquad (2.13)$$

We note that the combination of Fourier transform and inversion formula can be written in the form

$$f(x) = \frac{C_1 C_2}{2\pi} \int_{-\infty}^{\infty} \left[\frac{1}{C_1} \int_{-\infty}^{\infty} f(\xi)\, e^{-iC_2\omega'\xi}\, d\xi \right] e^{iC_2\omega' x}\, d\omega' \qquad (2.14)$$

where C_1 is an arbitrary constant (like C in the case of Fourier series) and C_2 is a second arbitrary constant by which we can change the frequency variable: $\omega = C_2\omega'$. We have chosen $C_1 = C_2 = 1$. Other popular choices in the literature are $(C_1, C_2) = (\sqrt{2\pi}, \pm 1)$, $(C_1, C_2) = (2\pi, \pm 1)$, and $(C_1, C_2) = (1, \pm 2\pi)$.

The convolution of "nice" functions on the line is defined by

$$(f_1 * f_2)(x) = \int_{-\infty}^{\infty} f_1(\xi)\, f_2(x - \xi)\, d\xi.$$

This operation produces a new well-behaved function from two old ones. For instance, we see that

$$\begin{aligned}
\|f_1 * f_2\|_1 &= \int_{-\infty}^{\infty} \left| \int_{-\infty}^{\infty} f_1(\xi)\, f_2(x - \xi)\, d\xi \right| dx \\
&\leq \int_{-\infty}^{\infty} \int_{-\infty}^{\infty} |f_1(\xi)| \cdot |f_2(x - \xi)|\, d\xi dx \\
&= \|f_1\|_1 \cdot \|f_2\|_1,
\end{aligned}$$

and when f_1 and f_2 are both differentiable, so too is $f_1 * f_2$ from the product rule.

Sometimes it is useful to consider the convolution of two shifted functions $(S_{x_i} f_i)(x) = f_i(x - x_i)$, resulting in

$$((S_{x_1} f_1) * (S_{x_2} f_2))(x) \doteq \int_{-\infty}^{\infty} f_1(\xi - x_1) f_2(x - \xi - x_2) \, d\xi$$
$$= (f_1 * f_2)(x - x_1 - x_2)$$
$$= (S_{x_1 + x_2}(f_1 * f_2))(x).$$

In many engineering books $((S_{x_1} f_1) * (S_{x_2} f_2))(x)$ is written as $f_1(x - x_1) * f_2(x - x_2)$. While this is okay in the context of the applications in which it is used, it leads to problems in the noncommutative context. And so we refrain from using this notation.

The Fourier transform has the property that

$$\mathcal{F}(f_1 * f_2)(\omega) = \hat{f}_1(\omega)\hat{f}_2(\omega).$$

This follows from the property $e^{i\omega(x_1 + x_2)} = e^{i\omega x_1} \cdot e^{i\omega x_2}$. More explicitly,

$$\mathcal{F}(f_1 * f_2)(\omega) = \int_{-\infty}^{\infty} \left(\int_{-\infty}^{\infty} f_1(\xi) f_2(x - \xi) \, d\xi \right) e^{-i\omega x} \, dx.$$

By changing variables $y = x - \xi$, and observing that the limits of integration are invariant to finite shifts, this is rewritten as

$$\int_{-\infty}^{\infty} \int_{-\infty}^{\infty} f_1(\xi) f_2(y) e^{-i\omega(y+\xi)} \, dy d\xi = \left(\int_{-\infty}^{\infty} f_1(\xi) e^{-i\omega\xi} \, d\xi \right) \left(\int_{-\infty}^{\infty} f_2(y) e^{-i\omega y} \, dy \right)$$
$$= \hat{f}_1(\omega)\hat{f}_2(\omega).$$

Using exactly the same argument as in the previous section, we observe the Parseval equality

$$\int_{-\infty}^{\infty} |f(x)|^2 dx = \frac{1}{2\pi} \int_{-\infty}^{\infty} |\hat{f}(\omega)|^2 d\omega$$

and its generalization

$$\int_{-\infty}^{\infty} f_1(x)\overline{f_2(x)} \, dx = \frac{1}{2\pi} \int_{-\infty}^{\infty} \hat{f}_1(\omega)\overline{\hat{f}_2(\omega)} \, d\omega.$$

In complete analogy with the Fourier series, a number of symmetry and operational properties of the Fourier transform result from symmetries and operations performed on the original functions. The following is a summary of the most common operational properties:

$$f(x) = \pm f(-x) \implies \hat{f}(\omega) = \pm \hat{f}(-\omega) \tag{2.15}$$

$$f(x) = \overline{f(x)} \implies \hat{f}(\omega) = \overline{\hat{f}(-\omega)} \tag{2.16}$$

$$\frac{df}{dx} \implies i\omega \hat{f}(\omega) \tag{2.17}$$

$$f(x - x_0) \implies e^{-i\omega x_0} \hat{f}(\omega) \tag{2.18}$$

$$f(ax) \implies \frac{1}{|a|} \hat{f}(\omega/a) \tag{2.19}$$

$$(f_1 * f_2)(x) \implies \hat{f}_1(\omega)\hat{f}_2(\omega) \tag{2.20}$$

2.2.1 Relationship Between the Fourier Series and the Fourier Transform

In this section we examine the relationship between the Fourier series on the circle and the Fourier transform on the real line.[9] Let $f_\mathbb{R}(x)$ denote a complex-valued function on the line, i.e., $f_\mathbb{R} : \mathbb{R} \to \mathbb{C}$, and let $f_{\mathbb{S}^1}(\theta)$ denote a function on the circle, i.e., $f_\mathbb{R} : \mathbb{S}^1 \to \mathbb{C}$. The Fourier transform of a function on the real line can be related to the Fourier series of a related function on the circle by defining

$$\hat{f}_{\mathbb{R},\mathbb{S}^1}(\omega) \doteq \hat{f}_\mathbb{R}(\omega) \cdot \sum_{k=-\infty}^{\infty} \delta(\omega - k)$$

and observing that

$$(\mathcal{F}^{-1}(\hat{f}_{\mathbb{R},\mathbb{S}^1}))(x) = \frac{1}{2\pi} \int_{-\infty}^{\infty} \hat{f}_\mathbb{R}(\omega) \cdot \sum_{k=-\infty}^{\infty} \delta(\omega - k)\, e^{i\omega x}\, d\omega$$

$$= \frac{1}{2\pi} \sum_{k=-\infty}^{\infty} \hat{f}_\mathbb{R}(k)\, e^{ikx}$$

$$= f_{\mathbb{S}^1}(x).$$

In real space these two functions are related by "wrapping $f_\mathbb{R}(x)$ around the circle" as

$$f_{\mathbb{S}^1}(\theta) = \sum_{n=-\infty}^{\infty} f_\mathbb{R}(\theta - 2\pi n). \tag{2.21}$$

This construction is particularly common for defining Gaussian distributions on the circle from Gaussians on the real line, as reviewed in [5].

2.2.2 Multi-Dimensional Fourier Transforms and Fourier Series

The n-dimensional Euclidean space endowed with an origin and vector and scalar multiplication can be identified with the n-fold *Cartesian product* $\mathbb{R}^n \doteq \mathbb{R} \times \cdots \times \mathbb{R}$, and similarly, the n-*torus* is defined as $\mathbb{T}^n \doteq \mathbb{T} \times \cdots \times \mathbb{T}$ where $\mathbb{T} = \mathbb{S}^1$. Elements of these spaces are denoted as column vectors $\mathbf{x} = [x_1, x_2, ..., x_n]^T \in \mathbb{R}^n$ and $\boldsymbol{\theta} = [\theta_1, \theta_2, ..., \theta_n]^T \in \mathbb{T}^n$. (Mathematically, $\mathbb{R}^n$ is a vector space but $\mathbb{T}$ is not, and by 'column vector' we mean here nothing more than an array of numbers.)

The multidimensional Fourier transform and Fourier series generalize in essentially the same way, and are related to each other in analogy with how they are related in the one-dimensional case. Therefore, it suffices to examine the properties of only the Fourier transform. The Fourier transform of any function in $(\mathcal{L}^1 \cap \mathcal{L}^2)(\mathbb{R}^n)$ can be defined as[10]

$$\hat{f}(\boldsymbol{\omega}) \doteq \frac{1}{(2\pi)^{n/2}} \int_{\mathbb{R}^n} f(\mathbf{x})\, e^{-i\boldsymbol{\omega}\cdot\mathbf{x}}\, d\mathbf{x} \tag{2.22}$$

---

[9]The circle is denoted as $\mathbb{S}^1$, and more generally a sphere in N-dimensional Euclidean space is denoted as $\mathbb{S}^{N-1}$.

[10]Here we have split the factor of $1/(2\pi)^n$ evenly between the Fourier transform and the inversion formula.

where $d\mathbf{x} = dx_1 dx_2 \cdots dx_n$ is the differential volume element in n-dimensional space constructed as a product for the differential volume elements along each of the orthogonal coordinate axes.[11] Moreover, $\int_{\mathbb{R}^n}$ is shorthand for $\int_{x_1 \in \mathbb{R}} \int_{x_2 \in \mathbb{R}} \cdots \int_{x_n \in \mathbb{R}}$, an integral over n independent copies of the real line. The dot product of vectors is defined as

$$\boldsymbol{\omega} \cdot \mathbf{x} \doteq \omega_1 x_1 + \omega_2 x_2 + \cdots + \omega_n x_n.$$

It is the structure of this operation that relates the Fourier transform in multiple dimensions to that in a single dimension since the integral in (2.22) can be written as

$$\int_{\mathbb{R}^{n-1}} \left(\int_{\mathbb{R}} f(x_1, ..., x_{n-1}, x_n)\, e^{-i\omega_n x_n}\, dx_n \right) \exp[-i(\omega_1 x_1 + \cdots + \omega_{n-1} x_{n-1})]\, dx_1 \cdots dx_{n-1}.$$

This separability property means that for each fixed value of $[x_1, ..., x_{n-1}]^T \in \mathbb{R}^{n-1}$, the one-dimensional Fourier transform in the variable x_n can be computed. Then with x_{n-1} taken as the variable, and $x_1, ..., x_{n-2}, \omega_n$ held fixed, the Fourier transform in x_{n-1} can be computed. This process can continue until all of the x_i's are visited.

In a similar way, with $d\boldsymbol{\omega} \doteq d\omega_1 \cdots d\omega_n$, the multi-dimensional Fourier inversion formula

$$f(\mathbf{x}) = \frac{1}{(2\pi)^{n/2}} \int_{\mathbb{R}^n} \hat{f}(\boldsymbol{\omega})\, e^{i\boldsymbol{\omega} \cdot \mathbf{x}}\, d\boldsymbol{\omega} \tag{2.23}$$

can be proved using the one-dimensional inversion formula n times, each time holding $n - 1$ variables fixed, as in the discussion for the Fourier transform above.

As with the one-dimensional case, it is sometimes more convenient to denote $\hat{f}(\boldsymbol{\omega})$ as $(\mathcal{F}(f))(\boldsymbol{\omega})$ and to write the reconstruction formula as $f(\mathbf{x}) = (\mathcal{F}^1(\hat{f}))(\mathbf{x})$. The multi-dimensional convolution

$$(f_1 * f_2)(\mathbf{x}) \doteq \int_{\mathbb{R}^n} f_1(\boldsymbol{\xi})\, f_2(\mathbf{x} - \boldsymbol{\xi})\, d\boldsymbol{\xi} \tag{2.24}$$

can also be separated into N one-dimensional convolutions, and the convolution theorem in the one-dimensional case extends to the multi-dimensional case as

$$\widehat{(f_1 * f_2)}(\boldsymbol{\omega}) = \hat{f}_1(\boldsymbol{\omega})\, \hat{f}_2(\boldsymbol{\omega}). \tag{2.25}$$

2.3 Using the Fourier Transform to Solve PDEs and Integral Equations

The Fourier transform is a powerful tool when used to solve, or simplify, linear partial differential equations (PDEs) with constant coefficients. Several examples are given in Subsections 2.3.1–2.3.4. The usefulness of the Fourier transform in this context is a direct result of the operational properties and convolution theorem discussed earlier in this chapter. The convolution theorem, Parseval equality, and operational properties are also useful tools in solving certain kinds of integral equations. Often such equations do not admit exact solutions, and Fourier techniques provide the tools for regularization (approximate solution) of such ill-posed problems.

---

[11] Another notation in use for $d\mathbf{x}$ is dx^n.

2.3.1 Diffusion Equations

A one-dimensional linear diffusion equation with constant coefficients is one of the form

$$\frac{\partial u}{\partial t} = a\frac{\partial u}{\partial x} + b\frac{\partial^2 u}{\partial x^2} \qquad (2.26)$$

where $a \in \mathbb{R}$ is called the drift coefficient and $b \in \mathbb{R}_{>0}$ is called the diffusion coefficient. When modeling diffusion phenomena in an infinite medium, it is common to solve the above diffusion equation for $u(x,t)$ subject to given initial conditions $u(x,0) = f(x)$. We note that (2.26) is a special case of the Fokker-Planck equation which we will examine in great detail in Chapter 16. When the drift coefficient is zero, the diffusion equation is called the *heat equation*.

Taking the Fourier transform of $u(x,t)$ for each value of t (i.e., treating time as a constant for the moment and x as the independent variable), we write $\hat{u}(\omega,t)$. Applying the Fourier transform to both sides of (2.26), and the initial conditions, we observe that for each fixed frequency ω we get a linear first-order ordinary differential equation in the Fourier transform with initial conditions:

$$\frac{d\hat{u}}{dt} = (ia\omega - b\omega^2)\hat{u} \quad \text{with} \quad \hat{u}(\omega,0) = \hat{f}(\omega).$$

The solution to this initial value problem is of the form

$$\hat{u}(\omega,t) = \hat{f}(\omega)\, e^{(ia\omega - b\omega^2)t}.$$

Application of the inverse Fourier transform yields a solution. The above expression for $\hat{u}(\omega,t)$ is a Gaussian with phase factor, and on inversion this becomes a shifted Gaussian:

$$\mathcal{F}^{-1}\left(e^{ia t\omega}e^{-b\omega^2 t}\right) = \frac{1}{\sqrt{4\pi b t}}\exp\left(-\frac{(x+at)^2}{4bt}\right).$$

The convolution theorem then allows us to write

$$u(x,t) = \frac{1}{\sqrt{4\pi b t}}\int_{-\infty}^{\infty} f(\xi)\exp\left(-\frac{(x+at-\xi)^2}{4bt}\right)d\xi. \qquad (2.27)$$

2.3.2 The Schrödinger Equation

In quantum mechanics, the Schrödinger equation for a free particle (i.e., when no potential field is present) is

$$\frac{\partial u}{\partial t} = ic^2\frac{\partial^2 u}{\partial x^2}. \qquad (2.28)$$

The corresponding initial and boundary conditions are $u(x,0) = f(x)$ and $u(-\infty,t) = u(\infty,t) = 0$. At first sight this looks much like the heat equation. Only now, u is a complex-valued function, and $c^2 = \hbar/2m$ where m is the mass of the particle and $\hbar = 1.05457 \times 10^{-27}\,\text{erg}\cdot\text{sec} = 1.05457 \times 10^{-34}\,\text{J}\cdot\text{sec}$ is Planck's constant divided by 2π ($\hbar = h/2\pi$).

The solution of this equation progresses much like before. Taking the Fourier transform of both sides yields

$$\frac{d\hat{u}}{dt} = -ic^2\omega^2\hat{u}$$

for each fixed value of ω. The Fourier transform is then written in the form

$$\hat{u}(\omega, t) = \hat{f}(\omega) e^{-ic^n \omega^a t}.$$

Applying the Fourier inversion formula yields the solution in the form of the integral

$$u(x,t) = \frac{1}{2\pi} \int_{-\infty}^{\infty} \hat{f}(\omega) e^{-ic^2 \omega^2 t} e^{i\omega x} \, d\omega. \tag{2.29}$$

While this is not a closed-form solution, it does tell us important information about the behavior of the solution. For instance, if $\|f\|_2^2 = 1$, then from the above equation

$$\|u\|_2^2 = \int_{-\infty}^{\infty} u(x,t) \overline{u(x,t)} \, dx = 1.$$

This follows from direct substitution of (2.29) into the above equation as:

$$\|u\|_2^2 = \int_{-\infty}^{\infty} \left(\frac{1}{2\pi} \int_{-\infty}^{\infty} \hat{f}(\omega) e^{-ic^2 \omega^2 t} e^{i\omega x} \, d\omega \right) \left(\frac{1}{2\pi} \int_{-\infty}^{\infty} \overline{\hat{f}(\nu)} e^{+ic^2 \nu^2 t} e^{-i\nu x} \, d\nu \right) \, dx.$$

Reversing the order of integration so that integration with respect to x is performed first, a closed-form solution of the x-integral is $2\pi\delta(\omega - \nu)$. Integration over ν then results in

$$\|u\|_2^2 = \frac{1}{2\pi} \int_{-\infty}^{\infty} \hat{f}(\omega) \overline{\hat{f}(\omega)} \, d\omega = \|f\|_2^2.$$

2.3.3 The Wave Equation

The one-dimensional wave equation is

$$\frac{\partial^2 u}{\partial t^2} = c^2 \frac{\partial^2 u}{\partial x^2}. \tag{2.30}$$

The corresponding initial and boundary conditions are $u(x,0) = f(x)$ and $u(-\infty, t) = u(\infty, t) = 0$. This equation is often used to model a long, taut string with uniform mass density per unit length ρ and tension T. The constant c is related to these physical quantities as $c^2 = T/\rho$.

As in the previous sections, the problem is to find $u(x,t)$. We begin to solve this problem by applying the Fourier transform to the equation, initial conditions, and boundary conditions to yield

$$\frac{d^2 \hat{u}}{dt^2} = -c^2 \omega^2 \hat{u} \quad \hat{u}(\omega, 0) = \hat{f}(\omega) \quad \hat{u}(0, t) = 0.$$

Again, ω is treated as a constant in this stage of the solution process. Clearly the solution to this ordinary differential equation (the simple harmonic oscillator) for the given initial conditions is of the form

$$\hat{u}(\omega, t) = \hat{f}(\omega) \cos c\omega t.$$

Using the identity $\cos \theta = (e^{i\theta} + e^{-i\theta})/2$ together with the operational property for shifts allows us to write

$$\hat{f}(\omega) = \hat{f}(\omega)(e^{-ic\omega t} + e^{ic\omega t})/2,$$

which after application of the inverse Fourier transform gives

$$f(x) = \frac{1}{2}\{f(x - ct) + f(x + ct)\}. \tag{2.31}$$

This is known as d'Alembert's solution of the wave equation.

2.3.4 The Telegraph Equation

The telegraph (or transmission-line) equation models the flow of electrical current in a long cable. It is expressed as [12, 29, 31]

$$\frac{\partial^2 u}{\partial t^2} + 2b\frac{\partial u}{\partial t} + au = c^2\frac{\partial^2 u}{\partial x^2}. \tag{2.32}$$

The constants a, b and c can be written in terms of the following cable properties (all per unit length): resistance, R, inductance, L, capacitance, C, and leakage, S. Explicitly,

$$b = \frac{1}{2}(R/L + S/C), \quad c^2 = 1/CL, \quad a = RS/CL.$$

It is straightforward to see that application of the Fourier transform yields

$$\frac{\partial^2 \hat{u}}{\partial t^2} + 2b\frac{\partial \hat{u}}{\partial t} + (a + c^2\omega^2)\hat{u} = 0 \tag{2.33}$$

which becomes an ordinary differential equation for each fixed value of ω. Initial conditions of the form $u(x,0) = f_1(x)$ and $\partial u/\partial t(x,0) = f_2(x)$ transform to $\hat{u}(\omega,0) = \hat{f}_1(\omega)$ and $\partial \hat{u}/\partial t(\omega,0) = \hat{f}_2(\omega)$.

Since for each fixed value of ω, (2.33) is a linear ordinary differential equation with constant coefficients, we seek solutions of the form $\hat{u} = e^{\lambda t}$ and find that the solutions of the corresponding characteristic equation

$$\lambda^2 + 2b\lambda + (a + c^2\omega^2) = 0$$

are of the form

$$\lambda = -b \pm \sqrt{b^2 - c^2\omega^2 - a}.$$

From the physical meaning of the parameters a and b, it must be that $b^2 - a \geq 0$. Hence we have two cases:

Case 1: $b^2 - a = 0$, $c \neq 0$

In this case the solution for $\hat{u}$ is of the form

$$\hat{u}(\omega,t) = e^{-bt}\{C_1(\omega)\,e^{i\omega ct} + C_2(\omega)\,e^{-i\omega ct}\},$$

where $C_i(\omega)$ for $i = 1,2$ are determined by the initial conditions by solving the equations

$$C_1(\omega) + C_2(\omega) = \hat{f}_1(\omega)$$

and

$$C_1(\omega)[i\omega c - b] - C_2(\omega)[b + i\omega c] = \hat{f}_2(\omega)$$

simultaneously.

Case 2: $b^2 - a > 0$

In this case there are three subcases, corresponding to the sign of $b^2 - c^2\omega^2 - a$. Hence we write

$$\hat{u}(\omega,t) = \begin{cases} e^{-bt}\{C_1(\omega) + C_2(\omega)t\}; & b^2 - c^2\omega^2 - a = 0 \\[2mm] e^{-bt}\{C_1(\omega)\exp(it\sqrt{c^2\omega^2 - b^2 + a}) + C_2(\omega)\exp(-it\sqrt{c^2\omega^2 - b^2 + a}); & b^2 - c^2\omega^2 - a < 0 \\[2mm] e^{-bt}\{C_1(\omega)\exp(t\sqrt{b^2 - c^2\omega^2 - a}) + C_2(\omega)\exp(-t\sqrt{b^2 - c^2\omega^2 - a}); & b^2 - c^2\omega^2 - a > 0 \end{cases}$$

The constants $C_1(\omega)$ and $C_2(\omega)$ are again solved for given initial conditions represented in Fourier space.

Application of the Fourier inversion formula to the resulting Fourier transforms provides an integral representation of the solution.

2.3.5 Integral Equations with Convolution Kernels

Often when taking a physical measurement, the quantity which is of interest is not found directly, and is obscured by a filter or measurement device of some sort. Examples of this are considered in Chapter 13 in the context of the tomography/tomotherapy problem.

In general, an *inverse problem*[12] is one in which the desired quantity (or function) is inside of an integral and the values of that integral are given (or measured). Extracting the desired function is the goal in solving the inverse problem.

Fourier analysis is a valuable tool in solving a particular kind of inverse problem which is posed as a Fredholm integral equation with a convolution kernel:[13]

$$g(x) = cu(x) + \int_{-\infty}^{\infty} f(x - \xi)\, u(\xi)\, d\xi. \tag{2.34}$$

Here all the functions are real and c is either 1 or 0 since all other real coefficients can be absorbed in the definitions of the given "nice" functions $g(x)$ and $f(x)$. Our goal is, of course, to solve for $u(x)$.

Clearly the Fourier transform is a useful tool in this context because it yields

$$\hat{g}(\omega) = (c + \hat{f}(\omega))\, \hat{u}(\omega).$$

If $c + \hat{f}(\omega) \neq 0$ for all values of ω, then

$$\hat{u}(\omega) = \frac{\hat{g}(\omega)}{c + \hat{f}(\omega)}.$$

Applying the Fourier inversion formula then gives

$$u(x) = (\mathcal{F}^{-1}(\hat{u}))(x).$$

In the case when $c + \hat{f}(\omega) = 0$ for one or more values of ω a *regularization* technique is required. Essentially, regularization is any method for generating approximate solutions where the singularity associated with division by zero is "smoothed out." One popular method is zeroth order *Tikhonov regularization* [20].

In this method, one seeks a solution to the modified problem of minimization of the functional[14]

$$F = \|cu + f * u - g\|_2^2 + \epsilon^2 \|u\|_2^2$$

where ϵ is a small real parameter. In the limit as $\epsilon \to 0$, the solution to this modified problem approaches the exact solution to (2.34).

The regularized solution is generated using Fourier techniques by first using the Parseval equality to write

$$F = \frac{1}{2\pi}\{\|(c + \hat{f})\hat{u} - \hat{g}\|_2^2 + \epsilon^2 \|\hat{u}\|_2^2\}.$$

Then treating $\hat{u}$ as a complex variable,

---

[12]The subject of inverse problems is more general than the solution of integral equations, but all of the inverse problems encountered in this book are integral equations.

[13]See Appendix E.5 for more on this topic.

[14]Higher order Tikhonov regularization methods add derivative terms to the functional. For instance, first order would add a term such as $\nu^2 \|du/dx\|^2$ for a chosen parameter $\nu^2 \ll 1$ in an attempt to generate smoother solutions.

$$2\pi \frac{\partial F}{\partial \bar{\hat{u}}} = [(c + \hat{f})\hat{u} - \hat{g}](c + \overline{\hat{f}}) + \epsilon^2 \hat{u} = 0.$$

Solving for $\hat{u}$ and applying the Fourier inversion formula, the regularized solution (which now depends on the choice of ϵ) is

$$u(x, \epsilon) = (\mathcal{F}^{-1}(\hat{u}))(x, \epsilon) \text{ where } \hat{u}(\omega, \epsilon) = \frac{\hat{g}(\omega)(c + \overline{\hat{f}(\omega)})}{|c + \hat{f}(\omega)|^2 + \epsilon^2}.$$

We conclude this section by stating that the examples considered here are only a few of the kinds of partial differential and integral equations that are solvable using classical Fourier methods. The interested reader is referred to [2, 3, 31, 38] for further reading.

2.4 Fourier Optics

In this section, we review how the behavior of light in some situations is described naturally using the Fourier transform, and conversely, how a combination of light and lenses can be used as a tool to calculate the Fourier transforms of certain kinds of functions.

Light is an electromagnetic wave, and is thus described by the equations governing electromagnetic phenomena. The electromagnetic state in a homogeneous isotropic medium (including a vacuum) without sources is described by Maxwell's equations [22]:

$$\nabla \times \mathbf{E} = -\mu \frac{\partial \mathbf{H}}{\partial t}$$

$$\nabla \times \mathbf{H} = \epsilon \frac{\partial \mathbf{E}}{\partial t}$$

$$\nabla \cdot \mathbf{E} = 0$$

$$\nabla \cdot \mathbf{H} = 0,$$

where $\mathbf{E} = \mathbf{E}(\mathbf{x}, t)$ is the electric field vector, and $\mathbf{H} = \mathbf{H}(\mathbf{x}, t)$ is the magnetic field vector. $\mathbf{x} \in \mathbb{R}^3$ is the position in the medium, ∇ is the gradient with respect to $\mathbf{x}$, $\nabla\times$ is the curl operation, and $\nabla\cdot$ is the divergence. The constants μ and ϵ are called *permeability* and *permittivity* in the medium, respectively. In the case of a vacuum, these are denoted μ_0 and ϵ_0, and the speed of light in a vacuum is given by $c = (\mu_0 \epsilon_0)^{-\frac{1}{2}} \approx 3 \times 10^8 \text{m/sec}$.

Observe that for an arbitrary vector-valued $\mathbf{K}$ that is at least twice differentiable in each component of $\mathbf{x}$,

$$\nabla \times (\nabla \times \mathbf{K}) = \nabla(\nabla \cdot \mathbf{K}) - \nabla^2(\mathbf{K}) \tag{2.35}$$

where

$$\nabla^2 = \nabla \cdot \nabla = \sum_{i=1}^{3} \frac{\partial^2}{\partial x_i^2}$$

is the Laplacian.

Taking the curl of both sides of the first two of Maxwell's equations, and using (2.35), we can reduce the first two of Maxwell's equations in a vacuum to the form

$$\nabla^2 \mathbf{E} = \frac{1}{c^2} \frac{\partial^2 \mathbf{E}}{\partial t^2}. \qquad \nabla^2 \mathbf{H} = \frac{1}{c^2} \frac{\partial^2 \mathbf{H}}{\partial t^2}. \tag{2.36}$$

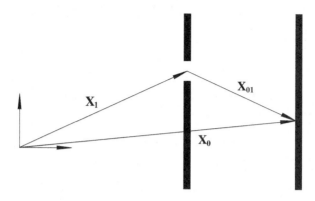

Fig. 2.1. Diffraction of Light Traveling through an Aperture

This is because for the case when $\mathbf{K}$ is $\mathbf{E}$ or $\mathbf{H}$, we see that (2.35) simplifies from the last two of Maxwell's equations: $\nabla \cdot \mathbf{E} = \nabla \cdot \mathbf{H} = 0$. Let us now define the function $u(\mathbf{x}, t)$ to represent any component of the electric or magnetic field. The function $u(\mathbf{x}, t)$ must then satisfy the scalar *wave equation*:

$$\nabla^2 u = \frac{1}{c^2} \frac{\partial^2 u}{\partial t^2}.$$

Solutions of this equation are of the form:

$$u(\mathbf{x}, t) = U(\mathbf{x}) \, e^{-i\omega t}.$$

Substitution of this solution into the wave equation yields the Helmholtz equation:

$$(\nabla^2 + k^2)U = 0$$

where $k = \omega/c = 2\pi/\lambda$ is called the wave number, and λ is the wave length of the light.

A particular solution to the Helmholtz equation for the case of a spherical point source at $\mathbf{x} = \mathbf{0}$ is given by[15]

$$U(\mathbf{x}) = \frac{e^{ik\|\mathbf{x}\|}}{\|\mathbf{x}\|}.$$

We now can consider what happens to light from this point source as it passes through an opening, or aperture, in an otherwise opaque screen, as illustrated in Figure 2.1. According to the Huygens-Fresnel principle, the observable light on the other side of the screen is the same as if a continuum of spherical sources had been placed at the opening. Only now, the intensity of these "secondary" spherical waves is modulated by a factor of the form $\cos(\mathbf{n} \cdot \mathbf{x})/i\lambda$, where $\mathbf{n}$ is the normal to the screen. We will not go into the details of the physics behind this modulation. Discussions can be found in [15, 17].

What is important is that the amplitude of the light at a point on the opposite side of the screen from the source, $U(\mathbf{x}_0)$, can be found from the amplitude at the aperture,

---

[15]Here $\|\mathbf{x}\| \doteq \sqrt{\mathbf{x} \cdot \mathbf{x}}$ where $\cdot$ is the dot product.

$U(\mathbf{x}_1)$, by adding the contributions from all of the secondary waves. This is performed with the superposition integral:

$$U(\mathbf{x}_0) = \int_{\mathbb{R}^2} \frac{1}{i\lambda} \frac{e^{ik\|\mathbf{x}_{01}\|}}{\|\mathbf{x}_{01}\|} \cos(\mathbf{n} \cdot \mathbf{x}_{01})\, U(\mathbf{x}_1)\, d\mathbf{x}_1, \qquad (2.37)$$

which is valid when the size of the aperture is small in comparison to the distance of the source from the aperture. The notation $\mathbf{x}_{01} = \mathbf{x}_0 - \mathbf{x}_1$ is shorthand for the relative position vector of point 0 with respect to point 1. The integral over the aperture area is written as being over the whole plane because $U(\mathbf{x}_1)$ has compact support (i.e., the aperture area) and $d\mathbf{x}_1 = dx_1 dy_1$ represents a differential area element.

Fresnel Approximation:

Assuming that the point $\mathbf{x}_0$ is far enough from the aperture, and its first and second components are small compared to its third, we have the following approximations:

$$\cos(\mathbf{n} \cdot \mathbf{x}_{01}) \approx 1 \qquad \text{and} \qquad \|\mathbf{x}_{01}\| \approx z \left[1 + \frac{1}{2z^2} \left((x_0 - x_1)^2 + (y_0 - y_1)^2 \right) \right], \quad (2.38)$$

where z is the component of $\mathbf{x}_{01}$ in the direction normal to the plane of the aperture.

The second approximation above is acceptable for the denominator of the integrand in (2.37) provided

$$z^2 \gg (x_0 - x_1)^2 + (y_0 - y_1)^2.$$

This just ensures that the Taylor series expansion truncated at the first term is an accurate approximation. However, it can lead to a poor result when substituted into the numerator of the integrand in (2.37) because $k = 2\pi/\lambda$, and the wavelength of visible light is extremely small (and hence its reciprocal is huge).

To ensure that $e^{ik|\mathbf{x}_{01}|}$ can be approximated well with the substitution (2.38), we assume that the stronger condition

$$z^3 \gg \frac{\pi}{4\lambda} [(x_0 - x_1)^2 + (y_0 - y_1)^2]_{max}^2 \qquad (2.39)$$

is satisfied.

All of these approximations (and particularly (2.39)) are called the *Fresnel approximations*. Assuming these are valid, the superposition integral (2.37) becomes

$$U(x_0, y_0) = \frac{e^{ikz}}{i\lambda z} \int_{\mathbb{R}^2} U(x_1, y_1) \exp\left(i\frac{k}{2z}[(x_0 - x_1)^2 + (y_0 - y_1)^2] \right) dx_1 dy_1.$$

This may further be expanded to show that

$$U(x_0, y_0) = \frac{e^{ikz}}{i\lambda z} \exp\left[i\frac{k}{2z}(x_0^2 + y_0^2) \right] \times$$

$$\int_{\mathbb{R}^2} U(x_1, y_1) \exp\left(i\frac{k}{2z}[x_1^2 + y_1^2] \right) \exp\left(-i\frac{2\pi}{\lambda z}[x_0 x_1 + y_0 y_1] \right) dx_1 dy_1.$$

Note that if the constants in front of the integral are disregarded, this is the 2D Fourier transform of $U(x_1, y_1)\exp\left(i\frac{k}{2z}[x_1^2 + y_1^2] \right)$ evaluated at $\boldsymbol{\omega} = \frac{2\pi}{\lambda z}\mathbf{x}_0$.

Fraunhofer Approximation:

If the more restrictive condition

$$z \gg \frac{k(x_1^2 + y_1^2)_{max}}{2} \qquad (2.40)$$

is valid, then $\exp\left(i\frac{k}{2z}[x_1^2 + y_1^2]\right) \approx 1$ and the superposition integral (2.37) is reduced further to

$$U(x_0, y_0) = \frac{e^{ikz}}{i\lambda z}\exp\left[i\frac{k}{2z}(x_0^2 + y_0^2)\right]\int_{\mathbb{R}^2} U(x_1, y_1)\exp\left(-i\frac{2\pi}{\lambda z}[x_0 x_1 + y_0 y_1]\right)dx_1 dy_1.$$

This shows that light traveling through an aperture is in fact performing a Fourier transform with modified phase and amplitude (due to the constant in front of the integral).

In practice the approximation in (2.40) is extremely limiting. However, the same effect can be observed by using lenses, and the physical phenomena discussed in this section can be used as an analog method for computing Fourier transforms (see e.g. [26]).

2.5 Discrete and Fast Fourier Transforms

As we have seen, many physical phenomena involve linear partial differential equations and/or wave propagation in continuous space. The Fourier transform is a natural tool for obtaining analytical solutions to describe these phenomena. However, when it comes time to perform calculations on a digital computer, discretization and truncation of these continuous descriptions of the world are required. In the subsections that follow, we present the general definitions and properties of the discrete Fourier transform (or DFT), show how it can be used to exactly reconstruct and convolve band-limited functions sampled at finite points, and sketch the basic ideas of the fast (discrete) Fourier transform (or FFT).

2.5.1 The Discrete Fourier Transform: Definitions and Properties

A digital computer cannot numerically perform the continuous integrations required in the definitions of the forward and inverse Fourier transform. However, a computer can calculate very well what is called the *discrete Fourier transform*, or DFT, of a function that is defined on a finite number, N, of equally spaced points. These points can be viewed as being on the unit circle, corresponding to the vertices of a regular N-gon. If the points are labeled x_0, ..., x_{N-1}, then the values of the discrete function, which are generally complex, are $f_j = f(x_j)$ for $j \in [0, N-1]$. In this context, the function f is then equivalent to the discrete set of values $\{f_1, ..., f_{N-1}\}$.

The DFT of such a function is defined as[16]

$$\hat{f}_k = \mathcal{F}_k(\{f_1, ..., f_{N-1}\}) = \frac{1}{N}\sum_{j=0}^{N-1} f_j e^{-i2\pi jk/N}. \qquad (2.41)$$

---

[16]As with the Fourier transform and Fourier series, a variety of normalizations can be found in the literature, the most common of which are $1/N$ and 1.

This can be viewed as an $N-$point approximation in the integral defining the Fourier coefficients $\hat{f}(k)$, where the sample points are $x_j = jL/N$, $f_j = f(x_j)$ and dx is approximated as $\Delta x = L/N$. Or it can simply be viewed as the definition of the Fourier transform of a function whose arguments take values in a discrete set.

As with the Fourier series and transform, the DFT inherits its most interesting properties from the exponential function. In this case, we have

$$\left[e^{-i2\pi/N}\right]^N = 1.$$

Using this property and the definition of the DFT it is easy to show that $\hat{f}_{k+N} = \hat{f}_k$. The following completeness relation can also be shown

$$\frac{1}{N} \sum_{k=0}^{N-1} e^{i2\pi(n-m)k/N} = \delta_{n,m} \qquad (2.42)$$

by observing that for a geometric sum with $r \neq 1$ and $|r| \leq 1$,

$$\sum_{k=0}^{N-1} r^k = \frac{1 - r^N}{1 - r}.$$

Setting $r = e^{i2\pi(n-m)/N}$ for $n \neq m$ has the property that $r^N = 1$, and so the numerator in the above equation is zero in this case. When $n = m$, all the exponentials in the sum reduce to the number 1, and so summing N times and dividing by N yields 1.

Equipped with (2.42), we observe that

$$\sum_{k=0}^{N-1} \hat{f}_k e^{i2\pi jk/N} = \sum_{k=0}^{N-1} \left(\frac{1}{N} \sum_{n=0}^{N-1} f_n e^{-i2\pi nk/N} \right) e^{i2\pi jk/N}$$

$$= \sum_{n=0}^{N-1} f_n \left(\frac{1}{N} \sum_{k=0}^{N-1} e^{i2\pi(j-n)k/N} \right)$$

$$= \sum_{n=0}^{N-1} f_n \delta_{j,n}$$

and thus the DFT inversion formula

$$f_j = \sum_{k=0}^{N-1} \hat{f}_k e^{i2\pi jk/N}. \qquad (2.43)$$

This is much like the Fourier series expansion of a band-limited function $f(x)$ sampled at $x = x_j$. Only in the case of the Fourier series, the sum contains $2B + 1$ terms (which is always an odd number), and in the case of the DFT, N is usually taken to be a power of 2 (which is always an even number). This discrepancy between the DFT and sampled band-limited Fourier series is rectified if the constraint $\hat{f}(B) = \hat{f}(-B)$ is imposed on the definition of a band-limited Fourier series. Then there are $N = 2B$ free parameters in both.

The convolution of two discrete functions f_j and g_j for $j = 0, ..., N-1$ is defined as

$$(f \star g)_j \doteq \frac{1}{N} \sum_{k=0}^{N-1} f_k g_{j-k},$$

and the corresponding convolution theorem is

$$\mathcal{F}(f \star g) = \hat{f}_j \hat{g}_j.$$

The proofs for Parseval's equality,

$$\frac{1}{N} \sum_{j=0}^{N-1} |f_j|^2 = \sum_{k=0}^{N-1} |\hat{f}_k|^2,$$

and the more general relationship

$$\frac{1}{N} \sum_{j=0}^{N-1} f_j \overline{g_j} = \sum_{k=0}^{N-1} \hat{f}_k \overline{\hat{g}_k},$$

follow in an analogous way to those for the continuous Fourier series.

2.5.2 The Fast Fourier Transform and Convolution

Direct evaluation of the DFT using (2.41) can be achieved with N multiplications and $N-1$ additions for *each* of the N values of k. Since for $k = 0$ and $j = 0$ the multiplications reduce to a product of a number with $e^0 = 1$, these need not be calculated explicitly, and a total of $(N-1)^2$ multiplications and $N(N-1)$ additions are sufficient. The values $e^{i2\pi jk/N}$ can be stored in advance and need not be calculated. Thus, $\mathcal{O}(N^2)$ arithmetic operations are sufficient to compute the DFT, and because of the symmetry of the definition, the inverse DFT is computed in this order as well.[17] Likewise, the discrete convolution of two functions with values at N points can be computed in $N-1$ additions and N multiplications for each value of k in the convolution product $(f \star g)_k$, and so N^2 multiplications and $N(N-1)$ additions are used for all values of k. Thus computation of the convolution product can also be performed in $\mathcal{O}(N^2)$ computations.

The *Fast Fourier transform* (or FFT) refers to a class of algorithms, building on the result of Cooley and Tukey from the 1960s [6], that allow the DFT to be computed in $\mathcal{O}(N \log_2 N)$ computations. By the convolution theorem, this means that convolutions can be computed in this order as well. For large values of N, the relative amount of savings is dramatic, since the rate at which the function $N/\log N$ increases continues to increase with N.

The core of this somewhat remarkable result actually goes back two hundred years! It is based on the fact that

$$\left(e^{-i2\pi jk/N} \right)^N = 1.$$

Gauss[18] is purported to have used DFT and FFT ideas, and some reports indicate that others may have even preceded him [24]. In the modern era, the result is attributed to the 1965 paper of Cooley and Tukey, though many elements of the FFT had been known previously [24, 7]. Over the past few decades, a number of different FFT algorithms have been developed, but they all have the same $\mathcal{O}(N \log N)$ computational complexity

---

[17]See Appendix A for the meaning of the $\mathcal{O}(\cdot)$ notation.

[18]Carl Friedrich Gauss (1777–1855), one of the greatest mathematicians of all time, was responsible for the divergence theorem, and many results in number theory and differential geometry. His unpublished work on the DFT and FFT was not collected until after his death [16].

(though the notation $\mathcal{O}(\cdot)$ can hide quite significant differences in the actual time the algorithms take to execute). For an examination of how the FFT became a ubiquitous computational tool see [8].

The core idea of the FFT is simple enough. The number N is restricted to be of the form $N = K^m$ for positive integers K and m. Then we can rewrite

$$\hat{f}_k = \frac{1}{N} \sum_{j=0}^{K^m-1} f_j e^{-i2\pi jk/K^m} = \frac{1}{N} \sum_{l=0}^{K-1} \left[\sum_{j=0}^{K^{m-1}-1} f_{K \cdot j + l} (e^{-i2\pi/K^{m-1}})^{jk} \right] (e^{i2\pi/K^m})^{lk}.$$

Each of the inner summations for each of the K values of l can be viewed as independent Fourier transforms for sequences with K^{m-1} points, and thus can be computed for all values of k in order $(N/K)^2$ time using the DFT. To compute all of the inner sums and to combine them all together to generate $\hat{f}_k$ for all values of k requires on the order of $(N/K)^2 \cdot K = N^2/K$ computations. Thus, splitting the sum once reduces the computation time by a factor of K. In fact, we can keep doing this to each of the remaining smaller sequences, until we get K^m sequences each with $K^0 = 1$ point. In this case there are $m = \log_K N$ levels to the recursion, each requiring $\mathcal{O}(N)$ multiplications and additions, so the total computational requirements are $\mathcal{O}(N \log_K N)$. Usually the most convenient choice is $K = 2$, and the trickiest part of the implementation of an FFT algorithm is the bookkeeping. Since a number of good books and FFT software packages are available (see e.g., [4, 11, 18, 28, 36, 37, 38]) we will not discuss the particular methods in any detail. We note also that in recent year, efforts devoted to FFTs for nonuniformly spaced data have been investigated [1].

2.5.3 Multi-Dimensional FFTs

The same basic ideas regarding separation of coordinates that were used in the multi-dimensional Fourier transform hold for Fourier series and FFTs as well. From a computational perspective, this means that the $\mathcal{O}(N \log N)$ performance of the singe-degree-of-freedom FFT can be generalized to the multi-dimensional case as $\mathcal{O}(N^n \log N^n) = \mathcal{O}(N^n \log N)$ where n is the dimension of the space (treated as a constant) and samples are selected on an $N \times N \times \cdots \times N$ Cartesian grid. It is worth noting that separability does not depend on the orthogonality of the Cartesian coordinate axes. For example, in the field of crystallography, FFTs are commonly computed for data presented in nonorthogonal coordinates. Moreover, so-called nonuniform FFTs developed for multi-dimensional data that does not fall onto any regular grid has become a popular topic in recent years.

2.5.4 Sampling, Aliasing, and Filtering

It is well known in many engineering disciplines that an arbitrary signal (function) cannot be reconstructed from a finite number of samples. Any attempt to do so leads to a phenomenon known as *aliasing*. One physical example of aliasing is when the wheel of an automobile in a movie appear to be turning slowly clockwise when the wheels are actually rotating rapidly counterclockwise. This is because the sampling rate of the movie camera is not fast enough to capture the true motion of the wheel. In general aliasing results because the amount of information in an arbitrary signal is infinite, and cannot be captured with a finite amount of collected data.

However, a signal (or function) on the circle with Fourier coefficients $\hat{f}(k)$ for $k \in \mathbb{Z}$ can be *band-pass filtered* to band-limit B by essentially throwing away (or zeroing) all Fourier coefficients for which $|k| > B$. In this way, $2B+1$ Fourier coefficients completely define the signal.

Since the Fourier coefficients enter linearly in the expansion of a function in a Fourier series, we can write at each sample time

$$f_B(\theta(t_k)) = \sum_{j=-B}^{B} \hat{f}_j e^{ij\theta(t_k)}$$

for each of $2B+1$ sample values of t_k. In principle, all that is then required to calculate the $2B+1$ independent Fourier coefficients from $2B+1$ samples of the function is to invert a $(2B+1) \times (2B+1)$ dimensional matrix with k, j^{th} entry $e^{ij\theta(t_k)}$ (though the computation is not actually preformed as a matrix inversion). However, in practice two issues arise: (1) real filters only approximate band-pass filters; and (2) measurements always contain some error. Hence it is safest to sample at a rate above the minimal rate to obtain a more robust estimate of a filtered signal.

The DFT can be viewed as a real-space-sampled version of a band-limited Fourier series with the additional constraint that

$$\hat{f}(B) = \overline{\hat{f}(-B)}.$$

This constraint is imposed because a DFT is typically sampled at $2B$ points where B is a power of 2, whereas there are $2B+1$ degrees of freedom in a Fourier series with bandlimit B.

Sampling and Reconstruction of Functions on the Line

Similar ideas apply for functions on the line. In this context a band-limited function is one whose Fourier transform vanishes outside of a finite range of frequencies. The formal criteria for sampling and reconstruction of functions on the line are often collectively referred to as the *Shannon sampling theorem*.[19] If samples are taken at no less than twice the band-limit frequency, the band-limited function can be exactly reconstruct. In the case of the line, this is referred to as the *Nyquist criterion*[20]

Clear descriptions of the Shannon sampling theorem can be found in [25, 30]. Our proof proceeds along similar lines as follows. Assume that $f \in \mathcal{L}^2(\mathbb{R})$ is a band-limited function with band-limit Ω. That is, $\hat{f}(\omega) = 0$ for $|\omega| > \Omega$. We will view f as a function of time, t. The product of $f(t)$ and an infinite sum of equally spaced shifted delta functions (called a periodic pulse train) yields

$$f_s(t) = f(t) \sum_{n=-\infty}^{\infty} \delta(t - nT) = \sum_{n=-\infty}^{\infty} f(nT)\,\delta(t - nT)$$

where T is the sample interval. Since the impulse train is T-periodic, it can be viewed as a single delta function on the circle of circumference T, and can be expanded in a Fourier series as

---

[19]First proved in the context of communication theory by Shannon (see [35] pp. 160–162)

[20]Nyquist reported on the importance of this sample frequency in the context of the telegraph [29].

$$\sum_{n=-\infty}^{\infty} \delta(t - nT) = \sum_{n=-\infty}^{\infty} e^{2\pi nit/T}.$$

Taking the Fourier transform of both sides of this equality yields

$$\mathcal{F}\left(\sum_{n=-\infty}^{\infty} \delta(t - nT)\right) = \sum_{n=-\infty}^{\infty} \mathcal{F}(e^{2\pi nit/T}) = 2\pi \sum_{n=-\infty}^{\infty} \delta(\omega - 2\pi n/T).$$

Since the product of functions in the time domain corresponds to a convolution in the frequency domain multiplied by a factor of $1/2\pi$,

$$\hat{f}_s(\omega) = \sum_{n=-\infty}^{\infty} \hat{f}(\omega) * \delta(\omega - 2\pi n/T) = \sum_{n=-\infty}^{\infty} \hat{f}(\omega - 2\pi n/T).$$

This is a periodic function with period T, and hence can be expanded in a Fourier series. As long as $2\pi/T \geq 2\Omega$, $\hat{f}_s(\omega)$ (which is $\hat{f}(\omega)$ "wrapped around the circle") will be identical to $\hat{f}(\omega)$ for $|\omega| < \Omega$. If $\pi/T < \Omega$, the wrapped version of the function will overlap itself. This phenomenon, called aliasing, results in the destruction of some information in the original function, and makes it impossible to perform an exact reconstruction.

Since $\hat{f}_s(\omega)$ is $2\pi/T$-periodic, and is identical to $\hat{f}(\omega)$ for $|\omega| < \Omega$, it can be expanded in a Fourier series as

$$\hat{f}_s(\omega) = \sum_{n=-\infty}^{\infty} \left(\frac{T}{2\pi} \int_{-\Omega}^{\Omega} \hat{f}(\sigma) e^{-inT\sigma} \, d\sigma\right) e^{inT\omega} \tag{2.44}$$

where σ is the variable of integration.

For the band-limited function $f(t)$, the Fourier inversion formula is written as

$$f(t) = \frac{1}{2\pi} \int_{-\Omega}^{\Omega} \hat{f}(\omega) e^{i\omega t} \, d\omega. \tag{2.45}$$

It then follows that the Fourier coefficients (quantities in parentheses) in (2.44) are equal to $T \cdot f(-nT)$ for each $n \in \mathbb{Z}$. Hence we can rewrite (2.45) as

$$f(t) = \frac{1}{2\pi} \int_{-\Omega}^{\Omega} \left[\sum_{n=-\infty}^{\infty} \frac{\pi}{\Omega} f(-nT) e^{inT\omega}\right] e^{i\omega t} \, d\omega.$$

Changing the order of integration and summation, and making the substitution $n \to -n$, results in Shannon's reconstruction:

$$f(t) = \sum_{n=-\infty}^{\infty} f(nT) \frac{\sin \Omega(t - nT)}{\Omega(t - nT)}. \tag{2.46}$$

In analogy with the way that the band-limited approximation of the delta function on the circle (or, equivalently, a 2π-periodic train of delta functions on the real line) can be written in terms of the Dirichlet kernel, the band-limited approximation of the non-periodic delta function can be written in terms of the (unnormalized) *sinc function* as

$$\delta_\Omega(t) \doteq \frac{1}{2\pi} \int_{-\Omega}^{\Omega} e^{i\omega t} \, d\omega = \frac{\Omega}{\pi} \text{sinc}(\Omega t)$$

where[21]

$$\text{sinc}(x) \doteq \frac{\sin x}{x}.$$

When $f(t)$ is sampled at the Nyquist rate $(T = \pi/\Omega)$ then (2.46) becomes

$$f(t) = \sum_{n=-\infty}^{\infty} f(n\pi/\Omega) \frac{\sin(\Omega t - n\pi)}{(\Omega t - n\pi)} = \sum_{n=-\infty}^{\infty} f(n\pi/\Omega) \operatorname{sinc}(\Omega t - n\pi).$$

Often this is written in terms of the band-limit B where $\Omega = 2\pi B$, in which case the use of $\underline{\text{sinc}}(x)$ becomes convenient.

In addition to communication and information theory, the sampling theorem is of general usefulness in linear systems theory and control (see, e.g., [34]). In the applications we will explore, it suffices to view functions on the line with finite support as functions on the circle provided the function is scaled properly, i.e., so that a sufficient portion of the circle maps to function values that are zero.

Commutativity of Sampling and Convolution

We now make one final note regarding the connection between classical continuous and discrete Fourier analysis. Let us assume that two band-limited functions on the unit circle $(L = 2\pi)$ with the same band limit, B, are sampled at the points $\theta_k = 2\pi k/N$ where $N \geq 2B + 1$. We now prove a point that is often used, but not always stated, in texts on Fourier analysis and numerical methods. Namely, the discrete convolution of evenly sampled band-limited functions on the circle yields exactly the same result as first convolving these functions, and then sampling.

To begin, consider band-limited functions of the form

$$f(x) = \sum_{k=-B}^{B} \hat{f}(k) e^{ikx} \tag{2.47}$$

where

$$\hat{f}(n) = \frac{1}{2\pi} \int_{0}^{2\pi} f(x) e^{-inx} \, dx.$$

As we discussed in the previous subsection, if we sample at $N \geq 2B + 1$ points, then enough information exists to reconstruct the function from its samples. We denote the set of sampled values as $S[f]$, with $S[f]_n$ denoting the single sample $f(2\pi n/N)$.

The convolution of two band-limited functions of the form of (2.47) is

$$(f_1 * f_2)(x) = \sum_{k=-B}^{B} \hat{f}_1(k) \hat{f}_2(k) e^{ikx}.$$

Sampling at points $x = 2\pi n/N$ for $n \in [0, N-1]$ yields the set of sampled values $S[f_1 * f_2]$.

Performing the convolution of the two sets of discrete values $S[f_1]$ and $S[f_2]$ yields

---

[21]The normalized sinc function is defined as $\underline{\text{sinc}}(x) \doteq \text{sinc}(\pi x)$, and hence has a factor of π both inside the sin in the numerator and in the denominator. The unnormalized sinc function used here does not have the factor of π in either location.

$$
(S[f_1] \star S[f_2])_j = \frac{1}{N} \sum_{k=0}^{N-1} S[f_1]_k S[f_2]_{j-k}
$$

$$
= \frac{1}{N} \sum_{k=0}^{N-1} \left(\sum_{l=-B}^{B} \hat{f}_1(l)\, e^{2\pi i l k/N} \right) \left(\sum_{m=-B}^{B} \hat{f}_2(m)\, e^{2\pi i m(j-k)/N} \right)
$$

$$
= \frac{1}{N} \sum_{l=-B}^{B} \sum_{m=-B}^{B} \hat{f}_1(l)\hat{f}_2(m) \left(e^{i2\pi m j/N} \sum_{k=0}^{N-1} e^{2\pi i k(l-m)/N} \right)
$$

$$
= \sum_{l=-B}^{B} \hat{f}_1(l)\hat{f}_2(l)\, e^{2\pi i j l/N}
$$

$$
= S[f_1 * f_2]_j.
$$

This follows from (2.42).

Hence, we write

$$
S[f_1 * f_2] = S[f_1] \star S[f_2]. \tag{2.48}
$$

2.6 Summary and Recent Developments

Classical Fourier analysis is ubiquitous in engineering and science. Its applications range from optics and X-ray crystallography, to the solution of linear partial differential equations with constant coefficients, and the theory of sound and vibration. We know of no discipline in the physical sciences and engineering which does not make use of either Fourier series, the Fourier transform, or the DFT/FFT. For a complete overview of applications and the history of Fourier analysis see [27].

Recent developments in classical Fourier analysis include two areas. The first is FFTs for non-equally-spaced samples, as described in [10, 13, 19, 32, 33] and references therein. The second is new kinds of FFTs with even faster performance in the case of sparse data (i.e., when many of the f_j's are zero) [21], which is related to a growing area of research called *compressed sensing* [9].

Classical Fourier analysis and the novel algorithms alluded to above are built on the arithmetic of scalar addition and multiplication (which are commutative operations). A relatively modern area of mathematics referred to as *noncommutative harmonic analysis* is unknown to the vast majority of engineers and computer scientists. In fact it is unknown to most scientists outside of theoretical physics.

In subsequent chapters we explain the basic theory behind noncommutative harmonic analysis in a way which we hope is much more accessible to various engineering and scientific communities than the pure mathematics books on the subject. We then illustrate a number of concrete applications and numerical implementations from our own research in the second half of this book.

References

1. Bagchi, S., Mitra, S., *The Nonuniform Discrete Fourier Transform and Its Applications in Signal Processing*, Kluwer Academic Publishers, Boston, 1999.
2. Benedetto, J.J., *Harmonic Analysis and Applications*, CRC Press, Boca Raton, 1997.
3. Bracewell, R.N., *The Fourier Transform and Its Applications*, 2nd ed., McGraw-Hill, New York, 1986.
4. Burrus, C.S., Parks, T.W., *DFT/FFT and Convolution Algorithms*, John Wiley and Sons, New York, 1985.
5. Chirikjian, G.S., *Stochastic Models, Information Theory, and Lie Groups, Vol 1.*, Birkhäuser, Boston, 2009.
6. Cooley, J.W., Tukey, J. "An Algorithm for the Machine Calculation of Complex Fourier Series," *Mathematics of Computation*, 19(90): 297 – 301, April 1965.
7. Cooley, J.W., "The Re-Discovery of the Fast Fourier Transform Algorithm," *Mikrochimica Acta III*, 93(1): 33 – 45, 1987.
8. Cooley, J.W., "How the FFT Gained Acceptance," *Proceedings of the ACM Conference on the History of Scientific and Numeric Computation*, pp. 133 – 140, Princeton, NJ, May 13 – 15, 1987.
9. Donoho, D.L., "Compressed Sensing," *IEEE Transactions on Information Theory*, 52(4): 1289 – 1306, 2006.
10. Dutt, A., Rokhlin, V., "Fast Fourier Transforms for Nonequispaced Data," *SIAM Journal on Scientific Computing*, 14(6): 1368 - 1393, Nov. 1993.
11. Elliott, D.F., Rao, K.R., *Fast Transforms: Algorithms, Analyses, Applications*, Academic Press, New York, London, 1982.
12. Farlow, S.J., *Partial Differential Equations for Scientists and Engineers*, Dover, New York, 1993.
13. Fessler, J. A., Sutton, B. P., "Nonuniform Fast Fourier Transforms Using Min-Max Interpolation," *IEEE Transactions on Signal Processing*, 51(2): 560 – 574, 2003.
14. Fourier, J.B.J., *Théorie Analytique de la Chaleur*, F. Didot, Paris, 1822.
15. Fowles, G.R., *Introduction to Modern Optics*, 2nd ed., Dover, New York, 1975.
16. Gauß, C.F., "Nachlass, Theoria Interpolationis Methodo Nova Tractata," in *Carl Friedrich Gauß Werke, Band 3*, Königliche Gesellschaft der Wissenschaften: Göttingen, pp. 265 – 330, 1866.
17. Goodman, J.W., *Introduction to Fourier Optics*, McGraw-Hill, New York, 1968.
18. Gray, R.M., Goodman, J.W., *Fourier Transforms: An Introduction for Engineers*, Kluwer Academic Publishers, Boston, 1995.
19. Greengard, L., Lee, J. Y., "Accelerating the Nonuniform Fast Fourier Transform," *SIAM Review*, 46(3): 443 – 454, 2004.
20. Groetsch, C. W., *The Theory of Tikhonov Regularization for Fredholm Equations of the First Kind*, Pitman, Boston, 1984.

21. Hassanieh, H., Indyk, P., Katabi, D., Price, E. "Simple and Practical Algorithm for Sparse Fourier Transform," in *Proceedings of the Twenty-Third Annual ACM-SIAM Symposium on Discrete Algorithms*, pp. 1183–1194, Jan. 2012.

22. Hayt, W.H. Jr., *Engineering Electromagnetics*, 5th ed., McGraw-Hill Book Company, New York, 1989.

23. Heaviside, O., "On Operators in Physical Mathematics," *Proceedings of the Royal Society*, 52: 504–529, 1893 and 54: 105–143, 1894.

24. Heideman, M.T., Johnson, D.H., Burrus, C.S., "Gauss and the History of the Fast Fourier Transform," *Archive for History of Exact Sciences*, 34(3): 265–277, 1985.

25. Hsu, H.P., *Applied Fourier Analysis*, Harcourt Brace Jovanovich College Outline Series, New York, 1984.

26. Karim, M.A., Awwal, A.A.S., *Optical Computing, An Introduction*, Wiley, New York, 1992.

27. Körner, T.W., *Fourier Analysis*, Cambridge University Press, 1988 (reprinted 1993).

28. Nussbaumer, H.J., *Fast Fourier Transform and Convolution Algorithms*, Springer-Verlag, New York, 1982.

29. Nyquist, H., "Certain Topics in Telegraph Transmission Theory," *AIEE Transactions*, 47(2): 617–644, April 1928.

30. Oppenheim, A.V., Willsky, A.S., *Signals and Systems*, Prentice-Hall, Inc., Englewood Cliffs, NJ, 1983.

31. Pinsky, M.A., *Introduction to Partial Differential Equations with Applications*, McGraw-Hill Book Company, New York, 1984.

32. Potts, D., Steidl G., Tasche, M., "Fast Fourier Transforms for Nonequispaced Data: A Tutorial," in *Modern Sampling Theory: Mathematics and Application*, J. J. B. P. Ferreira, ed., pp. 253-274, Birkhäuser, Boston, MA, 2000.

33. Press, W. H., Rybicki, G. B., "Fast Algorithm for Spectral Analysis of Unevenly Sampled Data," *Astrophysical Journal*, 338: 227-280, March 1989.

34. Rugh, W.J., et al., *Johns Hopkins University Signals, Systems, and Control Demonstrations*, http://www.jhu.edu/~signals/, 1996-present.

35. Sloane, N.J.A., Wyner, A.D., eds., *Claude Elwood Shannon : Collected Papers*, IEEE Press, New York, 1993.

36. Tolimieri, R., An, M., Lu, C., *Algorithms for Discrete Fourier Transform and Convolution*, 2nd ed. Springer-Verlag, New York, 1997.

37. Van Loan, C., *Computational Frameworks for the Fast Fourier Transform*, SIAM, Philadelphia, 1992.

38. Walker, J.S., *Fast Fourier Transforms*, 2nd ed., CRC Press, Boca Raton, FL, 1996.

3

Sturm-Liouville Expansions, Discrete Polynomial Transforms, and Wavelets

This chapter reviews some generalizations of Fourier analysis. We begin with Sturm-Liouville theory, which provides a tool for constructing so-called "special functions" that are orthogonal with respect to a given weight function on an interval. We also examine modern orthogonal polynomials and wavelet expansions.

3.1 Sturm-Liouville Theory

Consider the closed interval $a \le x \le b$ and real-valued differentiable functions $s(x)$, $w(x)$, and $q(x)$ defined on this interval, where $s(x) > 0$ and $w(x) > 0$ on the open interval $a < x < b$. The values of these functions may be positive at $x = a$ and $x = b$, or they may be zero. It has been known since the mid-nineteenth century that the solutions of differential equations of the form

$$\frac{d}{dx}\left(s(x)\frac{dy}{dx}\right) + (\lambda w(x) - q(x))y = 0 \qquad (3.1)$$

for different *eigenvalues*, λ, and appropriate boundary conditions produce sets of orthogonal *eigenfunctions* $y(x)$. Each eigenfunction-eigenvalue pair is denoted as $(y_i(x), \lambda_i)$ for $i = 1, 2, \dots$.

3.1.1 Orthogonality of Eigenfunctions

To observe the orthogonality of solutions, multiply (3.1) evaluated at the solution $(y_i(x), \lambda_i)$ by the eigenfunction $y_j(x)$ and then integrate:

$$\int_a^b y_j\,(sy_i')'dx + \int_a^b y_j\,(\lambda_i w - q)\,y_i\,dx = 0.$$

The notation y' is used in this context to denote dy/dx. The first integral in the above equation is written using integration by parts as

$$\int_a^b y_j\,(sy_i')'dx = y_j\,s\,y_i'\Big|_a^b - \int_a^b y_j'\,s\,y_i'dx.$$

Hence,

$$y_j \, s \, y_i' \Big|_a^b - \int_a^b y_j' s \, y_i' dx + \int_a^b y_j \left(\lambda_i w - q \right) y_i \, dx = 0.$$

If we were to repeat the same procedure with the roles of (y_i, λ_i) and (y_j, λ_j) reversed, then

$$y_i \, s \, y_j' \Big|_a^b - \int_a^b y_i' s \, y_j' dx + \int_a^b y_i \left(\lambda_j w - q \right) y_j \, dx = 0.$$

Subtracting the second equation from the first, we see that if

$$y_i \, s \, y_j' \Big|_a^b - y_j \, s \, y_i' \Big|_a^b = 0 \qquad (3.2)$$

then

$$\left(\lambda_i - \lambda_j \right) \int_a^b y_i(x) \, y_j(x) \, w(x) \, dx = 0. \qquad (3.3)$$

In other words, y_i and y_j are orthogonal with respect to the weight function w if $\lambda_i \neq \lambda_j$. If $s(a){\cdot}w(a) > 0$ and $s(b){\cdot}w(b) > 0$, then this is called a *regular* Sturm-Liouville problem. Otherwise it is called a *singular* Sturm-Liouville problem.[1]

Ways for $y_i s y_j' |_a^b - y_j s y_i' |_a^b = 0$ to be satisfied include the following boundary conditions for (3.1):

$$y'(a) \cos \alpha - y(a) \sin \alpha = 0 \quad \text{and} \quad y'(b) \cos \beta - y(b) \sin \beta = 0$$

for any $\alpha, \beta \in [0, 2\pi]$ (called *separable* boundary conditions);

$$y(a) = y(b) \quad \text{and} \quad y'(a) = y'(b) \quad \text{when} \quad s(a) = s(b)$$

(called *periodic* boundary conditions); and for a singular Sturm-Liouville problem with $s(a) = s(b) = 0$ the condition (3.2) is automatically satisfied.

One of the key features of Sturm-Liouville theory is that any "nice" function $f(x)$ defined on $a \leq x \leq b$ that satisfies the boundary conditions in (3.2) can be expanded in a series of eigenfunctions [12]:

$$f(x) = \sum_{n=0}^{\infty} c_n y_n(x) \qquad (3.4)$$

where

$$c_n = \frac{\int_a^b f(x) \, y_n(x) \, w(x) \, dx}{\int_a^b y_n^2(x) \, w(x) \, dx}.$$

In this context we say $f(x)$ is band-limited with band-limit B if the sum in (3.4) is truncated at $n = B - 1$. The existence and uniqueness of the eigenfunctions up to an arbitrary scale factor is addressed in a number of works (see, e.g., [70, 91]).

What is sometimes not stated is the relationship between different normalizations. Using the notation

$$(y_i, y_j)_w = \int_a^b y_i(x) \, y_j(x) \, w(x) \, dx,$$

orthogonality is stated as

$$(y_i, y_j)_w = C_i \delta_{ij}$$

---

[1]Named after Jacques Charles François Sturm (1803-1855) and Joseph Liouville (1809-1882)

where C_i is a constant. We can define orthonormal eigenfunctions as $\tilde{y}_i = y_i/\sqrt{C_i}$ so that

$$(\tilde{y}_i, \tilde{y}_j)_w = \delta_{ij}.$$

However, when a and b are finite, it is common for the first eigenfunction, y_0, to be constant, and to be chosen as $y_0 = 1$. It is desirable to retain this normalization of the first eigenfunction. This can be achieved together with orthonormality if we normalize the weighting function as:

$$\tilde{w}(x) = w(x)/ \int_a^b w(x)\, dx$$

(the functions s and q must be rescaled as well) and then normalize the eigenfunctions a second time as

$$\tilde{\tilde{y}}_i = \tilde{y}_i \sqrt{C_0} = y_i \sqrt{C_0/C_i}.$$

In this way, when $y_0 = 1$ so is $\tilde{\tilde{y}}_0$. Since by definition in the case when $y_0 = 1$ we have

$$(1, 1)_w = \int_a^b w(x)\, dx = C_0,$$

it follows that

$$(\tilde{\tilde{y}}_i, \tilde{\tilde{y}}_j)_{\tilde{w}} = C_0(\tilde{y}_i, \tilde{y}_j)_{\tilde{w}} = (\tilde{y}_i, \tilde{y}_j)_w = \delta_{ij}.$$

Hence the two conditions of orthonormality of all the eigenfunctions and normalization of the first eigenfunction to unity are simultaneously achievable.

Having said this, we will use whatever normalization is the most convenient in any given context, with the knowledge that the eigenfunctions *can* be normalized in the variety of ways described above.

Clearly, the Fourier series expansion of functions on the circle is one example of a solution to a Sturm-Liouville problem with $s(x) = w(x) = 1$, $q(x) = 0$, $\lambda = n^2$ for $n \in \mathbb{Z}$, $[a, b] = [0, 2\pi]$, and periodic boundary conditions.

3.1.2 Completeness of Eigenfunctions

Let

$$f_B(x) = \sum_{n=0}^{B-1} c_n y_n(x) \tag{3.5}$$

be a band-limited approximation of a real-valued function $f(x)$ with eigenfunction expansion (3.4). The square of the norm of a function in this context is defined as

$$\|f\|^2 \doteq \int_a^b [f(x)]^2 w(x)\, dx.$$

The sequence of functions $f_1, f_2, ..., f_B$ is called *mean-square convergent* if

$$\lim_{B \to \infty} \|f_B - f\|^2 = 0. \tag{3.6}$$

The set of functions $\{y_n(x)\}$ is called *complete in a set of functions S* if for any $f \in S$ it is possible to approximate f to arbitrary precision in the mean-squared sense with a band-limited series of the form (3.5). That is, $\{y_n(x)\}$ is complete in S if

$$\|f_B - f\| < \epsilon.$$

for any $\epsilon > 0$, $f \in S$, and sufficiently large B. Here $\| \cdot \|$ is shorthand for $\| \cdot \|_2$.

When $\{y_n(x)\}$ is orthonormal and complete in $\mathcal{L}^2([a,b], w(x)\,dx)$, expanding

$$\|f_B - f\|^2 = \|f_B\|^2 + \|f\|^2 - 2\int_a^b f(x)\,f_B(x)\,w(x)\,dx,$$

and substituting (3.5) into the above expanded form of (3.6) yields *Bessel's inequality*:

$$\sum_{n=0}^{B-1} c_n^2 \le \|f\|^2.$$

As $B \to \infty$, this becomes *Parseval's equality*:

$$\sum_{n=0}^{\infty} c_n^2 = \|f\|^2.$$

Parseval's equality holds for eigenfunctions of Sturm-Liouville equations (i.e., the eigenfunctions form a complete orthonormal set in $\mathcal{L}^2([a,b], w(x)\,dx)$). To our knowledge, only case-by-case proofs of completeness were known for the century following the development of the theory by Sturm and Liouville.

A paper by Birkoff and Rota [12] from the latter half of the twentieth century provides a general proof of the completeness of eigenfunctions of Sturm-Liouville equations. The essential point of this proof is that given one orthonormal basis $\{\phi_n\}$ for $\mathcal{L}^2([a,b], w(x)\,dx)$, then another infinite sequence of orthonormal functions $\{\psi_n\}$ for $n = 0, 1, 2, \dots$ will also be a basis if

$$\sum_{n=0}^{\infty} \|\psi_n - \phi_n\|^2 < \infty.$$

Hence, starting with a known orthonormal basis (such as the trigonometric functions $\{\sqrt{2/\pi}\cos nx\}$ for $[a,b] = [0,\pi]$ and $w(x) = 1$) and relating the asymptotic behavior of other eigenfunctions on the same interval to this basis provides a way to prove completeness.

The interested reader can see [12] for details. Throughout the remainder of the text, the completeness of eigenfunctions is assumed.

If $\{y_i(x)\,|\,i = 0, 1, 2, \dots\}$ is a set of orthonormal functions on the interval $[a,b]$ with respect to the weight $w(x)$, this means that

$$\int_a^b y_n(x)\,\overline{y_m(x)}\,w(x)\,dx = \delta_{mn}.$$

It follows from this that $y_i \in \mathcal{L}^2([a,b], w\,dx)$. If in addition the set $\{y_i(x)\,|\,i = 0, 1, 2, \dots\}$ is *complete* in $\mathcal{L}^2([a,b], w(x)\,dx)$ then this means that any function $f \in \mathcal{L}^2([a,b], w(x)\,dx)$ can be approximated as

$$f_N(x) = \sum_{n=0}^{N} a_n\,y_n(x) \tag{3.7}$$

and as $N \to \infty$ the $\mathcal{L}^2$ error between f_N and f decays to zero. In this book, as in many applications in engineering and science, the distinction between f_∞ and f can be neglected and we say $f_\infty = f$, where here equality is not a pointwise statement, but rather is interpreted in the weaker sense that[2]

$$\lim_{N\to\infty} \int_a^b |f_N(x) - f(x)|^2\, w(x)\, dx = 0.$$

From orthogonality, it follows that

$$a_n = \int_a^b f(z)\, \overline{y_n(z)}\, w(z)\, dz. \tag{3.8}$$

Then substituting (3.8) into (3.7) gives

$$f(x) = \int_a^b f(z)\, w(z) \left\{ \sum_{n=0}^{\infty} y_n(x)\, \overline{y_n(z)} \right\} dz.$$

This is equivalent to saying that[3]

$$\boxed{\sum_{n=0}^{\infty} y_n(x)\overline{y_n(z)} = \frac{\delta(x-z)}{w(x)}.} \tag{3.9}$$

This is called a *completeness relation*. . It is like an orthogonality relation with the domain of the function $\tilde{y}_x(n) \doteq y_n(x)$ now being the nonnegative integers and the space of functions now being indexed by the variables $x, z \in (a, b)$. We have already seen completeness relations in the context of classical Fourier expansions, and will encounter them in many different forms throughout the remainder of the book.

3.1.3 Generating Functions and Rodrigues Formulas

As we will see with the examples in Sections 3.2–3.4, the existence of simple closed-form *generating functions*, $g(x, t)$, that can be expanded in the form

$$g(x, t) = \sum_{n=0}^{\infty} \alpha_n y_n(x) t^n \tag{3.10}$$

can be quite useful in computing integrals involving orthonormal functions $y_n(x)$. Usually α_n is taken to be either 1 or $1/n!$, and possibly including a factor of $(-1)^n$. The existence of a simple generating function together with orthonormality of $\{y_n(x)\}$ means that

$$\int_a^b g(x, r) g(x, s) w(x)\, dx = \sum_{n=0}^{\infty} (rs)^n \alpha_n^2. \tag{3.11}$$

This equation can be used in the opposite direction as well. That is, suppose that a given set of functions $\{y_n(x)\}$ satisfies (3.10). If we do not know in advance that they

---

[2] If additional conditions are imposed on f, such as continuity, then this weak sense of equality leads to the stronger sense of pointwise equality.

[3] Purists in some areas of mathematics and probability theory do not believe in using Dirac delta functions, as they lead to some contradictions in logic. For them, this statement will be interpreted as nonsense. But in the applications considered throughout this book, completeness relations of this form will arise in multiple different contexts, and will be quite useful.

are solutions to a Sturm-Liouville problem, and hence do not have prior knowledge about their orthonormality, we can assess this using the above expression. Namely, if the integral on the left of (3.11) is computed and given the name $f(r,s)$, and if $f(r,s)$ is expanded in a bivariate Taylor series in r and s that matches the form on the right side for all values of r and s and some α_n, then a localization argument can be used to establish the orthonormality of $\{y_n(x)\}$ independent of Sturm-Liouville theory. Generating functions can also be used in some instances to derive expressions involving integrals of products of more than two functions in the set $\{y_n(x)\}$, or products of these functions with others that are not in the set.

In addition to producing orthogonality relations and other integrals, generating functions are also an important tool for producing recurrence relations. These are very important when developing "fast" numerical transforms.

Another related concept is that of a *Rodrigues formula*. By taking n partial derivatives of (3.10) with respect to t we find that

$$\left.\frac{\partial^n g}{\partial t^n}\right|_{t=0} = \alpha_n\, n!\, y_n(x).$$

Often the expression on the left can be rearranged to give a useful way to evaluate $y_n(x)$, in which case the above equation is called a Rodrigues formula.

Sections 3.2 – 3.4 provide numerous examples of these concepts. But first, we examine properties of orthogonal functions sampled at discrete values of their argument.

3.1.4 Sampling Band-Limited Sturm-Liouville Expansions

Let $y_k(x)$ for $k = 1, 2, \dots$ be normalized eigenfunctions that solve a Sturm-Liouville problem. Rewriting (3.5), recall that the function

$$f(x) = \sum_{k=0}^{B-1} c_k y_k(x) \tag{3.12}$$

is said to have band-limit B. If any such band-limited function is sampled at B distinct points $a < x_0 < x_1 < \cdots < x_{B-1} < b$, then (3.12) implies

$$f_j = \sum_{k=0}^{B-1} Y_{jk} c_k$$

where $f_j = f(x_j)$ and $Y_{jk} = y_k(x_j)$ are the entries of a $B \times B$ matrix Y. Assuming that the points $\{x_j\}$ can be chosen so that $\det(Y) \neq 0$ (where Y is the matrix with elements Y_{jk}), and the matrix Y^{-1} can be calculated exactly, then it is possible to calculate

$$c_j = \int_a^b f(x)\, y_j(x)\, w(x)\, dx = \sum_{k=0}^{B-1} Y_{jk}^{-1} f_k. \tag{3.13}$$

That is, when $f(x)$ is band-limited and $\det(Y) \neq 0$, the integral defining the coefficients $\{c_j\}$ can be computed exactly as a finite sum.

At first glance it might appear that the above sums defining f_j and c_j require $\mathcal{O}(B^2)$ arithmetic operations for each value of j (when Y^{-1} is precomputed), and hence $\mathcal{O}(B^3)$ in total.[4] However, more efficient algorithms exist. When the functions $\{y_k\}$ satisfy the

---

[4] See Appendix A for an explanation of the $\mathcal{O}(\cdot)$ notation.

properties described below, the previous computation can be performed *much* more efficiently.

For all of the Sturm-Liouville systems considered in this book, the functions $\{y_k\}$ satisfy *three-term recurrence relations* of the form

$$y_{k+1}(x) = C_1(x,k)\, y_k(x) + C_2(x,k)\, y_{k-1}(x). \qquad (3.14)$$

In the special case when $\{y_k\}$ is a set of polynomials that solve a Sturm-Liouville problem, the coefficient $C_1(x,k)$ is of the form $a_k x + b_k$, and C_2 is a constant that depends on k (see Section 3.5.1). These recurrence relations allow for much more efficient evaluation of the above sums [5, 24]. Polynomials are also convenient because a function that is band-limited in one system of complete orthonormal polynomials will be band-limited with the same band-limit in any such system.

The use of recurrence relations for the fast evaluation of a weighted sum of orthogonal polynomials has a long history (see [37, 38, 41, 42, 59, 106] and references therein), as do algorithms for the conversion from one polynomial basis to another (see, e.g., [100, 101]).

Many different paths lead to the theory of orthogonal polynomials. The path that we have taken is Sturm-Liouville theory. This theory is applicable to orthogonal functions in general, many of which are polynomials. Another path is through the theory of generating functions, which can be used to prove orthogonality relations without using differential equations. A third path is through Gram-Schmidt orthogonalization, which is particularly useful for constructing polynomials that are orthogonal to a given weighting function because, by definition, polynomials have a finite number of coefficients that can be uniquely determined by orthogonalization.

Appendix A reviews computational tools used in connection with polynomials that are independent of recurrence relations. The subsequent sections of this chapter provide concrete examples of solutions to Sturm-Liouville problems, as well as complete orthonormal systems of functions that are not generated as the eigenfunctions of such expansions.

3.2 Legendre Polynomials and Associated Legendre Functions

3.2.1 Legendre Polynomials

Legendre polynomials,[5] $P_l(x)$, are defined as solutions of the singular Sturm-Liouville equation

$$((1-x^2)y')' + l(l+1)y = 0,$$

or equivalently

$$(1-x^2)y'' - 2xy' + l(l+1)y = 0,$$

for $x \in [-1,1]$ where $l = 0,1,2,....$ The Legendre polynomials can be calculated analytically using the *Rodrigues formula*

$$P_l(x) = \frac{1}{2^l l!} \frac{d^l}{dx^l} \left[(x^2 - 1)^l \right],$$

or the integral

---

[5]Named after Adrien Marie Legendre (1752-1833)

$$P_l(x) = \frac{1}{2\pi} \int_0^{2\pi} \left(x + i\sqrt{1 - x^2} \cos \phi \right)^l d\phi.$$

In general $P_l(-x) = (-1)^l P_l(x)$. The first few Legendre polynomials are $P_0(x) = 1$, $P_1(x) = x$, $P_2(x) = (3x^2 - 1)/2$. All the others can be calculated using the recurrence relation:

$$(l + 1)P_{l+1}(x) - (2l + 1)x P_l(x) + l\, P_{l-1}(x) = 0.$$

This is sometimes called *Bonner's recursion.*

Moreover, the derivatives of Lengendre polynomials satisfy the conditions [102]

$$P'_{l+1}(x) - x P'_l(x) = (l + 1)P_l(x) \quad \text{and} \quad P'_{l+1}(x) - P'_{l-1}(x) = (2l + 1)P_l(x).$$

By the general Sturm-Liouville theory, the Legendre polynomials satisfy the orthogonality conditions

$$\int_{-1}^1 P_{l_1}(x) P_{l_2}(x)\, dx = \frac{2}{2l_1 + 1}\delta_{l_1, l_2}.$$

Note that either l_1 or l_2 can be used in the denominator of the normalization constant in the above equation because δ_{l_1, l_2} is only nonzero when $l_1 = l_2$.

As explained in [102], the Legendre polynomials were originally introduced as the coefficients in the Taylor series expansion

$$(1 - 2\rho \cos \theta + \rho^2)^{-\frac{1}{2}} = \sum_{l=0}^{\infty} \rho^l P_l(\cos \theta) \tag{3.15}$$

which appears in astronomical calculations, and the above equation can be viewed as a generating function. Another place where the Legendre polynomials also appear is in the expansion

$$\frac{1 - \rho^2}{(1 - 2\rho \cos \theta + \rho^2)^{\frac{3}{2}}} = \sum_{l=0}^{\infty} (2l + 1)\rho^l P_l(\cos \theta). \tag{3.16}$$

Finally, before moving to the next topic, it is worth noting that fast numerical evaluation of Legendre expansions remains of interest today [119].

3.2.2 Associated Legendre Functions

The *associated Legendre functions*, $P_l^m(x)$, are solutions of the equation

$$((1 - x^2)y')' + \left(l(l + 1) - \frac{m^2}{1 - x^2} \right) y = 0,$$

which is equivalent to

$$(1 - x^2)y'' - 2xy' + \left(l(l + 1) - \frac{m^2}{1 - x^2} \right) y = 0.$$

Here $l = 0, 1, 2, \dots$ and $0 \le m \le l$.

They can be computed using the *Rodrigues formula*

$$P_l^m(x) \doteq (-1)^m \frac{(1 - x^2)^{m/2}}{2^l l!} \frac{d^{l+m}}{dx^{l+m}} \left[(x^2 - 1)^l \right]. \tag{3.17}$$

These associated Legendre functions can calculated from the Legendre polynomials as

$$P_l^m(x) = (-1)^m(1-x^2)^{m/2}\frac{d^m}{dx^m}P_l(x),$$

and so their generating function can be computed easily from (3.15) with $x = \cos\theta$. $P_l^m(x)$ also can be computed using Laplace's formula

$$P_l^m(x) = \frac{(l+m)!}{\pi\,l!}i^m\int_0^\pi \left(x+i\sqrt{1-x^2}\cos\phi\right)^l \cos m\phi\,d\phi$$

$$= \frac{(l+m)!}{2\pi\,l!}i^m\int_0^{2\pi} \left(x+i\sqrt{1-x^2}\cos\phi\right)^l e^{im\phi}\,d\phi. \qquad (3.18)$$

We note that some texts such as [7, 102] define the associated Legendre functions without the $(-1)^m$ factor in (3.17), which is called the *Condon-Shortley phase*. We denote this alternative version as

$$\hat{P}_l^m(x) \doteq (-1)^m P_l^m(x). \qquad (3.19)$$

We also note that some sources refer to $P_l^m(x)$ as associated Legendre polynomials, when in fact they are not all polynomials. For example, $P_1^1(x) = -(1-x^2)^{1/2}$.

The associated Legendre functions (and their derivatives) satisfy a number of recurrence relations including

$$(l-m+1)P_{l+1}^m(x) - (2l+1)xP_l^m(x) + (l+m)P_{l-1}^m(x) = 0, \qquad (3.20)$$

and are extended to negative values of the superscript as

$$P_l^{-m}(x) \doteq (-1)^m\frac{(l-m)!}{(l+m)!}P_l^m(x)$$

for all $0 \le m \le l$. The two expressions given above hold for both P_l^m and $\hat{P}_l^m$, as do the following [7]:

$$P_l(x) = P_l^0(x) \quad\text{and}\quad P_l^m(-x) = (-1)^{l+m}P_l^m(x)$$

The orthogonality of the associated Legendre functions is

$$\int_{-1}^1 P_{l_1}^m(x)P_{l_2}^m(x)\,dx = \frac{2}{2l_1+1}\frac{(l_1+m)!}{(l_1-m)!}\delta_{l_1,l_2}.$$

For any fixed $m \in [-l,l]$, the associated Legendre functions form an orthonormal basis for $\mathcal{L}^2([-1,1])$. That is, any "nice" function on the interval $-1 \le x \le 1$ can be expanded in a series of normalized associated Legendre functions as

$$f(x) = \sum_{l=|m|}^\infty \hat{f}(l,m)\tilde{P}_l^m(x)$$

where

$$\tilde{P}_l^m(x) = \sqrt{\frac{2l+1}{2}\frac{(l-m)!}{(l+m)!}}P_l^m(x).$$

Due to the orthonormality of these functions, we have

$$\hat{f}(l,m) = \int_{-1}^{1} f(x)\tilde{P}_l^m(x)\,dx.$$

The coefficient $\hat{f}(l,m)$ is called the associated Legendre transform. When using the above normalization

$$\tilde{P}_l^{-m}(x) = (-1)^m \tilde{P}_l^m(x) \tag{3.21}$$

and this too is independent of whether or not the Condon-Shortley phase is used.

3.2.3 Fast Legendre and Associated Legendre Transforms

Using recurrence relations to efficiently evaluate a weighted sum of Legendre polynomials is an idea that has been explored for some time (see e.g., [33, 96, 118] and references therein). Over the past two decades there has been some very interesting work on the development of sampling and fast, exact, discrete Legendre (and associated Legendre) transforms. These algorithms, developed by Driscoll, Healy, Rockmore, and collaborators [34, 35, 55], can, in principle, compute $\hat{f}(l,m)$ exactly for each fixed $m \geq 0$ and all values of l in the range $m \leq l < B$, a finite band-limit, in $\mathcal{O}(B(\log B)^2)$ operations by sampling at $2B$ special points.

This is work that can be found in the literature, and we will not review the derivations for these results. The key idea is that the recurrence relations in (3.20) provide a tool for performing fast implementations. As a special case, when $m = 0$, the Legendre polynomial transform can be computed in $\mathcal{O}(B(\log B)^2)$ operations as well.

In addition to these theoretically exact algorithms, several recent works develop fast approximations that claim $\mathcal{O}(B \log B)$ speed and very good accuracy [31, 60]. Other approximations that are very fast and accurate in practice use the Fast Multipole Method [44] and the relationship between inverse power laws and Legendre polynomials in (3.15) to develop fast numerical algorithms for computing Legendre function transforms [49].

Throughout the text we will use asymptotic complexity bounds established for exact computation of problems using exact arithmetic. In practice, numerical stability issues and the tradeoff between numerical error and running time must be taken into account. For the Legendre transform, this has been addressed in [4].

3.3 Jacobi, Gegenbauer, and Chebyshev Polynomials

The Legendre polynomials form one complete orthogonal basis for functions on the interval $[-1,1]$. Several other important examples of polynomials that satisfy

$$(p_k, p_m) = \int_{-1}^{1} p_k(x)\,p_m(x)\,w(x)\,dx = C_k\,\delta_{k,m} \tag{3.22}$$

for some normalization C_k and weight $w(x)$ are reviewed here. One very broad class of orthogonal polynomials is the *Jacobi polynomials*, $\{P_n^{(\alpha,\beta)}(x)\}$, which, for each fixed $\alpha, \beta > -1$ and $n = 0, 1, 2, ...$, form an orthogonal basis with respect to the weight

$$w(x) = (1-x)^\alpha (1+x)^\beta.$$

The Jacobi polynomials are calculated using the Rodrigues formula

$$P_n^{(\alpha,\beta)}(x) = \frac{(-1)^n}{2^n n!}(1-x)^{-\alpha}(1+x)^{-\beta}\frac{d^n}{dx^n}\left[(1-x)^{\alpha+n}(1+x)^{\beta+n}\right]. \tag{3.23}$$

In the case when $\alpha = \beta = \lambda - \frac{1}{2}$, the set of functions $\{P_n^{(\lambda)}(x)\}$ for fixed λ and $n = 0, 1, 2, \ldots$ are called the *Gegenbauer polynomials*. If $\lambda = 0$, the Gegenbauer polynomials reduce to the Chebyshev (or Tschebysheff) polynomials of the second kind, which are defined as

$$T_n(x) \doteq \cos(n \cos^{-1} x) = \frac{1}{2}[(x + \sqrt{x^2 - 1})^n + (x - \sqrt{x^2 - 1})^n]$$

for $0 \leq \cos^{-1} x \leq \pi$. They have generating function

$$\frac{1 - xz}{1 - 2xz + z^2} = \sum_{n=0}^{\infty} T_n(x) z^n$$

and Rodrigues formula

$$T_n(x) = \frac{(-1)^n \sqrt{\pi}(1 - x^2)^{\frac{1}{2}}}{2^n \Gamma(n + 1/2)} \frac{d^n}{dx^n}[(1 - x^2)^{n - \frac{1}{2}}],$$

and satisfy the simple recurrence relations

$$T_{n+1}(x) = 2x T_n(x) - T_{n-1}(x)$$

with $T_0(x) = 1$ and $T_1(x) = x$. They are solutions of the equation

$$(1 - x^2)y'' - xy' + n^2 y = 0$$

where $n = 0, 1, 2, \ldots$. It can be shown that $T_n(-x) = (-1)^n T_n(x)$ and

$$\int_{-1}^{1} T_k(x) T_m(x) \frac{dx}{\sqrt{1 - x^2}} = F_m \, \delta_{k,m}$$

where

$$F_m = \begin{cases} \frac{\pi}{2} & m \neq 0 \\ \pi & m = 0 \end{cases}.$$

See [98] for more properties. Some computational aspects are addressed in [24].

A band-limited function in this context is one for which

$$f(x) = \sum_{n=0}^{B-1} c_n T_n(x)$$

for finite B where

$$c_n = \frac{1}{F_n} \int_{-1}^{1} f(x) T_n(x) \frac{dx}{\sqrt{1 - x^2}}.$$

We note that by setting $x = \cos y$, the function may be evaluated as $f(x) = f(\cos y) = \tilde{f}(y)$. Then the band-limited expansion above becomes

$$\tilde{f}(y) = \sum_{n=0}^{B-1} c_n \cos ny \qquad (3.24)$$

where

$$c_n = \frac{1}{F_n} \int_0^\pi \tilde{f}(y) \cos ny \, dy.$$

Sampling this integral at the points $y = \pi j / B$ for $j = 0, 1, 2, ..., B - 1$ produces the exact result

$$c_n = \frac{1}{B} \sum_{j=0}^{B-1} \tilde{f}(\pi j / B) \cos(\pi n j / B).$$

This *discrete cosine transformation* (DCT) can be viewed as a variant of the DFT. As such, the above equation together with (3.24) sampled at $y = \pi j / B$ can be computed for all $j = 0, ..., B - 1$ in $\mathcal{O}(B \log B)$ computations. Therefore, a fast discrete (sampled) Chebyshev transform results as well.

Chebyshev polynomials play an important role in approximation methods in numerical analysis [14, 71]. And in recent years a number of works have addressed fast transforms for the Jacobi and Gegenbauer (ultraspherical) polynomials [18, 66, 115].

3.4 Hermite and Laguerre Polynomials

Hermite polynomials[6] are solutions to the differential equation

$$y'' - 2xy' + 2ny = 0$$

or equivalently,

$$(e^{-x^2} y')' + 2n e^{-x^2} y = 0,$$

for $n = 0, 1, 2, ...$ and $x \in \mathbb{R}$. They are generated by the Rodrigues formula

$$H_n(x) = (-1)^n e^{x^2} \frac{d^n}{dx^n} (e^{-x^2}),$$

are the coefficients in the expansion[7]

$$\exp(-z^2 + 2xz) = \sum_{n=0}^\infty H_n(x) \frac{z^n}{n!},$$

and they satisfy the recurrence formulas

$$H_{n+1}(x) = 2x H_n(x) - 2n H_{n-1}(x) \quad \text{and} \quad H_n'(x) = 2n H_{n-1}(x).$$

In general $H_n(-x) = (-1)^n H_n(x)$. The first few Hermite polynomials are $H_0(x) = 1$, $H_1(x) = 2x$ and $H_2(x) = 4x^2 - 2$. The Hermite polynomials satisfy the orthogonality conditions

$$\int_{-\infty}^\infty H_m(x) H_n(x) \, e^{-x^2} \, dx = 2^n n! \sqrt{\pi} \delta_{m,n}.$$

The *Hermite functions* are defined as

$$h_n(x) \doteq \frac{1}{2^{n/2} \sqrt{n! \sqrt{\pi}}} H_n(x) \, e^{-\frac{x^2}{2}} \tag{3.25}$$

---

[6]Named after Charles Hermite (1822-1901)

[7]Sometimes in the literature Hermite polynomials are defined by the alternative generating function $\exp(-z^2 - 2xz)$ but with the same expansion on the right side. We denote the Hermite polynomials resulting from this alternative phase convention as $\bar{H}_n(x) \doteq (-1)^n H_n(x)$.

for $n = 0, 1, 2, \dots$. Note that $h_0(x)$ is a Gaussian scaled so that $\|h_0^2\|_1 = 1$.

The set $\{h_n(x)\}$ for $n = 0, 1, 2\dots$ forms a complete orthonormal basis for the set of square-integrable functions on the line, $\mathcal{L}^2(\mathbb{R})$, (with unit weighting function) and can be calculated using the recurrence relations

$$\sqrt{n+1}\,h_{n+1}(x) = \sqrt{2}\,h_n(x) - \sqrt{n}\,h_{n-1}(x)$$

which can be used to construct fast Hermite-function transforms [72] The Hermite functions are also very special because they are eigenfunctions of the Fourier transform. That is, the Fourier transform of a Hermite function is a scalar multiple of the same Hermite function evaluated at the frequency parameter. More specifically, it can be shown that [110]:

$$\int_{-\infty}^{\infty} e^{-i\omega x} h_n(x)\,dx = \sqrt{2\pi}(-i)^n h_n(\omega).$$

Hence, a function expanded in a series of Hermite functions in the spatial domain will have a Fourier transform that is expanded in a series of Hermite functions in the frequency domain.

The *Laguerre polynomials*[8] are solutions of the differential equation

$$xy'' + (1 - x)y' + ny = 0,$$

or

$$(xe^{-x}y')' + ne^{-x}y = 0,$$

for $x \in \mathbb{R}_{\geq 0}$. The unnormalized Laguerre polynomials satisfy the Rodrigues formula

$$L_n(x) = e^x \frac{d^n}{dx^n}(x^n e^{-x}).$$

The first few are $L_0(x) = 1$, $L_1(x) = 1 - x$, $L_2(x) = x^2 - 4x + 2$, and the rest can be generated using the recurrence formulas

$$L_{n+1}(x) = (2n + 1 - x)L_n(x) - n^2 L_{n-1}(x);$$

$$L_n'(x) = nL_{n-1}'(x) - nL_{n-1}(x);$$

$$xL_n'(x) = nL_n(x) - n^2 L_{n-1}(x).$$

The unnormalized Laguerre polynomials satisfy the orthogonality conditions

$$\int_0^{\infty} L_m(x)L_n(x)\,e^{-x}\,dx = (n!)^2 \delta_{m,n}.$$

For this reason the normalized Laguerre polynomials

$$\tilde{L}_n(x) \doteq \frac{1}{n!}L_n(x) = \sum_{k=0}^{n} \frac{(-1)^k}{k!}\binom{n}{k}x^k \tag{3.26}$$

are commonly used.

In analogy with the Legendre polynomials, we can define *associated Laguerre polynomials* as

---

[8]Named after Edmond Laguerre (1834-1886)

$$L_n^m(x) \doteq \frac{d^m}{dx^m}(L_n(x))$$

for $0 \le m \le n$. They satisfy the equation

$$xy'' + (m + 1 - x)y' + (n - m)y = 0,$$

and they satisfy the orthogonality conditions

$$\int_0^\infty L_n^m(x)L_p^m(x) \, e^{-x} x^m dx = \frac{(n!)^3}{(n-m)!}\delta_{p,n}.$$

Often in the literature the alternate definitions

$$\tilde{L}_n^m(x) = (-1)^m \frac{d^m}{dx^m}(\tilde{L}_{n+m}(x)) = \frac{(-1)^m}{(n+m)!}L_{n+m}^m(x) \tag{3.27}$$

are used, in which case the recurrence relations

$$(n+1)\tilde{L}_{n+1}^m(x) = (2n + m + 1 - x)\tilde{L}_n^m(x) - (n+m)\tilde{L}_{n-1}^m(x).$$

and

$$\tilde{L}_n^m(x) = \tilde{L}_n^{m+1}(x) - \tilde{L}_{n-1}^{m+1}(x)$$

result, as well as others involving their derivatives. These recursions can be started with $\tilde{L}_0^m(x) = 1$ and $\tilde{L}_{-1}^m(x) = 1$.

Corresponding changes to the orthogonality relation gives

$$\int_0^\infty \tilde{L}_n^m(x)\tilde{L}_p^m(x) \, e^{-x} x^m dx = \frac{(n+m)!}{n!}\delta_{p,n}.$$

This version of the associated Laguerre polynomials can be generated by [102]

$$(1 - z)^{-(m+1)} \exp\left(\frac{-xz}{1-z}\right) = \sum_{n=0}^\infty z^n \tilde{L}_n^m(x) \tag{3.28}$$

and satisfy the Rodrigues formula

$$\tilde{L}_n^m(x) = \frac{x^{-m} e^x}{n!} \frac{d^n}{dx^n}(x^{m+n} e^{-x}).$$

Moreover, if we use (3.28) to expand the definition of $\tilde{L}_n^m(x)$ to allow m to take non-integer values, then they can be related to $H_n(x)$ as [102, 108]

$$H_{2n}(x) = (-1)^n \, 2^{2n} \, n! \, \tilde{L}_n^{-1/2}(x^2) \quad \text{and} \quad H_{2n+1}(x) = (-1)^n \, 2^{2n+1} \, n! \, x\tilde{L}_n^{-1/2}(x^2).$$

3.5 Quadrature Rules and Discrete Polynomial Transforms

We saw earlier that exact fast discrete transforms exist for the Fourier series, Legendre transform, and Chebyshev series. A natural question to ask at this point is whether or not fast discrete polynomial transforms exist for the other polynomials orthogonal on the interval $[-1, 1]$. (We shall not consider discrete transforms on an infinite domain such as those corresponding to the Hermite and Laguerre polynomials, and we restrict the discussion to functions on $[-1,1]$ because any closed finite interval can be mapped to $[-1, 1]$ and vice versa).

We first need a systematic way to exactly convert integrals of band-limited functions into sums:

$$\int_{-1}^{1} f(x)\, w(x)\, dx = \sum_{i=0}^{n} f(x_i)\, w_i. \qquad (3.29)$$

Equation (3.29) can be viewed as a special case of (3.13). In this context, a band-limited polynomial of band-limit $B + 1$ is one of the form

$$f(x) = a_B x^B + a_{B-1} x^{B-1} + \cdots + a_1 x + a_0.$$

The set of all polynomials with band-limit $B + 1$ (or smaller) is denoted $\mathbb{P}_B$.

Any choice of the set of ordered pairs $\{(x_i, w_i) | i = 0, ..., n < \infty\}$ for which (3.29) holds for a given (fixed) weighting function $w(x)$ and *all* band-limited functions of band-limit $B+1$ is called a *quadrature rule*. An optimal quadrature rule is one which minimizes the required number of points, n, while satisfying (3.29) for a given B. On the other hand, this may require locating the points $x_0, ..., x_n$ in irregular ways. Therefore, we should expect a tradeoff between regular location of points, and the minimal number of points required for exact computation.

We now make the discussion concrete by reviewing some of the most common quadrature rules. For modern generalizations see [77]. Two of the most commonly used elementary rules are the *trapezoid rule* and *Simpson's rule*. These are exact for $w(x) = 1$ and small values of B. The even more elementary *centered difference rule* for numerical integration is only guaranteed to be exact when the integrand is a constant. The trapezoid and Simpson's rules are both based on evenly spaced samples and are respectively

$$\int_{-1}^{1} f(x)\, dx = \frac{1}{n}\left[f(-1) + 2\sum_{k=1}^{n-1} f(-1 + 2k/n) + f(1)\right]$$

and

$$\int_{-1}^{1} f(x)\, dx = \frac{2}{3n}\left[f(-1) + 4\sum_{k=1}^{n/2} f(-1 + 2(2k-1)/n) + 2\sum_{k=1}^{n/2-1} f(-1 + 4k/n) + f(1)\right].$$

In the case of Simpson's rule, n is assumed even. While both of these rules usually behave well numerically for polynomial functions f and large enough values of n, the trapezoid rule provides an *exact* answer only when $f(x) = a_1 x + a_0$, and Simpson's rule provides an exact answer only when $f(x) = a_3 x^3 + a_2 x^2 + a_1 x + a_0$ for finite values of a_i. In the case of the trapezoid rule it is assumed that $n \geq 1$, while for Simpson's rule $n \geq 2$.

Clearly these rules will not suffice for our purposes. The optimal quadrature rule (in the sense of requiring the fewest sample points) is presented in Subsection 3.5. But first we review an important general property of orthogonal polynomials.

3.5.1 Recurrence Relations for Orthogonal Polynomials

All of the orthogonal polynomials encountered thus far have had three-term recurrence relations of the form

$$p_n(x) = (a_n x + b_n)p_{n-1}(x) + c_n p_{n-2}(x) \tag{3.30}$$

for $n = 1, 2, 3, \dots$ and $p_{-1}(x) \doteq 0$. It can be shown that for appropriate choices of a_n, b_n, and c_n, (3.30) is *always* true for orthogonal polynomials [29, 107].

To prove this inductively, assume $p_0, \dots, p_n$ are orthogonal to each other, and are generated by recurrence. By defining

$$p_{n+1} = \left(\frac{x}{(p_n, p_n)^{\frac{1}{2}}} - \frac{(xp_n, p_n)}{(p_n, p_n)^{\frac{3}{2}}} \right) p_n - \frac{(p_n, p_n)^{\frac{1}{2}}}{(p_{n-1}, p_{n-1})^{\frac{1}{2}}} p_{n-1}, \tag{3.31}$$

we seek to show that $(p_{n+1}, p_k) = 0$ for all $k = 0, \dots, n$ given that $(p_r, p_s) = 0$ for all $r \neq s \in [0, n]$. Here $(\cdot, \cdot)$ is defined as in (3.22).

Since $xp_j \in \mathbb{P}_{j+1}$ for $p_j \in \mathbb{P}_j$ we can write $xp_j = \sum_{k=0}^{j+1} \alpha_k p_k(x)$ for some set of constants $\{\alpha_k\}$. For all $k \leq j + 1 < n$, it follows from $(p_n, p_k) = 0$ that $(xp_n, p_j) = (p_n, xp_j) = 0$. We therefore conclude that $(p_{n+1}, p_j) = 0$ for all $j = 0, \dots, n - 2$.

Similarly, it can be shown that

$$(p_{n+1}, p_{n-1}) = \frac{(xp_n, p_{n-1})}{(p_n, p_n)^{\frac{1}{2}}} - (p_n, p_n)^{\frac{1}{2}} (p_{n-1}, p_{n-1})^{\frac{1}{2}},$$

which reduces to zero because

$$(xp_n, p_{n-1}) = (p_n, xp_{n-1}) = (p_n, p_n)(p_{n-1}, p_{n-1})^{\frac{1}{2}}. \tag{3.32}$$

To see this, evaluate (3.31) at $n - 1$ in place of n:

$$p_n = \left(\frac{x}{(p_{n-1}, p_{n-1})^{\frac{1}{2}}} - \frac{(xp_{n-1}, p_{n-1})}{(p_{n-1}, p_{n-1})^{\frac{3}{2}}} \right) p_{n-1} - \frac{(p_{n-1}, p_{n-1})^{\frac{1}{2}}}{(p_{n-2}, p_{n-2})^{\frac{1}{2}}} p_{n-2}.$$

Next, take the inner product with p_n and rearrange terms to result in (3.32). The equality $(p_{n+1}, p_n) = 0$ is then verified by inspection.

Thus far we have shown that $(p_{n+1}, p_k) = 0$ for $k = 0, \dots, n$ under the inductive hypothesis that $p_0, \dots, p_n$ are orthogonal and generated by recurrence. By starting with $p_{-1}(x) = 0$, $p_0(x) = C$, and calculating $p_1(x)$ from (3.31), it can be shown that $(p_0, p_1) = 0$. This justifies our hypothesis up to $n = 1$, and this justification is carried to arbitrary n by the discussion above.

Evaluating (3.31) with $n - 1$ instead of n and comparing with (3.30) gives

$$a_n = \frac{1}{(p_{n-1}, p_{n-1})^{\frac{1}{2}}}; \quad b_n = -\frac{(xp_{n-1}, p_{n-1})}{(p_{n-1}, p_{n-1})^{\frac{3}{2}}}; \quad c_n = -\frac{(p_{n-1}, p_{n-1})^{\frac{1}{2}}}{(p_{n-2}, p_{n-2})^{\frac{1}{2}}}.$$

With appropriate normalization, p_i can be assumed to be orthonormal. We can also set

$$p_{-1}(x) = 0 \quad \text{and} \quad p_0(x) = 1 \tag{3.33}$$

by normalizing $w(x)$. In this case, the coefficients a_n, b_n, and c_n are related to the leading coefficients $p_n(x) = k_n x^n + s_n x^{n-1} + \cdots$ as [29, 108]

$$a_n = \frac{k_n}{k_{n-1}}; \quad b_n = -a_n \left(\frac{s_n}{k_n} - \frac{s_{n-1}}{k_{n-1}} \right); \quad c_n = -a_n \frac{k_{n-2}}{k_{n-1}}. \tag{3.34}$$

The expressions above for a_n and b_n follow from expanding p_n, p_{n-1} and p_{n-2} term by term in (3.30) and equating coefficients of powers x^n and x^{n-1}. The expression for c_n follows by observing that

$$(p_n, p_{n-2}) = a_n(x p_{n-1}, p_{n-2}) + c_n$$

and

$$(x p_{n-1}, p_{n-2}) = (p_{n-1}, x p_{n-2}) = (p_{n-1}, k_{n-2} x^{n-1} + s_{n-2} x^{n-2} + \cdots) = (p_{n-1}, k_{n-2} x^{n-1})$$

since p_{n-1} is orthogonal to all polynomials in $\mathbb{P}_{n-2}$. Likewise,

$$(p_{n-1}, k_{n-2} x^{n-1}) = \frac{k_{n-2}}{k_{n-1}} (p_{n-1}, k_{n-1} x^{n-1}) = \frac{k_{n-2}}{k_{n-1}} (p_{n-1}, p_{n-1}) = \frac{k_{n-2}}{k_{n-1}}$$

since p_{n-1} is normalized.

We note that since (3.30) holds, so too does the *Christoffel-Darboux formula* [107, 108]:

$$\sum_{i=0}^{n} p_i(x) p_i(y) = \frac{k_n}{k_{n+1}} \frac{p_{n+1}(x) p_n(y) - p_n(x) p_{n+1}(y)}{x - y}. \tag{3.35}$$

This follows by first using the substitution

$$p_{n+1}(x) = (a_{n+1} x + b_{n+1}) p_n(x) + c_{n+1} p_{n-1}(x)$$

to write

$$p_{n+1}(x) p_n(y) - p_n(x) p_{n+1}(y) =$$

$$a_{n+1}(x - y) p_n(x) p_n(y) - c_{n+1}(p_n(x) p_{n-1}(y) - p_{n-1}(x) p_n(y)).$$

Then, recursively expanding the term in parenthesis on the right, and using (3.34) to rewrite the resulting coefficients, we arrive at (3.35). As $y \to x$, a straightforward application of l'Hôpital's rule to (3.35) yields

$$\sum_{i=0}^{n} (p_i(x))^2 = \frac{k_n}{k_{n+1}} \left[p_n(x) p'_{n+1}(x) - p_{n+1}(x) p'_n(x) \right]. \tag{3.36}$$

3.5.2 Gaussian Quadrature

Let $p_k(x)$ for $k = 0, 1, 2, \ldots$ be a complete system of orthonormal polynomials in $\mathcal{L}^2([-1, 1], w(x)\, dx)$. Let $x_0, \ldots, x_n$ be the roots of p_{n+1}, and let $w_0, \ldots, w_n$ be the solution of the system of equations

$$\sum_{j=0}^{n} (x_j)^k w_j = \int_{-1}^{1} x^k w(x)\, dx \tag{3.37}$$

for $k = 0, \ldots, n$. Then [21, 40]:

$$\sum_{j=0}^{n} p_k(x_j) w_j = \int_{-1}^{1} p_k(x) w(x)\, dx$$

for any $p_k \in \mathbb{P}_{2n+1}$, or more generally

$$\sum_{j=0}^{n} f(x_j) w_j = \int_{-1}^{1} f(x) w(x)\, dx \tag{3.38}$$

for any $f \in \mathbb{P}_{2n+1}$. In order to verify (3.38), it is convenient to define the *Lagrange fundamental polynomials* for any set of distinct points $\xi_0, ..., \xi_n$ and $k \in \{0, 1, 2, ..., n\}$ as

$$l_k^n(x) = \frac{(x - \xi_0)(x - \xi_1) \cdots (x - \xi_{k-1})(x - \xi_{k+1}) \cdots (x - \xi_n)}{(\xi_k - \xi_0)(\xi_k - \xi_1) \cdots (\xi_k - \xi_{k-1})(\xi_k - \xi_{k+1}) \cdots (\xi_k - \xi_n)}.$$

From this definition $l_k^n(x) \in \mathbb{P}_n$ and $l_k^n(\xi_j) = \delta_{kj}$. The Lagrange fundamental polynomials can be written in the alternate form

$$l_k^n(x) = \frac{l^{n+1}(x)}{(x - \xi_k)(l^{n+1})'(\xi_k)}$$

where

$$l^{n+1}(x) = \prod_{k=0}^{n} (x - \xi_k) \in \mathbb{P}_{n+1}.$$

It follows from these definitions that

$$\sum_{k=0}^{n} l_k^n(x) = 1$$

and the function

$$\mathcal{L}_n(x) = \sum_{k=0}^{n} f_k l_k^n(x) \in \mathbb{P}_n$$

has the property $\mathcal{L}_n(\xi_j) = f_j$ for $j = 0, ..., n$. $\mathcal{L}_n(x)$ is called a *Lagrange interpolation polynomial.*

In the special case when $\xi_i = x_i$ for $i = 0, ..., n$, where x_i satisfy $p_{n+1}(x_i) = 0$, it follows that $l^{n+1}(x) = c_{n+1}p_{n+1}(x)$ for some nonzero constant $c_{n+1} \in \mathbb{R}$ because every polynomial in $\mathbb{P}_{n+1}$ is determined up to a constant by its zeros. In this case we write

$$\mathcal{L}_n(x) = \sum_{k=0}^{n} f(x_k) \frac{p_{n+1}(x)}{(x - x_k)p'_{n+1}(x_k)} = \sum_{k=0}^{n} f(x_k) l_k^n(x).$$

For any $f(x) \in \mathbb{P}_{2n+1}$ it then can be shown that [108]

$$f(x) - \mathcal{L}_n(x) = p_{n+1}(x)r(x)$$

for some $r(x) \in \mathbb{P}_n$. Therefore,

$$\int_{-1}^{1} f(x) w(x)\, dx = \int_{-1}^{1} \mathcal{L}_n(x) w(x)\, dx + \int_{-1}^{1} p_{n+1}(x) r(x) w(x)\, dx$$

$$= \int_{-1}^{1} \mathcal{L}_n(x) w(x)\, dx + 0 \qquad (3.39)$$

$$= \sum_{k=0}^{n} f(x_k) \int_{-1}^{1} l_k^n(x) w(x)\, dx.$$

This is (3.38) with

$$w_k = \int_{-1}^{1} l_k^n(x) w(x)\, dx. \qquad (3.40)$$

It can be shown (see [107]) that these weights satisfy (3.37).

Given (3.38), it follows that any polynomial of the form $f(x) = \sum_{k=0}^{2n+1} c_k\, p_k(x)$ can be integrated exactly with this rule. In particular, the product $p_j(x)\, p_k(x) \in \mathbb{P}_{2n+1}$ is true when $j, k \leq n$, and so

$$\sum_{k=0}^{n} p_i(x_k)\, p_j(x_k)\, w_k = \int_{-1}^{1} p_i(x)\, p_j(x)\, w(x)\, dx = \delta_{ij} \tag{3.41}$$

(the second equality holding by the definition of orthonormality). This means that orthonormality of discrete polynomials follows naturally from the orthonormality of continuous ones when using Gaussian quadrature.

Consider a band-limited polynomial function with band-limit $B + 1$:

$$f(x) = \sum_{k=0}^{B} c_k p_k(x).$$

When sampled at the $n + 1$ Gaussian quadrature points $x_0, ..., x_n$ this becomes

$$f(x_i) = \sum_{k=0}^{B} c_k p_k(x_i). \tag{3.42}$$

We can therefore write

$$\sum_{i=0}^{n} f(x_i)\, p_j(x_i)\, w_i = \sum_{i=0}^{n}\left(\sum_{k=0}^{B} c_k\, p_k(x_i)\right) p_j(x_i)\, w_i = \sum_{k=0}^{B} c_k \left(\sum_{i=0}^{n} p_k(x_i)\, p_j(x_i)\, w_i\right).$$

The last term in parenthesis can be simplified to δ_{jk} using (3.41) when $B \leq n + 1$. Taking $n = B$ then ensures that the coefficients corresponding to the discrete polynomial expansion in (3.42) are calculated as

$$c_j = \sum_{i=0}^{n} f(x_i)\, p_j(x_i)\, w_i. \tag{3.43}$$

Hence, when using Gaussian quadrature, the number of sample points is equal to the bandlimit. This comes, however, at the price of not having the freedom to choose the location of sample points.

Using the Christoffel-Darboux formula (3.35), it can be shown (see [30, 71, 107]) that the weights that solve (3.37) are determined as

$$w_i = -\frac{k_{n+2}}{k_{n+1}} \frac{1}{p_{n+2}(x_i) p'_{n+1}(x_i)} = \frac{k_{n+1}}{k_n} \frac{1}{p_n(x_i) p'_{n+1}(x_i)} = \frac{1}{\sum_{j=0}^{n} (p_j(x_i))^2}.$$

where k_n is the leading coefficient of $p_n(x) = k_n x^n + \dots$. This assumes that the set of functions $\{p_j(x)\}$ is orthonormal and $w(x)$ is normalized such that $p_0(x) = 1$. To verify the first two equalities, substitute $y = x_i$ into (3.35), integrate with respect to $w(x)\, dx$ over $[-1, 1]$, and simplify with (3.40). Combining either of the first two equalities with (3.36), and evaluating at $x = x_i$ yields the last equality.

3.5.3 Fast Polynomial Transforms

It is often convenient to specify the sample points in advance. The sample points in the cases of the DFT and discrete Legendre transform are examples of this. In general, when

$n + 1$ distinct sample points are chosen, (3.29) will still hold, but instead of holding for all $f \in \mathbb{P}_{2n+1}$, it will only hold for $f \in \mathbb{P}_n$. This makes sense because there are $n + 1$ free weights determined by the $n + 1$ equations resulting from (3.37) for $k = 0, ..., n$.

The implication of this for discrete polynomial transforms is that when the set of sample points are fixed a priori, the number must be related to the band-limit as $n \geq 2B$. Thus, by choosing $n = 2B$, (3.42)–(3.43) remain true. We already saw this for the discrete Legendre transform. A natural choice is to evenly space the sample points. This is the assumption in the following theorem due to Driscoll, Healy, and Rockmore.

Theorem 3.1. *[35]: Let $B = 2^K$ for positive integer K, and $\{\phi_k(x) \mid k = 0, ..., B-1\}$ be a set of functions satisfying (3.30) and (3.33). Then for any set of positive weights $w_0, ..., w_{n-1}$ (where $n = 2B$)*

$$\hat{f}(k) = \sum_{j=0}^{n-1} f(j)\, \phi_k(j)\, w_j$$

can be computed (exactly in exact arithmetic) for all $k \in \{0, ..., B-1\}$ in $\mathcal{O}(B(\log B)^2)$ arithmetic operations.

Because of the similarity of form between (3.42) and (3.43), this theorem indicates the existence of fast algorithms for both the computation of c_j for all $j = 0, ..., n-1$, as well as the fast reconstruction of $f(x_i)$ for all i.

For points not evenly spaced, the following is useful:

Theorem 3.2. *[80] Let $B = 2^K$ and $\{p_0, ..., p_{B-1}\}$ be a set of real orthogonal polynomials. Then for any sequence of real numbers, $f_0, ..., f_{B-1}$, the discrete polynomial transform defined by the sums*

$$\hat{f}(k) = \frac{1}{B} \sum_{j=0}^{B-1} f_j\, p_k \left(\cos^{-1}[(2j+1)\pi/2B] \right)$$

can be calculated exactly for all values of $k \in \{0, ..., B-1\}$ in $\mathcal{O}(B(\log B)^2)$ exact arithmetic operations.

Both of the above theorems depend on the fast polynomial evaluation and interpolation algorithms reviewed in Appendix A and assume exact arithmetic. In practice, modifications to account for numerical errors are required. Other results on fast discrete polynomial transforms can be found in [67, 92].

The following section reviews the Bessel functions, for which we know of no exact fast discrete transform. This is problematic when considering the fast numerical evaluation of the Fourier transform in polar coordinates discussed in Chapter 4, and will require us to interpolate back and forth between polar and Cartesian coordinates in order to benefit from FFT methods. This step, while usually numerically acceptable, does not preserve the exactness of the results.

3.6 Bessel and Spherical Bessel Functions

Bessel functions[9] are solutions, $(y_i(x), \lambda_i)$ to the equation

---

[9]Named after Friedrich Wilhelm Bessel (1784-1846)

$$y'' + (d-1)\frac{y'}{x} + \left(\lambda - \frac{m^2}{x^2}\right)y = 0. \qquad (3.44)$$

This equation can be written in the form of the singular Sturm-Liouville equation

$$(x^{d-1}y')' + (\lambda x^{d-1} - m^2 x^{d-3})y = 0.$$

The parameter d is called the dimension and m is called the angular frequency or order.

One feature of the above equations is that the alternate definition of the independent variable as $\xi = x\sqrt{\lambda}$ and corresponding function $z(\xi) = y(x)$ results in the equation

$$\frac{d^2 z}{d\xi^2} + (d-1)\frac{1}{\xi}\frac{dz}{d\xi} + \left(1 - \frac{m^2}{\xi^2}\right)z = 0, \qquad (3.45)$$

which is independent of λ.

This follows from the chain rule

$$\frac{d}{dx} = \frac{d}{d\xi}\frac{d\xi}{dx} = \sqrt{\lambda}\frac{d}{d\xi}.$$

Hence, in the absence of boundary conditions, we can take $\lambda = 1$ and consider solutions to (3.45) without loss of generality since the solution to (3.44) will then be $y(x) = z(x\sqrt{\lambda})$.

Detailed treatments of the Bessel functions can be found in [13, 116]. Material in the remainder of this section summarizes without proof the most relevant definitions and properties associated with Bessel functions.

3.6.1 Bessel Functions

Consider first the most common case when $d = 2$. For each fixed $m \in \{0, 1, 2, ...\}$ there is a solution corresponding to $\lambda = 1$ of the form [7, 13]:

$$J_m(x) \doteq \frac{x^m}{2^m}\sum_{n=0}^{\infty}\frac{(-1)^n}{2^{2n}n!(m+n)!}x^{2n}. \qquad (3.46)$$

and are referred to as "Bessel functions of the first kind." The definition

$$J_{-m}(x) \doteq (-1)^m J_m(x),$$

extends the index of the Bessel functions from the nonnegative integers to all the integers.

It can be shown that an equivalent expression for these Bessel functions of the first kind is

$$J_m(x) = \frac{1}{2\pi}\int_0^{2\pi}\cos(m\theta - x\sin\theta)\,d\theta. \qquad (3.47)$$

And the Bessel functions can be generated as the coefficients in the Laurent series expansion [1]

$$e^{\frac{1}{2}x(z-z^{-1})} = \sum_{m=-\infty}^{\infty}z^m J_m(x)$$

for $z \in \mathbb{C} - \{0\}$.

From the above equation it is possible to write the *Jacobi-Anger expansions* [19]

$$e^{ipr\sin\theta} = \sum_{m=-\infty}^{\infty} J_m(pr)\, e^{im\theta} \quad \text{and} \quad e^{ipr\cos\theta} = \sum_{m=-\infty}^{\infty} i^m J_m(pr)\, e^{im\theta} \qquad (3.48)$$

For the first of these, we see that $J_m(x)$ is the mth Fourier coefficient of the function $e^{ix\sin\theta}$.

By taking derivatives of the expression for $J_m(x)$ in (3.47) it can be shown that

$$J_m'(x) = \frac{1}{2}[J_{m-1}(x) - J_{m+1}(x)] \quad \text{and} \quad [x^m J_m(x)]' = x^m J_{m-1}(x).$$

And when $m \neq 0$ they satisfy the recurrence formula

$$J_m(x) = \frac{x}{2m}[J_{m-1}(x) + J_{m+1}(x)].$$

The Bessel functions also satisfy certain "addition formulas" including [19]

$$J_n(x+y) = \sum_{m=-\infty}^{\infty} J_{n-m}(x) J_m(y).$$

And by combining some results presented in [6],

$$\sum_{m=-\infty}^{\infty} J_m(x) J_{n+m}(x) = \delta_{n,0}.$$

The Bessel functions also satisfy the normalization conditions [1, 116]

$$\sum_{m=-\infty}^{\infty} J_m(x) = 1 \quad \text{and} \quad \int_0^{\infty} J_m(x)\, dx = 1$$

and other integrals including [1, 13]

$$\int_0^{\infty} e^{-ax^2} J_m(bx)\, x^{m+1}\, dx = \frac{b^m}{(2a)^{m+1}} e^{-b^2/4a}. \qquad (3.49)$$

The Bessel functions, $\{J_m(x)\}$, for any *fixed* $m = 0,1,2,\ldots$ also satisfy several kinds of orthogonality conditions. Namely, if $\{\rho_n^m = \rho_n^m(\beta)\}$ for $n = 1,2,\ldots$ are nonnegative solutions of the equation

$$\rho_n^m J_m'(\rho_n^m)\cos\beta + J_m(\rho_n^m)\sin\beta = 0 \qquad (3.50)$$

for some fixed $m \geq 0$ and some fixed $0 \leq \beta \leq \pi/2$, then [91, 116]

$$\int_0^1 J_m(\rho_{n_1}^m x) J_m(\rho_{n_2}^m x)\, x\, dx = C_{m,n_1} \delta_{n_1,n_2}$$

for the constant

$$C_{m,n} = \begin{cases} \frac{1}{2} J_{m+1}^2(\rho_n^m) & \text{for } \beta = \pi/2 \\[2mm] \frac{((\rho_n^m)^2 - m^2 + \tan^2\beta) J_m^2(\rho_n^m)}{2(\rho_n^m)^2} & \text{for } 0 \leq \beta < \pi/2. \end{cases}$$

Equation (3.50) results from a separable boundary condition on $y_n(x) \doteq J_m(\rho_n^m x)$ (again for fixed m) at $x = 1$. The scaling of the dependent variable used here is consistent with the eigenvalue $\lambda_n = (\rho_n^m)^2$. From Sturm-Liouville theory it follows that the functions y_{n_1} and y_{n_2} corresponding to $\lambda_{n_1} \neq \lambda_{n_2}$ must be orthogonal as shown above.

The above implies that functions on the interval $[0,1]$ that are square-integrable with respect to $x\, dx$ can be expanded in a *Fourier-Bessel series*[10] for any fixed $m \in \{0,1,2,\ldots\}$

---

[10] Also called *Dini's series*

as

$$f(x) = \sum_{n=1}^{\infty} c_{n,m} J_m(\rho_n^m x),$$ (3.51)

where

$$c_{n,m} = \frac{1}{C_{m,n}} \int_0^1 f(x) J_m(\rho_n^m x) \, x \, dx.$$

In analogy with the way the concept of the Fourier series of functions on the circle is extended to the Fourier transform of functions on the line, (3.51) can be extended to a transform based on Bessel functions called the mth *order Hankel Transform*:[11]

$$\hat{f}_m(p) = \int_0^{\infty} f(x) J_m(px) \, x \, dx,$$ (3.52)

which has an inverse of the same form:

$$f(x) = \int_0^{\infty} \hat{f}_m(p) J_m(px) \, p \, dp.$$ (3.53)

The use of the "hat" notation here should not be confused with the same notation used for the Fourier transform. The context will make clear which transform is being used. The orthogonality and completeness relations that allow the inversion formula (3.53) to work are

$$\int_0^{\infty} J_m(px) J_m(p'x) \, x \, dx = \frac{1}{p} \delta(p - p')$$

and

$$\int_0^{\infty} J_m(px) J_m(px') \, p \, dp = \frac{1}{x} \delta(x - x').$$

Interestingly, because of the symmetry of the arguments, these formulas are essentially one and the same.

3.6.2 Spherical Bessel Functions

For the case when $d = 3$, (3.44) becomes

$$y'' + \frac{2}{x} y' + \left(\lambda - \frac{l(l+1)}{x^2} \right) y = 0$$

where m^2 is replaced with $l(l+1)$. The reason for this becomes clear in Chapter 4 in the context of orthogonal expansions in spherical coordinates. If we let $y = x^{-\frac{1}{2}} z$, then

$$z'' + \frac{1}{x} z' + \left(\lambda - \frac{(m')^2}{x^2} \right) z = 0,$$

and hence $z(x) = c J_{m'}(x)$ for some normalizing constant c where $(m')^2 = l(l+1) + \frac{1}{4} = (l + \frac{1}{2})^2$. When $\lambda = 1$, solutions $y(x)$ are then of the form

$$j_l(x) \doteq \left(\frac{\pi}{2x} \right)^{\frac{1}{2}} J_{l+\frac{1}{2}}(x)$$ (3.54)

---

[11] Named after Hermann Hankel (1839-1873)

for $l = 0, 1, 2, ...$ Therefore, the series representation for $j_l(x)$ is obtained by combining (3.46) with $m = l + 1/2$ and (3.54).

Unlike the functions $J_m(x)$ for integer m, the *spherical Bessel functions* $j_l(x)$ can always be written as finite combinations of elementary trigonometric and rational functions for finite integers l. For instance, $j_0(x) = (1/x) \sin x$, $j_1(x) = (1/x^2) \sin x - (1/x) \cos x$, and $j_2(x) = (3/x^3 - 1/x) \sin x - (3/x^2) \cos x$.

By defining $j_{-1}(x) \doteq (1/x) \cos x$, a generating function for the spherical Bessel functions is [1]

$$\frac{\cos \sqrt{x^2 - 2xt}}{x} = \sum_{l=0}^{\infty} \frac{t^l}{l!} j_{l-1}(x)$$

and they have the Rodrigues formula (which in this case is called Rayleigh's formula) [7]

$$j_l(x) = (-1)^l x^l \left(\frac{1}{x} \frac{d}{dx} \right)^l \left(\frac{\sin x}{x} \right)$$

and satisfy the recurrence relations [7]

$$j_{l-1}(x) + j_{l+1}(x) = \frac{2l+1}{x} j_l(x) \qquad l\, j_{l-1}(x) - (l+1) j_{l+1}(x) = (2l+1) j_l'(x) \quad (3.55)$$

and

$$[x^{l+1} j_l(x)]' = x^{l+1} j_{l-1}(x) \qquad [x^{-l} j_l(x)]' = x^{-l} j_{l+1}(x). \quad (3.56)$$

Here again, as a special case of Sturm-Liouville theory, the functions $y_n(x) = j_l(\rho_n^l x)$ for any fixed $l \in \{0, 1, 2, ...\}$ are orthogonal on the finite domain $0 \le x \le 1$ where $\rho_n^l = \rho_n^l(\beta)$ is determined by the separable boundary condition

$$y_n'(1) \cos \beta + y_n(1) \sin \beta = 0. \quad (3.57)$$

For $0 \le \beta \le \pi/2$, this orthogonality is [91]

$$\int_0^1 j_l(\rho_{n_1}^l x)\, j_l(\rho_{n_2}^l x)\, x^2 dx = D_{n_1,l}\, \delta_{n_1,n_2}$$

where

$$D_{n,l} = \begin{cases} \frac{\pi}{4} (J_{l+1/2}')^2 (\rho_n^l) & \text{for } \beta = \pi/2 \\ \pi(\rho_n^2 + \tan^2 \beta - (m')^2) J_{l+1/2}^2 (\rho_n^l) & \text{for } 0 \le \beta < \pi/2, \end{cases}$$

$m' = l + \frac{1}{2}$, and series analogous to (3.51) can be constructed.

When $\beta = \pi/2$, $\{\rho_n^l\}$ are simply the solutions to $j_l(\rho_n^l x) = 0$.

A continuous transform on $\mathbb{R}_{\ge 0}$ exists. It is the *spherical Hankel transform* [7]:

$$\hat{f}_l(p) = \int_0^\infty f(x) j_l(px) x^2 dx, \quad (3.58)$$

which has the inverse

$$f(x) = \frac{2}{\pi} \int_0^\infty \hat{f}_l(p) j_l(px) p^2 dp. \quad (3.59)$$

The corresponding completeness/orthogonality relation is [7]

$$\int_0^\infty j_l(px) j_l(p'x) x^2 dx = \frac{\pi}{2p^2} \delta(p - p').$$

3.6.3 The Bessel Polynomials

The *Bessel polynomials* [69], $y_n(x)$, are the functions that satisfy the Sturm-Liouville equation

$$(x^2 e^{-2/x} y_n')' = n(n+1)\, e^{-2/x} y_n.$$

This is equivalent to the equation

$$x^2 y_n'' + 2(x+1) y_n' = n(n+1) y_n.$$

These polynomials satisfy the three-term recurrence relation

$$y_{n+1} = (2n+1)x y_n + y_{n-1},$$

and the first few are of the form $y_0(x) = 1$, $y_1(x) = 1 + x$, $y_2(x) = 1 + 3x + 3x^2$. They can be generated with the Rodrigues formula

$$y_n(x) = \frac{e^{2/x}}{2^n} \frac{d^n}{dx^n}(x^{2n} e^{-2/x}).$$

The orthogonality relation for these polynomials is [69]:

$$\frac{1}{2\pi i} \oint_{\mathbb{S}^1} y_m(z) y_n(z)\, e^{-2/z}\, dz = (-1)^{n+1} \frac{2}{2n+1} \delta_{m,n}$$

where $\mathbb{S}^1$ is the unit circle in the complex plane centered at the origin. (This can be replaced with any other closed curve containing the origin.) Choosing $\mathbb{S}^1$ and $z = e^{i\theta}$,

$$\frac{1}{2\pi} \int_0^{2\pi} y_m(e^{i\theta}) y_n(e^{i\theta}) \exp(i\theta - 2e^{-i\theta})\, d\theta = (-1)^{n+1} \frac{2}{2n+1} \delta_{m,n}.$$

These polynomials are connected to the spherical Bessel functions through the relationship [69]

$$j_n(x) = (1/2x)[i^{-n-1} e^{ix} y_n(-1/ix) + i^{n+1} e^{-ix} y_n(1/ix)]. \tag{3.60}$$

For further reading on Bessel polynomials see [47].

3.6.4 Fast Numerical Transforms

In many applications it is desirable to compute either the Hankel or the spherical-Hankel transform as efficiently and accurately as possible (see, e.g., [2, 22, 26, 36, 48, 53, 58, 61, 62, 73, 74, 75, 78, 84, 85, 90, 97, 103, 104]). Unfortunately, it seems that unlike the FFT for the exact computation of Fourier series sampled at a finite number of evenly spaced points, a fast, exact Hankel transform is yet to be formulated. A number of efficient and accurate (though not exact) numerical approximations have been reported in the literature. A review can be found in [27].

We are aware of four main categories of fast numerical Hankel transform algorithms. These are: (1) approximation of a compactly supported function in a Fourier-Bessel series and use of approximate quadrature rules; (2) conversion of the Hankel transform to the Fourier transform by a change of variables and interpolation; (3) expansion of a function in a band-limited series of Laguerre polynomials and closed-form analytical computation of Hankel transform; and (4) approximation of Bessel functions using asymptotic expansions. Below we review these methods in some detail.

Sampling the Fourier-Bessel Series

If we assume that the independent variable is normalized so that $f(x) = 0$ for $x \geq 1$, then the Hankel transform becomes the equation for the coefficients of the the Fourier-Bessel (Dini) series where $\beta = \pi/2$ and $J_n(x_m) = 0$ in (3.50).

Sampling the integral at quadrature points results in a sum. Since the Bessel functions are not polynomials, this will not result in an exact quadrature. Using (3.13) is also possible, but this is slow. Unfortunately, the numerical stability is poor when using the recurrence relations for Bessel functions [41, 43, 118], and we know of no exact fast algorithm.

Converting to the Fourier Transform

A number of methods use tricks that convert the Hankel transform to a form where the Fourier transform and/or the convolution theorem can be used.

Candel [15, 16, 17] uses the relationship

$$e^{ir \sin \theta} = \sum_{k=-\infty}^{+\infty} e^{ik\theta} J_k(r)$$

with $r = px$ to write

$$\int_0^\infty e^{ipx \sin \theta} f(x)\, x\, dx = \sum_{k=-\infty}^{+\infty} e^{ik\theta} \hat{f}_k(p)$$

where $\hat{f}_k(p)$ denotes the kth-order Hankel transform of $f(x)$. The left side of the above is the Fourier transform of the function

$$F(x) = \begin{cases} xf(x) & \text{for } x > 0 \\ 0 & \text{for } x \leq 0 \end{cases}$$

evaluated at frequency $p \sin \theta$. The right side indicates that the kth order Hankel transform of $f(x)$ is the kth Fourier series coefficient of the expansion on the right. Multiplying both sides by $e^{-ik\theta}$, integrating over $0 \leq \theta < 2\pi$ and dividing by 2π isolates $\hat{f}_k(p)$. Discretizing the above steps provides a numerical procedure for approximate calculation of the Hankel transforms. A similar result is obtained by using the projection-slice and back-projection theorems [87, 88] discussed in Chapter 13.

Another trick is to change variables so that the independent variables are of the form [105, 109]:

$$x = x_0 e^{\alpha \xi} \qquad p = p_0 e^{\alpha \eta} \tag{3.61}$$

where x_0, p_0, and α are constants that are chosen for good numerical performance [105]. Substituting (3.61) into the kth order Hankel transform results in the expression

$$\tilde{g}(\eta) = \int_{-\infty}^\infty \tilde{f}(\xi) \tilde{h}(\xi + \eta)\, d\xi \tag{3.62}$$

where $\tilde{f}(\xi) = xf(x)$, $\tilde{g}(\eta) = pg(p)$, and $\tilde{h}(\xi+\eta) = \alpha xp J_k(xp)$. Since (3.62) is of the form of a correlation (a convolution with a sign difference), the FFT can be used when $\tilde{f}$ and $\tilde{h}$ are approximated as band-limited functions. The drawbacks of this method are that the band-limited approximations are not exact, and the use of unequally (exponentially) spaced sample points in the original domain is awkward.

Using the Laguerre Series Expansion

The method described below was developed in [20]. Recall that the associated Laguerre polynomials are an orthogonal basis for $\mathcal{L}^2(\mathbb{R}_{\geq 0}, x^m e^{-x} \, dx)$, and when the normalization in (3.27) is used,

$$\int_0^\infty \tilde{L}_n^m(x)\tilde{L}_p^m(x)x^m e^{-x} \, dx = \frac{(n+m)!}{n!}\delta_{p,n}.$$

The *associated-Laguerre functions* (also called *Gauss-Laguerre functions*) are defined as

$$\varphi_{n,m}(x) \doteq N_{n,m}x^{m/2}\tilde{L}_n^m(x)\,e^{-x/2},$$

where $N_{n,m}$ is the normalization so that

$$\int_0^\infty \varphi_{n,m}(x)\,\varphi_{p,m}(x)\,dx = \delta_{n,p}.$$

These functions constitute an orthonormal basis for $\mathcal{L}^2(\mathbb{R}_{\geq 0})$. They have the property [20, 108][12]

$$\frac{1}{2}\int_0^\infty J_m(\sqrt{zw})\,\varphi_{n,m}(z)\,dz = (-1)^n\varphi_{n,m}(w)$$

or equivalently, letting $z = x^2$ and $w = p^2$

$$\int_0^\infty x^m\tilde{L}_n^m(x^2)\,e^{-x^2/2}J_m(px)\,x\,dx = (-1)^n p^m\tilde{L}_n^m(p^2)\,e^{-p^2/2}. \tag{3.63}$$

Therefore if $f(x) \in \mathcal{L}^2(\mathbb{R}_{\geq 0})$ is expanded in a Gauss-Laguerre series as

$$f(x) = \sum_{m=0}^\infty C_n^m\varphi_{n,m}(x^2)$$

where

$$C_n^m = \int_0^\infty f(x)\varphi_{n,m}(x^2)\,d(x^2),$$

then

$$\hat{f}_m(p) = \int_0^\infty f(x)J_m(px)\,x\,dx = \sum_{m=0}^\infty (-1)^n C_n^m\varphi_{n,m}(p^2).$$

This is also a Gauss-Laguerre series. And if f has band limit B (i.e., $C_n^m = 0$ for all $m \geq B$) then $\hat{f}_m(p)$ will also have band-limit B.

The approximation required in this method is to calculate the coefficients C_n^m numerically. A recursive numerical procedure for doing this is explained in [20].

For functions that are not band-limited, it is sometimes useful to introduce a scale parameter α so that $x \to \alpha x$ to minimize the number of terms in the series expansion for a given error criterion.

Asymptotic Expansions

In this approximate method, the asymptotic form of Bessel functions for large values of their argument is used to gain simplifications. We shall not consider this method in detail, and refer the reader to [89] for descriptions of this method.

---

[12]This relationship was originally derived by Sonine in 1880, and can be obtained by taking derivatives of (3.49) with respect to a and b and using the recurrence relations or generating functions for $\tilde{L}_n^m(x)$ and $J_m(px)$.

3.7 Hypergeometric Series and Gamma Functions

The generalized *hypergeometric series* is defined as

$$
{}_pF_q(\alpha_1, \alpha_2, ..., \alpha_p; \rho_1, \rho_2, ..., \rho_q; z) \doteq \sum_{k=0}^{\infty} \frac{z^k}{k!} \left[\frac{\prod_{h=1}^{p}(\alpha_h)_k}{\prod_{l=1}^{q}(\rho_l)_k} \right]. \tag{3.64}
$$

where for $k \in \mathbb{Z}^+$ and for $a = \alpha$ or ρ,

$$
(a)_k \doteq \prod_{j=0}^{k-1}(a+j) = a(a+1)\cdots(a+k-1) \quad \text{and} \quad (a)_0 \doteq 1.
$$

The term $(a)_n$ is called the *(ascending or rising) Pochhammer symbol*. The series in (3.64) terminates if one of the α_i is zero or a negative integer.

Using the above notation, the *binomial coefficient* is written as

$$
C_n^k = \binom{n}{k} = \frac{n!}{k!(n-k)!} = \frac{(-1)^k(-n)_k}{k!}.
$$

The compact notation

$$
{}_pF_q(\alpha_1, \alpha_2, ..., \alpha_p; \rho_1, \rho_2, ..., \rho_q; z) = {}_pF_q \left(\begin{matrix} \alpha_1, \alpha_2, ..., \alpha_p \\ \rho_1, \rho_2, ..., \rho_q \end{matrix} \middle| z \right)
$$

is often used. The most commonly encountered values are $(p, q) = (2, 1)$ and $(p, q) = (1, 1)$:

$$
{}_2F_1(a, b; c; z) = {}_2F_1 \left(\begin{matrix} a, b \\ c \end{matrix} \middle| z \right) = \sum_{k=0}^{\infty} \frac{(a)_k(b)_k}{(c)_k} \frac{z^k}{k!}
$$

$$
{}_1F_1(a; c; z) = {}_1F_1 \left(\begin{matrix} a \\ c \end{matrix} \middle| z \right) = \sum_{k=0}^{\infty} \frac{(a)_k}{(c)_k} \frac{z^k}{k!}.
$$

The former is called the *hypergeometric function*, and the latter is called the *confluent hypergeometric function*.

The *gamma function* is defined for complex argument z when $Re(z) > 0$ as [76]

$$
\Gamma(z) = \int_0^{\infty} e^{-t} t^{z-1} \, dt.
$$

It satisfies the difference equation

$$
\Gamma(z+1) = z\Gamma(z),
$$

and for positive integer argument

$$
\Gamma(n) = (n-1)!
$$

The gamma function is related to the coefficients in the hypergeometric series as

$$
(a)_k = \frac{\Gamma(a+k)}{\Gamma(a)}.
$$

The hypergeometric and confluent hypergeometric functions can be written as the integrals [112]:

$$_2F_1(a, b; c; z) = \frac{\Gamma(c)}{\Gamma(a)\Gamma(c-a)} \int_0^1 (1 - tz)^{-b} t^{a-1} (1-t)^{c-a-1} \, dt$$

and

$$_1F_1(a; c; z) = \frac{\Gamma(c)}{\Gamma(a)\Gamma(c-a)} \int_0^1 e^{zt} t^{a-1} (1-t)^{c-a-1} \, dt.$$

The classical functions of mathematical physics such as Bessel functions and orthogonal polynomials can be written in terms of hypergeometric functions and gamma functions. Below are some of these expressions. Many more such relationships can be found in [1, 76, 112], including

$$e^z = {}_0F_0(z);$$

$$(1+z)^a = {}_1F_0(-a; -z) \quad \text{for} \quad |z| < 1;$$

$$z^{-1} \ln(1+z) = {}_2F_1(1, 1; 2; -z) \quad \text{for} \quad |z| < 1;$$

$$z^{-1} \text{Erf}(z) \doteq z^{-1} \int_0^z e^{-x^2} dx = {}_1F_1(\frac{1}{2}; \frac{3}{2}; -z^2);$$

$$J_\nu(z) = \frac{(z/2)^\nu e^{\pm iz}}{\Gamma(\nu+1)} \, {}_1F_1(\frac{1}{2} + \nu; 1 + 2\nu; \mp 2iz);$$

$$P_\nu^\mu(z) = \frac{[(z+1)/(z-1)]^{\mu/2}}{\Gamma(1-\mu)} \, {}_2F_1(-\nu, \nu+1; 1-\mu; (1-z)/2);$$

$$P_n^{(\alpha,\beta)}(z) = \frac{\Gamma(n+\alpha+1)}{n!\Gamma(\alpha+1)} \, {}_2F_1(-n, n+\alpha+\beta+1; \alpha+1; (1-z)/2);$$

$$\tilde{L}_n^\alpha(z) = \frac{\Gamma(n+\alpha+1)}{n!\Gamma(\alpha+1)} \, {}_1F_1(-n; \alpha+1; z).$$

The hypergeometric functions are convenient for defining the polynomials in the subsections that follow.

3.7.1 Hahn Polynomials

Whereas quadrature rules induce discrete orthogonality relations from continuous ones, a number of twentieth-century polynomials are defined with discrete orthogonality in mind from the beginning.

The *Hahn polynomials* [51, 64] are defined for $\alpha > -1$ and $\beta > -1$ and for positive integer N as

$$Q_n(x; \alpha, \beta, N) = {}_3F_2(-n, -x, n+\alpha+\beta+1; \alpha+1, -N+1; 1). \quad (3.65)$$

For fixed values of α and β, Q_n is a polynomial in the variable x that satisfies two discrete orthogonalities. These are

$$\sum_{x=0}^{N-1} Q_n(x;\alpha,\beta,N)\,Q_m(x;\alpha,\beta,N)\,\mu(x;\alpha,\beta,N) = \delta_{m,n}/\nu_n(\alpha,\beta,N)$$

and

$$\sum_{n=0}^{N-1} Q_n(x;\alpha,\beta,N)\,Q_n(y;\alpha,\beta,N)\,\nu_n(\alpha,\beta,N) = \delta_{x,y}/\mu(x;\alpha,\beta,N)$$

where

$$\mu(x;\alpha,\beta,N) = \frac{\left(\begin{matrix}\alpha+x\\x\end{matrix}\right)\left(\begin{matrix}\beta+N-1-x\\N-1-x\end{matrix}\right)}{\left(\begin{matrix}N+\alpha+\beta\\N-1\end{matrix}\right)}$$

and

$$\nu_n(\alpha,\beta,N) = \frac{\left(\begin{matrix}N-1\\n\end{matrix}\right)}{\left(\begin{matrix}N+\alpha+\beta+n\\n\end{matrix}\right)} \times \frac{\Gamma(\beta+1)}{\Gamma(\alpha+1)\Gamma(\alpha+\beta+1)} \times$$

$$\frac{\Gamma(n+\alpha+1)\Gamma(n+\alpha+\beta+1)}{\Gamma(n+\beta+1)\Gamma(n+1)} \times \frac{(2n+\alpha+\beta+1)}{(\alpha+\beta+1)}.$$

The Hahn polynomials satisfy a three-term recurrence relation and a difference equation that can be found in [64].

3.7.2 Charlier, Krawtchouk, and Meixner Polynomials

In this subsection, we review three modern polynomials with discrete orthogonalities.
The functions

$$C_n(x,a) \doteq \frac{\Gamma(x+1)(-a)^{-n}}{\Gamma(x-n+1)}\,{}_1F_1(-n;x-n+1;a)$$

are called *Charlier polynomials*. When the argument x is chosen from the numbers $0,1,2,\ldots$ the symmetry

$$C_n(x,a) = C_x(n,a) \tag{3.66}$$

is observed. They satisfy two discrete orthogonalities:

$$\sum_{x=0}^{\infty} \frac{a^x}{x!} C_k(x,a) C_p(x,a) = k!\,e^a a^{-k}\delta_{kp}$$

and

$$\sum_{n=0}^{\infty} \frac{a^n}{n!} C_n(x,a) C_n(y,a) = x!\,e^a a^{-x}\delta_{xy}.$$

The second orthogonality results from the first orthogonality and the symmetry (3.66).
The *Krawtchouk polynomials* of degree s and argument x are defined as

$$K_s(x;p;N) \doteq \frac{(-1)^s}{p^s C_N^s} P_s^{(x-s,N-s-x)}(1+2p) = {}_2F_1(-x;-s;-N;p^{-1}).$$

They satisfy the discrete orthogonality [112]

$$\sum_{x=0}^{N} C_N^x p^x (1-p)^{N-x} K_s(x; p; N) K_q(x; p; N) = \frac{1}{C_N^s} \left(\frac{1-p}{p}\right)^s \delta_{sq}$$

(where C_N^x is the binomial coefficient) and the symmetry relation

$$K_s(x; p; N) = K_x(s; p; N).$$

Combining the above two equations, we find the second orthogonality:

$$\sum_{s=0}^{N} C_N^s p^s (1-p)^{N-s} K_s(x; p; N) K_s(y; p; N) = \frac{1}{C_N^x} \left(\frac{1-p}{p}\right)^x \delta_{xy}.$$

The *Meixner polynomials* [81, 82, 112] are defined as

$$M_s(x; \gamma; c) \doteq s! P_s^{(\gamma-1, -s-x-\gamma)} \left(\frac{2}{c} - 1\right) \doteq \frac{\Gamma(\gamma+s)}{\Gamma(\gamma)} {}_1F_1(-x; -s; \gamma; 1-1/c)$$

for $0 < c < 1$ and $\gamma > 0$. The symmetry relation

$$M_s(x; \gamma; c) = \frac{\Gamma(\gamma+s)}{\Gamma(\gamma+x)} M_x(s; \gamma; c)$$

holds, as does the discrete orthogonality relation

$$\sum_{x=0}^{\infty} \frac{c^x (\gamma+x-1)!}{x!} M_s(x; \gamma; c) M_q(x; \gamma; c) = s!(\gamma+s-1)! c^{-s} (1-c)^{-\gamma} \delta_{sq}.$$

Similar to the Charlier and Krawtchouk polynomials, a second orthogonality results from the symmetry relation:

$$\sum_{s=0}^{\infty} \frac{c^s (\gamma+s-1)!}{s!} M_s(x; \gamma; c) M_s(y; \gamma; c) = x!(\gamma+x-1)! c^{-x} (1-c)^{-\gamma} \delta_{xy}$$

3.7.3 Zernike Polynomials

The *Zernike polynomials* [120] are defined as

$$Z_n^m(x) \doteq \sum_{s=0}^{(n-|m|)/2} (-1)^s \frac{(n-s)!}{s! \left(\frac{n+|m|}{2} - s\right)! \left(\frac{n-|m|}{2} - s\right)!} x^{n-2s} \tag{3.67}$$

for $0 \le m \le n$ with $n \equiv m \mod 2$. They satisfy the differential equation

$$x(1-x^2)y'' + (1-3x^2)y' + [n(n+2)x - m^2/x]y = 0,$$

and can be written in terms of the hypergeometric function as

$$Z_n^m(x) = (-1)^{\frac{n-m}{2}} \left(\frac{\frac{1}{2}(n+m)}{m}\right) x^m {}_2F_1 \left(\frac{n+m+1}{2}, \frac{m-n}{2}; m+1, x^2\right).$$

They are orthogonal on the interval $[0, 1]$ with respect to the weight $w(x) = x$:

$$\int_0^1 Z_n^m(x)\, Z_{n'}^m(x)\, x\, dx = \frac{1}{2n+2}\delta_{n,n'}.$$

They satisfy the Rodrigues formula

$$Z_n^m(x) = \frac{x^{-m}}{\left(\frac{n-m}{2}\right)!}\left(\frac{d}{d(x^2)}\right)^{\frac{n-m}{2}}\left[x^{n+m}(x^2-1)^{\frac{n-m}{2}}\right]$$

and can be related to the Jacobi polynomials as

$$Z_n^m(x) = x^m\frac{P_{\frac{n-m}{2}}^{(0,m)}(2x^2-1)}{P_{\frac{n-m}{2}}^{(0,m)}(1)}.$$

The normalization is chosen so that $Z_n^m(1) = 1$. The first few are $Z_0^0(x) = 1$, $Z_1^1(x) = x$, $Z_2^0(x) = 2x^2 - 1$, $Z_2^2(x) = x^2$, $Z_3^1(x) = 3x^2 - 2x$.

Zernike showed that

$$\int_0^1 Z_n^m(x)\, J_m(px)\, x\, dx = (-1)^{\frac{n-m}{2}}p^{-1}J_{n+1}(p).$$

In Chapter 4 we show how Zernike polynomials can be used as part of multi-dimensional expansions.

3.7.4 Other Polynomials with Continuous and Discrete Orthogonalities

A set of functions that generalizes the Zernike polynomials has been developed by T. Koornwinder (see [68] and references therein) which are orthogonal on the interval $[0,1]$ with respect to the weight $(1-x^2)^\alpha$ for $\alpha > -1$.

A collection of polynomials of complex argument (called *Wilson polynomials*) defined as

$$\mathcal{W}_n(z^2; a, b, c, d) \doteq (a+b)_n(a+c)_n(a+d)_n\; {}_4F_3\left(\begin{array}{c}-n,\, a+b+c+d+n-1,\, a-z,\, a+z\\ a+b,\, a+c,\, a+d\end{array}\bigg| 1\right)$$

and satisfying certain orthogonalities can be found in [112, 117].

Sets of polynomials with discrete orthogonality related to topics covered in Chapter 9 can be found in [9, 95].

3.8 Piecewise Constant Orthogonal Functions

Sturm-Liouville theory is over a century old. The recent advances in discrete polynomial transforms reviewed in Section 3.5 have deep historical roots as well. In contrast, the focus of this section is on non-polynomial orthonormal expansions and transforms developed in the twentieth century. In particular, we review the Haar and Walsh functions as well as the basics of wavelets.

In the twentieth century, a number of new classes of non-polynomial orthogonal expansions on intervals and the real line were developed. One such class consists of functions that are piecewise constant over intervals. Three examples of this class are the Haar, Rademacher, and Walsh functions. All of these functions are defined for the independent variable $x \in [0, 1]$, are indexed by the number $m = 0, 1, 2, \dots$. They return values in the set $\{0, \pm C_m\}$ for a real constant C_m. They are all real-valued and orthonormal with respect to the inner product

$$(f_1, f_2) = \int_0^1 f_1(x)\, f_2(x)\, dx.$$

3.8.1 Haar Functions

The *Haar function* $\mathrm{har}(0, 1, x)$ is defined as [50]

$$\mathrm{har}(0, 1, x) \doteq \begin{cases} 1 & \text{for } x \in (0, 1/2) \\ -1 & \text{for } x \in (1/2, 1) \ . \\ 0 & \text{elsewhere} \end{cases}$$

Higher order Haar functions are defined as

$$\mathrm{har}(r, m, x) \doteq \begin{cases} 2^{r/2} & \text{for } \frac{m-1}{2^r} < x < \frac{m-1/2}{2^r} \\ -2^{r/2} & \text{for } \frac{m-1/2}{2^r} < x < \frac{m}{2^r} \\ 0 & \text{elsewhere} \end{cases} \ .$$

Here $1 \le m \le 2^r$, and the only Haar function not defined by the above rule is $\mathrm{har}(0, 0, x) \doteq 1$.

We note that the Haar system of functions is orthonormal and complete, and the definition of $\mathrm{har}(r, m, x)$ for $r = 0, 1, 2, \dots$ and $m = 1, \dots, 2^r$ can be written in terms of the recurrence relations:

$$\mathrm{har}(r, 1, x) = \sqrt{2}\, \mathrm{har}(r-1, 1, 2x)$$

and

$$\mathrm{har}(r, m, x) = \mathrm{har}(r, m-1, x - 2^{-r}).$$

The Haar functions form a complete orthonormal system, and so any function $f(x) \in \mathcal{L}^2([0, 1])$ can be expanded in a Haar series as

$$f(x) = \sum_{n=0}^{\infty} h_n\, \mathrm{har}(n, x)$$

where

$$h_n = \int_0^1 f(x)\, \mathrm{har}(n, x)\, dx.$$

If $f(x)$ is piecewise constant over $N = 2^r$ evenly spaced sub-intervals of $[0, 1]$ then the *finite Haar transform* pair results from truncating the sum and evaluating the integral above:

$$h_n = \frac{1}{N} \sum_{i=0}^{N-1} f(i/N)\, \mathrm{har}(n, i/N).$$

and

$$f(i/n) = \sum_{n=0}^{N-1} h_n\, \mathrm{har}(n, i/N).$$

A fast algorithm exists for evaluating the finite Haar transform. It can be performed in $\mathcal{O}(N)$ arithmetic operations [99].

3.8.2 Rademacher Functions

The *Rademacher function* [94] $\operatorname{rad}(m, x)$ is defined for $x \in [0, 1]$ and returns values ± 1 at all but a finite number of points. The Rademacher functions are generated using the equality

$$\operatorname{rad}(m, x) = \operatorname{rad}(1, 2^{m-1} x),$$

or equivalently, the one-term recurrence relation

$$\operatorname{rad}(m, x) = \operatorname{rad}(m - 1, 2x),$$

where

$$\operatorname{rad}(1, x) = \begin{cases} 1 & \text{for } x \in (0, 1/2) \\ -1 & \text{for } x \in (1/2, 1) \\ 0 & \text{elsewhere} \end{cases} .$$

The exception to the above rule for calculating $\operatorname{rad}(m, x)$ is $\operatorname{rad}(0, x) = 1$. They observe the periodicity [3]:

$$\operatorname{rad}(m, t + n/2^{m-1}) = \operatorname{rad}(m, t)$$

for $m = 1, 2, \ldots$ and $n = \pm 1, \pm 2, \ldots$. These periodicities are apparent when writing the Rademacher functions in the form

$$\operatorname{rad}(m, x) = \operatorname{sign}[\sin 2^m \pi x].$$

While the Rademacher functions are orthonormal, they are not complete in the sense that an arbitrary $f \in \mathcal{L}^2([0, 1])$ cannot be approximated to any desired accuracy in the $\mathcal{L}^2$ sense.

3.8.3 Walsh Functions and Transforms

In order to understand Walsh functions [113], we first need to introduce some notation for binary numbers. The value of the sth bit of an r-bit binary number b is denoted $b_s \in \{0, 1\}$ for $s = 0, 1, \ldots, r - 1$. Here $s = 0$ and $s = r - 1$ correspond to the most and least significant bit, respectively. Hence, in the binary number system we would write

$$b = b_0 b_1 \cdots b_{r-1} = \sum_{s=0}^{r-1} b_s 2^{r-1-s}.$$

Given the integers $n, k \in [0, 2^r - 1]$, their s^{th} bit values when written as binary numbers are n_s and k_s. We define the coefficient [65]:

$$W_k(n/2^r) = (-1)^{\sum_{s=0}^{r-1} k_{r-1-s} n_s} = \exp\left(\pi i \sum_{s=0}^{r-1} k_{r-1-s} n_s \right).$$

This can be related to the Rademacher functions as

$$W_k(n/2^r) = \prod_{s=0}^{r-1} (\operatorname{rad}_{s+1}(n/2^r))^{k_{r-1-s}}.$$

It is also convenient for all $x, a, b \in \mathbb{R}$ with $a < b$ to define the *window function*

$$\text{win}(x, a, b) = \begin{cases} 1 & \text{for } x \in (a, b) \\ 0 & \text{for } x \notin [a, b] \\ \frac{1}{2} & \text{for } x \in \{a, b\} \end{cases}.$$

This function has the property

$$\text{win}(x, c \cdot a, c \cdot b) = \text{win}(x/c, a, b)$$

for any $c \in \mathbb{R}_{>0}$.

The *Walsh function* $\text{wal}(k, x)$ is defined on the interval $[0, 1]$ as

$$\text{wal}(k, x) = \sum_{n=0}^{2^r - 1} W_k(n/2^r) \cdot \text{win}(x \cdot 2^r, n, n+1). \tag{3.68}$$

When $x = m/N$ where $N = 2^r$ for some integer $0 < m < N$ we see that

$$\text{wal}(k, m/N) = W_k(m/N)\,\delta_{m,n} = W_k(n/N)\,\delta_{n,m} = \text{wal}(m, k/N).$$

In addition to the orthonormality

$$(\text{wal}(m, x), \text{wal}(n, x)) = \delta_{m,n},$$

the Walsh functions satisfy the discrete orthogonality [11]:

$$\sum_{i=0}^{N-1} \text{wal}(m, i/N)\,\text{wal}(n, i/N) = N\delta_{m,n}$$

where $N = 2^r$.

The Walsh functions form a complete orthonormal system, and so any function $f(x) \in \mathcal{L}^2([0, 1])$ can be expanded in a Walsh series as

$$f(x) = \sum_{n=0}^{\infty} w_n\,\text{wal}(n, x)$$

where

$$w_n = \int_0^1 f(x)\,\text{wal}(n, x)\,dx.$$

If f is piecewise constant over $N = 2^r$ evenly spaced sub-intervals of $[0, 1]$ then the *finite Walsh transform* pair results from truncating the sum and evaluating the integral above:

$$w_n = \frac{1}{N} \sum_{i=0}^{N-1} f(i/N)\,\text{wal}(n, i/N) = \frac{1}{N} \sum_{i=0}^{N-1} f(i/N)\,W_k(n/N)$$

and

$$f(i/n) = \sum_{n=0}^{N-1} w_n\,\text{wal}(n, i/N) = \sum_{n=0}^{N-1} w_n\,W_k(n/N).$$

Walsh functions can be generated by a recurrence relation, and fast algorithms exist for evaluating these [11, 54]. One of these algorithms is called the *Walsh-Hadamard transform*, and can be performed in $\mathcal{O}(N \log_2 N)$ arithmetic operations.

See [3], [11], and [65] for modern reviews of applications and algorithms for fast implementations.

3.9 Wavelets

3.9.1 Continuous Wavelet Transforms

The basic reason for the popularity of wavelets since the mid 1980's is that certain functions can be expressed very efficiently (in the sense of requiring few terms for good approximation) as a weighted sum (or integral) of scaled, shifted, and modulated versions of a single function ("mother wavelet") with certain properties. The concept of wavelets initially entered the recent literature as a technique for the analysis of seismic measurements as described by J. Morlet et al [86]. The connection between the concept of wavelets and group representations (the latter of which is a large part of Chapter 8) was established shortly thereafter by Grossmann, Morlet and Paul [45, 46]. I. Daubechies (see, e.g., [28]), Y. Meyer (see, e.g., [83]), S. Mallat (see, e.g., [79]) and others contributed further to a rigorous theory behind the concept of wavelets, connected the idea to other techniques of classical harmonic analysis, and popularized the concept. Since then, there has been an explosion of work in both the theory and applications of wavelets. While wavelets can be superior to other expansions (in the sense or requiring fewer terms to achieve the same least-squares error), when a high degree of self-similarity exists in the signal or function being decomposed, they do have some drawbacks. These include poor performance under traditional operations such as convolution and differentiation.

The three basic operations upon which wavelet analysis is based are scale (dilation), translation (shift) and modulation of functions. These are applied to a function f as:

$$(\sigma_s f)(x) = s^{-\frac{1}{2}} f(x/s) \tag{3.69}$$

$$(\tau_t f)(x) = f(x - t) \tag{3.70}$$

$$(\mu_m f)(x) = e^{-imx} f(x), \tag{3.71}$$

respectively. These are *unitary operations* in the sense that $U \in \{\sigma_s, \tau_t, \mu_m\}$ preserves the inner product of two complex-valued functions:

$$(U f_1, U f_2) = (f_1, f_2) \tag{3.72}$$

where

$$(f_1, f_2) = \int_{-\infty}^{\infty} \overline{f_1(x)} f_2(x) \, dx.$$

As a special case, we see

$$\|U f\|^2 = \|f\|^2$$

where $\|f\|^2 = (f, f)$. We note that operators unitary with respect to (f_1, f_2) are unitary with respect to $\overline{(f_1, f_2)}$ also (and vice versa).

Continuous wavelet transforms are based on combinations of continuous translational and scale changes to a primitive function $\varphi \in \mathcal{L}^2(\mathbb{R})$. This is in contrast to the continuous Fourier transform which is based on continuous modulations of the function $\varphi(x) = 1$.

Continuous Scale-Translation-Based Wavelet Transforms

Let $\varphi \in \mathcal{L}^2(\mathbb{R})$ and $\hat{\varphi}$ be its Fourier transform. φ is chosen so that

$$0 < \int_0^{\infty} |\hat{\varphi}(\pm\omega)|^2 \frac{d\omega}{|\omega|} = C_{\pm} < \infty \tag{3.73}$$

(for reasons that will become clear shortly). It follows that

$$0 < \int_{-\infty}^{\infty} |\hat{\varphi}(\pm\omega)|^2 \frac{d\omega}{|\omega|} = C = C_+ + C_- < \infty.$$

A typical (but not the only) choice is

$$\varphi(x) = \frac{2}{\pi^{1/4}\sqrt{3}}(1 - x^2)\,e^{-x^2/2}.$$

Let

$$\varphi_{s,t}^p(x) = |s|^{-p}\varphi\left(\frac{x - t}{s}\right). \tag{3.74}$$

The range of values for scale changes and translations for the independent variable is $s \in \mathbb{R} - \{0\}$ and $t \in \mathbb{R}$. The number $p \in \mathbb{R}_{>0}$ is chosen differently in different papers on wavelets. We follow the notation in [63] and leave p undetermined for the moment.

Since $\varphi \in \mathcal{L}^2(\mathbb{R})$, so is $\varphi_{s,t}^p$ since

$$\|\varphi_{s,t}^p\|^2 = |s|^{-2p}\int_{-\infty}^{\infty}\left|\varphi\left(\frac{x - t}{s}\right)\right|^2 dx = |s|^{1-2p}\|\varphi\|^2.$$

(The choice $p = 1/2$ is common, and natural given the above equation but we will not fix the value of p).

The function $\varphi_{s,t}^p$ is called a *wavelet*, and $\varphi_{1,0}^p = \varphi$ is called the *mother wavelet*. The Fourier transform of $\varphi_{s,t}^p$ is

$$\hat{\varphi}_{s,t}^p(\omega) = \mathcal{F}(\varphi_{s,t}^p) = \int_{-\infty}^{\infty} |s|^{-p}\varphi\left(\frac{x - t}{s}\right)e^{-i\omega x}\,dx = |s|^{1-p}e^{-i\omega t}\hat{\varphi}(\omega s).$$

(The choice $p = 1$ is also natural to simplify the above equation.)

The *continuous wavelet transform* of the function $f \in \mathcal{L}^2(\mathbb{R})$ associated with the wavelet $\varphi_{s,t}^p$ is

$$\tilde{f}_p(s,t) = (\varphi_{s,t}^p, f) = \int_{-\infty}^{\infty} \overline{\varphi_{s,t}^p(x)}f(x)\,dx = \overline{\varphi_{-s,0}^p(t)} * f(t). \tag{3.75}$$

The derivation of the inverse of the continuous wavelet transform procedes as follows. Applying the Fourier transform to both sides of (3.75) and using the convolution theorem, we see that

$$\mathcal{F}\left(\overline{\varphi_{-s,0}^p(t)} * f(t)\right) = \mathcal{F}\left(\overline{\varphi_{-s,0}^p(t)}\right)\hat{f}(\omega).$$

Explicitly,

$$\mathcal{F}\left(\overline{\varphi_{-s,0}^p(t)}\right) = \int_{-\infty}^{\infty} |s|^{-p}\overline{\varphi(-t/s)}e^{-i\omega t}\,dt$$

$$= \int_{\mathrm{sgn}(s)\cdot\infty}^{-\mathrm{sgn}(s)\cdot\infty} |s|^{-p}\overline{\varphi(\tau)}e^{-i\omega t}(-s\,dt)$$

$$= |s|^{1-p}\overline{\hat{\varphi}(-s\omega)}.$$

Therefore, $f(x)$ is extracted from (3.75) by first observing that

$$\int_0^{\infty} \mathcal{F}(\tilde{f}_p(s,t))\hat{\varphi}(s\omega)s^{p-2}\,ds = \int_0^{\infty} \mathcal{F}\left(\overline{\varphi_{-s,0}^p(t)} * f(t)\right)\hat{\varphi}(s\omega)s^{p-2}\,ds$$

$$= C_+\hat{f}(\omega)$$

where C_+ is defined in (3.73). Dividing by C_+ and writing the result using the relationships given above,

$$
\begin{aligned}
\hat{f}(\omega) &= \frac{1}{C_+} \int_0^\infty \left(\int_{-\infty}^\infty e^{-i\omega t} \hat{\varphi}(s\omega) \tilde{f}(s,t)\, dt \right) s^{p-2}\, ds \\
&= \frac{1}{C_+} \int_0^\infty \left(\int_{-\infty}^\infty \hat{\varphi}_{s,t}^p(\omega) \tilde{f}(s,t)\, dt \right) s^{2p-3}\, ds.
\end{aligned}
$$

Applying the inverse Fourier transform to both sides yields

$$
f(x) = \frac{1}{C_+} \int_{-\infty}^\infty \int_0^\infty \varphi_{s,t}^p(x) \tilde{f}(s,t) s^{2p-3}\, ds\, dt.
$$

Sometimes it is more natural to integrate over negative as well as positive values of s. Since $s = 0$ constitutes a set of measure zero (i.e., a set of no area) in the (s,t) plane, it need not be excluded from the integral defining the inverse wavelet transform, and we write:

$$
f(x) = \frac{1}{C} \int_{\mathbb{R}^2} \varphi_{s,t}^p(x)\, \tilde{f}(s,t)\, |s|^{2p-3} ds\, dt. \tag{3.76}
$$

From the above equation, another "natural" choice for the value of p is $3/2$.

As a second example of a suitable function φ, consider the normalized sinc function:

$$
\varphi(x) \doteq \mathrm{sinc}(x) = \frac{\sin \pi x}{\pi x}.
$$

This is also called the *Shannon wavelet*, and its orthogonality and completeness as a wavelet system follow from the properties expressed in our discussion of the Shannon sampling theorem in Chapter 2. In particular, Shannon observed the orthogonality

$$
\int_{-\infty}^\infty \mathrm{sinc}(2Wx - m)\, \mathrm{sinc}(2Wx - n)\, dx = \frac{1}{2W} \delta_{m,n}.
$$

Continuous Modulation-Translation-Based Wavelet Transforms

In 1946, D. Gabor [39] introduced the idea of expanding functions in terms of translated and modulated versions of a single window-like function. Continuous transforms based on this idea can be found in the literature (see [10] and references therein). To begin, assume that a complex-valued function $g \in \mathcal{L}^2(\mathbb{R})$ is given, and has the additional property that it is either supported on a closed interval, or at least decreases rapidly to zero outside of such an interval. Further, assume that $g(x)$ is normalized so that

$$
\|g\|^2 = (g,g) = \int_{-\infty}^\infty |g(x)|^2 dx = 1.
$$

Then the *continuous Gabor transform* (or *complex spectrogram* [10]) of a function $f \in \mathcal{L}^2(\mathbb{R})$ is defined as

$$
\tilde{f}_g(t,\omega) \doteq \int_{\mathbb{R}} f(x)\, \overline{g(x-t)}\, e^{-i\omega x}\, dx. \tag{3.77}
$$

If we multiply by $g(x-t)$ and integrate both sides over $t \in \mathbb{R}$, the result is

$$
\int_{\mathbb{R}} \tilde{f}_g(t,\omega)\, g(x-t)\, dt = \int_{\mathbb{R}} f(x)\, \|g(x-t)\|^2\, e^{-i\omega x}\, dx = \|g\|^2 \int_{\mathbb{R}} f(x)\, e^{-i\omega x}\, dx.
$$

In the above manipulations, we used the fact that the value of $\|g(x-t)\|^2$ is independent of t. Since $\|g\|^2 = 1$, all that remains is the Fourier transform on f. Hence, application of the inverse Fourier transform recovers f. Combining all the steps, we obtain the inversion formula:

$$f(x) = \int_{\mathbb{R}^2} \tilde{f}_g(t,\omega)\, g(x-t)\, e^{i\omega x}\, d\omega\, dt. \tag{3.78}$$

3.9.2 Discrete Wavelet Transforms

In analogy with complete series expansions on the real line (like the Hermite functions), orthonormal wavelet series expansions (as opposed to transforms) are possible. As with the continuous wavelet transform where translation-scale and translation-modulation combinations are possible, we find these combinations for the discrete wavelet transform.

Discrete Scale-Translation Wavelet Series

Let $\psi \in \mathcal{L}^2(\mathbb{R})$ be normalized so that $\|\psi\|_2 = 1$, and let $\hat{\psi}$ be its Fourier transform. We seek conditions on the function ψ so that the set of functions

$$\psi_{j,k}(x) = 2^{j/2}\psi(2^j x - k) \tag{3.79}$$

for all $j,k \in \mathbb{Z}$ is orthonormal and complete. Following [57], and using the notation

$$(\psi,\phi) = \int_{\mathbb{R}} \psi\,\overline{\phi}\,dx,$$

a straightforward change of variables $x = 2^{-n}(y+m)$ shows that

$$(\psi_{j,k},\psi_{n,m}) = \int_{\mathbb{R}} 2^{j/2}\psi(2^j x - k)\cdot 2^{n/2}\overline{\psi(2^n x - m)}\,dx \tag{3.80}$$

$$= \int_{\mathbb{R}} 2^{(j-n)/2}\overline{\psi(2^{(j-n)}y - (k - 2^{(j-n)}m))}\cdot \psi(y)\,dy \tag{3.81}$$

$$= (\psi_{l,p},\psi) \tag{3.82}$$

where $l = j - n$ and $p = k - 2^{j-n}m$. A similar calculation shows that

$$(\psi_{j,k},\psi_{j,l}) = (\psi_{0,k},\psi_{0,l}).$$

The orthonormality condition $(\psi_{j,k},\psi_{j,l}) = \delta_{k,l}$ can therefore be examined in the context of the simpler expression $(\psi_{0,k},\psi_{0,l}) = (\psi,\psi_{0,l-k}) = \delta_{k,l} = \delta_{0,l-k}$. It is convenient to define $r = l - k$. Again following [57], necessary and sufficient conditions for this orthogonality to hold are:

$$\delta_{0,r} = \int_{\mathbb{R}} \psi(x)\,\overline{\psi(x-r)}\,dx = \frac{1}{2\pi}\int_{\mathbb{R}} \|\hat{\psi}\|^2 e^{ir\xi}\,d\xi.$$

We have used the shift operational property and Parseval's equality above. Since the integrand on the right is oscillatory, it makes sense to break the integration up as:

$$\frac{1}{2\pi}\sum_{r=-\infty}^{+\infty}\int_{2r\pi}^{2(r+1)\pi} |\hat{\psi}(\xi)|^2 e^{ir\xi}\,d\xi = \frac{1}{2\pi}\int_0^{2\pi}\left(\sum_{r=-\infty}^{+\infty} |\hat{\psi}(\xi + 2r\pi)|^2\right) e^{ir\xi}\,d\xi.$$

This last integral is equal to $\delta_{0,r}$ when the term in parenthesis satisfies

$$\sum_{r\in\mathbb{Z}}|\hat{\psi}(\xi+2r\pi)|^2 = 1. \tag{3.83}$$

A similar calculation follows from the condition $(\psi, \psi_{j,k}) = 0$. Namely, observing that the Fourier transform of (3.79) is

$$\mathcal{F}(\psi_{j,k}) = 2^{-j/2}\hat{\psi}(2^{-j}\xi)\,e^{-ik\xi/2^j},$$

Parseval's equality together with the change of scale in frequency $\nu = 2^{-j}\xi$ yields [57]

$$0 = (\psi, \psi_{j,k}) = \frac{1}{2\pi}\int_{\mathbb{R}} 2^{j/2}\hat{\psi}(2^j\nu)\overline{\hat{\psi}(\nu)}e^{ik\nu}\,d\nu$$

for $j,k \neq 0$. Breaking up the integral into a sum of integrals over intervals of length 2π as before and extracting the factor multiplying $e^{ik\nu}$, we see that the above equation holds when

$$\sum_{k\in\mathbb{Z}}\hat{\psi}(2^j(\nu+2k\pi))\overline{\hat{\psi}(\nu+2k\pi)} = 0. \tag{3.84}$$

The *discrete wavelet transform* (or discrete wavelet coefficient j,k) is defined as

$$\tilde{f}_{j,k} \doteq (\psi_{j,k}, f).$$

It can be shown [57] that the conditions

$$\sum_{j\in\mathbb{Z}}|\hat{\psi}(2^j\xi)|^2 = 1 \tag{3.85}$$

and

$$\sum_{j=0}^{\infty}\hat{\psi}(2^j\xi)\overline{\hat{\psi}(2^j(\xi+2m\pi))} = 0 \tag{3.86}$$

for all $\xi \in \mathbb{R}$ (except possibly for a set of measure zero) and all odd integers $m \in 2\mathbb{Z}+1$ ensure the completeness of the set of functions $\{\psi_{j,k}|j,k \in \mathbb{Z}\}$. It may also be shown that the orthonormality conditions (3.83) and (3.84) follow from (3.85) and (3.86), and so these conditions completely characterize functions ψ that give rise to systems of complete orthonormal wavelets.

When these conditions hold, the Parseval (Plancherel) equality

$$\sum_{j,k\in\mathbb{Z}}|(\psi_{j,k}, f)|^2 = \|f\|_2^2,$$

and the inversion formula

$$f = \sum_{j,k\in\mathbb{Z}}(\psi_{j,k}, f)\psi_{j,k}$$

do as well.

A concrete example of a discrete wavelet basis is the *Haar system*.

Discrete Modulation-Translation Wavelet Series

Discrete wavelet transforms based on modulations and translations (instead of dilations and translations) are possible. For example, the system

$$g_{m,n}(x) = e^{2\pi imx}g(x-n)$$

for $m,n \in \mathbb{Z}$ introduced by D. Gabor in 1946 can be shown to be an orthonormal basis for certain classes of $g \in \mathcal{L}^2(\mathbb{R})$. See [39, 57] for a discussion.

For further reading on wavelets more generally, see [23, 25, 114].

3.10 Summary

In this chapter we reviewed a number of orthogonal expansions on the real line and various intervals. The topics ranged in chronology from Sturm-Liouville theory and the nineteenth-century functions of mathematical physics, to the beginning of the twentieth century (Haar, Walsh, Zernike, etc.). Our discussion of orthonormal wavelets and associated transforms is work that can be found in the literature of the late 1980s, and the discussion of fast polynomial transforms dates to the late 1990s. This continues to be an area of active investigation. For example, in very recent work, classical asymptotic expansions of special functions have been used to compute fast, accurate, and numerically stable Legendre [52] and Hankel [111] transforms.

Of course we have not considered all possible orthogonal functions and transforms. Our goal here was to provide the necessary background for the reader to put the developments of the later chapters of this book in perspective. For those readers interested in more exhaustive treatments of orthogonal functions and transforms, the books [8, 32, 93] may be useful, and for a review of wavelets see [56].

References

1. Abramowitz, M., Stegun, I.A., eds., *Handbook of Mathematical Functions*, Dover edition, 1972.
2. Agnesi, A., Reali, G.C., Patrini, G., Tomaselli, A., "Numerical Evaluation of the Hankel Transform – Remarks," *Journal of the Optical Society of America A-Optics, Image Science and Vision*, 10(9): 1872 – 1874, Sept. 1993.
3. Ahmed, N., Rao, K.R., *Orthogonal Transforms for Digital Signal Processing*, Springer-Verlag, New York, 1975.
4. Alpert, B., Rokhlin, V., "A Fast Algorithm for the Evaluation of Legendre Expansions," *SIAM J. Scientific and Statistical Computing*, 12(1): 158 – 179, 1991.
5. Amos, D.E., Burgmeier, J.W., "Computation with Three-Term Linear Nonhomogeneous Recurrence Relations," *SIAM Review*, 15(2): 335 – 351, 1973.
6. Andrews, G.E., Askey, R., Roy, R., *Special Functions*, Encyclopedia of Mathematics and Its Applications, Vol. 71, Cambridge University Press, 1999.
7. Arfken, G.B., Weber, H.J., *Mathematical Methods for Physicists*, 7th ed., Academic Press, San Diego, 2010.
8. Askey, R., *Orthogonal Polynomials and Special Functions*, SIAM, Philadelphia, 1975.
9. Askey, R., Wilson, J., "Set of Orthogonal Polynomials that Generalize the Racah Coefficients or 6-j Symbols," *SIAM Journal on Mathematical Analysis*, 10(5): 1008 – 1016, 1979.
10. Bastiaans, M.J., "A Sampling Theorem for the Complex Spectrogram, and Gabor's Expansion of a Signal in Gaussian Elementary Signals," *Optical Engineering*, 20(4): 594-598, July/August 1981.
11. Beauchamp, K.G., *Applications of Walsh and Related Functions*, Academic Press, New York, 1984.
12. Birkhoff, G., Rota, G.-C., "On the Completeness of Sturm-Liouville Expansions," *American Mathematical Monthly*, 67(9): 835 – 841, Nov. 1960.
13. Bowman, F., *Introduction to Bessel Functions*, Dover Publications, Inc., New York, 2010.
14. Boyd, J.P., *Chebyshev and Fourier Spectral Methods*, 2nd revised ed., Dover Publications, Mineola, NY, 2001.
15. Candel, S.M., "An Algorithm for the Fourier-Bessel Transform," *Computer Physics Communications*, 23(4): 343 – 353, 1981.
16. Candel, S.M., "Simultaneous Calculation of Fourier-Bessel Transforms up to Order n," *Journal of Computational Physics*, 44(2): 243 – 261, 1981.
17. Candel, S.M., "Dual Algorithms for Fast Calculation of the Fourier-Bessel Transform," *IEEE Transactions on Acoustics Speech and Signal Processing*, 29(5): 963 – 972, 1981.
18. Cantero, M.J., Iserles, A., "On Rapid Computation of Expansions in Ultraspherical Polynomials," *SIAM Journal on Numerical Analysis*, 50(1): 307-327, 2012.
19. Cantrell, C.D., *Modern Mathematical Methods for Physicists and Engineers*, Cambridge University Press, 2000.

20. Cavanagh, E., Cook, B.D., "Numerical Evaluation of Hankel Transforms via Gaussian-Laguerre Polynomial Expansions," *IEEE Transactions on Acoustics Speech and Signal Processing*, 27(4): 361–366, 1979.
21. Canuto, C., Hussaini, M.Y., Quarteroni, A., Zang, T.A., *Spectral Methods in Fluid Dynamics*, Springer-Verlag, Berlin, 1988.
22. Christensen, N.B., "The Fast Hankel Transform–Comment," *Geophysical Prospecting*, 44(3): 469–471, May 1996.
23. Chui, C.K., *An Introduction to Wavelets*, Academic Press, Boston, 1992.
24. Clenshaw, C.W., "A Note on the Summation of Chebyshev Series," *Mathematics of Computation*, 9(51): 118–120, 1955.
25. Cody, M.A., "The Fast Wavelet Transform," *Dr. Dobb's Journal*, 16(4): 16–28, April 1992.
26. Corbató, F.J., Uretsky, J.L., "Generation of Spherical Bessel Functions in Digital Computers," *Journal of the ACM*, 6(3): 366–375, 1959.
27. Cree, M.J., Bones, P.J., "Algorithms to Numerically Evaluate the Hankel Transform," *Computers & Mathematics with Applications*, 26(1): 1–12, July 1993.
28. Daubechies, I., "Orthonormal Bases of Compactly Supported Wavelets," *Communications on Pure and Applied Mathematics*, 41(7): 909–996, 1988.
29. Davis, P.J., *Interpolation and Approximation*, Ginn (Blaisdell), Boston, 1963. (Dover Edition, 2014)
30. Davis, P.J., Rabinowitz, P., *Methods of Numerical Integration*, Academic Press, New York, 1975. (Dover Edition, 2007).
31. De Micheli, E., Viano, G.A., "A New and Efficient Method for the Computation of Legendre Coefficients," arXiv:1106.0463, 2011.
32. Debnath, L., Bhatta, D., *Integral Transforms and Their Applications*, 2nd ed., Chapman and Hall/CRC, Boca Raton, 2006.
33. Deprit, A., "Note on the Summation of Legendre Series," *Celestial Mechanics*, 20(4): 319–323, 1979.
34. Driscoll, J.R., Healy, D., "Computing Fourier Transforms and Convolutions on the 2-Sphere," *Advances in Applied Mathematics* 15(2): 202–250, 1994.
35. Driscoll, J.R., Healy, D., Rockmore, D., "Fast Discrete Polynomial Transforms with Applications to Data Analysis for Distance Transitive Graphs," *SIAM Journal on Computing*, 26(4): 1066-1099, 1997.
36. Ferrari, J.A., "Fast Hankel Transform of Order Zero," *Journal of the Optical Society of America A-Optics, Image Science and Vision*, 12(8): 1812–1813, Aug. 1995.
37. Fino, B.J., Algazi, V.R., "A Unified Treatment of Discrete Fast Transforms," *SIAM J. Computing*, 6(4): 700–717, 1977.
38. Forsythe, G.E., "Generation and Use of Orthogonal Polynomials for Data Fitting with a Digital Computer," *Journal of the Society for Industrial and Applied Mathematics*, 5(2): 74–88, 1957.
39. Gabor, D., "Theory of Communication," *Journal of the Institution of Electrical Engineers-Part III: Radio and Communication Engineering*, 93:429–457, 1946.
40. Gauß, C.F., "Methodus nova integralium valores per approximationem inveniendi," in *Carl Friedrich Gauß Werke, Band 3*, Königliche Gesellschaft der Wissenschaften: Göttingen, pp. 163-196, 1866.
41. Gautschi, W., "Computational Aspects of Three-Term Recurrence Relations," *SIAM Review*, 9(1): 24–82, 1967.
42. Gautschi, W., "Minimal Solutions of Three-Term Recurrence Relations and Orthogonal Polynomials," *Mathematics of Computation*, 36(154): 547–554, 1981.
43. Goldstein, M., Thaler, R.M., "Recurrence Techniques for the Calculation of Bessel Functions," *Mathematical Tables and other Aids to Computation*, 13(66): 102–108, April 1959.
44. Greengard, L., Rokhlin, V., "A Fast Algorithm for Particle Simulations," *Journal of Computational Physics*, 73(2): 325–348, 1987.
45. Grossmann, A., Morlet, J., Paul, T., "Transforms Associated to Square Integrable Group Representations. I. General Results," *Journal of Mathematical Physics*, 26(10): 2473–2479, 1985.

46. Grossmann, A., Morlet, J., Paul, T., "Transforms Associated to Square Integrable Group Representations. II. Examples," *Annales de l'Institut Henri Poincaré-Physique théorique*, 45(3): 293 – 309, 1986.

47. Grosswald, E., *Bessel Polynomials*, Lecture Notes in Mathematics No.698, Springer-Verlag, New York, 1978.

48. Gueron, S., "Methods for Fast Computation of Integral Transforms," *Journal of Computational Physics*, 110(1): 164 – 170, Jan. 1994.

49. Gumerov, N.A., Duraiswami, R., *Fast Multipole Methods for the Helmholtz equation in Three Dimensions*, Elsevier, Amsterdam, 2005.

50. Haar, A., "Zur Theorie der Orthogonalfunktionensysteme," *Mathematische Annalen*, 69: 331 – 371, 1910.

51. Hahn, W., "Über Orthogonalpolynome, die q-Differenzengleichungen genügen," *Mathematische Nachrichten*, 2: 4 – 34, 1949.

52. Hale, N., Townsend, A., "A Fast FFT-Based Discrete Legendre Transform," *IMA Journal on Numerical Analysis*, published online Nov. 2015.

53. Hansen, E.W., "Fast Hankel Transform Algorithm," *IEEE Transactions on Acoustics Speech and Signal Processing*, 33(3): 666 – 671, 1985.

54. Harmuth, H.F., *Transmission of Information by Orthogonal Functions*, 2nd ed., Springer-Verlag, Berlin, 1972.

55. Healy Jr., D.M., Rockmore, D.N., Kostelec, P.J., Moore, S., "FFTs for the 2-Sphere – Improvements and Variations," *Journal of Fourier Analysis and Applications*, 9(4): 341 – 385, 2003.

56. Heil, C.E., Walnut, D.F., "Continuous and Discrete Wavelet Transforms," *SIAM Review*, 31(4): 628 – 666, 1989.

57. Hernández, E., Weiss, G., *A First Course on Wavelets*, CRC Press, Boca Raton, 1996.

58. Higgins, W.E., Munson, D.C., "An Algorithm for Computing General Integer-Order Hankel-Transforms," *IEEE Transactions on Acoustics Speech and Signal Processing*, 35(1): 86 – 97, Jan. 1987.

59. Higham, N.J., "Fast Solution of Vandermonde-like Systems Involving Orthogonal Polynomials," *IMA Journal on Numerical Analysis*, 8(4): 473 – 486, 1988.

60. Iserles, A., "A Fast and Simple Algorithm for the Computation of Legendre Coefficients," *Numerische Mathematik*, 117(3): 529 - 553, 2011.

61. Jerri, A.J., "Towards a Discrete Hankel Transform and Its Applications," *Applicable Analysis*, 7(2): 97 – 109, 1978.

62. Johansen, H.K., Sorensen, K., "Fast Hankel Transforms," *Geophysical Prospecting*, 27(4): 876 – 901, 1979.

63. Kaiser, G., *A Friendly Guide to Wavelets*, Birkhäuser, Boston, 1994 (reprinted in 2011).

64. Karlin, S., McGregor, J.L., "The Hahn Polynomials, Formulas and an Application," *Scripta Mathematica*, 26(1): 33 – 46, Nov. 1961.

65. Karpovsky, M.G., *Finite Orthogonal Series in the Design of Digital Devices*, Halsted Press, John Wiley and Sons, Newy York, 1975.

66. Keiner, J., "Computing with Expansions in Gegenbauer Polynomials," *SIAM Journal on Scientific Computing*, 31(3): 2151 - 2171, 2009.

67. Keiner, J., *Fast Polynomial Transforms*, Logos Verlag Berlin GmbH., 2011.

68. Koornwinder, T., "The Addition Formula for Laguerre Polynomials," *SIAM Journal on Mathematical Analysis*, 8(3): 535 – 540, 1977.

69. Krall, H.L., Frink, O., "A New Class of Orthogonal Polynomials: The Bessel Polynomials," *Transactions of the American Mathematical Society*, 65: 100 – 115, 1949.

70. Kreyszig, E., *Advanced Engineering Mathematics*, 10th edition, John Wiley & Sons, New York, 2011.

71. Krylov, V.I., *Approximate Calculation of Integrals*, (A. H. Stroud, trans.), MacMillan, New York, 1962. (Dover edition, 2005).

72. Leibon, G., Rockmore, D.N., Park, W., Taintor, R., Chirikjian, G.S., "A Fast Hermite Transform," *Theoretical Computer Science*, 409(2): 211 – 228, 2008.

73. Lemoine, D., "The Discrete Bessel Transform Algorithm," *Journal of Chemical Physics*, 101(5): 3936 – 3944, Sept. 1994.
74. Liu, Q.H., Chew, W.C., "Applications of the Conjugate-Gradient Fast Fourier Hankel Transfer Method with an Improved Fast Hankel Transform Algorithm," *Radio Science* 29(4): 1009 – 1022, July-August 1994.
75. Liu, Q.H., "Applications of the Conjugate-Gradient Fast Hankel Fourier Transfer Method with an Improved Fast Hankel Transform Algorithm," *Radio Science* 30(2): 479, March-April 1995.
76. Luke, Y.L., *Mathematical Functions and Their Approximations*, Acadmic Press, 1975.
77. Ma, J., Rokhlin, V., Wandzura, S., "Generalized Gaussian Quadrature Rules for Systems of Arbitrary Functions," *SIAM J. Numerical Analysis*, 33(3): 971 – 996, June 1996.
78. Magni, V., Cerullo, G., De Silvestri, S., "High-Accuracy Fast Hankel Transform for Optical Beam Propagation," *Journal of the Optical Society of America A – Optics, Image Science and Vision*, 9(11): 2031 – 2033, Nov. 1992.
79. Mallat, S., "Multiresolution Approximations and Wavelet Orthonormal Bases of $\mathcal{L}^2(\mathbb{R})$," *Transactions of the American Mathematical Society*, 315(1): 69 – 87, 1989.
80. Maslen, D.K., Rockmore, D.N., "Generalized FFTs – A Survey of Some Recent Results," *DIMACS Series in Discrete Mathematics and Theoretical Computer Science*, 28: 183 – 237, 1997.
81. Meixner, J., "Orthogonale Polynomsysteme mit einer besonderen Gestalt der erzeugenden Funktion," *Journal of the London Mathematical Society*, 9(1): 6 – 13, 1934.
82. Meixner, J., "Symmetric Systems of Orthogonal Polynomials," *Archive for Rational Mechanics and Analysis*, 44(1): 69 – 75, 1972.
83. Meyer, Y., *Ondelettes et opérateurs I: ondelettes*, Hermann, 1989.
84. Mohsen, A.A., Hashish, E.A., "The Fast Hankel Transform," *Geophysical Prospecting*, 42(2): 131 – 139, Feb. 1994.
85. Mook, D.R., "An Algorithm for the Numerical Evaluation of the Hankel and Abel Transforms," *IEEE Transactions on Acoustics, Speech, and Signal Processing*, 31(4): 979 – 985, 1983.
86. Morlet, J., Arens, G., Fourgeau, I., Giard, D., "Wave Propagation and Sampling Theory," *Geophysics*, 47(2): 203 – 236, 1982.
87. Oppenheim, A.V., Frisk, G.V., Martinez, D.R., "Algorithm for Numerical Evaluation of Hankel Transform," *Proceedings of the IEEE*, 66(2): 264 – 265, 1978.
88. Oppenheim, A.V., Frisk, G.V., Martinez, D.R., "Computation of the Hankel Transform Using Projections," *Journal of the Acoustical Society of America*, 68(2): 523 – 529, Aug. 1980.
89. Orszag, S.A., "Fast Eigenfunction Transforms," in *Science and Computers, Advances in Mathematics Supplementary Studies*, 10: 23 – 30, Academic Press, 1986.
90. Piessens, R., Branders, M., "Modified Clenshaw-Curtis Method for the Computation of Bessel-function Integrals, " *BIT*, 23(3): 370 – 381, 1983.
91. Pinsky, M.A., *Introduction to Partial Differential Equations with Applications*, McGraw-Hill Book Company, New York, 1984.
92. Potts, D., Steidl, G., Tasche, M., "Fast Algorithms for Discrete Polynomial Transforms," *Mathematics of Computation*, 67(224): 1577 – 1590, Oct. 1998.
93. Poularikas, A.D., (ed.), *The Transforms and Applications Handbook*, 3rd ed., CRC Press, 2010.
94. Rademacher, H., "Einige Sätze von allgemeinen Orthogonalfunktionen," *Mathematische Annalen*, 87(1): 122 – 138, 1922.
95. Rao, K.S., Santhanam, T.S., Gustafson, R.A., "Racah Polynomials and a 3-term Recurrence Relation for the Racah Coefficients," *Journal of Physics A-Mathematical and General*, 20(10): 3041 – 3045, July 1987.
96. Renault, O., "A New Algorithm for Computing Orthogonal Polynomials," *Journal of Computational and Applied Mathematics*, 75(2): 231 – 248, Nov. 1996.
97. Rijo, L., "'The Fast Hankel Transform' - Comment," *Geophysical Prospecting*, 44(3): 473 – 477, May 1996.

98. Rivlin, T.J., *Chebyshev Polynomials*, 2nd edition, John Wiley and Sons, New York, 1990.

99. Roeser, P.R., Jernigan, M.E., "Fast Haar Transform Algorithms," *IEEE Transactions on Computers*, 31(2): 175 – 177, 1982.

100. Salzer, H.E., "A Recurrence Scheme for Converting from One Orthogonal Expansion into Another," *Communications of the ACM*, 16(11): 705 – 707, 1973.

101. Salzer, H.E., "Computing Interpolation Series into Chebyshev Series," *Mathematics of Computation*, 30(134): 295 – 302, 1976.

102. Sansone, G., *Orthogonal Functions*, Wiley Interscience, New York, 1958. (Revised Dover Edition, 2012).

103. Secada, J.D., "Numerical Evaluation of the Hankel Transform," *Computer Physics Communications*, 116(2-3): 278 – 294, Feb. 1999.

104. Sharafeddin, O.A., Ferrel Bowen, H., Kouri, D.J., Hoffman, D.K., "Numerical Evaluation of Spherical Bessel Transforms via Fast Fourier Transforms," *Journal of Computational Physics*, 100(2): 294 – 296, June 1992.

105. Siegman, A.E., "Quasi Fast Hankel Transform," *Optics Letters*, 1(1): 13 – 15, 1977.

106. Smith, F.J., "An Algorithm for Summing Orthogonal Polynomial Series and Their Derivatives with Application to Curve Fitting and Interpolation," *Mathematics of Computation*, 19(89): 33 – 36, 1965.

107. Stroud, A.H., Secrest, D., *Gaussian Quadrature Formulas*, Prentice-Hall, Inc., Englewood Cliffs, N.J., 1996.

108. Szegö, G., *Orthogonal Polynomials*, American Mathematical Society Colloquium Publications, Vol. 23, AMS, Providence RI, 1939 (4th ed., 1975).

109. Talman, J.D., "Numerical Fourier and Bessel Transforms in Logarithmic Variables," *Journal of Computational Physics*, 29(1): 35 – 48, 1978.

110. Thangavelu, S., *Lectures on Hermite and Laguerre Expansions*, Princeton University Press, 1993.

111. Townsend, A., "A Fast Analysis-Based Discrete Hankel Transform Using Asymptotic Expansions," *SIAM Journal on Numerical Analysis*, 53(4): 1897 - 1917, 2015.

112. Vilenkin, N.J., Klimyk, A.U., *Representation of Lie Group and Special Functions*, Vol. 1-3, Kluwer Academic Publishers, The Netherlands, 1991.

113. Walsh, J.L., "A Closed Set of Orthogonal Functions," *American Journal of Mathematics*, 45: 5 – 24, 1923.

114. Walter, G.G., *Wavelets and Other Orthogonal Systems with Applications*, CRC Press, Boca Raton, 1994.

115. Wang, H., Huybrechs, D., "Fast and Accurate Computation of Jacobi Expansion Coefficients of Analytic Functions," arXiv preprint arXiv:1404.2463. (2014).

116. Watson, G.N., *A Treatise on the Theory of Bessel Functions*, Cambridge University Press, 1995 edition.

117. Wilson, J.A., "Some Hypergeometric Orthogonal Polynomials," *SIAM Journal on Mathematical Analysis*, 11(4): 690 – 701, 1980.

118. Wimp, J., *Computation with Recurrence Relations*, Pitman Press, Boston, 1984.

119. Xiang, S., "On Fast Algorithms for the Evaluation of Legendre Coefficients," *Applied Mathematics Letters*, 26(2): 194 – 200, Feb. 2013.

120. Zernike, F., "Beugungstheorie des Schneidenverfahrens und seiner verbesserten Form, der Phasenkontrastmethode," *Physica*, 1: 689 – 704, 1934.

Orthogonal Expansions in Curvilinear Coordinates

In this chapter we review orthogonal expansions for " complex-valued functions on $\mathbb{R}^N$," i.e., functions of the form $f : \mathbb{R}^N \to \mathbb{C}$. In particular, when $\mathbf{x} \in \mathbb{R}^N$ is expressed in curvilinear coordinates, then $f(\mathbf{x})$ also depends on those coordinates. This chapter therefore reviews orthogonal expansions in curvilinear coordinates and associated computations. We also examine orthogonal expansions on surfaces in $\mathbb{R}^3$ and higher dimensions.

4.1 Introduction to Curvilinear Coordinates and Surface Parameterizations

While Cartesian coordinates are the most common methods for identifying positions in space, they are by no means the only ones. In general, positions in space are parameterized by an array of parameters $(u_1, ..., u_N)$, which we will denote as $\mathbf{u} \in D \subset \mathbb{R}^N$, i.e., D is a region (or domain) of parameter values in a different copy of $\mathbb{R}^N$ than that in which $\mathbf{x}$ resides. Then $\mathbf{x} = \tilde{\mathbf{x}}(\mathbf{u})$, where $\mathbf{x} \in \mathbb{R}^N$, $\mathbf{u} \in D$, and $\tilde{\mathbf{x}} : D \to \mathbb{R}^N$.

The relationship between differential elements of the arclength, $dL = (dL/dt)\, dt$, of any curve $\mathbf{x}(t) = \tilde{\mathbf{x}}(\mathbf{u}(t))$ in the different curvilinear coordinates is related to those in Cartesian coordinates as:

$$\left(\frac{dL}{dt} \right)^2 = \frac{d\mathbf{x}}{dt} \cdot \frac{d\mathbf{x}}{dt} = \left(J(\mathbf{u}) \frac{d\mathbf{u}}{dt} \right)^T \left(J(\mathbf{u}) \frac{d\mathbf{u}}{dt} \right) = \frac{d\mathbf{u}}{dt} \cdot \left(G(\mathbf{u}) \frac{d\mathbf{u}}{dt} \right),$$

where

$$G(\mathbf{u}) = J^T(\mathbf{u}) J(\mathbf{u}),$$

and

$$J(\mathbf{u}) = \left(\tfrac{\partial \mathbf{x}}{\partial u_1}, \, ... \, , \tfrac{\partial \mathbf{x}}{\partial u_N} \right).$$

G is called the *metric tensor* and J is called the *Jacobian matrix*.

If for some reason the units used to measure distance in each of the coordinate directions are different from each other, this must be rectified if some meaningful measure of arc length and volume are to be defined. In such cases, we have

$$G(\mathbf{u}) = J^T(\mathbf{u}) W J(\mathbf{u}), \tag{4.1}$$

where W is a diagonal matrix with elements of the form $w_{ij} = w_i \delta_{ij}$. The constant weights w_i are measures of distance or ratios of distance units used in each coordinate direction. For instance, if x_1 through x_{N-1} are measured in centimeters and x_N is measured in inches, then $w_1 = \ldots = w_{N-1} = 1$ and $w_N = (2.54)^2$ will provide the required uniformity.

The volume of an infinitesimal element in $\mathbb{R}^N$ written in the curvilinear coordinates $\mathbf{u}$ is

$$
\begin{aligned}
dx^N &= \sqrt{\det(G(\mathbf{u}))}\, du^N \\
&= w_1 \cdots w_N \sqrt{\det(J^T(\mathbf{u})J(\mathbf{u}))}\, du^N \\
&= w_1 \cdots w_N |\det(J(\mathbf{u}))| du^N.
\end{aligned}
$$

We use the notation $da^N = da_1 da_2 \ldots da_N$ for any $\mathbf{a} \in \mathbb{R}^N$.

In the special case when

$$
\sqrt{\det(G(\mathbf{u}))} = f_1(u_1) \cdots f_N(u_N),
$$

we can construct orthogonal expansions in curvilinear coordinates as N separate one-dimensional problems (one in each parameter) provided the weighting functions $f_i(u_i)$ all satisfy the conditions imposed by Sturm-Liouville theory.

4.2 Parameterizations of the Unit Circle, Semi-Circle, and Planar Rotations

The unit circle, $\mathbb{S}^1$, is the set of all points in $\mathbb{R}^2$ that are of unit distance from the origin:

$$
x_1^2 + x_2^2 = 1.
$$

A parameterization of $\mathbb{S}^1$ that is very natural (in the sense that uniform increments of the curve parameter correspond to equal increments of arc length) is

$$
\mathbf{x}(\theta) = \begin{pmatrix} \cos\theta \\ \sin\theta \end{pmatrix}. \tag{4.2}
$$

Here $\theta \in [0, 2\pi)$ is the counter-clockwise measured angle that the vector $\mathbf{x}$ makes with the x_1-axis. In this parameterization of the circle, the Fourier series is a natural way to expand functions $f(\mathbf{x}(\theta))$.

Rotations in the plane are identified with the circle by mapping the angle θ to a particular 2×2 rotation matrix

$$
R(\theta) = \begin{pmatrix} \cos\theta & -\sin\theta \\ \sin\theta & \cos\theta \end{pmatrix}. \tag{4.3}
$$

If we identify points in the plane $\mathbb{R}^2$ with points in the complex plane $\mathbb{C}$ by the rule

$$
\mathbf{x} \longleftrightarrow x_1 + ix_2,
$$

then the effect of the rotation matrix (4.3) applied to a vector $\mathbf{x}$ is equivalent to the product $e^{i\theta}(x_1 + ix_2)$:

$$
R(\theta)\mathbf{x} \longleftrightarrow e^{i\theta}(x_1 + ix_2).
$$

Formulations that are natural from a geometric perspective can sometimes be inefficient from a computational perspective. In the case of the circle and rotation parameterizations, it is computationally attractive to write

$$\mathbf{x}(z) = \begin{pmatrix} \frac{1-z^2}{1+z^2} \\ \frac{2z}{1+z^2} \end{pmatrix} \quad \text{and} \quad R(z) = \begin{pmatrix} \frac{1-z^2}{1+z^2} & \frac{-2z}{1+z^2} \\ \frac{2z}{1+z^2} & \frac{1-z^2}{1+z^2} \end{pmatrix}, \tag{4.4}$$

where

$$z = \tan \frac{\theta}{2} \quad \text{and} \quad \theta = 2\tan^{-1} z.$$

The reason why this parameterization is attractive is that the evaluation of (4.4) requires only the most basic arithmetic operations (addition, subtraction, multiplication, and division), without the need to compute trigonometric functions. In computer graphics, these descriptions (called "algebraic" or "rational") are often preferred to the transcendental (trigonometric) descriptions.

One computationally attractive way to implement rotations is as a succession of shears. A linear shear transformation of the plane that maps vertical lines into vertical lines (preserving their x_1 value) is of the form $\mathbf{x}' = A\mathbf{x}$ where

$$A = \begin{pmatrix} 1 & 0 \\ \beta & 1 \end{pmatrix},$$

and $\beta \in \mathbb{R}$ describes the amount of shear. A linear shear transformation of the plane that maps horizontal lines into horizontal lines (preserving their x_2 value) is of the form

$$A = \begin{pmatrix} 1 & \alpha \\ 0 & 1 \end{pmatrix}.$$

It can be shown that the arbitrary planar rotation (4.3) can be implemented as concatenated shears [41]:

$$\begin{pmatrix} \cos\theta & -\sin\theta \\ \sin\theta & \cos\theta \end{pmatrix} = \begin{pmatrix} 1 & \alpha(\theta) \\ 0 & 1 \end{pmatrix} \begin{pmatrix} 1 & 0 \\ \beta(\theta) & 1 \end{pmatrix} \begin{pmatrix} 1 & \alpha(\theta) \\ 0 & 1 \end{pmatrix}$$

where

$$\alpha(\theta) = -\tan\frac{\theta}{2} \quad \text{and} \quad \beta(\theta) = \sin\theta.$$

In rational form, this becomes

$$\begin{pmatrix} \frac{1-z^2}{1+z^2} & \frac{-2z}{1+z^2} \\ \frac{2z}{1+z^2} & \frac{1-z^2}{1+z^2} \end{pmatrix} = \begin{pmatrix} 1 & -z \\ 0 & 1 \end{pmatrix} \begin{pmatrix} 1 & 0 \\ 2z/(1+z^2) & 1 \end{pmatrix} \begin{pmatrix} 1 & -z \\ 0 & 1 \end{pmatrix},$$

which is a particularly useful tool for the rotation of images stored in pixelized form since the operations are performed on whole rows and columns of pixels.

In principle, an infinite number of different parameterizations of the circle and planar rotations can be generated by an appropriate choice of a monotonically increasing function $\theta(t)$ with $\theta(-\pi) = -\pi$ and $\theta(\pi) = \pi$. Geometrically, one class of parameterizations result from *stereographic projection* between the circle and line. This is illustrated in Figure 4.1.

We refer to the horizontal line bisecting the circle as the *equatorial line* and use the variable x to denote the distance measured along this line from the center of the circle (with positive sense to the right). All lines passing through the south pole (called *projectors*) assign a unique point on the circle to the equatorial line and vice versa

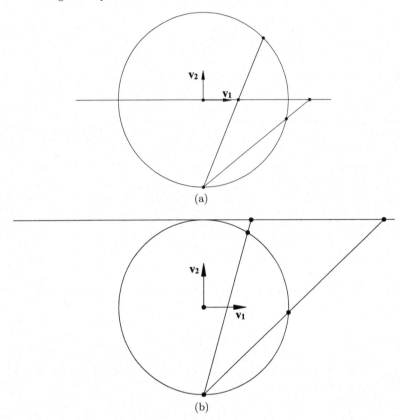

(a)

(b)

Fig. 4.1. (a) Projection onto the Equatorial Line; (b) Projection onto the Tangent to the North Pole

(except the south pole). Then, from elementary geometry[1], the relationship between the point $(v_1, v_2)^T \in \mathbb{S}^1$ and $x \in \mathbb{R}$ is

$$v_1 = \frac{2x}{1 + x^2} \quad v_2 = \frac{1 - x^2}{1 + x^2}; \quad x = \frac{1 - v_2}{v_1} = \frac{v_1}{1 + v_2}. \tag{4.5}$$

In the above expressions for x as a function of v_1 and v_2, the first one breaks down when $v_1 = 0$, and the second breaks down when $v_2 = -1$.

This stereographic projection is the same as the rational description in (4.4) when the north pole is associated with $\theta = 0$ and x is made positive, in the sense of counterclockwise rotation.

Of course, this is not the only kind of projection possible between the line and circle. Figure 4.1(b) illustrates a projection where a line tangent to the north pole replaces the role of the equatorial line. Denoting the distance along this line as measured from the north pole (positive to the right) as y, the relationship between $(v_1, v_2)^T \in \mathbb{S}^1$ and $y \in \mathbb{R}$ is

---

[1]This kind of projection, together with planar and spherical trigonometry is attributed to Hipparchus (circa 150 BC) [51].

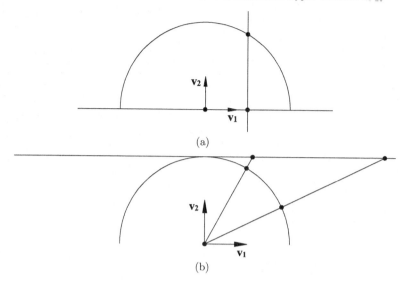

Fig. 4.2. (a) Vertical Projection; (b) Projection Through the Center.

$$v_1 = \frac{4y}{4 + y^2} \qquad v_2 = \frac{4 - y^2}{4 + y^2}; \qquad y = \frac{2(1 - v_2)}{v_1} = \frac{2v_1}{1 + v_2}. \qquad (4.6)$$

The above expressions for y as a function of v_1 and v_2 break down when $v_1 = 0$ and $v_2 = -1$, respectively.

We note that when using projectors through the south pole, a similar relationship can be established between the circle and any line lying between the tangent to the north pole and equatorial line and parallel to both.

The upper semi-circle ($\mathbb{S}^1$ with $x_2 \geq 0$ and denoted here as $(\mathbb{S}^1)^+$) can be parameterized using any of the above circle parameterizations. We may also project the upper semi-circle to the line in additional ways not possible for the whole circle. For instance, points can be projected straight up and down yielding the relationships

$$(v_1, v_2) \to v_1; \qquad x \to (x, \sqrt{1 - x^2}).$$

And a projection between the tangent to the north pole and the circle through its center yields

$$(v_1, v_2) \to \frac{v_1}{v_2}; \qquad x \to (x/\sqrt{1 + x^2}, 1/\sqrt{1 + x^2}).$$

These projections are shown in Figure 4.2.

The extensions of these simple examples to higher dimensions will be valuable in Chapter 5 in the context of parameterizing rotations in three dimensions.

4.3 *M*-Dimensional Hyper-Surfaces in $\mathbb{R}^N$

In the case when we have an M-dimensional hyper-surface $\mathcal{S} \subset \mathbb{R}^N$, the parameterization follows in exactly the same way as for curvilinear coordinates, only now

$\mathbf{u} \in D \subset \mathbb{R}^M$ where $M < N$. We note that $\mathbf{u}$ is not a vector in the physical sense, but rather an array of m real numbers.

The Jacobian matrix $J(\mathbf{u})$ is no longer square, and we can no longer compute $|\det(J(\mathbf{u}))|$. Thus, in the calculation of a hyper-surface volume element $ds(\mathbf{u})$, we usually write

$$ds(\mathbf{u}) = \sqrt{\det(G(\mathbf{u}))}\, du^M. \tag{4.7}$$

The volume element for an M dimensional hyper-surface in $\mathbb{R}^N$ can also be computed without introducing the metric tensor G. This follows since the columns of J form a basis for the tangent hyper-plane at any point on the hyper-surface. By using the Gram-Schmidt orthogonalization procedure (see Appendix C), an orthonormal basis for the hyper-plane is constructed from the columns of J. We denote the elements of this basis as $\mathbf{v}_1, ..., \mathbf{v}_M$. Then an $M \times M$ matrix with entry i, j corresponding to the projection of $\partial\mathbf{x}/\partial u_i$ onto $\mathbf{v}_j$ can be constructed. The determinant of this matrix then yields the same result as $\sqrt{\det(G(\mathbf{u}))}$. This is perhaps the most direct generalization of the use of the triple product in three dimensions for computing volume in higher dimensions.

In the special case when $N = 3$ and $M = 2$, both of the above approaches are the same as

$$ds(\mathbf{u}) = \left| \frac{\partial\mathbf{x}}{\partial u_1} \times \frac{\partial\mathbf{x}}{\partial u_2} \right| du_1 du_2.$$

Given a parameterized M-dimensional hyper-surface, $\mathbf{x}(\mathbf{u}) \in \mathbb{R}^N$, with metric $G(\mathbf{u})$, the delta function is defined as

$$\delta_S(\mathbf{u}, \mathbf{v}) \doteq \frac{\delta^M(\mathbf{u} - \mathbf{v})}{\sqrt{\det(G(\mathbf{u}))}} \tag{4.8}$$

when $\det(G(\mathbf{v})) \neq 0$.

In this way

$$\int_S f(\mathbf{u})\, \delta_S(\mathbf{u}, \mathbf{v})\, ds(\mathbf{u}) = \int_{D \subset \mathbb{R}^N} f(\mathbf{u})\, \delta^M(\mathbf{u} - \mathbf{v})\, du_1 \cdots du_M = f(\mathbf{v}).$$

Here $\delta^M(\mathbf{u} - \mathbf{v}) = \delta(u_1 - v_1) \cdots \delta(u_M - v_M)$.

The *curvature* of an M-dimensional surface in $\mathbb{R}^N$ is obtained by performing certain operations on the entries of G, which are denoted as g_{ij}. As a way of reducing notational complexity, mathematicians and physicists often denote the entries of G^{-1} by g^{ij}. The *Christoffel symbols* (of the second kind) are then defined as [1, 15, 38]:

$$\Gamma^i_{jk} \doteq \frac{1}{2} \sum_{l=1}^{M} g^{il} \left(\frac{\partial g_{lj}}{\partial u_k} + \frac{\partial g_{lk}}{\partial u_j} - \frac{\partial g_{jk}}{\partial u_l} \right). \tag{4.9}$$

They are symmetrical in the lower two indices,

$$\Gamma^i_{jk} = \Gamma^i_{kj},$$

and have many other properties derived in [15].

In the case of a two-dimensional surface in $\mathbb{R}^3$, the above definition of the Christoffel symbols follows from the expression

$$\frac{\partial^2 \mathbf{x}}{\partial u_i \partial u_j} = L_{ij}\mathbf{n} + \sum_{k=1}^{2} \Gamma^k_{ij} \frac{\partial\mathbf{x}}{\partial u_k}$$

where

$$\mathbf{n} = \frac{\frac{\partial \mathbf{x}}{\partial u_1} \times \frac{\partial \mathbf{x}}{\partial u_2}}{\left\| \frac{\partial \mathbf{x}}{\partial u_1} \times \frac{\partial \mathbf{x}}{\partial u_2} \right\|}$$

is the unit normal to the surface.

It follows that

$$L_{ij} = \frac{\partial^2 \mathbf{x}}{\partial u_i \partial u_j} \cdot \mathbf{n},$$

where $\cdot$ is the usual dot product in $\mathbb{R}^3$. If L is the matrix with coefficients L_{ij}, then the *Gaussian curvature* of the surface at $\mathbf{x}(u_1, u_2)$ is

$$K = \det(L)/\det(G) = \frac{1}{\sqrt{\det(G)}} \mathbf{n} \cdot \left(\frac{\partial \mathbf{n}}{\partial u_1} \times \frac{\partial \mathbf{n}}{\partial u_2} \right). \tag{4.10}$$

The *Riemannian curvature tensor*[2] is defined as [15, 38]

$$R^i{}_{jkl} \doteq -\frac{\partial \Gamma^i_{jk}}{\partial u_l} + \frac{\partial \Gamma^i_{jl}}{\partial u_k} + \sum_{m=1}^{M} (-\Gamma^m_{jk}\Gamma^i_{ml} + \Gamma^m_{jl}\Gamma^i_{mk}). \tag{4.11}$$

The *Ricci* and *scalar curvature* formulas are defined relative to this as [1, 38]

$$R_{jl} = \sum_{i=1}^{M} R^i{}_{jil} \quad \text{and} \quad R = \sum_{j,l=1}^{M} g^{jl} R_{jl},$$

respectively. The scalar curvature for $\mathbb{R}^N$ is zero. This is independent of what curvilinear coordinate system is used. For a sphere in any dimension, the scalar curvature is constant. In the case when $M = 2$ and $N = 3$, the scalar curvature is twice the *Gaussian* curvature.

As an example of these definitions, consider the parametric equations for the 2-torus embedded in $\mathbb{R}^3$,

$$\mathbf{x}(\phi, \theta) = \begin{pmatrix} (R + r\cos\theta)\cos\phi \\ (R + r\cos\theta)\sin\phi \\ r\sin\theta \end{pmatrix},$$

where $u_1 = \phi$ and $u_2 = \theta$.

The metric tensor is written in this parameterization as

$$G(\phi, \theta) = \begin{pmatrix} (R + r\cos\theta)^2 & 0 \\ 0 & r^2 \end{pmatrix}.$$

We find that

$$\Gamma^1_{11} = \Gamma^1_{22} = \Gamma^2_{12} = \Gamma^2_{21} = \Gamma^2_{22} = 0,$$

$$\Gamma^1_{12} = \Gamma^1_{21} = -\frac{r\sin\theta}{R + r\cos\theta},$$

and

$$\Gamma^2_{11} = \frac{(R + r\cos\theta)\sin\theta}{r}.$$

We also calculate

---

[2]Named after Bernhard Riemann (1826-1866)

$$R^2_{121} = -R^2_{112} = \frac{R\cos\theta}{r} + \cos^2\theta,$$

$$R^1_{212} = -R^1_{221} = \frac{r\cos\theta}{R + r\cos\theta},$$

with all other $R^i_{jkl} = 0$.

In contrast, it is possible to define a 2-torus as a surface in $\mathbb{R}^4$ by the equations

$$\mathbf{y}(\phi,\theta) = \begin{pmatrix} R\cos\phi \\ R\sin\phi \\ r\cos\theta \\ r\sin\theta \end{pmatrix}.$$

In this representation of the torus, the metric tensor is constant (it only depends on R and r), and so $\Gamma^i_{jk} = 0$ and $R^i_{jkl} = 0$ for all values of indices. As a result, the scalar curvature is also zero, and therefore this is called the *flat torus*, or a *flat embedding* of the abstract 2-torus. More generally, an "embedding" of an abstract manifold (see Appendix G) is a realization of that manifold as a subset of a Euclidean space, with that subset also satisfying the conditions that it is a manifold. (For example, it cannot cross itself anywhere.)

In an effort to extend the concepts of length, volume, and curvature to the most general scenarios, mathematicians toward the end of the nineteenth century formulated the geometry of higher dimensional curved spaces without regard to whether or not these spaces fit in a higher dimensional Euclidean space.[3] Roughly speaking, the only properties that were found to be important were: (1) that the curved space "locally looks like" $\mathbb{R}^M$; (2) that it is possible to construct a set of overlapping patches in the curved space in such a way that coordinates can be imposed on each patch; and (3) that the description of points in the overlapping patches are consistent with respect to the different parameterizations that are defined in the overlapping regions. In mathematics, the study of curved spaces is called differential geometry. Smooth curved spaces for which all points look locally like Euclidean space are called *manifolds*.[4] Each manifold can be covered with a collection of overlapping *patches*, in which a local *coordinate chart* relates a point in the manifold to one in a copy of $\mathbb{R}^M$. The collection of all of the coordinate charts is called an *atlas*. The inverse of a coordinate chart gives the parameterization of a patch in the manifold. Precise definitions of these terms can be found in any of a number of excellent books on differential geometry, and a short review is provided in Appendix G. It is assumed from the beginning that all of the manifolds of interest in this book can be embedded in $\mathbb{R}^N$ for sufficiently large N. Hence, we need not use formal methods of coordinate-free differential geometry. Rather, it suffices to use the methods described above for surfaces in higher dimensions, with the one modification described below.

We associate matrices $X \in \mathbb{R}^{N \times N}$ with vectors $\mathbf{x} \in \mathbb{R}^{N^2}$ by simply taking each column of the matrix and stacking sequentially the columns one on top of the other, until an N^2-dimensional vector results. The M^2 elements of the matrix $G(\mathbf{u})$ defined in (4.1) (which is called a *Riemannian metric tensor*) can be written in this case as

$$g_{ij} = \left(\frac{\partial \mathbf{x}}{\partial u_i}\right)^T W \frac{\partial \mathbf{x}}{\partial u_j} = \operatorname{tr}\left(\frac{\partial X}{\partial u_i} W_0 \frac{\partial X^T}{\partial u_j}\right), \qquad (4.12)$$

---

[3] Albert Einstein used these mathematical techniques as a tool to express the general theory of relativity.

[4] See Appendix G for a precise definition.

where $\mathbf{u} \in \mathbb{R}^M$ and $\mathbf{x}(\mathbf{u}) \in \mathbb{R}^{N^2}$ are the vectors associated with the matrices $X(\mathbf{u}) \in \mathbb{R}^{N \times N}$, and the matrix W_0 is defined relative to W in a way that makes the equality hold. As always, $\text{tr}(\cdot)$ denotes the trace of a matrix.

Note that it is always the case that $g_{ij} = g_{ji}$. In the special case when $G(\mathbf{u})$ is diagonal, the coordinates $\mathbf{u}$ are called *orthogonal* curvilinear coordinates.

The topics in this section originate from the field of mathematics known as *differential geometry* . For an accessible introduction to differential geometry, see [55], and for more advanced concepts of differential geometry applied to physics see [23].

4.4 Gradients, Divergence, and the Laplacian

In $\mathbb{R}^N$, the concept of the gradient of a scalar function of Cartesian coordinates $f(x_1, ..., x_N)$ is defined as the vector ∇f, where $\nabla \doteq \sum_{i=1}^{N} \mathbf{e}_i \partial / \partial x_i$. Recall that $\mathbf{e}_i$ is the ith natural basis vector for $\mathbb{R}^N$ and $(\mathbf{e}_i)_j = \delta_{ij}$. The divergence of a vector field $\mathbf{F} = \sum_{i=1}^{N} \mathbf{e}_i F_i$ is $\nabla \cdot \mathbf{F} \doteq \sum_{i=1}^{N} \partial F / \partial x_i$. The Laplacian[5] of a scalar function is $\nabla^2 f \doteq \nabla \cdot \nabla f = \sum_{i=1}^{N} \partial^2 f / \partial x_i^2$.

The concepts of gradient, divergence, and the Laplacian of smooth functions, $\tilde{f}(\mathbf{x}(\mathbf{u})) \doteq f(\mathbf{u})$, on a hyper-surface $\mathcal{S} \subset \mathbb{R}^N$ with points $\mathbf{x}(\mathbf{u}) \in \mathcal{S}$, are naturally generalized using the definitions given earlier in this section. Before continuing, we state that our notation of using subscripts to denote components is different than in physics, where subscripts and superscripts have special meaning. Basis vectors for the tangent hyper-plane (which are also vectors in $\mathbb{R}^N$) are

$$\mathbf{v}_i \doteq \frac{\partial \mathbf{x}}{\partial u_i} = J \mathbf{e}_i.$$

These are generally not orthogonal with respect to the natural inner product in $\mathbb{R}^N$ because for $W = \mathbb{I}$,

$$\mathbf{v}_i \cdot \mathbf{v}_j = g_{ij}.$$

However, when G is diagonal,

$$\mathbf{e}_{u_i} = \mathbf{v}_i / (g_{ii})^{\frac{1}{2}} \qquad (4.13)$$

are orthonormal with respect to this inner product.

The gradient of f is

$$\text{grad}(f) \doteq \sum_{i=1}^{M} \text{grad}(f)_i \mathbf{v}_i$$

where

$$\text{grad}(f)_i = \sum_{j=1}^{M} g^{ij} \frac{\partial f}{\partial u_j},$$

and again the shorthand g^{ij} denotes $(G^{-1})_{ij}$. The divergence of a vector field $\mathbf{F} = \sum_{i=1}^{M} F_i \mathbf{v}_i$ on a hyper-surface is defined as

$$\text{div}(\mathbf{F}) \doteq \sum_{i=1}^{M} \frac{1}{\sqrt{\det G}} \frac{\partial}{\partial u_i} (\sqrt{\det G} F_i).$$

---

[5]Named after Pierre Simon Marquis de Laplace (1749-1827)

The gradient and divergence are the "dual" (or adjoint) of each other in the following sense. For real-valued functions f and g that take their argument in $\mathcal{S}$, define the inner product

$$(f,g) = \int_{\mathcal{S}} f(\mathbf{x})g(\mathbf{x})\,ds(\mathbf{u}) = \int_{D \subset \mathbb{R}^M} f(\mathbf{x}(\mathbf{u}))g(\mathbf{x}(\mathbf{u}))\sqrt{\det G(\mathbf{u})}\,du_1\cdots du_M.$$

Here we have assumed for convenience that the whole surface can be considered as one coordinate patch, and either $\mathcal{S}$ is closed and bounded (and $f(\mathbf{x}(\mathbf{u}))$ and $g(\mathbf{x}(\mathbf{u}))$ therefore have compact support), or $\mathcal{S}$ has infinite extent (and f and g decay rapidly to zero outside of a compact region).

If $\mathbf{F}$ and $\mathbf{K}$ are two real vector fields on $\mathcal{S}$ with components F_i and K_i, we define

$$\langle \mathbf{F}, \mathbf{K} \rangle = \int_{\mathcal{S}} \mathbf{F}(\mathbf{x}) \cdot \mathbf{K}(\mathbf{x})\,ds(\mathbf{u}).$$

It follows that

$$\langle \mathbf{F}, \mathbf{K} \rangle = \Big\langle \sum_{i=1}^{m} F_i \mathbf{v}_i, \sum_{j=1}^{m} K_j \mathbf{v}_j \Big\rangle = \sum_{i,j=1}^{m} \langle F_i \mathbf{v}_i, K_j \mathbf{v}_j \rangle.$$

Equivalently,

$$\langle \mathbf{F}, \mathbf{K} \rangle = \sum_{i,j=1}^{m} \int_{D \subset \mathbb{R}^M} F_i g_{ij} K_j \sqrt{\det G}\,du_1\cdots du_M.$$

Then

$$\langle \mathrm{grad}(f), \mathbf{F} \rangle = \sum_{i,j,k=1}^{M} \int_{D \subset \mathbb{R}^M} g^{ki} \frac{\partial f}{\partial u_i} g_{kj} F_j \sqrt{\det G(\mathbf{u})}\,du_1\cdots du_M.$$

Using the fact that

$$\sum_{k=1}^{M} g_{kj} g^{ki} = \delta_{ij}$$

we write

$$\langle \mathrm{grad}(f), \mathbf{F} \rangle = \sum_{i=1}^{M} \int_{\mathbb{R}^M} \frac{\partial f}{\partial u_i} F_i \sqrt{\det G(\mathbf{u})}\,du_1\cdots du_M.$$

Rearranging terms and using integration by parts,

$$\sum_{i=1}^{M} \int_{D \subset \mathbb{R}^M} \frac{\partial f}{\partial u_i} F_i \sqrt{\det G(\mathbf{u})}\,du_1\cdots du_M = -\sum_{i=1}^{M} \int_{D \subset \mathbb{R}^M} f \frac{\partial}{\partial u_i}\Big(F_i \sqrt{\det G(\mathbf{u})}\Big)\,du_1\cdots du_M.$$

In the above equation we assume that the surface terms in the integration by parts vanish. Dividing the integrand by $\sqrt{\det G(\mathbf{u})}$ and multiplying by the same factor to restore the integration measure $ds(\mathbf{u}) = \sqrt{\det G(\mathbf{u})}\,du_1\cdots du_M$, we observe the duality

$$\langle \mathrm{grad}(f), \mathbf{F} \rangle = -(f, \mathrm{div}(\mathbf{F})). \tag{4.14}$$

The Laplacian (or *Laplace Beltrami operator*) of a smooth real-valued function is defined as the divergence of the gradient:

$$\text{div}(\text{grad} f) = \sum_{i=1}^{M} \frac{1}{\sqrt{\det G}} \frac{\partial}{\partial u_i} \left(\sqrt{\det G} \sum_{j=1}^{M} g^{ij} \frac{\partial f}{\partial u_j} \right)$$

$$= \sum_{i,j=1}^{M} g^{ij} \left(\frac{\partial^2 f}{\partial u_i \partial u_j} - \sum_{k=1}^{M} \Gamma_{ij}^{k} \frac{\partial f}{\partial u_k} \right). \tag{4.15}$$

In the case of $\mathbb{R}^N$ with Cartesian coordinates, $\mathbf{v}_i = \mathbf{e}_i$, $g^{ij} = g_{ij} = \delta_{ij}$, and $\Gamma_{ij}^{k} = 0$ for all i, j, k.

In general, the eigenfunctions of the Laplacian on a surface (or Riemannian manifold) form a complete orthonormal basis for the set of square-integrable, complex-valued functions on that surface [34, 36, 44].

4.5 Examples of Curvilinear Coordinates

In this section we illustrate the general definitions given previously with polar and spherical coordinates. See [6, 11] as references. For examples of other curvilinear coordinate systems see [40].

4.5.1 Polar Coordinates

Consider the polar (cylindrical) coordinates

$$x_1 = r \cos \phi \tag{4.16}$$
$$x_2 = r \sin \phi \tag{4.17}$$
$$x_3 = z. \tag{4.18}$$

Here $u_1 = r$, $u_2 = \phi$, and $u_3 = z$. In this case the Jacobian matrix is

$$J(r, \phi, z) = [\mathbf{v}_r, \mathbf{v}_\phi, \mathbf{v}_z] = \left[\frac{\partial \mathbf{x}}{\partial r}, \frac{\partial \mathbf{x}}{\partial \phi}, \frac{\partial \mathbf{x}}{\partial z} \right] = \begin{pmatrix} \cos \phi & -r \sin \phi & 0 \\ \sin \phi & r \cos \phi & 0 \\ 0 & 0 & 1 \end{pmatrix}.$$

The metric tensor is

$$G(r, \phi, z) = \begin{pmatrix} 1 & 0 & 0 \\ 0 & r^2 & 0 \\ 0 & 0 & 1 \end{pmatrix}.$$

According to (4.13),

$$\mathbf{v}_1 = \mathbf{e}_r = \begin{pmatrix} \cos \phi \\ \sin \phi \\ 0 \end{pmatrix}; \quad \mathbf{v}_2/r = \mathbf{e}_\phi = \begin{pmatrix} -\sin \phi \\ \cos \phi \\ 0 \end{pmatrix}; \quad \mathbf{v}_3 = \mathbf{e}_z = \begin{pmatrix} 0 \\ 0 \\ 1 \end{pmatrix}$$

and the vector $\mathbf{F}$ is expanded as

$$\mathbf{F} = F_1 \mathbf{v}_1 + F_2 \mathbf{v}_2 + F_3 \mathbf{v}_3 = F_r \mathbf{e}_r + F_\phi \mathbf{e}_\phi + F_z \mathbf{e}_z.$$

The above expression serves as a definition. That is, $F_1 = F_r$, $F_2 = F_\phi/r$, and $F_3 = F_z$.

By the general definitions given previously, the gradient, divergence, and Laplacian are given in polar (cylindrical) coordinates as

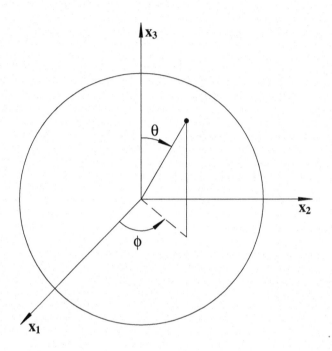

Fig. 4.3. Spherical Coordinates.

$$\mathrm{grad} f = \frac{\partial f}{\partial r} \mathbf{v}_1 + \frac{1}{r^2} \frac{\partial f}{\partial \phi} \mathbf{v}_2 + \frac{\partial f}{\partial z} \mathbf{v}_3 = \frac{\partial f}{\partial r} \mathbf{e}_r + \frac{1}{r} \frac{\partial f}{\partial \phi} \mathbf{e}_\phi + \frac{\partial f}{\partial z} \mathbf{e}_z,$$

$$\mathrm{div} \mathbf{F} = \frac{1}{r} \frac{\partial (r F_1)}{\partial r} + \frac{\partial F_2}{\partial \phi} + \frac{\partial F_3}{\partial z} = \frac{1}{r} \frac{\partial (r F_r)}{\partial r} + \frac{1}{r} \frac{\partial F_\phi}{\partial \phi} + \frac{\partial F_z}{\partial z},$$

and

$$\mathrm{div}(\mathrm{grad} f) = \frac{1}{r} \frac{\partial}{\partial r} \left(r \frac{\partial f}{\partial r} \right) + \frac{1}{r^2} \frac{\partial^2 f}{\partial \phi^2} + \frac{\partial f^2}{\partial z^2}.$$

We will see shortly that the eigenfunctions of the Laplacian in polar coordinates involve Bessel functions, and for fixed ϕ and z, the eigenfunction problem reduces to the Sturm-Liouville equation defining the Bessel functions.

4.5.2 Spherical Coordinates

Consider the orthogonal curvilinear coordinates:[6]

$$x_1 = r \cos \phi \sin \theta$$
$$x_2 = r \sin \phi \sin \theta \tag{4.19}$$
$$x_3 = r \cos \theta.$$

These are the familiar spherical coordinates, where $0 \leq \theta \leq \pi$ and $0 \leq \phi \leq 2\pi$. r is the radial distance from the origin, θ is called the polar angle (or colatitude), and ϕ is called azimuthal angle (or longitude). These are denoted in Figure 4.3.

In this case, the Jacobian matrix with respect to coordinates $(u_1, u_2, u_3) = (r, \phi, \theta)$ is

$$J(r, \phi, \theta) = \left[\frac{\partial \mathbf{x}}{\partial r}, \frac{\partial \mathbf{x}}{\partial \phi}, \frac{\partial \mathbf{x}}{\partial \theta} \right] = \begin{pmatrix} \cos \phi \sin \theta & -r \sin \phi \sin \theta & r \cos \phi \cos \theta \\ \sin \phi \sin \theta & r \cos \phi \sin \theta & r \sin \phi \cos \theta \\ \cos \theta & 0 & -r \sin \theta \end{pmatrix}.$$

The metric tensor is

$$G(r, \phi, \theta) = \begin{pmatrix} 1 & 0 & 0 \\ 0 & r^2 \sin^2 \theta & 0 \\ 0 & 0 & r^2 \end{pmatrix}.$$

By the general definitions given previously, the gradient, divergence, and Laplacian in spherical coordinates are

$$\mathrm{grad} f = \frac{\partial f}{\partial r} \mathbf{e}_r + \frac{1}{r} \frac{\partial f}{\partial \theta} \mathbf{e}_\theta + \frac{1}{r \sin \theta} \frac{\partial f}{\partial \phi} \mathbf{e}_\phi, \tag{4.20}$$

$$\mathrm{div} \mathbf{F} = \frac{1}{r^2} \frac{\partial (r^2 F_r)}{\partial r} + \frac{1}{r \sin \theta} \frac{\partial F_\phi}{\partial \phi} + \frac{1}{r \sin \theta} \frac{\partial (F_\theta \sin \theta)}{\partial \theta}, \tag{4.21}$$

and

$$\mathrm{div}(\mathrm{grad} f) = \frac{1}{r^2} \frac{\partial}{\partial r} \left(r^2 \frac{\partial f}{\partial r} \right) + \frac{1}{r^2 \sin^2 \theta} \frac{\partial^2 f}{\partial \phi^2} + \frac{1}{r^2 \sin \theta} \frac{\partial}{\partial \theta} \left(\sin \theta \frac{\partial f}{\partial \theta} \right), \tag{4.22}$$

where

$$\mathbf{e}_r = \begin{pmatrix} \cos \phi \sin \theta \\ \sin \phi \sin \theta \\ \cos \theta \end{pmatrix}; \quad \mathbf{e}_\phi = \begin{pmatrix} -\sin \phi \\ \cos \phi \\ 0 \end{pmatrix}; \quad \mathbf{e}_\theta = \begin{pmatrix} \cos \phi \cos \theta \\ \sin \phi \cos \theta \\ -\sin \theta \end{pmatrix}$$

and the vector $\mathbf{F} = \sum_{i=1}^3 F_i \mathbf{e}_i$ is expanded as $\mathbf{F} = F_r \mathbf{e}_r + F_\phi \mathbf{e}_\phi + F_\theta \mathbf{e}_\theta$.

If we fix the value of radius as $r \equiv 1$ before taking derivatives, then all of the above derivatives with respect to r vanish. In this case, (ϕ, θ) are the only variables, the above definitions give the equations for grad, div, and Laplacian for the unit sphere.

The Dirac delta function for the sphere centered as $\mathbf{u}_0 = \mathbf{u}(\phi_0, \theta_0)$ can be calculated using (E.4) as

$$\delta_{\mathbb{S}^2}(\mathbf{u}, \mathbf{u}_0) = \begin{cases} \delta(\theta - \theta_0)\, \delta(\phi - \phi_0)/\sin \theta & \theta_0 \notin \{0, \pi\} \\ \delta(\theta - \theta_0)/(2\pi \sin \theta) & \theta_0 \in \{0, \pi\} \end{cases}. \tag{4.23}$$

---

[6]Often in engineering the angles ϕ and θ in Figure 4.3 are reversed (see, e.g., [30]). We use the notation more common in physics and mathematics to be consistent with the classical references (see, e.g., [6]).

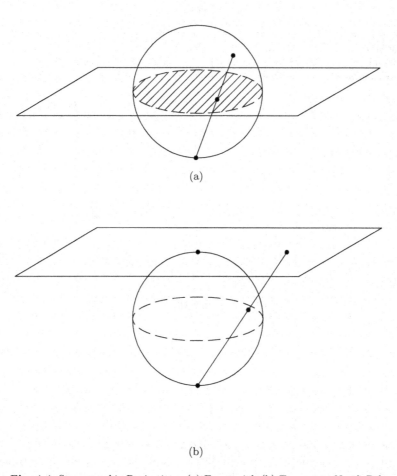

(a)

(b)

Fig. 4.4. Stereographic Projections: (a) Equatorial; (b) Tangent to North Pole

4.5.3 Stereographic Projection

Stereographic projections from the unit sphere to the plane can be defined in an analogous manner to those described earlier in the case of the projection of the circle onto the line.

In the particular case of the projection when the projecting line originates at the south pole and the plane intersects the sphere at the equator (Figure 4.4(a)), the relationship between planar coordinates (x_1, x_2) and $\mathbf{v} \in \mathbb{S}^2$ is

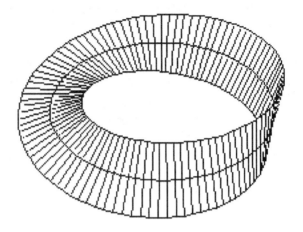

Fig. 4.5. The Möbius Band

$$x = x_1 + ix_2 = \frac{v_1 + iv_2}{1 + v_3}. \tag{4.24}$$

The inverse relationship is

$$v_1 = \frac{2x_1}{1 + x_1^2 + x_2^2} \qquad v_2 = \frac{2x_2}{1 + x_1^2 + x_2^2} \qquad v_3 = \frac{1 - x_1^2 - x_2^2}{1 + x_1^2 + x_2^2}. \tag{4.25}$$

It can be verified that this projection has the following properties [13]: (1) circles on the sphere are mapped to circles or lines in the plane; and (2) angles between tangents to intersecting curves on the sphere are preserved under projection to the plane. The other stereographic projection shown in Figure 4.4(b) is less common.

Using the general theory, it is possible to formulate the gradient, divergence, and Laplacian of functions on the sphere using the projected coordinates, though this is much less common than the (θ, ϕ) coordinates discussed earlier.

4.5.4 The Möbius Band

The Möbius band is a rectangular strip, the ends of which are turned through an angle of $N\pi$ relative to each other and attached, where N is an odd integer. Figure 4.5 illustrates the $N = 1$ case.

In the general case, such a band can be parameterized with the variables $0 \le \theta \le 2\pi$ and $-r_0 < r < r_0$ for the constants $r_0 \le 1$ and $N \in \mathbb{Z}$ as

$$\mathbf{x}(\theta, r) = \mathbf{n}(\theta) + r\mathbf{u}(\theta)$$

where

$$\mathbf{n}(\theta) = [\cos\theta, \sin\theta, 0]^T$$

is the position of the center circle, and

$$\mathbf{u}(\theta) = \begin{pmatrix} \cos\theta \sin N\theta/2 \\ \sin\theta \sin N\theta/2 \\ \cos N\theta/2 \end{pmatrix}$$

is the unit tangent to the surface in the direction orthogonal to $\mathbf{n}$.

We note that the values of the parameters (r,θ) and $((-1)^N r, \theta + 2\pi)$ define the same point on the strip:

$$\mathbf{x}(r,\theta) = \mathbf{x}((-1)^N r, \theta + 2\pi).$$

However, when we compute the Jacobian

$$J(\theta,r) = \left[\frac{\partial \mathbf{x}}{\partial\theta}, \frac{\partial \mathbf{x}}{\partial r} \right],$$

we find that

$$J(\theta,r) = J((-1)^N r, \theta + 2\pi) \begin{pmatrix} 1 & 0 \\ 0 & (-1)^N \end{pmatrix}.$$

This suggests that for odd N there is not a well-defined *orientation* for the Möbius band, since the direction of the normal $\partial \mathbf{x}/\partial\theta \times \partial \mathbf{x}/\partial r$ differs depending on whether it is evaluated at the parameters (r,θ) or $((-1)^N r, \theta + 2\pi)$, even though these correspond to the same point.

In order to clarify what is meant by the word orientation, the following definitions are helpful.

A *Gauss map* is a function that assigns to each point on a surface a unit normal vector. Since this may be viewed as a mapping between the surface and the unit sphere, $\mathbb{S}^2$, it is also called a *normal spherical image*.

A surface (2-manifold) is *orientable* if the Gauss map is continuous at all points.

The orientation of any orientable surface is fixed by specifying the orientation of the normal at any point on the surface (e.g. up or down, in or out).

The Möbius band is a standard example of a manifold that is not *orientable*. Throughout the rest of the text, we will deal exclusively with orientable manifolds for which well-defined normals and concepts of volume and integration exist.

4.6 Topology of Surfaces and Regular Tessellations

Topology is a field of mathematics which considers only those properties of manifolds (or more general spaces) that remain invariant under certain kinds of invertible transformations. Topological properties include connectedness and compactness.

Path connectedness means that it is possible to form a path between any two points in the object, space or surface under consideration, with the whole path being contained in the object. We say the object is *simply connected* if any closed path in the object can be continuously "shrunk" to a point, while always remaining in the object. *Compactness* means that the object is closed and bounded.

An object can have more than one component, each of which is connected (such as a hyperboloid of two sheets or the set of all orthogonal matrices of dimension n), in which case the whole object is not connected; an object can be simply connected (such as the sphere, the surface of a cube, or the plane); or an object can be path-connected (but not simply connected) such as the torus or the set of rotations $SO(3)$.

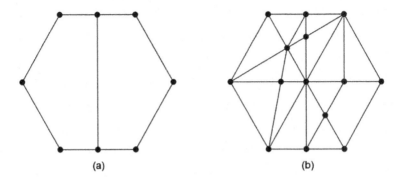

Fig. 4.6. Successive Tessellations of a Polygon

Even though the plane and sphere are both simply connected, the former is non-compact while the latter is compact. Therefore, it is not possible to find an invertible mapping between the sphere and plane (recall that stereographic projection has a singularity). On the other hand, the sphere and surface of the cube can be invertibly mapped to each other (for instance, by lining up their centers and projecting along all lines passing through this point) and are considered to be the same from the topological perspective.

In Subsection 4.6.1 we provide the oldest and perhaps the most illustrative example of a topological property of surfaces. In Subsection 4.6.2 we apply this to the determination of decompositions of the sphere and plane into congruent regular components.

4.6.1 The Euler Characteristic and Gauss-Bonnet Theorem

Consider a polygon in the plane. Figure 4.6(a) shows what happens if we divide it in half by introducing a line segment (edge) terminating at two points (vertices). The remaining figures show additional introductions of edges and vertices. Enclosed within some set of edges is always an area (called a face). If we count the number of vertices (V), edges (E), and faces (F), we find that no matter how we tessellate the polygon, the formula $V - E + F = 1$ holds.

In the plane with no boundary, if we start forming a larger and larger collection of polygons we observe the condition $V - E + F = 2$ where the whole plane minus the collection of inscribed shapes is counted as one face.

Likewise, consider a rectangular strip. If we imagine tessellating the strip in such a way that is consistent with the two ends being twisted by π radians relative to each other and glued, we observe the result $V - E + F = 0$. If without twisting, but gluing parallel edges together, the same formula holds. This is the case of a torus.

The quantity

$$V - E + F = \chi(\mathcal{S}) \tag{4.26}$$

is called the *Euler characteristic* of the surface $\mathcal{S}$. See [16] for a more detailed discussion. Euler discovered it more than two hundred years ago. From the above arguments, we see $\chi(\mathbb{R}^2) = 2$, $\chi(T^2) = \chi(Mo) = 0$. Here T^2 and Mo denote the 2-torus and Möbius strip, respectively. Similar arguments for the sphere show that $\chi(\mathbb{S}^2) = 2$.

Another remarkable result is that the total curvature of a surface (curvature integrated over the whole surface) is a topological (as opposed to geometrical) property. The

Gauss-Bonnet theorem says that the total curvature is related to the Euler characteristic as

$$\int_{\mathcal{S}} K \, ds = 2\pi \chi(\mathcal{S}) \tag{4.27}$$

where K is the Gaussian curvature defined in (4.10) and $ds = \sqrt{\det G} \, du_1 du_2$ is the area element for the surface. This is easy to see by direct calculation for the sphere or the torus.

To see that it really is invariant under smooth topological transformations, start with the surface parameterized as $\mathbf{x}(u_1, u_2)$, and apply an arbitrary invertible transformation $\mathbf{f} : \mathbb{R}^3 \to \mathbb{R}^3$ that is smooth with a smooth inverse. The surface then becomes $\mathbf{f}(\mathbf{x}(u_1, u_2))$. Going through the calculation of the volume element and curvature of this new surface, we find that the integral of total curvature remains unchanged.

In the next section we use the Euler characteristic to determine constraints that regular tessellations of the plane and sphere must obey.

4.6.2 Regular Tessellations of the Plane and Sphere

A *regular tessellation* is a division of a surface (or space) into congruent regular polygons (polyhedra in the three-dimensional case, and polytopes in higher dimensional cases). Our current discussion is restricted to tessellations of surfaces, in particular the plane and sphere.

The Case of the Plane

For a regular planar N-gon, the exterior angle between adjacent edges will be $\theta = 2\pi/N$. The interior angle will therefore be $\pi - \theta$.

Suppose M such N-gons meet at each vertex. Then it must be the case that

$$M \cdot (\pi - \theta) = 2\pi.$$

Dividing by π, this is rewritten as

$$M \cdot (1 - 2/N) = 2,$$

or solving for N in terms of M,

$$N = 2M/(M - 2).$$

Since M and N must both be positive integers, and $N \geq 3$, and $M - 2$ does not divide $2M$ without remainder for $M > 6$, we see immediately that

$$(M, N) \in \{(3, 6), (4, 4), (6, 3)\}.$$

These correspond to hexagonal, square, and equilateral triangular tessellations, respectively.

The Case of the Sphere

Given a regular tessellation, if m faces meet at each vertex and each face is a regular n-gon (polygon with n edges), then the tessellation (as well as the constituent polygons) is said to be of type $\{n, m\}$. It follows that for this type of tessellation,

$$V \cdot m = 2E = n \cdot F. \tag{4.28}$$

The above equation simply says that each edge joins two vertices and each edge lies at the interface of two faces.

Substituting (4.28) into (4.26) for the case of a sphere, we have

$$2E/m - E + 2E/n = 2,$$

or

$$\frac{2n - mn + 2m}{2mn} = \frac{1}{E}.$$

Since $m, n, E > 0$, we have $2n - mn + 2m > 0$, or equivalently

$$(n - 2)(m - 2) < 4.$$

For positive integer values of m and n, it is easy to see that a necessary (though not sufficient) condition for this equation to hold is that both the numbers m and n must be less than six. Direct enumeration of all pairs $\{n, m\}$ weeded out by conditions such as that every face must be surrounded by at least three edges and vertices, yields

$$\{3, 3\}, \{4, 3\}, \{3, 4\}, \{5, 3\}, \{3, 5\}.$$

These correspond to the tetrahedron, cube, octahedron, dodecahedron, and icosahedron.

4.7 Orthogonal Expansions and Transforms in Curvilinear Coordinates

The definition of orthogonal functions in curvilinear coordinates (and on curved surfaces in higher dimensions) is analogous in many ways to Sturm-Liouville theory. Given an M-dimensional hyper-surface S, or a finite-volume section of the surface $\Omega \subset S$, orthogonal real-valued functions are those for which

$$(\phi_i, \phi_j) \doteq \int_\Omega \phi_i(\mathbf{u}) \phi_j(\mathbf{u}) \sqrt{\det(G(\mathbf{u}))} \, du^M = \delta_{ij}.$$

Hence, the square root of the determinant of the Riemannian metric tensor serves as a weighting function. If this determinant can be written as the product of functions of only one variable, then the classical Sturm-Liouville theory can be used to expand functions on the surface in an orthonormal basis.

A complete orthogonal set of functions on any orientable hyper-surface can be generated as eigenfunctions of the Laplacian operator. Sometimes in geometry, the Laplacian is defined as $-\nabla^2$ rather than ∇^2. Instead of adopting that notation we will use the Laplacian defined earlier in this chapter, and include the negative sign when needed. For example, the (negative) Laplacian of functions on the circle is defined as $-d^2/d\theta^2$. Eigenfunctions of the (negative) Laplacian are those which satisfy $-d^2 f/d\theta^2 = \lambda f$ and the periodic boundary conditions $f(0) = f(2\pi)$ and $f'(0) = f'(2\pi)$. The result is the set of functions $f_n(\theta) = e^{in\theta}$ for $n \in \mathbb{Z}$. These functions form the starting point for classical Fourier analysis.

More generally, it has been observed [48] that the eigenfunctions of the Laplacian operator on a smooth compact Riemannian manifold also form an orthogonal basis for functions that are square integrable on the manifold. The integration measure is the one defined relative to the Riemannian metric in (4.7).

4.7.1 Expansions on the Euclidean Plane in Polar Coordinates

In this section two different kinds of orthogonal decompositions of functions on the plane are described. The first of these is a direct conversion of the Fourier transform in Cartesian coordinates into a series involving Bessel functions in the radial direction and Fourier series in the angular coordinate. The second is a Hermite-function expansion in Cartesian coordinates that transforms nicely into a series involving associated Laguerre functions in the radial direction and Fourier series in the angular coordinate.

The 2-D Fourier Transform in Polar Coordinates

The Fourier transform in two dimensions is written in Cartesian coordinates as

$$\hat{f}(\mathbf{p}) = \hat{f}(p_1, p_2) = \frac{1}{2\pi} \int_{-\infty}^{\infty} \int_{-\infty}^{\infty} f(r_1, r_2) \, e^{-i(p_1 r_1 + p_2 r_2)} \, dr_1 dr_2 = \frac{1}{2\pi} \int_{\mathbb{R}^2} f(\mathbf{r}) \, e^{-i\mathbf{p} \cdot \mathbf{r}} \, d\mathbf{r}.$$

Changing to polar coordinates in both the spatial and Fourier domains,

$$r_1 = r \cos\theta \qquad r_2 = r \sin\theta$$

and

$$p_1 = p \cos\phi \qquad p_2 = p \sin\phi,$$

then the original function is $f(r_1, r_2) = \tilde{f}(r, \theta)$, and using (3.47) the Fourier transform is rewritten as

$$\hat{\tilde{f}}(p, \phi) = \frac{1}{2\pi} \int_0^{2\pi} \int_0^{\infty} \tilde{f}(r, \theta) \, e^{-ipr \cos(\theta - \phi)} r \, dr \, d\theta$$

$$= \frac{1}{2\pi} \sum_{n=-\infty}^{\infty} i^{-n} e^{in\phi} \int_0^{2\pi} \int_0^{\infty} \tilde{f}(r, \theta) J_n(pr) \, e^{-in\theta} r \, dr \, d\theta. \qquad (4.29)$$

If we make the expansion

$$\tilde{f}(r, \theta) = \sum_{n=-\infty}^{\infty} f_n(r) \, e^{in\theta},$$

then

$$\hat{\tilde{f}}(p, \phi) = \sum_{n=-\infty}^{\infty} i^{-n} e^{in\phi} \hat{f}_n(p)$$

where $\hat{f}_n(p)$ is the Hankel transform of $f_n(r)$. Multiplying both sides of the above expression by $(2\pi)^{-1} e^{-im\phi}$ and integrating and then setting $m = n$ gives

$$\hat{f}_n(p) = \frac{i^n}{2\pi} \int_0^{2\pi} \hat{\tilde{f}}(p, \phi) \, e^{-in\phi} \, d\phi.$$

And applying the (inverse) Hankel transform,

$$f_n(r) = \frac{i^n}{2\pi} \int_0^{2\pi} \int_0^\infty \hat{\tilde{f}}(p, \phi) \, e^{-in\phi} J_n(pr) p \, dp \, d\phi.$$

Then substituting this into the expression for $\tilde{f}(r, \theta)$ results in

$$\tilde{f}(r, \theta) = \frac{1}{2\pi} \sum_{n=-\infty}^{\infty} i^n e^{in\theta} \int_0^{2\pi} \int_0^\infty \hat{\tilde{f}}(p, \phi) \, e^{-in\phi} J_n(pr) p \, dp \, d\phi, \tag{4.30}$$

which is the Fourier reconstruction formula (or inverse Fourier transform) in polar coordinates.

We also note the expansion

$$e^{i\mathbf{p} \cdot \mathbf{r}} = \sum_{n=-\infty}^{\infty} i^n J_n(pr) \, e^{in\theta} e^{-in\phi} \tag{4.31}$$

and its complex conjugate

$$e^{-i\mathbf{p} \cdot \mathbf{r}} = \sum_{n=-\infty}^{\infty} i^{-n} J_n(pr) \, e^{-in\theta} e^{in\phi}$$

which could have been substituted into the Fourier transform and its inverse in Cartesian coordinates to derive (4.29) and (4.30).

Hermite and Fourier-Laguerre Expansions

The classical Fourier expansions in Cartesian and polar coordinates are important in many applications. However, from the computational perspective they have the problematic property that a band-limited expansion in Cartesian coordinates will not be band-limited in polar coordinates, and vice versa. However, it is possible to define alternative expansions that are simultaneously band-limited in both coordinate systems. In particular, we can expand a function in the band-limited, bivariate, Hermite expansion

$$f(x, y) = \sum_{m=0}^{N} \sum_{n=0}^{N-m} \hat{f}_{m,n} h_m(x) h_n(y) \tag{4.32}$$

where $h_n(x)$ is defined in (3.25). The expansion in (4.32) can be converted into polar coordinates to result in $f(r\cos\theta, r\sin\theta)$. Interestingly, this can be expressed as another finite double sum that decomposes in the r and θ directions rather than x and y. The r dependence can be decomposed into associated Laguerre functions, and the θ dependence can be expressed as a Fourier series [43, 50]. The polar-coordinate version of (4.32) are known in the cosmology literature as "shapelets" [32, 46] and are used to describe the shape of galaxies. And they have potential applications in image processing, optics, and tomography as well [25, 29, 43].

Moreover, expansions of the form in (4.32) have the nice property under planar rotations $(x, y) \rightarrow (x\cos\alpha + y\sin\alpha, -x\sin\alpha + y\cos\alpha)$ the bandlimit is preserved. The Hermite-function coefficients transform in an invertible linear way that depends on α, $\hat{f}_{m,n} \rightarrow \hat{f}_{m,n}(\alpha)$, as described in [42].

4.7.2 Orthogonal Expansions on the Unit Disk

The open unit disk is the region in the Euclidean plane defined by the condition on Cartesian coordinates $x^2 + y^2 < 1$, or in polar coordinates $r \in [0,1)$. Let us denote this space as $\mathbb{B}^2$, the "unit ball" in $\mathbb{R}^2$. The closure of this space is $\overline{\mathbb{B}^2} = \mathbb{B}^2 \cup \mathbb{S}^1$. For most of the calculations we will do, the distinction between $\overline{\mathbb{B}^2}$ and $\mathbb{B}^2$ will be unimportant.

In Chapter 3 we reviewed two orthogonal function expansions on the unit interval. Since for the unit disk we have $r \in [0,1)$ with measure rdr, either Bessel functions or Zernike polynomials can be used to capture the radial dependence of any well-behaved function on the disk. A Fourier series expansion can be used in the angular variable. The combination results in an orthogonal function expansions for the unit disk.

2D Fourier-Bessel Expansions

By combining the Bessel series expansion in (3.51) for the radial direction, and a Fourier series in the angular coordinate, functions on the unit disk can be expanded as

$$f(r,\phi) = \sum_{m=-\infty}^{\infty} \sum_{n=1}^{\infty} c_{m,n} J_m(\rho_n^m r) \, e^{im\phi}. \qquad (4.33)$$

The coefficients in this expansion are

$$c_{m,n} = \frac{1}{2\pi C_{m,n}} \int_0^{2\pi} \int_0^1 f(r,\phi) J_m(\rho_n^m r) r dr d\phi$$

where ρ_n^m and $C_{m,n}$ were defined in association with (3.50).

Note that unlike the expansion in Subsection 4.7.1, where the radial frequency variable p was continuous, both the radial and angular frequency variables are discrete in the 2D Fourier-Bessel expansion in (4.33).

Fourier-Zernike Expansions

The Zernike polynomials in (3.67) are often used in combination with the functions $e^{\pm im\theta}$ to form the orthogonal basis elements

$$V_n^{\pm m}(r,\theta) = Z_n^m(r) \, e^{\pm im\theta} \qquad (4.34)$$

for the set of square-integrable functions on the disc of unit radius, i.e., $\mathcal{L}^2(\mathbb{B}^2)$. For examples of Zernike polynomials being applied in the field of pattern analysis, see, for example, [28] (and references therein).

4.7.3 Orthogonal Expansions on the Sphere

When restricted to the surface of a sphere, the result of orthogonal expansions in spherical coordinates is the spherical harmonics. A more general theory of orthogonal expansions on higher dimensional manifolds follows in the same way.

The unit sphere in $\mathbb{R}^3$ is a two-dimensional surface denoted as $\mathbb{S}^2$ defined by the constraint $x_1^2 + x_2^2 + x_3^2 = 1$. The spherical coordinates (θ, ϕ) are the most common.

Other parameterizations include stereographic projection [13] and other angular measurements. In the (θ, ϕ) parameterization, integration of a function on the sphere is performed as

$$\int_{\mathbb{S}^2} f(\mathbf{u})\, d(\mathbf{u}) \doteq c \int_{\theta=0}^{\pi} \int_{\phi=0}^{2\pi} f(\theta,\phi) \sin\theta \, d\phi \, d\theta \qquad (4.35)$$

where $\mathbf{u} = \mathbf{u}(\theta,\phi)$ is the standard parameterization of the unit sphere obtained by evaluating (4.20) at $r = 1$, and c is a constant that is chosen to be either 1 or 4π depending on the context. Henceforth we choose $c = 1$ unless otherwise stated.

Using the spherical coordinates introduced in the previous section, we can expand functions on the sphere into a series approximation much like Fourier series. The volume element can be viewed as the product of the volume elements for $\mathbb{S}^1$ and $[0,\pi]$ (with weight $\sin\theta$). Functions on the sphere are then viewed as functions on $\mathbb{S}^1 \times [0,\pi]$ with the weighting factor $w(\theta,\phi) = w_1(\theta)w_2(\phi) = \sin\theta \cdot 1$. This allows us to use Sturm-Liouville theory to construct orthonormal functions in these two domains separately. We already know that an orthogonal basis for $\mathcal{L}^2(\mathbb{S}^1)$ is given by $\{e^{im\phi}\}$ for $m \in \mathbb{Z}$. Likewise for $\mathcal{L}^2([-1,1], dx)$, we have the Legendre polynomials $P_l(x)$. It follows that by the change of coordinates $x = \cos\theta$, the functions $\sqrt{(2l+1)/2}P_l(\theta)$ form an orthonormal basis for $\mathcal{L}^2([0,\pi], d\theta \sin\theta)$, and the collection of all products $\{\sqrt{(2l+1)/4\pi}e^{im\phi}P_l(\theta)\}$ for $l = 0,1,2...$ and $m \in \mathbb{Z}$ form a complete orthonormal set of functions on the sphere. This is, however, not the only choice. In fact, it is much more common to choose the *associated Legendre functions* $\{P_l^m(\theta)\}$ for any fixed $m \in \mathbb{Z}$ satisfying $|m| \leq l$ as the basis for $\mathcal{L}^2([0,\pi], d\theta \sin\theta)$. These functions were defined in Subsection 3.2. The orthonormal basis that is most commonly used to expand functions on the sphere has elements of the form

$$Y_l^m(\theta,\phi) = \sqrt{\frac{(2l+1)(l-m)!}{4\pi(l+m)!}}P_l^m(\cos\theta)\,e^{im\phi}. \qquad (4.36)$$

These are called *spherical harmonics*[7] for reasons we will see shortly. They form a basis for describing square-integrable functions on the sphere, and obey the following addition theorem and Unsöld normalization conditions:

$$P_l(\mathbf{u} \cdot \mathbf{u}') = \frac{4\pi}{2l+1}\sum_{m=-l}^{l} Y_l^m(\mathbf{u})\overline{Y_l^m(\mathbf{u}')} \implies \frac{4\pi}{2l+1}\sum_{m=-l}^{l}|Y_l^m(\mathbf{u})|^2 = 1. \qquad (4.37)$$

They satisfy the orthogonality and completeness relations

$$\int_{\mathbb{S}^2} Y_l^m(\mathbf{u})\overline{Y_{l'}^{m'}(\mathbf{u})}\, d\mathbf{u} = \delta_{m,m'}\delta_{l,l'} \qquad (4.38)$$

and

$$\sum_{l=0}^{\infty}\sum_{m=-l}^{l} Y_l^m(\mathbf{u})\overline{Y_l^m(\mathbf{u}')} = \delta_{\mathbb{S}^2}(\mathbf{u},\mathbf{u}'). \qquad (4.39)$$

The Dirac delta function for the sphere, $\delta_{\mathbb{S}^2}(\mathbf{u},\mathbf{u}')$, is derived using the methods in Appendix E and can be written in components as in (4.23). When $\theta \in (0,\pi)$, this can also be written as

$$\delta_{\mathbb{S}^2}(\mathbf{u},\mathbf{u}') = \delta(\cos\theta - \cos\theta')\,\delta(\phi - \phi')$$

where

---

[7]Two variants on this form are common. Often a factor of $(-1)^m$ is present, which we have absorbed in the definition of the associated Legendre functions (3.18). Also, it is common to define harmonics $\tilde{Y}_l^m$ with the property $\overline{\tilde{Y}_l^m} = \tilde{Y}_l^{-m}$. In contrast $\overline{Y_l^m} = (-1)^m Y_l^{-m}$.

$$\delta(\cos\theta - \cos\theta') = \frac{\delta(\theta - \theta')}{\sin\theta}.$$

It can be shown that the addition theorem for spherical harmonics in (4.37) together with (3.16) in the limit as $\rho \to 1$ from below yields (4.39), as discussed in [45].

Any function in $\mathcal{L}^2(\mathbb{S}^2)$ can be expanded in a (spherical) *Fourier series* as

$$f(\mathbf{u}) = \sum_{l=0}^{\infty}\sum_{m=-l}^{l}\hat{f}(l,m)Y_l^m(\mathbf{u}) \quad \text{where} \quad \hat{f}(l,m) = \int_{\mathbb{S}^2}f(\mathbf{u})\overline{Y_l^m(\mathbf{u})}\,d\mathbf{u}. \quad (4.40)$$

Here the convention in (4.35) is used with $c = 1$.

Each of the coefficients $\hat{f}(l,m)$ is called the (spherical) *Fourier transform* of $f(\theta, \phi)$, whereas the collection $\{\hat{f}(l,m)\}$ is called the spectrum. In order to avoid a proliferation of names of various transforms, we will henceforth refer to these simply as the Fourier series, transform and spectrum for functions on the sphere.

The spherical harmonics, $Y_l^m(\theta, \phi)$, are the most commonly used expansion on the sphere because they are eigenfunctions of the Laplacian operator defined below.

Spherical Harmonics as Eigenfunctions of the Laplacian

The gradient of a differentiable function on the sphere results from setting $r = 1$ in the expression for the gradient in spherical coordinates in (4.20):

$$\text{grad}(f) = \frac{\partial f}{\partial\theta}\mathbf{e}_\theta + \frac{1}{\sin\theta}\frac{\partial f}{\partial\phi}\mathbf{e}_\phi.$$

Likewise, the divergence of a differentiable vector field $\mathbf{F}(\theta,\phi) = F_\theta\mathbf{e}_\theta + F_\phi\mathbf{e}_\phi$ on the sphere results from setting $r = 1$ in (4.21):

$$\text{div}(\mathbf{F}) = \frac{1}{\sin\theta}\frac{\partial(F_\theta\sin\theta)}{\partial\theta} + \frac{1}{\sin\theta}\frac{\partial F_\phi}{\partial\phi}.$$

The Laplacian of a smooth function on the sphere is given by

$$\text{div}(\text{grad}f) = \frac{1}{\sin\theta}\frac{\partial}{\partial\theta}\left(\sin\theta\frac{\partial f}{\partial\theta}\right) + \frac{1}{\sin^2\theta}\frac{\partial^2 f}{\partial\phi^2}.$$

In analogy with Sturm-Liouville theory, the eigenfunctions of the (negative) Laplacian operator are defined as

$$-\text{div}(\text{grad}f) = \lambda f$$

for eigenvalues λ. The boundary conditions resulting from the spherical geometry are periodic in ϕ: $f(\theta, \phi + 2\pi) = f(\theta, \phi)$. One possible set of eigenfunctions are the spherical harmonics defined in (4.36), with $\lambda = l(l+1)$.

FFTs for the Sphere

Recall that the DFT is a regularly sampled version of the Fourier transform of a continuous band-limited function. In the case of the sphere, all regular samplings are very coarse, e.g., the vertices of the platonic solids. Regular sampling in ϕ and θ is appealing, but yields closely packed sample points near the poles of the sphere, and sparsely packed points near the equator. In [14], Driscoll and Healy formulated an efficient technique for calculating the Fourier transform of band-limited functions on the sphere based on this sampling. Namely, if $\hat{f}(l,m)$ is the Fourier transform of $f(\theta,\phi)$, and $\hat{f}(l,m) = 0$ for all $l \geq B$ (where $B = 2^K$ is the band limit), then it is possible to exactly calculate the whole spectrum of $f(\theta,\phi)$ by the discrete computations:

$$\hat{f}(l,m) = \frac{\sqrt{\pi}}{B} \sum_{j=0}^{2B-1} \sum_{k=0}^{2B-1} a_j f(\theta_j, \phi_k) \overline{Y_l^m(\theta_j, \phi_k)}, \qquad (4.41)$$

where $\theta_j = \pi j/2B$, $\phi_k = \pi k/B$ and a_j are pre-computed constants defined in [14, 19]. The total number of samples in this case is $N = (2B)^2$.

Using the recursion relations for Legendre polynomials, and writing the discrete spherical Fourier transform as

$$\hat{f}(l,m) = \frac{1}{2B} \sqrt{\frac{(2l+1)(l-m)!}{(l+m)!}} \sum_{j=0}^{2B-1} a_j P_l^m(\cos\theta_j) \left(\sum_{k=0}^{2B-1} e^{-im\phi_k} f(\theta_j, \phi_k) \right), \quad (4.42)$$

it is possible to use both the usual FFT and the fast Legendre transform to realize an FFT of functions on the sphere. There are $N = \mathcal{O}(B^2)$ sample points, and the whole spectrum of f can be computed in $\mathcal{O}(N(\log N)^2)$ operations. Using uniform sampling in $x = \cos\theta$ rather than uniform sampling in θ also results in an $\mathcal{O}(N(\log N)^2)$ transform as a result of the general theory reviewed in Subsection 3.5. The inversion formula can be computed at the N sample points in $\mathcal{O}(N(\log N)^2)$ as well [19]. We note that when the discrete Legendre transform in θ is used in the traditional way (without the fast implementation) together with the FFT in ϕ, an $\mathcal{O}(B^3) = \mathcal{O}(N^{3/2})$ algorithm results, and this is what is often used in practice, as in [20].

In contrast to this algorithm, which is exact in exact arithmetic, a number of approximate numerical methods for computing spherical Fourier transforms have been proposed [12, 24, 37]. While from a theoretical perspective these algorithms do not provide exact results, they have been shown to work well in practice.

4.7.4 The 3-D Fourier Transform in Spherical Coordinates

In analogy with the expansion (4.31) in the planar case, we make the expansion [18, 35]

$$\exp(i\mathbf{p} \cdot \mathbf{r}) = 4\pi \sum_{l=0}^{\infty} \sum_{k=-l}^{l} i^l j_l(pr) Y_l^k(\theta_\mathbf{r}, \phi_\mathbf{r}) \overline{Y_l^k(\theta_\mathbf{p}, \phi_\mathbf{p})} \ . \qquad (4.43)$$

The direct Fourier transform is written in spherical coordinates as:

$$\hat{f}(p, \theta_p, \phi_p) = \sum_{l=0}^{\infty} \sum_{k=-l}^{l} \hat{f}_l^k(p) Y_l^k(\theta_p, \phi_p)$$

(the inverse spherical Fourier transform), where

$$\hat{f}_l^k(p) = 4\pi i^l f_l^k(p)$$

and

$$f_l^k(p) = \int_0^\infty A_l^k(r) j_l(pr) r^2 \, dr$$

(the spherical Hankel transform).

The functions $A_l^k(r)$ are defined as

$$A_l^k(r) \doteq \int_0^\pi \int_0^{2\pi} f(r, \theta_r, \phi_r) \overline{Y_l^k(\theta_r, \phi_r)} \sin\theta_r \, d\phi_r \, d\theta_r$$

(the direct spherical Fourier transform). Note that the complex conjugate has been interchanged between $Y_l^k(p)$ and $Y_l^k(r)$ in the expansion (4.43) of a plane wave in spherical harmonics.

The inverse Fourier transform is written as

$$\hat{f}(r, \theta_r, \phi_r) = \frac{2}{\pi} \sum_{l=0}^\infty \sum_{k=-l}^l g_l^k(r) Y_l^k(\theta_r, \phi_r)$$

where

$$g_l^k(r) = \int_0^\infty B_l^k(p) j_l(pr) p^2 \, dp$$

(the spherical Hankel transform), and the functions $B_l^k(p)$ are defined as

$$B_l^k(p) \doteq \int_0^\pi \int_0^{2\pi} \int_0^\infty f(r, \theta_r, \phi_r) \overline{Y_l^k(\theta_r, \phi_r)} r^2 \sin\theta_r \, dr \, d\phi_r \, d\theta_r.$$

This sort of expansion is classical, but remains of interest today. See, for example, [7].

We mention that, in analogy with the planar case, band-limited tri-variate Hermite expansions can be constructed. And these can be evaluated in spherical coordinates and expanded in a series akin to the Laguerre-Fourier decomposition, but with spherical harmonics taking the place of the Fourier basis for the angular variables.

4.7.5 Orthogonal Expansions on the Unit Ball

In analogy with the case of the disk, orthogonal expansions on the unit ball, $\mathbb{B}^3$, can be synthesized from those defined for the unit radial interval with measure $r^2 dr$, and those defined for the sphere. Two of the most common of these expansions use spherical harmonics for the angular variables, and either the spherical Bessel functions or a version of the Zernike polynomials in the radial direction. Here we review only the spherical-harmonic-spherical-Bessel expansion. In this expansion, we have

$$f(r, \mathbf{u}) = \sum_{n=1}^\infty \sum_{l=0}^\infty \sum_{m=-l}^l c_{m,n,l} \, j_l(\rho_n^l r) \, Y_l^m(\mathbf{u}) \qquad (4.44)$$

where now ρ_n^l denotes solutions of (3.57).

4.8 Multi-Dimensional Convolutions on Euclidean Space

This section reviews some basic, yet rarely articulated, aspects of multi-dimensional convolutions, and relates these properties to geometric operations and Sturm-Liouville expansions on compact domains in $\mathbb{R}^n$.

4.8.1 Convolution and Minkowski Sum

A convex body[8], $C \subset \mathbb{R}^n$, is a set of points such that for any $\mathbf{x}, \mathbf{y} \in C$ then all of the points $\lambda \mathbf{x} + (1 - \lambda)\mathbf{y}$ are also in C for all $\lambda \in (0, 1)$.

Let $V(C)$ denote the volume of a convex body C. Re-scaling C as

$$s \cdot C \doteq \{s\mathbf{x} \,|\, \mathbf{x} \in C \subset \mathbb{R}^n, s \in \mathbb{R}_{>0}\}$$

clearly does not change its convexity. However, the volume of the rescaled body will be $V(s \cdot C) = |s|^n V(C)$. We use absolute value signs here because it is also possible to scale by negative numbers. For example, $-C \doteq (-1) \cdot C$ denotes an inversion of the body through the origin. If $C = -C$ then we call C *centrally symmetric*.

Let us denote a translated copy of the body C by vector $\mathbf{t} \in \mathbb{R}^n$ as

$$\mathbf{t} + C \doteq \{\mathbf{x} + \mathbf{t} \,|\, \mathbf{x} \in C\}.$$

Then $V(\mathbf{t} + C) = V(C)$. Likewise, if $A \in \mathbb{R}^{n \times n}$ with $\det A > 0$, define

$$A \cdot C \doteq \{A\mathbf{x} \,|\, \mathbf{x} \in C\}.$$

It follows that $V(A \cdot C) = |\det A| \cdot V(C)$. A consequence of these statements is the well-known fact that rigid-body motions do not change the volume of a body.

Given two convex bodies, C_1 and C_2, the *Minkowski sum* of the bodies is defined as

$$C_1 + C_2 \doteq \{\mathbf{x}_1 + \mathbf{x}_2 \,|\, \mathbf{x}_1 \in C_1, \mathbf{x}_2 \in C_2\}. \tag{4.45}$$

This results in a new convex body. The Minkowski sum of some bodies can be computed simply. For example, the Minkowski sum of two solid balls, both centered at the origin, with radii r_1 and r_2 is a ball centered at the origin with radius $r_1 + r_2$. And the Minkowski sum of two solid boxes oriented to be parallel with each other is also easy. Surprisingly, until recently no closed-form solution for the Minkowski sum of solid ellipsoids with arbitrary aspect ratios and orientations was reported [56].

If C is convex then $C + C = 2C$, and if C in addition is centrally symmetric, then $C + (-C) = 2C$, which is a warning that regular arithmetic does not always apply for Minkowski sums.

A close relationship exists between convolution and the Minkowski sum. Let $I_C : \mathbb{R}^n \to \{0, 1\}$ denote the set indicator function for body C. That is,

$$I_C(\mathbf{x}) \doteq \begin{cases} 1 & \text{if } \mathbf{x} \in C \\ 0 & \text{if } \mathbf{x} \notin C \end{cases}. \tag{4.46}$$

If $f_i : \mathbb{R}^n \to \mathbb{R}$ is any function that takes positive values on the body C_i, and zero values otherwise, then

$$I_{C_1 + C_2}(\mathbf{x}) = H[(f_1 * f_2)(\mathbf{x})] \tag{4.47}$$

where $H[\cdot]$ denotes the *Heaviside step function* defined on $H : \mathbb{R}_{\geq 0} \to \{0, 1\}$ with $H(0) = 0$ and $H(x > 0) = 1$.

Two famous inequalities are associated with convolution and the Minkowski sum, respectively. The first of these is the *entropy power inequality*, or EPI, which holds when the constraints $f_i(\mathbf{x}) \geq 0$ and $\int_{\mathbb{R}^n} f_i(\mathbf{x}) \, d\mathbf{x} = 1$ are enforced:[9]

---

[8]We use C for "corpus" since B would be too similar to $\mathbb{B}^n$ which we reserve for a special convex body: the unit ball in $\mathbb{R}^n$.

[9]This inequality holds for more general probability densities as well, but our goal here is to illustrate similarities between these inequalities.

$$\exp\left(\frac{2}{n}S(f_1 * f_2)\right) \geq \exp\left(\frac{2}{n}S(f_1)\right) + \exp\left(\frac{2}{n}S(f_2)\right) \qquad (4.48)$$

where entropy is defined as

$$S(f_i) \doteq -\int_{\mathbb{R}^n} f_i(\mathbf{x}) \log_e f_i(\mathbf{x}) \, d\mathbf{x}.$$

The second is the *Brunn-Minkowski inequality* which states that

$$[V(C_1 + C_2)]^{1/n} \geq [V(C_1)]^{1/n} + [V(C_2)]^{1/n}. \qquad (4.49)$$

These inequalities look especially similar when we take $f_i(\mathbf{x}) = I_{C_i}(\mathbf{x})/V(C_i)$, in which case $S(f_i) = [V(C_i)]^{2/n}$. Since the maximum entropy distribution on $C_1 + C_2$ is $I_{C_1+C_2}(\mathbf{x})/V(C_1+C_2)$, we find that $[V(C_1+C_2)]^{2/n} \geq \exp\left(\frac{2}{n}S(f_1 * f_2)\right)$ and so from the entropy power inequality

$$[V(C_1 + C_2)]^{2/n} \geq [V(C_1)]^{2/n} + [V(C_2)]^{2/n}.$$

But this is weaker than the Brunn-Minkowski. And similarly, the Brunn-Minkowski inequality can produce inequalities that "look like" weaker versions of the EPI, but these inequalities seem not to be derivable from each other.

4.8.2 Windowed Convolution and Zero Padding

Let $i \in \{1,2\}$ and consider functions $f_i : \mathbb{R}^d \to \mathbb{C}$ that are supported on the open convex centrally-symmetric body $\frac{1}{2}C \in \mathbb{R}^d$. (In most cases we will be interested in functions that take values in $\mathbb{R}_{\geq 0}$, but the results presented here are valid more generally.) From the discussion in the previous subsection, since $f_i(\mathbf{x}) = 0$ when $\mathbf{x} \notin \frac{1}{2}C$, it follows that $f_1 * f_2$ will have support in C.

Moreover, obviously for functions f_i supported on $\frac{1}{2}C$, they will also be supported on C, and so

$$f_i(\mathbf{x}) = f_i(\mathbf{x}) \cdot I_C(\mathbf{x}) \quad \text{and} \quad (f_1 * f_2)(\mathbf{x}) = (f_1 * f_2)(\mathbf{x}) \cdot I_C(\mathbf{x}) \qquad (4.50)$$

for all $\mathbf{x} \in \mathbb{R}^d$. From the multi-dimensional version of the convolution theorem, the convolution of two functions in real space is their product in Fourier space. And in the Euclidean setting this convolution theorem can be used in reverse because real space and Fourier space "look like each other" as is evidenced by (2.22) and (2.23).[10] Therefore, the fact that products of functions in real space correspond to the convolution of their Fourier transforms in Fourier space, together with (4.50), imposes the following conditions:

$$\hat{f}_i(\boldsymbol{\omega}) = (\hat{f}_i * \hat{I}_C)(\boldsymbol{\omega}) \quad \text{and} \quad \hat{f}_1(\boldsymbol{\omega})\hat{f}_2(\boldsymbol{\omega}) = ((\hat{f}_1 \cdot \hat{f}_2) * \hat{I}_C)(\boldsymbol{\omega}). \qquad (4.51)$$

Now consider functions $\tilde{f}_i$ that are defined to be identical to f_i on C, but in contrast to f_i, we allow $\tilde{f}_i$ to take on arbitrary values outside of C. That is,

$$f_i(\mathbf{x}) = \tilde{f}_i(\mathbf{x})I_C(\mathbf{x}). \qquad (4.52)$$

---

[10] This is not true for the Fourier series expansion of functions on the torus, $\mathbb{T}^d$, and it is not true in the group-theoretic setting that will follow in Chapter 8 and beyond.

We define a "windowed convolution" operation for these functions as

$$(\tilde{f}_1 \circledast \tilde{f}_2)(\mathbf{x}) \doteq ((I_C \cdot \tilde{f}_1) * \tilde{f}_2)(\mathbf{x}) = \int_C \tilde{f}_1(\boldsymbol{\xi}) \tilde{f}_2(\mathbf{x} - \boldsymbol{\xi}) \, d\boldsymbol{\xi}. \tag{4.53}$$

The significance of this is that for any $\mathbf{x}$ inside the body C, (4.53) will be identical to $(f_1 * f_2)(\mathbf{x})$ because the range of the integral does not allow the product of $\tilde{f}_1(\boldsymbol{\xi})$ and $\tilde{f}_2(\mathbf{x} - \boldsymbol{\xi})$ to contribute any values from outside of C. However, when $\mathbf{x} \notin C$, it can be that $(\tilde{f}_1 \circledast \tilde{f}_2)(\mathbf{x}) \neq (f_1 * f_2)(\mathbf{x}) = 0$. Therefore, windowing again gives

$$\boxed{I_C(\mathbf{x}) \cdot (\tilde{f}_1 \circledast \tilde{f}_2)(\mathbf{x}) = (f_1 * f_2)(\mathbf{x})} \tag{4.54}$$

for all $\mathbf{x} \in \mathbb{R}^d$.

If $\tilde{f}_i \in (\mathcal{L}^1 \cap \mathcal{L}^2)(\mathbb{R}^d)$, and hence $\hat{\tilde{f}}_i \in (\mathcal{L}^1 \cap \mathcal{L}^2)(\mathbb{R}^d)$ also, then

$$f_i(\mathbf{x}) = \tilde{f}_i(\mathbf{x}) \cdot I_C(\mathbf{x}) \implies \hat{f}_i(\boldsymbol{\omega}) = \widehat{(I_C \cdot \tilde{f}_i)}(\boldsymbol{\omega}) = (\hat{I}_C * \hat{\tilde{f}}_i)(\boldsymbol{\omega}).$$

Taking the Fourier transform of both sides of the defining equality in (4.53) then gives

$$\widehat{(\tilde{f}_1 \circledast \tilde{f}_2)}(\boldsymbol{\omega}) = (\hat{I}_C * \hat{\tilde{f}}_1)(\boldsymbol{\omega}) \cdot \hat{\tilde{f}}_2(\boldsymbol{\omega}). \tag{4.55}$$

In contrast, taking the Fourier transform of (4.54) and using the defining equality in (4.53) gives

$$(\hat{I}_C * \widehat{(\tilde{f}_1 \circledast \tilde{f}_2)})(\boldsymbol{\omega}) = (\hat{I}_C * [(\hat{I}_C * \hat{\tilde{f}}_1) \cdot \hat{\tilde{f}}_2])(\boldsymbol{\omega}) = \hat{\tilde{f}}_1(\boldsymbol{\omega}) \cdot \hat{\tilde{f}}_2(\boldsymbol{\omega}). \tag{4.56}$$

4.8.3 Convolution Theorem for Zero-Padded Functions on Balls

Many of the equalities formulated in the previous section assume the existence of the Fourier transform of $\tilde{f}_i$. An exception to this is (4.54). Here we examine a class of $\tilde{f}_i$'s for which the Fourier transform does not exist. In particular, from (4.33) and (4.44), it is clear how functions can be expanded on the unit disk and unit ball, but the basis functions in these expansions decay very slowly as the radius tends to infinity, and hence $\tilde{f}_i \notin (\mathcal{L}^1 \cap \mathcal{L}^2)(\mathbb{R}^d)$.

However, since it is possible to view (4.33) as a sampled version of (4.30) with Dirac delta functions in frequency-space, these results can be combined with (4.54) to yield a convolution theorem for zero-padded functions on balls. We have not seen this in the literature and write the result as a theorem.

This will build on the Fourier transform pair

$$\hat{f}(\boldsymbol{\omega}) = \frac{1}{(2\pi)^{d/2}} \int_{\mathbb{R}^d} f(\mathbf{x}) e^{-i\boldsymbol{\omega}\cdot\mathbf{x}} \, d\mathbf{x} \qquad f(\mathbf{x}) = \frac{1}{(2\pi)^{d/2}} \int_{\mathbb{R}^d} \hat{f}(\boldsymbol{\omega}) e^{i\boldsymbol{\omega}\cdot\mathbf{x}} \, d\boldsymbol{\omega} \tag{4.57}$$

and the convolution property

$$\widehat{(f_1 * f_2)}(\boldsymbol{\omega}) = \hat{f}_1(\boldsymbol{\omega}) \hat{f}_2(\boldsymbol{\omega}). \tag{4.58}$$

Let $d = 2$ or 3 and let f_1 and f_2 each be functions on the unit ball $\mathbb{B}^d$ that take nonzero values only on $\frac{1}{2}\mathbb{B}^d$. Let $\tilde{f}_i$ denote an approximation of f_i (of the same sort as in the previous section) and expand this in a polar/spherical-coordinate expansion of the form

$$\tilde{f}_i(r\mathbf{u}) = \sum_{\mathbf{m},n,l} C^{(i)}_{\mathbf{m},n,l} \, \mathcal{J}_l(\rho_n^l r) \, \mathcal{Y}_l^{\mathbf{m}}(\mathbf{u}) \tag{4.59}$$

where $r \in [0,1]$, $\mathbf{u} \in \mathbb{S}^{d-1}$, $\mathcal{J}_l(\cdot)$ denotes a generalization of the Bessel function for $\mathbb{R}^d$, ρ_n^l are solutions to satisfy appropriate boundary conditions (e.g., $\mathcal{J}_l(\rho_n^l) = 0$), and $\mathcal{Y}_l^{\mathbf{m}}(\mathbf{u})$ are spherical harmonics for $\mathbb{S}^{d-1}$. Then, equipped with orthogonality relations of the form

$$\int_0^1 \mathcal{J}_l(\rho_n^l r) \, \mathcal{J}_l(\rho_{n'}^l r) \, r^{n-1} \, dr = c_{n,l} \, \delta_{n,n'} \tag{4.60}$$

and

$$\int_{\mathbf{u} \in \mathbb{S}^{d-1}} \mathcal{Y}_{k'}^{\mathbf{m}'}(\mathbf{u}) \, \overline{\mathcal{Y}_k^{\mathbf{m}}(\mathbf{u})} \, d\mathbf{u} = \delta_{\mathbf{m}',\mathbf{m}} \delta_{k,k'} \tag{4.61}$$

we can recover the coefficients in (4.59) from a given f_i as

$$C^{(i)}_{\mathbf{m},n,l} = \frac{1}{c_{n,l}} \int_{\mathbb{S}^{d-1}} \int_0^1 f_i(r\mathbf{u}) \, \mathcal{J}_l(\rho_n^l r) \, \overline{\mathcal{Y}_l^{\mathbf{m}}(\mathbf{u})} \, r^{n-1} \, dr \, d\mathbf{u}. \tag{4.62}$$

Moreover, given a generalized Jacobi-Anger (or spherical wave) expansion of the form

$$e^{i\boldsymbol{\omega}\cdot\mathbf{x}} = \sum_{k=0}^{\infty} \sum_{\mathbf{m}} \alpha_k \, \mathcal{J}_k(pr) \, \mathcal{Y}_k^{\mathbf{m}}(\mathbf{u}) \, \overline{\mathcal{Y}_k^{\mathbf{m}}(\mathbf{v})} \tag{4.63}$$

and the completeness relation

$$\int_0^{\infty} \mathcal{J}_k(p' r) \mathcal{J}_k(pr) \, r^{n-1} \, dr = \beta_k \frac{\delta(p - p')}{p^{n-1}} \tag{4.64}$$

where α_k and β_k are constants that depend on the dimension, d, and $\mathbf{m}$ is a vector of indices (the length of which depends on d), then the following theorem results.

Theorem 4.1. *Convolution Theorem for Functions on Balls. Let $i \in \{1,2\}$ and $f_i : \mathbb{R}^d \to \mathbb{C}$ be smooth functions with support in the open ball $\frac{1}{2}\mathbb{B}^d$, and let $\tilde{f}_i : \mathbb{R}^d \to \mathbb{C}$ be the approximation of these functions of the form in (4.59) that satisfy (4.52) with $C = \mathbb{B}^d$. Then for all $r < 1$*

$$(\tilde{f}_1 \circledast \tilde{f}_2)(r\mathbf{u}) = \widetilde{(f_1 * f_2)}(r\mathbf{u}) \tag{4.65}$$

and

$$\widetilde{(f_1 * f_2)}(r\mathbf{u}) = \sum_{\mathbf{m},n,l} C^{(1*2)}_{\mathbf{m},n,l} \, \mathcal{J}_l(\rho_n^l r) \, \mathcal{Y}_l^{\mathbf{m}}(\mathbf{u}) \tag{4.66}$$

where

$$C^{(1*2)}_{\mathbf{m},n,l} = \frac{\alpha_l \beta_l}{(2\pi)^{d/2} c_{n,l}} \int_{\mathbf{v} \in \mathbb{S}^{d-1}} \hat{f}_1(\rho_n^l \mathbf{v}) \, \hat{f}_2(\rho_n^l \mathbf{v}) \, \overline{\mathcal{Y}_l^{\mathbf{m}}(\mathbf{v})} \, d\mathbf{v}. \tag{4.67}$$

Proof. The equality in (4.65), which can also be written with a $\circledast$ in place of $*$ on the right-hand side, follows from the fact that f_i and $\tilde{f}_i$ are identical when evaluated in $\mathbb{B}$ and hence (4.52) and (4.54) apply.

To prove the above formula for $C^{(1*2)}_{\mathbf{m},n,l}$, we note that even if the Fourier transform of $\tilde{f}_i$ and $\tilde{f}_1 \circledast \tilde{f}_2$ may not exist, it does exist for f_i and $f_1 * f_2$. If we use spherical coordinates for $\mathbb{R}^d$ and let

$$\mathbf{x} = r\mathbf{u} \quad \text{and} \quad \boldsymbol{\omega} = p\mathbf{v}, \tag{4.68}$$

then for any $h \in \mathcal{L}^1(\mathbb{R}^d)$,

$$\int_{\mathbb{R}^d} h(\mathbf{x})\, d\mathbf{x} = \int_{\mathbf{u} \in \mathbb{S}^{d-1}} \int_{r \in \mathbb{R}_{\geq 0}} h(r\mathbf{u}) r^{d-1}\, dr\, d\mathbf{u}$$

and similarly for integration in Fourier space.

Then using (4.63) the Fourier inversion formula in (4.57) can be written as

$$f_i(r\mathbf{u}) = \frac{1}{(2\pi)^{d/2}} \int_{\mathbf{v} \in \mathbb{S}^{d-1}} \int_{p \in \mathbb{R}_{\geq 0}} \hat{f}_i(p\mathbf{v}) \sum_{k=0}^{\infty} \sum_{\mathbf{m}} \alpha_k \mathcal{J}_k(pr) \mathcal{Y}_k^{\mathbf{m}}(\mathbf{u}) \overline{\mathcal{Y}_k^{\mathbf{m}}(\mathbf{v})} p^{d-1}\, dp\, d\mathbf{v}.$$

(4.69)

The coefficients of $\tilde{f}_i(r\mathbf{u})$ are obtained by observing that in the unit ball $\tilde{f}_i(r\mathbf{u}) = f_i(r\mathbf{u})$ and so (4.62) can be used. Moreover, since f_i is defined to have support inside of the unit ball, the integral over r can be extended to infinity. Then substituting (4.69) into (4.62) and using (4.61) and (4.64) with $p' = \rho_n^l$ gives

$$C_{\mathbf{m},n,l}^{(i)} = \frac{\alpha_l \beta_l}{(2\pi)^{d/2} c_{n,l}} \int_{\mathbf{v} \in \mathbb{S}^{d-1}} \hat{f}_i(\rho_n^l \mathbf{v}) \overline{\mathcal{Y}_l^{\mathbf{m}}(\mathbf{v})}\, d\mathbf{v}. \tag{4.70}$$

Then, if the support of $f_1 * f_2$ is also inside of the unit ball, the the same reasoning together with the convolution theorem (4.58) can be used to write (4.67).

4.9 Continuous Wavelet Transforms on the Plane and $\mathbb{R}^n$

The idea of describing a function as a weighted sum (or integral) of translated and scaled (or modulated) versions of a given function extends to domains other than the real line. In this section, we review the work of a number of researchers who have examined continuous wavelet transforms in multi-dimensional Cartesian space.

4.9.1 Multi-Dimensional Gabor Transform

The continuous Gabor transform for $\mathbb{R}^n$ is defined in exact analogy as in the one-dimensional case:

$$G_f(\mathbf{b}, \omega) = \int_{\mathbb{R}^n} f(\mathbf{x}) \overline{g_{(\mathbf{b},\omega)}(\mathbf{x})}\, dx_1 \cdots dx_n$$

where

$$g_{(\mathbf{b},\omega)} = e^{i\omega \cdot (\mathbf{x} - \mathbf{b})} g(\mathbf{x} - \mathbf{b}).$$

Using the same calculations as in the one-dimensional case with the n-dimensional Fourier transform replacing the one-dimensional version, the inversion formula

$$f(\mathbf{x}) = \frac{1}{(2\pi)^n \|g\|^2} \int_{\mathbb{R}^{2n}} G_f(\mathbf{b}, \omega) e^{i\omega \cdot (\mathbf{x} - \mathbf{b})} g(\mathbf{x} - \mathbf{b})\, db_1 \cdots db_n d\omega_1 \cdots d\omega_n.$$

4.9.2 Multi-Dimensional Continuous Scale-Translation Wavelet Transforms

In a series of papers, the continuous wavelet transform based on translations and dilations has been extended to $\mathbb{R}^n$ by considering dilations and rigid-body motions (translations and rotations) [3, 39]. We review here only the planar case (the higher dimensional
cases follow analogously, provided one knows how to describe rotations in $\mathbb{R}^n$).

Let $\varphi \in \mathcal{L}^1(\mathbb{R}^2) \cap \mathcal{L}^2(\mathbb{R}^2)$ and its 2-D Fourier transform, $\hat{\varphi}(\omega)$ satisfy the admissibility
condition

$$0 < k_\varphi = 2\pi \int_{\mathbb{R}^2} |\hat{\varphi}(\omega)|^2 \frac{d\omega_1 d\omega_2}{\|\omega\|^2} < \infty.$$

Then the *scale-Euclidean* wavelet transform is defined as

$$T_f(\mathbf{b}, a, R(\theta)) \doteq \int_{\mathbb{R}^2} f(\mathbf{x}) \overline{\varphi_{(\mathbf{b},a,R(\theta))}(x)}\, dx_1 dx_2$$

where

$$\varphi_{(\mathbf{b},a,R(\theta))}(x) \doteq \frac{1}{a} \varphi\left(R^{-1}(\theta)(\mathbf{x} - \mathbf{b})/a\right)$$

for any $R(\theta) \in SO(2)$.

The inversion follows in a similar way as for the one-dimensional case and is written
explicitly as [39]:

$$f(\mathbf{x}) = \frac{1}{k_\varphi} \int_{\theta=0}^{2\pi} \int_{\mathbf{b}\in\mathbb{R}^2} \int_{a\in\mathbb{R}_{>0}} T_f(\mathbf{b}, a, R(\theta)) \varphi_{\mathbf{b},a,R(\theta)}(x) \frac{da}{a^3}\, db_1 db_2 d\theta.$$

We also note that continuous wavelet transforms based on rigid-body motion, modulation, and dilation have also been presented in the literature. These are called *Weil-
Heisenberg wavelets* [26].

4.10 Gabor Wavelets on the Circle and Sphere

In a sense, the basis $\{e^{in\theta}\}$ is a wavelet basis, and Fourier series expansions are Gabor
wavelet expansions with mother wavelet $\varphi = 1$. Each $e^{in\theta}$ for $n \neq 0$ can also be viewed
as a scaled version of $e^{i\theta}$. The set of Haar functions can also be used to expand functions
on the circle, and so this is another natural wavelet series expansion on a domain other
than the real line.

Here we explore a number of ways to extend the concept of modulation on the
circle and sphere. Together with the appropriate definition of a shift (which in this
case is simply a rotation), a number of different Gabor-like transforms can be defined.
As always, we are considering only transforms of functions on the sphere that are well
behaved (continuous, and hence square integrable).

4.10.1 The Modulation-Shift Wavelet Transform for the Circle

We examine two kinds of modulation-shift wavelets in this subsection. We refer to these
as Gabor and Torresani wavelets, respectively.

Gabor Wavelets for the Circle

Given a mother wavelet $\psi \in \mathcal{L}^2(\mathbb{S}^1)$, we define

$$\psi_{n,\alpha}(\theta) = e^{in\theta}\psi(\theta - \alpha).$$

Then the Gabor transform of a function $f \in \mathcal{L}^2(\mathbb{S}^1)$ is

$$\tilde{f}(n,\alpha) = \int_0^{2\pi} f(\theta)\overline{\psi_{n,\alpha}(\theta)}\, d\theta.$$

The function $f(\theta)$ can be recovered from $\tilde{f}(n,\alpha)$ with the inverse transform

$$f(\theta) = \frac{1}{2\pi\|\psi\|_2^2} \sum_{n=-\infty}^{\infty} \int_0^{2\pi} \tilde{f}(n,\alpha)\psi_{n,\alpha}(\theta)\, d\alpha,$$

where $\|\psi\|^2 = (\psi, \psi)$. This follows by direct substitution and using the formula for the delta function on the circle:

$$\sum_{n=-\infty}^{\infty} e^{in\theta} = 2\pi\delta(\theta).$$

The Torresani Wavelets for the Circle

This subsection describes a transform introduced in [53]. The starting point is a mother wavelet $\psi(\theta)$ with support on the circle in the interval $|\theta| \leq \pi/2$ and which satisfies the admissibility condition

$$0 \neq c_\psi = 2\pi \int_{-\pi/2}^{\pi/2} \frac{|\psi(\gamma)|^2}{\cos\gamma}\, d\gamma < \infty. \tag{4.71}$$

The corresponding wavelet transform is

$$\tilde{f}(\alpha,p) = \int_0^{2\pi} f(\theta)\overline{\psi_{\alpha,p}(\theta)}\, d\theta \tag{4.72}$$

where

$$\psi_{\alpha,p}(\theta) = e^{ip\sin(\theta-\alpha)}\psi(\theta - \alpha)$$

is a modulated and shifted version of $\psi(\theta)$ with p and α specifying the amount of modulation and rotation, respectively.

The inversion formula is

$$f(\theta) = \frac{1}{c_\psi} \int_0^{2\pi} \int_{-\infty}^{\infty} \tilde{f}(\alpha,p)\psi_{\alpha,p}(\theta)\, dp\, d\alpha. \tag{4.73}$$

In order to see that this does in fact reproduce the function f from the transform $\tilde{f}$, we substitute

$$\int_0^{2\pi} \int_{-\infty}^{\infty} \left(\int_0^{2\pi} f(\theta)\overline{\psi_{\alpha,p}(\theta)}\psi_{\alpha,p}(\gamma)\, d\theta \right) dp\, d\alpha =$$

$$\int_0^{2\pi} \int_0^{2\pi} f(\theta)\frac{\delta(\gamma-\theta)}{\cos(\theta-\alpha)}\overline{\psi(\theta-\alpha)}\psi(\gamma-\alpha)\, d\alpha\, d\theta = c_\psi f(\gamma). \tag{4.74}$$

In the left-most term in (4.74), the integral over p is performed first, resulting in the delta function:

$$\int_{-\infty}^{\infty} \exp(ip(\sin(\gamma - \alpha) - \sin(\theta - \alpha))\, dp = 2\pi\delta(\sin(\gamma - \alpha) - \sin(\theta - \alpha)) = \frac{2\pi\delta(\gamma - \theta)}{\cos(\theta - \alpha)}.$$

The final step in (4.74) is the observation that integration on the circle is invariant under shifts.

4.10.2 The Modulation-Shift Wavelet Transform for the Sphere

In this subsection we examine Gabor and Torresani wavelets for the sphere that are analogous to the definitions for the circle.

Gabor Wavelets for the Sphere

The concepts of modulation and translation are generalized to the sphere as multiplication by a spherical harmonic and rotation, respectively. Given a mother wavelet function $\psi \in \mathcal{L}^2(\mathbb{S}^2)$, the spherical Gabor wavelets are then

$$\psi_l^m(\mathbf{u}(\theta, \phi), R) = Y_l^m(\mathbf{u}(\theta, \phi))\psi(R^{-1}\mathbf{u}(\theta, \phi))$$

where R is a 3×3 rotation matrix (see Chapter 5). The spherical Gabor transform is then

$$\tilde{f}_l^m(R) = \int_{\mathbb{S}^2} f(\mathbf{u})\overline{\psi_l^m(\mathbf{u}(\theta, \phi), R)}\, d(\mathbf{u}). \tag{4.75}$$

In order to determine an inversion formula to recover f from $\tilde{f}_l^m(R)$, we need to know how to integrate over the set of all rotations This is addressed in Chapters 5 and 9. We therefore postpone the discussion of the inversion formula until Chapter 9.

Torresani Wavelets for the Sphere

Here a mother wavelet $\psi(\mathbf{u})$, with support on the northern hemisphere, is used to generate a family of wavelets of the form

$$\psi(\mathbf{u}; \mathbf{p}, R) = e^{i(R\mathbf{p})\cdot\mathbf{u}}\psi(R^{-1}\mathbf{u})$$

where R is a rotation matrix and $\mathbf{p} = [p_1, p_2, 0]^T \in \mathbb{R}^3$. The transform is

$$\tilde{f}(\mathbf{p}, R) = \int_{\mathbb{S}^2} f(\mathbf{u})\overline{\psi(\mathbf{u}; \mathbf{p}, R)}\, d(\mathbf{u}). \tag{4.76}$$

As with the spherical Gabor transform, the inversion formula corresponding to (4.76) requires properties of integrals over the set of rotations. Such topics are addressed in Chapters 5 and 9. See [53] for the inversion formula.

4.11 Scale-Translation Wavelet Transforms for the Circle and Sphere

Whereas modulation of functions is a natural operation on the circle and sphere, the dilation of functions on these compact manifolds is somewhat problematic. This is because we are immediately confronted with the problem of what to do with the "extra" part of the dilated version of a function $\psi(\theta)$ that is "pushed outside" of the finite domain $[0, 2\pi]$ or $[0, 2\pi] \times [0, \pi]$ for the circle and sphere, respectively.

The two techniques for handling this problem are *periodization* and *stereographic projection*. Periodization is essentially a way to "wrap" a function on the line (plane) around the circle (sphere). Stereographic projection maps all points of finite value from the line (plane) onto all but one point on the circle (sphere), and vice versa. Hence these approaches convert a function on the circle (sphere) to one on the line (plane), perform the scale change, and then map the function back to the circle (sphere). An appoach to wavelets on the sphere based on diffusion is presented in Chapter 9.

4.11.1 The Circle

Continuous dilation-translation-based wavelet transforms are perhaps less natural than the Gabor wavelets because of the issue of what to do when the support of the dilated function is wider than 2π. To this end, Holschneider [21] defined a continuous wavelet transform for the circle based on dilations and translations by "wrapping" the "leftover part" around the circle. The technical word for this wrapping is *periodization*. Here, we shall review the formulation in [21].

Given a function $\varphi \in \mathcal{L}^2(\mathbb{R})$ that satisfies the same conditions as a mother wavelet for the continuous wavelet transform on the line, the dilated, translated, and periodized version of φ (with period 2π) is denoted

$$[\varphi_{t,a}(x)]_{2\pi} = \sum_{n \in \mathbb{Z}} \frac{1}{a} \varphi \left(\frac{x - t + 2\pi n}{a} \right). \tag{4.77}$$

The periodization operation

$$\varphi \to [\varphi_{0,1}(x)]_{2\pi} = p_{2\pi}(\varphi)$$

commutes with modulations and translations:

$$(p_{2\pi}(\mu_m f))(x) = (\mu_m(p_{2\pi} f))(x)$$

$$(p_{2\pi}(\tau_t f))(x) = (\tau_t(p_{2\pi} f))(x)$$

Recall that

$$(\sigma_s f)(x) = s^{-\frac{1}{2}} f(x/s), \quad (\tau_t f)(x) = f(x - t) \quad \text{and} \quad (\mu_m f)(x) = e^{-imx} f(x),$$

where $s \in \mathbb{R}_{>0}$, $t \in \mathbb{R}$ and $m \in \mathbb{Z}$.

The continuous wavelet transform of a complex-valued function on the circle, $f(x)$, is defined as

$$[\tilde{f}]_{2\pi}(t, a) \doteq \int_0^{2\pi} f(x) \overline{[\varphi_{t,a}(x)]_{2\pi}} \, dx. \tag{4.78}$$

The function f is recovered from $[\tilde{f}]_{2\pi}$ as [21]

$$f(x) = \int_{\mathbb{S}^1 \times \mathbb{R}_{>0}} [\tilde{f}]_{2\pi}(t, a)[\varphi_{t,a}(x)]_{2\pi} \frac{da\,dt}{a}.$$

Our discussion here has been for a wavelet transform for the circle using periodization. Stereographic projection of the continuous translation-scale wavelet transform for the line onto the circle can also be used, and this is discussed in [3]. It follows in much the same way as the case for the sphere discussed in the next subsection.

4.11.2 The Sphere

Here we review the results of [4, 5] where the mother wavelet is assumed to satisfy the admissibility conditions

$$\int_0^{2\pi} \int_0^{\pi} \frac{\psi(\theta, \phi)}{1 + \cos \theta} \sin \theta \, d\theta \, d\phi = 0 \quad \text{and} \quad \int_0^{2\pi} \psi(\theta, \phi) \, d\phi \neq 0.$$

The spherical wavelet transform defined using such mother wavelets is based on stereographic projection of the continuous translation-scale wavelet transform for the plane.

The function $\psi(\theta, \phi)$ is projected onto the plane using stereographic projection. The resulting function on the plane is dilated by a and then projected back. The dilation operator defined in this way, together with a rotation operator, produces the family of wavelets

$$\psi(\mathbf{u}(\theta, \phi); a, R) = \frac{2a}{[(a^2 - 1)(R e_3 \cdot \mathbf{u}) + (a^2 + 1)]} \psi((R^{-1}\mathbf{u}(\theta, \phi))_{1/a})$$

where the notation $(\cdot)_{1/a}$ means

$$(\mathbf{u}(\theta, \phi))_{1/a} = \mathbf{u}(\theta_{1/a}, \phi)$$

where

$$\tan \frac{\theta_{1/a}}{2} = \frac{1}{a} \tan \frac{\theta}{2}.$$

The associated wavelet transform is

$$\tilde{f}(a, R) = \int_0^{2\pi} \int_0^{\pi} f(\mathbf{u}(\theta, \phi)) \overline{\psi(\mathbf{u}(\theta, \phi); a, R)} \sin \theta \, d\theta \, d\phi. \qquad (4.79)$$

Again, in order to invert, we need techniques that are presented in subsequent chapters. See [4, 5] for the inversion formula. For other approaches to spherical wavelets, see [22].

4.11.3 Discrete Wavelets and Other Expansions on the Sphere

A number of other wavelet-like expansions on spheres have been presented in the recent literature.

Discrete wavelet series expansions on the sphere have been presented [9]. At the core of this approach is the use of a subset of the Haar functions on the product of intervals $[0, 1] \times [0, 1]$ that satisfy certain boundary conditions together with a mapping

$$[0, 1] \times [0, 1] \leftrightarrow \mathbb{S}^2.$$

In this way, functions on the interior and boundary of the unit square are used to describe functions on the sphere.

An expansion based on pulse functions with support in successively smaller subdivisions of the sphere is presented in [49]. In that work, the triangular regions generated by the projection of an icosahedron onto the sphere serve as the starting point.

Other methods for describing functions on spheres based on splines can be found in the literature (see, e.g., [2, 8, 10]). For a complete review of many different methods for expanding functions on spheres, see [17].

4.12 Summary

In this chapter we examined various expansions of functions on curved surfaces (and manifolds) with emphasis on spheres. Some of the methods we reviewed here have been known for well over a hundred years, while others have been introduced in the last few decades. Indeed, since the publication of the original version of this book, a number of interesting new sampling schemes for spheres [20, 33, 57] and fast spherical-harmonic transforms have been developed, including [27, 31, 47, 52], which are not covered here. Although spherical harmonics are by no means new, they continue to be applied in new and interesting ways, e.g., [54].

Knowledge of the basic methods that have been discussed in this chapter provides a context for our discussion of Fourier expansions on functions of rotation and motion in later chapters. Chapters 5 and 6 explain rotations and rigid-body motions in great detail. In addition to being useful in their own right, these chapters provide concrete examples to give later chapters context.

References

1. Abbena, E., Salamon, S., Gray, A., *Modern Differential Geometry of Curves and Surfaces with Mathematica*, 3rd ed., Chapman and Hall/CRC, 2006.
2. Alfred, P., Neamtu, M., Schumaker, L.L., "Bernstein-Bézier Polynomials on Spheres and Sphere-like Surfaces," *Computer Aided Geometric Design*, 13(4): 333 – 349, 1996.
3. Antoine, J.-P., Carrette, P., Murenzi, R., Piette, B., "Image Analysis with Two-Dimensional Continuous Wavelet Transform," *Signal Processing*, 31(3): 241 – 272, April 1993.
4. Antoine, J.-P., Vandergheynst, P., "Wavelets on the 2-Sphere and Related Manifolds," *Reports on Mathematical Physics*, 43(1-2): 13 – 24, 1999.
5. Antoine, J.-P., Vandergheynst, P., "Wavelets on the *n*-Sphere and Related Manifolds," *Journal of Mathematical Physics*, 3(8): 3987 – 4008, Aug. 1998.
6. Arfken, G.B., Weber, H.J., *Mathematical Methods for Physicists*, 7th ed., Academic Press, San Diego, 2010.
7. Baddour, N., "Operational and Convolution Properties of Three-Dimensional Fourier Transforms in Spherical Polar Coordinates," *Journal of the Optical Society of America - Series A*, 27(10): 2144 – 2155, 2010.
8. Bülow, T., "Multiscale Image Processing on the Sphere," In *Pattern Recognition* (pp. 609 – 617). Springer, Berlin/Heidelberg, 2002.
9. Dahlke, S., Dahmen, W., Weinreich, I., "Multiresolution Analysis and Wavelets on $\mathbb{S}^2$ and $\mathbb{S}^3$," *Numerical Functional Analysis and Optimization*, 16(1–2): 19 – 41, 1995.
10. Dahlke, S., Maass, P., "A Continuous Wavelet Transform on Tangent Bundles of Spheres," *Journal of Fourier Analysis and Applications*, 2: 379 – 396, 1996.
11. Davis, H.F., Snider, A.D., *Introduction to Vector Analysis*, 4th edition, Allyn and Bacon, Boston, 1979.
12. Dilts, G.A., "Computation of Spherical Harmonic Expansion Coefficients via FFTs," *Journal of Computational Physics*, 57(3): 439 – 453, 1985.
13. Donnay, J.D.H., *Spherical Trigonometry: After the Cesàro Method*, Interscience Publishers, Inc., New York, 1945. (reprinted, Church Press, 2007).
14. Driscoll, J.R., Healy, D.M. Jr., "Computing Fourier Transforms and Convolutions on the 2-Sphere," *Advances in Applied Mathematics*, 15(2): 202 – 250, 1994.
15. Eisenhart, L.P., *Riemannian Geometry*, Princeton University Press, 1925 (reprinted 1997).
16. Firby, P.A., Gardiner, C.F., *Surface Topology*, 3rd ed., Woodhead Publishing, Philadelphia, PA, 2011.
17. Freeden, W., Gervens, T., Schreiner, M., *Constructive Approximation on the Sphere, With Applications to Geomathematics*, Clarendon Press, Oxford 1998.
18. Friedman, B., Russek, J., "Addition Theorems for Spherical Waves," *Quarterly of Applied Mathematics*, 12: 13 – 23, 1954
19. Healy Jr., D.M., Rockmore, D.N., Kostelec, P.J., Moore, S., "FFTs for the 2-Sphere – Improvements and Variations," *Journal of Fourier Analysis and Applications*, 9(4): 341 – 385, 2003.

20. Gorski, K.M., Hivon, E., Banday, A.J., Wandelt, B.D., Hansen, F.K., Reinecke, M., Bartelmann, M., "HEALPix: A Framework for High-Resolution Discretization and Fast Analysis of Data Distributed on The Sphere," *The Astrophysical Journal*, 622(2): 759, 2005.

21. Holschneider, M., "Wavelet Analysis on the Circle," *Journal of Mathematical Physics*, 31(1): 39–44, Jan. 1990.

22. Holschneider, M., "Continuous Wavelet Transforms on the Sphere," *Journal of Mathematical Physics*, 37(8): 4156–4165, Aug. 1996.

23. Isham, C.J., *Modern Differential Geometry for Physicists*, World Scientific Publishing, Singapore, 1989.

24. Jakob-Chien, R., Alpert, B.K., "A Fast Spherical Filter with Uniform Resolution," *Journal of Computational Physics*, 136(2): 580–584, 1997.

25. Julier, S.J., Uhlmann, J.K., "A Consistent, Debiased Method for Converting Between Polar and Cartesian Coordinate Systems, in *AeroSense'97*, pp.110-121, Society of Photo-Optical Instrumentation Engineers (SPIE), 1997.

26. Kalisa, C., Torrésani, B., "*N*-dimensional Affine Weyl-Heisenberg Wavelets," *Annales de l'Institut Henri Poincaré - Physique théorique*, 59(2): 201–236, 1993.

27. Keiner, J., Potts, D., "Fast Evaluation of Quadrature Formulae on the Sphere," *Mathematics of Computation*, 77(261): 397–419, 2008.

28. Kim, W.-Y., Kim, Y.-S., "Robust Rotation Angle Estimator," *IEEE Transactions on Pattern Analysis and Machine Intelligence*, 21(8): 768–773, Aug. 1999.

29. Kimel, I., Elias, L.R., "Relations Between Hermite and Laguerre Gaussian Modes," *IEEE Journal of Quantum Electronics*, 29(9): 2562-2567, Sept. 1993.

30. Kreyszig, E., *Advanced Engineering Mathematics*, 10th edition, John Wiley & Sons, New York, 2011.

31. Kunis, S., Potts, D., "Fast Spherical Fourier Algorithms," *Journal of Computational and Applied Mathematics*, 161(1): 75–98, 2003.

32. Massey, R., Refregier, A., "Polar Shapelets," *Monthly Notices of the Royal Astronomical Society*, 363(1): 197-210, 2005.

33. McEwen, J.D., Wiaux, Y., "A Novel Sampling Theorem on the Sphere," *IEEE Transactions on Signal Processing*, 59(12): 5876–5887, 2011.

34. McKean, H.P., Singer, I.M., "Curvature and the Eigenvalues of the Laplacian," *Journal of Differential Geometry*, 1(1): 43–69, 1967.

35. Miller, W., Jr., *Lie Theory and Special Functions*, Academic Press, New York, 1968;

36. Minakshisundaram, S., Pleijel, Å., "Some Properties of the Eigenfunctions of the Laplace Operator on Riemannian Manifolds," *Canadian Journal of Mathematics*, 1:242–256, 1949.

37. Mohlenkamp, M.J., "A Fast Transform for Spherical Harmonics," *Journal of Fourier Analysis and Applications*, 5(2-3): 159–184, 1999.

38. Morgan, F., *Riemannian Geometry : A Beginner's Guide*, Jones and Bartlett Publishers, Boston, 1993.

39. Murenzi, R., "Wavelet Transforms Associated to the *n*-Dimensional Euclidean Group with Dilations: Signal in More Than One Dimension," in *Wavelets: Time-Frequency Methods and Phase Space* (J.M. Combes, A. Grossmann, and Ph. Tchamitchian eds.), Springer-Verlag, New York, 1989.

40. Neutsch, W., *Coordinates*, Walter de Gruyter and Company, Berlin, 1996

41. Paeth, A.W., "A Fast Algorithm for General Raster Rotation," pp. 179–195 in *Graphics Gems* (A.S. Glassner ed.), Academic Press, Boston, 1990.

42. Park, W., Leibon, G., Rockmore, D.N., Chirikjian, G.S., "Accurate Image Rotation using Hermite Expansions," *IEEE Transactions on Image Processing*, 18(9): 1988–2003. 2009

43. Park, W., Chirikjian, G.S.,. "Interconversion Between Truncated Cartesian and Polar Expansions of Images," *IEEE Transactions on Image Processing*, 16(8): 1946–1955, 2007.

44. Patodi, V.K., "Curvature and the Eigenforms of the Laplace Operator," *Journal of Differential Geometry*, 5: 233–249, 1971.

45. Prosperetti, A., *Advanced Mathematics for Applications*, Cambridge University Press, 2001.

46. Refregier, A., "ShapeletsI. A Method for Image Analysis," *Monthly Notices of the Royal Astronomical Society*, 338(1): 35–47, 2003.

47. Rokhlin, V., Tygert, M., "Fast Algorithms for Spherical Harmonic Expansions," *SIAM Journal on Scientific Computing*, 27(6): 1903–1928, 2006.
48. Rosenberg, S., *The Laplacian on a Riemannian Manifold : An Introduction to Analysis on Manifolds,* (London Mathematical Society Student Texts , No 31), Cambridge University Press, 1997.
49. Schroeder, P., Swelden, W., "Spherical Wavelets: Texture Processing," Tech. Report, U. South Carolina, 1995.
50. Skibbe, H., Reisert, M., Schmidt, T., Brox, T., Ronneberger, O., Burkhardt, H., "Fast Rotation Invariant 3D Feature Computation Utilizing Efficient Local Neighborhood Operators," *IEEE Transactions on Pattern Analysis and Machine Intelligence*, 34(8): 1563–1575, 2012.
51. Sohon, F.W., *The Stereographic Projection*, Chemical Publishing Co., Inc., Brooklyn, N.Y, 1941 (Reprinted by Literary Licensing, LLC, 2013)
52. Suda, R., Takami, M., "A Fast Spherical Harmonics Transform Algorithm," *Mathematics of Computation*, 71(238): 703–715, 2002.
53. Torrésani, B., "Position-Frequency Analysis for Signals Defined on Spheres," *Signal Processing*, 43(3): 341–346, 1995.
54. Tunwattanapong, B., Fyffe, G., Graham, P., Busch, J., Yu, X., Ghosh, A., Debevec, P., "Acquiring Reflectance and Shape from Continuous Spherical Harmonic Illumination," *ACM Transactions on Graphics*, 32(4): 109, 2013.
55. Willmore, T.J., *An Introduction to Differential Geometry*, Dover, 2012.
56. Yan, Y., Chirikjian, G.S., "Closed-Form Characterization of the Minkowski Sum and Difference of Two Ellipsoids," *Geometriae Dedicata*, 177(1): 103–128, 2015.
57. Yershova, A., LaValle, S.M., "Deterministic Sampling Methods for Spheres and *SO*(3)," in *Proceedings 2004 IEEE International Conference on Robotics and Automation, ICRA'04*, Vol. 4, pp. 3974–3980, 2004.

Rotations in Three Dimensions

In this chapter, we introduce rigid-body rotations in three-dimensional Euclidean space. We show that the set of all such rotations can be viewed as a three-dimensional manifold embedded in Euclidean space, and that this manifold can be related to the hyper-sphere in four-dimensional Euclidean space with antipodal points identified, or "glued." We present numerous parameterizations of rotations and examine the relationship between angular velocity and rotation parameters in detail.

5.1 Deformations of Nonrigid Objects

In order to fully understand the constraints that rigidity imposes, we begin our discussion with motion of nonrigid media. A general motion (or deformation) of a continuous medium (including a rigid body) is a transformation of the form

$$\mathbf{x} = \mathbf{x}(\mathbf{X}, t)$$

where $\mathbf{X} \in \mathbb{R}^3$ is a vector that parameterizes the position of any material point in the medium at some reference time $t = t_0$. In contrast, $\mathbf{x} \in \mathbb{R}^3$ is the position of the same material point at time t. By definition, we then have $\mathbf{x}(\mathbf{X}, t_0) = \mathbf{X}$. The orthonormal basis $\{\mathbf{e}_1, \mathbf{e}_2, \mathbf{e}_3\}$ is used as a frame of reference fixed in space, and the vectors $\mathbf{x}$ and $\mathbf{X}$ are expressed in components as $x_i = \mathbf{x} \cdot \mathbf{e}_i$ and $X_i = \mathbf{X} \cdot \mathbf{e}_i$ for $i = 1, 2, 3$. In the case of a non-rigid continuous medium, many possible transformations exist. For instance, it may be possible to shear the medium in one direction:

$$\begin{aligned}
x_1 &= X_1 + k(t - t_0)X_2 \quad k \in \mathbb{R} \\
x_2 &= X_2 \\
x_3 &= X_3.
\end{aligned} \tag{5.1}$$

Or, we could stretch the medium along one axis:

$$\begin{aligned}
x_1 &= e^{ct}X_1 \\
x_2 &= e^{-ct}X_2 \\
x_3 &= X_3
\end{aligned} \tag{5.2}$$

for $c \in \mathbb{R}$. These deformations are depicted in Figure 5.1.

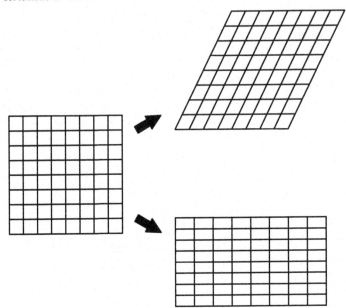

Fig. 5.1. Geometric Interpretation of Shear and Stretch

Researchers in the mechanics of solid media make great use of this kind of description of motion, which is called a *referential* or *Lagrangian* description. A subset of such motions that preserves the volume of any closed surface in $\mathbb{R}^3$ has been applied by one of the authors as a design tool in CAD and geometric modeling [5].

In contrast, one of the main aims of solid mechanics is to understand the relationship between forces applied to a continuous medium and the resulting deformation. We address applications of noncommutative harmonic analysis in solid mechanics in Chapter 18, which deals with the particular case of materials whose macroscopic strength properties depend on the orientational distribution of microscopic particles. An essential first step is to fully understand how orientations, or rotations, are quantified.

5.2 Rigid-Body Rotations

Spatial rigid-body rotations are defined as motions that preserve the distance between points in a body before and after the motion and leave one point fixed under the motion. By definition a motion must be physically realizable, and so reflections are not allowed. If $\mathbf{X}_1$ and $\mathbf{X}_2$ are any two points in a body before a rigid motion, then $\mathbf{x}_1$, and $\mathbf{x}_2$ are the corresponding points after rotation, and

$$d(\mathbf{x}_1, \mathbf{x}_2) = d(\mathbf{X}_1, \mathbf{X}_2)$$

where

$$d(\mathbf{x}, \mathbf{y}) = \|\mathbf{x} - \mathbf{y}\| = \sqrt{(x_1 - y_1)^2 + (x_2 - y_2)^2 + (x_3 - y_3)^2}$$

is the *Euclidean* distance. Note that rotations do not generally preserve other measures of distance between points.

By appropriately choosing our frame of reference in space, it is possible to make the pivot point (the point which does not move under rotation) the origin. Therefore $\mathbf{x}(\mathbf{0}, t) = \mathbf{0}$. With this choice, it can be shown that a necessary condition for a motion to be a rotation is

$$\mathbf{x}(\mathbf{X}, t) = A(t)\mathbf{X},$$

where $A(t) \in \mathbb{R}^{3 \times 3}$ is a time-dependent matrix. Note that this is *not* a sufficient condition, as the stretch and shear we saw earlier in (5.2) and (5.3) also can be written in this form.

Constraints on the form of $A(t)$ arise from the distance-preserving properties of rotations. If $\mathbf{X}_1$ and $\mathbf{X}_2$ are vectors defined in the frame of reference attached to the pivot, then the triangle with sides of length $\|\mathbf{X}_1\|$, $\|\mathbf{X}_2\|$, and $\|\mathbf{X}_1 - \mathbf{X}_2\|$ is congruent to the triangle with sides of length $\|\mathbf{x}_1\|$, $\|\mathbf{x}_2\|$, and $\|\mathbf{x}_1 - \mathbf{x}_2\|$. Hence, the angles between the vectors $\mathbf{x}_1$ and $\mathbf{x}_2$ must be the same as the angle between $\mathbf{X}_1$ and $\mathbf{X}_2$. In general $\mathbf{x} \cdot \mathbf{y} = \|\mathbf{x}\| \|\mathbf{y}\| \cos \theta$ where θ is the angle between $\mathbf{x}$ and $\mathbf{y}$. Since $\|\mathbf{x}_i\| = \|\mathbf{X}_i\|$ in our case, it follows that

$$\mathbf{x}_1 \cdot \mathbf{x}_2 = \mathbf{X}_1 \cdot \mathbf{X}_2.$$

Observing that $\mathbf{x} \cdot \mathbf{y} = \mathbf{x}^T \mathbf{y}$ and since $\mathbf{x}_i = A\mathbf{X}_i$, we see that

$$(A\mathbf{X}_1)^T (A\mathbf{X}_2) = \mathbf{X}_1^T \mathbf{X}_2. \tag{5.3}$$

Moving everything to the left side of the equation, and using the transpose rule for matrix vector multiplication, (5.3) is rewritten as

$$\mathbf{X}_1^T (A^T A - \mathbb{I})\mathbf{X}_2 = 0.$$

Since $\mathbf{X}_1$ and $\mathbf{X}_2$ where arbitrary points to begin with, this holds *for all* possible choices. The only way this can hold is if

$$A^T A = \mathbb{I} \tag{5.4}$$

where $\mathbb{I}$ is the 3×3 identity matrix. This means that $A^{-1} = A^T$.

An easy way to see this is to choose $\mathbf{X}_1 = \mathbf{e}_i$ and $\mathbf{X}_2 = \mathbf{e}_j$ for $i, j \in \{1, 2, 3\}$. This forces all the components of the matrix $A^T A - \mathbb{I}$ to be zero.

Equation (5.4) says that a rotation matrix is one whose inverse is its transpose. Taking the determinant of both sides of this equation yields $(\det A)^2 = 1$. There are two possible ways to satisfy this: $\det A = \pm 1$. The case $\det A = -1$ is a reflection and is not physically realizable in the sense that a rigid body cannot be reflected (only its image can be). A rotation is what remains:

$$\det A = +1. \tag{5.5}$$

Thus, a rotation matrix A is one which satisfies both (5.4) and (5.5). The set of all real matrices satisfying both (5.4) and (5.5) is called the set of *special orthogonal*[1] matrices. In general, the set of all $N \times N$ special orthogonal matrices is called $SO(N)$, and the set of all rotations in three-dimensional space is refered to as $SO(3)$.

We note that neither of the conditions in the definiton is enough by itself. Equation (5.4) is satisfied by reflections as well as rotations, and (5.5) is satisfied by a matrix such as

$$S(t) = \begin{pmatrix} 1 & k(t - t_0) & 0 \\ 0 & 1 & 0 \\ 0 & 0 & 1 \end{pmatrix},$$

---

[1]Also called *proper orthogonal*

which corresponds to the shear deformation $\mathbf{x} = S(t)\mathbf{X}$ in (5.1). The equations defining a rotation impose constraints on the behavior of the nine elements of a rotation matrix. Equation (5.4) imposes six scalar constraints. To see this, first write $A = [\mathbf{a}_1, \mathbf{a}_2, \mathbf{a}_3]$, and rewrite (5.4) as the nine constraints $\mathbf{a}_i \cdot \mathbf{a}_j = \delta_{i,j}$ for $i, j \in \{1, 2, 3\}$. Since the order of the dot product does not matter, $\mathbf{a}_i \cdot \mathbf{a}_j = \mathbf{a}_j \cdot \mathbf{a}_i$, and so three of the nine constraints are redundant. Six independent constraints remain. These six constraints and (5.5) are equivalent to saying that

$$A = [\mathbf{u}, \mathbf{v}, \mathbf{u} \times \mathbf{v}] \text{ where } \mathbf{u} \cdot \mathbf{u} = \mathbf{v} \cdot \mathbf{v} = 1 \text{ and } \mathbf{u} \cdot \mathbf{v} = 0 \tag{5.6}$$

where $\times$ denotes the vector cross product.

Hence, the six components of $(\mathbf{u}, \mathbf{v})$, together with three constraints that cause $\mathbf{u}$ and $\mathbf{v}$ to be mutually orthogonal unit vectors means that (5.5) does not reduce the number of degrees of freedom to less than three for spatial rotation.

In the special case of rotation about a fixed axis by an angle ϕ, the rotation only has one degree of freedom. In particular, counter-clockwise rotations about the $\mathbf{e}_1$, $\mathbf{e}_2$, and $\mathbf{e}_3$ axes are denoted as $R_1(\phi)$, $R_2(\phi)$, and $R_3(\phi)$, and respectively take the forms

$$\begin{pmatrix} 1 & 0 & 0 \\ 0 & \cos\phi & -\sin\phi \\ 0 & \sin\phi & \cos\phi \end{pmatrix}; \begin{pmatrix} \cos\phi & 0 & \sin\phi \\ 0 & 1 & 0 \\ -\sin\phi & 0 & \cos\phi \end{pmatrix}; \begin{pmatrix} \cos\phi & -\sin\phi & 0 \\ \sin\phi & \cos\phi & 0 \\ 0 & 0 & 1 \end{pmatrix}. \tag{5.7}$$

The range of ϕ in these matrices can be taken to be either $[0, 2\pi]$ or $[-\pi, \pi]$, with the opposite ends of these intervals identified so that $R_i(0) = R_i(2\pi)$ or $R_i(-\pi) = R_i(\pi)$.

5.2.1 Eigenvalues and Eigenvectors of Rotation Matrices

Recall that an eigenvalue/eigenvector pair $(\lambda, \mathbf{x})$ for any square matrix A is defined as the solution to the equation

$$A\mathbf{x} = \lambda\mathbf{x}.$$

Since a rotation matrix consists of only real entries, complex conjugation of both sides of this equation yields $\overline{A\mathbf{x}} = A\overline{\mathbf{x}} = \overline{\lambda}\overline{\mathbf{x}}$. We can therefore equate the dot product of vectors on the left and right sides of these equations as

$$(A\mathbf{x}) \cdot (A\overline{\mathbf{x}}) = (\lambda\mathbf{x}) \cdot (\overline{\lambda}\overline{\mathbf{x}}). \tag{5.8}$$

Using the transpose rule on the left of (5.8) together with (5.4) we have

$$(A\mathbf{x}) \cdot (A\overline{\mathbf{x}}) = \mathbf{x} \cdot (A^T A\overline{\mathbf{x}}) = \mathbf{x} \cdot \overline{\mathbf{x}} = \|\mathbf{x}\|^2.$$

On the right of (5.8) things simplify to $\lambda\overline{\lambda}\|\mathbf{x}\|^2$. Since $A^T A = \mathbb{I}$ and $\|\mathbf{x}\| \neq 0$, we have

$$\lambda\overline{\lambda} = 1.$$

This can be satisfied for the eigenvalues $\lambda_1 = 1$ and $\lambda_{2,3} = e^{\pm i\phi}$ for $\phi \in [0, \pi]$. Here we limit the range to half of what it was in conjunction with the discussion of the matrices in (5.7) because switching the roles of λ_2 and λ_3 is like extending the range of ϕ to $[-\pi, \pi]$. Since the trace of a matrix is equal to the sum of its eigenvalues, we have in this case

$$\text{tr}(A) = 1 + e^{i\phi} + e^{-i\phi} = 1 + 2\cos\phi,$$

or equivalently, the value of ϕ can be computed from the expression

$$\cos\phi = \frac{\mathrm{tr}(A) - 1}{2}.$$

The eigenvalue $\lambda = 1$ has a corresponding unit eigenvector $\mathbf{n}$, which is unchanged under application of the rotation matrix. Furthermore, $A\mathbf{n} = \mathbf{n}$ implies $A^T\mathbf{n} = \mathbf{n}$ and $A^m\mathbf{n} = \mathbf{n}$ for any integer power m. The eigenvalues $e^{\pm i\phi}$ have corresponding eigenvectors $\mathbf{c}_\pm$. By taking the complex conjugate of both sides of the equation $A\mathbf{c}_+ = e^{i\phi}\mathbf{c}_+$, it is clear that $\mathbf{c}_- = \overline{\mathbf{c}_+}$. Taking the dot product of the eigenvector/eigenvalue equations corresponding to $(1, \mathbf{n})$ and $(e^{\pm i\phi}, \mathbf{c}_\pm)$ and using (5.4) we see that

$$(A\mathbf{n}) \cdot (A\mathbf{c}_\pm) = e^{\pm i\phi}\mathbf{n} \cdot \mathbf{c}_\pm,$$

and so

$$\mathbf{n} \cdot \mathbf{c}_\pm = 0$$

when $e^{\pm i\phi} \neq 1$. From the complex vectors $\mathbf{c}_\pm$, can we construct real vectors

$$\mathbf{c}_1 = (\mathbf{c}_+ + \mathbf{c}_-)/2 \quad \text{and} \quad \mathbf{c}_2 = i(\mathbf{c}_+ - \mathbf{c}_-)/2$$

that are both orthogonal to $\mathbf{n}$. And from these, Graham-Schmidt orthogonalization can be used to construct new vectors in the plane spanned by $\mathbf{c}_1$ and $\mathbf{c}_2$ that are mutually orthogonal.

For further reading see [2, 3, 24]. For surveys on different parameterizations of rotation see [30, 32, 34].

5.2.2 Relationships between Rotation and Skew-Symmetric Matrices

There is a deep relationship between rotation matrices and *skew-symmetric* matrices. Recall that a real matrix is skew-symmetric if its transpose is its negative. Any 3×3 skew symmetric matrix, $S = -S^T$, can be written as

$$S = \begin{pmatrix} 0 & -s_3 & s_2 \\ s_3 & 0 & -s_1 \\ -s_2 & s_1 & 0 \end{pmatrix}, \tag{5.9}$$

where s_1, s_2, and s_3 can be viewed as the components of a vector $\mathbf{s} \in \mathbb{R}^3$, called the *dual vector* of S. The set of all 3×3 skew-symmetric matrices is denoted $so(3)$. This set is closed under the *matrix commutator*

$$[S_1, S_2] \doteq S_1 S_2 - S_2 S_1. \tag{5.10}$$

The product of a skew-symmetric matrix with an arbitrary vector is the same as the cross product of the dual vector with the same arbitrary vector:

$$S\mathbf{x} = \mathbf{s} \times \mathbf{x}. \tag{5.11}$$

We use the notation $S^\vee = \mathbf{s}$ and $S = \hat{\mathbf{s}}$ to express this relationship throughout the rest of the book.

It is easy to check that the eigenvalues of a 3×3 skew-symmetric matrix satisfy

$$-\det(S - \lambda\mathbb{I}) = \lambda^3 + (\mathbf{s} \cdot \mathbf{s})\lambda = (\lambda^2 + \|\mathbf{s}\|^2)\lambda = 0$$

and so the eigenvalues of S are $0, \pm i\|\mathbf{s}\|$.

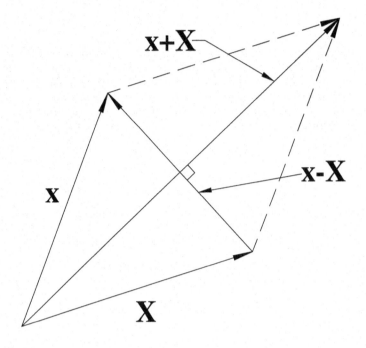

Fig. 5.2. Orthogonality of $\mathbf{x} + \mathbf{X}$ and $\mathbf{x} - \mathbf{X}$

It is also useful to know that the eigenvectors and eigenvalues of real symmetric matrices $M = M^T$ are real, and it is always possible to construct eigenvectors of M which form an orthonormal set. It follows that for any real symmetric matrix we can write

$$M = Q\Lambda Q^T; \quad MQ = Q\Lambda; \quad \Lambda = Q^T M Q \tag{5.12}$$

where Λ is a real diagonal matrix with nonzero entries corresponding to the eigenvalues of M, and Q is a rotation matrix whose columns are the eigenvectors of M. A proof of this statement is provided in Appendix D.

Cayley's Formula

Let $\mathbf{X}$ be an arbitrary vector in $\mathbb{R}^3$, and let $\mathbf{x} = A\mathbf{X}$ be a rotated version of this vector. Then it is clear from Figure 5.2.2 that the vectors $\mathbf{x} + \mathbf{X}$ and $\mathbf{x} - \mathbf{X}$ are orthogonal, and so we write

$$(\mathbf{x} + \mathbf{X}) \cdot (\mathbf{x} - \mathbf{X}) = 0. \tag{5.13}$$

Since by definition $\mathbf{x} = A\mathbf{X}$, it follows that

$$\mathbf{x} + \mathbf{X} = (A + \mathbb{I})\mathbf{X} \quad \text{and} \quad \mathbf{x} - \mathbf{X} = (A - \mathbb{I})\mathbf{X}. \tag{5.14}$$

For all $\phi \neq \pi$, the eigenvalues of a rotation matrix are not equal to -1, and so $A + \mathbb{I}$ is invertible. We will disregard the case when $\phi = \pi$ for the moment and examine this special case more closely later. Thus we write $\mathbf{X} = (A + \mathbb{I})^{-1}(\mathbf{x} + \mathbf{X})$, and substitute back into the second equation in (5.14) to yield

$$\mathbf{x} - \mathbf{X} = (A - \mathbb{I})(A + \mathbb{I})^{-1}(\mathbf{x} + \mathbf{X}).$$

From this statement and the condition in (5.13), we see that the matrix

$$S = (A - \mathbb{I})(A + \mathbb{I})^{-1} \tag{5.15}$$

must satisfy

$$\mathbf{y}^T S \mathbf{y} = 0 \tag{5.16}$$

where $\mathbf{y} = \mathbf{x} + \mathbf{X}$. Since $\mathbf{X}$ was arbitrary to begin with, so is $\mathbf{y}$.

We now prove that S in (5.16) must be skew-symmetric. As with any matrix with real entries, we can make the following decomposition: $S = S_{sym} + S_{skew}$ where $S_{sym} = \frac{1}{2}(S + S^T)$ and $S_{skew} = \frac{1}{2}(S - S^T)$ are respectively the symmetric and skew-symmetric parts of S. Hence

$$\mathbf{y}^T S \mathbf{y} = \mathbf{y}^T S_{sym} \mathbf{y} + \mathbf{y}^T S_{skew} \mathbf{y}.$$

The second term on the right hand side of the above equation must be zero because of the vector identity $\mathbf{y} \cdot (\mathbf{s}_{skew} \times \mathbf{y}) = 0$. As for the first term, since $\mathbf{y}$ is arbitrary, it is always possible to choose $\mathbf{y} = Q\mathbf{e}_i$ for any $i = 1, 2, 3$, in which case from (5.12) we have $\mathbf{y}^T S_{sym} \mathbf{y} = \lambda_i$. Therefore $\mathbf{y}^T S_{sym} \mathbf{y}$ can only be zero for all choices of $\mathbf{y}$ if $\lambda_i = 0$ for $i = 1, 2, 3$, which means that $S_{sym} = \mathbb{O}$. This means that S satisfying (5.16) must be skew symmetric. The equation

$$S = (A - \mathbb{I})(A + \mathbb{I})^{-1}$$

can be inverted to find A by multiplying by $A + \mathbb{I}$ on the right and recollecting terms to get

$$\mathbb{I} + S = (\mathbb{I} - S)A.$$

Since the number 1 is not an eigenvalue of S, $\mathbb{I} - S$ is invertible, and we arrive at *Cayley's formula* [4] for a rotation matrix:

$$A = (\mathbb{I} - S)^{-1}(\mathbb{I} + S). \tag{5.17}$$

This is the first of several examples we will see that demonstrate the relationship between skew-symmetric matrices and rotations. Cayley's formula provides one technique for *parameterizing* rotations. Each vector $\mathbf{s}$ contains the information required to specify a rotation, and given a rotation, a unique $\mathbf{s}$ can be extracted from the rotation matrix using (5.15).

Expanding out (5.17) in components, we find that

$$A = \frac{1}{1 + \|\mathbf{s}\|^2} \begin{pmatrix} 1 + s_1^2 - s_2^2 - s_3^2 & 2(s_1 s_2 - s_3) & 2(s_1 s_3 + s_2) \\ 2(s_1 s_2 + s_3) & 1 - s_1^2 + s_2^2 - s_3^2 & 2(s_2 s_3 - s_1) \\ 2(s_1 s_3 - s_2) & 2(s_2 s_3 + s_1) & 1 - s_1^2 - s_2^2 + s_3^2 \end{pmatrix} \tag{5.18}$$

From this expression it is clear that this description of rotation breaks down, for example, for rotation matrices of the form:

$$R_1(\pi) = \begin{pmatrix} 1 & 0 & 0 \\ 0 & -1 & 0 \\ 0 & 0 & -1 \end{pmatrix}; \quad R_2(\pi) = \begin{pmatrix} -1 & 0 & 0 \\ 0 & 1 & 0 \\ 0 & 0 & -1 \end{pmatrix}; \quad R_3(\pi) = \begin{pmatrix} -1 & 0 & 0 \\ 0 & -1 & 0 \\ 0 & 0 & 1 \end{pmatrix}.$$
$$\tag{5.19}$$

These are the degenerate cases because there are no nonnegative values of s_1^2, s_2^2, s_3^2 in (5.18) that will allow A to attain these values. All of the matrices above, which are of the form $R_i(\pi)$ for $i = 1, 2, 3$, are similar to each other under the appropriate similarity transformation $Q R_i(\pi) Q^T$ for $Q \in SO(3)$. More generally, any rotation matrix similar to these will also not lend itself to description using Cayley's formula. These all correspond to the case $\phi = \pi$ discussed earlier.

Transformation of Cross Products under Rotation

Given two vectors in a rotated frame of reference, $\mathbf{x}$ and $\mathbf{y}$, their cross product in the same frame of reference will be $\mathbf{x} \times \mathbf{y}$. In the unrotated (inertial) frame, this result will appear as $A(\mathbf{x} \times \mathbf{y})$ where A is the rotation that performs motion from the unrotated to rotated frame (and hence, converts vectors described in the rotated frame to their description in the unrotated frame). On the other hand, the vectors that appear as $\mathbf{x}$ and $\mathbf{y}$ in the rotated frame will be viewed as $A\mathbf{x}$ and $A\mathbf{y}$ in the inertial frame. Taking their cross product must then give

$$(A\mathbf{x}) \times (A\mathbf{y}) = A(\mathbf{x} \times \mathbf{y}). \qquad (5.20)$$

The geometrical argument used above may be verified by algebraic manipulations.

We note also that if $S \in so(3)$ and $S\mathbf{y} = \mathbf{s} \times \mathbf{y}$, i.e., $S^\vee = \mathbf{s}$, then

$$(ASA^T)^\vee = A\mathbf{s} \quad \text{and} \quad (ASA^T)\mathbf{y} = (A\mathbf{s}) \times \mathbf{y}. \qquad (5.21)$$

That is, the skew-symmetric matrix corresponding to a rotated vector is the similarity transformed version of the skew-symmetric matrix corresponding to the unrotated vector.

Euler's Theorem

Euler's theorem [9, 10][2]: states the general displacement of a rigid body with one point fixed is a rotation about some axis. That is not to say that every rotational *motion* $A(t) \in SO(3)$ for $t \in \mathbb{R}$ is a rotation about a fixed axis. Rather, the *result* of any rotational motion between two times $t = t_1$ and $t = t_2$ is a rotational *displacement*. This resulting displacement can be described as a rotation about an axis fixed in space that depends on the times t_1 and t_2.

By geometric construction, we can find the relationship between the axis of rotation, $\mathbf{n}$, and the angle of rotation about that axis, θ, which together describe an arbitrary spatial displacement. Observing an arbitrary point $\mathbf{X}$ in the body before rotation we break this into components as follows: $(\mathbf{X} \cdot \mathbf{n})\mathbf{n}$ is the component along the axis of rotation, and $\mathbf{X} - (\mathbf{X} \cdot \mathbf{n})\mathbf{n}$ is the component orthogonal to the axis of rotation. The plane containing the point $(\mathbf{X} \cdot \mathbf{n})\mathbf{n}$ and orthogonal to the axis of rotation is spanned by the two vectors $\mathbf{X} - (\mathbf{X} \cdot \mathbf{n})\mathbf{n}$ and $\mathbf{X} \times \mathbf{n}$. (See Figure 5.3.) These vectors are orthogonal to each other and have the same magnitude. Rotation about $\mathbf{n}$ by θ may be viewed as a planar rotation in this plane, and so the rotated version of the point $\mathbf{X}$ is

$$\mathbf{x} = (\mathbf{X} \cdot \mathbf{n})\mathbf{n} + (\mathbf{X} - (\mathbf{X} \cdot \mathbf{n})\mathbf{n}) \cos\theta + (\mathbf{n} \times \mathbf{X}) \sin\theta. \qquad (5.22)$$

---

[2]Leonard Euler (1707 - 1783) was one of the most prolific mathematicians in history. This is but one of the many theorems and formulae attributed to him.

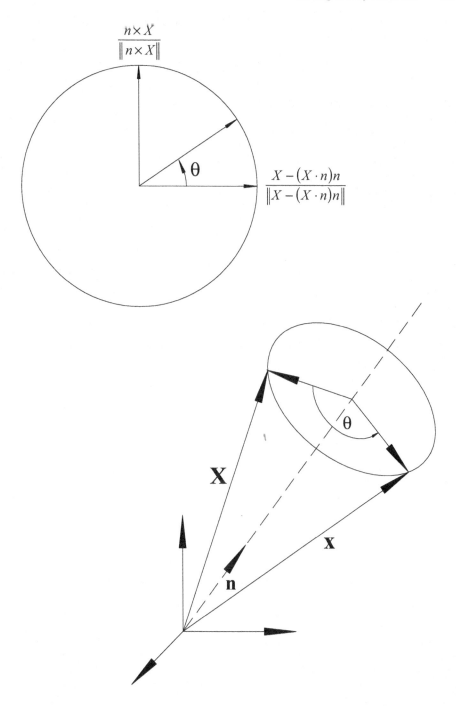

$$\frac{n \times X}{\|n \times X\|}$$

$$\frac{X - (X \cdot n)n}{\|X - (X \cdot n)n\|}$$

θ

X

x

n

Fig. 5.3. The Axis and Angle of Rotation: (top) Looking Down the Axis of Rotation; (bottom) View Perpendicular to the Axis of Rotation

To each unit vector $\mathbf{n}$ describing the axis of rotation, we assign a skew-symmetric matrix N. Then (5.22) is written as[3]

$$\mathbf{x} = A(\mathbf{n}, \theta)\mathbf{X} \quad \text{where} \quad A(\mathbf{n}, \theta) = \mathbb{I} + \sin\theta\, N + (1 - \cos\theta)N^2. \tag{5.23}$$

If we were to allow $\mathbf{n} \in \mathbb{S}^2$ and $\theta \in (-\pi, \pi]$, this would "double cover" $SO(3)$ because exactly two choices of pairs $(\theta, \mathbf{n})$ can correspond to the same rotation since $A(\mathbf{n}, \theta) = A(-\mathbf{n}, -\theta)$. There are several common choices for restricting the range of values that $\mathbf{n}$ and θ can take so as to make them an "almost" unique description. For example, $\mathbf{n}$ can be allowed to roam over the whole unit sphere, $\mathbf{n} \in \mathbb{S}^2$, while restricting $\theta \in [0, \pi]$, as we did earlier in this chapter. We can take the further step of identifying each vector on the sphere of radius π with its antipodal point, $\pi\mathbf{n} \leftrightarrow -\pi\mathbf{n}$, since these correspond to the same rotations. Alternatively, we can allow $\theta \in (-\pi, \pi]$ (which can be mapped to the unit circle) and restrict $\mathbf{n}$ to one hemisphere with half open and half closed boundary. Both of these descriptions have slight degeneracies. For example, $\mathbf{n}$ can be arbitrary when $\theta = 0$. But for our purposes, descriptions that break down at sets of measure zero will not affect our calculations. In different scenarios one or the other of the above choices for the range of θ and $\mathbf{n}$ will be more convenient.

Equations (5.22) and (5.23) are often called *Rodrigues'* equations for a rotational displacement, and (θ, n_1, n_2, n_3) is a four-parameter description of rotation called the *Rodrigues parameters* (after [29]).

Given any rotation matrix, these parameters can be extracted by observing that

$$A^T(\mathbf{n}, \theta) = A(\mathbf{n}, -\theta) = \mathbb{I} - \sin\theta\, N + (1 - \cos\theta)N^2,$$

and so

$$2\sin\theta N = A - A^T.$$

Then $\mathbf{n}$ is the normalized dual vector of $N = (A - A^T)/(2\sin\theta)$, and θ is determined by solving $2\sin\theta = \|(A - A^T)^\vee\|$.

5.2.3 The Matrix Exponential

The result of Euler's theorem can be viewed in another way using the concept of a matrix exponential. Recall from introductory calculus that the Taylor series expansion of the scalar exponential function is

$$e^x = 1 + \sum_{k=1}^{\infty} \frac{x^k}{k!}.$$

The matrix exponential is the same formula evaluated at a square matrix, X,

$$e^X = \mathbb{I} + \sum_{k=1}^{\infty} \frac{X^k}{k!}.$$

X is an arbitrary square matrix (not to be confused with the vector $\mathbf{X}$ used at the beginning of this chapter). Sometimes this is denoted as $\exp(X)$. In the case when $X = \theta N \in so(3)$ where $N = -N^T$ is the matrix whose dual is the unit vector $\mathbf{n}$, it can be shown that $N^2 = \mathbf{n}\mathbf{n}^T - \mathbb{I}$ and all higher powers of N can be related to either N or N^2 as

---

[3]An alternative notation that will be used is $\text{rot}[\mathbf{n}, \theta]$.

$$N^{2k+1} = (-1)^k N \quad \text{and} \quad N^{2k} = (-1)^{k+1} N^2. \tag{5.24}$$

The first few terms in the Taylor series of $e^{\theta N}$ are then expressed as

$$e^{\theta N} = \mathbb{I} + (\theta - \theta^3/3! + ...)N + (\theta^2/2! - \theta^4/4! + ...)N^2.$$

Hence (5.23) results, and we can write any rotational displacement as

$$A(\mathbf{n}, \theta) = e^{\theta N}.$$

This form clearly illustrates that $(\theta, \mathbf{n})$ and $(-\theta, -\mathbf{n})$ correspond to the same rotation.
Since $\theta = \|\mathbf{x}\|$ where $\mathbf{x} = X^\vee$ and $N = X/\|\mathbf{x}\|$, the following alternative form is also useful:

$$e^X = \mathbb{I} + \frac{\sin \|\mathbf{x}\|}{\|\mathbf{x}\|} X + \frac{(1 - \cos \|\mathbf{x}\|)}{\|\mathbf{x}\|^2} X^2. \tag{5.25}$$

In this light, the rotations in (5.7) can be expressed as

$$R_i(\theta) = \exp(\theta E_i) \quad \text{where} \quad E_i = \hat{\mathbf{e}}_i. \tag{5.26}$$

We now consider the special rotation matrix, $R(\mathbf{a}, \mathbf{b})$, which most directly transforms a unit vector $\mathbf{a}$ into the unit vector $\mathbf{b}$:

$$\mathbf{b} = R(\mathbf{a}, \mathbf{b})\mathbf{a}.$$

Let $\theta_{ab} \in [0, \pi]$ denote the angle of rotation measured counterclockwise from $\mathbf{a}$ to $\mathbf{b}$ around the axis defined by $\mathbf{a} \times \mathbf{b}$. Then

$$R(\mathbf{a}, \mathbf{b}) = \exp\left(\frac{\theta_{ab}}{\sin \theta_{ab}} \widehat{\mathbf{a} \times \mathbf{b}}\right) = \mathbb{I} + \widehat{\mathbf{a} \times \mathbf{b}} + \frac{(1 - \mathbf{a} \cdot \mathbf{b})}{\|\mathbf{a} \times \mathbf{b}\|^2} \left(\widehat{\mathbf{a} \times \mathbf{b}}\right)^2. \tag{5.27}$$

This follows easily from the facts that $\|\mathbf{a} \times \mathbf{b}\| = \sin \theta_{ab}$ and $\mathbf{a} \cdot \mathbf{b} = \cos \theta_{ab}$.
The matrix logarithm is defined as the inverse of the exponential:

$$X = \log(\exp X).$$

When $A = \exp X$ is a rotation, the logarithm is explicitly calculated as

$$\log(A) = \frac{1}{2} \frac{\theta(A)}{\sin \theta(A)} (A - A^T)$$

where

$$\theta(A) = \cos^{-1}\left(\frac{\text{tr}(A) - 1}{2}\right)$$

is the angle of rotation. From this definition it follows that for $\theta = 0$ we have $\log(\mathbb{I}) = \mathbb{O}$. The above definitions break down when $\theta = \pi$.

5.3 Rules for Composing Rotations

Consider three frames of reference A, B, and C, which all have the same origin. The vectors $\mathbf{x}^A$, $\mathbf{x}^B$, $\mathbf{x}^C$ represent *the same* arbitrary point in space, $\mathbf{x}$, as it is viewed in the three different reference frames, all centered at the origin. With respect to some common frame fixed in space with axes defined by $\{\mathbf{e}_1, \mathbf{e}_2, \mathbf{e}_3\}$ where $(\mathbf{e}_i)_j = \delta_{ij}$, the rotation matrices describing the basis vectors of the frames A, B, and C are

$$R_A = [\mathbf{e}_1^A, \mathbf{e}_2^A, \mathbf{e}_3^A] \quad R_B = [\mathbf{e}_1^B, \mathbf{e}_2^B, \mathbf{e}_3^B] \quad R_C = [\mathbf{e}_1^C, \mathbf{e}_2^C, \mathbf{e}_3^C],$$

where the vectors $\mathbf{e}_i^A$, $\mathbf{e}_i^B$, and $\mathbf{e}_i^C$ are unit vectors along the ith axis of frame A, B, or C. The "absolute" coordinates of the vector $\mathbf{x}$ are then given by

$$\mathbf{x} = R_A \mathbf{x}^A = R_B \mathbf{x}^B = R_C \mathbf{x}^C.$$

In this notation, which is often used in the field of robotics (see, e.g., [7, 26]), there is effectively a "cancellation" of indices along the upper right to lower left diagonal.

Given the rotation matrices R_A, R_B, and R_C, it is possible to define rotations of one frame *relative* to another by observing that, for instance, $R_A \mathbf{x}^A = R_B \mathbf{x}^B$ implies $\mathbf{x}^A = (R_A)^{-1} R_B \mathbf{x}^B$. Therefore, given any vector $\mathbf{x}^B$ as it looks in B, we can find how it looks in A, $\mathbf{x}^A$, by performing the transformation:

$$\mathbf{x}^A = R_B^A \mathbf{x}^B \quad \text{where} \quad R_B^A = (R_A)^{-1} R_B. \tag{5.28}$$

It follows from substituting the analogous expression $\mathbf{x}^B = R_C^B \mathbf{x}^C$ into $\mathbf{x}^A = R_B^A \mathbf{x}^B$ that concatenation of rotations is calculated as

$$\mathbf{x}^A = R_C^A \mathbf{x}^C \quad \text{where} \quad R_C^A = R_B^A R_C^B. \tag{5.29}$$

Again there is effectively a cancellation of indices, and this propogates through for any number of relative rotations. Note that the order of multiplication is critical.

In addition to changes of basis, rotation matrices can be viewed as descriptions of motion. Multiplication of a rotation matrix Q (which represents a frame of reference) by a rotation matrix R (representing motion) on the left, RQ, has the effect of moving Q by R relative to the *base frame*. Multiplying by the same rotation matrix on the right, QR, has the effect of moving by R relative to the the frame Q.

To demonstrate the difference, consider again the result of Euler's theorem. Let us define a frame of reference $Q = [\mathbf{a}, \mathbf{b}, \mathbf{n}]$ where $\mathbf{a}$ and $\mathbf{b}$ are unit vectors orthogonal to each other, and $\mathbf{a} \times \mathbf{b} = \mathbf{n}$. By first rotating from the identity $\mathbb{I} = [\mathbf{e}_1, \mathbf{e}_2, \mathbf{e}_3]$ fixed in space to Q and then rotating relative to Q by $R_3(\theta)$, results in $Q R_3(\theta)$. On the other hand, a rotation about the vector $\mathbf{e}_3{}^Q = \mathbf{n}$ as viewed in the fixed frame is $A(\mathbf{n}, \theta)$. Hence, shifting the frame of reference Q by multiplying on the left by $A(\mathbf{n}, \theta)$ has the same effect as $QR_3(\theta)$, and so we write

$$A(\mathbf{n}, \theta)Q = QR_3(\theta) \quad \text{or} \quad A(\mathbf{n}, \theta) = QR_3(\theta)Q^T. \tag{5.30}$$

This is yet another way to view the expression of $A(\mathbf{n}, \theta)$ given previously in (5.23). We note that in component form this is expressed as

$$A(\theta, \mathbf{n}) = \begin{pmatrix} n_1^2 v\theta + c\theta & n_2 n_1 v\theta - n_3 s\theta & n_3 n_1 v\theta + n_2 s\theta \\ n_1 n_2 v\theta + n_3 s\theta & n_2^2 v\theta + c\theta & n_3 n_2 v\theta - n_1 s\theta \\ n_1 n_3 v\theta - n_2 s\theta & n_2 n_3 v\theta + n_1 s\theta & n_3^2 v\theta + c\theta \end{pmatrix} \tag{5.31}$$

where $s\theta = \sin\theta$, $c\theta = \cos\theta$, and $v\theta = 1 - \cos\theta$.

Note that $\mathbf{a}$ and $\mathbf{b}$ do not appear in the final expression. There is nothing magical about $\mathbf{e}_3$, and we could have used the same construction using any other basis vector, $\mathbf{e}_i$, and we would get the same result so long as $\mathbf{n}$ is in the ith column of Q.

5.4 Parameterizations of Rotation

In this section, several parameterizations other than the axis-angle description of rotation are reviewed.

5.4.1 Euler Parameters

Euler parameters, $\{u_1, u_2, u_3, u_4\}$, are defined relative to the axis-angle parameters $\mathbf{n}, \theta$ as:

$$\mathbf{u} \doteq \mathbf{n}\sin(\theta/2) \qquad u_4 \doteq \cos(\theta/2) \tag{5.32}$$

where $\mathbf{u} = (u_1, u_2, u_3)^T$. With this definition, and using the fact that $\mathbf{n} \cdot \mathbf{n} = 1$, we see that

$$u_1^2 + u_2^2 + u_3^2 + u_4^2 = 1,$$

and so a natural relationship clearly exists between $SO(3)$ and the 3–sphere, $\mathbb{S}^3$. In particular, because $-\pi \le \theta \le \pi$ and $\cos(\theta/2) \ge 0$, u_4 is limited to the "upper half space" of $\mathbb{R}^4$ and so we associate $SO(3)$ with the "upper hemisphere" of $\mathbb{S}^3$ corresponding to $u_4 \ge 0$.

The rotation matrix when using Euler parameters takes the form

$$A_E(u_1, u_2, u_3, u_4) = \begin{pmatrix} u_1^2 - u_2^2 - u_3^2 + u_4^2 & 2(u_1u_2 - u_3u_4) & 2(u_3u_1 + u_2u_4) \\ 2(u_1u_2 + u_3u_4) & u_2^2 - u_3^2 - u_1^2 + u_4^2 & 2(u_2u_3 - u_1u_4) \\ 2(u_3u_1 - u_2u_4) & 2(u_2u_3 + u_1u_4) & u_3^2 - u_1^2 - u_2^2 + u_4^2 \end{pmatrix}. \tag{5.33}$$

The Euler parameters can be extracted from any rotation matrix A as:

$$u_4 = \frac{1}{2}(1 + \operatorname{tr}(A))^{\frac{1}{2}}$$

$$u_1 = \frac{a_{32} - a_{23}}{4u_4}$$

$$u_2 = \frac{a_{13} - a_{31}}{4u_4}$$

$$u_3 = \frac{a_{21} - a_{12}}{4u_4}$$

(where a_{ij} is the $(i, j)^{th}$ entry of A) so long as $u_4 \ne 0$. In this special case, we have

$$A_E(u_1, u_2, u_3, 0) = \begin{pmatrix} u_1^2 - u_2^2 - u_3^2 & 2u_1u_2 & 2u_3u_1 \\ 2u_1u_2 & u_2^2 - u_3^2 - u_1^2 & 2u_2u_3 \\ 2u_3u_1 & 2u_2u_3 & u_3^2 - u_1^2 - u_2^2 \end{pmatrix},$$

and we can usually find the remaining parameters using the formula

$$\mathbf{u} = \frac{1}{\sqrt{(a_{12}a_{13})^2 + (a_{12}a_{23})^2 + (a_{13}a_{23})^2}} \begin{pmatrix} a_{13}a_{12} \\ a_{12}a_{23} \\ a_{13}a_{23} \end{pmatrix}.$$

However, this too can be degenerate if $a_{12} = a_{23} = 0$ or $a_{13} = a_{23} = 0$ or $a_{12} = a_{13} = 0$. These cases correspond to the rotations about natural basis vectors in (5.7), and can be detected in advance by counting the number of zeros in the matrix.

5.4.2 Cayley/Rodrigues Parameters

If we set $u_4 = 1$ and normalize the resulting matrix $A_E(u_1, u_2, u_3, 1)$ by $1/(1 + u_1^2 + u_2^2 + u_3^2)$, then the three parameters u_1, u_2, u_3 can be viewed as the elements of the vector $\mathbf{s}$ corresponding to the skew-symmetric matrix in Cayley's formula, (5.17). In this case, u_1, u_2, u_3 are called *Cayley parameters*. As with all 3−parameter descriptions of rotation, the Cayley parameters are singular at some points.

Rodrigues parameters can also be defined relative to the Euler and axis-angle parameterization. They are defined as

$$\mathbf{r} = \mathbf{n} \tan \theta/2 = \mathbf{u}/u_4.$$

$\mathbf{r}$ is often called the *Rodrigues vector* or the *Gibbs vector*.

In terms of the exponential coordinates in (5.25), $\mathbf{x}$, the following relationship holds:

$$\mathbf{r} = \frac{\mathbf{x}}{\|\mathbf{x}\|} \tan \left(\frac{\|\mathbf{x}\|}{2} \right) \qquad \leftrightarrow \qquad \mathbf{x} = 2 \tan^{-1}(\|\mathbf{r}\|) \frac{\mathbf{r}}{\|\mathbf{r}\|}.$$

The resulting rotation matrix is the same as that for the Cayley parameters (and so the Cayley and Rodrigiues parameters are equivalent):

$$A_R(\mathbf{r}) = \frac{1}{1 + r_1^2 + r_2^2 + r_3^2} \begin{pmatrix} 1 + r_1^2 - r_2^2 - r_3^2 & 2(r_1 r_2 - r_3) & 2(r_3 r_1 + r_2) \\ 2(r_1 r_2 + r_3) & 1 + r_2^2 - r_3^2 - r_1^2 & 2(r_2 r_3 - r_1) \\ 2(r_3 r_1 - r_2) & 2(r_2 r_3 + r_1) & 1 + r_3^2 - r_1^2 - r_2^2 \end{pmatrix}. \quad (5.34)$$

We note that

$$A_R(\mathbf{r}) = \frac{(1 - \|\mathbf{r}\|^2)\mathbb{I} + 2\mathbf{r}\mathbf{r}^T + [\hat{\mathbf{r}}]}{1 + \|\mathbf{r}\|^2}$$

where $[\hat{\mathbf{r}}]$ is the skew-symmetric matrix with

$$\mathbf{r} = [\hat{\mathbf{r}}]^\vee.$$

When

$$A_R(\mathbf{r}) A_R(\mathbf{r}') = A_R(\mathbf{r}''),$$

the Rodrigues vector $\mathbf{r}''$ is calculated in terms of $\mathbf{r}$ and $\mathbf{r}'$ as

$$\mathbf{r}'' = \frac{\mathbf{r} + \mathbf{r}' - \mathbf{r}' \times \mathbf{r}}{1 - \mathbf{r} \cdot \mathbf{r}'}.$$

The Rodrigues parameters can be extracted from any rotation matrix as

$$r_1 = \frac{a_{32} - a_{23}}{1 + \operatorname{tr}(A)}$$

$$r_2 = \frac{a_{13} - a_{31}}{1 + \operatorname{tr}(A)}$$

$$r_3 = \frac{a_{21} - a_{12}}{1 + \operatorname{tr}(A)}$$

so long as $\operatorname{tr}(A) \neq -1$. This is the singularity in this description of rotation.

As explained by Kane et al. [17], when the position of two points $(\mathbf{X}_1, \mathbf{X}_2)$ in the rigid body are known both prior to rotation and after rotation $(\mathbf{x}_1, \mathbf{x}_2)$, then the relationship

$$\mathbf{x}_i - \mathbf{X}_i = \mathbf{r} \times (\mathbf{x}_i + \mathbf{X}_i)$$

(which follows from our discussion of Cayley's formula) is used to extract the Rodrigues vector as

$$\mathbf{r} = -\frac{(\mathbf{x}_1 - \mathbf{X}_1) \times (\mathbf{x}_2 - \mathbf{X}_2)}{(\mathbf{x}_1 + \mathbf{X}_1) \cdot (\mathbf{x}_2 - \mathbf{X}_2)}. \tag{5.35}$$

This formula follows by observing that

$$(\mathbf{x}_1 - \mathbf{X}_1) \times (\mathbf{x}_2 - \mathbf{X}_2) = (\mathbf{r} \times (\mathbf{x}_1 + \mathbf{X}_1)) \times (\mathbf{x}_2 - \mathbf{X}_2)$$

and expanding the right side using the vector identity

$$(\mathbf{a} \times \mathbf{b}) \times \mathbf{c} = (\mathbf{a} \cdot \mathbf{c})\mathbf{b} - (\mathbf{b} \cdot \mathbf{c})\mathbf{a}$$

to isolate $\mathbf{r}$.

5.4.3 Cartesian Coordinates in $\mathbb{R}^4$

Identifying $SO(3)$ with the upper hemisphere in $\mathbb{R}^4$, produces the Cartesian-coordinate description of the surface as

$$x_1 = u_1$$
$$x_2 = u_2$$
$$x_3 = u_3$$
$$x_4 = \sqrt{1 - u_1^2 - u_2^2 - u_3^2}.$$

The corresponding rotation matrix is (5.33) evaluated with these coordinates as

$$A_C(u_1, u_2, u_3) = A_E(u_1, u_2, u_3, \sqrt{1 - u_1^2 - u_2^2 - u_3^2}).$$

5.4.4 Spherical Coordinates

Parameterizing rotations using spherical coordinates is essentially the same as parameterizing $\mathbb{S}^3$ in spherical coordinates. We associate the set of all rotations with the upper hemisphere, which is parameterized as

$$x_1 = \cos \lambda \sin \nu \sin \theta/2$$
$$x_2 = \sin \lambda \sin \nu \sin \theta/2$$
$$x_3 = \cos \nu \sin \theta/2$$
$$x_4 = \cos \theta/2.$$

λ and ν are the polar and azimuthal angles for the axis of rotation, and θ is the angle of rotation.

The corresponding rotation matrix results from the spherical-coordinate parameterization of the vector of the axis of rotation, $\mathbf{n} = \mathbf{n}(\nu, \lambda)$ as in (5.37), and is given by

$$A_S(\lambda, \nu, \theta) = \begin{pmatrix} v\theta\, s^2\nu\, c^2\lambda + c\theta & v\theta\, s^2\nu\, c\lambda\, s\lambda - s\theta\, c\nu & v\theta\, s\nu\, c\nu\, c\lambda + s\theta\, s\nu\, s\lambda \\ v\theta\, s^2\nu\, c\lambda\, s\lambda + s\theta\, c\nu & v\theta\, s^2\nu\, s^2\lambda + c\theta & v\theta\, s\nu\, c\nu\, s\lambda - s\theta\, s\nu\, c\lambda \\ v\theta\, s\nu\, c\nu\, c\lambda - s\theta\, s\nu\, s\lambda & v\theta\, s\nu\, c\nu\, s\lambda + s\theta\, s\nu\, c\lambda & v\theta\, c^2\nu + c\theta \end{pmatrix},$$

$$\tag{5.36}$$

where $s\theta = \sin\theta$, $c\theta = \cos\theta$, and $v\theta = 1 - \cos\theta$. It is clear from (5.36) that (θ, ν, λ) and $(-\theta, \pi - \nu, \lambda \pm \pi)$ correspond to the same rotation.

5.4.5 Parameterization of Rotation as a Solid Ball in $\mathbb{R}^3$

Another way to view the set of rotations is as a solid ball in $\mathbb{R}^3$ with radius π. In this description, each concentric spherical shell within the ball represents a rotation by an angle which is the radius of the shell. That is, given any point in the solid ball,

$$\mathbf{x}(\theta, \nu, \lambda) = \theta\, \mathbf{n}(\nu, \lambda) \quad \text{where} \quad \mathbf{n}(\nu, \lambda) = \begin{pmatrix} \sin\nu \cos\lambda \\ \sin\nu \sin\lambda \\ \cos\nu \end{pmatrix}. \tag{5.37}$$

The ranges of these angles are $0 \le \theta, \nu \le \pi$, and either $0 \le \lambda \le 2\pi$ or $-\pi \le \lambda \le \pi$. In this model, antipodal points on the surface of the ball represent the same rotations.

Exponentiating the skew symmetric matrix corresponding to $\mathbf{x}(\theta, \nu, \lambda)$ results in the matrix in (5.36). Similarly, the parameters (x_1, x_2, x_3) defining the matrix exponential (5.25) can be viewed as Cartesian coordinates to the interior of this same ball.

5.4.6 Euler Angles

Euler angles are by far the most widely known parameterization of rotation. They are generated by three successive rotations about independent axes. Three of the most common choices are the ZXZ, ZYZ, and ZYX Euler angles. We will denote these as

$$A_{ZXZ}(\alpha, \beta, \gamma) = R_3(\alpha)R_1(\beta)R_3(\gamma) \tag{5.38}$$

$$A_{ZYZ}(\alpha, \beta, \gamma) = R_3(\alpha)R_2(\beta)R_3(\gamma) \tag{5.39}$$

$$A_{ZYX}(\alpha, \beta, \gamma) = R_3(\alpha)R_2(\beta)R_1(\gamma) \tag{5.40}$$

Of these, the ZXZ and ZYZ Euler angles are the most common, and the corresponding matrices $A_{ZXZ}(\alpha, \beta, \gamma)$ and $A_{ZYZ}(\alpha, \beta, \gamma)$ are, respectively,

$$\begin{pmatrix} \cos\gamma\cos\alpha - \sin\gamma\sin\alpha\cos\beta & -\sin\gamma\cos\alpha - \cos\gamma\sin\alpha\cos\beta & \sin\beta\sin\alpha \\ \cos\gamma\sin\alpha + \sin\gamma\cos\alpha\cos\beta & -\sin\gamma\sin\alpha + \cos\gamma\cos\alpha\cos\beta & -\sin\beta\cos\alpha \\ \sin\beta\sin\gamma & \sin\beta\cos\gamma & \cos\beta \end{pmatrix},$$

and

$$\begin{pmatrix} \cos\gamma\cos\alpha\cos\beta - \sin\gamma\sin\alpha & -\sin\gamma\cos\alpha\cos\beta - \cos\gamma\sin\alpha & \sin\beta\cos\alpha \\ \sin\alpha\cos\gamma\cos\beta + \sin\gamma\cos\alpha & -\sin\gamma\sin\alpha\cos\beta + \cos\gamma\cos\alpha & \sin\beta\sin\alpha \\ -\sin\beta\cos\gamma & \sin\beta\sin\gamma & \cos\beta \end{pmatrix}.$$

The ranges of angles for these choices are $0 \le \alpha \le 2\pi$, $0 \le \beta \le \pi$, and $0 \le \gamma \le 2\pi$. When ZYZ Euler angles are used,

$$R_3(\alpha)R_2(\beta)R_3(\gamma) = R_3(\alpha)(R_3(\pi/2)R_1(\beta)R_3(-\pi/2))R_3(\gamma)$$
$$= R_3(\alpha + \pi/2)R_1(\beta)R_3(-\pi/2 + \gamma),$$

and so

$$R_{ZYZ}(\alpha, \beta, \gamma) = R_{ZXZ}(\alpha + \pi/2, \beta, \gamma - \pi/2).$$

In recent times, the Euler angles have been considered by some to be undesirable as compared to the axis-angle or Euler parameters because of the singularities that they (and all) three-parameter descriptions possess. However, for our purposes in later chapters, where integral quantities that are not sensitive to singularities are of interest, the Euler angles will serve well.

5.4.7 Parameterization Based on Stereographic Projection

The Cayley/Rodrigues parameters can be viewed geometrically as points in $\mathbb{R}^3$ that are mapped onto half of the unit hypersphere in $\mathbb{R}^4$ via the four-dimensional analog of the two-dimensional stereographic projection shown in Figure 4.1(a). This is not the only way stereographic projection can be used to generate parameterizations of rotation. The subsequent subsubsections describe parameterizations developed in the spacecraft attitude literature based on stereographic projection.

Modified Rodrigues Parameters

If the stereographic projection in $\mathbb{R}^4$ analogous to the one described in Figure 4.1(b) is used, then the Euler parameters $\{u_i\}$ are mapped to modified Rodrigues parameters [23, 31]

$$\sigma_i = \frac{u_i}{1 + u_4} = n_i \tan(\theta/4)$$

for $i = 1, 2, 3$. The benefit of this is that the singularity at $\theta = \pm\pi$ is pushed to $\theta = \pm2\pi$. The corresponding rotation matrix is [31]

$$R(\boldsymbol{\sigma}) = \frac{1}{(1 + \|\boldsymbol{\sigma}\|^2)^2} \times$$

$$\begin{pmatrix} 4(\sigma_1^2 - \sigma_2^2 - \sigma_3^2) + \Sigma^2 & 8\sigma_1\sigma_2 - 4\sigma_3\Sigma & 8\sigma_3\sigma_1 + 4\sigma_2\Sigma \\ 8\sigma_1\sigma_2 + 4\sigma_3\Sigma & 4(-\sigma_1^2 + \sigma_2^2 - \sigma_3^2) + \Sigma^2 & 8\sigma_2\sigma_3 - 4\sigma_1\Sigma \\ 8\sigma_3\sigma_1 - 4\sigma_2\Sigma & 8\sigma_2\sigma_3 + 4\sigma_1\Sigma & 4(-\sigma_1^2 - \sigma_2^2 + \sigma_3^2) + \Sigma^2 \end{pmatrix} \quad (5.41)$$

where $\Sigma = 1 - \|\boldsymbol{\sigma}\|^2$.

The Parameterization of Tsiotras and Longuski

Consider the concatenation of the transformation $R(\mathbf{e}_3, \mathbf{v})$ (the rotation that moves $\mathbf{e}_3$ to $\mathbf{v}$) and $R_3(\varphi) = e^{\varphi \hat{\mathbf{e}}_3}$ (the rotation about the local z axis). The result is a four-parameter description of rotation with matrix

$$R(\mathbf{e}_3, \mathbf{v}) R_3(\varphi) = \begin{pmatrix} \frac{v_3 \cos\varphi - v_1 v_2 \sin\varphi + (v_2^2 + v_3^2)\cos\varphi}{1 + v_3} & -\frac{v_3 \sin\varphi + v_1 v_2 \cos\varphi + (v_2^2 + v_3^2)\sin\varphi}{1 + v_3} & v_1 \\ \frac{v_3 \sin\varphi - v_1 v_2 \cos\varphi + (v_1^2 + v_3^2)\sin\varphi}{1 + v_3} & \frac{v_3 \cos\varphi + v_1 v_2 \sin\varphi + (v_1^2 + v_3^2)\cos\varphi}{1 + v_3} & v_2 \\ -v_2 \sin\varphi - v_1 \cos\varphi & v_1 \sin\varphi - v_2 \cos\varphi & v_3 \end{pmatrix}$$

$$(5.42)$$

where $\|\mathbf{v}\| = 1$.

Given

$$A(\mathbf{v}, \varphi) = R(\mathbf{e}_3, \mathbf{v}) e^{\varphi \hat{\mathbf{e}}_3} \qquad (5.43)$$

it follows that

$$A\mathbf{e}_3 = R(\mathbf{e}_3, \mathbf{v}) \mathbf{e}_3 = \mathbf{v}.$$

Substituting this back into (5.43) and isolating $e^{\varphi \hat{\mathbf{e}}_3}$ gives

$$R(A\mathbf{e}_3, \mathbf{e}_3) A = e^{\varphi \hat{\mathbf{e}}_3}.$$

Then φ can be computed as

$$\varphi = \mathbf{e}_3 \cdot (\log(R(A\mathbf{e}_3, \mathbf{e}_3)\, A))^\vee .$$

This process recovers $(\mathbf{v}, \varphi)$ from any A provided that $\varphi(A) \in (-\pi, \pi)$.

Tsiotras and Longuski [35] have used stereographic projection (4.24) of the $\|\mathbf{v}\| = 1$ sphere to the plane (viewed as the complex plane $\mathbb{C}$). In this way, two parameters, $w_1 = x_2$ and $w_2 = -x_1$, are used to parameterize the unit vector $\mathbf{v}$. This allows us to write [35]

$$R(w, \varphi) = \frac{1}{1 + |w|^2} \begin{pmatrix} Re[(1 + w^2)\,e^{-i\varphi}] & Im[(1 + w^2)\,e^{-i\varphi}] & -2Im(w) \\ Im[(1 - \overline{w}^2)\,e^{i\varphi}] & Re[(1 - \overline{w}^2)\,e^{i\varphi}] & 2Re(w) \\ 2Im(we^{-i\varphi}) & -2Re(we^{-i\varphi}) & 1 - |w|^2 \end{pmatrix}. \tag{5.44}$$

Modified Axis-Angle Parameterization

The four-parameter axis-angle parameterization $R(\mathbf{n}, \theta)$ resulting from Euler's theorem and written as $\exp[\theta N]$ was reduced to a three dimensional parameterization earlier in this chapter by expressing $\mathbf{n}$ in spherical coordinates. A different reduction to three parameters is achieved by sterographic projection of the sphere $\|\mathbf{n}\| = 1$ onto the plane in the same way discussed in the previous subsubsection. We then write

$$R(\mathbf{n}(w), \theta) = \frac{1}{(k_+(w))^2} \times$$

$$\begin{pmatrix} 4w_2^2 v\theta + (k_+(w))^2 c\theta & -4w_1 w_2 v\theta - (1 - |w|^4)s\theta & -2w_2(k_-(w))v\theta + 2w_1(k_+(w))s\theta \\ -4w_1 w_2 v\theta + (1 - |w|^4)s\theta & 4w_1^2 v\theta + (k_+(w))^2 c\theta & 2w_1(k_-(w))v\theta + 2w_2(k_+(w))s\theta \\ -2w_2(k_-(w))v\theta - 2w_1(k_+(w))s\theta & 2w_1(k_-(w))v\theta - 2w_2(k_+(w))s\theta & (k_-(w))^2 v\theta + (k_+(w))^2 c\theta \end{pmatrix}.$$

where $k_+(w) = 1 + |w|^2$ and $k_-(w) = 1 - |w|^2$. We can go further and make the substitution

$$\cos\theta = \frac{1 - z^2}{1 + z^2} \qquad \sin\theta = \frac{2z}{1 + z^2}$$

so that $R(\mathbf{n}(w), \theta(z))$ is a rational expression.

5.4.8 Parameterizations Based on the Hopf Fibration

Another recent parameterization is motivated by the Hopf fibration of the three-sphere [36] in which unit quaternions are parameterized as

$$(q_1, q_2, q_3, q_4) = \left(\cos\frac{\theta}{2}\cos\frac{\psi}{2}, \cos\frac{\theta}{2}\sin\frac{\psi}{2}, \sin\frac{\theta}{2}\cos\left(\phi + \frac{\psi}{2}\right), \sin\frac{\theta}{2}\sin\left(\phi + \frac{\psi}{2}\right) \right).$$

This parameterization has been shown to have good properties with regard to sampling as measured in terms of dispersion and discrepancy. But since it becomes singular at $\theta = \pi$, for the purpose of biomolecular docking applications there remains motivation to look for alternatives.

5.4.9 Koh-Ananthasuresh Fully Reversible Rotation Sequences

In a series of papers [18, 19, 20], a paradigm for reorientation of micro satellites is established wherein small flippers rotate in a sequence followed by reverse rotations of the same sequence. Due to nonholonomic effects associated with the conservation of angular momentum, the final orientation of the micro satellite after the flippers return to their original angles is altered. By iterating this process, the overall orientation of the satellite can be changed as desired. Each sequence of reversible rotations is of the form

$$R(\phi_1, \phi_2, \phi_3) = R_1(\phi_1) R_2(\phi_2) R_3(\phi_3) R_1(-\phi_1) R_2(-\phi_2) R_3(-\phi_3) \qquad (5.45)$$

where $R_i(\phi_i)$ is a rotation around body-fixed principal axis i. The angle ϕ_i is not the angle through which the flipper rotates, but rather is the angle through which the micro satellite rotates as a result of the action of the flipper. This angle is related to the angle of the flipper via concervation of angular momentum in a simple way that depends on the moments of inertia of the flippers and the main body of the satellite when the flippers are aligned with principal axese. The resulting rotation matrix in (5.45) describes the result after one sequence of fully reversed flipper rotations, and can be viewed as a new parameterization of rotations.

5.5 Infinitesimal Rotations, Angular Velocity, and Integration

It is clear from Euler's theorem that when $|\theta| \ll \pi$, a rotation matrix reduces to the form

$$A_E(\theta, \mathbf{n}) = \mathbb{I} + \theta N.$$

This means that for small rotation angles θ_1 and θ_2, rotations commute:

$$A_E(\theta_1, \mathbf{n}_1) A_E(\theta_2, \mathbf{n}_2) = \mathbb{I} + \theta_1 N_1 + \theta_2 N_2 = A_E(\theta_2, \mathbf{n}_2) A_E(\theta_1, \mathbf{n}_1).$$

Given two frames of reference, one of which is fixed in space and the other of which is rotating relative to it, a rotation matrix describing the orientation of the rotating frame as seen in the fixed frame is written as $R = R(t)$ at each time t. The concepts of small rotations and angular velocity can be connected as follows. Observe that if $\mathbf{x}_0$ is a fixed (constant) position vector in the rotating frame of reference, then the position of the same point as seen in a frame of reference fixed in space with the same origin as the rotating frame is related to this as

$$\mathbf{x} = R\mathbf{x}_0 \iff \mathbf{x}_0 = R^T \mathbf{x}.$$

The velocity as seen in the frame fixed in space is then

$$\mathbf{v} = \dot{\mathbf{x}} = \dot{R}\mathbf{x}_0 = \dot{R}R^T \mathbf{x}. \qquad (5.46)$$

Observing that since R is a rotation matrix,

$$\frac{d}{dt}\left(RR^T\right) = \frac{d}{dt}(\mathbb{I}) = \mathbb{O},$$

and using the product rule from Calculus and the transpose rule for matrix multiplication, we write

$$\dot{R}R^T = -R\dot{R}^T = -(\dot{R}R^T)^T.$$

Due to the skew-symmetry of this matrix, we can rewrite (5.46) in the form most familiar to engineers and physicists:

$$\mathbf{v} = \boldsymbol{\omega}_L \times \mathbf{x},$$

where $\boldsymbol{\omega}_L = (\dot{R}R^T)^\vee$. The vector $\boldsymbol{\omega}_L$ is the angular velocity as seen in the space-fixed frame of reference (i.e., the frame in which the moving frame appears to have orientation given by R). The subscript L is not conventional; It denotes that the derivative of R appears on the "left" side of R^T. In contrast, the angular velocity as seen in the rotating frame of reference (where the orientation of the moving frame appears to have orientation given by the identity rotation) is the dual vector of $R^T \dot{R}$, which is also a skew-symmetric matrix. This is denoted as $\boldsymbol{\omega}_R$. Therefore we have

$$\boldsymbol{\omega}_R = (R^T \dot{R})^\vee = (R^T(\dot{R}R^T)R)^\vee = R^T \boldsymbol{\omega}_L.$$

Another way to write this is

$$\boldsymbol{\omega}_L = R\boldsymbol{\omega}_R.$$

In other words, the angular velocity as seen in the frame of reference fixed in space is obtained from the angular velocity as seen in the rotating frame in the same way in which the absolute position is obtained from the relative position.

5.5.1 Jacobians Associated with Parameterized Rotations

When a time-varying rotation matrix is parameterized as

$$R(t) = A(q_1(t), q_2(t), q_3(t)) = A(\mathbf{q}(t)),$$

then by the chain rule from calculus,

$$\dot{R} = \frac{\partial A}{\partial q_1}\dot{q}_1 + \frac{\partial A}{\partial q_2}\dot{q}_2 + \frac{\partial A}{\partial q_3}\dot{q}_3.$$

Multiplying on the right by R^T and extracting the dual vector from both sides, we find that

$$\boldsymbol{\omega}_L = J_L(A(\mathbf{q}))\dot{\mathbf{q}} \tag{5.47}$$

where

$$J_L(A(\mathbf{q})) = \left[\left(\frac{\partial A}{\partial q_1}A^T\right)^\vee, \left(\frac{\partial A}{\partial q_2}A^T\right)^\vee, \left(\frac{\partial A}{\partial q_3}A^T\right)^\vee\right].$$

Similarly,

$$\boldsymbol{\omega}_R = J_R(A(\mathbf{q}))\dot{\mathbf{q}} \tag{5.48}$$

where

$$J_R(A(\mathbf{q})) = \left[\left(A^T\frac{\partial A}{\partial q_1}\right)^\vee, \left(A^T\frac{\partial A}{\partial q_2}\right)^\vee, \left(A^T\frac{\partial A}{\partial q_3}\right)^\vee\right].$$

These two Jacobian matrices are related as

$$J_L = AJ_R. \tag{5.49}$$

It is easy to verify from the above expressions that for an arbitrary constant rotation $R_0 \in SO(3)$:

$$J_L(R_0 A(\mathbf{q})) = R_0 J_L(A(\mathbf{q})); \qquad J_L(A(\mathbf{q})R_0) = J_L(A(\mathbf{q}))$$

$$J_R(R_0 A(\mathbf{q})) = J_R(A(\mathbf{q})); \qquad J_R(A(\mathbf{q})R_0) = R_0^T J_R(A(\mathbf{q})).$$

In the following subsections, we provide the explicit forms for the Jacobians of several of the parameterizations discussed earlier in this chapter. As a notational shorthand, we will use $J_L(\mathbf{q})$ in place of $J_L(A(\mathbf{q}))$, and similarly for J_R.

The Jacobians for ZXZ Euler Angles

In this subsection we explicitly calculate the Jacobian matrices J_L and J_R for the ZXZ Euler angles. In this case, $A(\alpha, \beta, \gamma) = R_3(\alpha) R_1(\beta) R_3(\gamma)$, and the skew-symmetric matrices whose dual vectors form the columns of the Jacobian matrix J_L are given as:[4]

$$\frac{\partial A}{\partial \alpha} A^T = (R'_3(\alpha) R_1(\beta) R_3(\gamma))(R_3(-\gamma) R_1(-\beta) R_3(-\alpha)) = R'_3(\alpha) R_3(-\alpha),$$

$$\frac{\partial A}{\partial \beta} A^T = (R_3(\alpha) R'_1(\beta) R_3(\gamma))(R_3(-\gamma) R_1(-\beta) R_3(-\alpha))$$
$$= R_3(\alpha)(R'_1(\beta) R_1(-\beta)) R_3(-\alpha),$$

$$\frac{\partial A}{\partial \gamma} A^T = R_3(\alpha) R_1(\beta)(R'_3(\gamma))(R_3(-\gamma)) R_1(-\beta) R_3(-\alpha).$$

Noting that $(R'_i R_i^T)^\vee = \mathbf{e}_i$ regardless of the value of the parameter, and using the rule $(RXR^T)^\vee = R(X)^\vee$, we find that

$$J_L(\alpha, \beta, \gamma) = [\mathbf{e}_3, R_3(\alpha)\mathbf{e}_1, R_3(\alpha) R_1(\beta)\mathbf{e}_3].$$

This is written explicitly as

$$J_L(\alpha, \beta, \gamma) = \begin{pmatrix} 0 & \cos\alpha & \sin\alpha\sin\beta \\ 0 & \sin\alpha & -\cos\alpha\sin\beta \\ 1 & 0 & \cos\beta \end{pmatrix}.$$

The Jacobian J_R can be derived similarly, or we can calculate it easily from J_L as

$$J_R = A^T J_L = [R_3(-\gamma) R_1(-\beta)\mathbf{e}_3, R_3(-\gamma)\mathbf{e}_1, \mathbf{e}_3].$$

Explicitly, this is

$$J_R = \begin{pmatrix} \sin\gamma\sin\beta & \cos\gamma & 0 \\ \cos\gamma\sin\beta & -\sin\gamma & 0 \\ \cos\beta & 0 & 1 \end{pmatrix}.$$

These Jacobian matrices are important in understanding the differential operators and integration measure used in Chapters 8 and 9. These in turn provide the tools for defining orthogonal expansions of functions of rotation-valued arguments.

It is easy to see that

$$J_L^{-1} = \begin{pmatrix} -\cot\beta\sin\alpha & \cos\alpha\cot\beta & 1 \\ \cos\alpha & \sin\alpha & 0 \\ \csc\beta\sin\alpha & -\cos\alpha\csc\beta & 0 \end{pmatrix}.$$

---

[4]For one-parameter rotations we use the notation $'$ to denote differentiation with respect to the parameter.

$$J_R^{-1} = \begin{pmatrix} \csc\beta\sin\gamma & \cos\gamma\csc\beta & 0 \\ \cos\gamma & -\sin\gamma & 0 \\ -\cot\beta\sin\gamma & -\cos\gamma\cot\beta & 1 \end{pmatrix}. \tag{5.50}$$

$$\det(J_L) = \det(J_R) = -\sin\beta.$$

The negative sign here is irrelevant since only $|\det(J_L)|$ and $|\det(J_R)|$ appear in calculations, much like the Jacobian determinants used in Chapter 4.

The Jacobians for the Matrix Exponential

Relatively simple analytical expressions have been derived by Park [27] for the 3×3 Jacobian matrix J_L and its inverse when rotations are parameterized as in (5.25). These expressions are

$$J_L(\mathbf{x}) = \mathbb{I} + \frac{1-\cos\|\mathbf{x}\|}{\|\mathbf{x}\|^2}X + \frac{\|\mathbf{x}\| - \sin\|\mathbf{x}\|}{\|\mathbf{x}\|^3}X^2 \tag{5.51}$$

and

$$J_L^{-1}(\mathbf{x}) = \mathbb{I} - \frac{1}{2}X + \left(\frac{1}{\|\mathbf{x}\|^2} - \frac{1+\cos\|\mathbf{x}\|}{2\|\mathbf{x}\|\sin\|\mathbf{x}\|}\right)X^2.$$

The corresponding Jacobian J_L and its inverse are then calculated as $J_L = AJ_R$ and $J_L^{-1} = J_R^{-1}A^T$ to yield

$$J_R(\mathbf{x}) = \mathbb{I} - \frac{1-\cos\|\mathbf{x}\|}{\|\mathbf{x}\|^2}X + \frac{\|\mathbf{x}\| - \sin\|\mathbf{x}\|}{\|\mathbf{x}\|^3}X^2$$

and

$$J_R^{-1}(\mathbf{x}) = \mathbb{I} + \frac{1}{2}X + \left(\frac{1}{\|\mathbf{x}\|^2} - \frac{1+\cos\|\mathbf{x}\|}{2\|\mathbf{x}\|\sin\|\mathbf{x}\|}\right)X^2.$$

Note that unlike the Euler angles,

$$J_L = J_R^T.$$

The determinants are

$$\det(J_L) = \det(J_R) = \frac{2(1-\cos\|\mathbf{x}\|)}{\|\mathbf{x}\|^2}.$$

The Jacobians for the Cayley/Rodrigues Parameters

We note that

$$J_R = \frac{2}{1+r_1^2+r_2^2+r_3^2}\begin{pmatrix} 1 & r_3 & -r_2 \\ -r_3 & 1 & r_1 \\ r_2 & -r_1 & 1 \end{pmatrix}$$

and

$$J_R^{-1} = \frac{1}{2}\begin{pmatrix} 1+r_1^2 & r_1r_2-r_3 & r_1r_3+r_2 \\ r_2r_1+r_3 & 1+r_2^2 & r_2r_3-r_1 \\ r_3r_1-r_2 & r_3r_2+r_1 & 1+r_3^2 \end{pmatrix}.$$

We also observe the interesting property of this parameterization:

$$J_L = J_R^T,$$

and so
$$J_L^{-1} = (J_R^T)^{-1} = (J_R^{-1})^T.$$

The Jacobian determinant is
$$\det(J_L) = \det(J_R) = \frac{8}{(1 + r_1^2 + r_2^2 + r_3^2)^2}.$$

The Jacobians for Spherical Coordinates

By ordering columns with variables in the order θ, λ, ν, the Jacobians in spherical coordinates are

$$J_L = \begin{pmatrix} c\lambda s\nu & -c\lambda s^2(\theta/2)s(2\nu) - s\lambda s\nu s\theta & -2s^2(\theta/2)s\lambda + c\lambda c\nu s\theta \\ s\lambda s\nu & -s\lambda s(2\nu)s^2(\theta/2) + c\lambda s\nu s\theta & 2c\lambda s^2(\theta/2) + c\nu s\lambda s\theta \\ c\nu & 2s^2\nu s^2(\theta/2) & -s\nu s\theta \end{pmatrix}$$

and

$$J_R = \begin{pmatrix} c\lambda s\nu & c\lambda s(2\nu)s^2(\theta/2) - s\lambda s\nu s\theta & 2s\lambda s^2(\theta/2) + c\lambda c\nu s\theta \\ s\lambda s\nu & s\lambda s(2\nu)s^2(\theta/2) + c\lambda s\nu s\theta & -2s^2(\theta/2)c\lambda + c\nu s\lambda s\theta \\ c\nu & -2s^2\nu s^2(\theta/2) & -s\nu s\theta \end{pmatrix}.$$

The Jacobian determinant is then
$$\det(J_L) = \det(J_R) = -4s^2(\theta/2)s\nu.$$

Note that while transforming $(\theta, \lambda, \nu) \to (-\theta, \lambda \pm \pi, \pi - \nu)$ changes the sign of the first and third columns of the Jacobians, the sign of the Jacobian determinant does not change. Such a transformation can be useful when θ is taken over the doubled range $[-\pi, \pi]$.

The Jacobians for the Modified Rodrigues Parameters

The Jacobians for the parameterization have some interesting properties. First we observe that
$$J_L = \frac{4}{1 + \|\sigma\|^2} A_R(\boldsymbol{\sigma}) =$$

$$\frac{4}{(1 + \|\sigma\|^2)^2} \begin{pmatrix} 1 + \sigma_1^2 - \sigma_2^2 - \sigma_3^2 & 2(\sigma_1\sigma_2 - \sigma_3) & 2(\sigma_1\sigma_3 + \sigma_2) \\ 2(\sigma_2\sigma_1 + \sigma_3) & 1 - \sigma_1^2 + \sigma_2^2 - \sigma_3^2 & 2(\sigma_2\sigma_3 - \sigma_1) \\ 2(\sigma_3\sigma_1 - \sigma_2) & 2(\sigma_3\sigma_2 + \sigma_1) & 1 - \sigma_1^2 - \sigma_2^2 + \sigma_3^2 \end{pmatrix},$$

where A_R is the Rodrigues rotation matrix in (5.34), evaluated at $\boldsymbol{\sigma}$ instead of $\mathbf{r}$. Since this Jacobian is a scalar multiple of a rotation matrix (as was observed in [31]), it follows that its inverse is easy to calculate:

$$J_L^{-1} = \frac{1}{4} \begin{pmatrix} 1 + \sigma_1^2 - \sigma_2^2 - \sigma_3^2 & 2(\sigma_1\sigma_2 + \sigma_3) & 2(\sigma_1\sigma_3 - \sigma_2) \\ 2(\sigma_2\sigma_1 - \sigma_3) & 1 - \sigma_1^2 + \sigma_2^2 - \sigma_3^2 & 2(\sigma_2\sigma_3 + \sigma_1) \\ 2(\sigma_3\sigma_1 + \sigma_2) & 2(\sigma_3\sigma_2 - \sigma_1) & 1 - \sigma_1^2 - \sigma_2^2 + \sigma_3^2 \end{pmatrix}.$$

As with the Rodrigues parameters, we have
$$J_R = J_L^T,$$

and so
$$J_R^{-1} = (J_L^{-1})^T.$$

The Jacobian determinants are
$$\det(J_L) = \det(J_R) = \frac{64}{(1 + \|\boldsymbol{\sigma}\|^2)^3}.$$

The Jacobians for the Euler Parameters

The relationship between the time derivatives of the four Euler parameters and the angular velocity $\boldsymbol{\omega}_L$ can be expressed using a 4×4 Jacobian matrix J_L as
$$\begin{pmatrix} \boldsymbol{\omega} \\ 0 \end{pmatrix} = J_L(\mathbf{u})\dot{\mathbf{u}}$$

or
$$\dot{\mathbf{u}} = J_L^{-1}(\mathbf{u})[\boldsymbol{\omega}^T, 0]^T.$$

Here the last row of J_L corresponds to the constraint $\|\mathbf{u}\|^2 = 1$. The Jacobian matrix J_L is explicitly
$$J_L = 2 \begin{pmatrix} u_4 & u_3 & -u_2 & -u_1 \\ -u_3 & u_4 & u_1 & -u_2 \\ u_2 & -u_1 & u_4 & -u_3 \\ u_1 & u_2 & u_3 & u_4 \end{pmatrix}$$

(from [31] with first row and column moved) and its inverse is
$$J_L^{-1} = \frac{1}{2} \begin{pmatrix} u_4 & -u_3 & u_2 & u_1 \\ u_3 & u_4 & -u_1 & u_2 \\ -u_2 & u_1 & u_4 & u_3 \\ -u_1 & -u_2 & -u_3 & u_4 \end{pmatrix}.$$

5.5.2 Rigid-Body Mechanics

In the case of a rigid body, the angular momentum $\mathbf{L}$ has a special form. This is because the absolute position to the ith particle in a rigid body constructed from n particles can be decomposed into a component due to the position of the center of mass and a component due to rotation about the center of mass. That is,
$$\mathbf{x}_i = \mathbf{x}_{cm} + R\mathbf{y}_i \qquad (5.52)$$

where $\mathbf{y}_i$ is the position of the ith particle as measured in a frame attached to the center of mass of the rigid body. The angular momentum is then:
$$\mathbf{L} = \sum_{i=1}^{n} m_i(\mathbf{x}_i \times \dot{\mathbf{x}}_i) = \sum_{i=1}^{n} m_i \left((\mathbf{x}_{cm} + R\mathbf{y}_i) \times (\dot{\mathbf{x}}_{cm} + \dot{R}\mathbf{y}_i) \right).$$

Expanding this, we get:
$$\mathbf{L} = \sum_{i=1}^{n} m_i \left(\mathbf{x}_{cm} \times \dot{\mathbf{x}}_{cm} + \mathbf{x}_{cm} \times \dot{R}\mathbf{y}_i + R\mathbf{y}_i \times \dot{\mathbf{x}}_{cm} + R\mathbf{y}_i \times \dot{R}\mathbf{y}_i \right).$$

Now, by the definition of the center of mass, multiplying (5.52) by m_i and summing both sides, we find that $\sum_{i=1}^{n} m_i R \mathbf{y}_i = \mathbf{0}$, which means that $\sum_{i=1}^{n} m_i \mathbf{y}_i = \mathbf{0}$ (because the rotation matrix can come out of the summation sign and we can multiply both sides of the equation by its inverse on the left).

This means that all terms with a sum of this form will drop out, and so the angular momentum expression simplifies to:

$$\mathbf{L} = \left(\sum_{i=1}^{n} m_i \right) (\mathbf{x}_{cm} \times \dot{\mathbf{x}}_{cm}) + \sum_{i=1}^{n} m_i \left(R\mathbf{y}_i \times \dot{R\mathbf{y}_i} \right),$$

which is a component due to the motion of the center of mass and a component due to rotation about the center of mass. In cases where the center of mass is fixed or travels in a straight line (with constant or variable velocity) passing through the origin of a coordinate system, the first part of this expression will be zero.

In order to describe the second term in the angular momentum expression more easily, we can define the position of the ith particle relative to the center of mass (defined in a frame parallel to the inertial/fixed frame) as $\mathbf{z}_i = R\mathbf{y}_i$. Then

$$R\mathbf{y}_i \times \dot{R\mathbf{y}_i} = \mathbf{z}_i \times \dot{R}R^T \mathbf{z}_i = \mathbf{z}_i \times (\boldsymbol{\omega}_L \times \mathbf{z}_i).$$

Using the identity $\mathbf{a} \times (\mathbf{b} \times \mathbf{a}) = (\mathbf{a} \cdot \mathbf{a})\mathbf{b} - (\mathbf{a} \cdot \mathbf{b})\mathbf{a}$, for the case when $\mathbf{x}_{cm} \times \dot{\mathbf{x}}_{cm} = \mathbf{0}$, we can write

$$\mathbf{L}_L = \sum_{i=1}^{n} m_i \left((\mathbf{z}_i \cdot \mathbf{z}_i)\boldsymbol{\omega} - (\mathbf{z}_i \cdot \boldsymbol{\omega})\mathbf{z}_i \right) = I_L \boldsymbol{\omega}_L,$$

where

$$I_L \doteq \sum_{i=1}^{n} m_i \left((\mathbf{z}_i \cdot \mathbf{z}_i)\mathbb{I} - \mathbf{z}_i \mathbf{z}_i^T \right).$$

Note that the angular momentum vector, just like any vector, can be represented in any frame. Therefore, if instead of representing it in the fixed frame we represent it in the rotated frame as $\mathbf{L}_R$, then $\mathbf{L}_L = R\mathbf{L}_R$. Likewise, recall that then angular velocity can be represented in any frame, and so $\boldsymbol{\omega}_L = R\boldsymbol{\omega}_R$. Therefore, we can write

$$\mathbf{L}_R = R^T \mathbf{L}_L = R^T I_L R \boldsymbol{\omega}_R = I_R \boldsymbol{\omega}_R,$$

where

$$I_R = \sum_{i=1}^{n} m_i \left((\mathbf{y}_i \cdot \mathbf{y}_i)\mathbb{I} - \mathbf{y}_i \mathbf{y}_i^T \right).$$

In other words, the moment of inertia matrix is different when represented in different frames. It is a constant matrix I_R, when represented in the frame fixed to the body, and time-dependent in any frame which moves relative to the body. Any physical quantity which can be represented by a matrix $A \in \mathbb{R}^{3 \times 3}$ which transforms under any rotation $Q \in SO(3)$ as $A_L = QA_R Q^T$ is called a 2nd order tensor. Vectors are often called 1st order tensors because they transform under the rule $\mathbf{a}_L = Q\mathbf{a}_R$ under rotations.

Parallel Axis Theorem

For a rigid body with one fixed point, we can attach a frame at that point which is fixed in space and a second frame at that point which is fixed in the body. The vector to the center of mass is then $\mathbf{x}_{cm} = R\mathbf{y}_{cm}$, where $\mathbf{y}_{cm}$ is a *constant* vector. In this case, we can write:

$$\left(\sum_{i=1}^{n} m_i\right)(\mathbf{x}_{cm} \times \dot{\mathbf{x}}_{cm}) = \left(\sum_{i=1}^{n} m_i\right)\left((R\mathbf{y}_{cm}) \times (\dot{R}\mathbf{y}_{cm})\right)$$

$$= \left(\sum_{i=1}^{n} m_i\right) R(\mathbf{y}_{cm} \times (R^T \dot{R}\mathbf{y}_{cm}))$$

$$= M \cdot R\left(\mathbf{y}_{cm} \times (R^T \dot{R}\mathbf{y}_{cm})\right)$$

$$= R\left(M(\mathbf{y}_{cm}) \times (\boldsymbol{\omega}_R \times \mathbf{y}_{cm})\right).$$

where the rule $(Q\mathbf{a}) \times (Q\mathbf{b}) = Q(\mathbf{a} \times \mathbf{b})$ has been used (for $Q \in SO(3)$) and $M \doteq \sum_{i=1}^{n} m_i$ is the total mass.

This can be simplified in one final step as

$$R\left(M(\mathbf{y}_{cm}) \times (\boldsymbol{\omega}_R \times \mathbf{y}_{cm})\right) = R\left(M(\mathbf{y}_{cm} \cdot \mathbf{y}_{cm}\mathbb{I} - \mathbf{y}_{cm}\mathbf{y}_{cm}^T)\right)\boldsymbol{\omega}_R.$$

In other words, if we define a constant matrix

$$P_R = (\mathbf{y}_{cm} \cdot \mathbf{y}_{cm})\mathbb{I} - \mathbf{y}_{cm}\mathbf{y}_{cm}^T,$$

then the angular momentum vector described in the rotating frame with origin at the fixed point is

$$\mathbf{L}_R = (I_R + MP_R)\boldsymbol{\omega}_R.$$

It is often easier to write these two matrices together as one matrix:

$$I'_R = I_R + MP_R.$$

This is the *parallel axis theorem*. It says that if a rigid body has one fixed point, and that point is not the center of mass, then the angular momentum vector can still be written as $\mathbf{L}_R = I'_R\boldsymbol{\omega}_R$, or $\mathbf{L}_L = I'_L\boldsymbol{\omega}_L$.

Euler's Equations of Motion

We can use this to derive compact equations of motion for any rigid body with one point fixed, e.g., a spatial pendulum, or spinning top. Since the rate of change of angular momentum is equal to the applied moment when both vectors are described in an inertial reference frame,

$$\mathbf{N}_L = \frac{d\mathbf{L}_L}{dt}.$$

However,

$$\frac{d\mathbf{L}_L}{dt} = \frac{d}{dt}(R\mathbf{L}_R) = \dot{R}\mathbf{L}_R + R\dot{\mathbf{L}}_R,$$

and we can define $\mathbf{N}_R = R^T\mathbf{N}_L$ to be the resultant moment represented in the rotated frame. And so

$$\mathbf{N}_R = R^T\mathbf{N}_L = R^T\dot{R}\mathbf{L}_R + \dot{\mathbf{L}}_R,$$

Recognizing that $\boldsymbol{\omega}_R = (R^T\dot{R})^\vee$, and substituting $\mathbf{L}_R = I'_R\boldsymbol{\omega}_R$, we get:

$$\mathbf{N}_R = \boldsymbol{\omega}_R \times (I'_R\boldsymbol{\omega}_R) + I'_R\dot{\boldsymbol{\omega}}_R,$$

This is called *Euler's Equation of Motion*. In particular, if I'_R is diagonal with components I_1, I_2, and I_3, (which can always be achieved by choosing the body-fixed frame as the eigenvectors of the moment of inertia matrix), $\boldsymbol{\omega}_R$ has components ω_1, ω_2, ω_3, and $\mathbf{N}_R$ has components N_1, N_2, N_3, then these equations have the form

$$I_1\dot{\omega}_1 + (I_3 - I_2)\omega_2\omega_3 = N_1$$
$$I_2\dot{\omega}_2 + (I_1 - I_3)\omega_3\omega_1 = N_2$$
$$I_3\dot{\omega}_3 + (I_2 - I_1)\omega_1\omega_2 = N_3.$$

Kinetic Energy of a Rotating Rigid Body

The kinetic energy of a rigid body consists of a part due to translation of the center of mass, and a part due to rotation about the center of mass:

$$T = T_{trans} + T_{rot}.$$

The rotational kinetic energy is usually given by

$$T_{rot} = \frac{1}{2}\boldsymbol{\omega}_R^T I_R \boldsymbol{\omega}_R$$

where I_R and $\boldsymbol{\omega}_R$ are the moment of inertia tensor and angular velocity, respectively, as seen in the frame of reference attached to the rigid body at its mass center. The same equation for T_{rot} could be used without the subscripts on I and $\boldsymbol{\omega}$. However, for convenience the frame fixed in the body is usually chosen because in this frame I_R is a constant matrix. For a mass distribution that is continuous over a finite body, we write

$$I_R = \text{tr}(\mathcal{I}_R)\,\mathbb{I} - \mathcal{I}_R$$

where

$$\mathcal{I}_R = \int_{\mathbb{R}^3} \mathbf{z}\mathbf{z}^T \rho(\mathbf{z})\, dz_1 dz_2 dz_3$$

and $\mathbf{z}$ is defined with respect to the frame attached to the rigid body at its center of mass.

An equivalent expression for rotational kinetic energy in terms of the rotation matrix R is

$$T_{rot} = \frac{1}{2}\text{tr}(\dot{R}\mathcal{I}_R\dot{R}^T).$$

5.6 Other Methods for Describing Rotations in Three-Space

5.6.1 Quaternions

Quaternions[5] were developed in an attempt to generalize the concept of the complex numbers from $2D$ to $3D$ in the following sense. Each complex number $x = x_1 + ix_2$ is associated with a position in the plane $\mathbf{x} = (x_1, x_2)^T \in \mathbb{R}^2$. While it does not make sense to talk about dividing by vectors, it is perfectly legitimate to find for any complex number $x \neq 0$, its inverse x^{-1} such that $xx^{-1} = x^{-1}x = 1$. This is given explicitly as

---

[5]Developed by Sir William Rowan Hamilton (1805-1865)

$x^{-1} = (x_1 - ix_2)/(x_1^2 + x_2^2)$ and is easily verified by writing the rule for multiplication of complex numbers as $xy = (x_1y_1 - x_2y_2) + i(x_1y_2 + x_2y_1)$. It is interesting to note that the operation on vectors $\mathbf{x}, \mathbf{y} \in \mathbb{R}^2$ defined as

$$\mathbf{x} \wedge \mathbf{y} \doteq (x_1y_1 - x_2y_2)\mathbf{e}_1 + (x_1y_2 + x_2y_1)\mathbf{e}_2$$

endows the pair $(\mathbb{R}^2, \wedge)$ with the properties of an algebra (see Appendix C.3). Rotations in the complex plane are achieved as $x' = e^{i\theta}x$ for $\theta \in [0, 2\pi]$. This yields exactly the same result as the matrix-vector operation

$$\begin{pmatrix} x_1' \\ x_2' \end{pmatrix} = \begin{pmatrix} \cos\theta & -\sin\theta \\ \sin\theta & \cos\theta \end{pmatrix} \begin{pmatrix} x_1 \\ x_2 \end{pmatrix}.$$

Hamilton attempted for over ten years to find a spatial analog of the complex numbers, i.e., one of the form $x = x_1 + ix_2 + jx_3$ where x^{-1} is well defined. It is rumoured that during this time his son would often ask, "Father, have you learned how to divide vectors?"[12]. Needless to say, progress was slow until Hamilton had the inspiration to expand his thinking to four dimensions. In analogy with the complex numbers which have the basis $\{1, i\}$ satisfying $i^2 = -1$, the *quaternions* (or hyper-complex numbers) have a basis $\{1, \tilde{\mathbf{i}}, \tilde{\mathbf{j}}, \tilde{\mathbf{k}}\}$ satisfying the conditions

$$\tilde{\mathbf{i}}^2 = \tilde{\mathbf{j}}^2 = \tilde{\mathbf{k}}^2 = -1 \tag{5.53}$$

$$\tilde{\mathbf{i}}\tilde{\mathbf{j}} = -\tilde{\mathbf{j}}\tilde{\mathbf{i}} = \tilde{\mathbf{k}} \tag{5.54}$$

$$\tilde{\mathbf{k}}\tilde{\mathbf{i}} = -\tilde{\mathbf{i}}\tilde{\mathbf{k}} = \tilde{\mathbf{j}} \tag{5.55}$$

$$\tilde{\mathbf{j}}\tilde{\mathbf{k}} = -\tilde{\mathbf{k}}\tilde{\mathbf{j}} = \tilde{\mathbf{i}}. \tag{5.56}$$

Multiplication by 1 on either side of $\tilde{\mathbf{i}}$, $\tilde{\mathbf{j}}$, or $\tilde{\mathbf{k}}$ leaves it unchanged. The quaternions form an algebra over $\mathbb{R}^4$, when viewing the multiplication of two quaternions $p = p_4 + \tilde{\mathbf{i}}p_1 + \tilde{\mathbf{j}}p_2 + \tilde{\mathbf{k}}p_3$ and $q = q_4 + \tilde{\mathbf{i}}q_1 + \tilde{\mathbf{j}}q_2 + \tilde{\mathbf{k}}q_3$ as an operation on vectors in $\mathbb{R}^4$ with components (q_1, q_2, q_3, q_4) and (p_1, p_2, p_3, p_4). They also almost satisfy the definition of a *field* (see Appendix C for definition), except for the fact that $pq \neq qp$ in general, and so the quaternions are referred to as a *skew field*. Often a quaternion q is written as the pair $(q_4, \mathbf{q})$ where $\mathbf{q} = \tilde{\mathbf{i}}q_1 + \tilde{\mathbf{j}}q_2 + \tilde{\mathbf{k}}q_3$ is called a "pure quaternion" or the "vector part" of q, and q_4 is called the "real part" of q. Using the rules in (5.53) - (5.56), the product of two quaternions is written as

$$(q_4, \mathbf{q})(p_4, \mathbf{p}) = (q_4p_4 - \mathbf{q} \cdot \mathbf{p}, q_4\mathbf{p} + p_4\mathbf{q} + \mathbf{q} \times \mathbf{p})$$

where $\cdot$ and $\times$ are the scalar and cross products for $\mathbb{R}^3$.

In analogy with the complex numbers, the conjugate of a quaternion $q = (q_4, \mathbf{q})$ is $q^* = (q_4, -\mathbf{q})$. The norm (or modulus) of a quaternion is the square root of

$$|q|^2 = qq^* = q^*q = q_1^2 + q_2^2 + q_3^2 + q_4^2.$$

The inverse of a quaternion is

$$q^{-1} = \frac{q^*}{qq^*}.$$

In addition to being an algebra (and hence a vector space and an Abelian group under addition), the quaternions form a group under multiplication, with the small modification that $q = 0$ be excluded.[6] *Unit quaternions* are those satisfying $|q| = 1$. Henceforth

---

[6]See Chapter 7 for an introduction to group theory.

we use the notation u to denote unit quaternions instead of q. As with rotations, the unit quaternions can be viewed as lying on the surface of $\mathbb{S}^3$. In fact, the transformation,

$$x' = uxu^* \tag{5.57}$$

for $u = u_4 + \tilde{i}u_1 + \tilde{j}u_2 + \tilde{k}u_3$ implements exactly the same rotation of the vector $\mathbf{x} = (x_1, x_2, x_3)^T$ as multiplication by the matrix $A(u_1, u_2, u_3, u_3)$ of Euler parameters in (5.33). The unit quaternion corresponding to rotation about $\mathbf{n}$ by angle θ is $q = (\cos\theta/2, \mathbf{n}\sin\theta/2)$. Whereas the set of rotations is viewed as the upper hemisphere of $\mathbb{S}^3$, two unit quaternions, u and $-u$, represent the same rotation. If u is in the upper hemisphere, then $-u$ is in the lower one, and the set of all quaternions is viewed as the whole of $\mathbb{S}^3$ and is called the *quaternion sphere*.

For further reading on quaternions, see Hamilton's original paper [13], or books on kinematics such as [3, 24].

As a final note to this subsection, we observe that generalizations of the Fourier transform based on unit quaternions, instead of unimodular complex numbers, have been investigated in recent years. The motivation goes something like this: $e^{i\phi} = \cos\phi + i\sin\phi$ and satisfies $(e^{i\phi})^k = \cos k\phi + i\sin k\phi$. A unit quaternion $u = (\cos\phi, \mathbf{n}\sin\phi)$ satisfies the analogous expression $u^k = (\cos k\phi, \mathbf{n}\sin k\phi)$. For real and complex-valued functions, the Fourier transform is defined based on the properties of $e^{i\phi}$, so an analogous Fourier transform for quaternion-valued functions should follow based on the properties of unit quaternions. Several recent research papers and books in the field of "hyper-complex analysis" have defined this kind of generalization of the Fourier transform [12, 22, 33].

To our knowledge, there is no associated generalization of the convolution theorem. It is also difficult to imagine that one is possible, given the noncommutative nature of multiplication of quaternion-valued functions. In later chapters we will explore noncommutative harmonic analysis, which is a generalization of Fourier analysis when the *argument* of a function is a quantity such as a rotation or quaternion and the *value* of the function is real or complex.

5.6.2 Expressing Rotations as 4×4 Matrices

The rules for quaternion multiplication reviewed in the previous subsection can be observed in several kinds of matrices. For instance, we can define

$$\mathbb{I}_4 \doteq \begin{pmatrix} 1 & 0 & 0 & 0 \\ 0 & 1 & 0 & 0 \\ 0 & 0 & 1 & 0 \\ 0 & 0 & 0 & 1 \end{pmatrix} ; \quad \tilde{i}_4 \doteq \begin{pmatrix} 0 & -1 & 0 & 0 \\ 1 & 0 & 0 & 0 \\ 0 & 0 & 0 & 1 \\ 0 & 0 & -1 & 0 \end{pmatrix} ;$$

$$\tilde{j}_4 \doteq \begin{pmatrix} 0 & 0 & -1 & 0 \\ 0 & 0 & 0 & -1 \\ 1 & 0 & 0 & 0 \\ 0 & 1 & 0 & 0 \end{pmatrix} ; \quad \tilde{k}_4 \doteq \begin{pmatrix} 0 & 0 & 0 & 1 \\ 0 & 0 & -1 & 0 \\ 0 & 1 & 0 & 0 \\ -1 & 0 & 0 & 0 \end{pmatrix} ; \tag{5.58}$$

It is easy to see that $\tilde{i}_4\tilde{i}_4 = -\mathbb{I}_4$, $\tilde{i}_4\tilde{j}_4 = \tilde{k}_4$, etc. This choice is not unique. For example, if the matrices we have assigned to the symbols $\tilde{i}_4$, $\tilde{j}_4$, $\tilde{k}_4$ are instead assigned to a cyclic reordering of these symbols, the quaternion multiplication rules will still hold. This is also the case if the assignment of these matrices is made to an acyclic reordering of the symbols, provided that certain signs are changed. For example, $i_4 = \tilde{k}_4$, $j_4 = -\tilde{j}_4$, $k_4 = \tilde{i}_4$ is another choice that is found in the literature.

Using the assignment given above, the rotation encoded by a quaternion can be expressed as

$$Q(u_1, u_2, u_3, u_4) = u_4 \mathbb{I}_4 + u_1 \tilde{i}_4 + u_2 \tilde{j}_4 + u_3 \tilde{k}_4$$

$$= \begin{pmatrix} u_4 & -u_1 & -u_2 & u_3 \\ u_1 & u_4 & -u_3 & -u_2 \\ u_2 & u_3 & u_4 & u_1 \\ -u_3 & u_2 & -u_1 & u_4 \end{pmatrix}. \tag{5.59}$$

Note that since $u_1^2 + u_2^2 + u_3^2 + u_4^2 = 1$, all of the columns are orthonormal, and $\det Q = +1$. Therefore, $Q \in SO(4)$. However, four parameters are not sufficient to parameterize $SO(4)$, and hence it is *not* true that all 4×4 rotation matrices are of the form above.

Quaternion multiplication is implemented in this notation as 4×4 real-matrix multiplication, with the action of rotation of a vector $\mathbf{x} = (x_1, x_2, x_3)^T \in \mathbb{R}^3$ implemented as

$$X' = QXQ^* \tag{5.60}$$

where

$$X = x_1 \tilde{i}_4 + x_2 \tilde{j}_4 + x_3 \tilde{k}_4 = \begin{pmatrix} 0 & -x_1 & -x_2 & x_3 \\ x_1 & 0 & -x_3 & -x_2 \\ x_2 & x_3 & 0 & x_1 \\ -x_3 & x_2 & -x_1 & 0 \end{pmatrix},$$

and

$$Q^*(u_1, u_2, u_3, u_4) = \begin{pmatrix} u_4 & u_1 & u_2 & -u_3 \\ -u_1 & u_4 & u_3 & u_2 \\ -u_2 & -u_3 & u_4 & -u_1 \\ u_3 & -u_2 & u_1 & u_4 \end{pmatrix} = Q(-u_1, -u_2, -u_3, u_4).$$

In the present context, Q^* is simply Q^T, and (5.60) is somewhat like (5.21). More generally, $*$ is the Hermitian conjugate. One difference is that in the 4×4 description, taking the determinant of both sides yields $\|\mathbf{x}'\|^4 = \|\mathbf{x}\|^4$, which is a check that length is preserved, whereas in the 3×3 case the determinant is zero, since 3×3 skew-symmetric matrices are singular. However, in both cases, $\mathrm{tr}(XX^T)$ (which is a multiple of $\|\mathbf{x}\|^2$) is preserved under the transformation.

Finally, we note that the axis-angle description of $3D$-rotations using 4×4 matrices takes the form

$$Q(n_1 \sin \theta/2, n_2 \sin \theta/2, n_3 \sin \theta/2, \cos \theta/2) = e^{\frac{\theta}{2}N} = \cos \frac{\theta}{2} \mathbb{I}_4 + \sin \frac{\theta}{2} N \tag{5.61}$$

where

$$N = n_1 \tilde{i}_4 + n_2 \tilde{j}_4 + n_3 \tilde{k}_4$$

and $\mathbf{n} \cdot \mathbf{n} = 1$.

5.6.3 Special Unitary 2×2 Matrices: $SU(2)$

The rules of quaternion multiplication can also be realized with matrices of the form:

$$\mathbb{I}_2 \doteq \begin{pmatrix} 1 & 0 \\ 0 & 1 \end{pmatrix}; \quad \tilde{i}_2 \doteq \begin{pmatrix} 0 & -1 \\ 1 & 0 \end{pmatrix};$$

$$\tilde{j}_2 \doteq \begin{pmatrix} -i & 0 \\ 0 & i \end{pmatrix}; \quad \tilde{k}_2 \doteq \begin{pmatrix} 0 & -i \\ -i & 0 \end{pmatrix}. \tag{5.62}$$

As with the case of the 4×4 matrices in the previous subsection, this assignment of matrices to the symbols $\tilde{i}_2$, $\tilde{j}_2$, $\tilde{k}_2$ is not unique. Within the convention defined in (5.63), 2×2 matrices of the form

$$U(u_1, u_2, u_3, u_4) = u_4 \mathbb{I}_2 + u_1 \tilde{i}_2 + u_2 \tilde{j}_2 + u_3 \tilde{k}_2$$

$$= \begin{pmatrix} u_4 - iu_2 & -u_1 - iu_3 \\ u_1 - iu_3 & u_4 + iu_2 \end{pmatrix} \tag{5.63}$$

encode rotations. Composition of rotations is achieved by multiplication of these complex matrices, and rotation of a vector $\mathbf{x} = (x_1, x_2, x_3)^T \in \mathbb{R}^3$ is achieved as

$$X' = UXU^*$$

where

$$X = \begin{pmatrix} -ix_2 & -x_1 - ix_3 \\ x_1 - ix_3 & ix_2 \end{pmatrix} = -X^* \tag{5.64}$$

is a traceless *skew-Hermitian* matrix corresponding to the pure quaternion representing $\mathbf{x}$. It is easy to verify that

$$\det(X) = \|\mathbf{x}\|^2 = \frac{1}{2}\operatorname{tr}(XX^*)$$

is preserved under the transformation, as should be the case for a rotation.

In physics it is common to describe rotations using the *Pauli spin matrices*[7]:

$$\sigma_1 = i \cdot \tilde{k}_2 = \begin{pmatrix} 0 & 1 \\ 1 & 0 \end{pmatrix};$$

$$\sigma_2 = i \cdot \tilde{i}_2 = \begin{pmatrix} 0 & -i \\ i & 0 \end{pmatrix};$$

$$\sigma_3 = i \cdot \tilde{j}_2 = \begin{pmatrix} 1 & 0 \\ 0 & -1 \end{pmatrix};$$

In this notation, a position in space is represented as a traceless *Hermitian* matrix:

$$Y = y_1 \sigma_1^T + y_2 \sigma_2^T + y_3 \sigma_3^T = \begin{pmatrix} y_3 & y_1 + iy_2 \\ y_1 - iy_2 & -y_3 \end{pmatrix},$$

(here T denotes transpose), and a rotation is given by a matrix of the form:

---

[7]Named after the theoretical physicist Wolfgang Pauli (1900-1958)

$$U(a,b) = \begin{pmatrix} a & b \\ -\overline{b} & \overline{a} \end{pmatrix}$$

where $a = a_1 + ia_2$, $b = b_1 + ib_2$, and $a\overline{a} + b\overline{b} = 1$. These are called the *Cayley-Klein* parameters [11], which are essentially a variant of the Euler parameters when rotation is performed as

$$Y' = UYU^*. \tag{5.65}$$

In the most general case, the rotation matrix $R(a,b) \in SO(3)$ to which the special unitary matrices $\pm U(a,b)$ correspond can be found using the formula

$$U(a,b) \left(\sum_{j=1}^{3} y_j \sigma_j^T \right) U^*(a,b) = \sum_{j=1}^{3} (R(a,b)\mathbf{y})_j \, \sigma_j^T \tag{5.66}$$

which must hold for an arbitrary vector $\mathbf{y} = [y_1, y_2, y_3]^T \in \mathbb{R}^3$.

Explicitly, the rotation matrix can be written as

$$R(a,b) = \begin{pmatrix} \frac{a^2 - b^2 + \overline{a}^2 - \overline{b}^2}{2} & \frac{-i(\overline{a}^2 + \overline{b}^2 - a^2 - b^2)}{2} & -(\overline{a}\overline{b} + ab) \\ \frac{-i(a^2 - b^2 - \overline{a}^2 + \overline{b}^2)}{2} & \frac{(\overline{a}^2 + \overline{b}^2 + a^2 + b^2)}{2} & i(-\overline{a}\overline{b} + ab) \\ \overline{a}b + a\overline{b} & i(-\overline{a}b + a\overline{b}) & a\overline{a} - b\overline{b} \end{pmatrix}. \tag{5.67}$$

It can be shown by direct calculation for particular cases that rotations of $\mathbf{y} \in \mathbb{R}^3$ about $\mathbf{e}_1$, $\mathbf{e}_2$, $\mathbf{e}_3$ are achieved by substituting the following matrices in for U in (5.65):

$$U_1(\phi) = \begin{pmatrix} \cos(\phi/2) & i\sin(\phi/2) \\ i\sin(\phi/2) & \cos(\phi/2) \end{pmatrix} = \mathbb{I}\cos(\phi/2) + i\sigma_1^T \sin(\phi/2);$$

$$U_2(\phi) = \begin{pmatrix} \cos\phi/2 & -\sin\phi/2 \\ \sin\phi/2 & \cos\phi/2 \end{pmatrix} = \mathbb{I}\cos(\phi/2) + i\sigma_2^T \sin(\phi/2);$$

$$U_3(\phi) = \begin{pmatrix} e^{i\phi/2} & 0 \\ 0 & e^{-i\phi/2} \end{pmatrix} = \mathbb{I}\cos(\phi/2) + i\sigma_3^T \sin(\phi/2).$$

It is also easy to show by direct calculation that for any 2×2 skew-Hermitian matrix T, that $A = e^{iT}$ is a 2×2 unitary matrix of the form

$$e^{iT} = \mathbb{I}_2 \cos \|\mathbf{t}\| + iT \frac{\sin \|\mathbf{t}\|}{\|\mathbf{t}\|}.$$

Likewise, the axis-angle description becomes

$$e^{i\theta N/2} = \mathbb{I}_2 \cos \frac{\theta}{2} + iN \sin \frac{\theta}{2}$$

where $N = T/\|\mathbf{t}\|$.

In particular,

$$U_k(\phi) = e^{i(\phi/2)\sigma_k^T} \tag{5.68}$$

for $k = 1, 2, 3$.

Finally, we note that other conventions for the definitions of the Pauli spin matrices lead to similar-looking results. For instance, Miller [25] uses $\tilde{\sigma}_i = U_1(\pi)\sigma_i^T U_1(-\pi)$. Substitution of this into (5.67) results in a matrix $R'(a,b)$ which, in our notation, corresponds to the unitary matrices $U'(a,b) = U_1(-\pi)U(a,b)U_1(\pi)$ and is calculated relative to ours by the similarity transformation $R'(a,b) = R_1(-\pi)R(a,b)R_1(\pi)$.

5.7 Rotations in 3D as Bilinear Transformations in the Complex Plane

A *bilinear* (or *fractional linear*) transformation is one of the form

$$x' = \frac{ax + b}{cx + d} \tag{5.69}$$

where $x, a, b, c, d \in \mathbb{C}$. When $ad - bc = 1$, this is called a *Möbius transformation*.

Henceforth the real part of these variables will be denoted with a subscript 1 and the complex part with a subscript 2, e.g., $a = a_1 + ia_2$. Expanding this expression and separating real and imaginary parts, we find

$$x_1' = \frac{(a_1x_1 - a_2x_2 + b_1)(c_1x_1 - c_2x_2 + d_1) + (b_2 + a_2x_1 + a_1x_2)(d_2 + c_2x_1 + c_1x_2)}{(c_1x_1 - c_2x_2 + d_1)^2 + (d_2 + c_2x_1 + c_1x_2)^2} \tag{5.70}$$

and

$$x_2' = \frac{(b_2 + a_2x_1 + a_1x_2)(c_1x_1 - c_2x_2 + d_1) - (a_1x_1 - a_2x_2 + b_1)(d_2 + c_2x_1 + c_1x_2)}{(c_1x_1 - c_2x_2 + d_1)^2 + (d_2 + c_2x_1 + c_1x_2)^2}. \tag{5.71}$$

Clearly this is a ratio of two second order polynomials in the variables x_1 and x_2.

Refering back to the equations of stereographic projection, $(4.24)-(4.25)$, we write

$$\mathbf{x} = \mathbf{f}(\mathbf{v}) \quad \text{and} \quad \mathbf{v} = \mathbf{g}(\mathbf{x})$$

where $\mathbf{v} = [v_1, v_2, v_3]^T$ and $\mathbf{x} = [x_1, x_2]^T$.

It can be verified that the composed operations

$$\mathbf{x}' = \mathbf{f}(R\mathbf{g}(\mathbf{x}))$$

also results in a ratio of two second order polynomials in the variables x_1 and x_2, where $R \in \mathbb{R}^{3 \times 3}$ is a matrix independent of $\mathbf{x}$ and $\mathbf{v}$. Expanding this expression and equating the components of $\mathbf{x}'$ defined in this way with those defined in (5.70- 5.71) imposes a system of constraint equations. When $|a|^2 + |b|^2 = 1$, and

$$x' = \frac{ax + b}{-\bar{b}x + \bar{a}} \tag{5.72}$$

(which is a special kind of Möbius transformation), the matrix R may be solved as

$$R = \begin{pmatrix} a_1^2 - a_2^2 - b_1^2 + b_2^2 & -2(a_1a_2 + b_1b_2) & 2(a_1b_1 - a_2b_2) \\ 2(a_1a_2 - b_1b_2) & a_1^2 - a_2^2 + b_1^2 - b_2^2 & 2(a_1b_2 + a_2b_1) \\ -2(a_1b_1 + a_2b_2) & -2(a_1b_2 - a_2b_1) & a_1^2 + a_2^2 - b_1^2 - b_2^2 \end{pmatrix}.$$

This is clearly related to the rotation matrices with Cayley-Klein parameters and Euler parameters.

5.8 Metrics on Rotations

In the literature, several "metrics" for computing distance between rotations have been presented. Before examining these, we review the standard techniques for measuring distance between positions, and the properties that a proper distance function should observe.

Definitions and Examples of Metrics

Definition: Given a set $\mathcal{S}$, a *metric* is a positive real-valued function, $d : \mathcal{S} \times \mathcal{S} \to \mathbb{R}$, which has the following properties for all $A, B, C \in \mathcal{S}$:

$$d(A, B) \geq 0 \quad \text{and} \quad d(A, B) = 0 \Longleftrightarrow A = B$$

$$d(A, B) = d(B, A) \tag{5.73}$$

$$d(A, B) + d(B, C) \geq d(A, C).$$

We refer to these as positive definiteness, symmetry and the triangle inequality, respectively. The pair $(\mathcal{S}, d)$ is called a *metric space*.

Note that we will be dealing with metrics on *sets of points* (e.g., measuring distance between any two points in $\mathbb{R}^N$) *and sets of sets of points* (e.g., measuring distance between any two sets of points in $\mathbb{R}^N$). When the distinction is required, we will use S to represent a set of points and $\mathcal{S}$ to represent a set of sets. It will be clear from the context when this distinction needs to be made.

The most common examples of metrics on $\mathbb{R}^N$ are of the form

$$d_p(\mathbf{x}, \mathbf{y}) = \sqrt[p]{\sum_{i=1}^{N} |x_i - y_i|^p} \qquad \text{for } p \in \{1, 2, \dots, \infty\},$$

and for the case $p = 1$ the root sign disappears.

The above metrics are usually denoted $d_p(\mathbf{x}, \mathbf{y}) = \|\mathbf{x} - \mathbf{y}\|_p$ where $\|\mathbf{x}\|_p$ is the p-norm of $\mathbf{x} \in \mathbb{R}^N$. Note that $d_1(\mathbf{x}, \mathbf{y})$ is the Manhattan or "taxicab" metric, $d_2(\mathbf{x}, \mathbf{y})$ is the Euclidean metric, and $d_\infty(\mathbf{x}, \mathbf{y}) = \max_i |x_i - y_i|$ is the infinity metric. Clearly, evaluation of each of these metrics requires $\mathcal{O}(N)$ computations for fixed p.

The following metric (called the *discrete* or *trivial* metric) can also be thought of as a degenerate case of $d_p(\cdot, \cdot)$ defined above (for $p = 0$ if we allow the nonstandard definition $0^0 \doteq 0$) when $\mathcal{S} \subset \mathbb{R}^N$,

$$d_0(A, B) = \begin{cases} 0 & A = B \\ 1 & A \neq B. \end{cases}$$

However, this metric is much more general. In fact, this metric can be defined on *any* set. Clearly it satisfies the metric properties independent of the structure of $\mathbb{R}^N$.

Below we present a result regarding combining metrics to form new ones.

Theorem 5.1. *Given metric spaces $(\mathcal{S}, d_1)$ and $(\mathcal{S}, d_2)$, the following are metric spaces: (a) $(\mathcal{S}, \alpha_1 d_1 + \alpha_2 d_2)$ for $\alpha_1, \alpha_2 \in \mathbb{R}_{>0}$; (b) $(\mathcal{S}, \max(d_1, d_2))$ where $\max(\cdot, \cdot)$ takes the value of the greater of the two arguments; (c) $(\mathcal{S}, f(d_1))$ where $f \in C^2(\mathbb{R}_{\geq 0})$ with $f(0) = 0$, $f'(x) > 0$, and $f''(x) \leq 0$ for all $x \in \mathbb{R}_{\geq 0}$.*

Proof. Let $A, B, C \in \mathcal{S}$.

(a) Clearly the properties of scalar multiplication and addition cause positive definiteness, symmetry, and the triangle inequality to be preserved when $\alpha_1, \alpha_2 \in \mathbb{R}_{>0}$.

(b) Again, positive definiteness and symmetry follow trivially. The triangle inequality follows because $d_1(A, C) \leq d_1(A, B) + d_1(B, C)$ and $d_2(A, C) \leq d_2(A, B) + d_2(B, C)$, and so

$$\max(d_1(A, C), d_2(A, C)) \leq \max(d_1(A, B) + d_1(B, C), d_2(A, B) + d_2(B, C))$$
$$\leq \max(d_1(A, B), d_2(A, B)) + \max(d_1(B, C), d_2(B, C)).$$

(c) Symmetry and positive definiteness hold from the fact that d is a metric and $f(d(A, A)) = f(0) = 0$, and since $f(x)$ always increases with increasing x, if $f(x) = 0$ this implied that $x = 0$. The triangle inequality holds because

$$f(d(A, B)) + f(d(B, C)) \geq f(d(A, B) + d(B, C)) \geq f(d(A, C))$$

where the first inequality is due to the fact that $f''(x) \leq 0$ and the second is due to the fact that $f'(x) > 0$ and $f(x) > 0$ for $x > 0$ (which follows from $f'(0) > 0$ and the other conditions imposed on $f(x)$ in the statement of the theorem).

In fact, the conditions in case (c) can be relaxed since $f(x)$ need not even be differentiable as long as it is positive and strictly nondecreasing with strictly nonincreasing slope. Concrete examples of acceptable functions $f(x)$ are $f(x) = \log(1 + x)$, $f(x) = x/(1 + x)$, $f(x) = \min(x, T)$ for some fixed threshold $T \in \mathbb{R}_{>0}$, and $f(x) = x^r$ for $0 < r < 1$.

5.8.1 Metrics on Rotations Viewed as Points in $\mathbb{R}^4$

Using the description of rotations as unit vectors in $\mathbb{R}^4$, the vector 2-norm of the previous subsection can be used to define [28]

$$d(q, r) = \sqrt{(q_1 - r_1)^2 + (q_2 - r_2)^2 + (q_3 - r_3)^2 + (q_4 - r_4)^2}.$$

Since the length of a vector calculated using the 2-norm is invariant under rotations, it follows that for any unit quaternion s,

$$d(s\,q, s\,r) \doteq d(R(s)q, R(s)r) = d(q, r).$$

5.8.2 Metrics Resulting from Matrix Norms

When rotations are described using 3×3 matrices, an intuitive metric takes advantage of the matrix 2-norm:
$$d(R_1, R_2) = \|R_1 - R_2\| \tag{5.74}$$
where

$$\|A\| = \sqrt{\text{tr}(AA^T)} = \left(\sum_{i,j=1}^{3} a_{ij}^2 \right)^{\frac{1}{2}}.$$

This satisfies the general properties of a metric, and it is easy to see that

$$d(R_1 R, R_2 R) = d(R_1, R_2).$$

Because of the general property $\text{tr}(RAR^T) = \text{tr}(A)$ for any $A \in \mathbb{R}^{3 \times 3}$ and $R \in SO(3)$, it follows that

$$d(RR_1, RR_2) = d(R_1, R_2).$$

These properties are called *right* and *left* invariance, respectively. They are useful because they allow us to write the distance between orientations as a function of the relative rotatio from one orientation to the other, without regard to the order:

$$d(R_1, R_2) = d(\mathbb{I}, R_1^{-1} R_2) = d(\mathbb{I}, R_2^{-1} R_1).$$

This is similar to the result when the norm of the difference of two unit quaternions is considered.

By definition, we can write (5.74) in the different form

$$d(R_1, R_2) = \sqrt{\text{tr}[(R_1 - R_2)(R_1 - R_2)^T]} = \sqrt{6 - 2\text{tr}(R_1^T R_2)},$$

(or equivalently $d(R_1, R_2) = \sqrt{6 - 2\text{tr}(R_2^T R_1)}$). Since $\text{tr}(R) = 2\cos\theta + 1$, where $R = e^{\theta V}$ for some $V \in so(3)$, we can write

$$d(R_1, R_2) = 2\sqrt{1 - \cos\theta_{1,2}}$$

where $|\theta_{1,2}| \leq \pi$ is the relative angle of rotation.

5.8.3 Geometry-Based Metrics

The metrics discussed thus far are, in a sense, not natural because they do not take advantage of the geometric structure of the space of the rotations. For example, if the space of rotations is identified with the upper hemisphere of a unit sphere in $\mathbb{R}^4$, then the above metrics measure the straight-line distance between points on the sphere, rather than along great arcs on its surface.

Using the unit-quaternion notation, the distance as measured on the unit hypersphere is given by the angle between the corresponding rotations described as unit vectors:

$$d(q, r) = \cos^{-1}(q_1 r_1 + q_2 r_2 + q_3 r_3 + q_4 r_4).$$

When q is the north pole, this gives

$$d(1, r) = \theta/2$$

where θ is the angle of rotation.

Likewise, we can simply define the distance between orientations to be the angle of rotation from one to the other:

$$d(R_1, R_2) = |\theta_{1,2}|,$$

where the angle of rotation is extracted from a rotation as

$$\theta(R) = \|(\log(R))^{\vee}\|.$$

5.8.4 Metrics Based on Dynamics

In some situations, such as spacecraft attitude control[8], the geometric definition of distance between orientations needs to be augmented so as to reflect the effort (power or fuel consumption) required to get to a final orientation from an initial one. In this case the inertial properties of the satellite should enter in the definition of distance. For instance, we could define distance between two orientations based on the minimum of the integral of kinetic energy required to move a rigid body from one orientation to another (where both orientations are static). This is written mathematically as

---

[8]See [15] for an introduction to this subject.

$$d(R_1, R_2) = \sqrt{\min_{R(t)\in SO(3)} \frac{1}{2} \int_{t_1}^{t_2} \mathrm{tr}(\dot{R}\,\mathcal{I}_R\,\dot{R}^T)\,dt} \tag{5.75}$$

subject to the boundary conditions $R(t_1) = R_1$ and $R(t_2) = R_2$. This is a variational calculus problem that can be handled using the methods described in Appendix F (and their generalizations). Problems like this arise in planning spacecraft attitude maneuvers [16]. The numerical calculation of $d(R_1, R_2)$ in (5.75) involves solving Euler's equations of motion together with the kinematic constraint relating angular velocity and the rotation matrix, subject to the boundary conditions stated above. In practice, a numerical "shooting" technique can be used.

5.9 Integration and Convolution of Rotation-Dependent Functions

In many of the applications that follow, starting in Chapter 12, we will be considering functions that take rotations as their arguments and return real values. It is natural to ask how the standard concepts of differentiation and integration defined in the context of curvilinear coordinates are extended to the case of the set of rotations.

We present an informal discussion now, with a more formal treatment following in Chapter 7. To begin, recall that the Jacobian matrix relates the rate of change of the parameters used to describe the rotation to the angular velocity of a rigid body:

$$\boldsymbol{\omega} = J(R(\mathbf{q}))\,\dot{\mathbf{q}}.$$

Depending on whether body or spatial coordinates are used, the angular velocity and Jacobian would be subscripted with R or L, respectively. The angular velocity can be viewed as existing in the tangent to the set of rotations, in analogy with the way a velocity vector of a particle moving on a surface moves instantaneously in the tangent plane to the surface. In analogy with the way area (or volume) is defined for a surface, volume for the set of rotations is defined relative to a metric tensor $G = J^T I_0 J$ for some positive definite I_0 by defining the volume element $\sqrt{\det(G)}\,dq_1 dq_2 dq_3$. Since J is 3×3, this reduces to a positive multiple of $|\det(J)|dq_1 dq_2 dq_3$. Hence we write the integral over the set of rotations as

$$\int_{SO(3)} f(R)\,d(R) = \frac{1}{V} \int_{\mathbf{q}\in Q} f(R(\mathbf{q}))\,|\det(J(R(\mathbf{q})))|\,dq_1 dq_2 dq_3.$$

In this equation it does not matter if J_R or J_L is used, since they are related by a rotation matrix, and their determinants are therefore the same. Q denotes the parameter space (which can be viewed as a subset of $\mathbb{R}^3$) and

$$V = \int_{\mathbf{q}\in Q} |\det(J(R(\mathbf{q})))|\,dq_1 dq_2 dq_3$$

is the volume as measured in this parameterization. The integral of a function on $SO(3)$ is normalized by the volume V so that the value of the integral does not depend on the parameterization or the choice of I, and

$$\int_{SO(3)} d(R) = 1.$$

5.9.1 Integration Measure for $SO(3)$ in Euler Angles and Exponential Coordinates

As a concrete example, we have for the case of ZXZ Euler angles:

$$\int_{SO(3)} f(R)\,d(R) = \frac{1}{8\pi^2} \int_0^{2\pi} \int_0^\pi \int_0^{2\pi} f(\alpha, \beta, \gamma) \sin\beta \, d\alpha d\beta d\gamma. \qquad (5.76)$$

For the other 3-parameter descriptions we have

$$d(R) = \frac{1 - \cos\|\mathbf{x}\|}{4\pi^2 \|\mathbf{x}\|^2} \, dx_1 dx_2 dx_2 = \frac{dr_1 dr_2 dr_3}{\pi^2 (1 + \|\mathbf{r}\|^2)^2} \qquad (5.77)$$

$$(5.78)$$

$$= \frac{1}{4\pi^2} s^2(\theta/2) s\nu \, d\theta d\lambda d\nu = \frac{4d\sigma_1 d\sigma_2 d\sigma_3}{\pi^2 (1 + \|\boldsymbol{\sigma}\|^2)^3}. \qquad (5.79)$$

The bounds of integration for these variables are: $\mathbf{r}, \boldsymbol{\sigma} \in \mathbb{R}^3$, $\mathbf{x}$ inside the ball of radius π in $\mathbb{R}^3$, and $-\pi \le \theta \le \pi$, $0 \le \nu \le \pi$, $0 \le \lambda \le 2\pi$. Choosing θ in the range covers $SO(3)$ twice. Alternatively, if $\theta \in [0, \pi]$ then the factor of $1/(4\pi^2)$ becomes $1/(2\pi^2)$.

The four-parameter descriptions will not be used in the context of integration in other chapters, but for completeness we write

$$d(R) = \frac{1}{8\pi^2} \delta(\|\mathbf{u}\|^2 - 1) \, du_1 du_2 du_3 du_4$$

$$= \frac{1}{8\pi^2} (1 - \cos\theta)\, \delta(\|\mathbf{n}\|^2 - 1) \, dn_1 dn_2 dn_3 d\theta$$

where $\delta(\cdot)$ is the Dirac delta function, and the bounds of integration are $\mathbf{u} \in \mathbb{R}^4$ and $(\mathbf{n}, \theta) \in \mathbb{R}^3 \times [0, 2\pi]$.

This definition of the integral of a function over the set of rotations has some useful properties. For instance, the integral of a function is invariant under "shifts" from the left and right by an arbitrary rotation. It is clear from the change of variables $A = R_1 R R_2$ that

$$\int_{SO(3)} f(R_1 R R_2)\,d(R) = \int_{SO(3)} f(A)\,d(R_1^{-1} A R_2^{-1}).$$

Using the fact that

$$|J(R_1^{-1} A R_2^{-1})| = |J(A)|$$

allows us to write

$$\int_{SO(3)} f(R_1 R R_2)\,d(R) = \int_{\mathbf{q} \in Q} f(A(\mathbf{q}))\, |\det(J(R_1^{-1} A(\mathbf{q}) R_2))| \, dq_1 dq_2 dq_3$$

$$= \int_{\mathbf{q} \in Q \subset \mathbb{R}^3} f(A(\mathbf{q}))\, |\det(J(A(\mathbf{q}))| \, dq_1 dq_2 dq_3 = \int_{SO(3)} f(A)\,d(A).$$

5.9.2 Integration Measure for $SO(3)$ Using the Parameterization $R(\mathbf{u}, \mathbf{v})\, e^{\theta \hat{\mathbf{u}}}$

Let $\mathbf{u} \in \mathbb{S}^2$ be a fixed vector, and let $(\mathbf{v}, \theta) \in \mathbb{S}^2 \times (-\pi, \pi)$ be variable. The volume element for $SO(3)$ expressed in the parameterization

$$Q = R(\mathbf{u}, \mathbf{v}) \, e^{\theta \hat{\mathbf{u}}} \tag{5.80}$$

is then, to within a normalizing constant,

$$dQ = d\mathbf{v} d\theta \,. \tag{5.81}$$

This result is derived as follows. First observe that

$$R(\mathbf{u}, \mathbf{e}_3) \, Q \, R(\mathbf{e}_3, \mathbf{u}) = R(\mathbf{e}_3, R(\mathbf{u}, \mathbf{e}_3)\mathbf{v}) \, e^{\theta E_3} \,.$$

This simplifies the problem to one of computing the volume element defined by θ and $\mathbf{v}'(\nu, \lambda) = R(\mathbf{u}, \mathbf{e}_3)\mathbf{v}$ where ν and λ are angles in unit sphere such that $d\mathbf{v}' = \sin \nu \, d\nu \, d\lambda$ ($\nu \in [0, \pi]$ and $\lambda \in [0, 2\pi]$).

If we can show that the volume element corresponding to $R(\mathbf{e}_3, \mathbf{v}') \, e^{\theta E_3}$ is $d\mathbf{v}' d\theta$, then the volume element corresponding to $R(\mathbf{u}, \mathbf{v}) \, e^{\theta \hat{\mathbf{u}}}$ will be $d\mathbf{v} d\theta$. Or, equivalently,

$$\int_{-\pi}^{\pi} \int_{\mathbb{S}^2} f\left(R(\mathbf{e}_3, \mathbf{v}') \, e^{\theta E_3}\right) d\mathbf{v}' d\theta = \int_{-\pi}^{\pi} \int_{\mathbb{S}^2} f\left(R(\mathbf{e}_3, R(\mathbf{u}, \mathbf{e}_3)\mathbf{v}) \, e^{\theta E_3}\right) d\mathbf{v} d\theta$$

$$= \int_{-\pi}^{\pi} \int_{\mathbb{S}^2} f\left(R(\mathbf{u}, \mathbf{v}) \, e^{\theta \hat{\mathbf{u}}}\right) d\mathbf{v} d\theta \,.$$

This is a consequence of two properties: (1) the invariance of integration over the sphere due to rotational shifts,

$$\int_{\mathbb{S}^2} F(R^T \mathbf{v}') \, d\mathbf{v}' = \int_{\mathbb{S}^2} F(\mathbf{v}) \, d(R\mathbf{v}) = \int_{\mathbb{S}^2} F(\mathbf{v}) \, d\mathbf{v} \,;$$

and (2) the bi-invariance of the Haar measure for $SO(3)$,

$$\int_{SO(3)} f(A \, Q \, A^T) \, dQ = \int_{SO(3)} f(Q) \, dQ \,.$$

In the current context, $\mathbf{v}' = R(\mathbf{u}, \mathbf{e}_3)\mathbf{v}$ and $A = R(\mathbf{u}, \mathbf{e}_3)$.

Therefore, (5.81) can be proven without loss of generality if it can be shown to be true when $\mathbf{u} = \mathbf{e}_3$. In this case, we have

$$R(\mathbf{e}_3, \mathbf{v}) = \begin{pmatrix} 1 - \frac{v_1^2}{1+v_3} & \frac{-v_1 v_2}{1+v_3} & v_1 \\ \frac{-v_1 v_2}{1+v_3} & 1 - \frac{v_2^2}{1+v_3} & v_2 \\ -v_1 & -v_2 & v_3 \end{pmatrix} \tag{5.82}$$

and its inverse is

$$R(\mathbf{v}, \mathbf{e}_3) = \begin{pmatrix} 1 - \frac{v_1^2}{1+v_3} & \frac{-v_1 v_2}{1+v_3} & -v_1 \\ \frac{-v_1 v_2}{1+v_3} & 1 - \frac{v_2^2}{1+v_3} & -v_2 \\ v_1 & v_2 & v_3 \end{pmatrix} \,. \tag{5.83}$$

If $\mathbf{v}$ is parameterized as $\mathbf{v}(\nu, \lambda)$ (i.e., spherical coordinates), the Jacobian is computed as

$$J_r(A(\mathbf{q})) = \left[\left(A^T \frac{\partial A}{\partial \nu} \right)^\vee, \left(A^T \frac{\partial A}{\partial \lambda} \right)^\vee, \left(A^T \frac{\partial A}{\partial \theta} \right)^\vee \right] \tag{5.84}$$

with $\mathbf{q} = [\nu, \lambda, \theta]^T$. Note that $J_l = A J_r$. Hence let us calculate J_r.

Specifically regarding J_r, we can use

$$A^T \frac{\partial A}{\partial q_i} = e^{-\theta \hat{e}_3} R(\mathbf{v}, \mathbf{e}_3) \frac{\partial}{\partial q_i} \left(R(\mathbf{e}_3, \mathbf{v}) e^{\theta \hat{e}_3} \right) \tag{5.85}$$

to calculate each column of the Jacobian matrix. Then we can easily see that

$$\left(A^T \frac{\partial A}{\partial \theta} \right)^\vee = \mathbf{e}_3. \tag{5.86}$$

Other column elements can be calculated by using the following relation

$$\left(e^{\theta \hat{e}_3} R(\mathbf{v}, \mathbf{e}_3) \frac{\partial}{\partial q_i} R(\mathbf{e}_3, \mathbf{v}) e^{\theta \hat{e}_3} \right)^\vee = e^{-\theta \hat{e}_3} \left(R(\mathbf{v}, \mathbf{e}_3) \frac{\partial R(\mathbf{e}_3, \mathbf{v})}{\partial q_i} \right)^\vee \tag{5.87}$$

Specifically,

$$\left(R(\mathbf{v}, \mathbf{e}_3) \frac{\partial R(\mathbf{e}_3, \mathbf{v})}{\partial \nu} \right)^\vee = \begin{pmatrix} -\sin\lambda \\ \cos\lambda \\ 0 \end{pmatrix} \tag{5.88}$$

and

$$\left(R(\mathbf{v}, \mathbf{e}_3) \frac{\partial R(\mathbf{e}_3, \mathbf{v})}{\partial \lambda} \right)^\vee = \begin{pmatrix} -\cos\lambda \sin\nu \\ -\sin\lambda \sin\nu \\ -2\sin^2 \frac{\nu}{2} \end{pmatrix} \tag{5.89}$$

Since $e^{-\theta \hat{e}_3} = R_3(-\theta)$, it follows that

$$J_r = \begin{pmatrix} \sin(\theta - \lambda) & -\cos(\theta - \lambda)\sin\nu & 0 \\ \cos(\theta - \lambda) & \sin(\theta - \lambda)\sin\nu & 0 \\ 0 & -2\sin^2 \frac{\nu}{2} & 1 \end{pmatrix}. \tag{5.90}$$

Then $|\det J_r| = \sin\nu$, which leads to the normalized Haar measure in these coordinates being

$$\int_{SO(3)} f(Q)\,dQ = \frac{1}{8\pi^2} \int_0^{2\pi} \int_0^\pi \int_0^{2\pi} f\left(R(\mathbf{e}_3, \mathbf{v}(\nu, \lambda)) e^{\theta E_3} \right) \sin\nu \, d\lambda d\nu d\theta$$

$$= \int_{\mathbb{S}^1} \int_{\mathbb{S}^2} f\left(R(\mathbf{e}_3, \mathbf{v}(\nu, \lambda)) e^{\theta E_3} \right) d\mathbf{v}\, d\theta.$$

Therefore,

$$\int_{SO(3)} f(Q)\,dQ = \int_{\mathbb{S}^1} \int_{\mathbb{S}^2} f\left(R(\mathbf{u}, \mathbf{v}(\nu, \lambda)) e^{\theta \hat{u}} \right) d\mathbf{v}\, d\theta. \tag{5.91}$$

5.9.3 Convolution of Functions on $SO(3)$

Given the ability to integrate functions on the set of rotations, a natural question to ask is if the concept of convolution also generalizes. The answer is most definitely yes, provided the types of functions being convolved are "nice" in the sense that

$$\int_{SO(3)} |f(R)|^2 d(R) < \infty.$$

In analogy with the case of functions on the circle or line, the set of such square-integrable functions is denoted as $\mathcal{L}^2(SO(3))$. Any two functions $f_1, f_2 \in \mathcal{L}^2(SO(3))$ are convolved as

$$(f_1 * f_2)(R) = \int_{SO(3)} f_1(Q) f_2(Q^{-1}R) \, d(Q). \tag{5.92}$$

It is easy to see with the change of variables $Q \to RQ^{-1}$ that

$$(f_1 * f_2)(R) = \int_{SO(3)} f_1(RQ^{-1}) f_2(Q) \, d(Q).$$

While these two forms are always equal, it is rarely the case that the convolution of functions on $SO(3)$ commute. That is, in general

$$(f_1 * f_2)(R) \neq (f_2 * f_1)(R).$$

However, for certain special functions, convolutions will commute.

5.10 Summary

In this chapter we presented a variety of different ways to describe rotations in three-dimensional space. These methods include 3×3 special orthogonal matrices parameterized with three or four variables, 2×2 special unitary matrices, unit quaternions, and 4×4 special orthogonal matrices. The relationships between these descriptions of rotation were also examined. In addition we addressed topics such as how to define distance between two rotations, and how to integrate of function of rotation (or orientation) over all rotations. For other perspectives on some of these topics see [1, 6, 8, 14, 21].

References

1. Altmann, S.L., *Rotations, Quaternions, and Double Groups*, Dover, 2005.
2. Angeles, J., *Rational Kinematics*, Springer-Verlag, New York, 1988.
3. Bottema, O., Roth, B., *Theoretical Kinematics*, Dover Publications, Inc., New York, reprinted 1990.
4. Cayley, A., "On the Motion of Rotation of a Solid Body," *Cambridge Mathematics Journal*, 3:224–232, 1843.
5. Chirikjian, G.S., "Closed-Form Primitives for Generating Locally Volume Preserving Deformations." *ASME Journal of Mechanical Design*, 117(3): 347–354, Sept. 1995.
6. Conway, J.H., Smith, D., *On Quaternions and Octonions*, A K Peters Ltd., Natick, MA, 2003.
7. Craig, J.J., *Introduction to Robotics, Mechanics and Control*, Addison-Wesley Publishing Company, Reading Mass., 1986.
8. Du Val, P., *Homographies, Quaternions, and Rotations*, Oxford University Press, 1964.
9. Euler, L., "Du Mouvement de Rotation des Corps Solides Autour d'un Axe Variable," *Mémoires de l'Académie des Sciences de Berlin*, 14: 154–193, 1758.
10. Euler, L., "Nova Methodus Motum Corporum Rigidorum Determinandi," *Novii Comentarii AcademiæScientiarum Petropolitanæ*, 20: 208–238, 1775/76.
11. Goldstein, H., *Classical Mechanics*, 2nd ed., Addison-Wesley Pub. Co., Reading, Mass. 1980.
12. Gürlebeck, K., Sprössig, W., *Quaternionic and Clifford Calculus for Physicists and Engineers*, John Wiley and Sons, New York, 1997.
13. Hamilton, Sir W.R., " On a New Species of Imaginary Quantities Connected with a Theory of Quaternions," *Proceedings of the Royal Irish Academy*, 2: 424–434, Nov. 1843.
14. Hanson, A., *Visualizing Quaternions* (The Morgan Kaufmann Series in Interactive 3D Technology), Elsevier, 2006.
15. Hughes, P.C., *Spacecraft Attitude Dynamics*, John Wiley and Sons, New York, 1986.
16. Junkins, J.L., Turner, J.D., *Optimal Spacecraft Rotational Maneuvers*, Elsevier, New York, 1986.
17. Kane, T.R., Levinson, D.A., *Dynamics: Theory and Applications*, McGraw-Hall Book Company, New York, 1985.
18. Koh, S. K., Ostrowski, J. P., and Ananthasuresh, G. K., "Control of Micro-Satellite Orientation Using Bounded-Input, Fully-Reversed MEMS Actuators," *International Journal of Robotics Research*, 21(5–6): 591–605, 2002.
19. Koh, S. K., Ananthasuresh, G. K., and Croke, C., "Analysis of Fully-Reversed Sequences of Non-Commutative Free-Body Rotations," *Journal of Mechanical Design*, 126(4): 609–616, 2004.
20. Koh, S., Chirikjian, G.S., Ananthasuresh, G.K., "A Jacobian-Based Algorithm for Planning Attitude Maneuvers Using Forward and Reverse Rotations," *ASME Journal of Computational and Nonlinear Dynamics*, 4(1): 1–12, Article 011012, Jan. 2009.

21. Kuipers, J.B., *Quaternions and Rotation Sequences*, Princeton University Press, 2002.
22. Li, C., McIntosh, A., Qian, T., "Clifford Algebras, Fourier Transforms, and Singular Convolution Operators on Lipschitz Surfaces," *Revista Matemática Iberoamericana*, 10(3): 665 – 721, 1994.
23. Marandi, S.R., Modi, V.J., "A Preferred Coordinate System and the Associated Orientation Representation in Attitude Dynamics," *Acta Astronautica*, 15(11): 833 – 843, 1987.
24. McCarthy, J.M., *Introduction to Theoretical Kinematics*, MIT Press, 1990. (revised and available as an iBook as of 2013.)
25. Miller, W., Jr., *Lie Theory and Special Functions*, Academic Press, New York, 1968.
26. Murray, R.M., Li, Z., Sastry, S.S., *A Mathematical Introduction to Robotic Manipulation*, CRC Press, Boca Raton, 1994.
27. Park, F.C., *The Optimal Kinematic Design of Mechanisms*, Ph.D. Thesis, Division of Engineering and Applied Sciences, Harvard University, Cambridge, MA 1991.
28. Ravani, B., Roth, B., "Motion Synthesis Using Kinematic Mapping," *ASME Journal of Mechanisms, Transmissions and Automation in Design*, 105(3): 460 – 467, Sept. 1983.
29. Rodrigues, O., "Des lois géométriques qui régissent les déplacements d'un système solide dans l'espace, et de la variation des coordonnées provenant de ces déplacements considérés independamment des causes qui peuvent les produire" *J. Mathématique Pures et Appliquées*, 5: 380 – 440, 1840.
30. Rooney, J., "A Survey of Representations of Spatial Rotation about a Fixed Point," *Environment and Planning B*, 4(2): 185 – 210, 1977.
31. Schaub, H., Tsiotras, P., Junkins, J.L., "Principal Rotation Representations of Proper $N \times N$ Orthogonal Matrices," *International Journal of Engineering Science*, 33(15): 2277 – 2295, Dec. 1995.
32. Shuster, M.D., "A Survey of Attitude Representations," *Journal of the Astronautical Sciences* 41(4): 439 – 517, 1993.
33. Sommen, F., "Plane Waves, Biregular Functions and Hypercomplex Fourier Analysis," *Proceedings of the 13th Winter School on Abstract Analysis*, Jan. 1985
34. Stuelpnagel, J.H., "On the Parameterization of the Three-Dimensional Rotation Group," *SIAM Review*, 6(4): 422 – 430, 1964.
35. Tsiotras, P., Longuski, J.M., "A New Parameterization of the Attitude Kinematics," *The Journal of the Astronautical Sciences*, 43(3): 243 – 262, July–Sept., 1995.
36. Yershova, A., Jain, S., LaValle, S., Mitchell, J.C., "Generating Uniform Incremental Grids on SO(3) Using the Hopf Fibration," *The International Journal of Robotics Research*, 29(7): 801 - 812, 2010.

Rigid-Body Motion

This chapter is concerned with characterizing the most general motions of a rigid body. Thus far we have considered spatial rotations, which form the subset of rigid-body motions which leave one point fixed. Clearly these are not the only rigid-body motions since a translation of the form $\mathbf{x}' = \mathbf{x} + \mathbf{b}$ preserves distance between points: $\|\mathbf{x}'_1 - \mathbf{x}'_2\|_p = \|\mathbf{x}_1 - \mathbf{x}_2\|_p$. Note that unlike rotations, translations leave many more general measures of distance invariant than $d_2(\mathbf{x}, \mathbf{y}) \doteq \|\mathbf{x} - \mathbf{y}\|_2$, although $\| \cdot \|_2$ is the one of greatest interest.

The following statements address what comprises the complete set of rigid-body motions.

Theorem 6.1. *(Chasles) [12]:*[1] *(1) Every motion of a rigid body can be considered as a translation in space and a rotation about a point; (2) Every spatial displacement of a rigid body can be equivalently affected by a single rotation about an axis and translation along the same axis.*

In modern notation, (1) is expressed by saying that every point $\mathbf{x}$ in a rigid body may be moved as

$$\mathbf{x}' = R\mathbf{x} + \mathbf{b} \tag{6.1}$$

where $R \in SO(3)$ is a rotation matrix, and $\mathbf{b} \in \mathbb{R}^3$ is a translation vector.

The pair $g = (R, \mathbf{b}) \in SO(3) \times \mathbb{R}^3$ describes both motion of a rigid body *and* the relationship between frames fixed in space and in the body. Furthermore, motions characterized by a pair $(R, \mathbf{b})$ could either describe the behavior of a rigid body or of a deformable object undergoing a rigid-body motion during the time interval for which this description is valid.

6.1 Composition of Motions

Consider a rigid-body motion which moves a frame originally coincident with the "natural" frame $(\mathbb{I}, \mathbf{0})$ to $(R_1, \mathbf{b}_1)$. Now consider a relative motion of the frame $(R_2, \mathbf{b}_2)$ with respect to the frame $(R_1, \mathbf{b}_1)$. That is, given any vector $\mathbf{x}$ defined in the terminal frame, it will look like $\mathbf{x}' = R_2\mathbf{x} + \mathbf{b}_2$ in the frame $(R_1, \mathbf{b}_1)$. Then the same vector will appear in the natural frame as

---

[1] Named after Michel Chasles (1793-1880)

$$\mathbf{x}'' = R_1(R_2\mathbf{x} + \mathbf{b}_2) + \mathbf{b}_1 = R_1 R_2 \mathbf{x} + R_1 \mathbf{b}_2 + \mathbf{b}_1.$$

The net effect of composing the two motions (or changes of reference frame) is equivalent to the definition

$$(R_3, \mathbf{b}_3) = (R_1, \mathbf{b}_1) \circ (R_2, \mathbf{b}_2) \doteq (R_1 R_2, R_1 \mathbf{b}_2 + \mathbf{b}_1). \tag{6.2}$$

From this expression, we can calculate the motion $(R_2, \mathbf{b}_2)$ which for any $(R_1, \mathbf{b}_1)$ will return the floating frame to the natural frame. All that is required is to solve $R_1 R_2 = \mathbb{I}$ and $R_1 \mathbf{b}_2 + \mathbf{b}_1 = \mathbf{0}$ for the variables R_2 and $\mathbf{b}_2$ and given R_1 and $\mathbf{b}_1$. The result is $R_2 = R_1^T$ and $\mathbf{b}_2 = -R_1^T \mathbf{b}_1$. Thus we denote the inverse of a motion as

$$(R, \mathbf{b})^{-1} = (R^T, -R^T \mathbf{b}). \tag{6.3}$$

This inverse, when composed either on the left or the right side of $(R, \mathbf{b})$, yields $(\mathbb{I}, \mathbf{0})$.

The set of all pairs $(R, \mathbf{b})$ together with the operation $\circ$ is denoted as $SE(3)$, which is shorthand for "special Euclidean transformations of three-dimensional space." The word "special" refers to the fact that $\det R = +1$.

Note that every rigid-body motion (element of $SE(3)$) can be decomposed into a pure translation followed by a pure rotation as

$$(R, \mathbf{b}) = (\mathbb{I}, \mathbf{b}) \circ (R, \mathbf{0}),$$

and every translation *conjugated*[2] by a rotation yields a translation:

$$(R, \mathbf{0}) \circ (\mathbb{I}, \mathbf{b}) \circ (R^T, \mathbf{0}) = (\mathbb{I}, R\mathbf{b}). \tag{6.4}$$

6.2 Homogeneous Transformation Matrices

It is of great convenience in many fields to *represent*[3] each rigid-body motion with a transformation matrix instead of a pair of the form $(R, \mathbf{b})$ and to use matrix multiplication in place of a composition rule.

This is achieved by assigning to each pair $(R, \mathbf{b})$ a unique 4×4 matrix

$$H(R, \mathbf{b}) = \begin{pmatrix} R & \mathbf{b} \\ \mathbf{0}^T & 1 \end{pmatrix}. \tag{6.5}$$

This is called a homogeneous transformation matrix, or simply a *homogeneous transform*. It is easy to see by the rules of matrix multiplication and the composition rule for rigid-body motions that

$$H((R_1, \mathbf{b}_1) \circ (R_2, \mathbf{b}_2)) = H(R_1, \mathbf{b}_1) H(R_2, \mathbf{b}_2).$$

Likewise, the inverse of a homogeneous transformation matrix represents the inverse of a motion:

$$H((R, \mathbf{b})^{-1}) = [H(R, \mathbf{b})]^{-1}.$$

In this notation, vectors in $\mathbb{R}^3$ are augmented by appending a "1" to form a vector

---

[2]Conjugation of a motion $g = (R, \mathbf{x})$ by a motion $h = (Q, \mathbf{y})$ is defined as $h \circ g \circ h^{-1}$.

[3]The word "representation" will take on a very precise and important mathematical meaning in subsequent chapters. For now the usual meaning of the word will suffice.

$$\mathbf{X} = \begin{pmatrix} \mathbf{x} \\ 1 \end{pmatrix}.$$

The following are then equivalent expressions:

$$\mathbf{Y} = H(R, \mathbf{b})\mathbf{X} \qquad \Longleftrightarrow \qquad \mathbf{y} = R\mathbf{x} + \mathbf{b}.$$

In the case of planar motion (viewed as a subset of spatial motion), we have

$$H(R(\mathbf{e}_3, \theta), \mathbf{b}) = \begin{pmatrix} \cos\theta & -\sin\theta & 0 & b_1 \\ \sin\theta & \cos\theta & 0 & b_2 \\ 0 & 0 & 1 & 0 \\ 0 & 0 & 0 & 1 \end{pmatrix}.$$

These are a lot of "0's" and "1's" to be carrying around, so it is common to reduce the transformation matrices for planar motion by removing the third row and column to yield

$$H(\theta, b_1, b_2) = \begin{pmatrix} \cos\theta & -\sin\theta & b_1 \\ \sin\theta & \cos\theta & b_2 \\ 0 & 0 & 1 \end{pmatrix}. \tag{6.6}$$

There is usually no confusion in distinguishing 3×3 and 4×4 transformation matrices, since the former is used exclusively in the case of planar motions, and the latter is used in the spatial case.

In analogy with rotations, we can express planar motions using the matrix exponential parameterization

$$\exp\begin{pmatrix} 0 & -\theta & v_1 \\ \theta & 0 & v_2 \\ 0 & 0 & 0 \end{pmatrix} = \begin{pmatrix} \cos\theta & -\sin\theta & [v_2(-1+\cos\theta)+v_1\sin\theta]/\theta \\ \sin\theta & \cos\theta & [v_1(1-\cos\theta)+v_2\sin\theta]/\theta \\ 0 & 0 & 1 \end{pmatrix}.$$

The logarithm of a planar motion is defined so that

$$\log\exp\begin{pmatrix} 0 & -\theta & v_1 \\ \theta & 0 & v_2 \\ 0 & 0 & 0 \end{pmatrix} = \begin{pmatrix} 0 & -\theta & v_1 \\ \theta & 0 & v_2 \\ 0 & 0 & 0 \end{pmatrix}.$$

As $\theta \to 0$, this expression is not singular from a mathematical perspective (since l'Hôpital's rule provides a clear limit), but care must be taken in numerical computations since there is a division by zero.

Alternative complex-number descriptions of planar motion exist as well, i.e., it is easy to see that the matrices

$$K_1(\theta, b_1, b_2) = \begin{pmatrix} e^{i\theta} & b_1 + ib_2 \\ 0 & 1 \end{pmatrix}$$

or

$$K_2(\theta, b_1, b_2) = \begin{pmatrix} e^{i\theta} & 0 & (b_2 - ib_1)/2 \\ 0 & e^{-i\theta} & (-b_2 - ib_1)/2 \\ 0 & 0 & 1 \end{pmatrix}$$

contain the same information as $H(\theta, b_1, b_2)$. It is easy to verify that the product of these matrices with matrices of the same kind results in matrices containing the information of the corresponding composed motions.

6.3 Screw Motions

The axis in the second part of Chasles' theorem is called the *screw axis*. It is a line in space about which a rotation is performed and along which a translation is performed.[4] As with any line in space, it is specified completely by a direction $\mathbf{n} \in \mathbb{S}^2$ and the position of any point $\mathbf{p}$ on the line. Given $(\mathbf{p}, \mathbf{n})$, a line is parameterized as

$$\mathbf{L}(t) = \mathbf{p} + t\mathbf{n}, \quad \forall \quad t \in \mathbb{R}.$$

Since there are an infinite number of vectors $\mathbf{p}$ on the line that can be chosen, the one which is "most natural" is that which has the smallest magnitude. This is the vector originating at the origin of the coordinate system and terminating at the line to which it intersects orthogonally. Hence the condition $\mathbf{p} \cdot \mathbf{n} = 0$ is satisfied. Since $\mathbf{n}$ is a unit vector and $\mathbf{p}$ satisfies a constraint equation, a line is uniquely specified by only four parameters. Often instead of the pair of line coordinates $(\mathbf{n}, \mathbf{p})$, the pair $(\mathbf{n}, \mathbf{p} \times \mathbf{n})$ is used to describe a line because this implicitly incorporates the constraint $\mathbf{p} \cdot \mathbf{n} = 0$. That is, when $\mathbf{p} \cdot \mathbf{n} = 0$, $\mathbf{p}$ can be reconstructed as $\mathbf{p} = \mathbf{n} \times (\mathbf{p} \times \mathbf{n})$, and it is clear that for unit $\mathbf{n}$, that the pair $(\mathbf{n}, \mathbf{p} \times \mathbf{n})$ has four degrees of freedom. Such a description of lines is called the *Plücker coordinates*. For more on this subject, and kinematics in general, see [2, 7, 8, 9, 34, 35, 39, 50, 63, 66, 68, 69, 70, 71].

Given an arbitrary point $\mathbf{x}$ in a rigid body, the transformed position of the same point after translation by d units along a screw axis with direction specified by $\mathbf{n}$ is $\mathbf{x}' = \mathbf{x} + d\mathbf{n}$. Rotation about the same screw axis is given as $\mathbf{x}'' = \mathbf{p} + e^{\theta N}(\mathbf{x}' - \mathbf{p})$.

Since $e^{\theta N}\mathbf{n} = \mathbf{n}$, it does not matter if translation along a screw axis is performed before or after rotation. Either way, $\mathbf{x}'' = \mathbf{p} + e^{\theta N}(\mathbf{x} - \mathbf{p}) + d\mathbf{n}$.

Another way to view this is that the homogeneous transforms

$$\text{trans}(\mathbf{n}, d) = \begin{pmatrix} \mathbb{I} & d\mathbf{n} \\ \mathbf{0}^T & 1 \end{pmatrix}$$

and

$$\text{rot}(\mathbf{n}, \mathbf{p}, \theta) = \begin{pmatrix} e^{\theta N} & (\mathbb{I} - e^{\theta N})\mathbf{p} \\ \mathbf{0}^T & 1 \end{pmatrix}$$

commute, and the homogeneous transform for a general rigid-body motion along screw axis $(\mathbf{n}, \mathbf{p})$ is given as

$$H(\mathbf{n}, \mathbf{p}, \theta, d) \doteq \text{rot}(\mathbf{n}, \mathbf{p}, \theta)\text{trans}(\mathbf{n}, d) = \text{trans}(\mathbf{n}, d)\text{rot}(\mathbf{n}, \mathbf{p}, \theta) = \begin{pmatrix} e^{\theta N} & (\mathbb{I} - e^{\theta N})\mathbf{p} + d\mathbf{n} \\ \mathbf{0}^T & 1 \end{pmatrix}.$$
$$(6.7)$$

A natural question to ask at this point is how the screw axis parameters $(\mathbf{n}, \mathbf{p})$ and motion parameters (θ, d) can be extracted from a given rigid displacement $(R, \mathbf{b})$. Since we have already done this for pure rotations, half the problem is already solved, i.e., $\mathbf{n}$ and θ are calculated from $R = e^{\theta N}$ as described in Chapter 5. What remains is to find for given R, $\mathbf{n}$, and $\mathbf{b}$ the variables $\mathbf{p}$ and d satisfying

$$(\mathbb{I} - R)\mathbf{p} + d\mathbf{n} = \mathbf{b} \quad \text{and} \quad \mathbf{p} \cdot \mathbf{n} = 0.$$

This is achieved by first taking the dot product of the left of the above equations with $\mathbf{n}$ and observing that $\mathbf{n} \cdot e^{\theta N}\mathbf{p} = \mathbf{n} \cdot \mathbf{p} = 0$, and so

---

[4]The theory of screws was developed by Sir Robert Stawell Ball (1840-1913) [2].

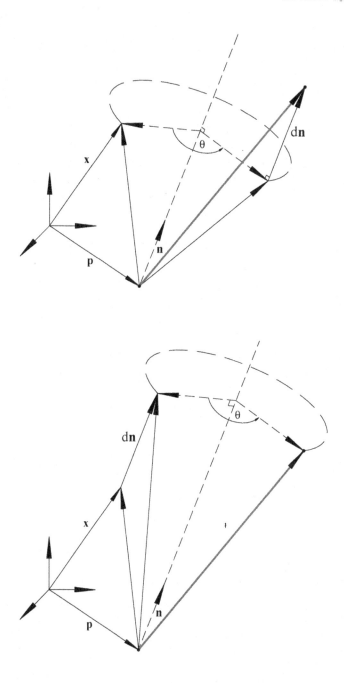

Fig. 6.1. A General Screw Transformation: (top) Rotation Followed by Translation; (bottom) Translation Followed by Rotation. The Net Effect is the Same.

$$d = \mathbf{b} \cdot \mathbf{n}.$$

We then can write

$$(\mathbb{I} - e^{\theta N})\mathbf{p} = \mathbf{b} - (\mathbf{b} \cdot \mathbf{n})\mathbf{n}. \tag{6.8}$$

Next for any $\mathbf{n} \in \mathbb{S}^2 - \{\mathbf{e}_3, -\mathbf{e}_3\}$, we introduce the unit vector

$$\mathbf{u} = \frac{1}{\sqrt{n_1^2 + n_2^2}} \begin{pmatrix} -n_2 \\ n_1 \\ 0 \end{pmatrix}, \tag{6.9}$$

where n_1 and n_2 are the first two components of $\mathbf{n}$. This is simply a choice we make that has the property $\mathbf{u} \cdot \mathbf{n} = 0$ and $\|\mathbf{u}\| = 1$. In this way, $\{\mathbf{n}, \mathbf{u}, \mathbf{n} \times \mathbf{u}\}$ forms a right-handed coordinate system. We can then write $\mathbf{p}$ in the form

$$\mathbf{p} = c_1 \mathbf{u} + c_2 (\mathbf{n} \times \mathbf{u}).$$

This can be substituted into (6.8) and projected onto $\mathbf{u}$ and $\mathbf{n} \times \mathbf{u}$ to yield the set of linear equations

$$\begin{pmatrix} 1 - \cos\theta & \sin\theta \\ -\sin\theta & 1 - \cos\theta \end{pmatrix} \begin{pmatrix} c_1 \\ c_2 \end{pmatrix} = \begin{pmatrix} \mathbf{b} \cdot \mathbf{u} \\ \mathbf{b} \cdot (\mathbf{n} \times \mathbf{u}) \end{pmatrix}, \tag{6.10}$$

which is solved for c_1 and c_2, provided $\theta \neq 0$. In this special case, any $\mathbf{p}$ in the plane normal to $\mathbf{n}$ will do.

A very important fact related to (6.7) is that for any $H_0 \in SE(3)$,

$$H_0 \, H(\mathbf{n}, \mathbf{p}, \theta, d) \, H_0^{-1} = H(\mathbf{n}', \mathbf{p}', \theta, d) \tag{6.11}$$

where $\mathbf{n}'$ and $\mathbf{p}'$ depend on H_0, but θ and d do not. In other words, θ and d are invariant under conjugation by arbitrary motions. $SE(2)$ and $SO(3)$ only have θ as an invariant. In these cases all of the matrix invariants only depend on θ as well. However, even though all of the *matrix invariants* of $H \in SE(3)$ depend only on θ, $SE(3)$ has the additional *group invariant* d which is independent of the eigenvalues of H. These invariants will be used in several ways in later chapters.

We have seen that every rigid body displacement can be viewed as a screw motion (translation along a line and rotation about the same line). Every rigid-body displacement may be viewed in other ways as well. For instance, given that every homogeneous transform can be written in the form of (6.7), the framed helix (helix with frames of reference attached at each point) that results from taking any real power of a homogeneous transform is

$$[H(\mathbf{n}, \mathbf{p}, \theta, d)]^\alpha = H(\mathbf{n}, \mathbf{p}, \alpha \cdot \theta, \alpha \cdot d).$$

Therefore, every rigid body in $\mathbb{R}^3$ is connected to the identity transformation by such a framed helix. Since a helix has constant curvature and torsion (and therefore has the same intrinsic geometry at every point), every helical path in $\mathbb{R}^3$ passing through the origin is given as a rotated version of the standard form of the helix that is given later in this chapter (see (6.27)).

In the planar case this helix degenerates to a circle, which for pure rotation shrinks to a point, and for pure translation expands to become a line.

For more theoretical issues in rigid-body kinematics, see [21, 22, 46].

0.4 Parameterization of Motions and Associated Jacobians

Any rigid-body motion can be parameterized by first parameterizing rotations in the ways that have already been discussed in Chapter 5, and then parameterizing translation in Cartesian, polar, spherical, or any other convenient curvilinear coordinate system. Any rigid-body motion in space can also be parameterized according to its six screw parameters: $(\mathbf{n}, \mathbf{p}, \theta, d)$ where $\mathbf{n} \cdot \mathbf{n} = 1$ and $\mathbf{n} \cdot \mathbf{p} = 0$. After describing this parameterization, we examine the relationship between differential changes in parameters and the corresponding infinitesimal spatial motions. Two other ways of describing rigid body motions in space which do not use homogeneous transforms are addressed in the final sections of this chapter.

6.4.1 The Matrix Exponential

In analogy with the way skew-symmetric matrices are exponentiated to result in rotations, we can exponentiate scalar multiples of "screw matrices" of the form

$$\Sigma = \begin{pmatrix} N & h\mathbf{n} \\ \mathbf{0}^T & 0 \end{pmatrix}$$

to generate rigid body motions. Here $N = -N^T$, $\mathbf{n} = N^{\vee}$, and $\mathbf{n} \cdot \mathbf{n} = 1$, and $h \in \mathbb{R}$ is a parameter called the *pitch* of the screw displacement. In analogy with a physical screw, the pitch of a screw displacement is the ratio of translation along the screw axis to the rotation about it, i.e., $h = d/\theta$.

The idea of exponentiating screw matrices of a more general form than those above to parameterize rigid-body motions was introduced into the engineering literature by Brockett [10]. It is easy to observe that for $m \geq 2$ [53]

$$\Sigma^m = \begin{pmatrix} N^m & \mathbf{0} \\ \mathbf{0}^T & 0 \end{pmatrix}.$$

Therefore, computing the exponential

$$e^{\theta \Sigma} = 1 + \theta \Sigma + \frac{1}{2!} \theta^2 \Sigma^2 + \cdots$$

results in

$$e^{\theta \Sigma} = \begin{pmatrix} e^{\theta N} & h\theta\mathbf{n} \\ \mathbf{0}^T & 1 \end{pmatrix}. \tag{6.12}$$

While this certainly represents a rigid-body motion, it does not parameterize all such motions. In analogy with rotations, where a general rotation $A(\mathbf{n}, \theta)$ is generated by similarity transformation of a rotation about $\mathbf{e}_3$, the above matrix exponential can be generalized to describe arbitrary rigid-body motions using a similarity transformation of the form

$$B e^{\theta \Sigma} B^{-1} = e^{\theta B \Sigma B^{-1}}.$$

Since the rotation part of the homogeneous transform in (6.12) is completely general, it is only the translational part that needs to be "fixed." This can be achieved by using the choice [53]

$$B = \begin{pmatrix} \mathbb{I} & \mathbf{n} \times \mathbf{r} \\ \mathbf{0}^T & 1 \end{pmatrix}$$

with $\mathbf{r} \cdot \mathbf{n} = 0$. This results in

$$B(\theta \Sigma)B^{-1} = \theta \begin{pmatrix} N & \mathbf{r} + h\mathbf{n} \\ \mathbf{0}^T & 0 \end{pmatrix},$$

which when exponentiated yields

$$B \left[\exp \theta \begin{pmatrix} N & h\mathbf{n} \\ \mathbf{0}^T & 0 \end{pmatrix} \right] B^{-1} = \exp \theta \begin{pmatrix} N & \mathbf{r} + h\mathbf{n} \\ \mathbf{0}^T & 0 \end{pmatrix}$$

$$= \begin{pmatrix} e^{\theta N} & (\mathbb{I} - e^{\theta N})(\mathbf{n} \times \mathbf{r}) + h\theta\mathbf{n} \\ \mathbf{0}^T & 1 \end{pmatrix},$$

where h is defined by $h\theta = d$. Letting $\mathbf{v} = \mathbf{r} + h\mathbf{n}$ and observing that $\mathbf{n} \times \mathbf{r} = \mathbf{n} \times \mathbf{v}$ and $h = \mathbf{n} \cdot \mathbf{v}$ then gives the following general expression for the $SE(3)$ exponential map:

$$\exp \theta \begin{pmatrix} N & \mathbf{v} \\ \mathbf{0}^T & 0 \end{pmatrix} = \begin{pmatrix} e^{\theta N} & (\mathbb{I} - e^{\theta N})(\mathbf{n} \times \mathbf{v}) + \theta\mathbf{n}\mathbf{n}^T\mathbf{v} \\ \mathbf{0}^T & 1 \end{pmatrix}. \tag{6.13}$$

In a similar way to how $\vee$ and $\widehat{}$ were defined to relate elements of $so(3)$ and $\mathbb{R}^3$, for any $\mathbf{n}, \mathbf{v} \in \mathbb{R}^3$ we can define

$$\begin{pmatrix} N & \mathbf{v} \\ \mathbf{0}^T & 0 \end{pmatrix}^{\vee} \doteq \begin{pmatrix} \mathbf{n} \\ \mathbf{v} \end{pmatrix} \quad \text{and} \quad \widehat{\begin{pmatrix} \mathbf{n} \\ \mathbf{v} \end{pmatrix}} = \begin{pmatrix} N & \mathbf{v} \\ \mathbf{0}^T & 0 \end{pmatrix}.$$

Though the same symbols are used to mean different things for pure rotations and for full spatial motions, the context makes the meaning clear.

The expression for the $SE(3)$ exponential map in (6.13) can be written concisely using ideas from [55] as [13]

$$\boxed{\exp(\theta \hat{\xi}) = \mathbb{I} + \theta\hat{\xi} + (1 - \cos\theta)\hat{\xi}^2 + (\theta - \sin\theta)\hat{\xi}^3} \tag{6.14}$$

where $\xi = [\mathbf{n}^T, \mathbf{v}^T]^T$.

If $\theta \in [0, \pi)$, it is possible to invert the exponential map, thereby defining log. By using the constraint $\mathbf{r} \cdot \mathbf{n} = 0$ it is possible to uniquely solve

$$(\mathbb{I} - e^{\theta N})(\mathbf{n} \times \mathbf{r}) + h\theta\mathbf{n} = \mathbf{b} \tag{6.15}$$

for $\mathbf{r}$ and h for any given $\mathbf{b}$, θ, and $\mathbf{n}$ by performing a construction like the one in Section 6.3. In the current case, we write $\mathbf{r} = r_1\mathbf{n} + r_2\mathbf{u} + r_3(\mathbf{n} \times \mathbf{u})$ where $h = \mathbf{b} \cdot \mathbf{n}/\theta$. The constraint $\mathbf{r} \cdot \mathbf{n} = 0$ forces $r_1 = 0$. There is no loss of generality in imposing this constraint because $h\theta$ is the component of $\mathbf{b}$ in the $\mathbf{n}$ direction.

Substituting into (6.15) and projecting onto $\mathbf{u}$ and $\mathbf{n} \times \mathbf{u}$ yields

$$\begin{pmatrix} \sin\theta & \cos\theta - 1 \\ 1 - \cos\theta & \sin\theta \end{pmatrix} \begin{pmatrix} r_2 \\ r_3 \end{pmatrix} = \begin{pmatrix} \mathbf{b} \cdot \mathbf{u} \\ \mathbf{b} \cdot (\mathbf{n} \times \mathbf{u}) \end{pmatrix}, \tag{6.16}$$

provided $\theta \neq 0$. Note that the matrix in (6.16) is different than that in (6.10) due to the $\mathbf{n} \times$ term in (6.15) which is not present in (6.8). In the special case when $\theta \to 0$, the motion approaches being a pure translation, when $h \to \infty$. Letting $h = \|\mathbf{b}\|/\theta$ and $\mathbf{r} = \mathbf{b}/\|\mathbf{b}\|$ solves the problem in this special case. If h is finite and $\theta \to 0$, then the identity (null) motion results. Thus, (6.13) is a parameterization of the entire set of rigid-body motions.

6.4.2 Infinitesimal Motions

For "small" motions the matrix exponential description is approximated well when truncated at the first two terms:

$$\exp\left[\begin{pmatrix} \Omega & \mathbf{v} \\ \mathbf{0}^T & 0 \end{pmatrix}\Delta t\right] \approx \mathbb{I} + \begin{pmatrix} \Omega & \mathbf{v} \\ \mathbf{0}^T & 0 \end{pmatrix}\Delta t. \tag{6.17}$$

Here $\Omega = -\Omega^T$ and $\Omega^\vee = \boldsymbol{\omega}$ describe the rotational part of the displacement. Since the second term in (6.17) consists mostly of zeros, it is common to extract the information necessary to describe the motion as

$$\begin{pmatrix} \Omega & \mathbf{v} \\ \mathbf{0}^T & 0 \end{pmatrix}^\vee = \begin{pmatrix} \boldsymbol{\omega} \\ \mathbf{v} \end{pmatrix}.$$

This six-dimensional vector is called an *infinitesimal* screw motion or *infinitesimal twist*. This $\vee$ operation, which produced a 6-dimensional vector from a 4×4 "infinitesimal screw matrix" of the form above, is not to be confused with $\vee$ applied to a 3×3 skew-symmetric matrix. The meaning of $\vee$ (and its inverse, $\hat{}$) will always be clear from the object to which it is applied.

Given a homogeneous transform

$$H(\mathbf{q}) = \begin{pmatrix} R(\mathbf{q}) & \mathbf{b}(\mathbf{q}) \\ \mathbf{0}^T & 0 \end{pmatrix}$$

parameterized with $(q_1, ..., q_6)$, which we write as a vector $\mathbf{q} \in \mathbb{R}^6$, we can express the homogeneous transform corresponding to a slightly changed set of parameters as the truncated Taylor series

$$H(\mathbf{q} + \delta\mathbf{q}) = H(\mathbf{q}) + \sum_{i=1}^{6} \Delta q_i \frac{\partial H}{\partial q_i}(\mathbf{q}).$$

This result can be shifted to the identity transformation by multiplying on the right or left by H^{-1} to define an equivalent relative infinitesimal motion. In this case we write

$$\begin{pmatrix} \boldsymbol{\omega}_L \\ \mathbf{v}_L \end{pmatrix} = \mathcal{J}_L(\mathbf{q})\dot{\mathbf{q}} \quad \text{where} \quad \mathcal{J}_L(\mathbf{q}) = \left[\left(\frac{\partial H}{\partial q_1}H^{-1}\right)^\vee, \cdots, \left(\frac{\partial H}{\partial q_6}H^{-1}\right)^\vee\right] \tag{6.18}$$

where $\boldsymbol{\omega}_L$ is defined as before in the case of pure rotation and

$$\mathbf{v}_L = -\boldsymbol{\omega}_L \times \mathbf{b} + \dot{\mathbf{b}}. \tag{6.19}$$

Similarly,

$$\begin{pmatrix} \boldsymbol{\omega}_R \\ \mathbf{v}_R \end{pmatrix} = \mathcal{J}_R(\mathbf{q})\dot{\mathbf{q}} \quad \text{where} \quad \mathcal{J}_R(\mathbf{q}) = \left[\left(H^{-1}\frac{\partial H}{\partial q_1}\right)^\vee, \cdots, \left(H^{-1}\frac{\partial H}{\partial q_6}\right)^\vee\right]. \tag{6.20}$$

Here

$$\mathbf{v}_R = R^T\dot{\mathbf{b}}.$$

The left and right Jacobian matrices are related as

$$\mathcal{J}_L(H) = [Ad(H)]\mathcal{J}_R(H). \tag{6.21}$$

where the matrix $[Ad(H)]$ is called the *adjoint*[5] and is written as [50, 53]

$$[Ad(H)] = \begin{pmatrix} R & \mathbb{O} \\ BR & R \end{pmatrix}. \tag{6.22}$$

The matrix B is skew-symmetric, and $B^\vee = \mathbf{b}$.

Jacobians when Rotations and Translations are Parameterized Separately

When the rotations are parameterized as $R = R(q_1, q_2, q_3)$ and the translations are parameterized using Cartesian coordinates $\mathbf{b}(q_4, q_5, q_6) = [q_4, q_5, q_6]^T$, we find that

$$\mathcal{J}_R = \begin{pmatrix} J_R & \mathbb{O} \\ \mathbb{O} & R^T \end{pmatrix} \quad \text{and} \quad \mathcal{J}_L = \begin{pmatrix} J_L & \mathbb{O} \\ BJ_L & \mathbb{I} \end{pmatrix}, \tag{6.23}$$

where J_L and J_R are the left and right Jacobians for the case of rotation.

6.5 Integration over Rigid-Body Motions

In general, the volume element with which to integrate over motions will be of the form[6]

$$d(H) = c \cdot |\det(\mathcal{J})| \, dq_1 \cdots dq_N$$

where $N = 3$ in the case of planar motion, and $N = 6$ in the case of spatial motion, and q_i are the parameters used to describe the motion. Here $\mathcal{J}$ is either $\mathcal{J}_L$ or $\mathcal{J}_R$; it does not matter which one is used. In both cases, $\det(\mathcal{J})$ reduces to a product of Jacobian determinants for the rotations and translations. When translations are described using Cartesian coordinates, $\det(\mathcal{J}) = \det(J)$ (the Jacobian determinant for rotations).

In particular, for the planar case we get to within an arbitrary constant factor

$$d(H(x_1, x_2, \theta)) = \frac{1}{2\pi} \, dx_1 dx_2 d\theta,$$

which is essentially the same as the volume element for $\mathbb{R}^3$. It is trivial to show that this is left and right invariant by direct calculation.

For the spatial case, we see the invariance of the volume element as follows. Right invariance follows from the fact that for any constant homogeneous transform

$$H_0 = \begin{pmatrix} R_0 & \mathbf{b}_0 \\ \mathbf{0}^T & 1 \end{pmatrix},$$

$$\mathcal{J}_L(HH_0) = \left[\left(\frac{\partial H}{\partial q_1} H_0 (HH_0)^{-1} \right)^\vee \cdots \left(\frac{\partial H}{\partial q_6} H_0 (HH_0)^{-1} \right)^\vee \right].$$

Since $(HH_0)^{-1} = H_0^{-1} H^{-1}$, and $H_0 H_0^{-1} = 1$, we have that $\mathcal{J}_L(HH_0) = \mathcal{J}_L(H)$.

The left invariance follows from the fact that

---

[5]We will be seeing this in a more general context in Chapter 7.

[6]The notation $d(H)$ used here will be simplified later as dH. The parentheses can be viewed as training wheels. The constant c is arbitrary. Sometimes it is chosen as $c = 1$, and sometimes it is chosen to be $1/Vol(SO(n))$.

$$\mathcal{J}_L(H_0 H) = \left[\left(H_0 \frac{\partial H}{\partial q_1} H^{-1} H_0^{-1} \right)^\vee \cdots \left(H_0 \frac{\partial H}{\partial q_6} H^{-1} H_0^{-1} \right)^\vee \right],$$

where

$$\left(H_0 \frac{\partial H}{\partial q_i} H^{-1} H_0^{-1} \right)^\vee = [Ad(H_0)] \left(\frac{\partial H}{\partial q_i} H^{-1} \right)^\vee.$$

Therefore,

$$\mathcal{J}_L(H_0 H) = [Ad(H_0)] \mathcal{J}_L(H).$$

But since $\det(Ad(H_0)) = 1$,

$$\det(J_L(H_0 H)) = \det(J_L(H)).$$

Explicitly, the volume element in the case of $SE(3)$ is found when the rotation matrix is parameterized using ZXZ or ZYZ Euler-Angles (α, β, γ) as

$$d(H(x_1, x_2, x_3, \alpha, \beta, \gamma)) = \frac{1}{8\pi^2} \sin \beta d\alpha d\beta d\gamma dx_1 dx_2 dx_3,$$

which is the product of the volume elements for $\mathbb{R}^3$ ($d\mathbf{x} = dx_1 dx_2 dx_3$), and for $SO(3)$ ($d(R) = \frac{1}{8\pi^2} \sin \beta d\alpha d\beta d\gamma$). Since $\beta \in [0, \pi]$, this is positive, except at the two points $\beta = 0$ and $\beta = \pi$, which constitute a set of zero measure and therefore does not contribute to the integral of singularity free functions.

If the above derivation is too mathematical, there is a physically intuitive way to show the invariance of the integration measure. Starting with the definition

$$d(H) = d(R)d(\mathbf{x}),$$

the left shift by the constant $H_0 = (R_0, \mathbf{b}_0)$ gives

$$d(H_0 H) = d(R_0 R)d(R_0 \mathbf{x} + \mathbf{b}_0) = d(R)d(R_0 \mathbf{x} + \mathbf{b}_0) = d(R)d(\mathbf{x}). \qquad (6.24)$$

The first equality follows from the invariance of the $SO(3)$ volume element. The second equality follows from the invariance of integration of functions of spatial position under rigid-body motions. That is, whenever $\int_{\mathbb{R}^N} f(\mathbf{x}) \, d\mathbf{x}$ exists, it is the case that

$$\int_{\mathbb{R}^N} f(R_0 \mathbf{x} + \mathbf{b}_0) \, d\mathbf{x} = \int_{\mathbb{R}^N} f(\mathbf{x}) \, d\mathbf{x}$$

for any $R_0 \in SO(3)$ and $\mathbf{b}_0 \in \mathbb{R}^3$ that do not depend on $\mathbf{x}$. Likewise, for right shifts,

$$d(H H_0) = d(R R_0)d(R \mathbf{b}_0 + \mathbf{x}) = d(R)d(R \mathbf{b}_0 + \mathbf{x}) = d(R)d(\mathbf{x}). \qquad (6.25)$$

The first equality again follows because of the invariance of the $SO(3)$ integration measure under shifts. And while $R\mathbf{b}_0$ is not constant, it is independent of $\mathbf{x}$, and so the same reasoning as before applies.

We note that unlike the case of the integration measure for the rotations, there is no unique way to scale the integration measure for rigid-body motions. This is due to the fact that translational motions can extend to infinity, and so it does not make sense to set the volume of the whole set of motions to unity as it did in the case of pure rotations. In the coming chapters, we will use different normalizations depending on the context so that equations take their simplest form.

Our motivation for studying the mathematical techniques reviewed in this section is that all of the problems stated in Chapter 1 involve convolutions of functions of rigid-body motion. The concept of convolution for rigid-body motions is exactly analogous to that discussed in the previous chapter for the case of rotations. For other applications of integration over rigid-body motions, see [57].

6.6 Assigning Frames to Curves and Serial Chains

In a variety of applications it is important to assign frames of reference (described using homogeneous transforms) to points along a curve or cascade of rigid links. Several methods for doing this, and their resulting properties, are explored in the sections that follow.

6.6.1 The Frenet-Serret Apparatus

To every smooth curve $\mathbf{x}(s) \in \mathbb{R}^3$, it is possible to define two scalar functions, termed *curvature* and *torsion*, which completely specify the shape of the curve without regard to how it is positioned and oriented in space. That is, curvature and torsion are intrinsic properties of the curve, and are the same for $\mathbf{x}(s)$ and $\mathbf{x}'(s) = R\mathbf{x} + \mathbf{b}$ for any $R \in SO(3)$ and $\mathbf{b} \in \mathbb{R}^3$. The most natural curve parameterization is by arclength s. The *tangent* of a curve $\mathbf{x}(s)$ parameterized by arclength is:

$$\mathbf{t}(s) = \frac{d\mathbf{x}}{ds}.$$

This is a unit vector satisfying $\mathbf{t}(s) \cdot \mathbf{t}(s) = 1$ because, by the definition of arclength, the equality

$$s = \int_0^s \|\mathbf{t}(\sigma)\| d\sigma \tag{6.26}$$

can only hold *for all* values of $s \in \mathbb{R}$ if the integrand is equal to unity. By defining the *normal* vector as

$$\mathbf{n} = \frac{1}{\kappa(s)} \frac{d\mathbf{t}}{ds}$$

when $\kappa(s) = \|d\mathbf{t}/ds\| \neq 0$, we observe that $\frac{d}{ds}(\mathbf{t}(s) \cdot \mathbf{t}(s)) = \frac{d}{ds}(1) = 0$ implies $\mathbf{t}(s) \cdot \frac{d\mathbf{t}}{ds} = 0$ and so

$$\mathbf{t}(s) \cdot \mathbf{n}(s) = 0.$$

Thus the tangent and normal vectors define two orthonormal vectors in three-dimensional space at each s for which $\kappa(s) \neq 0$. A right-handed frame of reference can be defined in these cases by defining a third vector, termed the *binormal*, as

$$\mathbf{b}(s) = \mathbf{t}(s) \times \mathbf{n}(s).$$

The frames of reference given by the positions $\mathbf{x}(s)$ and orientations $[\mathbf{t}(s), \mathbf{n}(s), \mathbf{b}(s)] \in SO(3)$ for all values of s parameterizing the curve are called the *Frenet frames* attached to the curve.

The *torsion* of the curve is defined as

$$\tau(s) \doteq -\frac{d\mathbf{b}(s)}{ds} \cdot \mathbf{n}(s)$$

and is a measure of how much the curve bends out of the $(\mathbf{t}, \mathbf{n})$- plane at each s.

A *right-circular helix* is a curve with constant curvature and torsion. Helices can be parameterized as

$$\mathbf{x}(s) = (r \cos as, r \sin as, has)^T \tag{6.27}$$

where $a = (r^2 + h^2)^{-1/2}$, and r and h are constants.

The information contained in the collection of Frenet frames, the curvature, and the torsion, is termed the Frenet-Serret apparatus.[7] Since the curvature and torsion completely specify the intrinsic geometry of a curve, it should be no surprise that the way the Frenet frames change along the curve depends on these functions. In particular, it can be shown that

$$\frac{d}{ds}\begin{pmatrix} \mathbf{t}(s) \\ \mathbf{n}(s) \\ \mathbf{b}(s) \end{pmatrix} = \begin{pmatrix} 0 & \kappa(s) & 0 \\ -\kappa(s) & 0 & \tau(s) \\ 0 & -\tau(s) & 0 \end{pmatrix}\begin{pmatrix} \mathbf{t}(s) \\ \mathbf{n}(s) \\ \mathbf{b}(s) \end{pmatrix}. \tag{6.28}$$

The vectors $\mathbf{t}$, $\mathbf{n}$, and $\mathbf{b}$ are treated like scalars when performing the matrix-vector multiplication on the right hand side of (6.28). This can be written in the different form

$$\frac{d}{ds}[\mathbf{t}(s), \mathbf{n}(s), \mathbf{b}(s)] = [\kappa(s)\mathbf{n}(s), -\kappa(s)\mathbf{t}(s) + \tau(s)\mathbf{b}(s), -\tau(s)\mathbf{n}(s)]$$
$$= -[\mathbf{t}(s), \mathbf{n}(s), \mathbf{b}(s)]\Lambda,$$

or

$$\frac{dQ_{FS}}{ds} = -Q_{FS}\Lambda$$

where Λ is the skew-symmetric matrix in (6.28) and $Q_{FS} = [\mathbf{t}(s), \mathbf{n}(s), \mathbf{b}(s)]$. The Frenet frame at each value of arclength is then $(Q_{FS}(s), \mathbf{x}(s))$.

6.6.2 Frames of Least Variation

The classical Frenet frames are only one of a number of techniques that can be used to assign frames of reference to space curves as explained by Bishop [3]. We now examine ways to assign frames which vary as little as possible along a curve segment. Two variants of this idea are presented here. First we examine how the Frenet frames should be "twisted" along the tangent to the curve for each value of arclength so as to result in a set of frames of least variation subject either to end constraints or or no end constraints. Next, we examine how frames of reference should be distributed along the length of the curve in an optimal way. This constitutes a reparameterization of a space curve. Finally, we combine the optimal set of frames and reparameterization. The tools from variational calculus used throughout this section are reviewed in Appendix F.

Optimally Twisting Frames

In order to determine the twist about the tangent of the Frenet frames which yields a minimally varying set of frames with orientation $Q(s)$ with $Q(s)\mathbf{e}_1 = \mathbf{t}(s)$, we seek an arclength-dependent angle, $\theta(s)$, such that

$$Q = \text{rot}[\mathbf{t}, \theta]\,Q_{FS} = Q_{FS}\,\text{rot}[\mathbf{e}_1, \theta]$$

minimizes the functional

$$J = \frac{1}{2}\int_0^1 \text{tr}\left(\frac{dQ}{ds}\frac{dQ}{ds}^T\right) ds. \tag{6.29}$$

Explicitly,

$$\frac{1}{2}\text{tr}\left(\frac{dQ}{ds}\frac{dQ}{ds}^T\right) = \kappa^2 + \tau^2 + 2\frac{d\theta}{ds}\mathbf{e}_1 \cdot \boldsymbol{\omega}_{FS} + \left(\frac{d\theta}{ds}\right)^2.$$

---

[7]Published independently by Frenet (1852) and Serret (1851)

Here we have used the definition

$$\boldsymbol{\omega}_{FS} = (-\Lambda)^\vee = \begin{pmatrix} \tau \\ 0 \\ \kappa \end{pmatrix},$$

and the properties of the trace, including the fact that for skew-symmetric matrices, Ω_1 and Ω_2,

$$\frac{1}{2}\mathrm{tr}\,(\Omega_1 \Omega_2) = \boldsymbol{\omega}_1 \cdot \boldsymbol{\omega}_2$$

where $\boldsymbol{\omega}_i = (\Omega_i)^\vee$. Clearly,

$$\frac{1}{2}\,\mathrm{tr}\left(\frac{dQ}{ds}\frac{dQ^T}{ds}\right) = \kappa^2 + \left(\tau + \frac{d\theta}{ds}\right)^2, \tag{6.30}$$

and the cost functional is minimized when

$$\frac{d\theta}{ds} = -\tau.$$

If, in addition, a specified twist relative to the Frenet frames is required at the end points, then the optimal solution is obtained by substituting (6.30) into the Euler-Lagrange equations with θ as the generalized coordinate. The solution is then

$$\theta(s) = c_1 + c_2 s - \int_0^s \tau(\sigma)\,d\sigma$$

where c_1 and c_2 are determined by the end conditions $\theta(0) = \theta_0$ and $\theta(1) = \theta_1$.

Optimal Reparameterization for Least Variation

In the optimal reparameterization problem, we seek to replace the arclength, s, with a curve parameter t along a unit length of curve such that $s(t)$ satisfies $s(0) = 0$ and $s(1) = 1$ and the cost functional

$$J = \int_0^1 \left\{ \frac{1}{2} r^2\,\mathrm{tr}\left(\frac{dQ(s(t))}{dt}\frac{dQ^T(s(t))}{dt}\right) + \left(\frac{ds}{dt}\right)^2 \right\} dt$$

is minimized. Here r is a length constant introduced to define the trade off between the cost of bending and extending. The integrand in this problem is clearly of the form $g(s)(s')^2$ where $s' = ds/dt$ and

$$g(s) = \frac{1}{2} r^2\,\mathrm{tr}\left(\frac{dQ}{ds}\frac{dQ^T}{ds}\right) + 1,$$

and so the optimal solution to this reparemetrization is that given in (F.8) of Appendix F with $y = s$ and $x = t$.

An Alternative to the Intrinsic Approach

Instead of describing curve properties (and twist about a curve) based on the Frenet frames, it is often convenient to have a description that does not degenerate when $\kappa = 0$. One way to do this is to consider arclength-parameterized curves that evolve as

$$\mathbf{x}(s) = \int_0^s Q(\sigma)\mathbf{e}_3 d\sigma.$$

Here $Q(s) \in SO(3)$ is the orientation of the curve frame at s. When $Q(s) = R_{ZXZ}(\alpha(s), \beta(s), \gamma(s))$, then α and β describe the orientation of the unit tangent vector, and γ specifies the twist of the frame about the tangent. The least varying twist (in the sense of minimizing the cost functional 6.29) subject to end constraints is of the form

$$\gamma(s) = c_1 + c_2 s - \int_0^s \cos\beta(\sigma)\alpha'(\sigma)\,d\sigma,$$

where α' is the derivative of α and c_1 and c_2 provide freedom to match the end constraints.

6.6.3 Global Properties of Closed Curves

We now present, without proof, some theorems relating the integrals of curvature and torsion of "nice" closed curves in $\mathbb{R}^3$ to global topological properties. The curves are all assumed to be smooth and self avoiding. The curve parameter, s, is taken to be arclength, and an integral with a little circle at its center denotes integration around the whole whole closed curve.

Theorem 6.2. *(Fenchel)[31]:*

$$\oint \kappa(s)\,ds \geq 2\pi \tag{6.31}$$

with equality holding only for some kinds of planar $(\tau(s) = 0)$ curves.

In contrast to the above theorem, we observe that for any closed smooth planar curve

$$\oint k(s)\,ds \cong 0 \bmod 2\pi$$

where $k(s)$ is the *signed curvature* of the curve, such that $|k(s)| = \kappa(s)$ with $k(s) > 0$ for counterclockwise bending and $k(s) < 0$ for clockwise bending.

Furthermore, any planar curve can be parameterized, up to a rigid-body displacement, as

$$\mathbf{x}(s) = \begin{pmatrix} \int_0^s \cos\left(\int_0^\sigma k(\nu)\,d\nu\right) d\sigma \\ \int_0^s \sin\left(\int_0^\sigma k(\nu)\,d\nu\right) d\sigma \end{pmatrix}. \tag{6.32}$$

Theorem 6.3. *(Fary-Milnor) [28, 52]: For closed space curves forming a knot*

$$\oint \kappa(s)\,ds \geq 4\pi. \tag{6.33}$$

Theorem 6.4. *(see [51, 64, 65]): If a closed curve is contained in the surface of the sphere $\mathbb{S}^2$, then*

$$\oint \tau(s)\,ds = \oint \frac{\tau(s)}{\kappa(s)}\,ds = 0. \tag{6.34}$$

Given two closed curves, $\mathbf{x}_1(s)$ and $\mathbf{x}_2(s)$, then the *Gauss integral* is a functional defined as

$$G(\mathbf{x}_1, \mathbf{x}_2) \doteq \frac{1}{4\pi} \oint_{C_1} ds_1 \oint_{C_2} ds_2 [\dot{\mathbf{x}}_1(s_1) \times \dot{\mathbf{x}}_2(s_2)] \cdot \frac{\mathbf{x}_1(s_1) - \mathbf{x}_2(s_2)}{\|\mathbf{x}_1(s_1) - \mathbf{x}_2(s_2)\|^3} \tag{6.35}$$

where ˙ is shorthand for d/ds.

Gauss showed that this integral is a topological invariant in the sense that its value only depends on the degree to which the curves intertwine. It is also called the *linking number* of the two curves, and the notation $Lw = G(\mathbf{x}_1, \mathbf{x}_2)$ is common.

Suppose we are given a closed curve of unit length, $\mathbf{x}(s)$. Then a *ribbon* (or strip) associated with this backbone curve is any smoothly evolving set of line segments of fixed length $2r$ for $0 \le s \le 1$ with centers at $\mathbf{x}(s)$, such that the line segments point in a direction in the plane normal to $\dot{\mathbf{x}}(s)$, and such that the tips of the line segments trace out closed curves. The tips of the ribbon can be described using the Frenet-Serret apparatus as the two curves

$$\mathbf{x}_\pm(s) = \mathbf{s} \pm r\mathbf{v}(s)$$

where

$$\mathbf{v}(s) = \mathbf{n}(s)\cos\theta(s) + \mathbf{b}(s)\sin\theta(s)$$

and $\theta(0) = \theta(1)$.

When r is sufficiently small, it is useful to represent the linking number into the sum of two quantities: the *writhe* (or writhing number), denoted as Wr, and the *twist* (or twisting number), denoted as Tw. It has been shown [11, 60, 61, 75] that the linking number of $\mathbf{x}$ and $\mathbf{x}_+$ is

$$Lw(\mathbf{x}, \mathbf{x} + r\mathbf{v}) = Wr(\mathbf{x}) + Tw(\mathbf{x}, \mathbf{v}) \tag{6.36}$$

(or more simply $Lw = Wr + Tw$) where

$$Wr = \frac{1}{4\pi} \oint ds_1 \oint ds_2 [\dot{\mathbf{x}}(s_1) \times \dot{\mathbf{x}}(s_2)] \cdot \frac{\mathbf{x}(s_1) - \mathbf{x}(s_2)}{\|\mathbf{x}(s_1) - \mathbf{x}(s_2)\|^3}$$

and

$$Tw = \frac{1}{2\pi} \oint \dot{\mathbf{x}}(s) \cdot [\mathbf{v}(s) \times \dot{\mathbf{v}}(s)] / \|\dot{\mathbf{x}}(s)\| ds.$$

We note that in the case of an arc-length-parameterized curve $\|\dot{\mathbf{x}}(s)\| = 1$, and when $\theta(s) = 0$ then

$$Tw = \frac{1}{2\pi} \oint \tau(s) \, ds.$$

When $\theta(s) = \alpha(s)$ (the optimal twist distribution), then $Tw = 0$. For a simple (non-self-intersecting) closed curve in the plane or on the surface of a sphere, it has been shown [32] that $Wr = 0$.

For any fixed unit vector $\mathbf{u}$ not parallel to the tangent to the curve $\mathbf{x}(s)$ for any value of s, the *directional writhing number* is defined as [32]:

$$Wr(\mathbf{x}, \mathbf{u}) \doteq Lk(\mathbf{x}, \mathbf{x} + \epsilon\mathbf{u})$$

where for sufficiently small ϵ, the value of the directional writhing number is independent of ϵ. The writhe can be calculated from the directional writhing number by integrating over all directions not parallel to the tangent of $\mathbf{x}$. This amounts to integration over the sphere (except at the set of measure zero where the tangent traces out a curve on the surface of the sphere) and so [32]

$$Wr(\mathbf{x}) = \int_{\mathbb{S}^2} Wr(\mathbf{x}, \mathbf{u}) \, d\mathbf{u}.$$

(Here the integral is normalized so that $\int_{\mathbb{S}^2} d\mathbf{u} = 1$.)

We mention these relationships because they play an important role in the study of DNA topology [72]. Chapter 17 discusses statistical mechanics of macromolecules in some detail.

6.6.4 Frames Attached to Serial Linkages

It is interesting to note that while the Frenet-Serret apparatus has been known for over a hundred and fifty years, methods for assigning frames of reference to chains with discrete links is a far newer problem. Two examples of this are the the Denavit-Hartenberg framework in robotics, and the analogous formulation in polymer science and biophysics. We provide a short review of these formulations below. See references [19], [20], and [30] for detailed explainations as to how frames are uniquely attached to serial chain robot arms and molecules.

The Denavit-Hartenberg Parameters in Robotics

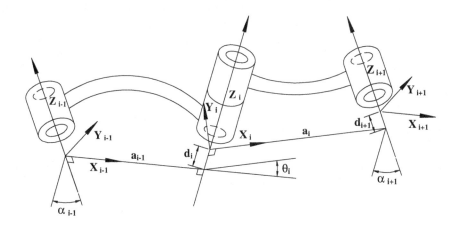

Fig. 6.2. The Denavit-Hartenberg Frames

The Denavit-Hartenberg (D-H) framework is a method for assigning frames of reference to a serial robot arm constructed of sequential rotary joints connected with rigid links. If the robot arm is imagined at any fixed time, the axes about which the joints turn are viewed as lines in space. In the most general case, these lines will be skew, and in degenerate cases they can be parallel or intersect.

In the D-H framework (Figure 6.2), a frame of reference is assigned to each link of the robot at the joint where it meets the previous link. The z-axis of the ith D-H frame points along the ith joint axis. Since a robot arm is usually attached to a base, there is no ambiguity in terms of which of the two ($\pm$) directions along the joint axis should be chosen, i.e., the "up" direction for the first joint is chosen. Since the $(i+1)^{st}$ joint axis in space will generally be skew relative to axis i, a unique x-axis is assigned to frame i, by defining it to be the unit vector pointing in the direction of the shortest line segment from axis i to axis $i+1$. This segment intersects both axes orthogonally. In addition to completely defining the relative orientation of the ith frame relative to the $(i-1)^{st}$, it also provides the relative position of the origin of this frame.

The D-H parameters, which completely specify this model, are:

- The distance from joint axis i to axis $i+1$ as measured along their mutual normal. This distance is denoted as a_i.

- The angle between the projection of joint axes i and $i+1$ in the plane of their common normal. The sense of this angle is measured counterclockwise around their mutual normal originating at axis i and terminating at axis $i+1$. This angle is denotes α_i.
- The distance between where the common normal of joint axes $i-1$ and i, and that of joint axes i and $i+1$ intersect joint axis i, as measured along joint axis i. This is denoted as d_i.
- The angle between the common normal of joint axes $i-1$ and i, and the common normal of joint axes i and $i+1$. This is denoted as θ_i, and has positive sense when rotation about axis i is counterclockwise.

Hence, given all the parameters $\{a_i, \alpha_i, d_i, \theta_i\}$ for all the links in the robot, together with how the base of the robot is situated in space, the geometry of the arm at any fixed time can be completely specified. Generally, θ_i is the only parameter that depends on time.

In order to solve the *forward kinematics* problem, which is to find the position and orientation of the distal end of the arm relative to the base, the homogeneous transformations of the relative displacements from one D-H frame to another are multiplied sequentially. This is written as

$$H_N^0 = H_1^0 H_2^1 \cdots H_N^{N-1}.$$

It is clear from Figure 6.2 that the the relative transformation, H_i^{i-1}, from frame $i-1$ to frame i is performed by first rotating about the x-axis of frame $i-1$ by α_{i-1}, then translating along this same axis by a_{i-1}. Next we rotate about the z-axis of frame i by θ_i and translate along the same axis by d_i. Since all these transformations are relative, they are multiplied sequentially on the right as rotations (and translations) about (and along) natural basis vectors. Furthermore, since the rotations and translations appear as two screw motions (translations and rotations along the same axis), we write

$$H_i^{i-1} = \text{Screw}(\mathbf{e}_1, a_{i-1}, \alpha_{i-1})\text{Screw}(\mathbf{e}_3, d_i, \theta_i),$$

where in this context

$$\text{Screw}(\mathbf{v}, c, \gamma) = \text{rot}(\mathbf{v}, \mathbf{0}, \gamma)\text{trans}(\mathbf{v}, c).$$

Explicitly,

$$H_i^{i-1}(a_{i-1}, \alpha_{i-1}, d_i, \theta_i) =$$

$$\begin{pmatrix} \cos\theta_i & -\sin\theta_i & 0 & a_{i-1} \\ \sin\theta_i \cos\alpha_{i-1} & \cos\theta_i \cos\alpha_{i-1} & -\sin\alpha_{i-1} & -d_i \sin\alpha_{i-1} \\ \sin\theta_i \sin\alpha_{i-1} & \cos\theta_i \sin\alpha_{i-1} & \cos\alpha_{i-1} & d_i \cos\alpha_{i-1} \\ 0 & 0 & 0 & 1 \end{pmatrix}.$$

Here we have used the conventions in [19]. For other alternatives, see [67].

Polymer Kinematics

A polymer is a long chain of repeated chemical units, where the bonds between units can be modeled as a rotary joint. The $(i+1)^{st}$ joint axis intersects axis i at a distance l_i from where joint axis i is intersected by joint axis $i-1$. Polymer chains can be modeled using the D-H framework (as a degenerate case where the joint axes intersect). In polymer science, the joint (bond) axes are labeled as the x-axis, and the angle between joint axis i and $i + 1$ is denoted as θ_i. The z-axis is defined normal to the plane of joint axes i and $i + 1$, and the angle of rotation about bond i is denoted as ϕ_i. The corresponding homogeneous transformation matrix for one such convention is

$$H_{i+1}^i = \text{Screw}(\mathbf{e}_1, l_i, \phi_i)\text{Screw}(\mathbf{e}_3, 0, \theta_i).$$

Explicitly,

$$H_{i+1}^i(l_i, \phi_i, 0, \theta_i) = \begin{pmatrix} \cos\theta_i & -\sin\theta_i & 0 & l_i \\ \sin\theta_i\cos\phi_i & \cos\theta_i\cos\phi_i & -\sin\phi_i & 0 \\ \sin\theta_i\sin\phi_i & \cos\theta_i\sin\phi_i & \cos\phi_i & 0 \\ 0 & 0 & 0 & 1 \end{pmatrix}. \tag{6.37}$$

This assumes a convention for the sense of angles and axes consistent with the D-H framework outlined in the previous subsection. The above convention is the same (to within a shifting of axis labels by one place) as those presented in [26, 54].

Other conventions for the signs of angles and directions of the axes in the context of polymers are explained in [30] (p. 387), [48] (p. 112). In order to define angles and coordinates so as to be consistent with [30], we can transform (6.37) by setting $\phi_i \to \phi_i - \pi$ and performs a similarity transformation with respect to the matrix

$$S = \text{rot}(\mathbf{e}_1, \pi) = \begin{pmatrix} 1 & 0 & 0 & 0 \\ 0 & -1 & 0 & 0 \\ 0 & 0 & -1 & 0 \\ 0 & 0 & 0 & 1 \end{pmatrix}.$$

Therefore,

$$\mathcal{H}_i = SH_{i+1}^i(l_i, \phi_i - \pi, 0, \theta_i)S^{-1} = \begin{pmatrix} \cos\theta_i & \sin\theta_i & 0 & l_i \\ \sin\theta_i\cos\phi_i & -\cos\theta_i\cos\phi_i & \sin\phi_i & 0 \\ \sin\theta_i\sin\phi_i & -\cos\theta_i\sin\phi_i & -\cos\phi_i & 0 \\ 0 & 0 & 0 & 1 \end{pmatrix}.$$

The notation

$$\mathcal{H}_i = \begin{pmatrix} T_i & \mathbf{l}_i \\ \mathbf{0}^T & 1 \end{pmatrix}$$

is used in Chapter 17.

The angle ϕ_i is called the ith *torsion angle*, and the angle θ_i is related to the ith *bond angle* as $\pi - \theta_i$.

Several relationships exist between continuous curves and discrete chains with zero offset (such as polymer chains). Consider the planar N-link chain shown in Figure 6.3 where the total length is normalized so that $\sum_{i=1}^{N} L_i = 1$.

The end position and orientation of this chain with respect to its base are given as

$$\mathbf{x}_{end} = \begin{pmatrix} \sum_{i=1}^{N} L_i \cos\left(\sum_{j=1}^{i}\theta_j\right) \\ \sum_{i=1}^{N} L_i \sin\left(\sum_{j=1}^{i}\theta_j\right) \end{pmatrix}$$

and

$$\theta_{end} = \sum_{i=1}^{N}\theta_i.$$

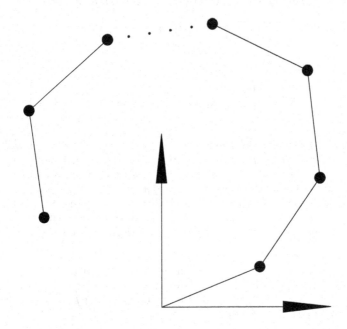

Fig. 6.3. A Planar N-Link Serial Chain

This example can be modeled using either the DH or polymer frames as a degenerate case. What is perhaps less obvious is that it can also be described as the limiting case of a curve with signed curvature

$$k(s) = \sum_{i=1}^{N} \theta_i \delta \left(s - \sum_{j=0}^{i-1} L_j \right)$$

where by definition $L_0 = 0$. It can be verified that substitution of this signed curvature into (6.32) results in $\mathbf{x}_{end} = \mathbf{x}(1)$ and $\theta_{end} = \int_0^1 k(s) \, ds$. From another perspective, as $N \to \infty$, and as $\theta_i \to 0$, the discrete linkage becomes a smooth curve with $\theta_i = (1/N) \cdot k(i/N)$. Both of these observed relationships between discrete chains and continuous curves have been useful in the kinematics and motion planning of snake-like ("hyper-redundant") manipulator arms [14].

In the case of a polymer, a similar extension of the Frenet-Serret apparatus is

$$\kappa(s) = \sum_{i=1}^{N} |\theta_i| \delta \left(s - \sum_{j=0}^{i-1} L_j \right),$$

and

$$|\tau(s)| = \sum_{i=1}^{N} |\phi_i| \delta \left(s - \sum_{j=0}^{i-1} L_j \right).$$

Hence, relationships such as the Fary-Milnor theorem provide the necessary conditions for a discrete linkage to form a knot

$$\sum_{i=1}^{N} |\theta_i| \geq 4\pi,$$

or if all the links have the same bond angle, θ_0, then the minimal number of links that a chain must posess before it can possibly form a knot is

$$N = 1 + \text{int}(4\pi/\theta_0)$$

(where $\text{int}(x)$ is the greatest integer less than or equal to x).

6.7 Dual Numbers

In Chapter 5 rotations were described in a number of equivalent ways including 3×3 orthogonal matrices, 2×2 unitary matrices, and unit quaternions. Each of these reductions in dimension requires a more sophisticated number system, i.e., real numbers to complex numbers to quaternions.

In analogy with this, the "dual numbers" are used in kinematics so that rigid-body motions can be expressed more compactly than with 4×4 matrices (see, e.g., [29]). In analogy with the imaginary number $i^2 = -1$, a dual number, $\hat{\epsilon}$, is defined so that $\hat{\epsilon} \neq 0$ and $\hat{\epsilon}^2 = 0$. Of course no real number satisfies these conditions, but taking these as defining properties of $\hat{\epsilon}$ allows us to define a new arithmetic of dual numbers. For instance, calculating the product of two numbers with real and dual parts results in:

$$(a_1 + \hat{\epsilon}a_2)(b_1 + \hat{\epsilon}b_2) = a_1b_1 + \hat{\epsilon}(b_1a_2 + a_1b_2).$$

The term $\hat{\epsilon}^2(a_2b_2)$ that would appear in usual arithmetic disappears by definition of $\hat{\epsilon}$. Note that scalar multiplication using dual numbers is equivalent in some sense to multiplying 2×2 matrices with special structure:

$$\begin{pmatrix} a_1 & a_2 \\ 0 & a_1 \end{pmatrix} \begin{pmatrix} b_1 & b_2 \\ 0 & b_1 \end{pmatrix} = \begin{pmatrix} a_1b_1 & b_1a_2 + a_1b_2 \\ 0 & a_1b_1 \end{pmatrix}.$$

We can "dualize" real, complex, or quaternion-valued vectors and matrices by adding another of the same kind of quantity multiplied by a dual number.

We can also discuss real-valued functions of a dual-number-valued argument. For instance, if $f(x)$ is a smooth function and hence has a Taylor series that exists and is convergent for all $x \in \mathbb{R}$, then we can evaluate the same function at $\hat{x} = x_0 + \hat{\epsilon}x_1$ for $x_0, x_1 \in \mathbb{R}$ by using its Taylor series about x_0:

$$f(\hat{x}) = f(x_0 + \hat{\epsilon}x_1) = f(x_0) + \hat{\epsilon}x_1 f'(x_0) + (\hat{\epsilon}x_1)^2 f''(x_0)/2 + \cdots$$
$$= f(x_0) + \hat{\epsilon}x_1 f'(x_0). \tag{6.38}$$

The above equality holds because $\hat{\epsilon}^2 = 0$, and so all higher powers of $\hat{\epsilon}$ are also zero. As we will see later, an example of this with particular importance in kinematics is

$$\sin\hat{\theta} = \sin\theta + \hat{\epsilon}\,d\cos\theta \qquad (6.39)$$

$$\cos\hat{\theta} = \cos\theta - \hat{\epsilon}\,d\sin\theta \qquad (6.40)$$

where $\hat{\theta} = \theta + \hat{\epsilon}\,d$.

By using the concept of dual numbers, we can describe a rigid-body motion in space using a 3×3 matrix of real and dual numbers, or a 2×2 matrix of complex and dual numbers, or by using a dual quaternion. Each of these different ways of describing motion in space is examined below.

6.7.1 Dual Orthogonal Matrices

A dual orthogonal matrix is constructed to contain all the information concerning a rigid-body displacement in a 3×3 matrix instead of a homogeneous transform. In particular, given a rigid-body displacement $(R, \mathbf{b})$, the corresponding dual orthogonal matrix is

$$\hat{R} = R + \hat{\epsilon}BR$$

where B is defined to be the skew-symmetric matrix satisfying $B\mathbf{x} = \mathbf{b} \times \mathbf{x}$ for any $\mathbf{x} \in \mathbb{R}^3$. It is clear from the usual rules of matrix multiplication, the fact that $B^T = -B$, and the properties of dual-number arithmetic, that

$$\hat{R}\hat{R}^T = (R + \hat{\epsilon}BR)(R^T - \hat{\epsilon}R^T B) = \mathbb{I}.$$

Furthermore, the product of dual orthogonal matrices results in a dual orthogonal matrix as:

$$\hat{R}_1\hat{R}_2 = R_1 R_2 + \hat{\epsilon}(R_1 B_2 R_1^T + B_1)R_1 R_2.$$

Since $R_1 B_2 R_1^T$ is the skew-symmetric matrix corresponding to the vector $R_1\mathbf{b}_2$, we see that the skew-symmetric matrix $R_1 B_2 R_1^T + B_1$ corresponds to the translation $R_1\mathbf{b}_2 + \mathbf{b}_1$ resulting from concatenated rigid motions. Hence the product of dual orthogonal matrices captures the the property of composition of general rigid-body motions.

6.7.2 Dual Unitary Matrices

If one desires to express motions using 2×2 matrices with entries that are both complex and dual numbers, this can be achieved by using dual unitary matrices defined as

$$\hat{U} \doteq U + \frac{1}{2}\hat{\epsilon}BU$$

where

$$B = \begin{pmatrix} -ix_2 & -x_1 - ix_3 \\ x_1 - ix_3 & ix_2 \end{pmatrix}$$

is a matrix containing all the translational information, and $U \in SU(2)$ contains all of the rotational information. Composition of motions is achieved by the multiplication $\hat{U}_1\hat{U}_2$.

6.7.3 Dual Quaternions

Starting with 4×4 homogeneous transform matrices to represent rigid-body displacements, we reduced the dimensions to 3×3 by introducing dual numbers, and to 2×2 by, in addition, introducing complex numbers. It should therefore be no suprise that the lowest dimensional representation of rigid-body motion uses the most complicated combination of elements - dual numbers and quaternions.

The dual unit quaternions are written as [23, 59]

$$\hat{u} = u + \frac{1}{2}\hat{\epsilon}bu$$

where $b = b_1\tilde{i} + b_2\tilde{j} + b_3\tilde{k}$ is the vector quaternion representing translation vector $\mathbf{b}$ in the displacement $(R, \mathbf{b})$.

As with the case of pure rotation, where there was a one-to-one correspondence between elements of $SU(2)$ and unit quaternions, there is also a one-to-one correspondence between dual special orthogonal matrices and dual unit quaternions. It follows then that

$$\hat{u}\hat{u}^* = 1.$$

6.8 Approximating Rigid-Body Motions in $\mathbb{R}^N$ as Rotations in $\mathbb{R}^{N+1}$

In many ways it is more convenient to deal with pure rotations than the combination of rotations and translations that compose rigid-body motions. The set of rotations have a well-defined measure of distance that is invariant under shifts from the left and right, and the set of rotations does not extend to infinity in any direction. Neither of these are characteristics of the set of rigid-body motions.

This fact has lead some researchers to investigate approximating rigid-body motions with pure rotations. To our knowledge, this concept was introduced into the engineering literature by McCarthy [49]. More generally, the concept of "contraction" of a set of transformations of infinite extent to one of finite extent has been known for many years in the physics literature (see, e.g., [37, 76]).

Methods for performing this contraction are geometrically intuitive. We have already seen in previous chapters how stereographic projection can be used to map points on the sphere to points in the plane, and vice versa. One way to approximate a rigid-body motion with a pure rotation is to examine how an arbitrary moving point in the plane is mapped to the sphere. The rotation that approximates the motion is then calculated as that which rotates the image of the projected points on the sphere. Given a homogeneous transform, H, another approach is to calculate the closest rotation using the polar decomposition as $R = H(H^T H)^{-\frac{1}{2}}$. Both of these methods for contracting the set of rigid-body motions to the set of rotations are examined in detail for the planar and spatial cases in the following subsections.

6.8.1 Planar Motions

The set of planar rigid-body motions is parameterized with three variables (x, y, θ). We seek an explicit mapping between planar motions and three-dimensional rotations.

McCarthy's Approach

Any rigid-body transformation can be decomposed into the product of a translation followed by a rotation about the x_3-axis. Any rotation in space can be described by first rotating about the x_2 axis by an angle ϵ, then about the x_1 axis by $-\nu$, followed by a rotation about the x_3 axis by θ. These two scenarios are written as

$$R = R_2(\epsilon)R_1(-\nu)R_3(\theta)$$

and

$$H = \begin{pmatrix} 1 & 0 & b_1/L \\ 0 & 1 & b_2/L \\ 0 & 0 & 1 \end{pmatrix} \begin{pmatrix} \cos\theta & -\sin\theta & 0 \\ \sin\theta & \cos\theta & 0 \\ 0 & 0 & 1 \end{pmatrix},$$

where L is the measure of length used to normalize the translations.

The approximation

$$R \approx H$$

is then valid with $\mathcal{O}(1/L^2)$ error when the plane of the rigid motion is taken to be the one that kisses the sphere at the north pole when one sets

$$\tan\epsilon = b_1/L \quad \text{and} \quad \tan\nu = b_2/L.$$

This idea, as presented in [49].

We note that rigid-body motions in the plane can also be approximated as pure rotations in $\mathbb{R}^3$ using stereographic projection or the polar decomposition.

6.8.2 Spatial Motions

The set of spatial rigid-body motions is parameterized with six variables (three for translation and three for rotation). We seek an explicit mapping between spatial motions and rotations in four dimensions. Since in general a rotation in $\mathbb{R}^N$ is described with $N(N-1)/2$ parameters, it should be no surprise that rotations in $\mathbb{R}^4$ require the same number of parameters as rigid-body motion in three-dimensional space.

Before going into the details of the mapping between $SE(3)$ and $SO(4)$ (the set of rotations in $\mathbb{R}^4$), we first review some important features of $SO(4)$

Rotations in $\mathbb{R}^4$

Rotations in four-space can be parameterized in ways very similar to those reviewed in Chapter 5. For instance, any $SO(4)$-rotation can be described as the exponential of a 4×4 skew-symmetric matrix [25, 33]:

$$M(\mathbf{z}, \mathbf{w}) = \begin{pmatrix} 0 & -z_3 & z_2 & w_1 \\ z_3 & 0 & -z_1 & w_2 \\ -z_2 & z_1 & 0 & w_3 \\ -w_1 & -w_2 & -w_3 & 0 \end{pmatrix}, \tag{6.41}$$

$$R = \exp M(\mathbf{z}, \mathbf{w}). \tag{6.42}$$

Following [33], two new matrices M^+ and M^- are formed as

$$M^\pm = (M(\mathbf{z}, \mathbf{w}) \pm M(\mathbf{w}, \mathbf{z}))/2.$$

These matrices have the form

$$M^+(\mathbf{s}) = \begin{pmatrix} 0 & -s_3 & s_2 & s_1 \\ s_3 & 0 & -s_1 & s_2 \\ -s_2 & s_1 & 0 & s_3 \\ -s_1 & -s_2 & -s_3 & 0 \end{pmatrix} = \|\mathbf{s}\| M^+(\mathbf{s}').$$

and

$$M^-(\mathbf{s}) = \begin{pmatrix} 0 & -s_3 & s_2 & -s_1 \\ s_3 & 0 & -s_1 & -s_2 \\ -s_2 & s_1 & 0 & -s_3 \\ s_1 & s_2 & s_3 & 0 \end{pmatrix} = \|\mathbf{s}\| M^-(\mathbf{s}').$$

where $\mathbf{s} = \|\mathbf{s}\|\mathbf{s}'$ and $\mathbf{s}' \in \mathbb{S}^2$.

Vectors $\mathbf{s}_\pm$ are defined such that

$$M(\mathbf{z}, \mathbf{w}) = M^+(\mathbf{s}_+) + M^-(\mathbf{s}_-).$$

The reason for performing this decomposition is that

$$M^+(\mathbf{s})M^-(\mathbf{t}) = M^-(\mathbf{s})M^+(\mathbf{t})$$

for any $\mathbf{s}, \mathbf{t} \in \mathbb{R}^3$. Since in general the exponential of commuting matrices can be written as the product of each matrix exponentiated separately, we can rewrite (6.42) as

$$R = \exp M^+(\mathbf{s}_+) \exp M^-(\mathbf{s}_-).$$

This is convenient because each of these simpler matrices has the closed form

$$\exp M^\pm(\mathbf{s}) = \cos \|\mathbf{s}\| \mathbb{I}_4 + \sin \|\mathbf{s}\| M^\pm(\mathbf{s}'). \tag{6.43}$$

It has been observed that $\exp M^+(\mathbf{s})$ and $\exp M^-(\mathbf{s})$ each correspond to unit quaternions describing rotation in three dimensions, with axis of rotation $\mathbf{s}$, and angle $\pm 2\|\mathbf{s}\|$, with sign corresponding to $M^\pm$. Hence the name *biquaternions* [18] is sometimes used.

This becomes particularly clear when the basis $i_4 = \tilde{k}_4$, $j_4 = -\tilde{j}_4$, $k_4 = \tilde{i}_4$ is used. Then given a point $\mathbf{x} \in \mathbb{R}^3$, three dimensional rotations are implemented as

$$M^+(\mathbf{x}') = \exp M^+(\mathbf{s})M^+(\mathbf{x}) \exp M^+(-\mathbf{s}).$$

An interesting and well-known result is that by defining the 4×4 matrix

$$X = x_4 \mathbb{I} + M^+(\mathbf{x}),$$

rotations in four dimensions are achieved as

$$X' = \exp M^+(\mathbf{s}) X \exp M^-(\mathbf{t}). \tag{6.44}$$

The fact that this is a rotation is easy to verify because

$$\det(X) = (x_1^2 + x_2^2 + x_3^2 + x_4^2)^2$$

and

$$\operatorname{tr}(XX^T) = 4(x_1^2 + x_2^2 + x_3^2 + x_4^2)$$

are preserved under the transformation.

The transformation of the vector $\tilde{\mathbf{x}} = [x_1, x_2, x_3, x_4]^T$ can also be written as

$$\tilde{\mathbf{x}}' = \exp M^+(\mathbf{s}) \exp M^-(\mathbf{t})\tilde{\mathbf{x}}, \tag{6.45}$$

and so it is clear that this transformation is both orientation and length preserving, and hence must be a rotation. Equation (6.45) follows from the relationship

$$X' = \exp M^\pm(\mathbf{s}) X \exp M^\pm(-\mathbf{s}) \iff \tilde{\mathbf{x}}' = \exp M^\pm(\mathbf{s})\tilde{\mathbf{x}},$$

and the commutativity of $M^+(\mathbf{s})$ and $M^-(\mathbf{t})$.

If instead of 4×4 real rotation matrices one prefers 2×2 special unitary matrices or unit quaternions, the equations analogous to (6.43) and (6.44) follow naturally with the elements of $SU(2)$ or the unit quaternions corresponding to $M^\pm$ and

$$X = \begin{pmatrix} x_4 - ix_2 & -x_1 - ix_3 \\ x_1 - ix_3 & x_4 + ix_2 \end{pmatrix}$$

or $x = x_1 i + x_2 j + x_3 k + x_4$.

Since both of the pairs $(\exp M^+(\mathbf{s}), \exp M^-(\mathbf{t}))$ and $(-\exp M^+(\mathbf{s}), -\exp M^-(\mathbf{t}))$ describe the same rotation in $\mathbb{R}^4$, the biquaternions and $SU(2) \times SU(2)$ (which is notation for two independent copies of $SU(2)$) both have two-to-one mappings to $SO(4)$. The biquaternions and $SU(2) \times SU(2)$ are both called double covers of $SO(4)$, in the sense that there is a 2-to-1 mapping from those spaces onto $SO(4)$.

Contraction of $SE(3)$ to $SO(4)$

We now consider the approximation of rigid-body motions in three space with rotations in four dimensions. The same three approaches presented in Section 6.8 are applicable here.

McCarthy's Approach

To begin, McCarthy's approach using Euler angles requires us to extend the concept of Euler angles to four dimensions. In four dimensions, rotations are no longer "about an axis k," but rather are rotations "in the plane of axes i and j." We denote such rotations as $R_{ij}(\theta)$, with the convention that $R_{ij}(\pi/2)$ is the rotation that takes axis i into axis j, leaving the other two axes unchanged. It then follows that

$$R_{12}(\theta) = \begin{pmatrix} \cos\theta & -\sin\theta & 0 & 0 \\ \sin\theta & \cos\theta & 0 & 0 \\ 0 & 0 & 1 & 0 \\ 0 & 0 & 0 & 1 \end{pmatrix} \quad R_{31}(\theta) = \begin{pmatrix} \cos\theta & 0 & \sin\theta & 0 \\ 0 & 1 & 0 & 0 \\ -\sin\theta & 0 & \cos\theta & 0 \\ 0 & 0 & 0 & 1 \end{pmatrix}$$

$$R_{23}(\theta) = \begin{pmatrix} 1 & 0 & 0 & 0 \\ 0 & \cos\theta & -\sin\theta & 0 \\ 0 & \sin\theta & \cos\theta & 0 \\ 0 & 0 & 0 & 1 \end{pmatrix} \quad R_{42}(\theta) = \begin{pmatrix} 1 & 0 & 0 & 0 \\ 0 & \cos\theta & 0 & \sin\theta \\ 0 & 0 & 1 & 0 \\ 0 & -\sin\theta & 0 & \cos\theta \end{pmatrix}$$

$$R_{34}(\theta) = \begin{pmatrix} 1 & 0 & 0 & 0 \\ 0 & 1 & 0 & 0 \\ 0 & 0 & \cos\theta & -\sin\theta \\ 0 & 0 & \sin\theta & \cos\theta \end{pmatrix} \quad R_{14}(\theta) = \begin{pmatrix} \cos\theta & 0 & 0 & -\sin\theta \\ 0 & 1 & 0 & 0 \\ 0 & 0 & 1 & 0 \\ \sin\theta & 0 & 0 & \cos\theta \end{pmatrix}$$

One of the many ways rotations in $\mathbb{R}^4$ can be parameterized using Euler angles is as follows [25]. Start with any of the usual Euler angles, e.g.:

$$K(\alpha,\beta,\gamma) = R_{12}(\alpha)R_{23}(\beta)R_{12}(\gamma) = \begin{pmatrix} A(\alpha,\beta,\gamma) & \mathbf{0} \\ \mathbf{0}^T & 1 \end{pmatrix}.$$

This particular choice is a 3×3 rotation matrix corresponding to ZXZ Euler angles embedded in a 4×4 matrix.

Recall that any rigid-body motion with translation measured in units of L can be decomposed into a translation followed by a rotation, as

$$H = \begin{pmatrix} \mathbb{I}_{3\times 3} & \mathbf{b}/L \\ \mathbf{0}^T & 1 \end{pmatrix} \begin{pmatrix} A(\alpha,\beta,\gamma) & \mathbf{0} \\ \mathbf{0}^T & 1 \end{pmatrix}.$$

We therefore seek a second 4×4 rotation matrix, J, to approximate the translation. so that

$$R = J(\epsilon,\nu,\eta)K(\alpha,\beta,\gamma) \approx H.$$

One possible choice is

$$J(\epsilon,\nu,\eta) = R_{43}(\epsilon)R_{42}(\nu)R_{41}(\eta) = \begin{pmatrix} c\epsilon & 0 & 0 & s\epsilon \\ -s\nu s\epsilon & c\nu & 0 & s\nu c\epsilon \\ -s\eta c\nu s\epsilon & -s\eta s\nu c\eta & c\eta & s\eta c\nu c\epsilon \\ -c\eta c\nu s\epsilon & -s\nu c\eta & -s\eta & c\eta c\nu c\epsilon \end{pmatrix}$$

where $c\theta$ and $s\theta$ are shorthand for $\cos\theta$ and $\sin\theta$.

In the limit as $1/L \to 0$, the approximation $H \approx R$ has $\mathcal{O}(1/L^2)$ error with

$$\tan\epsilon = b_1/L \qquad \tan\nu = b_2/L \qquad \tan\eta = b_3/L.$$

We note that rigid-body motions in three-dimensional space can also be approximated as pure rotations in $\mathbb{R}^4$ using stereographic projection or the polar decomposition of a homogeneous transform.

6.9 Metrics on Motion

In this section we present two general methods for generating metrics on the set of rigid-body motions. The basic idea behind the two classes of metrics presented here is to use metrics on $\mathbb{R}^N$ and on function spaces on the set of motions to induce metrics on the set of motions itself. This formulation in this section follows those in [17, 27, 40, 47]. Applications of metrics on motion are discussed in [58, 62, 77].

6.9.1 Metrics on Motion Induced by Metrics on $\mathbb{R}^N$

Perhaps the most straightforward way to define metrics on the set of motions is to take advantage of the well-known metrics on $\mathbb{R}^N$. Namely, if $\rho(\mathbf{x})$ is a continuous real-valued function on $\mathbb{R}^N$ which satisfies the properties

$$0 \leq \rho(\mathbf{x}) \quad \text{and} \quad 0 < \int_{\mathbb{R}^N} \|\mathbf{x}\|^m \rho(\mathbf{x})\, dx_1...\, dx_N < \infty,$$

for any $m \geq 0$, then

$$d(g_1, g_2) = \int_{\mathbb{R}^N} \|g_1 \cdot \mathbf{x} - g_2 \cdot \mathbf{x}\| \, \rho(\mathbf{x}) \, dx_1 \ldots dx_N$$

is a metric when $\|\cdot\|$ is any norm for vectors in $\mathbb{R}^N$ (in particular, the p-norm is denoted $\|\cdot\|_p$). Here $g = (A, \mathbf{b})$ is an affine transformation with corresponding homogeneous transform $H(A, \mathbf{b})$. The fact that this is a metric was observed in [47] for the case of rigid-body motion, i.e., when $A \in SO(N)$. We use the notation $g \cdot \mathbf{x} = A\mathbf{x} + \mathbf{b}$.

Natural choices for $\rho(\mathbf{x})$ are either the mass density of the object, or a function which is equal to one on the object and zero otherwise. However, it is also possible to choose a function such as $\rho(\mathbf{x}) = e^{-a^2 \mathbf{x} \cdot \mathbf{x}}$ (for any $a \in \mathbb{R}_{>0}$) which is positive everywhere yet decreases rapidly enough for $d(g_1, g_2)$ to be finite.

The fact that this is a metric on the the set of rigid-body transformations is observed as follows. The symmetry property $d(g_1, g_2) = d(g_2, g_1)$ results from the symmetry of vector addition and the properties of vector norms. The triangle inequality also follows from properties of norms. Positive definiteness of this metric follows from the fact that for rigid-body transformations

$$\|g_1 \cdot \mathbf{x} - g_2 \cdot \mathbf{x}\| = \|(A_1 - A_2)\mathbf{x} + (\mathbf{b}_1 - \mathbf{b}_2)\|,$$

and because of the positive definiteness of $\|\cdot\|$, the only time this quantity can be zero is when

$$(A_1 - A_2)\mathbf{x} = \mathbf{b}_2 - \mathbf{b}_1.$$

If $(A_1 - A_2)$ is invertible, this only happens at one point, i.e., $\mathbf{x} = (A_1 - A_2)^{-1}(\mathbf{b}_2 - \mathbf{b}_1)$.

In any case, the set of all $\mathbf{x}$ for which this equation is satisfied will have dimension less than N when $g_1 \neq g_2$, and so the value of the integrand at these points does not contribute to the integral. Thus, because $\rho(\mathbf{x})$ satisfies the properties listed above, and $\|g_1 \cdot \mathbf{x} - g_2 \cdot \mathbf{x}\| > 0$ for $g_1 \neq g_2$ except on a set of measure zero, the integral in the definition of the metric must satisfy $d(g_1, g_2) > 0$ unless $g_1 = g_2$, in which case $d(g_1, g_1) = 0$.

While this does satisfy the properties of a metric and *could* be used in applications such as CAD and robot design and path planning problems, it has the significant drawback that the integral of a pth root (or absolute value) must be taken. This means that numerical computations are required. For practical problems, devoting computer power to the computation of the metric detracts significantly from other aspects of the application in which the metric is being used. Therefore, this is not a practical metric. On the other hand, we can modify this approach slightly so as to generate metrics which are calculated in closed form. This yields tremendous computational advantages.

Namely, we observe that

$$d^{(p)}(g_1, g_2) = \sqrt[p]{\int_{\mathbb{R}^N} \|g_1 \cdot \mathbf{x} - g_2 \cdot \mathbf{x}\|_p^p \, \rho(\mathbf{x}) \, dx_1 \ldots dx_N}$$

is a metric. Clearly, this is symmetric and positive definite for all of the same reasons as the metric presented earlier. In order to prove the triangle inequality, we must use Minkowski's inequality. That is, if $a_1, a_2, \ldots, a_n$ and $b_1, b_2, \ldots, b_n$ are non-negative real numbers and $p > 1$, then

$$\left(\sum_{k=1}^{n} (a_k + b_k)^p \right)^{\frac{1}{p}} \leq \left(\sum_{k=1}^{n} a_k^p \right)^{\frac{1}{p}} + \left(\sum_{k=1}^{n} b_k^p \right)^{\frac{1}{p}}. \tag{6.46}$$

In our case, $a = \|g_1 \cdot \mathbf{x} - g_2 \cdot \mathbf{x}\|_p [\rho(\mathbf{x})]^{\frac{1}{p}}$, $b = \|g_2 \cdot \mathbf{x} - g_3 \cdot \mathbf{x}\|_p [\rho(\mathbf{x})]^{\frac{1}{p}}$, summation is replaced by integration, and because $\|g_1 \cdot \mathbf{x} - g_2 \cdot \mathbf{x}\|_p + \|g_2 \cdot \mathbf{x} - g_3 \cdot \mathbf{x}\|_p \geq \|g_1 \cdot \mathbf{x} - g_3 \cdot \mathbf{x}\|_p$, then

$$d^{(p)}(g_1, g_2) + d^{(p)}(g_2, g_3) \geq \sqrt[p]{\int_{\mathbb{R}^N} (\|g_1 \cdot \mathbf{x} - g_2 \cdot \mathbf{x}\|_p + \|g_2 \cdot \mathbf{x} - g_3 \cdot \mathbf{x}\|_p)^p \rho(\mathbf{x}) \, dx_1 \dots dx_N}$$

$$\geq \sqrt[p]{\int_{\mathbb{R}^N} \|g_1 \cdot \mathbf{x} - g_3 \cdot \mathbf{x}\|_p^p \rho(\mathbf{x}) \, dx_1 \dots dx_N}$$

$$= d^{(p)}(g_1, g_3).$$

From the above argument, we can conclude that $d^p(\cdot, \cdot)$ is a metric. Likewise, it is easy to see that

$$d^{(p')}(g_1, g_2) = \sqrt[p]{\frac{\int_{\mathbb{R}^N} \|g_1 \cdot \mathbf{x} - g_2 \cdot \mathbf{x}\|_p^p \rho(\mathbf{x}) \, dx_1 \dots dx_N}{\int_{\mathbb{R}^N} \rho(\mathbf{x}) \, dx_1 \dots dx_N}}$$

is also a metric, since division by a positive real constant has no effect on metric properties.

The obvious benefit of using $d^{(p)}(\cdot, \cdot)$ or $d^{(p')}(\cdot, \cdot)$ is that the pth root is now outside of the integral, and so the integral can be calculated in closed form.

A particularly useful case is when $p = 2$. In this case it is easy to see that all of the metrics presented in this section satisfy the property

$$d(h \circ g_1, h \circ g_2) = d(g_1, g_2)$$

where $h, g_1, g_2 \in SE(N)$.

This is because if $h = (R, \mathbf{b}) \in SE(N)$, then

$$(h \circ g_i) \cdot \mathbf{x} = h \cdot (g_i \cdot \mathbf{x})$$

and

$$\|h \cdot (g_1 \cdot \mathbf{x}) - h \cdot (g_2 \cdot \mathbf{x})\|_2 = \|R[g_1 \cdot \mathbf{x}] + \mathbf{b} - R[g_2 \cdot \mathbf{x}] - \mathbf{b}\|_2 = \|g_1 \cdot \mathbf{x} - g_2 \cdot \mathbf{x}\|_2$$

It is also interesting to note that there is a relationship between this kind of metric for $SE(N)$ and the Hilbert-Schmidt norm of $N \times N$ matrices. That is, for $g \in SE(N)$

$$d^{(2)}(g, e) = \sqrt{\int_{\mathbb{R}^N} \|g \cdot \mathbf{x} - \mathbf{x}\|_2^2 \rho(\mathbf{x}) \, d\mathbf{x}}$$

is the same as a weighted norm

$$\|g - e\|_W = \sqrt{\operatorname{tr}((g - e)^T W (g - e))},$$

where $W = W^T \in \mathbb{R}^{4 \times 4}$ and g and e are expressed as 4×4 homogeneous transformations matrices. To show that this is true, we begin our argument by expanding as follows.

Letting $g = (R, \mathbf{b})$, then

$$\|g \cdot \mathbf{x} - \mathbf{x}\|_2^2 = (R\mathbf{x} + \mathbf{b} - \mathbf{x})^T (R\mathbf{x} + \mathbf{b} - \mathbf{x})$$
$$= \operatorname{tr}((R\mathbf{x} + \mathbf{b} - \mathbf{x})(R\mathbf{x} + \mathbf{b} - \mathbf{x})^T)$$
$$= \operatorname{tr}((R\mathbf{x} + \mathbf{b} - \mathbf{x})(\mathbf{x}^T R^T + \mathbf{b}^T - \mathbf{x}^T))$$
$$= \operatorname{tr}(R\mathbf{x}\mathbf{x}^T R^T + R\mathbf{x}\mathbf{b}^T - R\mathbf{x}\mathbf{x}^T + \mathbf{b}\mathbf{x}^T R^T + \mathbf{b}\mathbf{b}^T - \mathbf{b}\mathbf{x}^T - \mathbf{x}\mathbf{x}^T R^T - \mathbf{x}\mathbf{b}^T + \mathbf{x}\mathbf{x}^T)$$
$$= \operatorname{tr}(\mathbf{x}\mathbf{x}^T + R\mathbf{x}\mathbf{b}^T - R\mathbf{x}\mathbf{x}^T + R\mathbf{x}\mathbf{b}^T + \mathbf{b}\mathbf{b}^T - \mathbf{x}\mathbf{b}^T - R\mathbf{x}\mathbf{x}^T - \mathbf{x}\mathbf{b}^T + \mathbf{x}\mathbf{x}^T)$$
$$= 2\operatorname{tr}(\mathbf{x}\mathbf{x}^T + R\mathbf{x}\mathbf{b}^T - R\mathbf{x}\mathbf{x}^T - \mathbf{x}\mathbf{b}^T + \frac{1}{2}\mathbf{b}\mathbf{b}^T).$$

Note that

$$\int_{\mathbb{R}^N} \operatorname{tr}\left[F(\mathbf{x})\right] \rho(\mathbf{x}) \, d\mathbf{x} = \operatorname{tr}\left[\int_{\mathbb{R}^N} F(\mathbf{x})\rho(\mathbf{x}) \, d\mathbf{x}\right],$$

where $F(\mathbf{x})$ is any matrix function. Under the assumption that

$$\int_{\mathbb{R}^N} \mathbf{x}\,\rho(\mathbf{x}) \, d\mathbf{x} = \mathbf{0},$$

we then have

$$\|g \cdot \mathbf{x} - \mathbf{x}\|_2^2 = 2\operatorname{tr}\left[(\mathbb{I} - R)\mathbf{x}\mathbf{x}^T + \frac{1}{2}\mathbf{b}\mathbf{b}^T\right],$$

and the type 1 metric is:

$$\left(d^{(2)}(g, e)\right)^2 = 2\operatorname{tr}\left[(\mathbb{I} - R)\int_{\mathbb{R}^N} \mathbf{x}\mathbf{x}^T\rho(\mathbf{x}) \, d\mathbf{x}\right] + \mathbf{b} \cdot \mathbf{b}\int_{\mathbb{R}^N} \rho(\mathbf{x}) \, d\mathbf{x}$$

or

$$d^{(2)}(g, e) = \sqrt{2\operatorname{tr}[(\mathbb{I} - R)J] + \mathbf{b} \cdot \mathbf{b}M} \qquad (6.47)$$

where $\mathbb{I}$ is the 3×3 identity matrix, $M = \int_{\mathbb{R}^N} \rho(\mathbf{x}) \, d\mathbf{x}$ is the mass and $J = \int_{\mathbb{R}^N} \mathbf{x}\mathbf{x}^T \rho(\mathbf{x}) \, d\mathbf{x}$ has a simple relationship with the moment of inertia matrix of the rigid body:

$$I = \int_{\mathbb{R}^N} \left((\mathbf{x}^T\mathbf{x})\mathbb{I} - \mathbf{x}\mathbf{x}^T\right)\rho(\mathbf{x}) \, d\mathbf{x} = \operatorname{tr}(J)\mathbb{I} - J.$$

Now we compare this to the weighted norm of $\|g - e\|_W$, defined by $\|g - e\|_W^2 = \operatorname{tr}((g - e)W(g - e)^T)$, where $W = \begin{pmatrix} W_{3\times3} & \mathbf{0} \\ \mathbf{0}^T & w_{44} \end{pmatrix}$:

$$\|g - e\|_W^2 = \operatorname{tr}((g - e)W(g - e)^T)$$
$$= \operatorname{tr}((g - e)^T(g - e)W)$$
$$= \operatorname{tr}\left(\begin{pmatrix} R^T - \mathbb{I} & \mathbf{0} \\ \mathbf{b}^T & 0 \end{pmatrix}\begin{pmatrix} R - \mathbb{I} & \mathbf{b} \\ \mathbf{0}^T & 0 \end{pmatrix}\begin{pmatrix} W_{3\times3} & \mathbf{0} \\ \mathbf{0}^T & w_{44} \end{pmatrix}\right)$$
$$= \operatorname{tr}\left(\begin{pmatrix} (R^T - \mathbb{I})(R - \mathbb{I}) & (R^T - \mathbb{I})\mathbf{b} \\ \mathbf{b}^T(R - \mathbb{I}) & \mathbf{b}^T\mathbf{b} \end{pmatrix}\begin{pmatrix} W_{3\times3} & \mathbf{0} \\ \mathbf{0}^T & w_{44} \end{pmatrix}\right)$$
$$= \operatorname{tr}\begin{pmatrix} (R^T - \mathbb{I})(R - \mathbb{I})W_{3\times3} & (R^T - \mathbb{I})\mathbf{b}w_{44} \\ \mathbf{b}^T(R - \mathbb{I})W_{3\times3} & \mathbf{b}^T\mathbf{b}w_{44} \end{pmatrix}$$
$$= \operatorname{tr}((R^T - \mathbb{I})(R - \mathbb{I})W_{3\times3}) + w_{44}\mathbf{b}^T\mathbf{b}$$
$$= 2\operatorname{tr}((\mathbb{I} - R)W_{3\times3}) + w_{44}\mathbf{b}^T\mathbf{b}.$$

It is exactly in the same form as (6.47), with $J = W_{3\times3}$ and $M = w_{44}$. Therefore, we conclude that

$$d^{(2)}(g,e) = \|g - e\|_W.$$

Furthermore, we can prove that $d(g_1, g_2) = \|g_1 - g_2\|_W$. First note that

$$g_1 - g_2 = \begin{pmatrix} R_1 - R_2 & \mathbf{b_1} - \mathbf{b_2} \\ \mathbf{0}^T & 0 \end{pmatrix}.$$

Therefore,

$$\|g_1 - g_2\|_W^2 =$$

$$\operatorname{tr}\left(\begin{pmatrix} R_1^T - R_2^T & \mathbf{0} \\ \mathbf{b_1}^T - \mathbf{b_2}^T & 0 \end{pmatrix} \begin{pmatrix} R_1 - R_2 & \mathbf{b_1} - \mathbf{b_2} \\ \mathbf{0}^T & 0 \end{pmatrix} \begin{pmatrix} W_{3\times3} & \mathbf{0} \\ \mathbf{0}^T & w_{44} \end{pmatrix} \right)$$

$$= \operatorname{tr}\left(\begin{pmatrix} (R_1^T - R_2^T)(R_1 - R_2) & (R_1^T - R_2^T)(\mathbf{b_1} - \mathbf{b_2}) \\ (\mathbf{b_1}^T - \mathbf{b_2}^T)(R_1 - R_2) & (\mathbf{b_1}^T - \mathbf{b_2}^T)(\mathbf{b_1} - \mathbf{b_2}) \end{pmatrix} \begin{pmatrix} W_{3\times3} & \mathbf{0} \\ \mathbf{0}^T & w_{44} \end{pmatrix} \right)$$

$$= \operatorname{tr}((R_1^T - R_2^T)(R_1 - R_2)W_{3\times3} + w_{44}(\mathbf{b_1} - \mathbf{b_2})^T(\mathbf{b_1} - \mathbf{b_2})$$

$$= \operatorname{tr}((2\mathbb{I} - R_1^T R_2 - R_2^T R_1)W_{3\times3}) + w_{44}(\mathbf{b_1} - \mathbf{b_2})^T(\mathbf{b_1} - \mathbf{b_2}).$$

Note that

$$\operatorname{tr}(R_2^T R_1 W_{3\times3}) = \operatorname{tr}(W_{3\times3} R_1^T R_2) = \operatorname{tr}(R_1^T R_2 W_{3\times3}).$$

We get

$$\|g_1 - g_2\|_W^2 = 2\operatorname{tr}((\mathbb{I} - R_1^T R_2)W_{3\times3}) + w_{44}\|\mathbf{b_1} - \mathbf{b_2}\|^2.$$

On the other hand, we already know that

$$d(g_2^{-1} \circ g_1, e) = d(g_1^{-1} \circ g_2, e) = d(g_1, g_2).$$

Noting that

$$g_1^{-1} \circ g_2 = \begin{pmatrix} R_1^T & -R_1^T \mathbf{b_1} \\ \mathbf{0}^T & 1 \end{pmatrix} \begin{pmatrix} R_2 & \mathbf{b_2} \\ \mathbf{0}^T & 1 \end{pmatrix} = \begin{pmatrix} R_1^T R_2 & R_1^T(\mathbf{b_2} - \mathbf{b_1}) \\ \mathbf{0}^T & 1 \end{pmatrix},$$

we can get

$$\left(d^{(2)}(g_1^{-1} \circ g_2, e) \right)^2 = \|g_1^{-1} \circ g_2 - e\|_W^2$$

$$= 2\operatorname{tr}((\mathbb{I} - R_1^T R_2)W_{3\times3} + w_{44}\|R_1^T(\mathbf{b_2} - \mathbf{b_1})\|^2$$

$$= 2\operatorname{tr}((\mathbb{I} - R_1^T R_2)W_{3\times3} + w_{44}\|(\mathbf{b_2} - \mathbf{b_1})\|^2.$$

This is exactly identical to $\|g_1 - g_2\|_W^2$. Therefore we conclude that for $g_1, g_2 \in SE(N)$

$$d^{(2)}(g_1, g_2) = \|g_1 - g_2\|_W, \tag{6.48}$$

where $W = \begin{pmatrix} J & \mathbf{0} \\ \mathbf{0}^T & M \end{pmatrix}$.

This property of these metrics is very convenient since we can use many well-developed theories of matrix norms, and the only explicit integration that is required to compute the metric is the computation of moments of inertia (which are already tabulated for most common engineering shapes).

6.9.2 Metrics on $SE(N)$ Induced by Norms on $\mathcal{L}^2(SE(N))$

It is always possible to define a continuous real-valued function $f : SE(N) \to \mathbb{R}$ that decays rapidly outside of a neighborhood of the identity motion. Furthermore, if $f \in \mathcal{L}^p(SE(N))$, by definition it is possible to integrate f^p over all motions. In this case, the measure of the function,

$$\mu(f) = \int_G f(g)\,d(g),$$

is finite where $d(g)$ is an integration measure on the set of rigid-body motions discussed earlier, and g denotes a motion.

Given this background, it is possible to define left and right invariant metrics on the set of motions in the following way: Let $f(g)$ be a continuous nonperiodic p-integrable function (i.e., $\mu(|f|^p)$ is finite). Then the following are metrics:

$$d_L^{(p)}(g_1, g_2) = \left(\int_G |f(g_1^{-1} \circ g) - f(g_2^{-1} \circ g)|^p \, d(g) \right)^{\frac{1}{p}},$$

$$d_R^{(p)}(g_1, g_2) = \left(\int_G |f(g \circ g_1) - f(g \circ g_2)|^p \, d(g) \right)^{\frac{1}{p}}.$$

The fact that these are metrics follow easily. The triangle inequality holds from the Minkowski inequality, symmetry holds from the symmetry of scalar addition, and positive definiteness follows from the fact that we choose $f(g)$ to be a nonperiodic continuous function. That is, we choose $f(g)$ such that the equalities $f(g) = f(g_1 \circ g)$ and $f(g) = f(g \circ g_1)$ do not hold for $g_1 \neq e$ except on sets of measure zero. Thus, there is no way for the integral of the difference of shifted versions of this function to be zero other than when $g_1 = g_2$, where it must be zero.

We prove the left invariance of $d_L^{(p)}(g_1, g_2)$ below. The proof for right invariance for $d_R^{(p)}(g_1, g_2)$ follows analogously.

$$d_L^{(p)}(h \circ g_1, h \circ g_2) = \left(\int_G |f((h \circ g_1)^{-1} \circ g) - f((h \circ g_2)^{-1} \circ g)|^p \, d(g) \right)^{\frac{1}{p}}$$

$$= \left(\int_G |f((g_1^{-1} \circ h^{-1}) \circ g) - f((g_2^{-1} \circ h^{-1}) \circ g)|^p \, d(g) \right)^{\frac{1}{p}}$$

$$= \left(\int_G |f(g_1^{-1} \circ (h^{-1} \circ g)) - f(g_2^{-1} \circ (h^{-1} \circ g))|^p \, d(g) \right)^{\frac{1}{p}}.$$

Because of the left invariance of the integration measure, we then have

$$= \left(\int_G |f(g_1^{-1} \circ g') - f(g_2^{-1} \circ g')|^p \, d(h \circ g') \right)^{\frac{1}{p}}$$

$$= d_L^{(p)}(g_1, g_2),$$

where the change of variables $g' = h^{-1} \circ g$ has been made.

If it were possible to find a square-integrable function with the property $f(g \circ h) = f(h \circ g)$ for $g, h \in SE(N)$ then it would be possible to define a bi-invariant metric on $SE(N)$. This is clear as follows, by illustrating the right invariance of a left-invariant metric.

$$d_L^{(p)}(g_1 \circ h, g_2 \circ h) = \left(\int_G |f((g_1 \circ h)^{-1} \circ g) - f((g_2 \circ h)^{-1} \circ g)|^p \, d(g) \right)^{\frac{1}{p}}$$

$$= \left(\int_G |f(h^{-1} \circ g_1^{-1} \circ g) - f(h^{-1} \circ g_2^{-1} \circ g)|^p \, d(g) \right)^{\frac{1}{p}}$$

For a class function $f(g_1 \circ (g_2 \circ g_3)) = f((g_2 \circ g_3) \circ g_1)$, for any $g_1 \in G$, and so

$$d_L^{(p)}(g_1 \circ h, g_2 \circ h) = \left(\int_G |f(g_1^{-1} \circ g \circ h^{-1}) - f(g_2^{-1} \circ g \circ h^{-1})|^p \, d(g) \right)^{\frac{1}{p}}$$

$$= \left(\int_G |f(g_1^{-1} \circ g') - f(g_2^{-1} \circ g')|^p \, d(h \circ g') \right)^{\frac{1}{p}}$$

$$= d_L^{(p)}(g_1, g_2),$$

where the substitution $g' = g \circ h^{-1}$ has been made and the right invariance of the integration has been assumed.

Unfortunately, for $SE(N)$ no such functions exist [43]. However, the construction above can be used to generate invariant metrics for $SO(N)$, where functions of the form $f(RQ) = f(QR)$ do exist.

This is consistent with results reported in the literature [56]. However, it does not rule out the existence of anomalous metrics such as the trivial metric, which can be generated for $p = 1$ when $f(g)$ is a delta function on $SE(N)$.

As a practical matter, $p = 1$ is a difficult case to work with since the integration must be performed numerically. Likewise, $p > 2$ does not offer computational benefits, and so we concentrate on the case $p = 2$. In this case we get:

$$d_L^{(2)}(g_1, g_2) = \sqrt{\frac{1}{2} \int_G |f(g_1^{-1} \circ g) - f(g_2^{-1} \circ g)|^2 \, d(g)}.$$

Note that introducing the factor of $1/2$ does not change the fact that this is a metric. Expanding the square, we see

$$2 \left(d_L^{(2)}(g_1, g_2) \right)^2 = \int_G f^2(g_1^{-1} \circ g) \, d(g) + \int_G f^2(g_2^{-1} \circ g) \, d(g) - 2 \int_G f(g_1^{-1} \circ g) f(g_2^{-1} \circ g) \, d(g).$$

$$(6.49)$$

Because of the left invariance of the measure, the first two integrals are equal. Furthermore, if we define $f^*(g) = f(g^{-1})$, then the last term may be written as a convolution. That is,

$$d_L^{(2)}(g_1, g_2) = \sqrt{\|f\|_2^2 - (f * f^*)(g_1^{-1} \circ g_2)}, \qquad (6.50)$$

where

$$\|f\|_2^2 = \int_G f^2(g) \, d(g),$$

and convolution on the set of rigid-body motions was defined earlier.

Since the maximum value of $(f * f^*)(g)$ occurs at $g = e$ and has the value $\|f\|^2$, we see that $d_L(g_1, g_1) = 0$, as must be the case for it to be a metric. The left invariance is clearly evident when written in the form of (6.50), since $(h \circ g_1)^{-1} \circ (h \circ g_2) = g_1^{-1} \circ (h^{-1} \circ h) \circ g_2$

$= g_1^{-1} \circ g_2$. We also recognize that unlike the class of metrics in the previous section, this one has a bounded value. That is

$$\max_{g_1, g_2 \in G} \sqrt{\|f\|_2^2 - (f * f^*)(g_1^{-1} \circ g_2)} \leq \|f\|_2.$$

By restricting the choice of f to symmetric functions, i.e., $f(g) = f(g^{-1})$, then this metric takes the form

$$d_L^{(2)}(g_1, g_2) = d_L^{(2)}(e, g_1^{-1} \circ g_2) = \sqrt{\|f\|_2^2 - (f * f)(g_1^{-1} \circ g_2)}.$$

One reason why the left invariance of metrics is useful is because as a practical matter it can be more convenient to calculate $d_L^{(2)}(e, g)$ and evaluate it at $g = g_1^{-1} \circ g_2$ than to calculate $d_L^{(2)}(g_1, g_2)$ directly.

6.9.3 Park's Metric

Distance metrics for $SE(3)$ can be constructed from those for $\mathbb{R}^3$ and $SO(3)$. If we denote $d_{SO(3)}(R_1, R_2)$ to be any metric on $SO(3)$ and $d_{\mathbb{R}^3}(\mathbf{b}_1, \mathbf{b}_2)$ to be any metric on $\mathbb{R}^3$, then the following will be metrics on $SE(3)$ for $g_i = (R_i, \mathbf{b}_i) \in SE(3)$:

$$d_{SE(3)}^{(1)}(g_1, g_2) = L \cdot d_{SO(3)}(R_1, R_2) + d_{\mathbb{R}^3}(\mathbf{b}_1, \mathbf{b}_2);$$

and

$$d_{SE(3)}^{(2)}(g_1, g_2) = \sqrt{L^2 \cdot (d_{SO(3)}(R_1, R_2))^2 + (d_{\mathbb{R}^3}(\mathbf{b}_1, \mathbf{b}_2))^2}.$$

Here L is a measure of length that makes the units of orientational and translational displacements compatible. If the metric is normalized a priori by some meaningful length, then we can choose $L = 1$. In the special case when $d_{SO(3)}(R_1, R_2) = \theta(R_1^{-1} R_2)$, the metric $d_{SE(3)}^{(1)}(g_1, g_2)$ above was introduced into the mechanical design community by Park [56].

6.9.4 Kinetic Energy Metric

A metric can be constructed for $SE(3)$ that is analogous to the one in (5.75). Namely, we can define

$$d(g_1, g_2) = \sqrt{\min_{g(t) \in SE(3)} \frac{1}{2} \int_{t_1}^{t_2} \left[M \dot{\mathbf{b}} \cdot \dot{\mathbf{b}} + \mathrm{tr}(\dot{R} J \dot{R}^T) \right] dt} \tag{6.51}$$

subject to the boundary conditions $g(t_1) = g_1$ and $g(t_2) = g_2$. For a nonspherical rigid body with mass M and moment of inertia in the body-fixed frame J (as in (6.47)), the path $g(t)$ that satisfies the above conditions must usually be generated numerically. This makes it less attractive than some of the other metrics discussed earlier.

6.9.5 Metrics on $SE(3)$ from Metrics on $SO(4)$

In Section 6.8.2 we reviewed how rigid-body motions in three dimensions can be approximated as rotations in four dimensions, and how rotations in four dimensions decompose into two copies of rotations (unit quaternions) in three dimensions. This fact has been used by Etzel and McCarthy [24, 25] to define approximate metrics on $SE(3)$. We now generalize this formulation and review their results in this context.

Any given pair of 4×4 rotation matrices $(\exp M^+(\mathbf{s}), \exp M^-(\mathbf{t}))$ can be mapped to a pair of 3×3 rotation matrices $(\exp S, \exp -T)$. We have metrics between rotations in three dimensions that can be easily extended to define distance between pairs of pairs of matrices. For instance, given the two pairs (R_1, Q_1) and (R_2, Q_2), it is easy to see that

$$D_1((R_1, Q_1), (R_2, Q_2)) = d(R_1, R_2) + d(Q_1, Q_2)$$

and

$$D_2((R_1, Q_1), (R_2, Q_2)) = ([d(R_1, R_2)]^2 + [d(Q_1, Q_2)]^2)^{\frac{1}{2}}$$

satisfy the metric properties whenever $d(R, Q)$ does. Here $d(R, Q)$ can be any of the metrics for $SO(3)$ discussed earlier. In addition, it is clear that this metrics $D_i(\cdot, \cdot)$ for $SO(4)$ inherit the bi-invarance of $d(\cdot, \cdot)$. That is, given any $R_0, Q_0 \in SO(3)$,

$$D_i(R_0 R, Q_0 Q) = D_i(R R_0, Q Q_0) = D_i(R, Q).$$

Likewise for right shifts.

The idea of approximating rigid-body motions in three-dimensional space with rotations in four-dimensional space has a long history, with some of the ideas expressed in [37].

6.9.6 Partial Bi-Invariance of $SE(3)$ Metrics

Since the publication of the original version of this book, a number of other metrics have been introduced. These include some based on infinitesimal differential-geometric approaches [45]. Others have been developed by computer scientists with an eye towards robot motion planning [1, 42]. And a third set of approaches are based on polar and singular value decompositions [44]. For a more comprehensive review of the recent literature, see [15]. Closely related to the concept of metrics is that of geodesics, and recent work in this area includes [36].

Regardless, one simple fact remains the same: nontrivial bi-invariant metrics for $SE(3)$ are not possible. However, many $SE(3)$ metrics are fully invariant with respect to shifts on one side, and *partially* invariant with respect to shifts on the other. For example, both of the metrics in Section 6.9.3, and the one in (6.48) (when W is defined with J as a multiple of the identity matrix), are fully left invariant *and, in addition,* invariant under right shifts by pure rotations. This "partial" bi-invariance opens up interesting opportunities as discussed in [16]

6.10 Summary

In this chapter we have reviewed a number of different ways to describe rigid-body motions, to measure distance between motions, and to assign frames of reference to curves and discrete linkages. Applications of these techniques from rigid-body kinematics can be found in a number of areas including manufacturing, robotics, computer-aided design (CAD), polymer science and mechanics. In CAD, problems of interest include the design of curves and surfaces with an associated set of attached frames (see, e.g., [38, 41, 73, 74]) and references therein). In manufacturing, the geometry of swept volumes (see, e.g., [4, 5, 6]) is closely related to the problem of path generation for machine tools. In Chapters 12, 17 and 18 we will examine problems in robotics, polymer science and mechanics where knowleges of rigid-body kinematics is useful.

References

1. Amato, N. M., Bayazit, O. B., Dale, L. K., Jones, C., Vallejo, D., "Choosing Good Distance Metrics and Local Planners for Probabilistic Roadmap Methods," *Proceedings of the 1998 IEEE International Conference on Robotics and Automation. ICRA '98*, (Vol. 1, pp. 630 – 637), 1998.
2. Ball, R.S., *A Treatise on the Theory of Screws*, Cambridge University press, 1900.
3. Bishop, R., "There is More Than One Way to Frame a Curve," *American Mathematical Monthly*, 82: 246 – 251, 1975.
4. Blackmore, D., Leu, M.C., Shih, F., "Analysis and Modelling of Deformed Swept Volumes," *Computer-Aided Design*, 26(4): 315 – 326, April 1994.
5. Blackmore, D., Leu, M.C., Wang, L.P., Jiang, H., "Swept Volume: A Retrospective and Prospective View," *Neural, Parallel and Scientific Computations*, 5(1–2): 81 – 102, 1997.
6. Blackmore, D., Leu, M.C., Wang, L.P., "The Sweep-Envelope Differential Equation Algorithm and Its Application to NC Machining Verification," *Computer-Aided Design*, 29(9): 629 – 637, Sept. 1997.
7. Blaschke, W., "Euklidische Kinematik und nichteuklidische Geometrie I,II," *Zeitschrift für Mathematik und Physik*, 60: 61 – 91, 203 – 204, 1911.
8. Blaschke, W. Müller, H.R., *Ebene Kinematik*, Verlag von R. Oldenbourg, München, 1956.
9. Bottema, O., Roth, B., *Theoretical Kinematics*, Dover Publications, Inc., New York, 1979.
10. Brockett, R.W., "Robotic Manipulators and the Product of Exponentials Formula," in *Mathematical Theory of Networks and Systems* (A. Fuhrman, ed.), pp. 120 – 129, Springer-Verlag, 1984.
11. Călugăreanu, G., "L'integrale de Gauss et l'analyse des noeuds tridimensionnels," *Revue de Mathématiques Pures et Appliquées*, 4(5): 5 – 20, 1959.
12. Chasles, M., "Note sur les propriétés générales du système de deux corps semblables entre eux et placés d'une manière quelconque dans l'espace; et sur le déplacement fini ou infiniment petit d'un corps solide libre." *Férussac, Bulletin des Sciences Mathématiques*, 14: 321 – 326, 1830.
13. Chen, G., Wang, H., Lin, Z., "Determination of Identifiable Parameters in Robot Calibration Based on the POE Formula," *IEEE Transactions on Robotics*, 2014.
14. Chirikjian, G.S., *Theory and Applications of Hyper-redundant Robotic Manipulators*, Ph.D. Dissertation, School of Engineering and Applied Science, California Institute of Technology, May 1992.
15. Chirikjian, G.S., *Stochastic Models, Information Theory, and Lie Groups: Volumes I + II*, Birkhäuser, Boston, 2009/2012.
16. Chirikjian, G.S., "Partial Bi-Invariance of $SE(3)$ Metrics," *Journal of Computing and Information Science in Engineering* 15(1): 011008, 2015.
17. Chirikjian, G.S., Zhou, S., "Metrics on Motion and Deformation of Solid Models," *Journal of Mechanical Design*, 120(2): 252 – 261, June 1998.

18. Clifford, W.K., "Preliminary Sketch of Biquaternions," in *Mathematical Papers* (R. Tucker ed.), MacMillan, London, 1882.

19. Craig, J.J., *Introduction to Robotics, Mechanics and Control*, Addison-Wesley Publishing Company, Reading Mass., 1986. (3rd ed., Prentice Hall, 2004.)

20. Denavit, J., Hartenberg, R.S., "A Kinematic Notation for Lower-Pair Mechanisms Based on Matrices," *Journal of Applied Mechanics*, 22: 215 – 221, June 1955.

21. Donelan, P. S., Gibson, C. G., "First-Order Invariants of Euclidean Motions," *Acta Applicandae Mathematicae*, 24(3): 233 – 251, 1991.

22. Donelan, P. S., Gibson, C. G., "On the Hierarchy of Screw Systems," *Acta Applicandae Mathematicae*, 32(3): 267 – 296, 1993.

23. Dooley, J., McCarthy, J.M., "Spatial Rigid Body Dynamics Using Dual Quaternion Components," *Proceedings of the 1991 IEEE International Conference on Robotics and Automation*, pp. 90-95, Sacramento, California, April 1991.

24. Etzel, K.R., *Biquaternion Theory and Applications to Spatial Kinematics*, M.S. Thesis, University of California, Irvine, Spring, 1996.

25. Etzel, K. R., McCarthy, J. M., "Spatial Motion Interpolation in an Image Space of $SO(4)$," *Proceedings of 1996 ASME Design Engineering Technical Conference and Computers in Engineering Conference*, August 18-22, 1996, Irvine, California

26. Eyring, H., "The Resultant Electric Moment of Complex Molecules," *Physical Review*, 39(4): 746 – 748, 1932.

27. Fanghella, P., Galletti, C., "Metric Relations and Displacement Groups in Mechanism and Robot Kinematic," *Journal of Mechanical Design, Transactions of the ASME*, 117(3): 470 – 478, Sept. 1995.

28. Fary, I., "Sur la courbure totale d'une courbe gauche faisant un noeud," *Bulletin de la Société Mathématique de France*, 77: 128 – 138, 1949.

29. Fischer, I.S., *Dual-Number Methods in Kinematics, Statics, and Dynamics*, CRC Press, Boca Raton, 1999.

30. Flory, P.J., *Statistical Mechanics of Chain Molecules*, Wiley Interscience Publishers, New York, 1969.

31. Fenchel, W., "Über Krümmung und Windung geschlossener Raumkurven," *Mathematische Annalen*, 101(1): 238 – 252, 1929.

32. Fuller, F.B., "The Writhing Number of a Space Curve," *Proc. Nat. Acad. Sci. USA*, 68(4): 815 – 819, April 1971.

33. Ge, Q.J., "On the Matrix Algebra Realization of the Theory of Biquaternions," *Proc. of the 1994 ASME Mechanisms Conference*, DE-Vol. 70, pp. 425 – 432, 1994.

34. Grünwald, J., "Ein Abbildungsprinzip, welches die ebene Geometrie und Kinematik mit der räumlichen Geometrie verknüpft," *Sitzungsbericht der kaiserlichen Akademie Wissenschaften. Wien*, 120: 677 – 741, 1911.

35. Hervé, J.M., "Analyse Structurelle des Mécanismes par Groupe des Déplacements," *Mechanisms and Machine Theory*, 13(4): 437 – 450, 1978.

36. Holm, D.D., Noakes, L., Vankerschaver, J., "Relative Geodesics in the Special Euclidean Group," *Proceedings of the Royal Society A: Mathematical, Physical and Engineering Science*, 469(2158), Article 20130297, 2013.

37. Inonu, E., Wigner, E.P., "On the Contraction of Groups and Their Representatations," *Proceedings of the National Academy of Sciences*, 39(6): 510 – 524, June 1953.

38. Kallay, M., and Ravani, B., "Optimal Twist Vectors as a Tool for Interpolating a Network of Curves with a Minimum Energy Surface," *Computer Aided Geometric Design*, 7(6): 465 – 473, 1990.

39. Karger, A., Novák, J., *Space Kinematics and Lie Groups*, Gordon and Breach Science Publishers, New York, 1985.

40. Kazerounian, K., Rastegar, J., "Object Norms: A Class of Coordinate and Metric Independent Norms for Displacement," *Flexible Mechanisms, Dynamics, and Analysis*, ASME DE-Vol.47, pp. 271–275, 1992.

41. Klok, F., "Two Moving Coordinate Frames for Sweeping Along a 3D Trajectory," *Computer Aided Geometric Design*, 3(3): 217 – 229, 1986.

42. Kuffner, J.J., "Effective Sampling and Distance Metrics for 3D Rigid Body Path Planning," *Proceedings of the 2004 IEEE International Conference on Robotics and Automation, ICRA'04,* (Vol. 4, pp. 3993–3998), April 2004.

43. Kyatkin, A.B., Chirikjian, G.S., "Regularized Solutions of a Nonlinear Convolution Equation on the Euclidean Group," *Acta Applicandae Mathematica,* 53(1): 89–123, 1998.

44. Larochelle, P.M., Murray, A.P., Angeles, J., "A Distance Metric for Finite Sets of Rigid-Body Displacements via the Polar Decomposition," *Journal of Mechanical Design,* 129(8): 883–886, 2007.

45. Lin, Q., Burdick, J.W., "Objective and Frame-Invariant Kinematic Metric Functions for Rigid Bodies," *The International Journal of Robotics Research,* 19(6): 612–625, 2000.

46. Martinez, J.M.R., "Representation of the Euclidean Group and Kinematic Mappings," *Proc. 9th World Congress on the Theory of Machines and Mechanisms,* Vol. 2, pp. 1594–1600, 1996.

47. Martinez, J.M.R., Duffy, J., "On the Metrics of Rigid Body Displacement for Infinite and Finite Bodies," *ASME Journal of Mechanical Design,* 117(1): 41–47, March 1995.

48. Mattice, W.L., Suter, U.W., *Conformational Theory of Large Molecules: The Rotational Isomeric State Model in Macromolecular Systems,* John Wiley and Sons, New York, 1994.

49. McCarthy, J.M., "Planar and Spatial Rigid Motion as Special Cases of Spherical and 3-Spherical Motion," *Journal of Mechanisms, Transmissions, and Automation in Design,* 105: 569–575, Sept. 1983.

50. McCarthy, J.M., *An Introduction to Theoretical Kinematics* MIT Press, Cambridge, Mass. 1990. (revised version published as an iBook in 2013).

51. Millman, R.S., Parker, G.D., *Elements of Differential Geometry,* Prentice-Hall Inc., Englewood Cliffs, NJ 1977.

52. Milnor, J., "On the Total Curvature of Knots," *Annals of Mathematics,* 52(2): 248–257, 1950.

53. Murray, R.M., Li, Z., Sastry, S.S., *A Mathematical Introduction to Robotic Manipulation,* CRC Press, Boca Raton, 1994.

54. Oka, S., "Zur Theorie der statistischen Molekülgestalt hochpolymerer Kettenmoleküle unter Berücksichtigung der Behinderung der freien Drehbarkeit," *Proc. Physico-Mathematical Society of Japan,* 24: 657–672, 1942.

55. Park, F.C., "Computational Aspects of the Product-of-Exponentials Formula for Robot Kinematics," *IEEE Transactions on Automatic Control,* 39(3): 643-647, March 1994.

56. Park, F.C., "Distance Metrics on the Rigid-Body Motions with Applications to Mechanism Design," *Journal of Mechanical Design, Transactions of the ASME,* 117(1): 48–54, March 1995.

57. Park, F.C., Brockett, R.W., "Kinematic Dexterity of Robotic Mechanisms," *The International Journal of Robotics Resaerch,* 13(1): 1–15, Feb. 1994.

58. Park, F.C., Ravani, B., "Bézier Curves on Riemannian Manifolds and Lie Groups with Kinematics Applications, " *ASME Journal of Mechanical Design* 117(1): 36–40, March 1995.

59. Pham, H.-L., Perdereau, V., Adorno, B.V., Fraisse, P., "Position and Orientation Control of Robot Manipulators Using Dual Quaternion Feedback," *Proceedings of the 2010 IEEE/RSJ International Conference on Intelligent Robots and Systems,* pp. 658-663, Taipei, Taiwan, October 1822 2010.

60. Pohl, W.F., "Some Integral Formulas for Space Curves and Their Generalizations," *American Journal of Mathematics,* 90:1321–1345, 1968.

61. Pohl, W.F., "The Self-Linking Number of a Closed Space Curve," *Journal of Mathematics and Mechanics,* 17(10): 975–985, 1968.

62. Ravani, B., Roth, B., "Motion Synthesis Using Kinematic Mapping," *ASME Journal of Mechanisms, Transmissions and Automation in Design,* 105(3): 460–467, Sept. 1983.

63. Rooney, J., "A Comparison of Representations of General Spatial Screw Displacements," *Environment and Planning B,* 5: 45–88, 1978.

64. Segre, B., "Una nuova caratterizzazione della sfera," *Atti della Accademia Nazionale del Lincei, Rendiconti,* 3: 420–422, 1947.

65. Segre, B., "Sulla torsione integrale delle curve chiuse sghembe," *Atti della Accademia Nazionale del Lincei, Rendiconti,* 3:422–426, 1947.
66. Selig, J.M., *Geometrical Fundamentals of Robotics,* 2nd ed., Springer, New York, 2005.
67. Spong, M., Hutchinson, S., Vidyasagar, M., *Robot Modeling and Control,* John Wiley and Sons, New York, 2006.
68. Stéphanos, M.C., "Mémoire sur la représentation des homographies binaires par des points de l'espace avec application à l'étude des rotations sphériques," *Mathematische Annalen,* 22(3): 299–367, 1883.
69. Study, E., *Geometrie der Dynamen,* Teubner Verlag, Leipzig, 1903.
70. Study, E., "Grundlagen und Ziele der analytischen Kinematik," *Sitzungsberichte der Berliner Mathematischen Gesellschaft,* 104: 36–60, 1912.
71. Tsai, L.-W., *Robot Analysis: The Mechanics of Serial and Parallel Manipulators,* John Wiley and Sons, New York, 1999.
72. Vologodskii, A., *Topology and Physics of Circular DNA,* CRC Press, Boca Raton, 1992.
73. Wang, W., Jüttler, B., Zheng, D., Liu, Y., "Computation of Rotation Minimizing Frames," *ACM Transactions on Graphics,* 27(1): Article 2, 2008.
74. Wesselink, W., Veltkamp, R.C., "Interactive Design of Constrained Variational Curves," *Computer Aided Geometric Design,* 12(5): 533–546, August 1995.
75. White, J.H., "Self-Linking and the Gauss Integral in Higher Dimensions," *American Journal of Mathematics,* 91: 693–728, 1969.
76. Wigner, E., "On Unitary Representations of the Inhomogeneous Lorentz Group," *Annals of Mathematics,* 40(1): 149–204, Jan. 1939.
77. Žefran, M., Kumar, V., Croke, C., "Metrics and Connections for Rigid-Body Kinematics," *The International Journal of Robotics Research,* 18(2): 243–258, Feb. 1999.

7

Group Theory

In this chapter we illustrate how some ideas introduced previously in the context of pure rotations and full rigid-body motions generalize to other kinds of transformations. We review the concept of a "group" including finite groups and Lie groups, and introduce methods for performing concrete calculations with these mathematical objects. Then, with this new perspective, we return to rotations and rigid-body motions.

7.1 Introduction

Group theory is a mathematical generalization of the study of symmetry. Before delving into the numerous definitions and fundamental results of group theory, we begin this chapter with a discussion of symmetries, or more precisely, operations that preserve symmetries, in geometrical objects and equations. For in-depth treatments of group theory from the perspective of algebra see [1, 2, 6, 15, 16, 20, 21, 23]; from the perspective of geometry see [13, 10, 25, 35, 43, 44, 48]; for applications in chemistry and physics see [5, 7, 14, 19, 24]; and for other of its applications see [3, 8, 22, 29, 32, 39] and the other references cited throughout this chapter.

7.1.1 Motivational Examples

Consider the set of numbers $\{0, 1, 2\}$ with addition "modulo 3." This means that when two numbers are added together, and the result is greater than 3, then 3 is subtracted from the result until a number in $\{0, 1, 2\}$ is obtained. For instance, $1 + 1 \equiv 2 \,(\mathrm{mod}\,3)$, whereas $1 + 2 \equiv 0 \,(\mathrm{mod}\,3)$. Modular arithmetic is common to our everyday experience when dealing with time. For instance, when working with a 12-hour clock, telling time is essentially an exercise in modulo 12 arithmetic (e.g., if it is 7 o'clock now, then seven hours later it is $7 + 7 \equiv 2 \,(\mathrm{mod}\,12)$). The only difference between telling time and modulo 12 arithemtic is that we say "12 o'clock" instead of "0 o'clock."

We can make the following addition table for $\{0, 1, 2\}$ with the operation $+_{[3]}$ (addition modulo 3)[1] that reads

$+_{[3]}$	0 1 2
0	0 1 2
1	1 2 0
2	2 0 1

---

[1] We use both $+_{[3]}$ and $+\,(\mathrm{mod}\,3)$ to mean addition modulo 3.

If $k \in \{0,1,2\}$, a similar table results for the multiplication of the numbers $W_k = e^{2\pi i k/3}$ reads

$$
\begin{array}{c|ccc}
\cdot & W_0 & W_1 & W_2 \\
\hline
W_0 & W_0 & W_1 & W_2 \\
W_1 & W_1 & W_2 & W_0 \\
W_2 & W_2 & W_0 & W_1
\end{array}.
$$

Hence $(\{0,1,2\}, + (\mathrm{mod}\,3))$ and $(\{W_0, W_1, W_2\}, \cdot)$ are in some sense equivalent. This sense is made more precise in Section 7.2.4.

Now consider the equilateral triangle shown in Figure 7.1. We attach labels a, b, c to the vertices and 1, 2, 3 to the positions in the plane to which the vertices correspond at the beginning of our thought experiment. The relative position of a with respect to the center point 0 is denoted with the vector $\mathbf{x}_{0a}$. Next consider all spatial rotations that move the triangle from its initial orientation to one in which the alphabetically labeled vertices are again matched (though possibly in a different way) to the numbered spaces. In fact, there are six kinds of motion. We can: (1) do nothing and leave the correspondence $a \to 1$, $b \to 2$, $c \to 3$; (2) we can rotate counterclockwise by $2\pi/3$ radians around $\mathbf{x}_{0a} \times \mathbf{x}_{0b}$ (the cross product of vectors $\mathbf{x}_{0a}$ and $\mathbf{x}_{0b}$) resulting in the new matching $a \to 2$, $b \to 3$, $c \to 1$; (3) we can rotate counterclockwise by $4\pi/3$ (or equivalently clockwise by $-2\pi/3$) around $\mathbf{x}_{0a} \times \mathbf{x}_{0b}$ resulting in $a \to 3$, $b \to 1$, $c \to 2$; (4) We can rotate the original triangle by an angle π around $\mathbf{x}_{0a}$ resulting in the correspondence $a \to 1$, $b \to 3$, $c \to 2$; (5) We can rotate the original triangle by an angle π around $\mathbf{x}_{0b}$, resulting in the correspondence $a \to 3$, $b \to 2$, $c \to 1$; (6) We can rotate the original triangle by an angle π around $\mathbf{x}_{0c}$, resulting in the correspondence $a \to 2$, $b \to 1$, $c \to 3$. We label the above set of operations as $g_0, ..., g_5$ where $g_0 = e$ is called the identity (or "do nothing") operation. We note that while there are an infinite number of motions that will take a triangle back into itself in this way, they are all of the "same kind" as one of the six mentioned above, provided we look only at where the motion starts and ends.

The natural issue to consider is what happens if one of these operations is followed by another. We denote this composition of relative motion as $g_i \circ g_j$ if first we perform g_j then g_i.[2] For instance, $g_2 \circ g_1$ is first a clockwise rotation by $2\pi/3$ followed by a clockwise rotation by $4\pi/3$. The result is a rotation by 2π which is the same as if no motion had been performed. Thus $g_2 \circ g_1 = e$. Likewise we can do this for all combinations of elements. If these elements truly represent all possible operations that map the vertices to the underlying positions in the plane, then the composition of such operations must result in another one of these operations. In general, a *Latin square* is a table indicating how the composition of operations results in new ones. In the present example it can be constructed as:

$$
\begin{array}{c|cccccc}
\circ & e & g_1 & g_2 & g_3 & g_4 & g_5 \\
\hline
e & e & g_1 & g_2 & g_3 & g_4 & g_5 \\
g_1 & g_1 & g_2 & e & g_4 & g_5 & g_3 \\
g_2 & g_2 & e & g_1 & g_5 & g_3 & g_4 \\
g_3 & g_3 & g_5 & g_4 & e & g_2 & g_1 \\
g_4 & g_4 & g_3 & g_5 & g_1 & e & g_2 \\
g_5 & g_5 & g_4 & g_3 & g_2 & g_1 & e
\end{array}. \tag{7.1}
$$

---

[2]If we were using rotation matrices to describe these relative motions, they would be multiplied in the opposite order.

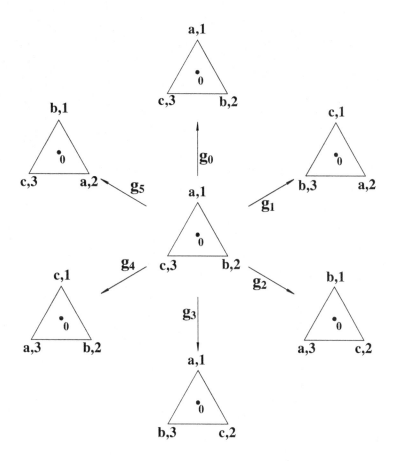

Fig. 7.1. Symmetry Operations of the Equilateral Triangle

This table is constructed by calculating $g_i \circ g_j$ where i indexes the row and j indexes the column. If we consider only the block

$$
\begin{array}{c|ccc}
\circ & e & g_1 & g_2 \\
\hline
e & e & g_1 & g_2 \\
g_1 & g_1 & g_2 & e \\
g_2 & g_2 & e & g_1
\end{array}
\tag{7.2}
$$

corresponding to rotations in the plane, the set of transformations C_3 results. More generally, C_n is the set of planar rotational symmetry operations of a regular n-gon.

Any finite set of transformations for which a Latin square results (i.e., with each transformation appearing once and only once in each row and column) is called a *finite group* when, in addition, the associative law $g_i \circ (g_j \circ g_k) = (g_i \circ g_j) \circ g_k$ holds. The Latin square for a finite group is called a *Cayley table*. The fact that each element appears once and only once implies that for each operation we can find a unique operation, which when composed, will result in the identity. Thus, each element in a group has an inverse.

We now consider a variant of the example shown in Figure 7.1. Consider the same basic set of transformations, except now everything is defined with respect to the labels fixed in space 1, 2, and 3. Whereas g_1 was a rotation about $\mathbf{x}_{0a} \times \mathbf{x}_{0b}$, we will now consider the transformation $\hat{g}_1$ that performs the rotation by the same angle as g_1, only now around the vector $\mathbf{x}_{01} \times \mathbf{x}_{02}$. Likewise, $\hat{g}_i$ is the version of g_i with space-fixed labels replacing the labels fixed in the moving triangle. The two operations $\hat{g}_i$ and g_i are the same when the moving and fixed triangles initially coincide, but differ after the triangle has been displaced.

In the center triangle in Figure 7.1, a is coincident with 1 and b is coincident with 2 so it appears that g_1 and $\hat{g}_1$ are the same. However, this is not so when we apply these transformations to the other triangles in the figure. Working through all the products, we find

$$
\begin{array}{c|cccccc}
\circ & e & \hat{g}_1 & \hat{g}_2 & \hat{g}_3 & \hat{g}_4 & \hat{g}_5 \\
\hline
e & e & \hat{g}_1 & \hat{g}_2 & \hat{g}_3 & \hat{g}_4 & \hat{g}_5 \\
\hat{g}_1 & \hat{g}_1 & \hat{g}_2 & e & \hat{g}_5 & \hat{g}_3 & \hat{g}_4 \\
\hat{g}_2 & \hat{g}_2 & e & \hat{g}_1 & \hat{g}_4 & \hat{g}_5 & \hat{g}_3 \\
\hat{g}_3 & \hat{g}_3 & \hat{g}_4 & \hat{g}_5 & e & \hat{g}_1 & \hat{g}_2 \\
\hat{g}_4 & \hat{g}_4 & \hat{g}_5 & \hat{g}_3 & \hat{g}_2 & e & \hat{g}_1 \\
\hat{g}_5 & \hat{g}_5 & \hat{g}_3 & \hat{g}_4 & \hat{g}_1 & \hat{g}_2 & e \\
\end{array}
\tag{7.3}
$$

Hence we see that in general $g_i \circ g_j \neq \hat{g}_i \circ \hat{g}_j$. However, there is a relationship between these two sets of operations, which is the same relationship between absolute and relative motions discussed in Chapter 6. The operations g_i are defined relative to the moving triangle, whereas $\hat{g}_i$ are defined relative to the fixed spatial triangle. It follows that each relative motion is generated by a similarity-like transformation of absolute motions using prior absolute motions:

$$
g_i \circ g_j = (\hat{g}_j \circ \hat{g}_i \circ \hat{g}_j^{-1}) \circ \hat{g}_j = \hat{g}_j \circ \hat{g}_i.
\tag{7.4}
$$

While the set of symmetry operations $\{e, g_1, g_2, g_3, g_4, g_5\}$ in the above example is finite, this need not be the case in general. For example, consider an infinite lattice where each cell is an $L \times L$ square. We observe that a rotation by $\pi/2$ about the center of any cell and translations along the x and y directions by one lattice unit take the whole lattice back into itself.

We can express any lattice transformation as:

$$
x'(j, k, l) = x \cos(l\pi/2) - y \sin(l\pi/2) + L \cdot j
$$

$$
y'(j, k, l) = x \sin(l\pi/2) + y \cos(l\pi/2) + L \cdot k
$$

where $l \in [0, 3]$, $j, k \in \mathbb{Z}$ and (x, y) and (x', y') are the coordinates of cell centers. Figure 7.2 illustrates the composition of this kind of transformation. Since $\mathbb{Z}$ is infinite, so too is this set of transformations, which is used in the field of crystallography. It is easy to show that the composition of two such transformations results in another one of the same kind, and that each can be inverted. The associative law holds as well.

Thus far we have seen two sets of operations (one finite and one infinite) which reflect discrete symmetries. The concept of a group is, however, not limited to discrete symmetry operations. Consider, for example, the set of all rigid-body transformations that leave a sphere unchanged in shape or location. This is nothing more than the set

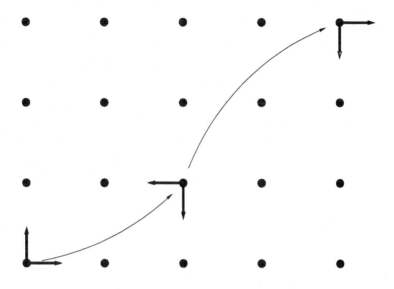

Fig. 7.2. Symmetry Operations of the Square Lattice

of rotations, $SO(3)$, studied in Chapter 5. It has an infinite number of elements in the same way that a solid ball in $\mathbb{R}^3$ consists of an infinite number of points. That is, it is a continuous set of transformations. The full set of rigid-body motions, $SE(3)$, is also a continuous set of transformations, but unlike $SO(3)$, it is not bounded. $SO(3)$ is an example of something called a compact Lie group, whereas $SE(3)$ is a noncompact Lie group (see Appendix G).

As a fifth motivational example of transformations which leave something unchanged, consider Newton's first and second laws expressed in the form of the equation

$$\mathbf{F} = m\frac{d^2\mathbf{x}}{dt^2}. \tag{7.5}$$

This is *the* fundamental equation of all of non-relativistic mechanics, and it forms the foundation for most of mechanical, civil, and aerospace engineering. It describes the position, $\mathbf{x}$, of a particle of mass m subjected to a force $\mathbf{F}$, when all quantities are viewed in an inertial (non accelerating) frame of reference. It has been known for centuries that this equation works independent of what inertial frame of reference is used. This can be observed by changing spatial coordinates

$$\mathbf{x}^1 = R_1\mathbf{x}^0 + \mathbf{b}_1 + \mathbf{v}_1 t_0,$$

where R_1 is a constant (time-independent) rotation matrix relating the orientation of the two frames, and $\mathbf{b}_1 + \mathbf{v}_1 t_0$ is the vector relating the position of the origin of reference frame 1 relative to reference frame 0 ($\mathbf{b}_1$ and $\mathbf{v}_1$ are constant vectors). By observing that forces transform from one frame to another as $\mathbf{F}^1 = R_1\mathbf{F}^0$, it is easy to see that

$$\mathbf{F}^0 = m\frac{d^2\mathbf{x}^0}{dt_0^2} \quad \Longleftrightarrow \quad \mathbf{F}^1 = m\frac{d^2\mathbf{x}^1}{dt_1^2}$$

since rotations can be inverted. Furthermore, in classical mechanics, the choice of when to begin measuring time is arbitrary, and so the transformation $t_1 = t_0 + a_1$ leaves Newton's laws unchanged as well. All together this can be written as

$$\begin{pmatrix} \mathbf{x}^1 \\ t_1 \\ 1 \end{pmatrix} = \begin{pmatrix} R_1 & \mathbf{v}_1 & \mathbf{b}_1 \\ \mathbf{0}^T & 1 & a_1 \\ \mathbf{0}^T & 0 & 1 \end{pmatrix} \begin{pmatrix} \mathbf{x}^0 \\ t_0 \\ 1 \end{pmatrix}. \tag{7.6}$$

A second change of reference frames relating $\mathbf{x}^2$ to $\mathbf{x}^1$ can be performed in exactly the same way. The composition of these changes results in the relationship between reference frames 0 and 2. It can be verified by multiplication of the matrices in (7.6) that this composition of transformations results in a transformation of the same kind. These invertible transformations reflect the so-called *Galilean invariance* of mechanics. The set of all such *Galilean transformations* contains both the set of rotations and rigid body motions.

Finally, we note that all of the examples considered thus far can be represented as invertible matrices with special structure. The set of all nonsingular $N \times N$-dimensional matrices with real (or complex) entries is denoted as $GL(N, \mathbb{R})$ (or $GL(N, \mathbb{C})$).

With these examples we hope the reader has gained some intuition into sets of transformations and why such transformations can be important in the physical sciences and engineering. In the subsequent sections of this chapter we formalize the concept and properties of groups in general. It is not critical that everything in these sections be crystal clear after the first reading. However, for readers unfamiliar with the basic concepts of group theory, it is important to have at least seen this material before proceeding to the following chapters.

7.1.2 General Terminology

A review of terminology, notation, and basic results from the mathematics literature are presented in this section. (See, e.g., [6, 16].) We begin with a few standard definitions, which are abbreviated so that the reader gets a feeling for the underlying concepts without the overhead required for complete rigor. It is assumed that the reader is familiar with the concepts of a set and function.

Definition 7.1. A *(closed) binary operation*, $\circ$, is a law of composition that produces an element of a set from two elements of the same set. More precisely, let G be a set and $g_1, g_2 \in G$ be arbitrary elements. Then $(g_1, g_2) \rightarrow g_1 \circ g_2 \in G$.

Definition 7.2. A *group* is a set G together with a (closed) binary operation $\circ$ such that for any elements $g, g_1, g_2, g_3 \in G$ the following properties hold:

- $g_1 \circ (g_2 \circ g_3) = (g_1 \circ g_2) \circ g_3$.
- There exists an element $e \in G$ such that $e \circ g = g \circ e = g$.
- For every element $g \in G$ there is an element $g^{-1} \in G$ such that $g^{-1} \circ g = g \circ g^{-1} = e$.

The first of the above properties is called associativity; the element e is called the identity of G; and g^{-1} is called the inverse of $g \in G$. In order to distinguish between the group and the underlying set, the former is denoted $(G, \circ)$, unless the operation is understood from the context, in which case G refers to both the set and the group. In the special case when for every two elements $g_1, g_2 \in G$, it is true that $g_1 \circ g_2 = g_2 \circ g_1$, the group is called *commutative* (or *Abelian*[3]), otherwise it is called *noncommutative* (or *non-Abelian*).

We have already seen five examples of noncommutative groups (all the sets of transformations described in Subsection 7.1.1 of this chapter), as well as four commutative ones that are central to standard Fourier analysis: $(\mathbb{R}, +)$ (real numbers with addition); $(T, + \,(\mathrm{mod}\,2\pi))$ (a circle with circular addition); $(\mathbb{Z}, +)$ (the integers with addition); $C_N = (Z_N, + \,(\mathrm{mod}\,N))$ (the integers $0, ..., N-1$ with addition modulo N, or equivalently, planar rotations that preserve a regular N-gon centered at the origin).

We define

$$h_g = g \circ h \circ g^{-1} \quad \text{and} \quad h^g = g^{-1} \circ h \circ g. \tag{7.7}$$

In various texts both h_g and h^g are called *conjugation* of the element h by g. When we refer to conjugation, we will specify which of the above definitions is being used. Of course they are related as $h^g = h_{g^{-1}}$. Note however that

$$(h^{g_1})^{g_2} = g_2^{-1} \circ (g_1^{-1} \circ h \circ g_1) \circ g_2 = (g_1 \circ g_2)^{-1} \circ h \circ (g_1 \circ g_2) = h^{g_1 \circ g_2}$$

whereas $(h_{g_1})_{g_2} = h_{g_2 \circ g_1}$.

For groups that are Abelian, conjugation leaves an element unchanged: $h_g = h^g = h$. For noncommutative groups (which is more often the case in group theory), conjugation usually does not result in the same element again. However, it is always true that conjugation of the product of elements is the same as the product of the conjugations:

$$h_1^g \circ h_2^g = (h_1 \circ h_2)^g \quad \text{and} \quad (h_1)_g \circ (h_2)_g = (h_1 \circ h_2)_g.$$

A *subgroup* is a subset of a group ($H \subseteq G$) which is itself a group such that it is closed under the group operation of G, $e \in H$ and $h^{-1} \in H$ whenever $h \in H$. The notation for this is $H \leq G$. One kind of subgroup, called a *conjugate subgroup*, is generated by conjugating all of the elements of an arbitrary subgroup with a fixed element g of the group. This is denoted as $gHg^{-1} = \{g \circ h \circ g^{-1} | h \in H\}$ for a single $g \in G$. If $H_1, H_2 \leq G$ and $gH_1g^{-1} = H_2$ for a $g \in G$, then H_1 and H_2 are said to be conjugate to each other. A subgroup $N \leq G$ which is conjugate to itself so that $gNg^{-1} = N$ for *all* $g \in G$ is called a *normal* subgroup of G. In this case, we use the notation $N \trianglelefteq G$. For sets H and G such that $H \subseteq G$ and $H \neq G$ we write $H \subset G$. H is then called a proper subset of G. Likewise, for groups if $H \leq G$ and $H \neq G$, we write $H < G$ (H is a proper subgroup of G), and for a normal subgroup we write $N \triangleleft G$ when $N \neq G$ (N is a proper normal subgroup of G).

Example 7.1: The group C_3 is both an Abelian and normal proper subgroup of the group of symmetry operations of the equilateral triangle (We will prove this shortly).

A *transformation group* $(G, \circ)$ is a group that acts on a set S in such a way that $g \cdot x \in S$ is defined for all $x \in S$ and $g \in G$ and has the properties:

$$e \cdot x = x \quad \text{and} \quad (g_1 \circ g_2) \cdot x = g_1 \cdot (g_2 \cdot x) \in S$$

for all $x \in S$ and $e, g_1, g_2 \in G$. The operation $\cdot$ defines the *action* of G on S. When such an action exists, we sometimes use the word "action" at the element level and say that $g \in G$ acts on $x \in S$ as $g \cdot x$.

If any two elements of $x_1, x_2 \in S$ can be related as $x_2 = g \cdot x_1$ for some $g \in G$, then G is said to act *transitively* on S.

---

[3]Named after Niels Henrick Abel (1802-1829)

Example 7.2: The group of rotations is a transformation group that acts transitively on the sphere, because any point on the sphere can be moved to any other point using an appropriate rotation. The group of rotations $SO(N)$ also acts on $\mathbb{R}^N$, though not transitively because a point on any sphere centered at the origin remains on that sphere after a rotation.

Example 7.3: The group of rigid-body motions $SE(N)$ is a transformation group that acts transitively on $\mathbb{R}^N$.

A group $(G, \circ)$ is called a *topological* group if the mapping $(g_1, g_2) \to g_1 \circ g_2^{-1}$ is continuous. Continuity is defined in the context of an appropriate topology (as defined in the Appendix B).

Definition 7.3. A Hausdorff space[4] X is called *locally compact* if for each $x \in X$ and every open set U containing x there exists an open set W such that $Cl(W)$ is compact and $x \in W \subseteq Cl(W) \subseteq U$. A *locally compact group* is a group for which the underlying set is locally compact [18, 31].

Definition 7.4. The *Cartesian product* of two sets H and G is the set $H \times G$ which consists of all ordered pairs of the form (h, g) for all $g \in G$ and $h \in H$. That is, $(h, g) \in H \times G$, whereas $(g, h) \in G \times H$. The *direct product* of two groups $(H, \hat{\circ})$ and $(G, \circ)$ is the group $(P, \odot) = (H, \hat{\circ}) \times (G, \circ)$ such that $P = H \times G$, and for any two elements $p_1 = (h_1, g_1)$ and $p_2 = (h_2, g_2) \in P$, the group operation is defined as

$$p_1 \odot p_2 \doteq (h_1 \hat{\circ} h_2, g_1 \circ g_2).$$

Often the group operation is suppressed and the expression $P = H \times G$ is used to denote both the direct product as well as the Cartesian product.

Definition 7.5. Let the group $(G, \circ)$ be a transformation group that acts on the set H where $(H, +)$ is an Abelian group. Then the *semi-direct product* of $(H, +)$ and $(G, \circ)$ is the group $(P, \hat{\circ}) = (H, +) \rtimes (G, \circ)$ such that $P = H \times G$, and for any two elements $p_1 = (h_1, g_1)$ and $p_2 = (h_2, g_2) \in P$ the group operation is defined as

$$p_1 \hat{\circ} p_2 \doteq (g_1 \cdot h_2 + h_1, g_1 \circ g_2).$$

Often in group theory the operations are suppressed and we simply write

$$P = H \rtimes G$$

when it is clear from the context that P refers to the group rather than the underlying set. Therefore, while it is true that as a set $(h, g) \in H \times G$, as a group $(h, g) \in H \rtimes G$.

A normal subgroup of $H \rtimes G$ which is "essentially the same" as H is[5]

$$\tilde{H} \doteq H \rtimes \{e\}$$

where e is the identity element of G. Usually the distinction between $\tilde{H}$ and H is ignored in group theory, and we write

---

[4]See Appendix B.
[5]This concept of "essentially the same" will be formalized and generalized later as the idea of *isomorphism*.

$$H \lhd (H \rtimes G).$$

Note that the triangles in this expression point in the same direction. It is also possible to define $G \ltimes H \doteq H \rtimes G$ so that the triangles can align in the other direction as

$$(G \ltimes H) \rhd H.$$

This is nothing deep, but sometimes makes it easier to write chains of equalities or subgroup relationships that would be difficult to express otherwise. It also allows us the flexibility to write $(h, g) \in H \rtimes G$ or $(g, h) \in G \ltimes H$.

As an example of a semi-direct product, we observe that

$$SE(N) = \mathbb{R}^N \rtimes SO(N) = SO(N) \ltimes \mathbb{R}^N.$$

We now examine a more exotic (though geometrically intuitive) example of a groups.

Example 7.4: Consider the ribbon defined in Chapter 6. A group of deformations that acts on ribbons to produce new ribbons was defined in [26]. This group consists of a set of ordered triplets of the form $g = (\mathbf{a}, \alpha(s), M(s))$ where $\mathbf{a} \in \mathbb{R}^3$ is a position vector, and for $s \in [0, 1]$, the smooth bounded functions $M(s) \in SO(3)$ and $\alpha(s) \in \mathbb{R}$ are defined. The group law is written as

$$(\mathbf{a}_1, \alpha_1(s), M_1(s)) \circ (\mathbf{a}_2, \alpha_2(s), M_2(s)) = (\mathbf{a}_1 + \mathbf{a}_2, \alpha_1(s) + \alpha_2(s), M_1(s)M_2(s)) \quad (7.8)$$

for all $s \in [0, 1]$.

A ribbon is specified by the ordered pair $p = (\mathbf{x}(s), \mathbf{n}(s))$, where $\mathbf{x}(s)$ is the backbone curve and $\mathbf{n}(s)$ is the unit normal to the backbone curve for each value of s. The action of the group defined in (7.8) on a ribbon is defined as

$$g \cdot p \doteq (\mathbf{a}, \alpha(s), M(s)) \cdot (\mathbf{x}(s), \mathbf{n}(s))$$

$$= \left(\mathbf{x}(0) + \mathbf{a} + \int_0^s e^{\alpha(\sigma)} M(\sigma) \mathbf{x}'(\sigma) \, d\sigma, M(s) \mathbf{n}(s) \right)$$

where $\mathbf{x}'(\sigma)$ is the derivative of $\mathbf{x}(\sigma)$. It is easy to confirm by direct substitution that

$$g_1 \cdot (g_2 \cdot p) = (g_1 \circ g_2) \cdot p.$$

7.2 Finite Groups

In this section, we examine *finite groups*, i.e., groups for which the underlying sets have a finite number of elements. We begin by enumerating all of the possible groups with 2, 3, and 4 elements and then introduce the concepts of a permutation and matrix representation.

7.2.1 Multiplication Tables

Since the group composition operation (also called group multiplication) assigns to each ordered pair of group elements a new group element, we can always describe a finite group in the form of a multiplication table. Entries of the form $g_j \circ g_k$ are inserted in this table, where g_k determines in which column and g_j determines in which row the result of the product is placed. This convention is illustrated as:

$$
\begin{array}{c|ccccc}
\circ & e & \cdots & g_k & \cdots & g_n \\
\hline
e & e & \cdots & g_k & \cdots & g_n \\
\vdots & \vdots & \cdots & \vdots & \cdots & \vdots \\
g_j & g_j & \cdots & g_j \circ g_k & \cdots & g_j \circ g_n \\
\vdots & \vdots & \cdots & \vdots & \ddots & \vdots \\
g_n & g_n & \cdots & g_n \circ g_k & \cdots & g_n \circ g_n
\end{array}
$$

The number of elements in a finite group G is called the *order* of G and is denoted as $|G|$.

We consider below the simplest abstract groups, which are those of order 2, 3 and 4. If $|G| = 1$, things are not interesting because the only element must be the identity element: $G = (\{e\}, \circ)$. If $|G| = 2$, things are not much more interesting. Here we have

$$
\begin{array}{c|cc}
\circ & e & a \\
\hline
e & e & a \\
a & a & e
\end{array}.
$$

Note that the form of this table is exactly the same as the form of those for the groups $(\{1, -1\}, \cdot)$ and $\mathbb{Z}_2 = (\{0, 1\}, + \,(\mathrm{mod}\,2))$. Groups with tables that have the same structure are abstractly the same group, even though they may arise in completely different situations and the group elements may be labeled differently. Later this concept will be formalized with the definition of *isomorphism*.

Considering the case when $|G| = 3$ we see that the table must look like

$$
\begin{array}{c|ccc}
\circ & e & a & b \\
\hline
e & e & a & b \\
a & a & b & e \\
b & b & e & a
\end{array}.
$$

By definition of the identity, the first row and column must be of the form above. Less obvious, but also required, is that $a \circ a = b$ must hold for this group. If this product had resulted in e or a it would not be possible to fill in the rest of the table while observing the constraint that every element appear in each row and column only once.

Moving on to the case when $|G| = 4$, we see that this is the first example where the table is not completely determined by the number of elements. The following three Latin squares can be formed:

$$
\begin{array}{c|cccc}
\circ & e & a & b & c \\
\hline
e & e & a & b & c \\
a & a & e & c & b \\
b & b & c & e & a \\
c & c & b & a & e
\end{array}
\tag{7.9}
$$

$$
\begin{array}{c|cccc}
\circ & e & a & b & c \\
\hline
e & e & a & b & c \\
a & a & e & c & b \\
b & b & c & a & e \\
c & c & b & e & a
\end{array}
\tag{7.10}
$$

$$
\begin{array}{c|cccc}
\circ & e & a & b & c \\
\hline
e & e & a & b & c \\
a & a & b & c & e \\
b & b & c & e & a \\
c & c & e & a & b
\end{array}
\tag{7.11}
$$

We observe right away by looking at the tables in (7.9) and (7.10) that the corresponding sets of transformations cannot be equivalent under renaming of elements because the number of times the identity element, e, appears on the diagonal is different. On the other hand it may not be clear that these two exhaust the list of all possible *nonisomorphic* tables. For the tables in (7.9) and (7.10), it is not difficult to show that the choice of element in the $(2, 2)$ entry in the table, together with how the 2×2 block in the lower right corner of the table is arranged, completely determines the rest of the table. An exhaustive check for associativity confirms that these are the Cayley tables of two groups of order 4.

Regarding the third table above, if the b in the $(2, 2)$ entry is replaced with c, the table is completely determined as

$$
\begin{array}{c|cccc}
\circ & e & a & b & c \\
\hline
e & e & a & b & c \\
a & a & c & e & b \\
b & b & e & c & a \\
c & c & b & a & e
\end{array}.
$$

However, this table is "the same as" the table in (7.11) because it just represents a reordering (or renaming) of elements. In particular, renaming $e \to e$, $a \to a$, $b \to c$, and $c \to b$ shows that these two tables are essentially the same, and so it is redundant to consider them both. Likewise (7.10) and (7.11) are isomorphic under the mapping $e \to e$, $a \to b$, $b \to a$, $c \to c$.

At this point it may be tempting to believe that every Latin square is the Cayley table for a group. However, this is not true. For instance,

$$
\begin{array}{c|ccccc}
\circ & e & a & b & c & d \\
\hline
e & e & a & b & c & d \\
a & a & e & d & b & c \\
b & b & c & a & d & e \\
c & c & d & e & a & b \\
d & d & b & c & e & a
\end{array}
$$

is a valid Latin square, but $(a \circ b) \circ c \neq a \circ (b \circ c)$, and so the associative property fails to hold.

We observe from the symmetry about the major diagonal of all of the tables in (7.9)-(7.11) that $g_j \circ g_k = g_k \circ g_j$, and therefore all of the operations defined by these tables are commutative. It may be tempting to think that all finite groups are Abelian, but this is not true, as can be observed from the Cayley table for the group of symmetry operations of the equilateral triangle.

While it is true that we can always construct a symmetric group table of any dimension corresponding to an Abelian group and certain theorems of group theory provide results when a group of a given order must be Abelian (such as if $|G|$ is a prime number or the square of a prime number), the groups of most interest in applications are often those that are not commutative.

For a finite group G, we can define the *order of an element* $g \in G$ as the smallest positive number r for which $g^r = e$, where $g^r = g \circ \cdots \circ g$ (r times), and by definition

$g^0 = e$. The cyclic group C_n has the property that $|C_n| = n$, and every element of C_n can be written as g^k for some $k \in \{0, ..., n-1\}$ and some $g \in C_n$ where $g^n = e$.

7.2.2 Permutations and Matrices

The group of permutations of n letters (also called the *symmetric group*) is denoted as S_n. It is a finite group containing $n!$ elements. The elements of S_n can be arranged in any order, and for any fixed arrangement we label the elements of S_n as σ_{i-1} for $i = 1, ..., n!$. We denote an arbitrary element $\sigma \in S_n$ as

$$\sigma = \begin{pmatrix} 1 & 2 & \dots & n \\ \sigma(1) & \sigma(2) & \dots & \sigma(n) \end{pmatrix}.$$

Changing the order of the columns in the above element does not change the element. Therefore, in addition to the above expression,

$$\sigma = \begin{pmatrix} 2 & 1 & \dots & n \\ \sigma(2) & \sigma(1) & \dots & \sigma(n) \end{pmatrix} = \begin{pmatrix} n & 2 & \dots & 1 \\ \sigma(n) & \sigma(2) & \dots & \sigma(1) \end{pmatrix}$$

where dots denote those columns not explicitly listed.

As an example of a permutation group, the elements of S_3 are

$$\sigma_0 = \begin{pmatrix} 1\,2\,3 \\ 1\,2\,3 \end{pmatrix}; \quad \sigma_1 = \begin{pmatrix} 1\,2\,3 \\ 2\,3\,1 \end{pmatrix}; \quad \sigma_2 = \begin{pmatrix} 1\,2\,3 \\ 3\,1\,2 \end{pmatrix};$$

$$\sigma_3 = \begin{pmatrix} 1\,2\,3 \\ 2\,1\,3 \end{pmatrix}; \quad \sigma_4 = \begin{pmatrix} 1\,2\,3 \\ 3\,2\,1 \end{pmatrix}; \quad \sigma_5 = \begin{pmatrix} 1\,2\,3 \\ 1\,3\,2 \end{pmatrix}.$$

More than one convention is in use for what these numbers mean and how the product of two permutations is defined. This is much like the way the product of symmetry operations of the equilateral triangle can be generated as $g_i \circ g_j$ or $\hat{g}_i \hat{g}_j$. The convention we use can be interpreted in the following way. Reading the expression $\sigma_i \circ \sigma_j$ from right to left we say "each k is sent to $\sigma_i(\sigma_j(k))$."

Using this convention we calculate

$$\sigma_2 \circ \sigma_3 = \begin{pmatrix} 1\,2\,3 \\ 3\,1\,2 \end{pmatrix} \circ \begin{pmatrix} 1\,2\,3 \\ 2\,1\,3 \end{pmatrix} = \begin{pmatrix} 1\,2\,3 \\ 1\,3\,2 \end{pmatrix} = \sigma_5.$$

The computation explicitly is $\sigma_2(\sigma_3(1)) = \sigma_2(2) = 1$, $\sigma_2(\sigma_3(2)) = \sigma_2(1) = 3$, $\sigma_2(\sigma_3(3)) = \sigma_2(3) = 2$. However, in one's mind it is convenient to simply read from upper right to lower left yielding the sequence $1 \to 2 \to 1$, $2 \to 1 \to 3$, $3 \to 3 \to 2$, with the first and last numbers of each composition kept as columns of the new permutation.

With this composition rule, the group table is

$\circ$	σ_0	σ_1	σ_2	σ_3	σ_4	σ_5
σ_0	σ_0	σ_1	σ_2	σ_3	σ_4	σ_5
σ_1	σ_1	σ_2	σ_0	σ_4	σ_5	σ_3
σ_2	σ_2	σ_0	σ_1	σ_5	σ_3	σ_4
σ_3	σ_3	σ_5	σ_4	σ_0	σ_2	σ_1
σ_4	σ_4	σ_3	σ_5	σ_1	σ_0	σ_2
σ_5	σ_5	σ_4	σ_3	σ_2	σ_1	σ_0

We note that the table for the symmetry operations on the equilateral triangle is the same as the table for the group of permutations of three objects. Permutations are an important example of finite groups. An important theorem of group theory (Cayley's theorem [15]) states that *every* finite group is isomorphic (i.e., the same to within an arbitrary relabling) to a subgroup of a permutation group.

We can assign to each permutation of n objects, σ, an invertible $n \times n$ matrix, $D(\sigma)$, that has the property

$$D(\sigma_i \circ \sigma_j) = D(\sigma_i)D(\sigma_j). \qquad (7.12)$$

The mapping $\sigma \to D(\sigma)$ is called a *matrix representation* of S_n. The matrices $D(\sigma)$ are constructed in a straightforward way: if the permutation assigns $i \to j$, put a 1 in the j, i entry of the matrix. Otherwise insert a zero.

For example, under the mapping D, the six elements of S_3 have matrix representations [1]:

$$D(\sigma_0) = \begin{pmatrix} 1 & 0 & 0 \\ 0 & 1 & 0 \\ 0 & 0 & 1 \end{pmatrix}; \quad D(\sigma_1) = \begin{pmatrix} 0 & 0 & 1 \\ 1 & 0 & 0 \\ 0 & 1 & 0 \end{pmatrix}; \quad D(\sigma_2) = \begin{pmatrix} 0 & 1 & 0 \\ 0 & 0 & 1 \\ 1 & 0 & 0 \end{pmatrix};$$

$$D(\sigma_3) = \begin{pmatrix} 0 & 1 & 0 \\ 1 & 0 & 0 \\ 0 & 0 & 1 \end{pmatrix}; \quad D(\sigma_4) = \begin{pmatrix} 0 & 0 & 1 \\ 0 & 1 & 0 \\ 1 & 0 & 0 \end{pmatrix}; \quad D(\sigma_5) = \begin{pmatrix} 1 & 0 & 0 \\ 0 & 0 & 1 \\ 0 & 1 & 0 \end{pmatrix}.$$

We observe by direct calculation that (7.12) holds.

Given a vector $\mathbf{x} = [x_1, ..., x_n]^T \in \mathbb{R}^n$ and a permutation $\sigma \in S_n$, we find that

$$D(\sigma)\mathbf{x} = \begin{pmatrix} x_{\sigma^{-1}(1)} \\ \vdots \\ x_{\sigma^{-1}(n)} \end{pmatrix} \quad \text{and} \quad \mathbf{x}^T D(\sigma) = \begin{pmatrix} x_{\sigma(1)} \\ \vdots \\ x_{\sigma(n)} \end{pmatrix}^T$$

both define actions of S_n on $\mathbb{R}^n$. For example, if $n = 3$, then

$$D(\sigma_1)\mathbf{x} = \begin{pmatrix} 0 & 0 & 1 \\ 1 & 0 & 0 \\ 0 & 1 & 0 \end{pmatrix} \begin{pmatrix} x_1 \\ x_2 \\ x_3 \end{pmatrix} = \begin{pmatrix} x_3 \\ x_1 \\ x_2 \end{pmatrix}.$$

Under this action we see that the component x_i is *sent to place* $\sigma_1(i)$, and the component in the ith place after the permutation is $x_{\sigma_1^{-1}(i)}$.

Since these matrix representations can be generated for any permutation, and every finite group is a subgroup of a permutation group, it follows that every finite group can be represented with square matrices composed of ones and zeros with exactly a single one in each row and column. These are called *permutation matrices*.

We now examine a somewhat more exotic example of a finite group constructed from simpler groups.

Example 7.5: Given a finite group $(G, \circ)$, the Cartesian product of the set G with itself n times is denoted as G^n. The *wreath product* of G and S_n is the group $G \wr S_n = (G^n \times S_n, \diamond)$ where the product of two elements in $G^n \times S_n$ is defined as:

$$(h_1, ..., h_n; \sigma) \diamond (g_1, ..., g_n; \pi) \doteq (h_1 \circ g_{\sigma^{-1}(1)}, ..., h_n \circ g_{\sigma^{-1}(n)}; \sigma\pi).$$

As an example of a wreath product, consider the *hyper-octahedral group* $\mathbb{Z}_2 \wr S_n$. This group has been used to model the statistics of changes in circular DNA caused by

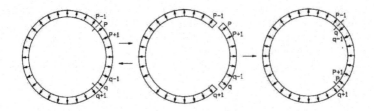

Fig. 7.3. The Action of $\mathbb{Z}_2 \wr S_n$ on Closed-Loop DNA

random removal and reattachment of segments [36]. See Figure 7.3 for an illustration of this process on circular DNA consisting of n base pairs. All possible operations defined by two cuts, a flip and reattachment into a single loop may be viewed as elements of $\mathbb{Z}_2 \wr S_n$, as are the compositions of two such operations. The number of base pairs in the circle is preserved under these operations, but the relative proximity of base pairs gets mixed up. Mathematically, this is an action of $\mathbb{Z}_2 \wr S_n$ on the circle divided into n segments.

7.2.3 Cosets, Fundamental Domains, and Orbits

In this section we review basic concepts concerning the decomposition of a group into cosets and the decomposition of a set on which a group acts into orbits.

Coset Spaces and Quotient Groups

Given a subgroup $H \leq G$, and any element $g \in G$, the *left coset* gH is defined as

$$gH \doteq \{g \circ h | h \in H\}.$$

Similarly, the right coset Hg is defined as

$$Hg \doteq \{h \circ g | h \in H\}.$$

In the special case when $g \in H$, the corresponding left and right cosets are equal to H. More generally for all $g \in G$, $g \in gH$ and $g_1 H = g_2 H$ if and only if $g_2^{-1} \circ g_1 \in H$. Likewise for right cosets $Hg_1 = Hg_2$ if and only if $g_1 \circ g_2^{-1} \in H$.

Any group is divided into disjoint left (right) cosets, and the statement "g_1 and g_2 are in the same left (right) coset" is an equivalence relation. This can be written explicitly for the case of left cosets as

$$g_1 \sim g_2 \quad \Leftrightarrow \quad g_1^{-1} \circ g_2 \in H. \tag{7.13}$$

Since H is a subgroup (and hence is itself a group), it is easy to verify that

$$g^{-1} \circ g = e \in H, \tag{7.14}$$

$$g_1^{-1} \circ g_2 \in H \quad \Rightarrow \quad (g_1^{-1} \circ g_2)^{-1} \in H, \tag{7.15}$$

and

$$g_1^{-1} \circ g_2 \in H, \quad g_2^{-1} \circ g_3 \in H \quad \Rightarrow \quad (g_1^{-1} \circ g_2) \circ (g_2^{-1} \circ g_3) = g_1^{-1} \circ g_3 \in H. \qquad (7.16)$$

Hence $\sim$ as defined in (7.13) is an equivalence relation since (7.14) says $g \sim g$, (7.15) says $g_1 \sim g_2$ implies $g_2 \sim g_1$, and (7.16) says that $g_1 \sim g_2$ and $g_2 \sim g_3$ implies $g_1 \sim g_3$. An analogous argument holds for right cosets with the equivalence relation defined as

$$g_1 \sim g_2 \Leftrightarrow g_1 \circ g_2^{-1} \in H$$

instead of (7.13). Sometimes instead of $\sim$, the notation

$$g_1 \equiv g_2 \bmod H$$

is used to denote either of the equivalence relations above when left or right has been specified in advance.

An important property of gH and Hg is that they have the same number of elements as H. Since the group is divided into disjoint cosets, each with the same number of elements, it follows that the number of cosets must divide without remainder the number of elements in the group. The set of all left(or right) cosets is called the left(or right) *coset space*, and is denoted as G/H (or $H\backslash G$). We can equate the order of a coset space with the order of the corresponding group and subgroup as

$$|G/H| = |H\backslash G| = |G|/|H|.$$

This result is called *Lagrange's theorem* [6]. Since the order of G/H and $H\backslash G$ are the same, it is sometimes convenient to use the notation

$$[G : H] \doteq |G|/|H|.$$

In analogy with the way a coset is defined, the conjugate of a subgroup H for a given $g \in G$ is defined as

$$gHg^{-1} \doteq \{g \circ h \circ g^{-1} | h \in H\}.$$

Recall that a subgroup $N \le G$ is called *normal* if and only if $gNg^{-1} \subseteq N$ for all $g \in G$. This is equivalent to the conditions $g^{-1}Ng \subseteq N$, and so we also write $gNg^{-1} = N$ and $gN = Ng$ for all $g \in G$.

When N is a normal subgroup of G, we use the notation $N \trianglelefteq G$, and when it is both a proper and normal subgroup we write $N \triangleleft G$.

Theorem 7.6. *If $N \trianglelefteq G$, then the coset space G/N together with the binary operation $(g_1 N)(g_2 N) = (g_1 \circ g_2)N$ is a group (called the quotient group).*

Proof. Note that for the operation defined in this way to make sense, the result should not depend on the particular choices of g_1 and g_2. Instead, it should only depend on which coset in G/N the elements g_1 and g_2 belong to. That is, if $g_1 N = h_1 N$ and $g_2 N = h_2 N$, with $g_i \ne h_i$, it should nevertheless be the case that the products are the same: $(g_1 \circ g_2)N = (h_1 \circ h_2)N$. This is equivalent to showing that $(g_1 \circ g_2)^{-1} \circ (h_1 \circ h_2) \in N$ whenever $g_1^{-1} \circ h_1 \in N$ and $g_2^{-1} \circ h_2 \in N$. Expanding out the product we see that

$$\begin{aligned}
(g_1 \circ g_2)^{-1} \circ (h_1 \circ h_2) &= g_2^{-1} \circ g_1^{-1} \circ h_1 \circ h_2 \\
&= g_2^{-1} \circ (g_1^{-1} \circ h_1) \circ h_2 \qquad (7.17) \\
&= (g_2^{-1} \circ h_2) \circ h_2^{-1} \circ (g_1^{-1} \circ h_1) \circ h_2.
\end{aligned}$$

Since $g_1^{-1} \circ h_1 \in N$ and N is normal, $h_2^{-1} \circ (g_1^{-1} \circ h_1) \circ h_2 \in N$ also. Since $g_2^{-1} \circ h_2 \in N$ and N is closed under the group operation, it follows that the whole product in (7.18) is in N.

With this definition of product, it follows that the identity of G/N is $eN = N$, and the inverse of $gN \in G/N$ is $g^{-1}N \in G/N$. The associative law follows from the way the product is defined and the fact that G is a group.

Fundamental Domains

Associated with any coset space, we can establish a rule for choosing exactly one element per coset to form a a special subset of G called a *fundamental domain*. For concreteness, we consider right cosets here, but the same concept applies equally well to left cosets.

Let $H < G$ and $s : H\backslash G \to G$ denote the rule for selecting coset representatives. That is, $s(Hg) \in G$ is a distinguished element that is chosen from Hg. A fundamental domain $F_{H\backslash G} \subset G$ is then defined as

$$F_{H\backslash G} \doteq \bigcup_{Hg \in H\backslash G} \{s(Hg)\}.$$

This fundamental domain is a set (not a group). It has the property that

$$F_{H\backslash G} \cap Hg = s(Hg). \tag{7.18}$$

If $h \in H - \{e\}$ and we act on $F_{H\backslash G}$ on the left by H, then we get a translated version of the fundamental domain, $h \cdot F_{H\backslash G}$. We observe that

$$F_{H\backslash G} \cap h \cdot F_{H\backslash G} = \emptyset$$

because, on the one hand, a nontrivial shift by h cannot leave $s(Hg)$ fixed, and, on the other hand, cosets are disjoint, and so $h \circ s(Hg) \neq s(Hg')$ for all $g' \in G$. Moreover, we can reconstruct G from shifted copies of $F_{H\backslash G}$ as

$$G = \bigcup_{h \in H} h \cdot F_{H\backslash G}.$$

Thus we can partition G into translates of fundamental domains, which is very different than partitioning G into cosets.

If G has a nontrivial left-invariant distance function associated with it, then a natural way to define the fundamental domain is as a *Voronoi cell*

$$F_{H\backslash G} \doteq \{g \in G \,|\, d(e,g) < d(g,h) \; \forall h \in H - \{e\}\}. \tag{7.19}$$

By defining $F_{H\backslash G}$ like this, then we can work backwards from (7.18) to define the selection rule $s : H\backslash G \to G$.

These ideas extend beyond the context of finite groups. For example if G is a group such as $SO(3)$ or $SE(2)$ it is possible to construct the interiors of Voronoi cells corresponding to discrete subgroups, H, as described in [45, 46].

Orbits and Stabilizers

Recall that a set X on which a group G acts is called a G-set. When the group G acts transitively on the set X, every two elements of the set $x_1, x_2 \in X$ are related as $x_2 = g \cdot x_1$ for some $g \in G$. When G acts on X but *not* transitively, then X is divided into multiple equivalence classes by G. The equivalence class containing $x \in X$ is called the *orbit* of x and is formally defined as

$$Orb(x) \doteq \{g \cdot x | g \in G\}.$$

The set of all orbits is denoted[6] as X/G, and it follows that

$$X = \bigcup_{\sigma \in X/G} \sigma \quad \text{and} \quad |X| = \sum_{i=1}^{|X/G|} |Orb(x_i)|,$$

where x_i is a representative of the ith orbit. The hierarchy established by these definitions is $x_i \in Orb(x_i) \in X/G$. When G acts transitively on X, there is only one orbit, and this is the whole of X. In this special case we write $X/G = X = Orb(x)$ for any x.

The subset of all elements of G that leave a particular element $x \in X$ fixed is called a *stabilizer, stability subgroup, little group*, or *isotropy subgroup* of x. It is formally defined as

$$G_x \doteq \{g \in G | g \cdot x = x\}.$$

The fact that G_x is a subgroup of G follows easily since $e \cdot x = x$ by definition, $x = g \cdot x$ implies

$$g^{-1} \cdot x = g^{-1} \cdot (g \cdot x) = (g^{-1} \circ g) \cdot x = e \cdot x = x,$$

and for any $g_1, g_2 \in G_x$

$$(g_1 \circ g_2) \cdot x = g_1 \cdot (g_2 \cdot x) = g_1 \cdot x = x.$$

Hence G_x contains the identity, $g^{-1} \in G_x$ for every $g \in G_x$, and G_x is closed under the group operation. We therefore write $G_x \leq G$.

We have seen thus far that G divides X into disjoint orbits. Isotropy subgroups of G are defined by the property that they leave elements of X fixed. The next theorem relates these two phenomena.

Theorem 7.7. *If G is a finite group and X is a finite G-set, then for each $x \in X$*

$$|Orb(x)| = |G/G_x|.$$

Proof. Define the mapping $m : Orb(x) \to G/G_x$ as $m(g \cdot x) = g\,G_x$. This is a mapping that takes in elements of the orbit $g \cdot x \in Orb(x)$ and returns the coset $g\,G_x \in G/G_x$. This mapping is well-defined since $g_1 \cdot x = g_2 \cdot x$ implies $g_2^{-1} \circ g_1 \in G_x$, and hence $g_1 G_x = g_2 G_x$. If we can show that this mapping is bijective, the number of elements in $Orb(x)$ and G/G_x must be the same. Surjectivity follows immediately from the definition of m. To see that m is injective, we observe that $g_1 G_x = g_2 G_x$ implies $g_2^{-1} \circ g_1 \in G_x$, and thus $x = (g_2^{-1} \circ g_1) \cdot x$, or equivalently $g_2 \cdot x = g_1 \cdot x$.

---

[6]An alternative notation (which actually makes a lot more sense because G acts on X on the left) is $G\backslash X$. But for consistency with the majority of the literature, and so as not to confuse the reader who may be consulting other works, we will use the notation X/G.

Another interesting property of isotropy groups is expressed below.

Theorem 7.8. *Given a group G acting on a set X, the conjugate of any isotropy group G_x by $g \in G$ for any $x \in X$ is an isotropy group, and in particular*

$$g\, G_x\, g^{-1} = G_{g \cdot x}. \tag{7.20}$$

Proof. $h \in G_x$ means $h \cdot x = x$. Likewise, $k \in G_{g \cdot x}$ means $k \cdot (g \cdot x) = g \cdot x$. Let $k = g \circ h \circ g^{-1}$. Then

$$k \cdot (g \cdot x) = (k \circ g) \cdot x = (g \circ h) \cdot x = g \cdot (h \cdot x) = g \cdot x.$$

Hence $k = g \circ h \circ g^{-1} \in G_{g \cdot x}$ whenever $h \in G_x$, and so we write $gG_xg^{-1} \subseteq G_{g \cdot x}$. On the other hand, *any* $k \in G_{g \cdot x}$ satisfies $k \cdot (g \cdot x) = g \cdot x$. We can write this as

$$g^{-1} \cdot (k \cdot (g \cdot x)) = g^{-1} \cdot (g \cdot x)$$

or

$$(g^{-1} \circ k \circ g) \cdot x = e \cdot x = x.$$

In other words, $g^{-1} \circ k \circ g = h \in G_x$. From this it follows that all $k \in G_{g \cdot x}$ must be of the form $k = g \circ h \circ g^{-1}$ for some $h \in G_x$, and so we write $gG_xg^{-1} \supseteq G_{g \cdot x}$.

7.2.4 Mappings

Definition 7.9. A *homomorphism* is a mapping from one group *into* another $h : (G, \circ) \to (H, \hat{\circ})$ such that[7]

$$h(g_1 \circ g_2) = h(g_1) \,\hat{\circ}\, h(g_2).$$

The word "into" refers to the fact that the values $h(g)$ for *all* $g \in G$ must be contained in H, but it is possible that elements of H exist for which there are no counterparts in G. This is illustrated in Figure 7.4.

It follows immediately from this definition that $h(g) = h(g \circ e) = h(g) \,\hat{\circ}\, h(e)$, and so $h(e)$ is the identity in H. Likewise, $h(e) = h(g \circ g^{-1}) = h(g) \,\hat{\circ}\, h(g^{-1})$ and so $(h(g))^{-1} = h(g^{-1})$. Thus a homomorphism $h : G \to H$ maps inverses of elements in G to the inverses of their counterparts in H, and the identity of G is mapped to the identity in H.

In general, a homomorphism will map more elements of G to the identity of H than just the identity. The set of all $g \in G$ for which $h(g) = h(e)$ is called the *kernel* of the homomorphism and is denoted as $\text{Ker}(h)$. It is easy to see from the definition of homomorphism that if $g_1, g_2 \in \text{Ker}(h)$, then so are their inverses and products. Thus $\text{Ker}(h)$ is a subgroup of G, and moreover it is a normal subgroup because given any $g \in \text{Ker}(h)$ and $g_1 \in G$,

$$h(g_1^{-1} \circ g \circ g_1) = (h(g_1))^{-1} \hat{\circ}\, h(g) \,\hat{\circ}\, h(g_1) = (h(g_1))^{-1} \hat{\circ}\, h(g_1) = h(e).$$

That is, conjugation of $g \in \text{Ker}(h)$ by any $g_1 \in G$ results in another element in $\text{Ker}(h)$, and so $\text{Ker}(h)$ is a normal subgroup of G and so we write $\text{Ker}(h) \trianglelefteq G$.

---

[7]Our notational choice of using $h(\cdot)$ as a mapping and $h(g)$ as an element of H is to emphasize that some part of H is, in a sense, parameterized by G.

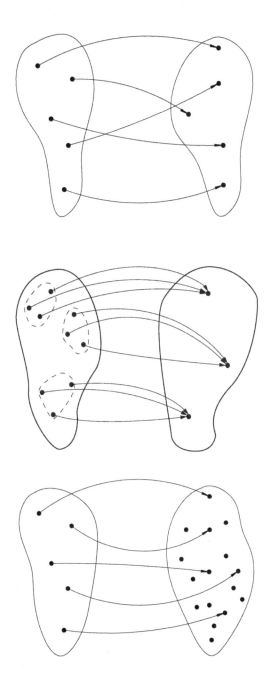

Fig. 7.4. Illustration of Homomorphisms: (top) Isomorphism; (middle) Epimorphism (points inside dashed lines constitute a fiber); (bottom) Monomorphism

In general, a homomorphism $h : G \to H$ will map all the elements of G to some subset of H. This subset is called the *image* of the homomorphism, and we write $\text{Im}(h) \subseteq H$. More specifically, since a homomorphism maps the identity of G to the identity of H, and inverses in G map to inverses in H, and since for any $g_1, g_2 \in G$ we have $h(g_1) \hat{\circ} h(g_2) = h(g_1 \circ g_2) \in \text{Im}(h)$, it follows that $\text{Im}(h)$ is a subgroup of H, and we write $\text{Im}(h) \le H$. (We note that even more generally if $K \le G$, then the image of K in H under the homomorphism h is also a subgroup of H.) A homomorphism that is surjective[8] is called an *epimorphism*. A homomorphism that is injective is called a *monomorphism*. A one-to-one homomorphism of G onto H is called an *isomorphism*, and when such an isomorphism exists between groups, the groups are called *isomorphic* to each other. If H and G are isomorphic, we write $H \cong G$. An isomorphism of a group G onto itself is called an *automorphism*. Conjugation of all elements in a group by one fixed element is an example of an automorphism. For a finite group, an automorphism is essentially a relabeling of elements. Isomorphic groups are fundamentally the same group expressed in different ways.

Example 7.6 (five examples): The two groups $(\{0, 1, 2\}, + \pmod 3)$ and $(\{W_0, W_1, W_2\}, \cdot)$ from Subsection 7.1.1 are isomorphic. The group $(\{e^{i\theta}\}, \cdot)$ and the set of all 2×2 rotation matrices

$$\begin{pmatrix} \cos\theta & -\sin\theta \\ \sin\theta & \cos\theta \end{pmatrix}$$

under matrix multiplication for $0 \le \theta < 2\pi$ are isomorphic. Given a fixed matrix $A \in GL(N, \mathbb{R})$, $\mathbf{x} \to A\mathbf{x}$ for all $\mathbf{x} \in \mathbb{R}^N$ defines an automorphism of $(\mathbb{R}^N, +)$ onto itself. The group of real numbers under addition, and the set of upper triangular 2×2 matrices of the form

$$\begin{pmatrix} 1 & x \\ 0 & 1 \end{pmatrix}$$

with operation of matrix multiplication are isomorphic. The mapping $\sigma \to D(\sigma)$ defining matrix representations of permutations in Section 7.2.2 is an isomorphism.

The Homomorphism Theorem

There are a number of interesting relationships between the concepts of homomorphism and quotient groups. For our purposes the following will be of the greatest importance.

Theorem 7.10. *Let G and H be groups and $h : G \to H$ be a homomorphism. Then $G/Ker(h) \cong Im(h)$.*

Proof. Define $\theta : G/Ker(h) \to H$ as

$$\theta(\eta(g)) = h(g)$$

for all $g \in G$ where $\eta : G \to G/Ker(h)$ is defined as

$$\eta(g) \doteq gKer(h). \tag{7.21}$$

(In general, the function $\eta : G \to G/K$ that identifies $g \in G$ with $gK \in G/K$ is called the "natural projection" or "natural map" from G to G/K.) We need to show that: (1) θ does not depend on the choice of coset representative; (2) θ is a homomorphism; (3) θ is bijective.

---

[8]See Appendix B for a review of definitions from set theory.

(1) The independence of the choice of coset representative is observed as follows. Let $g_1, g_2 \in G$. Then given $g_1 Ker(h) = g_2 Ker(h)$, it must be that $g_1 = g_2 \circ k$ for some $k \in Ker(h)$, and so we have

$$\theta(g_1 Ker(h)) = h(g_1) = h(g_2 \circ k)$$
$$= h(g_2) \,\hat{\circ}\, h(k) = h(g_2) \,\hat{\circ}\, h(e)$$
$$= h(g_2) = \theta(g_2 Ker(h)).$$

(2) Clearly θ is a homomorphism because

$$\theta(g_1 Ker(h)) \,\hat{\circ}\, \theta(g_2 Ker(h)) = h(g_1) \,\hat{\circ}\, h(g_2) = \theta((g_1 \circ g_2) Ker(h)).$$

(3) We now show that θ is both surjective and injective. By definition, $\theta(gKer(h)) = h(g)$. Running over all $g \in G$ yields the whole of $Im(h)$, and so $\theta : G/Ker(h) \to Im(h)$ is surjective. Injectivity in this context means that $h(g_1) = h(g_2)$ implies $\theta(g_1 Ker(h)) = \theta(g_2 Ker(h))$ for all $g_1, g_2 \in G$. When $h(g_1) = h(g_2)$, we can write

$$h(e) = [h(g_2)]^{-1} \,\hat{\circ}\, h(g_1) = h(g_2^{-1}) \,\hat{\circ}\, h(g_1) = h(g_2^{-1} \circ g_1),$$

which means that $g_2^{-1} \circ g_1 \in Ker(h)$. Therefore it must follow that $g_1 Ker(h) = g_2 Ker(h)$.

The above theorem is sometimes called the "Fundamental Theorem of Homomorphisms" or the "First Isomorphism Theorem." There are in fact a number of other isomorphism theorems that will not be discussed here.

G-Morphisms

Let X and Y be two sets on which a group G acts. To distinguish between the two different kinds of actions, denote $g \cdot x \in X$ and $g \bullet y \in Y$ for all $x \in X$ and $y \in Y$.

A *G-morphism* is a mapping $f : X \to Y$ with the property

$$f(g \cdot x) = g \bullet f(x). \tag{7.22}$$

This property is often called *G-equivariance*.

We now review three important examples of G-morphisms found in [38].

Example 7.7: Suppose that $g \bullet y = y$ for all $y \in Y$ and $g \in G$. Then $f : X \to Y$ with the property

$$f(g \cdot x) = f(x)$$

for all $x \in X$ is a G-morphism that is constant on each orbit $\sigma \in X/G$.

Example 7.8 Let Y be the set of all subgroups of G. That is, $H \in Y \Leftrightarrow H \leq G$. Define the action $g \bullet H = gHg^{-1}$. Then the mapping $f : X \to Y$ defined as $f(x) = G_x$ (the stabilizer of x) satisfies the equivariance property because

$$f(g \cdot x) = G_{g \cdot x} = g \, G_x \, g^{-1} = g \bullet f(x).$$

Example 7.9 Let M be a set and $g \odot m \in M$ for all $m \in M$ and $g \in G$ define an action. Let X be the Cartesian product $X = G \times M$, and let G act on X as

$$g \cdot (h, m) = (g \circ h \circ g^{-1}, g \odot m)$$

for $g \in G$ and $(h, m) \in X$. Following [38], if $Z \subset G \times M$ is defined as

$$Z \doteq \{(h, m) | h \odot m = m\},$$

then it can be verified that $g \cdot (h, m) \in Z$ and $g^{-1} \cdot (h, m) \in Z$ and so

$$gZ \doteq \{g \cdot (h, m) | (h, m) \in Z\} = Z.$$

Therefore, the mapping $f : Z \rightarrow Z$ defined as $f(h, m) \doteq g \cdot (h, m)$ is bijective.

It can then be shown that both $\theta : Z \rightarrow M$ defined as $\theta(h, m) \doteq m$, and $\tau : Z \rightarrow G$ defined as $\tau(h, m) \doteq h$ are G-morphisms.

The inverse images (or *fibers*) of θ and τ are the subsets of Z that map to any given $m \in M$ and $h \in G$, respectively. Recall that for any mapping $f : X \rightarrow Y$, the inverse image $f^{-1}(y)$ is the set of all $x \in X$ such that $f(x) = y \in Y$. It can be shown that the inverse images of the above G-morphisms are

$$\theta^{-1}(m) = \{(h, m) | h \in G_m\}$$

and

$$\tau^{-1}(h) = \{(h, m) | m \in F(h)\}$$

where

$$F(h) = \{m | h \odot m = m\} \subseteq M.$$

Counting Formulas

The G-morphisms τ and θ of the previous section provide a tool for counting the number of elements in the set Z. We can evaluate $|Z|$ by first counting how many elements are in $\tau^{-1}(h)$ for a given h, then sum over all h:

$$|Z| = \sum_{m \in F(h)} |\tau^{-1}(h)|,$$

or we can count the elements of Z as

$$|Z| = \sum_{h \in G_m} |\theta^{-1}(m)|.$$

Both of these equations can be viewed as special cases of (B.2) in Appendix B with the function taking the value 1 on every element of the set Z, and where membership in a fiber is the equivalence relation.

We exclude the subset $\{e\} \times M$ from Z, and define

$$P = \bigcup_{h \in G - \{e\}} F(h).$$

We can then perform the counting computations [38]:

$$|Z - \{e\} \times M| = \sum_{h \in G - \{e\}} |F(h)|$$

and

$$|Z - \{e\} \times M| = \sum_{m \in P} (|G_m| - 1) = \sum_{Orb(m) \in P/G} \frac{|G|}{|G_m|} (|G_m| - 1)$$

where in the sum on the right hand side it is understood that m is an arbitrary representative of $Orb(m)$. The last equality holds because $|G_{h \odot m}| = |G_m|$ is constant on each orbit $Orb(m)$, and from the Orbit-Stabilizer Theorem and Lagrange's Theorem

$$|Orb(m)| = |G/G_m| = |G|/|G_m|.$$

In Section 7.3.5 the formula

$$\sum_{h \in G - \{e\}} |F(h)| = \sum_{Orb(m) \in P/G} \frac{|G|}{|G_m|} (|G_m| - 1) \tag{7.23}$$

is used to classify the finite subgroups of $SO(3)$.

7.2.5 Conjugacy Classes, Class Functions, and Class Products

In this section we summarize important results that can be found in many texts on group theory including [19, 47].

Conjugacy Classes and Conjugate Subgroups

Two elements $a, b \in G$ are said to be *conjugate* to each other if $a = g^{-1} \circ b \circ g$ for some $g \in G$. Conjugacy of two elements is an equivalence relation because the following properties hold: $a = e^{-1} \circ a \circ e$ (reflexivity); $a = g^{-1} \circ b \circ g$ implies $b = g \circ a \circ g^{-1}$ (symmetry); and $a = g_1^{-1} \circ b \circ g_1$ and $b = g_2^{-1} \circ c \circ g_2$ implies $a = (g_2 \circ g_1)^{-1} \circ c \circ (g_2 \circ g_1)$ (transitivity).

If we conjugate all the elements of a subgroup $H \leq G$ with respect to a particular $g \in G$, the result is also a subgroup, called a *conjugate subgroup*, denoted $g^{-1}Hg$. It follows that $g^{-1}Hg$ is a subgroup because: $e \in H$ and hence $e = g^{-1} \circ e \circ g = e \in g^{-1}Hg$; for any $h, h^{-1} \in H$ we have $(g^{-1} \circ h \circ g)^{-1} = g^{-1} \circ h^{-1} \circ g \in g^{-1}Hg$; and for any $h_1, h_2 \in H$ we have $(g^{-1} \circ h_1 \circ g) \circ (g^{-1} \circ h_1 \circ g) = g^{-1} \circ (h_1 \circ h_2) \circ g \in g^{-1}Hg$.

If instead of fixing an element g and conjugating a whole subgroup by g, we choose a fixed element $b \in G$ and calculate $g^{-1} \circ b \circ g$ for *all* $g \in G$, the result is a set of elements in G called the *conjugacy class* (or simply *class* for short) containing b. All elements in a class are conjugate to every other element in the same class. An element of the group can only belong to one class. That is, the group is divided (or partitioned) into disjoint classes. This follows from the more general fact that an equivalence relation (of which conjugacy is one example) partitions a set into disjoint subsets. If C_i denotes the ith conjugacy class and $|C_i|$ denotes the number of group elements in this class, then

$$\sum_{i=1}^{\alpha} |C_i| = |G| \tag{7.24}$$

where α is the number of classes. Equation (7.24) is called the *class equation*. Clearly, $\alpha \leq |G|$ since $|C_i|$ are positive integers. Only in the case when G is Abelian is $|C_i| = 1$ for all i, and hence this is the only case where $\alpha = |G|$. By denoting C_1 as the class containing the identity element, we see that $|C_1| = 1$, i.e., it contains *only* the identity element, since $g^{-1} \circ e \circ g = e$ for all $g \in G$.

Class Functions and Class Sums

An important concept that we will see time and time again throughout this book is the *class function*. This is a function $\mathcal{C} : G \to \mathbb{C}$ which is constant on conjugacy classes. That is, its value is the same for all elements of each conjugacy class. This means that

$$\mathcal{C}(g) = \mathcal{C}(h^{-1} \circ g \circ h) \qquad \text{or} \qquad \mathcal{C}(h \circ g) = \mathcal{C}(g \circ h)$$

for all values of $g, h \in G$. The above two equations are completely equivalent because the change of variables $g \to h \circ g$ is invertible.

A very special kind of class function on a discrete (and not necessarily finite) group is the delta function:

$$\delta(h) \doteq \begin{cases} 1 \text{ for } h = e \\ 0 \text{ for } h \neq e \end{cases}.$$

The "shifted" version of this function: $\delta_{g_1}(g) = \delta(g_1^{-1} \circ g)$ for any $g_1 \in G$, while not a class function, has some useful properties that will be used shortly.

Let $f : G \to \mathbb{C}$ and define the operation

$$(L(g_1)f)(g) \doteq f(g_1^{-1} \circ g). \tag{7.25}$$

Using the shorthand $f_{g_1}(g)$ to denote the function $(L(g_1)f)(g)$, we see that

$$(L(g_1)L(g_2)f)(g) = (L(g_1)f_{g_2})(g) = f_{g_2}(g_1^{-1} \circ g) = f(g_2^{-1} \circ g_1^{-1} \circ g) = (L(g_1 \circ g_2)f)(g). \tag{7.26}$$

The equalities above can be a great source of confusion when seen for the first time. The temptation is to switch the order of g_1^{-1} and g_2^{-1}.

Using shifted δ-functions, we can define the *class sum function* as

$$\mathcal{C}_i(g) = \sum_{a \in C_i} \delta(a^{-1} \circ g). \tag{7.27}$$

This is a class function with the property

$$\mathcal{C}_i(g) = \begin{cases} 1 \text{ for } g \in C_i \\ 0 \text{ for } g \notin C_i \end{cases}.$$

It is therefore the characteristic function (see Appendix B) of the set C_i. Clearly, any class function can be written as

$$\mathcal{C} = \sum_{i=1}^{\alpha} \gamma_i \mathcal{C}_i.$$

Central to the extension of Fourier analysis to the case of finite groups in Chapter 8 is that the definition of convolution generalizes in a natural way:

$$(f_1 * f_2)(g) = \sum_{h \in G} f_1(h) \, f_2(h^{-1} \circ g) = \sum_{h \in G} f_1(g \circ h^{-1}) \, f_2(h).$$

The second equality above results from the invertible change of variables $h \to g \circ h^{-1}$.

Here f_1 and f_2 are complex-valued functions with group-valued argument, i.e., $f_i : G \to \mathbb{C}$. In general, $f_1 * f_2 \neq f_2 * f_1$. However, if either f_1 or f_2 is a class function, we have

$$(f * C)(g) = \sum_{h \in G} f(h)C(h^{-1} \circ g)$$

$$= \sum_{h \in G} f(h)C(g \circ h^{-1})$$

$$= \sum_{h \in G} C(g \circ h^{-1})f(h)$$

$$= (C * f)(g).$$

Furthermore, since $\delta_{g_1}(g)$ is zero unless $g = g_1$, it is easy to see that

$$(\delta_{g_1} * \delta_{g_2})(g) = \sum_{h \in G} \delta(g_1^{-1} \circ h)\,\delta(g_2^{-1} \circ h^{-1} \circ g)$$

$$= \delta(g_2^{-1} \circ g_1^{-1} \circ g) = \delta((g_1 \circ g_2)^{-1} \circ g)$$

$$= \delta_{g_1 \circ g_2}(g).$$

This result can be abbreviated as

$$\delta_{g_1} * \delta_{g_2} = \delta_{g_1 \circ g_2}$$

by suppressing the arguments of the functions. Hence,

$$\delta_{g_1} * \delta_{g_1^{-1}} = \delta.$$

This means that for a class function, C,

$$\delta_{g_1} * C * \delta_{g_1^{-1}} = \delta_{g_1} * (C * \delta_{g_1^{-1}}) = \delta_{g_1} * (\delta_{g_1^{-1}} * C) = (\delta_{g_1} * \delta_{g_1^{-1}}) * C = C.$$

Likewise, we find that for an arbitrary function on the group, $f(g)$,

$$(\delta_{g_1} * f * \delta_{g_1^{-1}})(g) = f(g_1^{-1} \circ g \circ g_1),$$

and so the *only* functions for which $\delta_{g_1} * f * \delta_{g_1^{-1}} = f$ for all $g_1 \in G$ are class functions.

Products of Class Sum Functions for Finite Groups

Given two class sum functions C_i and C_j (which, being class functions, must satisfy $\delta_h * C_i * \delta_{h^{-1}} = C_i$ and $\delta_h * C_j * \delta_{h^{-1}} = C_j$), we can write

$$C_i * C_j = (\delta_h * C_i * \delta_{h^{-1}}) * (\delta_h * C_j * \delta_{h^{-1}})$$

$$= \delta_h * C_i * (\delta_{h^{-1}} * \delta_h) * C_j * \delta_{h^{-1}}$$

$$= \delta_h * C_i * C_j * \delta_{h^{-1}}.$$

Therefore, the convolution product of two class sum functions must be a class function. Since all class functions are expandable as a linear combination of class sum functions we have

$$C_i * C_j = \sum_{k=1}^{\alpha} c_{ij}^k C_k, \tag{7.28}$$

where c_{ij}^k are called *class constants*. Recall that α is the number of conjugacy classes in the group.

Since the convolution of class functions (and hence class sum functions) commutes, it follows that $c_{ij}^k = c_{ji}^k$.

In analogy with the Cayley table, we can construct a class product table that illustrates the convolution of class sum functions. Note that this table must be symmetric about the diagonal.

$$
\begin{array}{c|ccccc}
* & \mathcal{C}_1 & \cdots & \mathcal{C}_k & \cdots & \mathcal{C}_n \\
\hline
\mathcal{C}_1 & \mathcal{C}_1 & \cdots & \mathcal{C}_k & \cdots & \mathcal{C}_n \\
\vdots & \vdots & \cdots & \vdots & \cdots & \vdots \\
\mathcal{C}_j & \mathcal{C}_j & \cdots & \mathcal{C}_j * \mathcal{C}_k & \cdots & \mathcal{C}_j * \mathcal{C}_n \\
\vdots & \vdots & \cdots & \vdots & \ddots & \vdots \\
\mathcal{C}_n & \mathcal{C}_n & \cdots & \mathcal{C}_n * \mathcal{C}_k & \cdots & \mathcal{C}_n * \mathcal{C}_n
\end{array}
$$

The $(1,1)$ entry in the table is $\mathcal{C}_1 * \mathcal{C}_1 = \mathcal{C}_1$ because $\mathcal{C}_1$ is the conjugacy class $\{e\}$, and hence $\mathcal{C}_1 = \delta$, which yields itself when convolved with itself. Likewise, $\mathcal{C}_1 * \mathcal{C}_i = \mathcal{C}_i * \mathcal{C}_1 = \mathcal{C}_i$, and so $c_{i,1}^1 = \delta_{i,1}$. More generally, the coefficient c_{ij}^1 is found as follows. First we expand out

$$
\begin{aligned}
(\mathcal{C}_i * \mathcal{C}_j)(g) &= \sum_{h \in G} \left(\sum_{a \in C_i} \delta(a^{-1} \circ h) \right) \left(\sum_{b \in C_j} \delta(b^{-1} \circ h^{-1} \circ g) \right) \\
&= \sum_{a \in C_i} \sum_{b \in C_j} \left(\sum_{h \in G} \delta(a^{-1} \circ h)\, \delta(b^{-1} \circ h^{-1} \circ g) \right).
\end{aligned}
$$

The last term in parenthesis is only nonzero when $h = a$, and so the inner summation vanishes and

$$
(\mathcal{C}_i * \mathcal{C}_j)(g) = \sum_{a \in C_i} \sum_{b \in C_j} \delta((a \circ b)^{-1} \circ g).
$$

The coefficient c_{ij}^1 can be viewed as the number of times the summand above reduces to $\delta(g)$. $c_{ij}^1 \neq 0$ only when it is possible to write $b = a^{-1}$. If C is the class containing c, then we denote C^{-1} to be the class containing c^{-1}. (We have not defined a product of conjugacy classes, so C^{-1} should not be interpreted as the inverse of C, but rather only as the set containing the inverses of all the elements of C.) Usually $C \neq C^{-1}$ but it is always true that $|C| = |C^{-1}|$. Using this notation, $c_{ij}^1 \neq 0$ only when $C_i = C_j^{-1}$. In this case, $c_{ij}^1 = |C_i|$ since we can go through all $a \in C_i$, and for each of them we can find $b = a^{-1} \in C_i^{-1}$. Therefore,

$$
c_{ij}^1 = |C_i| \delta_{C_i, C_j^{-1}} = \begin{cases} |C_i| & \text{for } C_i = C_j^{-1} \\ 0 & \text{for } C_i \neq C_j^{-1} \end{cases}.
$$

Double Cosets and Their Relationship to Conjugacy Classes

We now review the concept of double cosets. This concept is closely related to that of cosets as described in Section 7.2.3. At the end of this subsection, we examine a relationship between double cosets and conjugacy classes.

Let $H < G$ and $K < G$. Then for any $g \in G$, the set

$$
HgK \doteq \{h \circ g \circ k \mid h \in H, k \in K\} \tag{7.29}
$$

is called the *double coset* of H and K, and any $g' \in HgK$ (including $g' = g$) is called a *representative* of the double coset. Though a double coset representative often can be described with two or more different pairs (h_1, k_1) and (h_2, k_2) so that $g' = h_1 \circ g \circ k_1 = h_2 \circ g \circ k_2$, we only count g' once in HgK. Hence $|HgK| \le |G|$, and in general $|HgK| \ne |H| \cdot |K|$. If redundancies were not excluded in (7.29), we might expect the last inequality to be an equality. Having said this, we note that there are special cases when $|HgK| = |H| \cdot |K| = |G|$. For example, if every $g \in G$ can be uniquely decomposed as $g = h \circ k$ for $h \in H$ and $k \in K$ where $H \cap K = \{e\}$, then

$$H(h \circ k)K = H(h \circ e \circ k)K = (Hh)e(kK) = HeK = G.$$

In general, the set of all double cosets of H and K is denoted $H \backslash G / K$. Hence we have the hierarchy $g \in HgK \in H \backslash G / K$.

Theorem 7.11. *Membership in a double coset is an equivalence relation on G. That is, G is partitioned into disjoint double cosets, and for $H < G$ and $K < G$ either $Hg_1 K \cap Hg_2 K = \emptyset$ or $Hg_1 K = Hg_2 K$.*

Proof. Suppose $Hg_1 K$ and $Hg_2 K$ have an element in common. If

$$h_1 \circ g_1 \circ k_1 = h_2 \circ g_2 \circ k_2,$$

then

$$g_1 = (h_1^{-1} \circ h_2) \circ g_2 \circ (k_2 \circ k_1^{-1}).$$

The quantities in parenthesis are in H and K, respectively. Then

$$Hg_1 K = H(h \circ g_2 \circ k)K = (Hh)g_2(kK) = Hg_2 K.$$

From Theorem 7.11 it follows that for any finite group G, we can write

$$G = \bigcup_{i=1}^{|H \backslash G / K|} H\sigma_i K \quad \text{and} \quad |G| = \sum_{i=1}^{|H \backslash G / K|} |H\sigma_i K|$$

where σ_i is an arbitrary representative of the ith double coset. However, unlike single cosets where $|gH| = |H|$, we cannot assume that all the double cosets have the same size. In the special case mentioned above where we can make the unique decomposition $g = h \circ k$, the fact that double cosets must either be disjoint or identical means $H \backslash G / K = \{G\}$ and $|H \backslash G / K| = 1$.

Theorem 7.12. *Given $H < G$ and $K < G$ and $g_1, g_2 \in G$ such that $g_1 K \cap Hg_2 K \ne \emptyset$, then $g_1 K \subseteq Hg_2 K$.*

Proof. Let $g_1 \circ k \in g_1 K$. If $g_1 \circ k \in Hg_2 K$ then $g_1 \circ k = h_1 \circ g_2 \circ k_1$ for some $h_1 \in H$ and $k_1 \in K$. Then we can write $g_1 = h_1 \circ g_2 \circ k_1 \circ k^{-1}$. Hence for all $k_2 \in K$, we have $g_1 \circ k_2 = h_1 \circ g_2 \circ k_1 \circ k^{-1} \circ k_2 = h_1 \circ g_2 \circ k_3 \in HgK$ where $k_3 = k_1 \circ k^{-1} \circ k_2 \in K$, and so $g_1 K \subseteq Hg_2 K$.

We note without proof the important result of Frobenius that the number of single left cosets gK that appear in HgK is $\gamma_g = |H/(H \cap K_g)|$ where $K_g = gKg^{-1} < G$ is the subgroup K conjugated by g. Note that $H \cap K_g \le H$ and $H \cap K_g \le K_g$. This means that

$$HgK = \bigcup_{i=1}^{\gamma_g} \sigma_i K \tag{7.30}$$

where σ_i is a coset representative of $\sigma_i K \subseteq HgK$. An analogous result follows for right cosets. In the special case when H is a normal subgroup of G, we find $\gamma_g = 1$ for all $g \in G$, and so $HgH = gH$.

We now examine the relationship between double cosets and conjugacy classes. Recall that $G \times G$ denotes the direct product of G and G consisting of all pairs of the form (g, h) for $g, h \in G$ and with the product $(g_1, h_1)(g_2, h_2) \doteq (g_1 \circ g_2, h_1 \circ h_2)$. Let $G \cdot G$ denote the subset of $G \times G$ consisting of all pairs of the form $(g, g) \in G \times G$. Clearly $G \cdot G$ is a subgroup of $G \times G$.

There are some interesting features of the double coset space $(G \cdot G)\backslash(G \times G)/(G \cdot G)$.[9] For instance, without loss of generality, double coset representatives of $(G \cdot G)(g', g'')(G \cdot G)$ can always be written as (e, g). This follows because

$$(g_1, g_1)(g', g'')(g_2, g_2) = (g_1 \circ g' \circ g_2, g_1 \circ g'' \circ g_2) = (g_1, g_1)(e, g'' \circ (g')^{-1})(g' \circ g_2, g' \circ g_2).$$

Since $(g' \circ g_2, g' \circ g_2) \in G \cdot G$ for all $g' \in G$, we can always define $(e, g) = (e, g'' \circ (g')^{-1})$ to be the coset representative. Furthermore, for any $h \in G$, the representatives (e, g) and $(e, h \circ g \circ h^{-1})$ belong to the same double coset since

$$(g_1, g_1)(e, g)(g_2, g_2) = (g_1 \circ h^{-1}, g_1 \circ h^{-1})(e, h \circ g \circ h^{-1})(h \circ g_2, h \circ g_2).$$

This means that elements of the same conjugacy class of G yield representatives of the same double coset under the mapping $g \to (e, g)$. Conversely, if

$$(G \cdot G)(e, g)(G \cdot G) = (G \cdot G)(e, g')(G \cdot G),$$

then it must be possible to find $g, g', g_1, g_2, g_3, g_4 \in G$ such that

$$(g_1, g_1)(e, g)(g_2, g_2) = (g_3, g_3)(e, g')(g_4, g_4).$$

Expanding out the calculation on both sides and equating terms means $g_1 \circ g_2 = g_3 \circ g_4$ and $g_1 \circ g \circ g_2 = g_3 \circ g' \circ g_4$. Rewriting the first of these as $g_2 = g_1^{-1} \circ g_3 \circ g_4$ and substituting into $g' = g_3^{-1} \circ g_1 \circ g \circ g_2 \circ g_4^{-1}$ yields

$$g' = g_3^{-1} \circ g_1 \circ g \circ (g_1^{-1} \circ g_3 \circ g_4) \circ g_4^{-1} = (g_3^{-1} \circ g_1) \circ g \circ (g_3^{-1} \circ g_1)^{-1}.$$

This means that representatives (e, g) and (e, g') of the same double coset $(G \cdot G)(g_i, g_j)(G \cdot G)$ always map to elements of the same conjugacy class of G.

Written in a different way, we have the following result:

Theorem 7.13. *Given a finite group G, there is a bijective mapping between the set of double cosets $(G \cdot G)\backslash(G \times G)/(G \cdot G)$ and the set of conjugacy classes of G, and hence*

$$\alpha = |(G \cdot G)\backslash(G \times G)/(G \cdot G)|,$$

where α is the number of conjugacy classes of G.

---

[9]The parentheses around $G \times G$ and $G \cdot G$ are simply to avoid incorrectly reading $G \cdot G\backslash G \times G/G \cdot G$ as $G \cdot (G\backslash G) \times (G/G) \cdot G$.

Proof. Let $C(g)$ denote the conjugacy class containing g and $\mathcal{C}$ denote the set of all conjugacy classes (i.e., $C(g) \in \mathcal{C}$ for all $g \in G$). Let the function $f : \mathcal{C} \to (G \cdot G) \backslash (G \times G)/(G \cdot G)$ be defined as

$$f(C(g)) \doteq (G \cdot G)(e, g)(G \cdot G).$$

Then for any $h \in G$ the discussion above indicates that

$$f(C(g)) = f(C(h \circ g \circ h^{-1})) = (G \cdot G)(e, h \circ g \circ h^{-1})(G \cdot G) = (G \cdot G)(e, g)(G \cdot G),$$

and so f is defined consistently in the sense that all elements of the same class are mapped to the same double coset.

Since a double coset representative of $(G \cdot G)(g', g'')(G \cdot G)$ can always be chosen of the form (e, g) for some $g \in G$, and since

$$(G \cdot G)(e, g_1)(G \cdot G) = (G \cdot G)(e, g_2)(G \cdot G) \quad \Leftrightarrow \quad g_1 \in C(g_2),$$

it follows that f is a bijection between $\mathcal{C}$ and $(G \cdot G) \backslash (G \times G)/(G \cdot G)$, and therefore $|\mathcal{C}| = |(G \cdot G) \backslash (G \times G)/(G \cdot G)|$.

7.2.6 Examples of Definitions: Symmetry Operations on the Equilateral Triangle (Revisited)

Note that any element of the group of symmetry operations on the equilateral triangle can be written as products of powers of the elements g_1 and g_3. In such cases, we say that these elements are *generators* of the group. In the case when a whole group is generated by a single element, it is called a *cyclic* group. A cyclic group is always Abelian because all elements are of the form a^n, and $a^n \circ a^m = a^{n+m} = a^m \circ a^n$.

The proper subgroups can be observed from (7.1). They are $\{e\}$, $H_1 = \{e, g_1, g_2\}$, $H_2 = \{e, g_3\}$, $H_3 = \{e, g_4\}$, and $H_4 = \{e, g_5\}$. The conjugacy classes for this group are

$$C_1 = \{e\}; \quad C_2 = \{g_1, g_2\}; \quad C_3 = \{g_3, g_4, g_5\}.$$

The class multiplication constants are

$$c_{11}^1 = c_{21}^2 = c_{31}^3 = c_{12}^2 = c_{22}^2 = c_{13}^3 = 1;$$

$$c_{22}^1 = c_{32}^3 = c_{23}^3 = 2;$$

$$c_{33}^1 = c_{33}^2 = 3.$$

The subgroups H_2, H_3, and H_4 are conjugate to each other, e.g.,

$$g_1 H_2 g_1^{-1} = H_4; \quad g_1 H_3 g_1^{-1} = H_2; \quad g_1 H_4 g_1^{-1} = H_3;$$

H_1 is conjugate to itself:

$$g H_1 g^{-1} = H_1$$

for all $g \in G$. Hence, H_1 is a normal subgroup.

The left cosets are

$$H_1 = e H_1 = g_1 H_1 = g_2 H_1; \quad \{g_3, g_4, g_5\} = g_3 H_1 = g_4 H_1 = g_5 H_1;$$

$$H_2 = e H_2 = g_3 H_2; \quad \{g_1, g_4\} = g_1 H_2 = g_4 H_2; \quad \{g_2, g_5\} = g_2 H_2 = g_5 H_2.$$

$$H_3 = eH_3 = g_4H_3; \quad \{g_2, g_3\} = g_2H_3 = g_3H_3; \quad \{g_1, g_5\} = g_1H_3 = g_5H_3.$$

$$H_4 = eH_4 = g_5H_4; \quad \{g_1, g_3\} = g_1H_4 = g_3H_4; \quad \{g_2, g_4\} = g_2H_4 = g_4H_4.$$

In the present example we have

$$G/G = \{G\}$$

$$G/H_1 = \{\{e, g_1, g_2\}, \{g_3, g_4, g_5\}\}$$
$$G/H_2 = \{\{e, g_3\}, \{g_1, g_4\}, \{g_2, g_5\}\}$$
$$G/H_3 = \{\{e, g_4\}, \{g_2, g_3\}, \{g_1, g_5\}\}$$
$$G/H_4 = \{\{e, g_5\}, \{g_1, g_3\}, \{g_2, g_4\}\}.$$

The right cosets are

$$H_1 = H_1e = H_1g_1 = H_1g_2; \quad \{g_3, g_4, g_5\} = H_1g_3 = H_1g_4 = H_1g_5;$$

$$H_2 = H_2e = H_2g_3; \quad \{g_1, g_5\} = H_2g_1 = H_2g_5; \quad \{g_2, g_4\} = H_2g_2 = H_2g_4;$$
$$H_3 = H_3e = H_3g_4; \quad \{g_1, g_3\} = H_3g_1 = H_3g_3; \quad \{g_2, g_5\} = H_3g_2 = H_3g_5;$$
$$H_4 = H_4e = H_4g_5; \quad \{g_1, g_4\} = H_4g_1 = H_4g_4; \quad \{g_2, g_3\} = H_4g_2 = H_4g_3;$$

The corresponding coset spaces $H\backslash G$ are

$$G\backslash G = \{G\}$$

$$H_1\backslash G = \{\{e, g_1, g_2\}, \{g_3, g_4, g_5\}\}$$
$$H_2\backslash G = \{\{e, g_3\}, \{g_1, g_5\}, \{g_2, g_4\}\}$$
$$H_3\backslash G = \{\{e, g_4\}, \{g_1, g_3\}, \{g_2, g_5\}\}$$
$$H_4\backslash G = \{\{e, g_5\}, \{g_1, g_4\}, \{g_2, g_3\}\}.$$

Note that $G/H_1 = H_1\backslash G$, which follows from H_1 being a normal subgroup. The coset space G/H_1 (or $H_1\backslash G$) is therefore a group.

The Cayley table for G/H_1 is

$\circ$	H_1	gH_1
H_1	H_1	gH_1
gH_1	gH_1	H_1

.

Since all groups with two elements are isomorphic to each other, we can say immediately that $G/H_1 \cong H_2 \cong H_3 \cong H_4 \cong \mathbb{Z}_2$.

Examples of double cosets and double coset spaces associated with G are:

Example 7.10:

$$H_1gH_1 = gH_1$$

for all $g \in G$ (and hence $H_1\backslash G/H_1 = G/H_1$);

Example 7.11:

$$H_2g_3H_2 = H_2eH_2 = eH_2 = H_2$$

and

$$H_2g_1H_2 = H_2g_2H_2 = H_2g_4H_2 = H_2g_5H_2 = g_2H_2 \cup g_4H_2$$

(and hence $H_2\backslash G/H_2 = \{H_2, g_2H_2 \cup g_4H_2\}$ and $|H_2\backslash G/H_2| = 2$.). If $g \in H_2$ then $H_2gH_2 = gH_2 = H_2$ and $\gamma_g = 1$ since $H_2/(H_2 \cap H_2) = H_2/H_2 = \{e\}$. If $g \in g_4H_2 \cup g_2H_2$ then $H_2gH_2 = g_4H_2 \cup g_2H_2$ and $H_2/(H_2 \cap gH_2g^{-1}) = H_2/(H_2 \cap H_4) = H_2/\{e\} = H_2$. Hence in this case $\gamma_g = 2$. Then for any $\sigma_2 \in H_2$ and $\sigma_1 \in G - H_2$, we have $G = H_2\sigma_1H_2 \cup H_2\sigma_2H_2$.

Example 7.12:
$$H_1gH_2 = G$$
for all $g \in G$ and so $H_1\backslash G/H_2 = \{G\}$ and $|H_1\backslash G/H_2| = 1$.

Example 7.13:
$$H_3eH_2 = H_3g_1H_2 = H_3g_3H_2 = H_3g_4H_2 = \{e, g_1, g_3, g_4\} = g_1H_2 \cup eH_2,$$

We therefore write
$$H_3g_2H_2 = H_3g_5H_2 = \{g_2, g_5\} = g_2H_2$$
and so $H_3\backslash G/H_2 = \{g_1H_2 \cup eH_2, g_2H_2\}$ and $|H_3\backslash G/H_2| = 2$.

7.3 Lie Groups

In this section we examine a very important kind of group. These are the Lie groups.[10] These groups describe continuous symmetries such as those that arise in classical mechanical systems [28]. We begin the first subsection of this section with some motivation and an example. In the second subsection the formal definitions are provided. In previous chapters we already examined in great detail two very important Lie groups for engineering applications: the rotation group; and the Euclidean motion group. We revisit rotation and rigid-body motion with the tools of group theory.

7.3.1 An Intuitive Introduction to Lie Groups

From elementary calculus we know that for a constant $x \in \mathbb{C}$ and a complex-valued function $y(t)$ that if
$$\frac{dy}{dt} = xy; \qquad y(0) = y_0 \qquad \Rightarrow \qquad y(t) = e^{xt}y_0.$$

Likewise if $Y(t)$ and X take values in $\mathbb{C}^{N \times N}$, we have
$$\frac{dY}{dt} = XY; \qquad Y(0) = Y_0 \qquad \Rightarrow \qquad Y(t) = e^{tX}Y_0.$$

Here the matrix exponential is defined via the same power series that defines the scalar exponential function[11]
$$e^{tX} = \mathbb{I} + \sum_{n=1}^{\infty} \frac{t^n X^n}{n!}. \tag{7.31}$$

---

[10]Named after the Norwegian mathematician Marius Sophus Lie (1842-1899)

[11]In (7.31) we use $\mathbb{I}$ as shorthand for the $N \times N$ identity matrix, $\mathbb{I}_{N \times N}$, and later in this section we use $\mathbb{O}$ as shorthand for the $N \times N$ zero matrix $\mathbb{O}_{N \times N}$.

If we specify that $Y_0 = \mathbb{I}$, then $e^{0X} = \mathbb{I}$. In general, the exponential of a matrix with finite entries will be invertible because (see Appendix D)

$$\det(e^X) = e^{\text{tr}(X)} \neq 0.$$

By multiplying out the power series for $e^{t_1 X}$ and $e^{t_2 X}$, we also see that $e^{t_1 X} e^{t_2 X} = e^{(t_1 + t_2)X}$. Choosing $t_1 = t$ and $t_2 = -t$, the result of this product of exponentials is the identity matrix. Since matrix multiplication is associative, it follows that all the properties of an Abelian group are satisfied by the matrix exponential under the operation of matrix multiplication. Since $t \in \mathbb{R}$, this is an example of a one-dimensional group. This is one of the simplest examples of a Lie group. For an intuitive understanding at this point, it is sufficient to consider a Lie group to be one which is continuous in the same sense that $\mathbb{R}^N$ is. The group operation of a Lie group is also continuous in the sense that the product of slightly perturbed group elements are sent to an element "near" the product of the unperturbed elements. For a Lie group, the inverse of a perturbed element is also "close to" the inverse of the unperturbed one. These conditions are referred to as the continuity of the product and inversion of group elements.

If $X_1, X_2 \in \mathbb{C}^{N \times N}$, then in general $e^{X_1} e^{X_2} \neq e^{X_2} e^{X_1}$. A sufficient (though not necessary) condition for equality to hold is $X_1 X_2 = X_2 X_1$ (see Appendix D).

Now, if there are a set of linearly independent matrices $\{X_1, ..., X_n\}$, and we calculate $e^{t_i X_i}$, and it happens to be the case that the product of the resulting exponentials is closed under all possible multiplications and all values of $t_1, ..., t_n \in \mathbb{R}$, then the result is an n–dimensional continuous group.[12] When working with matrices $A_i \in \mathbb{C}^{N \times N}$, norms of differences provide a measure of distance $(d(A_i, A_j) = \|A_i - A_j\|)$ that can be used to verify continuity of the product and inversion of group elements. A Lie group that is also a set of matrices is called a *matrix Lie group*.

For example, given

$$X_1 = \begin{pmatrix} 0 & 1 & 0 \\ 0 & 0 & 0 \\ 0 & 0 & 0 \end{pmatrix} \quad X_2 = \begin{pmatrix} 0 & 0 & 1 \\ 0 & 0 & 0 \\ 0 & 0 & 0 \end{pmatrix} \quad X_3 = \begin{pmatrix} 0 & 0 & 0 \\ 0 & 0 & 1 \\ 0 & 0 & 0 \end{pmatrix}$$

the corresponding matrix exponentials are

$$e^{t_1 X_1} = \begin{pmatrix} 1 & t_1 & 0 \\ 0 & 1 & 0 \\ 0 & 0 & 1 \end{pmatrix} \quad e^{t_2 X_2} = \begin{pmatrix} 1 & 0 & t_2 \\ 0 & 1 & 0 \\ 0 & 0 & 1 \end{pmatrix} \quad e^{t_3 X_3} = \begin{pmatrix} 1 & 0 & 0 \\ 0 & 1 & t_3 \\ 0 & 0 & 1 \end{pmatrix}.$$

This choice of the X_i is special because they are *nilpotent* matrices, i.e., they satisfy $(X_i)^K = \mathbb{O}$ for some finite $K \in \mathbb{Z}^+$ (where in this case $K = 2$ and $N = 3$). It is also true that any weighted sum of the matrices X_i is nilpotent: $(\alpha X_1 + \beta X_2 + \gamma X_3)^3 = \mathbb{O}$. Hence, the matrix exponentials in this case have a very simple form. It is clear that no matter in what order we multiply the above matrix exponentials, the result will be one of the form

$$g = \begin{pmatrix} 1 & \alpha & \beta \\ 0 & 1 & \gamma \\ 0 & 0 & 1 \end{pmatrix}$$

for some value of $\alpha, \beta, \gamma \in \mathbb{R}$. In particular, we can make the Euler-angle-like factorization

---

[12]The number n (the dimension of the group) is generally different than N (the dimension of the matrix).

$$g = \begin{pmatrix} 1 & 0 & \beta \\ 0 & 1 & 0 \\ 0 & 0 & 1 \end{pmatrix} \begin{pmatrix} 1 & 0 & 0 \\ 0 & 1 & \gamma \\ 0 & 0 & 1 \end{pmatrix} \begin{pmatrix} 1 & \alpha & 0 \\ 0 & 1 & 0 \\ 0 & 0 & 1 \end{pmatrix}.$$

Furthermore, we see that

$$g_1 \circ g_2 = \begin{pmatrix} 1 & \alpha_1 & \beta_1 \\ 0 & 1 & \gamma_1 \\ 0 & 0 & 1 \end{pmatrix} \begin{pmatrix} 1 & \alpha_2 & \beta_2 \\ 0 & 1 & \gamma_2 \\ 0 & 0 & 1 \end{pmatrix} = \begin{pmatrix} 1 & \alpha_1 + \alpha_2 & \beta_1 + \beta_2 + \alpha_1\gamma_2 \\ 0 & 1 & \gamma_1 + \gamma_2 \\ 0 & 0 & 1 \end{pmatrix},$$

and so there is closure in the sense that the product of matrices of this form results in a matrix of the same form. It is also clear that the choice $\alpha_2 = -\alpha_1$, $\gamma_2 = -\gamma_1$, and $\beta_2 = -\beta_1 + \alpha_1\gamma_1$ means $g_2 = g_1^{-1}$ exists for every $g_1 \in G$. Hence this set is a group under matrix multiplication. This group is isomorphic to the 3-D *Heisenberg group* and 1-D *Galilei group*. It is an example of a more general class of groups called nilpotent Lie groups.

Every element of the Heisenberg group can be generated by the exponential

$$\exp \begin{pmatrix} 0 & \alpha' & \beta' \\ 0 & 0 & \gamma' \\ 0 & 0 & 0 \end{pmatrix} = \begin{pmatrix} 1 & \alpha' & \beta' + \alpha'\gamma'/2 \\ 0 & 1 & \gamma' \\ 0 & 0 & 1 \end{pmatrix}.$$

7.3.2 Rigorous Definitions

A *Lie group* $(G, \circ)$ is a group for which the set G is an analytic manifold, with the operations $(g_1, g_2) \rightarrow g_1 \circ g_2$ and $g \rightarrow g^{-1}$ being analytic (see Appendix G for definitions) [10, 34, 35, 41, 43]. The dimension of a Lie group is the dimension of the associated manifold G.

Since a metric (distance) function $d(\cdot, \cdot)$ exists for G, we have a way to measure "nearness," and hence continuity is defined in the usual way.

A *matrix Lie group* $(G, \circ)$ is a Lie group where G is a set of square matrices and the group operation is matrix multiplication. Another way to say this is that a matrix Lie group is a (closed) Lie subgroup of $GL(N, \mathbb{R})$ or $GL(N, \mathbb{C})$ for some $N \in \{1, 2, ...\}$. We will deal exclusively with matrix Lie groups.

Given a matrix Lie group, elements sufficiently close to the identity are written as $g(t) = e^{tX}$ for some $X \in \mathcal{G}$ (the Lie algebra of G) and t near 0. For matrix Lie groups, the corresponding Lie algebra is denoted with small letters. For example, the Lie algebras of the groups $GL(N, \mathbb{R})$, $SO(N)$ and $SE(N)$ are respectively denoted as $gl(N, \mathbb{R})$, $so(N)$ and $se(N)$.

In the case of the rotation and motion groups, the exponential mapping is surjective (see Chapters 5 and 6), so all group elements can be parameterized as matrix exponentials of Lie algebra elements.

The matrices $\frac{dg}{dt}g^{-1}$ and $g^{-1}\frac{dg}{dt}$ are respectively called the *tangent vectors* or *tangent elements* at g. Evaluated at $t = 0$, these vectors reduce to X.

The *adjoint* operator is defined as

$$Ad(g_1)X \doteq \frac{d}{dt}\left(g_1 e^{tX} g_1^{-1}\right)|_{t=0} = g_1 X g_1^{-1}. \tag{7.32}$$

This gives a homomorphism $Ad : G \rightarrow GL(\mathcal{G})$ from the group into the set of all invertible linear transformations of $\mathcal{G}$ onto itself. It is a homomorphism because

$$Ad(g_1)Ad(g_2)X = g_1(g_2 X g_2^{-1})g_1^{-1} = (g_1 g_2)X(g_1 g_2)^{-1} = Ad(g_1 g_2)X.$$

It is linear because

$$Ad(g)(c_1 X_1 + c_2 X_2) = g(c_1 X_1 + c_2 X_2)g^{-1} = c_1 g X_1 g^{-1} + c_2 g X_2 g^{-1}$$
$$= c_1 Ad(g)X_1 + c_2 Ad(g)X_2.$$

In the special case of a 1-parameter subgroup when $g = g(t)$ is an element close to the identity,[13] we can approximate $g(t) \approx \mathbb{I} + tX$ for small t. Then we get $Ad(\mathbb{I}+tX)Y = Y + t(XY - YX)$. The quantity

$$XY - YX = [X, Y] = \frac{d}{dt}(Ad(g(t))Y)|_{t=0} \doteq ad(X)Y \qquad (7.33)$$

is called the *Lie bracket* of the elements $X, Y \in \mathcal{G}$. The last equality in (7.33) defines $ad(X)$. We observe from this definition that

$$Ad(\exp tX) = \exp(t \cdot ad(X)). \qquad (7.34)$$

It is clear from the definition in (7.33) that the Lie bracket is linear in each entry:

$$[c_1 X_1 + c_2 X_2, Y] = c_1[X_1, Y] + c_2[X_2, Y]$$

and

$$[X, c_1 Y_1 + c_2 Y_2] = c_1[X, Y_1] + c_2[X, Y_2].$$

Furthermore, the Lie bracket is anti-symmetric:

$$[X, Y] = -[Y, X], \qquad (7.35)$$

and hence $[X, X] = 0$. Given a basis $\{X_1, ..., X_n\}$ for the Lie algebra $\mathcal{G}$, the Lie bracket of any two elements will result in a linear combination of all basis elements. This is written as

$$[X_i, X_j] = \sum_{k=1}^{n} C_{ij}^k X_k.$$

The constants C_{ij}^k are called the *structure constants* of the Lie algebra $\mathcal{G}$. Note that in contrast to the class multiplication constants c_{ij}^k that we saw in Section 7.2.5, the structure constants are antisymmetric: $C_{ij}^k = -C_{ji}^k$.

It can be checked that for any three elements of the Lie algebra, the *Jacobi identity* is satisfied:

$$[X_1, [X_2, X_3]] + [X_2, [X_3, X_1]] + [X_3, [X_1, X_2]] = 0. \qquad (7.36)$$

As a result of the Jacobi identity, $ad(X)$ satisfies

$$ad([X, Y]) = ad(X)ad(Y) - ad(Y)ad(X).$$

An important relationship called the *Baker-Campbell-Hausdorff formula* exists between the Lie bracket and matrix exponential (see [4, 9, 17]). Namely, the product of

---

[13] In the context of matrix Lie groups, distance is defined naturally as a matrix norm of the difference of two group elements.

two Lie group elements written as exponentials of Lie algebra elements can be expressed as

$$e^X e^Y = e^{Z(X,Y)} \tag{7.37}$$

where

$$Z(X,Y) = X + Y + \frac{1}{2}[X,Y] + \frac{1}{12}([X,[X,Y]] + [Y,[Y,X]]) + \cdots .$$

This expression is verified by expanding e^X, e^Y and $e^{Z(X,Y)}$ in Taylor series and comparing terms.

In summary, the matrix exponential provides a tool for parameterizing connected Lie groups. We already saw two ways of parameterizing the Heisenberg group using the matrix exponential. We can exponentiate each basis element of the Lie algebra and multiply the results (in any order), or we can exponentiate a weighted sum of all the basis elements. Both parameterization methods are suitable in more general contexts. We have already seen this in great detail in the context of the rotation and Euclidean motion groups.

7.3.3 Compact Lie Groups as Symmetric Spaces

The concept of a *symmetric space* has been studied extensively in modern differential geometry. Let M be a connected Riemannian manifold, p be an arbitrary point in M, and $\gamma(t) \in M$ be a geodesic curve for which $\gamma(0) = p$. (A geodesic curve segment in a Riemannian manifold can be thought of as the most taut curve segment connecting two points, and the distance between those points can be defined as the arclength along the shortest geodesic segment connecting them.[14]) An *isometry* (or isometric mapping) of a Riemannian manifold is a mapping $I : M \to M$ such that for a Riemannian distance function $d : M \times M \to \mathbb{R}_{\geq 0}$ it holds that

$$d(I(p_1), I(p_2)) = d(p_1, p_2) \tag{7.38}$$

for every $p_1, p_2 \in M$.

With this in mind, M is called a symmetric space if for *every* geodesic curve $\gamma(t)$ passing through p there is an isometric *involution*[15] operation $I_p : M \to M$ satisfying the condition

$$I_p(\gamma(t)) = \gamma(-t) \tag{7.39}$$

for every $p \in M$ in addition to I_p satisfying (7.38).

Compact Lie groups are an example of symmetric spaces. We can see this by first observing that for any fixed $X \in \mathcal{G}$, the curve

$$\gamma(t) \doteq p \circ \exp(tX)$$

is a geodesic passing through $\gamma(0) = p$. The metric used to define a geodesic in this context is a weighted log metric (akin to the ones we examined for $SO(3)$ and $SE(3)$ in Chapters 5 and 6) which has a corresponding distance function with the property

---

[14]For example, the geodesics for a sphere are the circles contained in the sphere that have the same radius as the sphere. Given two points on these "great circles," the distance between these points is measured along the shorter of the two arcs formed by removing the points, unless the points are antipodal, in which case the arcs have the same length.

[15]An involution is any bijective mapping with the property that it is its own inverse, e.g., $I_p(I_p(x)) = x$ for every $x \in M$.

$$d(e, \exp(tX) \doteq t \cdot \|X\|_W$$

where the symmetric positive definite weighting matrix W can be related to the metric tensor. For a compact Lie group this distance function is bi-invariant for an appropriate choice of W, and can be thought of as a generalization of the the metric for $SO(3)$ discussed in Chapter 5 of the form $d(g_1, g_2) = \|\log(g_1^{-1} \circ g_2)\|_W$, where in that case $W = c\mathbb{I}$.

Either by an appropriate choice of W, or by an averaging process described in [30], it is always possible to ensure for a compact Lie group, G, that this distance function is bi-invariant in the sense that

$$d(h \circ g_1, h \circ g_2) = d(g_1 \circ h, g_2 \circ h) = d(g_1, g_2)$$

for arbitrary $g_1, g_2, h \in G$.

When defining

$$I_p(g) \doteq p \circ g^{-1} \circ p$$

then clearly $I_p(I_p(g)) = g$ and [30]

$$I_p(\gamma(t)) = p \circ (p \circ \exp(tX))^{-1} \circ p = p \circ \exp(-tX) = \gamma(-t).$$

Moreover, from bi-invariance of the metric, it is not difficult to show that

$$d(p \circ \exp(tX), p \circ \exp(tY)) = d(p \circ \exp(-tX), p \circ \exp(-tY)).$$

Therefore, both (7.39) and (7.38) are satisfied.

Though we have focused on compact Lie groups, it is worth mentioning that they are not the only symmetric spaces. For example, Euclidean spaces with the operation of addition are noncompact Lie groups that are symmetric spaces (with isometry $I(\mathbf{x}) \doteq -\mathbf{x}$). And spheres are also symmetric spaces, which are compact Riemannian manifolds, but do not have a group structure.[16] In addition to being symmetric spaces, both of these examples are called *spaces of constant curvature* because at every point in these spaces their scalar curvature is the same.

7.3.4 Examples

Examples of Lie groups which act on $\mathbb{R}^N$ (N-dimensional Euclidean space) include:

1. The rotation group $SO(N)$, which consists of all $N \times N$ special orthogonal matrices R (i.e., $RR^T = I$, $\det R = +1$). As with all matrix groups, the group operation is matrix multiplication, and elements of $SO(N)$ act on $\mathbf{x} \in \mathbb{R}^N$ as $\mathbf{x}' = R\mathbf{x} \in \mathbb{R}^N$. $SO(N)$ is an $N(N-1)/2$-dimensional Lie group. The corresponding Lie algebra is denoted $so(N)$.

2. The group $SO(p,q)$ consists of all $N \times N$ matrices ($N = p+q$) with unit determinant that preserve the matrix

$$J(p,q) = \begin{pmatrix} -\mathbb{I}_{p \times p} & \mathbb{O}_{p \times q} \\ \mathbb{O}_{q \times p} & \mathbb{I}_{q \times q} \end{pmatrix}.$$

That is, $Q \in SO(p,q)$ satisfies

$$Q^T J(p,q) Q = J(p,q).$$

$SO(p,q)$ is an $N(N-1)/2$-dimensional Lie group. The corresponding Lie algebra is denoted $so(p,q)$.

---

[16]However, spheres are coset spaces.

0. The special Euclidean group (also called the Euclidean motion group, Euclidean group, or simply the motion group) $SE(N) = \mathbb{R}^N \rtimes SO(N)$, which consists of all pairs $g = (R, \mathbf{b})$, where $R \in SO(N)$ and $\mathbf{b} \in \mathbb{R}^N$, has the group law $g_1 \circ g_2 = (R_1 R_2, R_1 \mathbf{b}_2 + \mathbf{b}_1)$ and acts on $\mathbb{R}^N$ as $\mathbf{x}' = R\mathbf{x} + \mathbf{b}$. $SE(N)$ is the group of rigid body motions of N-dimensional Euclidean space and is itself an $N(N+1)/2$-dimensional Lie group. The corresponding Lie algebra is denoted $se(N)$. Discrete subgroups of $SE(N)$ include the crystallographic space groups, i.e., the symmetry groups of regular lattices.

4. The scale-Euclidean (or similitude) group, $SIM(N)$, which consists of all pairs $(e^a R, \mathbf{b})$, has the group law $g_1 \circ g_2 = (e^{a_1 + a_2} R_1 R_2, R_1 \mathbf{b}_2 + \mathbf{b}_1)$ and acts on $\mathbb{R}^N$ by translation, rotation and dilation as $\mathbf{x}' = e^a R\mathbf{x} + \mathbf{b}$. It is a $(1 + N(N+1)/2)$-dimensional Lie group, with corresponding Lie algebra $sim(N)$.

5. The real special linear group $SL(N, \mathbb{R})$, which consists of all $N \times N$ matrices L with $\det L = +1$ acts on $\mathbb{R}^N$ ($\mathbf{x}' = L\mathbf{x}$) by rotation, stretch, and shear in such a way that volume is preserved, and parallel lines remain parallel. It is an $(N^2 - 1)$-dimensional Lie group with corresponding Lie algebra $sl(N, \mathbb{R})$.

6. The special affine group of $\mathbb{R}^N$ (denoted $\mathrm{Aff}^+ \doteq \mathbb{R}^N \rtimes GL^+(N, \mathbb{R})$), which is the set of all pairs $g = (e^a L, \mathbf{b})$ acts on objects in $\mathbb{R}^N$ by translation, rotation, shear, stretch and dilation.[17] In short, this is the most general type of deformation of Euclidean space which transforms parallel lines into parallel lines. The special affine group is $N(N+1)$-dimensional.

7.3.5 Demonstration of Theorems with $SO(3)$

In this subsection we demonstrate many of the definitions and results of theorems using the rotation group $SO(3)$ and related groups, and sets on which these groups act.

The Baker-Campbell-Hausdorff Formula

For $SO(3)$ the Baker-Campbell-Hausdorff formula

$$e^X e^Y = e^{Z(X,Y)}$$

can be interpreted as a statement of what the axis and angle of rotation are for the composition of two rotations. That is, in the case of $SO(3)$

$$\exp(\theta_1 N_1) \exp(\theta_2 N_2) = \exp(\theta_3 N_3),$$

and we can find θ_3 and N_3 for given (θ_i, N_i) for $i = 1, 2$.

By expanding out

$$\exp(\theta_i N_i) = \mathbb{I} + \sin \theta_i N_i + (1 - \cos \theta_i) N_i^2$$

and matching terms, it can be shown that [37]:

$$\cos \frac{\theta_3}{2} = \cos \frac{\theta_1}{2} \cos \frac{\theta_2}{2} - (\mathbf{n}_1 \cdot \mathbf{n}_2) \sin \frac{\theta_1}{2} \sin \frac{\theta_2}{2},$$

from which $\theta_3 \in [0, \pi]$ can be found uniquely, and

---

[17]The notation $GL^+(N, \mathbb{R})$ stands for the subgroup of the *general linear group* consisting of $N \times N$ matrices with real entries and positive determinant.

$$N_3 = aN_1 + bN_2 + 2c[N_1, N_2]$$

where the equalities

$$a \sin \frac{\theta_3}{2} = \sin \frac{\theta_1}{2} \cos \frac{\theta_2}{2}$$

$$b \sin \frac{\theta_3}{2} = \cos \frac{\theta_1}{2} \sin \frac{\theta_2}{2}$$

$$c \sin \frac{\theta_3}{2} = \sin \frac{\theta_1}{2} \sin \frac{\theta_2}{2}$$

fully determine a, b, and c.

Isotropy Groups, Homogeneous Manifolds, and Orbits

$R \in SO(3)$ acts on $\mathbf{x} \in \mathbb{R}^3$ in a natural way with matrix multiplication. If we consider the set of all rotations that keep a particular $\mathbf{x} \in \mathbb{R}^3$ fixed, we find this set to be all rotation matrices of the form

$$\mathrm{rot}(\mathbf{x}/\|\mathbf{x}\|, \theta) = S(\mathbf{x}) R_3(\theta) S^{-1}(\mathbf{x})$$

for $\theta \in [0, 2\pi)$, where S is any rotation matrix with third column $\mathbf{x}/\|\mathbf{x}\|$. The set of all matrices of this form is the "little group" (stability subgroup, or isotropy subgroup) of $\mathbf{x}$. It is isomorphic to $SO(2)$ since each element in the little group is related to a planar rotation by conjugation. Since the little group of $\mathbf{x}$ and $\mathbf{x}/\|\mathbf{x}\|$ are the same, it suffices to consider the little groups of points on the unit sphere. Each point on $\mathbb{S}^2$ defines the axis of a little group, and every little group can be identified with a point in $\mathbb{S}^2$. Since every little group is isomorphic with $SO(2)$, we write

$$SO(3)/SO(2) \cong \mathbb{S}^2.$$

Hence the homogeneous space $SO(3)/SO(2)$ (set of all left cosets of $SO(2)$ in $SO(3)$) is identified with the unit sphere $\mathbb{S}^2$. Similarly for right cosets we write

$$SO(2)\backslash SO(3) \cong \mathbb{S}^2.$$

The unit vector specifying the axis of rotation of each little group serves as a name tag by which the little group is identified. We note also that

$$SU(2)/U(1) \cong \mathbb{S}^2/\mathbb{Z}_2.$$

$SO(3)$ divides $\mathbb{R}^3$ into concentric spherical orbits centered at the origin. Each orbit is specified by its radius, and so the set of all orbits is equated with the set of all radii:

$$\mathbb{R}^3/SO(3) = \mathbb{R}_{\geq 0}.$$

Here we use the notation $/$ in the sense of X/G where G is a group acting on X. Likewise, the orbits of $\mathbb{S}^2$ under the action of $SO(2)$ are parallel circles in the planes normal to the axis of rotation. This set of circles can be identified with either the set of points $[-1, 1]$ along the axis of rotation, or with the Euler angle $0 \leq \beta \leq \pi$. In the latter case we write

$$\mathbb{S}^2/SO(2) \cong [0, \pi].$$

Since $\mathbb{S}^2 \cong SU(2)\backslash SO(3)$, it follows that we can write the *double coset decomposition*

$$SO(2)\backslash SO(3)/SO(2) \cong [0, \pi].$$

Since there is a two-to-one surjective homomorphism from $SU(2)$ onto $SO(3)$, the homomorphism theorem says that

$$SU(2)/[\mathbb{I}_{2\times 2}, -\mathbb{I}_{2\times 2}] \cong SO(3).$$

This is an example of the quotient of a Lie group with respect to a finite normal subgroup being isomorphic to a Lie group of the same dimension as the original group. Another example of this is

$$SO(2) \cong \mathbb{R}/2\pi\mathbb{Z}.$$

Other homogeneous spaces that arise are

$$SO(N)/SO(N-1) \cong \mathbb{S}^{N-1}$$

and

$$SO(N,1)/SO(N) \cong H^N$$

(where H^N is a hyperboloid in $\mathbb{R}^{N+1}$).

Quotient groups that are of general interest are

$$GL(N,\mathbb{R})/SL(N,\mathbb{R}) \cong \mathbb{R} - \{0\}$$

and

$$GL^+(N,\mathbb{R})/SL(N,\mathbb{R}) \cong \mathbb{R}_{>0}.$$

These result from the homomorphism theorem with the homomorphisms $\det : GL(N,\mathbb{R}) \to \mathbb{R} - \{0\}$ and $\det : GL^+(N,\mathbb{R}) \to \mathbb{R}_{>0}$, respectively.

Conjugacy Classes and Class Functions

Every rotation can be parameterized with the exponential map as $A(\mathbf{n}, \theta) = \exp(\theta X)$, which is the same as a conjugated version of matrix $R_3(\theta)$, i.e., $A(\mathbf{n}, \theta) = Q\,R_3(\theta)\,Q^T$ where $Q\mathbf{e}_3 = \mathbf{n}$. It follows that functions of the form

$$\chi(\theta) = \mathrm{tr}(\exp(\theta X)) = 2\cos\theta + 1$$

are class functions for $SO(3)$. Since the cosine function is invertible when $\theta \in [0, \pi]$, it follows that θ itself is an invariant.

In fact, θ is the only part of a 3×3 rotation matrix that remains invariant under conjugation by all 3×3 rotations. This leads to the following theorem.

Theorem 7.14. *A function $f : SO(3) \to \mathbb{C}$ is a class function if and only if it is a function of the angle (and not the axis) of rotation. That is, $f(\exp(\theta N)) = \tilde{f}(\theta)$ for some function $\tilde{f} : [0, \pi] \to \mathbb{C}$.*

This follows easily from the fact that

$$R\exp(\theta N) R^T = \exp(\theta\, R N\, R^T).$$

Sometimes it is convenient to double the range of the parameters so that $(\theta, \mathbf{n}) \in \mathbb{S}^1 \times \mathbb{S}^2$. We can then identify a function $f' : \mathbb{S}^1 \times \mathbb{S}^2 \to \mathbb{C}$ with a function $f : SO(3) \to \mathbb{C}$ by forcing the equality

$$f(\exp(\theta N)) = f'(\theta, \mathbf{n}) \tag{7.40}$$

via the constraint

$$f'(\theta, \mathbf{n}) = f'(-\theta, -\mathbf{n}).$$

When the range of θ is expanded in this way from $[0, \pi]$ to $\mathbb{S}^1$, then f is a class function if and only if f' is an even function of θ whose value is independent of $\mathbf{n}$.

In contrast, a symmetric function on $SO(3)$ is one for which $f(R) = f(R^T)$ for all $R \in SO(3)$. Defining a function f' as in (7.40), then a symmetric function on $SO(3)$ must be even in both θ and $\mathbf{n}$ in the sense that $f'(-\theta, \mathbf{n}) = f'(\theta, \mathbf{n}) = f'(\theta, -\mathbf{n})$. But a symmetric function is not required to be independent of $\mathbf{n}$.

In the next section we change topics and examine finite subgroups of $SO(3)$. We will return to class functions for $SO(3)$ in Chapter 9.

Classification of Finite Subgroups

In this subsection we use (7.23) to determine conditions that any finite subgroup $G < SO(3)$ must satisfy. Other approaches that yield the same result can be found in [1, 12]. To be consistent with the literature, we simplify the form of the equations and write $n = |G|$, $r = |P/G|$ and $n_i = |G_m|$, where m is in the ith orbit. Since every rotation other than the identity keeps two (antipodal) points fixed, it follows in the case of a finite rotation that

$$\sum_{h \in G - \{e\}} |F(h)| = 2(|G| - 1).$$

Then (7.23) is rewritten as [29, 38]

$$2(n - 1) = \sum_{i=1}^{r} \frac{n}{n_i}(n_i - 1),$$

or equivalently

$$2 - \frac{2}{n} = r - \sum_{i=1}^{r} \frac{1}{n_i}. \tag{7.41}$$

Since n, r, and n_i must all be positive integers, the set of possible solutions is very small. In fact, it can be shown that (7.41) can only hold when $r = 2$ or $r = 3$. When $r = 2$ the condition $n_1 = n_2 = n$ is the only way in which (7.41) is satisfied. This is nothing more than the finite subgroup of rotations in a plane by angles $2\pi/n$ (called the *cyclic group*) denoted as C_n. When $r = 3$, there are three possibilities for the values $(n_1, n_2, n_3; n)$ (where the numbers are arranged so that $n_1 \leq n_2 \leq n_3 < n$). These are: the *dihedral groups* for any integer $k \geq 2$

$$D_k : \quad (2, 2, k; 2k);$$

the *tetrahedral* group

$$T_{12} : \quad (2, 3, 3; 12);$$

the *cubo-octahedral* group

$$O_{24} : \quad (2, 3, 4; 24);$$

the *icosa-dodecahedral* group

$$I_{60}: \quad (2,3,5;60).$$

D_k is the group of discrete spatial rotations that bring a regular k-gon back into itself. It includes C_k rotations about the normal to the face of the polygon through its center, as well as rotations by π around the axes through the center and each vertex of the polygon. When k is even, rotations by π around the axes passing through the faces and center of the polygon are also in D_k.

T_{12}, O_{24}, and I_{60} are the groups of rotational symmetry operations that preserve the platonic solids for which they are named. An alternate derivation of these results can be found in [12].

The platonic solids arise in nature in several contexts. For instance, the simplest viruses (i.e., those with the smallest genetic code) have evolved so that they only have one kind of protein in their protective shell. In order to best approximate a sphere (which is the surface that has the maximum volume to surface area) the protein molecules that these viruses produce fold into sections of the viral shell that self-assembles into an icosahedron [11, 27, 33].

7.3.6 Calculating Jacobians

Given a finite-dimensional matrix Lie group, an orthonormal basis for the Lie algebra can always be found when an appropriate inner product is defined. Such a basis can be constructed by the Gram-Schmidt orthogonalization procedure starting with any Lie algebra basis (see Appendix C).

Abstractly, an inner product on a Lie algebra satisfies all of the usual conditions defined in Appendix C. Orthonormality of a basis $\{X_1, ..., X_n\}$ means that

$$(X_i, X_j) = \delta_{ij} \tag{7.42}$$

for all values of i and j.

Let G be a matrix Lie group with $\{X_i\}$ being an orthonormal basis for the corresponding Lie algebra $\mathcal{G}$. Then when $Y = \sum_{i=1}^{n} y_i X_i$ and $Z = \sum_{i=1}^{n} z_i X_i$, it follows that

$$(Y, Z) = \sum_{i=1}^{n} y_i z_i.$$

In other words, we can make a correspondence between Lie algebra elements and vectors in $\mathbb{R}^n$ as $Y \leftrightarrow \mathbf{y}$ and also make a correspondence between the inner products as $(Y, Z) \leftrightarrow \mathbf{y} \cdot \mathbf{z}$. This can all be described in terms of a $\vee$ operation akin to that in Chapters 5 and 6 by setting

$$\mathbf{e}_i \doteq X_i^{\vee} \text{ and } (X_i, X_j) = X_i^{\vee} \cdot X_j^{\vee}. \tag{7.43}$$

A natural question to ask is "How can the inner product be constructed in the first place?" One answer is that given a metric tensor, the inner product can be defined relative to it. But this introduces the question of how to choose the matrix tensor. Rather than doing this, we take an alternative approach. Namely, choose a basis for $\mathcal{G}$, call it $\{X_i\}$, and use (7.42) to define the inner product. This is essentially equivalent to choosing a metric.

Sometimes it is convenient to have an equation to compute the inner product operation rather than a list of abstract properties. For an n-dimensional matrix Lie algebra represented by $N \times N$ matrices with $n \leq N$, we can define an inner product between elements of the Lie algebra as

$$(X, Y) = \frac{1}{2}\text{Re}[\text{tr}(XWY^*)] \qquad (7.44)$$

where W is a Hermitian weighting matrix with positive eigenvalues, and $\text{Re}[z]$ is the real part of the complex number z. In the case of a real matrix Lie group, the Lie algebra will consist of real matrices, and so

$$(X, Y) = \frac{1}{2}\text{tr}(XWY^T),$$

where W is a real symmetric positive definite matrix. Usually, but not always, W will be a scalar multiple of the identity. The condition $n \leq N$ ensures that there is enough freedom in W to satisfy (7.42). In some special cases (7.44) will still work when $n > N$, but it is not guaranteed in general.

It is easy to see that (7.44) follows the general definition of an inner product on a *real* vector space given in the Appendix C (in the mathematician's sense rather than the physicist's) because even though the Lie algebra basis elements X_i may be complex, all the scalars that multiply these basis elements will be real.

Given an orthonormal basis $\{X_1, ..., X_n\}$ for the Lie algebra, we can project the left and right tangent operators onto this basis to yield elements of the right- and left-Jacobian matrices

$$(J_R)_{ij} = \left(X_i, g^{-1}\frac{\partial g}{\partial x_j}\right) \quad \text{and} \quad (J_L)_{ij} = \left(X_i, \frac{\partial g}{\partial x_j}g^{-1}\right). \qquad (7.45)$$

where $g = g(x_1, ..., x_n)$. Since a Jacobian is a function of coordinates, and those coordinates specify g, sometimes it will be convenient to think of each entry of the Jacobian as a function of g.

Note that $J_R(h \circ g) = J_R(g)$ and $J_L(g \circ h) = J_L(g)$. Here $\{x_1, ..., x_n\}$ is the set of local coordinates used to parameterize a neighborhood of the group around the identity. For the groups considered below, these parameterizations extend over the whole group with singularities of measure zero.

The Jacobian matrix entries given in (7.45) can be related to the $\vee$ operation by the expressions

$$(J_R)_{ij} = \mathbf{e}_i^T \left(g^{-1}\frac{\partial g}{\partial x_j}\right)^{\vee}$$

and

$$J_R \mathbf{e}_j = \left(g^{-1}\frac{\partial g}{\partial x_j}\right)^{\vee} = \sum_{k=1}^{n}(J_R)_{kj}\,\mathbf{e}_k.$$

The analogous expressions hold for J_L as well.

The above right- and left-Jacobians are natural extensions of the Jacobians presented in Chapters 4, 5, and 6 with an appropriate concept of inner product of vectors in the tangent space to the group defined at every group element.

When $|\text{det}J_R(g)| = |\text{det}J_L(g)|$ for all $g \in G$ then G is called a *unimodular* Lie group.

For an n-dimensional matrix Lie group, we denote two special kinds of vector fields as

$$V_L(g) = \sum_{i=1}^{n} v_i X_i g \quad \text{and} \quad V_R(g) = \sum_{i=1}^{n} v_i g X_i.$$

The subscripts of V denote on which side the Lie algebra basis element X_i appears. Here $X_i g$ and $g X_i$ are simply matrix products, and $\{v_i\}$ are real numbers.[18] For a vector field

---

[18] We restrict the discussion to real vector fields.

$V(g)$ on a matrix Lie group (which need not be left- or right-invariant), the left- and right-shift operations are defined as

$$L(h)V(g) \doteq h\,V(g) \quad \text{and} \quad R(h)V(g) \doteq V(g)\,h$$

where $h \in G$. Then it is clear that V_R is *left invariant* and V_L is *right invariant* in the sense that

$$L(h)V_R(g) = V_R(h \circ g) \quad \text{and} \quad R(h)V_L(g) = V_L(g \circ h).$$

This means that there are left- and right-invariant ways to extend the inner product $(\cdot, \cdot)$ on the Lie algebra over the whole group. Namely, for all $Y, Z \in \mathcal{G}$, we can define right and left inner products respectively as

$$(gY, gZ)_g^R \doteq (Y, Z) \quad \text{and} \quad (Yg, Zg)_g^L \doteq (Y, Z)$$

for any $g \in G$. In this way, the inner product of two invariant vector fields $Y_R(g)$ and $Z_R(g)$ (or $Y_L(g)$ and $Z_L(g)$) yields

$$((Y_R(g), Z_R(g))_g^R = ((Y_L(g), Z_L(g))_g^L = (Y, Z).$$

In this notation, the (i, j) entry of the left-Jacobian is $\partial g / \partial x_j$ projected on the ith basis element for the tangent space to G at g as:

$$(J_L)_{ij} = \left(X_i g, \frac{\partial g}{\partial x_j} \right)_g^L = \left(X_i, \frac{\partial g}{\partial x_j} g^{-1} \right).$$

Likewise,

$$(J_R)_{ij} = \left(gX_i, \frac{\partial g}{\partial x_j} \right)_g^R = \left(X_i, g^{-1} \frac{\partial g}{\partial x_j} \right).$$

$SE(2)$ Revisited

According to the formulation of Chapter 5, $SE(2)$ can be parameterized as

$$g(x_1, x_2, \theta) = \exp(x_1 X_1 + x_2 X_2) \exp(\theta X_3) = \begin{pmatrix} \cos\theta & -\sin\theta & x_1 \\ \sin\theta & \cos\theta & x_2 \\ 0 & 0 & 1 \end{pmatrix}$$

where

$$X_1 = \begin{pmatrix} 0 & 0 & 1 \\ 0 & 0 & 0 \\ 0 & 0 & 0 \end{pmatrix}; \quad X_2 = \begin{pmatrix} 0 & 0 & 0 \\ 0 & 0 & 1 \\ 0 & 0 & 0 \end{pmatrix}; \quad X_3 = \begin{pmatrix} 0 & -1 & 0 \\ 1 & 0 & 0 \\ 0 & 0 & 0 \end{pmatrix}.$$

Using the weighting matrix

$$W = \begin{pmatrix} 1 & 0 & 0 \\ 0 & 1 & 0 \\ 0 & 0 & 2 \end{pmatrix},$$

we observe that $(X_i, X_j) = \delta_{ij}$.

The Jacobians for this parameterization, basis, and weighting matrix are then of the form

$$J_R = \begin{pmatrix} \cos\theta & \sin\theta & 0 \\ -\sin\theta & \cos\theta & 0 \\ 0 & 0 & 1 \end{pmatrix}$$

and

$$J_L = \begin{pmatrix} 1 & 0 & x_2 \\ 0 & 1 & -x_1 \\ 0 & 0 & 1 \end{pmatrix}.$$

The J_L and J_R calculated in this way are identical to those calculated using the technique of Chapter 6. Note that

$$\det(J_L) = \det(J_R) = 1.$$

We now examine the Jacobians for the exponential parameterization

$$g(v_1, v_2, \alpha) = \exp(v_1 X_1 + v_2 X_2 + \alpha X_3) = \exp \begin{pmatrix} 0 & -\alpha & v_1 \\ \alpha & 0 & v_2 \\ 0 & 0 & 0 \end{pmatrix}.$$

The closed-form expression for this was given previously. The corresponding Jacobians are

$$J_R = \begin{pmatrix} \frac{\sin\alpha}{\alpha} & \frac{\cos\alpha-1}{\sin\alpha} & 0 \\ \frac{1-\cos\alpha}{\alpha} & \frac{\sin\alpha}{\alpha} & 0 \\ \frac{\alpha v_1 - v_2 + v_2\cos\alpha - v_1\sin\alpha}{\alpha^2} & \frac{v_1 + \alpha v_2 - v_1\cos\alpha - v_2\sin\alpha}{\alpha^2} & 1 \end{pmatrix}^T$$

and

$$J_L = \begin{pmatrix} \frac{\sin\alpha}{\alpha} & \frac{1-\cos\alpha}{\sin\alpha} & 0 \\ \frac{\cos\alpha-1}{\alpha} & \frac{\sin\alpha}{\alpha} & 0 \\ \frac{\alpha v_1 + v_2 - v_2\cos\alpha - v_1\sin\alpha}{\alpha^2} & \frac{-v_1 + \alpha v_2 + v_1\cos\alpha - v_2\sin\alpha}{\alpha^2} & 1 \end{pmatrix}^T.$$

It follows that

$$\det(J_L) = \det(J_R) = \frac{-2(\cos\alpha - 1)}{\alpha^2}.$$

Jacobians for the 'ax + b' group

After having examined rotations and general rigid-body motions in some detail, one may be tempted to believe that bi-invariant integration measures always exist. We now provide an example that demonstrates otherwise.

The so-called "$ax + b$ group," or affine group of the line, may be viewed as the set of all matrices of the form

$$g(a, b) = \begin{pmatrix} a & b \\ 0 & 1 \end{pmatrix}.$$

It acts on the real line as

$$\begin{pmatrix} x' \\ 1 \end{pmatrix} = g(a, b) \begin{pmatrix} x \\ 1 \end{pmatrix}.$$

That is, $x' = ax + b$ (hence the name).

An orthonormal basis for the Lie algebra of this group is

$$X_1 = \begin{pmatrix} 1 & 0 \\ 0 & 0 \end{pmatrix}; \qquad X_2 = \begin{pmatrix} 0 & 1 \\ 0 & 0 \end{pmatrix}.$$

The weighting matrix W for which this basis is orthonormal is

$$W = \begin{pmatrix} 2 & 0 \\ 0 & 2 \end{pmatrix}.$$

A straightforward calculation shows that

$$J_R = \begin{pmatrix} 1/a & 0 \\ 0 & 1/a \end{pmatrix}; \qquad J_L = \begin{pmatrix} 1/a & 0 \\ -b/a & 1 \end{pmatrix}.$$

Hence,

$$\det(J_R) = \frac{1}{a^2} \neq \det(J_L) = \frac{1}{a}.$$

Jacobians for $GL(2, \mathbb{R})$

Elements of $GL(2, \mathbb{R})$ are invertible 2×2 real matrices:

$$g(x_1, x_2, x_3, x_4) = \begin{pmatrix} x_1 & x_2 \\ x_3 & x_4 \end{pmatrix}.$$

An orthonormal basis for the Lie algebra $gl(2, \mathbb{R})$ is

$$X_1 = \begin{pmatrix} 1 & 0 \\ 0 & 0 \end{pmatrix}; \qquad X_2 = \begin{pmatrix} 0 & 1 \\ 0 & 0 \end{pmatrix}; \qquad X_3 = \begin{pmatrix} 0 & 0 \\ 1 & 0 \end{pmatrix}; \qquad X_4 = \begin{pmatrix} 0 & 0 \\ 0 & 1 \end{pmatrix}.$$

The weighting matrix for which these basis elements are orthonormal is $W = 2\mathbb{I}_{2 \times 2}$. The Jacobians in this parameterization, basis, and weighting are

$$J_R = \frac{1}{\det g} \begin{pmatrix} x_4 & 0 & -x_2 & 0 \\ 0 & x_4 & 0 & -x_2 \\ -x_3 & 0 & x_1 & 0 \\ 0 & -x_3 & 0 & x_1 \end{pmatrix}$$

and

$$J_L = \frac{1}{\det g} \begin{pmatrix} x_4 & -x_3 & 0 & 0 \\ -x_2 & x_1 & 0 & 0 \\ 0 & 0 & x_4 & -x_3 \\ 0 & 0 & -x_2 & x_1 \end{pmatrix}.$$

The determinants are

$$\det(J_L) = \det(J_R) = \frac{1}{|\det g|^2}.$$

Jacobians for $SL(2, \mathbb{R})$

A basis for the Lie algebra $sl(2, \mathbb{R})$ is

$$X_1 = \begin{pmatrix} 0 & -1 \\ 1 & 0 \end{pmatrix}; \qquad X_2 = \begin{pmatrix} 1 & 0 \\ 0 & -1 \end{pmatrix}; \qquad X_3 = \begin{pmatrix} 0 & 1 \\ 1 & 0 \end{pmatrix}.$$

This basis is orthonormal with respect to the weighting matrix $W = \mathbb{I}_{2 \times 2}$. The *Iwasawa decomposition* allows us to write an arbitrary $g \in SL(2, \mathbb{R})$ in the form [40]

$$g = u_1(\theta) u_2(t) u_3(\xi)$$

where

$$u_1(\theta) = \exp(\theta X_1) = \begin{pmatrix} \cos\theta & -\sin\theta \\ \sin\theta & \cos\theta \end{pmatrix};$$

$$u_2(t) = \exp(tX_2) = \begin{pmatrix} e^t & 0 \\ 0 & e^{-t} \end{pmatrix};$$

$$u_3(\xi) = \exp(\frac{\xi}{2}(X_3 - X_1)) = \begin{pmatrix} 1 & \xi \\ 0 & 1 \end{pmatrix}.$$

The corresponding right and left Jacobians are

$$J_R(\theta, t, \xi) = \frac{1}{2} \begin{pmatrix} e^{-2t} + e^{2t}(1 + \xi^2) & -2e^{2t}\xi & e^{2t} - e^{-2t}(1 + e^{4t}\xi^2) \\ -2\xi & 2 & 2\xi \\ -1 & 0 & 1 \end{pmatrix}^T$$

and

$$J_L(\theta, t, \xi) = \frac{1}{2} \begin{pmatrix} 2 & 0 & 0 \\ 0 & 2\cos 2\theta & 2\sin 2\theta \\ -e^{2t} & -e^{2t}\sin 2\theta & e^{2t}\cos 2\theta \end{pmatrix}^T.$$

It is easy to verify that

$$\det(J_R(\theta, t, \xi)) = \det(J_L(\theta, t, \xi)) = \frac{1}{2}e^{2t}.$$

Hence, $SL(2, \mathbb{R})$ is *unimodular* (which means the determinants of the left and right Jacobians are the same).

In the matrix exponential parameterization

$$g(a, b, c) = \exp \begin{pmatrix} c & -a+b \\ a+b & -c \end{pmatrix} = \begin{pmatrix} \cosh x + \frac{c}{x}\sinh x & \frac{(a-b)e^{-x}(-1+e^{2x})}{2x} \\ \frac{(a+b)e^{-x}(-1+e^{2x})}{2x} & \cosh x - \frac{c}{x}\sinh x \end{pmatrix}$$

(where $x = \sqrt{-a^2 + b^2 + c^2}$), it can be verified that

$$\det(J_L) = \det(J_R) = -\frac{\sinh^2 x}{x^2}.$$

Jacobians for the Scale-Euclidean Group

A basis for the Lie algebra of the scale-Euclidean group[19] $SIM(2)$ is

$$X_1 = \begin{pmatrix} 0 & 0 & 1 \\ 0 & 0 & 0 \\ 0 & 0 & 0 \end{pmatrix}; \quad X_2 = \begin{pmatrix} 0 & 0 & 0 \\ 0 & 0 & 1 \\ 0 & 0 & 0 \end{pmatrix};$$

$$X_3 = \begin{pmatrix} 0 & -1 & 0 \\ 1 & 0 & 0 \\ 0 & 0 & 0 \end{pmatrix}; \quad X_4 = \begin{pmatrix} 1 & 0 & 0 \\ 0 & 1 & 0 \\ 0 & 0 & 0 \end{pmatrix}.$$

This basis is orthonormal with respect to the weighting matrix

$$W = \begin{pmatrix} 1 & 0 & 0 \\ 0 & 1 & 0 \\ 0 & 0 & 2 \end{pmatrix}.$$

---

[19] Also called the similitude group.

The parameterization

$$g(x_1, x_2, \theta, a) = \exp(x_1 X_1 + x_2 X_2) \exp(\theta X_3 + a X_4)$$

$$= \begin{pmatrix} e^a \cos\theta & -e^a \sin\theta & x_1 \\ e^a \sin\theta & e^a \cos\theta & x_2 \\ 0 & 0 & 1 \end{pmatrix}$$

extends over the whole group.

The Jacobians are

$$J_R = \begin{pmatrix} e^{-a} \cos\theta & -e^{-a} \sin\theta & 0 & 0 \\ e^{-a} \sin\theta & e^{-a} \cos\theta & 0 & 0 \\ 0 & 0 & 1 & 0 \\ 0 & 0 & 0 & 1 \end{pmatrix}^T$$

and

$$J_L = \begin{pmatrix} 1 & 0 & 0 & 0 \\ 0 & 1 & 0 & 0 \\ x_2 & -x_1 & 1 & 0 \\ -x_1 & -x_2 & 0 & 1 \end{pmatrix}^T$$

Note that

$$\det(J_R) = e^{-2a} \neq \det(J_L) = 1.$$

As a general rule, subgroups of the affine group with elements of the form

$$g = \begin{pmatrix} A & \mathbf{b} \\ \mathbf{0}^T & 1 \end{pmatrix}$$

will have left and right Jacobians whose determinants are different unless $\det(A) = 1$.

Parameterization and Jacobians for $SL(2, \mathbb{C})$ and $SO(3, 1)$

We now illustrate the construction of the Jacobians for a group of complex matrices.

$SL(N, \mathbb{C})$ is the group of all $N \times N$ complex matrices with unit determinant. It follows that the matrix exponential of any traceless complex matrix results in an element of $SL(N, \mathbb{C})$.

The case of $n = 2$ plays a very important role in physics, and as we shall see, it may play a role in engineering applications as well.

A general element of the Lie algebra $sl(2, \mathbb{C})$ is of the form

$$X = \begin{pmatrix} x_1 + ix_2 & x_3 + ix_4 \\ x_5 + ix_6 & -x_1 - ix_2 \end{pmatrix},$$

and exponentiating matrices of this form results in an element of $SL(2, \mathbb{C})$:

$$\exp X = \begin{pmatrix} \cosh x + (x_1 + ix_2)(\sinh x)/x & (x_3 + ix_4)(\sinh x)/x \\ (x_5 + ix_6)(\sinh x)/x & \cosh x - (x_1 + ix_2)(\sinh x)/x \end{pmatrix}$$

$$= \mathbb{I}_{2 \times 2} \cosh x + X \frac{\sinh x}{x}$$

where

$$x = i\sqrt{\det(X)} = \sqrt{(x_1 + ix_2)^2 + (x_3 + ix_4)(x_5 + ix_6)}.$$

The basis elements for the Lie algebra $sl(2, \mathbb{C})$ may be taken as

$$X_1 = \begin{pmatrix} 0 & i \\ i & 0 \end{pmatrix}; \quad X_2 = \begin{pmatrix} 0 & -1 \\ 1 & 0 \end{pmatrix}; \quad X_3 = \begin{pmatrix} i & 0 \\ 0 & -i \end{pmatrix};$$

$$X_4 = \begin{pmatrix} 0 & 1 \\ 1 & 0 \end{pmatrix}; \quad X_5 = \begin{pmatrix} 0 & i \\ -i & 0 \end{pmatrix}; \quad X_6 = \begin{pmatrix} 1 & 0 \\ 0 & -1 \end{pmatrix}.$$

Three of the above basis elements are exactly the same as those for $su(2)$, indicating that $SU(2)$ is a subgroup of $SL(2, \mathbb{C})$. This is an example where the inner product

$$(X, Y) = \frac{1}{2} \mathrm{Re}(\mathrm{tr}(XY^*))$$

for the Lie algebra makes $\{X_i\}$ an orthonormal basis.

$SL(2, \mathbb{C})$ is closely related to the *proper Lorentz group* $SO(3, 1)$. The Lorentz group is the group of linear transformations $A \in \mathbb{R}^{4 \times 4}$ that preserves the quadratic form

$$Q(\mathbf{x}) = \pm(x_1^2 + x_2^2 + x_3^2 - x_4^2).$$

That is,

$$Q(A\mathbf{x}) = Q(\mathbf{x}).$$

The *proper Lorentz group* satisfies the further constraint

$$\det(A) = +1.$$

The first three dimensions have the physical meaning of spatial directions, and $x_4 = ct$ is temporal.

More generally, the matrix group consisting of $(p+q) \times (p+q)$ dimensional matrices that preserve the quadratic form

$$Q(\mathbf{x}) = \sum_{i=1}^{p} x_i^2 - \sum_{j=p+1}^{p+q} x_j^2$$

and have determinant $+1$ is denoted as $SO(p, q)$. The two most important examples are $SO(N) = SO(N, 0)$ and the proper Lorentz group, $SO(3, 1)$.

The basis elements for the Lie algebra of the proper Lorentz group are

$$X_1 = \begin{pmatrix} 0 & 0 & 0 & 0 \\ 0 & 0 & -1 & 0 \\ 0 & 1 & 0 & 0 \\ 0 & 0 & 0 & 0 \end{pmatrix}; \quad X_2 = \begin{pmatrix} 0 & 0 & 1 & 0 \\ 0 & 0 & 0 & 0 \\ -1 & 0 & 0 & 0 \\ 0 & 0 & 0 & 0 \end{pmatrix}; \quad X_3 = \begin{pmatrix} 0 & -1 & 0 & 0 \\ 1 & 0 & 0 & 0 \\ 0 & 0 & 0 & 0 \\ 0 & 0 & 0 & 0 \end{pmatrix};$$

$$X_4 = \begin{pmatrix} 0 & 0 & 0 & 1 \\ 0 & 0 & 0 & 0 \\ 0 & 0 & 0 & 0 \\ 1 & 0 & 0 & 0 \end{pmatrix}; \quad X_5 = \begin{pmatrix} 0 & 0 & 0 & 0 \\ 0 & 0 & 0 & 1 \\ 0 & 0 & 0 & 0 \\ 0 & 1 & 0 & 0 \end{pmatrix}; \quad X_6 = \begin{pmatrix} 0 & 0 & 0 & 0 \\ 0 & 0 & 0 & 0 \\ 0 & 0 & 0 & 1 \\ 0 & 0 & 1 & 0 \end{pmatrix};$$

These basis elements are orthonormal with respect to the inner product $(X, Y) = \frac{1}{2} \mathrm{tr}(XY^T)$. The first three basis elements are isomorphic to those for $so(3)$, indicating that spatial rotations form a subgroup of proper Lorentz transformations.

Just as $SU(2)$ is a double cover of $SO(3)$, $SL(2,\mathbb{C})$ is a double cover for the Lorentz group, $SO(3,1)$. Using group-theoretic notation, this is stated as

$$SU(2)/\mathbb{Z}_2 \cong SO(3) \quad \text{and} \quad SL(2,\mathbb{C})/\mathbb{Z}_2 \cong SO(3,1)$$

where $\mathbb{Z}_2 \cong \{\mathbb{I}, -\mathbb{I}\}$. In other words, there are homomorphisms $SU(2) \to SO(3)$ and $SL(2,\mathbb{C}) \to SO(3,1)$ with kernel $\mathbb{Z}_2$, meaning that these are two-to-one homomorphisms.

There is a one-to-one correspondence between the basis elements of the Lie algebras $so(3)$ and $su(2)$, and similarly for the Lie algebras $so(3,1)$ and $sl(2,\mathbb{C})$. These correspondences preserve the Lie bracket. It follows that the Jacobians and adjoint matrices are the same when such correspondences exist.

It can be shown that in any parameterization $\det(J_R) = \det(J_L)$, or equivalently $\det[Ad(g)] = 1$ for both of these groups.

7.3.7 The Killing Form

A bilinear form, $B(X,Y)$ for $X,Y \in \mathcal{G}$ is said to be Ad-invariant if

$$B(X,Y) = B(Ad(g)X, Ad(g)Y)$$

for any $g \in G$. In the case of real matrix Lie groups (and the corresponding Lie algebras), which are the ones of most interest in engineering applications, a symmetric ($B(X,Y) = B(Y,X)$) and invariant bilinear form, called the *Killing form* [20] is defined as

$$B(X,Y) \doteq \text{tr}(ad(X)ad(Y)). \tag{7.46}$$

It can be shown that this form is written as [42]

$$B(X,Y) = \lambda\text{tr}(XY) + \mu(\text{tr}X)(\text{tr}Y)$$

for some constant real numbers μ and λ that are group dependent.

The Killing form is important in the context of harmonic analysis because the Fourier transform and inversion formula can be defined for large classes of groups that are defined by the behavior of their Killing form. The classification of Lie groups according to the properties of the Killing form is based on *Cartan's Criteria* [21] [34, 41]. For example, a Lie group is called *nilpotent* if $B(X,Y) = 0$ for all $X,Y \in \mathcal{G}$. If for all $X,Y,Z \in \mathcal{G}$ the equality $B(X,[Y,Z]) = 0$ holds, then this is equivalent to the corresponding Lie group being *solvable*. In contrast, *semi-simple* Lie groups are those for which $B(X,Y)$ is nondegenerate (i.e., the determinant of the $n \times n$ matrix with elements $B(X_i, X_j)$ is nonzero where $\{X_1, ..., X_n\}$ ($n \geq 2$) is a basis for $\mathcal{G}$). For example, the Heisenberg groups are nilpotent, the rotation groups are semi-simple, and groups of rigid-body motions are solvable.

7.3.8 The Matrices of $Ad(G)$, $ad(X)$, and $B(X,Y)$

The formal coordinate-independent definitions of the adjoints $Ad(g)$ and $ad(X)$ and the Killing form $B(X,Y)$ are central to the theory of Lie groups. And while such definitions

---

[20] Named after Wilhelm Karl Joseph Killing (1847-1923)

[21] Named after the mathematician Elie Cartan (1869-1951) who made major contributions to the theory of Lie groups

are sufficient for mathematicians to prove many fundamental properties, it is useful for the engineer interested in group theory to have such concepts illustrated with matrices.

As with all linear operators, $Ad(g)$ and $ad(X)$ are expressed as matrices using an appropriate inner product and concrete basis for the Lie algebra. In particular, with the inner product defined earlier for Lie algebras we have

$$[Ad(g)]_{ij} = (X_i, Ad(g)X_j) = (X_i, gX_jg^{-1}) \tag{7.47}$$

and

$$[ad(X)]_{ij} = (X_i, ad(X)X_j) = (X_i, [X, X_j]). \tag{7.48}$$

Another way to view these is that if we define the linear operation $\vee$ such that it converts Lie algebra basis elements X_i to elements of the natural basis element $\mathbf{e}_i \in \mathbb{R}^n$,

$$(X_i)^{\vee} = \mathbf{e}_i,$$

then the matrix with elements given in (7.47) will be

$$[Ad(g)] = [(gX_1g^{-1})^{\vee}, ..., (gX_ng^{-1})^{\vee}].$$

It is this matrix that relates left and right Jacobians. Using the $\vee$ notation, we can write

$$J_L = \left[\left(\frac{\partial g}{\partial x_1} g^{-1} \right)^{\vee}, ..., \left(\frac{\partial g}{\partial x_n} g^{-1} \right)^{\vee} \right]$$

and

$$J_R = \left[\left(g^{-1} \frac{\partial g}{\partial x_1} \right)^{\vee}, ..., \left(g^{-1} \frac{\partial g}{\partial x_n} \right)^{\vee} \right].$$

Since

$$\left(g \left(g^{-1} \frac{\partial g}{\partial x_1} \right) g^{-1} \right)^{\vee} = \left(\frac{\partial g}{\partial x_1} g^{-1} \right)^{\vee}$$

it follows that

$$J_L = [Ad(g)]J_R.$$

Hence, if the Jacobians are known, we can write

$$[Ad(g)] = J_L J_R^{-1}.$$

The matrix with elements given in (7.48) will be

$$[ad(X)] = [([X, X_1])^{\vee}, ..., ([X, X_n])^{\vee}].$$

This then gives a concrete tool with which to calculate the $n \times n$ matrix with entries

$$[B]_{ij} = B(X_i, X_j) = \operatorname{tr}([ad(X_i)][ad(X_j)]).$$

B is then degenerate if and only if

$$\det([B]) = 0.$$

If $\det([B]) \neq 0$ the Lie algebra is called *semi-simple*. If $[B]_{ij} = 0$ for all i, j, the Lie algebra is called *nilpotent*.

We now illustrate these concepts with some concrete examples.

Relationship Between $ad(X)$, $B(X,Y)$, and the Structure Constants

Recall that the structure constants of a real Lie algebra are defined by

$$[X_i, X_j] = \sum_{k=1}^{n} C_{ij}^k X_k .$$

From the anti-symmetry of the Lie bracket (7.35) and the Jacobi identity (7.36), respectively, we see that

$$C_{ij}^k = -C_{ji}^k$$

and

$$\sum_{j=1}^{n} (C_{ij}^l C_{km}^j + C_{mj}^l C_{ik}^j + C_{kj}^l C_{mi}^j) = 0.$$

The matrix entries of $[ad(X_k)]_{ij}$ are related to the structure constants as

$$[ad(X_k)]_{ij} = (X_i, [X_k, X_j]) = \left(X_i, \sum_{m=1}^{n} C_{kj}^m X_m \right).$$

For a real Lie algebra, the inner product (X,Y) is linear in the second argument, and so

$$[ad(X_k)]_{ij} = \sum_{m=1}^{n} (X_i, X_m) C_{kj}^m = C_{kj}^i.$$

Then

$$B(X_i, X_j) = \mathrm{tr}(ad(X_i)ad(X_j))$$

$$= \sum_{m=1}^{n} \sum_{r=1}^{n} [ad(X_i)]_{mr}[ad(X_j)]_{rm}$$

$$= \sum_{m=1}^{n} \sum_{r=1}^{n} C_{ir}^m C_{jm}^r.$$

The $ax + b$ group

$$gX_1g^{-1} = \begin{pmatrix} a & b \\ 0 & 1 \end{pmatrix} \begin{pmatrix} 1 & 0 \\ 0 & 0 \end{pmatrix} \begin{pmatrix} 1/a & -b/a \\ 0 & 1 \end{pmatrix} = \begin{pmatrix} 1 & -b \\ 0 & 0 \end{pmatrix} = X_1 - bX_2.$$

$$gX_2g^{-1} = \begin{pmatrix} a & b \\ 0 & 1 \end{pmatrix} \begin{pmatrix} 0 & 1 \\ 0 & 0 \end{pmatrix} \begin{pmatrix} 1/a & -b/a \\ 0 & 1 \end{pmatrix} = \begin{pmatrix} 0 & a \\ 0 & 0 \end{pmatrix} = aX_2.$$

Therefore,

$$[Ad(g)] = \begin{pmatrix} 1 & 0 \\ -b & a \end{pmatrix}.$$

Similarly, the calculation

$$[X, X_1] = \begin{pmatrix} x & y \\ 0 & 0 \end{pmatrix} \begin{pmatrix} 1 & 0 \\ 0 & 0 \end{pmatrix} - \begin{pmatrix} 1 & 0 \\ 0 & 0 \end{pmatrix} \begin{pmatrix} x & y \\ 0 & 0 \end{pmatrix} = \begin{pmatrix} 0 & -y \\ 0 & 0 \end{pmatrix}$$

and

$$[X, X_2] = \begin{pmatrix} x & y \\ 0 & 0 \end{pmatrix} \begin{pmatrix} 0 & 1 \\ 0 & 0 \end{pmatrix} - \begin{pmatrix} 0 & 1 \\ 0 & 0 \end{pmatrix} \begin{pmatrix} x & y \\ 0 & 0 \end{pmatrix} = \begin{pmatrix} 0 & x \\ 0 & 0 \end{pmatrix},$$

and so

$$[ad(X)] = \begin{pmatrix} 0 & 0 \\ -y & x \end{pmatrix}.$$

Substitution into the definition shows that

$$[B] = \begin{pmatrix} 1 & 0 \\ 0 & 0 \end{pmatrix}.$$

Clearly this is degenerate, and the $ax + b$ group is not semi-simple.

$SO(3)$

When the inner product is normalized so that $(X_i, X_j) = \delta_{ij}$,

$$[Ad(g)] = J_L J_R^{-1},$$

where J_R and J_L are given in Chapter 4 in a variety of parameterizations.
    A straightforward calculation shows

$$[ad(X)] = X$$

and

$$[B] = -2\,\mathbb{I}_{3 \times 3}.$$

Hence $SO(3)$ is semi-simple.

$SE(2)$

Substitution into the definitions yields

$$[Ad(g)] = \begin{pmatrix} \cos\theta & -\sin\theta & x_2 \\ \sin\theta & \cos\theta & -x_1 \\ 0 & 0 & 1 \end{pmatrix};$$

And when

$$X = \begin{pmatrix} 0 & -\alpha & v_1 \\ \alpha & 0 & v_2 \\ 0 & 0 & 0 \end{pmatrix},$$

$$[ad(X)] = \begin{pmatrix} 0 & -\alpha & v_2 \\ \alpha & 0 & -v_1 \\ 0 & 0 & 0 \end{pmatrix},$$

and

$$[B] = \begin{pmatrix} 0 & 0 & 0 \\ 0 & 0 & 0 \\ 0 & 0 & -2 \end{pmatrix}.$$

This is clearly degenerate, and $SE(2)$ is therefore not semi-simple. (Neither is $SE(3)$.)
    These are enough examples to illustrate the computation for real Lie groups. We note that $GL(N, \mathbb{R})$ and $SL(N, \mathbb{R})$ are both semi simple.

7.4 Summary

In this chapter we presented a comprehensive introduction to group theory. We considered both finite and Lie groups. Many good books already exist on group theory (for example, see the reference list below). Our goal here was to present group theory in a manner that builds on the mathematical knowledge of engineers and applied scientists. This was particularly so in the context of Lie groups. There our treatment of matrix Lie groups used basic linear algebra. Concepts such as the adjoint and Killing form were presented in such a way that concrete matrices could be computed (rather than the usual treatment of these concepts as abstract linear operators in mathematics books).

The techniques and terminology of this chapter provide the foundations for Chapter 8 and the remainder of the book.

References

1. Artin, M., *Algebra*, Prentice Hall, Upper Saddle River, New Jersey, 1991.
2. Alperin, J.L., Bell, R.B., *Groups and Representations*, Springer-Verlag, New York, 1995.
3. Baglivo, J., Graver, J.E., *Incidence and Symmetry in Design and Architecture*, Cambridge University Press, 1983.
4. Baker, H.F., "Alternants and Continuous Groups," *Proceedings of the London Mathematical Society* (Second Series), 3(1): 24–47, 1905.
5. Barut, A.O., Rączka, R., *Theory of Group Representations and Applications*, World Scientific, 1986.
6. Birkhoff, G., MacLane, S., *A Survey of Modern Algebra*, 4th ed., Macmillan Publishing Co., Inc., New York, 1977.
7. Bishop, D.M., *Group Theory and Chemistry*, Dover, New York, 1973.
8. Bradley, C.J., Cracknell, A.P., *The Mathematical Theory of Symmetry in Solids*, Clarendon Press, Oxford, 1972.
9. Campbell, J.E., "On a Law of Combination of Operators," *Proceedings of the London Mathematical Society* (Series 1), 29(1): 14–32, 1897.
10. Gilmore, R., *Lie Groups, Lie Algebras, and Some of Their Applications*, Dover, 2005.
11. Casper, D.L.D., Klug, A., "Physical Principles in the Construction of Regular Viruses," *Cold Spring Harbor Symposium on Quantitative Biology*, 27: 1–24, 1962.
12. Coxeter, H.S.M., *Regular Polytopes*, 2nd ed., MacMillan, New York, 1963.
13. Eisenhart, L.P., *Continuous Groups of Transformations*, Dover Publications Inc., New York, 1961.
14. Gel'fand, I.M., Minlos, R.A., Shapiro, Z.Ya., *Representations of the Rotation and Lorentz Groups and Their Applications*, Macmillan, New York, 1963.
15. Humphreys, J.F., *A Course in Group Theory*, Oxford University Press, 1996.
16. Herstein, I.N., *Topics in Algebra*, 2nd edition, John Wiley and Sons, New York, 1975.
17. Hausdorff, F., "Die symbolische Exponentialformel in der Gruppentheorie," *Berichte der Sächsischen Akademie der Wissenschaften*, Leipzig, 58: 19–48, 1906.
18. Husain, T., *Introduction to Topological Groups*, W.B. Sanders Company, Philadelphia, 1966.
19. Inui, T., Tanabe, Y., Onodera, Y., *Group Theory and Its Applications in Physics*, 2nd ed., Springer-Verlag, 1996.
20. Isaacs, I.M., *Character Theory of Finite Groups*, Dover, 1976.
21. James, G., Liebeck, M., *Representations and Characters of Groups*, Cambridge University Press, 1993.
22. Janssen, T., *Crystallographic Groups*, North-Holland/American Elsevier, New York, 1973.
23. Johnston, B.L., Richman, F.R., *Numbers and Symmetry: An Introduction to Algebra*, CRC Press, Boca Raton, FL, 1997.
24. Kettle, S.F.A., *Symmetry and Structure: Readable Group Theory for Chemists*, Wiley, New York, 1995.

25. Klein, F., *Lectures on the Icosahedron*, Kegan Paul, London, 1913.
26. Lawton, W., Raghavan, R., Viswanathan, R., "Ribbons and Groups: A Thin Rod Theory for Catheters and Proteins," *Journal of Physics A: Mathematical and General*, 32(9): 1709 – 1735, March 1999.
27. Liljas, L., "The Structure of Spherical Viruses," *Progress in Biophysics and Molecular Biology*, 48(1): 1 – 36, 1986.
28. Marsden, J.E., Ratiu, T.S., *Introduction to Mechanics and Symmetry*, 2nd ed., Springer, 1999.
29. Miller, W., *Symmetry Groups and Their Applications*, Academic Press, New York, 1972.
30. Milnor, J., *Morse Theory*, (Based on the notes of M. Spivak and R. Wells), Princeton University Press, Princeton, NJ, 1963.
31. Montgomery, D., Zippin, L., *Topological Transformation Groups*, Interscience Publishers Inc., New York, 1955.
32. Naimark, M.A., *Linear Representations of the Lorentz Group*, Macmillan, New York, 1964.
33. Rossmann, M.G., Johnson, J.E., "Icosahedral RNA Virus Structure," *Annual Review of Biochemistry*, 58(1): 533 – 73, 1989.
34. Sagle, A.A., Walde, R.E., *Introduction to Lie Groups and Lie Algebras*, Academic Press, Orlando, FL, 1973.
35. Sattinger, D.H., Weaver, O.L., *Lie Groups and Algebras with Applications to Physics, Geometry, and Mechanics*, Springer-Verlag, New York, 1986.
36. Schoolfield, C.H., *Random Walks on Wreath Products of Groups and Markov Chains on Related Homogeneous Spaces*, Ph.D. Dissertation, Dept. of Mathematical Sciences, Johns Hopkins University, Baltimore, MD, May 1998.
37. Selig, J.M., *Geometrical Methods in Robotics*, Springer Monographs in Computer Science, New York, 1996. (2nd edition, 2005)
38. Sternberg, S., *Group Theory and Physics*, Cambridge University Press, New York, 1994.
39. Sudarshan, E.C.G., Mukunda, N., *Classical Dynamics: A Modern Perspective*, Wiley-Interscience, New York, 1974.
40. Sugiura, M., *Unitary Representations and Harmonic Analysis*, 2nd ed., North-Holland, Amsterdam, 1990.
41. Varadarajan, V.S., *Lie Groups, Lie Algebras, and Their Representations*, Springer-Verlag, New York 1984.
42. Vilenkin, N.J., Klimyk, A.U., *Representation of Lie Groups and Special Functions*, Vols. 1-3, Kluwer Academic Publ., Dordrecht, Holland 1991.
43. Warner, F.W., *Foundations of Differentiable Manifolds and Lie Groups*, Springer, New York, 1983.
44. Weyl, H., *The Classical Groups*, Princeton University Press, Princeton, NJ, 1946.
45. Yan, Y., Chirikjian, G.S., "Almost-Uniform Sampling of Rotations for Conformational Searches in Robotics and Structural Biology," *Proceedings of ICRA 2012*, pp. 4254 – 4259, Minneapolis, MN, 14–18 May 2012.
46. Yan, Y., Chirikjian, G.S., "Voronoi Cells in Lie Groups and Coset Decompositions: Implications for Optimization, Integration, and Fourier Analysis," *Proceedings of the 52nd IEEE Conference on Decision and Control*, pp. 1137 – 1143, Firenze, Italy, December 10–13, 2013.
47. Zassenhaus, H.J., *The Theory of Groups*, Vandenhoeck and Ruprecht, 1958 (Dover edition, 2011)
48. Želobenko, D.P., *Compact Lie Groups and Their Representations*, Translations of Mathematical Monographs, American Mathematical Society, Providence, RI, 1973.

8

Harmonic Analysis on Groups

In this chapter we review Fourier analysis of functions on certain kinds of noncommutative groups. We begin with the simplest case of finite groups. Then compact Lie groups are considered. Finally, methods to handle a wide range of noncompact noncommutative unimodular groups are discussed. To this end, we review in some depth what it means to differentiate and integrate functions on Lie groups. The rotation and motion groups are discussed in detail in Chapters 9 and 10.

In their most general form, the results of this chapter are as follows. Given functions $f_i(g)$ for $i = 1, 2$ which are square integrable with respect to an invariant integration measure μ on a unimodular group $(G, \circ)$, i.e.,[1]

$$\mu(|f_i|^2) = \int_G |f_i(g)|^2 \, d\mu(g) < \infty$$

and

$$\mu(L(h)f_i) = \mu(R(h)f_i) = \mu(f_i) \quad \forall \ h \in G$$

where $L(h)$ and $R(h)$ are the left and right shift operators $(L(h)f_i)(g) = f_i(h^{-1} \circ g)$ and $(R(h)f_i)(g) = f_i(g \circ h)$.

We can define the convolution product

$$(f_1 * f_2)(g) \doteq \int_G f_1(h) \, f_2(h^{-1} \circ g) \, d\mu(h).$$

And for broad classes of unimodular groups we can define the Fourier transform as

$$\hat{f}(p) \doteq \int_G f(g) U(g^{-1}, \lambda) \, d\mu(g) \tag{8.1}$$

where $U(\cdot, \lambda)$ is a unitary matrix function akin to $e^{i\omega x}$ (called an irreducible matrix representation) for each value of the parameter λ, which is like a frequency parameter. These matrices have the homomorphism property

$$U(g_1 \circ g_2, \lambda) = U(g_1, \lambda)U(g_2, \lambda). \tag{8.2}$$

The Fourier transform defined in this way has corresponding inversion, convolution, and Parseval theorems

---

[1] When there is no ambiguity about which measure is being used, $d\mu(g)$ will be abbreviated as $d(g)$ or dg.

$$f(g) = \int_{\hat{G}} \mathrm{tr}[\hat{f}(\lambda)U(g,\lambda)]d\nu(\lambda), \qquad (8.3)$$

$$(\widehat{f_1 * f_2})(\lambda) = \hat{f}_2(\lambda)\hat{f}_1(\lambda) \qquad (8.4)$$

and

$$\int_G |f(g)|^2 \, d(g) = \int_{\hat{G}} \|\hat{f}(\lambda)\|^2 d\nu(\lambda). \qquad (8.5)$$

Here $\| \cdot \|$ is the Hilbert-Schmidt norm, $\hat{G}$ (which is defined on a case-by-case basis, and is the space of all λ values) is called the *unitary dual* of the group G, and ν is an appropriately chosen integration measure (in a generalized sense) on $\hat{G}$. In this chapter we examine the structure of $\hat{G}$ and how to construct the measures $d\mu(g)$ and $d\nu(\lambda)$.

As with the classical Fourier transform, sometimes we use the alternative notation $\mathcal{F}(f)$ in place of $\hat{f}$.

In addition to the condition of square integrability, we will assume functions to be as well-behaved as required for a given context. For instance, when discussing differentiation of functions on Lie groups with elements parameterized as $g(x_1, ..., x_n)$, we will assume all the partial derivatives of $f(g(x_1, ..., x_n))$ in all the parameters x_i exist. When considering noncompact groups, we will assume that functions $f(g)$ are in $\mathcal{L}^p(G)$ for all $p \in \mathbb{Z}^+$ and rapidly decreasing in the sense that

$$\int_G |f(g(x_1, ..., x_n))| |x_1^{p_1} \cdots x_n^{p_n} \, d(g(x_1, ..., x_n)) < \infty$$

for all $p_i \in \mathbb{Z}^+$.

In the case of finite groups, all measures are counting measures, and these integrals are interpreted (to within a constant) as sums. In the case of unimodular Lie groups, integration by parts provides a tool to derive operational properties that convert differential operators acting on functions to algebraic operations on the Fourier transforms of functions, much like classical Fourier analysis.

For historical perspective and abstract aspects of noncommutative harmonic analysis see [4, 20, 24, 29, 35, 52, 60, 74, 80, 81, 92]. For applications to other fields including physics and probability theory, see [28, 34, 88].

8.1 Fourier Transforms for Finite Groups

Let $(G, \circ)$ be a finite group with $|G| = n$ elements, $g_1, ..., g_n$. Let $f(g_i)$ be a function which assigns a complex value to each group element, i.e., $f : G \to \mathbb{C}$. In analogy with the standard (Abelian) convolution product and discrete Fourier transform that we saw in Chapter 2, the convolution product of two complex-valued functions on G is written as:

$$(f_1 * f_2)(g) \doteq \sum_{h \in G} f_1(h) \, f_2(h^{-1} \circ g) = \sum_{i=1}^{|G|} f_1(g_i) \, f_2(g_i^{-1} \circ g). \qquad (8.6)$$

Since we are summing over h, we can make the change of variables $k = h^{-1} \circ g$ and write the convolution in the alternate form:

$$(f_1 * f_2)(g) = \sum_{k \in G} f_1(g \circ k^{-1}) \, f_2(k).$$

However, unlike the case of Abelian (in particular, additive) groups where $\circ = +$, in the more general noncommutative context we can not switch the order of elements on either side of the group operation. This means that in general, the convolution of two arbitrary functions on G does not commute: $(f_1 * f_2)(g) \neq (f_2 * f_1)(g)$. There are, of course, exceptions to this. For example, if $f_1(g)$ is a scalar multiple of $f_2(g)$ or if one or both functions are class functions [2], then the convolution will commute.

A number of problems can be formulated as convolutions on finite groups. These include generating the statistics of random walks, as well as coding theory. See [71] for an overview. In some applications, instead of computing the convolution, one is interested in the inverse problem of solving the convolution equation

$$(f_1 * f_2)(g) = f_3(g)$$

where f_1 and f_3 are given and f_2 must be found. This problem is often refered to as *deconvolution*.

For a finite group with $|G| = n$ elements, the convolution (8.6) is performed in $\mathcal{O}(n^2)$ arithmetic operations when the definition above is used. The deconvolution can be performed as a matrix inversion in the following way. Assume f_1 and f_3 are given, and define $A_{ij} = f_1(g_i \circ g_j^{-1})$, $x_j = f_2(g_j)$ and $b_i = f_3(g_i)$. Then, when $A = [A_{ij}]$ is invertible, inversion of the matrix equation

$$A\mathbf{x} = \mathbf{b}$$

will solve the deconvolution problem. However, it requires $\mathcal{O}(n^\gamma)$ computations where $2 \leq \gamma \leq 3$ is the exponent required for matrix inversion and multiplication. (See Appendix D for a discussion.) If Gaussian elimination is used, $\gamma = 3$. Thus, it is desirable from a computational perspective to generalize the concept of the Fourier transform, and particularly the FFT, as a tool for efficiently computing convolutions and solving deconvolution problems. In order to do this, a generalization of the unitary function $u(x, \omega) = \exp(i\omega x)$ is required. The two essential properties of this function are

$$\exp(i\omega(x_1 + x_2)) = \exp(i\omega x_1) \cdot \exp(i\omega x_2)$$

and

$$\overline{\exp(i\omega x)}\exp(i\omega x) = 1.$$

The concept of a *unitary group representation* discussed in the following subsection is precisely what is required to define the Fourier transform of a function on a finite group.

In essence, if we are able to find a particular set of homomorphisms from G into a set of unitary matrices $\{U^\lambda(g)\}$ indexed by λ (i.e., for each λ, $U^\lambda(g)$ is a unitary matrix for all $g \in G$), then for functions $f : G \to \mathbb{C}$, we can define the matrix function[3]

$$\mathcal{T}^\lambda(f) = \tilde{f}^\lambda = \sum_{g \in G} f(g)U^\lambda(g). \tag{8.7}$$

Since $U^\lambda(g \circ h) = U^\lambda(g)U^\lambda(h)$, it can be shown (and will be shown later) that this transformation has the following properties under left and right shifts and convolutions:

$$\mathcal{T}^\lambda(L(h)f) = U^\lambda(h)\tilde{f}^\lambda \quad \text{and} \quad \mathcal{T}^\lambda(R(h)f) = \tilde{f}^\lambda U^\lambda(h^{-1}) \tag{8.8}$$

---

[2]Recall that a class function has the property $f(g \circ h) = f(h \circ g)$.

[3]Additional conditions will be required on $U^\lambda(g)$ before discrete analogs of (8.3) and (8.5) will hold, and so we use the different notations $U^\lambda(g)$ and $U(g, \lambda)$ on purpose.

and

$$\mathcal{T}^\lambda(f_1 * f_2) = \tilde{f}_1^\lambda \tilde{f}_2^\lambda. \tag{8.9}$$

Recall from Chapter 7 that $(L(h)f)(g) = f(h^{-1} \circ g)$ and $(R(h)f)(g) = f(g \circ h)$.

8.1.1 Representations of Finite Groups

An N-dimensional matrix representation of a group G is a homomorphism from the group into a set of invertible matrices, $D : G \to GL(N, \mathbb{C})$. Following from the definition of a homomorphism, $D(e) = \mathbb{I}_{N \times N}$, $D(g_1 \circ g_2) = D(g_1)D(g_2)$, and $D(g^{-1}) = D^{-1}(g)$. Here $\mathbb{I}_{N \times N}$ is the $N \times N$ dimensional identity matrix, and the product of two representations is ordinary matrix multiplication.

If the mapping $D : G \to D(G)$ is bijective, then the representation D is called *faithful*. Two representations D and D' are said to be *equivalent* if they are similar as matrices, i.e., if $SD(g) = D'(g)S$ for some $S \in GL(N, \mathbb{C})$ and all $g \in G$ where S is independent of g. In this case, the notation $D \cong D'$ is used. A representation is called *reducible* if it is equivalent to the direct sum of other representations. In other words, if a representation can be block-diagonalized, it is reducible. Thus, by an appropriate similarity transformation, any finite dimensional representation matrix can be decomposed by successive similarity transformations until the resulting blocks can no longer be reduced into smaller ones. These smallest blocks are called *irreducible* representations and form the foundations upon which all other representation matrices can be built. It is common to use the notation

$$B = A_1 \oplus A_2 = \begin{pmatrix} A_1 & \mathbb{O} \\ \mathbb{O} & A_2 \end{pmatrix}$$

for block diagonal matrices. B is said to be the *direct sum* of A_1 and A_2. If instead of two blocks, B consists of n blocks on the diagonal, the following notation is used:

$$B = \sum_{m=1}^{n} \bigoplus A_m = \begin{pmatrix} A_1 & \mathbb{O} & \mathbb{O} & \cdots & \mathbb{O} \\ \mathbb{O} & A_2 & \mathbb{O} & \cdots & \vdots \\ \mathbb{O} & \mathbb{O} & \ddots & \mathbb{O} & \vdots \\ \vdots & \vdots & \mathbb{O} & \ddots & \mathbb{O} \\ \mathbb{O} & \cdots & \cdots & \mathbb{O} & A_n \end{pmatrix}.$$

In the case when $A_1 = A_2 = \cdots = A_n = A$, we say that $B = \bigoplus n A$. If not for the $\bigoplus$ this would be easily confused with scalar multiplication $n \cdot A$, but this will not be an issue here.

Representations can be considered as linear operators which act on functions on a group. Depending on what basis is used, these operators are expressed as different representation matrices that are equivalent under similarity transformation. However, from a computational perspective the choice of basis can have a significant impact.

The two standard representations which act on functions on a group are called the *left-regular* and *right-regular* representations. They are defined respectively as:

$$(L(g)f)(h) \doteq f(g^{-1} \circ h) \qquad (R(g)f)(h) \doteq f(h \circ g) \tag{8.10}$$

for all $g, h \in G$. The shorthand $f_h^L(g)$ and $f_h^R(g)$ is used for these, respectively.

The fact that these are representations can be observed as follows:

$$(L(g_1)(L(g_2)f))(h) = (L(g_1)f_{g_2}^L)(h) = f_{g_2}^L(g_1^{-1} \circ h) =$$

$$f(g_2^{-1} \circ (g_1^{-1} \circ h)) = f((g_1 \circ g_2)^{-1} \circ h) = (L(g_1 \circ g_2)f)(h),$$

and

$$(R(g_1)(R(g_2)f))(h) = (R(g_1)f_{g_2}^R)(h) = f_{g_2}^R(h \circ g_1) =$$

$$f((h \circ g_1) \circ g_2) = f(h \circ (g_1 \circ g_2)) = (R(g_1 \circ g_2)f)(h).$$

Note also that left and right shifts commute with each other, but left (right) shifts do not generally commute with other left (right) shifts.

Matrix elements of these representations can be expressed using the inner product[4]

$$(f_1, f_2) = \sum_{g \in G} f_1(g)\overline{f_2(g)}.$$

Shifted versions of the delta function

$$\delta(h) = \begin{cases} 1 \text{ for } h = e \\ 0 \text{ for } h \neq e \end{cases}$$

for all $h \in G$ can be used as a basis for all functions on the finite group G. That is, any function $f(g)$ can be expressed as $f(g) = \sum_{h \in G} f(h)\delta_g(h)$ where $\delta_g(h) = \delta(g^{-1} \circ h)$ is a left-shifted version of $\delta(h)$. (Actually, $\delta(h)$ is a special function in the sense that $\delta(g^{-1} \circ h) = \delta(h \circ g^{-1})$, and $\delta(g^{-1}) = \delta(g)$, and so it doesn't matter if the shifting is from the left or right, and also $\delta_g(h) = \delta_h(g)$.). An $N \times N$ matrix representation of G is then defined element by element by computing

$$T_{ij}(g) = (\delta_{g_i}, T(g)\delta_{g_j})$$

where the operator T can be either of the shift operators L or R. Because of the properties of δ functions, this matrix representation has the property that $T(e) = \mathbb{I}_{N \times N}$, and the diagonal elements of $T(g)$ for $g \neq e$ are all zero. In the case when $T = L$, this can be seen by carrying out the computation:

$$L_{ij}(g) = \sum_{h \in G} \delta(g_i^{-1} \circ h)\delta(g_j^{-1} \circ g^{-1} \circ h) = \delta(g_j^{-1} \circ g^{-1} \circ g_i) = \delta(g \circ (g_j \circ g_i^{-1})),$$

which reduces to $L_{ii}(g) = \delta(g)$ for diagonal elements. A similar computation follows for $R(g)$.

In either case, taking the trace of $T(g)$ defines the *character* of T:

$$\chi(g) = \text{tr}(T(g)) = \sum_{i=1}^{|G|} \delta(g) = \begin{cases} |G| \text{ for } g = e \\ 0 \text{ for } g \neq e. \end{cases} \tag{8.11}$$

For all $g_i \in G$, let $\pi_j(g_i)$ be a $d_j \times d_j$ irreducible representation of G (in contrast to T, which may be reducible). Thus,

$$\pi_j(g_k \circ g_l) = \pi_j(g_k)\pi_j(g_l),$$

---

[4]Here the physicists convention is used whereas the complex conjugate of this (f_1, f_2) is more common in mathematics. The difference between these is immaterial for our purposes.

and

$$\pi_j(e) = \mathbb{I}_{d_j \times d_j} \qquad \pi_j^*(g) = \pi_j(g^{-1}).$$

(Recall that $*$ denotes complex conjugate transpose, also called the Hermitian conjugate, of a matrix.)

Let n_j be the number of times $\pi_j(g)$ appears in the decomposition of $T(g)$. Then, since the trace is invariant under similarity transformation, it follows from (8.11) that

$$\chi(g) = \sum_{k=1}^{N} n_k \chi_k(g) = \begin{cases} |G| & \text{for } g = e \\ 0 & \text{for } g \neq e \end{cases} \qquad (8.12)$$

where $\chi_k(g) = \text{tr}(\pi_k(g))$, and N is the number of inequivalent irreducible unitary representations (IURs).

Theorem 8.1. *If G is a finite group, $g \in G$, and $D(g)$ is a matrix representation, then a unitary representation equivalent to $D(g)$ results from the similarity transformation*

$$U(g) = M^{-1}D(g)M \quad \text{where} \quad M = \left(\sum_{g \in G} D^*(g)D(g) \right)^{\frac{1}{2}}.$$

Proof. The matrix square root in the above expression is well defined and results in a Hermitian matrix ($M^* = M$) since M^2 is a positive definite Hermitian matrix (see Appendix D). Furthermore, by the definition of a representation and the general property $(X_1 X_2)^* = X_2^* X_1^*$, we have for an arbitrary $h \in G$

$$D^*(h)M^2 D(h) = \sum_{g \in G} D^*(h)D^*(g)D(g)D(h) = \sum_{g \in G} D^*(g \circ h)D(g \circ h) = M^2.$$

Multiplication on the left and right by M^{-1} shows that the product of $MD(h)M^{-1} = U(h)$ with its hermitian conjugate is the identity matrix.

Theorem 8.2. *(Schur's Lemma) [78] If $\pi_{\lambda_1}(g)$ and $\pi_{\lambda_2}(g)$ are irreducible matrix representations of a group G (that act on complex vector spaces $V_1 \cong \mathbb{C}^{d_{\lambda_1}}$ and $V_2 \cong \mathbb{C}^{d_{\lambda_2}}$ respectively) and*

$$\pi_{\lambda_2}(g)X = X\pi_{\lambda_1}(g)$$

for all $g \in G$, then X must either be the zero matrix or an invertible matrix. Furthermore, $X = \mathbb{O}$ when $\pi_{\lambda_1}(g) \not\cong \pi_{\lambda_2}(g)$, X is invertible when $\pi_{\lambda_1}(g) \cong \pi_{\lambda_2}(g)$, and $X = c\mathbb{I}$ for some $c \in \mathbb{C}$ when $\pi_{\lambda_1}(g) = \pi_{\lambda_2}(g)$.

Proof. Case 1: $\pi_{\lambda_1}(g) = \pi_{\lambda_2}(g)$. Let $(\mathbf{x}, v)$ be an eigenvector/eigenvalue pair for the matrix X, i.e., a solution to the equation $X\mathbf{x} = v\mathbf{x}$. If $\pi_{\lambda_1}(g)X = X\pi_{\lambda_1}(g)$, it follows that

$$X\pi_{\lambda_1}(g)\mathbf{x} = \pi_{\lambda_1}(g)X\mathbf{x} = v\pi_{\lambda_1}(g)\mathbf{x},$$

and thus $\mathbf{x}'(g) = \pi_{\lambda_1}(g)\mathbf{x}$ is also an eigenvector of X. Thus, the space of all eigenvectors with eigenvalue v (which is a subspace of V_1) is invariant under the action of representations $\pi_{\lambda_1}(g)$. But by definition $\pi_{\lambda_1}(g)$ is irreducible, and so the vectors $\mathbf{x} \neq \mathbf{0}$ must span the *whole* vector space V_1. Hence, the only way $X\mathbf{x} = v\mathbf{x}$ can hold *for all* $\mathbf{x} \in V_1$ is if $X = v\mathbb{I}$.

Case 2: $\pi_{\lambda_1}(g) \ncong \pi_{\lambda_2}(g)$. If V_1 is d_1-dimensional and V_2 is d_2-dimensional, then X must have dimensions $d_2 \times d_1$. Since $d_1 \neq d_2$, without loss of generality let us choose $d_2 < d_1$. Then we can always find a nontrivial d_1-dimensional vector $\mathbf{n}$ such that $X\mathbf{n} = \mathbf{0}$ and the set of all such vectors (called the nullspace of X) is a vector subspace of V_1. It follows that

$$X\pi_{\lambda_1}(g)\,\mathbf{n} = \pi_{\lambda_2}(g)X\mathbf{n} = \mathbf{0}$$

and so $\pi_{\lambda_1}(g)\mathbf{n}$ is also in the nullspace. But by definition, no vector subspace of V_1 other than $\{\mathbf{0}\} \subset V_1$ and V_1 itself can be invariant under $\pi_{\lambda_1}(g)$ since it is irreducible. Thus, since $\mathbf{n} \neq \mathbf{0}$ the nullspace of X must equal V_1. The only way this can be so is if $X = \mathbb{O}$.

The next subsection reviews the very important, and somewhat surprising, fact that each π_j appears in the decomposition of the (left or right) regular representation T exactly d_j times and that the number of inequivalent irreducible representations of G is the same as the number of conjugacy classes in G. We therefore write

$$T(g) \cong \sum_{j=1}^{\alpha} \bigoplus d_j \pi_j(g) \tag{8.13}$$

where α is the number of conjugacy classes of G. (Note that here we use $\pi_j(g)$ rather than $\pi_{\lambda_j}(g)$ because the sum is over specific IURs, whereas Schur's Lemma refers to arbitrary ones.)

On first reading, (8.13) can be taken as fact, and the next subsection can be skipped. It can be read at a later time to gain full understanding.

We now present a theorem which illustrates the orthogonality of matrix elements of IURs.

Theorem 8.3. *(Orthogonality of IUR Matrix Elements) [22, 37] Given inequivalent irreducible unitary representation matrices $\pi_{\lambda_1} = [\pi_{ij}^{\lambda_1}]$ and $\pi_{\lambda_2} = [\pi_{ij}^{\lambda_2}]$, their elements satisfy the orthogonality relations*

$$\sum_{g \in G} \pi_{ij}^{\lambda_1}(g)\overline{\pi_{kl}^{\lambda_2}(g)} = \frac{|G|}{d_{\lambda_1}}\delta_{\lambda_1, \lambda_2}\delta_{i,k}\delta_{j,l}. \tag{8.14}$$

Proof. For any two irreducible representations π_{λ_1} and π_{λ_2}, we can define the matrix

$$A = \sum_{g \in G} \pi_{\lambda_1}(g)E\,\pi_{\lambda_2}(g^{-1}) \tag{8.15}$$

for any matrix $E \in \mathbb{C}^{d_{\lambda_1} \times d_{\lambda_2}}$. Due to the invariance of summation over the group under shifts and homomorphism property of representations,

$$\pi_{\lambda_1}(h)A\,\pi_{\lambda_2}(h^{-1}) = \sum_{g \in G} \pi_{\lambda_1}(h)\pi_{\lambda_1}(g)E\,\pi_{\lambda_2}(g^{-1})\pi_{\lambda_2}(h^{-1})$$

$$= \sum_{g \in G} \pi_{\lambda_1}(h \circ g)E\,\pi_{\lambda_2}((h \circ g)^{-1}) = A,$$

or equivalently,

$$\pi_{\lambda_1}(h)A = A\,\pi_{\lambda_2}(h).$$

Thus, from Schur's Lemma,

$$A = a\delta_{\lambda_1, \lambda_2} \mathbb{I}_{d_{\lambda_1} \times d_{\lambda_2}},$$

where $d_\lambda = \dim(\pi_\lambda(g))$ and $\mathbb{I}_{d_{\lambda_1} \times d_{\lambda_2}}$ when $\pi_{\lambda_1}(g) \ncong \pi_{\lambda_2}(g)$ is the rectangular matrix with $(i,j)^{th}$ element $\delta_{i,j}$ and all others zero. When $\lambda_1 = \lambda_2$ it becomes an identity matrix.

Following [37, 22], we choose $E = E^{j,l}$ to be the matrix with elements $E_{p,q}^{j,l} = \delta_{p,j}\delta_{l,q}$. Making this substitution into (8.15) and evaluating the $(i,j)^{th}$ element of the result yields

$$a_{j,l}\delta_{i,k}\delta_{\lambda_1, \lambda_2} = \sum_{p,q}\sum_{g \in G} \pi_{i,p}^{\lambda_1}(g)\,\delta_{p,j}\delta_{l,q}\pi_{q,k}^{\lambda_2}(g^{-1}) = \sum_{g \in G} \pi_{i,j}^{\lambda_1}(g)\pi_{l,k}^{\lambda_2}(g^{-1}). \qquad (8.16)$$

The constant $a_{j,l}$ is determined by taking the trace of both sides when $\lambda_1 = \lambda_2 = \lambda$. The result is $a_{j,l} = (|G|/d_\lambda)\,\delta_{j,l}$. Finally, using the unitarity of the representations we see that (8.14) holds.

The result of this theorem will be very useful in the proofs that follow.

8.1.2 Characters of Finite Groups

In this section we prove two fundamental results used in the inversion formula and Plancherel/Parseval equality for finite groups: (1) the number of irreducible representations in the decomposition of the regular representation is the same as the number of conjugacy classes in the group; and (2) the number of times that an irreducible representation appears in the decomposition of the regular representation is equal to the dimension of the irreducible representation.

A fundamental concept used in the proofs of these statements is that of a *character* of the irreducible representation $\pi_j(g)$,

$$\chi_j(g) \doteq \operatorname{tr}[\pi_j(g)].$$

The characters are all examples of *class* functions (i.e., $\chi_j(h^{-1} \circ g \circ h) = \chi_j(g)$), since they have the property that they are constant on conjugacy classes of the group. This property results directly from the facts that π_j is a representation (and therefore a homomorphism) and the invariance of the trace under similarity transformation. In fact, it can be shown that any class function can be expanded as a weighted sum of characters.

The proofs of statements (1) and (2) above follow from orthogonality properties of characters, which themselves need to be proved. Our proofs follow those given in [22, 37].

Theorem 8.4. *(First Orthogonality of Characters)*

$$\sum_{g \in G} \chi_{\lambda_1}(g)\overline{\chi_{\lambda_2}(g)} = |G|\delta_{\lambda_1, \lambda_2} \qquad (8.17)$$

for finite groups. This can be written in the equivalent form

$$\sum_{k=1}^{\alpha} |C_k|\chi_{\lambda_1}(g_{C_k})\overline{\chi_{\lambda_2}(g_{C_k})} = |G|\delta_{\lambda_1, \lambda_2} \qquad (8.18)$$

where g_{C_k} is any representative element of C_k (the kth conjugacy class), i.e., $g_{C_k} \in C_k \subset G$, and $|C_k|$ is the number of elements in C_k. It follows that $\sum_{k=1}^{\alpha} |C_k| = |G|$.

Proof. Equation (8.17) results from evaluating (8.14) at $i = j$ and $k = l$ and summing over all i and k. It can be written in the alternate form

$$\sum_{k=1}^{\alpha} \sum_{g \in C_k} \chi_{\lambda_1}(g) \overline{\chi_{\lambda_2}(g)} = |G| \delta_{\lambda_1, \lambda_2},$$

where the summation over the group is decomposed into a summation over classes and summation within each class. Because characters are class functions, and hence constant on conjugacy classes, we have

$$\sum_{g \in C_k} \chi_{\lambda_1}(g) \overline{\chi_{\lambda_2}(g)} = |C_k| \chi_{\lambda_1}(g_{C_k}) \overline{\chi_{\lambda_2}(g_{C_k})}$$

Theorem 8.5. *(Second Orthogonality of Characters)*

$$\sum_{\lambda=1}^{N} \chi_\lambda(g_{C_i}) \overline{\chi_\lambda(g_{C_j})} = \frac{|G|}{|C_i|} \delta_{i,j} \tag{8.19}$$

where $g_{C_i} \in C_i$ and $g_{C_j} \in C_j$ are class representatives, and N is the number of inequivalent irreducible representations of G.

Proof. Recall that the class sum function $\mathcal{C}_i$ is defined as $\mathcal{C}_i(g) \doteq \sum_{h \in C_i} \delta(h^{-1} \circ g)$. It then follows using the definition (8.7) that

$$\tilde{C}_i^\lambda = \sum_{g \in G} \mathcal{C}_i(g) U^\lambda(g) = \sum_{g \in G} \left(\sum_{h \in C_i} \delta(h^{-1} \circ g) \right) U^\lambda(g) = \sum_{h \in C_i} U^\lambda(h).$$

Since by definition every conjugacy class is closed under conjugation, we know that

$$\mathcal{C}(h \circ g) = \mathcal{C}(g \circ h). \tag{8.20}$$

Application of (8.7) and (8.8) to (8.20) results in

$$U^\lambda(h) \tilde{C}_i^\lambda = \tilde{C}_i^\lambda U^\lambda(h)$$

for all $h \in G$. Taking $U^\lambda = \pi_\lambda$ to be irreducible, it follows from Schur's lemma that

$$\tilde{C}_i^\lambda = \kappa_{i,\lambda} \mathbb{I}_{d_\lambda \times d_\lambda} \tag{8.21}$$

for some scalar $\kappa_{i,\lambda}$. We find the value of this constant by taking the trace of both sides of (8.21) and observing that

$$\kappa_{i,\lambda} \cdot d_\lambda = \operatorname{tr} \left(\sum_{h \in C_i} \pi_\lambda(h) \right) = |C_i| \operatorname{tr}(\pi_\lambda(h)) = |C_i| \chi_\lambda(g_{C_i})$$

for any $g_{C_i} \in C_i$. Hence

$$\tilde{C}_i^\lambda = \frac{|C_i|}{d_\lambda} \chi_\lambda(g_{C_i}) \mathbb{I}_{d_\lambda \times d_\lambda}. \tag{8.22}$$

The equality

$$\tilde{C}_i^\lambda \tilde{C}_j^\lambda = \sum_{k=1}^{\alpha} c_{ij}^k \tilde{C}_k^\lambda$$

follows from (7.28) and (8.9). Using (8.22), this is rewritten as

$$|C_i||C_j|\chi_\lambda(g_{C_i})\chi_\lambda(g_{C_j}) = d_\lambda \sum_{k=1}^{\alpha} c_{ij}^k \chi_\lambda(g_{C_k})$$

Dividing by $|C_i||C_j|$, summing over λ, and changing the order of summations yields

$$\sum_{\lambda=1}^{N} \chi_\lambda(g_{C_i})\chi_\lambda(g_{C_j}) = \frac{1}{|C_i||C_j|}\sum_{k=1}^{\alpha} c_{ij}^k \sum_{\lambda=1}^{N} d_\lambda \chi_\lambda(g_{C_k}) = \frac{1}{|C_i||C_j|}\sum_{k=1}^{\alpha} c_{ij}^k |G|\delta(g_{C_k}).$$

The last equality follows from (8.12). Since $k = 1$ is the class containing the identity, and we know from Subsection 7.2.5 that $c_{ij}^1 = |C_i|\delta_{C_i,C_j^{-1}}$, it follows that

$$\frac{1}{|C_i||C_j|}\sum_{k=1}^{\alpha} c_{ij}^k |G|\delta(g_{C_k}) = \frac{|G|}{|C_i||C_j|}c_{ij}^1 = \frac{|G|}{|C_j|}\delta_{C_i,C_j^{-1}}.$$

This is written as (8.19) by observing from the unitarity of π_λ that

$$\chi_\lambda(g_{C_j^{-1}}) = \overline{\chi_\lambda(g_{C_j})},$$

and so switching the roles of C_j and C_j^{-1} in the above equations yields

$$\sum_{\lambda=1}^{N} \chi_\lambda(g_{C_i})\overline{\chi_\lambda(g_{C_j})} = \frac{|G|}{|C_i|}\delta_{C_i,C_j}.$$

Recognizing that $\delta_{C_i,C_j} = \delta_{i,j}$ completes the proof.

Theorem 8.6. *The number of irreducible representations of a finite group is equal to the number of its conjugacy classes.*

Proof. Following [37], we can view $v_i^\lambda = \sqrt{|C_i|}\chi_\lambda(g_{C_i})$ for $i = 1, ..., \alpha$ as the components of an α-dimensional vector, $\mathbf{v}^\lambda \in \mathbb{C}^\alpha$. In this interpretation, the first orthogonality of characters then says that all such vectors are orthogonal to each other (when using the standard inner product on $\mathbb{C}^\alpha$). But by definition λ can have N values, each corresponding to a different irreducible representation. Thus it must be the case that $N \leq \alpha$ because N orthogonal vectors cannot exist in an α-dimensional vector space if $N > \alpha$.

Now let $w_i^C = \chi_i(g_C)$ be the components of an N-dimensional vector $\mathbf{w}^C \in \mathbb{C}^N$. (Note that for each component of the vector $\mathbf{w}^C$, the class is the same and the character function varies, whereas for the components of $\mathbf{v}^\lambda$, the character function is the same, but the class varies.) The second orthogonality of characters says that this set of vectors is orthogonal in $\mathbb{C}^N$. The number of such vectors, α, must then obey $\alpha \leq N$.

The two results $N \leq \alpha$ and $\alpha \leq N$ can only hold if $\alpha = N$.

Theorem 8.7. *The number of times each irreducible representation of a finite group appears in the decomposition of the regular representation is equal to its dimension.*

Proof. The regular representation is decomposed into the direct sum of irreducible ones as

$$T(g) = S^{-1} \left(\sum_{\lambda=1}^{N} \bigoplus n_\lambda \pi_\lambda(g) \right) S$$

for some invertible matrix S. Taking the trace of both sides yields

$$\chi(g) = \sum_{\lambda=1}^{N} n_\lambda \chi_\lambda(g).$$

Multiplying both sides by $\overline{\chi_{\lambda'}(g)}$ and summing over G we get

$$\sum_{g \in G} \chi(g) \overline{\chi_{\lambda'}(g)} = \sum_{\lambda=1}^{N} n_\lambda \sum_{g \in G} \chi_\lambda(g) \overline{\chi_{\lambda'}(g)} = n_{\lambda'} |G|.$$

For finite groups, we already saw in (8.12) that $\chi(g) = |G|\, \delta(g)$, and so

$$\sum_{g \in G} \chi(g) \overline{\chi_{\lambda'}(g)} = |G|\, \overline{\chi_{\lambda'}(e)} = |G|\, \mathrm{tr}\left(\mathbb{I}_{d_{\lambda'} \times d_{\lambda'}} \right) = |G|\, d_{\lambda'}.$$

Thus $n_{\lambda'} = d_{\lambda'}$.

It follows from this that every Abelian group has exactly $|G|$ characters, and each of them appears only once.

8.1.3 Fourier Transform, Inversion, and Convolution Theorem

Since the trace of a matrix is invariant under similarity transformation, the trace of both sides of (8.13) yields the equality

$$\mathrm{tr}(T(g)) = \sum_{j=1}^{\alpha} d_j \chi_j(g). \tag{8.23}$$

Evaluating both sides of (8.23) at $g = e$ (where the representations are $d_j \times d_j$ identity matrices), we get *Burnside's formula*:

$$|G| = \sum_{j=1}^{\alpha} d_j^2. \tag{8.24}$$

(Recall that $|G| = n$ is the number of group elements.) Evaluating (8.23) at $g \neq e$ yields

$$0 = \sum_{j=1}^{\alpha} d_j \chi_j(g), \qquad \forall\, g \neq e. \tag{8.25}$$

With these facts, we are now ready to define the Fourier transform of a function on a finite group and the corresponding inversion formula.

Definition 8.8. The *Fourier transform* of a complex-valued function $f : G \to \mathbb{C}$ at the representation π_j is the matrix-valued function with representation-valued argument defined as

$$\hat{f}(\pi_j) \doteq \sum_{g \in G} f(g)\pi_j(g^{-1}). \tag{8.26}$$

We refer to $\hat{f}(\pi_j)$ as the Fourier transform *at* the representation π_j, whereas the whole collection of matrices $\{\hat{f}(\pi_j)\}$ for $j = 1, ..., \alpha$ is called the *spectrum* of the function f. Note that unlike $\mathcal{T}^\lambda(f)$ defined in (8.7), the representations are now irreducible, and the Fourier transform is defined with $\pi_j(g^{-1})$ instead of $\pi_j(g)$.

Theorem 8.9. *The following inversion formula reproduces f from its spectrum:*

$$f(g) = \frac{1}{|G|} \sum_{j=1}^{\alpha} d_j \operatorname{tr}[\hat{f}(\pi_j)\pi_j(g)], \tag{8.27}$$

where the sum is taken over all inequivalent irreducible representations.

Proof. By definition of the Fourier transform and properties of the representations π_j,

$$\hat{f}(\pi_j)\pi_j(g) = \sum_{h \in G} f(h)\pi_j(h^{-1})\pi_j(g)$$

$$= \sum_{h \in G} f(h)\pi_j(h^{-1} \circ g)$$

$$= f(g)\,\mathbb{I}_{d_j \times d_j} + \sum_{h \neq g} f(h)\pi_j(h^{-1} \circ g).$$

Taking the trace of both sides,

$$\operatorname{tr}[\hat{f}(\pi_j)\pi_j(g)] = d_j f(g) + \sum_{h \neq g} f(h)\chi_j(h^{-1} \circ g).$$

Multiplying both sides by d_j and summing over all j yields

$$\sum_{j=1}^{\alpha} d_j \operatorname{tr}[\hat{f}(\pi_j)\pi_j(g)] = \left(\sum_{j=1}^{\alpha} d_j^2\right) f(g) + \sum_{h \neq g} f(g) \sum_{j=1}^{\alpha} d_j \chi_j(h^{-1} \circ g).$$

Using Burnside's formula (8.24) and (8.25), we see that

$$\sum_{j=1}^{\alpha} d_j \operatorname{tr}[\hat{f}(\pi_j)\pi_j(g)] = |G| f(g)$$

which, after division of both sides by $|G|$, results in the inversion formula.

While not directly stated in the above proof, the inversion formula works because of the orthogonality (8.14) and the completeness of the set of IURs.

Remark: Note that while we use irreducible representations that are unitary, because of the properties of the trace, the formulation would still work even if the representations were not unitary.

Theorem 8.10. *The Fourier transform of the convolution of two functions on G is the matrix product of the Fourier transform matrices:*

$$(\widehat{f_1 * f_2})(\pi_j) = \hat{f}_2(\pi_j)\hat{f}_1(\pi_j),$$

where the order of the products matters.

Proof. By definition we have

$$(\widehat{f_1 * f_2})(\pi_j) = \sum_{g\in G}\left(\sum_{h\in G} f_1(h)\,f_2(h^{-1}\circ g)\right)\pi_j(g^{-1}).$$

Making the change of variables $k = h^{-1}\circ g$ and substituting $g = h\circ k$, the above expression becomes

$$\sum_{k\in G}\sum_{h\in G} f_1(h)\,f_2(k)\pi_j(k^{-1}\circ h^{-1}).$$

Using the homomorphism property of π_j and the commutativity of scalar-matrix multiplication and summation, this can be split into

$$\left(\sum_{k\in G} f_2(k)\pi_j(k^{-1})\right)\left(\sum_{h\in G} f_1(h)\pi_j(h^{-1})\right) = \hat{f}_2(\pi_j)\,\hat{f}_1(\pi_j).$$

Remark: The reversal of order of the Fourier transform matrices relative to the order of convolution of functions is a result of the $\pi_j(g^{-1})$ in the definition of the Fourier transform. Often, the Fourier transform is defined using $\pi_j(g)$ in the definition instead of $\pi_j(g^{-1})$. This keeps the order of the product of the Fourier transform matrices the same as the order of convolution of functions in the convolution theorem. However, for the inversion formula to work, $\pi_j(g)$ must then be replaced by $\pi_j(g^{-1})$. As a matter of personal preference, and consistency with the way in which the standard Fourier series and transform were defined in Chapter 2, we view the inversion formula as an expansion of a function $f(g)$ in harmonics (matrix elements of the IURs) instead of their complex conjugates.

Theorem 8.11. *The generalization of the Parseval equality (called the Plancherel theorem in the context of group theory) is written for a finite group as*

$$\sum_{g\in G} |f(g)|^2 = \frac{1}{|G|}\sum_{j=1}^{\alpha} d_j \|\hat{f}(\pi_j)\|^2,$$

where $\|\cdot\|$ denotes the Hilbert-Schmidt norm of a matrix. A more general form of this equality is

$$\sum_{g\in G} f_1(g^{-1})\,f_2(g) = \frac{1}{|G|}\sum_{j=1}^{\alpha} d_j\,\mathrm{tr}[\hat{f}_1(\pi_j)\hat{f}_2(\pi_j)].$$

Proof. Applying the inversion formula to the result of the convolution theorem we find that

$$(f_1 * f_2)(g) = \frac{1}{|G|}\sum_{j=1}^{\alpha} d_j\,\mathrm{tr}[\hat{f}_2(\pi_j)\hat{f}_1(\pi_j)\pi_j(g)].$$

Evaluating this result at $g = e$ results in the second statement above. In the particular case when $f_1(g) = \overline{f(g^{-1})}$ and $f_2(g) = f(g)$, this reduces to the first statement.

In the special case when f is a class function, we can take $\hat{f}(\pi_i) = (f_i/d_i)\mathbb{I}_{d_i \times d_i}$ and the general formulation is simplified. Any class function can be expanded as a finite linear combination of characters (which forms a complete orthogonal basis for the space of all square integrable class functions):

$$f(g) = \sum_{i=1}^{\alpha} f_i \chi_i(g)$$

where

$$f_i = \frac{1}{|G|} \sum_{g \in G} f(g)\overline{\chi_i(g)}.$$

Furthermore, the Plancherel equality for class functions reduces to

$$\sum_{g \in G} |f(g)|^2 = \frac{1}{|G|} \sum_{i=1}^{\alpha} |f_i|^2.$$

In analogy with the way class functions are constant on classes, we can define functions which are constant on cosets. For instance, a function satisfying $f(g) = f(g \circ h)$ for all $h \in H < G$ is constant on all left cosets, and is hence viewed as a function on the coset space G/H. Using the property of such functions, the Fourier transform is written as

$$\hat{f}(\pi_j) = \sum_{g \in G} f(g \circ h)(\pi_j(g))^* = \pi_j(h) \sum_{k \in G} f(k)(\pi_j(k))^*,$$

where the change of variables $k = g \circ h$ has been made. Summing both sides over H, and dividing by $|H|$ gives:

$$\hat{f}(\pi_j) = \Delta(\pi_j)\hat{f}(\pi_j)$$

where

$$\Delta(\pi_j) = \frac{1}{|H|} \sum_{h \in H} \pi_j(h)$$

is a matrix which constrains the structure of $\hat{f}(\pi_j)$. For some values of j, $\Delta(\pi_j) = 0$, hence forcing $\hat{f}(\pi_j)$ to be zero. For others it will have a structure which causes only parts of $\hat{f}(\pi_j)$ to be zero. A similar construction follows for functions $f(g) = f(h \circ g)$ for all $h \in H < G$ which are constant on right cosets, and hence are functions on $H\backslash G$.

The material presented in this section is well known to mathematicians who concentrate on finite groups. References which explain this material from different perspectives include [22, 26, 38, 65]. Concrete representations of crystallographic space groups can be found in [48, 49]. A number of interesting applications involving convolution and Fourier analysis on finite groups have appeared in recent years, including [51, 47, 82, 100].

We note also that fast Fourier transform and inversion algorithms for broad classes of finite groups have been developed in the recent literature [14, 18, 61, 71]. This is discussed in the next subsection.

8.1.4 Fast Fourier Transforms for Finite Groups

At the cores of the FFT reviewed at the end of Chapter 2 and fast polynomial transforms reviewed in Chapter 3 were recurrence formulae used to decompose the sums defining discrete transforms. This has been extended to the context of finite groups in a series of papers by Diaconis and Rockmore [18], Rockmore [71], Maslen [61], and Maslen and Rockmore [62]. We shall not go into all of the technical detail of fast Fourier transforms for groups but will rather only examine the key enabling concepts.

Given a function $F : S \to V$, where S is a finite set and V is a vector space, we can make the decomposition

$$\sum_{s \in S} F(s) = \sum_{[s] \in S/\sim} \left(\sum_{x \in [s]} F(x) \right) \tag{8.28}$$

for any equivalence relation $\sim$. (See Appendix B for an explanation of terms.)

In the context of arbitrary groups, two natural equivalence classes are immediately apparent. These are conjugacy classes and cosets. In the case of conjugacy classes, (8.28) can be written as

$$\sum_{g \in G} F(g) = \sum_{i=1}^{\alpha} \sum_{g \in C_i} F(g).$$

The decomposition when the equivalence classes are chosen to be cosets is

$$\sum_{g \in G} F(g) = \sum_{\sigma \in G/H} \left(\sum_{s \in \sigma} F(s) \right). \tag{8.29}$$

Since each coset is of the form $\sigma = gH$, it follows that $|\sigma| = |H|$ and so $|G| = |G/H| \cdot |H|$, and the number of cosets is $|G/H| = |G|/|H|$.

Let $s_\sigma \in \sigma$ for each $\sigma \in G/H$. s_σ is called a *coset representative*. With the observation that

$$\sum_{s \in \sigma} F(s) = \sum_{h \in H} F(s_\sigma \circ h),$$

we rewrite (8.29) as

$$\sum_{g \in G} F(g) = \sum_{\sigma \in G/H} \left(\sum_{h \in H} F(s_\sigma \circ h) \right). \tag{8.30}$$

This is a key step in establishing a recursive computation of the Fourier transform for groups. However, this in and of itself does not yield a fast evaluation of an arbitrary sum of the form $\sum_{g \in G} F(g)$. The second essential feature is that the Fourier transform of a function on a finite group is special because

$$F(g) = f(g)\pi(g^{-1})$$

where $\pi(g)$ is an irreducible matrix representation of the group, and therefore $\pi(g_1 \circ g_2) = \pi(g_1)\pi(g_2)$. This allows us to write (8.26) as

$$\sum_{g \in G} f(g)\pi(g^{-1}) = \sum_{\sigma \in G/H} \left(\sum_{h \in H} f_\sigma(h)\pi(h^{-1}) \right) \pi(s_\sigma^{-1}) \tag{8.31}$$

where $f_\sigma(h) = f(s_\sigma \circ h)$. In general since the restriction of an irreducible representation to a subgroup is reducible, we can write it as a direct sum of the form

$$\pi(h) \cong \sum_k \bigoplus c_k \pi_k^H(h). \tag{8.32}$$

When using a special kind of basis (called "H-adapted"), the equivalence sign in (8.32) becomes an equality. The representations $\pi_k^H(h)$ are irreducible, and c_k is the number of times $\pi_k^H(h)$ appears in the decomposition of $\pi(h)$. Since $\sum_k c_k \dim(\pi_k^H) = \dim(\pi)$, it follows that it is more efficient to compute the Fourier transforms of all the functions $f_\sigma(h)$ in the subgroup H using the irreducible representations as $\sum_{h \in H} f_\sigma(h) \pi_k^H(h^{-1})$ for all $\sigma \in G/H$ and all values of k, and recompose the Fourier transform on the whole group using (8.31) and (8.32).

Given a "tower of subgroups"

$$G_n < \cdots < G_2 < G_1 < G, \tag{8.33}$$

it is possible to write (8.30) as

$$\sum_{g \in G} F(g) = \sum_{\sigma_1 \in G/G_1} \sum_{\sigma_2 \in G_1/G_2} \cdots \sum_{\sigma_n \in G_{n-1}/G_n} \sum_{h \in G_n} F(s_{\sigma_1} \circ \cdots \circ s_{\sigma_n} \circ h) \tag{8.34}$$

and recursively decompose the representations $\pi(g)$ into smaller and smaller ones as they are restricted to smaller and smaller subgroups.

The computational performance of this procedure is generally better than the $\mathcal{O}(|G|^2)$ computations required for direct evaluation of the whole spectrum of a function on a finite group. For some classes of groups $\mathcal{O}(|G| \log|G|)$ or $\mathcal{O}(|G|(\log|G|)^2)$ performance can be achieved. In general, $\mathcal{O}(|G|^{\frac{3}{2}})$ performance is possible.[5] The computational complexity depends on the structure of the group, and in particular, on how many subgroups are cascaded in the tower in (8.33). Below is a specific case from the literature.

Theorem 8.12. *[5] Let $G = G_1 \times G_2 \times \cdots \times G_k$ be the direct product of finite groups $G_1, ..., G_k$. Then the Fourier transform and its inverse can be computed in $\mathcal{O}(|G| \sum_{i=1}^k |G_i|)$ arithmetic operations.*

A number of works have addressed various aspects of the complexity of computation of the Fourier transform and convolution of functions on finite groups. These include [6, 12] in the commutative case, and [5, 9, 39, 73] in the noncommutative context. An efficient Fourier transform for wreath product groups is derived in [72]. Applications of fast Fourier transforms for finite groups are discussed in [40, 41, 42, 43, 87, 90].

8.2 Differentiation and Integration of Functions on Lie Groups

The concept of a Fourier transform pair extends to certain kinds of Lie groups in interesting ways, and these transforms have operational properties akin to those in classical Fourier analysis. Before defining the Fourier transform we must first know how to integrate on Lie groups, and to realize operational properties we must first have an appropriate concept of derivative. These foundational concepts are developed in this section.

---

[5]For the $\mathcal{O}(\cdot)$ symbol to make sense in this context, the groups under consideration must be members of infinite classes so that $|G|$ can become arbitrarily large and asymptotic complexity estimates have meaning.

8.2.1 Derivatives, Gradients, and Laplacians on Lie Groups

Given a function $f : G \to \mathbb{C}$ with $g \in G$ parameterized as $g(x_1, ..., x_n)$ and $f(g(x_1, ..., x_n))$ differentiable in all the parameters x_i, we define for any element of the Lie algebra, X:

$$\left(X^L f\right)(g) \doteq \frac{d}{dt} f(\exp(-tX) \circ g)\Big|_{t=0} = \lim_{t \to 0} \frac{f((\exp(-tX) \circ g) - f(g)}{t}. \tag{8.35}$$

Similarly, let

$$\left(X^R f\right)(g) \doteq \frac{d}{dt} f(g \circ \exp(tX))\Big|_{t=0} = \lim_{t \to 0} \frac{f(g \circ \exp(tX)) - f(g)}{t}. \tag{8.36}$$

The superscripts L and R denote on which side of the argument the infinitesimal operation is applied. If the above limits exist, then we say $f \in \mathcal{C}^1(G)$. And by extension, we can define $\mathcal{C}^n(G)$ to be the class of functions for which any combination of n such derivatives can be applied to result in a continuous function.

It follows from the fact that left and right shifts do not interfere with each other that

$$(L(g_0)\left(X^R f\right))(g) = \left(X^R f\right)(g_0^{-1} \circ g) = \left(X^R(L(g_0)f)\right)(g)$$

and

$$((R(g_0)X^L f)(g) = \left(X^L f\right)(g \circ g_0) = \left(X^L(R(g_0)f)\right)(g).$$

We note in passing that the (right) Taylor series about $g \in G$ of an an analytic function on a Lie group G is expanded as [91]

$$f(g \circ \exp tX) = \sum_{k=0}^{\infty} \frac{d^k}{dt^k} f(g \circ \exp tX)\Big|_{t=0} \frac{t^k}{k!}. \tag{8.37}$$

The left Taylor series follows analogously.

Particular examples of the operators X^L and X^R are X_i^R and X_i^L where X_i is a basis element for the Lie algebra.

These operators inherit the Lie bracket properties from $\mathcal{G}$:

$$[X, Y]^L = -[X^L, Y^L] \quad \text{and} \quad [X, Y]^R = -[X^R, Y^R] \tag{8.38}$$

More sophisticated differential operators can be constructed from these basic ones in a straight forward way. For instance, the left and right gradient vectors at the identity can be defined as

$$\text{grad}_e^R f \doteq \sum_{i=1}^{N} X_i \left(X_i^R f\right)(g)\Big|_{g=e} \quad \text{and} \quad \text{grad}_e^L f \doteq \sum_{i=1}^{N} X_i \left(X_i^L f\right)(g)\Big|_{g=e}.$$

These gradients are two ways to describe the same element of the Lie algebra $\mathcal{G}$, i.e., $\text{grad}_e^R f = \text{grad}_e^L f$. We can likewise generate images of these gradients at any tangent to the group by defining:

$$\text{grad}_g^R f = \sum_{i=1}^{N} gX_i \left(X_i^R f\right)(g) \quad \text{and} \quad \text{grad}_g^L f = \sum_{i=1}^{N} X_i g \left(X_i^L f\right)(g).$$

The products gX_i and $X_i g$ are matrix products, or more generally can be written as $d/dt(g \circ \exp(tX_i))|_{t=0}$ and $d/dt(\exp(tX_i) \circ g)|_{t=0}$.

The collection of all grad_g^R (resp. grad_g^L) can be viewed as a vector field.

We can also define the left and right divergence of a vector field, $Y(g) = \sum_{i=1}^N Y_i(g)X_i(g)$, on the group as

$$\mathrm{div}_L(Y) = (\mathrm{grad}_g^L, Y(g))_g^L \quad \text{and} \quad \mathrm{div}_R(Y) = (\mathrm{grad}_g^R, Y(g))_g^R,$$

where appropriate left or right inner products on the tangent space of G at g are used.

Given a right-invariant vector field $Y^L = \sum_{i=1}^N y_i X_i g$ where y_i are constants, we can also define differential operators of the form

$$Y^L f = \mathrm{div}_L(Yf) = \sum_{i=1}^N y_i X_i{}^L f,$$

and analogously for the left-invariant version. These are analogues of the directional derivatives.

In analogy with the Abelian case, the Laplacian operator is defined as the divergence of gradient vectors.

Using the orthogonality of Lie algebra basis elements, we have

$$\mathrm{div}_L(\mathrm{grad}_L(f)) = \sum_{i=1}^N (X_i{}^L)^2 f \tag{8.39}$$

$$\mathrm{div}_R(\mathrm{grad}_R(f)) = \sum_{i=1}^N (X_i{}^R)^2 f. \tag{8.40}$$

Hence, whenever $\sum_{i=1}^N (X_i{}^L)^2 = \sum_{i=1}^N (X_i{}^R)^2$, the Laplacian for the group is written without subscripts as $\mathrm{div}(\mathrm{grad}(f))$.

8.2.2 Integration Measures on Lie Groups and Their Homogeneous Spaces

In this section we review integration on Lie groups and on their left and right coset spaces. Obviously, we could not go very far in extending concepts from classical Fourier analysis to the noncommutative case without some concept of integration. Thus, it is important to get the basic idea of the material in this section before proceeding.

Haar Measures and Shifted Functions

Let G be a locally compact group (and in particular, a matrix Lie group). On G, two natural integration measures exist. These are called the *left* and *right Haar* measures.[31, 67][6]

The measures μ_L and μ_R respectively have the properties:

$$\mu_L(f) = \int_G f(g)\, d_L(g) = \int_G (L(a)f)(g)\, d_L(g) = \int_G f(a^{-1} \circ g)\, d_L(g)$$

and

---

[6]Haar measure, named after Alfréd Haar (1885-1933), is a term used for abstract locally compact groups. In the particular case of Lie groups, the Haar measure is sometimes called a *Hurwitz measure*, named after Adolf Hurwitz (1859-1919).

$$\mu_R(f) = \int_G f(g)\,d_R(g) = \int_G (R(a)f)(g)\,d_R(g) = \int_G f(g \circ a)\,d_R(g)$$

for any $a \in G$, where we will assume $f(g)$ to be a "nice" function. $f(g)$ is assumed to be "nice" in the sense that $f(g) \in \mathcal{L}^p(G, d_{L,R}) \cap \mathcal{C}^\infty(G)$ for all $p \in \mathbb{Z}^p$ (where $d_{L,R}$ stands for either d_L or d_R), and in addition is infinitely differentiable in each coordinate parameter.

The measures μ_L and μ_R are invariant under left, $L(a)$, and right, $R(a)$, shifts respectively. However, if the left Haar measure is evaluated under right shifts, and the right Haar measure is evaluated under left shifts, we find that

$$\mu_L(f) = \Delta_L(a) \int_G f(g \circ a)\,d_L(g)$$

and

$$\mu_R(f) = \Delta_R(a) \int_G f(a^{-1} \circ g)\,d_R(g),$$

where the scale factors $\Delta_L(g)$ and $\Delta_R(g)$ are respectively called the left and right *modular* functions. In general, these functions are related to each other as: $\Delta_L(g) = \Delta_R(g^{-1})$, and these functions are both homomorphisms from G into the multiplicative group of complex numbers (excluding zero) under scalar multiplication.

A group for which $\Delta_L(g) = \Delta_R(g) = 1$ is called *unimodular*. For such groups, we can set $d(g) = d_L(g) = d_R(g)$. Integration of "nice" functions on such groups is, by definition, shift invariant. In addition, integration on unimodular groups is invariant under inversions of the argument:

$$\int_G f(g^{-1})\,d(g) = \int_G f(g)\,d(g).$$

All but two of the groups on which we will integrate in this book (including $(\mathbb{R}, +)$, all compact groups, the motion groups and all direct products of the aforementioned) are unimodular. The exceptions are the affine and scale-Euclidean groups, which are of importance in image analysis and computer vision.

Integration on G/H and $H\backslash G$

Let G be a Lie group and H be a Lie subgroup of G. Recall that G/H and $H\backslash G$ respectively denote the spaces of all left and right cosets. Coset spaces are also referred to as quotient spaces, factor spaces, or homogeneous spaces. Integration and orthogonal expansions on such spaces is a field of study in its own right (see, e.g., [13, 86, 94, 97, 99]). In this section we review the fundamental concepts of invariant and quasi-invariant integration measures on G/H. The formulation follows in an exactly analogous manner for $H\backslash G$.

The phrase "invariant integration of functions on homogeneous spaces of groups" is by no means a standard one in the vocabulary of most engineers and scientists. However, the reader has, perhaps without knowing it, already seen in Chapters 2, 4 and 5 how to integrate functions on some of the simplest examples of homogeneous spaces: $\mathbb{R}/2\pi\mathbb{Z} \cong \mathbb{S}^1$ (the unit circle), $SO(3)/SO(2) \cong \mathbb{S}^2$ (the unit sphere), and $SE(3)/\mathbb{R}^3 \cong SO(3)$. Knowledge of the more general theory is important in the context of noncommutative harmonic analysis, because construction of the irreducible unitary representations of many groups (using something called the method of induced representations) is achieved using these ideas. In Chapter 10 we use this method to generate the IURs of the Euclidean motion group.

A differential volume element for a coset space, $d(gH)$, is called a G−invariant measure if for every "nice" function $\tilde{f} \in \mathcal{L}^1(G/H) \cap \mathcal{L}^2(G/H) \cap C^\infty(G/H)$,

$$\int_{G/H} \tilde{f}(a \cdot (gH)) \, d(gH) = \int_{G/H} \tilde{f}(gH) \, d(gH) \tag{8.41}$$

for all $a \in G$ where $a \cdot (gH) \doteq (a \circ g)H$. If H is discrete (i.e., a Lie group of dimension zero), then this integral can be computed conveniently in terms of integration over fundamental domains in the original group, G, as

$$\int_{G/H} \tilde{f}(gH) \, d(gH) = \int_{F_{G/H}} f(g) \, d(g)$$

where f is a function on the group which is constant on left cosets (i.e., $f(g) = f(g \circ h)$ for all $h \in H$).

For example, we can identify any unit vector with elements of $\mathbb{S}^2 \cong SO(3)/SO(2)$. In this case (8.41) is written as

$$\int_{\mathbb{S}^2} \tilde{f}(R^T \mathbf{u}) \, d\mathbf{u} = \int_{\mathbb{S}^2} \tilde{f}(\mathbf{u}) \, d\mathbf{u}$$

where $d\mathbf{u}$ is the usual integration measure on $\mathbb{S}^2$ and $R \in SO(3)$ is an arbitrary rotation.

Invariant measures of suitable functions on cosets exist whenever the modular function of G restricted to H is the same as the modular function of H [99]. Since we are most interested in unimodular groups (and their unimodular subgroups) this condition is automatically satisfied.

The existence of a G−invariant measure on G/H allows us to write [10, 33, 99]:

$$\int_G f(g) \, d(g) = \int_{F_{G/H}} \left(\int_H f(g \circ h) \, d(h) \right) d(g) \tag{8.42}$$

for any well-behaved $f : G \to \mathbb{C}$ where $d(g)$ and $d(h)$ are, respectively, the invariant integration measures on G and H.

Then, by defining

$$\tilde{f}(gH) \doteq \int_H f(g \circ h) \, d(h)$$

and using the notation in (8.41), we can rewrite (8.42) as

$$\int_G f(g) \, d(g) = \int_{G/H} \left(\int_H f(g \circ h) \, d(h) \right) d(gH) \tag{8.43}$$

where $g \in gH$ is taken to be the coset representative. In the special case when $f(g)$ is a left-coset function, (8.43) reduces to

$$\int_G f(g) \, d(g) = \int_{G/H} \tilde{f}(gH) \, d(gH)$$

where it is assumed that $d(h)$ is normalized so that $\mathrm{Vol}(H) = \int_H d(h) = 1$.

For example, let $SO(3)$ be parameterized with ZYZ Euler angles, $g = R_3(\alpha)R_2(\beta)R_3(\gamma)$. Taking $H \cong SO(2)$ to be the subgroup of all $R_3(\gamma)$, and identifying all matrices of the form $R_3(\alpha)R_2(\beta)$ with points on the unit sphere, (with (α, β) serving as spherical coordinates), we write

$$\int_{SO(3)} f(g)\,d(g) = \int_{\mathbb{S}^2} \left(\int_{SO(2)} f((R_3(\alpha)R_2(\beta))R_3(\gamma))\,d(R_3(\gamma)) \right) d\mathbf{u}(\alpha, \beta),$$

where $\mathbf{u}(\alpha, \beta) = R_3(\alpha)R_2(\beta)\mathbf{e}_3$. Even more explicitly, this is

$$\frac{1}{8\pi^2} \int_0^{2\pi} \int_0^{\pi} \int_0^{2\pi} f(R_3(\alpha)R_2(\beta)R_3(\gamma))\sin\beta\,d\alpha\,d\beta\,d\gamma =$$

$$\frac{1}{4\pi} \int_0^{\pi} \int_0^{2\pi} \left(\frac{1}{2\pi} \int_0^{2\pi} f((R_3(\alpha)R_2(\beta))R_3(\gamma))\,d\gamma \right) \sin\beta\,d\alpha\,d\beta.$$

If we are given a unimodular group G with unimodular subgroups K and H such that $K \leq H$, we can use the facts that

$$\int_G f(g)\,d(g) = \int_{G/H} \int_H f(g \circ h)\,d(h)\,d(gH),$$

and

$$\int_H f(h)\,d(h) = \int_{H/K} \int_K f(h \circ k)\,d(k)\,d(hK)$$

to decompose the integral of any nice function $f(g)$ as

$$\int_G f(g)\,d(g) = \int_{G/H} \int_{H/K} \int_K f(g \circ h \circ k)\,d(k)\,d(hK)\,d(gH). \tag{8.44}$$

For example, using the fact that any element of $SE(3)$ can be decomposed as the product of translation and rotation subgroups, $g = T(\mathbf{a}) \circ A(\alpha, \beta, \gamma)$, the integral of a function over $SE(3)$,

$$\int_{SE(3)} f(g)\,d(g) = \int_{\mathbb{R}^3} \int_{SO(3)} f(\mathbf{a}, A)\,dA\,d\mathbf{a},$$

can be written as

$$\int_{SE(3)/SO(3)} \int_{SO(3)/SO(2)} \int_{SO(2)} f(T(\mathbf{a}) \circ (R_3(\alpha)R_2(\beta)) \circ R_3(\gamma))\,d(R_3(\gamma))\,d\mathbf{u}(\alpha, \beta)\,d\mathbf{a}.$$

We can also decompose the integral using the subgroups of $(\mathbb{R}^3, +)$, but this is less interesting since the decomposition of a function on $\mathbb{R}^3$ is just

$$\int_{\mathbb{R}^3} F(\mathbf{a})\,d\mathbf{a} = \int_{\mathbb{R}} \int_{\mathbb{R}} \int_{\mathbb{R}} F(a_1\mathbf{e}_1 + a_2\mathbf{e}_2 + a_3\mathbf{e}_3)\,da_1\,da_2\,da_3.$$

Another interesting thing to note (when certain conditions are met) is the decomposition of the integral of a function on a group in terms of two subgroups and a double coset space:

$$\int_G f(g)\,d(g) = \int_K \int_{K \backslash G/H} \int_H f(k \circ g \circ h)\,d(h)\,d(KgH)\,d(k).$$

A particular example of this is the integral over $SO(3)$, which can be written as

$$\int_{SO(3)} = \int_{SO(2)} \int_{SO(2)\backslash SO(3)/SO(2)} \int_{SO(2)}.$$

8.2.3 Constructing Invariant Integration Measures

In this subsection we explicitly construct integration measures on several Lie groups, and give a short explanation of how this can be done in general.

According to Lie theory there always *exists* both unique left-invariant and right-invariant integration measures (up to an arbitrary scaling) on arbitrary Lie groups. Usually the left- and right-invariant integration measures are different. For the case of compact Lie groups, a unique bi-invariant (left and right invariant) measure exists, and often the most natural scaling in this case is the one for which $\int_G d(g) = 1$. For the case of $SO(3)$ we constructed this bi-invariant measure in Chapter 5 based on the physically intuitive relationship between angular velocity and the rate of change of rotation parameters. The fact that $SE(3)$ has a bi-invariant integration measure that is the product of those for $SO(3)$ and $\mathbb{R}^3$ was demonstrated in Chapter 6.

When it is possible to view the manifold of a Lie group as a hyper-surface in $\mathbb{R}^N$, volume elements based on the Riemannian metric can be used as integration measures. Invariance under left or right shifts must then be established by an appropriate choice of scaling matrix (W_0 in (4.12)). The theory of differential forms can also be used to generate left(right)-invariant integration measures.

Left- and right-invariant integration measures are then found as

$$d_L(g) = |\det(J_R)|\, dx_1 \cdots dx_n \quad \text{and} \quad d_R(g) = |\det(J_L)|\, dx_1 \cdots dx_n.$$

Generally, $d_L(g) \neq d_R(g)$. When equality holds, the group is called *unimodular*. The reason for this will be explained in Subsection 8.2.4.

We already had a preview of this general theory in action for $SO(3)$ and $SE(3)$. In those cases, as with the examples below, it is possible to identify each orthonormal basis element X_i of the Lie algebra with a basis element $\mathbf{e}_i \in \mathbb{R}^n$ through the $\vee$ operation. Having introduced the concept of a Lie algebra, and an inner product on the Lie algebras of interest here, we need not use the $\vee$ operation in the context of the examples that follow.

In any coordinate patch, elements of an n-dimensional matrix Lie group can be parameterized as $g(x_1, ..., x_n)$. In a patch near the identity, two natural parameterizations are

$$g(x_1, ..., x_n) = \prod_{i=1}^n \exp(x_i X_i) \quad \text{and} \quad \hat{g}(x_1, ..., x_n) = \exp\left(\sum_{i=1}^n x_i X_i\right). \tag{8.45}$$

In practice these are often the only parameterizations we need. Usually, for noncommutative groups, these parameterizations do not yield the same result, but by a nonlinear transformation of coordinates they can be related as $\hat{g}(\mathbf{x}) = g(\mathbf{f}(\mathbf{x}))$. We have already seen this for the case of the Heisenberg group, the rotation group, and the Euclidean motion group.

For a matrix Lie group, the operation of partial differentiation with respect to each of the parameters x_i is well defined. Thus we can compute the left and right "tangent operators"

$$T_i^L(g) = \frac{\partial g}{\partial x_i} g^{-1} \quad \text{and} \quad T_i^R(g) = g^{-1}\frac{\partial g}{\partial x_i}.$$

They have the property that they convert the matrix group elements g into their derivatives:

$$T_i^L(g)g = \frac{\partial g}{\partial x_i} = gT_i^R(g).$$

As a result of the product rule for differentiation of matrices, we have $\partial(g \circ g_0)/\partial x_i = (\partial g/\partial x_i)g_0$, and likewise for left shifts. This means that

$$T_i^L(g \circ g_0) = T_i^L(g) \quad \text{and} \quad T_i^R(g_0 \circ g) = T_i^R(g).$$

If either of these operators is evaluated at $g = e$, and if either of the "natural" parameterizations in (8.45) is used, we see that the tangent operators reduce to Lie algebra basis vectors

$$T_i^L(g)\big|_{g=e} = T_i^R(g)\big|_{g=e} = X_i.$$

We can use the tangent operators T_i^L and T_i^R defined above to construct left or right-invariant integration measures on the group. For the moment, let us restrict the discussion to the left-invariant case. First associate with each tangent operator $T_i^L(g)$ an $n \times 1$ array (or "column vector"), the components of which are the projection of $T_i^L(g)$ onto the basis vectors $\{X_j\}$. This projection is performed using one of the inner products on the Lie algebra in Subsection 7.3.6. The array corresponding to $T_i^L(g)$ is denoted as $(T_i^L(g))^\vee$. A Jacobian matrix is generated by forming an $n \times n$ matrix whose columns are these arrays:

$$J_L(g) = \left[(T_1^L(g))^\vee, ..., (T_n^L(g))^\vee \right], \tag{8.46}$$

and similarly for J_R. When the basis elements X_i are chosen to be orthonormal with respect to the inner product (X_i, X_j), it is clear that $J_L(e) = \mathbb{I}_{n \times n}$.

An infinitesimal volume element located at the identity element of the group can be viewed as an n-dimensional box in the Lie algebra $\mathcal{G} = \mathcal{G}(e)$. It is then parameterized by $\sum_{i=1}^n y_i X_i$ where y_i ranges from 0 to dx_i, and its volume is $dx_1 ... dx_n$. We can view the volume at this element as $d_L(e) = |\det(J_R(e))| dx_1 ... dx_n$. The volume at $g \neq e$ is computed in a similar way. Using the first parameterization in (8.45), we have a volume element parameterized by $0 \leq y_i \leq dx_i$ for $i = 1, .., n$:

$$g(x_1 + y_1, ..., x_n + y_n) = \prod_{i=1}^n \exp((x_i + y_i)X_i).$$

Using the smallness of each dx_i, we write

$$g(x_1 + dx_1, ..., x_n + dx_n) = g(x_1, ..., x_n) \left(\mathbb{I}_{N \times N} + \sum_{i=1}^n T_i^R(g) \, dx_i \right)$$

$$= \left(\mathbb{I}_{N \times N} + \sum_{i=1}^n T_i^L(g) \, dx_i \right) g(x_1, ..., x_n).$$

Recall that g is an $N \times N$ matrix, whereas the dimension of the Lie group consisting of such matrices has dimension n.

Each of the tangent vectors provide a direction in the tangent space $\mathcal{G}(g)$ defining an edge of the differential box, which is now deformed. The volume of this deformed box is

$$d_L(g(x_1, ..., x_n)) = |\det(J_R(g))| \, dx_1 ... dx_n \tag{8.47}$$

because the Jacobian matrix $J_R(g)$ reflects how the coordinate system in $\mathcal{G}(g)$ is distorted relative to the one in the Lie algebra $\mathcal{G}(e)$. Due to the left invariance of $T_i^R(g)$, and hence $J_R(g)$, we have that $d_L(g_0 \circ g) = d_L(g)$. By a simple change of variables, this means integration of well-behaved functions with respect to this volume element is also left invariant:

$$\int_G f(g)\,d_L(g) = \int_G f(g)\,d_L(g_0 \circ g) = \int_G f(g_0^{-1} \circ g)\,d_L(g).$$

A completely analogous derivation results in a right-invariant integration measure.

Finally, it is worth noting that from a computational point of view, the first parameterization in (8.45) is often the most convenient to use in analytical and numerical computations of $T_i^L(g)$ and $T_i^R(g)$, and hence the corresponding Jacobians. This is because

$$\frac{\partial g}{\partial x_i} = g(x_1, 0, ..., 0)g(0, x_2, 0, ..., 0)\cdots\frac{\partial}{\partial x_i}g(0, ..., x_i, 0, ..., 0)\cdots g(0, ..., 0, x_n),$$

and

$$g^{-1}(x_1, ..., x_n) = g^{-1}(0, ..., 0, x_n)\cdots g(0, ..., x_i, 0, ..., 0)\cdots g^{-1}(x_1, 0, ..., 0)$$

and so there is considerable cancellation in the products of these matrices in the definition of $T_i^R(g)$ and $T_i^L(g)$.

8.2.4 The Relationship between Modular Functions and the Adjoint

We saw in the previous section that $d_L(g)$ and $d_R(g)$ are defined and are invariant under left and right shifts, respectively. But what happens if $d_L(g)$ is shifted from the right, or $d_R(g)$ is shifted from the left ?

We see $T_i^R(g \circ g_0) = g_0^{-1}T_i^R(g)g_0$ and $T_i^L(g_0 \circ g) = g_0 T_i^L(g)g_0^{-1}$. This means that $(T_i^R(g \circ g_0))^\vee = [Ad(g_0^{-1})](T_i^R(g))^\vee$ and $(T_i^L(g_0 \circ g))^\vee = [Ad(g_0)](T_i^L(g))^\vee$. These transformations of tangent vectors propogate through to the Jacobian, and we see that

$$d_L(g \circ g_0) = \Delta_L(g_0)\,d(g) \quad \text{and} \quad d_R(g_0 \circ g) = \Delta_R(g_0)\,d(g)$$

where

$$\Delta_L(g_0^{-1}) = |\det[Ad(g_0)]| = \Delta_R(g_0).$$

We will compute the adjoint for each of the groups of interest in this book, and hence the modular function.

Since

$$\Delta_L(g) \cdot \Delta_R(g) = 1,$$

it is common to define

$$\Delta_L(g) = \Delta(g) \quad \text{and} \quad \Delta_R(g) = \frac{1}{\Delta(g)}.$$

Then the following statements hold:

$$\int_G f(g^{-1})\,d(g) = \int_G f(g)\Delta(g^{-1})\,d(g)$$

$$\int_G f(g_0 \circ g)\,d_R(g) = \Delta(g_0)\int_G f(g)\,d_R(g)$$

$$\int_G f(g \circ g_0)\, d_L(g) = \Delta(g_0^{-1}) \int_G f(g)\, d_L(g)$$

The modular function is a continuous homomorphism $\Delta : G \to (\mathbb{R}_{>0}, \cdot)$. The homomorphism property follows from the fact that shifting twice on the right (first by g_2, then by g_1) gives

$$\Delta(g_1 \circ g_2) = \Delta(g_1) \cdot \Delta(g_2).$$

Continuity of the modular function follows from the fact that matrix Lie groups are analytic. On a compact Lie group, all continuous functions are bounded, i.e., it is not possible for a function with compact support to shoot to infinity anywhere and still maintain continuity. But by the homomorphism property, products of modular functions result after each right shift. The only way this can remain finite and positive after an arbitrary number of right shifts is if $\Delta(g) = 1$. Hence, all compact Lie groups are unimodular. As it turns out, other kinds of Lie groups of interest in engineering and the sciences are also unimodular, and this will prove to be of great importance in later chapters.

8.2.5 Examples of Volume Elements

For $SE(2)$ in the matrix exponential parameterization,

$$d(g) = c_1 dx_1 dx_2 d\theta = c_2 \frac{\sin^2 \alpha/2}{\alpha^2}\, dv_1 dv_2 d\alpha,$$

where c_1 and c_2 are scaling constants.

The bi-invariant integration measure for $GL(2, \mathbb{R})$ is

$$d(g) = \frac{1}{|\det g|^2}\, dx_1 dx_2 dx_3 dx_4.$$

We state without proof that more generally, $GL(N, \mathbb{R})$ is unimodular with

$$d(g) = \frac{1}{|\det g|^N}\, dx_1 \cdots dx_{N^2} \tag{8.48}$$

where

$$g = \begin{pmatrix} x_1 & x_2 & \cdots & x_N \\ x_{N+1} & x_{N+2} & \cdots & \vdots \\ \vdots & \vdots & \ddots & \vdots \\ x_{N^2-N+1} & x_{N^2-N+2} & \cdots & x_{N^2} \end{pmatrix}.$$

The dimension of $GL(N, \mathbb{R})$ as a Lie group is $n = N^2$.

One possible normalization for the volume element of $SL(2, \mathbb{R})$ (with integration over the $SO(2)$ subgroup set to unity and unit normalization for the two noncompact subgroups) is then

$$d(g(\theta, t, \xi)) = \frac{1}{2\pi} e^{2t}\, d\theta dt d\xi.$$

Using exactly the same argument as for $SE(2)$ presented in Chapter 6, we can show that the product $d_L(g(\theta, t, \xi)) \cdot d(\mathbf{x})$ for $\mathbf{x} \in \mathbb{R}^2$ is the bi-invariant volume element for $\mathbb{R}^2 \rtimes SL(2, \mathbb{R})$.

8.3 Harmonic Analysis on Lie Groups

The results presented for finite groups carry over in a straightforward way to many kinds of groups for which the invariant integration measures exist. In this section we discuss representations of Lie groups, and the decomposition of functions on certain Lie groups into a weighted sum of the matrix elements of the IURs.

8.3.1 Representations of Lie Groups

The representation theory of Lie groups has many physical applications. This theory evolved with the development of quantum mechanics, and today serves as a natural language in which to articulate the quantum theory of angular momentum. There is also a very close relationship between Lie group representations and the theory of special functions [19, 66, 93, 96]. It is this connection that allows for our concrete treatment of harmonic analysis on the rotation and motion groups in Chapters 9 and 10.

Let K denote either $\mathbb{R}^N$ or $\mathbb{C}^N$. Consider a transformation group $(G, \circ)$ that acts on vectors $\mathbf{x} \in K$ as $g \circ \mathbf{x} \in K$ for all $g \in G$. The *left quasi-regular representation* of G is the group $GL(\mathcal{L}^2(K))$ (the set of all linear transformations of $\mathcal{L}^2(K)$ with operation of composition). That is, it has elements $L(g)$ and group operation $L(g_1)L(g_2)$, such that the linear operators $L(g)$ act on scalar-valued functions $f(\mathbf{x}) \in \mathcal{L}^2(K)$ (the set of all square-integrable complex-valued functions on K) in the following way:[7]

$$(L(g)f)(\mathbf{x}) = f(g^{-1} \cdot \mathbf{x}).$$

Since $GL(\mathcal{L}^2(K))$ is a group of linear transformations, all that needs to be shown to prove that L is a representation of the group G is that $L : G \to GL(\mathcal{L}^2(K))$ is a homomorphism. This follows because

$$(L(g_1)(L(g_2)f))(\mathbf{x}) = (L(g_1)(L(g_2)f))(\mathbf{x})) = (L(g_1)f_{g_2})(\mathbf{x})$$

$$= f(g_2^{-1} \circ g_1^{-1} \circ \mathbf{x}) = f((g_1 \circ g_2)^{-1} \circ \mathbf{x}) = L(g_1 \circ g_2)f(\mathbf{x}).$$

Here we have used the notation $f_{g_2}(\mathbf{x}) = f(g_2^{-1} \circ \mathbf{x})$. In other words, $L(g_1 \circ g_2) = L(g_1)L(g_2)$. When G is a matrix group, g is interpreted as an $N \times N$ matrix, and we write $g \circ \mathbf{x} = g\mathbf{x}$

The *right quasi-regular representation* of a matrix Lie group G defined as an action on functions of row vectors as

$$(R(g)f)(\mathbf{x}^T) \doteq f(\mathbf{x}^T \cdot g)$$

is also a representation.

Matrices corresponding to these operators are generated by selecting an appropriate basis for the invariant subspaces of $\mathcal{L}^2(K)$. This leads to the irreducible matrix representations of the group. In the case of noncompact noncommutative groups like the Euclidean motion group, the invariant subspaces will have an infinite number of basis elements, and so the representation matrices will be infinite-dimensional.

---

[7]Often in the literature $L(g)f$ is denoted as $L_g f$.

8.3.2 Compact Lie Groups

The formulation developed previously in this chapter for finite groups can almost be copied word-for-word to formulate the Fourier pair for functions on compact Lie groups with invariant integration replacing summation over the group. Instead of doing this, we present the representation theory and Fourier expansions on compact groups in a different way. This formulation is also applicable to the finite case. We leave it to the reader to decide which is preferable as a tool to learn the subject.

As we saw in Subsection 8.2.2, every compact Lie group has a natural integration measure which is invariant to left and right shifts. Let $U(g, \lambda) = [U_{ij}(g, \lambda)]$ be the λth irreducible unitary representation matrix of the compact Lie group G. We use $U(g, \lambda)$ in this context rather than $\pi_\lambda(g)$ that we used when discussing finite groups because in the case of Lie groups, representations have additional properties. For example, $U(g, \lambda)$ are differentiable functions of their arguments, whereas there is not even a concept of continuity of arguments for $\pi_\lambda(g)$. And so using different symbols avoids inadvertently associating properties of $U(g, \lambda)$ with $\pi_\lambda(g)$.

In analogy with the Fourier transform on the circle and line, it makes sense to define

$$\hat{f}(\lambda) = \int_G f(g) U(g^{-1}, \lambda) \, d(g), \qquad (8.49)$$

or in component form

$$\hat{f}_{i,j}(\lambda) = \int_G f(g) U_{i,j}(g^{-1}, \lambda) \, d(g).$$

The collection of all λ values is denoted as $\hat{G}$ and is called the dual of the group G. Unlike the case of finite groups where $\hat{G} = \{1, ..., \alpha\}$, for the case of compact groups, $\hat{G}$ contains a countably infinite number of elements. The collection of Fourier transforms $\{\hat{f}(\lambda)\}$ for all $\lambda \in \hat{G}$ is called the spectrum of the function f.

The convolution of two square-integrable functions on a compact Lie group is defined as

$$(f_1 * f_2)(g) \doteq \int_G f_1(h) \, f_2(h^{-1} \circ g) \, d(h).$$

Since

$$U(g_1 \circ g_2, \lambda) = U(g_1, \lambda) U(g_2, \lambda)$$

and $d(g_1 \circ g) = d(g \circ g_1) = d(g)$ for any fixed $g_1 \in G$, the convolution theorem

$$(\widehat{f_1 * f_2})(\lambda) = \hat{f}_2(\lambda) \hat{f}_1(\lambda),$$

follows from

$$(\widehat{f_1 * f_2})(\lambda) = \int_G \left(\int_G f_1(h) \, f_2(h^{-1} \circ g) \, d(h) \right) U(g^{-1}, \lambda) \, d(g)$$

by making the change of variables $k = h^{-1} \circ g$ (and replacing all integrations over g by ones over k) and changing the order of integration, which results in

$$\int_G \int_G f_1(h) \, f_2(k) U((h \circ k)^{-1}, \lambda) \, d(k) \, d(h) =$$

$$\left(\int_G f_2(k) U(k^{-1}, \lambda) \, d(k) \right) \left(\int_G f_1(h) U(h^{-1}, \lambda) \, d(h) \right) = \hat{f}_2(\lambda) \hat{f}_1(\lambda).$$

The fundamental fact that allows for the reconstruction of a function from its spectrum is that every irreducible representation of a compact group is equivalent to $U(g, \lambda)$ for some value of λ, and $U(g, \lambda_1)$ and $U(g, \lambda_2)$ are not equivalent if $\lambda_1 \neq \lambda_2$. This result is Schur's Lemma for compact groups.

In analogy for the case of finite groups, that this says that if $U(g)$ and $V(g)$ are irreducible matrix representations of a compact group G with dimensions $d(\lambda_1) \times d(\lambda_1)$ and $d(\lambda_2) \times d(\lambda_2)$ respectively, and

$$AU(g) = V(g)A,$$

where the dimensions of the matrix A are $d(\lambda_2) \times d(\lambda_1)$, then if $U(g)$ is not equivalent to $V(g)$, all the entries in the matrix A must be zero. If $U(g) \cong V(g)$ then A is uniquely determined up to an arbitrary scaling, and if $U(g) = V(g)$ then $A = a \, \mathbb{I}_{d(\lambda_1) \times d(\lambda_1)}$ for some constant a.

In particular, we can define the matrix

$$A = \int_G U(g, \lambda_1) \, E \, U(g^{-1}, \lambda_2) \, d(g) \tag{8.50}$$

for any matrix E of compatible dimensions, and due to the invariance of the integration measure and homomorphism property of representations, we observe that

$$U(h, \lambda_1) \, A \, U(h^{-1}, \lambda_2) = \int_G U(h, \lambda_1) \, U(g, \lambda_1) \, EU(g^{-1}, \lambda_2) \, U(h^{-1}, \lambda_2) \, d(g)$$

$$= \int_G U(h \circ g, \lambda_1) \, E \, U((h \circ g)^{-1}, \lambda_2) \, d(g) = A,$$

or equivalently,

$$U(h, \lambda_1) \, A = A \, U(h, \lambda_2).$$

Thus, from Schur's Lemma,

$$A = a \, \delta_{\lambda_1, \lambda_2} \, \mathbb{I}_{d(\lambda_1) \times d(\lambda_2)},$$

where $d(\lambda) = \dim(U(g, \lambda))$ and $\mathbb{I}_{d(\lambda_1) \times d(\lambda_2)}$ when $U(h, \lambda_1) \not\cong U(h, \lambda_2)$ is the rectangular matrix with $(i, j)^{th}$ element $\delta_{i,j}$ and all others zero. When $\lambda_1 = \lambda_2$ it becomes an identity matrix.

Following Sugiura's notation [83], we choose $E = E^{j,l}$ to be the matrix with elements $E_{p,q}^{j,l} = \delta_{p,j} \delta_{l,q}$. Making this substitution into (8.50) and evaluating the $(i, j)^{th}$ element of the result yields

$$a_{j,l} \delta_{i,k} \delta_{\lambda_1, \lambda_2} = \sum_{p,q} \int_G U_{i,p}(g, \lambda_1) \, \delta_{p,j} \, \delta_{l,q} \, U_{q,k}(g^{-1}, \lambda_2) \, d(g). \tag{8.51}$$

The constant $a_{j,l}$ is determined by taking the trace of both sides when $\lambda_1 = \lambda_2 = \lambda$. The result is $a_{j,l} = \delta_{j,l}/d(\lambda)$. Finally, using the unitarity of the representations $U(g, \lambda)$, we see that the matrix elements of irreducible unitary representations of a compact group G satisfy

$$\delta_{j,l} \delta_{i,k} \delta_{\lambda_1, \lambda_2} = d(\lambda) \cdot \int_G U_{i,j}(g, \lambda_1) \, \overline{U_{k,l}(g, \lambda_2)} \, dg \tag{8.52}$$

where

$$\int_G dg = 1.$$

This means that if

$$f(g) = \sum_{\lambda \in \hat{G}} \sum_{i,j=1}^{d(\lambda)} c_{i,j}(\lambda) U_{i,j}(g,\lambda),$$

then the orthogonality dictates that

$$c_{i,j}(\lambda) = d(\lambda) \hat{f}(\lambda)_{i,j}.$$

In other words, the inversion formula

$$f(g) = \sum_{\lambda \in \hat{G}} d(\lambda) \operatorname{tr} \left(\hat{f}(\lambda) U(g,\lambda) \right) \tag{8.53}$$

holds, provided that the collection of matrix elements $U_{i,j}(g,\lambda)$ is *complete* in the set of square integrable functions on G. Fortunately, this is the case, as stated in the Peter-Weyl theorem below. As a result, we also have the Plancherel formulae:

$$\int_G f_1(g) \overline{f_2(g)} \, d(g) = \sum_{\lambda \in \hat{G}} d(\lambda) \operatorname{tr} \left(\hat{f}_1(\lambda) \, \hat{f}_2^*(\lambda) \right)$$

and

$$\int_G |f(g)|^2 \, d(g) = \sum_{\lambda \in \hat{G}} d(\lambda) \| \hat{f}(\lambda) \|^2.$$

Theorem 8.13. *(Peter-Weyl)*[8] *[69]: For a compact Lie group, G, the collection of functions $\{ \sqrt{d(\lambda)} U_{i,j}(g,\lambda) \}$ for all $\lambda \in \hat{G}$ and $1 \le i,j \le d(\lambda)$ form a complete orthonormal basis for $\mathcal{L}^2(G)$. The Hilbert space $\mathcal{L}^2(G)$ can be decomposed into orthogonal subspaces*

$$\mathcal{L}^2(g) = \sum_{\lambda \in \hat{G}} \bigoplus V_\lambda. \tag{8.54}$$

For each fixed value of λ, the functions $U_{i,j}(g,\lambda)$ form a basis for the subspace V_λ.

In analogy with the finite-group case, the left and right regular representations of the group are defined on the space of square integrable functions on the group $\mathcal{L}^2(G)$ as

$$(L(g)f)(h) = f(g^{-1} \circ h) \qquad \text{and} \qquad (R(g)f)(h) = f(h \circ g),$$

where $f \in \mathcal{L}^2(G)$. Both of these representations can be decomposed as

$$T(g) \cong \sum_{\lambda \in \hat{G}} \bigoplus d(\lambda) U(g,\lambda),$$

where $d(\lambda)$ is the dimension of $U(g,\lambda)$. Here $T(g)$ stands for either $L(g)$ or $R(g)$. Each of the irreducible representations $U(g,\lambda)$ acts only on the corresponding subspaces V_λ.

Finally, we note that in analogy with the case of finite groups, the Fourier series on a compact group reduce to a simpler form when the function is constant on cosets or conjugacy classes. These formulae are the same as the finite case with integration replacing summation over the group.

For further reading, see the classic work on compact Lie groups by Želobenko [101].

---

[8]Hermann Weyl (1885-1955) was a German mathematician who held a position in Switzerland for many years before moving to the US. He did work in many areas including group representation theory and its applications in physics. F. Peter was his student

8.3.3 Noncommutative Unimodular Groups in General

The striking similarity between the Fourier transform, inversion, and Plancherel formulae for finite and compact groups results in large part from the existence of invariant integration measures. A number of works have been devoted to abstract harmonic analysis on locally compact unimodular groups (e.g., [56, 63, 64, 76, 79]). See [70] for an introduction.

Bi-invariant integration measures do not exist for arbitrary Lie groups. However, they do exist for a large enough set of noncompact Lie groups for harmonic analysis on many such groups to have been addressed. Harmonic analysis for general classes of unimodular Lie groups was undertaken by Harish-Chandra[9] [32]. The classification of irreducible representations of a wide variety of real algebraic groups was pioneered by Langlands [53]. See [46, 50] for more on the representation theory of Lie groups.

Expansions of functions in terms of IURs have been made in the mathematics and physics literature for many groups including the Galilei group [54], the Lorentz group [7, 68, 75, 98], and the Heisenberg group [17, 77, 89]. We shall not review expansions on these groups here.

The noncompact noncommutative groups of greatest relevance to engineering problems is the Euclidean motion group. Harmonic analysis on this group is reviewed in great detail in Chapter 10.

We now review some general operational properties for unimodular Lie groups. See [85] for further reading.

Operational Properties for Unimodular Lie Groups

Recall that a unimodular group G is one which has a left and right invariant integration measure, and hence $\Delta(g) = 1$.

The first thing that is the same in the noncommutative context is that a differential operator acting on the convolution of two functions is the convolution of the derivative of one of the functions with the other, only now the kind of operator matters:

$$X_i^L(f_1 * f_2) = (X_i^L f_1) * f_2 \quad \text{and} \quad X_i^R(f_1 * f_2) = f_1 * (X_i^R f_2).$$

There are many other similarities between the "operational properties" of the Abelian and noncommutative cases. In order to see the similarities, another definition is required.

Given a $d(\lambda) \times d(\lambda)$ matrix representation $U(g, \lambda)$ of G we can define a representation of the Lie algebra $\mathcal{G}$ as

$$u(X, \lambda) \doteq \left. \frac{d}{dt} (U(\exp tX), \lambda) \right|_{t=0}. \tag{8.55}$$

This is called a representation of $\mathcal{G}$ because it inherits the Lie bracket from the differential operators X^L and X^R, i.e.,

$$u([X, Y], \lambda) = [u(X, \lambda), u(Y, \lambda)].$$

If $X = \sum_i x_i X_i$ is an arbitrary element of $\mathcal{G}$ and $\{X_i\}$ is a basis for $\mathcal{G}$, then

---

[9]Harish-Chandra (1923-1983) was a mathematics professor at Princeton.

$$u(X, \lambda) = \sum_i x_i\, u(X_i, \lambda).$$

The left and right derivatives are also linear in the sense that $(\sum_i x_i X_i)^L = \sum_i x_i X_i^L$ and $(\sum_i x_i X_i)^R = \sum_i x_i X_i^R$.

Furthermore, by substitution we see that

$$
\begin{aligned}
\widehat{X^L f}(\lambda) &= \int_G \frac{d}{dt} f(\exp(-tX) \circ g)\Big|_{t=0} U(g^{-1}, \lambda)\, d(g) \\
&= \frac{d}{dt}\left[\int_G f(g') U(g'^{-1} \circ \exp(-tX), \lambda)\, d(g') \right]_{t=0} \\
&= \left(\int_G f(g') U(g'^{-1}, \lambda)\, d(g') \right) \left[\frac{d}{dt} U(\exp(-tX), \lambda) \right]_{t=0} \\
&= -\hat{f}(\lambda) u(X, \lambda).
\end{aligned}
$$

Similarly, for the right derivative,

$$
\begin{aligned}
\widehat{X^R f}(\lambda) &= \int_G \frac{d}{dt} f(g \circ \exp(tX))\Big|_{t=0} U(g^{-1}, \lambda)\, d(g) \\
&= \frac{d}{dt}\left[\int_G f(g') U(\exp(tX) \circ g'^{-1}, \lambda)\, d(g') \right]_{t=0} \\
&= \left[\frac{d}{dt} U(\exp(tX), \lambda) \right]_{t=0} \left(\int_G f(g') U(g'^{-1}, \lambda)\, d(g') \right) \\
&= u(X, \lambda) \hat{f}(\lambda).
\end{aligned}
$$

These are summarized as

$$\boxed{\widehat{X^R f}(\lambda) = u(X_i, p) \hat{f}(\lambda) \quad \text{and} \quad \widehat{X^L f}(\lambda) = -\hat{f}(\lambda) u(X, \lambda).} \qquad (8.56)$$

Using the Lie bracket property of the Lie algebra representations, expanding the bracket, and using the above operational properties, we see that

$$\mathcal{F}([X^L, Y^L] f)(\lambda) = \int_G ([X^L, Y^L] f(g)) U(g^{-1}, \lambda)\, d(g) = \hat{f}(\lambda)[u(Y, \lambda), u(X, \lambda)]$$

and

$$\mathcal{F}([X^R, Y^R] f)(\lambda) = \int_G ([X^R, Y^R] f(g)) U(g^{-1}, \lambda)\, d(g) = [u(X, \lambda), u(Y, \lambda)] \hat{f}(\lambda).$$

Note that this is an operational property with no analog in classical (Abelian) Fourier analysis.

The Fourier transform of the Laplacian of a function is transformed as

$$\mathcal{F}(\mathrm{div}_R(\mathrm{grad}_R(f)))(\lambda) = \sum_{i=1}^N (u(X_i, \lambda))^2 \hat{f}(\lambda)$$

and

$$\mathcal{F}(\mathrm{div}_L(\mathrm{grad}_L(f)))(\lambda) = \sum_{i=1}^N \hat{f}(\lambda)(u(X_i, \lambda))^2.$$

For the examples we will examine (such as $SO(3)$ and $SU(2)$), the matrices $\sum_{i=1}^{N}(u(X_i, \lambda))^2$ are multiples of the identity matrix and the subscripts L and R can be dropped so that $\mathcal{F}(\text{div}(\text{grad}(f))) = -\tilde{\alpha}(\lambda)\hat{f}(\lambda)$ where $\tilde{\alpha}(\lambda)$ is a positive real-valued function. When this happens, the heat equation

$$\frac{\partial^2 f}{\partial t^2} = K^2 \, \text{div}(\text{grad}((f))$$

can be solved in closed form for $f(g, t)$ subject to initial conditions $f(g, 0) = \delta(g)$. From this fundamental solution, the solution for any initial conditions $f(g, 0) = \tilde{f}(g)$ can be generated by convolution.

The Adjoint of Differential Operators

Consider a unimodular Lie group with the inner product of complex-valued square-integrable functions defined as

$$(f_1, f_2) \doteq \int_G \overline{f_1(g)} f_2(g) \, d(g).$$

The properties resulting from invariant integration on G means that it is easy to calculate the *adjoint*, $(X_i^R)^*$, of the differential operators

$$(X_i^R f)(g) = \frac{d}{dt} \left(f(g \circ \exp[tX_i]) \right) \Big|_{t=0},$$

as defined by the equality

$$((X_i^R)^* f_1, f_2) = (f_1, X_i^R f_2).$$

To explicitly calculate the adjoint, observe that

$$(f_1, X_i^R f_2) = \int_G \overline{f_1(g)} \frac{d}{dt} f_2(g \circ \exp[tX_i]) \Big|_{t=0} d(g) = \frac{d}{dt} \int_G \overline{f_1(g)} f_2(g \circ \exp[tX_i]) \, d(g) \Big|_{t=0}.$$

Letting $g' = g \circ \exp[tX_i]$, substituting for g, and using the invariance of integration under shifts results in

$$\frac{d}{dt} \int_G \overline{f_1(g' \circ \exp[-tX_i])} f_2(g') d(g') \Big|_{t=0} = \int_G \frac{d}{dt} \overline{f_1(g' \circ \exp[-tX_i])} \Big|_{t=0} f_2(g') \, d(g')$$

$$= -\int_G \overline{(X_i^R f_1)(g')} f_2(g') \, d(g') = (-X_i^R f_1, f_2).$$

Hence, we conclude that since X_i^R is real,

$$(X_i^R)^* = -X_i^R.$$

It is often convenient to consider the self-adjoint form of an operator, and in this context it is easy to see that the operator defined as

$$Y_i^R \doteq -iX_i^R$$

satisfies the self-adjointness condition

$$(Y_i^R)^* = Y_i^R.$$

Before moving on to the construction of explicit representation matrices for $SO(3)$ and $SE(N)$ in Chapters 9 and 10, a very general technique for generating group representations from representations of subgroups is examined in the next section. This method is applicable both to discrete and Lie groups.

8.4 Induced Representations and Tests for Irreducibility

In this chapter we have assumed that a complete set of IURs for a group either have been given in advance or have been obtained by decomposing the left or right regular representations. In practice, other methods for generating IURs are more convenient than decomposition of the regular representation. One such method for generating representations of a subgroup H of a group G for which representations are already known is called *subduction*. This is nothing more than the restriction of the representation $D(g)$ of G to H. This is often denoted $D(h) = D \downarrow H$. If $D(h)$ is irreducible, then so too is $D(g)$. But in general, irreducibility of $D(g)$ does not guarantee irreducibility of $D(h)$.

A very general method for creating representations of groups from representations of subgroups exists. It is called the method of *induced representations*. Given a representation $D(h)$ for $h \in H < G$, the corresponding induced representation is denoted in the literature in a variety of ways including $D \uparrow G$, $D(H) \uparrow G$, $U = \text{ind}_H^G(D)$,or $U^D = \text{ind}_{H,D}^G$. It is this method that will be of importance to us when constructing representations of $SE(2)$ and $SE(3)$.

The method of induced representations for finite groups was developed by Frobenius [25], Weyl, and Clifford [15]. The extension and generalization of these ideas to locally compact groups (and in particular to Lie groups) was performed by Mackey [56]-[59]. The collection of methods for generating induced representations of a locally compact group from a normal subgroup $N \triangleleft G$ and subgroups of G/N is often called the "Mackey machine."

In general, these representations will be reducible. It is therefore useful to have a general test to determine when a representation is reducible. We discuss here both the method of induced representations and general tests for irreducibility. These methods are demonstrated in subsequent chapters with the rotation and motion groups.

8.4.1 Finite Groups

Induced Representations of Finite Groups

Let $\{D^\lambda\}$ for $\lambda = 1, ..., \alpha$ be a complete set of IUR matrices for a subgroup $H < G$ and $dim(D^\lambda) = d_\lambda$. Then the matrix elements of an induced representation of G with dimension $d_\lambda \cdot |G/H|$ are defined by the expression [2, 16, 37]

$$U_{i\mu,j\nu}^\lambda(g) = \delta_{ij}(g)D_{\mu\nu}^\lambda(g_i^{-1} \circ g \circ g_j) \tag{8.57}$$

where $g \in G$ and g_i and g_j are respectively arbitrary elements of the cosets $g_i H$ and $g_j H$, and

$$\delta_{ij}(g) \doteq \begin{cases} 1 \text{ for } g \circ g_j \in g_i H \\ 0 \text{ for } g \circ g_j \notin g_i H \end{cases}.$$

This guarantees that $U_{i\mu,j\nu}^\lambda(g) = 0$ unless $g_i^{-1} \circ g \circ g_j \in H$.

From the above definitions we verify that

$$\sum_{j=1}^{|G/H|} \sum_{\nu=1}^{d_\lambda} U_{i\mu,j\nu}^\lambda(g_1)U_{j\nu,k\eta}^\lambda(g_2) = \sum_{j=1}^{|G/H|} \sum_{\nu=1}^{d_\lambda} \delta_{ij}(g_1)D_{\mu\nu}^\lambda(g_i^{-1} \circ g \circ g_j)\, \delta_{jk}(g_2)D_{\nu\eta}^\lambda(g_j^{-1} \circ g \circ g_k)$$

$$= \sum_{j=1}^{|G/H|} \delta_{ij}(g_1)\, \delta_{jk}(g_2) \sum_{\nu=1}^{d_\lambda} D_{\mu\nu}^\lambda(g_i^{-1} \circ g \circ g_j)D_{\nu\eta}^\lambda(g_j^{-1} \circ g \circ g_k).$$

Since D^λ is a representation,

$$\sum_{\nu=1}^{d_\lambda} D_{\mu\nu}^\lambda(g_i^{-1} \circ g \circ g_j) D_{\nu\eta}^\lambda(g_j^{-1} \circ g \circ g_k) = D_{\mu\eta}^\lambda(g_i^{-1} \circ (g_1 \circ g_2) \circ g_k).$$

Likewise,

$$\sum_{j=1}^{|G/H|} \delta_{ij}(g_1)\, \delta_{jk}(g_2) = \delta_{ik}(g_1 \circ g_2),$$

since if $g_i^{-1} \circ g_1 \circ g_j \in H$ and $g_j^{-1} \circ g_2 \circ g_k \in H$, then the product

$$g_i^{-1} \circ g_1 \circ g_j \circ g_j^{-1} \circ g_2 \circ g_k = g_i^{-1} \circ g_1 \circ g_2 \circ g_k \in H.$$

Therefore,

$$\sum_{j=1}^{|G/H|} \sum_{\nu=1}^{d_\lambda} U_{i\mu,j\nu}^\lambda(g_1) U_{j\nu,k\eta}^\lambda(g_2) = U_{i\mu,k\eta}^\lambda(g_1 \circ g_2).$$

We note that the four indices j, ν, k, η may be contracted to two while preserving the homomorphism property. There is more than one way to do this. We can denote

$$\tilde{U}_{IJ}^\lambda(g) = U_{i\mu,j\nu}^\lambda(g)$$

where

$$I = i + |G/H| \cdot (\mu - 1), \quad J = j + |G/H| \cdot (\nu - 1)$$

or

$$I = d_\lambda \cdot (i - 1) + \mu, \quad J = d_\lambda \cdot (j - 1) + \nu.$$

Either way,

$$\tilde{U}_{IK}^\lambda(g_1 \circ g_2) = \sum_{J=1}^{|G/H| \cdot d_\lambda} \tilde{U}_{IJ}^\lambda(g_1) \tilde{U}_{JK}^\lambda(g_2).$$

In this way standard two-index matrix operations can be performed.

For applications of induced representations of finite groups in chemistry and solid-state physics see [2, 37].

Checking Irreducibility of Represesentations of Finite Groups

Theorem 8.14. *A representation $\pi(g)$ of a finite group G is irreducible if and only if the character $\chi(g) = \mathrm{tr}(\pi(g))$ satisfies*

$$\frac{1}{|G|} \sum_{g \in G} \chi(g)\overline{\chi(g)} = 1. \tag{8.58}$$

Proof. For an arbitrary representation of G we write

$$\pi(g) = S^{-1}(c_1 \pi_1 \oplus \cdots \oplus c_n \pi_n) S,$$

where S is an invertible matrix and c_i is the number of times π_i appears in the decomposition of π. Since the trace is invariant under similarity transformation,

$$\chi(g) = \sum_{i=1}^{n} c_i \chi_i(g),$$

where $\chi_i(g) = \text{tr}(\pi_i(g))$. Substituting into (8.58) and using the orthogonality of characters yields the condition

$$\sum_{i=1}^{n} c_i^2 = 1.$$

Since c_i is a nonnegative integer, this equality holds only if one of the c_i's is equal to one and all the others are equal to zero. On the other hand, if π is irreducible $c_i = \delta_{ij}$ for some $j \in \{1, ..., n\}$. Therefore, when π satisfies (8.58) it must be the case that $\pi \cong \pi_j$ for some j, and π is therefore irreducible.

8.4.2 Lie Groups

Induced Representations of Lie Groups

A formal definition of the induced representations $D(H) \uparrow G$, where $D(H)$ are representations of the subgroup H of group G, is given below. For more details see, for example, [11, 16, 30, 58].

Definition 8.15. Let V be a complex vector space and $\phi : G/H \to V$. The operator $U(g) \in D \uparrow G$ acts on V as

$$(U(g)\phi)(\sigma) \doteq D(s_\sigma^{-1} g \, s_{g^{-1}\sigma}) \, \phi(g^{-1}\sigma), \tag{8.59}$$

where s_σ is an arbitrary representative of the coset $\sigma \in G/H$, $g \in G$, and D is a representation of $H < G$.

We observe the homomorphism property by first defining $\phi_g(\sigma) = (U(g)\phi)(\sigma)$. Then

$$(U(g_1)U(g_2)\phi)(\sigma) = (U(g_1)\phi_{g_2})(\sigma) = D(s_\sigma^{-1} g_1 \, s_{g_1^{-1}\sigma}) \, \phi_{g_2}(g_1^{-1}\sigma) =$$

$$D(s_\sigma^{-1} g_1 \, s_{g_1^{-1}\sigma}) \, D(s_{g_1^{-1}\sigma}^{-1} g_2 \, s_{g_2^{-1}g_1^{-1}\sigma}) \, \phi(g_1^{-1}g_2^{-1}\sigma) =$$

$$D(s_\sigma^{-1} g_1 g_2 \, s_{(g_1 g_2)^{-1}\sigma}) \, \phi((g_1 g_2)^{-1}\sigma) = (U(g_1 g_2)\phi)(\sigma).$$

Since in general, the homogeneous space G/H will be a manifold of dimension $\dim(G/H) = \dim(G) - \dim(H)$, it follows that matrix elements of $U(g)$ are calculated as

$$U_{i,j}(g) = (e_i, U(g)e_j),$$

where $\{e_k\}$ for $k = 1, 2, ...$ is is a set of orthonormal eigenfunctions for $\mathcal{L}^2(G/H)$, and the inner product is defined as

$$(\phi_1, \phi_2) \doteq \int_{G/H} \overline{\phi_1(\sigma)}\phi_2(\sigma)d(\sigma)$$

with $\sigma \in G/H$ and $d(\sigma)$ denotes an integration measure for G/H.

Since this method will generally produce infinite-dimensional representation matrices, it is clear that in the case of compact Lie groups, these representations will generally be reducible. However, as we shall see, the infinite-dimensional representations generated in this way for certain noncompact noncommutative Lie groups will be irreducible.

The next subsection provides tests for determining when a representation is reducible.

Checking Irreducibility of Representations of Compact Lie Groups

Several means are available to quickly determine if a given representation of a compact Lie group is irreducible or not. Since in many ways the representation theory and harmonic analysis for compact Lie groups is essentially the same as for finite groups (with summation over the group replaced with invariant integration) the following theorem should not be surprising.

Theorem 8.16. *A representation $U(g)$ of a compact Lie group G with invariant integration measure $d(g)$ normalized so that $\int_G d(g) = 1$ is irreducible if and only if the character $\chi(g) = \mathrm{tr}(U(g))$ satisfies*

$$\int_G \chi(g)\overline{\chi(g)}\, d(g) = 1. \tag{8.60}$$

Proof. Same as for (8.58) with integration replacing summation.

Another test for irreducibility of compact Lie group representations is based solely on their continuous nature and is not simply a direct extension of concepts originating from the representation theory of finite groups. For a Lie group, we can evaluate any given representation $U(g)$ at a one-parameter subgroup generated by exponentiating an element of the Lie algebra as $U(\exp(tX_i))$. Expanding this matrix function of t in a Taylor series, the linear term in t will have the constant coefficient matrix

$$u(X_i) = \left. \frac{dU(\exp(tX_i))}{dt} \right|_{t=0}. \tag{8.61}$$

If U is reducible, then all of the $u(X_i)$ must simultaneously be block-diagonalizable by the same similarity transformation. If no such similarity transformation can be found, then we can conclude that U is irreducible.

Theorem 8.17. *[27, 68, 93] Given a Lie group G, associated Lie algebra $\mathcal{G}$, and finite-dimensional $u(X_i)$ as defined in (8.61), then*

$$U(\exp(tX_i)) = \exp[tu(X_i)], \tag{8.62}$$

where $X_i \in \mathcal{G}$. Furthermore, if the matrix exponential parameterization

$$g(x_1, ..., x_n) = \exp\left(\sum_{i=1}^{n} x_i X_i \right) \tag{8.63}$$

is surjective, then when the $u(X_i)$'s are not simultaneously block-diagonalizable,

$$U(g) = \exp\left(\sum_{i=1}^{n} x_i\, u(X_i) \right) \tag{8.64}$$

is an irreducible representation for all $g \in G$.

Proof. For the exponential parameterization (8.63), we observe that

$$g(tx_1, ..., tx_n) \circ g(\tau x_1, ..., \tau x_n) = g((t+\tau)x_1, ..., (t+\tau)x_n)$$

for all $t, \tau \in \mathbb{R}$, i.e., the set of all $g(tx_1, ..., tx_n)$ forms a one-dimensional (Abelian) subgroup of G for fixed values of x_i. From the definition of a representation it follows that

$$U(g((t+\tau)x_1, ..., (t+\tau)x_n)) = U(g(tx_1, ..., tx_n))U(g(\tau x_1, ..., \tau x_n)) \tag{8.65}$$

$$= U(g(\tau x_1, ..., \tau x_n))U(g(tx_1, ..., tx_n)).$$

Define

$$\tilde{U}(x_1, ..., x_n) = U(g(x_1, ..., x_n)).$$

Then differentiating (8.65) with respect to τ and setting $\tau = 0$,

$$\frac{d}{dt}\tilde{U}(tx_1, ..., tx_n) = \frac{d}{d\tau}\tilde{U}(\tau x_1, ..., \tau x_n)\bigg|_{\tau=0} \tilde{U}(tx_1, ..., tx_n).$$

But since infinitesimal operations commute, it follows from (8.61) that

$$\frac{d}{d\tau}\tilde{U}(\tau x_1, ..., \tau x_n)|_{\tau=0} = \sum_{i=1}^{n} x_i \, u(X_i).$$

We therefore have the matrix differential equation

$$\frac{d}{dt}\tilde{U}(tx_1, ..., tx_n) = \left(\sum_{i=1}^{n} x_i \, u(X_i)\right)\tilde{U}(tx_1, ..., tx_n)$$

subject to the initial conditions

$$\tilde{U}(0, ..., 0) = \mathbb{I}_{dim(U) \times dim(U)}.$$

The solution is therefore

$$\tilde{U}(tx_1, ..., tx_n) = \exp\left(t\sum_{i=1}^{n} x_i \, u(X_i)\right).$$

Evaluating at $t = 1$, we find (8.64) and setting all $x_j = 0$ except x_i, we find (8.62). The irreducibility of these representations follows from the assumed properties of $u(X_i)$.

Essentially the same argument can be posed in different terminology by considering what happens to subspaces of function spaces on which representation operators (not necessarily matrices) act. This technique, which we will demonstrate for $SU(2)$ in the next chapter, also can be used for noncompact noncommutative Lie groups (which generally have infinite-dimensional representation matrices). The concrete example of $SE(2)$ is presented in Chapter 10 to demonstrate this.

8.5 Wavelets on Groups

The concept of wavelet transforms discussed in Chapter 3 extends to the group-theoretical setting with appropriate concepts of translation, dilation, and modulation. We discuss the modulation-translation case in Subsection 8.5.1 and the dilation-translation case in Subsection 8.5.2.

8.5.1 The Gabor Transform on a Unimodular Group

Let G be a unimodular group and $\gamma \in \mathcal{L}^2(G)$ be normalized so that $\|\gamma\| = 1$. The matrix representations of G in a suitable basis are denoted as $U(g, \lambda)$ for $\lambda \in \hat{G}$ and $g \in G$.

In direct analogy with the Gabor transform on the real line and unit circle, we can define the transform

$$\tilde{f}_\gamma(h, \lambda) = \int_G f(g)\overline{\gamma(h^{-1} \circ g)}U(g^{-1}, \lambda)\, dg. \tag{8.66}$$

Here $h \in G$ is the shift, and $U(g^{-1}, \lambda)$ is the modulation.

This group-theoretical Gabor transform is inverted by following the same steps as in cases of the line and circle. We first multiply $\tilde{f}_\gamma(h, \lambda)$ by $\gamma(h^{-1} \circ g)$ and integrate over all shifts:

$$\int_G \tilde{f}_\gamma(h, \lambda)\gamma(h^{-1} \circ g)\, dh = \int_G \left(\int_G \overline{\gamma(h^{-1} \circ g)}\gamma(h^{-1} \circ g)\, dh \right) f(g)U(g^{-1}, \lambda)\, dg.$$

Using the invariance of integration under shifts and inversions, this simplifies to

$$\int_G \tilde{f}_\gamma(h, \lambda)\gamma(h^{-1} \circ g)\, dh = \|\gamma\|^2 \int_G f(g)U(g^{-1}, \lambda)\, dg.$$

Since $\|\gamma\|^2 = 1$, this is nothing more than the Fourier transform of f and the inverse Fourier transform is used to recover f. Hence, we write:

$$f(g) = \int_{\hat{G}} d\nu(\lambda)\mathrm{tr} \left(U(g, \lambda) \int_G \tilde{f}_\gamma(h, \lambda)\gamma(h^{-1} \circ g)\, dh \right) \tag{8.67}$$

8.5.2 Wavelet Transforms on Groups Based on Dilation and Translation

Let $\varphi(g; s) \in \mathcal{L}^2(G \times \mathbb{R}_{>0})$, and call $\varphi(g; 1)$ the mother wavelet. Then define

$$\varphi_{s,h}^p(g) = |s|^{-p}\varphi(h^{-1} \circ g; s).$$

At this point, the structure of $\varphi(g; s)$ will be left undetermined, as long as the Hermitian matrices

$$K_\lambda = \int_0^\infty \hat{\varphi}(\lambda, s)\hat{\varphi}^*(\lambda, s)|s|^{-2}\, ds \tag{8.68}$$

have nonzero eigenvalues for all $\lambda \in \hat{G}$.

We define the continuous wavelet transform on G as

$$\tilde{f}_p(s, h) = \int_G \overline{\varphi_{s,h}^p(g)}f(g)\, dg. \tag{8.69}$$

Let $\varphi_{-1}(g; s) = \varphi(g^{-1}; s)$. Then (8.69) can be written as

$$\tilde{f}_p(s, h) = |s|^{-p} \int_G f(g)\overline{\varphi_{-1}(g^{-1} \circ h, s)}\, dg = |s|^{-p}f * \overline{\varphi_{-1}}(h; s).$$

Applying the Fourier transform to $\overline{\varphi_{-1}}$ we see that

$$(\widehat{\overline{\varphi_{-1}}})(\lambda;s) = \int_G \overline{\varphi_{-1}(g;s)} U(g^{-1},\lambda)\,dg$$
$$= \int_G \overline{\varphi(g^{-1};s)} U(g^{-1},\lambda)\,dg$$
$$= \int_G \overline{\varphi(g;s)} U(g,\lambda)\,dg$$
$$= (\hat{\varphi}(\lambda;s))^*,$$

and hence by the convolution theorem

$$\mathcal{F}(\tilde{f}_p(s,h)) = |s|^{-p}\hat{\varphi}^*(\lambda,s)\hat{f}(\lambda).$$

Multiplying on the left by $s^{p-2}\hat{\varphi}$, integrating over all values of s, and inverting the resulting matrix K_λ, we see that

$$\mathcal{F}(f) = K_\lambda^{-1}\int_0^\infty \mathcal{F}(\tilde{f}_p(s,h))|s|^{p-2}\,ds.$$

Finally, the Fourier inversion theorem applied to the above equation reproduces f. Hence we conclude that when a suitable mother wavelet exists, the inverse of the continuous wavelet transform on a unimodular group is of the form

$$f(g) = \int_{\hat{G}} d(\lambda)\mathrm{tr}\left(K_\lambda^{-1}\int_0^\infty\int_G \tilde{f}_p(s,h)U(h^{-1},\lambda)\,dh|s|^{p-2}\,ds\right)U(g,\lambda). \tag{8.70}$$

A natural question to ask at this point is "How do we construct suitable functions $\varphi(g;s)$?" While the dilation operation is not natural in the context of functions on groups, the diffusion operator is very natural (at least in the context of the unimodular Lie groups of interest in this book). Let $\kappa(g;s)$ be a function with the property that

$$\kappa(g;s_1)*\kappa(g;s_2) = \kappa(g;s_1+s_2).$$

Then define

$$\varphi(g;s) = \phi(g)*\kappa(g;s).$$

We investigate conditions on ϕ in Chapter 9 when $G = SO(3)$. See [45, 1] for background on coherent states and their relationship to group-theoretic aspects of wavelet transforms.

In the time since the publication of the original version of this book, a number of works have developed the theory of wavelets on Lie groups (in particular, $SO(3)$) in the context of various applications. See, for example, [8, 21] and additional references given in Chapter 13.

8.6 Summary

This chapter provided a broad introduction to harmonic analysis on finite, compact and particular noncompact unimodular groups. In order to gain a fuller understanding of this subject, the classic books in the bibliography for this chapter should be consulted. For aspects of commutative and noncommutatative harmonic analysis not covered here, the interested reader should see [44, 55, 84] and [3, 23, 35, 36, 95], respectively. However, our treatment is complete enough for the engineer or scientist to use these techniques in applications. The somewhat abstract material in this chapter lays the foundations for concrete harmonic analysis on $SO(3)$ and $SE(3)$ in the two chapters to follow. Chapters 11-18 then examine computational issues and their impact on applications.

References

1. Ali, S.T., Antoine, J.-P., Gazeau, J.-P., *Coherent States, Wavelets and Their Generalizations*, Graduate Texts in Contemporary Physics, Springer, New York, 2000.
2. Altmann, S.L., *Induced Representations in Crystals and Molecules*, Academic Press, New York, 1977.
3. Applebaum, D., *Probability on Compact Lie Groups*, Springer, New York, 2014.
4. Arthur, J., "Harmonic Analysis and Group Representations," *Notices of the American Mathematical Society*, 47(1): 26-34, 2000.
5. Atkinson, M.D., "The Complexity of Group Algebra Computations," *Theoretical Computer Science*, 5(2): 205 – 209, 1977.
6. Apple, G., Wintz, P., "Calculation of Fourier Transforms on Finite Abelian Groups," *IEEE Transactions on Information Theory*, 16(2): 233 – 234, 1970.
7. Bargmann, V., "Irreducible Unitary Representations of the Lorentz Group," *Annals of Mathematics*, 48(3): 568 – 640, July 1947.
8. Bernstein, S., Ebert, S., "Wavelets on S^3 and $SO(3)$-Their Construction, Relation to Each Other and Radon Transform of Wavelets on $SO(3)$," *Mathematical Methods in the Applied Sciences*, 33(16): 1895 – 1909, 2010.
9. Beth, T., "On the Computational Complexity of the General Discrete Fourier Transform," *Theoretical Computer Science*, 51(3): 331 – 339, 1987.
10. Bröcker, T., tom Dieck, T., *Representations of Compact Lie Groups*, Springer, New York 2010.
11. Bruhat, F., "Sur les représentations induites de groupes de Lie," *Bulletin de la Société Mathématique de France*, 84: 97 – 205, 1956.
12. Cairns, T.W., "On the Fast Fourier Transform on Finite Abelian Groups," *IEEE Transactions on Computers*, 20: 569 – 571, 1971.
13. Cartan, É., "Sur la determination d'un système orthogonal complet dans un espace de Riemann symmetrique clos," *Rendiconti del Circolo Matematico di Palermo*, 53(1): 217 – 252, 1929.
14. Clausen, M., "Fast Generalized Fourier Transforms," *Theoretical Computer Science*, 67(1): 55 – 63, 1989.
15. Clifford, A.H., "Representations Induced in an Invariant Subgroup," *Annals of Mathematics*, 38:533 – 550, 1937.
16. Coleman, A.J., *Induced Representations with Applications to S_n and $GL(n)$*, Queen's University, Kingston, Ontario, 1966.
17. Corwin, L., Greenleaf, F.P., *Representations of Nilpotent Lie Groups and Their Applications Part 1: Basic Theory and Examples*, Cambridge Studies in Advanced Mathematics 18, Cambridge University Press, New York 1990.
18. Diaconis, P., Rockmore, D., "Efficient Computation of the Fourier Transform on Finite Groups," *Journal of the American Mathematical Society*, 3(2): 297 – 332, April 1990.

19. Dieudonné, J., *Special Functions and Linear Representations of Lie Groups*, American Mathematical Society, Providence, RI, 1980.

20. Dunkl, C.F., Ramirez, D.E., *Topics in Harmonic Analysis*, Appleton-Century-Crofts, Meredith Corporation, New York, 1971.

21. Ebert, S., Wirth, J., "Diffusion Wavelets on Groups and Homogeneous Spaces," *Proceedings of the Royal Society of Edinburgh*, Section A, 141(3): 497–520, 2011.

22. Fässler, A., Stiefel, E., *Group Theoretical Methods and Their Applications*, Birkhäuser, Boston, 1992.

23. Feinsilver, P., Schott, R., *Algebraic Structures and Operator Calculus: Vols I-III*, Springer, 1994 (Softcover 2013).

24. Folland, G.B., *A Course in Abstract Harmonic Analysis*, CRC Press, Boca Raton, FL, 1995.

25. Frobenius, F.G., "Über Relationen zwischen den Charakteren einer Gruppe und denen ihrer Untergruppen," *Sitzungsbericht der Preussischen Akademie der Wissenschaften*, pp. 501-515, 1898.

26. Fulton, W., Harris, J., *Representation Theory: A First Course*, Springer-Verlag, 1991.

27. Gelfand, I. M., Minlos, R.A., Shapiro, Z.Ya., *Representations of the Rotation and Lorentz Groups and Their Applications*, Macmillan, New York, 1963. (Martino Fine Books Edition, 2012).

28. Grenander, U., *Probabilities on Algebraic Structures*, Dover, 2008.

29. Gross, K.I., "Evolution of Noncommutative Harmonic Analysis," *American Mathematical Monthly*, 85(7): 525–548, 1978.

30. Gurarie, D., *Symmetry and Laplacians. Introduction to Harmonic Analysis, Group Representations and Applications*, Elsevier Science Publisher, The Netherlands, 1992. (Dover Edition, 2008).

31. Haar, A., "Der Maßbegriff in der Theorie der kontinuierlichen Gruppen," *Annals of Mathematics*, 34: 147–169, 1933.

32. Harish-Chandra, *Collected Papers*, Vol. I-IV, Springer-Verlag, 1984.

33. Helgason, S., *Differential Geometry, Lie Groups, and Symmetric Spaces*, American Mathematical Society, 2001.

34. Hermann, R., *Fourier Analysis on Groups and Partial Wave Analysis*, W.A. Benjamin, Inc., New York, 1969.

35. Hewitt, E., Ross, K.A., *Abstract Harmonic Analysis I, and II*, Springer-Verlag, Berlin, 1963 and 1970. (reprinted 1994).

36. Howe, R., and Tan, E.C., *Non-Abelian Harmonic Analysis*, Springer, 1992.

37. Inui, T., Tanabe, Y., Onodera, Y., *Group Theory and Its Applications in Physics*, 2nd ed., Springer-Verlag, 1996.

38. Janssen, T., *Crystallographic Groups*, North Holland/American Elsevier, New York, 1973.

39. Karpovsky, M.G., "Fast Fourier Transforms on Finite Non-Abelian Groups," *IEEE Transactions on Computers*, 26(10): 1028–1030, Oct. 1977.

40. Karpovsky, M.G., Trachtenberg, E.A., "Fourier Transform over Finite Groups for Error Detection and Error Correction in Computation Channels," *Information and Control*, 40(3): 335–358, 1979.

41. Karpovsky, M.G., Trachtenberg, E.A., "Some Optimization Problems for Convolution Systems over Finite Groups," *Information and Control*, 34(3): 227–247, 1977.

42. Karpovsky, M.G., "Error Detection in Digital Devices and Computer Programs with the Aid of Linear Recurrent Equations over Finite Commutative Groups," *IEEE Transactions on Computers*, 26(3): 208–218, March 1977.

43. Karpovsky, M.G., Trachtenberg, E.A., "Statistical and Computational Performance of a Class of Generalized Wiener Filters," *IEEE Transactions on Information Theory*, 32(2): 303–307, March 1986.

44. Katznelson, Y., *An Introduction to Harmonic Analysis*, Wiley, New York, 1968.

45. Klauder, J.R., Skagerstam, B.S., *Coherent States: Application in Physics and Mathematical Physics*, World Scientific, Singapore, 1985.

46. Knapp, A.W , Representation Theory of Semisimple Groups, An Overview Based on Examples, Princeton University Press, 1986
47. Koo, J.Y., Kim, P.T., "Asymptotic Minimax Bounds for Stochastic Deconvolution over Groups," IEEE Transactions on Information Theory, 54(1): 289–298, 2008
48. Koster, G.F., Space Groups and Their Representations, Academic Press, New York, 1957.
49. Kovalev, V., Representations of the Crystallographic Space Groups: Irreducible Representations, Induced Representations, and Corepresentations, edited by Harold T. Stokes and Dorian M. Hatch ; translated from the Russian by Glen C. Worthey. Gordon and Breach Scientific Publishers, 1993.
50. Kirillov, A.A., ed., Representation Theory and Noncommutative Harmonic Analysis I,II, Springer-Verlag, Berlin, 1988,1995.
51. Kondor, R., "The Skew Spectrum of Functions on Finite Groups and Their Homogeneous Spaces," arXiv preprint arXiv:0712.4259, 2007.
52. Kunze, R., "L_p Fourier Transforms on Locally Compact Unimodular Groups," Transactions of the American Mathematical Society, 89: 519–540, 1958.
53. Langlands, R.P., "On the Classification of Irreducible Representations of Real Algebraic Groups," mimeographed notes, Institute for Advanced Study, 1973.
54. Loebl, E.M., ed., Group Theory and Its Applications, Vols. I, II, III, Academic Press, New York, 1968.
55. Loomis, L.H., Introduction to Abstract Harmonic Analysis, Courier Dover Publications, 2011.
56. Mackey, G.W., "Induced Representations of Locally Compact Groups I," Annals of Mathematics, 55:101–139, 1952.
57. Mackey, G.W., "Unitary Representations of Group Extensions," Acta Mathematicae, 99: 265–311, 1958.
58. Mackey, G.W., Induced Representations of Groups and Quantum Mechanics, W. A. Benjamin, Inc., New York and Amsterdam, 1968.
59. Mackey, G.W., The Theory of Unitary Group Representations, The University of Chicago Press, Chicago, 1976.
60. Mackey, G.W., The Scope and History of Commutative and Noncommutative Harmonic Analysis, American Mathematical Society, 1992.
61. Maslen, D.K., Fast Transforms and Sampling for Compact Groups, Ph.D. Dissertation, Dept. of Mathematics, Harvard University, May 1993.
62. Maslen, D.K., Rockmore, D.N., "Generalized FFTs - a Survey of Some Recent Results," DIMACS Series in Discrete Mathematics and Theoretical Computer Science, Vol. 28, pp. 183–237, 1997.
63. Mautner, F.I., "Unitary Representations of Locally Compact Groups II," Annals of Mathematics, 52: 528–556, 1950.
64. Mautner, F.I., "Note on the Fourier Inversion Formula on Groups," Transactions of the American Mathematical Society, 78(2): 371–384, 1955.
65. Miller, W., Symmetry Groups and Their Applications, Academic Press, New York, 1972.
66. Miller, W., Lie Theory and Special Functions, Academic Press, New York, 1968;
67. Nachbin, L., The Haar Integral, Van Nostrand Co. Inc., Princeton, 1965.
68. Naimark, M.A., Linear Representations of the Lorentz Group, Macmillan, New York, 1964.
69. Peter, F., Weyl, H., "Die Vollständigkeit der primitiven Darstellungen einer geschlossenen kontinuierlichen Gruppe," Mathematische Annalen, 97: 735–755, 1927.
70. Robert, A., Introduction to the Representation Theory of Compact and Locally Compact Groups, London Mathematical Society Lecture Note Series 80, Cambridge University Press, 1983.
71. Rockmore, D.N., "Efficient Computation of Fourier Inversion for Finite Groups," Journal of the Association for Computing Machinery, 41(1): 31–66, Jan. 1994.
72. Rockmore, D.N., "Fast Fourier Transforms for Wreath Products," Applied and Computational Harmonic Analysis, 2(3): 279–292, 1995.

73. Roziner, T.D., Karpovsky, M.G., Trachtenberg, E.A., "Fast Fourier Transforms over Finite Groups by Multiprocessor Systems," *IEEE Transactions on Acoustics, Speech, and Signal Processing*, 38(2): 226–240, Feb. 1990.

74. Rudin, W., *Fourier Analysis on Groups*, Intersceince Publications, New York, 1962.

75. Rühl, W., *The Lorentz Group and Harmonic Analysis*, W. A. Benjamin, New York, 1970.

76. Saito, K., "On a Duality for Locally Compact Groups," *Tohoku Mathematical Journal*, 20(3): 355–367, 1968.

77. Schempp, W., *Harmonic Analysis on the Heisenberg Nilpotent Lie Group, with Applications to Signal Theory*, Pitman Research Notes in Mathematics Series, 147, Longman Scientific & Technical, London, 1986.

78. Schur, I., "Über die Darstellung der endlichen Gruppen durch gebrochene lineare Substitutionen," *J. für Math.*, 127: 20–50, 1904.

79. Segal, I.E., "An Extension of Plancherel's Formula to Separable Unimodular Groups," *Annals of Mathematics*, 52(2): 272–292, 1950.

80. Serre, J.-P., *Linear Representations of Finite Groups*, Springer-Verlag, New York, 1977.

81. Simon, B., *Representations of Finite and Compact Groups*, Graduate Studies in Mathematics, Vol. 10, American Mathematical Society, Providence RI, 1996.

82. Stankovic, R.S., Moraga, C., Astola, J. *Fourier Analysis on Finite Groups with Applications in Signal Processing and System Design*, John Wiley and Sons, New York, 2005.

83. Sugiura, M., *Unitary Representations and Harmonic Analysis*, 2nd ed., North-Holland, Amsterdam, 1990.

84. Stein, E.M., Weiss, G.L., *Introduction to Fourier Analysis on Euclidean Spaces*, Princeton University Press, 1971.

85. Taylor, M.E., *Noncommutative Harmonic Analysis*, Mathematical Surveys and Monographs, American Mathematical Society, Providence, RI, 1986.

86. Terras, A., *Harmonic Analysis on Symmetric Spaces and Applications I+II*, Springer-Verlag, New York, 1985/1988.

87. Terras, A., *Fourier Analysis on Finite Groups and Applications*, London Mathematical Society Student Texts 43, Cambridge University Press, 1999.

88. Terras, A., *Harmonic Analysis on Symmetric Spaces - Euclidean Space, the Sphere, and the Poincaré Upper Half-Plane*, Springer, 2nd ed., 2013.

89. Thangavelu, S., *Harmonic Analysis on the Heisenberg Group*, Birkhäuser, Boston, 1998.

90. Trachtenberg, E.A., Karpovsky, M.G., "Filtering in a Communication Channel by Fourier Transforms over Finite Groups," in *Spectral Techniques and Fault Detection* (M.G. Karpovsky, ed.), Academic Press, 1985.

91. Varadarajan, V.S., *Lie Groups, Lie Algebras, and Their Representations*, Springer-Verlag, New York 1984.

92. Varadarajan, V.S., *An Introduction to Harmonic Analysis on Semisimple Lie Groups*, Cambridge University Press, New York, 1989.

93. Vilenkin, N.J., Klimyk, A.U., *Representation of Lie Groups and Special Functions*, Vols. 1–3, Kluwer Academic Publ., Dordrecht, Holland 1991.

94. Wallach, N.R., *Harmonic Analysis on Homogeneous Spaces*, Marcel Dekker, Inc., New York, 1973.

95. Warner, G., *Harmonic Analysis on Semi-Simple Lie Groups I + II*, Springer-Verlag, Berlin, 1972.

96. Wawrzyńczyk, A., *Group Representations and Special Functions*, Mathematics and Its Applications (East European Series), D. Reidel Publishing Company, Dordrecht, Holland, 1984.

97. Weyl, H., "Harmonics on Homogeneous Manifolds," *Annals of Mathematics*, 35: 486–499, 1934.

98. Wigner, E., "On Unitary Representations of the Inhomogeneous Lorentz Group," *Annals of Mathematics*, 40(1): 149–204, Jan. 1939.

99. Williams, F.L., *Lectures on the Sprectum of $\mathcal{L}^2(\Gamma \backslash G)$*, Pitman Research Notes in Mathematics Series, 242, Longman Scientific & Technical, London, 1991.

100. Yazici, B., "Stochastic Deconvolution over Groups, *IEEE Transactions on Information Theory*, 50(3): 494 - 510, March 2004.

101. Želobenko, D.P., *Compact Lie Groups and Their Representations*, Translations of Mathematical Monographs, American Mathematical Society, Providence, RI, 1973.

Representation Theory and Operational Calculus for $SU(2)$ and $SO(3)$

In many of the chapters that follow, $SO(3)$ representations will be essential in applications. While the range of applications that we will explore is broad (covering topics from solid mechanics to rotational Brownian motion and satellite attitude estimation), we do not cover every possible application. For instance, group representations are well known as a descriptive tool in theoretical physics. In this chapter we briefly explain why this is so, but since the intent of this work is to inform a broader audience, we do not pursue this in depth. We mention also in passing that the representation theory of $SO(3)$ has been used in applications not covered in this book. One notable area is in the design of stereo sound systems [13].

The current chapter presents a coordinate-dependent introduction to the representation theory of the groups $SU(2)$ and $SO(3)$. The emphasis is on explicit calculation of matrix elements and how these quantities transform under certain differential operators.

9.1 Representations of $SU(2)$ and $SO(3)$

9.1.1 Irreducible Representations from Homogeneous Polynomials

Several methods are available for generating a complete set of irreducible unitary representations of $SO(3)$. See, for example, the classic references [11, 25, 40].

Perhaps the most straightforward computational technique is based on the recognition that the space of *analytic functions* (i.e., functions that can be expanded and approximated to arbitrary precision with a Taylor series) of the form $f(\mathbf{z}) = \tilde{f}(z_1, z_2)$ for all $\mathbf{z} \in \mathbb{C}^2$ can be decomposed into subspaces of *homogeneous polynomials* on which each of the IURs of $SU(2)$ acts [10, 25, 39, 40]. Since there is a two-to-one homomorphism from $SU(2)$ onto $SO(3)$, representations of $SO(3)$ are a subset of those for $SU(2)$.

A homogeneous polynomial of degree m is one of the form

$$P_m(z_1, z_2) = \sum_{k=0}^{m} c_k z_1^k z_2^{m-k}.$$

Here z_i^k is z_i to the power k. The set of all $P_m(z_1, z_2)$ is a subspace of the space of all analytic functions of the form $\tilde{f}(z_1, z_2)$. An inner product on this complex vector space can be defined as

$$\left(\sum_{k=0}^{m} a_k z_1^k z_2^{m-k}, \sum_{k=0}^{m} b_k z_1^k z_2^{m-k}\right) \doteq \sum_{k=0}^{m} k!(m-k)! a_k \overline{b_k}. \tag{9.1}$$

With respect to this inner product, the following are elements of an orthonormal basis for the set of all $P_m(z_1, z_2)$:

$$e_j^m(\mathbf{z}) = \tilde{e}_j^m(z_1, z_2) = \frac{z_1^{j+m} z_2^{j-m}}{[(j+m)!(j-m)!]^{1/2}}. \tag{9.2}$$

Here the range of indices is $j = 0, 1/2, 1, 3/2, 2, \dots$ and the superscripts $m \in \{-j, -j+1, \dots, j-1, j\}$ are half-integer values with unit increment. Substituting (9.2) into (9.1) verifies that

$$(e_j^m, e_k^m) = \delta_{jk}.$$

The matrix elements of IURs for $SU(2)$ are found explicitly as the coefficients in the expansion

$$e_j^n(A^{-1}\mathbf{z}) = \sum_{m=-j}^{j} e_j^m(\mathbf{z}) U_{mn}^j(A). \tag{9.3}$$

By defining the $(2j+1)$-dimensional column vector

$$\mathbf{e}_j(\mathbf{z}) \doteq [e_j^{-j}(\mathbf{z}), e_j^{-j+1}(\mathbf{z}), \dots, e_j^{j-1}(\mathbf{z}), e_j^{j}(\mathbf{z})]^T$$

then (9.3) can be written in the matrix forms

$$[\mathbf{e}_j(A^{-1}\mathbf{z})]^T = [\mathbf{e}_j(\mathbf{z})]^T U^j(A) \iff \mathbf{e}_j(A^{-1}\mathbf{z}) = [U^j(A)]^T \mathbf{e}_j(\mathbf{z}) \tag{9.4}$$

where, as usual, T denotes the transpose and $U^j(A) = [U_{mn}^j(A)]$ is a $(2j+1) \times (2j+1)$ matrix.

From (9.3) we see that, on the one hand, if $A \to A_1 A_2$ and $m \to p$ then

$$e_j^n((A_1 A_2)^{-1}\mathbf{z}) = \sum_{p=-j}^{j} e_j^p(\mathbf{z}) U_{pn}^j(A_1 A_2),$$

and on the other hand using (9.3) twice gives

$$e_j^n((A_1 A_2)^{-1}\mathbf{z}) = e_j^n(A_2^{-1}(A_1^{-1}\mathbf{z})) = \sum_{m=-j}^{j} e_j^m(A_1^{-1}\mathbf{z}) U_{mn}^j(A_2)$$

$$= \sum_{m=-j}^{j} \left(\sum_{p=-j}^{j} e_j^p(\mathbf{z}) U_{pm}^j(A_1)\right) U_{mn}^j(A_2)$$

$$= \sum_{p=-j}^{j} e_j^p(\mathbf{z}) \left(\sum_{m=-j}^{j} U_{pm}^j(A_1) U_{mn}^j(A_2)\right).$$

Equating the above expressions for $e_j^n((A_1 A_2)^{-1}\mathbf{z})$ and localizing shows that U^j has the key property of a matrix representation that

$$U_{pn}^j(A_1 A_2) = \sum_{m=-j}^{j} U_{pm}^j(A_1) U_{mn}^j(A_2) \iff U^j(A_1 A_2) = U^j(A_1) U^j(A_2). \tag{9.5}$$

Using the inner product (9.1) we write

$$U_{kn}^j(A) = (e_j^k(\mathbf{z}), e_j^n(A^{-1}\mathbf{z})).$$

Here we use the notation that the matrices $U^j(A)$ have elements $U_{kn}^j(A)$. In the notation of Chapter 8, the corresponding quantities would be respectively labeled $U(g,j)$ and $U_{kn}(g,j)$.

Discussions of the irreducibility and unitarity of these representations can be found in [40, 43]. Taking this for granted, we can use (9.3) as a way to enumerate the IURs of $SU(2)$.

We shall not be concerned with proving the irreducibility or unitarity of $U^j(A)$ at this point (which can be done using the tools of Chapter 8), but rather their explicit calculation. This is demonstrated below for $j = 0, 1/2, 1, 3/2, 2$. Taking

$$A = \begin{pmatrix} a & b \\ -\bar{b} & \bar{a} \end{pmatrix} \in SU(2),$$

(and so $a\bar{a} + b\bar{b} = 1$) we have

$$A^{-1} = \begin{pmatrix} \bar{a} & -b \\ \bar{b} & a \end{pmatrix}.$$

Consider (9.2) in the case when $j = 1/2$. Then

$$e_{1/2}^{-1/2}(\mathbf{z}) = z_2 \quad \text{and} \quad e_{1/2}^{1/2}(\mathbf{z}) = z_1.$$

Therefore we write

$$e_{1/2}^{-1/2}(A^{-1}\mathbf{z}) = \bar{b}z_1 + az_2 = ae_{1/2}^{-1/2}(\mathbf{z}) + \bar{b}e_{1/2}^{1/2}(\mathbf{z})$$

and

$$e_{1/2}^{1/2}(A^{-1}\mathbf{z}) = \bar{a}z_1 - bz_2 = -be_{1/2}^{-1/2}(\mathbf{z}) + \bar{a}e_{1/2}^{1/2}(\mathbf{z}).$$

From (9.3) we see that the IUR matrix $U^{1/2}$ is found from the coefficients in the above equation as

$$\begin{pmatrix} e_{1/2}^{-1/2}(A^{-1}\mathbf{z}) \\ e_{1/2}^{1/2}(A^{-1}\mathbf{z}) \end{pmatrix} = \begin{pmatrix} U_{-1/2,-1/2}^{1/2} & U_{-1/2,1/2}^{1/2} \\ U_{1/2,-1/2}^{1/2} & U_{1/2,1/2}^{1/2} \end{pmatrix}^T \begin{pmatrix} e_{1/2}^{-1/2}(\mathbf{z}) \\ e_{1/2}^{1/2}(\mathbf{z}) \end{pmatrix}.$$

Hence we write

$$U^{1/2}(A) = \begin{pmatrix} a & -b \\ \bar{b} & \bar{a} \end{pmatrix}.$$

This is a 2×2 unitary matrix related to A by similarity as

$$U^{1/2}(A) = T_{1/2}^{-1} A T_{1/2}$$

where

$$T_{1/2} = T_{1/2}^{-1} = \begin{pmatrix} 1 & 0 \\ 0 & -1 \end{pmatrix}.$$

For $j = 1$ we have

$$e_1^{-1}(\mathbf{z}) = z_2^2/\sqrt{2}; \quad e_1^0(\mathbf{z}) = z_1 z_2; \quad e_1^1(\mathbf{z}) = z_1^2/\sqrt{2}.$$

Then

$$e_1^{-1}(A^{-1}\mathbf{z}) = \frac{1}{\sqrt{2}}(\bar{b}z_1 + az_2)^2 = a^2 e_1^{-1}(\mathbf{z}) + \sqrt{2}\,\bar{b}a e_1^0(\mathbf{z}) + \bar{b}^2 e_1^1(\mathbf{z});$$

$$e_1^0(A^{-1}\mathbf{z}) = -\sqrt{2}\,ab e_1^{-1}(\mathbf{z}) + (\bar{a}a - b\bar{b}) e_1^0(\mathbf{z}) + \sqrt{2}\,\bar{a}\bar{b} e_1^1(\mathbf{z});$$

$$e_1^1(A^{-1}\mathbf{z}) = \frac{1}{\sqrt{2}}(\bar{a}z_1 - bz_2)^2 = b^2 e_1^{-1}(\mathbf{z}) - \sqrt{2}\,\bar{a}b e_1^0(\mathbf{z}) + \bar{a}^2 e_1^1(\mathbf{z}).$$

The coefficients are collected together as

$$U^1(A) = \begin{pmatrix} a^2 & -\sqrt{2}\,ab & b^2 \\ \sqrt{2}\,\bar{b}a & \bar{a}a - b\bar{b} & -\sqrt{2}\,\bar{a}b \\ \bar{b}^2 & \sqrt{2}\,\bar{a}\bar{b} & \bar{a}^2 \end{pmatrix}.$$

We note that this matrix can be related by similarity transformation to the rotation matrix $R(A) \in SO(3)$ defined by the Cayley-Klein parameters (i.e., $R(A) = R(a,b)$):

$$U^1(A) = T_1^{-1} R(A) T_1 \tag{9.6}$$

where

$$T_1 = \frac{-1}{\sqrt{2}} \begin{pmatrix} -1 & 0 & 1 \\ i & 0 & i \\ 0 & -\sqrt{2} & 0 \end{pmatrix}$$

is unitary.

The analogous calculations for $j = 3/2$ and $j = 2$ respectively yield

$$U^{3/2}(A) = \begin{pmatrix} a^3 & -\sqrt{3}ba^2 & \sqrt{3}ab^2 & -b^3 \\ \sqrt{3}\,\bar{b}a^2 & a^2\bar{a} - 2b\bar{b}a & b^2\bar{b} - 2\bar{a}ab & \sqrt{3}\bar{a}b^2 \\ \sqrt{3}\,\bar{b}^2 a & 2\bar{b}a\bar{a} - b\bar{b}^2 & \bar{a}^2 a - 2\bar{a}b\bar{b} & -\sqrt{3}\bar{a}^2 b \\ \bar{b}^3 & \sqrt{3}\bar{a}\bar{b}^2 & \sqrt{3}\bar{a}^2\bar{b} & \bar{a}^3 \end{pmatrix}$$

and

$$U^2(A) = \begin{pmatrix} a^4 & -2a^3 b & \sqrt{6}a^2 b^2 & -2b^3 a & b^4 \\ 2\bar{b}a^3 & \bar{a}a^3 - 3b\bar{b}a^2 & \sqrt{6}(-\bar{a}a^2 b + b^2\bar{b}a) & 3\bar{a}ab^2 - b^3\bar{b} & -2\bar{a}ab^3 \\ \sqrt{6}\bar{b}^2 a^2 & \sqrt{6}(\bar{b}\bar{a}a^2 - \bar{b}^2 ba) & \bar{a}^2 a^2 - 4\bar{a}a b\bar{b} + b^2\bar{b}^2 & \sqrt{6}(-\bar{a}^2 ab + \bar{a}b^2\bar{b}) & \sqrt{6}\bar{a}^2 b^2 \\ 2\bar{b}^3 a & 3\bar{b}^2\bar{a}a - b\bar{b}^3 & \sqrt{6}(\bar{a}^2\bar{b}a - \bar{a}b\bar{b}^2) & \bar{a}^3 a - 3\bar{a}^2 b\bar{b} & -2\bar{a}^3 b \\ \bar{b}^4 & 2\bar{b}^3\bar{a} & \sqrt{6}\bar{a}^2\bar{b}^2 & 2\bar{a}^3\bar{b} & \bar{a}^4 \end{pmatrix}.$$

For values of $j \geq 3/2$, it follows from the definition of irreducibility that U^j is not equivalent to any direct sum of 1, A, and $R(A)$ since U^j cannot be block diagonalized.

The IURs presented above are one of an infinite number of choices that are all equivalent under similarity transformation. For instance, we might choose to enumerate basis elements from positive to negative superscripts instead of the other way around. Writing,

$$\begin{pmatrix} e_{1/2}^{1/2}(A^{-1}\mathbf{z}) \\ e_{1/2}^{-1/2}(A^{-1}\mathbf{z}) \end{pmatrix} = \begin{pmatrix} U_{1/2,1/2}^{1/2} & U_{1/2,-1/2}^{1/2} \\ U_{-1/2,1/2}^{1/2} & U_{-1/2,-1/2}^{1/2} \end{pmatrix}^T \begin{pmatrix} e_{1/2}^{1/2}(\mathbf{z}) \\ e_{1/2}^{-1/2}(\mathbf{z}) \end{pmatrix}$$

results in the an IUR equivalent to $U^{1/2}$:

$$S_{1/2}\, U^{1/2}\, S_{1/2}^{-1} = \begin{pmatrix} \bar{a} & \bar{b} \\ -b & a \end{pmatrix}$$

where $S_{1/2}$ is an invertible 2×2 matrix. Analogously for $j = 1$ we can write

$$S_1\, U^1\, S_1^{-1} = \begin{pmatrix} \bar{a}^2 & \sqrt{2}\bar{a}\bar{b} & \bar{b}^2 \\ -\sqrt{2}\bar{a}b & \bar{a}a - b\bar{b} & \sqrt{2}\bar{b}a \\ b^2 & -\sqrt{2}ab & a^2 \end{pmatrix}$$

where S_1 is an invertible 3×3 matrix. The same can be done for all U^j where S_j is the matrix which has the effect of reflecting both rows and columns of U^j about their centers.

Finally we note that the matrix elements U^l_{mn} can be calculated in this parameterization as [43]:

$$U^l_{mn}(A) = \frac{1}{2\pi} \left[\frac{(l-m)!(l+m)!}{(l-n)!(l+n)!} \right]^{\frac{1}{2}} \int_0^{2\pi} (ae^{i\phi} + \bar{b})^{l-n}(-be^{i\phi} + \bar{a})^{l+n} e^{i(m-l)\phi} \, d\phi. \quad (9.7)$$

9.1.2 The Adjoint and Tensor Representations of $SO(3)$

Let $X \in so(3)$ be an arbitrary 3×3 skew-symmetric matrix. Then for $R \in SO(3)$,

$$Ad(R)X = RXR^T$$

is the adjoint representation of $SO(3)$ with the property

$$Ad(R_1 R_2) = Ad(R_1)Ad(R_2). \quad (9.8)$$

In Chapter 5 we saw the rule

$$(RXR^T)^\vee = R(X^\vee).$$

It follows that the matrix of the adjoint representation is

$$[Ad(R)] = R.$$

From (9.6) we therefore observe that $Ad(R)$ is equivalent (under similarity transformation) to $U^1(R)$.

The homomorphism property (9.8) does not depend on X being in $so(3)$. We could have chosen $X \in \mathbb{R}^{3\times 3}$ arbitrarily, and the homomorphism property would still hold. In this more general context,

$$T(R)X = RXR^T$$

is called a *tensor representation* of $SO(3)$. In this context, let $(X)^\vee \in \mathbb{R}^9$ be the vector of independent numbers formed by sequentially stacking the columns of the matrix X. Then we can define a matrix $[T(R)]$ of the tensor representation $T(R)$ as

$$[T(R)](X)^\vee = (RXR^T)^\vee.$$

$[T(R)]$ is then a 9×9 matrix representation of $SO(3)$. As we shall see shortly, $[T(R)]$ (and hence $T(R)$) is a reducible representation.

It is well known that any matrix can be decomposed into a skew-symmetric part and a symmetric part. The symmetric part can further be decomposed into a part that is a constant multiple of the identity and a part that is traceless. Hence for arbitrary $X \in \mathbb{R}^{3 \times 3}$, we write

$$X = X_{skew} + X_{sym1} + X_{sym2}$$

where

$$X_{sym1} \doteq \frac{1}{3} \text{tr}(X) \mathbb{I}; \quad X_{skew} \doteq \frac{1}{2}(X - X^T); \quad X_{sym2} \doteq \frac{1}{2}(X + X^T) - \frac{1}{3} \text{tr}(X) \mathbb{I}.$$

Let the set of all matrices of the form X_{sym1} be called V_1, the set of all X_{skew} be called V_2 (which is the same as $so(3)$), and the set of all X_{sym2} be called V_3. Each of these spaces is invariant under similarity transformation by a rotation matrix. That is, if $Y \in V_i$ then $RYR^T \in V_i$. This is the very definition of reducibility of the representation $T(R)$ on the space $\mathbb{R}^{3 \times 3}$. We therefore write

$$\mathbb{R}^{3 \times 3} = V_1 \oplus V_2 \oplus V_3.$$

We can then define representations on each of these subspaces. Since $T_1(R)X_{sym1} = RX_{sym1}R^T = X_{sym1}$, it is clear that this is equivalent to the representation $U^0(R) = 1$. Since $X_{skew} \in so(3)$, it follows that the representation $T_2(R)$ acting on elements of V_2 is the same as the adjoint representation, which we already know is equivalent to $U^1(R)$. Finally, we note that $T_3(R)X_{sym2} = RX_{sym2}R^T$ is a five-dimensional representation. This can be shown to be irreducible and hence equivalent to $U^2(R)$. We therefore write

$$T \cong U^0 \oplus U^1 \oplus U^2.$$

Another way to say this is that there exists a matrix $S \in GL(9, \mathbb{C})$ such that the 9×9 matrix $[T(R)]$ is block-diagonalized as

$$[T(R)] = S^{-1} \begin{pmatrix} U^0(R) & \mathbb{O}_{1 \times 3} & \mathbb{O}_{1 \times 5} \\ \mathbb{O}_{3 \times 1} & U^1(R) & \mathbb{O}_{3 \times 5} \\ \mathbb{O}_{5 \times 1} & \mathbb{O}_{5 \times 3} & U^2(R) \end{pmatrix} S.$$

Explicitly, the matrix $[T(R)]$ is the *tensor (or Kronecker) product* of R with itself. That is,

$$[T(R)] = R \otimes R \tag{9.9}$$

where $A \otimes B$ is defined for any two matrices

$$A = \begin{pmatrix} a_{11} & a_{12} & \cdots & a_{1q} \\ a_{21} & a_{22} & \cdots & \vdots \\ \vdots & \vdots & \ddots & \vdots \\ a_{p1} & a_{p2} & \cdots & a_{pq} \end{pmatrix} \in \mathbb{R}^{p \times q}$$

and

$$B = \begin{pmatrix} b_{11} & b_{12} & \cdots & b_{1s} \\ b_{21} & b_{22} & \cdots & \vdots \\ \vdots & \vdots & \ddots & \vdots \\ b_{r1} & b_{r2} & \cdots & b_{rs} \end{pmatrix} \in \mathbb{R}^{r \times s}$$

as

$$A \otimes B = \begin{pmatrix} a_{11}B & a_{12}B & \cdots & a_{1q}B \\ a_{21}B & a_{22}B & \cdots & \vdots \\ \vdots & \vdots & \ddots & \vdots \\ a_{p1}B & a_{p2}B & \cdots & a_{pq}B \end{pmatrix} \in \mathbb{R}^{pr \times qs}.$$

From this definition, it is clear that if $C \in \mathbb{R}^{q \times l}$ and $D \in \mathbb{R}^{s \times t}$ for some l and t, then

$$(A \otimes B)(C \otimes D) = \begin{pmatrix} a_{11}B & a_{12}B & \cdots & a_{1q}B \\ a_{21}B & a_{22}B & \cdots & \vdots \\ \vdots & \vdots & \ddots & \vdots \\ a_{p1}B & a_{p2}B & \cdots & a_{pq}B \end{pmatrix} \begin{pmatrix} c_{11}D & c_{12}D & \cdots & c_{1l}D \\ c_{21}D & c_{22}D & \cdots & \vdots \\ \vdots & \vdots & \ddots & \vdots \\ c_{q1}D & c_{q2}D & \cdots & c_{ql}D \end{pmatrix}$$
$$= (AC) \otimes (BD). \tag{9.10}$$

If A and B are square, then

$$\mathrm{tr}(A \otimes B) = \mathrm{tr}(A) \cdot \mathrm{tr}(B)$$

and

$$\det(A \otimes B) = (\det A)^{dimB} \cdot (\det B)^{dimA} = \det(B \otimes A).$$

Note that this is in contrast to

$$\mathrm{tr}(A \oplus B) = \mathrm{tr}(A) + \mathrm{tr}(B)$$

and

$$\det(A \oplus B) = (\det A) \cdot (\det B)$$

for the direct sum.

When A, B, and X are square and of the same dimension, the tensor product plays an important role because

$$Y = AXB^T$$

is converted as[1]

$$Y^\vee = (B \otimes A)X^\vee, \tag{9.11}$$

or equivalently

$$(Y^T)^\vee = (A \otimes B)(X^T)^\vee.$$

Hence, (9.9) is a special case of (9.11).

We can generalize the concept of a tensor representation further, but postpone discussion of this until Chapter 18 where applications of the representation theory of $SO(3)$ in mechanics are discussed.

In the remainder of this chapter we examine the form of the representation matrices in various parameterizations and review harmonic analysis on $SU(2)$ and $SO(3)$ as concrete examples of the general theory presented in the previous chapter.

---

[1]In contrast to elsewhere in the book, in the context of Kroncker products, $\vee$ denotes the vector operation that converts a matrix into a long column vector by stacking columns of the matrix.

9.1.3 Irreducibility of the Representations $U_i(g)$ of $SU(2)$

We now demonstrate the irreducibility of the matrix representations $U_i(g)$ using three techniques: (1) By checking that the characters satisfy (8.60); (2) By observing that the infinitesimal matrices

$$u(X_i, l) = \frac{d}{dt} U^l(\exp t X_i)\Big|_{t=0}$$

are not all block-diagonalizable by the same matrix under similarity transformation; and (3) By observing how the operator corresponding to each $U_i(g)$ acts on the space of homogeneous polynomials.

Method 1: When $SU(2)$ is parameterized using Euler angles, the character functions $\chi_l(g) = \text{tr}(U^l(g))$ take the form (see (9.23) or [43] p. 359)

$$\chi_l(g) = \sum_{m=-l}^{l} e^{-im(\alpha+\gamma)} P^l_{mm}(\cos\beta) = \frac{\sin(l+\frac{1}{2})\theta}{\sin\frac{\theta}{2}},$$

where θ is the angle of rotation. Note that the range of θ can be taken as $[-\pi, \pi]$ or $[0, 2\pi]$. When $[-\pi, \pi]$ is selected, the class functions for $SU(2)$ are even functions of θ.

In (θ, λ, ν) coordinates, the normalized integral over $SU(2)$ is

$$\int_{SU(2)} f(g)\, dg = \frac{1}{4\pi^2} \cdot \int_0^{2\pi} \int_0^{\pi} \int_0^{2\pi} f(\theta, \lambda, \nu) \sin^2\frac{\theta}{2} \sin\nu\, d\lambda\, d\nu\, d\theta.$$

Then since for $l_1, l_2 \geq 0$

$$\int_0^{2\pi} \sin(l_1 + 1/2)\theta \, \sin(l_2 + 1/2)\theta\, d\theta = \pi \delta_{l_1, l_2},$$

it follows that

$$\int_{SU(2)} \chi_{l_1}(g) \overline{\chi_{l_2}(g)}\, dg = \delta_{l_1, l_2}.$$

(The same result can be obtained using Euler angles (α, β, γ) and the orthogonalities of $e^{im\phi}$ and the polynomials $P^l_{mn}(\cos\theta)$ in (9.28).)

When $l_1 = l_2$, the condition of Theorem 8.16 is met, and so the representations U^l are all irreducible.

Method 2:

A basis for the Lie algebra $su(2)$ is

$$X_1 = \frac{i}{2}\begin{pmatrix} 0 & 1 \\ 1 & 0 \end{pmatrix}; \quad X_2 = \frac{1}{2}\begin{pmatrix} 0 & -1 \\ 1 & 0 \end{pmatrix}; \quad X_3 = \frac{i}{2}\begin{pmatrix} 1 & 0 \\ 0 & -1 \end{pmatrix}.$$

The corresponding one-parameter subgroups are

$$g_1(t) = \exp(t X_1) = \begin{pmatrix} \cos\frac{t}{2} & i\sin\frac{t}{2} \\ i\sin\frac{t}{2} & \cos\frac{t}{2} \end{pmatrix}; \tag{9.12}$$

$$g_2(t) = \exp(t X_2) = \begin{pmatrix} \cos\frac{t}{2} & -\sin\frac{t}{2} \\ \sin\frac{t}{2} & \cos\frac{t}{2} \end{pmatrix}; \tag{9.13}$$

$$g_3(t) = \exp(tX_3) = \begin{pmatrix} e^{i\frac{t}{2}} & 0 \\ 0 & e^{-i\frac{t}{2}} \end{pmatrix}. \tag{9.14}$$

The matrices

$$u(X_i, l) = \left.\frac{dU^l(\exp(tX_i))}{dt}\right|_{t=0}$$

corresponding to infinitesimal motion for this case for $i = 1$ look like

$$u(X_1, \tfrac{1}{2}) = \begin{pmatrix} 0 & -\frac{i}{2} \\ -\frac{i}{2} & 0 \end{pmatrix};$$

$$u(X_1, 1) = \begin{pmatrix} 0 & -\frac{i}{\sqrt{2}} & 0 \\ -\frac{i}{\sqrt{2}} & 0 & -\frac{i}{\sqrt{2}} \\ 0 & -\frac{i}{\sqrt{2}} & 0 \end{pmatrix};$$

$$u(X_1, \tfrac{3}{2}) = \begin{pmatrix} 0 & -\frac{i\sqrt{3}}{2} & 0 & 0 \\ -\frac{i\sqrt{3}}{2} & 0 & -i & 0 \\ 0 & -i & 0 & -\frac{i\sqrt{3}}{2} \\ 0 & 0 & -\frac{i\sqrt{3}}{2} & 0 \end{pmatrix}.$$

For $i = 2$,

$$u(X_2, \tfrac{1}{2}) = \begin{pmatrix} 0 & \frac{1}{2} \\ -\frac{1}{2} & 0 \end{pmatrix};$$

$$u(X_2, 1) = \begin{pmatrix} 0 & \frac{1}{\sqrt{2}} & 0 \\ -\frac{1}{\sqrt{2}} & 0 & \frac{1}{\sqrt{2}} \\ 0 & -\frac{1}{\sqrt{2}} & 0 \end{pmatrix};$$

$$u(X_2, \tfrac{3}{2}) = \begin{pmatrix} 0 & \frac{\sqrt{3}}{2} & 0 & 0 \\ -\frac{\sqrt{3}}{2} & 0 & 1 & 0 \\ 0 & -1 & 0 & \frac{\sqrt{3}}{2} \\ 0 & 0 & -\frac{\sqrt{3}}{2} & 0 \end{pmatrix}.$$

For $i = 3$,

$$u(X_3, \tfrac{1}{2}) = \begin{pmatrix} \frac{i}{2} & 0 \\ 0 & -\frac{i}{2} \end{pmatrix};$$

$$u(X_3, 1) = \begin{pmatrix} i & 0 & 0 \\ 0 & 0 & 0 \\ 0 & 0 & -i \end{pmatrix};$$

$$u(X_3, \tfrac{3}{2}) = \begin{pmatrix} \frac{3i}{2} & 0 & 0 & 0 \\ 0 & \frac{i}{2} & 0 & 0 \\ 0 & 0 & -\frac{i}{2} & 0 \\ 0 & 0 & 0 & -\frac{3i}{2} \end{pmatrix}.$$

It can be verified that for all i=1, 2, 3, the above matrices (as well as those for all values of l) are not simultaneously block-diagonalizable [11].

Method 3:

For homogeneous polynomials of the form

$$f(z_1, z_2) = \sum_{n=-l}^{l} a_n z_1^{l-n} z_2^{l+n},$$

we observe that

$$f(z_1, z_2) = z_2^{2l} f(z_1/z_2, 1)$$

for $z_2 \neq 0$. It is convenient to define $\varphi(x) = f(x, 1)$. Hence, the operator $U^l(g)$ acting on the space of all homogeneous polynomials of degree l whose matrix elements are defined in (9.3) can be rewritten in the form [43]

$$\hat{U}^l(g)\varphi(z) = (\beta z + \overline{\alpha})^{2l} \varphi \left(\frac{\alpha z - \overline{\beta}}{\beta z + \overline{\alpha}} \right)$$

(with l taking the place of j, $\alpha = \overline{a}$ and $\beta = \overline{b}$). Evaluated at the one-parameter subgroups, this becomes

$$\hat{U}^l(\exp(tX_1))\varphi(x) = (ix \sin \frac{t}{2} + \cos \frac{t}{2})^{2l} \varphi \left(\frac{x \cos \frac{t}{2} + i \sin \frac{t}{2}}{ix \sin \frac{t}{2} + \cos \frac{t}{2}} \right);$$

$$\hat{U}^l(\exp(tX_2))\varphi(x) = (x \sin \frac{t}{2} + \cos \frac{t}{2})^{2l} \varphi \left(\frac{x \cos \frac{t}{2} - \sin \frac{t}{2}}{x \sin \frac{t}{2} + \cos \frac{t}{2}} \right);$$

$$\hat{U}^l(\exp(tX_3))\varphi(x) = e^{-itl} \varphi(xe^{it}).$$

Differential operators based on these representation operators are defined as

$$\hat{u}_i^l \doteq \frac{d\hat{U}^l(\exp(tX_i))}{dt} \bigg|_{t=0}.$$

Note that $\hat{u}_i^l$ are differential operators corresponding to the matrices $u(X_i, l)$. Explicitly we find

$$\hat{u}_1^l = ilx + \frac{i}{2}(1 - x^2)\frac{d}{dx};$$

$$\hat{u}_2^l = -lx + \frac{1}{2}(1 + x^2)\frac{d}{dx};$$

$$\hat{u}_3^l = i(x\frac{d}{dx} - l).$$

These operators act on the space of all polynomials of the form

$$\varphi(x) = \sum_{k=-l}^{l} a_k x^{l-k}$$

which can be equipped with an inner product $(\cdot, \cdot)$ defined such that

$$(\varphi, x^{l-k}) \doteq (l - k)!(l + k)!a_k.$$

Orthonormal basis elements for this space of polynomials are taken as

$$e_k^l(x) = \frac{x^{l-k}}{\sqrt{(l - k)!(l + k)!}}$$

for $-l \leq k \leq l$.

The irreducibility of the representation operators $U^l(g)$ can then be observed from the way the differential operators act on the basis elements. To this end, it is often convenient to use the operators [43]

$$\hat{H}^l_+ = i(\hat{u}^l_1 + i\hat{u}^l_2) = -\frac{d}{dx};$$

$$\hat{H}^l_- = i(\hat{u}^l_1 - i\hat{u}^l_2) = -2lx + x^2\frac{d}{dx};$$

$$\hat{H}^l_3 = i\hat{u}^l_3 = l - x\frac{d}{dx}.$$

We then observe that

$$\hat{H}^l_+ e^l_n(x) = -\sqrt{(l-n)(l+n+1)}e^l_{n+1}(x);$$

$$\hat{H}^l_- e^l_n(x) = -\sqrt{(l+n)(l-n+1)}e^l_{n-1}(x);$$

$$\hat{H}^l_3 e^l_n(x) = ne^l_n(x).$$

While subspaces of functions that consist of scalar multiples of $e^l_n(x)$ are stable under $\hat{H}^l_3$, it is clear that no subspaces (other than zero and the whole space) can be left invariant under $\hat{H}^l_+$, $\hat{H}^l_-$, and $\hat{H}^l_3$ simultaneously since $\hat{H}^l_+$ and $\hat{H}^l_+$ always "push" basis elements to "adjacent" subspaces. Hence the representation operators $\hat{U}(g)$ are irreducible, as must be their corresponding matrices.

9.2 Some Differential Geometry of $\mathbb{S}^3$ and $SU(2)$

The relationship between $SO(3)$, $SU(2)$ and $\mathbb{S}^3$ was established in Chapter 5. Here we calculate some geometric quantities using several different methods and show how they reduce to the same thing.

9.2.1 The Metric Tensor

The metric tensor for the group $SU(2)$ can be found using the parameterization of the rotation matrix in terms of Euler angles and in terms of coordinates of the unit sphere in $\mathbb{R}^4$ (see, for example, [11, 43]). For parameterization of the Euler angles $0 \leq \alpha < 2\pi$, $0 \leq \beta < \pi$, $0 \leq \gamma < 2\pi$ this tensor corresponds to the metric tensor on $SO(3)$.

The covariant metric tensor is

$$G = \begin{bmatrix} 1 & 0 & 0 \\ 0 & 1 & \cos\beta \\ 0 & \cos\beta & 1 \end{bmatrix} \tag{9.15}$$

where the element G_{11} corresponds to the $G_{\beta\beta}$ element. From a group-theoretic perspective, we recognize that

$$J^T_L J_L = J^T_R J_R = G.$$

From a purely geometric perspective, the vector

$$\mathbf{x}(\alpha, \beta, \gamma) = \begin{pmatrix} \cos\frac{\beta}{2}\cos\left(\frac{\alpha+\gamma}{2}\right) \\ \cos\frac{\beta}{2}\sin\left(\frac{\alpha+\gamma}{2}\right) \\ -\sin\frac{\beta}{2}\sin\left(\frac{\alpha-\gamma}{2}\right) \\ \sin\frac{\beta}{2}\cos\left(\frac{\alpha-\gamma}{2}\right) \end{pmatrix}$$

defines points on the unit sphere in $\mathbb{R}^4$. The Jacobian

$$J = \left[\frac{\partial\mathbf{x}}{\partial\beta}; \frac{\partial\mathbf{x}}{\partial\alpha}; \frac{\partial\mathbf{x}}{\partial\gamma}\right]$$

can be used to define the same metric tensor as

$$G = J^T W J$$

where $W = 4 \cdot \mathbb{I}_{4\times4}$.

Regardless of how G is computed we see that

$$G^{-1} = \begin{bmatrix} 1 & 0 & 0 \\ 0 & \frac{1}{\sin^2\beta} & -\frac{\cos\beta}{\sin^2\beta} \\ 0 & -\frac{\cos\beta}{\sin^2\beta} & \frac{1}{\sin^2\beta} \end{bmatrix}, \tag{9.16}$$

and for each differentiable $f(A)$ on $SO(3)$ we can define

$$\nabla_A f \doteq \left(\frac{\partial f}{\partial\beta}, \frac{\partial f}{\partial\alpha}, \frac{\partial f}{\partial\gamma}\right)^T.$$

This is *not* the differential-geometric gradient defined in Chapter 4, which in this notation would be

$$\mathrm{grad} f = \sum_{j=1}^{3} g^{ij}(\nabla_A f)_j \mathbf{v}_j$$

where

$$\mathbf{v}_j = \frac{\partial\mathbf{x}}{\partial\alpha_i}$$

with $\alpha_1 = \beta$, $\alpha_2 = \alpha$ and $\alpha_3 = \gamma$. The corresponding unit vectors are $\mathbf{e}_{\alpha_j}$.

It is easy to see that (for a real-valued function $f(A)$)

$$\int_{SO(3)} (\nabla_A f)^T G^{-1} \nabla_A f \, d(A) =$$

$$\frac{1}{8\pi^2}\int_{\gamma=0}^{2\pi}\int_{\beta=0}^{\pi}\int_{\alpha=0}^{2\pi}\left(\frac{\partial f}{\partial\beta}\frac{\partial f}{\partial\beta} + \frac{1}{\sin^2\beta}\frac{\partial f}{\partial\alpha}\frac{\partial f}{\partial\alpha} + \frac{1}{\sin^2\beta}\frac{\partial f}{\partial\gamma}\frac{\partial f}{\partial\gamma} - \frac{2\cos\beta}{\sin^2\beta}\frac{\partial f}{\partial\alpha}\frac{\partial f}{\partial\gamma}\right)\sin\beta \, d\alpha \, d\beta \, d\gamma.$$

It can be verified using integration by parts that

$$\int_{SO(3)} (\nabla_A f)^T G^{-1} \nabla_A f \, d(A) = \int_{SO(3)} (f, -\nabla_A^2 f) \, d(A) \tag{9.17}$$

where the Laplacian $\nabla_A^2 = \mathrm{div}(\mathrm{grad} f)$ is given in (9.22). We note that we can write a gradient vector in the unit vector notations as

$$\text{grad}(f) = \frac{\partial f}{\partial \beta}\mathbf{e}_\beta + \frac{1}{\sin \beta}\frac{\partial f}{\partial \alpha}\mathbf{e}_\alpha + \frac{1}{\sin \beta}\frac{\partial f}{\partial \gamma}\mathbf{e}_\gamma \ ,$$

where the unit vectors $\mathbf{e}_\beta$, $\mathbf{e}_\alpha$, $\mathbf{e}_\gamma$ satisfy the inner product relations

$$(\mathbf{e}_\beta, \mathbf{e}_\alpha) = (\mathbf{e}_\beta, \mathbf{e}_\gamma) = 0$$

$$(\mathbf{e}_\alpha, \mathbf{e}_\gamma) = (\mathbf{e}_\gamma, \mathbf{e}_\alpha) = \cos \beta \ .$$

In this notation, the relation (9.17) is written as

$$\int_{SO(3)} \|\text{grad}(f)\|^2 \, d(A) = \int_{SO(3)} (f, -\nabla_A^2 f) \, d(A).$$

We mention also a definition of the Dirac δ-function for the rotation group as a special case of (E.4):

$$\delta(A, A') = C \cdot \begin{cases} \frac{1}{\sin \beta} \delta(\alpha - \alpha')\,\delta(\beta - \beta')\,\delta(\gamma - \gamma') & \text{for } \sin \beta' \neq 0 \\[2ex] \frac{\delta(\beta - \beta')}{4\pi^2 \sin \beta} & \text{for } \sin \beta' = 0 \end{cases} \qquad (9.18)$$

where $C = 8\pi^2$ for the normalized integral and $C = 1$ for the unnormalized one.

The δ-function has the properties that

$$\int_{SO(3)} f(A)\,\delta(A, A')\,d(A) = f(A') \qquad (9.19)$$

and

$$\int_{SO(3)} \delta(A, A')\,d(A) = 1$$

The above definition is valid for either $SO(3)$ or a hemisphere in $\mathbb{R}^4$. Using the group operation of $SO(3)$, the Dirac delta function of two arguments is replaced with one of a single argument:

$$\delta(A, A') = \delta((A')^T A).$$

9.2.2 Differential Operators and Laplacian for $SO(3)$ in Spherical Coordinates

Spherical coordinates in $\mathbb{R}^4$ can be chosen as

$$\mathbf{r}(\theta_1, \theta_2, \theta_3) = \begin{bmatrix} r_1 \\ r_2 \\ r_3 \\ r_4 \end{bmatrix} = \begin{bmatrix} \cos \theta_1 \\ \sin \theta_1 \cos \theta_2 \\ \sin \theta_1 \sin \theta_2 \cos \theta_3 \\ \sin \theta_1 \sin \theta_2 \sin \theta_3 \end{bmatrix}$$

for $\theta_1 \in [0, \pi]$, $\theta_2 \in [0, \pi]$, $\theta_3 \in [0, 2\pi]$.

Using the four-dimensional matrix representation of rotation corresponding to $\mathbf{r}$ as in Chapter 5, i.e., by defining $R_{4\times4}(\mathbf{r})$ such that

$$R_{4\times4}(\mathbf{r})q = rq$$

(where r is the unit quaternion corresponding to $\mathbf{r} \in \mathbb{R}^4$ and rq is quaternion multiplication of r and q), we have:

$$R_{4\times4}(\mathbf{r}) = \begin{bmatrix} r_1 & -r_2 & -r_3 & -r_4 \\ r_2 & r_1 & -r_4 & r_3 \\ r_3 & r_4 & r_1 & -r_2 \\ r_4 & -r_3 & r_2 & r_1 \end{bmatrix}.$$

Hence, $R_{4\times4}(\mathbf{e}_1)$ is the identity matrix and corresponds to no rotation. If we consider infinitesimal motions on the sphere $\mathbb{S}^3$ in each of the remaining coordinate directions in the vicinity of the point $\mathbf{r} = \mathbf{e}_1$, the corresponding rotations will be of the form

$$R_{4\times4}(\mathbf{e}_1 + \epsilon\mathbf{e}_i) = \mathbb{I}_{4\times4} + \epsilon X_{i-1}$$

for $i = 2, 3, 4$ and $|\epsilon| \ll 1$. It is easy to see that $X_{i-1} = R_{4\times4}(\mathbf{e}_{i-1})$ and explicitly

$$X_1 = \begin{bmatrix} 0 & -1 & 0 & 0 \\ 1 & 0 & 0 & 0 \\ 0 & 0 & 0 & -1 \\ 0 & 0 & 1 & 0 \end{bmatrix}; \quad X_2 = \begin{bmatrix} 0 & 0 & -1 & 0 \\ 0 & 0 & 0 & 1 \\ 1 & 0 & 0 & 0 \\ 0 & -1 & 0 & 0 \end{bmatrix}; \quad X_3 = \begin{bmatrix} 0 & 0 & 0 & -1 \\ 0 & 0 & -1 & 0 \\ 0 & 1 & 0 & 0 \\ 1 & 0 & 0 & 0 \end{bmatrix}.$$

$$(9.20)$$

In analogy with the way the cross product in $\mathbb{R}^3$ takes unit basis vectors into unit basis vectors as $\mathbf{e}_1 \times \mathbf{e}_2 = \mathbf{e}_3$, $\mathbf{e}_2 \times \mathbf{e}_3 = \mathbf{e}_1$, $\mathbf{e}_3 \times \mathbf{e}_1 = \mathbf{e}_2$, we find that under the Lie bracket operation (which in this case we take to be half of matrix commutator),

$$[X_i, X_j] \doteq \frac{1}{2}(X_i X_j - X_j X_i),$$

that

$$[X_1, X_2] = X_3; \quad [X_2, X_3] = X_1; \quad [X_3, X_1] = X_2.$$

This indicates that the matrices X_i are representations of the basis elements of the Lie algebra $so(3)$.

These elements are not unique, as can be seen by the fact that

$$R_{4\times4}(\mathbf{e}_1 + \epsilon R(\mathbf{r})\mathbf{e}_i) = \mathbb{I}_{4\times4} + \epsilon R_{4\times4} X_{i-1} R_{4\times4}^T.$$

Hence any set of basis elements of the form $\tilde{X}_i = R_{4\times4} X_i R_{4\times4}^T$ is equally acceptable.

Differential motions on the surface of the sphere $\mathbb{S}^3$ can be used to define differential operators which act on real valued-functions of the form $f(\mathbf{r}(\theta_1, \theta_2, \theta_3))$ as:

$$D_i f = \frac{d}{dt} f(\exp(tX_i)\mathbf{r}(\theta_1, \theta_2, \theta_3))\Big|_{t=0}$$

for $i = 1, 2, 3$. Here

$$\exp(A) = \mathbb{I}_{4\times4} + \sum_{n=1}^{\infty} \frac{A^n}{n!}$$

is the matrix exponential for any $A \in \mathbb{R}^{4\times4}$.

After a little work, it can be shown that explicit form of these operators (in a basis, $\{\tilde{X}_i\}$ different than ours) is [47]

$$\tilde{D}_1 = -\cos\theta_2 \frac{\partial}{\partial\theta_1} + \sin\theta_2 \cot\theta_1 \frac{\partial}{\partial\theta_2} - \frac{\partial}{\partial\theta_3};$$

$$\tilde{D}_2 = -\sin\theta_2 \cos\theta_3 \frac{\partial}{\partial\theta_1} + (\sin\theta_3 - \cot\theta_1 \cos\theta_2 \cos\theta_3) \frac{\partial}{\partial\theta_2}$$

$$+ \left(\cot\theta_1 \frac{\sin\theta_3}{\sin\theta_2} + \cot\theta_2 \cos\theta_3 \right) \frac{\partial}{\partial\theta_3};$$

$$\tilde{D}_3 = -\sin\theta_2 \sin\theta_3 \frac{\partial}{\partial\theta_1} - (\cos\theta_3 + \cot\theta_1 \cos\theta_2 \sin\theta_3) \frac{\partial}{\partial\theta_2}$$

$$+ \left(-\cot\theta_1 \frac{\cos\theta_3}{\sin\theta_2} + \cot\theta_2 \sin\theta_3 \right) \frac{\partial}{\partial\theta_3}.$$

The Laplacian, which is invariant under the choice of $so(3)$ basis, is defined as

$$\nabla^2 \doteq (D_1)^2 + (D_2)^2 + (D_3)^2 = (\tilde{D}_1)^2 + (\tilde{D}_2)^2 + (\tilde{D}_3)^2 =$$

$$\frac{\partial^2}{\partial\theta_1^2} + 2\cot\theta_1 \frac{\partial}{\partial\theta_1} + \frac{\cot\theta_2}{\sin^2\theta_1} \frac{\partial}{\partial\theta_2} + \frac{1}{\sin^2\theta_1} \frac{\partial^2}{\partial\theta_2^2} + \frac{1}{\sin^2\theta_1 \sin^2\theta_2} \frac{\partial^2}{\partial\theta_3^2}. \tag{9.21}$$

9.3 Matrix Elements of $SU(2)$ Representations as Eigenfunctions of the Laplacian

Using the matrices g_k in $(9.12)-(9.14)$, elements of $SU(2)$ are parameterized with ZXZ Euler angles as

$$g(\alpha,\beta,\gamma) = g_3(\alpha)g_1(\beta)g_3(\gamma) = \begin{pmatrix} \cos\frac{\beta}{2} e^{i(\alpha+\gamma)/2} & i\sin\frac{\beta}{2} e^{i(\alpha-\gamma)/2} \\ -i\sin\frac{\beta}{2} e^{-i(\alpha-\gamma)/2} & \cos\frac{\beta}{2} e^{-i(\alpha+\gamma)/2} \end{pmatrix}.$$

Here $0 \le \alpha < 2\pi$, $-2\pi \le \gamma < 2\pi$, and $0 \le \beta \le \pi$. The normalized invariant integration measure written in terms of Euler angles is

$$d(g) = C \sin\beta d\alpha d\beta d\gamma,$$

and the constant C is set so that

$$\int_{SU(2)} d(g) = 1.$$

That is, $C = 1/16\pi^2$.

Identifying elements of $SU(2)$ with points on the sphere $\mathbb{S}^3$, the differential-geometric Laplacian operator is written in the Euler angles as

$$\nabla^2 = \frac{\partial^2}{\partial\beta^2} + \cot\beta \frac{\partial}{\partial\beta} + \frac{1}{\sin^2\beta} \left(\frac{\partial^2}{\partial\alpha^2} - 2\cos\beta \frac{\partial^2}{\partial\alpha\partial\gamma} + \frac{\partial^2}{\partial\gamma^2} \right). \tag{9.22}$$

This is the same as the "group-theoretic" Laplacian

$$\nabla^2 = (X_1^R)^2 + (X_2^R)^2 + (X_3^R)^2 = (X_1^L)^2 + (X_2^L)^2 + (X_3^L)^2.$$

The second equality above follows from the $Ad(g)$-invariance of the inner product on the Lie algebra $su(2)$. The explicit form of the operators X_i^L and X_i^R are given in Section 9.10.

The eigenfunctions $u(\alpha,\beta,\gamma)$ which satisfy

$$\nabla^2 u = \lambda u$$

can be found using the separation of variables $u(\alpha, \beta, \gamma) = u_1(\alpha)u_2(\gamma)u_3(\beta)$ which reduces this PDE to the three Sturm-Liouville problems:

$$u_1'' + m^2 u_1 = 0 \quad u_1(0) = u_1(2\pi) \quad u_1'(0) = u_1'(2\pi)$$

$$u_2'' + n^2 u_2 = 0 \quad u_2(-2\pi) = u_2(2\pi) \quad u_2'(-2\pi) = u_2'(2\pi)$$

$$(\sin \beta u_3')' + \left[l(l+1) \sin \beta - \frac{n^2 - 2mn \cos \beta + m^2}{\sin \beta} \right] u_3 = 0.$$

The above equation in $u_3(\beta)$ is transformed to a singular Sturm-Liouville problem in the variable $x = \cos \beta$, whose solution is written as $P_{mn}^l(x)$. That is,

$$\left[\frac{d}{dx} \left((1 - x^2) \frac{d}{dx} \right) - \frac{m^2 + n^2 - 2mnx}{1 - x^2} + l(l+1) \right] P_{mn}^l = 0.$$

The composite solution $u(\alpha, \beta, \gamma)$ for each set of l, m, n can be written as $U_{mn}^l(g(\alpha, \beta, \gamma))$. These are the matrix elements of the irreducible unitary representations of $SU(2)$ given by

$$U_{mn}^l(g(\alpha, \beta, \gamma)) = i^{m-n} e^{-i(m\alpha + n\gamma)} P_{mn}^l(\cos \beta). \tag{9.23}$$

The functions $P_{mn}^l(\cos \beta)$ are generalizations of the associated Legendre functions and are given by the Rodrigues formula [43]:

$$P_{mn}^l(x) = \frac{(-1)^{l-m}}{2^l} \left[\frac{(l+m)!}{(l-n)!(l+n)!(l-m)!} \right]^{\frac{1}{2}} \times$$

$$(1+x)^{-(m+n)/2}(1-x)^{(n-m)/2} \frac{d^{l-m}}{dx^{l-m}} [(1-x)^{l-n}(1+x)^{l+n}]. \tag{9.24}$$

They can also be calculated by the integral

$$P_{mn}^l(\cos \beta) = \frac{i^{n-m}}{2\pi} \left[\frac{(l-m)!(l+m)!}{(l-n)!(l+n)!} \right]^{\frac{1}{2}} \int_0^{2\pi} \left(\cos \frac{\beta}{2} e^{i\phi/2} + i \sin \frac{\beta}{2} e^{-i\phi/2} \right)^{l-n} \times$$

$$\left(\cos \frac{\beta}{2} e^{-i\phi/2} + i \sin \frac{\beta}{2} e^{i\phi/2} \right)^{l+n} e^{im\phi} \, d\phi, \tag{9.25}$$

or relative to the Jacobi polynomials as

$$P_{mn}^l(\cos \beta) = \left[\frac{(l-m)!(l+m)!}{(l-n)!(l+n)!} \right]^{\frac{1}{2}} \sin^{m-n} \frac{\beta}{2} \cos^{m+n} \frac{\beta}{2} P_{l-m}^{(m-n,m+n)}(\cos \beta).$$

These functions satisfy symmetry relations including

$$P_{mn}^l(x) = (-1)^{m+n} P_{nm}^l(x) \qquad P_{mn}^l(x) = (-1)^{m-n} P_{-m,-n}^l(x) \tag{9.26}$$

$$P_{mn}^l(x) = P_{-n,-m}^l(x) \qquad P_{mn}^l(x) = (-1)^{l+n} P_{-m,n}^l(-x). \tag{9.27}$$

The functions $P_{mn}^l(z)$ are also given as in [43]

$$P_{mn}^l(z) = \left[\frac{(l-n)!(l+m)!}{(l-m)!(l+n)!} \right]^{1/2} \frac{(1-z)^{\frac{m-n}{2}}(1+z)^{\frac{m+n}{2}}}{2^m(m-n)!} \cdot$$
$$ {}_2F_1\left(l+m+1, -l+m; m-n+1; \frac{1-z}{2}\right)$$

where $_2F_1$ denotes the hypergeometric function. This expression is valid for $m-n \geq 0$; the corresponding expressions for other possible values of m, n may be found using the properties of the hypergeometric functions and the symmetry properties of $P_{mn}^l(z)$. Other expressions for $P_{mn}^l(z)$ in terms of the Krawtchouk polynomials can be found in [43], and classical Fourier series expansions of $P_{mn}^l(\cos\theta)$ can be found in [6, 7].

They also satisfy certain recursion relations including [11, 42, 43]:[2]

$$\cos\beta P_{mn}^l = \frac{[(l^2-m^2)(l^2-n^2)]^{\frac{1}{2}}}{l(2l+1)} P_{mn}^{l-1} + \frac{mn}{l(l+1)} P_{mn}^l$$
$$ + \frac{[(l+1)^2-m^2]^{\frac{1}{2}}[(l+1)^2-n^2]^{\frac{1}{2}}}{(l+1)(2l+1)} P_{mn}^{l+1};$$

$$\frac{1}{2}c_n^l P_{m,n+1}^l + \frac{1}{2}c_{-n}^l P_{m,n-1}^l = \frac{m-n\cos\beta}{\sin\beta} P_{mn}^l;$$

$$-\frac{1}{2}c_m^l P_{m+1,n}^l - \frac{1}{2}c_{-m}^l P_{m-1,n}^l = \frac{n-m\cos\beta}{\sin\beta} P_{mn}^l;$$

where $c_n^l = \sqrt{(l-n)(l+n+1)}$. Note that $c_{-n}^l = c_{n-1}^l$.

Furthermore, they can be related to functions of classical physics such as the Legendre polynomials,

$$P_l(x) = P_{00}^l(x)$$

and the associated Legendre polynomials:

$$P_l^n(x) = \left[\frac{(l+n)!}{(l-n)!}\right]^{\frac{1}{2}} P_{-n,0}^l(x).$$

of a Sturm-Liouville problem, the following orthogonality relations hold:

$$\int_0^\pi P_{mn}^l(\cos\beta) P_{mn}^s(\cos\beta) \sin\beta \, d\beta = \frac{2}{2l+1}\delta_{ls}. \qquad (9.28)$$

These and other properties of the functions $P_{mn}^l(x)$ and relationships with other special functions are described in great detail in the three-volume set by Vilenkin and Klimyk [43].

It is easy to see that given the orthogonality properties of the functions $P_{mn}^l(\cos\beta)$ that the matrix elements U_{mn}^l form an orthogonal set of functions on $SU(2)$:

$$\int_{SU(2)} U_{mn}^l(g)\overline{U_{pq}^s(g)}\, d(g) = \frac{1}{2l+1}\delta_{ls}\delta_{mp}\delta_{nq}.$$

Hence, we expand functions on $SU(2)$ in a Fourier series as (see, e.g., [19, 20, 21]):

---

[2]Our notation is consistent with Vilenkin and Klimyk [43]. The functions that Gel'fand, et. al, [11], call P_{mn}^l differ from ours by a factor of i^{m-n}, which results in slightly different looking recurrence relations. Varshalovich et al. use ZYZ Euler angles. Their relations are different from ours by a factor of $(-1)^{m-n}$ for reasons which are explained in Section 9.4.

$$f(g) = \sum_{l=0,\frac{1}{2},1,\frac{3}{2},...} (2l+1) \sum_{m=-l}^{l} \sum_{n=-l}^{l} \hat{f}_{mn}^l U_{nm}^l(g) \tag{9.29}$$

where

$$\hat{f}_{mn}^l = \int_{SU(2)} f(g) U_{mn}^l(g^{-1}) \, d(g). \tag{9.30}$$

The homomorphism property $U^l(g_1 \circ g_2) = U^l(g_1)U^l(g_2)$ holds because of the derivation in Section 9.1.1, and $\{U^l\}$ for $l = 0, 1/2, 1, 3/2, ...$ is a complete set of irreducible unitary representations for $SU(2)$. Taking these facts for granted, the Plancherel equality and convolution theorem hold as special cases of the general theory.

Explicitly, the first few matrices $P^l(\cos\beta)$ with elements $P_{mn}^l(\cos\beta)$ are:

$$P^0 = P_{00}^0 = 1;$$

$$P^{\frac{1}{2}} = \begin{pmatrix} P_{-\frac{1}{2},-\frac{1}{2}}^{\frac{1}{2}} & P_{-\frac{1}{2},\frac{1}{2}}^{\frac{1}{2}} \\ P_{\frac{1}{2},-\frac{1}{2}}^{\frac{1}{2}} & P_{\frac{1}{2},\frac{1}{2}}^{\frac{1}{2}} \end{pmatrix} = \begin{pmatrix} \cos\frac{\beta}{2} & -\sin\frac{\beta}{2} \\ \sin\frac{\beta}{2} & \cos\frac{\beta}{2} \end{pmatrix};$$

$$P^1 = \begin{pmatrix} P_{-1,-1}^1 & P_{-1,0}^1 & P_{-1,1}^1 \\ P_{0,-1}^1 & P_{0,0}^1 & P_{0,1}^1 \\ P_{1,-1}^1 & P_{1,0}^1 & P_{1,1}^1 \end{pmatrix} = \begin{pmatrix} \frac{1+\cos\beta}{2} & -\frac{\sin\beta}{\sqrt{2}} & \frac{1-\cos\beta}{2} \\ \frac{\sin\beta}{\sqrt{2}} & \cos\beta & -\frac{\sin\beta}{\sqrt{2}} \\ \frac{1-\cos\beta}{2} & \frac{\sin\beta}{\sqrt{2}} & \frac{1+\cos\beta}{2} \end{pmatrix}.$$

$SU(2)$ representation matrices U^l are easily constructed as the matrix product of these matrices with diagonal ones which depend on α and γ.

Many more of the functions $P_{mn}^l(\cos\beta)$ can be found in [42], and all of them can be found up to any finite value of l by using the Rodrigues formula or recursion relations given earlier in this section. For instance, it is easy to verify that

$$P^2 = \begin{pmatrix} P_{-2,-2}^2 & P_{-2,-1}^2 & P_{-2,0}^2 & P_{-2,1}^2 & P_{-2,2}^2 \\ P_{-1,-2}^2 & P_{-1,-1}^2 & P_{-1,0}^2 & P_{-1,1}^2 & P_{-1,2}^2 \\ P_{0,-2}^2 & P_{0,-1}^2 & P_{0,0}^2 & P_{0,1}^2 & P_{0,2}^2 \\ P_{1,-2}^2 & P_{1,-1}^2 & P_{1,0}^2 & P_{1,1}^2 & P_{1,2}^2 \\ P_{2,-2}^2 & P_{2,-1}^2 & P_{2,0}^2 & P_{2,1}^2 & P_{2,2}^2 \end{pmatrix} = \begin{pmatrix} a & -b & c & -g & h \\ b & d & -e & k & -g \\ c & e & f & -e & c \\ g & k & e & d & -b \\ h & g & c & b & a \end{pmatrix}$$

where

$$a = \frac{(1+\cos\beta)^2}{4}$$

$$b = \frac{\sin\beta(1+\cos\beta)}{2}$$

$$c = \frac{1}{2}\sqrt{\frac{3}{2}}\sin^2\beta$$

$$d = \frac{2\cos^2\beta + \cos\beta - 1}{2}$$

$$e = \sqrt{\frac{3}{2}}\sin\beta\cos\beta$$

$$f = \frac{3\cos^2\beta - 1}{2}$$

$$g = \frac{(1 - \cos \beta) \sin \beta}{2}$$

$$h = \frac{(1 - \cos \beta)^2}{4}$$

$$k = \frac{-2 \cos^2 \beta + \cos \beta + 1}{2}.$$

9.4 *SO*(3) Matrix Representations in Various Parameterizations

In this section we consider the representations of $SO(3)$ when using several different parameterizations.

For $SO(3)$ the formulas look almost exactly the same as the $SU(2)$ case. Only now

$$\hat{f}^l_{mn} = \int_{SO(3)} f(A) U^l_{mn}(A^{-1}) \, dA, \tag{9.31}$$

and

$$\int_{SO(3)} U^l_{mn}(A) \overline{U^s_{pq}(A)} \, dA = \frac{1}{2l+1} \delta_{ls} \delta_{mp} \delta_{nq}. \tag{9.32}$$

where dA is scaled so that $\int_{SO(3)} dA = 1$. The Fourier series on $SO(3)$ has the form

$$f(A) = \sum_{l=0}^{\infty} (2l+1) \sum_{m=-l}^{l} \sum_{n=-l}^{l} \hat{f}^l_{mn} U^l_{nm}(A), \tag{9.33}$$

which results from the completeness relation

$$\sum_{l=0}^{\infty} (2l+1) \sum_{m=-l}^{l} \sum_{n=-l}^{l} U_{mn}(R^{-1}) U^l_{nm}(A) = \delta(R^{-1}A). \tag{9.34}$$

Since $\chi_l(R) \doteq \mathrm{tr}[U^l(R)]$, another way to write the above equation (with R in place of $R^{-1}A$) is

$$\sum_{l=0}^{\infty} (2l+1) \chi_l(R) = \delta(R). \tag{9.35}$$

This can be verified by direct calculation. Since

$$\chi_l(\exp(\theta N)) = \frac{\sin(l + 1/2)\theta}{\sin \theta/2},$$

all that is required is to show that when using this expression the left hand side of (9.35) satisfies

$$\int_{SO(3)} f(R) \, \delta(R) \, dR = f(\mathbb{I})$$

for any well-behaved function, $f(R)$.

Using the coordinates (θ, ν, λ) described in Chapter 5,

$$\int_{SO(3)} f(R) \, \delta(R_0^{-1} R) \, dR = \frac{1}{2\pi^2} \cdot \int_0^{\pi} \int_0^{\pi} \int_0^{2\pi} f(\exp(\theta N)) \, \delta(R_0^{-1} \exp(\theta N)) \sin^2 \frac{\theta}{2} \sin \nu \, d\lambda \, d\nu \, d\theta$$

where $N = N(\nu, \lambda)$ and $R_0 = \exp(\theta_0 N_0)$ and $1/(2\pi^2)$ is the normalizing constant so that $\int_{SO(3)} dR = 1$.

By using the rule in (E.4), the Dirac delta is then written in the coordinates (θ, ν, λ) as

$$\delta(R_0^{-1} R) = C \cdot \begin{cases} \frac{\delta(\theta - \theta_0)\,\delta(\nu - \nu_0)\delta(\lambda - \lambda_0)}{\sin^2 \frac{\theta}{2} \sin \nu} & \text{for} \quad \theta_0 \neq 0, \ \sin \nu_0 \neq 0 \\[2ex] \frac{\delta(\theta - \theta_0)\,\delta(\nu - \nu_0)}{2 \sin^2 \frac{\theta}{2}} & \text{for} \quad \theta_0 \neq 0, \ \sin \nu_0 = 0 \\[2ex] \frac{\delta(\theta - \theta_0)}{4\pi \sin^2 \frac{\theta}{2}} & \text{for} \quad \theta_0 = 0 \end{cases} \qquad (9.36)$$

where $C = 2\pi^2$ when dR is the normalized Haar measure, and $C = 1$ for the unnormalized case. This formula is derived in essentially the same way as the case of spherical coordinates in (E.6).

Using (9.36) in the third case, $\delta(R) = (\pi/2)\,\delta(\theta)/(\sin^2 \frac{\theta}{2})$, and so proving (9.35) is equivalent to showing that

$$\sum_{l=0}^{\infty} (2l + 1) \sin(l + 1/2)\theta \sin(\theta/2) = \frac{\pi}{2}\delta(\theta). \qquad (9.37)$$

To prove this, we use the identity

$$2 \sin^2 \frac{\theta}{2} = 1 - \cos \theta = 1 - \frac{1}{2}\left(e^{i\theta} + e^{-i\theta}\right).$$

Then (9.37) is written as

$$\left[1 - \frac{1}{2}\left(e^{i\theta} + e^{-i\theta}\right)\right] \sum_{l=0}^{\infty} (2l + 1) \sum_{m=-l}^{l} e^{-im\theta} = \pi\,\delta(\theta) \qquad (9.38)$$

The left-hand side is calculated as

$$\sum_{l=0}^{\infty} (2l + 1)\left[\sum_{m=-l}^{l} e^{-im\theta} - \frac{1}{2}\sum_{m=-l}^{l} e^{-i(m-1)\theta} - \frac{1}{2}\sum_{m=-l}^{l} e^{-i(m+1)\theta}\right]$$

$$= \sum_{l=0}^{\infty} (2l + 1)\left[\frac{1}{2}\left(e^{-il\theta} + e^{il\theta}\right) - \frac{1}{2}\left(e^{-i(l+1)\theta} + e^{i(l+1)\theta}\right)\right]$$

$$= \frac{1}{2}\sum_{l=0}^{\infty} \left\{(2l + 1)\left(e^{-il\theta} + e^{il\theta}\right) - (2l + 3)\left(e^{-i(l+1)\theta} + e^{i(l+1)\theta}\right)\right\} + \sum_{l=0}^{\infty} \left(e^{-i(l+1)\theta} + e^{i(l+1)\theta}\right)$$

$$= 1 + \sum_{n=1}^{\infty} \left(e^{-in\theta} + e^{in\theta}\right)$$

$$= \sum_{n=-\infty}^{\infty} e^{in\theta}. \qquad (9.39)$$

The right-hand side is

$$\pi\delta(\theta) = \frac{1}{2}\sum_{n=-\infty}^{\infty} e^{in\theta}. \qquad (9.40)$$

From Chapter 2, we know that an expression for $\delta(\theta)$ for the circle that satisfies

$$\int_0^{2\pi} f(\theta)\,\delta(\theta - \theta_0)\,d\theta = f(\theta_0)$$

is

$$\delta(\theta) = \frac{1}{2\pi} \sum_{n=-\infty}^{\infty} e^{in\theta}.$$

When restricting to $\theta \in [0, \pi]$ the scale factor of 2π becomes π. This means that (9.37) holds, and hence so does (9.35).

9.4.1 Euler Angles and Spherical Harmonics

The Fourier series for $SO(3)$ using ZXZ Euler angles is almost identical to that for $SU(2)$, with the following two exceptions: (1) only integer values of l appear in the Fourier series, and therefore only matrix representations with odd dimensions such as U^0, U^1, U^2, etc., need to be calculated; and (2) the integration in the Fourier transform is over half the range, with invariant integration measure $\frac{1}{8\pi^2}\sin\beta d\beta d\alpha d\gamma$. For $SO(3)$ we have $0 \le \gamma < 2\pi$, whereas for $SU(2)$ we have $-2\pi \le \gamma < 2\pi$.

In the literature, a range of equivalent (though frustratingly varied) formulae for matrix elements of the IURs of $SO(3)$ are provided. For instance, it is often convenient to consider the representations of $SO(3)$ Parameterized in ways other than ZXZ Euler angles. When ZYZ Euler angles are used,

$$R_3(\alpha)R_2(\beta)R_3(\gamma) = R_3(\alpha)(R_3(\pi/2)R_1(\beta)R_3(-\pi/2))R_3(\gamma)$$
$$= R_3(\alpha + \pi/2)R_1(\beta)R_3(\gamma - \pi/2),$$

and so

$$R_{ZYZ}(\alpha, \beta, \gamma) = R_{ZXZ}(\alpha + \pi/2, \beta, \gamma - \pi/2).$$

Hence, evaluating the matrix elements (9.23), we find

$$U_{mn}^l(R_{ZYZ}(\alpha, \beta, \gamma)) = e^{i(n-m)\pi/2}U_{mn}^l(R_{ZXZ}(\alpha, \beta, \gamma)) = e^{-im\alpha}P_{mn}^l(\cos\beta)\,e^{-in\gamma} \tag{9.41}$$

since the factor $e^{i(n-m)\pi/2} = i^{n-m}$ cancels with the i^{m-n} in (9.23). In the physics literature the similar expression

$$D_{mn}^l(R_{ZYZ}(\alpha, \beta, \gamma)) = e^{-im\alpha}\,d_{mn}^l(\cos\beta)\,e^{-in\gamma}$$

is common. Each $D_{mn}^l(R)$ is called a *Wigner D-function* and each $D^l(R) = [D_{mn}^l(R)]$ is called a *Wigner D-matrix*[3] [2, 3, 35, 42, 46], and $d_{mn}^l(\cos\beta) = (-1)^{m-n}P_{mn}^l(\cos\beta)$.

Since $U^l(R_{ZYZ}(\alpha, \beta, \gamma))$ and $U^l(R_{ZXZ}(\alpha, \beta, \gamma))$ are equivalent under similarity transformation by diagonal matrices with elements of the form $v_{mn} = i^m \delta_{m,n}$ (no sum over repeated indices) on their diagonals (which are unitary matrices), the distinction between representations parameterized using ZXZ and ZYZ is really unimportant, since a change of basis in the definition of the representations has the same effect as the change of parameterization. Likewise, two similarity transformations of this form relate $U^l(R_{ZYZ})$ to $D^l(R_{ZYZ})$. Hence, independent of whether ZXZ or ZYZ Euler angles

---

[3]Named after Eugene Wigner (1902-1995), a pioneer in the application of group theory in quantum mechanics, who shared the 1963 Nobel prize for Physics.

are used, or a factor of 1, i^{m-n}, i^{n-m}, $(-1)^{m-n}$ appears in the definition of matrix elements, they are all unitarily equivalent (i.e., equal under similarity transformation by a unitary matrix). Finally, another variant is that some texts use $e^{im\alpha}$ and $e^{in\gamma}$ in the definition of matrix elements instead of $e^{-im\alpha}$ and $e^{-in\gamma}$. This too is acceptable, since if U^l is a unitary representation than so is $\overline{U^l}$. While any of these choices is valid, for the sake of concreteness, we will stick with the ZXZ convention when using Euler angles and the factor i^{m-n} unless otherwise stated. Other choices will have slightly different-looking recurrence relations than those given in Section 9.3.

When ZYZ Euler angles are used, then $\mathbf{u}(\beta, \alpha) \doteq R_{ZYZ}(\alpha, \beta, \gamma)\mathbf{e}_3$ gives the standard parameterization of the unit sphere with $\theta = \beta$ and $\phi = \alpha$, i.e., (4.20) with $r = 1$. And similarly, when ZXZ Euler angles are used, then $[\mathbf{u}(\beta, \gamma)]^T = \mathbf{e}_3^T R_{ZXZ}(\alpha, \beta, \gamma)$, which of course is equivalent to $\mathbf{u}(\beta, \gamma) = R_{ZXZ}^T(\alpha, \beta, \gamma)\mathbf{e}_3$.

Using the notation $Y_l^m(\mathbf{u}(\theta, \phi)) = Y_l^m(\theta, \phi)$ to emphasize that a spherical harmonic is a function of the parameterized position on the sphere, the $SO(3)$ matrix elements can be related to the spherical harmonics using both of the above descriptions as

$$U_{m0}^l(R_{ZYZ}(\alpha, \beta, \gamma)) = e^{-im\alpha}P_{m0}^l(\cos\beta) = (-1)^m\sqrt{\frac{4\pi}{2l+1}}\overline{Y_l^m}(\mathbf{u}(\beta, \alpha)) \qquad (9.42)$$

and

$$U_{0n}^l(R_{ZXZ}(\alpha, \beta, \gamma)) = P_{0n}^l(\cos\beta)e^{-in\gamma} = \sqrt{\frac{4\pi}{2l+1}}\overline{Y_l^n}(\mathbf{u}(\beta, \gamma)). \qquad (9.43)$$

It follows from the homomorphism property of group representations that

$$U_{n0}^l(R_1 R_2) = \sum_{m=-l}^{l} U_{nm}^l(R_1)U_{m0}^l(R_2) \qquad (9.44)$$

and

$$U_{0n}^l(R_1 R_2) = \sum_{m=-l}^{l} U_{0m}^l(R_1)U_{mn}^l(R_2). \qquad (9.45)$$

In (9.44) let $R_1 = R_{ZYZ}(\alpha, \beta, \gamma)$ and let $R_2 = R_{ZYZ}(\phi, \theta, 0)$. Then substituting (9.42) and cancelling the common factor of $\sqrt{\frac{4\pi}{2l+1}}$ from both sides gives

$$(-1)^n\overline{Y_l^n}(R_{ZYZ}(\alpha, \beta, \gamma)\mathbf{u}(\theta, \phi)) = \sum_{m=-l}^{l} U_{nm}^l(R_{ZYZ}(\alpha, \beta, \gamma))(-1)^m\overline{Y_l^m}(\mathbf{u}(\theta, \phi))$$
$$(9.46)$$

Similarly, substituting (9.43) into (9.45),

$$\overline{Y_l^n}(R_{ZXZ}^T(\alpha, \beta, \gamma)\mathbf{u}(\theta, \phi)) = \sum_{m=-l}^{l} \overline{Y_l^m}(\mathbf{u}(\theta, \phi))U_{mn}^l(R_{ZXZ}(\alpha, \beta, \gamma)). \qquad (9.47)$$

Now for notational convenience we drop the arguments of the rotation matrices and unit vectors in these expressions.

Conjugating both sides of (9.46), and multiplying through by $(-1)^n$ gives

$$Y_l^n(R_{ZYZ}\mathbf{u}) = \sum_{m=-l}^{l} (-1)^{m-n}\overline{U_{nm}^l(R_{ZYZ})}Y_l^m(\mathbf{u}).$$

Substituting R_{ZYZ}^T for R_{ZYZ} and using the unitarity of the representations U^l gives

$$Y_l^n(R_{ZYZ}^T\mathbf{u}) = \sum_{m=-l}^{l} (-1)^{m-n} Y_l^m(\mathbf{u}) U_{mn}^l(R_{ZYZ}). \qquad (9.48)$$

There are several ways to make the factor of $(-1)^{m-n}$ disappear. These include: (1) Defining $D_{mn}^l(R_{ZYZ}) \doteq (-1)^{m-n} U_{mn}^l(R_{ZYZ})$ (the Wigner D-matrices); (2) Using the spherical harmonics $\tilde{Y}_l^m = (-1)^m Y_l^m$ in place of Y_l^m. Pursuing option 1, then

$$Y_l^n(R_{ZYZ}^T\mathbf{u}) = \sum_{m=-l}^{l} Y_l^m(\mathbf{u}) D_{mn}^l(R_{ZYZ}).$$

Returning to (9.47), and conjugating both sides gives

$$Y_l^n(R_{ZXZ}^T\mathbf{u}) = \sum_{m=-l}^{l} Y_l^m(\mathbf{u}) \overline{U_{mn}^l(R_{ZXZ})}.$$

Then switching $R_{ZXZ}^T \to R_{ZXZ}$ and using the unitary property gives

$$Y_l^n(R_{ZXZ}\mathbf{u}) = \sum_{m=-l}^{l} U_{nm}^l(R_{ZXZ}) Y_l^m(\mathbf{u}).$$

We can now do away with the crutches of coordinates and write

$$\boxed{Y_l^n(R^T\mathbf{u}) = \sum_{m=-l}^{l} Y_l^m(\mathbf{u}) D_{mn}^l(R) \ \text{ and } \ Y_l^n(R\mathbf{u}) = \sum_{m=-l}^{l} U_{nm}^l(R) Y_l^m(\mathbf{u}).} \qquad (9.49)$$

As a consistency check, we now examine what happens when two rotations are applied. On the one hand,

$$Y_l^n((R_1R_2)^T\mathbf{u}) = Y_l^n(R_2^T(R_1^T\mathbf{u})) = \sum_{m=-l}^{l} Y_l^m(R_1^T\mathbf{u}) D_{mn}^l(R_2)$$

$$= \sum_{m=-l}^{l} \left(\sum_{p=-l}^{l} Y_l^p(\mathbf{u}) D_{pm}^l(R_1) \right) D_{mn}^l(R_2)$$

$$= \sum_{p=-l}^{l} Y_l^p(\mathbf{u}) \left(\sum_{m=-l}^{l} D_{pm}^l(R_1) D_{mn}^l(R_2) \right)$$

$$= \sum_{p=-l}^{l} Y_l^p(\mathbf{u}) D_{pn}^l(R_1R_2).$$

A similar calculation gives

$$Y_l^n(R_1 R_2\, \mathbf{u}) = Y_l^n(R_1(R_2\,\mathbf{u})) = \sum_{m=-l}^{l} U_{nm}^l(R_1) Y_l^m(R_2\mathbf{u})$$

$$= \sum_{m=-l}^{l} U_{nm}^l(R_1) \left(\sum_{p=-l}^{l} U_{mp}^l(R_2) Y_l^p(\mathbf{u}) \right)$$

$$= \sum_{p=-l}^{l} \left(\sum_{m=-l}^{l} U_{nm}^l(R_1) U_{mp}^l(R_2) \right) Y_l^p(\mathbf{u})$$

$$= \sum_{p=-l}^{l} U_{np}^l(R_1 R_2) Y_l^p(\mathbf{u}).$$

Therefore, both equations in (9.49) are self consistent.

9.4.2 Axis-Angle and Exponential Parameterizations

We now review the form which the matrix elements U_{mn}^l take when rotations are parameterized as $R(\theta, \nu, \lambda) = \exp(\theta N(\nu, \lambda))$ where $N\mathbf{x} = \mathbf{n} \times \mathbf{x}$, and $\mathbf{n}$ is a unit vector defining the axis of rotation. The ZXZ Euler angles are related to the angle of rotation, θ, and polar and azimuthal angles (ν, λ) of $\mathbf{n}$ as in (5.36). Making appropriate substitutions and using trigonometric rules, we find [42]

$$U_{mn}^l(R(\theta, \nu, \lambda)) = i^{m-n} e^{-i(m-n)\lambda} \left(\frac{1 - i\tan\theta/2\cos\nu}{\sqrt{1 + \tan^2\theta/2\cos^2\nu}} \right)^{m+n} P_{mn}^l(x)$$

where x satisfies

$$\sin x/2 = \sin\theta/2 \sin\nu.$$

In particular,

$$U^1 = \begin{pmatrix} U_{1,1}^1 & U_{1,0}^1 & U_{1,-1}^1 \\ U_{0,1}^1 & U_{0,0}^1 & U_{0,-1}^1 \\ U_{-1,1}^1 & U_{-1,0}^1 & U_{-1,-1}^1 \end{pmatrix} = \begin{pmatrix} a^2 & ab & -c^2 \\ -a\bar{b} & 1 - 2c\bar{c} & \bar{a}b \\ -(\bar{c})^2 & -\overline{ab} & (\bar{a})^2 \end{pmatrix}$$

where

$$a = \cos(\theta/2) - i\sin(\theta/2)\cos\nu$$

$$b = -i\sqrt{2}\sin(\theta/2)\sin\nu\, e^{-i\lambda}$$

$$c = \sin(\theta/2)\sin\nu\, e^{-i\lambda}.$$

In general,

$$\chi_l(\theta) = \mathrm{tr}(U^l(R(\theta, \nu, \lambda))) = \frac{\sin(l + \frac{1}{2})\theta}{\sin\theta/2} = \sum_{m=-l}^{l} e^{-im\theta},$$

and it is easy to verify from the above matrix U^1 that

$$\chi_1(\theta) = 1 + 2\cos\theta = \frac{\sin 3\theta/2}{\sin\theta/2}.$$

The closely related functions

$$\chi_k^l(\theta) \doteq i^k \sum_{m=-l}^{l} C(l, m; k, 0 \,|\, l, m)\, e^{-im\theta},$$

where $\chi_0^l(\theta) = \chi_l(\theta)$, play an important role in the quantum theory of angular momentum, and the Clebsch-Gordan coefficients $C(l_1, m_1; l_2, m_2 \,|\, l_3, m_3)$ are defined in Section 9.9.

An alternative expression for matrix elements of IURs of $SO(3)$ in terms of the axis-angle parameters $R(\theta, \nu, \lambda) = \exp(\theta N(\nu, \lambda))$ (modified from [42]) is[4]

$$U_{mn}^l(R(\theta, \nu, \lambda)) \cong \frac{\sqrt{4\pi}}{2l+1} \sum_{k=0}^{2l} (-i)^k \sqrt{2k+1}\, C(l, m; k, n-m \,|\, l, n)\, \chi_k^l(\theta)\, Y_k^{n-m}(\mathbf{n}(\nu, \lambda))$$

where $Y_k^{n-m} = 0$ when $|n-m| > k$, and $\mathbf{n}(\nu, \lambda)$ is defined as in (5.37).

The differential operators considered in the case of the Euler angles can be written explicitly in this parameterization as well. For instance, the $SO(3)$-Laplacian is written as

$$\nabla^2 = \frac{\partial^2}{\partial\theta^2} + \cot\theta/2 \frac{\partial}{\partial\theta} + \frac{1}{4\sin^2\theta/2}\left(\frac{\partial^2}{\partial\nu^2} + \cot\nu\frac{\partial}{\partial\nu} + \frac{1}{\sin^2\nu}\frac{\partial^2}{\partial\lambda^2}\right) \tag{9.50}$$

where the term in parentheses is the Laplacian for the two-sphere. Regardless of the parameterization, the effect on the matrix elements U_{mn}^l is the same.

Moses [22, 23, 24] derived the matrix elements for the parameterization

$$R(\theta_1, \theta_2, \theta_3) = \exp\begin{pmatrix} 0 & -\theta_3 & \theta_2 \\ \theta_3 & 0 & -\theta_1 \\ -\theta_2 & \theta_1 & 0 \end{pmatrix}.$$

These matrix elements are of the form

$$U_{mn}^l(R(\theta_1, \theta_2, \theta_3)) = (-1)^{2l+m+n}\left[\frac{(l-m)!}{(l+m)!(l-n)!(l+n)!}\right]^{\frac{1}{2}} (\sin\theta/2)^{m-n}\left(\frac{-\theta_1 + i\theta_2}{\theta}\right)^{m-n} \times$$
$$\left(\cos\theta/2 - i\frac{\theta_3}{\theta}\sin\theta/2\right)^{m+n} P_{l-m}^{(m-n,m+n)}\left((1-\theta_3^2/\theta^2)\cos\theta + \theta_3^2/\theta^2\right)$$

where $\theta = \sqrt{\theta_1^2 + \theta_2^2 + \theta_3^2}$ and $P_n^{(\alpha,\beta)}(x)$ are the Jacobi polynomials.

In particular,

$$U^1 = \begin{pmatrix} U_{1,1}^1 & U_{1,0}^1 & U_{1,-1}^1 \\ U_{0,1}^1 & U_{0,0}^1 & U_{0,-1}^1 \\ U_{-1,1}^1 & U_{-1,0}^1 & U_{-1,-1}^1 \end{pmatrix} = \begin{pmatrix} c^2 & -\sqrt{2}a\bar{b}c & a^2(\bar{b})^2 \\ \sqrt{2}abc & -a^2|b|^2 + |c|^2 & -\sqrt{2}a\bar{b}c \\ a^2 b^2 & \sqrt{2}ab\bar{c} & (\bar{c})^2 \end{pmatrix}$$

where

$$a = \sin\theta/2; \quad b = \frac{-\theta_1 - i\theta_2}{\theta}; \quad c = \cos\theta/2 - i\frac{\theta_3}{\theta}\sin\theta/2.$$

---

[4]Here $\cong$ is used because this result is unitary-equivalent to the expression for $U_{mn}^l(R(\theta, \nu, \lambda))$ given above, but may differ in terms of exact equality for a variety of reasons including the choice of convention for spherical harmonics.

9.5 Sampling and FFT for $SO(3)$ and $SU(2)$

Sampling and fast Fourier transform techniques for the rotation group have been developed [14, 17, 18]. Essentially, the double coset decomposition of $SO(3)$ (or $SU(2)$) corresponding to ZXZ Euler angles yields matrix elements of the IUR matrices that lend themselves to fast transforms in each coordinate (Euler angle).

Using the notational simplification

$$f(g(\alpha,\beta,\gamma)) = f(\alpha,\beta,\gamma),$$

(9.29) and (9.30) are written in band-limited form explicitly as

$$f(\alpha,\beta,\gamma) = \sum_{l=0,\frac{1}{2},1,\frac{3}{2},\dots<B} (2l+1) \sum_{m=-l}^{l} \sum_{n=-l}^{l} \hat{f}_{mn}^{l} i^{m-n} e^{-i(m\alpha+n\gamma)} P_{mn}^{l}(\cos\beta) \quad (9.51)$$

where

$$\hat{f}_{mn}^{l} = \frac{1}{16\pi^2} \int_{\beta=0}^{\pi} \int_{\gamma=-2\pi}^{2\pi} \int_{\alpha=0}^{2\pi} f(\alpha,\beta,\gamma) i^{m-n} e^{i(n\alpha+m\gamma)} P_{nm}^{l}(\cos\beta) \sin\beta \, d\alpha \, d\gamma \, d\beta \;.$$

$$(9.52)$$

In the above formulas, we have written this for the $SU(2)$ case, and the $SO(3)$ case follows easily.

The whole spectrum (collection of Fourier transform matrix elements in (9.52)) can be calculated fast in principle by using the classical FFT over α and γ and fast polynomial transforms over β. By using a quadrature rule, (9.51) can be sampled in each coordinate at $\mathcal{O}(B)$ values to exactly compute the integral. The whole of $SU(2)$ is then sampled at $N = \mathcal{O}(B^3)$ points. Explicitly, we first calculate

$$\tilde{f}_n(\beta,\gamma) = \int_0^{2\pi} f(\alpha,\beta,\gamma) \, e^{in\alpha} \, d\alpha$$

for all $n \in \{-B, \dots, B\}$ and all sample values of β and γ. This requires $\mathcal{O}(B^2 \cdot B \log B)$ operations. Then we calculate

$$\tilde{\tilde{f}}_{mn}(\beta) = \int_{-2\pi}^{2\pi} \tilde{f}_n(\beta,\gamma) e^{im\gamma} \, d\gamma$$

in $\mathcal{O}(B^2 \cdot B \log B)$ operations ($\mathcal{O}(B \log B)$ for each value of n and β). Finally,

$$\hat{f}_{mn}^{l} = \frac{1}{16\pi^2} i^{m-n} \int_0^{\pi} \tilde{\tilde{f}}_{mn}(\beta) P_{nm}^{l}(\cos\beta) \sin\beta \, d\beta$$

is calculated in $\mathcal{O}(B^2 \cdot B(\log B)^2)$ operations ($\mathcal{O}(B(\log B)^2)$ for each value of m and n).

Since the limiting calculation is the $\mathcal{O}(B(\log B)^2)$ required for the fast polynomial transform in the variable β, the whole procedure is $\mathcal{O}(N(\log N)^2)$ for all values of m, n, l up to the band-limit. For an alternative numerical approach see [34]. If the expansion in β is done directly (i.e., using $\mathcal{O}(B^2)$ operations instead of using an $\mathcal{O}(B(\log B)^2)$ fast polynomial transform), then the $SU(2)$ Fourier transforms for all m, m, l up to the band-limit can be performed in $\mathcal{O}(B^4) = \mathcal{O}(N^{4/3})$ arithmetic operations.

The cost of reconstructing a function on $SU(2)$ from its spectrum is on the same order as computing the whole spectrum.

Since convolution of functions on $SU(2)$ with band-limit B requires the multiplication of matrices of dimensions $(2l+1) \times (2l+1)$ for $l = 0, ..., B$, the cost of convolution will be

$$\mathcal{O}\left(\sum_{l=0}^{B}(2l+1)^\gamma\right) = \mathcal{O}(B^{\gamma+1}).$$

Hence, when Gaussian elimination is used ($\gamma = 3$), it is clear that the order of computation of the Fourier transforms will be no greater than the cost of convolution (even when the $\mathcal{O}(B^4)$ version is used).

The only difference between the computation for $SU(2)$ and $SO(3)$ is the range of integration, values of l used, and the normalization of the volume element.

The topic of FFTs on the rotation group and fast numerically stable recursive computation of IURs remains a topic of current research [12, 29]. One can imagine FFTs for $SU(2)$ and $SO(3)$ in parameterizations other than Euler angles. To our knowledge, this is an open problem and initial progress along these lines is reported in [48].

9.6 Inverting Gabor Wavelet Transforms for the Sphere

In this section we examine an inversion formula for one of the spherical wavelet transforms given in Chapter 4. This inversion formula uses properties of the rotation group such as inversion under shifts and the completeness of the IUR matrix elements of $SO(3)$.

The concepts of modulation and translation are generalized to the sphere as multiplication by a spherical harmonic and rotation, respectively.

Recall from Chapter 4 that the spherical Gabor transform is defined as (4.75)

$$\tilde{f}_l^m(R) \doteq \int_{\mathbb{S}^2} f(\mathbf{u})\overline{\psi_l^m(\mathbf{u}, R)}\, d\mathbf{u}.$$

Here the mother wavelet is the function $\psi \in \mathcal{L}^2(\mathbb{S}^2)$, and $\psi_l^m(\mathbf{u}, R)$ is defined as

$$\psi_l^m(\mathbf{u}, R) \doteq Y_l^m(\mathbf{u})\psi(R^{-1}\mathbf{u})$$

where $R \in SO(3)$.

In analogy with the case of the circle, we calculate

$$\sum_{l=0}^{\infty}\sum_{m=-l}^{l}\int_{SO(3)} \tilde{f}_l^m(R)\psi_l^m(\mathbf{u}, R)\, dR =$$

$$\int_{SO(3)}\int_{\mathbb{S}^2} f(\mathbf{v})\left(\sum_{l=0}^{\infty}\sum_{m=-l}^{l} Y_l^m(\mathbf{u})\overline{Y_l^m(\mathbf{v})}\right)\psi(R^{-1}\mathbf{u})\overline{\psi(R^{-1}\mathbf{v})}\, d\mathbf{v}\, dR.$$

The term in parenthesis above is rewritten as

$$\sum_{l=0}^{\infty}\sum_{m=-l}^{l} Y_l^m(\mathbf{u})\overline{Y_l^m(\mathbf{v})} = \delta_{\mathbb{S}^2}(\mathbf{u}, \mathbf{v})$$

where $\delta_{\mathbb{S}^2}(\cdot, \cdot)$ is the Dirac delta function for the sphere. This means that

$$\sum_{l=0}^{\infty}\sum_{m=-l}^{l}\int_{SO(3)}\tilde{f}_l^m(R)\psi_l^m(\mathbf{u},R)\,dR = f(\mathbf{u})\int_{SO(3)}\overline{\psi(A^{-1}\mathbf{u})}\psi(A^{-1}\mathbf{u})\,dA.$$

All we have done is use the properties of the Dirac delta and changed the name of the variable of integration for the $SO(3)$ integral. The above integral over $SO(3)$ can be simplified by observing that if we write $\mathbf{u} = R(\mathbf{e}_3, \mathbf{u})\mathbf{e}_3$, and use the invariance of integration on $SO(3)$ under shifts then

$$\int_{SO(3)}\overline{\psi(A^{-1}\mathbf{u})}\psi(A^{-1}\mathbf{u})\,dA =$$

$$\int_{SO(3)}\overline{\psi((R(\mathbf{e}_3,\mathbf{u})^{-1}A)^{-1}\mathbf{e}_3)}\psi((R(\mathbf{e}_3,\mathbf{u})^{-1}A)^{-1}\mathbf{e}_3)\,dA =$$

$$\int_{SO(3)}\overline{\psi(A^{-1}\mathbf{e}_3)}\psi(A^{-1}\mathbf{e}_3)\,dA.$$

But $\psi(A^{-1}\mathbf{e}_3)$ is a function of only two of the Euler angles, and constant with respect to the third. This is because $\mathrm{rot}[\mathbf{e}_3, \theta]\mathbf{e}_3 = \mathbf{e}_3$. Therefore

$$\int_{SO(3)}\overline{\psi(A^{-1}\mathbf{u})}\psi(A^{-1}\mathbf{u})\,dA = \int_{\mathbb{S}^2}\overline{\psi(\mathbf{u}')}\psi(\mathbf{u}')\,d\mathbf{u}' = \|\psi\|_2^2.$$

Hence we write the inversion formula [41]

$$f(\mathbf{u}) = \frac{1}{\|\psi\|_2^2}\sum_{l=0}^{\infty}\sum_{m=-l}^{l}\int_{SO(3)}\tilde{f}_l^m(R)\psi_l^m(\mathbf{u},R)\,dR. \tag{9.53}$$

9.7 Helicity Representations

In this section we construct the representations of the little group $H_{\mathbf{u}'} < SO(3)$. Recall that $H_{\mathbf{u}'}$ is defined as the group which leaves the point $\mathbf{u}' \in \mathbb{S}^2$ fixed. Representations of $H_{\mathbf{u}'}$ are important in the constructions of Chapter 10.

   To calculate the representations of $H_{\mathbf{u}'}$ explicitly, we first choose a particular coset representative $\mathbf{u}' \doteq \mathbf{e}_3 \in \mathbb{S}^2 \cong SO(3)/SO(2)$. The vector $\mathbf{u}'$ is invariant with respect to rotations from the $SO(2)$ subgroup of $SO(3)$, and for this particular choice of $\mathbf{u}'$ we not only have $H_{\mathbf{u}'} \cong SO(2)$, but rather $H_{\mathbf{u}'} = SO(2)$. That is, the general statement

$$\mathrm{rot}[\mathbf{u}, \theta]\,\mathbf{u} = \mathbf{u} \tag{9.54}$$

reduces to $R_3(\theta)\,\mathbf{u}' = \mathbf{u}'$ for this choice of coset representative.
   For each $\mathbf{u} \in \mathbb{S}^2$ we can select one $R_{\mathbf{u}} \in SO(3)$ such that

$$R_{\mathbf{u}}\,\mathbf{u}' = \mathbf{u}.$$

These matrices are elements of cosets in the coset space $SO(3)/SO(2)$.
   Explicitly, this rotation matrix is the one which has an axis pointing in the direction defined by $\mathbf{u}' \times \mathbf{u}$ and has a rotation angle whose sin is $\|\mathbf{u}' \times \mathbf{u}\|$. That is,

$$R_{\mathbf{u}} = R(\mathbf{u}', \mathbf{u})$$

where $R(\mathbf{u}', \mathbf{u})$ is defined as in (5.27).

For any $A \in SO(3)$ it follows from the definition of $R_{\mathbf{u}}$ that

$$R_{A^{-1}\mathbf{u}}\,\mathbf{u}' = A^{-1}\mathbf{u}.$$

Multiplying both sides by A, making the replacement $\mathbf{u} = R_{\mathbf{u}}\,\mathbf{u}'$ on the right-hand-side, and multiplying both sides by $R_{\mathbf{u}}^{-1}$ means

$$(R_{\mathbf{u}}^{-1}\,A\,R_{A^{-1}\mathbf{u}})\,\mathbf{u}' = \mathbf{u}'.$$

Therefore, $Q(\mathbf{u}, A) = (R_{\mathbf{u}}^{-1}\,A\,R_{A^{-1}\mathbf{u}}) \in H_{\mathbf{u}'}$. It follows that

$$Q(\mathbf{u}, A)\,Q(A^{-1}\mathbf{u}, A^{-1}B) = Q(\mathbf{u}, B). \tag{9.55}$$

This expression has significance in quantum mechanics [45].

The representations of $H_{\mathbf{u}'}$ can be taken to be of the form

$$\Delta_s: \quad \phi \to e^{is\phi} \ ; \ 0 \le \phi \le 2\pi;$$

and $s = 0, \pm 1, \pm 2, \dots$. Here $\phi = \theta(Q(\mathbf{u}, A))$ is the angle of rotation of the matrix $Q(\mathbf{u}, A)$. The representations Δ_s form the usual Fourier basis for $\mathbb{S}^1 \cong SO(2) \cong H_{\mathbf{u}'}$.

We now derive the form of $Q(\mathbf{u}, A)$ explicitly. At first sight this would appear to be a complicated function of $\mathbf{u}$ and A. We show that this is in fact not the case.

We begin by observing that

$$R_{A^{-1}\mathbf{u}} = R(\mathbf{u}', A^{-1}\mathbf{u}).$$

Using the general cross-product rules in Chapter 5 (Section 2.2.2), we find that

$$\mathbf{u}' \times (A^{-1}\mathbf{u}) = A^{-1}[(A\mathbf{u}') \times \mathbf{u})]$$

and[5]

$$[A^{-1}\{(A\mathbf{u}') \times \mathbf{u}\}]^{\wedge} = A^{-1}[(A\mathbf{u}') \times \mathbf{u}]^{\wedge}A.$$

Since conjugation commutes with the matrix exponential, it follows from the first equality in (5.27) that

$$R_{A^{-1}\mathbf{u}} = A^{-1}R(A\mathbf{u}', \mathbf{u})A.$$

Substituting this into the definition of $Q(\mathbf{u}, A)$, and using the fact that

$$R_{\mathbf{u}}^{-1} = [R(\mathbf{u}', \mathbf{u})]^{-1} = R(\mathbf{u}, \mathbf{u}'),$$

we find the equation

$$Q(\mathbf{u}, A) = R(\mathbf{u}, \mathbf{u}')R(A\mathbf{u}', \mathbf{u})A. \tag{9.56}$$

9.8 Induced Representations

It is possible to construct representations of $SO(3)$ by using the method of induced representations together with the knowledge of $SO(2)$ representations. The method of induction yields representations of the form

---

[5]Recall that $\wedge$ is the inverse of $\vee$. In the current context $\wedge$ converts a 3-vector to a 3×3 skew-symmetric matrix.

$$(\mathcal{U}^s(A)\varphi)(\mathbf{u}) = \Delta_s(R_{\mathbf{u}}^{-1} A R_{A^{-1}\mathbf{u}})\, \varphi(A^{-1}\mathbf{u}), \tag{9.57}$$

where $A \in SO(3)$, Δ_s are the helicity representations of $H_{\mathbf{u}'} \cong SO(2)$ and $s = 0, \pm 1, \pm 2, \ldots$ discussed in Section 9.7.

It can be shown that the $\Delta_s(Q(\mathbf{u}, A))$ factor (in formula (9.57)) can be expressed as

$$\Delta_s(Q(\mathbf{u}', A)) = e^{is(\alpha+\gamma)}, \tag{9.58}$$

where α and γ are Euler angles of rotation around the z-axis, for the vector $\mathbf{u}' = (0,0,1)$ and arbitrary rotation $A \in SO(3)$ (see [40]).

Using this fact, the following basis functions for $\mathbb{S}^2$ can be defined:

$$h_{m\,s}^l(\beta, \alpha) = \Delta_s(Q(\mathbf{u}', P))(-1)^{(l-s)} \sqrt{\frac{2l+1}{4\pi}}\, \tilde{U}_{m,-s}^l(P) \tag{9.59}$$

where $\tilde{U}_{nm}^l$ are defined relative to (9.23) as

$$\tilde{U}_{mn}^l = i^{m-n} U_{mn}^l.$$

This then differs by a factor of $(-1)^{n-m}$ from the Wigner functions (which are defined in ZYZ Euler angles). Again α, β, γ are ZXZ Euler angles of $P \in SO(3)$. We note that (9.59) does not depend on γ.

For $s = 0$, $h_{m\,s}^l$ is the same as Y_l^m to within a constant factor. From the orthogonalities of the IURs of $SO(3)$ we observe for fixed s that

$$(h_{m'\,s}^{l'}, h_{m\,s}^l) = \delta_{mm'}\delta_{ll'}$$

where $(\cdot, \cdot)$ is the usual inner product for $\mathcal{L}^2(\mathbb{S}^2)$.

Under the rotation A, these functions are transformed as

$$(\mathcal{U}^s(A)\, h_{m\,s}^l)(\mathbf{u}) = \Delta_s(Q(\mathbf{u}, A))\, h_{m\,s}^l(A^{-1}\mathbf{u}) \tag{9.60}$$

where $\mathbf{u} = P\hat{u}$ is found by rotation P (for arbitrary γ) of $\mathbf{u}'$. This transformation law can be written as (see [40])

$$(\mathcal{U}^s(A)\, h_{m\,s}^l)(P\mathbf{u}') = \Delta_s(Q(P\mathbf{u}', A))\,(-1)^{(l-s)}\sqrt{\frac{2l+1}{4\pi}}\cdot$$
$$\Delta_s(Q(\mathbf{u}', A^{-1}P))\overline{\tilde{U}_{m,-s}^l(A^{-1}P)}. \tag{9.61}$$

Using the multiplication law (9.55) and the property

$$\Delta_s(Q(\mathbf{u}', A))^{-1} = \Delta_s(Q(\mathbf{u}', A^{-1}))$$

we can see that

$$\Delta_s(Q(P\mathbf{u}', A))\Delta_s(Q(\mathbf{u}', A^{-1}P)) = \Delta_s(Q(\mathbf{u}', P))\Delta_s(Q(\mathbf{u}', P^{-1}A))\cdot$$
$$\Delta_s(Q(\mathbf{u}', A^{-1}P)) = \Delta_s(Q(\mathbf{u}', P)) \tag{9.62}$$

The group property, the unitarity and (9.62) allow us to write a transformation law as

$$(\mathcal{U}^s(A)\, h_{m\,s}^l)(\mathbf{u}) = \Delta_s(Q(\mathbf{u}', P))\,(-1)^{(l-s)}\sqrt{\frac{2l+1}{4\pi}}\sum_{k=-l}^{l}\overline{\tilde{U}_{mk}^l(A^{-1})}\,\overline{\tilde{U}_{k,-s}^l(P)} =$$

$$\sum_{k=-l}^{l}\tilde{U}_{km}^l(A)\Delta_s(Q(\mathbf{u}', P))\,(-1)^{(l-s)}\sqrt{\frac{2l+1}{4\pi}}\,\overline{\tilde{U}_{k,-s}^l(P)} = \sum_{k=-l}^{l}\tilde{U}_{km}^l(A)\, h_{k\,s}^l(\mathbf{u}). \tag{9.63}$$

The matrix elements of this representation are

$$(h_{m'\,s}^{l'}, \mathcal{U}^s(A)h_{m\,s}^l) = \tilde{U}_{m'm}^l\delta_{ll'}.$$

9.9 The Clebsch-Gordan Coefficients and Wigner 3jm Symbols

In the case of the group $SO(2)$, the representations are all one dimensional, and they are of the form $u^l(\theta) = e^{il\theta}$. It is easy to see that the product of two $SO(2)$ representations is

$$u^{l_1}(\theta)u^{l_2}(\theta) = u^{l_1+l_2}(\theta) = \sum_{l=-\infty}^{\infty} \delta_{l,l_1+l_2} u^l(\theta).$$

Since the product of two matrix elements of IURs of a group is a function on the group, we can ask the more general question, "If the product of two matrix elements is expressed in a Fourier series on the group, what do the Fourier coefficients (or transforms) of this product look like ?"

In fact, it can be shown [42] [6] that in the case when $SU(2)$ and $SO(3)$ are parameterized using ZYZ Euler angles and the matrix elements in (9.23) are used, then

$$U_{m_1,n_1}^{l_1}(\alpha,\beta,\gamma)U_{m_2,n_2}^{l_2}(\alpha,\beta,\gamma) = \sum_{l=|l_1-l_2|}^{l_1+l_2} \sum_{m,n=-l}^{l} \mathcal{C}(l_1,l_2;m_1,m_2;n_1,n_2;l,n,m)U_{m,n}^l(\alpha,\beta,\gamma).$$

$$(9.64)$$

It also can be shown that the decomposition

$$\mathcal{C}(l_1,l_2;m_1,m_2;n_1,n_2;l,n,m) = C(l_1,m_1;l_2,m_2|l,m)C(l_1,n_1;l_2,n_2|l,n)$$

holds. The coefficients $C(l_1,m_1;l_2,m_2|l,m)$ are called the *Clebsch-Gordan coefficients* (CGCs).

It is easy to see from this definition that by taking $n_1 = m_1$, $n_2 = m_2$, and $n = m$ that the magnitude of these coefficients is calculated as

$$|C(l_1,m_1;l_2,m_2|l,m)| = \sqrt{(2j+1)(\mathcal{F}(U_{m_1,m_1}^{l_1}U_{m_2,m_2}^{l_2}))_{mm}}$$

where $\mathcal{F}(\cdot)_{mm}$ denotes the $(m,m)^{th}$ element of the $SO(3)$ Fourier transform. However, this only defines the CGCs to within a sign, which is fixed by the convention [42]

$$C(l_1,m_1;l_2,m_2|l,m)C(l_1,l_1;l_2,-l_2|l,l_1-l_2) = (2j+1)(\mathcal{F}(U_{m_1,l_1}^{l_1}U_{m_2,-l_2}^{l_2}))_{m,l_1-l_2}$$

and

$$C(l_1,l_1;l_2,-l_2|l,l_1-l_2) > 0.$$

The CGCs can be explicitly calculated using the integral [42]

$$C(l_1,m_1;l_2,m_2 \mid l_3,m_3) = \frac{(-1)^{l_1-l_3+m_2}}{2^{l_1+l_2+l_3+1}} \times$$

$$\left[\frac{(l_3+m_3)!(l_1+l_2-l_3)!(l_1+l_2+l_3+1)!(2l_3+1)}{(l_1-m_1)!(l_1+m_1)!(l_2-m_2)!(l_2+m_2)!(l_3-m_3)!(l_2+l_3-l_1)!(l_1+l_3-l_2)!}\right]^{\frac{1}{2}} \times$$

$$\int_{-1}^{1} (1-x)^{l_1-m_1}(1+x)^{l_2-m_2} \frac{d^{l_3-m_3}}{dx^{l_3-m_3}}\left[(1-x)^{l_2+l_3-l_1}(1+x)^{l_1+l_3-l_2}\right] dx. \quad (9.65)$$

They can be generated recursively and possess a number of symmetries including:

---

[6] Our choice of matrix elements U_{mn}^l is different than these by a factor of $(-1)^{m-n}$.

$$C(l_1, m_1; l_2, m_2 \mid l_3, m_3) = (-1)^{(m_2 - m_3)}(-1)^{(l_1 + l_2 + l_3)}\sqrt{\frac{2l_3 + 1}{2l_2 + 1}} \cdot$$

$$C(l_1, m_1; l_3, -m_3 \mid l_2, -m_2). \tag{9.66}$$

The CGCs are related to the *Wigner 3jm symbols* often used in the physics literature as [42]:

$$\begin{pmatrix} j_1 & j_2 & j \\ m_1 & m_2 & m \end{pmatrix} = \frac{(-1)^{m+j+2j_1}}{\sqrt{2j+1}} C(j_1, -m_1; j_2, -m_2 \mid j, m) \tag{9.67}$$

and

$$C(j_1, m_1; j_2, m_2 \mid j, m) = (-1)^{m+j_1-j_2}\sqrt{2j+1} \begin{pmatrix} j_1 & j_2 & j \\ m_1 & m_2 & -m \end{pmatrix}. \tag{9.68}$$

The CGCs play an important role in quantum physics in describing the relationship between spherical tensors and angular momentum via the *Wigner-Eckart Theorem*. In physics, the Racah coefficients [30] and related coefficients called the Wigner 6jm and 9jm coefficients also arise in the context of similar applications. A number of works have considered computational aspects of computing these coefficients using recurrence relations. See, for example, [4, 9, 15, 32, 33, 36, 37, 38, 44, 49].

9.10 Differential Operators for $SO(3)$

Let $A \in SO(3)$ be an arbitrary rotation, and $f(A)$ be a function which assigns a complex number to each value of A. In analogy with the definition of the partial derivative (or directional derivative) of a complex-valued function of $\mathbb{R}^N$-valued argument, we can define differential operators which act on functions of rotation-valued argument. Only now, there are two choices corresponding to whether the operation is applied to the right or left. These take the form

$$(X_{\mathbf{n}}^L f)(A) = \lim_{\epsilon \to 0} \frac{1}{\epsilon} [f(\mathrm{rot}[\mathbf{n}, -\epsilon] \cdot A) - f(A)] = \frac{df(\mathrm{rot}[\mathbf{n}, -t] \cdot A)}{dt}\bigg|_{t=0} \tag{9.69}$$

and

$$(X_{\mathbf{n}}^R f)(A) = \lim_{\epsilon \to 0} \frac{1}{\epsilon} [f(A \cdot \mathrm{rot}[\mathbf{n}, \epsilon]). - f(A)] = \frac{df(A \cdot \mathrm{rot}[\mathbf{n}, t])}{dt}\bigg|_{t=0}. \tag{9.70}$$

In the case of the left operator, $-\epsilon$ is used above to be consistent with notations that will follow in subsequent chapters. Note that for small motions,

$$\mathrm{rot}[\mathbf{n}, \theta] \approx \mathbb{I} + \theta N = \mathbb{I} + \theta(n_1 X_1 + n_2 X_2 + n_3 X_3)$$

where

$$X_1 = \begin{pmatrix} 0 & 0 & 0 \\ 0 & 0 & -1 \\ 0 & 1 & 0 \end{pmatrix}; \quad X_2 = \begin{pmatrix} 0 & 0 & 1 \\ 0 & 0 & 0 \\ -1 & 0 & 0 \end{pmatrix}; \quad X_3 = \begin{pmatrix} 0 & -1 & 0 \\ 1 & 0 & 0 \\ 0 & 0 & 0 \end{pmatrix}.$$

We now find the explicit forms of the operators $X_{\mathbf{n}}^L$ and $X_{\mathbf{n}}^R$ in any 3-parameter description of rotation $A = A(q_1, q_2, q_3)$. Expanding in a Taylor series, we write

$$X_{\mathbf{n}}^R f = \sum_{i=1}^{3} \frac{\partial f}{\partial q_i} \frac{\partial q_i^R}{\partial \epsilon}\bigg|_{\epsilon=0}$$

where $\{q_i^R\}$ are the parameters such that $A(q_1, q_2, q_3)\text{rot}[\mathbf{n}, \epsilon] = A(q_1^R, q_2^R, q_3^R)$.

The coefficients $\left.\frac{\partial q_i^R}{\partial \epsilon}\right|_{\epsilon=0}$ are determined by observing two different-looking, though equivalent, ways of writing $A \cdot \text{rot}[\mathbf{n}, \epsilon]$ for small ϵ:

$$A + \epsilon A N \approx A \cdot \text{rot}[\mathbf{n}, \epsilon] \approx A + \epsilon \sum_{i=1}^{3} \frac{\partial A}{\partial q_i} \left.\frac{\partial q_i^R}{\partial \epsilon}\right|_{\epsilon=0}.$$

We then have that

$$N = \sum_{i=1}^{3} A^T \frac{\partial A}{\partial q_i} \left.\frac{\partial q_i^R}{\partial \epsilon}\right|_{\epsilon=0},$$

or

$$\mathbf{n} = N^\vee = \sum_{i=1}^{3} \left(A^T \frac{\partial A}{\partial q_i}\right)^\vee \left.\frac{\partial q_i^R}{\partial \epsilon}\right|_{\epsilon=0},$$

which is written as $\mathbf{n} = J_R \left.\frac{\partial \mathbf{q}^R}{\partial \epsilon}\right|_{\epsilon=0}$. This allows us to solve for

$$\left.\frac{\partial \mathbf{q}^R}{\partial \epsilon}\right|_{\epsilon=0} = J_R^{-1}\mathbf{n}.$$

J_R is the "body" Jacobian calculated in (5.50) for the ZXZ Euler angles. Its inverse is

$$J_R^{-1} = \begin{pmatrix} \sin\gamma/\sin\beta & \cos\gamma/\sin\beta & 0 \\ \cos\gamma & -\sin\gamma & 0 \\ -\cot\beta\sin\gamma & -\cot\beta\cos\gamma & 1 \end{pmatrix}.$$

Making the shorthand notation $X_{\mathbf{e}_i}^R = X_i^R$, we then write for the ZXZ Euler angles

$$X_1^R = -\cot\beta\sin\gamma\frac{\partial}{\partial\gamma} + \frac{\sin\gamma}{\sin\beta}\frac{\partial}{\partial\alpha} + \cos\gamma\frac{\partial}{\partial\beta};$$

$$X_2^R = -\cot\beta\cos\gamma\frac{\partial}{\partial\gamma} + \frac{\cos\gamma}{\sin\beta}\frac{\partial}{\partial\alpha} - \sin\gamma\frac{\partial}{\partial\beta};$$

$$X_3^R = \frac{\partial}{\partial\gamma}.$$

For the ZYZ Euler-angles, these same operators take the form

$$X_1^R = \cot\beta\cos\gamma\frac{\partial}{\partial\gamma} - \frac{\cos\gamma}{\sin\beta}\frac{\partial}{\partial\alpha} + \sin\gamma\frac{\partial}{\partial\beta};$$

$$X_2^R = -\cot\beta\sin\gamma\frac{\partial}{\partial\gamma} + \frac{\sin\gamma}{\sin\beta}\frac{\partial}{\partial\alpha} + \cos\gamma\frac{\partial}{\partial\beta};$$

$$X_3^R = \frac{\partial}{\partial\gamma}.$$

In analogy with the directional derivative in $\mathbb{R}^3$, we have $X_\mathbf{n}^R = \sum_{i=1}^{3} n_i X_i^R$.

The operators $X_{\mathbf{e}_i}^L = X_i^L$ can be derived in a completely analogous way using the spatial Jacobian instead of the body Jacobian. Or they can be derived from $X_{\mathbf{e}_i}^R$ by observing that

$$\text{rot}[\mathbf{n}, \epsilon]A = A(A^T\text{rot}[\mathbf{n}, \epsilon]A) = A\,\text{rot}[A^T\mathbf{n}, \epsilon].$$

This means that

$$X_{\mathbf{e}_i}^L = X_{-A^T \mathbf{e}_i}^R.$$

Explicitly, in terms of ZXZ Euler angles,

$$J_R^{-1} A^T = \begin{pmatrix} -\sin\alpha\cot\beta & \cos\alpha\cot\beta & 1 \\ \cos\alpha & \sin\alpha & 0 \\ \sin\alpha/\sin\beta & -\cos\alpha/\sin\beta & 0 \end{pmatrix},$$

and so

$$X_1^L = \sin\alpha\cot\beta\frac{\partial}{\partial\alpha} - \cos\alpha\frac{\partial}{\partial\beta} - \sin\alpha/\sin\beta\frac{\partial}{\partial\gamma}$$

$$X_2^L = -\cos\alpha\cot\beta\frac{\partial}{\partial\alpha} - \sin\alpha\frac{\partial}{\partial\beta} + \cos\alpha/\sin\beta\frac{\partial}{\partial\gamma}$$

$$X_3^L = -\frac{\partial}{\partial\alpha}.$$

When ZYZ Euler angles are used, these operators take the form

$$X_1^L = \cos\alpha\cot\beta\frac{\partial}{\partial\alpha} + \sin\alpha\frac{\partial}{\partial\beta} - \cos\alpha/\sin\beta\frac{\partial}{\partial\gamma}$$

$$X_2^L = \sin\alpha\cot\beta\frac{\partial}{\partial\alpha} - \cos\alpha\frac{\partial}{\partial\beta} - \sin\alpha/\sin\beta\frac{\partial}{\partial\gamma}$$

$$X_3^L = -\frac{\partial}{\partial\alpha}.$$

9.11 Operational Properties

We now examine the effect of the operators X_i^R and X_i^L on the matrix elements U_{mn}^l. Due to the homomorphism property of representations, we write

$$(X_i^R U_{mn}^l)(A) = \lim_{\epsilon\to 0}\frac{1}{\epsilon}\left[\sum_{k=-l}^{l} U_{mk}^l(A)U_{kn}^l(\mathrm{rot}[\mathbf{e}_i,\epsilon]) - U_{mn}^l(A)\right] \qquad (9.71)$$

and

$$(X_i^L U_{mn}^l)(A) = \lim_{\epsilon\to 0}\frac{1}{\epsilon}\left[\sum_{k=-l}^{l} U_{mk}^l(\mathrm{rot}[\mathbf{e}_i,-\epsilon])U_{kn}^l(A) - U_{mn}^l(A)\right]. \qquad (9.72)$$

Expanding the integral in (9.25) for $\beta = \epsilon$, we find that

$$U_{mn}^l(\mathrm{rot}(\mathbf{e}_1,\epsilon)) = \delta_{mn} + \frac{i}{2}c_{-n}^l\epsilon\delta_{m+1,n} + \frac{i}{2}c_n^l\epsilon\delta_{m-1,n} + \mathcal{O}(\epsilon^2).$$

Again we use the definition $c_n^l = \sqrt{(l-n)(l+n+1)}$. Using the fact that

$$U_{mn}^l(\mathrm{rot}(\mathbf{e}_2,\epsilon)) = i^{n-m}U_{mn}^l(\mathrm{rot}[\mathbf{e}_1,\epsilon])$$

allows us to write

$$U^l_{mn}(\text{rot}[\mathbf{e}_2, \epsilon]) = \delta_{mn} - \frac{1}{2}c^l_{-n}\epsilon\delta_{m+1,n} + \frac{1}{2}c^l_n\epsilon\delta_{m-1,n} + \mathcal{O}(\epsilon^2).$$

Substituting these into (9.71) and (9.72), we find that

$$X^R_1 U^l_{mn} = \frac{1}{2}ic^l_{-n}U^l_{m,n-1} + \frac{1}{2}ic^l_n U^l_{m,n+1};$$

$$X^R_2 U^l_{mn} = -\frac{1}{2}c^l_{-n}U^l_{m,n-1} + \frac{1}{2}c^l_n U^l_{m,n+1};$$

$$X^L_1 U^l_{mn} = -\frac{1}{2}ic^l_{-m-1}U^l_{m+1,n} - \frac{1}{2}ic^l_{m-1}U^l_{m-1,n};$$

$$X^L_2 U^l_{mn} = \frac{1}{2}c^l_{-m-1}U^l_{m+1,n} - \frac{1}{2}c^l_{m-1}U^l_{m-1,n}.$$

The operators X^R_3 and X^L_3 can be applied directly to the closed-form formula (9.23) to result in

$$X^R_3 U^l_{mn} = -inU^l_{mn};$$

$$X^L_3 U^l_{mn} = imU^l_{mn}.$$

By repeated application of these rules we find

$$(X^R_1)^2 U^l_{mn} = -\frac{1}{4}c^l_{-n}c^l_{-n+1}U^l_{m,n-2} - \frac{1}{4}(c^l_{-n}c^l_{n-1} + c^l_n c^l_{-n-1})U^l_{mn} - \frac{1}{4}c^l_n c^l_{n+1}U^l_{m,n+2};$$

$$(X^R_2)^2 U^l_{mn} = \frac{1}{4}c^l_{-n}c^l_{-n+1}U^l_{m,n-2} - \frac{1}{4}(c^l_{-n}c^l_{n-1} + c^l_n c^l_{-n-1})U^l_{mn} + \frac{1}{4}c^l_n c^l_{n+1}U^l_{m,n+2};$$

$$(X^R_3)^2 U^l_{mn} = -n^2 U^l_{mn};$$

$$X^R_1 X^R_2 U^l_{m,n} = -\frac{i}{4}c^l_{-n}c^l_{-n+1}U^l_{m,n-2} + \frac{i}{4}(-c^l_{-n}c^l_{n-1} + c^l_n c^l_{-n-1})U^l_{m,n} + \frac{i}{4}c^l_n c^l_{n+1}U^l_{m,n+2};$$

$$X^R_2 X^R_1 U^l_{m,n} = -\frac{i}{4}c^l_{-n}c^l_{-n+1}U^l_{m,n-2} + \frac{i}{4}(c^l_{-n}c^l_{n-1} - c^l_n c^l_{-n-1})U^l_{m,n} + \frac{i}{4}c^l_n c^l_{n+1}U^l_{m,n+2}.$$

$$X^R_1 X^R_3 U^l_{m,n} = \frac{n}{2}(c^l_{-n}U^l_{m,n-1} + c^l_n U^l_{m,n+1});$$

$$X^R_3 X^R_1 U^l_{m,n} = \frac{n-1}{2}c^l_{-n}U^l_{m,n-1} + \frac{n+1}{2}c^l_n U^l_{m,n+1};$$

$$X^R_3 X^R_2 U^l_{m,n} = \frac{i(n-1)}{2}c^l_{-n}U^l_{m,n-1} - \frac{i(n+1)}{2}c^l_n U^l_{m,n+1};$$

$$X^R_2 X^R_3 U^l_{m,n} = \frac{in}{2}(c^l_{-n}U^l_{m,n-1} - c^l_n U^l_{m,n+1}).$$

Since $c^l_{-n}c^l_{n-1} + c^l_n c^l_{-n-1} = 2(l^2-n^2) + 2l$, we find that for arbitrary constant scalars D_1 and D_2,

$$[D_1((X^R_1)^2 + (X^R_2)^2) + D_2(X^R_3)^2]U^l_{mn} = -D_1[l(l+1) + n^2(D_2/D_1 - 1)]U^l_{mn}. \quad (9.73)$$

The Laplacian[7] is given by

$$\nabla^2 = (X^R_1)^2 + (X^R_2)^2 + (X^R_3)^2.$$

---

[7] The negative of this Laplacian is often denoted in physics as $\mathbf{J}^2$, and in mathematics it is called the Casimir operator.

This is exactly the same result from the differential-geometric definition of the Laplacian given in (9.21). The Laplacian commutes with all the operators mentioned above, as well as with left and right shifts, and can also be written as

$$\nabla^2 = (X_1^L)^2 + (X_2^L)^2 + (X_3^L)^2.$$

When $D_1 = D_2 = 1$ in Equation 9.73 the Laplacian results, and

$$\nabla^2 U_{mn}^l = -l(l+1)U_{mn}^l$$

as it must from its differential-geometric definition.

Other differential and algebraic operators also transform matrix elements into matrix elements. One of the many such operators found in [42] is:

$$\sin\beta\frac{\partial}{\partial\beta}\left(U_{mn}^l\right) = -\frac{(l+1)[(l^2-m^2)(l^2-n^2)]^{\frac{1}{2}}}{l(2l+1)}U_{mn}^{l-1} - \frac{mn}{l(l+1)}U_{mn}^l$$
$$+\frac{l[(l+1)^2-m^2]^{\frac{1}{2}}[(l+1)^2-n^2]^{\frac{1}{2}}}{(l+1)(2l+1)}U_{mn}^{l+1} \qquad (9.74)$$

Note that unlike the operators X_i^R and X_i^L, the operator $\sin\beta\frac{\partial}{\partial\beta}$ has the effect of shifting the value of l while leaving m and n unchanged.

9.12 Classification of Quantum States and Representations of $SU(2)$

Conservation laws summarize natural phenomena observed in physical experiments. Each conservation law states the fact that some physical quantities are left invariant under the action of some group of transformations. Important groups of transformations include translations, rotations and reflections. These groups lead to the conservation of momentum, angular momentum and parity, respectively.

The invariance of equations of motion under a group of transformations leads to the classification of quantum states as eigenvalues of representations of a group of transformations. Let us write the Schrödinger equation

$$i\frac{\partial\psi}{\partial t} = \hat{H}\psi$$

where $\hat{H}$ is an operator representing the Hamiltonian of the system and ψ is a wave function. Let us assume that the equation is invariant under some transformation $\hat{A}$. $\hat{A}$ is an operator and acts on wave functions as $(\hat{A}\psi)(x) = \psi(A^{-1} \cdot x)$, where A is an element of the group of transformations G. We note that the transformation law assumes that the operators $\hat{A}$ are representations of the group G that act on the space of wave functions $\psi(x)$. Invariance of Schrödinger's equation means that the transformed wave function $\psi' = \hat{A}\psi$ must still satisfy the Schrödinger equation

$$i\frac{\partial\psi'}{\partial t} = \hat{H}\psi'.$$

If we assume that the transformation is independent of time, then we can write

$$i\hat{A}\frac{\partial\psi}{\partial t} = \hat{H}\hat{A}\psi.$$

We observe that the Schrödinger equation is invariant if the Hamiltonian commutes with the transformation operator A:

$$\hat{H}\hat{A} = \hat{A}\hat{H}.$$

According to Schur's lemma, if the operator $\hat{H}$ commutes with all the irreducible representation operators of the group G, then $\hat{H}$ must be proportional to the identity operator. Thus, for the direct sum of different irreducible representations of G, the Hamiltonian can be represented in diagonal form. Therefore, the Hamiltonian has the simplest form in the basis where the wave functions are the basis functions of representations of the group of transformations G. If we assume that the quantum system is invariant with respect to the rotation group $SO(3)$, this leads to the classification of quantum states according to quantum numbers l, m, which enumerate different basis functions of representations of $SO(3)$. Wave functions (quantum states) can be considered as a sum over different irreducible representation of $SO(3)$, enumerated by $l = 0, 1, 2, ...$. Thus, for spherically summetric system (such as the hydrogen atom) the wave functions can be represented in the form

$$\psi(r, \theta, \phi) = \sum_{l=0}^{\infty} R_l(r, n) Y_l^m(\theta, \phi)$$

where $Y_l^m(\theta, \phi)$ are spherical harmonics, and $R_l(r, n)$ represents a radial dependence (n is an additional radial quantum number). The number l which enumerates irreducible representation of $SO(3)$ has the meaning of angular momentum of the system, and $m = -l, ..., l$ has the meaning of the projection of angular momentum on a fixed axis.

We note that in quantum mechanics the state with $l = 0$ is called the s state, $l = 1$ is the p state, $l = 2$ is the d state and so on. Because $l = 0, 1, 2, ...$, angular momentum can have only discrete values. This property is called *quantization* of angular momentum.

If X_3 is the generator of $SO(3)$ for rotations around the z-axis, it acts on quantum spherical states as

$$X_3 \, \psi_{l,m} = m \, \psi_{l,m}$$

because of the corresponding property of spherical harmonics. The operator $X^2 = X_1^2 + X_2^2 + X_3^2$ acts as

$$X^2 \, \psi_{l,m} = -l(l+1) \, \psi_{l,m}.$$

If we assume that the system has several non-interacting particles, then the Hamiltonian should be invariant under the combined transformation which is a direct product of representations of $SO(3)$ for each of the individual particles (and the corresponding spherical dependence on $\theta_1, \phi_1, \theta_2.\phi_2$ can be written as a product of spherical harmonics $Y_{l_1}^{m_1}(\theta_1, \phi_1)$ and $Y_{l_2}^{m_2}(\theta_2, \phi_2)$). The direct product of irreducible representations of $SO(3)$ can be written as a sum over irreducible representations of $SO(3)$. This leads to the summation rules for angular momentum. The direct product of irreducible representations A_{l_1} and A_{l_2} of $SO(3)$, enumerated by l_1 and l_2 can be written as a sum over irreducible representations A_l, where $l = |l_1 - l_2|, ..., l_1 + l_2$.

Because of this, the sum of two angular momenta l_1 and l_2 may have values from $l = |l_1 - l_2|$ to $l = l_1 + l_2$. The projections of momenta m_1 and m_2 are added as ordinary numbers $m = m_1 + m_2$ because the Clebsch-Gordan coefficients $C(l_1, m_1; l_2, m_2 | l, m)$ are not equal to zero only for $m_1 + m_1 = m$.

So far we have discussed physical systems which act as scalars under transformation, i.e. the ψ functions are scalar-valued functions. Many important physical systems can be described as vector-, spinor- or even tensor-valued functions. Vector-valued functions can be described as functions which transform under a representation operator $\hat{A}$ of $SO(3)$ as

$$(\hat{A}\psi_i)(x) = \sum_j A_{ij}\psi_j(A^{-1} \cdot x)$$

where $A \in SO(3)$. Because $A \in SO(3)$ is equivalent to an $l = 1$ representation of $SO(3)$, these functions describe particles with $l = 1$ internal momentum. We mention that transformations of ψ functions in the internal space of ψ describe the *internal* angular momentum of particles. Since the particles of interest in non-relativistic quantum mechanics are electrons (which have non-integer internal angular momentum), vector functions are rarely used in quantum mechanics (more general vector functions are used in relativistic quantum mechanics to describe photons).

Particles with non-integer internal angular momentum (called spin) are described by spinor-valued functions. A spinor-valued function transforms under the action of $SU(2)$ (corresponding to the rotation matrix A) on the space of functions containing ψ_i (i is an index describing internal degrees of freedom). The transformation law for spinor functions is

$$(\hat{A}\psi_i)(x) = \sum_j A'_{ij}\psi_j(A^{-1} \cdot x)$$

where $A \in SO(3)$ and $A' \in SU(2)$. The matrix $A' = \cos(\theta/2) + i(\sigma \cdot \mathbf{n})\sin(\theta/2)$ corresponds to the rotation matrix A parameterized with $(\mathbf{n}, \theta)$ where the unit vector $\mathbf{n}$ defines the axis of rotation, and θ is the angle of rotation. Matrices σ_i (Pauli matrices) are proportional to the basis elements of the Lie algebra $su(2)$:

$$\sigma_1 = \begin{pmatrix} 0 & 1 \\ 1 & 0 \end{pmatrix}; \quad \sigma_2 = \begin{pmatrix} 0 & -i \\ i & 0 \end{pmatrix}; \quad \sigma_3 = \begin{pmatrix} 1 & 0 \\ 0 & -1 \end{pmatrix}.$$

Because the $SU(2)$ matrix in the fundamental representation corresponds to the $l = 1/2$ representation of $SU(2)$, spinor functions describe particles with internal half-integer spin.

See [5, 8, 26, 28, 35, 42] for more in-depth treatments of the relationship between the representation theory of $SU(2)$ and quantum mechanics.

9.13 Representations of $SO(4)$ and $SU(2) \times SU(2)$

We have already seen that rotations in $\mathbb{R}^4$ can be described using two independent special unitary matrices as

$$Y = AXB^*$$

where

$$X = \begin{pmatrix} x_4 - ix_2 & -x_1 - ix_3 \\ x_1 - ix_3 & x_4 + ix_2 \end{pmatrix}.$$

Furthermore, the pair $(A, B) \in SU(2) \times SU(2)$ is redundant in the sense that $(-A, -B)$ describes the same rotation as (A, B). In group-theoretic language, $SU(2) \times SU(2)$ is the double covering group of $SO(4)$ in the same way that $SU(2)$ is the double cover of $SO(3)$.

This is useful to know, since it means that we need not construct the IURs of $SO(4)$ from scratch. The property (9.10) of the tensor product indicates that if $U^j(A)$ and $U^{j'}(B)$ are representations of $SU(2)$, then

$$\mathcal{U}^{j,j'}(A,B) = U^j(A) \otimes U^{j'}(B)$$

is a representation of $SU(2) \times SU(2)$. It can be shown that the set of all representations of the form $\mathcal{U}^{j,j'}(A,B)$ is complete, and each representation is unitary [40]. Furthermore, the complete set of representations for $SO(4)$ is just the subset of those for $SU(2) \times SU(2)$ for which $j + j'$ is an integer.

We note that due to the close relationship between representations of $SO(4)$ and those for $SU(2)$, the sampling and FFT techniques discussed earlier can be applied in this case as well. This and the relationship between $SO(4)$ and $SE(3)$ would seem to lead to a promising way to perform fast approximate FFTs for $SE(3)$. The difficulty is that the stereographic projection relating $SO(4)$ and $SE(3)$ is only accurate for small rotations. This means that probability density functions on $SO(4)$ used to approximate PDFs on $SE(3)$ would require an extremely large band limit in order to achieve any reasonable accuracy. That is, the uncertainty principle will not allow functions that simultaneously have a small band limit and small compact support. Therefore, we must consider the representation theory of $SE(N)$ for $N = 2$ and $N = 3$ and seek fast transforms in order that some of the problems posed in subsequent chapters can be solved using efficient numerical techniques.

9.14 Summary

In this chapter we presented a concrete overview of the representation theory of $SO(3)$ and $SU(2)$ together with harmonic analysis analysis on these groups. Irreducible unitary representation matrices were given in terms of several parameterizations. It was shown how differential operators acting on functions of these groups can be defined in a concrete way, and how the matrix elements of IURs behave under these operators. FFTs for $SO(3)$ result from recurrence relations of the Wigner D-functions when IURs are expressed in Euler angles. The results of this chapter are used in Chapter 10 to define the Fourier transform and operational calculus for $SE(3)$, and in Chapters 16 and 17 in the context of applications involving pure rotations. In addition, the representation theory of rotation groups has been applied recently in new and interesting ways not discussed in this book, including multi-resolution (or wavelet) analysis [16, 31], motion estimation [27], and the extraction of atomic potentials from experimental data [1].

References

1. Bartók, A.P., Kondor, R., Csányi, G., "On Representing Chemical Environments," *Physical Review B*, 87(18): 184115, 2013.
2. Biedenharn, L.C., Louck, J.D., *Angular Momentum in Quantum Physics*, Encyclopedia of Mathematics and Its Applications, Vol. 8, Cambridge University Press, 1985. (paperback version 2009).
3. Biedenharn, L.C., Louck, J.D., *Racah-Wigner Algebra in Quantum Theory*, Encyclopedia of Mathematics and Its Applications, Vol. 9, Cambridge University Press, 1985. (paperback version 2009).
4. Bretz, V., "Improved Method for Calculation of Angular-Momentum Coupling Coefficients," *Acta Physica Academiae Scientiarum Hungarucae*, 40(4): 255 – 259, 1976.
5. Brink, D.M., Satchler, G.R., *Angular Momentum*, 3rd ed., Oxford University Press, 2003.
6. Bunge, H.J., "Über eine Fourier-Entwicklung verallgemeinerter Kugelfunktionen," *Monatsberichte der deutschen Akademie der Wissenschaften*, 9: 652 – 658, 1967.
7. Bunge, H.J., "Calculation of the Fourier Coefficients of the Generalized Spherical Functions," *Kristall Techn.*, 9: 939 – 963, 1974.
8. Edmonds, A. R., *Angular Momentum in Quantum Mechanics*, Princeton University Press, Princeton, N.J., 1996.
9. Fack, V., Van der Jeugt, J., Rao, K.S., "Parallel Computation of Recoupling Coefficeints Using Transputers," *Computer Physics Communications*, 71(3): 285 – 304, Sept. 1992.
10. Fässler, A., Stiefel, E., *Group Theoretical Methods and Their Applications*, (Baoswan Dzung Wong, trans.), Birkhäuser, Boston, 1992.
11. Gelfand, I. M., Minlos, R.A., Shapiro, Z.Ya., *Representations of the Rotation and Lorentz Groups and Their Applications*, Macmillan, New York, 1963.
12. Gumerov, N.A., Duraiswami, R., "Recursive Computation of Spherical Harmonic Rotation Coefcients of Large Degree," http://arxiv.org/pdf/1403.7698, 2014.
13. Hannabuss, K., "Sound and Symmetry," *The Mathematical Intelligencer*, 19(4): 16 – 20, Fall 1997.
14. Kostelec, P.J., Rockmore, D.N., "FFTs on the Rotation Group," *Journal of Fourier Analysis and Applications*, 14(2): 145 – 179, 2008.
15. Larson, E.G., Li, M.S., Larson, G.C., "Some Comments on the Electrostatic Potential of a Molecule," *International Journal of Quantum Chemistry*, Suppl. 26: 181 – 205, 1992.
16. Mallat, S., "Group Invariant Scattering," *Communications on Pure and Applied Mathematics*, 65(10): 1331 – 1398, 2012.
17. Maslen, D.K., *Fast Transforms and Sampling for Compact Groups*, Ph.D. Dissertation, Dept. of Mathematics, Harvard University, May 1993.
18. Maslen, D.K., Rockmore, D.N., "Generalized FFTs - A Survey of Some Recent Results," *DIMACS Series in Discrete Mathematics and Theoretical Computer Science*, 28: 183 – 237, 1997.

19. Mayer, R.A., "Fourier Series of Differentiable Functions on $SU(2)$," *Duke Mathematical Journal*, 34: 549–554, 1967.
20. Mayer, R, A., "Summation of Fourier Series on Compact Groups," *American Journal of Mathematics*, 89: 661–692, 1967.
21. Mayer, R.A., "Localization for Fourier Series on $SU(2)$," *Transactions of the American Mathematical Society*, 130: 414–424, 1968.
22. Moses, H.E., "Irreducible Representations of the Rotation Group in Terms of Euler's Theorem," *Il Nuovo Cimento*, 40(4): 1120-1138, Dec. 1965.
23. Moses, H.E., "Irreducible Representations of the Rotation Group in Terms of the Axis and Angle of Rotation," *Annals of Physics*, 37: 224–226, 1966.
24. Moses, H.E., "Irreducible Representations of the Rotation Group in Terms of the Axis and Angle of Rotation: II. Orthogonality Relations Between Matrix Elements and Representations of Rotations in the Parameter Space," *Annals of Physics*, 42: 343–346, 1967.
25. Naimark, M.A., *Linear Representations of the Lorentz Group*, Macmillan, New York, 1964.
26. Normand, J.-M., *A Lie Group: Rotations in Quantum Mechanics*, North-Holland Publishing Company, Amsterdam, 1980.
27. Osteen, P.R., Owens, J.L., Kessens, C.C., "Online Egomotion Estimation of RGB-D Sensors Using Spherical Harmonics," *Proceedings of the 2012 IEEE International Conference on Robotics and Automation*, pp. 1679-1684, St. Paul, Minnesota, May, 2012.
28. Petrashen, M.I., Trifonov, E.D., *Applications of Group Theory in Quantum Mechanics*, Dover, 2009.
29. Potts, D., Prestin, J., Vollrath, A., "A Fast Algorithm for Nonequispaced Fourier Transforms on the Rotation Group," *Numerical Algorithms*, 52(3): 355-384, 2009.
30. Racah, G., "Theory of Complex Spectra. I, II and III," *Physical Review* 61: 186–197, 1942; 62: 438–462, 1942; 63: 367–382, 1943.
31. Rahman, I.U., Drori, I., Stodden, V.C., Donoho, D.L., Schröder, P., "Multiscale Representations for Manifold-Valued Data," *Multiscale Modeling and Simulation*, 4(4): 1201–1232, 2005.
32. Rao, K.S., Venkatesh, K., "New Fortran Programs for Angular-Momentum Coefficients," *Computer Physics Communications* 15(3–4): 227–235, 1978.
33. Rao, K.S., Rajeswari, V., Chiu, C.B., "A New Fortran Program for the 9-J Angular-Momentum Coefficient," *Computer Physics Communications*, 56(2): 231–248, Dec. 1989.
34. Risbo, T., "Fourier Transform Summation of Legendre Series and D-Functions," *Journal of Geodesy*, 70(7): 383–396, April 1996.
35. Rose, M.E., *Elementary Theory of Angular Momentum*, John Wiley and Sons, New York, 1957 (Dover Edition, 1995).
36. Schulten, K., Gordon, R.G., "Recursive Evaluation of 3J and 6J Coefficients," *Computer Physics Communications*, 11(2): 269–278, 1976.
37. Scott, N.S., Milligan, P., Riley, H.W.C., "The Parallel Computation of Racah Coefficients Using Transputers," *Computer Physics Communications*, 46(1): 83–98, July 1987.
38. Sherborne, B.S., Stedman, G.E., "Recursive Generation of Cartesian Angular-Momentum Coupling Trees for $SO(3)$," *Computer Physics Communications*, 59(2): 417–428, June 1990.
39. Sugiura, M., *Unitary Representations and Harmonic Analysis*, 2nd ed., North-Holland, Amsterdam, 1990.
40. Talman, J., *Special Functions*, W. A. Benjamin, Inc., Amsterdam, 1968.
41. Torrésani, B., "Position-Frequency Analysis for Signals Defined on Spheres," *Signal Processing*, 43:341–346, 1995.
42. Varshalovich, D.A., Moskalev, A.N., Khersonskii, V.K., *Quantum Theory of Angular Momentum*, World Scientific, Singapore, 1988.
43. Vilenkin, N.J.. Klimyk, A.U., *Representation of Lie Groups and Special Functions*, Vols. 1-3, Kluwer Academic Publ., Dordrecht, Holland 1991.
44. Wei, L.Q., "Unified Approach for Exact Calculation of Angular Momentum Coupling and Recoupling Coefficients," *Computer Physics Communications* 120(2–3): 222–230, Aug. 1999.

45. Wightman, A.S., "On the Localizability of Quantum Mechanical Systems," *Reviews in Modern Physics*, 34: 845–872, 1962.

46. Wigner, E.P., *Group Theory and its Applications to the Quantum Mechanics of Atomic Spectra*, Academic Press, New York, 1959.

47. Willsky, A.S., "Dynamical Systems Defined on Groups: Structural Properties and Estimation," Ph.D. Dissertation, MIT, 1973.

48. Yan, Y., Chirikjian, G., "Voronoi Cells in Lie Groups and Coset Decompositions: Implications for Optimization, Integration, and Fourier Analysis," in *Proceedings of the 52nd Annual IEEE Conference on Decision and Control*, pp. 1137–1143, Dec., 2013.

49. Zhao, D.Q., Zare, R.N., "Numerical Computation of 9-J Symbols," *Molecular Physics*, 65(5): 1263–1268, Dec. 1988.

Harmonic Analysis on the Euclidean Motion Groups

10.1 Introduction

The Euclidean motion group[1], $SE(N)$, is the semidirect product of $\mathbb{R}^N$ with the special orthogonal group, $SO(N)$. That is, $SE(N) = \mathbb{R}^N \rtimes SO(N) = SO(N) \ltimes \mathbb{R}^N$. We denote elements $g \in SE(N)$ as $g = (\mathbf{a}, A)$ or $g = (A, \mathbf{a})$ where $A \in SO(N)$ and $\mathbf{a} \in \mathbb{R}^N$. For any $g = (\mathbf{a}, A)$ and $h = (\mathbf{r}, R) \in SE(N)$, the group law is written as $g \circ h = (\mathbf{a} + A\mathbf{r}, AR)$, and $g^{-1} = (-A^T \mathbf{a}, A^T)$. Alternately, we can represent any element of $SE(N)$ as an $(N+1) \times (N+1)$ homogeneous transformation matrix of the form:

$$H(g) = \begin{pmatrix} A & \mathbf{a} \\ \mathbf{0}^T & 1 \end{pmatrix}$$

Clearly, $H(g)H(h) = H(g \circ h)$ and $H(g^{-1}) = H^{-1}(g)$, and the mapping $g \to H(g)$ is an isomorphism between $SE(N)$ and the set of homogeneous transformation matrices.

$SE(N)$ does not fit neatly into the most common classes of Lie groups of interest to mathematicians. That is, it is not compact, not semi-simple, not nilpotent, and not reductive. However, $SE(2)$ is a solvable Lie group, and general methods for constructing unitary representations of solvable Lie groups have been known for some time (see, e.g., [2, 20, 25]). $SE(N)$ for $N > 2$ is not solvable in general, but it is always a semi-direct product, and therefore other methods for building representations exist. In the past 50 years, the representation theory and harmonic analysis for the Euclidean groups have been developed in the pure mathematics and mathematical physics literature. The study of matrix elements of irreducible unitary representation of $SE(3)$ was initiated by N.Vilenkin [43] in 1957 (some particular matrix elements are also given in [44]). The most complete study of $\widetilde{SE}(3)$ (the universal covering group of $SE(3)$) with application to the harmonic analysis was given by W. Miller in [21]. The representations of $SE(3)$ were also studied in [11, 17, 26, 39].

However, despite the considerable progress in mathematical developments of the representation theory of $SE(3)$, these achievements have not yet been widely incorporated in engineering and applied fields. In subsequent chapters we try to fill this gap. In this chapter, we review the representation theory of $SE(3)$, derive the matrix elements of the irreducible unitary representations and define the Fourier transform for $SE(3)$.

---

[1] Recall from Chapter 6 that the notation $SE(N)$ comes from the terminology Special Euclidean group of N-dimensional space.

We derive new symmetry and operational properties of the Fourier transform and give explicit examples of Fourier transforms of functions on the motion group.

10.2 Irreducible Unitary Representations of $SE(2)$

Each element of $SE(2)$ is parameterized in either rectangular or polar coordinates as:

$$g(a_1, a_2, \theta) = \begin{pmatrix} \cos\theta & -\sin\theta & a_1 \\ \sin\theta & \cos\theta & a_2 \\ 0 & 0 & 1 \end{pmatrix}$$

or

$$g(a, \phi, \theta) = \begin{pmatrix} \cos\theta & -\sin\theta & a\cos\phi \\ \sin\theta & \cos\theta & a\sin\phi \\ 0 & 0 & 1 \end{pmatrix},$$

where $a = \|\mathbf{a}\|$.

A unitary representation of $SE(2)$ (see [24, 37, 39, 42, 44] for general definition) is defined by the unitary operator

$$(U(g,p)\tilde{\varphi})(\mathbf{x}) \doteq e^{-ip(\mathbf{a}\cdot\mathbf{x})}\tilde{\varphi}(A^T\mathbf{x}) \doteq \tilde{\varphi}_g(\mathbf{x}) \tag{10.1}$$

for each $g = (\mathbf{a}, A) = g(a_1, a_2, \theta) \in SE(2)$. Here $p \in \mathbb{R}^+$, and $\mathbf{x} \cdot \mathbf{y} = x_1 y_1 + x_2 y_2$. The vector $\mathbf{x}$ is a unit vector ($\mathbf{x} \cdot \mathbf{x} = 1$).

By definition, group representations observe the homomorphism property, which in this case is seen as follows:

$$\begin{aligned}
(U(g,p)U(h,p)\tilde{\varphi})(\mathbf{x}) &= (U(g,p)(U(h,p)\tilde{\varphi}))(\mathbf{x}) \\
&= (U(g,p)\tilde{\varphi}_h)(\mathbf{x}) = e^{-ip(\mathbf{a}\cdot\mathbf{x})}\tilde{\varphi}_h(A^T\mathbf{x}) \\
&= e^{-ip(\mathbf{a}\cdot\mathbf{x})}e^{-ip(\mathbf{r}\cdot(A^T\mathbf{x}))}\tilde{\varphi}(R^T A^T\mathbf{x}) \\
&= e^{-ip(\mathbf{a}+A\mathbf{r})\cdot\mathbf{x}}\tilde{\varphi}((AR)^T\mathbf{x}) \\
&= (U(g \circ h, p)\tilde{\varphi})(\mathbf{x}).
\end{aligned}$$

Since $\mathbf{x}$ is a unit vector, the function $\tilde{\varphi}(\mathbf{x}) = \tilde{\varphi}(\cos\psi, \sin\psi) \equiv \varphi(\psi)$ is a function on the unit circle. Henceforth we will not distinguish between $\tilde{\varphi}$ and φ.

10.2.1 Matrix Elements of IURs of $SE(2)$

Any function $\varphi(\psi) \in \mathcal{L}^2(\mathbb{S}^1)$ can be expressed as a weighted sum (Fourier series) of orthonormal basis functions as $\varphi(\psi) = \sum_{n\in\mathbb{Z}} c_n e^{in\psi}$. Likewise, the matrix elements of the operator $U(g,p)$ are expressed in this basis as

$$u_{mn}(g,p) = (e^{im\psi}, U(g,p)\,e^{in\psi}) = \frac{1}{2\pi}\int_0^{2\pi} e^{-im\psi}e^{-i(a_1 p\cos\psi + a_2 p\sin\psi)}e^{in(\psi-\theta)}\,d\psi \tag{10.2}$$

$\forall\, m, n \in \mathbb{Z}$, where the inner product $(\cdot, \cdot)$ is defined as

$$(\varphi_1, \varphi_2) \doteq \frac{1}{2\pi}\int_0^{2\pi} \overline{\varphi_1(\psi)}\varphi_2(\psi)\,d\psi.$$

It is easy to see that $(U(g,p)\varphi_1, U(g,p)\varphi_2) = (\varphi_1, \varphi_2)$ and that $U(g,p)$ is therefore unitary with respect to this inner product.

A number of works including [24], [39], and [42] have shown that the matrix elements of this representation are expressed as

$$u_{mn}(g(a,\phi,\theta),p) = i^{n-m} e^{-i[n\theta+(m-n)\phi]} J_{n-m}(p\,a) \tag{10.3}$$

where $J_\nu(x)$ is the νth order Bessel function.

From this expression, and the fact that $U(g,p)$ is a unitary representation, we have that

$$u_{mn}(g^{-1}(a,\phi,\theta),p) = u_{mn}^{-1}(g(a,\phi,\theta),p) =$$
$$\overline{u_{nm}(g(a,\phi,\theta),p)} = i^{n-m} e^{i[m\theta+(n-m)\phi]} J_{m-n}(pa). \tag{10.4}$$

Henceforth no distinction will be made between the operator $U(g,p)$ and the corresponding infinite-dimensional matrix with elements $u_{mn}(g,p)$.

Symmetry properties.   The matrix elements are related by the symmetries:

$$\overline{u_{mn}(g,p)} = (-1)^{m-n} u_{-m,-n}(g,p), \tag{10.5}$$

$$u_{mn}(g(-a,\phi,\theta),p) \doteq u_{mn}(g(a,\phi\pm\pi,\theta),p) = (-1)^{m-n} u_{m,n}(g(a,\phi,\theta),p) \tag{10.6}$$

and

$$(-1)^{m-n} u_{m,n}(g(a,\phi-\theta,-\theta),p) = \overline{u_{nm}(g(a,\phi,\theta),p)}. \tag{10.7}$$

The equality in (10.7) follows from (10.4) and (10.6).

10.2.2 Irreducibility of the Representations $U(g,p)$ of $SE(2)$

Since $SE(2)$ is neither compact nor commutative, the representation matrices will be infinite-dimensional. Therefore, it will be more convenient to show the irreducibility of the operator $U(g,p)$ rather than the corresponding matrix. To this end, we examine one parameter subgroups generated by exponentiating linearly independent basis elements of the Lie algebra $se(2)$. Using the basis

$$\tilde{X}_1 = \begin{pmatrix} 0 & 0 & 1 \\ 0 & 0 & 0 \\ 0 & 0 & 0 \end{pmatrix}; \quad \tilde{X}_2 = \begin{pmatrix} 0 & 0 & 0 \\ 0 & 0 & 1 \\ 0 & 0 & 0 \end{pmatrix}; \quad \tilde{X}_3 = \begin{pmatrix} 0 & -1 & 0 \\ 1 & 0 & 0 \\ 0 & 0 & 0 \end{pmatrix};$$

we find

$$g_1(t) \doteq \exp(t\tilde{X}_1) = \begin{pmatrix} 1 & 0 & t \\ 0 & 1 & 0 \\ 0 & 0 & 1 \end{pmatrix};$$

$$g_2(t) \doteq \exp(t\tilde{X}_2) = \begin{pmatrix} 1 & 0 & 0 \\ 0 & 1 & t \\ 0 & 0 & 1 \end{pmatrix};$$

$$g_3(t) \doteq \exp(t\tilde{X}_3) = \begin{pmatrix} \cos t & -\sin t & 0 \\ \sin t & \cos t & 0 \\ 0 & 0 & 1 \end{pmatrix}.$$

The corresponding differential operators $\tilde{X}_i^R$ (in polar coordinates) are

$$\tilde{X}_1^R = \cos(\theta - \phi)\frac{\partial}{\partial a} + \frac{\sin(\theta - \phi)}{a}\frac{\partial}{\partial \phi}$$

$$\tilde{X}_2^R = -\sin(\theta - \phi)\frac{\partial}{\partial a} + \frac{\cos(\theta - \phi)}{a}\frac{\partial}{\partial \phi}$$

$$\tilde{X}_3^R = \frac{\partial}{\partial \theta}.$$

The operators $\tilde{X}_i^L$ are

$$\tilde{X}_1^L = \cos\phi\frac{\partial}{\partial a} - \frac{\sin\phi}{a}\frac{\partial}{\partial \phi}$$

$$\tilde{X}_2^L = \sin\phi\frac{\partial}{\partial a} + \frac{\cos\phi}{a}\frac{\partial}{\partial \phi}$$

$$\tilde{X}_3^L = \frac{\partial}{\partial \theta} + \frac{\partial}{\partial \phi}.$$

These are completely analogous to the operators X_i^R and X_i^L for $SO(3)$. The tilde is to distinguish the $SE(2)$ case from the $SO(3)$ case. These operators act on functions on the group.

It follows from (10.1) that

$$U(g_1(t), p)\varphi(\psi) = e^{-ipt\cos\psi}\varphi(\psi);$$
$$U(g_2(t), p)\varphi(\psi) = e^{-ipt\sin\psi}\varphi(\psi);$$
$$U(g_3(t), p)\varphi(\psi) = \varphi(\psi - t).$$

Differentiating with respect to t and setting $t = 0$, we define the operators

$$\hat{X}_i(p)\varphi(\psi) = \left.\frac{dU(\exp(tX_i), p)\varphi(\psi)}{dt}\right|_{t=0}.$$

Explicitly,

$$\hat{X}_1(p)\varphi(\psi) = -ip\cos\psi\,\varphi(\psi);$$
$$\hat{X}_2(p)\varphi(\psi) = -ip\sin\psi\,\varphi(\psi);$$
$$\hat{X}_3(p)\varphi(\psi) = -\frac{d\varphi}{d\psi}.$$

In analogy with $SU(2)$, we define the operators

$$\hat{Y}_+(p) = \hat{X}_1(p) + i\hat{X}_2(p); \quad \hat{Y}_-(p) = \hat{X}_1(p) - i\hat{X}_2(p); \quad \hat{Y}_3(p) = \hat{X}_3(p).$$

Since $\varphi \in \mathcal{L}^2(\mathbb{S}^1)$, basis elements are of the form $e^{ik\psi}$. These basis elements are transformed by the operators $\hat{Y}_\pm$ and $\hat{Y}_3$ as [42]:

$$\hat{Y}_+(p)\, e^{ik\psi} = -ipe^{i(k+1)\psi}; \quad \hat{Y}_-(p)\, e^{ik\psi} = -ipe^{i(k-1)\psi}; \quad \hat{Y}_3(p)\, e^{ik\psi} = -ike^{ik\psi}.$$

As in the case of $SU(2)$, $\hat{Y}_+$ and $\hat{Y}_+$ always "push" basis elements to "adjacent" subspaces. Since no subspaces are left invariant by $\hat{Y}_\pm$, the representation operators $U(g, p)$ must be irreducible.

10.3 The Fourier Transform for $SE(2)$

Given the representations from the previous section, we are ready for the following definition:

Definition 10.1. *[37]* The Fourier transform of a function $f \in \mathcal{L}^2(G)$ (where $G = SE(2)$) is

$$\hat{f}(p) \doteq \int_G f(g)U(g^{-1},p)\,d(g)$$

with the volume element normalized as

$$d(g) = \frac{1}{4\pi^2}\,da_1\,da_2\,d\theta = \frac{1}{4\pi^2}\,a\,da\,d\phi\,d\theta$$

and

$$\int_G = \int_{\theta=0}^{2\pi}\int_{a_2=-\infty}^{\infty}\int_{a_1=-\infty}^{\infty} = \int_{\theta=0}^{2\pi}\int_{\phi=0}^{2\pi}\int_{a=0}^{\infty}.$$

Theorem 10.2. *The corresponding convolution theorem is*

$$\widehat{(f_1 * f_2)}(p) = \hat{f}_2(p)\hat{f}_1(p),$$

the Parseval/Phancherel equality is

$$\int_G |f(g)|^2\,d(g) = \int_0^\infty \|\hat{f}(p)\|^2\,p\,dp,$$

and the reconstruction formula is

$$f(g) = \int_0^\infty \mathrm{tr}(\hat{f}(p)U(g,p))\,p\,dp.$$

Proof. See Subsections 10.4.1, 10.4.4, and 10.4.2, respectively.

Sometimes the notation $\mathcal{F}(f) = \hat{f}$ and $\mathcal{F}^{-1}(\hat{f}) = f$ are useful to denote the operation of converting a function on $SE(2)$ to its Fourier transform, and to convert the Fourier transform back to the function. As with the Fourier transform of functions on $\mathbb{R}^N$,

$$\mathcal{F}\mathcal{F}^{-1}(\hat{f}) = \hat{f} \qquad \mathcal{F}^{-1}\mathcal{F}(f) = f,$$

and so we write symbolically that

$$\mathcal{F}\mathcal{F}^{-1} = \mathcal{F}^{-1}\mathcal{F} = id$$

where *id* is the identity operator. A proof that these identities hold is given in [37]. The fact that the inverse transform works depends on $\{U(g,p)\}$ being a complete set of irreducible representations, and the fact that each $U(g,p)$ is unitary allows us to write $U(g^{-1},p) = U^*(g,p)$ instead of computing the inverse of an infinite-dimensional matrix. In Subsection 10.4.2 we show that the inversion formula works.

The matrix elements of the transform can be calculated using the matrix elements of $U(g,p)$ defined in (10.3) as:

$$\hat{f}_{mn}(p) = (e^{im\psi}, \hat{f}(p)\,e^{in\psi}) = \int_G f(g)u_{mn}(g^{-1},p)\,d(g).$$

Likewise, the inverse transform can be written in terms of elements as:

$$f(g) = \sum_{n,m\in\mathbb{Z}}\int_0^\infty \hat{f}_{mn}(p)u_{nm}(g,p)\,p\,dp.$$

10.4 Properties of Convolution and Fourier Transforms of Functions on $SE(2)$

In Subsection 10.4.1, it is shown that the Fourier transform defined in Section 10.3 possesses the convolution property in an analogous way to the usual Fourier transform. In Subsection 10.4.2, the inversion formula is proved. In Subsection 10.4.4, Parseval's inequality is proved. In Subsection 10.4.5, some of the operational properties of the Fourier transform pair for functions on $SE(2)$ are derived.

10.4.1 The Convolution Theorem

Let us assume that there are real scalar-valued functions $f_1(\cdot)$, $f_2(\cdot) \in \mathcal{L}^2(G)$ where $G = SE(2)$. Recall that one of the most powerful properties of the Fourier transform of functions on $\mathbb{R}^N$ is that the Fourier transform of the convolution of two functions is the product of the Fourier transforms of the functions. This property extends to the concept of a Fourier transform for functions on $SE(2)$. The proof of this fact presented here follows that of Sugiura [37].

Given that

$$(f_1 * f_2)(g) = \int_G f_1(h) \, f_2(h^{-1} \circ g) \, d(h),$$

we get,

$$\widehat{(f_1 * f_2)}(p) = \int_G (f_1 * f_2)(g) U(g^{-1}, p) \, d(g) =$$

$$\int_G \left(\int_G f_1(h) \, f_2(h^{-1} \circ g) \, d(h) \right) U(g^{-1}, p) \, d(g). \qquad (10.8)$$

Switching the order of integration, we get:

$$\widehat{(f_1 * f_2)}(p) = \int_G \left(\int_G f_2(h^{-1} \circ g) U(g^{-1}, p) \, d(g) \right) f_1(h) \, d(h). \qquad (10.9)$$

Since $d(\cdot)$ is left and right invariant

$$\int_G f(h \circ g) \, d(g) = \int_G f(g \circ h) \, d(g) = \int_G f(g^{-1}) \, d(g) = \int_G f(g) \, d(g)$$

for any function $F \in \mathcal{L}^2(SE(2))$. These facts allow us to write the inner integral in (10.9) as

$$\int_G f_2(h^{-1} \circ (h \circ g)) U((h \circ g)^{-1}, p) \, d(g) = \int_G f_2(g) U(g^{-1} \circ h^{-1}, p) \, d(g).$$

Since $U(g, p)$ is a representation of $SE(2)$,

$$U(g^{-1} \circ h^{-1}, p) = U(g^{-1}, p) U(h^{-1}, p),$$

and so

$$\widehat{(f_1 * f_2)}(p) = \int_G \left(\int_G f_2(g) U(g^{-1}, p) U(h^{-1}, p) \, d(g) \right) f_1(h) \, d(h)$$

$$= \left(\int_G f_2(g) U(g^{-1}, p) \, d(g) \right) \left(\int_G f_1(h) U(h^{-1}, p) \, d(h) \right)$$

$$= \hat{f}_2(p) \hat{f}_1(p).$$

The order in which $U(g^{-1}, p)$ and $U(h^{-1}, p)$ appear is important because they are representations of $SE(2)$, which is not a commutative group. As seen above, the fact that $U(g, p)$ is a representation is critical for the separation required for the convolution theorem to hold. The noncommutative nature of $SE(2)$ expressed in $U(g, p)$ is responsible for the reversed order of the product of the Fourier transforms. Some authors define the Fourier transform of functions on groups differently so that the order of the product of the transform is the same as the order of the convolved functions. We use the definition of Fourier transform given in Section 10.3 because it is the most analogous to the standard Fourier transform.

10.4.2 Proof of the Inversion Formula

We now present a proof that the Fourier inversion formula for functions on $SE(2)$ actually works. This proof is coordinate-dependent to avoid the introduction of additional mathematical machinery. The proof presented here is a variant of one found in [42].

Functions $f \in \mathcal{L}^2(SE(2))$ can be expressed in a series of the form

$$f(g) = f(a, \phi, \theta) = \sum_{j,k \in \mathbb{Z}} F_{jk}(a) \, e^{-ij\phi} e^{-ik\theta}$$

where $g = g(a, \phi, \theta) \in SE(2)$.

The matrix elements of the Fourier transform of this function (as defined in Section 10.3) are

$$\hat{f}_{mn} = \int_G f(g) u_{mn}(g^{-1}, p) \, d(g) =$$

$$\frac{1}{(2\pi)^2} \int_{-\pi}^{\pi} \int_{-\pi}^{\pi} \int_0^{\infty} \left(\sum_{j,k \in \mathbb{Z}} F_{jk}(a) \, e^{-ij\phi} e^{-ik\theta} \right) \left(i^{n-m} e^{i[m\theta + (n-m)\phi]} J_{m-n}(pa) \right) a \, da \, d\phi \, d\theta.$$

Rearranging the integrals, this is rewritten as

$$\sum_{j,k \in \mathbb{Z}} \left(i^{n-m} \int_0^{\infty} F_{jk}(a) J_{m-n}(pa) a \, da \right) \left(\frac{1}{2\pi} \int_{-\pi}^{\pi} e^{-ij\phi} e^{i(n-m)\phi} \, d\phi \right) \left(\frac{1}{2\pi} \int_{-\pi}^{\pi} e^{-ik\theta} e^{im\theta} \, d\theta \right)$$

$$= \sum_{j,k \in \mathbb{Z}} \delta_{k,m} \delta_{j,n-m} \left(i^{n-m} \int_0^{\infty} F_{jk}(a) J_{m-n}(pa) a \, da \right) = i^{n-m} \int_0^{\infty} F_{n-m,m}(a) J_{m-n}(pa) a \, da.$$

$\delta_{p,q}$ is equal to 1 if $p = q$ and zero otherwise, and there is no summation implied by the repeated indices in the last term in the above integral.

The fact that this Fourier transform matrix with elements $\hat{f}_{mn}$ is inverted using the inversion formula to reconstruct $f(g)$ is seen as follows:

$$f(g) = \int_0^{\infty} \operatorname{tr}(\hat{f}(p) U(g, p)) \, p \, dp = \sum_{m,n \in \mathbb{Z}} \int_0^{\infty} \hat{f}_{mn} u_{nm}(g, p) \, p \, dp =$$

$$\sum_{m,n \in \mathbb{Z}} \int_0^{\infty} \left(i^{n-m} \int_{a=0}^{\infty} F_{n-m,m}(a) J_{m-n}(pa) a \, da \right) \left(i^{m-n} e^{-i[m\theta + (n-m)\phi]} J_{m-n}(pa) \right) p \, dp =$$

$$\sum_{m,n \in \mathbb{Z}} e^{-i[m\theta + (n-m)\phi]} \int_0^{\infty} \left(\int_{a=0}^{\infty} F_{n-m,m}(a) J_{m-n}(pa) a \, da \right) J_{m-n}(pa) \, p \, dp.$$

From here, the inversion is exactly the same as the Hankel transform pair (3.52)-(3.53):

$$\hat{\phi}(p) = \int_0^\infty \phi(a) J_\nu(pa) a\, da,$$

then

$$\phi(a) = \int_0^\infty \hat{\phi}(p) J_\nu(pa)\, p\, dp.$$

In our case, $\hat{\phi}(p) = \hat{F}_{n-m,m}(p)$, $\phi(a) = F_{n-m,m}(a)$, and $\nu = m - n$. This means that

$$\sum_{m,n\in\mathbb{Z}} \int_0^\infty \hat{f}_{mn} u_{nm}(g,p)\, p\, dp = \sum_{m,n\in\mathbb{Z}} e^{-i[m\theta+(n-m)\phi]} F_{n-m,m}(a)$$

$$= \sum_{j,k\in\mathbb{Z}} F_{jk}(a)\, e^{-ij\phi} e^{-ik\theta}$$

$$= f(g).$$

This last step was simply a renaming of variables: $k = m$, and $n - m = j$. Since the sums over n and m are over $\mathbb{Z}$, these shifts do not change the range of summation.

10.4.3 Proof of $SE(2)$-Completeness Formula without Using a Basis

Introducing the Fourier basis $\{e^{in\theta}\}$ converts the $SE(2)$-IUR operators into infinite-dimensional matrices with entries involving Bessel functions. This can be a valuable tool for concrete calculations in applications. However, the proof of the inversion formula and Parseval's equality can be obtained more directly than in the derivation we presented previously by establishing a completeness formula. This can be achieved without introducing a basis as was done in a very elegant way in [37]. This leads to simpler calculations than those that depend on introducing a basis, and the same methodology will be useful later when considering the $SE(3)$ case, which is more complicated.

Let $\varphi \in \mathcal{L}^2(\mathbb{S}^1)$ and let $U(g,p)$ be the IUR operator in (10.1). Then

$$(U(g,p)\,\varphi)(\psi) = e^{-ip\,\mathbf{a}\cdot\mathbf{u}(\psi)}\varphi(\psi - \theta)$$

where $g = (R(\theta), \mathbf{a})$, $\psi \in [0, 2\pi)$ parameterizes $\mathbb{S}^1$, and $\mathbf{u}(\psi) = [\cos\psi, \sin\psi]^T$. Then

$$(U(g^{-1},p)\,\varphi)(\psi) = e^{ip\,(R^T(\theta)\mathbf{a})\cdot\mathbf{u}(\psi)}\varphi(\psi + \theta) = e^{ip\,\mathbf{a}\cdot\mathbf{u}(\psi+\theta)}\varphi(\psi + \theta).$$

As an operator, the $SE(2)$-Fourier transform is written as

$$(\hat{f}(p)\,\varphi)(\psi) = \int_{SE(2)} f(g)(U(g^{-1},p)\,\varphi)(\psi)\, dg$$

$$= \frac{1}{(2\pi)^2} \int_0^{2\pi} \int_{\mathbb{R}^2} f(R(\theta),\mathbf{a})\, e^{ip\,\mathbf{a}\cdot\mathbf{u}(\psi+\theta)}\varphi(\psi + \theta)\, d\mathbf{a}\, d\theta$$

$$= \frac{1}{2\pi} \int_0^{2\pi} \left[\frac{1}{2\pi} \int_{\mathbb{R}^2} f(R(\theta),\mathbf{a})\, e^{ip\,\mathbf{a}\cdot\mathbf{u}(\psi+\theta)}\, d\mathbf{a}\right] \varphi(\psi + \theta)\, d\theta$$

$$= \frac{1}{2\pi} \int_0^{2\pi} \tilde{f}(R(\theta), -p\,\mathbf{u}(\psi + \theta))\, \varphi(\psi + \theta)\, d\theta$$

where $\tilde{f}$ denotes the classical $\mathbb{R}^2$ Fourier transform on the translational variables for each fixed value of θ according to the normalization convention in Subsection 2.2.2.

Defining $g' = (R(\theta'), \mathbf{a}')$ and applying the operator $U(g', p)$ gives:

$$(U(g', p)(\hat{f}(p)\,\varphi))(\psi) = \frac{1}{2\pi} e^{-ip\,\mathbf{a}'\cdot\mathbf{u}(\psi)} \int_0^{2\pi} \tilde{f}(R(\theta), -p\,\mathbf{u}(\psi + \theta - \theta'))\,\varphi(\psi + \theta - \theta')\,d\theta\,.$$

Integrating both sides with respect to the measure $p\,dp$, using the associativity of operator composition,

$$(U(g', p)(\hat{f}(p)\,\varphi))(\psi) = (U(g', p)\hat{f}(p)\,\varphi)(\psi)\,,$$

and making the change of variables,

$$\theta = \psi' - \psi + \theta' \text{ and } d\theta = d\psi'\,,$$

then gives

$$\int_0^\infty (U(g', p)\hat{f}(p)\,\varphi)(\psi)\,p\,dp =$$

$$\int_0^{2\pi} \left[\frac{1}{2\pi} \int_0^\infty e^{-ip\,\mathbf{a}'\cdot\mathbf{u}(\psi)}\,\tilde{f}(R(\theta), -p\,\mathbf{u}(\psi + \theta - \theta'))\,\varphi(\psi + \theta - \theta')\,p\,dp \right]\,d\theta$$

$$= \int_0^{2\pi} \left[\frac{1}{2\pi} \int_0^\infty e^{-ip\,\mathbf{a}'\cdot\mathbf{u}(\psi)}\,\tilde{f}(R(\psi' - \psi + \theta'), -p\,\mathbf{u}(\psi'))\,\varphi(\psi')\,p\,dp \right]\,d\psi'\,.$$

This means that

$$\int_0^\infty (U(g', p)\hat{f}(p)\,\varphi)(\psi)\,p\,dp = \int_0^{2\pi} K(\psi, \psi')\,\varphi(\psi')\,d\psi'$$

where

$$K(\psi, \psi') = \frac{1}{2\pi} \int_0^\infty e^{-ip\,\mathbf{a}'\cdot\mathbf{u}(\psi)}\,\tilde{f}(R(\psi' - \psi + \theta'), -p\,\mathbf{u}(\psi'))\,p\,dp\,.$$

Therefore, defining $\mathbf{p} = -p\,\mathbf{u}(\psi')$, and using Theorem E.1 from Appendix E.5, gives

$$\int_0^\infty \mathrm{tr}\left[U(g', p)\hat{f}(p) \right]\,p\,dp = \int_0^{2\pi} K(\psi, \psi)\,d\psi$$

$$= \frac{1}{2\pi} \int_0^{2\pi} \int_0^\infty e^{-ip\,\mathbf{a}'\cdot\mathbf{u}(\psi)}\,\tilde{f}(R(\theta'), -p\,\mathbf{u}(\psi))\,p\,dp\,d\psi$$

$$= \frac{1}{2\pi} \int_{\mathbb{R}^2} \tilde{f}(R(\theta'), \mathbf{p})\,e^{i\,\mathbf{a}'\cdot\mathbf{p}}\,d\mathbf{p}$$

$$= f(R(\theta'), \mathbf{a}')$$

$$= f(g')\,.$$

Evaluating this expression with $f(g) = \delta(g)$ gives the completeness relation

$$\int_0^\infty \mathrm{tr}\,[U(g', p)]\,p\,dp = \delta(g')\,. \tag{10.10}$$

10.4.4 Parseval's Equality

We now present a proof of Parseval's equality (also called the Plancherel formula) so that we have a mechanism for regularizing integral equations on $SE(2)$ using the Fourier transform. The proof presented here is similar to the one found in [37].

We begin by defining

$$f^*(g) \doteq \overline{f(g^{-1})} \in \mathcal{L}^2(SE(2)), \qquad \forall \; f \in \mathcal{L}^2(SE(2)).$$

Then

$$h(g) = f * f^*(g) =$$

$$\int_G f(h) f^*(h^{-1} \circ g) \, d(h) = \int_G f(h) \overline{f(g^{-1} \circ h)} \, d(h).$$

Evaluating this at the identity element $g = e$,

$$h(e) = \int_G f(h) \overline{f(h)} \, d(h) = \int_G |f(g)|^2 \, d(g).$$

The function $h(g)$ can also be expressed via the inversion formula as

$$h(g) = \int_0^\infty \operatorname{tr}(\hat{h}(p) U(g,p)) \, p \, dp.$$

Evaluated at the identity element, $U(e,p)$ is the identity operator, and so:

$$h(e) = \int_0^\infty \operatorname{tr}(\hat{h}(p)) \, p \, dp = \int_0^\infty \operatorname{tr}(\hat{f}^*(p)\hat{f}(p)) \, p \, dp.$$

But

$$\mathcal{F}_{mn}(f^*) = \int_G \overline{f(g^{-1})} u_{mn}(g^{-1},p) \, d(g) = \int_G \overline{f(g)} u_{mn}(g,p) \, d(g).$$

This follows from the fact that $G = SE(2)$ is a unimodular group (possessing a volume element which is left and right invariant), and thus for any function $f \in \mathcal{L}^2(G)$

$$\int_G f(g^{-1}) \, d(g) = \int_G f(g) \, d(g).$$

From (10.4) we can then write:

$$\mathcal{F}_{mn}(f^*) = \int_G \overline{f(g)} u_{nm}(g^{-1},p) \, d(g) = \overline{\hat{f}}_{nm} = \overline{\hat{f}}_{mn}^T = (\hat{f}_{mn})^*,$$

i.e., $\mathcal{F}(f^*) = (\hat{f})^*$ is the complex conjugate transpose of the Fourier transform matrix $\hat{f}$. This means that

$$h(e) = \int_0^\infty \operatorname{tr}((\hat{f})^*(p)\hat{f}(p)) \, p \, dp = \int_0^\infty \|\hat{f}(p)\|^2 \, p \, dp.$$

Equating the two expressions for $h(e)$, we get Parseval's equality for $SE(2)$:

$$\int_G |f(g)|^2 \, d(g) = \int_0^\infty \|\hat{f}(p)\|^2 \, p \, dp.$$

10.4.5 Operational Properties

Given a differentiable function f (with derivative f') such that $f, f' \in \mathcal{L}^2(\mathbb{R})$, recall that the Fourier transform of f' is defined using the inversion formula in (2.10) by differentiating under the integral:

$$\frac{df}{dx} = \frac{1}{\sqrt{2\pi}} \frac{d}{dx} \int_{-\infty}^{\infty} \hat{f}(\omega) u(x, \omega) \, d\omega = \frac{1}{\sqrt{2\pi}} \int_{-\infty}^{\infty} \hat{f}(\omega) \frac{\partial u}{\partial x} \, d\omega. \tag{10.11}$$

Since $\frac{\partial u}{\partial x} = i\omega u(x, \omega)$, we get that the Fourier transform of df/dx is $i\omega \hat{f}(\omega)$ where $\hat{f}(\omega)$ is the Fourier transform of $f(x)$.

The same argument can be used to generate the Fourier transform matrices of derivatives of functions $f(g)$ where $g \in SE(2)$. For example, using the the inverse transform given in Section 10.3 we get that

$$a \frac{\partial f}{\partial a} = a \int_0^{\infty} \sum_{m,n \in \mathbb{Z}} \hat{f}_{nm} \frac{\partial u_{mn}}{\partial a} \, p \, dp. \tag{10.12}$$

Since $u_{mn}(g(a, \phi, \theta), p) = i^{n-m} e^{-i[n\theta + (m-n)\phi]} J_{n-m}(pa)$, we get that $\partial u_{mn}/\partial a = i^{n-m} e^{-i[n\theta + (m-n)\phi]} J'_{n-m}(pa)p$. Integrating the part of the expression in (10.12) which depends on p by parts:

$$a \int_0^{\infty} \hat{f}_{nm}(p) J'_{n-m}(pa) p^2 \, dp = - \int_0^{\infty} \frac{d}{dp} (p^2 \hat{f}_{nm}(p)) J_{n-m}(pa) \, dp$$

$$= - \int_0^{\infty} \left(2\hat{f}_{nm}(p) + p\hat{f}'_{nm}(p) \right) J_{n-m}(pa) \, p \, dp.$$

Here we have used the fact that

$$a J'_{n-m}(pa) = \frac{\partial}{\partial p} J_{n-m}(pa).$$

This means that

$$a \frac{\partial f}{\partial a} = - \int_0^{\infty} \sum_{m,n \in \mathbb{Z}} \left(2\hat{f}_{nm} + p\hat{f}'_{nm} \right) u_{mn} \, p \, dp,$$

and so

$$\mathcal{F}\left(a \frac{\partial f}{\partial a} \right) = -2\hat{f} - p \frac{d\hat{f}}{dp}. \tag{10.13}$$

We can then use Parseval's equality together with integration by parts to show that:

$$\int_G \left| a \frac{\partial f}{\partial a} \right|^2 d(g) = \int_0^{\infty} \left\| p \frac{d\hat{f}}{dp} \right\|^2 p \, dp.$$

We find that the transforms of the derivatives of $f(g)$ with respect to ϕ and θ are:

$$\mathcal{F}\left(\frac{\partial f}{\partial \theta} \right) = D\hat{f} \tag{10.14}$$

and

$$\mathcal{F}\left(\frac{\partial f}{\partial \phi} \right) = \hat{f}D - D\hat{f} \tag{10.15}$$

where $D_{mn} = -im\delta_{mn}$ (no sum) is a diagonal matrix. This comes from the fact that $\partial u_{mn}/\partial \theta = (-in)i^{n-m} e^{-i[n\theta + (m-n)\phi]} J_{n-m}(pa) = -inu_{mn}$ (no sum), and $\partial u_{mn}/\partial \phi = -i(m-n)i^{n-m} e^{-i[n\theta + (m-n)\phi]} J_{n-m}(pa) = -i(m-n)u_{mn}$ (no sum).

The transforms of other derivatives follow from these by composition.

10.5 Differential Operators for $SE(3)$

The left and right differential operators $\tilde{X}_i^L$ and $\tilde{X}_i^R$ for $i = 1, ..., 6$ acting on functions on $SE(3)$ are calculated much like they were for the case of $SO(3)$ and $SE(2)$.

For small translational (rotational) displacements from the identity along (about) the ith coordinate axis, the homogeneous transforms representing infinitessimal motions look like

$$H_i(\epsilon) \doteq \exp(\epsilon\tilde{X}_i) \approx \mathbb{I}_{4\times4} + \epsilon\tilde{X}_i$$

where

$$\tilde{X}_1 = \begin{pmatrix} 0 & 0 & 0 & 0 \\ 0 & 0 & -1 & 0 \\ 0 & 1 & 0 & 0 \\ 0 & 0 & 0 & 0 \end{pmatrix} ; \quad \tilde{X}_2 = \begin{pmatrix} 0 & 0 & 1 & 0 \\ 0 & 0 & 0 & 0 \\ -1 & 0 & 0 & 0 \\ 0 & 0 & 0 & 0 \end{pmatrix} ; \quad \tilde{X}_3 = \begin{pmatrix} 0 & -1 & 0 & 0 \\ 1 & 0 & 0 & 0 \\ 0 & 0 & 0 & 0 \\ 0 & 0 & 0 & 0 \end{pmatrix} ;$$

$$\tilde{X}_4 = \begin{pmatrix} 0 & 0 & 0 & 1 \\ 0 & 0 & 0 & 0 \\ 0 & 0 & 0 & 0 \\ 0 & 0 & 0 & 0 \end{pmatrix} ; \quad \tilde{X}_5 = \begin{pmatrix} 0 & 0 & 0 & 0 \\ 0 & 0 & 0 & 1 \\ 0 & 0 & 0 & 0 \\ 0 & 0 & 0 & 0 \end{pmatrix} ; \quad \tilde{X}_6 = \begin{pmatrix} 0 & 0 & 0 & 0 \\ 0 & 0 & 0 & 0 \\ 0 & 0 & 0 & 1 \\ 0 & 0 & 0 & 0 \end{pmatrix} .$$

In the literature sometimes $\tilde{X}_1$, $\tilde{X}_2$, $\tilde{X}_3$ are denoted J_1, J_2, J_3 and $\tilde{X}_4$, $\tilde{X}_5$, $\tilde{X}_6$ are denoted P_1, P_2, P_3 because they correspond to infinitesimal rotations and translations about the 1, 2, and 3 axes. We use $\tilde{X}_i$ instead of X_i to avoid confusion with the $SO(3)$ case.

The commutation relations

$$[J_i, J_j] = \epsilon_{ijk}J_k \qquad [J_i, P_j] = \epsilon_{ijk}P_k \qquad [P_i, P_j] = 0$$

describe the Lie algebra $se(3)$.

It is often convenient to write these in vector form as

$$(\tilde{X}_1)^\vee = \begin{pmatrix} 1 \\ 0 \\ 0 \\ 0 \\ 0 \\ 0 \end{pmatrix} ; \quad (\tilde{X}_2)^\vee = \begin{pmatrix} 0 \\ 1 \\ 0 \\ 0 \\ 0 \\ 0 \end{pmatrix} ; \quad (\tilde{X}_3)^\vee = \begin{pmatrix} 0 \\ 0 \\ 1 \\ 0 \\ 0 \\ 0 \end{pmatrix} ;$$

$$(\tilde{X}_4)^\vee = \begin{pmatrix} 0 \\ 0 \\ 0 \\ 1 \\ 0 \\ 0 \end{pmatrix} ; \quad (\tilde{X}_5)^\vee = \begin{pmatrix} 0 \\ 0 \\ 0 \\ 0 \\ 1 \\ 0 \end{pmatrix} ; \quad (\tilde{X}_6)^\vee = \begin{pmatrix} 0 \\ 0 \\ 0 \\ 0 \\ 0 \\ 1 \end{pmatrix} .$$

Given that elements of $SE(3)$ (viewed as homogeneous transforms) are parameterized as $H = H(\mathbf{q})$, the differential operators take the form

$$(\tilde{X}_i^R f)(H) = \lim_{\epsilon\to 0} \frac{1}{\epsilon}[f(H \circ H_i(\epsilon)) - f(H)] = \left.\frac{df(H \circ H_i(\epsilon))}{dt}\right|_{t=0} \tag{10.16}$$

$$(\tilde{X}_i^L f)(H) = \lim_{\epsilon \to 0} \frac{1}{\epsilon} \left[f(H_i^{-1}(\epsilon) \circ H) - f(H) \right] = \left. \frac{df(H_i^{-1}(\epsilon) \circ H)}{dt} \right|_{t=0}. \tag{10.17}$$

Since H and $H_i(\epsilon)$ are 4×4 matrices, we henceforth drop the "$\circ$" notation since it is understood as matrix multiplication.

In analogy with the $SO(3)$ case, we observe for the case of $\tilde{X}_i^R$ that

$$H + \epsilon H \tilde{X}_i = H H_i(\epsilon) = H + \epsilon \sum_{j=1}^{6} \frac{\partial H}{\partial q_j} \left. \frac{\partial q_j^{R,i}}{\partial \epsilon} \right|_{\epsilon=0}.$$

We then have that

$$\tilde{X}_i = \sum_{j=1}^{6} H^{-1} \frac{\partial H}{\partial q_j} \left. \frac{\partial q_j^{R,i}}{\partial \epsilon} \right|_{\epsilon=0},$$

or

$$(\tilde{X}_i)^{\vee} = \sum_{j=1}^{6} \left(H^{-1} \frac{\partial H}{\partial q_j} \right)^{\vee} \left. \frac{\partial q_j^{R,i}}{\partial \epsilon} \right|_{\epsilon=0},$$

which is written as $(\tilde{X}_i)^{\vee} = \mathcal{J}_R(\mathbf{q}) \left. \frac{dq^{R,i}}{d\epsilon} \right|_{\epsilon=0}$ where $\mathcal{J}_R$ is the $SE(3)$ right Jacobian defined in Chapter 6. This allows us to solve for

$$\left. \frac{d\mathbf{q}^{R,i}}{d\epsilon} \right|_{\epsilon=0} = \mathcal{J}_R^{-1} (\tilde{X}_i)^{\vee},$$

which is used to calculate

$$\tilde{X}_i^R f = \sum_{j=1}^{6} \frac{\partial f}{\partial q_j} \left. \frac{\partial q_j^{R,i}}{\partial \epsilon} \right|_{\epsilon=0}.$$

We use the tilde to distinguish between the full motion and rotation operators. For the case when the rotations are parameterized with ZXZ Euler angles α, β, γ, and translations are parameterized in Cartesian coordinates a_1, a_2, a_3,

$$\tilde{X}_i^R = \begin{cases} X_i^R & \text{for } i = 1, 2, 3 \\ (R^T \nabla_{\mathbf{a}})_{i-3} & \text{for } i = 4, 5, 6 \end{cases}. \tag{10.18}$$

where X_i^R is defined in Section 9.10, and $(\nabla_{\mathbf{a}})_i = \partial/\partial a_i$.

The operators $\tilde{X}_i^L$ are calculated from $\tilde{X}_i^R$ by observing that $HJH^{-1} = Ad(H)J$, and so

$$\tilde{X}_i^L = \begin{cases} X_i^L + \sum_{k=1}^{3} (\mathbf{a} \times \mathbf{e}_i) \cdot \mathbf{e}_k \partial/\partial a_k & \text{for } i = 1, 2, 3 \\ -\partial/\partial a_{i-3} & \text{for } i = 4, 5, 6 \end{cases}. \tag{10.19}$$

For a more abstract treatment of differential operators on $SE(N)$ (including combinations of the above operators that commute with left and right shifts) see [9, 40].

10.6 Irreducible Unitary Representations of $SE(3)$

We define unitary representations of $SE(3)$ (see [11, 21, 26, 39, 44] for discussions and definitions) in the following way.

We start to construct the representation of the motion group in the space of functions $\varphi(\mathbf{p}) \in \mathcal{L}^2(\hat{T})$, where $\hat{T}$ is a dual (frequency) space of the $\mathbb{R}^3$ subgroup. Functions $\varphi(\mathbf{p})$

correspond to the Fourier transforms of the functions $\varphi(\mathbf{r}) \in \mathcal{L}^2(T)$, where $T = \mathbb{R}^3$, are defined as

$$\varphi(\mathbf{p}) \doteq \frac{1}{(2\pi)^{3/2}} \int_T e^{-i\mathbf{p}\cdot\mathbf{r}} \varphi(\mathbf{r}) \, d\mathbf{r} \,. \tag{10.20}$$

In the current context, we use the same notation $\varphi(\mathbf{p})$ for the Fourier transform of the function as for the function $\varphi(\mathbf{r})$ itself in order to avoid cumbersome notation. The argument $\mathbf{p}$ or $\mathbf{r}$ is sufficient to specify which we are considering.

The rotation subgroup $SO(3)$ of the motion group acts on $\hat{T}$ by rotations, so $\hat{T}$ is divided into orbits S_p, where S_p are $\mathbb{S}^2$ spheres of radius $p = \|\mathbf{p}\|$. The translation operator acts on $\varphi(\mathbf{p})$ as

$$(U(\mathbf{a}, \mathbb{I})\varphi)(\mathbf{p}) = e^{-i\mathbf{p}\cdot\mathbf{a}} \varphi(\mathbf{p}). \tag{10.21}$$

Therefore, the irreducible representations of the motion group can be built on spaces $\varphi(\mathbf{p}) \in \mathcal{L}^2(S_p)$, with the inner product defined as

$$(\varphi_1, \varphi_2) \doteq \int_{\Theta=0}^{\pi} \int_{\Phi=0}^{2\pi} \overline{\varphi_1(\mathbf{p})} \, \varphi_2(\mathbf{p}) \sin \Theta \, d\Theta \, d\Phi \,, \tag{10.22}$$

where $\mathbf{p} = (p \sin\Theta \cos\Phi, p \sin\Theta \sin\Phi, p \cos\Theta)$, and $p > 0$, $0 \le \Theta \le \pi$, $0 \le \Phi \le 2\pi$.

The inner product (φ_1, φ_2) is invariant with respect to transformations

$$\varphi(\mathbf{p}) \to e^{i\alpha} \varphi(A^{-1}\mathbf{p}) \,, \tag{10.23}$$

where $A \in SO(3)$ and $0 \le \alpha \le 2\pi$.

The parameter α in (10.23) may, in general, depend on p and group element $A \in SO(3)$. In this case, different functions $\alpha_s(p, A)$ (where s enumerates the irreducible representations of $SO(2)$), which are nonlinear functions of group element A, correspond to different irreducible representations of the motion group. Functions $\varphi(\mathbf{p})$, thus, may have different *internal* properties with respect to rotations.

With the help of functions $\alpha_s(p, A)$ we can construct the representations of $G = SE(3) \simeq \hat{T} \rtimes SO(3)$ from representations of its subgroup $G' = \hat{T} \rtimes SO(2)$ using the method of induced representations. In our case (we disregard for the moment the translation group $\hat{T}$), $G = SO(3)$, $H = SO(2)$ and $\sigma = \mathbf{p} \in S_p \simeq SO(3)/SO(2)$.

To construct the representations of the motion group explicitly, we choose a particular vector $\mathbf{p}' = (0, 0, p)$ on each orbit S_p. The vector $\mathbf{p}'$ is invariant with respect to rotations from the $SO(2)$ subgroup of $SO(3)$

$$\Lambda \mathbf{p}' = \mathbf{p}' \,; \quad \Lambda \in H_{\mathbf{p}'} = SO(2), \tag{10.24}$$

where $H_{\mathbf{p}'}$ is a little group of $\mathbf{p}'$. For each $\mathbf{p} \in S_p$ we can find $R_{\mathbf{p}} \in SO(3)/SO(2)$, such that

$$R_{\mathbf{p}} \, \mathbf{p}' = \mathbf{p}.$$

Then for any $A \in SO(3)$,

$$(R_{\mathbf{p}}^{-1} A R_{A^{-1}\mathbf{p}}) \, \mathbf{p}' = \mathbf{p}'.$$

Therefore, $Q(\mathbf{p}, A) = (R_{\mathbf{p}}^{-1} A R_{A^{-1}\mathbf{p}}) \in H_{\mathbf{p}'}$. The representations of $H_{\mathbf{p}'}$ can be taken to be of the form

$$\Delta_s : \quad \phi \to e^{is\phi} \,, \quad 0 \le \phi \le 2\pi$$

for $s = 0, \pm 1, \pm 2, \ldots$.

Thus, we can construct the induced representation $(\hat{T} \rtimes \Delta_s(H_{\mathbf{p}'})) \uparrow SE(3)$ of the motion group from the representations of its subgroup $\hat{T} \vartriangleleft H_{\mathbf{p}'}$.

Definition 10.3 The unitary representations $U^s(\mathbf{a}, A)$ of $SE(3)$, which act on the space of functions $\varphi(\mathbf{p})$ with the inner product (10.22), are defined by

$$(U^s(\mathbf{a}, A)\varphi)(\mathbf{p}) = e^{-i\mathbf{p}\cdot\mathbf{a}} \, \Delta_s(R_{\mathbf{p}}^{-1} \, A \, R_{A^{-1}\mathbf{p}}) \, \varphi(A^{-1}\mathbf{p}) \,, \qquad (10.25)$$

where $A \in SO(3)$, Δ_s are representations of $H_{\mathbf{p}'}$ and $s = 0, \pm 1, \pm 2, \dots$.

An explicit expression for $\Delta_s(R_{\mathbf{p}}^{-1} \, A \, R_{A^{-1}\mathbf{p}})$ is given in [26] when $A = R(a, b)$ is parameterized as in (5.67), and $\mathbf{p}$ is expressed in spherical coordinates.

Each representation characterized by $p = \|\mathbf{p}\|$ and s is irreducible (they, however, become reducible if we restrict $SE(3)$ to $SO(3)$, i.e. when $a = \|\mathbf{a}\| = 0$). They are unitary because $(U^s(\mathbf{a}, A)\varphi_1, U^s(\mathbf{a}, A)\varphi_2) = (\varphi_1, \varphi_2)$.

We note that (10.25) can be written in the form

$$\boxed{(U^s(\mathbf{a}, A; p)\varphi)(\mathbf{u}) = e^{-ip\mathbf{u}\cdot\mathbf{a}} \, \Delta_s(R_{\mathbf{u}}^{-1} \, A \, R_{A^{-1}\mathbf{u}}) \, \varphi(A^{-1}\mathbf{u})} \,, \qquad (10.26)$$

where $\mathbf{p} = p\mathbf{u}$ and $\mathbf{u}$ is a unit vector. Here $\varphi(\cdot)$ is defined on the unit sphere.

Representations (10.26), which we denote below by $U^s(g, p)$, satisfy the homomorphism properties

$$U^s(g_1 \circ g_2, p) = U^s(g_1, p) \cdot U^s(g_2, p) \,,$$

where $\circ$ is the group operation. The corresponding multiplication law for the $Q(\mathbf{p}, A)$ factors is [45]

$$Q(\mathbf{p}, A) \, Q(A^{-1}\mathbf{p}, A^{-1}B) = Q(\mathbf{p}, B) \,. \qquad (10.27)$$

In analogy with the planar case, we define $g_i(t) = \exp(t\tilde{X}_i)$ for $i = 1, \dots, 6$ and evaluate the representation (10.26). The result for $k = 1, 2, 3$ is

$$U^s(g_k(t), p)\varphi(\mathbf{u}) = \Delta_s(R_{\mathbf{u}}^{-1} \, \mathrm{rot}[\mathbf{e}_k, t] \, R_{\mathrm{rot}[\mathbf{e}_k, -t]\mathbf{u}}) \, \varphi(\mathrm{rot}[\mathbf{e}_k, -t]\mathbf{u});$$

$$U^s(g_{k+3}(t), p)\varphi(\mathbf{u}) = e^{-ipt\mathbf{e}_k\cdot\mathbf{u}}\varphi(\mathbf{u}).$$

Differentiating with respect to t and setting $t = 0$, we define the operators

$$\hat{X}_i(p, s)\varphi(\mathbf{u}) \doteq \left. \frac{dU^s(\exp(t\tilde{X}_i), p)\varphi(\mathbf{u})}{dt} \right|_{t=0}$$

for $i = 1, \dots, 6$. Explicitly, with $\mathbf{u} = \mathbf{u}(\Theta, \Phi)$,

$$\hat{J}_1\varphi(\mathbf{u}) \doteq \hat{X}_1(p, s)\varphi(\mathbf{u}) = \left(\sin\Phi \frac{\partial}{\partial\Theta} - is\frac{\cos\Phi}{\sin\Theta} + \cos\Phi \cot\Theta \frac{\partial}{\partial\Phi} \right) \varphi(\mathbf{u})$$

$$\hat{J}_2\varphi(\mathbf{u}) \doteq \hat{X}_2(p, s)\varphi(\mathbf{u}) = \left(-\cos\Phi \frac{\partial}{\partial\Theta} - is\frac{\sin\Phi}{\sin\Theta} + \sin\Phi \cot\Theta \frac{\partial}{\partial\Phi} \right) \varphi(\mathbf{u})$$

$$\hat{J}_3\varphi(\mathbf{u}) \doteq \hat{X}_3(p, s)\varphi(\mathbf{u}) = -\frac{\partial}{\partial\Phi}\varphi(\mathbf{u})$$

$$\hat{P}_k\varphi(\mathbf{u}) \doteq \hat{X}_{k+3}(p, s)\varphi(\mathbf{u}) = -ip\,(\mathbf{e}_k \cdot \mathbf{u})\,\varphi(\mathbf{u}) \qquad (10.28)$$

for $k = 1, 2, 3$.

10.7 Matrix Elements

To obtain the matrix elements of the unitary representations, we use the group property

$$U^s(\mathbf{a}, A) = U^s(\mathbf{a}, \mathbb{I}) \cdot U^s(0, A). \tag{10.29}$$

The basis eigenfunctions of the irreducible representations (10.25) of $SE(3)$ can be enumerated by the integer numbers l, m (for each s and p). We note that the values $l(l+1)$, $m+s$, ps and $-p^2$ correspond to the eigenvalues of the operators $\hat{\mathbf{J}}^2, \hat{J}^3, \hat{\mathbf{P}} \cdot \hat{\mathbf{J}} = \sum_{i=1}^{3} \hat{P}_i \hat{J}_i$ and $\hat{\mathbf{P}} \cdot \hat{\mathbf{P}} = \sum_{i=1}^{3} \hat{P}_i \hat{P}_i$ (where $\hat{J}_i, \hat{P}_i, i = 1, 2, 3$ are rotation and translation operators defined in (10.28) and $\hat{J}^3 = i\hat{J}_3$, $\hat{P}^3 = i\hat{P}_3$) which can be diagonalized simultaneously (i.e. they commute). The restrictions for the l, m, s numbers are

$$l \geq |s|; \quad l \geq |m|.$$

The basis functions can be expressed in the form [21]

$$h_{ms}^l(\mathbf{u}(\Theta, \Phi)) = Q_{s,m}^l(\cos \Theta) \, e^{i(m+s)\Phi} \tag{10.30}$$

where

$$Q_{-s,m}^l(\cos \Theta) = (-1)^{l-s} \sqrt{\frac{2l+1}{4\pi}} \, P_{sm}^l(\cos \Theta), \tag{10.31}$$

and generalized Legendre polynomials $P_{ms}^l(\cos \Theta)$ are given as in Vilenkin [44].

It can be shown that these basis functions are transformed under the rotations $h_{ms}^l(\mathbf{u}) \rightarrow \Delta_s(Q(\mathbf{u}, A)) \, h_{ms}^l(A^{-1}\mathbf{u})$ as (see Chapter 9 for the proof):

$$(U^s(0, A) \, h_{ms}^l)(\mathbf{u}) = \sum_{n=-l}^{l} \tilde{U}_{nm}^l(A) h_{ns}^l(\mathbf{u}), \tag{10.32}$$

where the matrix elements $\tilde{U}_{nm}^l(A)$ are

$$\tilde{U}_{mn}^l(A) = e^{-im\alpha} \, (-1)^{n-m} \, P_{mn}^l(\cos \beta) \, e^{-in\gamma}, \tag{10.33}$$

where α, β, γ are ZXZ Euler angles of the rotation. We note that the rotation matrix elements do not depend on s.

The translation matrix elements are given by the integral [21]

$$(h_{m's}^{l'}, U^s(\mathbf{a}, \mathbb{I}) h_{ms}^l) = [l', m' \mid p, s \mid l, m](\mathbf{a}) =$$

$$\int_{\Theta=0}^{\pi} \int_{\Phi=0}^{2\pi} Q_{s,m'}^{l'}(\cos \Theta) \, e^{-i(m'+s)\Phi} \, e^{-i\mathbf{p} \cdot \mathbf{a}} \, Q_{s,m}^l(\cos \Theta) \, e^{i(m+s)\Phi} \sin \Theta \, d\Phi \, d\Theta. \tag{10.34}$$

These are written in closed form as [21]

$$[l', m' \mid p, s \mid l, m](\mathbf{a}) =$$

$$(4\pi)^{1/2} \sum_{k-|l'-l|}^{l'+l} i^k \sqrt{\frac{(2l'+1)(2k+1)}{(2l+1)}} \, j_k(p\,a) \, C(k, 0; l', s \mid l, s)$$

$$\cdot C(k, m-m'; l', m' \mid l, m) \, Y_k^{m-m'}(\theta, \phi), \tag{10.35}$$

where θ, ϕ are polar and azimuthal angles of $\mathbf{a}$, $C(k, m - m'; l', m' \mid l, m)$ are Clebsch-Gordan coefficients (see, for example, [14]).

Finally, using the group property (10.29), the matrix elements of the unitary representation $U^s(g, p)$ (10.25) (for $s = 0, \pm 1, \pm 2, ...$) are expressed as

$$U^s_{l',m';l,m}(\mathbf{a}, A; p) = \sum_{j=-l}^{l} [l', m' \mid p, s \mid l, j](\mathbf{a})\, \tilde{U}^l_{jm}(A). \qquad (10.36)$$

Because (10.35) contains only half-integer Bessel functions, all matrix elements can be expressed in terms of elementary functions. Below we have shown several matrix elements in explicit form (using the notation $a = \|\mathbf{a}\|$ and $p = \|\mathbf{p}\|$):

$$U^0_{0,0;0,0}(\mathbf{a}, A; p) = \frac{\sin(a\,p)}{a\,p}\,;$$

$$U^0_{1,0;0,0}(\mathbf{a}, A; p) = \frac{i\sqrt{3}\,\cos(\theta)\,(-\cos(a\,p) + \sin(a\,p)/(a\,p))}{a\,p}\,;$$

$$U^0_{1,-1;0,0}(\mathbf{a}, A; p) = \frac{-i\sqrt{\frac{3}{2}}\,e^{i\phi}\,(a\,p\,\cos(a\,p) - \sin(a\,p))\,\sin(\theta)}{a^2\,p^2}\,;$$

$$U^0_{1,1;0,0}(\mathbf{a}, A; p) = \frac{i\sqrt{\frac{3}{2}}\,e^{-i\phi}\,(a\,p\,\cos(a\,p) - \sin(a\,p))\,\sin(\theta)}{a^2\,p^2}\,;$$

$$U^0_{2,1;0,0}(\mathbf{a}, A; p) = -\frac{1}{a^3\,p^3}\sqrt{\frac{15}{8}}\,e^{-i\phi}\,(3\,a\,p\,\cos(a\,p) - 3\,\sin(a\,p) +$$
$$a^2\,p^2\,\sin(a\,p))\,\sin(2\,\theta)\,;$$

$$U^0_{1,-1;1,1}(\mathbf{a}, A; p) = \frac{3}{4\,a^3\,p^3}\,e^{-i\alpha-i\gamma}\,(2\,a\,e^{2i\alpha}\,p\,\cos(a\,p)\,(\sin\tfrac{\beta}{2})^2 -$$
$$6\,a\,e^{2i\alpha}\,p\,\cos(a\,p)\,(\cos\theta)^2\,(\sin\tfrac{\beta}{2})^2 - 2\,e^{2i\alpha}\,(\sin\tfrac{\beta}{2})^2\,\sin(a\,p) +$$
$$2\,a^2\,e^{2i\alpha}\,p^2\,(\sin\tfrac{\beta}{2})^2\,\sin(a\,p)\,(\sin\theta)^2 + 6\,e^{2i\alpha}\,(\cos\theta)^2\,(\sin\tfrac{\beta}{2})^2\,\sin(a\,p) -$$
$$6\,a\,e^{2i\phi}\,p\,(\cos\tfrac{\beta}{2})^2\,\cos(a\,p)\,(\sin\theta)^2 + 6\,e^{2i\phi}\,(\cos\tfrac{\beta}{2})^2\,\sin(a\,p)\,(\sin\theta)^2 -$$
$$2\,a^2\,e^{2i\phi}\,p^2\,(\cos\tfrac{\beta}{2})^2\,\sin(a\,p)\,(\sin\theta)^2 + 3\,a\,e^{i\phi+i\alpha}\,p\,\cos(a\,p)\,\sin(\beta)\,\sin(2\,\theta) -$$
$$3\,e^{i\phi+i\alpha}\,\sin(\beta)\,\sin(a\,p)\,\sin(2\,\theta) + a^2\,e^{i\phi+i\alpha}\,p^2\,\sin(\beta)\,\sin(a\,p)\,\sin(2\,\theta))\,, \qquad (10.37)$$

where $a = \|\mathbf{a}\|$, θ, ϕ are polar and azimuthal angles of $\mathbf{a}$, and α, β, γ are Euler angles of the $SO(3)$ rotation.

Symmetry properties. We note symmetry properties of the matrix elements. Using the property of the Clebsch-Gordan coefficients [14]

$$C(l_1, m_1; l_2, m_2 \mid l, m) = (-1)^{(l_1 + l_2 + l)}\,C(l_1, -m_1; l_2, -m_2 \mid l, -m)\,, \qquad (10.38)$$

(for integer l_1, l_2, l), the property

$$Y_l^{-m}(\theta, \phi) = (-1)^m \, \overline{Y}_l^m(\theta, \phi)$$

and the symmetry relation [44]

$$P_{mn}^l(x) = (-1)^{(m-n)} P_{-m,-n}^l(x),$$

it can be shown that

$$\overline{U_{l',m';l,m}^s}(\mathbf{a}, A; p) = (-1)^{(l'-l)}(-1)^{(m'-m)} U_{l',-m';l,-m}^s(\mathbf{a}, A; p). \tag{10.39}$$

Also, it can be observed from the transformation law (10.25) that the complex conjugate transformation is related with the $(-s)$ transformation evaluated at $(-\mathbf{a}, A)$. Explicitly,

$$\overline{U_{l',m';l,m}^{-s}}(-\mathbf{a}, A; p) = (-1)^{(m'-m)} U_{l',-m';l,-m}^s(\mathbf{a}, A; p). \tag{10.40}$$

We also list here a unitarity relation

$$U_{l',m';l,m}^s(-A^{-1}\mathbf{a}, A^{-1}; p) = \overline{U_{l,m;l',m'}^s(\mathbf{a}, A; p)}. \tag{10.41}$$

Orthogonality. Using the orthogonality of the rotation matrix elements $\tilde{U}_{mn}^l$ that follow from the presentation in Chapter 9, the integral expression (10.34) for the translation matrix elements, and the integral representation for the δ-function,

$$\int_{\mathbb{R}^3} e^{i(\mathbf{p}-\mathbf{p}')\cdot\mathbf{r}} \, d\mathbf{r} = (2\pi)^3 \, \delta(\mathbf{p} - \mathbf{p}'), \tag{10.42}$$

it can be shown that the $SE(3)$ matrix elements satisfy the orthogonality relation [21]

$$\int_{\mathbb{R}^3} d\mathbf{a} \int_{SO(3)} dA \, \overline{U_{l_1,m_1;j_1,k_1}^{s_1}}(\mathbf{a}, A; p_1) U_{l,m;j,k}^s(\mathbf{a}, A; p) =$$

$$= 2\pi^2 \, \delta_{l_1 l} \, \delta_{j_1 j} \, \delta_{m_1 m} \, \delta_{k_1 k} \, \delta_{s_1 s} \, \frac{\delta(p_1 - p)}{p^2} \tag{10.43}$$

where $d\mathbf{a} = a^2 \, da \, \sin\theta \, d\theta \, d\phi$ and $dA = \frac{1}{8\pi^2} \sin\beta \, d\alpha \, d\beta \, d\gamma$.

10.8 The Fourier Transform for $SE(3)$

Here we define the Fourier transform of functions $f(\mathbf{a}, A) \in \mathcal{L}^2(SE(3))$. The inner product of two such functions is given by

$$(f_1, f_2) \doteq \int_{\mathbb{R}^3} \int_{SO(3)} \overline{f_1(\mathbf{a}, A)} \, f_2(\mathbf{a}, A) \, dA \, d\mathbf{a}. \tag{10.44}$$

To define the Fourier transform for functions on $SE(3)$, we have to use a complete orthogonal basis for functions on this group. The completeness of matrix elements (10.36) depends in part on the completeness of the rotation matrix elements $\tilde{U}_{mn}^l(A)$ on $SO(3)$ [21]. The orthogonality relation for the matrix elements is given in (10.43). So, using the unitary representations $U(g, p)$ (10.25) (for $s = 0, \pm 1, \pm 2...$), we can define the Fourier transform of functions on the motion group.

Definition 10.4. For any absolutely- and square-integrable complex-valued function $f(\mathbf{a}, A)$ on $SE(3)$ we define the Fourier transform as

$$\hat{f}(p) \doteq \int_{SE(3)} f(g) U(g^{-1}, p) \, d(g)$$

where $g = (\mathbf{a}, A) \in SE(3)$ and $d(g) = dA \, d\mathbf{a}$ and

$$U(g; p) \doteq \bigoplus_{s \in \mathbb{Z}} U^s(g; p)$$

and similarly for $\hat{f}(p)$. The operation of converting a function to its Fourier transform is denoted as $\mathcal{F}(f)$.

The matrix elements of the transform are given in terms of matrix elements (10.36) as

$$\hat{f}^s_{l',m';l,m}(p) = \int_{\mathbb{R}^3} \int_{SO(3)} f(\mathbf{a}, A) \overline{U^s_{l,m;l',m'}(\mathbf{a}, A; p)} \, dA \, d\mathbf{a} \qquad (10.45)$$

where we have used the unitarity property.

Theorem 10.5. *The inverse Fourier transform is*

$$f(g) = \mathcal{F}^{-1}(\hat{f}(p)) = \frac{1}{2\pi^2} \int_0^\infty \operatorname{tr}(\hat{f}(p) U(g, p)) \, p^2 \, dp . \qquad (10.46)$$

and is written explicitly in terms of matrix elements as

$$f(\mathbf{a}, A) = \frac{1}{2\pi^2} \sum_{s=-\infty}^{\infty} \sum_{l'=|s|}^{\infty} \sum_{l=|s|}^{\infty} \sum_{m'=-l'}^{l'} \sum_{m=-l}^{l} \int_0^\infty p^2 \, dp \, \hat{f}^s_{l,m;l',m'}(p) U^s_{l',m';l,m}(\mathbf{a}, A; p) .$$

$$(10.47)$$

And the Fourier transform pair has the convolution, Plancherel, and symmetry properties listed below.

We note that we can use any unitary equivalent representation $T^* U(g, p) T$ (where T is a unitary transformation, which does not depend on g) to define the Fourier transform.

Convolution of functions. Recall that the convolution integral of functions $f_1, f_2 \in \mathcal{L}^2(SE(3))$ can be defined as

$$(f_1 * f_2)(g) \doteq \int_{SE(3)} f_1(h) \, f_2(h^{-1} \circ g) \, d(h) . \qquad (10.48)$$

One of the most powerful properties of the Fourier transform of functions on $\mathbb{R}^N$ is that the Fourier transform of the convolution of two functions is the product of the Fourier transform of the functions. This property persists also for the convolution of functions on the group, namely

$$\mathcal{F}(f_1 * f_2) = \mathcal{F}(f_2) \mathcal{F}(f_1) \qquad (10.49)$$

or, in the matrix form

$$(\mathcal{F}(f_1 * f_2))^s_{l',m';l,m}(p) = \sum_{j=|s|}^{\infty} \sum_{k=-j}^{j} (\hat{f}_2)^s_{l',m';j,k}(p)(\hat{f}_1)^s_{j,k;l,m}(p). \qquad (10.50)$$

Parseval/Plancherel equality.   The following equality holds

$$\int_{SE(3)} |f(\mathbf{a}, A)|^2 \, dA \, d\mathbf{a} =$$

$$\frac{1}{2\pi^2} \sum_{s=-\infty}^{\infty} \sum_{l'=|s|}^{\infty} \sum_{l=|s|}^{\infty} \sum_{m'=-l'}^{l'} \sum_{m=-l}^{l} \int_0^{\infty} |\hat{f}^s_{l',m';l,m}(p)|^2 \, p^2 \, dp =$$

$$= \frac{1}{2\pi^2} \int_0^{\infty} \| \hat{f}(p) \|^2 \, p^2 \, dp, \qquad (10.51)$$

where $\| \hat{f}(p) \|$ is the Hilbert-Schmidt norm of $\hat{f}(p)$, the square of which is

$$\| \hat{f}(p) \|^2 \doteq \sum_{s=-\infty}^{\infty} \sum_{l'=|s|}^{\infty} \sum_{l=|s|}^{\infty} \sum_{m'=-l'}^{l'} \sum_{m=-l}^{l} |\hat{f}^s_{l',m';l,m}(p)|^2.$$

A similar relation holds for the inner product

$$(f,g) = \int_{SE(3)} \overline{f(\mathbf{a}, A)} g(\mathbf{a}, A) \, dA \, d\mathbf{a} =$$

$$\frac{1}{2\pi^2} \sum_{s=-\infty}^{\infty} \sum_{l'=|s|}^{\infty} \sum_{l=|s|}^{\infty} \sum_{m'=-l'}^{l'} \sum_{m=-l}^{l} \int_0^{\infty} \overline{\hat{f}^s_{l',m';l,m}(p)} \hat{g}^s_{l',m';l,m}(p) \, p^2 \, dp =$$

$$= \frac{1}{2\pi^2} \int_0^{\infty} \mathrm{tr}(\hat{f}^*(p)\hat{g}(p)) \, p^2 \, dp, \qquad (10.52)$$

where $\hat{f}^*_{l,m;l',m'} = \overline{\hat{f}_{l',m';l,m}}$ is the Hermitian conjugate.

Symmetries.   For the real function $f(\mathbf{a}, A)$, we note a symmetry property of the Fourier transform, which follows from symmetry (10.39) of the matrix elements

$$\overline{\hat{f}^s_{l',m';l,m}(p)} = (-1)^{(l'-l)}(-1)^{(m'-m)} \hat{f}^s_{l',-m';l,-m}(p) \qquad (10.53)$$

Others analogous to those for $SE(2)$ also exist.

Contraction of indices.   It is convenient to rewrite the 4-index Fourier transform matrix element $\hat{f}^s_{l',m';l,m}(p)$ as a 2-index matrix $\hat{f}^s_{ij}(p)$. To satisfy the matrix product definition in (10.50) we arrange l, m indices in a row (we show an example for $s = 0$) 0 (for $l = 0$); $-1, 0, 1$ (for $l = 1$); $-2, -1, 0, 1, 2$ (for $l = 2$); ..., which corresponds to $j = 1, 2, 3, 4.....$

Explicitly

$$\hat{f}^s_{l',m';l,m}(p) = \hat{f}^s_{ij}(p), \qquad (10.54)$$

where $i = l'(l' + 1) + m' - s^2 + 1; j = l(l + 1) + m - s^2 + 1$.

In particular, any 4-index matrix such as $\hat{f}^0_{l',m';l,m}$ where $0 \leq l, l' \leq L$, $|m| \leq l$ and $|m'| \leq l'$ can be written as an equivalent $(L+1)^2 \times (L+1)^2$ matrix with two indices using the rule

$$l'(l'+1) + m' + 1 \to i; \quad l(l+1) + m + 1 \to j$$

where

$$\hat{f}^0_{i,j} \doteq \hat{f}^0_{l',m';l,m}.$$

In this way the product is preserved. That is,

$$\sum_{j=1}^{(L+1)^2} \hat{f}^0_{i,j} \hat{g}^0_{j,k} = \sum_{l=0}^{L} \sum_{m=-l}^{l} \hat{f}^0_{l',m';l,m} \hat{g}^0_{l,m;l'',m''}$$

where $k = l''(l''+1) + m'' + 1$. For instance, if $L = 1$, we write

$$\hat{f}^0 = \begin{pmatrix} \hat{f}^0_{0,0;0,0} & \hat{f}^0_{0,0;1,-1} & \hat{f}^0_{0,0;1,0} & \hat{f}^0_{0,0;1,1} \\ \hat{f}^0_{1,-1;0,0} & \hat{f}^0_{1,-1;1,-1} & \hat{f}^0_{1,-1;1,0} & \hat{f}^0_{1,-1;1,1} \\ \hat{f}^0_{1,0;0,0} & \hat{f}^0_{1,0;1,-1} & \hat{f}^0_{1,0;1,0} & \hat{f}^0_{1,0;1,1} \\ \hat{f}^0_{1,1;0,0} & \hat{f}^0_{1,1;1,-1} & \hat{f}^0_{1,1;1,0} & \hat{f}^0_{1,1;1,1} \end{pmatrix}. \tag{10.55}$$

For other approaches to the representation theory and harmonic analysis on the Euclidean motion groups see [1, 4, 7, 12, 13, 27, 28, 29, 30, 31, 32, 33, 34, 35, 36, 38, 41]. For applications of representations of motion groups with scale changes, see [22, 23].

10.9 Reconstruction Formula and Completeness for $SE(3)$

In this section, a proof is given for the Fourier inversion formula stated earlier.
Recall from (10.26) that the IUR operator for $SE(3)$ is

$$(U^s(\mathbf{a}, A; p)\varphi)(\mathbf{u}) = e^{-ip\,\mathbf{u}\cdot\mathbf{a}} \, \Delta_s(R_{\mathbf{u}}^{-1} A R_{A^{-1}\mathbf{u}}) \, \varphi(A^{-1}\mathbf{u}) \tag{10.56}$$

where $g = (\mathbf{a}, A) \in SE(3)$. This means that evaluating at g^{-1} rather than g gives

$$(U^s(-A^{-1}\mathbf{a}, A^{-1}; p)\varphi)(\mathbf{u}) = e^{ip\,\mathbf{u}\cdot(A^{-1}\mathbf{a})} \, \Delta_s(R_{\mathbf{u}}^{-1} A^{-1} R_{A\mathbf{u}}) \, \varphi(A\mathbf{u}),$$

and the associated Fourier transform operator is

$$(\hat{f}(p)\,\varphi)(\mathbf{u}) = \int_{SO(3)} \int_{\mathbb{R}^3} f(\mathbf{a}, A)(U^s(-A^{-1}\mathbf{a}, A^{-1}; p)\varphi)(\mathbf{u}) \, d\mathbf{a}\,dA =$$

$$\int_{SO(3)} \int_{\mathbb{R}^3} f(\mathbf{a}, A) \, e^{ip\,(A\mathbf{u})\cdot\mathbf{a}} \, \Delta_s(R_{\mathbf{u}}^{-1} A^{-1} R_{A\mathbf{u}}) \, \varphi(A\mathbf{u}) \, d\mathbf{a}\,dA,$$

where $A^{-1} = A^T$ is used to write $\mathbf{u} \cdot (A^{-1}\mathbf{a}) = (A\mathbf{u}) \cdot \mathbf{a}$.
Then

$$(U^s(g'; p)\,\hat{f}(p)\,\varphi)(\mathbf{u}) = e^{-ip\,\mathbf{u}\cdot\mathbf{a}'} \, \Delta_s(R_{\mathbf{u}}^{-1} A' R_{A'^{-1}\mathbf{u}}) \cdot$$

$$\int_{SO(3)} \int_{\mathbb{R}^3} f(\mathbf{a}, A) \, e^{ip\,(AA'^{-1}\mathbf{u})\cdot\mathbf{a}} \, \Delta_s(R_{A'^{-1}\mathbf{u}}^{-1} A^{-1} R_{AA'^{-1}\mathbf{u}}) \, \varphi(AA'^{-1}\mathbf{u}) \, d\mathbf{a}\,dA,$$

Bringing the Δ_s term inside the integral and using the simplification

$$\Delta_s(R_{\mathbf{u}}^{-1} A' R_{A'^{-1}\mathbf{u}}) \Delta_s(R_{A'^{-1}\mathbf{u}}^{-1} A^{-1} R_{AA'^{-1}\mathbf{u}}) = \Delta_s(R_{\mathbf{u}}^{-1} A' A^{-1} R_{AA'^{-1}\mathbf{u}})$$

then gives

$$(U^s(g';p)\,\hat{f}(p)\,\varphi)(\mathbf{u}) =$$

$$e^{-ip\,\mathbf{u}\cdot\mathbf{a}'} \int_{SO(3)} \left[\int_{\mathbb{R}^3} f(\mathbf{a}, A)\, e^{ip\,(AA'^{-1}\mathbf{u})\cdot\mathbf{a}}\, d\mathbf{a} \right] \Delta_s(R_{\mathbf{u}}^{-1} A' A^{-1} R_{AA'^{-1}\mathbf{u}})\, \varphi(AA'^{-1}\mathbf{u})\, dA\,.$$

The term in brackets is a scaled version of the classical Fourier transform in the translation variables with the rotational parts held fixed. Therefore,

$$(U^s(g';p)\,\hat{f}(p)\,\varphi)(\mathbf{u}) =$$

$$(2\pi)^{3/2} e^{-ip\,\mathbf{u}\cdot\mathbf{a}'} \int_{SO(3)} \tilde{f}(-p\,AA'^{-1}\mathbf{u}, A)\, \Delta_s(R_{\mathbf{u}}^{-1} A' A^{-1} R_{AA'^{-1}\mathbf{u}})\, \varphi(AA'^{-1}\mathbf{u})\, dA\,.$$

Let $Q = AA'^{-1}$ and $\mathbf{v} = Q\mathbf{u}$, and decompose Q as

$$Q = R(\mathbf{u}, \mathbf{v})\, e^{\theta\hat{\mathbf{u}}}\,,$$

where $R(\mathbf{u}, \mathbf{v})$ is the rotation matrix that most directly converts $\mathbf{u}$ into $\mathbf{v}$. That is, $R(\mathbf{u}, \mathbf{v}) = \exp(\alpha(\widehat{\mathbf{u} \times \mathbf{v}})/\|\mathbf{u} \times \mathbf{v}\|)$ where α is the angle between $\mathbf{u}$ and $\mathbf{v}$.

Using the methods of Chapter 5, the Haar measure for $SO(3)$ can be computed as (5.81)

$$dQ = d\theta\, d\mathbf{v} \qquad\qquad (10.57)$$

(where the normalizing factor of $1/8\pi^2$ is reflected in how the integral over $SO(3)$ is fibered into normalized integrals over $\mathbb{S}^2$ and $\mathbb{S}^1$ as in (5.91)).

This allows us to decompose the integral over $SO(3)$ in the definition of the $SE(3)$-Fourier transform as:

$$(U^s(g';p)\,\hat{f}^s(p)\,\varphi)(\mathbf{u}) =$$

$$(2\pi)^{3/2} e^{-ip\,\mathbf{u}\cdot\mathbf{a}'} \int_{SO(3)} \tilde{f}(-p\,\mathbf{v}, QA')\, \Delta_s(R_{\mathbf{u}}^{-1} Q^{-1} R_{\mathbf{v}})\, \varphi(\mathbf{v})\, dQ =$$

$$(2\pi)^{3/2} e^{-ip\,\mathbf{u}\cdot\mathbf{a}'} \int_{\mathbb{S}^2} \int_{\mathbb{S}^1} \tilde{f}(-p\,\mathbf{v}, R(\mathbf{u}, \mathbf{v})e^{\theta\hat{\mathbf{v}}} A')\, \Delta_s(R_{\mathbf{u}}^{-1} e^{-\theta\hat{\mathbf{u}}} R(\mathbf{v}, \mathbf{u}) R_{\mathbf{v}})\, \varphi(\mathbf{v})\, d\theta d\mathbf{v}\,.$$

The first equality above results from using the invariance of integration on $SO(3)$ with respect to shifts and the second results from (10.57).

Defining

$$K(\mathbf{u}, \mathbf{v}) \doteq (2\pi)^{3/2} e^{-ip\,\mathbf{u}\cdot\mathbf{a}'} \int_{\mathbb{S}^1} \tilde{f}(-p\,\mathbf{v}, R(\mathbf{u}, \mathbf{v})e^{\theta\hat{\mathbf{v}}} A')\, \Delta_s(e^{-\theta\hat{\mathbf{u}}} R(\mathbf{v}, \mathbf{u}) R_{\mathbf{v}})\, d\theta\,,$$

and using the fact that

$$R_{\mathbf{v}}^{-1} e^{-\theta\hat{\mathbf{v}}} R(\mathbf{v}, \mathbf{v}) R_{\mathbf{v}} = e^{-\theta\hat{\mathbf{e}}_3}$$

we can compute the trace (using Theorem E.1 from Appendix E.5) as

$$\operatorname{tr}\left[U^s(g';p)\,\hat{f}^s(p)\right] = \int_{\mathbb{S}^2} K(\mathbf{v}, \mathbf{v})\, d\mathbf{v}$$

$$= (2\pi)^{3/2} \int_{\mathbb{S}^2} e^{-ip\,\mathbf{v}\cdot\mathbf{a}'} \int_{\mathbb{S}^1} \tilde{f}(-p\,\mathbf{v}, e^{\theta\hat{\mathbf{v}}} A')\, \Delta_s(e^{-\theta\hat{\mathbf{e}}_3})\, d\theta\, d\mathbf{v}$$

because $R(\mathbf{v}, \mathbf{v}) = \mathbb{I}$. Moreover, recall that

$$\Delta_s \left(e^{\theta E_3} \right) = e^{is\theta}.$$

Multiplying by $p^2 dp$, integrating over p, and summing over s then gives[2]

$$\sum_{s=-\infty}^{\infty} \int_0^{\infty} \mathrm{tr} \left[U^s(g'; p) \, \hat{f}^s(p) \right] p^2 dp =$$

$$= (2\pi)^{1/2} \int_0^{\infty} \int_{\mathbb{S}^2} e^{-ip\,\mathbf{v}\cdot\mathbf{a}'} \int_0^{2\pi} \tilde{f}(-p\,\mathbf{v}, e^{\theta\hat{\mathbf{v}}} A') \left[\sum_{s=-\infty}^{\infty} e^{-is\theta} \right] d\theta \, d\mathbf{v} p^2 dp \, .$$

But

$$\sum_{s=-\infty}^{\infty} e^{-is\theta} = 2\pi \, \delta(\theta) \, ,$$

and so letting $\mathbf{p} \doteq -p\mathbf{v}$ and observing that for any $F \in \mathcal{L}^1(\mathbb{R}^3)$

$$\int_0^{\infty} \int_{\mathbb{S}^2} F(p\mathbf{v}) \, d\mathbf{v} p^2 dp = \frac{1}{4\pi} \int_{\mathbb{R}^3} F(\mathbf{p}) \, d\mathbf{p}$$

(because the integral over $\mathbb{S}^2$ is normalized) gives

$$\sum_{s=-\infty}^{\infty} \int_0^{\infty} \mathrm{tr} \left[U^s(g'; p) \, \hat{f}^s(p) \right] p^2 dp = 4\pi^2 \, (2\pi)^{-1/2} \int_0^{\infty} \int_{\mathbb{S}^2} e^{-ip\,\mathbf{v}\cdot\mathbf{a}'} \tilde{f}(-p\,\mathbf{v}, A') \, d\mathbf{v} p^2 dp$$

$$= \pi \, (2\pi)^{-1/2} \int_{\mathbb{R}^3} e^{i\mathbf{p}\cdot\mathbf{a}'} \tilde{f}(\mathbf{p}, A') \, d\mathbf{p}$$

$$= 2\pi^2 f(\mathbf{a}', A') \, .$$

The completeness relation for $SE(3)$ is therefore

$$\frac{1}{2\pi^2} \sum_{s=-\infty}^{\infty} \int_0^{\infty} \mathrm{tr} \left[U^s(g'; p) \right] p^2 dp = \delta(g') \, .$$

10.10 Operational Properties

The general formulation of operational properties for Lie group Fourier transforms described in Chapter 8 has been worked out in [5] for the particular case when $G = SE(3)$. Hence, evaluating (8.55) for $i = 1, ..., 6$ results in a number of operational properties. However, these are not the only operational properties. Certain others result from the particular parameterization used. For instance, there are operational properties associated with Euler angles and with derivatives with respect to $a = \|\mathbf{a}\|$. In the following subsections we explore these in detail.

---

[2]Here $\int_{\mathbb{S}^1}$ has been replaced with $(1/2\pi) \int_0^{2\pi}$.

10.10.1 Properties of Translation Differential Operators

It can be observed from the integral representation (10.34) and (10.36) that the matrix elements of the motion group satisfy the relation

$$\nabla_{\mathbf{a}}^2 U_{l',m';l,m}^s(\mathbf{a}, A; p) = (-p^2) U_{l',m';l,m}^s(\mathbf{a}, A; p) \tag{10.58}$$

where $\nabla_{\mathbf{a}}^2$ is the Laplacian with respect to $\mathbf{a}$. This equation leads to the Fourier transform property

$$\mathcal{F}(\nabla_{\mathbf{a}}^2 f(\mathbf{a}, A)) = (-p^2) \hat{f}(p) \tag{10.59}$$

For a differentiable function $f(\mathbf{a}, A)$ which decreases rapidly as $a = \|\mathbf{a}\| \to \infty$, and is less singular than $\frac{1}{a}$ at $a \to 0$, the following relation is valid:

$$\mathcal{F}\left(a \frac{\partial}{\partial a} f(\mathbf{a}, A)\right) = -p \frac{d\hat{f}(p)}{dp} - 3\hat{f}(p). \tag{10.60}$$

This can be proven using the equation

$$\frac{\partial}{\partial a} f(\mathbf{a}, A) = \frac{1}{2\pi^2} \int_0^\infty \operatorname{tr}\left(\hat{f}(p) \frac{\partial}{\partial a} U(g, p)\right) p^2 \, dp, \tag{10.61}$$

the fact that matrix elements of $U(g, p)$ are functions of the product (pa) (which can be seen from (10.35)), and integrating (10.61) by parts.

The more general property

$$\mathcal{F}\left(\frac{1}{a^{n-1}} \frac{\partial}{\partial a}(a^n f(\mathbf{a}, A))\right) = (n - 3)\hat{f}(p) - p \frac{d\hat{f}(p)}{dp}$$

follows from (10.60).

We mention also the related property

$$\int_{SE(3)} \left| a \frac{\partial}{\partial a} f(\mathbf{a}, A) \right|^2 dA \, d\mathbf{a} = \frac{1}{2\pi^2} \int_0^\infty \left\| p \frac{d\hat{f}(p)}{dp} \right\|^2 p^2 \, dp,$$

which follows from the Parseval equality (10.51) and (10.60).

The equality $i \frac{\partial}{\partial a_3} e^{-i\mathbf{p} \cdot \mathbf{a}} = p \cos\theta \, e^{-i\mathbf{p} \cdot \mathbf{a}} = -p\sqrt{\frac{4\pi}{3}} Q_{00}^1(\cos\Theta) e^{-i\mathbf{p} \cdot \mathbf{a}}$ (where Θ is the polar angle of $\mathbf{p}$, and $Q_{-s,m}^l(\cos\Theta)$ is defined in (10.31)) and the explicit expressions for the Clebsch-Gordan coefficients lead to the relation

$$i \frac{\partial}{\partial a_3} U_{l',m';l,m}^s(\mathbf{a}, A; p) =$$

$$-p \left(\frac{(l'^2 - m'^2)(l'^2 - s^2)}{(2l' + 1)(2l' - 1)l'^2}\right)^{1/2} U_{l'-1,m';l,m}^s(\mathbf{a}, A; p)$$

$$-p \frac{m' s}{l'(l' + 1)} U_{l',m';l,m}^s(\mathbf{a}, A; p)$$

$$-p \left(\frac{((l' + 1)^2 - m'^2)((l' + 1)^2 - s^2)}{(2l' + 1)(l' + 1)^2(2l' + 3)}\right)^{1/2} U_{l'+1,m';l,m}^s(\mathbf{a}, A; p).$$

We can get from this relation

$$\mathcal{F}\left(i\frac{\partial}{\partial a_3}f(\mathbf{a}, A)\right) = -p\left(\frac{((l+1)^2 - m^2)\,((l+1)^2 - s^2))}{(2l+1)(l+1)^2(2l+3)}\right)^{1/2}\hat{f}^s_{l',m';l+1,m}(p)$$

$$-p\,\frac{m\,s}{l(l+1)}\,\hat{f}^{\partial}_{l',m';l,m}(p)$$

$$-p\left(\frac{(l^2 - m^2)\,(l^2 - s^2)}{(2l+1)(2l-1)l^2}\right)^{1/2}\hat{f}^s_{l',m';l-1,m}(p)\,.$$

For operators $P^{\pm} = i\frac{\partial}{\partial a_1} \pm \frac{\partial}{\partial a_2}$, which act on the exponent as

$$P^{\pm}e^{-i\mathbf{p}\cdot\mathbf{a}} = \pm p\sqrt{\frac{8\pi}{3}}\,Q^1_{0,\pm1}(\cos\theta)\,e^{\mp i\phi}e^{-\mathbf{p}\cdot\mathbf{a}}$$

$(\theta, \phi$ are polar and azimuthal angles of $\mathbf{p}$), we have the relations

$$P^+\,U^s_{l',m';l,m}(\mathbf{a}, A; p) =$$

$$-p\left(\frac{(l'-1-m')(l'-m')\,(l'^2 - s^2)}{(2l'+1)(2l'-1)l'^2}\right)^{1/2}U^s_{l'-1,m'+1;l,m}(\mathbf{a}, A; p)$$

$$-p\,\frac{\sqrt{(l'-m')(l'+1+m')}\,s}{l'(l'+1)}\,U^s_{l',m'+1;l,m}(\mathbf{a}, A; p)$$

$$+p\left(\frac{(l'+1+m')\,(l'+2+m')((l'+1)^2 - s^2)}{(2l'+1)(l'+1)^2(2l'+3)}\right)^{1/2}U^s_{l'+1,m'+1;l,m}(\mathbf{a}, A; p)$$

and

$$P^-\,U^s_{l',m';l,m}(\mathbf{a}, A; p) =$$

$$-p\left(\frac{(l'-1+m')(l'+m')\,(l'^2 - s^2)}{(2l'+1)(2l'-1)l'^2}\right)^{1/2}U^s_{l'-1,m'-1;l,m}(\mathbf{a}, A; p)$$

$$-p\,\frac{\sqrt{(l'+m')(l'+1-m')}\,s}{l'(l'+1)}\,U^s_{l',m'-1;l,m}(\mathbf{a}, A; p)$$

$$-p\left(\frac{(l'+1-m')\,(l'+2-m')((l'+1)^2 - s^2)}{(2l'+1)(l'+1)^2(2l'+3)}\right)^{1/2}U^s_{l'+1,m'-1;l,m}(\mathbf{a}, A; p)\,.$$

These relations yield the operational properties

$$\mathcal{F}(P^+ f(\mathbf{a}, A)) =$$

$$-p\left(\frac{(l+1-m)\,(l+2-m)((l+1)^2 - s^2)}{(2l+1)(l+1)^2(2l+3)}\right)^{1/2}\hat{f}^s_{l',m';l+1,m-1}(p)$$

$$-p\,\frac{\sqrt{(l+m)(l+1-m)}\,s}{l(l+1)}\,\hat{f}^s_{l',m';l,m-1}(p)$$

$$+p\left(\frac{(l-1+m)(l+m)\,(l^2 - s^2)}{(2l+1)(2l-1)l^2}\right)^{1/2}\hat{f}^s_{l',m';l-1,m-1}(p)$$

and

$$\mathcal{F}(P^- f(\mathbf{a}, A)) =$$

$$p \left(\frac{(l+1+m)(l+2+m)((l+1)^2 - s^2)}{(2l+1)(l+1)^2(2l+3)} \right)^{1/2} \hat{f}^s_{l',m';l+1,m+1}(p)$$

$$- p \frac{\sqrt{(l-m)(l+1+m)} \, s}{l(l+1)} \hat{f}^s_{l',m';l,m+1}(p)$$

$$- p \left(\frac{(l-1-m)(l-m)(l^2 - s^2)}{(2l+1)(2l-1)l^2} \right)^{1/2} \hat{f}^s_{l',m';l-1,m+1}(p) \,.$$

We note that $\hat{f}^s_{l',m';l,m}(p) = 0$ if $|\, m'(m)\, | > l'(l)$ because of the definition of $\tilde{U}^l_{mn}$. Using integration by parts together with the property (10.59), it can be shown that

$$\int_{SE(3)} |\, \nabla_{\mathbf{a}} f(g)\, |^2 \; d(g) = \int_{SE(3)} (f(g), -\nabla_{\mathbf{a}}^2 f(g)) \; d(g) =$$

$$\frac{1}{2\pi^2} \int_0^\infty \mathrm{tr}(p^2 \, \hat{f}^*(p) \, \hat{f}(p)) \, p^2 \, dp \,. \tag{10.62}$$

10.10.2 Properties of Rotational Differential Operators

From the fact that [39, 44]

$$\nabla_A^2 \, \tilde{U}^l_{mn}(A) = (-l(l+1)) \, \tilde{U}^l_{mn}(A) \,, \tag{10.63}$$

where the Laplacian operator on $SO(3)$ is given by

$$\nabla_A^2 = \frac{1}{\sin\beta} \frac{\partial}{\partial\beta}(\sin\beta \frac{\partial}{\partial\beta}) + \frac{1}{\sin^2\beta} \left(\frac{\partial^2}{\partial\alpha^2} - 2\cos\beta \frac{\partial^2}{\partial\alpha\partial\gamma} + \frac{\partial^2}{\partial\gamma^2} \right) \tag{10.64}$$

(recall that α, β, γ are ZXZ Euler angles of A), it follows that

$$\mathcal{F}(\nabla_A^2 f(\mathbf{a}, A))^s_{l',m';l,m} = (-l'(l'+1)) \hat{f}^s_{l',m';l,m}(p) \,. \tag{10.65}$$

The straightforward relation

$$\mathcal{F}\left(i\frac{\partial}{\partial\gamma} f(\mathbf{a}, A) \right)^s_{l',m';l,m} = m' \, \hat{f}^s_{l',m';l,m}(p)$$

follows from (10.33) and (10.36) and Fourier transform definition (10.46).
    The equations [8]

$$J_- \tilde{U}^l_{mn}(A) = i\sqrt{(l+n)(l-n+1)} \, \tilde{U}^l_{m,n-1}(A)$$

$$J_+ \tilde{U}^l_{mn}(A) = -i\sqrt{(l-n)(l+n+1)} \, \tilde{U}^l_{m,n+1}(A) \,,$$

where

$$J_- = e^{i\gamma}(-\cot\beta \frac{\partial}{\partial\gamma} + \frac{1}{\sin\beta} \frac{\partial}{\partial\alpha} + i\frac{\partial}{\partial\beta})$$

and

$$J_+ = e^{-i\gamma}(\cot\beta \frac{\partial}{\partial\gamma} - \frac{1}{\sin\beta} \frac{\partial}{\partial\alpha} + i\frac{\partial}{\partial\beta})$$

lead to the relations

$$\mathcal{F}(J_- f(\mathbf{a}, A))^s_{l',m';l,m} = i\sqrt{(l'+m'+1)(l'-m')} \, \hat{f}^s_{l',m'+1;l,m}(p)$$

and

$$\mathcal{F}(J_+ f(\mathbf{a}, A))^s_{l',m';l,m} = -i\sqrt{(l'-m'+1)(l'+m')} \, \hat{f}^s_{l',m'-1;l,m}(p) \,.$$

10.10.3 Other Operational Properties

A number of other operational properties of the $SE(3)$ Fourier transform exist such as those given below. Yet others that are not listed here can be found in [5].

Scaling property. For $\tilde{f}(\mathbf{a}, A) = f(\frac{\mathbf{a}}{k}, A)$, where $k \neq 0$ is any real number, the following scaling property is valid:

$$\mathcal{F}(\tilde{f}(\mathbf{a}, A)) = k^3 \hat{f}(k\,p).$$

Shift property. We note also the shift property:

$$\mathcal{F}(f(g_1 \circ g \circ g_2^{-1})) = U(g_2^{-1}, p) \cdot \hat{f}(p) \cdot U(g_1, p) = U^*(g_2, p) \cdot \hat{f}(p) \cdot U(g_1, p),$$

where $g_1, g_2 \in SE(3)$, $U^*(g_2, p)$ is the Hermitian conjugate of $U(g_2, p)$ and $\hat{f}(p)$ is the Fourier transform of $f(g) = f(\mathbf{a}, A)$.

Properties of the δ-function. The δ-function $\delta(g, g')$ is defined in Appendix E. Using the Fourier transform definition and the property (9.19) of δ-function, it is easy to see that

$$\mathcal{F}(\delta(g, e)) = \mathbb{I},$$

where $\mathbb{I}$ is a unit operator (a unit matrix after contraction of indices according to (10.54) for each s). Using the convolution property (10.49) the Fourier transform of the convolution

$$(f * \delta)(g) = \int_{SE(3)} f(h)\, \delta(h^{-1} \circ g)\, d(h)$$

is $\mathcal{F}(f)$ (where we use the notation $\delta(g)$ for $\delta(g, e)$). It means that

$$(f * \delta)(g) = f(g) \tag{10.66}$$

Analogously,

$$(\delta * f)(g) = f(g). \tag{10.67}$$

We note that the convolution (10.67) can be written as

$$(\delta * f)(g) = \int_{SE(3)} f(h)\, \delta(g \circ h^{-1})\, d(h).$$

Comparing (9.19) and the convolution equations we have

$$\delta(g, g_1) = \delta(g_1^{-1} \circ g) = \delta(g \circ g_1^{-1}).$$

10.11 Analytical Examples

Here we give some examples which illustrate the Fourier transform and inverse Fourier transform and allow us to perform analytical calculations.

First, let us consider spherically symmetric functions

$$f(\mathbf{a}, A) = f(a)$$

that decrease sufficiently fast as $a \to \infty$ to be square integrable.

Because this function does not depend on the Euler angles of rotation, only $U^s_{l',m';0,0}(\mathbf{a}, A; p)$ elements contribute to the Fourier transform. Moreover, $s = 0$ because $|s| \le l, l'$. Finally, examining the expression (10.35) and using the fact that

$$\int_{\theta=0}^{\pi} \int_{\phi=0}^{2\pi} Y_l^m \sin\theta \, d\theta \, d\phi = \sqrt{4\pi} \, \delta_{l0} \, \delta_{m0} \,,$$

we see that only the element $U^0_{0,0;0,0}(\mathbf{a}, A; p)$ contributes to the Fourier transform.

For the function $f(a) = e^{-a}$ the Fourier transform gives

$$\hat{f}^0_{0,0;0,0}(p) = \int_{SE(3)} f(a) \overline{U^0_{0,0;0,0}(\mathbf{a}, A; p)} \, da \, dA =$$

$$4\pi \int_0^{\infty} e^{-a} \frac{\sin(pa)}{p} a \, da = \frac{8\pi}{(1+p^2)^2} \,, \tag{10.68}$$

where we have used the fact that $\int_{SO(3)} dA = 1$.

The inverse Fourier transform reproduces the original function

$$f(a) = \frac{1}{2\pi^2} \int_0^{\infty} \hat{f}^0_{0,0;0,0}(p) U^0_{0,0;0,0}(\mathbf{a}, A; p) \, p^2 \, dp =$$

$$\frac{1}{2\pi^2} \int \frac{8\pi}{(1+p^2)^2} \frac{\sin(pa)}{a} p \, dp = e^{-a} \,.$$

For the function $f(a) = e^{-a^2}$ we have for the Fourier transform $\hat{f}(p) = (\pi)^{3/2} e^{-p^2/4}$. The inverse Fourier transform gives the original function.

Another example is the function

$$f(\mathbf{a}, A) = e^{-a} \cos\theta \cos\beta$$

where θ is the polar angle of $\mathbf{a}$ and β is Euler angle (around the x-axis) of rotation A. Because $U^1_{00}(A) = \cos\beta$, it can be shown from (10.35) and (10.36) that only $\hat{f}^0_{1,0;0,0}(p)$; $\hat{f}^0_{1,0;1,0}(p)$; $\hat{f}^0_{1,0;2,0}(p)$; $\hat{f}^1_{1,0;1,0}(p)$; $\hat{f}^1_{1,0;2,0}(p)$; $\hat{f}^{-1}_{1,0;1,0}(p)$; $\hat{f}^{-1}_{1,0;2,0}(p)$ may give nonzero contributions. Direct computations show that the Fourier transform elements are (we show only nonzero matrix elements)

$$\hat{f}^0_{1,0;0,0}(p) = -\frac{8i\pi}{3\sqrt{3}} \frac{p}{(1+p^2)^2} \,;$$

$$\hat{f}^0_{1,0;2,0}(p) = -\frac{16i\pi}{\sqrt{135}} \frac{p}{(1+p^2)^2} \,;$$

$$\hat{f}^1_{1,0;2,0}(p) = -\frac{8i\pi}{3\sqrt{5}} \frac{p}{(1+p^2)^2} \,;$$

$$\hat{f}^{-1}_{1,0;2,0}(p) = -\frac{8i\pi}{3\sqrt{5}} \frac{p}{(1+p^2)^2} \,. \tag{10.69}$$

For the inverse Fourier transform we obtain the following expression for the trace in (10.46)

$$\mathrm{tr}(\hat{f}(p) U(g,p)) = 8\pi \cos\theta \cos\beta \frac{(\sin(pa) - pa \cos(pa))}{(1+p^2)^2 \, p \, a^2}$$

The p integration in (10.46) reproduces the original function.

We computed also the Fourier transform of the function

$$f(\mathbf{a}, A) = a^2 e^{-a} \cos\theta \cos\beta \ .$$

The nonzero matrix elements are

$$\hat{f}^0_{1,0;0,0}(p) = \frac{32i\pi}{3\sqrt{3}} \frac{p\,(p^2 - 5)}{(1 + p^2)^4} \ ;$$

$$\hat{f}^0_{1,0;2,0}(p) = \frac{64i\pi}{\sqrt{135}} \frac{p\,(p^2 - 5)}{(1 + p^2)^4} \ ;$$

$$\hat{f}^1_{1,0;2,0}(p) = \frac{32i\pi}{3\sqrt{5}} \frac{p\,(p^2 - 5)}{(1 + p^2)^4} \ ;$$

$$\hat{f}^{-1}_{1,0;2,0}(p) = \frac{32i\pi}{3\sqrt{5}} \frac{p\,(p^2 - 5)}{(1 + p^2)^4} \ . \tag{10.70}$$

The inverse Fourier transform gives the original function.

We chop all zero elements for $l(l') > 2$. Therefore, after contraction of four indices to two using (10.54), the Fourier transform of the examples (10.70) and (10.69) can be written as a block-diagonal matrix

$$\hat{F} = \begin{bmatrix} \boxed{\hat{F}_{-1}} & & \\ & \boxed{\hat{F}_0} & \\ & & \boxed{\hat{F}_1} \end{bmatrix} . \tag{10.71}$$

The nonzero blocks are 9×9 matrix $\hat{F}_0$ and two 8×8 matrices $\hat{F}_{-1}$, $\hat{F}_1$ (lower indices correspond to s index). Using (10.54), these matrices can be depicted as

$$\hat{F}_0 = \begin{bmatrix} 0 & \cdots & 0 & & 0 \\ \vdots & & & & \vdots \\ \hat{f}^0_{31} & \cdots & \hat{f}^0_{37} & & 0 \\ \vdots & & & & \vdots \\ 0 & \cdots & 0 & & 0 \end{bmatrix} \tag{10.72}$$

where $\hat{f}^0_{31} = \hat{f}^0_{1,0;0,0}(p)$, $\hat{f}^0_{37} = \hat{f}^0_{1,0;2,0}(p)$. The other matrices are

$$\hat{F}_{\pm 1} = \begin{bmatrix} 0 & \cdots & 0 & & 0 \\ 0 & \cdots & \hat{f}^{\pm 1}_{26} & & 0 \\ \vdots & & & & \vdots \\ 0 & \cdots & 0 & & 0 \end{bmatrix} \tag{10.73}$$

where $\hat{f}^{\pm 1}_{26} = \hat{f}^{\pm 1}_{1,0;2,0}(p)$.

10.12 Linear-Algebraic Properties of Fourier Transform Matrices for $SE(2)$ and $SE(3)$

In this section we state several theorems that characterize the structure of $SE(N)$ Fourier transform matrices for $N = 2$ and $N = 3$.

Theorem 10.6. *[18] Complex eigenvalues of matrices with symmetries of the form (10.5) and (10.53) appear in conjugate pairs.*

Proof. <u>Case 1</u>: $SE(2)$.
    The eigenvalue problem is written in component form as

$$\lambda x_m = \sum_n \hat{f}_{mn} x_n = (-1)^m \sum_n (-1)^n \overline{\hat{f}_{-m,-n}} \, x_n.$$

Multiplying both sides by $(-1)^m$ and regrouping terms we get

$$\sum_n (-1)^n \overline{\hat{f}_{-m,-n}} x_n = \sum_n \overline{\hat{f}_{-m,-n}} ((-1)^n x_n) = \lambda((-1)^m x_m).$$

Changing the variables (m,n) to $(-m,-n)$, and taking the complex conjugate of both sides,

$$\sum_n \hat{f}_{m,n} ((-1)^n \overline{x_{-n}}) = \overline{\lambda}((-1)^m \overline{x_{-m}}),$$

indicating that if (λ, x_m) is an eigenvalue/vector pair, then so is $(\overline{\lambda}, (-1)^m \overline{x_{-m}})$.
    <u>Case 2</u>: $SE(3)$.
    The eigenvalue problem is written for each s block as

$$\sum_{l,m} \hat{f}^s_{l',m';l,m} x^s_{l,m} = \lambda x^s_{l',m'},$$

where there is summation over l and m.
    Substitution using symmetry (10.53) yields

$$\sum_{l,m} (-1)^{(l'-l)} (-1)^{(m'-m)} \overline{\hat{f}^s_{l',-m';l,-m}} x^s_{l,m} =$$

$$(-1)^{(l'+m')} \sum_{l,m} \overline{\hat{f}^s_{l',-m';l,-m}} ((-1)^{(l+m)} x^s_{l,m}) = \lambda x^s_{l',m'}.$$

Multiplication on both sides by $(-1)^{(l'+m')}$, and complex conjugation yield:

$$\sum_{l,m} \hat{f}^s_{l',-m';l,-m} ((-1)^{(l+m)} \overline{x^s_{l,m}}) = \overline{\lambda}((-1)^{(l'+m')} \overline{x^s_{l',m'}}).$$

Finally, changing the dummy variables (m,m') to $(-m,-m')$, we get

$$\sum_{l,m} \hat{f}^s_{l',m';l,m} ((-1)^{(l-m)} \overline{x^s_{l,-m}}) = \overline{\lambda}((-1)^{(l'-m')} \overline{x^s_{l',-m'}}),$$

indicating that for every eigenvalue/vector pair $(\lambda, x^s_{l',m'})$, there is also a pair of the form $(\overline{\lambda}, (-1)^{(l'-m')} \overline{x^s_{l',-m'}})$.

While the Fourier transform is generally used for harmonic analysis of square-integrable functions in this book, there are a few notable exceptions. These are the generalized (singularity) functions (or distributions). These are discussed here, and used throughout the book.

The Dirac delta function for $\mathbb{R}^N$ is defined to have the following properties:

$$\int_{\mathbb{R}^N} \delta(\mathbf{r})\, d\mathbf{r} = 1\ , \qquad \int_{\mathbb{R}^N} f(\mathbf{r})\, \delta(\mathbf{x} - \mathbf{r})\, d\mathbf{r} = (f * \delta)(\mathbf{x}) = f(\mathbf{x}),$$

where $d\mathbf{r} = dr_1 dr_2 \ldots dr_N$. (An alternative notation for this is $d^N r$.)

The Dirac delta function on $SO(N)$ has the analogous properties:

$$\int_{SO(N)} \delta(\mathcal{R})\, d\mathcal{R} = 1\ , \qquad \int_{SO(N)} f(\mathcal{R})\, \delta(\mathcal{R}^T R)\, d\mathcal{R} = (f * \delta)(R) = f(R).$$

It follows directly from the invariance of integration under shifts and inversion of the arguments of functions on $\mathbb{R}^N$ and $SO(N)$ that

$$\delta(\mathbf{x} - \mathbf{r}) = \delta(\mathbf{r} - \mathbf{x}) \quad \text{and} \quad \delta(\mathcal{R}^T R) = \delta(R^T \mathcal{R}).$$

The delta function for $SE(N)$ is the product of these:

$$\delta(g) = \delta(\mathbf{r})\, \delta(A) \quad \text{for} \quad g = (\mathbf{r}, A) \in SE(N).$$

The integration over $SE(N)$ in the convolution integral can be rewritten as integration over position and orientation separately:

$$(f_1 * f_2)(\mathbf{x}, R) = \int_{SO(N)} \int_{\mathbb{R}^N} f_1(\boldsymbol{\xi}, \mathcal{R})\, f_2(\mathcal{R}^T(\mathbf{x} - \boldsymbol{\xi}), \mathcal{R}^T R)\, d\boldsymbol{\xi}\, d\mathcal{R}. \tag{10.74}$$

where $g = (\mathbf{x}, R)$ and $h = (\boldsymbol{\xi}, \mathcal{R})$ are elements of $SE(N)$.

Using this notation, it is easy to see that the Fourier transforms of functions such as $f(R)\,\delta(\mathbf{x})$ and $f(\mathbf{x})\,\delta(R)$ reduce to

$$\int_{SO(N)} f(R)\, U(\mathbf{0}, R^T; p)\, dR$$

and

$$\int_{\mathbb{R}^N} f(\mathbf{x})\, U(-\mathbf{x}, \mathbb{I}; p)\, d\mathbf{x}$$

respectively.

Theorem 10.7. *Functions $f(g) \in \mathcal{L}^2(SE(N))$ with Fourier transforms satisfying the condition of normality*

$$\hat{f}^*(p)\hat{f}(p) = \hat{f}(p)\hat{f}^*(p)$$

include:
a) *$f(g)$ which satisfy the condition*

$$\overline{f(g)} = \pm f(g^{-1}).$$

We call these functions symmetric (antisymmetric).
b) *$f(g)$ which is a class function, i.e.*

$$f(g) = f(h^{-1} \circ g \circ h)$$

for any g, $h \in G$.
c) *$f(g)$ with a Fourier transform matrix which is proportional to a unitary matrix.*

Proof. Cases (a) and (b) can be shown easily using the Fourier transform definition, the unitarity of $U(g,p)$, and the invariance of integration with $d(g)$ under shifts and inversion of the argument. In particular, it is easy to see that the Fourier transform of a symmetric (antisymmetric) function is a Hermitian (skew-Hermitian) matrix

$$\hat{F}^*(p) = \pm \hat{F}(p).$$

Case (c) follows directly from the definition of unitary matrices.

We can also find normal matrices which are not in these categories. For example, a function with the Fourier transform matrix $\hat{f}^*(p) = e^{ic(p)} \hat{f}(p)$ (where $c(p)$ is some function) is also normal. The next two theorems examine how broad the sets of class and symmetric/antisymmetric functions are for $SE(N)$.

Theorem 10.8. *For $N \in \{2,3\}$ there are no class functions in $\mathcal{L}^2(SE(N))$ for which* $\int_{SE(N)} |f(g)| \, d(g) > 0.$

Proof. Let $g = (R, \mathbf{t}) \in SE(3)$ with $R \neq \mathbb{I}$. From (6.11) the conjugation invariants for $SE(3)$ are $\theta = \|(\log R)^\vee\|$ and $d = \mathbf{t} \cdot (\log R)^\vee / \|(\log R)^\vee\|$. And any class function, $\chi(g)$, must be a function of one or both of these and no other parameters of the motion. That is, it must be the case that

$$\chi(g) = f\left(\|(\log R)^\vee\|, \mathbf{t} \cdot \frac{(\log R)^\vee}{\|(\log R)^\vee\|}\right).$$

Since d only measures distance along the single direction, $\mathbf{n} = (\log R)^\vee / \|(\log R)^\vee\|$, if we fix a value of R, it becomes clear that $\chi(R, \mathbf{t})$ can only vary in the direction parallel to $\mathbf{n}$ and must remain constant in the other two translational directions. Therefore, it does not have the freedom to decay in these directions, and hence cannot be square integrable. But if we force the conditions under which the (θ, d) invariants are defined to break down (which constitute a set of meausure zero), then other class functions that are square integrable are possible, but these do not satisfy the condition $\int_{SE(N)} |f(g)| \, d(g) > 0$.

In the $SE(2)$ case, the only invariant is θ and so any function that depends at all on translation cannot be class function for the same reasons as the $SE(3)$ case.

Examples: Consider the following examples: forcing R to be $\mathbb{I}$ means that the dependence on A is as a Dirac delta on $SO(N)$ and a class function of the form $\chi_1(g) \doteq f_1(\|\mathbf{t}\|) \delta(A)$ is possible. If $\chi_2(g) \doteq f_2(\theta)$ (either in the planar or spatial case), and $f_2(\theta) = f_2(-\theta)$, this is also a class function. But neither $\chi_i(g)$ is square integrable. Pathological functions that take nonzero values only on sets of measure zero of even lower dimension are also possible, such as a function $\chi_3(g) = 1$ if $g = e$ and zero otherwise, but they do not satisfy the second condition in the statement of theorem.

For a function to be a class function on $SE(N)$ necessary conditions are that $f(g_1^{-1} \circ g \circ g_1) = f(g)$ for $g_1 = (\mathbf{0}, A_1)$ and $f(g_2^{-1} \circ g \circ g_2) = f(g)$ for $g_2 = (\mathbf{r}_1, I)$. That is, the definition must hold for general automorphisms, and so it must hold for pure rotations and translations individually. Using the notation $f(g) = f(\mathbf{r}, A)$, these conditions are written as:

$$f(\mathbf{r}, A) = f(A_1^T \mathbf{r}, A_1^T A A_1) \tag{10.75}$$

and

$$f(\mathbf{r}, A) = f(\mathbf{r} + (A - 1)\mathbf{r}_1, A). \tag{10.76}$$

Equation (10.76) can only be true for arbitrary $A \neq I$ if it has no dependence on $\mathbf{r}$. This leads to:

<u>case 1</u>: $f(\mathbf{r}, A) = f_1(A)$.

If, however, $f(\mathbf{r}, A) = 0$ for all $A \neq I$ then (10.76) is also satisfied. The only way this can be true and $\int_{SE(N)} |f(g)| \, d(g) > 0$ is when:

<u>case 2</u>: $f(\mathbf{r}, A) = f_2(\mathbf{r}) \delta(A)$.

Neither of these functions are square integrable on $SE(N)$. While this completes the proof, it is interesting to note that in order to satisfy (10.75), $f_1(A)$ must be a class function on $SO(N)$, and $f_2(\mathbf{r}) = f_2(|\mathbf{r}|) = f_2(r)$.

From Chapter 6 we know that all class functions for $SO(3)$ are functions of the rotation angle, and so $f_1(A) = f_1(\theta)$.

Theorem 10.9. *There exist nontrivial square-integrable symmetric and antisymmetric functions on $SE(N)$.*

Proof. The proof is by construction. Let $f_i : \mathbb{R}^{N \times N} \times \mathbb{R}^N \to \mathbb{C}$ (or $\mathbb{R}$), for $i = 1, 2, 3, 4$, be square integrable functions on $\mathbb{R}^{N \times N \times N}$ such that

$$f_1((-1)^n B, (-1)^m \mathbf{y}) = f_1(B, \mathbf{y}),$$

$$f_2((-1)^n B, (-1)^m \mathbf{y}) = (-1)^m f_2(B, \mathbf{y}),$$

$$f_3((-1)^n B, (-1)^m \mathbf{y}) = (-1)^n f_3(B, \mathbf{y}),$$

$$f_4((-1)^n B, (-1)^m \mathbf{y}) = (-1)^{(n+m)} f_4(B, \mathbf{y}),$$

for all $B \in \mathbb{R}^{N \times N}$, $\mathbf{y} \in \mathbb{R}^N$, and $m, n \in \{0, 1\}$. Then it is easy to confirm by direct substitution of $g^{-1} = (A^T, -A^T \mathbf{r})$ for $g = (A, \mathbf{r}) \in SE(N)$ in the above functions that

$$f_1(A + A^T, A^{-1/2} \mathbf{r})$$

$$f_3(A + A^T, A^{-1/2} \mathbf{r})$$

$$f_1(A - A^T, A^{-1/2} \mathbf{r})$$

$$f_4(A - A^T, A^{-1/2} \mathbf{r})$$

are symmetric, and

$$f_2(A + A^T, A^{-1/2} \mathbf{r})$$

$$f_4(A + A^T, A^{-1/2} \mathbf{r})$$

$$f_3(A - A^T, A^{-1/2} \mathbf{r})$$

$$f_2(A - A^T, A^{-1/2} \mathbf{r})$$

are antisymmetric, and they are all square integrable by definition (the $A^{-1/2}$ is defined as rotation around the same axis as in A by half of the angle in opposite direction). For instance, if $f(g) = f(\mathbf{r}, A) = f_1(A + A^T, A^{-1/2} \mathbf{r})$, then $f(g^{-1}) = f(-A^T \mathbf{r}, A^T) = f_1(A^T + A, (A^T)^{-1/2}(-A^T \mathbf{r}))$. But since any matrix commutes with powers of itself, and $A^T = A^{-1}$, $(A^T)^{-1/2}(-A^T \mathbf{r}) = (A^{-1})^{-1/2}(-A^{-1} \mathbf{r}) = A^{1/2}(-A^{-1} \mathbf{r}) = -A^{-1/2} \mathbf{r}$. Clearly then, $f(g^{-1}) = f_1(A^T + A, -A^{-1/2} \mathbf{r}) = f(g)$ in this case. The other cases follow in the same way.

In the case of $SE(3)$, if the rotation matrix is parameterized using the axis and angle of rotation, $A = A(\theta, \boldsymbol{\omega})$, and if the direction of the axis of rotation is parameterized by polar and azimuthal angles, $\boldsymbol{\omega} = \boldsymbol{\omega}(\alpha_1, \alpha_2)$, then functions of the form $f(\theta, \alpha_1, \alpha_2; r)$ where $f(\pm \theta, \alpha_1, \alpha_2; r) = \pm f(\theta, \alpha_1, \alpha_2; r)$ are symmetric/antisymmetric.

10.13 Summary

In this chapter we described one possible definition of the irreducible unitary representations and Fourier transform for $SE(2)$ and $SE(3)$. Operational properties of the Fourier transform under differential operators acting on functions on these groups were investigated. These operational properties are important in the context of applications in Chapter 17 (see also [6]). They have also been used to formulate deconvolution problems, as in [19]. Chapter 11 addresses computational issues of the Fourier transforms for $SE(N)$ that are used in Chapters 12 and 13.

References

1. Arnal, D., Cortet, J.C., "Star Representation of $E(2)$," *Letters in Mathematical Physics*, 20: 141–149, 1990.
2. Auslander, L., Moore, C.C., *Unitary Representations of Solvable Lie Groups*, Memoirs of the American Mathematical Society, No. 62, 1966.
3. Ballesteros, A., Celeghini, E., Giachetti, R., Tarlini, M., "An R-Matrix Approach to the Quantization of the Euclidean Group $E(2)$," *Journal of Physics and Mathematics*, 26: 7495–7501, 1993.
4. Bohm, M., Junker, G., "Path Integration over the n-Dimensional Euclidean Group," *Journal of Mathematical Physics*, 30: 1195–1197, June 1989.
5. Chirikjian, G.S., *Stochastic Models, Information Theory, and Lie Groups: Vols. 1 + 2*, Birkhäuser, Boston, 2009/2011.
6. Chirikjian, G.S., Kyatkin, A.B., "An Operational Calculus for the Euclidean Motion Group with Applications in Robotics and Polymer Science," *Journal of Fourier Analysis and Applications* 6(6): 583–606, 2000.
7. Feinsilver, P., "Lie Algebras and Recurrence Relations IV: Representations of the Euclidean Group and Bessel Functions," *Acta Applicandae Mathematicae*, 43: 289–316, 1996.
8. Gel'fand, I.M., Minlos, R.A., Shapiro, Z. Ya., *Representations of the Rotation and Lorentz Groups and Their Applications*, Pergamon Press, New York, 1963.
9. Gonzalez, F.B., "Bi-invariant Differential Operators on the Euclidean Motion Group and Applications to Generalized Radon Transforms," *Achiv for Matematik*, 26(2): 191–204, Oct. 1988.
10. Gross, K.I., Kunze, R.A., "Bessel Functions and Representation Theory," *Journal of Functional Analysis*, 22(2): 73–105, 1976.
11. Gurarie, D., *Symmetry and Laplacians. Introduction to Harmonic Analysis, Group Representations and Applications*, Elsevier Science Publisher, The Netherlands, 1992. (Dover 2008)
12. Humi, M., "Representations and Invariant Equations of $E(3)$," *Journal of Mathematical Physics*, 28: 2807–2811, Dec. 1987.
13. Isham, C. J., Klauder, J.R., "Coherent States for n-Dimensional Euclidean Groups $E(n)$ and Their Application," *Journal of Mathematical Physics*, 32: 607–620, March 1991.
14. Jones, M.N., *Spherical Harmonics and Tensors for Classical Field Theory*, Research Studies Press Ltd., England, 1985.
15. Kopský, V., "Translation Normalizers of Euclidean Groups," *Journal of Mathematics and Physics*, 34: 1548–1556, April 1993.
16. Kopský, V., "Translational Normalizers of Euclidean Groups (II)," *Journal of Mathematics and Physics*, 34: 1557–1576, April 1993.
17. Kumahara, K., Okamoto, K., "An Analogue of the Paley-Wiener Theorem for the Euclidean Motion Group," *Osaka Journal of Mathematics*, 10: 77–92, 1973.

18. Kyatkin, A.B., Chirikjian, G.S., "Regularization of a Nonlinear Convolution Equation on the Euclidean Group," *Acta Applicandae Mathematicae*, 53: 89–123, August 1998.
19. Luo, Z.M., Kim, P.T., Kim, T.Y., Koo, J.Y., "Deconvolution on the Euclidean Motion Group SE(3)," *Inverse Problems*, 27(3): 035014, 2011.
20. Mackey, G.W., *Induced Representations of Groups and Quantum Mechanics*, W. A. Benjamin, Inc., New York and Amsterdam, 1968.
21. Miller, W., Jr., *Lie Theory and Special Functions*, Academic Press, New York, 1968; also see Miller, W. Jr., "Some Applications of the Representation Theory of the Euclidean Group in Three-Space," *Communications on Pure and Applied Mathematics*, 17: 527–540, 1964.
22. Moses, H.E., Quesada, A.F., "The Expansion of Physical Quantities in Terms of the Irreducible Representations of the Scale-Euclidean Group and Applications to the Construction of Scale-Invariant Correltation Functions Part I," *Archives of Rational Mechanics and Analysis*, 44(3): 217–248, 1972.
23. Moses, H.E., Quesada, A.F., "The Expansion of Physical Quantities in Terms of the Irreducible Representations of the Scale-Euclidean Group and Applications to the Construction of Scale-Invariant Correlation Functions Part II," *Archives of Rational Mechanics and Analysis*, 50: 194–236, 1973.
24. Orihara, A., "Bessel Functions and the Euclidean Motion Group," *Tohoku Mathematical Journal*, 13: 66–71, 1961.
25. Pukanszky, L., "Unitary Representations of Solvable Lie Groups," *Annales scientifiques de l'École normale supérieure*, 4(4): 457–608, 1971.
26. Rno, J. S., *Clebsch-Gordan Coefficients and Special Functions Related to the Euclidean Group in Three-Space*, University of Minnesota, Ph.D., 1973.
27. Rno, J.S., "Clebsch-Gordan Coefficients and Special Functions Related to the Euclidean Group in Three-Space," *Journal of Mathematical Physics*, 15(12): 2042–2047, Dec. 1974.
28. Rno, J.S., "Harmonic Analysis on the Euclidean Group in Three-Space, I, II," *Journal of Mathematical Physics*, 26: 675–677, 2186–2188, 1985.
29. Rozenblyum, A.V., Rozenblyum, L.V., "Orthogonal Polynomials of Several Variables Related to Representations of Groups of Euclidean Motion," *Differentsial'nye Uravneniya*, 22(11): 1972–1977, Nov. 1986.
30. Rozenblyum, A.V., "Representations of Lie Groups and Multidimensional Special Functions," *Acta Applicandae Mathematicae*, 29: 171–240, 1992.
31. Rubin, R.L., "Harmonic Analysis on the Group of Rigid Motions of the Euclidean Plane," *Studia Mathematica*, 62: 125–141, 1978.
32. Sakai, K., "Irreducible Unitary Representations of the Group of Motions in 3-Dimensional Euclidean Space and the Spherical Bessel Functions," (in japanese) *Review of the Marine Technical College*, 8: 93–115, 1964.
33. Sakai, K., "On the Representations of the Motion Group of n-Dimensional Euclidean Space. I," *Science Reports of the Kagoshima University*, 16: 25–33, 1967.
34. Sakai, K., "Some Remarks on Unitary Representations of the Euclidean Motion Group in Π_m-spaces," *Science Reports of the Kagoshima University*, 29: 13–26, 1980.
35. Sakai, K., "On Indecomposable Unitary Representations of the 2-Dimensional Euclidean Motion Group in Finite Dimensional Indefinite Inner Product Spaces I," *Science Reports of the Kagoshima University*, 29: 27–51, 1980.
36. Sakai, K., "On Indecomposable Unitary Representations of the 2-dimensional Euclidean Motion Group in Finite Dimensional Indefinite Inner Product Spaces II," *Science Reports of the Kagoshima University*, 30: 1–21, 1981.
37. Sugiura, M., *Unitary Representations and Harmonic Analysis*, 2nd ed., Elsevier Science Publisher, The Netherlands, 1990.
38. Symons, J., "Irreducible Representations of the Group of Movements of the Euclidean Plane," *The Journal of the Australian Mathematical Society*, 18: 78–96, Aug. 1974.
39. Talman, J., *Special Functions*, W. A. Benjamin, Inc., Amsterdam, 1968.

40. Takiff, S., "Invariant Polynomials on Lie Algebras of Inhomogeneous Unitary and Special Orthogonal Groups," *Transactions of the American Mathematical Society*, 170: 221–230, 1972.

41. Torres del Castillo, G. F., "Spin-weighted Cylindrical Harmonics and the Euclidean Group of the Plane," *Journal of Mathematical Physics*, 34: 3856–3862, Aug. 1993.

42. Vilenkin, N.J., "Bessel Functions and Representations of the Group of Euclidean Motions," *Uspehi Mat. Nauk.*, 11(3): 69–112, 1956 (in Russian).

43. Vilenkin, N.J., Akim, E.L., Levin, A.A., "The Matrix Elements of Irreducible Unitary Representations of the Group of Euclidean Three-Dimensional Space Motions and Their Properties," *Dokl. Akad. Nauk SSSR*, 112: 987–989, 1957 (in Russian).

44. Vilenkin, N.J., Klimyk, A.U., *Representation of Lie Group and Special Functions*, Vol. 1-3, Kluwer Academic Publishers, The Netherlands, 1991.

45. Wightman, A.S., "On the Localizability of Quantum Mechanical Systems," *Reviews in Modern Physics*, 34: 845–872, 1962.

11

Fast Fourier Transforms for Motion Groups

In this chapter we apply techniques from noncommutative harmonic analysis to the development of fast numerical algorithms for the computation of convolution integrals on motion groups.[1] In particular, we focus on the group of rigid-body motions in the plane and 3-space. Using IURs in operator form, we write the Fourier transform of functions on the motion group as an integral over the product space $SE(3) \times \mathbb{S}^2$. The integral form of the Fourier transform matrix elements allows us to apply Fast Fourier Transform (FFT) methods developed previously for $\mathbb{R}^3$, $\mathbb{S}^2$, and $SO(3)$ to speed up considerably the numerical computation of convolutions of functions on $SE(3)$. Such convolutions play an important role in a number of engineering disciplines. Numerical algorithms for the fast computation of the Fourier transform are given and the complexity of the numerical implementations are discussed. The Fourier transform on the 3D "semi-discrete motion group" (semi-direct product of the icosahedral group with the continuous translation group[2]) is also developed and the results of the numerical implementation are discussed. Finally, we examine alternative approaches to computing fast motion-group convolutions. These include the FFT for the direct product group $SO(3) \times \mathbb{R}^3$, and the use of reducible representations of $SE(3)$.

11.1 Preliminaries: Direct Convolution and the Cost of Interpolation

If we let $g = (\mathbf{r}, R)$ and $h = (\mathbf{a}, A)$ be elements of $SE(n)$, then the convolution

$$(f_1 * f_2)(h) = \int_{SE(n)} f_1(h) \, f_2(h^{-1} \circ g) \, d(h)$$

is written as

$$(f_1 * f_2)(\mathbf{r}, R) = \int_{SO(n)} \int_{\mathbb{R}^n} f_1(\mathbf{a}, A) \, f_2(A^T(\mathbf{r} - \mathbf{a}), A^T R) \, d\mathbf{a} \, dA. \tag{11.1}$$

If we compute this convolution integral by direct evaluation of each of N values of g and sum over N values of h, then a total of $\mathcal{O}(N^2)$ computations are required. Here

---

[1]Much of the material in this chapter was first presented in [18].

[2]Previously we called this simply the "discrete motion group" [17, 18], but to avoid confusion with chiral crystallographic groups, which have both discrete rotations and translations we now use the current name.

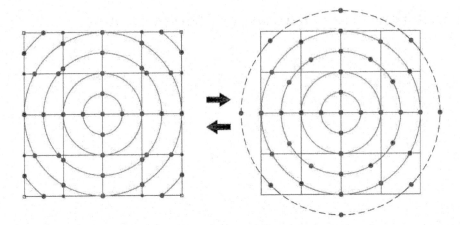

Fig. 11.1. Interpolation from Cartesian to Polar Grid. (Reprinted from Kyatkin, A.B., Chirikjian, G.S., "Algorithms for Fast Convolutions on Motion Groups," *Applied and Computational Harmonic Analysis*, 9: 220–241, 2000, with permission from Elsevier)

we assume there are N_r samples in the $\mathbb{R}^n$ part, and N_R samples in the $SO(n)$ part. If there are $\mathcal{O}(S)$ samples in each coordinate direction, this means that in the case of $SE(3)$ we get $N = \mathcal{O}(S^6)$, and $\mathcal{O}(S^{12})$ computations are required to perform a direct convolution. Hence, even $S \approx 10$ is prohibitive (at the time of the writing of this book).

If the functions f_i are given numerically with the $\mathbb{R}^n$ part given on a Cartesian lattice, the direct evaluation of (11.1) will require the interpolation of function values off grid. We therefore devote this first section of this chapter to issues related to the complexity of interpolation.

11.1.1 Fast and Accurate Fourier Interpolation

We can, without loss of any information, interpolate finite values of a periodic band-limited function on the plane (or, equivalently, a band-limited function on T^2) sampled at Cartesian grid points to points on a polar grid. This is shown in the left hand side of Figure 11.1, where function values are specified at each intersection of the straight lines. The first step in this procedure is the interpolation of the given values to each of the points marked with circular dots. Since each straight grid line (which can be viewed as an unwrapped circle) defines a one-dimensional Fourier series, this step amounts to the evaluation of a one-dimensional Fourier series at nonequally spaced points. Since band-limited Fourier series can be viewed as polynomials of the form $\sum_{n=0}^{N_r^{1/2}-1} a_n Z^n$ where Z is a complex exponential, each vertical and horizontal evaluation can be performed in $\mathcal{O}(N_r^{\frac{1}{2}}(\log N_r^{\frac{1}{2}})^2)$ [1, 3]. Since there are $\mathcal{O}(N_r^{\frac{1}{2}})$ such lines, the total procedure of interpolating values from the Cartesian to polar grid requires $\mathcal{O}(N_r^{\frac{1}{2}} \cdot N_r^{\frac{1}{2}}(\log N_r^{\frac{1}{2}})^2)$ $= \mathcal{O}(N_r(\log N_r)^2)$ computations. This procedure is an exact one in exact arithmetic. And since we are careful to choose a polar grid for which there are a sufficient number of intersections of circles with each Cartesian grid line, no information is lost.

Once the values have been interpolated to each of the concentric circles, they define a band-limited Fourier series on each circle that can be evaluated at equally spaced angles within each circle. This is again an $\mathcal{O}(N_r(\log N_r)^2)$ computation since there are $\mathcal{O}(N_r^{\frac{1}{2}})$ circles. It too is exact in the sense that it conserves the original information defined at the N_r grid points.

In order for the information of the function values to be recoverable to the Cartesian grid from the polar one, a sufficient number of circles extending outside of the original Cartesian grid must have values defined on them. In the example shown, there is only one circle that extends beyond the Cartesian grid.

To see that this procedure is reversible, first evaluate the Fourier series on each circle at the points on the circle that intersect the Cartesian grid. Then fit the band-limited Fourier series on each straight line that pass through the values at the points of intersection. This fitting of a Fourier series to irregularly spaced data on each line is performed with the same order of computations as the evaluation of Fourier series at irregular points [1, 3]. In order for there to be enough information to reconstruct the original band-limited Fourier series on the outer edges of the Cartesian grid, the information contained in the partially outlying circle is required. For finer grids, multiple such circles are required. In the current example, the corner values could not be determined from the polar grid unless the information in the circle extending outside of the polar grid is included. The values marked with an asterisk are found using a two-step process whereby function values on surrounding lines must be evaluated at Cartesian grid points before there is enough data available to exactly recover the values.

From the above discussion, we observe that interpolation back and forth between polar and Cartesian grids can be performed in $\mathcal{O}(N_r(\log N_r)^2)$ computations. This procedure of Fourier interpolation between Cartesian and polar grids is "exact" in the sense that in exact arithmetic it reversibly takes values from one kind of grid into another.

The number of function values on each circle enclosed in the Cartesian grid increases linearly with the radius. This ensures that the information in a band-limited function on T^2 is exactly transferred to the collection of band-limited functions on the circles of the polar grid using the procedure described above. Another way to view this is that a band-limited periodic function on the plane is set to zero outside of a finite window with dimensions equal to the period of the function. The version of this function sampled in polar coordinates defines a new function that, in principle, can be nonzero outside of this finite window. Given the standard assumptions in numerical Fourier analysis, the values on all circles not fully contained within the finite square grid will be made arbitrarily close to zero by appropriate choice of the band-limit and zero padding.

The circles partially enclosed in the Cartesian grid diminish in significance with radius. In the context of the example shown in Figure 11.1, we see that on circles from the center outward there are 1, 4, 4, 12, 12, 8 values. This means that the band-limit for functions of the angle ψ for each of these circles will be the numbers given in the previous sentence. At radius p, the band limit will be $\mathcal{O}(p)$, with $p \leq \mathcal{O}(N_r^{\frac{1}{2}})$.

Note that while the above procedure is exact at the specific discrete points specified, it is not exact at other points. However, sufficient accuracy can be guaranteed when inter-converting, as described in [2].

11.1.2 Approximate Spline Interpolation

In spline interpolation, only the N_s nearest grid points to the point at which the interpolated value is desired enter the calculation. This necessarily means that some information about the function is destroyed in the spline interpolation process. However, since the functions in question are all assumed to be band-limited Fourier series, they do not oscillate on the length scale of the distance between sample points. Hence, a polynomial spline of sufficiently high order will approximate well the local neighborhood of the point at which the interpolated value is to be determined. While this technique is an approximate one, it has been used with great success in compute tomography [6, 8, 27], image processing [7, 13, 20, 26, 32], and other fields [4, 36].

The benefit of this approach is that only $\mathcal{O}(N_s N_r) = \mathcal{O}(N_r)$ computations are required. The value of N_s is chosen for a given finite error threshold. The drawback is that unlike Fourier interpolation, it is not mathematically exact in exact arithmetic. For more discussion of splines see [30].

In the sections that follow, we use the notation $\epsilon(N_r)$ to denote the relative complexity of Fourier interpolation to spline interpolation. That is, for Fourier interpolation

$$\epsilon(N_r) = (\log N_r)^2,$$

whereas for spline interpolation

$$\epsilon(N_r) = N_s = \mathcal{O}(1).$$

11.2 A Fast Algorithm for the Fourier Transform on the 2D Motion Group

Numerical algorithms for Fourier transforms on the 2D motion group were developed in [19] (continuous motion group) and [17, 18] (semi-discrete motion group). For completeness we describe below a new algorithm for the continuous motion group Fourier transform using FFT methods. This algorithm is similar to the semi-discrete motion group algorithm used in [17] but uses $\exp(i\,m\,\Phi)$ as basis functions instead of pulse functions. We begin with the representation operators for the 2D motion group:

$$U(g,p)\tilde{\varphi}(\mathbf{x}) = e^{-ip(\mathbf{r}\cdot\mathbf{x})}\tilde{\varphi}(R^T\mathbf{x}), \qquad (11.2)$$

which are defined for each $g = (\mathbf{r}, R(\theta)) \in SE(2)$. Here $p \in \mathbb{R}^+$, and $\mathbf{x}\cdot\mathbf{y} = x_1 y_1 + x_2 y_2$. The vector $\mathbf{x}$ is a unit vector $(\mathbf{x}\cdot\mathbf{x} = 1)$, so $\tilde{\varphi}(\mathbf{x}) = \tilde{\varphi}(\cos\psi, \sin\psi) \equiv \varphi(\psi)$ is a function on the unit circle. Henceforth we will not distinguish between $\tilde{\varphi}$ and φ.

Any function $\varphi(\psi) \in \mathcal{L}^2([0, 2\pi))$ can be expressed as a weighted sum of orthonormal basis functions as $\varphi(\psi) = \sum_n a_n e^{in\psi}$. Likewise, the matrix elements of the operator $U(g,p)$ are expressed in this basis as [33]:

$$U_{mn}(g,p) = (e^{im\psi}, U(g,p)\,e^{in\psi}) = \frac{1}{2\pi}\int_0^{2\pi} e^{-im\psi}e^{-i(r_1 p\cos\psi + r_2 p\sin\psi)}e^{in(\psi-\theta)}\,d\psi$$

$$(11.3)$$

$\forall\ m,n \in \mathbb{Z}$. The inner product $(\cdot,\cdot)$ is defined as:

$$(\varphi_1, \varphi_2) \doteq \frac{1}{2\pi}\int_0^{2\pi} \overline{\varphi_1(\psi)}\varphi_2(\psi)\,d\psi.$$

It is easy to see that $(U(g,p)\varphi_1, U(g,p)\varphi_2) = (\varphi_1, \varphi_2)$, and that $U(g,p)$ is, therefore, unitary with respect to this inner product.

We can express the Fourier matrix elements as an integral:

$$\hat{f}_{mn}(p) = \int_{\mathbf{r}\in\mathbb{R}^2}\int_{\theta=0}^{2\pi}\int_{\psi=0}^{2\pi} f(\mathbf{r},\theta)\,e^{in\psi}e^{i(\mathbf{p}\cdot\mathbf{r})}e^{-im(\psi-\theta)}\,d^2r\,d\theta\,d\psi.$$

We compute the band-limited approximation of the Fourier transform for $|m|, |n| \leq S$ harmonics. Furthermore, we assume that it is computed at $N_p = \mathcal{O}(S)$ points along the p-interval, and we assume that the order of sampling points in an $\mathbb{R}^2$ region is of the order of S^2 ($N_r = \mathcal{O}(S^2)$) and the number of sampling points of orientation angle θ is $N_R = \mathcal{O}(S)$. In this way the total number of points sampled in $SE(2)$ is $N = \mathcal{O}(S^3)$, which is on the same order as the total number of sample points in the Fourier domain.

For estimates of complexity of numerical algorithms we introduce the following notations:

N_r	Number of samples on $\mathbb{R}^2$
N_R	Number of samples on $SO(2)$
N_p	Number of samples of p interval
N_u	Number of samples on $[0, 2\pi)$
N_F	Total number of harmonics

We assume that only a finite number of harmonics are required for an accurate approximation of the function $f(g)$. Hence only matrix elements in the range $|s| < S$ and $l, l' < L$ are computed. We assume also that $L = \mathcal{O}(S)$. We have made the following assumptions about the number of samples in term of S:

N_r	$\mathcal{O}(S^2)$
N_R	$\mathcal{O}(S)$
N_p	$\mathcal{O}(S)$
N_u	$\mathcal{O}(S)$
N_F	$\mathcal{O}(S^2)$

From these definitions, $N = N_r \cdot N_R = \mathcal{O}(S^3)$ and $N_p \cdot N_F = \mathcal{O}(N)$.
We can perform first $\mathbb{R}^2$ integration using the usual FFT

$$f_1(\mathbf{p}, \theta) = \int_{\mathbb{R}^2} f(\mathbf{r}, \theta) \, e^{i(\mathbf{p} \cdot \mathbf{r})} \, d^2 r$$

in $\mathcal{O}(N_R N_r \log N_r)$ computations.

Then, we perform interpolation to the polar coordinate mesh. For N_s-point spline interpolation this can be performed in $\mathcal{O}(N_s N_r N_R)$ computations. If Fourier interpolation is used this becomes $\mathcal{O}((\log N_r)^2 N_r N_R)$ computations.

The next step is to perform integration on $SO(2)$:

$$f_2^{(m)}(p, \psi) = \int_{SO(2)} f_1(p, \psi, \theta) \, e^{im\theta} \, d\theta.$$

This can be computed in $\mathcal{O}(N_r N_R \log N_R)$ computations. Then, ψ integration

$$\hat{f}_{mn}(p) = \int_0^{2\pi} \left[f_2^{(m)}(p, \psi) \, e^{-im\psi} \right] e^{in\psi} \, d\psi$$

can be computed in $\mathcal{O}(S N_p S \log S) = \mathcal{O}(N_r N_R \log N_R)$ computations.

We denote the total number of samples on $SE(2)$ as $N = N_r N_R = \mathcal{O}(S^3)$. Thus, the total number of arithmetic operations used to compute the direct Fourier transform is of $\mathcal{O}(N \log N) + \mathcal{O}(N\epsilon(N^{\frac{1}{2}}))$. When spline interpolation is used, the first term in this complexity estimate dominates for large values of N, whereas the second term dominates when Fourier interpolation is used.

The Fourier inversion formula for $SE(2)$ with matrix elements $U_{mn}(g, p)$ written in integral form is

$$f(g) = \frac{1}{2\pi} \int_0^\infty p \, dp \sum_{m,n=-\infty}^\infty \hat{f}_{nm}(p) \int_0^{2\pi} e^{i(n-m)\psi} e^{-i\mathbf{r} \cdot \mathbf{P}} e^{-in\theta} \, d\psi.$$

This can be rewritten as

$$f(g) = \frac{1}{2\pi} \sum_{n=-\infty}^\infty e^{-in\theta} \int_{\mathbb{R}^2} \tilde{f}_n(\mathbf{p}) \, e^{-i\mathbf{r} \cdot \mathbf{p}} \, d^2p$$

where

$$\tilde{f}_n(\mathbf{p}) = \sum_{m=-\infty}^\infty \hat{f}_{nm}(p) \, e^{i(n-m)\psi}.$$

If we assume that the above sums over m and n are trucated at $\pm S$ where $S = \mathcal{O}(N_R)$ and $\tilde{f}_n(\mathbf{p})$ is a band-limited function of $\mathbf{p}$ for each value of n, then it can be shown that the Fourier inversion can be performed in the same order of computations as the set of Fourier transforms. Note that this is faster than if the ψ-integration is performed first, even though that integration results in closed-form solutions for the matrix elements $U_{mn}(g, p)$.

The matrix product in the convolution can be performed in $\mathcal{O}(N^{(\gamma+1)/3})$ computations, where $2 \leq \gamma \leq 3$ is the cost of matrix multiplication.

11.2.1 Algorithms for Fast $SE(3)$ Convolutions Using FFTs

Here we describe an algorithm for computing 3D continuous motion group convolutions using FFTs. We use irreducible unitary representations of $SE(3)$ (as described in the Chapter 10) to calculate Fourier matrix elements in integral form.

To write the Fourier transform in matrix form we calculate the matrix elements of $U(\mathbf{r}, R; p)$ as

$$U_{l',m';l,m}^s(\mathbf{r}, R; p) = \int_{\mathbb{S}^2} \overline{h_{l'm'}^s(\mathbf{u})} \, (U(\mathbf{r}, R; p) h_{lm}^s)(\mathbf{u}) \, d\mathbf{u}$$

where $d\mathbf{u} = \sin\Theta d\Theta \, d\Phi$ and $h_{l'm'}^s(\mathbf{u}) = h_{l'm'}^s(\mathbf{u}(\Theta, \Phi))$ are generalized spherical harmonics defined in Chapter 9 and used in Chapter 10. Here $g = (\mathbf{r}, R) \in SE(3)$.

11.2.2 Direct Fourier Transform

We use basis eigenfunctions $h_{lm}^s(\mathbf{u})$ to write Fourier matrix elements in the integral form

$$\hat{f}_{l',m';l,m}^s(p) = \int_{\mathbf{u}\in\mathbb{S}^2} \int_{\mathbf{r}\in\mathbb{R}^3} \int_{R\in SO(3)} f(\mathbf{r}, R) \, h_{lm}^s(\mathbf{u}) \, e^{ip\,\mathbf{u}\cdot\mathbf{r}} \, \overline{\Delta_s(Q(\mathbf{u}, R)) h_{l'm'}^s(R^{-1}\mathbf{u})} \, d\mathbf{u} \, d^3r \, dR$$

where dR is the normalized invariant integration measure on $SO(3)$.

Recall from Chapter 10 that the basis functions can be expressed in the form [23, 24]

$$h^s_{lm}(\mathbf{u}(\Theta,\Phi)) = Q^l_{s,m}(\cos\Theta)\, e^{i(m+s)\Phi} \tag{11.4}$$

where

$$Q^l_{-s,m}(\cos\Theta) = (-1)^{l-s}\sqrt{\frac{2l+1}{4\pi}}\, P^l_{s\,m}(\cos\Theta)\ ,$$

and generalized Legendre polynomials $P^l_{m\,s}(\cos\Theta)$ are given as in Vilenkin [34].

Under the rotation R these functions are transformed as

$$(U^s(\mathbf{0},R;p)\,h^s_{lm})(\mathbf{u}) = \Delta_s(Q(\mathbf{u},R))\,h^s_{lm}(R^{-1}\mathbf{u}) = \sum_{n=-l}^{l} U^l_{n\,m}(R)h^s_{ln}(\mathbf{u})\ .$$

$U^l_{nm}(R)$ are matrix elements of $SO(3)$ representations

$$U^l_{m\,n}(A) = e^{-im\alpha}\,(-1)^{n-m}\,P^l_{m\,n}(\cos\beta)\,e^{-in\gamma}\ , \tag{11.5}$$

where α,β,γ are ZXZ Euler angles of the rotation and $P^l_{mn}(\cos\beta)$ is a generalization of the associated Legendre functions [33].

Thus, the Fourier transform matrix elements can be written in the form

$$\hat{f}^s_{l',m';l,m}(p) =$$

$$\int_{\mathbf{u}\in\mathbb{S}^2}\int_{\mathbf{r}\in\mathbb{R}^3}\int_{R\in SO(3)} f(\mathbf{r},R)\,h^s_{lm}(\mathbf{u})\,e^{ip\,\mathbf{u}\cdot\mathbf{r}} \sum_{n=-l'}^{l'} \overline{U^{l'}_{nm'}(R)}\overline{h^s_{l'n}(\mathbf{u})}\,d\mathbf{u}\,d^3r\,dR.$$

For estimates of complexity of numerical algorithms we introduce the following notations:

N_r	Number of samples on $\mathbb{R}^3$
N_R	Number of samples on $SO(3)$
N_p	Number of samples of p interval
N_u	Number of samples on S^2
N_F	Total number of harmonics

We assume that only a finite number of harmonics are required for an accurate approximation of the function $f(g)$. Hence only matrix elements in the range $|s| < S$ and $l, l' < L$ are computed. We assume also that $L = \mathcal{O}(S)$. We have made the following assumptions about the number of samples in term of S

N_r	$\mathcal{O}(S^3)$
N_R	$\mathcal{O}(S^3)$
N_p	$\mathcal{O}(S)$
N_u	$\mathcal{O}(S^2)$
N_F	$\mathcal{O}(S^5)$

From these definitions, $N = N_r \cdot N_R = \mathcal{O}(S^6)$ and $N_p \cdot N_F = \mathcal{O}(N)$.

Our algorithm for the numerical computation of the direct Fourier transform is as follows:

a) First, we compute

$$f_1(R, \mathbf{p}) = \int_{\mathbb{R}^3} f(\mathbf{r}, R) \, e^{i\mathbf{p}\cdot\mathbf{r}} \, d^3r$$

using FFTs for a Cartesian lattice in $\mathbb{R}^3$. This integral can be computed in $\mathcal{O}(N_r \log(N_r) N_R)$ computations. The resulting Fourier transform is computed on a rectangular grid. We need to perform interpolation to the spherical coordinate grid in order to compute values

$$f_1(R; p, \mathbf{u}) \, .$$

The complexity of this interpolation is discussed in Section 11.1. In general, $\mathcal{O}(N_r \epsilon(N_r))$ computations will be required for each different sampled rotation. For high precision numerical approximations, 3D spline interpolation technique can be used (see, e.g., [6, 30]). For N_s-point spline interpolation for all values of rotation, the complexity of interpolation is of $\mathcal{O}(N_s N_r N_R)$ computations, i.e., $\epsilon(N_r) = N_s$. Since spline interpolation only uses a small subset of the sample points, we assume $N_s = \mathcal{O}(1)$. For exactly reversible interpolation (in exact arithmetic) Fourier interpolation can be used. In this case $\epsilon(N_r) = \mathcal{O}((\log N_r)^2)$, for reasons that are explained in Section 11.1.

b) Then, we perform integration on $SO(3)$:

$$(f_2)_{nm'}^{l'}(p, \mathbf{u}) = \int_{SO(3)} f_1(R, p, \mathbf{u})\overline{U_{nm'}^{l'}(R)} \, dR \, .$$

This is the Fourier transform on the rotation group, computed for different values of p and $\mathbf{u}$. A fast Fourier transform technique for $SO(3)$ has been developed by Maslen and Rockmore which calculates the forward and inverse Fourier transform of band-limited functions on $SO(3)$ in $\mathcal{O}(N_R(\log N_R)^2)$ arithmetic operations for N_R sample points [22]. This can be applied to compute f_2 for all values of p and $\mathbf{u}$ and all indices in $\mathcal{O}(N_p N_u N_R (\log N_R)^2)$. We assume that $N_p N_u \approx N_r$, thus the order of computations is $\mathcal{O}(N_r N_R (\log N_R)^2)$.

c) Then, we can perform integrations on the unit sphere:

$$\hat{f}_{l',m';l,m}^s(p) = \sum_{n=-l'}^{l'} \int_{\mathbb{S}^2} (f_2)_{nm'}^{l'}(p, \mathbf{u}) h_{lm}^s(\mathbf{u}) \overline{h_{l'n}^s(\mathbf{u})} \, d\mathbf{u}. \tag{11.6}$$

Using expression (10.30) for the basis functions we write (11.6) in the form

$$\hat{f}_{l',m';l,m}^s(p) =$$

$$\sum_{n=-l'}^{l'} \int_{\mathbb{S}^2} (f_2)_{nm'}^{l'}(p, \mathbf{u}) Q_{s,m}^l(\cos\Theta) Q_{s,n}^{l'}(\cos\Theta) \, \exp(i(m-n)\Phi) \, d\Phi \sin\Theta d\Theta \, .$$

We can perform integration with respect to Φ using the FFT on $\mathbb{S}^1$:

$$(f_3)_{l',m',n;m}(p, \Theta) = \int_0^{2\pi} [(f_2)_{nm'}^{l'}(p, \Phi, \Theta) \, \exp(-in\Phi)] \, \exp(im\Phi) \, d\Phi \, .$$

Integrations can be performed in $\mathcal{O}(N_p N_\Theta S^3 N_\Phi \log N_\Phi)$, where $N_\Phi = \mathcal{O}(S)$ is the number of samplings on the Φ interval. Thus, the order of computations is $\mathcal{O}(N_r N_R \log N_R)$.

Then, we perform integration with respect to Θ:

$$(f_4)^s_{l',m';l,m,n}(p) = \int_0^\pi [(f_3)_{l',m',n;m}(p,\Theta)\,Q^{l'}_{s,n}(\cos\Theta)]\,Q^l_{s,m}(\cos\Theta)\sin\Theta d\Theta. \quad (11.7)$$

Using the fact that $Q^l_{s,m}(\cos\Theta)$ is $P^l_{-s,m}(\cos\Theta)$ (up to a constant coefficient), the integration (summation) for all but l fixed indices can be performed using the Driscoll and Healy technique [11, 12, 22] in $\mathcal{O}(S^5\,N_p\,N_\Theta\,(\log N_\Theta)^2)$, where $N_\Theta = \mathcal{O}(S)$ is the number of samples on the Θ interval. Thus, the computations is $\mathcal{O}(N_R^{5/3}\,N_r^{2/3}\,(\log N_R)^2)$. The Θ integration can be performed in $\mathcal{O}(S^6\,N_p\,N_\Theta) = \mathcal{O}(N_R^2\,N_r^{2/3})$ operations using plain integration.

Then, the matrix elements of the $SE(3)$-Fourier transform can be found by the summation

$$\hat{f}^s_{l',m';l,m}(p) = \sum_{n=-l'}^{l'} (f_4)^s_{l',m';l,m,n}(p)$$

which is of $\mathcal{O}(S^6\,N_p) = \mathcal{O}(N_R^2\,N_p)$.

Thus, the total computational cost of the direct Fourier transform is $\mathcal{O}(N_r\,N_R\,(\log(N_r) + (\log N_R)^2 + \epsilon(N_r)) + N_R^{5/3}\,N_r^{2/3}\,(\log N_R)^2)$. Under the assumption that $N_r = \mathcal{O}(N_R)$ and using the notation $N_r\,N_R = N$ (N is the total number of samples on $SE(3)$) we write the leading order terms as $\mathcal{O}(N^{7/6}\,(\log N)^2 + N\,\epsilon(N^{\frac{1}{2}}))$. For plain Θ integration (i.e., if the fast generalized Legendre transform is not used to evaluate (11.7)) the estimate becomes $\mathcal{O}(N^{4/3} + N\,\epsilon(N^{\frac{1}{2}}))$.

11.2.3 Inverse Fourier Transform

The inverse Fourier transform integral can be written as

$$f(\mathbf{r}, R) =$$

$$\frac{1}{2\pi^2} \int_{\mathbf{u}\in\mathbb{S}^2} \int_{p=0}^\infty \sum_{s,l,m,l',m'} \hat{f}^s_{l',m';l,m}(p)\,\overline{h^s_{lm}(\mathbf{u})}\,\exp(-ip\mathbf{u}\cdot\mathbf{r}) \sum_{n=-l'}^{l'} U^{l'}_{nm'}(R)h^s_{l'n}(\mathbf{u})\,p^2\,dp\,d\mathbf{u}.$$

where

$$\sum_{s,l,m,l',m'} = \sum_{s=-\infty}^{\infty}\sum_{l=|s|}^{\infty}\sum_{m=-l}^{l}\sum_{l'=|s|}^{\infty}\sum_{m'=-l'}^{l'}.$$

A band-limited approximation results when the restrictions $s \leq S$ and $l,l' \leq L = \mathcal{O}(S)$ are imposed.

We note that $p^2\,dpd\mathbf{u} = d^3p$. Our algorithm for the inverse Fourier transform is as follows:
a) We compute first

$$(g_1)^s_{l',m';n}(p,\mathbf{u}) = \sum_{l=|s|}^{L}\left[\sum_{m=-l}^{l}\hat{f}^s_{l',m';l,m}(p)\,\overline{h^s_{lm}(\mathbf{u})}\right]h^s_{l'n}(\mathbf{u}) \quad (11.8)$$

for fixed values of l', m', n, s. The summation can be performed in the following way. We perform first the summation in square brackets. Using the expression for basis functions (10.30) this summation can be written as

$$\exp(-is\Phi)\,(g_{11})^s_{l',m';l}(p,\Theta,\Phi)\ =$$

$$\exp(-is\Phi)\sum_{m=-l}^{l}\hat{f}^s_{l',m';l,m}(p)\,Q^l_{s,m}(\cos\Theta)\,\exp(-im\Phi).$$

Replacing the summation limits $|m| \le l$ by $|m| \le L = \mathcal{O}(S)$ (and assuming that the corresponding elements $\hat{f}^s_{l',m';l,m}(p)$ are zero for $|m| > l$ for given l) we can compute this sum using the one-dimensional FFT for fixed values of other indices. This can be computed in $\mathcal{O}(S^4\,N_p\,N_\Theta\,S\log S) = \mathcal{O}(N_R^{5/3}\,N_r^{2/3}\,\log N_R)$. We note that the term $\exp(-is\Phi)$ is canceled with the corresponding term from $h^s_{l'n}(\mathbf{u})$ in (11.8).

Then, we compute the summation

$$(g_{12})^s_{l',m'}(p,\Theta,\Phi)\ =\ \sum_{l=|s|}^{L}(g_{11})^s_{l',m';l}(p,\Theta,\Phi)$$

which can be performed in $\mathcal{O}(N_p\,N_\Theta\,N_\Phi\,S^4) = \mathcal{O}(N_r\,N_R^{4/3})$ computations. The product

$$(g_1)^s_{l',m';n}(p,\mathbf{u})\ =\ (g_{12})^s_{l',m'}(p,\Theta,\Phi)\,Q^{l'}_{s,n}(\cos\Theta)\,\exp(in\Phi)\qquad(11.9)$$

can be computed in the same amount of computations.

Thus, the sum in (11.8) can be performed in $\mathcal{O}(N^{7/6}\log N)$ computations. Then, we interpolate from a spherical coordinate grid to a rectangular grid. This requires $\mathcal{O}(N_r N_R^{4/3}\epsilon(N_r)) = \mathcal{O}(N^{7/6}\,\epsilon(N^{\frac{1}{2}}))$ operations.

Application of the Driscoll-Healy fast polynomial transform technique gives an additional savings in computation of sum (11.8). We define formally the $\hat{f}^s_{l',m';l,m}(p)$ matrix elements to be zero for $|m|,|s| > l$ and extend the limits of summation in (11.8) from $l = 0$ and from $m = -L$ to $m = L$. Then the summation with respect to l can be performed first as

$$(g_{11})^s_{l',m';m}(p,\Theta)\ =\ \sum_{l=0}^{L}\hat{f}^s_{l',m';l,m}(p)\,Q^l_{s,m}(\cos\Theta)$$

using the fast polynomial technique in $\mathcal{O}(S^4 N_p\,S(\log S)^2) = \mathcal{O}(N_R^{4/3}\,N_r^{2/3}\,(\log N_R)^2)$ computations. Then, the summation

$$(g_{12})^s_{l',m'}(p,\Theta,\Phi)\ =\ \sum_{m=-L}^{L}(g_{11})^s_{l',m';m}(p,\Theta)\,\exp(-im\Phi)$$

can be performed using the one-dimensional FFT in

$$\mathcal{O}(S^3\,N_p\,N_\Theta\,S\log S) = \mathcal{O}(N_R^{4/3}\,N_r^{2/3}\,\log N_R).$$

b) Next we compute integrals of the form

$$(g_2)^s_{l',m';n}(\mathbf{r})\ =\ \int_{\mathbb{R}^3}(g_1)^s_{l',m';n}(\mathbf{p})\,\exp(-i\mathbf{p}\cdot\mathbf{r})\,d^3p$$

using 3D FFTs. This requires $\mathcal{O}(S^4\,N_r\,\log(N_r)) = \mathcal{O}(N^{7/6}\log N)$ operations to compute.

c) The function $f(\mathbf{r}, R)$ can be recovered by the summation

$$f(\mathbf{r}, R) = \sum_{s=-S}^{S} \left[\sum_{l'=|s|}^{L} \sum_{m'=-l'}^{l'} \sum_{n=-l'}^{l'} U_{nm'}^{l'}(R) \, (g_2)_{l',m',n}^{s}(\mathbf{r}) \right].$$

The expression in square brackets is a set of inverse Fourier transforms on $SO(3)$ for fixed values of $\mathbf{r}$ and s. It takes $\mathcal{O}(N_r \, S \, N_R \, (\log N_R)^2) = \mathcal{O}(N^{7/6} \, (\log N)^2)$ operations to compute.

Thus, the total computational cost for the inverse Fourier transform is $\mathcal{O}(N^{7/6} \, ((\log N)^2 + N \, \epsilon(N^{\frac{1}{2}}))$.

Convolution of Functions.   The convolution integral

$$(f_1 * f_2)(g) = \int_{SE(3)} f_1(h) \, f_2(h^{-1} \circ g) \, d\mu(h) \tag{11.10}$$

can be written as a matrix product in Fourier space

$$(\mathcal{F}(f_1 * f_2))_{l',m';l,m}^{s}(p) = \sum_{j=|s|}^{\infty} \sum_{k=-j}^{j} (\hat{f}_2)_{l',m';j,k}^{s}(p) \, (\hat{f}_1)_{j,k;l,m}^{s}(p).$$

When $f_1(g)$ and $f_2(g)$ are band-limited in the sense defined earlier, the matrix product can be computed directly in $\mathcal{O}(N_p \, S^7) = \mathcal{O}(N_r^{1/3} \, N_R^{7/3}) = \mathcal{O}(N^{4/3})$ operations, which can be the largest time consuming computation. We note that a fast matrix multiplication algorithm can be applied for $n \times n = 2^m \times 2^m$ matrices, which is of the order of $n^{\log_2 7}$ instead of n^3 [28, 29, 31, 35]. Using this algorithm the matrix product can be computed in $\mathcal{O}(N^{(\log_2 7+1)/3})$ computations. Since fast matrix multiplications is an active research field in its own right, we characterize the order of computations for the convolution product as $\mathcal{O}(N^{(\gamma+1)/3})$, where $2 \leq \gamma \leq 3$ indicates the cost of matrix multiplication.

Therefore, the total order of computations of convolution is, at most, $\mathcal{O}(N^{(\gamma+1)/3}) + \mathcal{O}(N^{7/6} \, (\log N)^2) + \mathcal{O}(N^{7/6} \, \epsilon(N^{\frac{1}{2}}))$. Thus, when $\epsilon(N_r) \leq \mathcal{O}((\log N_r)^2)$, the algorithm described above provides very considerable savings compared to the direct integration in (11.10), which is of the order of $\mathcal{O}(N_r^2 \, N_R^2) = \mathcal{O}(N^2)$.

11.3 Harmonic Analysis on the Semi-Discrete Motion Group (SDMG) of the Plane

The subgroup of $SE(2)$ where $\theta = 2\pi i/N_R$ for $i = 0, ..., N_R - 1$ is called the *semi-discrete motion group* (or SDMG) of the plane [14]. While $SE(n)$ and crystallographic space groups have been studied extensively in the literature, the semi-discrete motion groups have received very little attention.

11.3.1 Irreducible Unitary Representations of the SDMG of the Plane

We define in this subsection matrix elements of the irreducible unitary representations (IURs) of the semi-discrete motion group $G = \mathbb{R}^2 \rtimes C_{N_R}$, where C_{N_R} is the N_R-element finite subgroup of $SO(2)$ (i.e. the group of rotational symmetries of a regular planar N_R-gon), and the notation $\rtimes$ means semi-direct product (which in the present context means nothing more than that group elements are expressed as homogeneous transforms).

The IURs $U(\mathbf{a}, A)$ of $SE(2)$ act on functions $f(\mathbf{u}) \in \mathcal{L}^2(S)$ (S is a unit circle, $\mathbf{u} = (\cos\Theta, \sin\Theta)^T$ is a vector to a point on the unit circle) with the inner product

$$(\varphi_1 \cdot \varphi_2) = \frac{1}{2\pi} \int_{\mathbb{S}^1} \overline{\varphi_1(\Theta)}\, \varphi_2(\Theta)\, d\Theta$$

where Θ is an angle on the unit circle. These operators are defined by the expression

$$(U_p(\mathbf{a}, A; p)\,\varphi)(\mathbf{u}) = e^{-ip\mathbf{u}\cdot\mathbf{a}}\,\varphi(A^{-1}\mathbf{u}) = e^{-i\mathbf{p}\cdot\mathbf{a}}\,\varphi(A^{-1}\cdot(\mathbf{p}/p)), \tag{11.11}$$

where $A \in SO(2)$, $p \in \mathbb{R}_{\geq 0}$, $\mathbf{p} = p\mathbf{u}$ is the vector to arbitrary points in the dual (frequency) space of $\mathbb{R}^2$ (p is its magnitude and $\mathbf{u}$ is its direction).

We choose a pulse orthonormal basis $\varphi_{N_R,n}(\mathbf{u})$ on $\mathbb{S}^1$, i. e. we subdivide the circle into identical segments F_n and choose the f-functions to satisfy the orthonormality relations

$$\frac{1}{2\pi} \int_{\mathbb{S}^1} \varphi_{N_R,n}(\mathbf{u})\, \varphi_{N_R,m}(\mathbf{u})\, d\Theta = \delta_{nm}.$$

We can choose the orthonormal functions as

$$\varphi_{N_R,n}(\mathbf{u}) = \begin{cases} (N_R)^{1/2} & \text{if } \mathbf{u} \in F_n \\ 0 & \text{otherwise} \end{cases}.$$

$n = 0, ..., N_R - 1$ enumerates different segments. We denote these pulse functions as δ-like functions $\varphi_{N_R,n}(\mathbf{u}) = (1/N_R)^{1/2}\delta_{N_R}(\mathbf{u}, \mathbf{u_n})$, where $\mathbf{u_n}$ is the vector to the center of the F_n segment.

The matrix elements in this basis are

$$U_{mn}(A, \mathbf{r}; p) = \frac{1}{2\pi} \int_S \varphi_{N_R,m}(\mathbf{u})\, e^{-ip\,\mathbf{u}\cdot\mathbf{r}}\, \varphi_{N_R,n}(A^{-1}\mathbf{u})\, d\Theta. \tag{11.12}$$

Using the "delta-function" notation, this can be written as

$$U_{mn}(A, \mathbf{r}; p) = \frac{1}{2\pi N_R} \int_S \delta_{N_R}(\mathbf{u}, \mathbf{u_m})\, e^{-ip\,\mathbf{u}\cdot\mathbf{r}}\, \delta_{N_R}(A^{-1}\mathbf{u}, \mathbf{u_n})\, d\Theta.$$

This integral can be approximated as

$$U_{mn}(A, \mathbf{r}; p) \approx 1/N_R\, e^{-ip\,\mathbf{u_m}\cdot\mathbf{r}}\, \delta_{N_R}(A^{-1}\mathbf{u_m}, \mathbf{u_n}). \tag{11.13}$$

We approximate delta-functions as

$$1/N_R\,\delta_{N_R}(A^{-1}\mathbf{u_m}, \mathbf{u_n}) = \delta_{A^{-1}\mathbf{u_m}, \mathbf{u_n}} = \begin{cases} 1 \text{ if } A^{-1}\mathbf{u_m} = \mathbf{u_n} \\ 0 \text{ otherwise} \end{cases}$$

which means that we restrict rotations to the rotations A_j from the finite subgroup C_{N_R} of $SO(2)$, and $A_j^{-1}\mathbf{u_m} = \mathbf{u_{m-j}} = \mathbf{u_n}$.

Thus, the matrix elements of the irreducible unitary representations of the semi-discrete motion subgroup $G = \mathbb{R}^2 \rtimes C_{N_R}$ are given as

$$U_{mn}(A_j, \mathbf{r}; p) = e^{-ip\,\mathbf{u_m}\cdot\mathbf{r}}\, \delta_{A_j^{-1}\mathbf{u_m}, \mathbf{u_n}} \tag{11.14}$$

where $\delta_{A_j^{-1}\mathbf{u_m}, \mathbf{u_n}} = \delta_{m-j,n}$ in this case.

11.3.2 Fourier Transforms for the SDMG of the Plane

Though motivated as an approximation, the matrix elements (11.14) are exact expressions for the matrix elements of the unitary representations of the semi-discrete motion group. However, this set of matrix elements (11.14) is incomplete. This means that the direct and inverse Fourier transforms, defined using these matrix elements, would reproduce the original function with $\mathcal{O}(1/N_R)$ error, i.e.,

$$\mathcal{F}^{-1}(\mathcal{F}(f(A_i, \mathbf{r})) = f(A_i, \mathbf{r})\left(1 + \mathcal{O}\left(\frac{1}{N_R}\right)\right).$$

The reason for this is that summing through all possible segments cannot replace integration over all possible angles on the circle. It is also clear that the additional continuous parameter which enumerates possible angles inside each segment on the circle must give the complete set of the matrix elements.

Thus, the matrix elements are modified as

$$U_{mn}(A_j, \mathbf{r}; p, \Phi) = e^{-ip\,\mathbf{u}_m^{\Phi}\cdot\mathbf{r}}\,\delta_{A_j^{-1}\mathbf{u}_m, \mathbf{u}_n} \tag{11.15}$$

where $\mathbf{u}_k^{\Phi}$ denotes the vector to the angle Θ on the unit circle on the interval

$$F_k = [2\pi\,k/N_R\,,\,2\pi\,(k+1)/N_R],$$

where $k = 0, ..., N_R - 1$ (Φ measures the angle on this segment).

The completeness relation

$$\sum_{m=0}^{N_R-1}\sum_{n=0}^{N_R-1}\int_0^{\infty}\int_0^{2\pi/N_R}\overline{U_{mn}(A_i, \mathbf{r}_1; p, \Phi)}\,U_{mn}(A_j, \mathbf{r}_2; p, \Phi)\,p\,dp\,d\Phi =$$

$$(2\pi)^2\,\delta^2(\mathbf{r}_1 - \mathbf{r}_2)\,\delta_{A_i, A_j} \tag{11.16}$$

is exact, because the integration is now over the whole space of $\mathbf{p}$ values. This can be proven using the integral representation of the δ-function

$$\frac{1}{(2\pi)^2}\int_{\mathbb{R}^2}e^{-i\mathbf{p}\cdot\mathbf{r}}\,d^2p = \delta^2(\mathbf{r}).$$

The orthogonality relation is written as

$$\sum_{i=0}^{N_R-1}\int_{\mathbb{R}^2}\overline{U_{mn}(A_i, \mathbf{r}; p, \Phi)}\,U_{m'n'}(A_i, \mathbf{r}; p', \Phi')\,d^2r = \tag{11.17}$$

$$(2\pi)^2\,\frac{\delta(p - p')}{p}\,\delta_{m,m'}\,\delta_{n,n'}\,\delta(\Phi - \Phi').$$

The direct Fourier transform is defined as

$$\hat{f}_{mn}(p, \Phi) \doteq \sum_{i=0}^{N_R-1}\int_{\mathbb{R}^2}f(A_i, \mathbf{r})\,U_{mn}^{-1}(A_i, \mathbf{r}; p, \Phi)\,d^2r. \tag{11.18}$$

The vector $\mathbf{u}_m^{\Phi}$, which is inside the segment F_m can be found by the rotation A_m (which transforms F_0 to F_m) from $\mathbf{u}_0^{\Phi}$, $\mathbf{u}_m^{\Phi} = A_m\,\mathbf{u}_0^{\Phi}$. The parameter Φ denotes the position inside the segment F_0.

The inverse Fourier transform is

$$\mathcal{F}^{-1}(\hat{f}) = \frac{1}{4\pi^2} \sum_{m=0}^{N_R-1} \sum_{n=0}^{N_R-1} \int_0^\infty \int_0^{2\pi/N_R} \hat{f}_{mn}(p, \Phi) \, U_{nm}(A_i, \mathbf{r}; p, \Phi) \, p \, dp \, d\Phi. \quad (11.19)$$

We note that this result is in agreement with [14].

We define a convolution on the semi-discrete motion group as

$$F(\mathbf{r}, A_j) = \frac{1}{2\pi N_R} \sum_{i=0}^{N_R-1} \int_{\mathbb{R}^2} f_1(\mathbf{a}, A_i) \, f_2(A_i^{-1}(\mathbf{r} - \mathbf{a}), A_{j-i}) \, d^2a. \quad (11.20)$$

(the normalization factor corresponds in the limit $N_R \to \infty$ to the normalization of the convolution on the continuous (Euclidean) motion group in [19]).

The Fourier transform of the convolution of functions on the semi-discrete motion group is just a product of the Fourier matrices with the corresponding normalization

$$\hat{F}_{mn}^\Phi(p) = \frac{A}{2\pi N_R} \sum_{k=0}^{N_R-1} (\hat{f}_2)_{mk}^\Phi(p) \, (\hat{f}_1)_{kn}^\Phi(p) \quad (11.21)$$

where A is the area of a compact region of $\mathbb{R}^2$ containing $\mathbf{r}$ where the FFT is computed. The functions must have support inside the area A, and are considered periodic outside of this region. The area factor arises because the discrete Fourier transform can be computed as an approximation of the continuous case using

$$r_1 \to \frac{L}{N_r} i; \quad p_1 \to \frac{2\pi}{L} i$$

(and analogously for the 2nd component). Here L is the length of the compact region in the x (y) direction. While the factor L is cancelled out of equations if we apply direct and inverse discrete Fourier transform to the function, it appears in the convolution defined in (11.20).

11.3.3 Efficiency of Computation of $SE(2)$ Convolution Integrals Using the SDMG Fourier Transform

In this section we show that using the Fourier transform on the semi-discrete motion group is a fast method to compute convolution integrals on the discrete motion group (we assume that the finite rotation group has N_R elements). Convolution on the semi-discrete motion group is an approximation to $SE(2)$ convolution. Particularly, we show that the convolution of a function $f(\mathbf{r}, A_i)$ sampled at $N = N_R \cdot N_r$ points (N_r is the number of samples in an $\mathbb{R}^2$ region) can be performed in $\mathcal{O}(N \log N_r) + \mathcal{O}(N N_R)$ operations instead of $\mathcal{O}(N_g^2)$ computations in the direct coordinate space integration using the "plain" integration in (11.20). The structure of matrix elements of (11.14) allows us to apply Fast Fourier Methods and reduce the amount of computations (without the application of FFT the amount of computations using the Fourier transform method is $\mathcal{O}(N^2/N_R)$).

First, we estimate the amount of computations to perform the direct and inverse Fourier transforms of $f(g)$. We assume that we restrict p values to a finite interval and sample it at N_p points, and sample the Φ values at N_Φ points. We also assume that the total number of harmonics $N_p N_\Phi N_R^2 = N = N_R N_r$.

Let us consider the direct Fourier transform (11.18). Each term i (for fixed A_i) gives one nonzero term in each row and column of the Fourier matrix $\hat{f}_{mn}^{\Phi}(p)$ (because only one element in each row and column of $U_{mn}^{-1}(g;p,\Phi)$ is nonzero). For each fixed i we can compute the usual FFT of $f(\mathbf{r};A_i)$, which can be computed in $\mathcal{O}(N_r\log(N_r))$ operations. The Fourier transform elements found by FFT are computed on a square grid of $\mathbf{p}$ values. We can, however, interpolate the Fourier elements computed on the grid to the Fourier elements computed at polar coordinates. The radial part p is determined by the length of $\mathbf{p}$, the angular part determines the indices m,Φ (other index n is determined uniquely for given A_i). We linearly interpolate values on a square grid to values on a polar grid. Such an interpolation can be performed in $\mathcal{O}(N_r)$ computations. Each term i in (11.18) can be computed in $\mathcal{O}(N_r\log(N_r))$ computations. The whole Fourier matrix can be calculated in $\mathcal{O}(N_R N_r\log(N_r))$ computations.

Again, one element from each row and column is used in computation of the inverse Fourier transform for each rotation element A_i. After inverse interpolation to Cartesian coordinates (which can be done in $\mathcal{O}(N_r)$ computations), the inverse Fourier integration can be performed in $\mathcal{O}(N_r\log(N_r))$ for each of the N_R nonzero matrix elements of U using the FFT. Thus, in $\mathcal{O}(N_R N_r\log(N_r))$ computations we reproduce the function for all A_i.

The matrix product of $\hat{f}_{mn}^{\Phi}(p)$ can be computed directly in $\mathcal{O}(N_R^3)$ computations, for each value of p and parameters Φ.[3] This means that the convolution (which is a matrix product of Fourier matrices) can be performed in $\mathcal{O}(N_R^3 N_p N_\Phi) = \mathcal{O}(N_R N)$ computations.

Therefore, the convolution of functions on the semi-discrete motion group can be performed in $\mathcal{O}(N\log N_r) + \mathcal{O}(N N_R)$ using Fourier methods on the semi-discrete motion group and the usual FFT, without assuming any special matrix multiplication technique.

It can be shown that without the application of the FFT the convolution can be performed using the Fourier transform in $\mathcal{O}(N^2/N_R)$, which is still faster than evaluating the direct integration.

11.4 Fourier Transform for the 3D SDMGs

In this section we develop fast approximate algorithms for computing the Fourier transform of functions on the discrete motion groups, G_{N_R}, which are the semi-direct product of the continuous translation group $\mathbf{T} = (\mathbb{R}^3, +)$ and a finite subgroup $I_{N_R} \subset SO(3)$ with N_R elements. I_{N_R} can be the icosahedral, the cubo-octahedral, or tetrahedral rotational symmetry groups. Our formulation is completely general, though in discussions of numerical implementations we focus on the icosahedral semi-discrete motion group, where the number of elements is $N_R = 60$. In Subsection 11.4.1 the mathematical formulation is presented and in Subsection 11.4.2 the computational complexity of implementing Fourier transforms and convolution of functions on G_{N_R} is discussed.

11.4.1 Mathematical Formulation

Instead of using spherical harmonics (10.30) as basis functions as was done when calculating matrix elements of the IURS of the continuous motion group, we now choose

---

[3]Estimates as fast as $\mathcal{O}(N_R^{2.38})$ have been reported (see [9] and references therein) which, depending on how they are implemented, have the potential to increase speed further.

pulse functions $\varphi_{N_R,n}(\mathbf{u})$ on $\mathbb{S}^2$, i.e., we subdivide the sphere into spherical regions and choose the φ-functions to satisfy the orthonormality relations

$$\int_{\mathbb{S}^2} \varphi_{N_R,n}(\mathbf{u})\,\varphi_{N_R,m}(\mathbf{u})\,d\mathbf{u} = \delta_{nm}$$

where $d\mathbf{u} = \sin\Theta\,d\Theta\,d\Phi$, and (Θ,Φ) are spherical coordinates.[4] For example, in the case of I_{60} we can subdivide the sphere into 20 equilateral triangles or 12 regular pentagons. These figures can be used as the support for pulse functions, but as we will see shortly, it is convenient to subdivide these regular figures so that 60 congruent (but irregular) regions, F_n, result. We then choose the orthonormal functions as

$$\varphi_{N_R,n}(\mathbf{u}) = \begin{cases} (\frac{N_R}{4\pi})^{1/2} & \text{if } \mathbf{u} \in F_n \\ 0 & \text{otherwise} \end{cases}.$$

Here $n = 1, ..., N_R$ enumerates the different congruent polygonal regions on the sphere. We denote these δ-like functions as $\varphi_{N_R,n}(\mathbf{u}) = (4\pi/N_R)^{1/2}\delta_{N_R}(\mathbf{u},\mathbf{u_n})$, where $\mathbf{u_n}$ is a vector from the center of sphere to a point in F_n.

The matrix elements can be found in this basis as

$$U_{mn}^s(A,\mathbf{r};p) = \int_{\mathbb{S}^2} \varphi_{N_R,m}(\mathbf{u})\,e^{-i\,p\,\mathbf{u}\cdot\mathbf{r}}\,\Delta_s(R_{\mathbf{u}}^{-1}\,A\,R_{A^{-1}\mathbf{u}})\,\varphi_{N_R,n}(A^{-1}\mathbf{u})\,d\mathbf{u}. \quad (11.22)$$

Using the "delta"-function notation, this integral can be written as

$$U_{mn}^s(A,\mathbf{r};p) = \frac{4\pi}{N_R}\int_{\mathbb{S}^2} \delta_{N_R}(\mathbf{u},\mathbf{u_m})\,e^{-i\,p\,\mathbf{u}\cdot\mathbf{r}}\,\Delta_s(R_{\mathbf{u}}^{-1}\,A\,R_{A^{-1}\mathbf{u}})\,\delta_{N_R}(A^{-1}\mathbf{u},\mathbf{u_n})\,d\mathbf{u}.$$

This integral can be approximated as

$$U_{mn}^s(A,\mathbf{r};p) \approx 4\pi/N_R\,e^{-i\,p\,\mathbf{u_m}\cdot\mathbf{r}}\,\Delta_s(R_{\mathbf{u_m}}^{-1}\,A\,R_{A^{-1}\mathbf{u_m}})\,\delta_{N_R}(A^{-1}\mathbf{u_m},\mathbf{u_n}). \quad (11.23)$$

We again approximate "delta"-functions as

$$4\pi/N_R\,\delta_{N_R}(A^{-1}\mathbf{u_m},\mathbf{u_n}) = \delta_{A^{-1}\mathbf{u_m},\mathbf{u_n}} = \begin{cases} 1 & \text{if } A^{-1}\mathbf{u_m} = \mathbf{u_n} \\ 0 & \text{otherwise} \end{cases}$$

which means that we discretize the rotation group, i.e. we restrict rotations to the rotations A_j from the finite subgroup I_{N_R} of $SO(3)$, and $A_j^{-1}\mathbf{u_m} = \mathbf{u_n}$.

Thus, matrix elements of irreducible unitary representations of the semi-discrete motion subgroup G_{N_R} are given as

$$U_{mn}^s(A_j,\mathbf{r};p) = e^{-i\,p\,\mathbf{u_m}\cdot\mathbf{r}}\,\Delta_s(R_{\mathbf{u_m}}^{-1}\,A_j\,R_{\mathbf{u_n}})\,\delta_{A_j^{-1}\mathbf{u_m},\mathbf{u_n}}. \quad (11.24)$$

Here s enumerates the representations of the little group, which is a finite subgroup $C_n \subset SO(2)$ in this case, and $s = 0, 1, ..., n-1$ for C_n. Although the expression (11.24) has been derived as an approximation of the continuous expression (11.22), the matrix elements (11.24) are exact expressions for the matrix elements of the irreducible unitary representations of G_{N_R}, i.e. the relation $U(g_1,p)\cdot U(g_2,p) = U(g_1 \circ g_2,p)$ holds.

We note $R_{\mathbf{u_m}}\mathbf{u_0} = A_m\mathbf{u_0} = \mathbf{u_m}$, where $\mathbf{u_0}$ is the center of an arbitrary chosen regular spherical polygon F_0.

---

[4]The set of functions $\{\varphi_{N_R,n}(\mathbf{u})\}$ is not complete in $\mathcal{L}^2(\mathbb{S}^2)$ and therefore is not a basis in which to expand matrix elements of the IURs, but we take this into account shortly.

However, due to the incompleteness of the set of functions $\{\varphi_{N_R,n}\}$, the Fourier transform matrix elements defined by (11.24) form an incomplete set of matrix elements. A complete set of elements are defined in the following way.

Let us allow the vector $\mathbf{u}_m^w$ to point to an arbitrary position, w, inside the spherical shape which forms the support for the pulse basis function $\varphi_{N_R,m}$.

It is then possible to verify that the following matrix elements form a complete set of matrix elements for G_{N_R}:

$$U_{mn}^s(A_j, \mathbf{r}; p, w) = e^{-i p\, \mathbf{u}_m^w \cdot \mathbf{r}}\, \Delta_s(R_{\mathbf{u}_m}^{-1}\, A_j\, R_{\mathbf{u}_n})\, \delta_{A_j^{-1}\mathbf{u}_m,\mathbf{u}_n}. \tag{11.25}$$

In particular, it is easy to check that $U(g_1; p, w) \cdot U(g_2; p, w) = U(g_1 \circ g_2; p, w)$.

As an example we consider the case when I_{N_R} is the Icosahedral subgroup of the $SO(3)$, which has $N_R = 60$ elements. This is the largest finite subgroup of $SO(3)$ [15]. If we subdivide the sphere into 20 equilateral spherical triangles (see Figure 11.2), this group has 6 axes of rotation of order 5 (i.e rotation $2\pi/5 \cdot n$, $n = 0, 1, 2, 3, 4$ around each axis) located at the triangle corners, 10 axes of order 3 located at the triangle centers and 15 axes of order 2 located in the middle point of each triangle side.

Different representations of this semi-discrete motion group can be classified according to different choices of little group C_n. This corresponds to choosing orthogonal pulse functions with differently shaped support. To illustrate the possible cases consider the tessellations of the sphere in Figure 11.2. In order to have a complete set of matrix elements for non-trivial little group, C_n, we have to consider all possible $s = 0, ..., n - 1$. The representations for different s can be viewed as blocks in a 60×60 block-diagonal (and hence reducible) representation matrix.

The possible choices for little group corresponding to the shapes illustrated in Figure 11.2 are:

1) Little group C_5: For 12 regular spherical pentagons, such as ECGJH, chosen as the support for spherical pulse functions we have 12×12 representations. The little group of $\mathbf{u}_0$ is C_5, thus $s = 0, 1, 2, 3, 4$. These representation matrices, each corresponding to an element of the little group enumerated by a value of s, can be viewed as blocks in 60×60 representation matrices of G.

2) Little group C_3: Twenty equilateral spherical triangles are chosen as pulse functions on $\mathbb{S}^2$, such as triangle ABO. We have 20 vectors $\mathbf{u}_m$ pointing to the centers of triangles, i. e. the representations and Fourier matrices consist of 20×20 nonzero blocks (for each fixed p and s). The little group of the arbitrary chosen vector $\mathbf{u}_0$ is C_3, thus we have three inequivalent representation of the little group for each value of p enumerated by $s = 0, 1, 2$. These representation can be combined as blocks to form 60×60 representation matrices of G.

3) Little group C_2: Thirty spherical parallelograms (such as OEAC) are chosen as the support of pulse functions and we have 30×30 representations matrices, the little group is C_2, and $s = 0, 1$.

4) Trivial little group: The 4-sided figure ODCF of Figure 11.2 can be chosen as the support for pulse functions. We have 60 such figures, the little group is trivial in this case since they possess no rotational symmetry. The representation matrices are then 60×60. We note that some other divisions of the sphere into 60 equal spherical figures

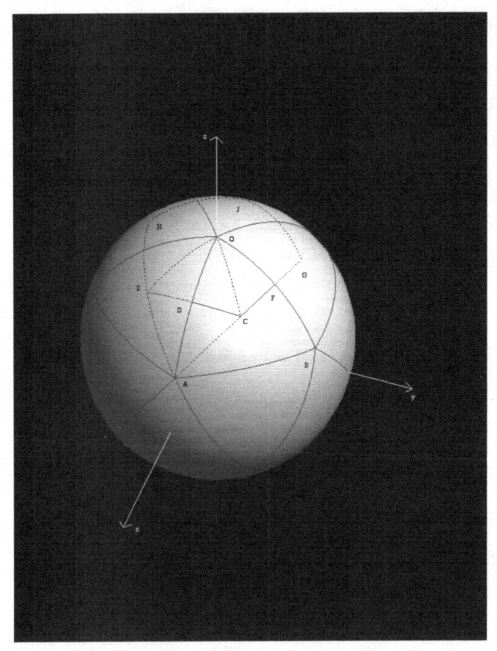

Fig. 11.2. Illustration of the Support of a Basis Function for the Semi-Discrete Motion Group. (Reprinted from Kyatkin, A.B., Chirikjian, G.S., "Algorithms for Fast Convolutions on Motion Groups," *Applied and Computational Harmonic Analysis*, 9: 220–241, 2000, with permission from Elsevier)

lead to equivalent representations (for example, the choice of the triangles ACO or ECO). These representations are irreducible.

We note that each row and column of the 60×60 representation matrices corresponding to the trivial little group contain only one non-zero element. The matrix elements of the representations are written in this case as

$$U_{mn}(A_j, \mathbf{r}; p, w) = e^{-i p \, \mathbf{u}_m^w \cdot \mathbf{r}} \delta_{A_j^{-1} \mathbf{u}_m, \mathbf{u}_n} \,. \tag{11.26}$$

Henceforth we restrict the discussion to this case.

The direct Fourier transform is defined as

$$\hat{f}_{mn}(p, w) \doteq \sum_{i=0}^{N_R-1} \int_{\mathbb{R}^3} f(A_i, \mathbf{r}) \, U_{mn}^{-1}(A_i, \mathbf{r}; p, w) \, d^3 r \tag{11.27}$$

where it depends now on w. The vector $\mathbf{u}_m^w$, which is inside the figure F_m can be found by the rotation A_m (which transforms F_0 to F_m) from $\mathbf{u}_0^w$, as $\mathbf{u}_m^w = A_m \, \mathbf{u}_0^w$. The parameter w, thus denotes the position inside the figure F_0.

The inverse Fourier transform is

$$\mathcal{F}^{-1}(\hat{f}) = \frac{1}{8\pi^3} \sum_{m=0}^{N_R-1} \sum_{n=0}^{N_R-1} \int_0^\infty \int_{F_0} \hat{f}_{mn}(p, w) \, U_{nm}(A_i, \mathbf{r}; p, w) \, p^2 \, dp \, d^2 w \tag{11.28}$$

(integration with respect to w is over the area of the basis figure F_0). The vector $\mathbf{u}_m^w$ is found by rotation from $\mathbf{u}_0^w$.

We choose the z-axis passing through the vertex O. The $\Phi = 0$ circular arc then contains side OD and the $\Phi = 2\pi/5$ arc contains side OF. With this choice, the border of the area F_0 for the 4-sided Figure ODCF of Figure 11.2 can be parameterized in terms of the spherical angles Θ, Φ as

$$\Theta(\Phi) = \arcsin \left[\frac{0.525732}{\sqrt{1 - 0.723605 \sin^2(\Phi')}} \right],$$

where

$$\Phi' = \Phi, \quad \text{if} \quad 0 \leq \Phi \leq \pi/5 \,,$$

and

$$\Phi' = 2\pi/5 - \Phi, \quad \text{for } \pi/5 \leq \Phi \leq 2\pi/5 \,.$$

This dependence is derived using the relationships between the angles of spherical triangles (e.g., the "sin"- and "cos"-theorems from spherical trigonometry).

Due to the completeness of matrix elements the application of direct and inverse Fourier transform reproduces function on the semi-discrete motion group G

$$\mathcal{F}^{-1}\mathcal{F}(f(g)) = f(g) \,.$$

The completeness of the set of IURs we have developed for the 3D semi-discrete motion group can be seen by first observing the integral representation for the δ-function in $\mathbb{R}^3$:

$$\int_{\mathbb{R}^3} e^{-i \mathbf{p} \cdot \mathbf{r}} \, d^3 p = (2\pi)^3 \delta(\mathbf{r}) \,.$$

Here integration is through the Fourier space, which is parameterized by $\mathbf{p}$.

The completeness relation

$$\sum_{m=0}^{N_R-1} \sum_{n=0}^{N_R-1} \int_0^\infty \int_{F_0} \overline{U_{mn}(A_i, \mathbf{r}_1; p, w)} \, U_{mn}(A_j, \mathbf{r}_2; p, w) \, p^2 \, dp \, d^2w =$$

$$(2\pi)^3 \, \delta^3(\mathbf{r}_1 - \mathbf{r}_2) \, \delta_{A_i, A_j} . \tag{11.29}$$

$(d^2w = \sin\Theta d\Theta \, d\Phi)$ then follows, because the integration is over the whole space of $\mathbf{p} = p\mathbf{u}$ values. Repeated integration along the boundary of the basis figures F_i gives zero contribution, because the functions are nonsingular and the integration measure along the boundaries is zero.

The orthogonality relation is written as

$$\sum_{i=0}^{N_R-1} \int_{\mathbb{R}^3} \overline{U_{mn}(A_i, \mathbf{r}; p, w)} \, U_{m'n'}(A_i, \mathbf{r}; p', w') \, d^3r =$$

$$(2\pi)^3 \, \frac{\delta(p - p')}{p^2} \, \delta_{m,m'} \, \delta_{n,n'} \, \delta^2(w - w') . \tag{11.30}$$

Using the orthogonality relations we can check that the convolution properties and Plancherel (Parseval) identity are exact for square integrable functions on the discrete motion group.

11.4.2 Computational Complexity

We now analyze the computational complexity of an algorithm for the numerical implementation of convolution of functions on the semi-discrete motion group (with the icosahedral group as the rotation subgroup). In particular, we show that the convolution of functions $f_1(\mathbf{r}, A_i)$ and $f_2(\mathbf{r}, A_i)$ sampled at $N = N_R \cdot N_r$ points (N_r is the number of samples in a region in $\mathbb{R}^3$ and N_R is the order of a finite rotation subgroup, which is 60 for the icosahedral group) can be performed in $\mathcal{O}(N \, (\epsilon(N_r) + \log N_r)) + \mathcal{O}(N \, N_R^{\gamma-2})$ operations (where again $2 \le \gamma \le 3$ is the exponent for matrix multiplication) instead of the $\mathcal{O}(N^2)$ computations required for the direct computation of convolution by discretization of the convolution integral and evaluation at each discrete value of translation and rotation. The structure of matrix elements of (11.26) allows us to apply Fast Fourier methods and reduce the amount of computations. However, even without the application of the FFT, the amount of computations required to compute convolutions using the group-theoretical Fourier transform method is $\mathcal{O}(N^2/N_R))$, which is a savings over brute-force discretization of the convolution integral.

First, we estimate the amount of computations to perform the direct and inverse Fourier transforms of $f(g)$. We assume that we restrict p values to a finite interval and sample it at N_p points, and sample the w region at N_w points. We also assume that the total number of harmonics $N_p N_w N_R^2 = N = N_R N_r$.

Let us consider the direct Fourier transform (11.27). Each term i (for fixed A_i) gives one nonzero term in each row and column of the Fourier matrix $\hat{f}_{mn}^w(p)$ (because only one element in each row and column of $U_{mn}^{-1}(g; p, w)$ is nonzero). For each fixed i we can compute the FFT of $f(\mathbf{r}; A_i)$, which can be computed in $\mathcal{O}(N_r \log(N_r))$ operations. The Fourier transform elements found by FFT are computed on a square grid of $\mathbf{p}$ values. We, however, need to interpolate the Fourier elements computed on the grid to the Fourier elements computed in polar (spherical) coordinates. The radial part p is determined by the length of $\mathbf{p}$, the angular part is determined by the indices m and w (the other index n is determined uniquely for given A_i). If we interpolate the values from the square grid to the values on the polar (spherical) grid, $\mathcal{O}(N_r\epsilon(N_r)N_R)$ computations are required. Each term i in (11.27) can be computed in $\mathcal{O}(N_r \log(N_r))$ computations. The whole Fourier matrix is found in $\mathcal{O}(N_R N_r \log(N_r))$ computations.

Again, one element from each row and column is used in the computation of the inverse Fourier transform for each rotation element A_i. After inverse interpolation to Cartesian coordinates (which can be done in $\mathcal{O}(N_r\epsilon(N_r))$ computations), the inverse Fourier integration can be performed in $\mathcal{O}(N_r \log(N_r))$ computation using the FFT. Thus, in $\mathcal{O}(N_R N_r \log(N_r))$ computations we reproduce the function for each A_i.

The matrix product of $\hat{f}_{mn}(p, w)$ can be computed in $\mathcal{O}(N^\gamma)$ computations for each value of p and parameters w. This means that the convolution (which is a matrix product of Fourier matrices) can be performed in $\mathcal{O}(N_\gamma^3 N_p N_w) = \mathcal{O}(N_R^{\gamma-2}N)$ computations.

Therefore, the convolution of functions on the semi-discrete motion group can be performed in $\mathcal{O}(N \cdot (\log N_r + \epsilon(N_r))) + \mathcal{O}(N N_R^{\gamma-2})$ using the Fourier methods on the motion group and FFT. When spline interpolation is used, the term with $\epsilon(N_r)$ is of sub-leading order, whereas this term is of leading order when Fourier interpolation is used.

11.5 Alternative Methods for Computing $SE(D)$ Convolutions

In this section we examine alternatives to calculating convolutions on $SE(D)$ using Fourier transforms based on IURs of $SE(D)$. First we consider a Fourier transform based on reducible unitary representations. Then we consider the Fourier transform for the direct product $\mathbb{R}^D \times SO(D)$ as a tool to compute $SE(D)$ convolutions. Finally, we consider contraction of $SE(D)$ to $SO(D+1)$.

11.5.1 An Alternative Motion-Group Fourier Transform Based on Reducible Representations

Let $g = (\mathbf{a}, A) \in SE(3)$, and consider the following operator that acts on functions $\varphi \in \mathcal{L}^2(SO(3))$:

$$\mathcal{U}(g, \mathbf{p})\varphi(R) = e^{i((R\mathbf{p})\cdot\mathbf{a})}\varphi(A^T R). \tag{11.31}$$

These representations were studied in [16, 25]. We observe that

$$\begin{aligned}
\mathcal{U}(g_1, \mathbf{p})\left[\mathcal{U}(g_2, \mathbf{p})\varphi(R)\right] &= \mathcal{U}(g_1, \mathbf{p})\left[e^{i((R\mathbf{p})\cdot\mathbf{a}_2)}\varphi(A_2^T R)\right] \\
&= e^{i((R\mathbf{p})\cdot\mathbf{a}_1)} \cdot e^{i((A_1^T R\mathbf{p})\cdot\mathbf{a}_2)}\varphi(A_2^T A_1^T R) \\
&= e^{i((R\mathbf{p})\cdot(\mathbf{a}_1 + A_1\mathbf{a}_2))}\varphi((A_1 A_2)^T R) \\
&= \mathcal{U}(g_1 \circ g_2, \mathbf{p})\varphi(R).
\end{aligned}$$

In this context, the matrix elements corresponding to $\mathcal{U}(g, \mathbf{p})$ are calculated as

$$\mathcal{U}_{m,n;m',n'}^{l,l'} = \sqrt{d_l d_{l'}} \int_{SO(3)} \overline{U_{m,n}^l(R)} e^{i(R\mathbf{p})\cdot\mathbf{a}} U_{m',n'}^{l'}(A^T R)\, dR \tag{11.32}$$

where $U_{m,n}^l$ are $SO(3)$ matrix elements, $d_l = 2l+1$ and $\{\sqrt{d_l}\, U_{m,n}^l\}$ is an orthonormal basis for $\mathcal{L}^2(SO(3))$. Using the homomorphism property, we see that

$$U_{m',n'}^{l'}(A^T R) = \sum_{r=-l'}^{l'} U_{m',r}^{l'}(A^T) U_{r,n'}^{l'}(R)$$

and

$$U_{m',r}^{l'}(A^T) = \overline{U_{r,m'}^{l'}(A)}.$$

This allows us to write

$$\mathcal{U}_{m,n;m',n'}^{l,l'}(g,\mathbf{p}) = \sqrt{d_l d_{l'}} \sum_{r=-l'}^{l'} \overline{U_{r,m'}^{l'}(A)} \int_{SO(3)} e^{i(R\mathbf{p})\cdot \mathbf{a}} U_{r,n'}^{l'}(R)\overline{U_{m,n}^{l}(R)}\, dR. \quad (11.33)$$

This can be computed in closed form using the expansion of $e^{i(R\mathbf{p})}$ in spherical coordinates, but for computational purposes it is sufficient to keep it as is.

The matrix elements of a Fourier transform based on these matrix elements are computed as

$$\hat{f}_{m,n;m',n'}^{l,l'}(\mathbf{p}) = \int_{SE(3)} f(g)\mathcal{U}_{m,n;m',n'}^{l,l'}(g^{-1},\mathbf{p})\, dg. \quad (11.34)$$

The reconstruction formula is written as

$$f(g) = \int_{\mathbb{R}^3} \operatorname{tr}(\hat{f}(\mathbf{p})\mathcal{U}(g,\mathbf{p}))\, d\mathbf{p} \quad (11.35)$$

$$= \sum_{l=0}^{\infty} \sum_{l'=0}^{\infty} \sum_{m=-l}^{l} \sum_{n=-l}^{l} \sum_{m'=-l'}^{l'} \sum_{n'=-l'}^{l'} \int_{\mathbb{R}^3} \hat{f}_{m,n;m',n'}^{l,l'}(\mathbf{p})\mathcal{U}_{m',n';m,n}^{l,l'}(g,\mathbf{p})\, d\mathbf{p}.$$

The associated completeness relation follows from those for $SO(3)$ and $\mathbb{R}^3$ as

$$\int_{SE(3)} \mathcal{U}_{m,n;m',n'}^{l,l'}(g,\mathbf{p})\overline{\mathcal{U}_{r,s;r',s'}^{k,k'}(g,\mathbf{p}')}\, dg = (2\pi)^3 \delta_{kl}\delta_{k'l'}\delta_{mr}\delta_{ns}\delta_{m'r'}\delta_{n's'}\delta(\mathbf{p}-\mathbf{p}').$$

Instead of a six-index Fourier transform we can define

$$\hat{f}_{ij} \doteq \hat{f}_{m,n;m',n'}^{l,l'}$$

where

$$i = \sum_{k=0}^{l}(2k-1)^2 + (2l+1)(l+n) + l + n$$

(and likewise for j).

The multiplication of Fourier transform matrices is usually a computational bottleneck when using the group Fourier transform to calculate convolutions. In the present case, if $N = N_r \cdot N_R$ sample points in $SE(3)$ are taken, and $l, l' \leq \mathcal{O}(N_R^{1/3})$, the two-index Fourier matrices will be $\mathcal{O}(N_R) \times \mathcal{O}(N_R)$. Mltiplication of these matrices will require $\mathcal{O}(N_R^\gamma)$ for each value of $\mathbf{p}$. In other words, $\mathcal{O}(N_R^\gamma \cdot N_r) = \mathcal{O}(N^{(1+\gamma)/2})$ computations are required. If matrices are multiplied with complexity $\gamma = 3$, then it is clear that this method will require the same order of computation as brute-force numerical integration of the convolution integral. Hence, we conclude that this method is not computationally advantageous.

11.6 Computing $SE(D)$ Convolutions Using the Fourier Transform for $\mathbb{R}^D \times SO(D)$

Given the pairs $g = (\mathbf{a}, A)$ and $h = (\mathbf{r}, R)$ where $\mathbf{a}, \mathbf{r} \in \mathbb{R}^D$ and $A, R \in SO(D)$, if no group law is specified, there is no natural definition of the Fourier transform. That is, g and h could be elements of the semi-direct product group $SE(D)$, or the direct product group $\mathbb{R}^D \times SO(D)$. And while the natural choice for the definition of a group Fourier transform is based on the IURs of that group, the similarities between $SE(D)$ and $\mathbb{R}^D \times SO(D)$ are worth exploiting because the IURs of the former are infinite-dimensional, while those of the latter are finite dimensional.

The group law for $\mathbb{R}^D \times SO(D)$ is simply

$$(\mathbf{a}, A) \circ (\mathbf{r}, R) = (\mathbf{a} + \mathbf{r}, AR).$$

The representations are the direct product of representations of $\mathbb{R}^D$ and $SO(D)$:

$$V^l(\mathbf{a}, A; \mathbf{p}) = e^{i\mathbf{a}\cdot\mathbf{p}} U^l(A)$$

where $U^l(A)$ are IURs of $SO(D)$. (We are only interested in the cases $D = 2$ and $D = 3$.) The corresponding Fourier transform of a well-behaved function is

$$\hat{f}^l(\mathbf{p}) = \mathcal{F}(f) = \int_{\mathbb{R}^D \times SO(D)} f(g) V^l(g^{-1}; \mathbf{p})\, d(g)$$

$$= \int_{SO(D)} \int_{\mathbb{R}^D} f(\mathbf{a}, A)\, e^{-i\mathbf{a}\cdot\mathbf{p}} U^l(A^{-1})\, d\mathbf{a}\, dA. \tag{11.36}$$

The computationally attractive nature of this Fourier transform is emphasized when written in the form

$$\hat{f}^l(\mathbf{p}) = \int_{SO(D)} \left[\int_{\mathbb{R}^D} f(\mathbf{a}, A)\, e^{-i\mathbf{a}\cdot\mathbf{p}}\, d\mathbf{a} \right] U^l(A^{-1})\, dA. \tag{11.37}$$

We compute these integrals by sampling at $N = N_r \cdot N_R$ points (where N_r is the number of samples in $\mathbb{R}^D$ and N_R is the number of samples in $SO(D)$). For $D = 2$, $N_r = \mathcal{O}(N^{2/3})$ and $N_R = \mathcal{O}(N^{1/3})$, while for $D = 3$, $N_r = \mathcal{O}(N^{1/2})$ and $N_R = \mathcal{O}(N^{1/2})$. The term in parenthesis in (11.37) is calculated in $\mathcal{O}(N_R \cdot N_r \log N_r)$ operations using the FFT for $\mathbb{R}^D$. We denote

$$\tilde{f}(\mathbf{p}, A) \doteq \int_{\mathbb{R}^D} f(\mathbf{a}, A)\, e^{-i\mathbf{a}\cdot\mathbf{p}}\, d\mathbf{a}.$$

Then (11.36) becomes

$$\hat{f}^l(\mathbf{p}) = \int_{SO(D)} \tilde{f}(\mathbf{p}, A) U^l(A^{-1})\, dA.$$

This can be computed in $\mathcal{O}(N_r \cdot N_R\, \eta(N_R))$ where $\eta(N_R) = \log N_R$ when $D = 2$, and $\eta(N_R) = (\log N_R)^2$ when $D = 3$. In either case, by adding the complexities of the two steps, (11.37) is computed for all values of l up to the band-limit and all sampled values of $\mathbf{p}$ in $\mathcal{O}(N\, \eta(N))$ computations.

The inverse Fourier transform for $\mathbb{R}^D \times SO(D)$ is computed as

$$f(\mathbf{a}, A) = \frac{1}{(2\pi)^D} \sum_{l=0}^{B-1} d_l \int_{\mathbb{R}^D} \mathrm{tr}(\hat{f}^l(\mathbf{p})\, e^{i\mathbf{p}\cdot\mathbf{a}} U^l(A))\, d\mathbf{p}. \tag{11.38}$$

This follows immediately from the Fourier inversion formulas for $\mathbb{R}^D$ and $SO(D)$, and can also be computed in $\mathcal{O}(N\,\eta(N))$ arithmetic operations. When $D = 2$ we have $d_l = 1$ and $B = \mathcal{O}(N_R)$, whereas for $D = 3$ we have $d_l = 2l + 1$ and $B = \mathcal{O}(N_R^{1/3})$.

We now examine the computational complexity of performing $SE(D)$ convolutions using the above fast Fourier transform pair for $\mathbb{R}^D \times SO(D)$. Recall that for $SE(D)$ the convolution of two well-behaved functions is

$$f_3(\mathbf{a}, A) = (f_1 * f_2)(\mathbf{a}, A) = \int_{SO(D)} \int_{\mathbb{R}^D} f_1(\mathbf{r}, R)\, f_2(R^{-1}(\mathbf{a} - \mathbf{r}), R^{-1}A)\, d\mathbf{r}\, dR.$$

The $\mathbb{R}^D \times SO(D)$ Fourier transform of this is

$$\hat{f}_3^l(\mathbf{p}) = \int_{SO(D)} \int_{\mathbb{R}^D} \left[\int_{SO(D)} \int_{\mathbb{R}^D} f_1(\mathbf{r}, R)\, f_2(R^{-1}(\mathbf{a} - \mathbf{r}), R^{-1}A)\, d\mathbf{r}\, dR \right] e^{-i\mathbf{a}\cdot\mathbf{p}} U^l(A^{-1})\, d\mathbf{a}\, dA.$$

$$(11.39)$$

Making the change of variables $(R^{-1}(\mathbf{a} - \mathbf{r}), R^{-1}A) = (\mathbf{b}, B)$, observing that $d\mathbf{a}\, dA = d\mathbf{b}\, dB$ and performing the outer integration in (11.39) first, we write

$$\int_{SO(D)} \int_{\mathbb{R}^D} f_2(\mathbf{b}, B)\, e^{-i(R\mathbf{b}+\mathbf{r})\cdot\mathbf{p}} U^l(B^{-1}R^{-1})\, d\mathbf{b}\, dB = \hat{f}_2^l(R^{-1}\mathbf{p})\, e^{-i\mathbf{p}\cdot\mathbf{r}} U^l(R^{-1}).$$

This means that

$$\hat{f}_3^l(\mathbf{p}) = \int_{SO(D)} \int_{\mathbb{R}^D} f_1(\mathbf{r}, R)\hat{f}_2^l(R^{-1}\mathbf{p})\, e^{-i\mathbf{p}\cdot\mathbf{r}} U^l(R^{-1})\, d\mathbf{r}\, dR.$$

This can be rewritten as

$$\hat{f}_3^l(\mathbf{p}) = \int_{SO(D)} \hat{f}_2^l(R^{-1}\mathbf{p})\tilde{f}_1(\mathbf{p}, R) U^l(R^{-1})\, dR \qquad (11.40)$$

where

$$\tilde{f}_1(\mathbf{p}, R) = \int_{\mathbb{R}^D} f_1(\mathbf{r}, R)\, e^{-i\mathbf{p}\cdot\mathbf{r}}\, d\mathbf{r}$$

is computed in $\mathcal{O}(N_R N_r \log N_r)$ operations.

Hence, the price of using the Fourier transform for $\mathbb{R}^D \times SO(D)$ instead of that for $SE(D)$ is that pointwise multiplication of Fourier matrices is replaced with the integral in (11.40). This integral is *not* just the Fourier transform on $SO(D)$ for each fixed value of $\mathbf{p}$ since

$$F^l(R, \mathbf{p}) \doteq \hat{f}_2^l(R^{-1}\mathbf{p})\tilde{f}_1(\mathbf{p}, R) \qquad (11.41)$$

depends on l.

When $D = 3$, $F^l(R, \mathbf{p})$ is a matrix, each element of which can be calculated in $\mathcal{O}(N_R \cdot N_r \epsilon(N_r))$ operations for all values of $\mathbf{p}$ and R where $\epsilon(\cdot)$ depends on what form of interpolation is used, as described at the beginning of this chapter. This calculation can be done for all $\mathcal{O}(N_R)$ matrix elements using $\mathcal{O}(N^{3/2}\epsilon(N^{1/2}))$ operations. Since

$$\sum_{l=0}^{B-1} d_l^\gamma = \mathcal{O}(B^{\gamma+1})$$

operations are required to multiply $F^l(R, \mathbf{p})$ and $U^l(R^{-1})$ for all values of $l < B$ and each value of $\mathbf{p}$ and R, it follows that a total of $\mathcal{O}(B^{\gamma+1} \cdot N_r \cdot N_R)$ arithmetic operations are required to multiply $F^l(R, \mathbf{p})$ and $U^l(R^{-1})$ for all values of l up to the band-limit, and all values of $\mathbf{p}$ and R at the sample points. When Gaussian elimination is used, $\gamma = 3$ and the order of computation to perform all the matrix multiplications becomes $\mathcal{O}(B^{(\gamma+1)} \cdot N_r \cdot N_R) = \mathcal{O}(N^{5/3})$ since $B = \mathcal{O}(N_R^{1/3}) = \mathcal{O}(N^{1/6})$.

When $D = 2$, the $SO(D)$ representations are one-dimensional and $\gamma = 0$. In this case, (11.41) is a scalar function of R and $\mathbf{p}$ that can be calculated for all values of $\mathbf{p}$, R and l in $\mathcal{O}(N_R^2 \cdot N_r \epsilon(N_r))$ operations. Then (11.40) can be computed for each value of $\mathbf{p}$ and l in $\mathcal{O}(B \cdot N_r \cdot N_R) = \mathcal{O}(N^{4/3})$ since $B = \mathcal{O}(N_R) = \mathcal{O}(N^{1/3})$.

Hence, it appears that for the $D = 2$ case, this method is on the same order of complexity as when using the $SE(D)$ Fourier transform, while for the $D = 3$ case it is slower. In addition, it must be noted that this technique will always be an approximate one. This is because

$$ f(\mathbf{a}, A) = \frac{1}{(2\pi)^D} \sum_{l=0}^{B-1} d_l \int_{\mathbb{R}^D} \mathrm{tr} \left(\hat{f}^l(\mathbf{p}) \, e^{i\mathbf{p} \cdot \mathbf{a}} U^l(A) \right) d\mathbf{p} $$

is *not* a band-limited function on $SE(D)$ even though it is a band-limited function on $\mathbb{R}^D \times SO(D)$. This can be observed directly by taking the $SE(D)$ Fourier transform. And even if two functions f_1 and f_2 are band-limited on $\mathbb{R}^D \times SO(D)$, if they are convolved on $SE(3)$ the result $f_1 * f_2$ will generally no longer be band-limited on $\mathbb{R}^D \times SO(D)$. This is because $\hat{f}_2^l(R^{-1}\mathbf{p})$ need not have a band-limited expansion in the $SO(D)$ harmonics for each fixed value of $\mathbf{p}$.

11.6.1 Contraction of $SE(D)$ to $SO(D+1)$

The methods for approximating rigid-body motions in D-dimensional space with rotations in $(D + 1)$-dimensional space that are discussed in Chapter 6 can be used as a method for performing fast approximate $SE(D)$ convolutions.

Using this idea, functions $f_i : SE(D) \to \mathbb{C}$ are replaced with functions of the form $f_i' : SO(D + 1) \to \mathbb{C}$ where the conversion from f_i to f_i' is achieved by mapping points from $SE(D)$ onto $SO(D+1)$, with the correspondence $f_i'(A) = f_i(g)$ established when $g \in SE(D)$ is mapped to $A \in SO(D+1)$.

The convolution $f_1' * f_2'$ can then be performed using the fast Fourier transform for $SO(D+1)$. In Chapter 9, the IURs of $SO(4)$ are shown to be closely related to those for $SU(2)$, and an $\mathcal{O}(N(\log N)^2)$ algorithm exists for $SO(4)$ Fourier transforms [21]. The result is then mapped back to $SE(D)$.

While this method has the best computational speed of any algorithm we have considered for computing group Fourier transforms, the drawback is that the mapping between $SE(D)$ and $SO(D + 1)$ results in a distortion so that

$$ (f_1' * f_i')(A) \neq (f_1 * f_2)(g). $$

11.7 Summary

Fast numerical algorithms for computing the convolution product of functions on motion groups were derived. These algorithms use the group-theoretic Fourier transform with the irreducible unitary representations written in operator form, and their matrix elements calculated numerically instead of analytically. This, together with interpolation between Cartesian and spherical coordinate grids, made it possible to use well-known FFTs for compactly supported functions on $\mathbb{R}^3$, and more recent FFTs for the sphere and rotation group. Using the Fourier transform based in IURs of $SE(3)$ appears to be more suitable than using alternatives based on other kinds of group Fourier transforms. Other alternatives can also be found in the literature (see, e.g., [10]).

References

1. Aho, A.V., Hopcroft, J.E., Ullman, J.D., *The Design and Analysis of Computer Algorithms*, Addison-Wesley Publishing Company, Reading Mass., 1974.
2. Averbuch, A., Coifman, R.R., Donoho, D.L., Elad, M., Israeli, M., "Fast and Accurate Polar Fourier Transform," *Applied and Computational Harmonic Analysis*, 21(2): 145–167, 2006.
3. Borodin, A., Munro, I., *The Computational Complexity of Algebraic and Numerical Problems*, Elsevier, New York, 1975.
4. Bucci, O.M., Gennarelli, C., Savarese, C., "Fast and Accurate Near-Field-Far-Field Transformation by Sampling Interpolation of Plane-Polar Measurements," *IEEE Transactions on Antennas and Propagation*, 39(1): 48–55, 1991.
5. Chirikjian, G.S., Kyatkin, A.B., "Algorithms for Fast Convolutions on Motion Groups," *Applied and Computational Harmonic Analysis*, 9(2): 220–241, 2000.
6. Choi, H., Munson, D.C., Jr., "Direct-Fourier Reconstruction in Tomography and Synthetic Aperture Radar," *International Journal of Imaging Systems and Technology*, 9(1): 1–13, 1998.
7. Danielsson, P.-E., Hammerin, M., "High-Accuracy Rotation of Images," *CVGIP: Graphical Models and Image Processing*, 54(4): 340–344, July 1992.
8. Deans, S.R., *The Radon Transform and Some of Its Applications*, John Wiley and Sons, New York, 1983. (Dover, 2007).
9. Diaconis, P., Rockmore, D., "Efficient Computation of the Fourier Transform on Finite Groups," *Journal of the American Mathematical Society*, 3(2): 297–332, April 1990.
10. Dooley, A.H., "A Nonabelian Version of the Shannon Sampling Theorem," *SIAM Journal on Mathematical Analysis*, 20(3): 624–633, May 1989.
11. Driscoll, J.R., Healy, D., "Computing Fourier Transforms and Convolutions on the 2-Sphere," *Advances in Applied Mathematics*, 15(2): 202–250, 1994.
12. Driscoll, J.R., Healy, D., Rockmore, D.N., "Fast Discrete Polynomial Transform with Applications to Data Analysis for Distance Transitive Graphs," *SIAM Journal on Computing*, 26(4): 1066–1099, 1997.
13. Fraser, D., Schowengerdt, R.A., "Avoidance of Additional Aliasing in Multipass Image Rotations," *IEEE Transactions on Image Processing*, 3(6): 721–735, Nov. 1994.
14. Gauthier, J. P., Bornard, G., Sibermann, M., "Motion and Pattern Analysis: Harmonic Analysis on Motion Groups and Their Homogeneous Spaces," *IEEE Transactions on Systems, Man, and Cybernetics*, 21(1): 159–172, 1991.
15. Gurarie, D., *Symmetry and Laplacians. Introduction to Harmonic Analysis, Group Representations and Applications*, Elsevier Science Publisher, The Netherlands, 1992. (Dover, 2008).
16. Kumahara, K., Okamoto, K., "An Analogue of the Paley-Wiener Theorem for the Euclidean Motion Group," *Osaka Journal of Mathematics*, 10(1): 77–92, 1973.
17. Kyatkin, A.B., Chirikjian, G.S., "Pattern Matching as a Correlation on the Discrete Motion Group, " *Computer Vision and Image Understanding*, 74(1): 22–35, April 1999.

18. Kyatkin, A.B., Chirikjian, G.S., "Algorithms for Fast Convolutions on Motion Groups, " *Applied and Computational Harmonic Analysis*, 9(2): 220 – 241, 2000.
19. Kyatkin, A.B., Chirikjian, G.S., "Synthesis of Binary Manipulators Using the Fourier Transform on the Euclidean Group," *ASME J. Mechanical Design*, 121(1): 9 – 14, March 1999.
20. Larkin, K.G., Oldfield, M.A., Klemm, H., "Fast Fourier Method for the Accurate Rotation of Sampled Images," *Optics Communications*, 139(1): 99 – 106, June 1997.
21. Maslen, D.K., *Fast Transforms and Sampling for Compact Groups*, Ph.D. Dissertation, Dept. of Mathematics, Harvard University, May 1993.
22. Maslen, D.K., Rockmore, D.N., "Generalized FFTs - A Survey of Some Recent Results," *DIMACS Series in Discrete Mathematics and Theoretical Computer Science*, Vol. 28, pp. 183 – 237, 1997.
23. Miller, W.,Jr., *Lie Theory and Special Functions*, Academic Press, New York, 1968;
24. Miller, W., "Some Applications of the Representation Theory of the Euclidean Group in Three-Space," *Communications on Pure and Applied Mathematics*, 17(4): 527 – 540, 1964.
25. Orihara, A., "Bessel Functions and the Euclidean Motion Group," *Tohoku Mathematical Journal*, 13(1): 66 – 71, 1961.
26. Paeth, A.W., "A Fast Algorithm for General Raster Rotation," in *Graphics Gems* (A.S. Glassner ed.), pp. 179 – 195, Academic Press, Boston, 1990.
27. Pan, S.X., Kak, A.C., "A Computational Study of Reconstruction Algorithms for Diffraction Tomography: Interpolation Versus Filtered Backpropagation," *IEEE Transactions on Acoustics, Speech, and Signal Processing*, 31(5): 1262 – 1275, Oct., 1983.
28. Pan, V., "How Can We Speed Up Matrix Multiplication," *SIAM Review*, 26(3): 393 – 416, 1984.
29. Pan, V., *How to Multiply Matrices Fast*, Springer-Verlag, Berlin, Heidelberg 1984.
30. Speath, H., *Two Dimensional Spline Interpolation Algorithms*, Wellesley, Massachusetts, AK Peters, 1995.
31. Strassen, V., "Gaussian Elimination is not Optimal," *Numerische Mathematik*, 13(4): 354 – 356, 1969.
32. Unser, M., Thévenaz, P., Yaroslavsky, L., "Convolution-Based Interpolation for Fast, High-Quality Rotation of Images," *IEEE Transactions on Image Processing*, 4(10): 1371 – 1381, October 1995.
33. Vilenkin, N.J., "Bessel Functions and Representations of the Group of Euclidean Motions," *Uspehi Mat. Nauk.*, 11: 69 – 112, 1956 (in Russian).
34. Vilenkin, N.J., Klimyk, A.U., *Representation of Lie Group and Special Functions*, Vol. 1 – 3, Kluwer Academic Publishers, The Netherlands, 1991.
35. Winograd, S., "A New Algorithm for Inner Products," *IEEE Transactions on Computers*, 17(7): 693 – 694, 1968.
36. Yaghjian, A.D., Woodworth, M.B., "Sampling in Plane-Polar Coordinates," *IEEE Transactions on Antennas and Propagation*, 44(5): 696 – 700, May 1996.

Robotics

12.1 A Brief Introduction to Robotics

Robotics is the study of machines that exhibit some degree of autonomy and flexibility in the tasks that they perform. There are two major kinds of robotic devices: (1) manipulators; and (2) mobile robots.

A manipulator is a robot arm and/or hand that is usually fixed to some kind of base. A manipulator can have a *serial chain* topology in which there are no loops formed by the links of the arm It is also possible to have a *parallel* or *platform* architecture in which one or more loops exist. A third possibility is a *tree-like* topology such as the case of a hand or cooperating serial manipulators. In any of these topologies, the actuators which drive the movement of the arm can be revolute (rotational) such as an electric motor, prismatic (translational) such as a hydraulic cylinder, or some combination of the two (e.g., a screw). The key computational issues in the use of manipulator arms in industrial or service environments all depend on the fast calculation of joint angles[1] which will place the functional end of the arm (called the *end effector*) at the desired position and/or orientation relative to its base. The set of all positions and orientations that an arm can reach is called its *workspace*. The determination of the joint angles that result in the desired end-effector state is called the *inverse kinematics* problem. The *forward kinematics* problem is the problem of finding the position and/or orientation of the end effector when the value of the joint angles is given. As a rule, the forward kinematics problem is very easy to solve for serial manipulators, and the inverse problem is more difficult. However, for parallel manipulators, the opposite is true. And in a sense, hybrid manipulators (in which two or more parallel structures are stacked) inherit the difficult aspects of both serial and parallel manipulators. The *workspace generation* problem is that of determining all positions and/or orientations that are reachable by a robotic arm (see, e.g., [2, 8, 43, 63, 64]). And a number of works have studied workspaces by sampling joint angles and analyzing the resulting distribution of reachable positions (see, e.g., [41, 68]).

Excellent introductions to the issues involved in the kinematics, dynamics, and control of manipulator arms can be found in the textbooks [20, 24, 53, 66, 70, 72]. For group-theoretic issues and configuration modeling in robotics see [6, 54, 55, 62, 67].

A mobile robot is a machine with wheels, tracks, legs or other means of propulsion that is intended to move from one location to another. We will concentrate on issues

---

[1] The term *joint angles* refers not only to angles but to any generalized actuator displacement such as stroke lengths of prismatic actuators.

in the motion planning of a single rigid-body robot. The most basic problem is that of navigating a mobile robot of known shape through an environment with known obstacles without regard to effects of measurement error, wheel slippage or the vehicle dynamics. This problem has been addressed extensively in the literature (see, e.g., [7, 18, 48]), and the approach presented here is one of a variety of acceptable techniques.

Combinations of manipulators and mobile robots, which can be considered the two most basic subsystems, can be arranged in a variety of ways. For example, one or more manipulator arms can be affixed atop a mobile platform which serves as a transport device for the arms. We do not examine combined manipulator and mobile platform systems. The reader interested in this area will find the following references useful [36, 22].

In this chapter, we examine manipulators and mobile robots using concepts from noncommutative harmonic analysis. In Section 12.2 a particular kind of manipulator (called a *binary* or *discretely-actuated* manipulator) is examined in detail, and the concept of a *density function* which describes where such an arm can reach is defined. It is shown in Section 12.3 why this information is useful in solving the inverse kinematics problem for discretely-actuated manipulators. Section 12.4 examines the symmetries of the density function that are inherited from geometrical symmetries of the arm. Section 12.5 then describes the design problem for binary arms, which involves the inversion of convolution equations on the Euclidean motion group. In Section 12.6 we turn to an issue which is of more general interest: the accumulation of error in serial chains (including manipulators). Finally, in Section 12.7 we illustrate how convolution-like integrals arise when considering the set of all positions and orientations that a rigid mobile robot can attain without intersecting obstacles.

12.2 The Density Function of a Discretely-Actuated Manipulator

The convolution product of real-valued functions on the Euclidean Group is used in this section as a computational tool. The primary application is the generation of discretely-actuated-manipulator workspaces, and determination of the density of reachable frames in any portion of the workspace. A discretely-actuated manipulator is an arm for which each of the actuators have a finite number of states. This includes manipulators driven by stepper motors, pneumatic cylinders or solenoids. In the case when the discrete-state actuators have only two states, the resulting arm is called a *binary manipulator*. Figure 12.1 shows a schematic of a 3-bit binary platform and a photo of a manipulator made out of this kind of 3-bit unit. This structure is a truss where all the vertices can be thought of as passive hinges, and each of the labeled legs is an actuator which changes length. The labels indicate the two states, with the shorter being labeled '0' and the longer labeled '1'. The three-dimensional version of this kind of platform (regardless of whether or not the actuators have discrete states) has six extensible legs and is called a Stewart/Gough platform.

Figure 12.2 shows a three-dimensional binary manipulator which is constructed of a series of Stewart/Gough platforms that are stacked, or cascaded, on top of each other. Since each actuator (which in this case is a pneumatic cylinder in parallel with a viscous dashpot) has two stable states, each platform has 2^6 states, and the whole arm has $(2^6)^6 = 2^{36}$ states. Binary arms are attractive because they require no feedback control, and are very inexpensive to construct. However, issues such as inverse kinematics become much more computationally challenging than in the case of continuous actuation.

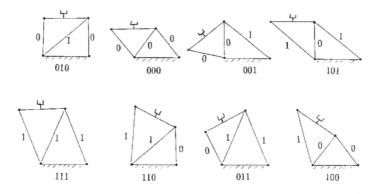

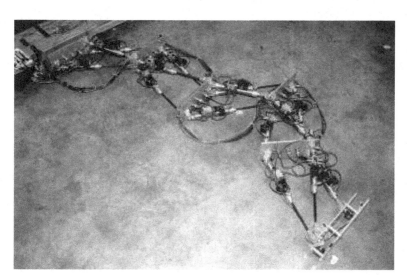

Fig. 12.1. A 3-Bit, 8-State Platform Manipulator: (top) Allowable States of One Module; (bottom) Photograph of Whole Manipulator with Five Cascaded Modules and 8^5 States. (Reprinted from Chirikjian, G.S., Ebert-Uphoff, I., "Numerical Convolution on the Euclidean Group with Applications to Workspace Generation," *IEEE Transactions on Robotics and Automation*, 14(1): 123–136, 1998, with permission from IEEE)

In general, if a discretely-actuated manipulator has P units (where, for example, each unit is a platform) and each unit has K states, then the arm will have K^P states. It will generally be desirable to avoid direct computation of the K^P different configurations of the arm when performing calculations like inverse kinematics (see Section 12.3). In fact, while the concept of discretely-actuated manipulators has been in the literature for more than 30 years (see [61, 65, 38]), it seems that the exponential complexity of the problem has been a major stumbling block. See [9, 10, 12, 13, 14, 15, 16, 44, 45, 49, 50, 51] for various approaches to circumventing this complexity.

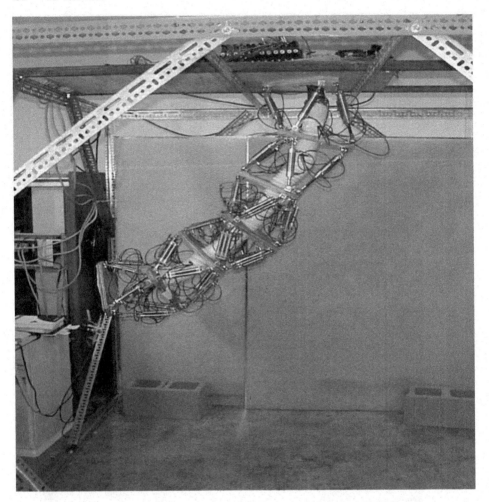

Fig. 12.2. A Three-Dimensional Manipulator with 2^{36} States. (Reprinted from Chirikjian, G.S., Ebert-Uphoff, I., "Numerical Convolution on the Euclidean Group with Applications to Workspace Generation," *IEEE Transactions on Robotics and Automation*, 14(1): 123–136, 1998, with permission from IEEE)

In the context of discrete actuation, the density of frames (number of frames per unit volume in $SE(3)$) in many ways replaces classical measures of dexterity used in robotics (see, e.g., [3, 5, 37, 39, 57, 75]) as a scalar function of importance defined over the workspace. This is because density in the neighborhood of a given frame is an indicator of how accurately a discretely actuated manipulator can reach that point/frame.

To compute the workspace density function using brute force is computationally intractable, e.g., it requires $O(K^P)$ evaluations of the forward kinematic equations for a manipulator with P actuated modules each with K states. In addition, an array storing the density of all volume elements in the workspace must be incremented $O(K^P)$ times if brute force computation is used.

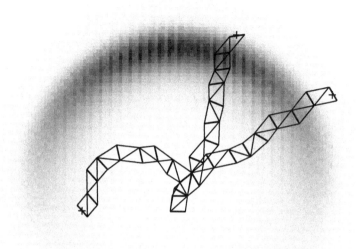

Fig. 12.3. A Discretely-Actuated Manipulator with 4^{30} States with Superimposed Density Function (Reprinted from Chirikjian, G.S., Ebert-Uphoff, I., "Numerical Convolution on the Euclidean Group with Applications to Workspace Generation," *IEEE Transactions on Robotics and Automation*, 14(1): 123 – 136, 1998, with permission from IEEE)

Figure 12.3 shows a schematic of the density of frames reachable by a discretely actuated variable geometry truss manipulator. If there are 30 actuated truss elements (ten modules) and each element has 4 states (and thus the whole manipulator has $4^{30} \approx 10^{18}$ states) the workspace density cannot simply be computed using brute force because this could take years using current computer technology. This combinatorial explosion is a major reason why discrete actuation is not commonly used, despite the fact that the concept is almost five decades old (see, e.g., [61, 65]). Having a representation of the density of reachable frames is important for performing inverse kinematics and design of discretely actuated manipulators. Using the concept of Euclidean-group convolution, an approximation of the workspace density can be achieved in $O(\log P)$ convolutions for macroscopically serial (hybrid) manipulators[2] composed of P identical modules. This reduces the computation time to minutes when convolutions are implemented in an efficient way.

The remainder of this section is organized as follows. In Subsection 12.2.1 we give the geometrical intuition behind Euclidean-group convolution. In Subsection 12.2.2 we show

---

[2]A serial cascade of modules where each module may be a serial or parallel kinematic structure.

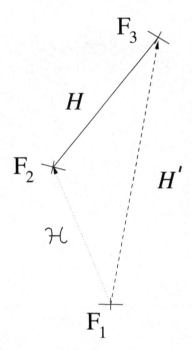

Fig. 12.4. Concatenation of Homogeneous Transformations. (Reprinted from Chirikjian, G.S., Ebert-Uphoff, I., "Numerical Convolution on the Euclidean Group with Applications to Workspace Generation," *IEEE Transactions on Robotics and Automation*, 14(1): 123–136, 1998, with permission from IEEE)

why this concept is important for workspace generation of discretely actuated manipulators. In Subsection 12.2.3 the computational benefit of this approach is explained. In Subsection 12.2.4 the two-dimensional case is considered explicitly. In Subsection 12.2.5 a direct numerical implementation of convolution is explained in detail for the planar case. Numerical results are generated in Subsection 12.2.6. Much of the material presented in this section originally appeared in [12]. The concept of using sweeping as a tool to generate workspaces was considered in [27].

12.2.1 Geometric Interpretation of Convolution of Functions on $SE(D)$

Suppose there are three frames in space, F_1, F_2 and F_3, as shown in Figure 12.4. The first frame can be viewed as fixed, the second frame as moving with respect to the first, and the third frame as moving with respect to the second. Let the homogeneous transform $\mathcal{H}$ describe the position and orientation of F_2 w.r.t. F_1, and H describe the position and orientation of F_3 w.r.t. F_2. Then the position and orientation of F_3 with respect to F_1 is $H' = \mathcal{H}H$. The position and orientation of F_3 with respect to F_2 can then be written as
$$H = \mathcal{H}^{-1}H'.$$

We can divide up $SE(D)$ into volume elements, or "voxels," of finite but small size. The volume of the voxel centered at $H \in SE(D)$ (for $D = 2$ or 3) is denoted $\Delta(H)$, and as the element size is chosen smaller and smaller it becomes closer to the differential volume element $d(H)$.

The motion of F_2 relative to F_1 and the motion of F_3 relative to F_2 can both be considered as elements of $SE(D)$, and no distinction is made between these motions and the transformation matrices $\mathcal{H}$ and H which represent these motions.

Assuming we move $\mathcal{H}$ and H through a finite number of different positions and orientations, let ρ_1 be a function that records how often the $\mathcal{H}$ frames appear in each voxel, divided by the voxel volume $\Delta(\mathcal{H})$. Likewise, let ρ_2 be the function describing how often the H frames appear in each voxel normalized by voxel volume.

To calculate how often the H' frames appear in each voxel in $SE(D)$ for all possible values of $\mathcal{H}$ and H, we can perform the following steps:

- Evaluate $\rho_1 = \rho_1(\mathcal{H})$ (frequency of occurrence of $\mathcal{H}$).
- Evaluate $\rho_2 = \rho_2(H) = \rho_2(\mathcal{H}^{-1}H')$ (frequency of occurrence of $H = \mathcal{H}^{-1}H'$).
- Weight (multiply) the left-shifted density histogram $\rho_2(\mathcal{H}^{-1}H')$ by the number of frames which are doing the shifting. This number is $\rho_1(\mathcal{H})\Delta(\mathcal{H})$ for each $\mathcal{H}$.[3]
- Sum (integrate) over all these contributions:

$$(\rho_1 * \rho_2)(H') = \int_{SE(D)} \rho_1(\mathcal{H})\rho_2(\mathcal{H}^{-1}H') \, d(\mathcal{H}).$$

As will be seen in subsequent sections, this approach yields an approximation to the density of H' frames which can be computed very efficiently. The number of H' frames in each voxel of $SE(D)$ can be calculated from this density as simply $(\rho_1 * \rho_2)(H')\Delta(H')$.

12.2.2 The Use of Convolution for Workspace Generation

In this subsection, we show how the concept of the convolution product of functions on $SE(D)$ can be applied to the generation of workspaces.

Let us consider a manipulator that consists of two mechanisms stacked on top of one another. For example, the two mechanisms can be in-parallel platform mechanisms or serial linkages. A frame F_1 is attached to the bottom of the first mechanism and a frame F_2 to its top, which also defines the bottom of the second mechanism. A third frame, F_3, defines the position and orientation of the top of the second mechanism. This leads back to the situation shown in Figure 12.4, where now $\mathcal{H}$ describes the homogeneous transformation corresponding to the lower mechanism, H the one for the upper mechanism, and H' the one for the whole manipulator.

If the manipulator is actuated discretely, then each mechanism only has a finite number of different states, which can be described by two finite sets, $\mathcal{S}_1$ and $\mathcal{S}_2$, which contain m_1 and m_2 elements respectively. The set of all homogeneous transformations that can be attained by the distal end of the manipulator when the base is fixed results from all possible combinations of these two sets,

$$\mathcal{S}' = \{H' = \mathcal{H}H : H \in \mathcal{S}_1, \mathcal{H} \in \mathcal{S}_2\},$$

---

[3]Note that the product $\rho_2(\mathcal{H}^{-1}H')\rho_1(\mathcal{H})\Delta(\mathcal{H})$ is then an approximation to the sum of histograms which would result by sweeping $\rho_2(H)$ by the homogeneous transforms $\mathcal{H}$ in the voxel with volume $\Delta(\mathcal{H})$. In the limiting case when the volume size becomes small, this approximation becomes better.

and hence consists of $m_1 \cdot m_2$ elements.

For a finite set of frames (elements of $SE(D)$) that is very large, it is useful to reduce the amount of data by approximating its information with a density function. This is done by dividing a bounded region in $SE(D)$ into small volume elements, counting how many reachable frames occupy each volume element, and dividing this number by the volume of each element. This density function describes the distribution of frames in the workspace. Figure 12.3 shows the integral of such a distribution over all orientations reachable by a manipulator. The result is a function of position in the plane (gray scale corresponds to density).

If the sets $\mathcal{S}_1$ and $\mathcal{S}_2$ are approximated with density functions $\rho_1(\cdot)$ and $\rho_2(\cdot)$ respectively, then the density function resulting from the convolution of the two is a density function for the whole manipulator:

$$\rho(H') = (\rho_1 * \rho_2)(H').$$

Furthermore, this reasoning can be applied to manipulators consisting of more than two mechanisms (modules) stacked on top of one another. For instance, if four modules are stacked, the density of the lower two is $\rho_1 * \rho_2$ and the density of the upper two is $\rho_3 * \rho_4$ by the reasoning given previously. Treating the lower two modules as one big module, and the upper two as one big module, the density of the collection of all four modules is $(\rho_1 * \rho_2) * (\rho_3 * \rho_4) = \rho_1 * \rho_2 * \rho_3 * \rho_4$.

If the possible configurations of a discretely-actuated manipulator consisting of P independent modules are described by sets $S_1, \ldots, S_P$ in the way described above, then the density of the whole manipulator is derived by multiple convolution as

$$\rho(H') = (\rho_1 * \rho_2 * \ldots * \rho_P)(H').$$

12.2.3 Computational Benefit of this Approach

In this section the computational complexity of the approach described in the preceding section is compared with brute force enumeration of manipulator states.

Suppose a manipulator consisting of P modules is considered and the number of homogeneous transformations in each set $\mathcal{S}_i$ is m_i. The explicit (brute force) calculation of all combinations of homogeneous transformations is of the same order as the number of all combinations, $\mathcal{O}(\prod_{i=1}^{P} m_i)$. If $m_1 = \ldots = m_P = K$, then this is an $\mathcal{O}(K^P)$ calculation.

In our approach, density functions are used to describe the frame distribution for each module. The frame distribution of the whole manipulator results by performing P convolutions. These calculations also depend on the dimension of $SE(D)$ for $D = 2, 3$. While we treat D as a constant because it does not change with the number of actuator states, it is worth noting that if a compact subset of $SE(D)$ is divided into $\mathcal{N}_j$ increments in each dimension, then $Q = \prod_{j=1}^{D(D+1)/2} \mathcal{N}_j$ voxels result. Treating Q as constant, the calculation of $\rho_i(\cdot)$ requires $\mathcal{O}(m_i)$ additions to increment the number of frames in each voxel. Since the voxels are uniform in size in the $SE(2)$ case, there is no need to explicitly divide by voxel volume (i.e., this normalization can be performed concurrently with convolution). Thus the calculation of $\rho_i(\cdot)$ is effectively $\mathcal{O}(m_i)$.

Consider the convolution of density functions of any two adjacent modules. The numerical approximation of the convolution integral evaluated at a single point in the support of $(\rho_i * \rho_{i+1})(\cdot)$ becomes a sum over all voxels in the support of $\rho_i(\cdot)$. This calculation must be performed for all voxels in the support of $(\rho_i * \rho_{i+1})(\cdot)$, and so the computations required to perform one convolution are $\mathcal{O}(\text{convolution}) = \mathcal{O}(Q_i \cdot Q_i^*)$, where Q_i and Q_i^* are respectively the number of voxels in the support of $\rho_i(\cdot)$ and $(\rho_i * \rho_{i+1})(\cdot)$. If the voxel size is kept constant after convolution, then $Q_i^* > Q_i$, because the workspace of any two adjacent modules is bigger than either one individually. Treating Q_i and D as constants, the calculation of P convolutions would then be polynomial in P. The order of this polynomial would depend on D. However, if the voxel size is rescaled after convolution, so that $Q_i^* \approx Q_i$, then each convolution is $\mathcal{O}(Q_i^2) = \mathcal{O}(1)$. Hence the total order of this approach is $\mathcal{O}(\sum_{i=1}^{P} m_i) + P \cdot \mathcal{O}(1) = \mathcal{O}(P)$ for this data storage strategy.

In the special case of a manipulator consisting of P identical modules the number of convolutions to be performed can be further reduced by using a different strategy. In this case the frame distribution $\rho(H')$ is calculated from P identical functions as a P-fold convolution:

$$\rho = \rho_1^{(P)} = \rho_1 * \rho_1 * \ldots * \rho_1.$$

Note, that the repeated convolution of a function with itself generates the following sequence of functions:

$$\rho^{(2)} = \rho_1 * \rho_1, \quad \rho^{(4)} = \rho^{(2)} * \rho^{(2)}, \quad \rho^{(8)} = \rho^{(4)} * \rho^{(4)}, \quad \text{etc.,}$$

i.e. it is possible to generate $\rho^{(2)}$ by one convolution, $\rho^{(4)}$ by two convolutions, and more generally $\rho^{(2^n)}$ by n convolutions. Thus, for a manipulator with P identical modules, approximately $\mathcal{O}(\log P)$ convolutions have to be performed, which is an $\mathcal{O}(\log P)$ calculation if the number of voxels is held constant after each convolution (i.e., if we allow voxel size to grow with each convolution).[4]

An estimate of the support of the convolution of two density functions can be obtained by first making a gross over estimate and doing a very crude (low resolution) convolution. Those voxels which have zero density after convolution can be discarded, and what is left over is a closer overestimate of the support of the convolved functions. This region is smaller than the original estimate, and voxel sizes can be scaled down to get the best resolution for the allowable memory.

12.2.4 Computation of the Convolution Product of Functions on $SE(2)$

In the two-dimensional case, the homogeneous transforms H' and $\mathcal{H}$ in the convolution integral can be parameterized as

$$H'(x, y, \theta) = \begin{pmatrix} \cos\theta & -\sin\theta & x \\ \sin\theta & \cos\theta & y \\ 0 & 0 & 1 \end{pmatrix}$$

and

$$\mathcal{H}(\xi, \eta, \alpha) = \begin{pmatrix} \cos\alpha & -\sin\alpha & \xi \\ \sin\alpha & \cos\alpha & \eta \\ 0 & 0 & 1 \end{pmatrix}.$$

We define a parameterized density function $\rho(x, y, \theta)$ by identifying

$$\rho(x, y, \theta) \equiv \rho(H'(x, y, \theta)),$$

---

[4] In this context the cost of performing a convolution is considered to be a constant, and we are interested in determining the computational cost as a function of the number of manipulator modules P.

which leads to an explicit form of the convolution product on $SE(2)$,

$$(\rho_1 * \rho_2)(x, y, \theta) = \int_{SE(2)} \rho_1(\mathcal{H})\rho_2(\mathcal{H}^{-1}H')\, d(\mathcal{H}) =$$

$$\int_{-\pi}^{\pi}\int_{-\infty}^{\infty}\int_{-\infty}^{\infty} \rho_1\Big(\xi, \eta, \alpha\Big)\rho_2\Big((x-\xi)c\alpha + (y-\eta)s\alpha, -(x-\xi)s\alpha + (y-\eta)c\alpha, \theta-\alpha\Big) d\xi d\eta d\alpha,$$

where $c\alpha = \cos\alpha$ and $s\alpha = \sin\alpha$.

In general if a subset of $SE(D)$ for $D = 2,3$ is parameterized with $D(D+1)/2$ variables $q_1, ..., q_{D(D+1)/2}$, then to within a constant[5]

$$d(\mathcal{H}(\mathbf{q})) = |\det(\mathcal{J})|dq_1 \cdots dq_{D(D+1)/2},$$

where $\mathcal{J}$ (taken to be either $\mathcal{J}_R$ or $\mathcal{J}_L$) is a $D(D+1)/2 \times D(D+1)/2$ Jacobian matrix of the parameterization. In the case when $D = 2$ the determinant of the Jacobian matrix is unity.

For the following derivation we assume that ρ_1 and ρ_2 are real-valued functions on $SE(D)$, that are nonzero and bounded everywhere, and have "compact support." That is, they vanish outside of a compact (closed and bounded) subset of $SE(2)$, which for simplicity is chosen of the form

$$[x_{min}^{(j)}, x_{max}^{(j)}] \times [y_{min}^{(j)}, y_{max}^{(j)}] \times [-\pi, \pi] \quad \text{for} \quad j = 1, 2.$$

The range of $x - y$ values is chosen to include the support of the workspace density of the concatenated modules.

The convolution product of two such functions on $SE(2)$ can be expressed in the form:

$$(f_1 * f_2)(x, y, \theta) = \int_{-\pi}^{\pi}\int_{y_{min}^{(1)}}^{y_{max}^{(1)}}\int_{x_{min}^{(1)}}^{x_{max}^{(1)}} \rho_1\Big(\xi, \eta, \alpha\Big)\rho_2\Big((x-\xi)\cos\alpha + (y-\eta)\sin\alpha,$$

$$-(x-\xi)\cos\alpha + (y-\eta)\sin\alpha, (\theta-\alpha)\bmod 2\pi\Big)d\xi d\eta d\alpha,$$

where $\bmod 2\pi$ is used here to mean that the difference $\theta - \alpha$ is taken in the range $-\pi$ to π.[6]

12.2.5 Workspace Generation for Planar Manipulators

In Chapter 8 we showed how the convolution product of functions on Lie groups is defined. In Subsection 12.2.2 we showed how convolution of functions on $SE(D)$ can be applied to workspace generation. This section describes the details of a numerical implementation which is based on a description of density as a piecewise constant histogram.

We start this section with a summary of the procedure for generating the workspace of a discretely actuated planar manipulator:

---

[5] For compact groups the constant is set so that $\int_G d(g) = 1$, but $SE(D)$ is not compact and so there is no unique way to scale the volume element.

[6] This is different than the standard definition which would put the result in $[0, 2\pi)$.

1. The manipulator is divided into P kinematically independent modules. The modules are numbered from 1 to P, starting at the base with module 1 and increasing up to the most distal module, module P. For each module one frame is attached to the base of the module and a second one to the top, where the next module is attached. Modules can have a parallel kinematic structure internally, but the modules are all cascaded in a serial way.

2. For each module, $(p = 1, \dots, P)$ the finite set $\mathcal{S}_p = \{H_{i_p}\}$ of all frames the top of module p can attain relative to the bottom is determined. That is, the position and orientation of the upper frame with respect to the lower frame is described by homogeneous transformations, $H_{i_p} \in \mathcal{S}_p \subset SE(2)$, while the module undergoes all possible discrete configurations.

3. A compact subset $\mathcal{C}_p \subset SE(2)$ is chosen that contains all of the discrete sets $\mathcal{S}_p$. The subset $\mathcal{C}_p$ is discretized and a piecewise constant density function/histogram ρ_p is calculated for each $\mathcal{S}_p$ to represent this information.

4. Finally, the discrete density functions are convolved in the order

$$\rho_W = \rho_1 * \rho_2 * \dots * \rho_P$$

to yield an approximation to the density function of the workspace.

The implementation of Steps 1 and 2 of the workspace generation procedure depends on the architecture of the manipulator. Generating these sets is a simple task if the manipulator can be separated into a sufficiently large number of kinematically independent modules of simple structure. We assume that the discrete sets $\mathcal{S}_p$ are calculated efficiently either numerically or in closed form for all modules of the manipulator.

For Step 3 and 4 we have to define discretized density functions used to represent each set $\mathcal{S}_p$. The discretization of the parameter space is described in the following subsection for the case $SE(2)$, and the subsection after that describes the resulting discrete form of the convolution.

Discretization of Parameter Space

For each set $\mathcal{S}_p$ arising in Step 3 of the procedure, the support of ρ_P in terms of the parameters (x, y, θ) is of the form $[x_{\min}^{(p)}, x_{\max}^{(p)}] \times [y_{\min}^{(p)}, y_{\max}^{(p)}] \times [-\pi, \pi]$. This set of parameters is divided into elements/voxels of equal size. Note, that in the following discussion the superscript "p" is dropped, unless we refer to a particular set $\mathcal{S}_p$. We choose the resolution in the x and y-directions to be identical, i.e. $\Delta x = \Delta y$, and explain below how to choose the resolution in the angular direction such that the resulting errors from inaccuracy in position and rotation are of the same order.

We denote by N_1, N_2 and M the number of discretizations in the x-, y- and θ-directions, respectively, i.e. $\Delta x = \frac{(x_{\max} - x_{\min})}{N_1}$, $\Delta y = \frac{(y_{\max} - y_{\min})}{N_2}$, $\Delta\theta = \frac{(2\pi - 0)}{M}$, and choose N_1 and N_2 such that $\Delta x = \Delta y$.[7] Each voxel is a volume element of the form $[x_{\min} + i\Delta x, x_{\min} + (i+1)\Delta x] \times [y_{\min} + j\Delta y, y_{\min} + (j+1)\Delta y] \times [-\pi + k\Delta\theta, -\pi + (k+1)\Delta\theta]$ with center coordinates $(x_i, y_j, \theta_k) = (x_{\min} + (i+0.5)\Delta x, y_{\min} + (j+0.5)\Delta y, -\pi + (k + 0.5)\Delta\theta)$.

To characterize the error resulting from discretization we consider a homogeneous transform $H \in SE(2)$ corresponding to some exact parameters: $H = H(x, y, \theta)$. H is

---

[7]To get exact equality it is usually necessary to slightly change one of the workspace boundaries, e.g. to slightly increase $y_{\max}$.

then compared to the homogeneous transformation $\hat{H}$ corresponding to the rounded coordinates: $\hat{H} = H(\hat{x}, \hat{y}, \hat{\theta})$. By definition (x, y, θ) differs from $(\hat{x}, \hat{y}, \hat{\theta})$ at most by $(\frac{\Delta x}{2}, \frac{\Delta y}{2}, \frac{\Delta \theta}{2})$.

If we apply H and $\hat{H}$ to any vector $\mathbf{v} \in \mathbb{R}^2$ and use $(R, \mathbf{b})$ and $(\hat{R}, \hat{\mathbf{b}})$ to denote the rotation and translation of H and $\hat{H}$ respectively (so that, for instance, $H \cdot \mathbf{v} = R\mathbf{v} + \mathbf{b}$), then the error $\|H \cdot \mathbf{v} - \hat{H} \cdot \mathbf{v}\|$ is bounded as

$$\|H \cdot \mathbf{v} - \hat{H} \cdot \mathbf{v}\| = \|(R\mathbf{v} + \mathbf{b}) - (\hat{R}\mathbf{v} + \hat{\mathbf{b}})\|$$
$$= \|(R\mathbf{v} - \hat{R}\mathbf{v}) + (\mathbf{b} - \hat{\mathbf{b}})\|$$
$$\leq \underbrace{\|R\mathbf{v} - \hat{R}\mathbf{v}\|}_{\text{rot. part}} + \underbrace{\|\mathbf{b} - \hat{\mathbf{b}}\|}_{\text{trans. part}}$$
$$\leq \frac{\Delta \theta}{2} \|\mathbf{v}\| + \left(\left(\frac{\Delta x}{2}\right)^2 + \left(\frac{\Delta y}{2}\right)^2\right)^{\frac{1}{2}}$$

If V is a set of vectors, then the maximal difference in displacement between transformed versions of a vector $\mathbf{v} \in V$ after transformation by H and $\hat{H}$ is bounded by

$$\max_{\mathbf{v} \in V} \|H \cdot \mathbf{v} - \hat{H} \cdot \mathbf{v}\| = \frac{\Delta \theta}{2} \max_{\mathbf{v} \in V} \|\mathbf{v}\| + \left(\left(\frac{\Delta x}{2}\right)^2 + \left(\frac{\Delta y}{2}\right)^2\right)^{\frac{1}{2}} \tag{12.1}$$

$$= \frac{\Delta \theta}{2} \max_{\mathbf{v} \in V} \|\mathbf{v}\| + \frac{\Delta x}{\sqrt{2}}. \tag{12.2}$$

The last equality holds because $\Delta x = \Delta y$.

In the case of convolution of two workspace densities we can use (12.1) to balance the error between the angular and translational part. We denote the smallest simply connected continuous regions containing the sets $\mathcal{S}_1$ and $\mathcal{S}_2$ as $\mathcal{C}_1$ and $\mathcal{C}_2$ respectively. To discretize $\mathcal{C}_1$ we first select a maximal acceptable error e (which results from a trade-off between memory and accuracy). The resolution parameters $\Delta x^{(1)}, \Delta y^{(1)}, \Delta \theta^{(1)}$ of $\mathcal{C}_1$ are then determined such that the two parts of the error are of the same order and add up to e, i.e.

$$\frac{\Delta \theta^{(1)}}{2} \max_{\mathbf{v} \in T(\mathcal{C}_2)} \|\mathbf{v}\| \overset{!}{=} \frac{\Delta x^{(1)}}{\sqrt{2}} \overset{!}{=} \frac{e}{2},$$

where $T(\mathcal{C}_2) \subset \mathbb{R}^2$ is the union of all projections of constant theta "slices" of $\mathcal{C}_2$ onto the $x - y$ plane.

This results in the choice

$$\Delta x^{(1)} = \Delta y^{(1)} = \frac{e}{\sqrt{2}}, \qquad \Delta \theta^{(1)} = \frac{e}{\displaystyle\max_{\mathbf{v} \in T(\mathcal{C}_2)} \|\mathbf{v}\|}. \tag{12.3}$$

The step sizes for the discretization of $\mathcal{C}_2$ are chosen analogously. The error bound in (12.1) was derived in [26].

Numerical Convolution of Histograms on $SE(2)$

Our goal is to store density functions in the form of piecewise constant histograms, i.e., we only want to store average values for each voxel. This section presents convolution in a form applicable to histograms on $SE(2)$. As a first step, the integral from the previous section

$$f_3(x,y,\theta) = \int_{x_{min}^{(1)}}^{x_{max}^{(1)}} \int_{y_{min}^{(1)}}^{y_{max}^{(1)}} \int_{-\pi}^{\pi} f_1(\xi,\eta,\alpha) f_2\Big((x-\xi)\cos\alpha + (y-\eta)\sin\alpha,$$

$$-(x-\xi)\sin\alpha + (y-\eta)\cos\alpha, (\theta-\alpha)\bmod 2\pi\Big)\, d\xi d\eta d\alpha$$

is approximated by a Riemann-Stieltjes sum:

$$(f_1 * f_2)(x,y,\theta) \approx \Delta\xi\Delta\eta\Delta\alpha \sum_{l=0}^{N_1} \sum_{m=0}^{N_2} \sum_{n=0}^{M} f_1(\xi_l,\eta_m,\alpha_n) f_2\Big((x-\xi_l)\cos\alpha_n + (y-\eta_m)\sin\alpha_n,$$

$$-(x-\xi_l)\sin\alpha_n + (y-\eta_m)\cos\alpha_n, (\theta-\alpha_n)\bmod 2\pi\Big).$$

Although the right hand side of the equation is approximated by a discrete sum, the function f_2 is required to exist for any values of arguments because its arguments generally do not coincide with points on the grid of any discretization. We therefore approximate the functions f_2 for any real-valued arguments by interpolation using function values at neighboring discrete points. Since this problem is frequently encountered in many applications, there exist many different strategies. The simplest strategy is that the function $f_2(x,y,\theta)$ is approximated by the value of $f_2(x_i,y_j,\theta_k)$ (i.e., at the closest point on the grid). Because this can lead to large round-off errors, we instead use linear interpolation. For each coordinate (x,y,θ) we find the indices i,j,k in the grid such that $x_i \leq x \leq x_{i+1}$, $y_j \leq y \leq y_{j+1}$, etc., and define the ratios

$$t = \frac{x-x_i}{\Delta x}, \qquad u = \frac{y-y_j}{\Delta y}, \qquad v = \frac{\theta-\theta_k}{\Delta\theta}.$$

The value $f_2(x,y,\theta)$ is then interpolated from the values at eight discrete points (8-point interpolation):

$$\begin{aligned} f_2(x,y,\theta) = &(1-t)(1-u)(1-v)\, f_2(x_i,y_j,\theta_k) + \\ &(t)(1-u)(1-v)f_2(x_{i+1},y_j,\theta_k) + \\ &(1-t)(u)(1-v)\, f_2(x_i,y_{j+1},\theta_k) + \\ &\dots + (t)(u)(v)\, f_2(x_{i+1},y_{j+1},\theta_{k+1}). \end{aligned}$$

While this constitutes an approximation, as we shall see in the next subsection, this produces acceptable results.

12.2.6 Numerical Results for Planar Workspace Generation

In this section we present numerical results for the generation of workspaces using the methods presented earlier in this chapter.[8]

The algorithm is implemented on a SUN SPARCstation 5, 110 MHz, in the C programming language. Figures were made using Mathematica version 3.0. The algorithm is applied to a version of the discretely actuated manipulator shown in Figure 12.3. The manipulator in Figure 12.3 consists of 10 modules composed of three legs each, where each leg has two bits (four states). Hence each module has $4^3 = 64$ discrete states. In our example we consider a manipulator consisting of only eight identical modules of this

---

[8]We thank Dr. Imme Ebert-Uphoff for generating the numerical results that appear here.

kind, resulting in a manipulator with $(64)^8 \approx 2.8 \cdot 10^{14}$ states. Unless specified otherwise, the width of each platform is chosen as $w = 0.2$ compared to the minimal and maximal actuator lengths of $q_1 = 0.15$ and $q_2 = 0.22$.

For this manipulator we calculate the workspace density corresponding to only the first two modules by brute force $(64^2 = 4096$ states), which we will refer to as W2 in the following. Since all modules are identical, convolution of this density with itself leads to the density of the four-module workspace, W4. Convolving this workspace again with itself leads to the workspace density of the whole manipulator, W8.

For the workspace of four modules it is possible to calculate the results using brute force $((64)^4 \approx 1.7 \cdot 10^7$ states). In the following we first compare the results of this approach to the results obtained from convolution and quantify the error. Afterwards we show results for the workspace of the 8-module manipulator, which cannot simply be calculated by brute force $(2.8 \cdot 10^{14}$ states).

Figure 12.5(a) is generated by calculating the histogram directly (brute force) and Figure 12.5(b) is generated by convolution of the density function corresponding to two modules with itself. In the brute force calculation we use linear (8-point) smoothing when incrementing voxels, because this makes the raw data less sensitive to small shifts in the way the grid is superimposed. eight-point interpolation is also used when evaluating the discrete density functions for the discrete convolution.

In each figure the z axis corresponds to the angle θ (orientation of end-effector), and the point density per voxel is represented by a gray scale (black representing very high density). The base of the manipulator lies at the origin of the coordinate system. To enhance differences in low density areas, we chose a nonlinear gray scale: Each density value is normalized to a value between 0 and 1 (divided by the largest density value in the drawing), and the fourth root of this value is displayed as gray value for W4 (the eighth root for W8).

Convolving the density array in Figure 12.5(a) with itself results in W8. The result is shown in Figure 12.6. Figures 12.7(a),(b) show the workspace of the same manipulator as in Figure 12.6, if the maximal actuator length is decreased to $q_2 = 0.2$ or increased to $q_2 = 0.25$, respectively.

Error Measures:
To quantify the error resulting from convolution for W4 (as compared to direct calculation) three different error measures are used. The first two measures compare the density resulting from the brute force approach, ρ, with the density from numerical convolution, $\tilde{\rho}$:

$$E_1 = \frac{\sum\limits_{i,j,k} |\rho(x_i, y_j, \theta_k) - \tilde{\rho}(x_i, y_j, \theta_k)|}{\sum\limits_{i,j,k} |\rho(x_i, y_j, \theta_k)|},$$

$$E_2 = \frac{\sum\limits_{i,j,k} |\rho(x_i, y_j, \theta_k) - \tilde{\rho}(x_i, y_j, \theta_k)|^2}{\sum\limits_{i,j,k} |\rho(x_i, y_j, \theta_k)|^2}.$$

The third measure is a function of the shapes of the workspaces (by shape we mean the set of all voxels with nonzero density), by counting the number of voxels which belong to one of the workspaces, but not to the other:

$$E_3 = \frac{\# \text{ for which } ((\rho > 0)\&(\tilde{\rho} = 0)) \text{ or } ((\rho = 0)\&(\tilde{\rho} > 0))}{\# \text{ for which } (\rho > 0)}.$$

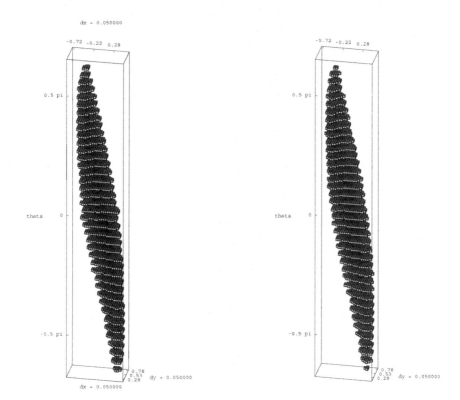

Fig. 12.5. Workspace Density for a 4-Module Manipulator: (a) Calculated by Brute Force; (b) Using Convolution. Scale is $\Delta x = \Delta y = 0.05$ and $\Delta\theta = \pi/30$. (Reprinted from Chirikjian, G.S., Ebert-Uphoff, I., "Numerical Convolution on the Euclidean Group with Applications to Workspace Generation," *IEEE Transactions on Robotics and Automation*, 14(1): 123–136, 1998, with permission from IEEE)

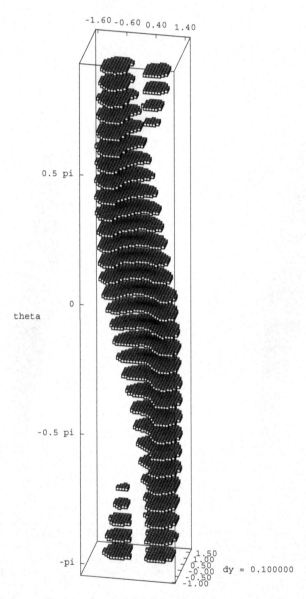

Fig. 12.6. Workspace Density for an 8-Module Manipulator Using Convolution. Scale is $\Delta x = \Delta y = 0.1$ and $\Delta \theta = \pi/15$. (Reprinted from Chirikjian, G.S., Ebert-Uphoff, I., "Numerical Convolution on the Euclidean Group with Applications to Workspace Generation," *IEEE Transactions on Robotics and Automation*, 14(1): 123 – 136, 1998, with permission from IEEE)

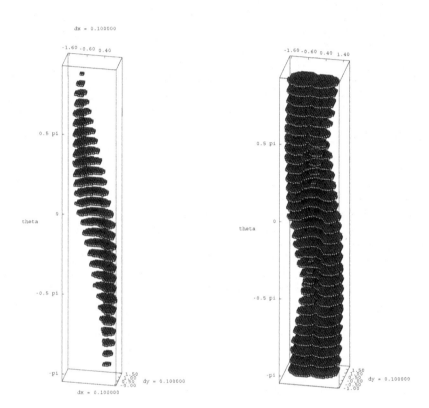

Fig. 12.7. Workspace Density for an 8-Module Manipulator with Kinematic Parameters: (a) $q_2 = 0.2$; (b) $q_2 = 0.25$. Scale is $\Delta x = \Delta y = 0.1$ and $\Delta \theta = \pi/15$. (Reprinted from Chirikjian, G.S., Ebert-Uphoff, I., "Numerical Convolution on the Euclidean Group with Applications to Workspace Generation," *IEEE Transactions on Robotics and Automation*, 14(1): 123–136, 1998, with permission from IEEE)

Tables 12.1 and 12.2 show error, memory, and time, for the approximation of the manipulator of 4 modules described above, if different discretizations are chosen for the representation of the workspace. In particular, the time listed is the net time to calculate workspace W4 by one numerical convolution of W2 with itself. The memory listed is the amount of memory needed to represent either W2 or W4 (whichever has more voxels). Since voxels are consolidated after convolution, it is possible for W4 to require fewer voxels than W2.

In all simulations the discretization is chosen as follows:
(a) choose Δx_2 for workspace W2 and Δx_4 for W4,
(b) calculate the angular discretization, $\Delta\theta_2$, of W2 according to Equation (12.3),
(c) define the scale factor f as $f = \frac{\Delta x_4}{\Delta x_2}$, and determine $\Delta\theta_4$ such that f is also the factor for the angular discretization, i.e. $\Delta\theta_4 = f \cdot \Delta\theta_2$. As a practical matter, integer scale factors are the easiest to work with. However it is possible to consolidate voxels using more general scale factors if interpolation is used.

Results and Analysis:

(1) For the case considered, W4 can be calculated from W2 with one convolution in less than 12 minutes with an error smaller than 10%. W8 is calculated from W4 with one convolution within another 9 minutes. (No error is reported, since comparison with brute force results is impossible for W8).

(2) The best results in this particular example were obtained using a factor of $f \approx 2$, i.e. if the resolution is twice as fine before convolution than after (see Table 12.1). In general, an appropriate scale factor can be estimated even if there is no brute force data against which to compare the results of convolution. This can be achieved by examining the relative error of sequences of density functions generated using convolution, and picking the scale factor for which the relative error is smallest.

(3) For a constant factor $f = 2$, the error decreases slowly with higher discretization (see Table 12.2).

(4) The result of the discrete convolution is closer to a brute-force calculation if 8-point interpolation is used to generate the brute force results, as compared to no interpolation. Table 12.3 lists the error between brute force (BF) and discrete (numerical) convolution (DC) if different numbers of points are used in the interpolation, together with the corresponding computation time for either method. (1-point means no interpolation at all, 2-point means interpolation only in the angle, 8-point means interpolation in all three coordinate axes of the grid.)

One would expect the error to reduce more or less monotonically with finer discretization. As can be seen in Table 12.2 this is true for the shape error, E_3. For error E_1 and E_2 the general tendency is to decrease for higher resolution, but this does not happen strictly monotonically. One likely reason for this is sensitivity of density to shifting of the grid by a tiny amount. This effect only appears if the density is distributed very unevenly and differs considerably in neighboring voxels. Hence we expect this effect also to disappear for manipulators of higher resolution, i.e. if the actuators have a higher resolution or a larger number of modules are considered as a single unit. This expectation is supported by the data in Table 12.3 which indicates a dramatic convergence of results obtained by brute force and convolution as the number of actuator states in each leg is doubled.

It is worth noting that for manipulators composed of P modules which each have a very large number of states, the $\mathcal{O}(P)$ or $\mathcal{O}(\log P)$ calculations required for P convolutions can be a significant savings over $\mathcal{O}(K^p)$. However, in the example presented here, the 64 frames reachable by each module are relatively sparse. This means that the actual time required to perform the $\log_2(8) = 3$ convolutions would far exceed the time required to generate the density function for half of the manipulator (which requires $(64)^{8/2}$ kinematic calculations) and perform one convolution. Thus in this case, $\mathcal{O}(K^{P/2})$ is better than $\mathcal{O}(\log P)$. In either case convolution plays a critical role in avoiding the $\mathcal{O}(K^P)$ calculations which cannot be performed in a reasonable amount of time.

Finally, we note that the purpose of this implementation is to show that the concept of numerical convolution on $SE(D)$ works and provides usable results. The error analysis resulting in (12.1) serves as a guide for balancing the number of discretizations in position and orientation. This direct numerical convolution also gives a sense for what kinds of errors are acceptable when using FFT algorithms. That is, if we know a priori that Euclidean group convolutions will not result in exact results, it becomes less important for the FFT to reproduce convolutions exactly. Some error is tolerable. For the spatial ($D = 3$) case, it is not possible to do brute force convolutions with the fine resolution we desire, and this indicates a need for $SE(3)$ FFT algorithms.

factor f	($\varDelta x_2$, $\varDelta\theta_2$)	($\varDelta x_4$, $\varDelta\theta_4$)	E_1	E_2	E_3	Max. Memory	Time
1	(0.0500, 0.101)	(0.050, 0.101)	18.95%	17.08%	96.42%	48 KB	2.4 min
2	(0.0250, 0.051)	(0.050, 0.101)	8.27%	9.64%	9.42%	442 KB	11.5 min
4	(0.0125, 0.025)	(0.050, 0.101)	21.82%	27.05%	30.18%	3546 KB	70.8 min

Table 12.1: Results for Fixed Discretization ($\varDelta x_4, \varDelta\theta_4$) and Varying Factor f. (Reprinted from Chirikjian, G.S., Ebert-Uphoff, I., "Numerical Convolution on the Euclidean Group with Applications to Workspace Generation," *IEEE Transactions on Robotics and Automation*, 14(1): 123 – 136, 1998, with permission from IEEE)

factor f	($\varDelta x_2$, $\varDelta\theta_2$)	($\varDelta x_4$, $\varDelta\theta_4$)	E_1	E_2	E_3	Max. Memory	Time
2	(0.0500, 0.101)	(0.1000, 0.203)	13.77%	15.37%	13.65%	54 KB	23 sec
2	(0.0375, 0.077)	(0.0750, 0.153)	7.48%	7.91%	13.58%	131 KB	91 sec
2	(0.0250, 0.051)	(0.0500, 0.101)	8.27%	9.64%	9.42%	442 KB	11.5 min
2	(0.0188, 0.038)	(0.0375, 0.076)	7.46%	9.05%	7.68%	1056 KB	47.1 min
2	(0.0125, 0.025)	(0.0250, 0.050)	5.20%	6.18%	6.47%	3546 KB	7hrs 11min

Table 12.2: Results for Fixed Factor $f = 2$ and Varying Discretization ($\varDelta x_4, \varDelta\theta_4$). (Reprinted from Chirikjian, G.S., Ebert-Uphoff, I., "Numerical Convolution on the Euclidean Group with Applications to Workspace Generation," *IEEE Transactions on Robotics and Automation*, 14(1): 123 – 136, 1998, with permission from IEEE)

12.3 Inverse Kinematics of Binary Manipulators: The Path-of-Probability Algorithm

The concept of the manipulator workspace density function is useful in solving the inverse kinematics problem for discretely-actuated manipulators with many states [26, 28]. If we were to try to solve the inverse kinematics problem by evaluating the manipulator forward kinematics for all K^P states, the computational cost would be prohibitive. If we have a cascade of manipulator workspace density functions corresponding to each

Four States per Actuator:

$(\Delta x_2$, $\Delta\theta_2)$	$(\Delta x_4$, $\Delta\theta_4)$	Brute Force	Disc. Conv.	E_1	E_2	E_3	T_{BF}	T_{DC}
(0.0200, 0.041)	(0.040, 0.082)	1-point	2-point	10.56%	13.26%	13.58%	4.8 min	12.7 min
(0.0200, 0.041)	(0.040, 0.082)	1-point	8-point	10.90%	14.56%	9.83%	4.8 min	20.8 min
(0.0200, 0.041)	(0.040, 0.082)	8-point	8-point	7.72%	9.58%	8.27%	12.1 min	33.9 min
(0.0250, 0.051)	(0.050, 0.101)	1-point	2-point	11.91%	15.24%	16.74%	4.8 min	4.3 min
(0.0250, 0.051)	(0.050, 0.101)	1-point	8-point	9.78%	12.58%	9.60%	4.8 min	7.2 min
(0.0250, 0.051)	(0.050, 0.101)	8-point	8-point	8.27%	9.64%	9.42%	16.8 min	11.5 min

Two States per Actuator:

$(\Delta x_2$, $\Delta\theta_2)$	$(\Delta x_4$, $\Delta\theta_4)$	brute force	disc. conv.	E_1	E_2	E_3	T_{BF}	T_{DC}
(0.0250, 0.051)	(0.050, 0.101)	1-point	2-point	151.22%	235.52%	75.43%	0.1 sec	3.4 min
(0.0250, 0.051)	(0.050, 0.101)	1-point	8-point	121.50%	158.80%	62.51%	0.3 sec	12.5 min
(0.0250, 0.051)	(0.050, 0.101)	8-point	8-point	37.70%	21.64%	21.64%	0.7 sec	20.3 min

Table 12.3: Results for Different Types of Interpolation and Different Number of States per Actuator. (Reprinted from Chirikjian, G.S., Ebert-Uphoff, I., "Numerical Convolution on the Euclidean Group with Applications to Workspace Generation," *IEEE Transactions on Robotics and Automation*, 14(1): 123–136, 1998, with permission from IEEE)

of P sections of the manipulator, this exponential complexity is reduced to a problem that is linear in P.

Let $\rho_1, ..., \rho_P$ be the workspace density functions for each of the P sections of the manipulator. Instead of computing densities from the base to the distal end of the manipulator, the calculation can be performed in reverse order. That is,

$$\rho^{(P)} \doteq \rho_P,$$

$$\rho^{(P-1)} \doteq \rho_{P-1} * \rho_P,$$

$$\rho^{(P-k)} \doteq \rho_{P-k} * \cdots * \rho_P,$$

$$\rho^{(1)} \doteq \rho_1 * \cdots * \rho_P.$$

$\rho^{(1)}$ is then the workspace density for the whole manipulator.

Now for $k \in \{1, ..., P\}$, let g_k be the transformation that relates the distal end of the kth segment to its own base. The position of the top of the kth segment relative to the base of the whole manipulator is then

$$g^{(k)} \doteq g_1 \circ g_2 \circ \cdots \circ g_k$$

and the position and orientation of the distal end of the manipulator relative to the distal end of the kth segment is

$$(g^{(k)})^{-1} \circ g^{(P)} = g_{k+1} \circ g_{k+2} \circ \cdots \circ g_P.$$

There are K possible states for each g_k. Using the information in the cascade of density functions $\rho^{(2)}, ..., \rho^{(P-1)}$, we can sequentially choose states of each section which, at each instance, maximize the probability density around the particular frame of reference that we seek to reach.

In other words, given that we want the end of the manipulator to reach $g_{des} \in SE(D)$, we start at the base of the manipulator and ask which state of segment 1 maximizes $\rho^{(2)}((g_{(1)})^{-1} \circ g_{des})$. After searching through all K possible values of g_1, and the optimal $g^{(1)} = g_1$ is fixed, we proceed up the manipulator one unit. That is, we next calculate $\rho^{(3)}((g^{(2)})^{-1} \circ g_{des})$. Since g_1 is fixed, K values of $g^{(2)} = g_1 \circ g_2$ are searched until the value of g_2 that maximizes is found. This procedure is performed by sequentially maximizing $\rho^{(k+1)}((g^{(k)})^{-1} \circ g_{des})$ for all $k \in \{1, 2, ..., P-1\}$. When $k = P$, the one out of K values of g_P that minimizes some measure of distance

$$C = d(g_{des}, g_1 \circ g_2 \circ \cdots \circ g_P)$$

is chosen.

The procedure we have described here is a way of specifying a state of the whole manipulator in $\mathcal{O}(P)$ arithmetic operations such that the distal end reaches g_{des} approximately. This is called the *Path-of-Probability* (or POP) *Algorithm*.

12.4 Symmetries in Workspace Density Functions

When a manipulator arm possesses geometrical symmetries, this will induce certain symmetries in the manipulator's density function. As an example, consider the planar 3-bit platform in Figure 12.1. This manipulator can be rotated by 180 degrees about its center through the axis pointing out of the plane of the Figure, and if all the legs have the same two length states, the result will be the same manipulator with the same eight states. The only difference is that the configurations will be labeled differently, e.g., 110 might become 011. Hence, performing the same operations on the corresponding density function (which in this case is a sum of eight Dirac delta functions) should, in some sense, preserve the density function as long as we take into account the fact that frames that were originally moving freely at the top are now fixed, and frames originally fixed at the base are now free to move. In the following subsection we quantify how such discrete symmetries in the arm induce symmetries in the density function. Then in Subsection 12.4.2 we show how continuous symmetries reduce the density function of a spatial manipulator to a form which does not depend on all six parameters of $SE(3)$.

12.4.1 Discrete Manipulator Symmetries

Let us now consider in greater detail the density function of the symmetric 3-bit planar binary platform manipulator. This is a platform where all the states are uniform, i.e., the zero (one) state corresponds to the same length for all of the legs, and so rotating this platform about its center by 180 degrees results in functionally the same device. The density function is of the form

$$f(g) = \sum_{i=0}^{1}\sum_{j=0}^{1}\sum_{k=0}^{1} \delta(g_{(ijk)}^{-1} \circ g)$$

where ijk is a binary number describing one of the eight frames.

Now imagine shifting each of the original eight configurations of the platform so that the top plate in the platform now resides where the fixed base was previously. If a delta function is placed at the frame attached to what used to be the base of the manipulator for each of the eight configurations we get the new density function

$$\tilde{f}(g) = \sum_{i=0}^{1}\sum_{j=0}^{1}\sum_{k=0}^{1} \delta((g_{(ijk)}^{-1})^{-1} \circ g).$$

The second inverse is introduced because instead of the top of the manipulator reaching $g_{(ijk)}$, the bottom is now reaching $g_{(ijk)}^{-1}$ since the top is fixed at the identity frame. In general for a group G, and any $h \in G$, the equality $(h^{-1})^{-1} = h$ holds. Also, the delta function has the special properties $\delta(g^{-1}) = \delta(g)$ and $\delta(h \circ g) = \delta(g \circ h)$. Applying these three rules we observe that

$$\tilde{f}(g) = \sum_{i=0}^{1}\sum_{j=0}^{1}\sum_{k=0}^{1} \delta(g_{(ijk)}^{-1} \circ g^{-1}) = f(g^{-1}). \tag{12.4}$$

This is true independent of any symmetry in the manipulator. Furthermore, the relationship $\tilde{f}(g) = f(g^{-1})$ holds for a cascade of platforms since the density function for the cascade will be a convolution, and it is easy to check that

$$(\tilde{f}_2 * \tilde{f}_1)(g) = (f_1 * f_2)(g^{-1}).$$

Hence, what holds for a single platform holds for a cascade due to the fact that $SE(n)$ is unimodular, and the resulting properties of convolution. The function $\tilde{f}(g)$ in (12.4) is shown in Figure 12.8.

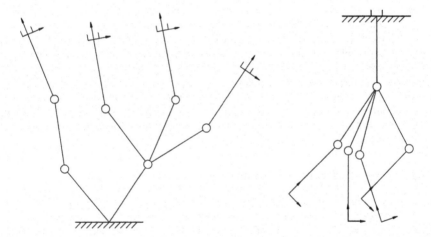

Fig. 12.8. Workspace Density of a Manipulator with Distal End Fixed and Proximal End Free

In the case when rotational symmetry about the center of the manipulator exists we can say more. In this case, we observe that if any of the original configurations of the platform had been rotated by 180 degrees about the e_1 axis pointing out of the page *at the center of its base*, the resulting configuration would coincide with one of the shifted configurations discussed earlier. However the frames attached to the platform would differ by an orientation of 180 degrees about e_1.

Let $f(g)$ be the sum of eight Dirac delta functions on $SE(2)$ which constitute the workspace of the 3-bit manipulator in Figure 12.1, which is depicted graphically in Figure 12.9 (left). Let h_0 be a 180 rotation about c_1 (the axis pointing out of the plane) in Figure 12.9. Then $h_0 = h_0^{-1}$. The version of the density function corresponding to the case when the original platform is rotated at its base (Figure 12.9 (middle)) is $f(h_0 \circ g)$. An additional rotation by h_0 of the frames reachable at the distal end of the platform

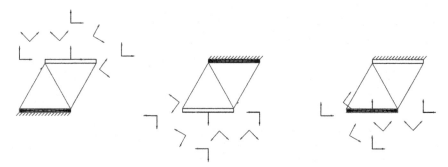

Fig. 12.9. Frames Reachable by a 3-Bit Manipulator with C_2 Rotational Symmetry: (left) Proximal End Fixed; (middle) Proximal End Rotated by 180 Degrees and Fixed; (right) Distal End Fixed

changes this to $f(h_0 \circ g \circ h_0)$. In the case of a manipulator with rotational symmetry about its center, this twice rotated density function will be the same as $\tilde{f}(g)$ (which is shown for this example platform in Figure 12.9 (right)). Hence, we write

$$f(g^{-1}) = f(h_0 \circ g \circ h_0). \tag{12.5}$$

Using the notation $g = (\mathbf{a}, A)$, this is written in the equivalent forms

$$f(-A^T\mathbf{a}, A^T) = f(-\mathbf{a}, A) \quad f(A^{\frac{1}{2}}\mathbf{a}, A) = f(A^{-\frac{1}{2}}\mathbf{a}, A^{-1}). \tag{12.6}$$

It makes intuitive sense that if $f(g)$ has this symmetry, then so too will $(f * f)(g)$ because the concatenation of two identical manipulator units with 180 degree rotational symmetry will also have this symmetry. This intuition is verified by observing that

$$(f * f)(g^{-1}) = \int_G f(k)\, f(k^{-1} \circ g^{-1})\, dk = \int_G f(h_0 \circ k^{-1} \circ h_0)\, f(h_0 \circ g \circ k \circ h_0)\, dk,$$

which, after the change of variables $k' = h_0 \circ g \circ k \circ h_0$, takes the form

$$\int_G f((k')^{-1} \circ (h_0 \circ g \circ h_0))\, f(k')\, dk' = (f * f)(h_0 \circ g \circ h_0).$$

Thus the intuition is correct. A direct consequence of this is that when generating the density function for a manipulator which is a concatenation of identical modules, we only need to calculate the density function for half of its support, since the other half can be reconstructed from (12.5) or (12.6).

Other discrete symmetries can exist as well. For instance, if a manipulator is composed of a cascade of binary Stewart/Gough platforms with three-fold symmetry about the $\mathbf{e}_3$ axis pointing from the center of the base plate through the center of the top one (as in Figure 12.2), the density function will reflect this symmetry with the constraint

$$f(g) = f(h_1 \circ g), \tag{12.7}$$

where h_1 is a rotation by $2\pi/3$ around the $\mathbf{e}_3$ axis pointing along the backbone of the manipulator in its reference state (where all bit values are 0).

These symmetries can be used to speed up the direct computation of convolution integrals. For instance, if $f_1(g)$ has the kind of symmetry in (12.7), then it follows that

$$(f_1 * f_2)(h_1 \circ g) = \int_G f_1(k)\, f_2(k^{-1} \circ h_1 \circ g)\, dk$$

$$= \int_G f_1(h_1 \circ r)\, f_2(r^{-1} \circ g)\, dr$$

$$= (f_1 * f_2)(g).$$

All that has been done here is to use the symmetry of $f_1(g)$ and the change of variables $r = h_1^{-1} \circ k$, and from this it is clear that $(f_1 * f_2)(g)$ inherits the symmetry of $f_1(g)$ (apparently regardless of any symmetries of $f_2(g)$).

More generally, if H is any cyclic subgroup of $G = SE(3)$ and h_1 is the generator of H with (12.7) holding, then the computation of the convolution integral can be sped up by a factor of $|H|$ by evaluating $f_1(g)$ and $(f_1 * f_2)(g)$ only at one representative of each of the cosets $\sigma \in H \backslash G$ instead of evaluating them for all $g \in G$. To see this, use

$$\int_G f(g)\, dg = \sum_{h \in H} \int_{\sigma \in H \backslash G} f(h \circ g_\sigma)\, d_{H \backslash G}(\sigma),$$

for any $g_\sigma \in \sigma$ where $G = SE(3)$ and H a finite subgroup of G, to rewrite the convolution integral

$$(f_1 * f_2)(g_2) = \int_G f_1(g_1)\, f_2(g_1^{-1} \circ g_2)\, dg_1$$

as

$$(f_1 * f_2)(g_2) = \sum_{h \in H} \int_{\sigma \in H \backslash G} f_1(h \circ g_\sigma)\, f_2((h \circ g_\sigma)^{-1} \circ g_2)\, d_{H \backslash G}(\sigma). \qquad (12.8)$$

Since $H \backslash G$ has the same dimension as G, and can be viewed as a measurable portion of G, it makes sense that $d_{H \backslash G}(\sigma) = d(g_\sigma)$ where $d(g) = dg$ is just the integration measure for $SE(3)$. And the only difference between integrating over G and $H \backslash G$ in this case is the bounds of integration for the parameters.

Using the fact that both $f_1(g)$ and $(f_1 * f_2)(g)$ have symmetry (12.7), we can calculate (12.8) at $g_2 = g_{\sigma'}$ where $\sigma' \in H \backslash G$ as

$$(f_1 * f_2)(g_{\sigma'}) = \int_{\sigma \in H \backslash G} f_1(g_\sigma) \left(\sum_{h \in H} f_2(g_\sigma^{-1} \circ h^{-1} \circ g_{\sigma'}) \right) d(g_\sigma).$$

The calculation of

$$f_2'(g_\sigma, g_{\sigma'}) = \sum_{h \in H} f_2(g_\sigma^{-1} \circ h^{-1} \circ g_{\sigma'})$$

can be performed in $\mathcal{O}(|H| \cdot (N/|H|)^2)$ computations where N is the number of discretizations of G.

Integrating over $H \backslash G$ is $|H|$ times faster than integrating over G. The combination of faster integration (due to a reduction in the domain of integration by a factor of $|H|$) and evaluation of the product on a domain reduced by a factor of $|H|$ leads to a speed-up by a factor of $|H|^2$ in all computations once $f_2'(g_\sigma, g_{\sigma'})$ is known. Hence, the limiting calculation is $f_2'(g_\sigma, g_{\sigma'})$, and a speed-up by a factor of $|H|$ is realized. This savings is not nearly as helpful as using the FFT for the motion group.

12.4.2 3D Manipulators with Continuous Symmetries: Simplified Density Functions

Consider a manipulator with a rotary actuator at the base as shown in Figure 12.10. As with all manipulators, the workspace can be viewed as the support of a workspace density function, $f(g)$. However, if position is described using polar or spherical coordinates, the density function for this manipulator will have no dependence on the azimuthal angle ϕ. Likewise, if a rotary actuator is placed at the distal end of the arm, it will have the effect or averaging over the γ Euler angle, and so instead of considering a density function on the six-dimensional space $\mathbb{R}^3 \times SO(3)$, we can instead deal with the four-dimensional space $\mathbb{R}^2 \times \mathbb{S}^2$. This leads to savings in the direct computation of convolutions for reasons analogous to those seen in the case of discrete symmetries.

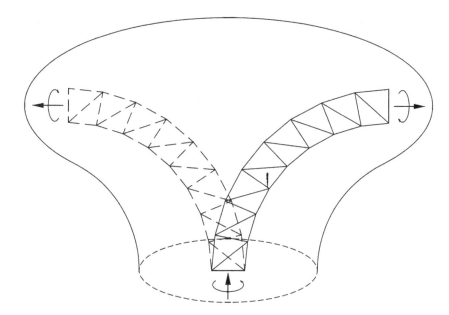

Fig. 12.10. A 3D Manipulator with Continuous Symmetries

It also forces many of the Fourier transform matrix elements to zero. From (10.35) it is clear for this case that only elements of the form

$$U_{l',m';l,0}^s(\mathbf{a}, A; p) = [l',m' \mid p,s \mid l,m'](\mathbf{a})\, \tilde{U}_{m'\,0}^l(A)$$

contribute to the Fourier expansion of functions on $SE(3)$ that are independent of ϕ and γ. The translation matrix elements take on the following simplified form for this special case:

$$[l',m' \mid p,s \mid l,m'](\mathbf{a}) =$$

$$\sqrt{\frac{(2l'+1)}{(2l+1)}} \cdot \sum_{k=|l'-l|}^{l'+l} i^k\,(2k+1)\,j_k(p\,a)\,C(k,0;l',s \mid l,s)\,C(k,0;l',m' \mid l,m')\,P_k(\cos\theta).$$

$$(12.9)$$

From the perspective of calculating workspaces using the $SE(3)$ Fourier transform and convolution theorem, the independence of workspace density functions of ϕ and γ gives special structure to the corresponding Fourier matrices. It is yet to be determined what computational advantages can be gained from this special structure. From the perspective of direct numerical convolution, these special density functions are functions of a smaller number of variables, and the convolutions can be performed more efficiently.

We note that if the manipulator consists of planar sections connected with rotary joints that allow the manipulator to twist out of the plane, then the workspace density function will be independent of α. This means that only $m' = 0$ elements will contribute to the Fourier expansion. While such a manipulator possesses the freedom to reach points in space, its workspace density function only has three degrees of freedom. In this case

$$U_{l',0;l,0}^s(\mathbf{a}, A; p) = [l', 0 \mid p, s \mid l, 0](\mathbf{a})\, P_l(\cos\beta).$$

By defining $U_{l',l}^s(r, \theta, \beta; p) \doteq U_{l',0;l,0}^s(\mathbf{a}, A; p)$ and $\hat{f}_{l,l'}^s(p) \doteq \hat{f}_{l,0;l',0}^s(p)$, the Fourier expansion for workspace density functions on $SE(3)$ with this high degree of symmetry reduces to

$$f(\mathbf{a}, A) = \frac{1}{2\pi^2} \sum_{s=-\infty}^{\infty} \sum_{l'=|s|}^{\infty} \sum_{l=|s|}^{\infty} \int_0^{\infty} p^2\, dp\, \hat{f}_{l,l'}^s(p) U_{l',l}^s(r, \theta, \beta; p)$$

where

$$\hat{f}_{l,l'}^s(p) = (2\pi)^3 \int_0^{\pi} \int_0^{\pi} \int_0^{\infty} f(r, \theta, \beta) \overline{U_{l',l}^s(r, \theta, \beta; p)} r^2 \sin\theta \sin\beta\, dr\, d\theta\, d\beta.$$

The convolution theorem becomes

$$(\mathcal{F}(f_1 * f_2))_{l',l}^s(p) = \sum_{j=|s|}^{\infty} (\hat{f}_2)_{l',j}^s(p)\, (\hat{f}_1)_{j,l}^s(p).$$

12.4.3 Another Efficient Case: Uniformly Tapered Manipulators

We have seen how the concept of convolution can reduce a problem which would take K^P evaluations of the forward kinematics to one which requires P convolutions. In the special case when all the units are the same, this is reduced further to $\log_2 P$ convolutions. When discrete or continuous symmetries exist, each of these convolutions can be computed more efficiently than when using direct integration. Fast Fourier transforms also provide a tool which will speed up the computation of each convolution. But by far the largest savings in computational time results when the K^P problem is reduced to a $\log_2 P$ problem. We therefore examine here a case, in addition to the manipulator with uniform modules, for which this dramatic savings is possible.

Consider the case when each module in a manipulator is a scaled version of each other, so the corresponding density function for the ith module is

$$f_i(\mathbf{a}, A) = f(\mathbf{a}/s_i, A)$$

where s_i is the scale factor for the ith module, and $f(\mathbf{a}, A)$ is the density function for the base module where $s_1 = 1$.

The convolution of two adjacent density functions of this kind is written as

$$(f_i * f_{i+1})(\mathbf{a}, A) = \int_{SO(n)} \int_{\mathbb{R}^n} f(\mathbf{r}/s_i, R) f(R^T((\mathbf{a} - \mathbf{r})/s_{i+1}), R^T A) \, d\mathbf{r} \, dR.$$

Making the change of variables $\mathbf{r}' = \mathbf{r}/s_i$, this integral is transformed to the following form when $s_i = (s_0)^i$ for some scalar s_0:

$$(s_i)^n \int_{SO(n)} \int_{\mathbb{R}^n} f(\mathbf{r}', R) f(R^T((\mathbf{a} - s_i \mathbf{r}')/s_{i+1}), R^T A) \, d\mathbf{r} \, dR = (s_i)^n (f * f_1)(\mathbf{a}/s_i, A).$$

This means that the density function for all pairs of modules for $i = 0, 2, 4, \ldots$ can be calculated with one convolution, and can be scaled appropriately to yield $f_{i,i+1} \doteq f_i * f_{i+1}$. Then, using the same logic, $f_{i,i+2}$ is calculated by convolving appropriately scaled versions of $f_{i,i+1}$ with itself. This too is found by performing a single convolution and rescaling the output.

12.5 Inverse Problems in Binary Manipulator Design

In the previous sections of this chapter we have observed that the density function for the workspace of a binary manipulator is generated by the convolution of the density functions for two halves:

$$(f_1 * f_2)(h) = \int_{SE(3)} f_1(g) f_2(g^{-1} \circ h) \, d(g) = f_3(h).$$

In addition to the *forward* problem, where $f_1(g)$ and $f_2(g)$ are given and we seek $f_3(g)$ for a manipulator with known geometry, certain *inverse* problems arise in the kinematic design of manipulators. One problem in the design of binary manipulators is to set kinematic parameters (e.g. actuator stroke length stops) so that a prescribed workspace density function is attained. This problem can be solved more easily if the manipulator is broken into subunits, and each is designed separately. Another problem is to design the distal end of the manipulator when the proximal end is already built so that the whole workspace density comes as close as possible to the desired one. The latter problem reduces to the solution of a linear convolution equation, i.e., find $f_2(g)$ for given $f_1(g)$ so that the workspace of the combination comes as close as possible to the specified function $f_3(g)$. A nonlinear convolution problem (find $f_1(g)$ so that $(f_1 * f_1)(g) = f_3(g)$ for given $f_3(g)$) arises when we seek to design the sections of the manipulator separately [11, 46, 47].

The most natural way to solve these problems is to use the Fourier transform of functions on $SE(3)$. Using the convolution theorem, the convolution equation can be written in the form

$$\hat{f}_2 \hat{f}_1 = \hat{f}_3 \,, \tag{12.10}$$

where $\hat{f}_i$ denotes the $SE(3)$ Fourier transform of function f_i.

In general, the Fourier transform can contain an infinite number of harmonics l, l' and block elements s. We assume that the contribution of the higher (rapidly oscillating) harmonics can be neglected, and truncate the Fourier transform at some $l = l'$ for each block and take all nonzero blocks for $\mid s \mid \leq l$. Thus, the problem can be reduced to the solution of the matrix equation (12.10). If the functions are "band-limited" (i. e. only a finite number of harmonics give contributions to the Fourier transform), the Fourier transform matrices are finite.

For nonsingular matrix $\hat{f}_1$ the inverse Fourier transform is used to generate the solution:

$$f_2(g) = \mathcal{F}^{-1}(\hat{f}_3(p)\hat{f}_1^{-1}(p)). \tag{12.11}$$

However, in practice, $\hat{f}_1(p)$ is usually singular for most or all values of p, and so a means of regularization is required.

This is a perfect application of the operational properties derived in Chapter 10. We can extend the Tikhonov regularization technique [31], used for integral equations of real-valued argument, to the case of $SE(3)$. That is, instead of solving the original problem, we seek an approximate solution which minimizes the cost function (this is a particular example of first order Tikhonov regularization):

$$C = \int_{SE(3)} (|(f_1 * f_2)(g) - f_3(g)|^2 + \epsilon |f_2(g)|^2 + \nu |\nabla_{\mathbf{a}} f_2(g)|^2 +$$

$$\eta (f_2(g), -\nabla_A^2 f_2(g))) \, d(g) \tag{12.12}$$

for small parameters ϵ, ν and η (higher order derivatives can be added for higher order regularization). Here $g = (\mathbf{a}, A)$.

Using the operational properties (10.62) and (10.65) together with the Plancherel equality (10.51), we can convert this cost function into an algebraic expression in the dual space, do algebraic manipulations, and convert back using the inverse transform.

The Fourier transform converts (12.12) into

$$C = \frac{1}{2\pi^2} \int_0^\infty c(\hat{f}_2^*(p), \hat{f}_2(p)) \, p^2 \, dp = \frac{1}{2\pi^2} \int_0^\infty (\| \hat{f}_2(p) \, \hat{f}_1(p) - \hat{f}_3(p) \|_2^2 +$$

$$\epsilon \| \hat{f}_2(p) \|_2^2 + \nu \| p \hat{f}_2(p) \|_2^2 + \eta \, Tr(\hat{f}_2^*(p) \, A \hat{f}_2(p)) \,) p^2 \, dp, \tag{12.13}$$

where $A_{l',m';l,m} = l'(l'+1) \, \delta_{l'l} \, \delta_{m'm}$.

The equation for $\hat{f}_2(p)$, which minimizes the functional C can be found by differentiating $c(\hat{f}_2^*(p), \hat{f}_2(p))$ with respect to $\hat{f}_2^*(p)$ (or $\hat{f}_2(p)$)

$$\frac{\partial c}{\partial \hat{f}_2^*} = 0,$$

(differentiation with respect to $\hat{f}_2(p)$ gives a Hermitian conjugate equation). This equation is written explicitly as

$$\hat{f}_2 (\hat{f}_1 \hat{f}_1^* + (\epsilon + \nu p^2) \mathbb{I}) + \eta A \hat{f}_2 = \hat{f}_3 \hat{f}_1^*, \tag{12.14}$$

where $\mathbb{I}$ is an appropriately dimensioned identity matrix. After truncating the Fourier transforms and contraction of indices according to (10.54) this equation is analogous to the matrix equation

$$\hat{f}_2 B + A \hat{f}_2 = D \tag{12.15}$$

for given matrices A, B, D. Methods for solving this equation can be found in [29, 4, 30].

We have to solve equation (12.14) for smaller and smaller values of the parameters ϵ, ν, η. When the solution starts to exhibit unpleasant behavior (the solution shows singular-like growth in some regions and starts to oscillate) the calculations must be stopped. "Physical" arguments can be used in the choice of the particular values of the parameters, i. e. we want to pay more attention to the restriction on the magnitude of derivatives of the solution (parameters ν) or to the restriction just on the absolute values of the solution (parameter ϵ). The solution for these values of the parameters ϵ, ν, η is an approximation to the solution of the convolution equation (10.4).

For more on this linear inverse problem, see [11]. For the nonlinear inverse problem, see [46].

12.6 Error Accumulation in Serial Linkages

Error accumulation in serial linkages and cascades of platform manipulators is another problem that is easily quantified as a group-theoretic convolution. Intuitively, the errors due to manufacturing inaccuracies, backlash and flexibility of the constituent components "add up" when traversing the length of an open kinematic chain. This intuitive notion is quantified below using the concepts of convolution of functions on the group of rigid body motions. But first, we review an approach that assumes infinitesimal (as opposed to finite) error.

12.6.1 Accumulation of Infinitesimal Spatial Errors

Consider two homogeneous transforms (positions and orientations) that are measured, and are hence known to some error. A natural question to ask is what the error of the concatenation of these transformations will look like. This question was addressed in detail in the case of planar motions by Smith and Cheeseman [69]. We now present a more general formulation of the same ideas.

Let H_1 be the "exact" proximal homogeneous transform, and let H_2 be the "exact" distal homogeneous transform. Relative to these transforms are measured homogeneous transforms $H_1(\mathbb{I} + \Sigma_1)$ and $H_2(\mathbb{I} + \Sigma_2)$. Here

$$\Sigma_i = \sum_{k=1}^{6} \epsilon_k^{(i)} X_k$$

is an element of the Lie algebra $se(3)$, and the $\epsilon_k^{(i)}$'s are error probability density functions. This model assumes the errors are very small. For larger errors, the discussion of errors in Section 12.6 is more appropriate.

While the exact concatenation of transformations is $H_1 H_2 = H_3$, when measurement error or uncertainty are introduced, it becomes

$$H_1(\mathbb{I} + \Sigma_1)H_2(\mathbb{I} + \Sigma_2) = H_3(\mathbb{I} + \Sigma_3).$$

We now determine Σ_3 from the given information.

Expanding the above products, and using the smallness of the error to neglect second order effects, we find that

$$H_3(\mathbb{I} + \Sigma_3) = H_1 H_2[\mathbb{I} + \Sigma_2 + H_2^{-1}\Sigma_1 H_2].$$

Using the 6-dimensional screw description of the small motions, we write

$$(\Sigma_3)^\vee = (\Sigma_2)^\vee + [Ad(H_2)]^{-1}(\Sigma_1)^\vee,$$

where by definition

$$[Ad(H_2)]^{-1}(\Sigma_1)^\vee = (H_2^{-1}\Sigma_1 H_2)^\vee.$$

Since the errors are small enough, they can be considered to evolve in the Lie algebra, $se(3)$, rather than on the group $SE(3)$, and the standard tools used in statistics on $\mathbb{R}^N$ are applicable here. For instance, we can calculate what the covariance of the resulting error is, given the covariances of the errors of the two constituent homogeneous transforms.

12.6.2 Model Formulation for Finite Error

Suppose we are given a manipulator consisting of two six degree-of-freedom sub-units. These units could be Stewart-Gough platforms or six serial links connected with revolute joints. One unit is stacked on top of the other one. The proximal unit will be able to reach each frame $h_1 \in SE(3)$ with some error. This error may be different for each different frames h_1. This is expressed mathematically as a real-valued function of $g \in SE(3)$ which has a peak in the neighborhood of h_1 and decays rapidly away from h_1. If the unit could reach h_1 exactly, this function would be a delta function. Explicitly this function may have many forms depending on what error model is used. However, it will always be the case that it is of the form $\rho_1(h_1, g_1)$ for $h_1, g_1 \in SE(3)$. That is, the error will be a function on $g_1 \in SE(3)$ for each frame h_1 that the top of the module tries to attain relative to its base. Likewise, the second module will have an error function $\rho_2(h_2, g_2)$ for $h_2, g_2 \in SE(3)$ that describes the distribution of frames around h_2 that might be reached when h_2 is the actual goal.

The error that results from the concatenation of two modules with errors $\rho_1(\cdot)$ and $\rho_2(\cdot)$ results from sweeping the error of the second module by that of the first. This is written mathematically as:

$$\rho(h_1 \circ h_2, g) = (\rho_1 \otimes \rho_2)(h_1 \circ h_2, g) \doteq \int_{SE(3)} \rho_1(h_1, g_1) \rho_2(h_2, g_1^{-1} \circ g) \, d(g_1). \quad (12.16)$$

In the case of no error, the multiplication of homogeneous transforms h_1 and h_2 as $h_1 \circ h_2$ represents the composite change in position and orientation from the base of the lower unit to the interface between units, and from the interface to the top of the upper unit. In the case of inexact kinematics, the error function for the upper unit is shifted by the lower unit $(\rho_2(h_2, g_1^{-1} \circ g))$, weighted by the error of the lower unit $(\rho_1(h_1, g_1))$ and integrated over the support of the error function of the lower unit (which is the same as integrating over all of $SE(3)$ since outside of the support of the error function the integral is zero). The result of this integration is by definition the error function around the frame $h_1 \circ h_2$, and this is denoted as $(\rho_1 \otimes \rho_2)(h_1 \circ h_2, g)$. We illustrated (12.16) in Figure 12.11.

12.6.3 Related Mathematical Issues

To test this formulation, consider the case of exact kinematics. In this case, the error functions have a very special form: they are Dirac delta functions on $SE(3)$. In complete analogy with the usual Dirac delta function on the real line, we have the properties

$$\int_{SE(3)} \delta(g) \, d(g) = 1 \qquad\qquad \delta(h^{-1} \circ g) = \delta(g^{-1} \circ h)$$

and

$$\int_{SE(3)} f(h) \, \delta(h^{-1} \circ g) \, d(h) = f(g).$$

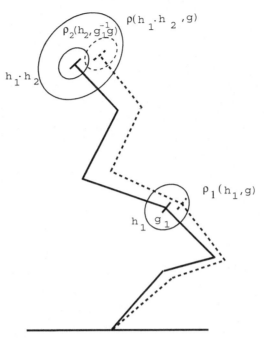

Fig. 12.11. Error Propagation in Serial Linkages. (Reprinted from Wang, Y., Chirikjian, G.S., "Error Propagation on the Euclidean Group with Applications to Manipulator Kinematics," *IEEE Transactions on Robotics*, 22(4): 591 – 602, 2006, with permission from IEEE)

Using these properties, the error functions for both units can be written as

$$\rho_i(h_i, g) = \delta(h_i^{-1} \circ g).$$

Then in this special case, (12.16) reduces to

$$
\begin{aligned}
(\delta \otimes \delta)(h_1 \circ h_2, g) &= \int_{SE(3)} \delta(h_1^{-1} \circ g_1)\, \delta(h_2^{-1} \circ g_1^{-1} \circ g)\, d(g_1) \\
&= \delta(h_2^{-1} \circ h_1^{-1} \circ g) \\
&= \delta((h_1 \circ h_2)^{-1} \circ g).
\end{aligned}
\tag{12.17}
$$

In other words, in the case of exact kinematics, we have exactly the result which is expected. Moreover, the delta function is a tool that also allows us to reexamine the general error propagation equation in (12.16) as a convolution of functions on the direct product group $SE(3) \times SE(3)$. Recall that the group law for the direct product of group with itself, $G \times G$, is simply $(h_1, g_1) \,\hat{\circ}\, (h_2, g_2) = (h_1 \circ h_2, g_1 \circ g_2)$ for $h_i, g_i \in G$. Likewise, the convolution of functions on the direct product is defined as

$$(\alpha * \beta)(h, g) \doteq \int_G \int_G \alpha(h_3, g_1) \beta(h_3^{-1} \circ h, g_1^{-1} \circ g)\, d(h_3)\, d(g_1) \tag{12.18}$$

By letting $\alpha(h, g) = \rho_1(h, g)\, \delta(h_1^{-1} \circ h)$ and $\beta = \rho_2(h, g)$ in the above equation and using the properties of the Dirac delta function, it is clear that error accumulation (12.16) can be written as a convolution on the direct product of $SE(3)$ with itself.

12.7 Mobile Robots: Generation of C-Space Obstacles

The question of computing the configuration-space obstacles of a mobile robot has been considered in a number of papers [32, 34, 52, 56]. Some of these algorithms [52] compute analytically the boundary of the regions of free configuration space for polygonal robots and obstacles. In this section we develop and implement a method which builds on work of [32, 34], which can be applied to both polygonal and non-polygonal shapes. We suggest faster implementation of this method to compute the configuration space, which also allows us to solve inverse problems, i.e. the design of a robot for given static obstacles and desired configuration space (or computation of the maximal size of the obstacles for a given robot and configuration space).

The mathematical formulation of the method is given below. We compute a function on the configuration space (the space of translations and rotations of the robot) which has nonzero values only in the regions where the robot hits the obstacle. The magnitude of this "density function" is the ratio of the overlapping volume (area) of the robot to the total volume (area), i. e. it changes from 0 to 1 in the overlapping regions. To calculate this value we compute the integral

$$c(\mathbf{x}, A) = \frac{\int_{\mathbb{R}^n} f_1(\mathbf{y})\, f_2(A^{-1}(\mathbf{y} - \mathbf{x}))\, d^n y}{\int_{\mathbb{R}^n} f_2(\mathbf{y})\, d^n y} \qquad (12.19)$$

where $f_{1,2}(\mathbf{x})$ are equal to 1 if the vector $\mathbf{x}$ is inside or on the boundary of obstacles (robot) and zero otherwise, $n = 2, 3$ for 2D (3D) coordinate space and $d^n y = dy_1 ... dy_n$ is the usual integration measure for $\mathbb{R}^n$. The function $c(\mathbf{x}, A)$ is normalized to have maximal value of 1, i.e., it is divided by the volume (area) of the robot, $\mathbf{x} \in \mathbb{R}^n$ and $A \in SO(n)$. The geometrical meaning of this function is that it is zero when the obstacle and robot do not intersect, and has increasing positive value as the area of intersection increases.

To compute this integral directly, or simply to check pixel by pixel, that the obstacle $f_1(\mathbf{y})$ and the rotated and translated robot $f_2(A^{-1}(\mathbf{y} - \mathbf{x}))$ do not overlap, we need to perform $O(N_R N_r^2)$ computations, where N_R is the number of sampled orientations and N_r is the number of sampled points in a bounded region of $\mathbb{R}^n$. For a large 3D array of values the computation of this integral by direct summation may be quite slow. The "overlap function" $c(\mathbf{x}, A)$, however, can be computed in $O(N_R N_r \log N_r)$ computations using Fourier methods on the semi-discrete motion group and FFT methods. We discuss applications of this method to mobile robot configuration space generation in detail in the following subsection.

12.7.1 Computing Configuration-Space Obstacles of a Mobile Robot

Here we apply the numerically implemented Fourier transform on the discrete motion group of the plane to the computation of the free configuration space of a rigid mobile robot moving among static 2D obstacles. To find the configuration space we compute the integral (12.19) using the above described Fourier methods for the semi-discrete motion group. Because the obstacles and the robot are functions only of Cartesian position

(i. e. to find (12.19) we need to compute the Fourier transform of $f_2(\mathbf{x})$ only for the fixed orientation) the direct Fourier transform is performed faster than for an arbitrary function on the motion group.

A function of position can be considered as a function on the semi-discrete motion group which does not depend on the orientation, i.e., $f(\mathbf{x}, A_i) = f(\mathbf{x})$. Then the "overlap" function (12.19) may be formally written as the integral

$$c(\mathbf{x}, A_j) = \frac{1}{N_R} \frac{\sum_{i=0}^{N_R-1} \int_{\mathbb{R}^2} f_1(\mathbf{y}, A_i) \, f_2(A_j^{-1}(\mathbf{y} - \mathbf{x}), A_j^{-1} A_i) \, d^2 y}{\int_{\mathbb{R}^2} f_2(\mathbf{y}) \, d^2 y}. \tag{12.20}$$

Because the functions are real, the integral in the numerator can be written as

$$\frac{1}{N_R} \sum_{i=0}^{N_R-1} \int_{\mathbb{R}^2} \overline{f_1(\mathbf{y}, A_i)} \, f_2(A_j^{-1}(\mathbf{y} - \mathbf{x}), A_j^{-1} A_i) \, d^2 y =$$

$$\int_{G_{N_R}} \overline{f_1(h)} \, f_2(g^{-1} h) \, d(h)$$

where we denote integration $d(h)$ over the semi-discrete motion group, G_{N_R}, to mean integration with respect to $\mathbb{R}^2$ and the summation through the A_i, and the group elements are of the form $g = (\mathbf{x}, A_j)$. Using the orthogonality properties of the Fourier matrix elements, this integral can be written as

$$\frac{1}{N_R} \sum_{q=0}^{N_R-1} \sum_{n=0}^{N_R-1} \int_0^\infty \int_v \sum_{m=0}^{N_R-1} \left(\overline{(\hat{f}_1)_{mn}} \, (\hat{f}_2)_{mq} \right) U_{qn}(g^{-1}; p, \phi) \, p \, dp \, d\phi = \tag{12.21}$$

$$\frac{1}{N_R} \sum_{q=0}^{N_R-1} \sum_{n=0}^{N_R-1} \int_0^\infty \int_v \sum_{m=0}^{N_R-1} \left(\overline{(\hat{f}_2)_{mq}} \, (\hat{f}_1)_{mn} \right) U_{nq}(g; p, \phi) \, p \, dp \, d\phi$$

where integration with respect to ϕ is integration on the circle in the interval $F_q = [2\pi q/N_R, 2\pi (q+1)/N_R]$. For the second expression we used the unitarity of the matrix elements and the fact that the expression is real (i.e., we take a complex conjugate of the integral). The matrices $(\hat{f}_{1,2})_{mn}$ are the Fourier transforms of the functions $f_{1,2}(\mathbf{x}, A_i)$.

Because functions $f_{1,2}(\mathbf{x}, A_i) = f_{1,2}(\mathbf{x})$ do not depend on the orientations A_i, matrix elements of the Fourier transforms in the same column are the same, i.e.,

$$(\hat{f}_{1,2})_{mn}' = (\hat{f}_{1,2})_{qn}$$

for any m, q. This can be observed from the expression

$$U(g^{-1}; p, \phi)_{mn} = e^{-i p \, \mathbf{u}_n^\phi \cdot \mathbf{r}} \, \delta_{A_i^{-1} \mathbf{u}_n, \mathbf{u}_m}$$

(the exponent depends only on the n-index), the definition of the direct semi-discrete-motion-group Fourier transform, and the fact that the functions do not depend on the orientation. Thus, we compute a row of the Fourier matrix for a particular orientation (for example $A_0 = \mathbb{I}$)

$$(\hat{f}_{1,2})_n = (\hat{f}_{1,2})_{nn}.$$

This can be done using the 2D FFT for the functions $f_{1,2}(\mathbf{x})$ and interpolating the Fourier values to points on a polar coordinate grid. This requires $O(N_r \log(N_r))$ computations. Thus, the "overlap" function (12.19) is written as

$$c(\mathbf{x}, A_j) = C \frac{\sum_q \sum_n \int_0^\infty \int_\phi \left(\hat{f}_{2q}\,\hat{f}_{1n}\right) U_{nq}(g; p, \phi)\,p\,dp\,d\phi}{\int_{\mathbb{R}^2} f_2(\mathbf{y})\,d^2y} \qquad (12.22)$$

where $C = \frac{1}{N_R}$. The product of the column $\hat{f}_2$ and the row $\hat{f}_1$ can be performed in $O(N_r\,N_R)$ computations, and the inverse transform can be performed in $(N_R\,N_r\,\log(N_r))$ computations. The normalization of the function $f_2(\mathbf{x})$ can be computed by direct integration in $O(N_r)$. Thus, the inverse transform is the largest time-consuming computation. We note that the direct Fourier transform is performed only for one orientation, this is N_R times faster (for discrete rotation group C_{N_R}) than to perform the FFT for each orientation as is done in [34], but the inverse transform is of the same order of computations. Thus, as N_R becomes large, there is a built-in factor of two speed increase using our method. We also mention that inverse problems, i.e., problems of finding the allowed shape of the robot for the given desired configuration space and the shape of the obstacles, can be solved using Fourier methods on the motion group, because it is reduced to a functional matrix equation (regularization methods for the solution of singular functional matrix equations are discussed in [46]).

When we discretize the coordinate regions and use the FFT to compute (12.19), the coefficient C in (12.22) must be defined as $C = A/N_R$, where A is the area of the compact region of $\mathbf{x}$ where the FFT is computed (the functions must have a support inside the area A, the functions are considered periodic outside of this region). The area factor arises because the discrete Fourier transform can be received as an approximation of the continuous case using

$$x_1 \to \frac{L}{N_r}i; \quad p_1 \to \frac{2\pi}{L}i\,,$$

(and analogously for the 2nd component). Here L is the length of the compact region in the x (y) direction. While the factor L is canceled out of equations if we apply direct and inverse discrete Fourier transform to the function, it appears in the convolution-like integral (12.19).

Thus, we compute the 2D FFT of $f_{1,2}(\mathbf{x})$, interpolate the Fourier elements to the polar grid, arrange them into the Fourier column and row, multiply column and row $\hat{f}_{mn} = \hat{f}_{2m}\,\hat{f}_{1n}$, and take the inverse Fourier transform (the corresponding elements $m = n - i$ from the Fourier matrix $\hat{f}_{mn}$ must be taken for each orientation A_i and interpolated back to the Cartesian grid to take the 2D inverse FFT).

We implemented the computation of (12.22) using the FFT in the C programming language. Time to compute $c(\mathbf{x}, A_j)$ was 30 sec (on a 250 MHz SGI workstation) for a 256×256 square grid in $\mathbb{R}^2$ for the C_{10} group ($N_R = 10$, and we subdivide each segment into $N_\phi = 20$ subsegments).

Because for small values of $c(\mathbf{x}, A_j)$ the function exhibits oscillations (due to finite discretization of the integration area in the Fourier transform), we depict the boundary of the configuration space, defined by the contour line where $c(\mathbf{x}, A_j) = \epsilon$. The smallest possible choice of ϵ was in our examples in the region $0.005 - 0.035$. To increase accuracy, we use the following method. We can increase the value of $f_1(\mathbf{x})$ in the region near the border of the robot. For a convex robot this can be done by scaling down the robot and by increasing the value of $f_1(\mathbf{x})$ between the scaled boundary of the robot and the original boundary. There is also another way to produce a rim for both concave and convex robots using the concept of an "offset curve." If the original boundary is the curve $\mathbf{x}(t)$, the offset curve is $\mathbf{O}_\delta(t) = \mathbf{x}(t) - \delta\mathbf{n}(t)$, where $\mathbf{n}(t)$ is the unit normal (pointing out) from the boundary curve.

While the change of the function values in the "rim" region does not change the shape and the size of the robot, the overlapping area of robot and obstacle is, effectively, smaller than ϵ for given ϵ. If the overlap is completely in the "rim" of the increased values near the boundary, the following estimate can be received for the intersection area ϵ' for given value of ϵ

$$\epsilon' = \epsilon/q \,,$$

where $q = \frac{V}{V-1/k^2\,(V-1)}$, (the function f_1 is equal to unity in the area scaled down by the factor k, and increased to the value V in the "rim" near the boundary of the robot). In addition, we can increase the size of the robot and depict the configuration space for the scaled robot. The area between the two configuration boundaries (of the scaled robot and of the original robot) is the "near-collision" area of the robot and the obstacle. This is an important region for motion planning. We note that the direct method (i.e direct integration of (12.19) or the direct check if the robot and the obstacle overlap) can be used to find a precise boundary in the "near-collision" area. This can be perform in $O(N^2\,A/A_{tot})$ computations, where A/A_{tot} is the ratio of the "near-collision" region to the total region of the coordinate space and $N = N_r N_R$. Because this ratio is generally small this gives very considerable (hundreds of times) saving in computations in the direct method.

We note that this method may not give good results if a very narrow and long object is attached to the robot (i.e. when the intersection area is not sensitive to the overlap). However, the value of $f_1(\mathbf{x})$ can be increased in these regions and this method may be useful for some of these shapes.

12.7.2 Examples of C-Space Obstacles

As an example we depict slices of configuration space in Figure 12.12. The obstacle has a polygonal shape (the Fourier method can be applied both for polygonal and for "smooth" obstacles. The results are better for "smoother" obstacle, i.e., the value of ϵ can be chosen smaller), the robot has an elliptic shape (the size and the orientation of the robot are depicted in the lower right corner on the figures). The solid line outside the obstacle depicts the boundary of the configuration space ($\epsilon \approx 0.030 - 0.035$). We increase the value of the function f_1 to 10 in the rim depicted in Figure 12.12 (between the original boundary and the boundary scaled down by factor $k = 1.1$). We also depict in the pictures the boundary (dashed line) which describes the configuration space of the robot when it is located completely inside of the obstacle ($g(\mathbf{x}, A) \approx 0.95 - 0.97$).

The figures also depict the corresponding boundaries for the scaled robot (enlarged by the factor 1.27). This is region where the exact position of the configuration space boundary can be found by direct integration.

After c-space obstacles are characterized, then efficient methods for constructing collision-free paths such as those in [35, 40] can be used.

12.8 Stochastic Nonholonomic Systems

Mobile robotic systems that locomote on the plane are typically nonholonomic due to constraints imposed by wheels corresponding to no-slip along the direction of the axel of each wheel. For slow-moving "kinematic" systems in which dynamics is unimportant, the governing equations are of the form

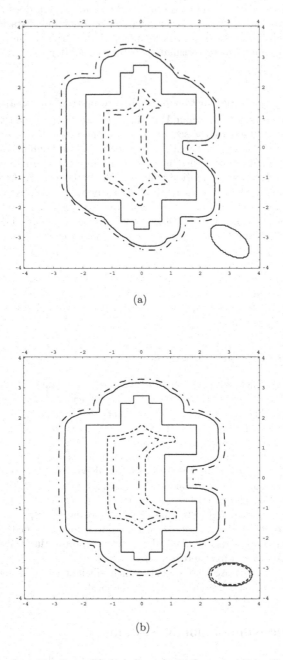

(a)

(b)

Fig. 12.12. Configuration-Space Obstacle Boundaries Generated Using the Fourier Transform for the Semi-Discrete Motion Group. (Reprinted from Kyatkin, A.B., Chirikjian, G.S., "Computation of Robot Configuration and Workspaces via the Fourier Transform on the Discrete-Motion Group," *International Journal of Robotics Research*, 18(6): 601–615, 1999, with permission from Sage Publications)

$$\left(g^{-1}\dot{g}\right)^{\vee} = C\,\dot{\boldsymbol{\phi}} \tag{12.23}$$

where $g \in SE(2)$, $\boldsymbol{\phi}$ is a column vector of wheel angles, and C is a constant matrix that describes how the wheels interact with the ground. For explicit examples of the matrix C for the kinematic cart and car with front-wheel steering see [76] and references therein. Similar models hold in 3D for underwater and aerial vehicles where the vector of wheel angle rates is replaced by thruster intensities, and in this case C is typically a constant matrix describing how the thrusters are positioned and oriented in the body frame of the vehicle.

Returning to the planar case, when the vector of actual wheel rates consists of a baseline commanded component, $\boldsymbol{\phi}_0(t)$, and in addition a stochastic component reflecting noise and uncertainty, then[9]

$$d\boldsymbol{\phi} = \boldsymbol{\phi}_0(t)\,dt + B(t)\,d\mathbf{W}$$

where $d\mathbf{W}$ is a vector of "uncorrelated unit-strength white noise" forcing functions and $\mathbf{W}$ is a "vector Wiener process." Here $B(t)$ is a weighting matrix. Substituting into (12.23) gives a stochastic nonholonomic model of the form

$$\left(g^{-1}\,dg\right)^{\vee} = \mathbf{h}(t)\,dt + B(t)\,d\mathbf{W} \tag{12.24}$$

where $\mathbf{h}(t) = C\,\boldsymbol{\phi}_0(t)$ and $H(t) = C\,B(t)$.

The field of *Simultaneous Localization and Mapping*, or SLAM, is built on such models. See [19, 71] for introductions to this field and pointers to the vast literature on this topic. In SLAM, often stochastic equations of the form (12.24) are simulated many times, resulting in an ensemble. That ensemble is then binned, and the resulting histogram, normalized to be a probability density function, looks essentially the same as the workspace density function described earlier in this chapter. The only difference is that that number of modules along a manipulator is replaced with the continuous time parameter. When we discuss semiflexible polymers in Chapter 17, in which DNA is described by a continuous elastic filament forced by Brownian motion, the same sorts of probabilities result again.

Equation (12.24) is a stochastic differential equation, the trajectories of which evolve on $G = SE(2)$. Corresponding to such an equation is a *Fokker-Planck equation* of the form [76]

$$\frac{\partial f}{\partial t} = -\sum_{i=1}^{n} h_i(t)\tilde{X}_i^R f + \frac{1}{2}\sum_{i,j=1}^{n} D_{ij}(t)\tilde{X}_i^R \tilde{X}_j^R f \tag{12.25}$$

where $D = HH^T$ and n is the dimension of G, i.e., $n = 3$ in the case when $G = SE(2)$.

This is a diffusion equation that describes the evolution of probability of the position and orientation of the robot. That is, $f : G \times \mathbb{R}_{\geq 0} \to \mathbb{R}_{\geq 0}$ is a probability density function. Typically the initial location of the robot is known and so $f(g, 0) = \delta(g)$ is a Dirac delta function on G. For a coordinate-dependent derivation of (12.25) see Chapter 16, and for coordinate-free derivations see [17, 60].

The relationships between equations of the form (12.25) and harmonic analysis is that the group Fourier transform can be applied to both sides, converting this to a system of ordinary differential equations that can be solved in the dual space $\hat{G}$, and then the solution can be obtained by applying the inverse Fourier transform. This solution

---

[9]The reason for writing it out in this form with $d\boldsymbol{\phi}$ instead of $\dot{\boldsymbol{\phi}}$ is that technically a stochastic trajectory is not differentiable. This is explained in detail in Chapters 15 and 16.

methodology works well in the case of long-time or large $\|D\|$. When $t \cdot \|D\|$ is small, alternative solution methodologies exisit, as outlined in [17, 60].

Three-dimensional versions of such models have been used to model the path that a flexible needle will take when inserted for medical treatment and steered from its base by pushing and twisting [1, 58, 59, 74], and this topic has become very popular in recent years.

Classic references to modeling uncertainty in robotics, artificial intelligence, and machine learning include [21, 23, 25, 33, 42]. Though these works do not use harmonic analysis as a tool, this opens up opportunities to do so.

12.9 Summary

In this chapter we saw how harmonic analysis on motion groups is a useful computational and analytical tool in the context of robotics problems. Such problems range from manipulator workspace generation and error accumulation in serial linkages to the computation of configuration-space obstacles of mobile robots. Moreover, the probability density describing the location of a nonholonomic robot at a particular time after it travels from a known starting position and orientation can be described by a diffusion equation on the special Euclidean group. This equation can be solved using methods of harmonic analysis. In Chapter 17, applications of motion-group harmonic analysis is applied in the context of polymer-chain statistical mechanics. This is very similar to the workspace generation problem. In Chapter 13, problems in image analysis and tomography are posed using motion-group harmonic analysis. These problems are in many ways like C-space obstacle generation.

References

1. Alterovitz, R., Lim, A., Goldberg, K., Chirikjian, G.S., Okamura, A.M., "Steering Flexible Needles Under Markov Motion Uncertainty," *Proceedings of the IEEE/RSJ Int. Conf. on Intelligent Robots and Systems*, pp. 120–125, Aug. 2005.
2. Alciatore, D.G., Ng, Chung-Ching D., "Determining Manipulator Workspace Boundaries Using the Monte Carlo Method and Least Square Segmentation," *ASME Robotics: Kinematics, Dynamics and Controls* DE-Vol. 72, pp.141–146, 1994.
3. Angeles, J., "The Design of Isotropic Manipulator Architectures in the Presence of Redundancies," *International Journal of Robotics Research*, 11(3): 196–201, June 1992.
4. Bartels, R.H., Stewart, G.W., "Solution of the Matrix Equation $AX + XB = C$," *Communications of the ACM*, 15(9): 820–826, 1972.
5. Basavaraj, U., Duffy, J., "End-Effector Motion Capabilities of Serial Manipulators," *International Journal of Robotics Research*, 12(2): 132–145, April 1993.
6. Blackmore, D., Leu, M.C., "Analysis of Swept Volume via Lie Groups and Differential Equations," *The International Journal of Robotics Research* 11(6): 516–537, Dec. 1992.
7. Canny, J.F., *The Complexity of Robot Motion Planning*, (1987 ACM Doctoral Dissertation Award) MIT Press, Cambridge MA, 1988.
8. Ceccarelli, M., Vinciguerra, A., "On the Workspace of General 4R Manipulators," *International Journal of Robotics Research*, 14(2): 152–160, April 1995.
9. Chirikjian, G.S.,"A Binary Paradigm for Robotic Manipulators," *Proceedings of the 1994 IEEE International Conference on Robotics and Automation*, pp. 3063–3069, San Diego, CA, May 1994.
10. Chirikjian, G.S., "Kinematic Synthesis of Mechanisms and Robotic Manipulators with Binary Actuators," *ASME Journal of Mechanical Design*, 117(4): 573–580, Dec. 1995.
11. Chirikjian, G.S. "Fredholm Integral Equations on the Euclidean Motion Group," *Inverse Problems* 12(5): 579–599, Oct. 1996.
12. Chirikjian, G.S., Ebert-Uphoff, I., "Numerical Convolution on the Euclidean Group with Applications to Workspace Generation," *IEEE Transactions on Robotics and Automation*, 14(1): 123–136, Feb. 1998.
13. Chirikjian, G.S., "Synthesis of Discretely Actuated Manipulator Workspaces Via Harmonic Analysis," in *Recent Advances in Robot Kinematics* (J. Lenarcic and V. Parenti-Castelli eds.), Kluwer Academic Publishers, 1996.
14. Chirikjian, G.S., "Group Theoretical Synthesis of Binary Manipulators," in *Ro.Man.Sy. '11: Theory anc Practice of Robots and Manipulators*, pp. 107–114, Springer, New York, 1997.
15. Chirikjian, G.S., Lees, D.S., "Inverse Kinematics of Binary Manipulators with Applications to Service Robotics," *Proceedings of IROS'95*, Pittsburgh, PA, Vol. 3, pp. 65–71, Aug. 1995.
16. Chirikjian, G.S., "Inverse Kinematics of Binary Manipulators Using a Continuum Model," *Journal of Intelligent and Robotic Systems*, 19(1): 5–22, 1997.

17. Chirikjian, G.S., *Stochastic Models, Information Theory, and Lie Groups: Vols. 1+2*, Birkhäuser, Boston, 2009/2011.

18. Choset, H., Burdick, J.W., "Sensor Based Motion Planning: The Hierarchical Generalized Voronoi Graph," *International Journal of Robotics Research*, 19(2): 96 – 125, 2000.

19. Choset, H., Lynch, K. M., Hutchinson, S., Kantor, G., Burgard, W., Kavraki, L.E., Thrun, S., *Principles of Robot Motion: Theory, Algorithms, and Implementations*, MIT Press, Boston, 2005.

20. Craig, J.J., *Introduction to Robotics, Mechanics and Control*, Addison-Wesley Publishing Company, Reading Mass., 1986.

21. Dellaert, F., Fox, D., Burgard, W., Thrun, S., "Monte Carlo Localization for Mobile Robots," *Proceedings of the 1999 IEEE International Conference on Robotics and Automation*, pp. 1322 – 1328, Detroit, Michigan, May 1999.

22. Desai, J., and Kumar, V., "Nonholonomic Motion Planning of Multiple Mobile Manipulators," *Proceedings of the IEEE International Conference on Robotics and Automation*, Vol. 4, pp. 3409 – 3414, Albuquerque, New Mexico, April 20–24, 1997.

23. Dissanayake, M.W.M.G., Newman, P., Clark, S., Durrant-Whyte, H.F., Csorba, M., "A Solution to the Simultaneous Localization and Map Building (SLAM) Problem," *IEEE Transactions on Robotics and Automation*, 17(3): 229 – 241, 2001.

24. Duffy, J., Crane, C.D. III, *Kinematic Analysis of Robot Manipulators*, Cambridge University Press, New York, 1998.

25. Durrant-Whyte, H.F., "Uncertain Geometry in Robotics," *IEEE Journal of Robotics and Automation*, 4(1): 23 – 31, 1988.

26. Ebert-Uphoff, I., "On the Development of Discretely-Actuated Hybrid-Serial-Parallel Manipulators," PhD Dissertation, Dept. of Mechanical Engineering, Johns Hopkins University, May 1997.

27. Ebert-Uphoff, I., Chirikjian, G.S.,"Efficient Workspace Generation for Binary Manipulators with Many Actuators," *Journal of Robotic Systems*, 12(6): 383 – 400, June 1995.

28. Ebert-Uphoff, I., Chirikjian, G.S., "Inverse Kinematics of Discretely Actuated Hyper-Redundant Manipulators Using Workspace Densities," *Proceedings of the 1996 IEEE International Conference on Robotics and Automation*, pp. 139 – 145, April 1996.

29. Gantmacher, F.R., *The Theory of Matrices*, Chelsea, New York, 1959.

30. Golub, G.H., Nash S., Van Loan, C., "A Hessenburg-Shur Method for the Problem $AX + XB = C$," *IEEE Transactions on Automatic Control*, 24(6): 909 – 913, 1979.

31. Groetsch, C.W., *The Theory of Tikhonov Regularization for Fredholm Equations of the First Kind*, Pitman, Boston, 1984.

32. Guibas, L., Ramshaw, L., Stolfi, J., "A Kinetic Framework for Computational Geometry," *Proceedings of the IEEE Symposium on Foundations of Computer Science*, pp. 100 – 111, 1983.

33. Kaelbling, L.P., Littman, M.L., Cassandra, A.R., "Planning and Acting in Partially Observable Stochastic Domains," *Artificial intelligence*, 101(1): 99 – 134, 1998.

34. Kavraki, L., "Computation of Configuration-Space Obstacles Using the Fast Fourier Transform," *IEEE Trans. on Robotics and Automation*, 11(3): 408 – 413, 1995.

35. Kavraki, L.E., Svestka, P., Latombe, J.C., Overmars, M.H., " Probabilistic Roadmaps for Path Planning in High-Dimensional Configuration Spaces," *IEEE Transactions on Robotics and Automation* 12(4): 566 – 580, Aug. 1996.

36. Khatib, O., "Mobile Manipulation: The Robotic Assistant," *Journal of Robotics and Autonomous Systems*, 26(2): 175 – 183, 1999.

37. Klein, C.A., Blaho, B.E., "Dexterity Measures for the Design and Control of Kinematically Redundant Manipulators," *International Journal of Robotics Research*, 6(2): 72 – 83, Summer 1987.

38. Koliskor, A., "The l-Coordinate Approach to the Industrial Robots Design," in *Information Control Problems in Manufacturing Technology 1986. Proceedings of the 5th IFAC/IFIP/IMACS/IFORS Conference*, pp. 225 – 232, Suzdal, USSR (Preprint).

39. Korein, J.U., "A Geometric Investigation of Reach," MIT Press, 1985.

40. Kuffner, J.J., LaValle, S.M., "RRT-Connect: An Efficient Approach to Single-Query Path Planning," *Proceedings of the IEEE International Conference on Robotics and Automation, ICRA'00*, Vol. 2. pp. 995–1001, San Francisco, CA, April, 2000.

41. Kumar, A., Waldron, K.J., "Numerical Plotting of Surfaces of Positioning Accuracy of Manipulators," *Mechanisms and Machine Theory*, 16(4): 361–368, 1980.

42. Kwon, J., Choi, M., Park, F.C., Chun, C., "Particle Filtering on the Euclidean Group: Framework and Applications," *Robotica*, 25(6): 725–737, 2007.

43. Kwon, S.-J., Youm, Y., Chung, K.C., "General Algorithm for Automatic Generation of the Workspace for *n*-link Planar Redundant Manipulators," *ASME Transactions* 116(3): 967–969, Sept. 1994.

44. Kyatkin, A.B., Chirikjian, G.S., "Fourier Methods on Groups: Applications in Robot Kinematics and Motion Planning," *Proc. 3rd Workshop on Algorithmic Foundations of Robotics*, Houston, Texas, March 5–8, 1998.

45. Kyatkin, A.B., Chirikjian, G.S., "Computation of Robot Configuration and Workspaces via the Fourier Transform on the Discrete-Motion Group," *International Journal of Robotics Research*, 18(6): 601–615, 1999.

46. Kyatkin, A.B., Chirikjian, G.S., "Regularized Solutions of a Nonlinear Convolution Equation on the Euclidean Group," *Acta Applicandae Mathematicae*, 53: 89–123, August 1998.

47. Kyatkin, A.B., Chirikjian, G.S., "Synthesis of Binary Manipulators Using the Fourier Transform on the Euclidean Group," *ASME Journal of Mechanical Design*, 121(1): 9–14, March 1999.

48. Latombe, J.-C., *Robot Motion Planning*, Kluwer Academic Publishers, Boston 1991.

49. Lees, D.S., Chirikjian, G.S. "An Efficient Trajectory Planning Method for Binary Manipulators," *ASME Mechanisms Conference*, 96-DETC/MECH-1161, (9 pages on CD ROM), August 1996.

50. Lees, D.S., Chirikjian, G.S. "An Efficient Method for Computing the Forward Kinematics of Binary Manipulators," *Proceedings of the IEEE International Conference of Robotics and Automation*, Minneapolis, MN, pp. 1012–1017, April, 1996.

51. Lees, D.S., Chirikjian, G.S. "A Combinatorial Approach to Trajectory Planning for Binary Manipulators," *Proceedings of the IEEE International Conference of Robotics and Automation*, Minneapolis, MN, April, pp. 2749–2754, 1996.

52. Lozano-Perez, T., "Spatial Planning: A Configuration Space Approach," *IEEE Transactions on Computers*, 32(2): 108–120, 1983.

53. Merlet, J.-P., *Les Robots Parallèles*, Traité des nouvelles Technologies, Série Robotique. Hermes, 1990.

54. Murray, R.M., Li, Z., Sastry, S.S., *A Mathematical Introduction to Robotic Manipulation*, CRC Press, Ann Arbor MI, 1994.

55. Nelaturi, S., Shapiro, V., "Configuration Products in Geometric Modeling," *Proceedings of the 2009 SIAM/ACM Joint Conference on Geometric and Physical Modeling* (pp. 247–258) October, 2009.

56. Newman, W., Branicky, M., "Real-Time Configuration Space Transforms for Obstacle Avoidance," *The International Journal of Robotics Research*, 10(6): 650–667, 1991.

57. Park, F.C., Brockett, R.W., "Kinematic Dexterity of Robotic Mechanisms," *The International Journal of Robotics Research*, 13(1): 1–15, 1994.

58. Park, W., Kim, J.S., Zhou, Y., Cowan, N.J., Okamura, A.M., Chirikjian, G.S., "Diffusion-Based Motion Planning for a Nonholonomic Flexible Needle Model," *Proceedings of the IEEE Int. Conf. on Robotics and Automation*, Barcelona, Spain, pp. 4600–4605, April 2005.

59. Park, W., Wang, Y., Chirikjian, G.S., "The Path-of-Probability Algorithm for Steering and Feedback Control of Flexible Needles," *The International Journal of Robotics Research*, 29(7): 813–830, 2010.

60. Park, W., Liu, Y., Zhou, Y., Moses, M., Chirikjian, G.S., "Kinematic State Estimation and Motion Planning for Stochastic Nonholonomic Systems Using the Exponential Map," *Robotica*, 26(4): 419–434, July-August 2008

61. Pieper, D.L., "The Kinematics of Manipulators under Computer Control," PhD Dissertation, Stanford University, Oct. 1968.
62. Popplestone, R.J., "Group Theory and Robotics," in *Robotics Research: The First International Symposium*, M. Brady and R. Paul, eds., MIT Press, Cambridge MA, 1984.
63. Rastegar, J., Deravi, P., "Methods to Determine Workspace, its Subspaces with Different Numbers of Configurations and All the Possible Configurations of a Manipulator," *Mechanisms and Machine Theory*, 22(4): 343–350, 1987.
64. Rastegar, J., Deravi, P., "The Effect of Joint Motion Constraints on the Workspace and Number of Configurations of Manipulators," *Mechanisms and Machine Theory*, 22(5): 401–409, 1987.
65. Roth, B., Rastegar, J., and Scheinman, V., "On the Design of Computer Controlled Manipulators," *First CISM-IFTMM Symp. on Theory and Practice of Robots and Manipulators*, pp. 93–113, 1973.
66. Sciavicco, L., Siciliano, B., *Modeling and Control of Robot Manipulators*, McGraw-Hill Inc., New York, 1996.
67. Selig, J.M., *Geometrical Methods in Robotics*, Springer, New York, 1996.
68. Sen, D., Mruthyunjaya, T.S., "A Discrete State Perspective of Manipulator Workspaces," *Mechanisms and Machine Theory*, 29(4): 591–605, 1994.
69. Smith, R.C., Cheeseman, P., "On the Representation and Estimation of Spatial Uncertainty," *The International Journal of Robotics Research*, 5(4): 56–68, 1986.
70. Spong, M.W., Hutchinson, S., Vidyasagar, M., *Robot Modeling and Control*, Wiley, New York, 2006.
71. Thrun, S., Burgard, W., Fox, D., *Probabilistic Robotics*, MIT Press, Cambridge, MA, 2005.
72. Tsai, L.-W., *Robot Analysis: The Mechanics of Serial and Parallel Manipulators*, John Wiley and Sons, New York, 1999.
73. Wang, Y., Chirikjian, G.S., "Error Propagation on the Euclidean Group with Applications to Manipulator Kinematics," *IEEE Transactions on Robotics*, 22(4): 591–602 August 2006.
74. Webster, R.J. III, Kim, J.-S., Cowan, N.J., Chirikjian, G.S., Okamura, A.M., "Nonholonomic Modeling of Needle Steering," *The International Journal of Robotics Research*, 25(5–6): 509–525, May-June 2006.
75. Yoshikawa, T., "Manipulability of Robotic Mechanisms," *International Journal of Robotics Research*, 4(2): 3–9, 1985. (Also see *Foundations of Robotics: Analysis and Control*, MIT Press, Cambridge MA, 1990.)
76. Zhou, Y., Chirikjian, G.S., "Probabilistic Models of Dead-Reckoning Error in Nonholonomic Mobile Robots," *ICRA'03*, Taipei, Taiwan, pp. 1594–1599, Sept., 2003.

Image Analysis and Tomography

In this chapter we formulate two kinds of problems in the language of noncommutative harmonic analysis: (1) the template matching problem from two-dimensional pattern recognition; (2) the forward and inverse tomography problems. Both (1) and (2) can be formulated as convolution equations on the group of rigid-body motions. In (1), the fast evaluation of $SE(2)$-convolutions is required. In (2) the accurate inversion of convolution equations with a particular kind of kernel is sought. In Section 13.1 we address template matching. In Section 13.2 we address the forward tomography (imaging) problem. In Section 13.3 we address the inverse tomography (radiotherapy planning) problem.

13.1 Image Analysis: Template Matching

For a given template object we want to find if this template object is present in a given image, and, if it is found, determine its position and orientation. We use a correlation method (see [52] and references therein) for this purpose, which is extended to include rotations and dilations of the template object in addition to translations.[1] Essentially, we translate, rotate and dilate the template object, overlap it with the image and compute an overlap area (weighted by the intensity value at each pixel) with the proper normalization.

The correlation method is implemented using the Fourier transform for the semi-discrete motion group. Fourier methods for the semi-discrete motion group also provide a fast method to distinguish "identical" images (up to possible translations and rotations of the image) from "different" ones.

The semi-discrete motion group (see Chapter 11) can be viewed as the set of matrices of the form

$$g = \begin{pmatrix} A & \mathbf{a} \\ \mathbf{0}^T & 1 \end{pmatrix}, \tag{13.1}$$

where

$$R = \begin{pmatrix} \cos 2\pi\, i/N_R & -\sin 2\pi\, i/N_R \\ \sin 2\pi\, i/N_R & \cos 2\pi\, i/N_R \end{pmatrix}, \tag{13.2}$$

---

[1]Much of the material in this section was first presented in [61].

for fixed natural number N_R and $i \in [0, N_R - 1]$, (A_i is replaced by an arbitrary element of $SO(2)$ for the continuous motion group $SE(2)$). The group law is simply matrix multiplication.

The problem of template matching is quite old, and has been approached in a number of different ways. Perhaps the most common (and oldest) approach is that of "matched filters" [97]. In this approach the Fourier transform of the image and template are taken and multiplied, and a peak is sought. This method can be implemented via digital computer, or by analog optical computation [56]. The drawback of this standard approach is that rotations are handled in a very awkward manner. Several works have considered rotation-invariant approaches (e.g., [3]). In such approaches, polar coordinates are used and images are expanded in a series of Zernike polynomials (see, e.g., [11]) or by using the Hankel transform. The problem with such approaches is that rotational invariance is usually gained at the expense of the translational invariance offered by the usual (Abelian) Fourier transform.

A number of works have considered using invariants of images for recognition (see, e.g., [1, 44] and references therein). When discussing invariants, one of the most natural analytical tools is group theory. In this Chapter we apply group theory (in particular noncommutative harmonic analysis) to the template matching problem. In particular, for a given function $f(\mathbf{x})$, the Euclidean-group Fourier transform (from Chapters 10 and 11) is a matrix function which has the property:

$$\mathcal{F}(f(A^T(\mathbf{x} - \mathbf{a}))) = \mathcal{F}(f(\mathbf{x}))U(A, \mathbf{a}),$$

where U is a unitary representation matrix that depends on rotation A and translation $\mathbf{a}$, and $\mathcal{F}$ denotes the non-Abelian Fourier transform. The above expression cannot be written as a matrix product for the usual Abelian Fourier transform for $A \neq \mathbb{I}_{2 \times 2}$, though it is completely analogous to the behavior of the Abelian Fourier transform applied to translated functions. In other words, noncommutative harmonic analysis provides a tool for translation *and* rotation invariant pattern matching. Furthermore, since U is unitary $||U\mathcal{F}(f)||_2 = ||\mathcal{F}(f)||_2$, and so this generalized Fourier transform provides a tool for generating a whole continuum of pattern invariants under rigid-body motion.

The connection between group theory and the theory of wavelets (which has become a very popular tool in image analysis) has been well established. In essence, expanding a function in a wavelet basis is achieved by starting with a mother wavelet and superposing affine-transformed versions of the mother wavelet to best approximate a given function. The interested reader is pointed to [2, 64, 77, 92] for further reading on the subject of wavelets, their applications in image analysis, and their connection with group theory.

The approach presented in this chapter is to use the non-Abelian Fourier transform and generalized concepts of convolution and correlation. This is very different than wavelet approaches, which have become very popular in the image analysis context in recent years. While wavelets typically allow one to efficiently approximate functions (or images), they have the drawback of not behaving well under operations such as convolution, which is the most natural tool in matched filtering.

In Subsection 13.1.1 we describe the correlation method. Subsection 13.1.2 shows why Fourier analysis on the semi-discrete motion group is a useful tool in this context. In Subsection 13.1.3 we describe the numerical implementation of the correlation method using the Fourier transform for the semi-discrete motion group.

13.1.1 Method for Pattern Recognition

In this section we extend the correlation method for pattern recognition [52] to include rotations and dilations (in addition to translations) as the allowed transformations of the image. To find if the template object is present in the image we take a section from the image and compare it with a rotated, translated, and dilated version of the template pattern. Taking a section from the image is equivalent to multiplication of the image by a "window" function, which is rotated, translated, and dilated the same way as the template pattern. Mathematically, the correlation function is written as

$$q(\mathbf{a}, A, k) =$$

$$\frac{\int_{\mathbb{R}^2} f_1(\mathbf{x}) W(A^{-1}(k\,\mathbf{x} - \mathbf{a})) f_2(A^{-1}(k\,\mathbf{x} - \mathbf{a}))\, d^2x}{[\int_{\mathbb{R}^2} (f_1(\mathbf{x}))^2 (W(A^{-1}(k\,\mathbf{x} - \mathbf{a})))^2 d^2x]^{1/2} [\int_{\mathbb{R}^2} (f_2((A^{-1}(k\,\mathbf{x} - \mathbf{a})))^2 d^2x]^{1/2}} \tag{13.3}$$

where $A \in SO(2)$, $\mathbf{a} \in \mathbb{R}^2$, $k \in \mathbb{R}_{>0}$ close to unity and $W(\mathbf{x})$ is a window function. $f_1(\mathbf{x})$ is the image function and $f_2(\mathbf{x})$ is the template function. For similar template pattern and window from the image the value of the correlation coefficient should be close to unity. We note that for $k = 1$ the integral

$$\int_{\mathbb{R}^2} (f_2(A^{-1}(k\,\mathbf{x} - \mathbf{a})))^2\, d^2x \tag{13.4}$$

is just the square of norm of function f_2

$$\int_{\mathbb{R}^2} (f_2(\mathbf{x}))^2\, d^2x.$$

According to the Cauchy-Schwarz inequality,

$$\int f_1(\mathbf{x})\, f_2(\mathbf{x})\, d^2x \le \left[\int (f_1(\mathbf{x}))^2\, d^2x \int (f_2(\mathbf{x}))^2\, d^2x \right]^{1/2},$$

the correlation coefficient (13.3) is always smaller or equal to one, and it is equal to one for identical pattern and windowed image. We note that the value of correlation coefficient does not change if we change overall intensity of the original image or the template object.

For the dilation coefficient $k = 1$, we observe that the correlation function $q = q(\mathbf{a}, A)$ is a function on the Euclidean motion group $SE(2)$. It appears that this group has not been used extensively in applications to the image processing, the authors are aware of only a few previous works using this group, (e.g., [42, 55, 65]).

Using Fourier methods on the motion group, we can compute the correlation coefficient in a much more efficient way than using direct integration. Indeed, the direct computation of integral (13.3) is very costly (we consider for simplicity the $k = 1$ case). For $N_r = N_x \cdot N_y$ samples of the image (and template) on an $N_x \times N_y$ rectangular grid, and for N_R samples of orientation, we need to perform $O(N_r^2 N_R)$ computations (and we need to compute the convolution-like integrals twice, in the denominator and numerator of (13.3)). For $N_r = 256 \times 256$ and $N_R = 60$, the computations require $5 \cdot 10^{11}$ operations, which requires a day of computer work on a 250 MHZ workstation. In this chapter we use advantages of Fourier methods for the semi-discrete motion group (i.e., subgroup of $SE(2)$, where the orientation angle has discrete values from the C_{N_R} subgroup of $SO(2)$, $\theta = 2\pi i/N_R$ for $i = 0, ..., N_R - 1$), and Fast Fourier Transform (FFT) methods [25, 37] to compute the correlation coefficient in $O(N_R N_r \log N_r)$ computations. In addition, Fourier methods for the semi-discrete motion group provide a very fast method for comparison of two images which are translated and rotated relative to each other.

In the next subsection we briefly discuss applications of Fourier methods for the semi-discrete motion group in the context of the template matching problem.

13.1.2 Application of the FFT on $SE(2)$ to the Correlation Method

The convolution-like integrals in the numerator and denominator of (13.3) can be formally written (for simplicity we consider first the $k = 1$ case, the case including dilations is considered in Section 13.1.3) as integrals

$$c(\mathbf{x}, A_j) = \frac{1}{N_R} \sum_{i=0}^{N_R-1} \int_{\mathbb{R}^2} f_1(\mathbf{y}, A_i) f_2(A_j^{-1}(\mathbf{y} - \mathbf{x}), A_j^{-1} A_i) d^2y \qquad (13.5)$$

where the functions $f_{1,2}$ are orientation-independent, i.e., $f_{1,2} = f_{1,2}(\mathbf{x})$. The correlation function, however, is a function on the semi-discrete motion group G_{N_R}, so we can use the Fourier transform on G_{N_R} to write this integral as a product of Fourier transforms.

Because the functions are real the integral in the numerator of (13.3) can be written as

$$\frac{1}{N_R} \sum_{i=0}^{N_R-1} \int_{\mathbb{R}^2} \overline{f_1(\mathbf{y}, A_i)} \, f_2(A_j^{-1}(\mathbf{y} - \mathbf{x}), A_j^{-1} A_i) d^2y =$$

$$\int_{G_{N_R}} \overline{f_1(h)} \, f_2(g^{-1} \circ h) \, d\mu(h) \qquad (13.6)$$

where we denote integration over the semi-discrete motion group to mean integration over $\mathbb{R}^2$ and summation through the A_i, and the group elements are of the form $g = (\mathbf{x}, A_j)$. Note that (13.6) is not a convolution of f_1 and f_2. Rather, it is a correlation.

Using the orthogonality and homomorphism properties of the Fourier matrix elements, (13.6) can be written as

$$\frac{1}{N_R} \sum_{q=0}^{N_R-1} \sum_{n=0}^{N_R-1} \int_0^{2\pi/N_R} \int_0^\infty \sum_m \left(\overline{(\hat{f}_1)_{mn}} \, (\hat{f}_2)_{mq} \right) U_{qn}(g^{-1}; p, \phi) \, p \, dp \, d\phi = \qquad (13.7)$$

$$\frac{1}{N_R} \sum_{q=0}^{N_R-1} \sum_{n=0}^{N_R-1} \int_0^{2\pi/N_R} \int_0^\infty \sum_m \left(\overline{(\hat{f}_2)_{mq}} \, (\hat{f}_1)_{mn} \right) U_{nq}(g; p, \phi) \, p \, dp \, d\phi$$

where ϕ is measured from $2\pi q/N_R$. For the second expression we used the unitarity of the matrix elements U_{mn} and the fact that the expression is real. The matrices $(\hat{f}_{1,2})_{mn}$ are the G_{N_R}-Fourier transforms of the functions $f_{1,2}(\mathbf{x}, A_i)$. We note that this integral is the inverse Fourier transform of $\hat{f}_2^* \cdot \hat{f}_1$, and thus the expression depends only on three indices.

Because functions $f_{1,2}(\mathbf{x}, A_i) = f_{1,2}(\mathbf{x})$ do not depend on the orientations A_i, matrix elements in the same column are the same, i.e.,[2]

$$(\hat{f}_{1,2})_{mn} = (\hat{f}_{1,2})_{qn}$$

---

[2] Here we use the shorthand $f_{1,2}$ to indicate that the same holds for f_1 and f_2.

for any $m, q \in [0, N_R - 1]$. This can be observed from the expression

$$U_{mn}(g^{-1}; p, \phi) = e^{-i\, p\, \mathbf{u}_n^{\phi} \cdot \mathbf{a}}\, \delta_{A_i^{-1}\mathbf{u}_n, \mathbf{u}_m}, \tag{13.8}$$

the definition of the forward G_{N_R}-Fourier transform, and the fact that the functions $f_1(\mathbf{x})$ and $f_2(\mathbf{x})$ do not depend on the orientation. (Note that $g = (\mathbf{a}, A_i) \in G_{N_R}$, and the exponent in (13.8) depends only on the n-index.) Thus, we compute a row of the Fourier matrix for a particular orientation (for example $A_0 = \mathbb{I}_{N_R \times N_R}$)

$$(\hat{f}_{1,2})_n \doteq (\hat{f}_{1,2})_{nn}. \tag{13.9}$$

This can be done using the 2D FFT for the functions $f_{1,2}(\mathbf{x})$ and interpolating the Fourier values to points on a polar coordinate grid. The value of p is determined by $|\mathbf{p}|$, the values of m and v are determined by the angular part of $\mathbf{p}$. This requires $O(N_r \log(N_r))$ computations. Thus, the integrals in (13.3) can be written as

$$c(\mathbf{x}, A_j) = C \sum_{q=0}^{N_R-1} \sum_{n=0}^{N_R-1} \int_0^{2\pi/N_R} \int_0^{\infty} \left(\overline{(\hat{f}_2)_q(p, \phi)}\, (\hat{f}_1)_n(p, \phi) \right) U_{nq}(g; p, \phi)\, p\, dp\, d\phi \tag{13.10}$$

where $C = \frac{1}{N_R}$.

We observe that the convolution-like integrals can be computed by taking the Fourier transform, computing the product of transforms and taking the inverse Fourier transform for the semi-discrete motion group.

Invariants of the Semi-Discrete Motion Group

Let us assume that it is desirable to compute properties of the image (object) which are invariant with respect to translations and rotations of the image. The Fourier transform for the semi-discrete motion group provides a very efficient tool to compute these invariants. Let us construct a function with values in $\mathbb{R}_{\geq 0}$

$$\eta(p; \phi) = \sum_{m=0}^{N_R-1} [\overline{\hat{f}_m(p; \phi)}\, \hat{f}_m(p; \phi)] \tag{13.11}$$

for each fixed $\phi = 0, ..., N_\phi - 1$, where $\hat{f}_m(p; \phi)$ is the Fourier transform for the semi-discrete motion group of $f(\mathbf{x})$. Then (13.11) is invariant with respect to rotations and translations of $f(\mathbf{x})$, i. e. $\eta(p; \phi)$ does not change if we compute (13.11) using the Fourier transform for the semi-discrete motion group for $f'(\mathbf{x}) = f(A^{-1}(\mathbf{x} - \mathbf{a}))$.

We note that for orientation-independent functions (i.e., for functions on $\mathbb{R}^2$) the Fourier transform elements $\hat{f}_m$ can be arranged as a matrix which has the same matrix elements in the same column

$$\hat{f}_{qm} = \hat{f}_{rm} = \hat{f}_m.$$

Then (13.11) can be written also as a trace

$$\eta(p; \phi) = \mathrm{tr}[\hat{f}^*(p; \phi)\, \hat{f}(p; \phi)] \tag{13.12}$$

where $\hat{f}^*(p; \phi)$ is the Hermitian conjugate matrix.

Using the general definition of a group Fourier transform given in Chapter 8, $\hat{f}_{qm}$ for $f(\mathbf{x})$ can be written as

$$\hat{f}_{qm}(p;\phi) = \int_{G_{N_R}} f(h)\, U_{qm}^{-1}(h;p,\phi)\, d\mu(h)$$

where the integral over G_{N_R} denotes integration with respect to $\mathbf{x}$ *and* summation through the elements of C_{N_R}, and $f(h) = f(\mathbf{x})$. The function $f'(\mathbf{x}) = f(A_i^{-1}(\mathbf{x} - \mathbf{a}))$ can be formally written as $f(g^{-1} \circ h)$, where $g = (\mathbf{a}, A_i) \in G_{N_R}$. Then $\hat{f}'_{qm}$ is written as

$$\hat{f}'_{qm}(p;\phi) = \int_{G_{N_R}} f(g^{-1} \circ h)\, U_{qm}^{-1}(h;p,\phi)\, d\mu(h).$$

Using the invariance of the integration measure we can write this integral as

$$\hat{f}'_{qm}(p;\phi) = \int_{G_{N_R}} f(h')\, U_{qm}^{-1}(g \circ h';p,\phi)\, d\mu(h').$$

Using the homomorphism properties of U we can write it as

$$\hat{f}'_{qm}(p;\phi) = \sum_{r=0}^{N_R-1} \left[\int_{G_{N_R}} f(h')\, U_{qr}^{-1}(h';p,\phi)\, d\mu(h') \right] \cdot U_{rm}^{-1}(g;p,\phi)$$

$$= \sum_{r=0}^{N_R-1} \hat{f}_{qr}(p,\phi)\, U_{rm}^{*}(g;p,\phi)$$

where we have used a unitarity property of U. Thus, the Fourier matrix is transformed under rotations and translations $g \in G_{N_R}$ as

$$\hat{f}'(p,\phi) = \hat{f}(p,\phi)\, U^{*}(g;p,\phi).$$

Using the cyclic property of $\mathrm{tr}(\cdot)$ and unitarity of U it is clear that

$$\mathrm{tr}[(\hat{f}')^{*}(p,\phi)\hat{f}'(p,\phi)] = \mathrm{tr}[U(g;p,\phi)\,\hat{f}^{*}(p,\phi)\,\hat{f}(p,\phi)\,U^{*}(g;p,\phi)] =$$

$$\mathrm{tr}[\hat{f}^{*}(p,\phi)\,\hat{f}(p,\phi)]\ ,$$

which proves the invariance of (13.11). We note that the invariant, written in the form (13.12) is valid also for orientation-dependent functions (i.e., for general functions on the semi-discrete motion group). The use of invariants for pattern recognition was suggested in [42].

Efficiency of Computation of Convolution-Like Integrals

As we mentioned before, the direct integration of (13.3) requires $O(N_r^2 N_R)$ computations for C_{N_R}, where N_r is the number of sampling points in an $\mathbb{R}^2$ region. Using the Fourier transform for the semi-discrete motion group we have to compute direct Fourier transforms for image and template, compute the matrix product (in our case it is a column-row product) of the Fourier transform, which describes the Fourier transform of the convolved functions, and then calculate the inverse Fourier transform.

The calculation of direct Fourier transform and the "matrix" (column - row) product is a fast computation. The direct Fourier transform for $f_{1,2}(\mathbf{x})$ can be computed using a usual two-dimensional FFT [37] in $O(N_r \log N_r)$ computations. The FFT gives, however, values of Fourier elements computed on the Cartesian square (rectangular) grid of $\mathbf{p}$ values. To obtain the Fourier transform elements $\hat{f}_m(p, \phi)$ on the semi-discrete motion group we have to interpolate values on the Cartesian grid to a polar coordinate grid, the p value is the magnitude of $\mathbf{p}$, the m and ϕ indices are determined by the angular part of $\mathbf{p}$ (thus, the constraint $N_p N_\phi N_R \approx N_r$ can be used). The linear interpolation requires $O(N_r)$ computations. The product of Fourier column $\hat{f}_m^T(p, \phi)$ and row $\hat{f}_n(p, \phi)$, which defines the matrix $\hat{F}_{mn}(p, \phi)$, can be performed in $O(N_R^2 N_p N_\phi) = O(N_R N_r)$ computations. We note that the trace in invariants (13.11) can be computed in $O(N_R N_p N_\phi) = O(N_r)$ (for all ϕ-values). Thus, the direct Fourier transform and the "matrix" product can be computed in $O(N_r \log N_r) + O(N_R N_r)$ computations.

The inverse Fourier transform calculation is a slower computation. One element from each row and column of $\hat{F}_{mn}(p, \phi)$ is used in computation of the inverse Fourier transform for each rotation element A_i. First, we interpolate the value of the Fourier transform on the square grid $N_r \times N_r$ of $\mathbf{p}$ to polar coordinates. The radial coordinate is $p = |\mathbf{p}|$, the polar angle is determined by the values of m and ϕ (the value of n is determined by m and the index of rotation i, $n = m + i$, thus we take $\hat{F}_{m,m+i}$ elements from the Fourier matrix to compute the inverse transform for fixed orientation A_i). After inverse interpolation to Cartesian coordinates (which can be done in $O(N_R N_r)$ computations), the inverse Fourier integration can be performed in $O(N_r \log(N_r))$ for each of the N nonzero matrix elements of U using the FFT. Thus, in $O(N_R N_r \log(N_r))$ computations we reproduce the function for all A_i. We note that the inverse Fourier transform computation is $O(N_R)$ (or $O(\log N_r)$, depending which is larger) times more time-consuming, because we reproduce a function on the semi-discrete motion group, rather than a function on $\mathbb{R}^2$.

Thus, the total required is $O(N_R N_r \log(N_r))$ computations, and these computations are, for the most part, calculation of the inverse Fourier transform. We also note that we have to perform calculations twice, to compute convolution-like integrals in denominator and numerator of (13.3).

For a comparison with standard methods, see [61].

13.1.3 Numerical Examples

Correlation Method Including Rotations and Translations

In this section we compute the correlation function (13.3) (for dilation $k = 1$) for some practical examples. We compute most examples for $N_r = 256 \times 256$ and $N_R = 60$ (C_{60} group), although the computing time for other arrays is also reported.

We consider the image depicted in Figure 13.1. This is a 256×256 array of gray values (256 gray levels of intensity for each pixel). We note that because our implementation of the direct and inverse Fourier transforms on the motion group use spline interpolations, the image after application of direct and inverse Fourier transforms for the semi-discrete motion group is not reproduced with the same quality.

We choose a template, depicted in Figure 13.2, which is a rotated (at angle $\theta = -\pi/3$) and translated pattern taken from the image. The arrow shown on the picture is used as a reference arrow to find the position and orientation of this template in the image. The correlation function depicted for the $\theta = \pi/3$ angle is depicted in Figure 13.3. The highest value of the correlation function is at the original position and orientation of the pattern in the image. We also find positions and orientations of local maxima in each of $m \times m$ subregions of the original image. For $m = 8$ the positions and orientations

Fig. 13.1. The Image - 256 × 256 Array of Gray Values. (Reprinted from Kyatkin, A.B., Chirikjian, G.S., "Pattern Matching as a Correlation on the Discrete Motion Group," *Computer Vision and Image Understanding*, 74(1): 22 – 35, 1999, with permission from Elsevier)

of local maxima in each of subregions are shown in Figure 13.4. The highest value is depicted by the arrow, which is rotated and translated from the arrow in Figure 13.2. Other local maxima (with a value of correlation which is greater than 0.85) are depicted by a white square, small line attached to the square shows orientation.

We note that the precise values of correlation at the locations of 64 maxima can be found by direct integration, and the Fourier method can be used as a fast filter

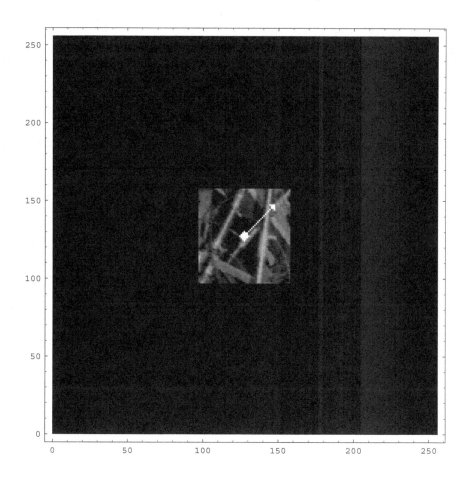

Fig. 13.2. The Template Pattern: The Arrow is Used as a Reference to Find Position and Orientation of This Pattern in the Image. (Reprinted from Kyatkin, A.B., Chirikjian, G.S., "Pattern Matching as a Correlation on the Discrete Motion Group," *Computer Vision and Image Understanding*, 74(1): 22 – 35, 1999, with permission from Elsevier)

method to find locations of these maxima. It is especially important to compute precise values in the case when the template object does not match exactly the pattern in the image. For example, for a template extracted from a filtered version of Figure 13.1 (which does not match exactly the corresponding pattern in Figure 13.1), the position of the absolute maximum found by the Fourier method (0.97) is located at the "wrong"

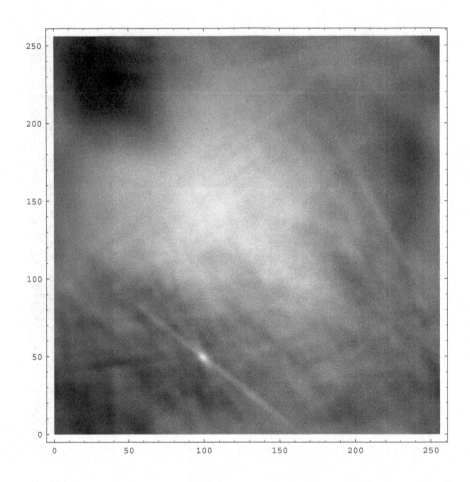

Fig. 13.3. The Correlation Function Depicted for the $\theta = \pi/3$ Orientation Angle. (Reprinted from Kyatkin, A.B., Chirikjian, G.S., "Pattern Matching as a Correlation on the Discrete Motion Group," *Computer Vision and Image Understanding*, 74(1): 22 – 35, 1999, with permission from Elsevier)

position, although the local maxima with the close value (0.91) of correlation coefficient is located at the precise position. Computation of precise values of correlation coefficients by direct integration at the locations of maxima finds the precise location of the pattern with correlation 0.954 (for identical template and pattern in the image correlation was 0.999).

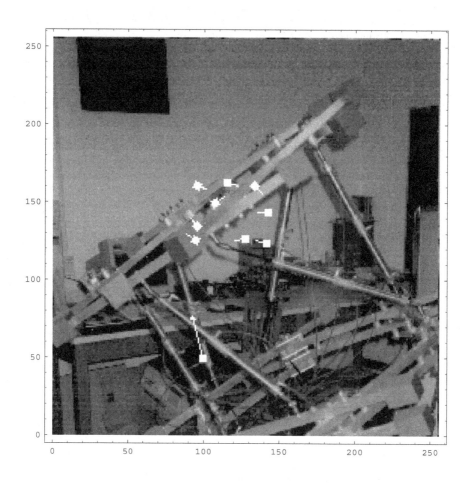

Fig. 13.4. The Image: Positions and Orientations of Absolute Maximum (Shown by Arrow) and Local Maxima of Correlation Function are Depicted. (Reprinted from Kyatkin, A.B., Chirikjian, G.S., "Pattern Matching as a Correlation on the Discrete Motion Group," *Computer Vision and Image Understanding*, 74(1): 22 – 35, 1999, with permission from Elsevier)

The approximate value of the correlation coefficient, which distinguishes the case when the template object is "found" from the case when it is "not-found" in the image, depends on the desired level of accuracy and general properties of the image (such as if it contains a background or it is a "black and white" contrast image). For images containing almost uniform background, like in the example discussed above, the level of "threshold" correlation must be high and can be set around 0.95.

The accuracy of the method in our examples was around 0.5 pixel. For example, for a rotation of $\pi/7$ (4.29 rotational pixels) and translations of 41.2 and 68.8 pixels of the template relative to the similar pattern in the image (they do not match exactly, i.e., the template taken from a filtered version of Figure 13.1), the found object was at the location 4 rotational pixels ($2\pi/15$) and translations of 41 and 69 pixels.

We note that the resolution of the Fourier method can be increased in some cases if we "precompute" the image to increase contrast. For images containing almost uniform background, application of the Laplacian operator (in Fourier space this is just a multiplication of Fourier transform by p^2) smoothed by a Gaussian weight function

$$\mathcal{F}(p) \to const\, p^2 \, \exp(-p^2/(2\sigma^2)) \, \mathcal{F}(p) \qquad (13.13)$$

(like in edge detection problems as in [18], but we do not compute "zero crossing" to get actual edges since we desire the overlap of the template and the image to be large), can increase the resolution. For example the template shown in Figure 13.5 produces a local maximum at the location of a similar (they do not match exactly) pattern in Figure 13.1. This still gives an absolute maximum at the "right" location when we compute precise values of correlation using direct integration at the locations of local maxima. Application of the Laplacian operator gives the image shown in Figure 13.6 for $\sigma^2 = p_{max}^2/12$ where $p_{max} = N_x/2 = N_y/2$ is the maximal p value, the $const$ in (13.13) was set to normalize the gray values of the image to the interval $[0, 1]$. When we use the Fourier method to find the position of the template (we also apply the Laplacian operator to the template), it produces an absolute maximum with correlation 0.71. The next closest local maximum has the value 0.63. Direct integration performed at the location of maxima gives the precise value of correlation 0.81. The next closest local maxima has a value 0.61. For the "black and white" contrast pictures the "threshold" correlation can be set around 0.8.

In the table below we listed the computing time of the method (given in minutes and seconds, on a 250 MHz SGI workstation), implemented in the C programming language. N_R is listed along the horizontal, the right column lists the time to compute correlation coefficients at 64 maxima using direct integration. The N_r array size is given along the vertical.

	$N_R = 60$	$N_R = 30$	$N_R = 10$	Dir. int.
$N_r = 256 \times 256$	4:06	2:11	0:48	0:55
$N_r = 128 \times 128$	1:12	0:36	0:16	0:13
$N_r = 64 \times 64$	0:25	0:12	0:04	0:03

Using the Invariants of the Motion Group to Compare Images

As we have shown before, function (13.11) is the same for images which are rotated and translated relative to each other. It can be used to compare images and determine if

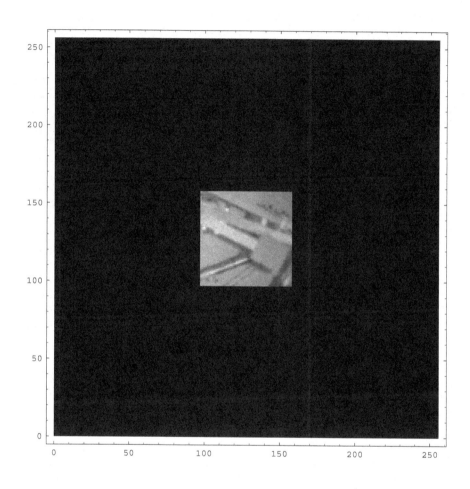

Fig. 13.5. The Template Object.

they are identical (similar) or not. Again, given image functions $f_{1,2}(\mathbf{x})$, we can compute the correlation coefficient of invariant functions $\eta_{1,2}(p, \phi)$ defined in (13.11). Then these can be compared as

$$\eta(\phi) = \frac{\int_0^\infty \eta_1(p, \phi)\, \eta_2(p, \phi) p\, dp}{(\int_0^\infty \eta_1(p, \phi)^2\, p\, dp)^{1/2} \, (\int_0^\infty \eta_2(p, \phi)^2\, p\, dp)^{1/2}}.$$

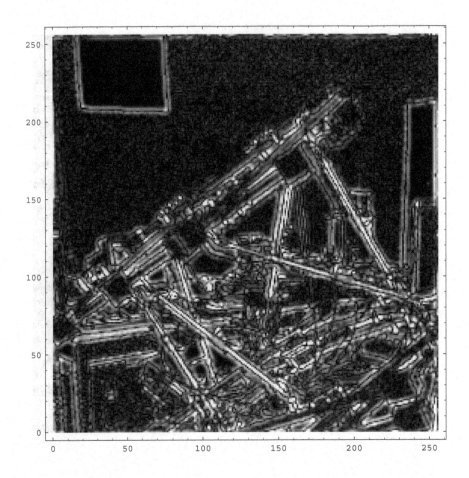

Fig. 13.6. The Image of Figure 13.1 After Application of "Smoothed" Laplacian Operator.

This is a fast computation of the order $O(N_p\,N_\phi) \approx O(N_r/N_R)$ (to compute N_ϕ coefficients) and it can be done using usual integration techniques. As we mentioned before, the direct Fourier transform can be computed in $O(N_r \log(N_r))$ arithmetic operations; computation of the sum in (13.11) can be done in $O(N_r)$ arithmetic operations.

We compare the images depicted in Figure 13.7 and Figure 13.8, which are just rotated and translated relative to each other. We use the value $\nu = (1.0 - \eta) \cdot 10^3$ to compare images, which is more convenient to use for η values which are close to 1.0. The

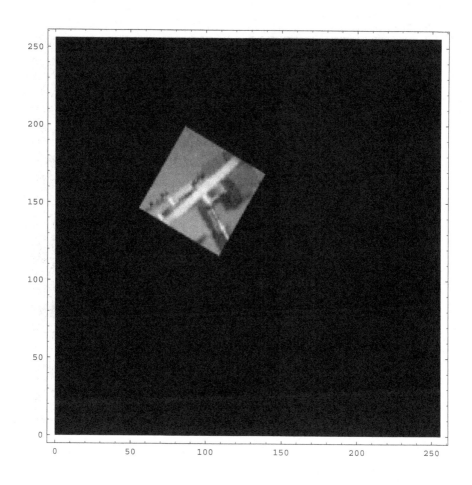

Fig. 13.7. A Window of the Original Image (Reprinted from Kyatkin, A.B., Chirikjian, G.S., "Pattern Matching as a Correlation on the Discrete Motion Group," *Computer Vision and Image Understanding*, 74(1): 22–35, 1999, with permission from Elsevier)

greater ν is, the worse the correlation is. In the table below we show ν for $\phi = 0, ..., 5$.

ϕ	0	1	2	3	4	5
ν	0.016	0.017	0.017	0.017	0.016	0.016

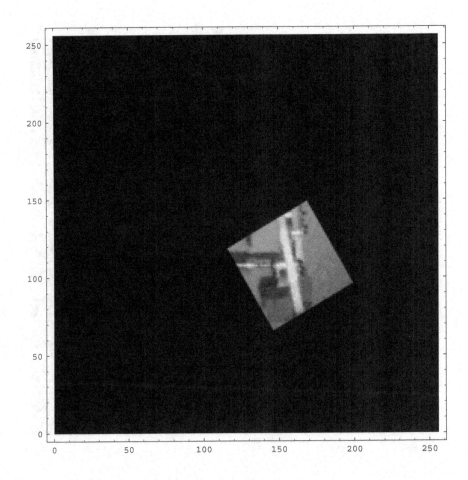

Fig. 13.8. A Rotated and Translated Version of the Image in Figure 13.7. (Reprinted from Kyatkin, A.B., Chirikjian, G.S., "Pattern Matching as a Correlation on the Discrete Motion Group," *Computer Vision and Image Understanding,* 74(1): 22–35, 1999, with permission from Elsevier)

We see that values are very close for different ϕ, thus we can use ν for any one of the ϕ values. The values are small which indicates very strong correlation.

The exact level to separate "identical" from "different" images depends on the array size and desired level of accuracy and must be set "by hand." For images of type depicted in Figures 13.7-13.8 the value $\nu \approx 0.5$ separates "identical" from "different" images. We note that when we compare images of different size, the ν value becomes large, if we compare Figure 13.1 and Figure 13.7 $\nu = 520$.

Time to compute direct Fourier transforms and correlations ν was around 3 sec on a 250 MHZ workstation.

Correlation Method Including Dilations

It can be shown that the convolution-like integral which includes dilations

$$c(\mathbf{x}, A_j, k) = \int_{\mathbb{R}^2} f_1(\mathbf{y}) f_2(A_j^{-1}(k\mathbf{y} - \mathbf{x})) d^2 y$$

can be written, in analogy with the expression (13.10) as

$$c(\mathbf{x}, A_j, k) = \frac{1}{N_R} \sum_{q=0}^{N_R-1} \sum_{n=0}^{N_R-1} \int_0^{2\pi/N_R} \int_0^{\infty} \left(\overline{(\hat{f}_2)_q(p, \phi)} \, (\hat{f}_1)_n(k\,p, \phi) \right) U_{nq}(g; p, \phi) \, p \, dp \, d\phi$$

$$(13.14)$$

The derivation of (13.14) is analogous to the derivation of (13.10). We also have to use the property of matrix elements of U:

$$U(k\,\mathbf{r}, A; p, \phi) = U(\mathbf{r}, A; k\,p, \phi)$$

which can be easily observed from the expression (13.8).

We also note that the integral (13.4) is equal to

$$\frac{1}{k^2} \int_{\mathbb{R}^2} (f_2(\mathbf{x}))^2 \, d^2 x.$$

Thus, to compute the correlation function in the case when dilations are allowed, we have to compute the direct Fourier transforms for $f_{1,2}(\mathbf{x})$ (which can be done in $O(N_r \log N_r)$ operations), multiply the Fourier transforms for each k from the discrete set of values N_k (which can be done in $O(N_k N_R N_r)$ operations), and take the inverse Fourier transform for the semi-discrete motion group for each k (which can be done in $O(N_k N_R N_r \log N_r)$ operations). In total, the computations are approximately N_k times more time-consuming.

We note that we also have to increase the number of subregions where we look for local maxima of correlation functions (because the number of local maxima is increased).

In Figure 13.9 we depict the template which is enlarged in size by the factor of 1.2, and compare with the similar pattern in the image (they are not identical). The scaling can be observed by comparing the pictures of Figure 13.2 and Figure 13.9. We computed the correlation function on a $256 \times 256 \times 60$ array for possible scaling (reduction) by factor 1.0, 1.1, 1.2 (computing time is around 12 minutes). Local maxima were found in a 12×12 grid of subregions of the image. The template object produces local maxima at the locations with similar pattern in the image (with correct scaling factor). We compute the correlation coefficient at the locations of local maxima. The "right" location and scaling factor gives the highest correlation, as depicted in Figure 13.10 (for correlation greater than 0.9 it is a single maximum).

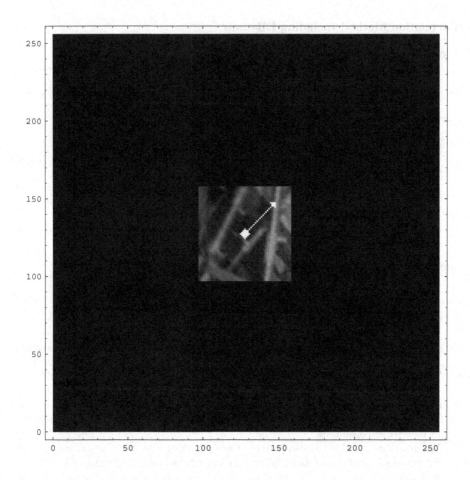

Fig. 13.9. The Template Object Enlarged by the Factor 1.2 Compared to the Similar Pattern in the Image.

13.2 The Radon Transform and Tomography

The Radon transform and its generalizations play an important role in fields as diverse as medical imaging [5, 30, 54, 93], electron microscopy, and radio astronomy [15]. The reconstruction of images from data collected in the form of a Radon transform is an inverse problem which has received much attention over the past twenty-five years.

Fig. 13.10. The Absolute Maximum Found by Direct Integration at the Location of Local Maxima.

New methods for the inversion of the Radon transform remain popular today (see, e.g., [49, 60, 91]).

In this section, it is shown how the Radon transform can be viewed as a convolution on the Euclidean motion group (group of rigid-body motions), where the convolution product is defined relative to the group operation. While this has been observed before in abstract settings (see, e.g., [43, 47, 84]), we go two steps further and shown: (1) how the Euclidean-group Fourier transform is used to rewrite the Radon transform as a matrix product; and (2) how the regularized inversion of the Radon transform is achieved with efficient and accurate numerical codes based on this group-theoretical approach. The key to this approach is the use of "fast" Fourier transforms for motion groups.

We show how this approach handles the case of the usual Radon transform, as well as the case of finite-width beams. We note that only one other work, [27], was found which deals with this kind of generalization of the Radon transform. In contrast, a number of works have dealt with the exponential/attenuated Radon transforms in the context of single photon emission tomography (see, e.g., [10, 22, 46, 75, 78, 96]) and their applications in radiation treatment planning [13] and inverse source problems [82]. While in principle exponential/attenuated Radon transforms can also be written as a matrix product upon application of the Euclidean group Fourier transform, we do not handle this case in this chapter for lack of an appropriate regularization technique.

In Subsection 13.2.1 the relationship between the Radon transform and motion-group convolutions is extablished. In Subsection 13.2.2 the derivation of the Radon transform for finite beam width is given. The application of Fourier transforms on motion groups to computed tomography algorithms is given in Subsection 13.2.3.

13.2.1 Radon Transforms as Motion-Group Convolutions

The Radon transform and its generalizations, such as the attenuated and exponential Radon transforms, play a central role in medical imaging and computed tomography (See e.g., [30] and references therein). This is because when an x-ray passes through an object it loses energy. Sensors on the other side of the object which measure the final energy in the rays effectively provide a means of determining the integral of the mass density of the object along the path of the x-ray. For recent works that address applications and techniques for inversion of the Radon transform see [16, 26, 36, 62].

In this section, we show how the Radon transform can be viewed as a convolution on the Euclidean motion group (group of rigid-body motions), where the convolution product is defined relative to the group operation. We then show how the Fourier transform of functions on the Euclidean motion group is used to rewrite the Radon transform as a matrix product. The regularized inversion of the Radon transform then reduces to matrix computations, followed by the inverse Fourier transform for the Euclidean motion group.

Given a function $f(\mathbf{x})$ on $\mathbb{R}^N$ which is integrable on each $(N-1)$-dimensional hyperplane in $\mathbb{R}^N$, the corresponding Radon transform is

$$\hat{f}(p, \boldsymbol{\xi}) = \mathcal{R}\{f(\mathbf{x})\} = \int_{\mathbb{R}^N} f(\mathbf{x})\, \delta(p - \boldsymbol{\xi} \cdot \mathbf{x})\, d\mathbf{x} \qquad (13.15)$$

where $d\mathbf{x} = dx_1...dx_N$ is the usual integration measure for $\mathbb{R}^N$, and $\delta(\cdot)$ is the Dirac delta function on the real line. In principle, if $f(\mathbf{x})$ is differentiable, and the value of Radon transform is known for all of these hyperplanes (each hyperplane in $\mathbb{R}^N$ is determined by the equation $\boldsymbol{\xi} \cdot \mathbf{x} = p$ where $\boldsymbol{\xi}$ is its unit normal), then the original function can be reconstructed using the inverse Radon transform.

The Radon transform has certain properties under affine transformations of the form: $\mathbf{x} \to A\mathbf{x} + \mathbf{b} \doteq a \circ \mathbf{x}$ where $A \in GL(N, \mathbb{R})$ (i.e., A is an invertible $N \times N$ matrix). Like the Euclidean motion group, the group of affine transformations of $\mathbb{R}^N$ can be expressed using homogeneous transform notation. While $a \circ \mathbf{x}$ denotes the action of the affine transformation a on $\mathbf{x} \in \mathbb{R}^N$, the same affine transform applied to a function on $\mathbb{R}^N$ yields

$$f(\mathbf{x}) \to f(a^{-1} \circ \mathbf{x}) = f(A^{-1}(\mathbf{x} - \mathbf{b})).$$

Applying the Radon transform to an affinely transformed function, we see that

$$\mathcal{R}\{f(A^{-1}(\mathbf{x} - \mathbf{b}))\} = \int_{\mathbb{R}^N} f(A^{-1}(\mathbf{x} - \mathbf{b})) \, \delta(p - \boldsymbol{\xi} \cdot \mathbf{x}) \, d\mathbf{x}.$$

Making the change of coordinates $\mathbf{y} = A^{-1}(\mathbf{x} - \mathbf{b})$, or equivalently $\mathbf{x} = A\mathbf{y} + \mathbf{b}$, we see that $d\mathbf{x} = |\det(A)| d\mathbf{y}$ and

$$\mathcal{R}\{f(A^{-1}(\mathbf{x} - \mathbf{b}))\} = |\det(A)| \int_{\mathbb{R}^N} f(\mathbf{y}) \, \delta \left(p - \boldsymbol{\xi} \cdot (A\mathbf{y} + \mathbf{b}) \right) \, d\mathbf{y}$$

$$= |\det(A)| \hat{f}(p - \boldsymbol{\xi} \cdot \mathbf{b}, A^T \boldsymbol{\xi}).$$

In the planar case we write explicitly

$$\hat{f}(p,\theta) = \hat{f}(p, \boldsymbol{\xi}(\theta)) = \int_{-\infty}^{\infty} \int_{-\infty}^{\infty} f(x_1, x_2) \, \delta(p + x_1 \sin\theta - x_2 \cos\theta) \, dx_1 dx_2.$$

Johann Radon proved in 1917 that for the case when $N = 2$, the following inversion formula holds:

$$f(r,\phi) = \frac{1}{2\pi^2} \int_0^\pi \int_{-\infty}^{\infty} \frac{\partial \hat{f}}{\partial p} \frac{dp}{r \sin(\phi - \theta) - p} \, d\theta$$

where (r, ϕ) are the polar coordinates for the corresponding Cartesian coordinates (x_1, x_2). For $N = 3$ we have

$$f(\mathbf{x}) = -\frac{1}{8\pi^2} \nabla^2 \left(\int_{\mathbb{S}^2} \hat{f}(\boldsymbol{\xi} \cdot \mathbf{x}, \boldsymbol{\xi}) \, ds(\boldsymbol{\xi}) \right)$$

where $\mathbb{S}^2$ is the unit sphere in three dimensions, ∇^2 is the Laplacian, and $ds(\boldsymbol{\xi})$ is the area element on this sphere (see, e.g., [30, 43, 47]).

Extensions of the inversion formula to other dimensions are also well known (though, as can be seen above the formulae, for even and odd dimensions the inversion formulae are different). Regardless, in many practical applications, it is not possible to simply use this inversion formula because data is sampled discretely (i.e., only a finite number of measurements are taken), and the original function $f(\mathbf{x})$ may in fact not be continuous, much less differentiable.

To complicate matters further, in many practical applications, it is not the Radon transform in (13.15), but rather a *generalized, attenuated,* or *exponential* Radon transform which descibes the physical scenario of interest [22, 75, 78, 96]. For example, instead of obtaining discrete measurements of $\hat{f}(p, \boldsymbol{\xi})$, in an experimental setting one might observe

$$\hat{f}_\mu(p, \boldsymbol{\xi}) = \int_{\mathbb{R}^2} f(\mathbf{x}) \exp\left(-\int_{0 \le \mathbf{x}' \cdot \boldsymbol{\xi}_\perp \le \mathbf{x} \cdot \boldsymbol{\xi}_\perp} \mu(\mathbf{x}') \, \delta(p - \mathbf{x}' \cdot \boldsymbol{\xi}) \, d\mathbf{x}' \right) \delta(p - \mathbf{x} \cdot \boldsymbol{\xi}) \, d\mathbf{x} \quad (13.16)$$

where $\boldsymbol{\xi}_\perp(\theta) = (-\sin\theta, \cos\theta)$ satisfies $\boldsymbol{\xi}_\perp \cdot \boldsymbol{\xi} = 0$. In the general case when $\mu(\mathbf{x})$ is completely unknown, this problem cannot be solved. In the cases when $\mu(\cdot)$ is a known nonlinear function of $\rho(\mathbf{x})$ and/or $\mathbf{x}$, the inversion is still a hopelessly difficult problem. While the Radon transform in (13.15) corresponds to the case when $\mu(\mathbf{x}) = 0$, in practice consideration of the case when $\mu(\mathbf{x}) = \mu_0 \neq 0$ is a better model for the real situation, which leaves the following linear inverse problem to solve for $f(\mathbf{x})$:

$$\hat{f}_{\mu_0}(p, \boldsymbol{\xi}) = \int_{\mathbb{R}^2} f(\mathbf{x}) \exp\left(-\mu_0 \mathbf{x} \cdot \boldsymbol{\xi}_\perp\right) \delta(p - \mathbf{x} \cdot \boldsymbol{\xi}) \, d\mathbf{x}. \tag{13.17}$$

Our main observation in this section is that the Radon transform its generalizations such as (13.17) can be viewed as a Euclidean-group convolution written in the form:

$$\int_{SE(N)} k(g_2 \circ g_1^{-1}) f(g_1) \, d(g_1) = (k * f)(g_2). \tag{13.18}$$

If we view a function on the motion group, $f(g)$, as a function on the Cartesian product of the translation and rotation subgroups, $f(\mathbf{r}, R)$, the convolution as written in (13.18) can be rewritten as

$$(f_2 * f_1)(\mathbf{r}, R) = \int_{SO(N)} \int_{\mathbb{R}^N} f_1(\mathbf{x}, Q) f_2(-RQ^T \mathbf{x} + \mathbf{r}, RQ^T) \, d\mathbf{x} dQ$$

where dQ is the invariant integration measure on the rotation group and $d\mathbf{x}$ is the usual differential volume element in $\mathbb{R}^N$. By making the choice

$$f_1(\mathbf{x}, Q) = f(\mathbf{x}) \, \delta(Q) \qquad f_2(\mathbf{x}, Q) = \delta(\mathbf{e}_1 \cdot \mathbf{x})$$

where $\delta(x_1)$ is the usual Dirac delta function on the x_1 coordinate axis in $\mathbb{R}^N$ and $\delta(Q)$ is the Dirac delta function on the rotation group which has the properties

$$\int_{SO(N)} \delta(Q) \, dQ = 1 \qquad \delta(Q) = \delta(Q^T) \qquad \int_{SO(N)} f(Q) \, \delta(Q^T R) \, dQ = f(R),$$

we see that

$$(f_1 * f_2)(\mathbf{q}, R) = \int_{\mathbb{R}^N} f(\mathbf{x}) \left[\int_{SO(N)} \delta(Q) \, \delta(\mathbf{e}_1 \cdot (\mathbf{r} - RQ^T \mathbf{x})) \, dQ \right] d\mathbf{x}$$

$$= \int_{\mathbb{R}^N} f(\mathbf{x}) \, \delta(\mathbf{e}_1 \cdot \mathbf{r} - \mathbf{x} \cdot (R^T \mathbf{e}_1)) \, d\mathbf{x} = \hat{f}(\mathbf{r} \cdot \mathbf{e}_1, R^T \mathbf{e}_1).$$

While the above choice of functions $f_1(\cdot)$ and $f_2(\cdot)$ yields the Radon transform, this choice is not the only one that will work. For example, recalling that the *conjugate* of a complex-valued function $f(g)$ on a group G is defined as $f^*(g) \doteq \overline{f(g^{-1})}$, we observe that

$$(f_1 * f_2^*)(g) = \int_G f_1(h) f_2(g^{-1} \circ h) d(h) = \int_G f_1(g \circ h) f_2(h) \, d(h).$$

This is useful in the current context, because when $G = SE(N)$ we can write

$$(f_1 * f_2^*)(\mathbf{r}, R) = \int_{SO(N)} \int_{\mathbb{R}^N} f_1(R\mathbf{x} + \mathbf{r}, RQ) f_2(\mathbf{x}, Q) \, d\mathbf{x} dQ.$$

Choosing $f_1(\cdot)$ and $f_2(\cdot)$ as follows:

$$f_1(\mathbf{r}, R) = \delta(\mathbf{r} \cdot \mathbf{e}_1) \qquad f_2(\mathbf{r}, R) = f(-\mathbf{r}) \tag{13.19}$$

(where $f(\cdot)$ is real-valued), then

$$(f_1 * f_2^*)(\mathbf{r}, R) = \text{Vol}(SO(N)) \cdot \int_{\mathbb{R}^N} \delta((R^t \mathbf{e}_1) \cdot \mathbf{x} + \mathbf{r} \cdot \mathbf{e}_1) f(-\mathbf{x}) \, dx.$$

By proper normalization of the volume element for $SO(N)$, $\text{Vol}(SO(N)) = 1$. Making the change of coordinates $\mathbf{x} \to -\mathbf{x}$ we see that the choice in (13.19) yields

$$(f_1 * f_2^*)(\mathbf{r}, R) = \hat{f}(\mathbf{r} \cdot \mathbf{e}_1, R^T \mathbf{e}_1). \tag{13.20}$$

Thus we have a choice in defining the functions f_1 and f_2 in such a way that their convolution results in the Radon transform. Similar manipulations can be performed for the $\mu_0 \neq 0$ case.

13.2.2 The Radon Transform for Finite Beam Width

We consider below the 2D computed tomography problem for the case when finite-width x-ray beams are used for taking parallel projections, i.e., when several detectors (that are each several pixels wide) are used for taking the parallel projection values. We assume that the beam width is Δ and, as in ordinary computed tomography approaches for parallel beams, the parallel projections are taken for angles $0 \leq \theta \leq \pi$ at distances ξ from the origin to the center line of the beam. In order to get an expression for the ratio of the incident and transmitted beam intensity we consider the $\theta = 0$ beam, shifted at distance $x = \xi$ from the origin. We denote $I(y)$ to be the total energy flow along the direction of propagation at coordinate y. The intensity of the beam $i(x, y)$ (the energy flow per unit length, in the direction of propagation) can be written as

$$I(y) = \int_{-\infty}^{\infty} i(x, y) \, dx$$

where the integral (the beam width can be infinite) denotes the integral along the cross-section of the beam, centered at $x = \xi$. Then, considering two close cross sections of the beam located at coordinates y and $y + dy$, taking the small "volume element" of length dy and cross-section (which is just a length in the planar case) dx and using the conservation of energy, the equation for $i(x, y)$ can be written as

$$(i(x, y + dy) - i(x, y)) \, dx = -(\mu(x, y) \, i(x, y)) \, dx \, dy$$

The right part of the equation shows the energy loss (mostly due to scattering and photoelectric absorption at small photon energies [53]) which is effectively taken into account by the linear attenuation coefficient $\mu(x, y)$. This equation is illustrated in Figure 13.11. Integrating with respect to x along the cross section of the beam, the equation can be written as

$$\int_{-\infty}^{\infty} \frac{\partial i(x, y)}{\partial y} \, dx = -\int_{-\infty}^{\infty} \mu(x, y) \, i(x, y) \, dx. \tag{13.21}$$

We write the intensity of the beam in the form

$$i(x, y) = j(y) \, f^\xi(x, y),$$

where $f^\xi(x, y)$ is a dimensionless function which describe the intensity "profile" along the cross-section of the beam, located at ξ.

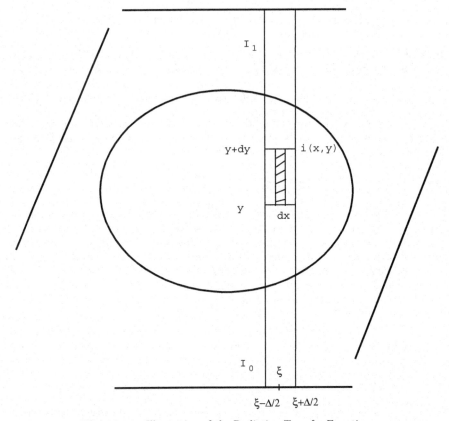

Fig. 13.11. Illustration of the Radiation Transfer Equation.

We assume also that the intensity beam profile is not changing along the propagation path of the beam, i.e., we are looking for solutions to (13.21) of the form

$$i(x, y) = j(y) f^\xi(x) .$$

We assume also that the profile of the beam located at ξ is just a shifted version of the beam located at $\xi = 0$

$$f^\xi(x) = f^0(x - \xi) = f(x - \xi) .$$

Then, (13.21) takes the form

$$\frac{dj(y)}{dy} \int_{-\infty}^{\infty} f(x - \xi) \, dx = - \left(\int_{-\infty}^{\infty} \mu(x, y) \, f(x - \xi) \, dx \right) j(y) .$$

This equation can be solved in the form

$$j(y) = j(y_0) \exp\left(-\frac{\left(\int_{y_0}^{y} \int_{-\infty}^{\infty} \mu(x, y) \, f(x - \xi) \, dx dy \right)}{\int_{-\infty}^{\infty} f(x) \, dx} \right) \tag{13.22}$$

where y_0 is the coordinate of the source and y is the detector coordinate. We used the invariance of integration of functions on the real line under shifts to simplify

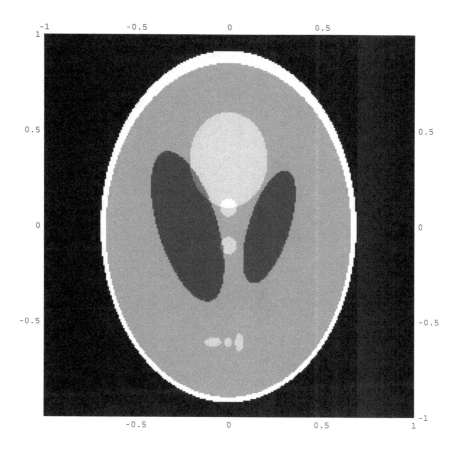

Fig. 13.12. The Shepp-Logan Image Phantom.

$$\int_{-\infty}^{\infty} f(x - \xi)\, dx \;=\; \int_{-\infty}^{\infty} f(x)\, dx \ .$$

The ratio of the total energy flow $I(y)/I(y_0)$ is the same as $j(y)/j_0$ in the assumption of constant beam profile. The logarithm of this ratio is the measured quantity (projection value) and it is equal to

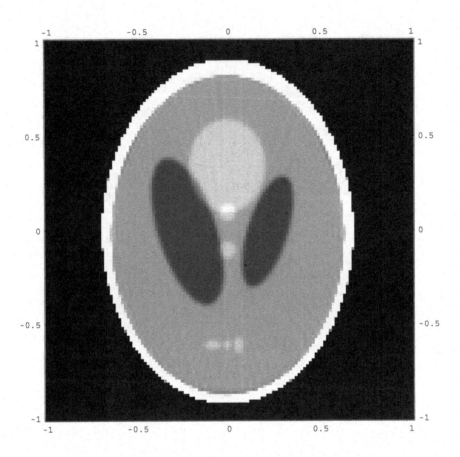

Fig. 13.13. The Shepp-Logan Image Reconstruction.

$$\rho(0,\xi) = -\log(I(y)/I(y_0)) = -\log(j(y)/j_0) = \frac{\left(\int_{y_0}^{y}\int_{-\infty}^{\infty}\mu(x,y)\,f(x-\xi)\,dx\,dy\right)}{\int_{-\infty}^{\infty}f(x)\,dx}.$$

We can view the beam profile function as a function on the plane, which does not depend on the y coordinate, i.e., $f(\mathbf{r}) = f(x)$. This also can be written using the scalar product as $f(\mathbf{r}) - f(\mathbf{e}_1 \cdot \mathbf{r})$, where $\mathbf{e}_1$ is the unit vector along the x direction. For the shift $\mathbf{a}$ the beam profile function centered at $\xi = (\mathbf{e}_1 \cdot \mathbf{a})$ can be written as

$$f(x - \xi) = f(\mathbf{e}_1 \cdot (\mathbf{r} - \mathbf{a})) \ .$$

For arbitrary orientation angle of the beam θ and shift ξ the beam profile function can be described by the rotated and shifted version of $f(\mathbf{r})$:

$$f(\mathbf{e}_1 \cdot (R^{-1}\mathbf{r} - \mathbf{a})) \ ,$$

where

$$R^{-1} = \begin{pmatrix} \cos\theta & \sin\theta \\ -\sin\theta & \cos\theta \end{pmatrix}, \tag{13.23}$$

describes the planar $SO(2)$ rotation. Then, the projection $\rho(\theta, \xi)$ can be written as

$$\rho(\theta, \xi) = \frac{\left(\int_{-\infty}^{\infty} \int_{-\infty}^{\infty} \mu(\mathbf{r}) f(\mathbf{e}_1 \cdot (R^{-1}\mathbf{r} - \mathbf{a})) \, d^2 r \right)}{\int_{-\infty}^{\infty} f(x) \, dx} \tag{13.24}$$

where we have use the relation

$$\int_{-\infty}^{\infty} f(x) \, dx = \int_{-\infty}^{\infty} f(\mathbf{e}_1 \cdot (R^{-1}\mathbf{r} - \mathbf{a})) \, d\mathbf{n}_\perp$$

for integration along the direction $\mathbf{n}_\perp$ perpendicular to the propagation path. Equation (13.24) also can be derived writing an energy transfer equation for the rotated beam.

Equation (13.24) can be written explicitly in (x, y) components as

$$\rho(\theta, \xi) = \frac{\left(\int_{-\infty}^{\infty} \int_{-\infty}^{\infty} \mu(x, y) f(x \cos\theta + y \sin\theta - \xi) \, dx \, dy \right)}{\int_{-\infty}^{\infty} f(x) \, dx} \ .$$

For $f(x) = \delta(x)$ this equation reproduces the ordinary Radon transform (see [54] and references therein).

13.2.3 Computed Tomography Algorithm on the Motion Group

We observe that the numerator of the finite-beam-width Radon transform equation in (13.24) can be viewed as a correlator on the 2D Euclidean motion group of the form

$$c(\mathbf{x}, R) = \int_{\mathbb{R}^2} f_1(\mathbf{y}) f_2(R\mathbf{y} + \mathbf{x}) \, d^2 y \tag{13.25}$$

where R is an $SO(2)$ rotation matrix and $\mathbf{x}$ is the translation vector. As with the template-matching problem, this is a correlation.

We observe that the correlation integral (13.25) can be written in Fourier space as[3]

$$A \, \overline{\hat{f}_n^{(1)}(p)} \, \hat{f}_m^{(2)}(p) = \hat{c}_{nm}(p) \ . \tag{13.26}$$

Here A is the area of the compact region in $\mathbb{R}^2$ where the FFT is computed.

Then, the solution for $\hat{f}_n^{(1)}$ can be written as

---

[3]In this section we use the notation $\hat{f}_n^{(i)}(p)$ to denote the semi-discrete-motion-group Fourier transform of a function $f_i(\mathbf{x})$ as in (13.9). Here $n \in [0, N_R - 1]$ where N_R is the number of orientations and A is a scalar constant.

$$\hat{f}_n^{(1)}(p) = \frac{\overline{\hat{c}_{nm}(p)}}{A\,\hat{f}_m^{(2)}(p)} \tag{13.27}$$

where m is some fixed value of the matrix index, and we also take $\phi = 0$. We note that in cases when $\hat{f}_m^{(2)}(p) = 0$ for some p, regularization is needed. We discuss regularization methods next.

For the case of the Radon transform the left part of (13.25) is a function of $\mathbf{e}_1 \cdot \mathbf{a} = \xi$ rather than a function of $\mathbf{a}$. This means that only the $\hat{c}_{n,0}(p)$ and $\hat{c}_{n,N_R/2}(p)$ Fourier elements are nonzero. The $m = N_R/2$ index corresponds to the $\phi = \pi$ slice of the Fourier transform. We assume for simplicity that N_R is an even number. For odd N_R the $\phi \neq 0$ index corresponding to $\phi = \pi$ has to be taken. This can be observed from the fact that

$$\frac{1}{(2\pi)^2} \int_{-\infty}^{\infty} \int_{-\infty}^{\infty} c(x)\exp(-ip_1 x)\,\exp(-ip_2\,y)\,dx\,dy \;=\; c(p_1)\,\delta(p_2)\;. \tag{13.28}$$

The Fourier transform on the group can be found (using the discrete version of (13.28)) by interpolating the values on the Cartesian grid to the polar grid. It is clear from (13.28) that only the values for $p_2 = 0$ (i.e., along the p_1 axis) are nonzero, which means that only $m = 0$ and $m = N_R/2$ components give contributions.

Taking the 2D Fourier transform of projection slice $c(x,\theta)$ corresponds to computation

$$\mathcal{F}(c(x,\theta)) \;=\; \hat{c}_i\,\delta_{k0} \tag{13.29}$$

where i,k correspond to the first and second component of the motion-group Fourier vector. The values in (13.29) correspond to $\hat{c}_{n0}(p)$ (for $i \geq 0$)(and $\hat{c}_{n+N_R/2,N_R/2}(p)$ for $i < 0$) motion group Fourier transform matrix elements, where $n = -j$ for $\theta = (2\pi/N_R)\,j$ rotation angle, and $p > 0$.

We compute for convenience the correlation

$$\eta(\theta,\xi) \;=\; \frac{\left(\int_{-\infty}^{\infty}\int_{-\infty}^{\infty} \mu(\mathbf{r})\,f(\mathbf{e}_1 \cdot (R\mathbf{r}+\mathbf{a}))\,d^2 r\right)}{\int_{-\infty}^{\infty} f(x)\,dx}\;, \tag{13.30}$$

which is related with measured quantity (13.24) as

$$\eta(\theta,\xi) \;=\; \rho(-\theta,-\xi)\;.$$

We note that due to the relation

$$\rho(\theta,\xi) \;=\; \rho(\theta+\pi,-\xi) \tag{13.31}$$

the relation

$$\eta(\theta,\xi) \;=\; \rho(-\theta+\pi,\xi) \tag{13.32}$$

is also valid.

The Radon Transform as a Particular Case of Correlation on the Motion Group

For the case of the ordinary Radon transform, i.e., for intensity profile function $f(x) = f_2(x) = \delta(x)$, the solution of the motion group convolution equation takes the form

$$\hat{f}_n^{(1)}(p) \;=\; \frac{\overline{\hat{\eta}_{n0}(p)}}{A\,\hat{f}_0^{(2)}(p)} \tag{13.33}$$

in Fourier space (the function $\eta(\theta, \xi)$ can be computed from the measured projections $\rho(\theta, \xi)$ via (13.32)), and the corresponding equation

$$\hat{f}^{(1)}_{n+N_R/2}(p) = \frac{\overline{\hat{\eta}_{n+N_R/2,N_R/2}(p)}}{A\,\overline{\hat{f}^{(2)}_{N_R/2}(p)}} \tag{13.34}$$

for index $N_R/2$ in the right part of the equation. Equation (13.34) is equivalent to considering $p < 0$ values. $\eta_n(p)$ are radial slices in Radon transform with finite width beam, and $f_{N_R/2}(p)$ is Fourier transform of the beam profile function. $f^1_n(p)$ corresponds to the slices in the ordinary Radon transform (i.e. it is the the Fourier transform of $f_1(r)$).

In fact, because for the rotation A_j the relations between indices m and n in $\hat{\eta}_{mn}$ is $m = n - j$, for the rotation θ ($\theta = 2\pi j/N_R$) we find that (13.33) takes the form

$$\hat{f}^{(1)}_n(p) = \frac{\overline{\hat{\eta}_{n0}(p)}}{L} \tag{13.35}$$

(and the corresponding equation for $N_R/2$ index), where $n = -j$ for the projection enumerated by j ($\theta = 2\pi j/N_R$ rotation angle), L is a linear dimension of the discrete Fourier transform. Here we use the fact that the discrete 2D Fourier transform of $\delta(x)$ can be written as

$$\hat{\delta} = \frac{1}{L}\,\delta_{k0}$$

where k corresponds to the second component (dual to y) of the Fourier vector (as we mentioned before it means that $\hat{f}^{(2)}_0(p) = 1/L$ for the motion group Fourier transform). The $\delta(x)$ is approximated as $(L/N_R)\,\delta_{i0}$ (i corresponds to the x component) on a discrete 2D grid with $L/N_R \times L/N_R$ pixel size.

Thus, (13.35) provides an algorithm for image reconstruction. Using $j = 0, ..., N_R/2$ rotation angles (for 0 and $N_R/2$ indices) we can reproduce the Fourier transform of $\hat{f}^{(1)}$ on the polar grid. Interpolation in Fourier space to the Cartesian grid and, then, the 2D FFT can be used. In this case a technique more complicated than linear interpolation has to be used, such as spline interpolation [24, 63, 94], or as described in [95]. Otherwise, the backprojection algorithm (see [54] and references therein) can be used, in which case linear interpolation in the spatial domain gives good results. Recall that the backprojection algorithm is based on the expression [54]

$$f_1(\mathbf{r}) = \int_0^\pi \left[\int_{-\infty}^\infty \hat{f}_1(p, \theta)|p|\exp(i2\pi p\,t)\,dp\right] d\theta \tag{13.36}$$

where $\hat{f}_1(p, \theta)$ is a radial slice of the usual Abelian Fourier transform of $f_1(\mathbf{r})$ where $\mathbf{r} \in \mathbb{R}^2$, and $t = x\cos\theta + y\sin\theta$. Due to the symmetry relation (13.31) the backprojection can be also computed through the $(\pi, 2\pi)$ interval

$$f_1(\mathbf{r}) = \int_\pi^{2\pi} \left[\int_{-\infty}^\infty \hat{f}_1(p, \theta)|p|\exp(i2\pi p\,t)\,dp\right] d\theta .$$

The Fourier slice $\hat{f}_1(p, \theta)$ at angle $\theta = 2\pi n/N_R$ corresponds to the motion group Fourier transform matrix elements

$$\hat{f}_1(p, \theta) = \hat{f}^{(1)}_n(p)$$

for $p \geq 0$, and

$$\hat{f}_1(p, \theta) = \hat{f}^{(1)}_{n+N_R/2}(|p|)$$

for $p < 0$.

We also note that for real function $f(\mathbf{x})$,

$$\overline{\hat{f}_n(p)} = \hat{f}_{n+N_R/2}(p)$$

which reflects the fact that $\hat{f}(\mathbf{p}) = \overline{\hat{f}(-\mathbf{p})}$ for the ordinary 2D transform of real function. For convenience we formally allow p to take $p < 0$ values and denote

$$\hat{f}^{(1)}_{n+N_R/2}(|p|) = \hat{f}^{(1)}_n(p)$$

for $p < 0$.

Thus, (13.35) provides the Fourier slice

$$\hat{f}^{(1)}_{-j}(p) = \hat{f}_1(p, -2\pi j/N_R)$$

which can be used in (13.36) for filtered backprojection.

Using motion group notation we write the backprojection algorithm as

$$f_1(\mathbf{r}) = 2\pi/N_R \sum_{j=0}^{N_R/2-1} \left[\int_{-\infty}^{\infty} \hat{f}^{(1)}_{-j}(p) |p| \exp(i2\pi p\, t)\, dp \right] \tag{13.37}$$

where $t = x\cos(2\pi j/N_R) - y\sin(2\pi j/N_R)$, and $\hat{f}^{(1)}_{-j}(p)$ are computed using (13.35).

We have shown in Figure 13.13 the backprojection reconstruction (13.36) for the Shepp and Logan "head phantom" [89] (depicted in Figure 13.12) using expression (13.37). The analytical expression for the projections and $N_R/2 = 60$ rotation angles were used. The image was located on a 128×128 grid, but the projection data were zero padded to a larger 256×256 grid to avoid the effect of periodic approximation to the continuous Fourier transform [54]. The $|p|$ multiplier in (13.36) was replaced by the Fourier transform of

$$h(0) = N_x N_x/4; \quad h(i) = 0; \text{ for even } i;$$

$$h(i) = -\frac{N_x N_x}{(\pi^2 i^2)} \text{ for odd } i$$

($N_x = 256$ and $-N_x/2 < i \leq N_x/2$) in order to improve accuracy, and low pass filtering was used to reduce high-frequency "noise" [54]. We have used the Gaussian filter

$$g(p) = \exp(-\frac{p^2}{2\sigma^2})$$

which is measured in units of maximal "momentum" $p_{max} = 2\pi N_R/(2L)$, $\sigma = c_1 p_{max}$. Using the notation $p = 2\pi i/L$, where $-N_R/2 < i \leq N_R/2$, we write the filter in the discrete form as

$$g(i) = \exp(-\sigma_1 \frac{i^2}{N_R^2}) \tag{13.38}$$

where $\sigma_1 = 2/c_1^2$. The reconstruction image depicted in Figure 13.12 corresponds to the $\sigma_1 = 4$ value, i. e. for $\sigma = p_{max}/\sqrt{2}$.

The Backprojection Algorithm for Finite-Width Beams

It is clear from the expression in (13.33) that the backprojection algorithm can be easily modified to include the finite beam case. For finite beams a non-constant $\hat{f}_0^{(2)}(p)$ has to be used in the denominator of (13.33) and (13.34) to correctly reproduce the left side of (13.35). The $\hat{f}_0^{(2)}(p)$ Fourier element is just the ordinary Fourier transform of the intensity profile function $f_2(x) = f(x)$.

Below we consider the case when the profile function is a step-like function

$$f(x) = \begin{cases} 1 \text{ if } |x| \le \Delta/2 \\ 0 \text{ otherwise} \end{cases}.$$

Then the denominator in (13.24) is just the width of the beam Δ. We assume that Δ is measured in an integer number of pixels. In this case the analytical expression for the projection values can be easily found so we can test the accuracy of the algorithms on an analytically defined model.

The analytical expressions for the projections of an ellipse

$$x^2/a^2 + y^2/b^2 = 1$$

with constant linear attenuation coefficient q are given by

$$\rho(\theta,\xi) = \frac{qab}{d(\theta)\,\Delta}\left((\xi+\Delta/2)\sqrt{d(\theta)-(\xi+\Delta/2)^2} - \right.$$

$$(\xi-\Delta/2)\sqrt{d(\theta)-(\xi-\Delta/2)^2}+$$

$$\left. d(\theta)\left(\arcsin\left(\frac{(\xi+\Delta/2)}{d(\theta)}\right) - \arcsin\left(\frac{(\xi-\Delta/2)}{d(\theta)}\right)\right)\right)$$

if $-\sqrt{d(\theta)} \le (\xi-\Delta/2)$ and $(\xi+\Delta/2) \le \sqrt{d(\theta)}$. Here $d(\theta) = a^2(\cos\theta)^2 + b^2(\sin\theta)^2$. For $(\xi-\Delta/2) \le \sqrt{d(\theta)} < (\xi+\Delta/2)$ we find the expression

$$\rho(\theta,\xi) = \frac{qab}{d(\theta)\,\Delta}(-(\xi-\Delta/2)\sqrt{d(\theta)-(\xi-\Delta/2)^2}$$

$$+d(\theta)(\pi/2 - \arcsin(\frac{(\xi-\Delta/2)}{d(\theta)}))\,.$$

For $(\xi-\Delta/2) < -\sqrt{d(\theta)} \le (\xi+\Delta/2)$ the expression is

$$\rho(\theta,\xi) = \frac{qab}{d(\theta)\,\Delta}((\xi+\Delta/2)\sqrt{d(\theta)-(\xi+\Delta/2)^2}+$$

$$d(\theta)(\arcsin(\frac{(\xi+\Delta/2)}{d(\theta)}) + \pi/2))\,.$$

The projection values are zero otherwise. We assume for simplicity that $\Delta < a$ and $\Delta < b$.

The Fourier transform of the profile functions is calculated as

$$f_0^{(2)}(i) = L\frac{\sin(\pi/L\,\Delta i)}{\pi\Delta i} \tag{13.39}$$

where $-N_R/2 < i < N_R/2$. Here L is a linear dimension of the image, $\delta = L/N_R$ is the pixel size.

Numerical Results

We observe that the expression (13.39) can be equal to zero for some values of i (i.e., for some values of p). For the odd number $N_1 = \Delta/\delta$ (for power-of-two number of pixels N_R) (13.39) is nonzero, while for even number of pixels regularization is required.

Recall that our goal is to invert (13.26) for given A, $\hat{f}_m^{(2)}(p)$ and $\hat{c}_{nm}(p)$. When $\hat{f}_m^{(2)}(p) \approx 0$ then (13.34) is not a valid expression, and we must seek regularized solutions. One common method of regularizing expressions of the form $\hat{f}_i \hat{g}_i = \hat{h}_i$ for known $\hat{g}_i$ and $\hat{h}_i$ is Tikhonov regularization [45]

$$\hat{f}_i = \frac{\hat{h}_i \overline{\hat{g}_i}}{\epsilon + \hat{g}_i \overline{\hat{g}_i}} .$$

However this technique does not yield reconstructions with acceptable accuracy in this context. Instead, we use spline interpolation to find $\hat{f}_i$ when $\hat{g}_i \approx 0$. Neighboring values of $\hat{f}$ can be found by (13.34), and the value at a singular point can be found using spline interpolation.

We used the spline interpolation technique of degree three [29] to get the Fourier value at the singular points. The Fourier transform was interpolated by a piecewise cubic function with continuous second order derivatives. A fixed number of interpolating points was used. For the 12-point interpolating interval the value of the Fourier transform at the singular point enumerated by i was interpolated as

$$\hat{f}_i = m_{i-1}/12 + m_{i+1}/12 + (\hat{f}_{i-1} - 2/3\, m_{i-1})/2 + (\hat{f}_{i+1} - 2/3\, m_{i+1})/2 ,$$

$-N_R/2 < i \leq N_R/2$ enumerates discrete p values, and values of the second order derivatives m_{i-1}, m_{i+1} were found from the solution of a system of linear equations with tridiagonal matrix [29]. The boundary conditions for the interpolating interval were specified, we used the natural splines (zero second order derivatives) and splines with the given slope $\hat{f}'$ at the ends of interval. Accuracy of the same order was achived for both of these boundary conditions.

In Figure 13.14 the Shepp-Logan-like computed simulation model is reconstructed using the ordinary backprojection algorithm, while the beam width is 5 pixels. We have listed the parameters of ellipses for simulations in Table 13.1.

Center x Coordinate	Center y Coordinate	Major Axis	Minor Axis	Rotation Angle	Refractive Index
0.0	0.0	0.92	0.69	$\pi/2$	2.0
0.0	-0.0184	0.874	0.6624	$\pi/2$	-0.98
0.22	0.0	0.31	0.11	1.25664	-0.02
-0.22	0.0	0.41	0.16	1.88496	-0.02
0.0	0.35	0.25	0.21	$\pi/2$	0.01
0.0	0.1	0.087	0.087	0.0	0.01
0.0	-0.1	0.087	0.087	0.0	0.01
-0.08	-0.605	0.09	0.087	0.0	0.01
0.08	-0.605	0.09	0.087	$\pi/2$	0.01

Table 13.1: Parameters Defining the Model

It is assumed that image is the sum of ellipses with the parameters given previously.

Figure 13.15 depicts the reconstructed image using algorithm (13.33) which takes into account the beam width. The Gaussian filter (13.38) with $\sigma_1 = 8$ was used.

The improvement in the image quality can be clearly observed.

Taking into account the beam width in the singular case also improves quality when the object size is comparable with the beam width. In Figure 13.16(a) we depicted the reconstructed projection of the Shepp-Logan-like image for a beam width of 24 pixels, reconstructed using the ordinary backprojection algorithm. Figure 13.16(b) depicts the reconstruction (singular case) which takes into account beam width ($\sigma_1 = 20$ was used). The image is reconstructed with better accuracy, even though the deviation of order 0.4% can be observed. The non-singular case (23 pixels width) reproduces the image with much better accuracy, see Figure 13.16(bottom) ($\sigma_1 = 12$ was used).

13.3 Inverse Tomography: Radiation Therapy Treatment Planning

Radiation therapy treatment planning (which also goes by other related names such as radiotherapy planning or radiation treatment planning) is the field concerned with delivering the best dose of radiation to a patient in order to damage or distroy a tumor using radiation, while causing no more than an acceptable amount of damage to surrounding healthy tissue. Radiation therapy treatment planning can be delivered either by external beams using a columnator, or by internal placement of radioactive material (either by ingestion or by a surgical procedure). When we refer to the radiation therapy treatment planning problem, we will mean radiation is delivered through external-beams. This problem has been studies extensively (see, e.g., [19, 20, 67, 68, 87, 88, 90] and references therein). For recent studies in medical physics and therapy planning see [14, 17, 28, 69, 70, 71, 79, 80, 81, 83, 101]. The book by Webb [100] serves as a very nice introduction to this topic.

In group-theoretic language, the problem can be stated as follows. A columnator produces a beam at position and orientation g. Associated with this beam is a certain normailzed profile (shape) that can be described as a function $b(\mathbf{x}, g)$ where $\mathbf{x} \in \mathbb{R}^3$. The form of the function $b(\mathbf{x}, g)$ (which describes the relative damage that the beam does to point $\mathbf{x}$ as compared to other positions in space when the beam originates from the origin of the frame g and is directed along an axis) depends on how tissue attenuates and/or distorts the cross sectional profile of the beam, and what the beam profile is as it exits the columnator.

In the highly idealized case when it is assumed that there is no attenuation or beam spreading, and the beam-profile is taken to have the same pin-point support for all positions and orientations of the columnator, then

$$b(\mathbf{x}, g) = b_0(g^{-1} \circ \mathbf{x})$$

where $b_0(\mathbf{x}) = \delta(x_1)\delta(x_2)$ is a function that has support on the x_3 axis, and constant values on this axis. In this idealized model, the question becomes how to specify the intensity, $i(g)$, of the beam for each position and orientation of the columnator so that the total dose is as close as possible to the desired one, $t(\mathbf{x})$. This is written as

$$t(\mathbf{x}) = \int_G i(g) b_0(g^{-1} \circ \mathbf{x}) \, dg. \tag{13.40}$$

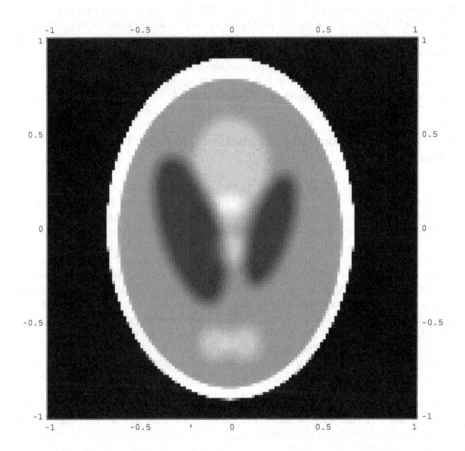

Fig. 13.14. Image Reconstruction for the Finite Beam Case Using the Ordinary Backprojection Algorithm

While our explanation of (13.40) was for a highly idealized function $b_0(\cdot)$, there are a number of ways in which $b_0(\cdot)$ can be made more realistic. For instance, in the reference frame $g = e$, the axis of the columnator is x_3, and the beam originates at $x_3 = 0$ and travels in the postive x_3 direction. The beam profile can be taken to be a

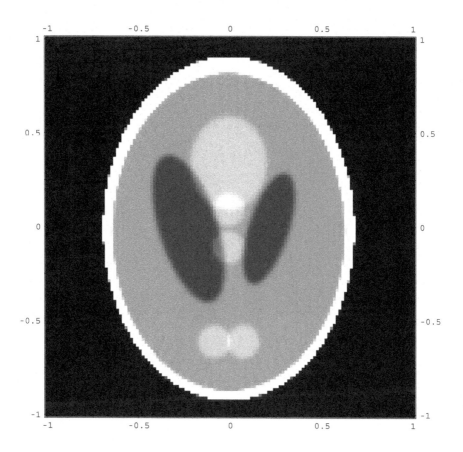

Fig. 13.15. Image Reconstruction which Takes into Account the Beam Width

Gaussian function with intensity that decreases with increasing value of x_3 and width that increases with x_3. That is,

$$b_0(\mathbf{x}) = c_1(x_3) \exp(-c_2(x_3)(x_1^2 + x_2^2))$$

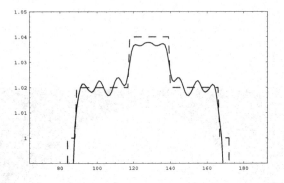

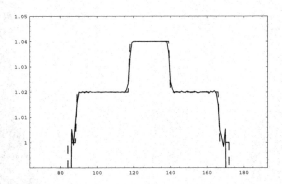

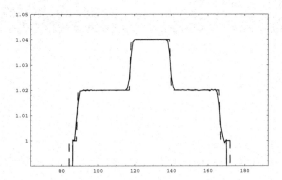

Fig. 13.16. (top) The Projection of Reconstructed Image Without Taking into Account Beam Width; (middle) The Projection of Reconstructed Image for Finite Width Beam (Singular Case); (bottom) The Projection of Reconstructed Image for Finite Width Beam (Non-Singular Case)

where $c_1(x)$ and $c_0(r_0)$ decrease with x_3. This is a model that has been used to describe the evolution of a profile of a beam of high energy photons through water. Other kinds of profiles have been developed for other kinds of beams (e.g., protons).

The point is, (13.40) can be used to describe a wide variety of superposition problems involving beams of radiation eminating from different frames of reference provided the attenuation throughtout the environment is constant. If $c_2(\cdot)$ is approximately constant and the attenuation constant within the patient is approximately constant, then the solution of (13.40) for $i(g)$ can be used as a first step in determining a treatment plan. The next step would be to compensate for the fact that the beam does not attenuate nearly as much in air as it does in the patient, and to modify $i(g)$ accordingly.

Since (13.40) can be written as a convolution on $G = SE(3)$, we can either use an FFT for motion groups to rapidly evaluate candidate functions $i(g)$ (treatment plans) or to attempt the regularized inversion of this convolution equation for $i(g)$ when $t(\mathbf{x})$ is given. This remains for future work.

13.4 Summary

In the first part of this chapter we posed the template matching problem in image analysis as a correlation on the motion group of the plane, and performed a fast implemenentation using the Fourier transform for the semi-discrete motion group. We also showed how a comparison of image invariants based on this group-theoretic Fourier transform yields a method to determine how alike two objects are, regardless of their position and orientation. Generalizations of these ideas have been used to do pose detection of 3D objects [50, 51].

In the second part of this chapter we showed how the Radon transform, and its generalization to the case of finite-width beams, can be viewed as a convolution (correlation) of functions on the group of rigid-body motions. By discretizing the rotational part of this group into equal increments, the semi-discrete motion group results. The Fourier transform of functions on this semi-discerete motion group is used as a tool to reduce the tomographic reconstruction problem to a scalar algebraic equation in a generalized Fourier space. Numerical results using this approach were presented and appear to be promising. Other applications of group-theoretic aspects of the Radon transform to omnidirectional vision can be found in [38].

In recent years, a number of interesting approaches to imaging that use harmonic analysis have been presented in the literature. For example, the representation theory of $SE(2)$ has been used in [31, 32, 33, 34, 39, 41, 59]. Harmonic analysis on $SO(3)$ has been used in robots equipped with fish-eye lenses that project images onto a sphere [72, 73, 74] to assist in localization. Moreover, in the field of crystallographic texture analysis the concept of Radon transforms on $SO(3)$ arise [7, 8, 9], and diffusion-based wavelets on groups akin to those discussed in Chapter 8 have been explored for solving the associated inverse Radon transform [21]. Texture analysis will be discussed in Chapter 18.

The use of group-theoretic methods for other image analysis and pattern matching remains a topic of active interest as is evidenced by publications such as [12, 35, 66, 99]. In addition, in the area of perception, learning, and machine intelligence other interesting group-theoretic problems arise such as those described in [4, 6, 40, 57, 58, 76, 98].

References

1. Abu-Mostafa, Y.S., Psaltis, D., "Recognition Aspects of Moment Invariants," *IEEE Transansactions on Pattern Analysis and Machine Intelligence*, 6(6): 698–706, 1984.
2. Antonini, M., Barlaud, M., Mathieu, P., Daubechies, I., "Image Coding Using Wavelet Transform," *IEEE Transactions on Image Processing*, 40(2): 205–220, 1992.
3. Arsenault, H.H., Hsu, Y.N., Chalasinska-Macukow, K., "Rotation-Invariant Pattern Recognition," *Optical Engineering*, 23(6): 705–709, 1984.
4. Barbieri, D., Citti, G., Cocci, G., Sarti, A., "A Cortical-Inspired Geometry for Contour Perception and Motion Integration," *Journal of Mathematical Imaging and Vision*, 49(3): 511–529, 2014.
5. Barrett, H.H., Swindell, W., *Radiological Imaging: The Theory of Image Formation, Detection and Processing*, Academic Press, New York, 1981.
6. Bayro-Corrochano, E., *Geometric Computing: for Wavelet Transforms, Robot Vision, Learning, Control and Action*, Springer, 2010.
7. Bernstein, S., Ebert, S., Pesenson, I.Z., "Generalized Splines for Radon Transform on Compact Lie Groups with Applications to Crystallography," *Journal of Fourier Analysis and Applications*, 19(1): 140–166, Feb. 2013.
8. Bernstein, S., Hielscher, R., Schaeben, H., "The Generalized Totally Geodesic Radon Transform and Its Application to Texture Analysis," *Mathematical Methods in the Applied Sciences*, 32(4): 379–394, 2009.
9. Bernstein, S., Schaeben, H., "A One-Dimensional Radon Transform on $SO(3)$ and Its Application to Texture Goniometry," *Mathematical Methods in the Applied Sciences*, 28(11): 1269–1289, 2005.
10. Beylkin, G., "The Inversion Problem and Applications of the Generalized Radon Transform," *Communications on Pure and Applied Mathematics*, 37(5): 579–599, 1984.
11. Bhatia, A.B., Wolf, E., "On the Circle Polynomials of Zernike and Related Orthogonal Sets," *Proceedings of the Cambridge Philosophical Society*, 50(1): 40–48, 1954.
12. Bigot, J., Loubes, J.M., Vimond, M., "Semiparametric Estimation of Shifts on Compact Lie Groups for Image Registration," *Probability Theory and Related Fields*, 152(3–4): 425–473, 2012.
13. Bortfeld, T.R., Boyer, A.L., "The Exponential Radon Transform and Projection Filtering in Radiotherapy Planning," *International Journal of Imaging Systems and Technology*, 6(1): 62–70, 1995.
14. Bortfeld, T., "An Analytical Approximation of the Bragg Curve for Therapeutic Proton Beams," *Medical Physics*, 24(12): 2024–2033, Dec. 1997.
15. Bracewell, R.N., Riddle, A.C., "Inversion of Fan-Beam Scans in Radio Astronomy," *Astrophysical Journal* 150: 427–434, 1967.
16. Brady, M.L., "A Fast Discrete Approximation Algorithm for the Radon Transform," *SIAM Journal of Computing*, 27(1): 107–119, Feb. 1998.

512   References

17. Brugmans, M.J.P., van der Horst, A., Lebesque, J.V., Mijnheer, B.J., "Dosimetric Verification of the 95% Isodose Surface for a Conformal Irradiation Technique," *Medical Physics*, 25(4): 424–434, April 1998.

18. Canny, J., "A Computational Approach to Edge Detection," *IEEE Transactions on Pattern Analysis and Machine Intelligence*, 8(6): 679–698, 1986.

19. Censor, Y., Altschuler, M.D., Powlis, W.D., "A Computational Solution of the Inverse Problem in Radiation-Therapy Treatment Planning," *Applied Mathematics and Computation*, 25(1): 57–87, 1988.

20. Censor, Y., Altschuler, M.D., Powlis, W.D., "On the Use of Cimmino's Simultaneous Projections Method for Computing a Solution of the Inverse Problem in Radiation Therapy Treatment Planning," *Inverse Problems*, 4(3): 607–623, 1988.

21. Cerejeiras, P., Ferreira, M., Kähler, U., Teschke, G., "Inversion of the Noisy Radon Transform on $SO(3)$ by Gabor Frames and Sparse Recovery Principles," *Applied and Computational Harmonic Analysis*, 31(3): 325–345, 2011.

22. Chang, L.-T., "Attenuation Correction and Incomplete Projection in Single Photon Emission Computed Tomography," *IEEE Transactions on Nuclear Science*, 26(2): 2780–2789, April 1979.

23. Chirikjian, G.S., "Fredholm Integral Equations on the Euclidean Motion Group." *Inverse Problems* 12(5): 579–599, Oct. 1996.

24. Choi, H., Munson, D.C., Jr. "Direct-Fourier Reconstruction in Tomography and Synthetic Aperture Radar," *International Journal of Imaging Systems and Technology*, 9(1): 1–13, 1998.

25. Cooley, J.W., Tukey, J., "An Algorithm for the Machine Calculation of Complex Fourier Series," *Mathematics of Computation*, 19(90): 297–301, 1965.

26. Cools, O.F., Herman, G.C., van der Weiden, R.M., Kets, F.B., "Fast Computation of the 3-D Radon Transform," *Geophysics*, 62(1): 362–364, Jan-Feb. 1997.

27. Cormack, A.M., "Sampling the Radon Transform with Beams of Finite Width," *Phys. Med. Biol.*, 23(6): 1141–1148, 1978.

28. Das, I.J., McGee, K.P., Cheng, C., "Electron-Beam Characteristics at Extended Treatment Distances," *Medical Physics*, 22(10): 1667–1674, Oct. 1995.

29. Davis, P.J., Rabinowitz, P., *Methods of Numerical Integration*, 2nd ed., Dover, 2007.

30. Deans, S.R., *The Radon Transform and some of Its Applications*, Dover, 2007.

31. Duits, R., van Almsick, M., Duits, M., Franken, E., Florack, L.M.J., "Image Processing via Shift-Twist Invariant Operations on Orientation Bundle Functions," in 7th *International Conference on Pattern Recognition and Image Analysis: New Information Technologies*, (Niemann Zhuralev et al. Geppener, Gurevich, editors), pp. 193-196, St. Petersburg, October 2004.

32. Duits, R., Felsberg, M., Granlund, G., ter Haar Romeny, B., "Image Analysis and Reconstruction Using a Wavelet Transform Constructed from a Reducible Representation of the Euclidean Motion Group," *International Journal of Computer Vision*, 72(1): 79–102, 2007.

33. Duits, R., Franken, E., "Left-invariant Parabolic Evolutions on $SE(2)$ and Contour Enhancement via Invertible Orientation Scores Part I: Linear left-invariant Diffusion Equations on SE(2)," *Quarterly of Applied Mathematics*, 68(2): 255–292, 2010.

34. Duits, R., Franken, E., "Left-invariant Parabolic Evolutions on $SE(2)$ and Contour Enhancement via Invertible Orientation Scores. Part II: Nonlinear Left-invariant Diffusions on Invertible Orientation Scores," *Quarterly of Applied Mathematics*, 68(2): 293–331, 2010.

35. Duits, R., Franken, E., "Left-invariant Diffusions on the Space of Positions and Orientations and Their Application to Crossing-Preserving Smoothing of HARDI Images," *International Journal of Computer Vision*, 92(3): 231–264, 2011.

36. Defrise, M., Clack, R., Townsend, D.W., "Image Reconstruction from Truncated, Two-Dimensional, Parallel Projections," *Inverse Problems*, 11(2): 287–313, 1995.

37. Elliott, D.F., Rao, K.R., *Fast Transforms: Algorithms, Analyses, Applications*, Academic Press, New York, London, 1982.

38. Falcón, L.E., Bayro-Corrochano, E., "Radon Transform and Harmonical Analysis Using Lines for 3D Rotation Estimation without Correspondences from Omnidirectional Vision," in *11th IEEE International Conference on Computer Vision (ICCV 2007)*, (pp. 1–6), October, 2007.

39. Ferraro, M., Caelli, T.M., "Lie Transformation Groups, Integral Transforms, and Invariant Pattern Recognition," *Spatial Vison*, 8(1): 33–44, 1994.

40. Földiák, P., "Learning Invariance from Transformation Sequences," *Neural Computation*, 3(2): 194–200, 1991.

41. Franken, E.M., *Enhancement of Crossing Elongated Structures in Images*, Ph.D. thesis, Eindhoven University of Technology, Department of Biomedical Engineering, 2008, Eindhoven, The Netherlands.

42. Gauthier, J.P., Bornard, G., Sibermann, M., "Motion and Pattern Analysis: Harmonic Analysis on Motion Groups and Their Homogeneous Spaces," *IEEE Transactions on Systems, Man, and Cybernetics*, 21(1): 159–172, 1991.

43. Gel'fand, I.M., Graev, M.I., Vilenkin, N.Ya., *Generalized Functions*, Academic Press, 1966.

44. Grenander, U., Miller, M.I., *Pattern theory: from representation to inference*, Oxford: Oxford University Press, 2007.

45. Groetsch, C.W., *The Theory of Tikhonov Regularization for Fredholm Equations of the First Kind*, Pitman, Boston, 1984.

46. Heike, U., "Single-Photon Emission Computed Tomography by Inverting the Attenuated Radon Transform with Least-Squares Collocation," *Inverse Problems*, 2(3): 307–330, 1986.

47. Helgason, S., *The Radon Transform*, Progress in Mathematics Vol. 5, Birkhäuser, 1980.

48. Hladky, R.K., Pauls, S.D., "Minimal Surfaces in the Roto-Translation Group with Applications to a Neuro-Biological Image Completion Model," *Journal of Mathematical Imaging and Vision*, 36(1): 1-27, 2010.

49. Holschneider, M., "Inverse Radon Transforms through Inverse Wavelet Transforms," *Inverse Problems* 7(6): 853–861, 1991.

50. Hoover, R.C., Maciejewski, A.A., Roberts, R.G., "Pose Detection of 3-D Objects Using Images Sampled on $SO(3)$, Spherical Harmonics, and Wigner-D Matrices," *Proceedings of the 2008 IEEE International Conference on Automation Science and Engineering (CASE 2008)*, pp. 47–52, Washington D.C., August, 2008.

51. Hoover, R.C., Maciejewski, A.A., Roberts, R.G., "Fast Eigenspace Decomposition of Images of Objects with Variation in Illumination and Pose," *IEEE Transactions on Systems, Man, and Cybernetics, Part B: Cybernetics*, 41(2): 318–329, 2011.

52. Jahne, B., *Spatio-Temporal Image Processing: Theory and Scientific Applications*, Springer-Verlag, Berlin, 1993.

53. Johns, H.E., Cunningham, J.R., *The Physics of Radiology*, 4th ed., Charles C Thomas Pub. Ltd., Springfield, Illinois, 1983.

54. Kak, A., Slaney, M., *Principles of Computerized Tomographic Imaging*, SIAM Press, 2001.

55. Kanatani, K., *Group-Theoretical Methods in Image Understanding*, Springer-Verlag, Berlin, Heidelberg, New York, 1990. (softcover edition, 2011).

56. Karim, M.A., Awwal, A.A.S., *Optical Computing: An Introduction*, Wiley, New York, 1992.

57. Knill, D.C., Kersten, D., Yuille, A., "Introduction: A Bayesian Formulation of Visual Perception," *Perception as Bayesian Inference*, pp. 1–21, D.C. Knill and W. Richards, eds., Cambridge University Press, 1996.

58. Kondor, R., *Group Theoretical Methods in Machine Learning*, PhD Dissertation, Columbia University, 2008.

59. Kondor, R., "A Complete Set of Rotationally and Translationally Invariant Features for Images," arXiv preprint cs/0701127

60. Kuchment, P., Lancaster, K., Mogilevskaya, L., "On Local Tomography," *Inverse Problems*, 11(3): 571–589, 1995.

61. Kyatkin, A.B., Chirikjian, G.S., "Pattern Matching as a Correlation on the Discrete Motion Group, " *Computer Vision and Image Understanding*, 74(1): 22–35, April 1999.

62. Lanzavecchia, S., Bellon, P.L., "Fast Computation of 3D Radon Transform via a Direct Fourier Method," *Bioinformatics*, 14(2): 212–216, 1998.

63. La Riviere, P.J., Pan, X., "Spline-based Inverse Radon Transform in Two and Three Dimensions," *IEEE Transactions on Nuclear Science*, 45(4): 2224–2231, August 1998.

64. Leduc, J.-P., "Spatio-temporal Wavelet Transforms for Digital Signal Analysis," *Signal Processing*, 60(1): 23–41, 1997.

65. Lenz, R., *Group Theoretical Methods in Image Processing*, Lecture Notes in Computer Science, Springer-Verlag, Berlin, Heidelberg, New York, 1990.

66. Li, L., Li, S., Abraham, A., Pan, J.S., "Geometrically Invariant Image Watermarking Using Polar Harmonic Transforms," *Information Sciences*, 199:1–19, 2012.

67. Lind, B., "Properties of an Algorithm for Solving the Inverse Problem in Radiation Therapy," *Inverse Problems*, 6(3): 415–426, 1990.

68. Lind, B., Brahme, A.,"Development of Treatment Techniques for Radiotherapy Optimization," *International Journal of Imaging Systems and Technology*, 6(1): 33–42, 1995.

69. Liu, H.H., Mackie, T.R., McCullough, E.C., "Correcting Kernal Tilting and Hardening in Convolution/Superposition Dose Calculations for Clinical Divergent and Polychromatic Photon Beams," *Medical Physics*, 24(11): 1729–1741, Nov. 1997.

70. Llacer, J., "Inverse Radiation Treatment Planning Using the Dynamically Penalized Likelihood Method," *Medical Physics*, 24(11): 1751–1764, Nov. 1997.

71. Lovelock, D.M.J., Chui, C.S., Mohan, R., "A Monte Carlo model of Photon Beams Used in Radiation Therapy," *Medical Physics*, 22(9): 1387–1394, Sept. 1995.

72. Makadia, A., Sorgi, L., Daniilidis, K., "Rotation Estimation from Spherical Images," in *Proceedings of the 17th International Conference on Pattern Recognition (ICPR 2004)*, Vol. 3, pp. 590–593, August, 2004.

73. Makadia, A., Daniilidis, K., "Rotation Recovery from Spherical Images without Correspondences," *IEEE Transactions on Pattern Analysis and Machine Intelligence*, 28(7): 1170–1175, 2006.

74. Makadia, A., Daniilidis, K.,"Spherical Correlation of Visual Representations for 3D Model Retrieval," *International Journal of Computer Vision*, 89(2-3): 193–210, 2010.

75. Markoe, A., "Fourier Inversion of the Attenuated X-Ray Transform," *SIAM Journal on Mathematical Analysis*, 15(4): 718–722, July 1984.

76. Miao, X., Rao, R.P.N., "Learning the Lie Groups of Visual Invariance," *Neural Computation*, 19(10): 2665–2693, 2007.

77. Murenzi, R., "Wavelet Transforms Associated to the n-Dimensional Euclidean Group with Dilations: Signals in More Than One Dimension," in *Wavelets: Time-Frequency Methods and Phase Space* (J.M. Combes, A. Grossmann, Ph. Tchamitchian, eds) pp. 239–246, 1990.

78. Natterer, F., "On the Inversion of the Attenuated Radon Transform," *Numerische Mathematik*, 32(4): 431–438, 1979.

79. Nizin, P.S., Mooij, R.B., "An Approximation of Central-Axis Absorbed Dose in Narrow Photon Beams," *Medical Physics*, 24(11): 1775–1780, Nov. 1997.

80. Olivares-Pla, M., Podgorsak, E.B., Pla, C., "Electron Arc Dose Distributions as a Function of Beam Energy," *Medical Physics*, 24(1): 127–132, Jan. 1997.

81. Ostapiak, O.Z., Zhu, Y., Van Dyk, J., "Refinements of the Finite-Size Pencil Beam Model of Three-Dimensional Photon Dose Calculations," *Medical Physics*, 24(5): 743–750, May 1997.

82. Panchenko, A.N., "Inverse Source Problem of Radiative Transfer: A Special Case of the Attenuated Radon Transform," *Inverse Problems*, 9(2): 321–337, 1993.

83. Perry, D., Wollin, M., Olch, A., Buffa, A., "Range Spectra in Electron Penetration Problems," *Medical Physics*, 25(1): 43–55, Jan. 1998.

84. Pintsov, D.A., "Invariant Pattern Recognition, Symmetry, and the Radon Transforms," *Journal of the Optical Society of America A*, 6(10): 1544–1554, Oct. 1989.

85. Prince, J.L., Links, J., *Medical Imaging Signals and Systems*, 2nd ed., Prentice Hall, 2014

86. Radon, J., "Über die Bestimmung von Funktionen durch ihre Integralwerte längs gewisser Mannigfaltigkeiten," *Berichte Sächsische Akademie der Wissenschaften. Leipzig*, 69: 262–267, 1917.

87. Raphael, C., "Mathematical Modelling of Objectives in Radiation Therapy Treatment Planning," *Physics in Medicine and Biology*, 37(6): 1293–1311, 1992.

88. Raphael, C., "Radiation Therapy Treatment Planning: An $\mathcal{L}^2$ Approach," *Applied Mathematics and Computation*, 52: 251 – 277, 1992.

89. Rosenfeld, A., Kak, A.C., *Digital Picture Processing*, Vols. 1+2, 2nd ed., Academic Press, New York, 1982.

90. Sandham, W.A., Yuan, Y., Durrani, T.S., "Conformal Therapy Using Maximum Entropy Optimization," *International Journal of Imaging Systems and Technology*, 6(1): 80 – 90, 1995.

91. Sahiner, B., Yagle, A.E., "Iterative Inversion of the Radon Transform," *IEEE Engineering in Medicine and Biology*, 15(5): 112 – 117, Sept./Oct. 1996.

92. Segman, J., Zeevi, Y., "Image Analysis by Wavelet-Type Transforms: Group Theoretical Approach," *Journal of Mathematical Imaging and Vision*, 3(1): 51 – 77, 1993.

93. Shepp, L.S., Logan, B.F., "The Fourier Reconstruction of a Head Section," *IEEE Transactions on Nuclear Science*, 21(3): 21 – 43, 1974.

94. Späth, H., *Two Dimensional Spline Interpolation Algorithms*, Wellesley, Massachusetts, AK Peters, 1995.

95. Stark, H., Woods, J.W., Paul, I., Hingorani, R., "Direct Fourier Reconstruction in Computed Tomography," *IEEE Transactions on Acoustics, Speech, and Signal Processing*, 29(2): 237 – 245, 1981.

96. Tretiak, O.J., Metz, C., "The Exponential Radon Transform," *SIAM Journal on Applied Mathematics*, 39(2): 341 – 354, 1980.

97. Turin, G.L., "An Introduction to Matched Filters," *IRE Transactions on Information Theory*, 6(3): 311 – 329, 1960.

98. Tuzel, O., Porikli, F., Meer, P., "Learning on Lie Groups for Invariant Detection and Tracking," in *Proceedings of the 2008 IEEE Conference on Computer Vision and Pattern Recognition (CVPR 2008)*, pp. 1 – 8, June, 2008.

99. Viola, P., Wells, W.M., III, "Alignment by Maximization of Mutual Information," *International Journal of Computer Vision*, 24(2): 137 – 154, 1997.

100. Webb, S., *The Physics of Three-Dimensional Radiation Therapy : Conformal Radiotherapy, Radiosurgery and Treatment Planning*, Institute of Physics Publishing, Medical Science Series, Bristol, UK, 1993.

101. Wong, E., Van Dyk, J., "Lateral Electron Transport in FFT Photon Dose Calculations," *Medical Physics*, 24(12): 1992 – 2000, Dec. 1997.

Statistical Pose Determination and Camera Calibration

In this chapter we examine the following problems: (1) finding the best rigid-body motion (pose) to fit to noisy measured data; and (2) performing camera calibration in vision systems mounted on robotic manipulators. In both cases it is assumed that only inexact measurements with known statistical properties are provided. We therefore begin this chapter with sections on basic probability theory and its extension to data on spheres, the rotation group, and the motion group.

First, Section 14.1 reviews basic definitions from probability theory. Section 14.2 discusses probability and statistics on the circle. Section 14.3 reviews the corresponding concepts for PDFs on groups. Section 14.4 introduces PDFs on $SO(3)$ that have special properties. Section 14.5 provides definitions of mean and variance of PDFs on rotation and motion groups. Section 14.6 addresses the problem of finding the best rigid-body transformation to fit a set of measured data. Finally, Section 14.7 discusses the problem of camera calibration and the statistical aspects of this problem.

14.1 Review of Basic Probability

Given a probability density function (PDF) with real argument, $\rho(x)$, the classical apparatus of probability theory is used to analyze $\rho(x)$ in terms of its moments. The expected value is the center of mass, or mean, $E(x) = \mu$ defined as the solution of

$$\int_{\mathbb{R}} (x - \mu)\rho(x)\, dx = 0. \tag{14.1}$$

Obviously, this gives

$$\mu = \int_{\mathbb{R}} x\rho(x)\, dx.$$

But, as we shall see later in this chapter, this form does not generalize in a convenient way to non-Euclidean spaces whereas (14.1) does.

The median value corresponding to $\rho(x)$ is the value $x = m$ for which there is as much "mass" on one side as on the other:

$$\int_{-\infty}^{m} \rho(x)\, dx = \int_{m}^{\infty} \rho(x)\, dx.$$

But since

$$\int_{-\infty}^{\infty} \rho(x)\,dx = 1,$$

this indicates that m satisfies the equation

$$\int_{-\infty}^{m} \rho(x)\,dx = \frac{1}{2}.$$

The median can also be viewed as the value of y that minimizes

$$F(y) = \int_{-\infty}^{\infty} |y - x|\rho(x)\,dx = -\int_{y}^{\infty}(y - x)\rho(x)\,dx + \int_{-\infty}^{y}(y - x)\rho(x)\,dx. \qquad (14.2)$$

It is easy to see by setting $dF/dy = 0$ and solving for y that $F(m)$ is the minimal value for any $\rho(x)$.

The nth moments about the expected value $E(x) = \mu$ for all integers $n \geq 0$ are defined as

$$M_n(\mu) \doteq \int_{\mathbb{R}} (x - \mu)^n \rho(x)\,dx. \qquad (14.3)$$

$M_0(\mu) = 1$ is the "mass" of the probability density. Setting $M_1(\mu) = 0$ is equivalent to the definition of μ in (14.1). $M_2 = \sigma^2$ is the variance, and σ is the standard deviation. Knowing all of the moments from $n = 2, ..., N$ for some finite N means that the important properties of the distribution $\rho(x)$ are known to some degree without regard to all of its details. As $N \to \infty$, the properties of $\rho(x)$ become more determined as more of the moments are known. Usually only the first few moments are of concern.

In analogy with the definition of the median as the value which minimizes the functional in (14.2), it is easy to show by direct calculation that $E(x) = \mu$ can be calculated as the value of y which minimizes the functional

$$M_2(y) = \int_{\mathbb{R}} (x - y)^2 \rho(x)\,dx. \qquad (14.4)$$

Hence, the value of y that minimizes $M_2(y)$ is μ, and as mentioned before, $M_2(\mu) = \sigma^2$. Since this is important we highlight it as

$$\mu = \min_{y} M_2(y) \text{ and } \sigma^2 = M_2(\mu). \qquad (14.5)$$

The moments of probability density functions have nice properties under the operation of convolution. Recall that on the line

$$(\rho_1 * \rho_2)(x) = \int_{-\infty}^{\infty} \rho_1(\xi)\rho_2(x - \xi)\,d\xi.$$

Since $\rho_1(x) \geq 0$ for all $x \in \mathbb{R}$, it follows from a change of variables $z = x - \xi$ that the mass of the convolution of two PDFs is unity:

$$\int_{-\infty}^{\infty} (\rho_1 * \rho_2)(x)\,dx = \left(\int_{-\infty}^{\infty} \rho_1(\xi)\,d\xi\right)\left(\int_{-\infty}^{\infty} \rho_2(z)\,dz\right) = 1 \cdot 1 = 1. \qquad (14.6)$$

That is, the convolution of two PDFs results in a PDF. Similarly, one has that the higher moments of convolutions of PDFs can be generated from the moments of the PDFs being convolved:

$$\mu_{1*2} = \int_{-\infty}^{\infty} x\,(\rho_1 * \rho_2)(x)\,dx = \int_{-\infty}^{\infty} x\rho_1(x)\,dx + \int_{-\infty}^{\infty} x\rho_2(x)\,dx = \mu_1 + \mu_2 \qquad (14.7)$$

and

$$
\begin{aligned}
(\sigma^2)_{1*2} &= \int_{-\infty}^{\infty} (x - \mu^{1*2})^2 (\rho_1 * \rho_2)(x)\,dx \\
&= \int_{-\infty}^{\infty} (x - \mu_1)^2 \rho_1(x)\,dx + \int_{-\infty}^{\infty} (x - \mu_2)^2 \rho_2(x)\,dx \\
&= (\sigma^2)_1 + (\sigma^2)_2.
\end{aligned}
\qquad (14.8)
$$

As with (14.6), we find (14.7) and (14.8) by substituting $z = x - \xi$ and using the invariance of integration under shifts.

In the n-dimensional case, the definitions of mean and variance generalize as

$$\int_{\mathbb{R}^n} (\mathbf{x} - \boldsymbol{\mu})\rho(\mathbf{x})\,d\mathbf{x} = \mathbf{0} \iff \boldsymbol{\mu} = \int_{\mathbb{R}^n} \mathbf{x}\rho(\mathbf{x})\,d\mathbf{x} \qquad (14.9)$$

and

$$\Sigma \doteq \int_{\mathbb{R}^n} (\mathbf{x} - \boldsymbol{\mu})(\mathbf{x} - \boldsymbol{\mu})^T \rho(\mathbf{x})\,d\mathbf{x}. \qquad (14.10)$$

$\Sigma = \Sigma^T \in \mathbb{R}^{n \times n}$ is called a *covariance matrix* and it has elements of the form

$$\sigma_{ij} = \int_{\mathbb{R}^n} (x_i - \mu_i)(x_j - \mu_j)\rho(\mathbf{x})\,d\mathbf{x}.$$

As with the one-dimensional case, it is not difficult to show that the mean and covariance of $(\rho_1 * \rho_2)(\mathbf{x})$ are related to those for $\rho_1(\mathbf{x})$ and $\rho_2(\mathbf{x})$ as

$$\boldsymbol{\mu}_{1*2} = \boldsymbol{\mu}_1 + \boldsymbol{\mu}_2 \text{ and } \Sigma_{1*2} = \Sigma_1 + \Sigma_2. \qquad (14.11)$$

Moreover, if in analogy with the 1D case we now define the matrix

$$M_2(\mathbf{y}) \doteq \int_{\mathbb{R}^n} (\mathbf{x} - \mathbf{y})(\mathbf{x} - \mathbf{y})^T \rho(\mathbf{x})\,d\mathbf{x}$$

then

$$\boldsymbol{\mu} = \min_{\mathbf{y}} \mathrm{tr}(M_2(\mathbf{y})) \text{ and } \Sigma = M_2(\boldsymbol{\mu}). \qquad (14.12)$$

14.1.1 Bayes' Rule

Let

$$P(x) = \int_{x}^{x+\Delta x} \rho(\xi)\,d\xi$$

be the probability that $\xi \in [x, x + \Delta x]$ corresponding to the PDF $\rho(x)$. Let $\rho(x, y)$ be the *joint probability* density describing the possibility that x and y occur simultaneously, and let

$$P(x, y) = \int_{x}^{x+\Delta x} \int_{y}^{y+\Delta y} \rho(\xi, \eta)\,d\eta d\xi.$$

The *conditional probability* that $\xi \in [x, x + \Delta x]$ given $\eta \in [y, y + \Delta y]$ is denoted $P(x|y)$, and is calculated as

$$P(x|y) = P(x,y)/P(y).$$

Similarly,

$$P(y|x) = P(x,y)/P(x).$$

Combining these two we get *Bayes' Rule:*[1]

$$P(x|y) = \frac{P(y|x)P(x)}{P(y)}. \tag{14.13}$$

In analogy with the definition of $P(x|y)$, let

$$\rho(x|y) = \rho(x,y)/\rho(y)$$

be the *conditional probability density function* of x given y. In the limit as Δx and Δy are allowed to approach zero, (14.13) is rewritten as

$$\rho(x|y)\Delta x = \frac{\rho(y|x)\Delta y \rho(x)\Delta x}{\rho(y)\Delta y}$$

which is simplified to

$$\rho(x|y) = \frac{\rho(y|x)\rho(x)}{\rho(y)}, \tag{14.14}$$

which is called *Bayes' rule for PDFs.*

14.1.2 The Gaussian Distribution

We have already seen the Gaussian distribution:

$$\rho_G(x; m, \sigma) = \frac{1}{\sqrt{2\pi}\sigma} e^{-(x-m)^2/2\sigma^2}$$

in earlier chapters. The expected value of this distribution is $\mu = m$ (the mean and median are the same), and its standard deviation is σ. The multi-dimensional Gaussian distribution is of the form

$$\rho_G(\mathbf{x}; \boldsymbol{\mu}, \Sigma) = \frac{\exp\left[-\frac{1}{2}\sum_{i,j=1}^n \sigma_{ij}^{-1}(x_i - m_i)(x_j - m_j)\right]}{[(2\pi)^n \det \Sigma]^{\frac{1}{2}}}$$

where $x_i = \mathbf{e}_i \cdot \mathbf{x}$, and μ_i are the components of the mean, σ_{ij}^{-1} are the entries of Σ^{-1}, and the normalization

$$\int_{\mathbb{R}^n} \rho_G(\mathbf{x}; \boldsymbol{\mu}, \Sigma) \, d\mathbf{x} = 1$$

is observed.

In the context of statistics, Gaussian distributions have some extremely useful properties. These include:

- Closure under convolution (i.e., the convolution of two Gaussians results in a Gaussian, with mean and covariance obeying (14.11)).

---

[1]Thomas Bayes (1702-1761) established the foundations of statistical inference.

- Closure under conditioning (i.e., the product of two Gaussians is a Gaussian (to within a scale factor), and when the quotient can be normalized to be a PDF, this PDF will be a Gaussian).
- The Central Limit Theorem (i.e., the convolution of a large number of well-behaved PDFs tends to the Gaussian distribution (see, e.g., [30] and references therein)).
- Gaussians are solutions to the heat equation with δ-function as initial conditions (see Chapter 2).
- The Fourier transform of a Gaussian is, to within a normalization factor, a Gaussian.
- The Gaussian is an even function of its argument.

In the following sections we show how the well-known definitions and properties reviewed here are generalized in ways unfamiliar to many engineers and scientists. But first we review some non-Gaussian PDFs.

14.1.3 Other Common Probability Density Functions on the Line

While the Gaussian distribution and its generalizations play an important role in this chapter and those that follow, a number of other probability density functions on the line are worth noting.

The *Cauchy* distribution has the density

$$\rho_C(x; m, a) = \frac{a}{(\pi a)^2 + (x - m)^2} \tag{14.15}$$

for $a \in \mathbb{R}_{>0}$ and mean $m \in \mathbb{R}$. This density has the property that it is closed under convolution, but it does not have many of the other nice properties of the Gaussian distribution listed in the previous section.

The *Laplace* distribution has the density

$$\rho_L(x; m, \sigma) = \frac{1}{\sqrt{2}\sigma} \exp\left(-\frac{\sqrt{2}|x - m|}{\sigma}\right).$$

The *rectangular* distribution has a density defined by the value $1/(b - a)$ on the interval $[a, b]$. For all values of $x \in \mathbb{R}$ outside of $[a, b]$ it takes the value zero. Clearly this is not closed under convolution. The result of the convolution of two rectangular distributions is one that is triangular.

In contrast, the sinc function defined in Chapter 2 has a Fourier transform that is (to within a constant) a rectangular distribution. The product of two rectangular distributions (in Fourier space) results in a rectangular distribution, and so sinc functions are closed under convolution. But a sinc function is not a PDF because it is not a strictly non-negative function.

14.2 Probability and Statistics on the Circle

We note that there are major differences between PDFs on the line and on the circle. Since the line is infinite in extent, it is not possible to define a PDF on the line that takes constant values. On the circle, the PDF $\rho(\theta) = 1/2\pi$ is perfectly valid. On the line, a PDF $\rho(x)$ always has a well-defined center of mass (the expected value of x). On the circle, it can be possible to have a single, multiple, or even an infinite number of "mean" values (e.g., for $\rho(\theta) = 1/2\pi$ all values can be called mean values). Of course, when the density function is concentrated on one small portion of the circle, the tools developed for statistics on the line can be used without much difficulty. However, for distributions that are more spread out, the tools introduced in Section 14.3 are required to define concepts of mean and variance.

Probability density functions for the circle can be generated by "wrapping" (or "folding") PDFs defined on the line. Given that $\rho(x)$ is a PDF on the line, the function

$$\rho_W(\theta) = \sum_{n=-\infty}^{\infty} \rho(\theta - 2\pi n) \tag{14.16}$$

will be a PDF on the unit circle (or equivalently, $SO(2)$).

It is easy to verify that

$$\int_0^{2\pi} \sum_{n=-\infty}^{\infty} \rho(\theta - 2\pi n)\, d\theta = \int_{-\infty}^{\infty} \rho(x)\, dx = 1.$$

Likewise, since $\rho(x) \geq 0$ for all $x \in \mathbb{R}$, it follows that $\rho_W(\theta) \geq 0$ for all $\theta \in [0, 2\pi]$. Also, if $\rho(x, a)$ is a member of a family of density functions on the line such that

$$\rho(x, a) * \rho(x, b) = \rho(x, c)$$

(closure under convolution), then the convolution of wrapped versions of these functions will be closed under convolution on the circle. We verify this below:

$$\rho_W(\theta, a) * \rho_W(\theta, b) = \int_0^{2\pi} \left(\sum_{n=-\infty}^{\infty} \rho(\xi - 2\pi n, a) \right) \left(\sum_{m=-\infty}^{\infty} \rho(\theta - \xi - 2\pi m, b) \right) d\xi$$

$$= \int_{-\infty}^{\infty} \rho(\xi, a) \left(\sum_{m=-\infty}^{\infty} \rho(\theta - \xi - 2\pi m, b) \right) d\xi$$

$$= \sum_{m=-\infty}^{\infty} \int_{-\infty}^{\infty} \rho(\xi, a) \rho(\theta - \xi - 2\pi m, b)\, d\xi$$

$$= (\rho(x, a) * \rho(x, b))_W = \rho_W(x, c).$$

In other words, the operation of convolution on the line and wrapping on the circle can be replaced with wrapping first then convolution on the circle, or vice versa. Hence, whenever functions are closed under convolution on the line, their wrapped versions will be closed under convolution on the circle.

If the PDF is very distributed, a Fourier series expansion will capture its shape well using few terms. If the PDF is very concentrated, then (14.16) can be truncated with good accuracy at small values of n.

For instance, the heat equation on the circle is

$$\frac{\partial f}{\partial t} = \frac{1}{2} K \nabla^2 f$$

where

$$\nabla^2 = \frac{\partial^2}{\partial \theta^2}.$$

It is well known that the Fourier series solution of this heat equation under the initial condition $f(\theta, 0) = \delta(\theta - 0)$ is of the form

$$f(\theta, t) = \frac{1}{2\pi} \sum_{n=-\infty}^{\infty} e^{-n^2 K t/2} e^{in\theta} = \frac{1}{2\pi} + \frac{1}{\pi} \sum_{n=1}^{\infty} e^{-n^2 K t/2} \cos n\theta. \tag{14.17}$$

when the integration measure on the circle is $d\theta$. Another well-known form of the solution to the heat equation on the circle is what results from "wrapping" the solution of the heat equation on the line,

$$\rho(x, t) = \frac{1}{\sqrt{2\pi K t}} e^{-x^2/2Kt},$$

around the circle. It can be shown that the solution in (14.17) is related to this as

$$f(\theta, t) = \sum_{n=-\infty}^{\infty} \rho(\theta - 2\pi n, t).$$

The density of the Cauchy distribution (14.15) can also be wrapped around the circle, the result of which can be expanded in the Fourier series [69]:

$$\rho_C(\theta, b) = (2\pi)^{-1} \left(1 + 2 \sum_{n=1}^{\infty} b^n \cos n\theta \right)$$

where $b = e^{-a}$. Since

$$\sum_{n=1}^{\infty} b^n e^{-in\theta} = \sum_{n=1}^{\infty} (be^{-i\theta})^n$$

is a geometric series, it follows from the fact that the real part of this series is $\sum_{n=1}^{\infty} b^n \cos n\theta$ that [69]

$$\rho_C(\theta, b) = \frac{1}{2\pi} \frac{1 - b^2}{1 + b^2 - 2b \cos \theta}$$

for $0 \le b \le 1$.

Other PDFs for the circle have been defined without wrapping PDFs on the line around the circle. For instance, the *von Mises* distribution has a PDF

$$\rho_{VM}(\theta; m, \kappa) = \frac{1}{2\pi I_0(\kappa)} e^{\kappa \cos(\theta - m)}$$

where m is the mean (or mode), κ dictates how spread out the density is ($\kappa = 0$ is the uniform density), and

$$I_0(\kappa) = \sum_{n=0}^{\infty} \frac{1}{(n!)^2} \left(\frac{\kappa}{2}\right)^{2r}$$

is the modified Bessel function of the first kind of order zero. Another PDF used in the context of statistics on the circle is the *cardioid* density

$$\rho_{CD}(\theta) = \frac{1}{2\pi}[1 + \frac{1}{2}\kappa \cos(\theta - m)].$$

This is a degenerate case of the von Mises density for small values of κ. See [69, 97] for a complete treatment. See [10, 40] for applications of orientational statistics in biology.

14.3 Metrics as a Tool for Statistics on Groups

In previous chapters we have seen a number of applications in which probability density functions (PDFs) on groups and metrics on groups arise separately. Now we consider how these two concepts can synergistically interact.

Clearly it would be useful to reduce the information in a probability density function on a group to the specification of its lowest moments . For example, prior to explicitly performing numerical convolutions on the Euclidean motion group (with or without the help of an FFT), it would be of value to know how moments of probability density functions interact under convolution. But how should this be done ? It is tempting to consider as a direct analogy with the Abelian case reviewed in the previous section, a definition such as $E(g) = \int_G g\rho(g)\, d(g)$ for the definition of expected value. However, this definition would really not make any sense because while integration of functions on a group is well defined, integration of group elements is not. In fact, even the product of a scalar and a group element is not well defined. Even when we can get around these problems, such as when G is a matrix group, the issue remains that $E(g)$ as defined above is not even a group element in general. And while there are ways to adjust this definition to make sense in some cases (e.g., see [97]), we will use a different definition in the following subsections.

14.3.1 Moments of Probability Density Functions on Groups and Their Homogeneous Spaces

In this section we view the difference $x - y$ as a signed measure of distance between points on the line (instead of as the group product $x + (-y)$). From this point of view, (14.4) is written as

$$M_2(y) = \int_{\mathbb{R}} d^2(x, y)\rho(x)\, dx$$

where $d(x, y) = |x - y|$, and $E(x)$ is the value of y which minimizes $M_2(y)$. Similarly, we can write (14.2) as

$$F(y) = \int_{\mathbb{R}} d(x, y)\rho(x)\, dx.$$

The straightforward generalization of these to PDFs on groups is that the expected value $E_d(g_1) = \mu^d \in G$ is the group element which minimizes the function

$$M_2(g_1) = \int_G [d(g_1, g_2)]^2 \rho(g_2)\, d(g_2). \tag{14.18}$$

Here $d(g_1, g_2)$ is a metric (not to be confused with the integration measure $d(g)$), and clearly the center of mass in this case depends on how this metric is defined. Hence $E_d(g)$ is called the d-mean, and the value $C(\mu^d)$ is what we will refer to as the d-variance. Similar concepts arise in Riemannian geometry, and analogous concepts of the mean are referred to as the *Karcher mean* after [61] or the *Fréchet mean* after [39]. Similarly, the group element $g_{med}^d \in G$ which minimizes the function

$$F(g_1) = \int_G d(g_1, g_2)\rho(g_2)\, d(g_2) \tag{14.19}$$

is called the d-median. These definitions are known in the theoretical statistics literature (see, e.g., [30]), but are not part of what is generally considered to be standard engineering mathematics. For other treatments of the concept of generalized medians see [23] and references therein.

In traditional statistics in $\mathbb{R}^N$, $N \times N$ covariance matrices with entries of the form

$$\sigma_{ij} = \int_{\mathbb{R}^N} (x_i - E[x_i])(x_j - E[x_j])\rho(\mathbf{x})\,d\mathbf{x}$$

play an important role. Several possible extensions of this concept in the context of the rotation and motion groups are clear when using the matrix exponential. Assuming an arbitrary element is of the form

$$g(\alpha_1, ..., \alpha_N) = \exp(\sum_{i=1}^{N} \alpha_i X_i)$$

where $\{X_i\}$ is a basis for the Lie algebra, then $g(0, ..., \alpha_i, ..., 0)$ is a one parameter subgroup of G,

$$g(0, ..., \alpha_i, ..., 0) \circ g(0, ..., \alpha_i', ..., 0) = g(0, ..., \alpha_i + \alpha_i', ..., 0).$$

The logarithm of a group element is defined as

$$\log g(\alpha_1, ..., \alpha_N) = \sum_{i=1}^{N} \alpha_i X_i.$$

The inner product of two basis elements X_i and X_j is

$$(X_i, X_j) = \frac{1}{2}\mathrm{tr}(X_i W X_j^T)$$

for an appropriate weighting matrix when the Lie algebra consists of real elements (see Chapter 7).

Then we can define

$$\sigma_{ij}^1 = \int_G \mathrm{sgn}[(\alpha_i - E[\alpha_i])(\alpha_j - E[\alpha_j])]d(g(0, ..., \alpha_j, ..., 0), g(0, ..., E[\alpha_j], ..., 0))\cdot$$

$$d(g(0, ..., \alpha_i, ..., 0), g(0, ..., E[\alpha_i], ..., 0))\rho(g)\,dg,$$

where we define $E[\alpha_i]$ as

$$E[\alpha_i] \doteq (\log(E[g]), X_i).$$

Using the left invariance of the metric, the exponential notation, and the shorthand $\Delta_i = \alpha_i - E[\alpha_i]$, we find

$$\sigma_{ij}^1 = \int_G \mathrm{sgn}[\Delta_i \Delta_j]d(e, \exp(\Delta_i X_i))\,d(e, \exp(\Delta_j X_j))\rho(g)\,dg.$$

A second possible extension of the concept of covariance arises by using the same definitions as above, but with $E[\alpha_i]$ defined as the value of α_i that minimizes

$$\int_G d^2(g(0, ..., 0, \alpha_i, 0, ..., 0))\rho(g)\,dg.$$

If this definition is used, we denote the corresponding covariance matrix as σ_{ij}^2.

A third possibility is to first left shift the PDF so that the mean value is at the identity, and then do the calculations as in the σ_{ij}^1 case, hence resulting in

$$\sigma_{ij}^3 = \int_G \text{sgn}[\alpha_i\alpha_j] d(e, \exp(\alpha_i X_i))\, d(e, \exp(\alpha_j X_j))\rho(E[g] \circ g)\, dg. \qquad (14.20)$$

In the original version of this book we used σ_{ij}^3 in (14.20) as the definition of covariance since it does not require the introduction of a choice for $E[\alpha_i]$. While there is no single correct choice, we have subsequently adopted different definitions of mean and covariance that have very convenient properties under convolution that will be used later. These are

$$\boxed{\int_G [\log(\mu^{-1} \circ g)]^\vee \rho(g)\, dg = \mathbf{0} \text{ and } \Sigma \doteq \int_G [\log(\mu^{-1} \circ g)]^\vee \left([\log(\mu^{-1} \circ g)]^\vee\right)^T \rho(g)\, dg.}$$
$$(14.21)$$

This definition of mean is analogous to the first one in (14.9) (but not the second one!) and this definition of covariance resembles that in (14.10).

In the case when $G = \mathbb{R}^N$ all the definitions of mean and covariance given above degenerate to the classical ones, and when limiting the discussion to particular kinds of unimodular Lie groups some of these definitions can be made to coincide. But in general they will all be different from each other.

Other issues relating to probability density functions on groups in general, and $SO(N)$ in particular, are addressed in [63, 66, 85]. The topics of orientational statistics [5, 12, 15, 31, 47, 58, 62, 82, 83], spherical regression [18, 19, 20, 21, 22, 84], and regression and deconvolution on the sphere and $SO(3)$ [48, 49, 50], are all relevant to the topics discussed here, since the sphere can be viewed as the coset space $SO(3)/SO(2)$. Applications of directional statistics are quite varied ranging from geophysical data analysis [21, 71] to the problems in estimation discussed in the next chapter. Therefore, the next section discusses some of the issues encountered in spherical statistics.

14.3.2 Statistics on the Homogeneous Space $SO(3)/SO(2)$

Analogous definitions hold for PDFs on homogeneous spaces G/H endowed with an integration measure invariant under G-shifts. The only difference is that integration is with respect to this measure (denoted as $d_{G/H}(g)$ or $d(gH)$ in Chapter 8) instead of $d_G(g)$, and the metric used is generated from a right-invariant metric on G as

$$d_{G/H}(g_1 H, g_2 H) \doteq \min_{h_1, h_2 \in H} d(g_1 \circ h_1, g_2 \circ h_2) = \min_{h \in H} d(g_1, g_2 \circ h).$$

For instance, a substantial amount of literature deals with statistics on the homogeneous space $SO(3)/SO(2) \cong \mathbb{S}^2$. Applications include the orientation of fibers in composite materials, the orientation of pebbles in soil, and a number of other interesting problems referred to in [97]. As with the circle, it is possible to have multiple (or even an infinite number) of means and median values of a PDF on the sphere.

In the case of statistics on the sphere, there appears to be two natural ways to formulate statistics. The first is to use the concepts outlined above, where $d\mathbf{u} = (1/4\pi) \sin\theta\, d\theta\, d\phi$ is the normalized integration measure, and $d(\mathbf{x}, \mathbf{y}) = |\cos^{-1}(\mathbf{x} \cdot \mathbf{y})|$ is the geodesic distance between two points $\mathbf{x}$ and $\mathbf{y}$ on the unit sphere $\mathbb{S}^2$ written as vectors in $\mathbb{R}^3$. The second approach is to work in $\mathbb{R}^3$ without trying to "stay on the sphere" as is done in [97]. In that approach, the average of a set of points on the sphere

Is defined as the normalized spatial average of the points. Normalization ensures that the result is "put back" on the sphere after the average is calculated. There are some conveniences to both approaches. The first approach has the benefit of being part of a more general theory, and hence the results of general theorems can be localized to problems on the sphere. The benefit of the second approach is that a number of application areas already use it and have developed highly specialized tools in this context. For instance, in the context of paleomagnetic data the *Fisher distribution* [35, 36]:

$$f_F(\theta) = f(\mathbf{x}(\theta, \phi)) = \frac{\kappa}{\sinh \kappa} e^{\kappa \mathbf{e}_3 \cdot \mathbf{x}(\theta, \phi)} = \frac{\kappa}{\sinh \kappa} e^{\kappa \cos \theta}. \qquad (14.22)$$

is used to capture the distribution of magnetic poles in rocks. Here $\mathbf{e}_3$ is the direction of the mode and $\mathbf{x}$ is parameterized as in (4.20). A rotation of this PDF by $R \in SO(3)$ changes $\mathbf{x}$ to $R^T \mathbf{x}$ in the PDF (which is the same as changing $\mathbf{e}_3$ to $R\mathbf{e}_3$ and leaving $\mathbf{x}$ unchanged).

Another PDF used commonly to fit statistical data on spheres is the *Bingham distribution* [12]:

$$b(\mathbf{x}) = \frac{e^{\mathbf{x}^T K \mathbf{x}}}{\int_{\mathbb{S}^2} e^{\mathbf{x}^T K \mathbf{x}} \, d\mathbf{x}}, \qquad (14.23)$$

where again $\mathbf{x} = \mathbf{x}(\theta, \phi) \in \mathbb{S}^2$ and $K \in \mathbb{R}^{3 \times 3}$.

Clearly, when one is interested in the orientational statistics of sets of objects without an axis of symmetry, statistics on $SO(3)$ becomes important for exactly the same reason as statistics on $\mathbb{S}^2$. The densities in (14.22) and (14.23) have natural extensions to the quaternion sphere $\mathbb{S}^3$, and densities on $\mathbb{S}^3$ with antipodal symmetry are ideal for statistics on $SO(3)$.

A natural metric between elements $R_1, R_2 \in SO(3)$ can be interpreted as the absolute value of the angle of rotation, θ, found by solving $e^{\theta N} = R_1^T R_2$, or by viewing rotations as points on the upper hemisphere of $\mathbb{S}^3$, and calculating $d(\mathbf{x}, \mathbf{y}) = |\cos^{-1}(\mathbf{x} \cdot \mathbf{y})|$ where now $\mathbf{x}, \mathbf{y} \in \mathbb{R}^4$ are unit vectors corresponding to rotations.

Questions regarding the average of a set of orientations or frames of reference are natural ones to ask in the field of robotics. In this context one is often interested in how repeatably a robot arm can reach the same position and orientation in space. If, for instance, the same pick-and-place task were performed several hundred times, there would be some distribution of actual frames of reference reached. Ideally, all the reached frames would correspond to the desired one. However in practice, flexibility in the arm, backlash in gears, and even thermal effects in the motors and sensors due to the use of the arm can mean that the arm will not reach the desired frame, but rather a cloud of frames around it. Clearly then, having a measure of the average frame (position and orientation) together with the variance provides tools for describing the repeatability of a robotic arm. The methods presented in this subsection provide the vocabulary for analysis in the context of this very applied problem.

14.4 PDFs on Rotation Groups with Special Properties

In this section we review PDFs on the rotation groups $SO(2)$ and $SO(3)$ that have the special property of being closed under convolution. Since $SO(2)$ is Abelian, convolutions and shifts naturally commute. However, some effort is required in the three-dimensional case to find class functions that commute and are closed under convolution.

14.4.1 The Folded Gaussian for One-Dimensional Rotations

The Gaussian function "spread on the circle" is what results from dividing up the real line into increments of 2π, shifting (in multiples of 2π) the Gaussian function to the interval $[-\pi, \pi]$, and adding up all the shifted values:

$$\psi(\theta; \beta) = \beta \sum_{k=-\infty}^{\infty} e^{-\pi\beta^2(\theta-2k\pi)^2}. \tag{14.24}$$

It is easy to check by direct calculation that this set of functions is closed under convolution, with

$$\psi(\theta; \beta) * \psi(\theta; \alpha) = \psi(\theta; \alpha')$$

where

$$\frac{1}{\alpha'^2} = \frac{1}{\alpha^2} + \frac{1}{\beta^2}. \tag{14.25}$$

The Fourier series description of the same functions requires relatively few terms when β is small, and many terms when β is large. In contrast, the description of the same function as a sum of shifted Gaussians can be truncated at one or two terms with great accuracy for large values of β. Hence, for all intents and purposes in engineering applications in which accurate sensors are used, the methods of classical probability theory on the line can be used for rotational processes in one dimension.

14.4.2 Gaussians for the Rotation Group of Three-Dimensional Space

In order to do statistics on $SO(3)$ it is useful to have a concept of Gaussian functions that has the properties enumerated at the end of Subsection 14.1.2.

The issue of closure under convolution is straight forward. As we have already seen, the convolution of any two band-limited functions on $SO(3)$ will be band limited, and the Fourier transform of the result will be the product of the Fourier transform matrices (in reverse order) of the original functions. In the special case when the Fourier transforms of two function f and h are of the form

$$\hat{f}^l = \exp A_l \qquad \hat{h}^l = \exp B_l$$

(where l enumerates IURs of $SO(3)$), then it is clear that the result of the convolution will be of the form $\exp B_l \exp A_l$. If for all values of l, A_l, and B_l commute ($[A_l, B_l] = \mathbb{O}_{(2l+1)\times(2l+1)}$), then

$$\mathcal{F}(f * h)_l = \exp(A_l + B_l) \qquad \text{and} \qquad f * h = h * f.$$

As a special case,

$$\rho(g, t_1) * \rho(g, t_2) = \rho(g, t_1 + t_2)$$

when

$$\hat{\rho}^l = \exp(B_l t).$$

In the special case when $B_l = b_l \mathbb{I}$ for all values of l, then $\rho(g, t)$ will be a class function for any value of t.

In Chapter 16, many of the solutions that result from rotational Brownian motion will have Fourier transforms of the form $\hat{f}^l(t) = \exp(A_l t)$, and hence $\hat{f}^l(t_1)\hat{f}^l(t_2) = \exp(A_l(t_1 + t_2))$. That is, solutions of certain PDEs on $SO(3)$ evaluated at different times will be closed under convolution.

We now consider a specific property of class functions under convolution.

Theorem 14.1. *The convolution of any function and a class function on a unimodular group is commutative, and the convolution of two class functions results in a class function.*

Proof. By the definition of convolution, and the bi-invariance of the integraton measure, a change of variables $k = h^{-1} \circ g$ allows one to write the convolution as

$$(f_1 * f_2)(g) = \int_G f_1(g \circ k^{-1}) f_2(k) \, d(k)$$

regardless of whether f_i are class functions or not. Assuming f_1 is a class function, then $f_1(g \circ k^{-1}) = f_1(k^{-1} \circ g)$, and since multiplication of scalar functions is commutative we have $f_1 * f_2 = f_2 * f_1$. The same follows if f_2 had been a class function instead of f_1.

To see that the convolution of two class functions is a class function, we observe that

$$(f_1 * f_2)(k^{-1} \circ g \circ k) = \int_G f_1(h) \, f_2(h^{-1} \circ k^{-1} \circ g \circ k) \, d(h).$$

If f_2 is a class function, then $f_2(h^{-1} \circ k^{-1} \circ g \circ k) = f_2(k \circ h^{-1} \circ k^{-1} \circ g)$. Making the change of variables $s^{-1} = k \circ h^{-1} \circ k^{-1}$, we find that

$$(f_1 * f_2)(k^{-1} \circ g \circ k) = \int_G f_1(k^{-1} \circ s \circ k) \, f_2(s^{-1} \circ g) \, d(s) = \int_G f_1(s) \, f_2(s^{-1} \circ g) \, d(s),$$

where the last equality follows when f_1 is a class function. Hence we have

$$(f_1 * f_2)(k^{-1} \circ g \circ k) = (f_1 * f_2)(g),$$

which means $f_1 * f_2$ is a class function.

In addition to closure under convolution, a much stronger condition is required for statistics on the rotation group to be of practical use in applications. Namely, the convolution of left-shifted versions of PDFs in a given family should be closed in the sense that they result in left-shifted PDFs in the same family, with resulting shift equal to the composition of the originals. That is, we desire the property:

$$f_1(g_1^{-1} \circ g) * f_2(g_2^{-1} \circ g) = (f_1 * f_2)((g_1 \circ g_2)^{-1} \circ g).$$

with f_1, f_2, and $f_1 * f_2$ in the same set. The next theorem describes when this is true.

Theorem 14.2. *The convolution of two left (right)-shifted functions on a non-Abelian unimodular group is equal to a left (right)-shifted version of the convolution of the unshifted functions if the original functions are class functions.*

Proof. Let $\tilde{f}_i(g) = f_i(g_i^{-1} \circ g)$. Then $\tilde{f}_2(h^{-1} \circ g) = f_2(g_2^{-1} \circ (h^{-1} \circ g))$ and

$$(\tilde{f}_1 * \tilde{f}_2)(g) = \int_G f_1(g_1^{-1} \circ h) \, f_2(g_2^{-1} \circ h^{-1} \circ g) \, d(h).$$

Since f_1 is a class function, then

$$f_1(g_1^{-1} \circ h) = f_1(h \circ g_1^{-1}).$$

Letting $s = h \circ g_1^{-1}$, and using the fact that f_2 is a class function, then

$$(\tilde{f}_1 * \tilde{f}_2)(g) = \int_G f_1(s)\, f_2(g_2^{-1} \circ g_1^{-1} \circ s^{-1} \circ g)\, d(s) = (f_1 * f_2)(g \circ g_2^{-1} \circ g_1^{-1}).$$

From Theorem 14.1, $f_1 * f_2$ is a class function, and so

$$(f_1 * f_2)(g \circ g_2^{-1} \circ g_1^{-1}) = (f_1 * f_2)((g_1 \circ g_2)^{-1} \circ g).$$

A similar proof follows for right shifts.

For $SO(3)$ we saw in Chapter 7 that all class functions are even functions of the angle of rotation. An analog of the folded normal distribution for the case of $SO(3)$ is [25]:

$$\varphi_\beta(A) = \varphi(\theta'; \beta) = \frac{\beta^3}{\sin \theta'} \sum_{k=-\infty}^{\infty} (\theta' - 2k\pi)\, e^{-\pi\beta^2(\theta' - 2k\pi)^2} \qquad (14.26)$$

for each fixed value of β where $\theta' = \theta/2$ and θ is the angle of rotation defined by the formula $A = e^{\theta N}$ and $\|N^\vee\| = 1$. It can be shown [25] that

$$(\varphi_\beta * \varphi_\alpha)(A) = (\varphi_\alpha * \varphi_\beta)(A) = \varphi_{\alpha'}(A)$$

where

$$\frac{1}{\alpha'^2} = \frac{1}{\alpha^2} + \frac{1}{\beta^2}. \qquad (14.27)$$

We note that since $\varphi(\theta; \beta)$ is closed under convolution *and* is an even function of only θ for each fixed β (and hence a class function), then the shifted versions of these functions are closed under convolution as well (from Theorem 14.2).

14.5 Mean and Variance for $SO(N)$ and $SE(N)$

We now demonstrate the definitions of mean and variance in the context of the rotation and motion groups. For the rotation groups there are a variety of metrics we can use (see Chapters 5 and 6). Here we use the metric

$$d_{SO(N)}(A, R) = \|A - R\|_2 = \sqrt{\sum_{i,j=1}^{N} (A_{ij} - R_{ij})^2}$$

for all $A, R \in SO(N)$. For $SE(N)$ we use the metric

$$d_{SE(N)}(g_1, g_2) = \sqrt{\|\mathbf{a}_1 - \mathbf{a}_2\|_2^2 + L^2\|A_1 - A_2\|_2^2}$$

where $g_i = (\mathbf{a}_i, A_i)$ and $L \in \mathbb{R}_{>0}$ is a length scale to put orientational and positional quantities in the same units.

14.5.1 Explicit Calculation for $SO(3)$

Using the metric $d_{SO(N)}$ described above, the mean (expected value) associated with a PDF $\rho \in \mathcal{L}^2(SO(N))$ is $A_{cm} \in SO(N)$ that minimizes the function

$$C(A_1) = \int_{SO(N)} \|A_1 - A_2\|_2^2 \rho(A_2)\, dA_2.$$

In order to minimize with respect to A_1, we can differentiate the modified cost function

$$C'(A_1) = C(A_1) + \operatorname{tr}(\Lambda(A_1^T A_1 - \mathbb{I}))$$

with respect to the elements of A_1 and the elements of the Lagrange multiplier matrix $\Lambda = \Lambda^T \in \mathbb{R}^{N \times N}$. The solution to this problem can be written in closed form.

We note that when $N = 3$ instead of using the elements of the matrix A_1 as the variables and using six constraint equations, we could use Euler angles as the variables and no constraints, or the Euler parameters (or axis-angle parameterization) and one constraint. When $N = 2$, it is natural to use the absolute values of the differences of angles of rotation as a metric instead of $d_{SO(2)}$. Then the calculation of the mean is much like the case of statistics on $\mathbb{R}$.

The $d_{SO(N)}$-variance of $\rho(A)$ is found by evaluating $C(A_{cm})$. We know of no simple exact closed-form formula for the $d_{SO(N)}$-mean of the convolution $(\rho_1 * \rho_2)(A)$ in terms of the $d_{SO(N)}$-means of ρ_1 and ρ_2, and likewise for $d_{SO(N)}$-variances. However, good approximations exist when the PDFs being convolved have small variances. Such approximations are used extensively in Chapter 19.

14.5.2 Explicit Calculation for $SE(2)$ and $SE(3)$

The mean (expected value) associated with a PDF $\rho \in \mathcal{L}^2(SE(N))$ is the pair $(\mathbf{a}_{cm}, A_{cm}) \in SE(N)$ that minimizes the function

$$C(\mathbf{a}_1, A_1) = \int_{\mathbb{R}^N} \int_{SO(N)} \{\|\mathbf{a}_1 - \mathbf{a}_2\|_2^2 + L\|A_1 - A_2\|_2^2\} \rho(\mathbf{a}_2, A_2)\, dA_2 d\mathbf{a}_2\,.$$

The minimization with respect to $\mathbf{a}_1$ follows exactly like the case of a function on $\mathbb{R}^N$, and we find the value to be

$$\mathbf{a}_{cm} = \int_{\mathbb{R}^N} \mathbf{a}_2 \left(\int_{SO(N)} \rho(\mathbf{a}_2, A_2)\, dA_2 \right) d\mathbf{a}_2. \tag{14.28}$$

In order to minimize with respect to A_1, we can differentiate the modified cost function

$$C'(\mathbf{a}_1, A_1) = C(\mathbf{a}_1, A_1) + \operatorname{tr}(\Lambda(A_1^T A_1 - \mathbb{I}))$$

with respect to the elements of A_1 and the elements of the Lagrange multiplier matrix $\Lambda = \Lambda^T \in \mathbb{R}^{N \times N}$. The solution to this problem is essentially like that for $SO(3)$.

As with the case of $SO(N)$, we know of no closed-form solution for the orientational means and variances of the convolution of two PDFs on $SE(N)$ in terms of the means and variances of the original PDFs. However, we are able to write the translational mean of the convolution of two PDFs in terms of the translational means of each PDF. In order to see this, evaluate (14.28) with the PDF $(\rho_1 * \rho_2)(\mathbf{a}, A)$. We denote the result as

$$\mathbf{a}_{cm}^{1*2} = \int_{\mathbb{R}^N} \mathbf{a} \left(\int_{SO(N)} (\rho_1 * \rho_2)(\mathbf{a}, A) \, dA \right) da.$$

Substitution of the explicit form of $(\rho_1 * \rho_2)(\mathbf{a}, A)$ into the above equation yields

$$\mathbf{a}_{cm}^{1*2} = \int_{\mathbb{R}^N} \mathbf{a} \int_{SE(3)} \rho_1(\mathbf{r}, R) \left(\int_{SO(N)} \rho_2(R^{-1}(\mathbf{a} - \mathbf{r}), R^{-1}A) \, dA \right) d\mathbf{r} dR da.$$

Due to the invariance of integration on $SO(N)$, we can simplify the quantity inside the parenthesis by defining

$$F_2(\mathbf{a}) \doteq \int_{SO(N)} \rho_2(\mathbf{a}, A) \, dA.$$

Making the substitution $\mathbf{x} = R^{-1}(\mathbf{a} - \mathbf{r})$, we then have

$$\mathbf{a}_{cm}^{1*2} = \int_{SE(3)} \int_{\mathbb{R}^N} (R\mathbf{x} + \mathbf{r}) \rho_1(\mathbf{r}, R) F_2(\mathbf{x}) \, d\mathbf{x} d\mathbf{r} dR.$$

Passing integrals through terms that are invariant under the integral and using the fact that

$$\int_{SE(3)} \rho_i(g) \, dg = 1,$$

we find

$$\mathbf{a}_{cm}^{1*2} = \mathbf{a}_{cm}^1 + M \mathbf{a}_{cm}^2 \tag{14.29}$$

where $\mathbf{a}_{cm}^i$ is the translational part of the mean of ρ_i, and

$$M = \int_{SO(N)} R \left(\int_{\mathbb{R}^N} \rho_1(\mathbf{r}, R) \, d\mathbf{r} \right) dR.$$

As an example of the usefulness of (14.29), consider the mean of the convolution of a PDF $\rho(\mathbf{a}, A)$ with itself n times. If $\mathbf{a}_{cm}$ is the translational mean of $\rho(\mathbf{a}, A)$, then the translational mean of $(\rho * \rho * \cdots * \rho)(\mathbf{a}, A)$ will be

$$\mathbf{a}_{cm}^{(n)} = \left(\mathbb{I} + \sum_{k=1}^{n} M^k \right) \mathbf{a}_{cm}.$$

But since

$$\left(\mathbb{I} + \sum_{k=1}^{n} M^k \right) (\mathbb{I} - M) = \left(\mathbb{I} - M^{n+1} \right),$$

it follows that if M has no eigenvalues equal to unity, then we can write

$$\mathbf{a}_{cm}^{(n)} = (\mathbb{I} - M)^{-1} \left(\mathbb{I} - M^{n+1} \right) \mathbf{a}_{cm}.$$

If all of the eigenvalues $|\lambda_i(M)| < 1$, then as $n \to \infty$ we have

$$\mathbf{a}_{cm}^{(n \to \infty)} \to (\mathbb{I} - M)^{-1} \mathbf{a}_{cm}.$$

In other words, for such PDFs on $SE(N)$, after an infinite number of convolutions, the translational mean will remain finite even though the translational mean of the original PDF is not zero. This does not happen with PDFs on $\mathbb{R}^N$.

14.6 Statistical Determination of a Rigid-Body Displacement

The problem of finding the rigid-body motion that will optimally match two sets of N points is of fundamental importance in a number of fields. Two variants on this static (time-independent) problem are described in the following subsections.

14.6.1 Pose Determination without A Priori Correspondence

In the first variant, the one-to-one correspondence between points in each of the sets is assumed not to be known and must be determined. For each rigid-body motion that one of the sets undergoes relative to the other there are $|\Pi_N| = N!$ possible correspondences. However, all $N!$ need not be investigated, and algorithms exist that will calculate the optimal assignment of two sets of points in $O(N^3)$ arithmetic operations. The *value* of the resulting optimal assignment is the sum of distances between points whose correspondence is established by the optimal assignment algorithm. A measure of the congruence of the two sets of points comes from determining the value of the optimal assignment as one of the sets of points is displaced over all rigid-body motions and a minimum is sought. Particular applications of this problem include the correspondence problem in computer vision and the motion planning of modular self-reconfigurable robots [77]. While this problem makes use of the group of permutations, we are not aware of applications of noncommutative harmonic analysis in this context.

In the case when two objects (or patterns) are given, and one wants to determine how alike the objects are, measures of distance between shape are required. See [77, 43] for methods of defining distance between shapes and finding the best fit of shapes under rigid-body motion. Likewise, if it is assumed that the two patterns are the same (to within a rigid-body motion) it can still be the case that the correspondence between points in the two patterns is not known. One way to establish a correspondence is to have a metric that distinguishes how different the collection of coordinates are for two spatial patterns. That is, if the two patterns A and B occupy the same exact points in space, the distance between the patterns should be zero: $D(A, B) = 0$. $D(\cdot, \cdot)$ should satisfy all the properties of a metric. There are many pattern metrics used in computational geometry and discrete mathematics. One such metric is to assume the two point patterns are replaced with a sum of Gaussian functions centered at the points, and the $\mathcal{L}^2$ distance between the functions is calculated. That is, if $A = \{\mathbf{a}_1, ..., \mathbf{a}_n\}$ and $B = \{\mathbf{b}_1, ..., \mathbf{b}_n\}$, we define

$$f_A(\mathbf{x}) = \sum_{i=1}^{n} \frac{\exp\left[-\frac{1}{2\sigma}\|\mathbf{x} - \mathbf{a_i}\|^2\right]}{(2\pi\sigma)^{\frac{n}{2}}},$$

and

$$D_1(A, B) = \sqrt{\int_{\mathbb{R}^N} |f_A(\mathbf{x}) - f_B(\mathbf{x})|^2 d\mathbf{x}}.$$

Another metric, called the *Hausdorff distance* is defined as follows. Let

$$h(A, B) \doteq \max_{a \in A} \min_{b \in B} d(a, b)$$

where $d(\cdot, \cdot)$ is any metric on $\mathbb{R}^N$. For example, the metric $d(\cdot, \cdot)$ could be Euclidean distance, or we could assign to each point a Gaussian function and calculate the $\mathcal{L}^2$ distance. It is well known that the triangle inequality holds for $h(A, B)$. For completeness, we illustrate this below:

$$h(A,C) = \max_{a \in A} \min_{c \in C} d(a,c) \leq \max_{a \in A} \min_{c \in C} (d(a,b') + d(b',c))$$
$$= \max_{a \in A} d(a,b') + \min_{c \in C} d(b',c) \quad \forall b' \in B.$$

This must be true for all $b' \in B$. Therefore it must be true for the choice $b' \in B$ such that $\max_{a \in A} d(a,b') = \max_{a \in A} \min_{b \in B} d(a,b)$. Therefore,

$$h(A,C) \leq \max_{a \in A} \min_{b \in B} d(a,b) + \min_{b \in B} \min_{c \in C} d(b,c) \leq h(A,B) + \max_{b \in B} \min_{c \in C} d(b,c)$$
$$= h(A,B) + h(B,C).$$

Since in general $h(A,B) \neq h(B,A)$ we can "repair" this by defining the undirected Hausdorff distance as

$$D_2(A,B) = \max(h(A,B), h(B,A)). \tag{14.30}$$

Another metric that has been shown to be useful in applications is the *optimal assignment metric* [77]. Again we begin with a measure of distance between points, $d(\cdot,\cdot)$. Then we define

$$D_3(A,B) = \min_{\pi \in \Pi_n} \sum_{i=1}^{n} d(\mathbf{a}_i, \mathbf{b}_{\pi(i)}) \tag{14.31}$$

where Π_n is the group of permutations on n letters.

Given a metric $D(\cdot,\cdot)$ that distinguishes between point patterns (e.g., any of the $D_i(\cdot,\cdot)$ mentioned above), the rigid-body transformation with best fit will be the one that minimizes the cost

$$C(g) = D(A, g \cdot B)$$

where $g \cdot B$ denotes each $\mathbf{b}_i \in B$ being moved as $g \cdot \mathbf{b}_i = R\mathbf{b}_i + \mathbf{r}$ where $g = (\mathbf{r}, R)$.

The optimal assignment metric (14.31) is particularly attractive from the perspective that in addition to generating the optimal rigid-body match, it also generates a correspondence between points in A and B. This correspondence is defined by the permutation (optimal assignment) that performs the minimization in (14.31).

14.6.2 Pose Determination with A Priori Correspondence

In the second variant of the pose determination problem, it is assumed that the correspondence between points has been established a priori, and one seeks to find the relative rigid-body motion that will best match the two sets of points in the sense of least squared error. In this subsection, we review the work presented in [8, 51, 52, 73, 94]. This is what we call the static rigid-body motion estimation problem with a priori correspondence. In the context of computer vision and image analysis many other related pose determination/estimation has been studied extensively. See e.g., [32, 45, 57, 76, 80, 86]. In particular, the problem of rotation/orientation estimation has been addressed in [42, 59, 75].

Consider two sets of N points (positions) in $\mathbb{R}^3$ denoted as $\{\mathbf{p}_i\}$ and $\{\mathbf{q}_i\}$ where the correspondence $\mathbf{q}_i \leftrightarrow \mathbf{p}_i$ is assumed for all $i = 1, ..., N$. The set $\{\mathbf{q}_i\}$ can be thought of as exact points originating from a database (e.g., the vertices of a polygonal object). The set $\{\mathbf{p}_i\}$ can be thought of as measurements (e.g., if the corners of an object are detected experimentally). The goal is to find the rigid-body motion $(\mathbf{a}, A)$ such that $\{\mathbf{q}_i\}$ is moved to fit in the "best" way to $\{\mathbf{p}_i\}$. One way to define the best fit is as the minimization of the weighted mean-square error

$$E^2(A, \mathbf{a}) = \sum_{i=1}^{N} \|\mathbf{p}_i - (A\mathbf{q}_i + \mathbf{a})\|^2_{W_i}. \tag{14.32}$$

Here we use the notation

$$\|\mathbf{x}\|^2_{W_i} = \mathbf{x}^T W_i \mathbf{x}.$$

The 3×3 weighting matrices W_i are chosen based on how noisy the measurement of $\mathbf{p}_i$ is assumed to be. If the measurement error of all the points is assumed to be the same, then $W_i = W$ for all $i = 1, ..., N$.

Assuming the errors in position measurements are Gaussian distributed, we define the *likelihood function* as

$$L(A, \mathbf{a}) = \sqrt{\det K} \exp(-\frac{1}{2}\epsilon^T K \epsilon)$$

where $K = K^T$ is positive definite and

$$\epsilon(A, \mathbf{a}) = \begin{pmatrix} \mathbf{p}_1 - (A\mathbf{q}_1 + \mathbf{a}) \\ \vdots \\ \mathbf{p}_N - (A\mathbf{q}_N + \mathbf{a}) \end{pmatrix} \in \mathbb{R}^{3N}$$

is a concatenated error vector.

When K is block diagonal of the form

$$K = \begin{pmatrix} W_1 & 0 & ... & 0 \\ 0 & W_2 & 0 & ... \\ 0 & ... & W_{N-1} & 0 \\ 0 & ... & 0 & W_N \end{pmatrix},$$

then

$$-2\log L = -\log\left(\prod_{i=1}^{N} \det W_i\right) + E^2(A, \mathbf{a})$$

where $E^2(A, \mathbf{a}) = \epsilon^T K \epsilon$.

In general the minimum of a function $f(x_1, ...x_n)$ and the minimum of $F(f(x_1, ...x_n))$ result in the same answer when $F(\cdot)$ is monotonically increasing. Hence the maximization of the likelihood function L is the same as the maximization of $\log L$. This in turn is the same as the minimization of the mean square error $E^2(A, \mathbf{a})$.

The problem of finding the optimal rotation matrix when $\mathbf{a} = \mathbf{0}$ and $W_i = w_i \mathbb{I}$ is addressed in a number of ways including those described in [17, 33, 95]. This is sometimes called the *(orthogonal) Procrustes problem* [32, 44, 88]. A very simple approach to solving this problem is outlined in [60], which we follow here. Namely, let

$$S = \sum_{i=1}^{n} w_i \, \mathbf{p}\mathbf{q}^T$$

and perform the singular-value decomposition (SVD) to give $S = U\Sigma V^T$ where Σ is the diagonal matrix of singular values ordered from greatest to smallest. Then the optimal rotation to solve the above Procrustes problem is simply

$$\boxed{A = U \, \text{diag}[1, ..., 1, \det(UV^T)] \, V^T.} \tag{14.33}$$

From this equation it is clear that

$$\det A = \det U \cdot \det(UV^T) \cdot \det V^T = |\det(UV^T)|^2 = 1.$$

Sometimes it is convenient to write S in the more general form

$$S = PKQ^T$$

where

$$P = M(\mathbf{p}) \doteq [\mathbf{p}_1, ..., \mathbf{p}_n]; \quad Q = M(\mathbf{q}) \doteq [\mathbf{q}_1, ..., \mathbf{q}_n].$$

As an alternative to the SVD, we can compute the optimal estimate A in two stages by first computing

$$\hat{A} = \left[\sum_{k=1}^{n} \mathbf{p}_k \mathbf{q}_k^T \right] \left[\sum_{i=1}^{n} \mathbf{q}_i \mathbf{q}_i^T \right]^{-1} \tag{14.34}$$

(which solves the unconstrained least squares problem) and then finding the closest $A \in SO(3)$ to $\hat{A} \in \mathbb{R}^{3 \times 3}$ in the sense of the Frobenius (Hilbert-Schmidt) norm. The solution is the orthogonal matrix in the polar decomposition of $\hat{A}$:

$$A = \hat{A}(\hat{A}^T \hat{A})^{-\frac{1}{2}}. \tag{14.35}$$

The benefit of (14.33) over (14.35) is that (14.33) is more stable and ensures that $\det A = +1$.

Nádas [73] considers the case when all the individual error vectors $\boldsymbol{\epsilon}_i = \mathbf{p}_i - (A\mathbf{q}_i + \mathbf{a})$ are assumed to be uncorrelated with each other, but have the same covariance matrices (hence $W_i = W$). That is, all the errors are assumed to be sampled from the same zero-mean normal distribution.

Define

$$M(\mathbf{x}) = [\mathbf{x}_1, ..., \mathbf{x}_N], \quad \mathbf{x}_{cm} = \frac{1}{N} \sum_{i=1}^{N} \mathbf{x}_i, \quad \text{and} \quad \mathbf{x}_i^{rel} = \mathbf{x}_i - \mathbf{x}_{cm}. \tag{14.36}$$

Using the definition (14.36) to define $\mathbf{p}^{rel}$ and $\mathbf{q}^{rel}$, the matrices $M(\mathbf{p}^{rel})$, $M(\mathbf{q}^{rel})$, and $M(\boldsymbol{\epsilon})$ then play an important role. In particular, the 3×3 matrix

$$C = M(\mathbf{p}^{rel})[M(\mathbf{q}^{rel})]^T \tag{14.37}$$

plays a special role. In the case considered by Nádas [73], it is straight forward to show that the optimal value of translation is determined by

$$\frac{\partial E^2}{\partial a_i} = 0 \quad \Rightarrow \quad \mathbf{a}_{opt} = \mathbf{p}_{cm} - WC(C^T W^2 C)^{-\frac{1}{2}} \mathbf{q}_{cm}.$$

This allows one to decouple the solution for the translational and rotational parts of the problem.

The solution for the optimal $A \in SO(3)$ can be approached in a number of ways. Perhaps the most intuitive is to formulate the problem as a constrained optimization problem, with the constraint of orthonormality of A imposed using Lagrange multipliers. This approach can be implemented in a variety of ways. For instance the six independent constraints given by the definition $AA^T = \mathbb{I}$ can be used (in which case the condition $\det(A) = 1$ must be checked after the fact), or if we define $A = [\mathbf{a}_1, \mathbf{a}_2, \mathbf{a}_3]$, then the six constraints

$$\mathbf{a}_1 \times \mathbf{a}_2 = \mathbf{a}_3, \quad \mathbf{a}_1 \cdot \mathbf{a}_1 = \mathbf{a}_2 \cdot \mathbf{a}_2 = 1, \quad \mathbf{a}_1 \cdot \mathbf{a}_2 = 0$$

ensure that a rotation will result. One could parameterize A with unit quaternions (Euler parameters) and minimize over these parameters subject to a single scalar constraint equation, or minimize over Euler angles subject to no constraint. Below we minimize over $A \in \mathbb{R}^{3 \times 3}$ subject to the constraint $AA^T = \mathbb{I}$.

After substitution of $\mathbf{a}_{opt}$ into the cost function, and ignoring the constant factor $\log(\prod_{i=1}^{N} \det W_i)$, we then have the constrained cost function

$$F = \sum_{i=1}^{N} \| \mathbf{p}_i^{rel} - A\mathbf{q}_i^{rel} \|_{W_i}^2 + \mathrm{tr}(\Lambda(A^T A - \mathbb{I})),$$

where Λ is a symmetric Lagrange multiplier matrix with elements Λ_{ij} (and therefore has six independent Lagrange multipliers). Setting

$$\frac{\partial F}{\partial A_{ij}} = 0 \quad \text{and} \quad \frac{\partial F}{\partial \Lambda_{ij}} = 0$$

and assuming $W_i = W$ results in the system of equations [73]

$$W(M(\mathbf{p}^{rel}) - AM(\mathbf{q}^{rel}))[M(\mathbf{q}^{rel})]^T + A\Lambda = 0 \quad \text{and} \quad A^T A = \mathbb{I}.$$

One can verify by inspection that when $W = w\mathbb{I}$,

$$A = WC(C^T W^2 C)^{-\frac{1}{2}} = C(C^T C)^{-\frac{1}{2}}$$

where the matrix C is defined in (14.37).

14.6.3 Problem Statement Using Harmonic Analysis

If $f(h)$ is the error probability density function when trying to fit rigid-body motion, then the measurements are

$$\int_{SE(N)} f(h)\, \delta_{\mathbf{q}_i}(h^{-1} \cdot \mathbf{x})\, dh = p_i(\mathbf{x}). \tag{14.38}$$

Here $\delta_{\mathbf{q}_i}(\mathbf{x}) \doteq \delta(\mathbf{x} - \mathbf{q}_i)$ and $p_i(\mathbf{x})$ is the error PDF for point i. If the measurements were exact, then we would have $p_i(\mathbf{x}) = \delta(\mathbf{x} - \mathbf{p}_i)$ and $f(h) = \delta(g^{-1} \circ h)$ where $g \in SE(N)$ is the motion for which $g \cdot \mathbf{q}_i = \mathbf{p}_i$ for all i.

In the non-ideal case of inexact measurements we can write the system of equations (14.38) for all sample measurements $i \in [1, M]$ as

$$f * \Delta = P$$

where

$$\Delta(\mathbf{x}) = [\delta_{\mathbf{q}_1}(\mathbf{x}), ..., \delta_{\mathbf{q}_M}(\mathbf{x})]$$

and

$$P(\mathbf{x}) = [p_1(\mathbf{x}), ..., p_M(\mathbf{x})],$$

and the convolution of f with each of these row vectors is defined to be the convolution with each element. The Fourier transform for $SE(N)$ can be used to reduce these convolutions to matrix products.

In practice, we will choose in advance an ansatz for $f(h)$ and seek to fit a few parameters to the data. In particular, we can define a Gaussian function on $SE(N)$ to be the products of Gaussians on $\mathbb{R}^N$ and $SO(N)$ centered around the respective identities. We denote

$$f(\mathbf{a}, A) = f_t(\mathbf{a}) \, f_r(A).$$

The convolution of two such functions results is a function of the same form when the translational Gaussian function is spherically symmetric.

Explicitly,

$$(f_1 * f_2)(\mathbf{a}, A) = \int_{SO(N)} \int_{\mathbb{R}^N} f_1(\mathbf{r}, R) \, f_2(R^{-1}(\mathbf{a} - \mathbf{r}), R^{-1}A) \, d\mathbf{r} dR.$$

Substitution of $f_i(\mathbf{a}, A) = f_{t_i}(\mathbf{a}) \, f_{r_i}(A)$ and using the spherical symmetry $f_{t_i}(Q\mathbf{a}) = f_{t_i}(\mathbf{a})$ for all $Q \in SO(3)$, it can be shown that

$$(f_1 * f_2)(\mathbf{a}, A) = (f_{t_1} * f_{t_2})(\mathbf{a}) \cdot (f_{r_1} * f_{r_2})(A).$$

On the right-hand side of the above equation, the convolutions are respectively those for $\mathbb{R}^N$ and $SO(N)$. Since the Gaussians for $\mathbb{R}^N$ and $SO(N)$ are closed under convolutions (in the sense that the convolution of two Gaussians is a Gaussian), it follows that the Gaussian defined for $SE(N)$ is also closed under convolution. Furthermore, if f_1 and f_2 are both Gaussians on $SE(N)$ then

$$(f_1 * f_2)(\mathbf{a}, A) = (f_2 * f_1)(\mathbf{a}, A).$$

However, while convolutions of such Gaussians commute with each other, they do not commute with arbitrary functions on $SE(N)$. That is, these Gaussians are not class functions.

14.7 Robot Sensor Calibration

This section reviews the robot sensor calibration problem solved by Shiu and Ahmad [90], Chou and Kamel [27], and Park and Martin [79]. We then extend the problem in a new direction, and show how noncommutative harmonic analysis may be applicable.

The problem is depicted in Figure 14.1 and is stated as follows. A robot arm which has already been calibrated has a frame of reference attached to its distal end. For any given movement of the arm away from a home position, let this distal frame be denoted as g_A. When the arm is in the home position, we take g_A to be the identity frame. We now attach to this distal end (which can be considered the wrist of the robot arm) a new sensor (e.g., camera, or hand with tactile sensors). It is assumed that the kinematic parameters describing the position and orientation of a frame of reference attached to the new sensors, g_X, relative to the wrist frame, g_A, is unknown. However, the relative displacement of the new sensor location to its own position and orientation when the robot arm is in the home position can be measured. Hence, the composition of motions $g_A \circ g_X$ results in the position and orientation of the sensor frame relative to the identity frame. Likewise, $g_X \circ g_B$ describes the concatenation of motions from the identity frame to sensor frame when the arm is in the home position, and then the motion of this frame. Hence the equality [27, 79, 90]:

$$g_A \circ g_X = g_X \circ g_B. \tag{14.39}$$

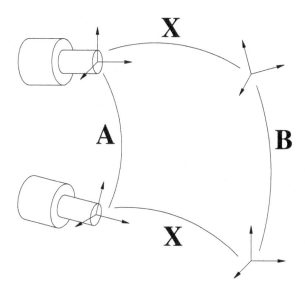

Fig. 14.1. How $AX = XB$ Arises

This problem arises in the fields of robotics and computer vision as mentioned above, and also in computer-integrated surgical systems [13, 14, 70, 81].

The solution to this problem for g_X in the case when it is assumed that one exact value of g_A and one exact value of g_B are known is underdetermined (i.e., there is a continuum of solutions) [90]. The case when multiple calibrating motions, (g_{A_i}, g_{B_i}) for $i = 1, ..., N$, are performed is addressed in [79]. We review both of these cases, and extend the analysis to answer the question: Given two sets of measurements, where each measurement of g_A and g_B are assumed to have known error distributions, how is this reflected in the distribution of error in g_X?

14.7.1 Euclidean-Group Invariants: Necessary and Sufficient Conditions for Unique Solution to the $AX = XB$ Problem

The problem of solving

$$A^i X = X B^i \tag{14.40}$$

for X when multiple corresponding pairs of A's and B's are presented has been examined in the context of many diverse applications over the past quarter century such as given in [8, 6, 9, 28, 29, 34, 56, 64, 68, 79]. The particular presentation in this section follows that in [2].

From screw theory it is known that any homogeneous transformation can be written as [24]:

$$H = \begin{pmatrix} e^{\theta N} & (\mathbb{I}_3 - e^{\theta N})\mathbf{p} + d\mathbf{n} \\ \mathbf{0}^T & 1 \end{pmatrix}$$

where $e^{\theta N}$ denotes the matrix exponential, $\mathbb{I}_n$ is the $n \times n$ identity matrix, and $\theta \in [0, \pi]$ is the angle of rotation.

$$N = \begin{pmatrix} 0 & -n_3 & n_2 \\ n_3 & 0 & -n_1 \\ -n_2 & n_1 & 0 \end{pmatrix}$$

where $\mathbf{n} = [\mathbf{n_1}, \mathbf{n_2}, \mathbf{n_3}]^T \in \mathbb{R}^3$ is the unit vector describing the axis of rotation, which connects the origin and any point on the unit sphere, and $\mathbf{p} \cdot \mathbf{n} = 0$. Together, $\{\theta, d, \mathbf{n}, \mathbf{p}\}$ define the Plücker coordinates of the screw motion.

If we write (14.40) as

$$A^i = XB^iX^{-1} \quad \text{where} \quad i \in \{1, 2\}, \tag{14.41}$$

then explicitly calculating and equating the matrix product gives two invariant relations,

$$\theta_{A^i} = \theta_{B^i} \quad d_{A^i} = d_{B^i} \tag{14.42}$$

where d_{A^i} and d_{B^i} are computed from A^i and B^i as in (14.41). Additionally, let

$$\mathbf{l}_{A^i}(t) = \mathbf{p}_{A^i} + t\mathbf{n}_{A^i} \quad \text{and} \quad \mathbf{l}_{B^i}(t) = \mathbf{p}_{B^i} + t\mathbf{n}_{B^i}$$

be the directed screw axis lines of A_i and B_i in three-dimensional Euclidean space. If the lines are not parallel or anti-parallel, i.e., if $\mathbf{n}_{A_i} \neq \pm\mathbf{n}_{B_i}$, then the distance between the two lines is given by

$$\Delta(\mathbf{l}_{A^{i_1}}, \mathbf{l}_{A^{i_2}}) = \frac{|[\mathbf{n}_{A^{i_1}}, \mathbf{n}_{A^{i_2}}, \mathbf{p}_{A^{i_2}} - \mathbf{p}_{A^{i_1}}]|}{\|\mathbf{n}_{A^{i_1}} \times \mathbf{n}_{A^{i_2}}\|} \tag{14.43}$$

where for any $\mathbf{a}, \mathbf{b}, \mathbf{c} \in \mathbb{R}^3$, the triple product is $[\mathbf{a}, \mathbf{b}, \mathbf{c}] \doteq \mathbf{a} \cdot (\mathbf{b} \times \mathbf{c})$. In the current context we can think of $i_1 = 1$ and $i_2 = 2$ but, in later discussion, i_1 and i_2 can represent more general values.

If in addition, $\Delta(\mathbf{l}_{A^{i_1}}, \mathbf{l}_{A^{i_2}}) \neq 0$, i.e., if the lines are skew, then the angle $\phi(\mathbf{l}_{A^{i_1}}, \mathbf{l}_{A^{i_2}}) \in [0, 2\pi)$ is uniquely specified by

$$\cos\phi(\mathbf{l}_{A^{i_1}}, \mathbf{l}_{A^{i_2}}) = \mathbf{n}_{A^{i_1}} \cdot \mathbf{n}_{A^{i_2}} \tag{14.44}$$

$$\sin\phi(\mathbf{l}_{A^{i_1}}, \mathbf{l}_{A^{i_2}}) = \Delta(\mathbf{l}_{A^{i_1}}, \mathbf{l}_{A^{i_2}})^{-1}[\mathbf{n}_{A^{i_1}}, \mathbf{n}_{A^{i_2}}, \mathbf{p}_{A^{i_2}} - \mathbf{p}_{A^{i_1}}].$$

Therefore, if $\theta_{A^{i_1}}, \theta_{A^{i_2}} \in (0, \pi)$ and $\phi(\mathbf{l}_{A^{i_1}}, \mathbf{l}_{A^{i_2}}) \notin \{0, \pi\}$, then a unique solution of (14.40) exists if and only if the following four conditions hold:

1. $\theta_{A^{i_1}} = \theta_{B^{i_1}}$ and $\theta_{A^{i_2}} = \theta_{B^{i_2}}$;
2. $d_{A^{i_1}} = d_{B^{i_1}}$ and $d_{A^{i_2}} = d_{B^{i_2}}$;
3. $\phi(\mathbf{l}_{A^{i_1}}, \mathbf{l}_{A^{i_2}}) = \phi(\mathbf{l}_{B^{i_1}}, \mathbf{l}_{B^{i_2}})$;
4. $\Delta(\mathbf{l}_{A^{i_1}}, \mathbf{l}_{A^{i_2}}) = \Delta(\mathbf{l}_{B^{i_1}}, \mathbf{l}_{B^{i_2}})$.

If these do not hold, then a solution will not be possible [2]. Fig. 14.2 illustrates the Plücker coordinates, and the parameters of the above four conditions for two arbitrary rigid-body motions.

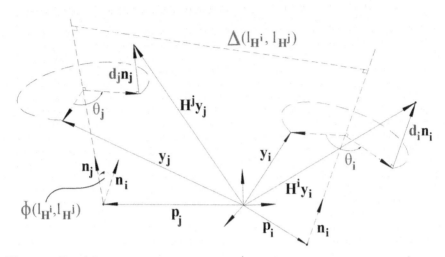

Fig. 14.2. Two Arbitrary Rigid-Body Motions: H^i which is shown acting on y_i and H^j which is shown acting on y_j. Their Plücker coordinates and the parameters of the four described conditions are also depicted. (Reprinted from Ackerman, M.K., Cheng, A., Shiffman, B., Boctor, E., Chirikjian, G., "Sensor Calibration with Unknown Correspondence: Solving $AX = XB$ Using Euclidean-Group Invariants," *Proceedings of the 2013 IEEE/RSJ International Conference on Intelligent Robots and Systems (IROS'13)*, (pp. 1308–1313), November 2013, with permission from IEEE)

14.7.2 Solution with Two Sets of Exact Measurements

Using homogeneous transform notation, let $g = (\mathbf{q}, Q) \in SE(3)$ be represented as

$$H(g) = \begin{pmatrix} Q & \mathbf{q} \\ \mathbf{0}^T & 1 \end{pmatrix}.$$

Then (14.39) is rewritten as

$$\begin{pmatrix} Q_A & \mathbf{q}_A \\ \mathbf{0}^T & 1 \end{pmatrix} \begin{pmatrix} Q_X & \mathbf{q}_X \\ \mathbf{0}^T & 1 \end{pmatrix} = \begin{pmatrix} Q_X & \mathbf{q}_X \\ \mathbf{0}^T & 1 \end{pmatrix} \begin{pmatrix} Q_B & \mathbf{q}_B \\ \mathbf{0}^T & 1 \end{pmatrix}.$$

Performing the matrix multiplication results in two equations of the form

$$Q_A Q_X = Q_X Q_B \tag{14.45}$$

and

$$Q_A \mathbf{q}_X + \mathbf{q}_A = Q_X \mathbf{q}_B + \mathbf{q}_X. \tag{14.46}$$

The strategy to solve (14.39) would appear to reduce to first solving (14.45), and then rearranging (14.46) so as to find acceptable values of $\mathbf{q}_X$:

$$(Q_A - \mathbb{I}_{3\times3})\mathbf{q}_X = Q_X \mathbf{q}_B - \mathbf{q}_A.$$

However, there are some problems with this naive approach. As pointed out in [27, 79, 90], there is a one-parameter set of solutions to (14.45), and the matrix $Q_A - \mathbb{I}_{3\times3}$ in general has rank 2. Hence, there are two unspecified degrees of freedom to the problem, and it cannot be solved uniquely unless additional measurements are taken.

This situation is rectified by considering two sets of exact measurements of the form in (14.39), i.e.,

$$g_{A_1} \circ g_X = g_X \circ g_{B_1} \quad \text{and} \quad g_{A_2} \circ g_X = g_X \circ g_{B_2}.$$

We now consider the technical details of the problem. To begin, we observe that (14.45) can be rewritten as

$$Q_A = Q_X Q_B Q_X^{-1}. \tag{14.47}$$

Regardless of what Q_X is, the trace is invariant under similarity transformation $(\text{tr}(Q_X Q_B Q_X^{-1}) = \text{tr}(Q_B))$, and so

$$\text{tr}(Q_A) = \text{tr}(Q_B). \tag{14.48}$$

This means that the angle of rotation of Q_A and Q_B must be the same. For the present discussion, we will assume that $\text{tr}(Q_A) \neq -1$ (and the same for Q_B).

Using the $\vee$ operation on both sides of (14.47), observing that $(Q_X Q_B Q_X^{-1})^\vee = Q_X (Q_B)^\vee$, and dividing both sides by the magnitude $\|(Q_A)^\vee\| = \|(Q_B)^\vee\|$ results in the equation

$$Q_X \mathbf{b} = \mathbf{a}$$

where $\mathbf{a}$ and $\mathbf{b}$ are the unit vectors pointing along the axes of rotation of Q_A and Q_B, respectively.

A particular solution to this equation for given $\mathbf{a}$ and $\mathbf{b}$ is

$$Q_Y = R(\mathbf{b}, \mathbf{a}), \tag{14.49}$$

which is the rotation defined in (5.27), that takes $\mathbf{b}$ into $\mathbf{a}$.

It is also clear that if Q_Y is a solution to (14.45), then so must be

$$Q_X = \text{rot}[\mathbf{a}, \theta_1] Q_Y \text{rot}[\mathbf{b}, \theta_2] \tag{14.50}$$

for arbitrary $\theta_1, \theta_2 \in [0, 2\pi]$. Using the notation $\exp(\theta \hat{\mathbf{a}}) = \text{rot}[\mathbf{a}, \theta]$, (14.50) is written as [2]

$$Q_X = Q_Y \exp\left(\theta_1 Q_Y^T \hat{\mathbf{a}} Q_Y\right) \text{rot}[\mathbf{b}, \theta_2]$$

where $(\hat{\mathbf{a}})^\vee = \mathbf{a}$. This is simplified to

$$Q_X = Q_Y \text{rot}[Q_Y^T \mathbf{a}, \theta_1] \text{rot}[\mathbf{b}, \theta_2] = Q_Y \text{rot}[\mathbf{b}, \theta_1 + \theta_2].$$

Hence, despite the appearance of (14.50), there is a one parameter set of solutions to the rotational part of the problem.

To solve for a unique Q_X, two sets of measurements (Q_{A_1}, Q_{B_1}) and (Q_{A_2}, Q_{B_2}) are required. For each we have the condition

$$Q_X \mathbf{b}_i = \mathbf{a}_i.$$

We also observe that

$$\mathbf{a}_1 \times \mathbf{a}_2 = (Q_X \mathbf{b}_1) \times (Q_X \mathbf{b}_2) = Q_X (\mathbf{b}_1 \times \mathbf{b}_2). \tag{14.51}$$

---

[2] Recall that for any $B, X \in \mathbb{R}^{N \times N}$ it holds that $B e^X B^{-1} = e^{BXB^{-1}}$.

Combining all this into one equation, we have

$$Q_X[\mathbf{b}_1, \mathbf{b}_2, \mathbf{b}_1 \times \mathbf{b}_2] = [\mathbf{a}_1, \mathbf{a}_2, \mathbf{a}_1 \times \mathbf{a}_2].$$

This is easily inverted to find Q_X provided $\mathbf{b}_1 \times \mathbf{b}_2 \neq \mathbf{0}$ (which, by (14.51) means $\mathbf{a}_1 \times \mathbf{a}_2 \neq \mathbf{0}$ as well).

Having solved the rotational part of the problem, the translation $\mathbf{q}_X$ is determined by simultaneously solving the two equations

$$(Q_{A_i} - \mathbb{I}_{3\times 3})\mathbf{q}_X = Q_X \mathbf{q}_{B_i} - \mathbf{q}_{A_i} \tag{14.52}$$

for $i = 1, 2$.

This can be done in closed form as follows. Fix the value of i and write

$$\mathbf{q}_X = x_i \mathbf{n}_i + y_i \mathbf{v}_i + z_i(\mathbf{n}_i \times \mathbf{v}_i)$$

where $\mathbf{n}_i$ is the unit vector pointing in the direction of the axis of rotation of Q_{A_i} and $\mathbf{v}_i = (n_1^2 + n_2^2)^{-\frac{1}{2}}[-n_2, n_1, 0]^T$. Then

$$Q_{A_i}\mathbf{v}_i = \mathbf{v}_i \cos\theta_i + (\mathbf{n}_i \times \mathbf{v}_i)\sin\theta_i$$
$$Q_{A_i}(\mathbf{n}_i \times \mathbf{v}_i) = -\mathbf{v}_i \sin\theta_i + (\mathbf{n}_i \times \mathbf{v}_i)\cos\theta_i.$$

Using this fact, and taking the dot product of (14.52) with $\mathbf{v}_i$ and $(\mathbf{n}_i \times \mathbf{v}_i)$, two scalar equations result that can be arranged as

$$\begin{pmatrix} \cos\theta_i - 1 & -\sin\theta_i \\ \sin\theta_i & \cos\theta_i - 1 \end{pmatrix} \begin{pmatrix} y_i \\ z_i \end{pmatrix} = \begin{pmatrix} (Q_X \mathbf{q}_{B_i} - \mathbf{q}_{A_i}) \cdot \mathbf{v}_i \\ (Q_X \mathbf{q}_{B_i} - \mathbf{q}_{A_i}) \cdot (\mathbf{n}_i \times \mathbf{v}_i) \end{pmatrix}. \tag{14.53}$$

For $\theta_i \neq 0$ we can invert this to find y_i and z_i for each $i = 1, 2$. What remains is to find x_1 and x_2 such that

$$\mathbf{q}_X = x_1 \mathbf{n}_1 + y_1 \mathbf{v}_1 + z_1(\mathbf{n}_1 \times \mathbf{v}_1) = x_2 \mathbf{n}_2 + y_2 \mathbf{v}_2 + z_2(\mathbf{n}_2 \times \mathbf{v}_2).$$

This is nothing more than the problem of finding where two lines (each parameterized by x_i) intersect. This problem can be solved by finding the values of x_1 and x_2 that minimize the square of the distance between points on the two lines. This leads to the equations

$$\begin{pmatrix} 1 & -\mathbf{n}_1 \cdot \mathbf{n}_2 \\ -\mathbf{n}_1 \cdot \mathbf{n}_2 & 1 \end{pmatrix} \begin{pmatrix} x_1 \\ x_2 \end{pmatrix} = \begin{pmatrix} \mathbf{n}_1 \cdot (y_2 \mathbf{v}_2 + z_2(\mathbf{n}_2 \times \mathbf{v}_2)) \\ \mathbf{n}_2 \cdot (y_1 \mathbf{v}_1 + z_1(\mathbf{n}_1 \times \mathbf{v}_1)) \end{pmatrix}.$$

We note that this solution is particularly simple if we can choose $\mathbf{n}_1$ and $\mathbf{n}_2$ such that $\mathbf{n}_1 \cdot \mathbf{n}_2 = 0$. Things simplify even more when $\mathbf{v}_1 = \mathbf{n}_2$ and $\mathbf{v}_2 = \mathbf{n}_1$, in which case $x_1 = y_2$ and $x_2 = y_1$.

The problem of solving for Q_X and $\mathbf{q}_X$ for $N > 2$ measurements is done in [27, 79].

14.7.3 The $AX = XB$ Problem as an $SE(3)$ Convolution Equation

This section reviews the results presented in [1, 2]. For other recent approaches and pointers to the literature on this topic, see [3, 4].

Notation and Mathematical Problem Formulation

Given a large set of pairs $(A_i, B_i) \in SE(3) \times SE(3)$ for $i = 1, ..., n$ that exactly satisfy the equation

$$A_i X = X B_i \tag{14.54}$$

numerous algorithms exist to find $X \in SE(3)$, as discussed earlier. Here we address a generalization of this problem in which the sets $\{A_i\}$ and $\{B_j\}$ are provided with elements written in any order and it is known that a correspondence exists between the elements of these sets such that (14.54) holds, but we do not know a priori this correspondence between each A_i and B_j. We seek to find X in this scenario.

The group of proper rigid-body motions, $SE(3)$, is a Lie group, and hence concepts of integration and convolution exist. If $H \in SE(3)$ is a generic 4×4 homogeneous transformation of the form in (13.1) where the rotation is parameterized in terms of Euler angles as $R = R_3(\alpha)R_1(\beta)R_3(\gamma)$ (where $R_i(\theta)$ is a counterclockwise rotation by θ around coordinate axis i) and the translation is $\mathbf{t} = [t_x, t_y, t_z]^T$, then the 'natural' integral of any rapidly decaying function is computed as

$$\int_{SE(3)} f(H)\, dH = \int_{\mathbb{R}^3} \int_{SO(3)} f(H(R, \mathbf{t}))\, dR\, d\mathbf{t}$$

where $dR = \sin\beta d\alpha d\beta d\gamma$ and $d\mathbf{t} = dt_x dt_y dt_z$, with $-\infty < t_x, t_y, t_z < \infty$ and $(\alpha, \beta, \gamma) \in [0, 2\pi] \times [0, \pi] \times [0, 2\pi]$. This integral is 'natural' in the sense that it is the unique one (up to scaling of the volume element by an arbitrary constant) such that

$$\int_{SE(3)} f(H)\, dH = \int_{SE(3)} f(H^{-1})\, dH = \int_{SE(3)} f(HH_0)\, dH = \int_{SE(3)} f(H_0 H)\, dH \tag{14.55}$$

for any fixed $H_0 \in SE(3)$. This choice of integral, being invariant under shifts on the left and on the right in the above equation, is called the bi-invariant, or Haar, measure. The above instantiation of the bi-invariant integral for $SE(3)$ using ZXZ Euler angles and Cartesian coordinates for translation is not unique. Any parameterization of $SE(3)$ will do.

If $\int_{SE(3)} |f(H)|^p dH < \infty$ then we say $f \in \mathcal{L}^p(SE(3))$. Most of our discussion will be limited to functions $f \in \mathcal{L}^1(SE(3)) \cap \mathcal{L}^2(SE(3))$, together with the special case of a Dirac delta function, which will be defined shortly. In this context, the convolution of two such functions is defined as

$$(f_1 * f_2)(H) \doteq \int_{SE(3)} f_1(\mathcal{H})\, f_2(\mathcal{H}^{-1}H)\, d\mathcal{H} \tag{14.56}$$

where $\mathcal{H} \in SE(3)$ is a dummy variable of integration.

A Dirac delta function can be defined for $SE(3)$ just like in the case of $\mathbb{R}^n$. It is defined by the properties $\int_{SE(3)} \delta(\mathcal{H})\, d\mathcal{H} = 1$ and $(f * \delta)(H) = f(\mathbb{I}_4)$ where $\mathbb{I}_4 = H(\mathbb{I}_3, \mathbf{0})$ is the 4×4 identity matrix (and the identity element of $SE(3)$), whereas $\mathbb{I}_3$ is the 3×3 identity. Intuitively a Dirac delta can be thought of as a function that has a spike with infinite height at the identity and vanishes everywhere else.

A shifted Dirac delta function can be defined as $\delta_A(H) \doteq \delta(A^{-1}H)$ which places the spike at $A \in SE(3)$. Note that the inverse operation, A^{-1}, must be applied to the argument of the function to move the spike from the identity to A.

Batch Solution

In this light we can think of (14.54) as the equation

$$(\delta_{A_i} * \delta_X)(H) = (\delta_X * \delta_{B_i})(H). \tag{14.57}$$

The addition of this mathematical terminology provides freedom to do something that we cannot do with (14.54). Namely, whereas the addition (as opposed to multiplication) of homogeneous transformation matrices is nonsensical, the addition of real-valued functions $f_1(H) + f_2(H)$ is a perfectly reasonable operation. And since convolution is a linear operation on functions, we can write all n instances of (14.57) into a single equation of the form

$$(f_A * \delta_X)(H) = (\delta_X * f_B)(H) \text{ where } f_A(H) = \frac{1}{n}\sum_{i=1}^{n}\delta(A_i^{-1}H) \tag{14.58}$$

and $f_B(H)$ is of a similar form computed from $\{B_j\}$.

When written in this way, it does not matter if we know the correspondence between each A_i and B_j or not. The above functions are normalized to be probability densities: $\int_{SE(3)} f_A(H)\,dH = \int_{SE(3)} f_B(H)\,dH = 1$. Of course, the functions $f_A(H)$ and $f_B(H)$ are not in $\mathcal{L}^2(SE(3))$, but for our purposes this will not be a problem.

Let us assume that the set of A_i's and the set of B_j's are each clumped closely together. In other words, given a measure of distance between reference frames, $d :$ $SE(3) \times SE(3) \to \mathbb{R}_{\geq 0}$, we have that $d(A_i, A_j), d(B_i, B_j) < \epsilon \ll 1$. This assumption can be made true for example, if we are using small relative motions between consecutive reference frames, regardless of whether the whole trajectory is long or not.

The convolution of "highly focused" distributions corresponding to closely clumped sets of reference frames have some interesting properties that we can exploit to solve for X. In particular, let the mean and covariance of a probability density $f(H)$ be defined by the conditions [96].

$$\int_{SE(3)} \log(M^{-1}H)\,f(H)\,dH = \mathbb{O} \text{ and}$$

$$\Sigma = \int_{SE(3)} \log^{\vee}(M^{-1}H)[\log^{\vee}(M^{-1}H)]^T f(H)\,dH. \tag{14.59}$$

where explicit formulas for the matrix logarithm, $\log(H)$, and its vectorized form, $\log^{\vee}(H)$, are given in [96]. The operation $\log(H)$ takes any element in $SE(3)$ (with rotational part that has an angle of rotation, θ, in the range $0 \leq \theta < \pi$) into the the corresponding unique element in the Lie algebra $se(3)$ such that $\exp(\log(H)) = H$, where $\exp(\cdot)$ is the matrix exponential. Since $SO(3)$ can be viewed as a solid three-dimensional ball of radius π with antipodal points identified, the exclusion of the 2D bounding sphere of radius π in $SO(3)$ defines a 5D set of measure zero in $SE(3)$ that has no effect on the computation of Σ in the above equation.

It is always the case that $\log(H) = \begin{pmatrix} \Omega & \mathbf{v} \\ \mathbf{0}^T & 0 \end{pmatrix}$ where $\Omega = -\Omega^T \in so(3)$. The map $\vee : se(3) \to \mathbb{R}^6$ is then composed with the log to give $\log^{\vee}(H) = [\boldsymbol{\omega}^T, \mathbf{v}^T]^T$ where $\boldsymbol{\omega} \in \mathbb{R}^3$ is the vector corresponding to Ω such that $\Omega\mathbf{x} = \boldsymbol{\omega} \times \mathbf{x}$ for any $\mathbf{x} \in \mathbb{R}^3$, where $\times$ is the vector cross product.

The definition of mean used above differs from that most often used in the literature when taking a Riemannian-geometric (rather than Lie-group) approach [38, 80, 91] or an extrinsic approach [41] which is of the form $M' = \arg\min_M \int_{SE(3)} [d(M, H)]^2 f(H) \, dH$ where $d(M, H)$ is a Riemannian distance function $(d(M, H) = \| \log(M^{-1} H) \|_W^2$, for example) and W is a weighting matrix related to the Riemannian metric tensor that is chosen. There are two reasons for our definition. First, in our definition there is no need to introduce a weighting matrix, and therefore we avoid coloring the result by an arbitrary choice. Second, in the context of robotics problems in which reference frames are attached to rigid links it is more natural in the following sense. If a single rigid link has a world frame attached to its base, and a reference frame attached to its distal end, and that distal reference frame is recorded at two different times as a joint at the base rotates, then the translation part of the average of these two reference frames should lie on the arc that joins the two. M will have this property, but M' will not. Having said this, if we were considering data on $SO(3)$ rather than $SE(3)$, and if $W = \mathbb{I}$ were chosen, the two definitions would become the same since for $SO(3)$ the distance (metric) function $d(R_1, R_2) \doteq \| \log(R_1^{-1} R) \|$ is bi-invariant and $Ad(R) = R$. But for $SE(3)$ the different definitions of mean are not equivalent because $Ad(H) \neq H$ and because there does not exist a bi-invariant metric (though there does exist a bi-invariant integration measure).

If $f(H)$ is of the form of $f_A(H)$ given above, then discrete versions of (14.59) are

$$\sum_{i=1}^n \log(M_A^{-1} A_i) = \mathbb{O} \quad \text{and} \quad \Sigma_A = \frac{1}{n} \sum_{i=1}^n \log^\vee(M_A^{-1} A_i)[\log^\vee(M_A^{-1} A_i)]^T. \qquad (14.60)$$

An iterative procedure for computing M_A was presented in [96] in which an initial estimate of the form $M_A^0 = \exp(\frac{1}{n} \sum_{i=1}^n \log(A_i))$ is chosen, and then a gradient descent procedure is used to update so as to minimize the cost $C(M) = \left\| \sum_{i=1}^n \log(M^{-1} A_i) \right\|^2$, and the minimum defines M_A.

It can be shown that if these quantities are computed for two highly focused functions, f_1 and f_2, then the same quantities for the convolution of these functions can be computed as [96]:

$$M_{1*2} = M_1 M_2 \quad \text{and} \quad \Sigma_{1*2} = Ad(M_2^{-1}) \Sigma_1 Ad^T(M_2^{-1}) + \Sigma_2 \qquad (14.61)$$

where $Ad(H) = \begin{pmatrix} R & \mathbb{O} \\ \hat{\mathbf{t}} R & R \end{pmatrix}$.

Here for any $\mathbf{a} \in \mathbb{R}^3$, $\hat{\mathbf{a}}$ is the skew-symmetric matrix such that $\hat{\mathbf{a}} \mathbf{b} = \mathbf{a} \times \mathbf{b}$. And by a slight abuse of notation, we use $\vee$ as the reverse map which gives $(\hat{\mathbf{a}})^\vee = \mathbf{a}$. The use of $\vee$ to denote maps from the Lie algebra $so(3)$ into $\mathbb{R}^3$ and from $se(3)$ into $\mathbb{R}^6$ should not be a source of confusion, as the version being used is defined by the argument to which it is applied.

The mean of $\delta_X(H)$ is $M_X = X$, and its covariance is the zero matrix. Therefore, (14.58) together with (14.61) gives two equations:

(a) $\boxed{M_A X = X M_B}$ and (b) $\boxed{Ad(X^{-1}) \Sigma_A Ad^T(X^{-1}) = \Sigma_B}$. $\qquad (14.62)$

These two equations can be solved in a similar way to how the two equations $A_1 X = X B_1$ and $A_2 X = X B_2$ are solved.

First, we seek the rotational component, R_X, of X. From (14.62a) we have that,

$$\mathbf{n}_{M_A} = R_X \mathbf{n}_{M_B} \tag{14.63}$$

where $\mathbf{n}_H$ is the direction of the screw axis of the homogeneous transform H [24].

If we decompose Σ_{M_A} and Σ_{M_B} into blocks as

$$\Sigma_i = \begin{pmatrix} \Sigma_i^1 & \Sigma_i^2 \\ \Sigma_i^3 & \Sigma_i^4 \end{pmatrix}$$

where $\Sigma_i^3 = (\Sigma_i^2)^T$, then we can take the first two blocks of (14.62b) and write

$$\Sigma_{M_B}^1 = R_X^T \Sigma_{M_A}^1 R_X \quad \text{and} \quad \Sigma_{M_B}^2 = R_X^T \Sigma_{M_A}^1 R_X(\widehat{R_X^T t_x}) + R_X^T \Sigma_{M_A}^2 R_X . \tag{14.64}$$

We can then find the eigendecomposition, $\Sigma_i = Q_i \Lambda Q_i^T$, where Q_i is the square matrix whose i^{th} column is the eigenvector of Σ_i and Λ is the diagonal matrix with corresponding eigenvalues as diagonal entries and write the first block equation of (14.64) as,

$$\Lambda = Q_{M_B}^T R_X^T Q_{M_A} \Lambda Q_{M_A}^T R_X Q_{M_B} = \mathcal{Q} \Lambda \mathcal{Q}^T \tag{14.65}$$

The set of $\mathcal{Q}$s that satisfy this equation is given as,

$$\mathcal{Q} = \left\{ \begin{pmatrix} 1 & 0 & 0 \\ 0 & 1 & 0 \\ 0 & 0 & 1 \end{pmatrix}, \begin{pmatrix} -1 & 0 & 0 \\ 0 & -1 & 0 \\ 0 & 0 & 1 \end{pmatrix}, \begin{pmatrix} -1 & 0 & 0 \\ 0 & 1 & 0 \\ 0 & 0 & -1 \end{pmatrix}, \begin{pmatrix} 1 & 0 & 0 \\ 0 & -1 & 0 \\ 0 & 0 & -1 \end{pmatrix} \right\}$$

with the simple condition that Q_i is constrained to be a rotation matrix. This means that the rotation component of X is given by,

$$R_x = Q_{M_A} \mathcal{Q} Q_{M_B}^T . \tag{14.66}$$

The correct solution, from the set of 4 possibilities of R_X (given (14.66)) can be found by applying (14.63) and choosing the one that minimizes $\|\mathbf{n}_{M_A} - R_X \mathbf{n}_{M_B}\|$. Once R_X is found in this way, $\mathbf{t}_X$ can be found easily from blocks the 2 and 4 of (14.62)(b).

14.8 Summary

In this chapter we have shown how statistical tools can be defined in a group-theoretic setting and applied in the context of pose determination and camera calibration. In addition, review material on statistics on the circle and spheres was presented, and we introduced the concept of a folded normal distribution for $SO(3)$, as well as concepts of mean and variance for $SO(N)$ and $SE(N)$. We computed the translational part of the mean of the convolution of two PDFs on $SE(N)$ in terms of the means of each function.

For other interesting approaches to solve for poses to match point sets when correspondence is known see [16, 37, 53, 60, 87, 88]. In recent years, the *Iterative Closest Point* (or ICP) method has become very popular for matching point sets without a priori correspondence [11, 26, 89, 98]. Other more recent alternatives are also popular including those in [72, 93]. And a number of novel applications in medical imaging have appeared including [54, 55, 65]. Calibrated ultrasound, which is an important application of $AX = XB$, is important for robotic surgical tasks such as those described in [13, 14, 70, 74, 81]. For discussions of other problems in computer vision that are not considered here, and a general introduction to this topic see [46, 67]. For other sensor calibration problems, such as laser scanners, see [78].

References

1. Ackerman, M.K., Chirikjian, G.S., "A Probabilistic Solution to the $AX = XB$ Problem: Sensor Calibration without Correspondence," in *Geometric Science of Information*, (pp. 693–701). Springer Berlin Heidelberg, 2013.
2. Ackerman, M.K., Cheng, A., Shiffman, B., Boctor, E., Chirikjian, G., "Sensor Calibration with Unknown Correspondence: Solving $AX = XB$ Using Euclidean-Group Invariants," *Proceedings of the 2013 IEEE/RSJ International Conference on Intelligent Robots and Systems (IROS'13)*, (pp. 1308–1313), November, 2013.
3. Ackerman, M. K., Cheng, A., Boctor, E., Chirikjian, G., "Online Ultrasound Sensor Calibration Using Gradient Descent on the Euclidean Group," *Proceedings of the 2014 IEEE International Conference on Robotics and Automation (ICRA'14)*, (pp. 4900–4905), May, 2014.
4. Ackerman, M.K., Cheng, A., Chirikjian, G., "An Information-Theoretic Approach to the Correspondence-Free $AX = XB$ Sensor Calibration Problem," *Proceedings of the 2014 IEEE International Conference on Robotics and Automation (ICRA'14)*, (pp. 4893–4899), May, 2014.
5. Arnold, K.J., *On Spherical Probability Distributions*, PhD Thesis, MIT, 1941.
6. Agrawal, M., "A Lie Algebraic Approach for Consistent Pose Registration for General Euclidean Motion," *Proceedings of the 2006 IEEE/RSJ International Conference on Intelligent Robots and Systems*, pp. 1891–1897, Beijing, China, October 9–15, 2006.
7. Andreff, N., Horaud, R., Espiau, B., "Robot Hand-Eye Calibration Using Structure-from-Motion," *The International Journal of Robotics Research*, 20(3): 228–248, 2001.
8. Arun, K.S., Huang, T.S., Blostein, S.D., "Least-Squares Fitting of Two 3-D Point Sets," *IEEE Transactions on Pattern Analysis and Machine Intelligence*, 9(5): 698–700, Sept. 1987.
9. Bai, S., Teo, M.Y., "Kinematic Calibration and Pose Measurement of a Medical Parallel Manipulator by Optical Position Sensors," *Journal of Robotic Systems*, 20(4): 201–209, 2003.
10. Batschelet, E., *Circular Statistics in Biology*, Academic Press, New York, 1981.
11. Besl, P.J., McKay, N.D., "A Method for Registration of 3-D Shapes," *IEEE Transactions on Pattern Analysis and Machine Intelligence*, 14(2): 239-256, Feb. 1992.
12. Bingham, C., "An Antipodally Symmetric Distribution on the Sphere," *The Annals of Statistics*, 2(6): 1201–1225, 1974; Also see "Distributions on the Sphere and Projective Plane," PhD Dissertation, Yale University, 1964.
13. Boctor, E.M., "Enabling Technologies For Ultrasound Imaging In Computer-Assisted Intervention," PhD Dissertation, Computer Science Department, Johns Hopkins University, 2006.
14. Boctor, E.M., Viswanathan, A., Choti, M.A., Taylor, R.H., Fichtinger, G., Hager, G.D., "A Novel Closed Form Solution for Ultrasound Calibration, *Proceedings of the IEEE International Symposium on Biomedical Imaging*, pp. 527–530, April, 2004.

15. Breitenberger, E., "Analogues of the Normal Distribution on the Circle and Sphere," *Biometrika*, 50: 81 – 88, 1963.

16. Breitenreicher, D., Schnörr, C., "Robust 3D Object Registration without Explicit Correspondence Using Geometric Integration," *Machine Vision and Applications*, 21(5): 601 – 611, 2010.

17. Brock, J.E., "Optimal Matrices Describing Linear Systems," *AIAA Journal*, 6(7): 1292 – 1296, July 1968.

18. Chang, T., "Spherical Regression," *Annals of Statistics*, 14(3): 907 – 924, 1986.

19. Chang, T., "On Statistical Properties of Estimated Rotations," *Journal of Geophysical Research*, 92(B7): 6319 – 6329, June 1987.

20. Chang, T., "Spherical Regression with Errors in Variables," *The Annals of Statistics*, 17(1): 293 – 306, 1989.

21. Chang, T., Stock, J., Molnar, P., "The Rotation Group in Plate Tectonics and the Representation of Uncertainties of Plate Reconstructions," *Geophysical Journal International*, 101(3): 649 – 661, 1990.

22. Chapman, G.R., Chen, G., Kim, P.T. "Assessing Geometric Integrity Through Spherical Regression Techniques," *Statistica Sinica*, 5: 173 – 220, 1995.

23. Chaudhury, K.N., Singer, A., "Non-Local Euclidean Medians," *IEEE Signal Processing Letters*, 19(11): 745 – 748, 2012.

24. Chen, H.H., "A Screw Motion Approach to Uniqueness Analysis of Head-Eye Geometry," *Proceedings of the 1991 IEEE Conference on Computer Vision and Pattern Recognition*, pp. 145 – 151, 1991.

25. Chételat, O., Chirikjian, G.S., "Sampling and Convolution on Motion Groups Using Generalized Gaussian Functions," *Electronic Journal of Computational Kinematics*, 1(1), 2002.

26. Chetverikov, D., Svirko, D., Stepanov, D., Krsek, P., "The Trimmed Iterative Closest Point Algorithm," *Proceedings of the 16th International Conference on Pattern Recognition*, (Vol. 3, pp. 545 – 548), 2002.

27. Chou, J.C.K., Kamel, M., "Finding the Position and Orientation of a Sensor on a Robot Manipulator Using Quaternions," *The International Journal of Robotics Research*, 10(3): 240 – 254, June 1991.

28. Dai, Y., Trumpf, J., Li, H., Barnes, N., Hartley, R., "Rotation Averaging with Application to Camera-Rig Calibration," H. Zha, R.-I. Taniguchi, and S. Maybank (Eds.): *ACCV 2009*, Part II, LNCS 5995, pp. 335 – 346, 2010.

29. Daniilidis, K., "Hand-Eye Calibration Using Dual Quaternions," *The International Journal of Robotics Research*, 18(3): 286 – 298, 1999.

30. Diaconis, P., *Group Representations in Probability and Statistics*, Lecture Notes-Mongraph Series, S.S. Gupta Series Editor, Institute of Mathamatical Statistics, Hayward, CA 1988.

31. Downs, T.D., "Orientation Statistics," *Biometrika*, 59(3): 665 – 676, 1972.

32. Eggert, D.W., Lorusso, A., Fischer, R.B., "Estimating 3-D Rigid Body Transformations: A Comparison of Four Major Algorithms," *Machine Vision and Applications*, 9(5–6): 272 – 290, 1997.

33. Farrell, J., Stuelpnagel, J., Wessner, R., Velman, J., Brook, J., "A Least Squares Estimate of Satellite Attitude (Grace Wahba)," *SIAM Review*, 8(3): 384 - 386, 1966.

34. Fassi, I., Legnani, G., "Hand to Sensor Calibration: A Geometrical Interpretation of the Matrix Equation $AX = XB$," *Journal of Robotic Systems*, 22(9): 497 – 506, 2005.

35. Fisher, R., Sir, "Dispersion on a Sphere," *Proceedings of the Royal Society*, Series A, 217(1130): 295 – 305, 1953.

36. Fisher, N.I., Lewis, T., Embleton, B.J., *Statistical Anaysis of Spherical Data*, Cambrige Univesity Press, 1987.

37. Fitzgibbon, A.W., "Robust Registration of 2D and 3D Point Sets," *Image and Vision Computing*, 21: 1145 - 1153, 2003.

38. Fletcher, P.T., Joshi, S., Lu, C., Pizer, S.M., "Gaussian Distributions on Lie Groups and Their Application to Statistical Shape Analysis," in *Information Processing in Medical Imaging* (pp. 450 – 462). Springer Berlin Heidelberg, 2003.

39. Fréchot, M., "Les éléments aleatoires de nature quelconque dans un espace distancié," *Ann. Inst. H. Poincaré*, 10: 215 - 310, 1948.

40. Giske, J., Huse, G., Fiksen, O., "Modelling Spatial Dynamics of Fish," *Reviews in Fish Biology and Fisheries*, 8: 57 - 91, 1998.

41. Glover, J., Bradski, G., Rusu, R.B., "Monte Carlo Pose Estimation with Quaternion Kernels and the Bingham Distribution," *Robotics: Science and Systems VII*, 2012 (online procedings).

42. Goryn, D., Hein, S., "On the Estimation of Rigid-Body Rotation from Noisy Data," *IEEE Transactions on Pattern Analysis and Machine Intelligence*, 17(12): 1219 - 1220, Dec. 1995.

43. Goodrich, M.T., Mitchell, J.S.B., Orletsky, M.W., "Approximate Geometric Pattern Matching Under Rigid Motions," *IEEE Transactions on Pattern Analysis and Machine Intelligence*, 21(4): 371 - 379, April 1999.

44. Gower, J.C., Dijksterhuis, G.B., *Procrustes Problems*, Oxford University Press, 2004.

45. Haralick, R.M., Joo, H., Lee, C.-N., Zhuang, X., Vaidya, V.G., Kim, M.B., "Pose Estimation from Corresponding Point Data," *IEEE Transactions on Systems, Man, and Cybernetics*, 19(6): 1426 - 1446, Nov/Dec. 1989.

46. Hartley, R.I., Zisserman, A., *Multiple View Geometry in Computer Vision*, 2nd edition, Cambridge University Press, 2003.

47. Hartman, P., Watson, G.S., " 'Normal' Distribution Functions on Spheres and the Modified Bessel Function," *The Annals of Probability*, 2(4): 593 - 607, 1974.

48. Healy, D.M., Jr., Hendriks, H., Kim, P.T., "Spherical Deconvolution," *Journal of Multivariate Analysis*, 67(1): 1 - 22, 1998.

49. Healy, D.M., Jr., Kim, P.T., "An Empirical Bayes Approach to Directional Data and Efficient Computation on the Sphere," *The Annals of Statistics*, 24(1): 232 - 254, 1996.

50. Hendriks, H. "Nonparametric Estimation of a Probability Density on a Riemannian Manifold Using Fourier Expansions," *The Annals of Statistics*, 18(2): 832 - 849, 1990.

51. Horn, B.K.P., "Closed-Form Solution of Absolute Orientation Using Unit Quaternions," *Journal of the Optical Society of America*, 4(4): 629 - 642, April 1987.

52. Horn, B.K.P., Hilden, H.M., Negahdaripour, S., "Closed-Form Solution of Absolute Orientation Using Orthonormal Matrices," *Journal of the Optical Society of America*, 5(7): 1127 - 1135, July 1988.

53. Horowitz, M.B., Matni, N., Burdick, J.W., "Convex Relaxations of $SE(2)$ and $SE(3)$ for Visual Pose Estimation," arXiv:1401.3700v2 [cs.CV] 6 Apr 2014.

54. Jain, A., Zhou, Y., Mustufa, T., Burdette, E.C., Chirikjian, G.S., Fichtinger, G., "Matching and Reconstruction of Brachytherapy Seeds Using the Hungarian Algorithm (MARSHAL)," *Medical Physics*, 32(11): 3475 - 3492, Nov. 2005.

55. Jain, A., Mustafa, T., Zhou, Y., Burdette, E.C ., Chirikjian, G.S., Fichtinger, G., "Robust Fluoroscope Tracking Fiducial," *Medical Physics* 32(10): 3185 - 3198, Oct. 2005.

56. Jordt, A., Siebel, N., Sommer, G., "Automatic High-Precision Self-Calibration of Camera-Robot Systems," *2009 IEEE International Conference on Robotics and Automation Kobe International Conference Center* pp. 1244 - 1249, May 2009.

57. Joseph, S.H., "Optimal Pose Estimation in Two and Three Dimensions," *Computer Vision and Image Understanding*, 73(2): 215 - 231, Feb. 1999.

58. Jupp, P.E., Mardia, K.V., "Maximum Likelihood Estimation for the Matrix von Mises-Fisher and Bingham Distributions," *Annals of Statistics*, 7(3): 599 - 606, 1979.

59. Kanatani, K., "Analysis of 3-D Rotation Fitting," *IEEE Transactions on Pattern Analysis and Machine Intelligence*, 16(5): 543 - 549, May 1994.

60. Kanatani, K., *Statistical Optimization for Geometric Computation*, Elsevier, 1996 (Dover Edition, 2005.)

61. Karcher, H., "Riemannian Center of Mass and Mollifier Smoothing," *Communications on Pure and Applied Mathematics* 30(5): 509 - 541, 1977.

62. Khatri, C.G., Mardia, K.V., "The von Mises-Fisher Matrix Distribution in Orientation Statistics," *Journal of the Royal Statistical Society*, Series B, 39(1): 95 - 106, 1977.

63. Kim, P.T., "Deconvolution Density Estimation on $SO(N)$," *Annals of Statistics* 26(3): 1083 - 1102, June 1998.

64. Kim, S., Jeong, M., Lee, J., Lee, J., Kim, K., You, B., Oh, S., "Robot Head-Eye Calibration Using the Minimum Variance Method," *Proceedings of the 2010 IEEE International Conference on Robotics and Biomimetics* pp. 1446–1451, Dec. 2010.

65. Lee, S., Fichtinger, G., Chirikjian, G.S., " Novel Algorithms for Robust Registration of Fiducials in CT and MRI," *Medical Physics*, 29(8): 1881–1891, Aug. 2002.

66. Lo, J.T.-Y., Ng, S.K., "Characterizing Fourier-Series Representation of Probability-Distributions on Compact Lie-Groups," *SIAM Journal on Applied Mathematics*, 48(1): 222–228, Feb. 1988.

67. Ma, Y., Košecká, J., Soatto, S., Sastry, S., *An Invitation to 3-D Vision: From Images to Models*, Springer, 2004.

68. Mair, E., Fleps, M., Suppa, M., Burschka, D., "Spatio-Temporal Initialization for IMU to Camera Registration," *Proceedings of 2011 IEEE ROBIO*, pp. 557–564, Dec. 2011.

69. Mardia, K.V., *Statistics of Directional Data*, Academic Press, New York, 1972.

70. Mercier, L., Langø, T., Lindseth, F., Collins, D.L., "A Review of Calibration Techniques for Freehand 3-D Ultrasound Systems, *Ultrasound in Medicine and Biology*, 31(2): 143–165, 2005.

71. Moran, P.A.P., "Quaternions, Haar Measure and the Estimation of a Paleomagnetic Rotation," in *Perspectives in Probability and Statistics*, edited by J. Gani, pp. 295–301, Academic Press, Orlando, FL, 1976.

72. Myronenko, A., Song, X., "Point Set Registration: Coherent Point Drift," *IEEE Transactions on Pattern Analysis and Machine Intelligence*, 32(12): 2262–2275, Dec. 2010.

73. Nádas, A., "Least Squares and Maximum Likelihood Estimates of Rigid Motion," *IBM Research Report* RC 6945 (#29783), Mathematics, Jan. 17, 1978.

74. Novotny, P.M., Stoll, J.A., Dupont, P.E., Howe, R.D., "Real-Time Visual Servoing of a Robot Using Three-Dimensional Ultrasound," *IEEE International Conference on Robotics and Automation (ICRA'07)*, pp. 2655–2660, 2007.

75. Ohta, N., Kanatani, K., "Optimal Estimation of Three-Dimensional Rotation and Reliability Evaluation," *IEICE Transactions on Information and Systems*, 81(11): 1247–1252, Nov. 1998.

76. Or, S.H., Luk, W.S., Wong, K.H., King, I., "An Efficient Iterative Pose Estimation Algorithm," *Image and Vision Computing*, 16(5): 353–362, April 1998.

77. Pamecha, A., Ebert-Uphoff, I., Chirikjian, G.S., "Useful Metrics for Modular Robot Motion Planning," *IEEE Transactions on Robotics and Automation*, 13(4): 531–545, August 1997.

78. Pandey, G., McBride, J., Savarese, S., Eustice, R., "Extrinsic Calibration of a 3D Laser Scanner and an Omnidirectional Camera," 7th *IFAC Symposium on Intelligent Autonomous Vehicles*, 7(1): 336–341, July 2010.

79. Park, F.C., Martin, B.J., "Robot Sensor Calibration: Solving $AX = XB$ on the Euclidean Group," *IEEE Transactions on Robotics and Automation*, 10(5): 717–721, Oct. 1994.

80. Pennec, X., Thirion, J.P., "A Framework for Uncertainty and Validation of 3-D Registration Methods Based on Points and Frames," *International Journal of Computer Vision*, 25(3): 203–229, Dec. 1997.

81. Poon, T., Rohling, R., "Comparison of Calibration Methods for Spatial Tracking of a 3-D Ultrasound Probe," *Ultrasound in Medicine and Biology* 31(8): 1095-1108, 2005.

82. Prentice, M.J., "Antipodally Symmetric Distributions for Orientation Statistics," *Journal of Statistical Planning and Inference*, 6(3): 205–214, 1982.

83. Prentice, M.J., "Orientation Statistics without Parametric Assumptions," *Journal of the Royal Statistical Society B*, 48(2): 214–222, 1986.

84. Rivest, L.-P., "Spherical Regression for Concentrated Fisher-von Mises Distributions," *The Annals of Statistics*, 17(1): 307–317, 1989.

85. Rosenthal, J.S., "Random Rotations - Characters and Random-Walks on $SO(N)$," *Annals of Probability*, 22(1): 398–423, Jan. 1994.

86. Rosin, P.L., "Robust Pose Estimation," *IEEE Transactions on Systems Man and Cybernetics Part B-Cybernetics*, 29(2): 297–303, April 1999.

87. Said, S., Courty, N., Le Dihan, N., Sangwine, S.J., "Exact Principal Geodesic Analysis for Data on $SO(3)$," *Proceedings of the 15th European Signal Processing Conference, EUSIPCO-2007* (pp. 1700–1705), 2007.

88. Schönemann, P.H., "A Generalized Solution of the Orthogonal Procrustes Problem," *Psychometrika*, 31(1): 1–10, 1966.

89. Sharp, G.C., Lee, S.W., Wehe, D.K., "ICP Registration Using Invariant Features," *IEEE Transactions on Pattern Analysis and Machine Intelligence*, 24(1): 90–102, 2002.

90. Shiu, Y.C., Ahmad, S., "Calibration of Wrist-Mounted Robotic Sensors by Solving Homogeneous Transform Equations of the Form $AX = XB$," *IEEE Transactions on Robotics and Automation*, 5(1): 16–29, Feb. 1989.

91. Tron, R., Vidal, R., Terzis, A., "Distributed Pose Averaging in Camera Networks via Consensus on $SE(3)$," in *Proceedings of the Second ACM/IEEE International Conference on Distributed Smart Cameras, ICDSC 2008*, pp. 1-10, Stanford, CA, Sept. 7-11, 2008.

92. Tsai, R., Lenz, R., "A New Technique for Fully Autonomous and Efficient 3D Robotics Hand/Eye Calibration," *IEEE Transactions on Robotics and Automation*, 5(3): 345–358, June 1989.

93. Tsin, Y., Kanade, T., "A Correlation-Based Approach to Robust Point Set Registration," in *Computer Vision-ECCV 2004* (pp. 558-569). Springer Berlin Heidelberg, 2004.

94. Umeyama, S., "Least-Squares Estimation of Transformation Parameters Between Two Point Patterns," *IEEE Transactions on Pattern Analysis and Machine Intelligence*, 13(4): 376–380, April 1991.

95. Wahba, G., "Problem 65-1, A Least Squares Estimate of Spacecraft Attitude," *SIAM Review*, 7(3): 409, 1965.

96. Wang, Y., Chirikjian, G.S., "Nonparametric Second-Order Theory of Error Propagation on the Euclidean Group," *International Journal of Robotics Research*, 27(11-12): 1258-1273, Nov/Dec. 2008. (Corrigendum: 33(10): 1413, Sept. 2014).

97. Watson, G.S., *Statistics on Spheres*, Wiley-Interscience Publications, New York, 1983.

98. Zhang, Z., "Iterative Point Matching for Registration of Free-Form Curves and Surfaces," *International Journal of Computer Vision*, 13(2): 119–152, 1994.

Stochastic Processes, Estimation, and Control

In this chapter we consider estimation of systems in motion, and control of certain kinds of dynamical systems. A dynamical system could be something as simple as the furnace in a building. In this case the controller might consist of a thermostat which measures the current temperature, compares to the desired, and either turns on or off the furnace to push the actual temperature toward the desired. The dynamical system of interest could also be something as complicated as a high performance airplane, a robot, or even the human body. The physical mechanism by which the system evolves over time may be completely different. The system could be purely electronic, electromechanical, chemical, or even physiological. And it might be desirable to control the rate of the system (as in chemical processes), or the position (as in a robot).

The field of feedback control seeks to modify existing dynamical systems in order for the modified system to evolve in time in a desired way. Pose determination is a prerequisite for the control of rigid bodies in the context of an imprecise world. Hence, the developments presented here build on those of Chapter 14.

In the first few sections of this chapter, we briefly review basic nonlinear models of mechanical systems, classical linear systems theory, recursive least-squares fitting, and stochastic processes, together with the associated control system design techniques. Many of the most popular stochastic control techniques use the properties of a Gaussian probability density function. We extend these ideas to recursive least-squares fitting to systems which have states in the rotation and motion groups. Finally, we extend the results of linear control and estimation theory to systems that evolve in these Lie groups.

15.1 Stability and Control of Nonlinear Systems

Regardless of the particular application, there are several broad categories of models into which most dynamical systems fall, and corresponding analytical/computational techniques for control system design have been developed for each of these categories. In the most general context, the "system" can be described with the mathematical model

$$\frac{d\mathbf{x}}{dt} = \mathbf{f}[\mathbf{x}, \mathbf{u}, \mathbf{w}, t]. \tag{15.1}$$

Here $\mathbf{x}(t)$ is the actual state of the system at time t. For a purely mechanical system, the state can be viewed as the set of all generalized coordinates that describe the geometry of the system and the rates of these generalized coordinates. $\mathbf{u}(t)$ is the control inputs to the system. For a mechanical system this may be a set of generalized forces which alter the behavior of the system. $\mathbf{w}(t)$ is some kind of external random disturbance acting on the system at time t. It is assumed within this model that the system itself is known exactly (i.e. that it has already been calibrated). If this were not the case, then the parameters characterizing the system would have to be identified. The area of *system identification* is a large area of research in itself, which is not addressed here.

In practice, one is often not able to directly measure the state vector $\mathbf{x}$. Instead, indirect measurements are observed:

$$\mathbf{y} = \mathbf{h}[\mathbf{x}, \mathbf{v}, t], \tag{15.2}$$

where $\mathbf{v}(t)$ is a vector describing measurement error at time t.

The goal behind *stochastic control* is to derive a *feedback law*:

$$\mathbf{u}(t) = \mathbf{g}(\mathbf{y}, \mathbf{w}(t), \mathbf{v}(t))$$

based on the observation $\mathbf{y}$ and the statistics of the system and measurement noise $\mathbf{w}$ and $\mathbf{v}$ so that, within acceptable error bounds, the system executes a desired behavior $\mathbf{x}_{des}(t)$. In contrast, in *deterministic control* it is assumed that there is no uncertainty, and $\mathbf{v}$ and $\mathbf{w}$ are zero. We discuss the basics of deterministic control below as background for the stochastic case (which is where noncommutative harmonic analysis is applicable).

In the study of systems it is natural to make several classifications. In addition to the labels "deterministic" or "stochastic," system models are treated differently depending on whether they are *linear* or *nonlinear*. The most developed theory is for linear systems (see, e.g., [85, 142] for complete treatments). Nonlinear systems come in many varieties. One important class of nonlinear systems for which globally stable control laws can be applied is classical finite-dimensional mechanical systems. The configuration of such a system is specified by n *generalized coordinates* that are often written in the form of an array $\mathbf{q} = [q_1, ..., q_n]^T$. These generalized coordinates specify the "shape" of the system as well as its location in space. The *state* of the system is the pair $(\mathbf{q}, \dot{\mathbf{q}})$.

For example, Figure 15.1 shows a two-link planar robot arm with revolute (hinge-like) joints. This is a two-degree-of-freedom mechanical system. The generalized coordinates can be taken to be the counterclockwise-measured angle that the first link makes with respect to the horizontal, and the second generalized coordinate can be taken to be the counterclockwise-measured angle that the second link makes with respect to the first, i.e., $\mathbf{q} = [q_1, q_2]^T$. These two angles completely specify the geometry of such an arm with fixed link lengths L_1 and L_2. If we assume that the mass in this arm is concentrated at the hinges, the positions to the masses with respect to a frame of reference fixed in space at the base of the arm are

$$\mathbf{x}_1 = \begin{pmatrix} L_1 \cos q_1 \\ L_1 \sin q_1 \end{pmatrix}$$

and

$$\mathbf{x}_2 = \begin{pmatrix} L_1 \cos q_1 + L_2 \cos(q_1 + q_2) \\ L_1 \sin q_1 + L_2 \sin(q_1 + q_2) \end{pmatrix}$$

where q_1 is the angle of the first link, and q_2 is the relative angle of the second link with respect to the first.

In general, the kinetic energy (energy of motion) of a finite-dimensional conservative mechanical system can be written as

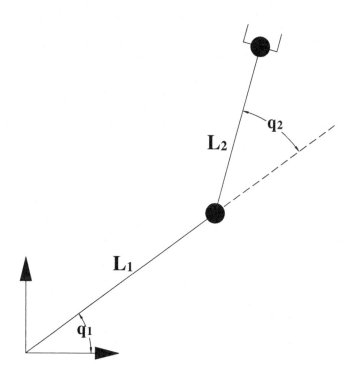

Fig. 15.1. A Planar Two-Link Revolute Manipulator

$$T = \frac{1}{2}\dot{\mathbf{q}}^T M(\mathbf{q})\dot{\mathbf{q}} \tag{15.3}$$

where $M(\mathbf{q})$ is a symmetric positive definite matrix (called the *mass matrix*) and the potential energy is typically a function $V = V(\mathbf{q})$. Henceforth when referring to mechanical systems we will mean finite-dimensional conservative systems.

For the above example of a two-link revolute planar manipulator arm (operating in the plane normal to gravity) $V = 0$ and

$$T = \frac{1}{2}m_1\dot{\mathbf{x}}_1 \cdot \dot{\mathbf{x}}_1 + \frac{1}{2}m_2\dot{\mathbf{x}}_2 \cdot \dot{\mathbf{x}}_2 \tag{15.4}$$

where m_1 and m_2 are the masses at the hinges. Explicitly, the kinetic energy is written as

$$T = \frac{1}{2}m_1 L_1^2 \dot{q}_1^2 + \frac{1}{2}m_2\{L_1^2\dot{q}_1^2 + (\dot{q}_1 + \dot{q}_2)^2 L_2^2 + 2\dot{q}_1(\dot{q}_1 + \dot{q}_2)L_1 L_2 \cos q_2\}.$$

We can rearrange this to read

$$T = \frac{1}{2}[\dot{q}_1 \; \dot{q}_2] \begin{pmatrix} (m_1 + m_2)L_1^2 + m_2(L_2^2 + 2L_1L_2 \cos q_2) & m_2(L_1L_2 \cos q_2 + L_2^2) \\ m_2(L_1L_2 \cos q_2 + L_2^2) & m_2 L_2^2 \end{pmatrix} \begin{bmatrix} \dot{q}_1 \\ \dot{q}_2 \end{bmatrix}.$$

The 2×2 matrix in this equation is the mass matrix $M(\mathbf{q})$ for this example.

Alternatively, a straight forward application of the chain rule in (15.4) allows us to rewrite the elements of this mass matrix as

$$M_{ij}(\mathbf{q}) = \sum_{k=1}^{2} m_k \frac{\partial \mathbf{x}_k}{\partial q_i} \cdot \frac{\partial \mathbf{x}_k}{\partial q_j}.$$

Clearly this is symmetric, and positive definiteness can be reasoned on the physical grounds that when the system is moving (and hence $\dot{\mathbf{q}} \neq \mathbf{0}$) its kinetic energy is nonzero.

Torques τ_i provided by motors at the joints cause the motion of the arm. When $V = 0$, the equations governing this motion are of the form

$$\frac{d}{dt}\left(\frac{\partial T}{\partial \dot{q}_i}\right) - \frac{\partial T}{\partial q_i} = \tau_i.$$

See Appendix F for a derivation of these Euler-Lagrange equations using variational methods. These equations of motion can be written in the form

$$M(\mathbf{q})\ddot{\mathbf{q}} + C(\mathbf{q}, \dot{\mathbf{q}})\dot{\mathbf{q}} = \boldsymbol{\tau}. \tag{15.5}$$

As an exercise the reader can generate the equations of motion for the two-link arm discussed above, and write them in the above form.

The second-order system of n equations in (15.5) is converted to a first-order system of $2n$ equations by defining the state vector $[\mathbf{q}^T, \mathbf{q}'^T]^T$ where $\mathbf{q}' = \dot{\mathbf{q}}$. That is,

$$\frac{d}{dt}\begin{bmatrix} \mathbf{q} \\ \mathbf{q}' \end{bmatrix} = \begin{pmatrix} \mathbb{O}_{n \times n} & \mathbb{I}_{n \times n} \\ \mathbb{O}_{n \times n} & -M^{-1}(\mathbf{q})C(\mathbf{q}, \mathbf{q}') \end{pmatrix} \begin{bmatrix} \mathbf{q} \\ \mathbf{q}' \end{bmatrix} + \begin{bmatrix} 0 \\ M^{-1}(\mathbf{q})\boldsymbol{\tau} \end{bmatrix}.$$

This system of equations is nonlinear as is evident from the fact that the above matrix relating $[\mathbf{q}^T, \mathbf{q}'^T]^T$ and its derivative depends on $[\mathbf{q}^T, \mathbf{q}'^T]^T$.

These equations of motion (as well as those for all conservative mechanical systems) exhibit the *passivity* condition when $\tau_i = 0$. That is, substitution of the solution to (15.5) into the expression

$$\dot{E} = \frac{d}{dt}(T + V)$$

yields

$$\dot{E} = 0.$$

In our case $V = 0$. This is just the condition that the total energy is conserved. For conservative mechanical systems with $V = 0$, the following proportional-derivative (PD) control law is guaranteed to drive the system to the desired $\mathbf{q}_d$ as $t \to \infty$:

$$\boldsymbol{\tau} = -K(\mathbf{q} - \mathbf{q}_d) - D\dot{\mathbf{q}}.$$

Here K and D are symmetric positive definite matrices.

The first term is like a linear spring with equilibrium at $\mathbf{q}_d$ and the second term is like a viscous damping. If we write $V' = \frac{1}{2}(\mathbf{q} - \mathbf{q}_d)^T K(\mathbf{q} - \mathbf{q}_d)$, and define an "artificial energy" as $E = T + V'$, then we see that

$$\dot{E} = -\dot{\mathbf{q}}^T D \dot{\mathbf{q}} \le 0$$

for all $(\mathbf{q}, \dot{\mathbf{q}}) \ne (\mathbf{0}, \mathbf{0})$, with equality holding only when $\dot{\mathbf{q}} = \mathbf{0}$. From physical arguments it is reasonable to assume that as $t \to \infty$ the kinetic energy dissipates to zero since $E \ge 0$ and $\dot{E} \le 0$. The only way $E \to 0$ is if $\dot{\mathbf{q}} \to \mathbf{0}$ and $\mathbf{q} \to \mathbf{q}_d$. What we have reasoned above can be formalized as an example of *Lyapunov stability theory*, and in particular, La Salle's Invariance Theorem.

More generally, control laws for which we can write

$$\tau_i = -\frac{\partial R}{\partial \dot{q}_i} - \frac{\partial V'}{\partial q_i}$$

(where $V'(\mathbf{q})$ has one critical point which is a global minimum at $\mathbf{q}_d$) will cause $(\mathbf{q}, \dot{\mathbf{q}}) \to (\mathbf{q}_d, \mathbf{0})$ as $t \to \infty$ when

$$R = \frac{1}{2} \dot{\mathbf{q}}^T D(\mathbf{q}) \dot{\mathbf{q}}$$

and $D(\mathbf{q})$ is a positive definite matrix. R is called a *Rayleigh dissipation function*, and in general

$$\frac{d}{dt}(T + V') = -R$$

means that as $t \to \infty$ then $T + V' \to 0$, and so $\mathbf{q} \to \mathbf{q}_d$.

Of course, this assumes that we can exactly measure the current state of the system $(\mathbf{q}, \dot{\mathbf{q}})$ at each time t and respond instantaneously. It also assumes that the actuators (in the case of the two-link robot these are motors) are able to provide these desired torques without any time delay.

While this PD control law is powerful in its generality, some natural questions arise. For instance: How should the gain matrices K and D be chosen so that $\mathbf{q}$ approaches $\mathbf{q}_d$ rapidly ? What if we want to track a desired trajectory $\mathbf{q}_d(t)$ instead of moving from point to point? How do we handle the issues of limited actuator capability, time delays, and uncertainty in the measurement of the system's state?

Some of these questions have been answered completely for highly restricted classes of systems. The next section reviews results for one of these classes: linear systems. Later in this chapter we come back to a class of nonlinear systems that can be viewed as evolving on Lie groups, and $SO(3)$ in particular.

15.2 Linear Systems Theory and Control

The linear form of (15.1) and (15.2) are respectively written as

$$\frac{d\mathbf{x}}{dt} = F(t)\mathbf{x} + G(t)\mathbf{u} + L(t)\mathbf{w} \tag{15.6}$$

and

$$\mathbf{y} = H(t)\mathbf{x} + \mathbf{v}. \tag{15.7}$$

If $G(t) = 0$ for all values of time t, the problem of determining how (15.6) and (15.7) evolve over time is the *linear estimation* problem. If there is no dynamics (i.e., $F(t)$, $G(t)$ and $L(t)$ are all zero in (15.6)) the determination of (15.7) for constant $H(t)$ is the *static linear estimation* problem. This was addressed in Chapter 14 in the context of least squares pose determination. Section 15.3 reviews the recursive computation of this solution.

On the other hand, if $\mathbf{v}$ and $\mathbf{w}$ are taken to be appropriately dimensioned zero vectors, the pair (15.6) and (15.7) represents a linear deterministic system. The stability, controllability, and observability of such systems has been treated extensively in the literature and a number of textbooks. See, e.g., [85, 142]. Given a deterministic linear system for which the state can be observed and for which a finite control effort can direct the system to a desired state, the question becomes what control law will achieve this result with the least effort.

In this deterministic context the *linear quadratic regulator* (LQR) provides an optimal solution. In the remainder of this section we explain the LQ regulator, with an eye towards its generalization to systems on Lie groups, and the stochastic analogs of the deterministic theory.

Given the system

$$\frac{d\mathbf{x}}{dt} = F(t)\mathbf{x} + G(t)\mathbf{u} \tag{15.8}$$

with exact observation of the state $\mathbf{x}$, the question becomes how to select the control input $\mathbf{u}$ optimally to drive $\mathbf{x}$ to $\mathbf{0}$ as $t \to T < \infty$. Optimality is defined in the LQR setting as the $\mathbf{u}$ that minimizes the cost functional [54, 106, 156]:

$$C = \frac{1}{2}\mathbf{x}^T(T)S\mathbf{x}(T) + \frac{1}{2}\int_0^T [\mathbf{x}^T(t)Q(t)\mathbf{x}(t) + \mathbf{u}^T(t)R(t)\mathbf{u}(t)]dt \tag{15.9}$$

where $t \in [0, T]$. The matrices S, $Q(t)$ and $R(t)$ are all symmetric and positive definite. The vector $\mathbf{x}(t)$ is defined so that $\mathbf{x}(T) = \mathbf{0}$ is the desired state.

It can be shown (see, e.g., [27]) that the optimal control law in this context is

$$\mathbf{u} = -\mathcal{C}\mathbf{x} \tag{15.10}$$

where $\mathcal{C}$ is the matrix function of time defined by

$$\mathcal{C} = R^{-1}G^T\mathcal{S}$$

where $\mathcal{S}$ is a matrix that satisfies the *Riccati equation*

$$\dot{\mathcal{S}} = -\mathcal{S}F - F^T\mathcal{S} + \mathcal{S}GR^{-1}G^T\mathcal{S} - Q$$

subject to the end constraint $\mathcal{S}(T) = S$.

In practice, control laws are implemented using digital computers by sampling continuous systems at discrete times. If τ' is the sample period with $N\tau' = T$, and if the control law keeps $\mathbf{u}$ constant during each sample period, then (15.8) is replaced with the approximation

$$\mathbf{x}_{k+1} = \mathcal{F}_k\mathbf{x}_k + \mathcal{G}_k\mathbf{u}_k \tag{15.11}$$

where $\mathbf{x}_k = \mathbf{x}(k\tau')$, $\mathbf{u}_k = \mathbf{u}(k\tau')$,

$$\mathcal{F}_k = \Phi((k+1)\tau', k\tau')$$

and

$$\mathcal{G}_k = \int_{k\tau'}^{(k+1)\tau'} \Phi((k+1)\tau', t)G(t)\, dt$$

where $\Phi(t, t_0)$ is the *state transition matrix*, which is the solution to the matrix differential equation

$$\frac{d}{dt}\Phi(t, t_0) = F(t)\Phi(t, t_0) \quad \text{with} \quad \Phi(t_0, t_0) = \mathbb{I}.$$

The matrix function $\Phi(t, t_0)$ has the property that solutions to $\dot{\mathbf{y}} = F(t)\mathbf{y}$ with $\mathbf{y}(t_0)$ given can be written as $\mathbf{y}(t) = \Phi(t, t_0)\mathbf{y}(t_0)$. In the special case when $F(t) = F_0$ (a constant matrix), we write

$$\Phi(t, t_0) = \exp((t - t_0)F_0).$$

The cost function (15.9) is modified in the case of digital control to

$$C = \frac{1}{2}\mathbf{x}_N^T S_N \mathbf{x}_N + \frac{1}{2}\tau' \sum_{k=0}^{N-1}[\mathbf{x}_k^T Q_k \mathbf{x}_k + \mathbf{u}_k^T R_k \mathbf{u}_k]$$

where $Q_k = Q(k\tau')$, $R_k = R(k\tau')$, and $S_N = S$. In what follows, we assume time is measured in units such that

$$\tau' = 1.$$

The optimal solution of this discretized problem is found by minimizing $\mathbf{x}_k^T Q_k \mathbf{x}_k + \mathbf{u}_k^T R_k \mathbf{u}_k$ at each value of k subject to the constraint (15.11). This is achieved recursively by defining [26, 40, 54, 106]:

$$S_k = \mathcal{F}_k^T[S_{k+1} - S_{k+1}\mathcal{G}_k(\mathcal{G}_k^T S_{k+1}\mathcal{G}_k + R_k)^{-1}\mathcal{G}_k^T S_{k+1}]\mathcal{F}_k + Q_k$$

and

$$\mathbf{u}_k = -(\mathcal{G}_k^T S_{k+1}\mathcal{G}_k + R_k)^{-1}\mathcal{G}_k^T S_{k+1}\mathcal{F}_k\mathbf{x}_k$$

for $k < N$. This provides an optimal digital control law for deterministic linear systems.

The next section shows how estimation of a static linear transformation subject to measurement noise can also be found recursively.

15.3 Recursive Estimation of a Static Quantity

Consider the linear observer equation

$$\mathbf{y} = H\mathbf{x} + \mathbf{v},$$

where $\mathbf{x} \in \mathbb{R}^n$ and $H \in \mathbb{R}^{k \times n}$, and $\mathbf{y}, \mathbf{v} \in \mathbb{R}^k$. $\mathbf{x}$ can be viewed as either a static vector with k fixed, or as a series of k scalar measurements taken over time. Here $\mathbf{v}$ represents the measurement noise.

We seek an estimate of $\mathbf{x}$ (denoted $\hat{\mathbf{x}}$) such that a quadratic cost function of the form

$$C_{\mathbf{x}}^{(1)}(\hat{\mathbf{x}}) = \frac{1}{2}(\mathbf{x} - \hat{\mathbf{x}})^T \mathcal{W}(\mathbf{x} - \hat{\mathbf{x}})$$

is minimized for a symmetric positive-definite weighting matrix $\mathcal{W}$. Since $\mathbf{x}$ is not known, we instead seek $\hat{\mathbf{x}}$ which minimizes a cost function of the observed quantity $\mathbf{y}$:

$$C_{\mathbf{y}}^{(2)}(\hat{\mathbf{x}}) = \frac{1}{2}(\mathbf{y} - H\hat{\mathbf{x}})^T \mathcal{W}(\mathbf{y} - H\hat{\mathbf{x}}).$$

Minimization of $C_{\mathbf{y}}^{(2)}(\hat{\mathbf{x}})$ with respect to $\hat{\mathbf{x}}$ yields

$$\hat{\mathbf{x}} = (H^T \mathcal{W}^{-1} H)^{-1} H^T \mathcal{W}^{-1} \mathbf{y}. \tag{15.12}$$

A suitable choice for the weighting matrix W that incorporates properties of noise into the formulation is the *measurement error covariance*:

$$W = E(\mathbf{v}\mathbf{v}^T).$$

In cases when k (the dimension of the noise and observation vectors) is very large, or if new observation data is rapidly streaming in, it makes sense to consider *recursive least-squares estimators*. If, for instance, the observation data is gathered sequentially over time, it often makes sense to calculate the best estimate at a given time for $k' < k$ observations, and recursively update the estimate thereafter instead of recalculating (15.12) for successively large values of k.

Let us assume that for some number of measurements, k_{i-1}, the estimation problem

$$\mathbf{y}_{i-1} = H_{i-1}\mathbf{x} + \mathbf{v}_{i-1}$$

has been solved to yield

$$\hat{\mathbf{x}}_{i-1} = (H_{i-1}^T W_{i-1}^{-1} H_{i-1})^{-1} H_{i-1}^T W_{i-1}^{-1} \mathbf{y}_{i-1}.$$

Now assume a new set of measurements,

$$\mathbf{y}_i = H_i\mathbf{x} + \mathbf{v}_i,$$

are made. The goal is to use the old information together with this newly observed information to arrive at the best estimate $\hat{\mathbf{x}}_2$. This is achieved by minimizing the quadratic cost function [156]

$$C_{\mathbf{y}_i}^{(3)}(\hat{\mathbf{x}}_i) = [(\mathbf{y}_{i-1} - H_{i-1}\hat{\mathbf{x}}_i)^T, (\mathbf{y}_i - H_i\hat{\mathbf{x}}_i)^T] \begin{pmatrix} W_{i-1}^{-1} & \mathbb{0}_{(i-1)\times(i)} \\ \mathbb{0}_{(i)\times(i-1)} & W_i^{-1} \end{pmatrix} \begin{bmatrix} \mathbf{y}_{i-1} - H_{i-1}\hat{\mathbf{x}}_i \\ \mathbf{y}_i - H_i\hat{\mathbf{x}}_i \end{bmatrix}.$$

The zero blocks appear in the weighting matrix because it is assumed that the noise is uncorrelated in the two different sequences of measurements.

The value of $\hat{\mathbf{x}}_i$ that minimizes $C_{\mathbf{y}_i}^{(3)}$ is of the form

$$\hat{\mathbf{x}}_i = (P_{i-1}^{-1} + H_i^T W_i^{-1} H_i)^{-1}(H_{i-1}^T W_{i-1}^{-1}\mathbf{y}_{i-1} + H_i^T W_i^{-1}\mathbf{y}_i) \qquad (15.13)$$

where (see [156] p. 310)

$$P_{i-1}^{-1} = H_{i-1}^T W_{i-1}^{-1} H_{i-1} = E[(\mathbf{x} - \hat{\mathbf{x}})(\mathbf{x} - \hat{\mathbf{x}})^T]$$

is the *residual covariance matrix*.

It is possible to calculate analytically (see, e.g., [54] p. 558)

$$(P_{i-1}^{-1} + H_i^T W_i^{-1} H_i)^{-1} = P_{i-1} - P_{i-1}H_i^T(H_i P_{i-1}H_i^T + W_i)^{-1}H_i P_{i-1}.$$

Making this substitution, and using the definitions of $\hat{\mathbf{x}}_{i-1}$ and the *recursive weighted least-squares estimator gain matrix*

$$K_i = P_{i-1}H_i^T(H_i P_{i-1}H_i^T + W_i)^{-1}, \qquad (15.14)$$

simplifies (15.13) to

$$\hat{\mathbf{x}}_i = \hat{\mathbf{x}}_{i-1} + K_i(\mathbf{z}_i - H_i\hat{\mathbf{x}}_{i-1}). \qquad (15.15)$$

The matrix P_k is updated recursively by observing that

$$P_i^{-1} = P_{i-1}^{-1} + H_i^T W_i^{-1} H_i.$$

For more on estimation (of dynamic as well as static quantities) see [88, 49, 87].

15.4 Determining a Rigid-Body Displacement: Recursive Estimation of a Static Orientation or Pose

In this section we consider recursive formulations of algorithms presented in Chapter 14 for least-squares fitting of rotations and rigid-body motions to measured data. Various iterative and recursive approaches to least-squares fitting of a rotation matrix have been introduced in the satellite attitude estimation and control literature over the years (see, e.g., [2, 7, 8, 9, 13, 32, 44, 45, 108, 131, 132, 138, 139, 172]).

A recursive formulation of the batch least-squares fitting procedure for finding the optimal rotation matrix to fit measured vector data was derived in [32]. A Kalman filtering technique was presented in [8].

One method to compute $A \in SO(3)$ that provides the best least-squares fit between two sets of measured vector data $\{\mathbf{p}_i\}$ and $\{\mathbf{q}_i\}$ related as $\mathbf{p}_i = A\mathbf{q}_i$ is to recursively calculate a series of matrices $\hat{A}_k$ with $\hat{A}_N = \hat{A}$ (where $\hat{A}$ is defined in (14.34)) and use (14.35) at the end of the process. This is addressed in [32]. The basic idea is that the unconstrained least-squares estimate based on the first k pairs of data vectors $\{\mathbf{p}_i, \mathbf{q}_i\}$ for $i = 1, ..., k$ can be written as

$$\hat{A}_k = [R_{k-1} + \mathbf{p}_k \mathbf{q}_k^T] S_k^{-1}$$

where

$$R_{k-1} = \sum_{i=1}^{k-1} \mathbf{p}_i \mathbf{q}_i^T$$

and

$$S_k = \sum_{i=1}^{k} \mathbf{q}_i \mathbf{q}_i^T.$$

A recursion can be started because S_k^{-1} can be written in terms of S_{k-1}^{-1}, $\mathbf{p}_k$ and $\mathbf{q}_k$. Explicitly,

$$S_k^{-1} = S_{k-1}^{-1} - [1 + \mathbf{q}_k^T S_{k-1}^{-1} \mathbf{q}_k]^{-1} \cdot S_{k-1}^{-1} \mathbf{q}_k \mathbf{q}_k^T S_{k-1}^{-1}.$$

The matrix $\hat{A}_k$ is then found recursively as

$$\hat{A}_k = \hat{A}_{k-1} + [1 + \mathbf{q}_k^T S_{k-1}^{-1} \mathbf{q}_k]^{-1} \cdot [\mathbf{p}_k - \hat{A}_{k-1} \mathbf{q}_k] \mathbf{q}_k^T S_{k-1}^{-1}.$$

Note that in the formulations of this subsection the group structure of rigid-body motions was not used. And further, the concepts of probability density functions on groups were not required. We show in later sections of this chapter how the group structure can be used in different kinds of estimation problems.

Before examining more sophisticated estimation problems, we need to have a language for describing noise in systems that leads to measurement errors. This is the subject of the next section.

15.5 Gaussian and Markov Processes

In this section we review basic stochastic processes. See [61, 63, 102] for physical introductions and [81, 104, 125, 157] for detailed mathematical treatments. Our brief review follows [94, 124, 168].

Let $p(\mathbf{x}, t)\, d\mathbf{x}$ denote the probability that the random process $\mathbf{X}(t)$ is contained in the d-dimensional voxel with volume $d\mathbf{x} = dx_1...dx_d$ centered at $\mathbf{x} \in \mathbb{R}^d$. Here the t inside of $p(\mathbf{x}, t)$ indexes this family of PDFs. That is, $p(\mathbf{x}, t)$ could have been denoted equally well as $p_t(\mathbf{x})$.

Similarly, let $p(\mathbf{x}_n, t_n; \mathbf{x}_{n-1}, t_{n-1}; ...\mathbf{x}_2, t_2; \mathbf{x}_1, t_1)\, d\mathbf{x}_n d\mathbf{x}_{n-1} \cdots d\mathbf{x}_2 d\mathbf{x}_1$ be the probability that for each time t_i, each $\mathbf{X}(t_i)$ is in the voxel centered at $\mathbf{x}_i$ for each $i = 1, ..., n$. The function $p(\mathbf{x}, t)$ is a probability density on $\mathbb{R}^d$ for each fixed value of time, t, whereas $p(\mathbf{x}_n, t_n; \mathbf{x}_{n-1}, t_{n-1}; ...; \mathbf{x}_2, t_2; \mathbf{x}_1, t_1)$ is a PDF on $\mathbb{R}^{d \cdot n} = \mathbb{R}^d \times \mathbb{R}^d \times \cdots \times \mathbb{R}^d$ for each fixed choice of $(t_1, ..., t_n)^T \in \mathbb{R}^n$. As a matter of notational convenience, we assume that the times t_i are ordered as $t_1 < t_2 < \cdots < t_n$ so that t_n is the most recent time.

By integrating the PDF $p(\mathbf{x}_n, t_n; \mathbf{x}_{n-1}, t_{n-1}; ...; \mathbf{x}_2, t_2; \mathbf{x}_1, t_1)$ over the last k copies of $\mathbb{R}^d$, we observe the general relationship

$$p(\mathbf{x}_n, t_n; ...; \mathbf{x}_{k+1}, t_{k+1}) = \int_{\mathbb{R}^d} \cdots \int_{\mathbb{R}^d} p(\mathbf{x}_n, t_n; ...; \mathbf{x}_1, t_1)\, d\mathbf{x}_k \cdots d\mathbf{x}_1. \qquad (15.16)$$

In a similar way, $p(\mathbf{x}_k, t_k; \mathbf{x}_{k-1}, t_{k-1}; ...; \mathbf{x}_1, t_1)$ is obtained by integrating the probability density $p(\mathbf{x}_n, t_n; \mathbf{x}_{n-1}, t_{n-1}; ...; \mathbf{x}_2, t_2; \mathbf{x}_1, t_1)$ over $\mathbf{x}_n$ through $\mathbf{x}_{k+1}$.

For the case when $d = 1$, a closed-form example of (15.16) is easily verified for the *Gaussian process*:

$$p(x_n, t_n; x_{n-1}, t_{n-1}; ...; x_1, t_1) = \frac{\exp\left[-\frac{1}{2}\sum_{i,j=1}^n C_{ij}^{-1}(x_i - m_i)(x_j - m_j)\right]}{[(2\pi)^n \det C]^{\frac{1}{2}}}$$

where C is an $n \times n$ covariance matrix with elements C_{ij} and $m_i = \langle X_i(t) \rangle$ are the components of the mean of $p(\cdot)$. The extension to the case of higher dimensions is straight forward.

The two PDFs $p(\mathbf{x}_k, t_k; \mathbf{x}_{k-1}, t_{k-1}; ...; \mathbf{x}_1, t_1)$ and $p(\mathbf{x}_n, t_n; \mathbf{x}_{n-1}, t_{n-1}; ...; \mathbf{x}_1, t_1)$ are also related by the conditional probability density function

$$p(\mathbf{x}_n, t_n; ...; \mathbf{x}_{k+1}, t_{k+1} \,|\, \mathbf{x}_k, t_k; ...; \mathbf{x}_1, t_1) \doteq \frac{p(\mathbf{x}_n, t_n; \mathbf{x}_{n-1}, t_{n-1}; ...; \mathbf{x}_1, t_1)}{p(\mathbf{x}_k, t_k; \mathbf{x}_{k-1}, t_{k-1}; ...; \mathbf{x}_1, t_1)}. \qquad (15.17)$$

This definition means that $p(\mathbf{x}_n, t_n; ...; \mathbf{x}_{k+1}, t_{k+1} \,|\, \mathbf{x}_k, t_k; ...; \mathbf{x}_1, t_1)\, d\mathbf{x}_n \cdots d\mathbf{x}_{k+1}$ is the probability that each $\mathbf{X}(t_i)$ is in the voxel centered at $\mathbf{x}_i$ at time t_i for all $i = n, ..., k+1$ under the condition that each $\mathbf{X}(t_j)$ is in the voxel centered at $\mathbf{x}_j$ at time t_j for $j = k, ..., 1$.

A direct consequence of the definition in (15.17) and the observation in (15.16) is that

$$\int_{\mathbb{R}^d} \cdots \int_{\mathbb{R}^d} p(\mathbf{x}_n, t_n; ...; \mathbf{x}_{k+1}, t_{k+1} \,|\, \mathbf{x}_k, t_k; ...; \mathbf{x}_1, t_1)\, d\mathbf{x}_n \cdots d\mathbf{x}_{k+1} = 1.$$

A *Markov process*[1] is one which satisfies the condition

$$p(\mathbf{x}_n, t_n \,|\, \mathbf{x}_{n-1}, t_{n-1}; \mathbf{x}_{n-2}, t_{n-2}; ...; \mathbf{x}_1, t_1) = p(\mathbf{x}_n, t_n \,|\, \mathbf{x}_{n-1}, t_{n-1}). \qquad (15.18)$$

That is, it is a process with memory limited to only the preceding step. Therefore, in this case we can limit the discussion to $n = 2$.

For a Markov process, the *Chapman-Kolmogorov equation*

---

[1]Named after the Russian mathematician Andrei Andreyevich Markov (1856-1922).

$$p(\mathbf{x}_2, t_2 \mid \mathbf{x}_1, t_1) = \int_{\mathbb{R}^d} p(\mathbf{x}_2, t_2 \mid \boldsymbol{\xi}, s) p(\boldsymbol{\xi}, s \mid \mathbf{x}_1, t_1) \, d\boldsymbol{\xi} \tag{15.19}$$

is satisfied where s is the value of time corresponding to $\boldsymbol{\xi} \in \mathbb{R}^d$. It also follows from the definition in (15.17) that

$$p(\mathbf{x}_2, t_2; \mathbf{x}_1, t_1) = p(\mathbf{x}_2, t_2 \mid \mathbf{x}_1, t_1) \, p(\mathbf{x}_1, t_1).$$

A *stationary* Markov process is one for which

$$p(\mathbf{x}_i, t_i \mid \mathbf{x}_{i-1}, t_{i-1}) = p(\mathbf{x}_i, t_i - t_{i-1} \mid \mathbf{x}_{i-1}, 0). \tag{15.20}$$

15.6 Wiener Processes and Stochastic Differential Equations

Let $X_i(t)$ denote the ith entry in a time-varying vector in d-dimensional Euclidean space. Consider the case when the behavior of $X_1(t), ..., X_d(t)$ is governed by the system of d stochastic differential equations

$$dX_i = h_i(X_1, ..., X_d, t) \, dt + \sum_{j=1}^{m} H_{ij}(X_1, ..., X_d, t) \, dW_j(t) \tag{15.21}$$

where $\mathbf{W}(t) = [W_1, ..., W_m]^T$ is a vector of independent *Wiener processes*. That is, each of the components W_j have zero ensemble (time) average, are taken to be zero at time zero, and are stationary and independent processes. Denoting an ensemble average as $\langle \cdot \rangle$, these properties are written as: (a) $\langle W_j(t) \rangle = 0$; (b) $W_j(0) = 0$; (c) $\langle (W_j(t_1 + t) - W_j(t_2 + t))^2 \rangle = \langle (W_j(t_1) - W_j(t_2))^2 \rangle$ for $t_1 + t \geq 0$ and $t_2 + t \geq 0$; (d) $\langle (W(t_i) - W(t_j))(W(t_k) - W(t_l)) \rangle = 0$ for $t_i > t_j \geq t_k > t_l \geq 0$.

From these defining properties, it is clear that for the Wiener process, $W_j(t)$, we have

$$\langle [W_j(t_1 + t_2)]^2 \rangle = \langle [W_j(t_1 + t_2) - W_j(t_1) + W_j(t_1) - W_j(0)]^2 \rangle$$

$$= \langle [W_j(t_1 + t_2) - W_j(t_1)]^2 + [W_j(t_1) - W_j(0)]^2 \rangle = \langle [W_j(t_1)]^2 \rangle + \langle [W_j(t_2)]^2 \rangle.$$

For the equality

$$\langle [W_j(t_1 + t_2)]^2 \rangle = \langle [W_j(t_1)]^2 \rangle + \langle [W_j(t_2)]^2 \rangle \tag{15.22}$$

to hold for all values of time t_1, t_2, it must be the case that [124]

$$\langle [W_j(t - s)]^2 \rangle = \sigma_j^2 |t - s|. \tag{15.23}$$

for some positive real number σ_j^2. This together with the absolute value signs ensures that $\langle [W_j(t - s)]^2 \rangle > 0$.

We calculate the correlation of a scalar-valued Wiener process with itself at two different times t and s with $0 \leq s \leq t$ as

$$\langle W_j(s) W_j(t) \rangle = \langle W_j(s)(W_j(s) + W_j(t) - W_j(s)) \rangle =$$

$$\langle [W_j(s)]^2 \rangle + \langle (W_j(s) - W_j(0))(W_j(t) - W_j(s)) \rangle = \sigma_j^2 s.$$

The notation dW_j is defined by

$$dW_j(t) = W_j(t + dt) - W_j(t).$$

Hence, from the definitions and discussion above,

$$\langle dW_j(t) \rangle = \langle W_j(t+dt) \rangle - \langle W_j(t) \rangle = 0$$

and

$$\langle [dW_j(t)]^2 \rangle = \langle (W_j(t+dt) - W_j(t))(W_j(t+dt) - W_j(t)) \rangle$$
$$= \langle [W_j(t+dt)]^2 \rangle - 2\langle W_j(t)W_j(t+dt) \rangle + \langle [W_j(t)]^2 \rangle$$
$$= \sigma_j^2(t+dt-2t+t) = \sigma_j^2 dt.$$

Finally, we note that for the m-dimensional Wiener process $\mathbf{W}(t)$, each component is uncorrelated with the others for all values of time. This is written together with what we already know from above as

$$\langle W_i(s)W_j(t)) \rangle = \sigma_j^2 \delta_{ij} \min(s,t)$$

and

$$\langle dW_i(t_i)\, dW_j(t_j) \rangle = \sigma_j^2 \delta_{ij} dt_j. \tag{15.24}$$

If at time t the stochastic variable $X_i(t)$ passes through the deterministic coordinate x_i for all $i = 1, ..., d$, then from the above properties and (15.21), and the linearity of ensemble averaging, we observe that[2]

$$\langle dX_i \rangle = \langle h_i(X_1, ..., X_d, t)\, dt \rangle + \sum_{j=1}^{m} \langle H_{ij}(X_1, ..., X_d, t)\, dW_j(t) \rangle$$

$$= \langle h_i(X_1, ..., X_d, t)\, dt \rangle + \sum_{j=1}^{m} \langle H_{ij}(X_1, ..., X_d, t) \rangle \langle dW_j(t) \rangle$$

$$= h_i(x_1, ..., x_d, t)\, dt \tag{15.25}$$

and

$$\langle dX_i dX_k \rangle = \Bigg\langle \left(h_i(X_1, ..., X_d, t)\, dt + \sum_{j=1}^{m} H_{ij}(X_1, ..., X_d, t)\, dW_j(t) \right) \left(h_k(X_1, ..., X_d, t)\, dt \right.$$

$$+ \left. \sum_{l=1}^{m} H_{kl}(X_1, ..., X_d, t)\, dW_l(t) \right) \Bigg\rangle$$

$$= \sum_{j=1}^{m}\sum_{l=1}^{m} \langle H_{ij}(X_1, ..., X_d, t)H_{kl}(X_1, ..., X_d, t) \rangle \langle dW_j(t)\, dW_l(t) \rangle + O((dt)^2).$$

Substitution of (15.24) into the equation above, and neglecting the higher order terms in dt results in

$$\langle dX_i dX_k \rangle = \sum_{j=1}^{m} \sigma_j^2 H_{ij}(x_1, ..., x_d, t)H_{kj}(x_1, ..., x_d, t)\, dt \tag{15.26}$$

where $H_{kj} = H_{jk}^T$. The above equation will be useful when deriving the Fokker-Planck equation in Chapter 16.

---

[2]The steps wherein $\langle H_{ij}(X_1, ..., X_d, t)\, dW_j(t) \rangle$ is replaced by $\langle H_{ij}(X_1, ..., X_d, t) \rangle \langle dW_j(t) \rangle$ and $\langle H_{ij}(X_1, ..., X_d, t)H_{kl}(X_1, ..., X_d, t)\, dW_j(t)\, dW_l(t) \rangle$ is replaced with $\langle H_{ij}(X_1, ..., X_d, t)H_{kl}(X_1, ..., X_d, t) \rangle \langle dW_j(t)\, dW_l(t)) \rangle$ are only valid when (15.21) is interpreted as an Itô SDE.

15.7 Stochastic Optimal Control for Linear Systems

For a linear system with Gaussian noise, we have

$$\frac{d\mathbf{x}}{dt} = F(t)\mathbf{x} + G(t)\mathbf{u} + L(t)\mathbf{w}. \tag{15.27}$$

where $\mathbf{w}dt = d\mathbf{W}$ is a zero-mean Gaussian white noise process. An observation with noise is of the form

$$\mathbf{y} = H(t)\mathbf{x} + \mathbf{v}. \tag{15.28}$$

We assume both $\mathbf{v}$ is also a zero-mean Gaussian white noise process that is uncorrelated with $\mathbf{w}$.

The LQG regulator is a control system based on the minimization of the expected value of the same cost function as in the deterministic case, but subject to the fact that measurements of the state are observed indirectly through (15.28). The main result of LQG control theory is that the *certainty-equivalence principle* holds (see, e.g., [54, 156]). That is, the best controller-estimator combination is the same as the optimal estimator of the uncontrolled system and the deterministic LQ regulator based on the mean values of the observed state.

The optimal estimator for a discretized system is called a *Kalman filter* [87]. For a continuous system it is the *Kalman-Bucy filter* [88]. This subject has received a great deal of attention over the past forty years, and the reader interested in learning more about this subject can draw on any of a number of sources. (see the reference list for this chapter). As it is a rather involved discussion to explain Kalman filtering, we will not address this subject here. Rather, we will examine the effect of noise in the system (15.27) on the quadratic cost function used in the deterministic controller case under the assumption of perfect state measurement.

Following [27, 156], the expected value of the quadratic cost function in the stochastic case is

$$C = E\left[\frac{1}{2}\mathbf{x}^T(T)S\mathbf{x}(T) + \frac{1}{2}\int_0^T [\mathbf{x}^T(t)Q(t)\mathbf{x}(t) + \mathbf{u}^T(t)R(t)\mathbf{u}(t)]dt\right]. \tag{15.29}$$

Using the control law (15.10) (which would be optimal if $\mathbf{x}$ were known exactly), then (15.27) becomes

$$\frac{d\mathbf{x}}{dt} = (F - GC)\mathbf{x} + L\mathbf{w}. \tag{15.30}$$

Define the matrix

$$X(t) \doteq E[(\mathbf{x}(t) - E[\mathbf{x}(t)])(\mathbf{x}(t) - E[\mathbf{x}(t)])^T].$$

Then a straightforward application of the product rule for differentiation gives

$$\frac{d}{dt}X(t) = E[\frac{d}{dt}(\mathbf{x}(t) - E[\mathbf{x}(t)])(\mathbf{x}(t) - E[\mathbf{x}(t)])^T] + E[(\mathbf{x}(t) - E[\mathbf{x}(t)])\frac{d}{dt}(\mathbf{x}(t) - E[\mathbf{x}(t)])^T].$$

Substitution of (15.30) into the right-hand-side of this equation together with appropriate changes of order in taking expectations gives

$$\dot{X} = (F - GC)X + X(F - GC)^T + LQ'L^T$$

where

$$E[\mathbf{w}(t)\mathbf{w}^T(\tau)] = Q'\delta(t - \tau).$$

This gives a deterministic set of equations that describe the evolution of covariances of the stochastic process $\mathbf{x}(t)$.

15.8 Deterministic Control on Lie Groups

The dynamics and control of rotating systems such as satellites has been addressed in a large number of works in the guidance and control literature (see, e.g., [79, 82, 89, 109, 147, 165, 173, 174, 179]). Group-theoretic and differential-geometric approaches to this problem have been explored in [12, 14, 15, 22, 23, 24, 31, 47, 84, 123, 163, 167, 171], and more recently in [96, 97, 100]. Similar problems have been studied in the context of underwater vehicles (see e.g., [53, 103]), and recently in the context of quad-rotor aerial vehicles [113].

In these works, it is assumed that the state of the system is known completely, and the control torques used to steer the system implement the control law exactly. Below we review this deterministic problem because a sub-optimal (though conceptually easy) solution to the stochastic version of this problem can be implemented by solving the estimation and control problems separately. This is analogous to the LQG case, except for the fact that optimality is not retained.

We begin by noting a difference between the conventions used throughout this book and those used in the attitude control literature. In attitude control, rotation matrices called *direction cosine matrices* describe a frame fixed in space relative to one fixed in a satellite. That is, whereas we use $R \in SO(3)$ to denote the orientation of a satellite with respect to a space-fixed frame, in the attitude control literature they would formulate everything in terms of $R' = R^T$. This means that when comparing the equations that follow with those in the literature, many things will appear to be "backwards." For example, if $\boldsymbol{\omega}$ is the angular velocity vector of the satellite with respect to a fixed frame (with components described in satellite-fixed coordinates) we would write

$$\boldsymbol{\omega} = (R^T \dot{R})^\vee, \tag{15.31}$$

or

$$\dot{R} = R\Omega \tag{15.32}$$

where $\boldsymbol{\omega} = (\Omega)^\vee$. In contrast, in the attitude control literature they may write everything in terms of R' as

$$\dot{R}' = \Omega' R'$$

where $\Omega' = -\Omega$.

15.8.1 Proportional Derivative Control on $SO(3)$ and $SE(3)$

There are many possible variants of the idea of PD control on the matrix Lie groups $SO(3)$ and $SE(3)$. When we make statements that hold for either of these groups we will refer to G rather than $SO(3)$ or $SE(3)$ in particular. Perhaps the conceptually simplest PD control scheme on G is to define an artificial potential based on any of the metrics discussed in chapters 5 and 6:

$$V'(g) = d^2(g, g_d)$$

where $g, g_d \in G$. A Rayleigh dissipation function of the form

$$R = \frac{1}{2}\operatorname{tr}(\dot{g}W\dot{g}^T)$$

for appropriate positive-definite symmetric damping matrix W can also be chosen. When doing this, the general principle behind PD control of mechanical systems ensures that controllers of this type will work.

Howovor there is some question as to whether or not such controllers are "natural" in the sense of utilizing all the structure that the properties of the group G has to offer. Such issues are addressed by Bullo and Murray [28]. We summarize some of their results below.

PD Control on $SO(3)$

Bullo and Murray [28] show that if $(\cdot, \cdot)$ is the usual Ad-invariant inner product on $so(3)$ (see Chapter 7) and if $\theta(t)$ is the angle of rotation of $R(\mathbf{n}, \theta)$ then

$$\frac{1}{2}\frac{d(\theta^2)}{dt} = (\log R, R^T \dot{R}) = (\log R, \dot{R}R^T),$$

and hence, given the system $\dot{R} = R\Omega$ where it is assumed that we have the ability to specify Ω, the control law

$$\Omega = -k_p \log R, \qquad k_p > 0$$

provides a feedback law that sends $R \to \mathbb{I}$ as $t \to \infty$. This is shown by using the Lyapunov function

$$\mathcal{V}(R) = \frac{1}{2}(\log R, \log R) = \frac{1}{2}\theta^2$$

and calculating

$$\dot{\mathcal{V}} = (\log R, -k_p \log R) = -2k_p\mathcal{V}$$

indicating that $\mathcal{V}$ is proportional to $e^{-2k_p t}$. Similarly, for a control torque that sends the system composed of Euler's equations of motion and the kinematic constraint $\dot{R} = R\Omega$ to the state $(R, \boldsymbol{\omega}) = (\mathbb{I}, \mathbf{0})$, Bullo and Murray [28] choose

$$\boldsymbol{\tau} = \boldsymbol{\omega} \times (I\boldsymbol{\omega}) - (k_p \log R - k_d R^T \dot{R})^\vee. \tag{15.33}$$

Here $k_p, k_d \in \mathbb{R}_{>0}$ are gains that provide an exponentially stable feedback law. We see this by constructing the Lyapunov function

$$V = \frac{1}{2}(\log R, \log R) + k_p(R^T \dot{R}, R^T \dot{R}).$$

Differentiation with respect to time and substitution of the controlled equations of motion results in $\dot{V} \leq 0$ in the usual way, with $\dot{V} = 0$ at a set of measure zero.

Of course this is not the only control law that will stabilize the system as $t \to \infty$. For instance, Koditschek [93] introduced the feedback law

$$\boldsymbol{\tau} = -k_p \log R - K_d(R^T \dot{R})^\vee$$

where $K_d = K_d^T$ is a positive definite matrix (not to be confused with the scalar k_d above). This control law is an example of the general approach described at the beginning of this subsection. In contrast (15.33) uses a feedforward term to cancel the nonlinearities, which strictly speaking no longer makes it a pure PD controller.

PD Control on $SE(3)$

Recall that if X is an arbitrary element of $se(3)$ it can be written as

$$X = \begin{pmatrix} \theta N & \mathbf{h} \\ \mathbf{0}^T & 0 \end{pmatrix}$$

where $N \in so(3)$ and $\|N^\vee\| = 1$. Then

$$\exp X = \begin{pmatrix} R(\mathbf{n}, \theta) & J_L(\theta\mathbf{n})\mathbf{h} \\ \mathbf{0}^T & 1 \end{pmatrix} \tag{15.34}$$

where $R(\mathbf{n}, \theta) = \exp \theta N$ is the axis-angle parameterization of $SO(3)$ and

$$J_L(\theta\mathbf{n}) = \mathbb{I} + \frac{1 - \cos\theta}{\theta^2} N + \left(\frac{1}{\theta^2} - \frac{\sin\theta}{\theta^3} \right) N^2$$

where $J_L(\mathbf{x})$ is the left Jacobian defined in (5.51) for the parameterization $R = \exp X$. Bullo and Murray observe that $J_L(\theta\mathbf{n})$ and $R(\mathbf{n}, \theta)$ are related by expressions such as [28]:

$$J_L(\theta\mathbf{n})R(\mathbf{n}, \theta) = R(\mathbf{n}, \theta)J_L(\theta\mathbf{n}) = 2J_L(2\theta\mathbf{n}) - J_L(\theta\mathbf{n})$$

and

$$\frac{d}{d\theta} J_L(\theta\mathbf{n}) = \frac{1}{\theta}(R(\mathbf{n}, \theta) - J_L(\theta\mathbf{n})).$$

These relationships are useful in constructing control systems on $SE(3)$. For instance, if $g \in SE(3)$ is parameterized as $g = \exp X$ consider the simple system

$$\frac{d}{dt} \exp X = (\exp X) \begin{pmatrix} \Omega & \mathbf{v} \\ \mathbf{0}^T & 0 \end{pmatrix}.$$

This is equivalent to the system

$$\mathcal{J}_R(\mathbf{x})\dot{\mathbf{x}} = \begin{pmatrix} \boldsymbol{\omega} \\ \mathbf{v} \end{pmatrix}$$

where $\mathbf{x} = (X)^\vee$ are exponential coordinates for $SE(3)$ and $\mathcal{J}_R(\mathbf{x})$ is the right Jacobian for $SE(3)$ using these coordinates.

Here $\boldsymbol{\omega} = \Omega^\vee$ and $\mathbf{v}$ are the quantities that are assumed to be specified by the controller. The above system can be controlled with the PD law [28]:

$$\begin{pmatrix} \Omega & \mathbf{v} \\ \mathbf{0}^T & 0 \end{pmatrix}^\vee = - \begin{pmatrix} k_\omega \mathbb{I}_{3\times3} & \mathbb{O}_{3\times3} \\ \mathbb{O}_{3\times3} & (k_\omega + k_v)\mathbb{I}_{3\times3} \end{pmatrix} (X)^\vee$$

where $g = \exp X$ as in (15.34). Bullo and Murray show that the closed-loop system with this control law can be written as

$$\frac{d}{dt}(\theta N) = -k_\omega \theta N$$

$$\frac{d\mathbf{h}}{dt} = -k_\omega \mathbf{h} - k_v [J_L(\theta\mathbf{n})]^{-T}\mathbf{h}.$$

(Here we use the notation $J^{-T} \doteq (J^{-1})^T = (J^T)^{-1}$.) Clearly the first of these equations indicates that $\theta N \to \mathbb{O}_{3\times3}$ exponentially as $t \to \infty$, and substitution of this information into the second equation indicates that $\mathbf{h} \to \mathbf{0}$ exponentially as $t \to \infty$ as well.

Similar control laws for second-order systems can also be devised. Our point in addressing this material is to show that the Lie-group approach to kinematics presented in earlier Chapters has direct applicability in deterministic control of mechanical systems.

15.8.2 Deterministic Optimal Control on $SU(3)$

A straightforward engineering way to state the optimal attitude control problem is as follows. We seek to find the control torques $\tau(t)$ (defined in satellite-fixed coordinates) that will bring a satellite from state (R_1, ω_1) at time t_1 to state (R_2, ω_2) at time t_2 while minimizing the cost

$$C = \int_{t_1}^{t_2} \{\omega^T K(t)\omega + \tau^T L(t)\tau\} dt$$

subject to the kinematic constraint (15.31) and the governing equation of motion

$$I\dot{\omega} + \omega \times (I\omega) = \tau. \tag{15.35}$$

Such problems have been addressed in the attitude control literature (see e.g., [83]).

In the geometric control literature, different variations on this problem have been addressed (see, e.g., [167]) To our knowledge the case when $K(t) = I$ (so that kinetic energy is the cost) and $L(t) \neq \mathbb{O}_{3\times3}$ does not have a closed-form solution. However, a number of numerical techniques exist for approaching this problem (see [83]).

However, a number of subcases with simple solutions do exist. For instance, if $I = a\mathbb{I}$, then (15.35) becomes a linear equation. Other cases when I is not a multiple of the identity also have solutions. When $L(t) = \mathbb{O}_{3\times3}$ and $K(t) = k(t)\mathbb{I}$, and assuming that angular velocity can be controlled exactly, an optimal solution will be [154, 155]:

$$\omega^* = \mathbf{c}/k(t)$$

where if $R_0 \doteq R_1^{-1} R_2$ then

$$\mathbf{c} = \frac{\theta(R_0)}{2(\sin\theta(R_0))\int_{t_1}^{t_2} 1/k(t)\,dt}(R_0 - R_0^T)^\vee.$$

This value of ω^* is substituted for ω in (15.35) to find the control torques.

15.8.3 Euler's Equations as a Double Bracket

In recent years, it has become fashionable to rewrite (15.35) in Lie-theoretic notation as [3]:

$$\dot{L} = [L, \Omega] \tag{15.36}$$

when $\tau = \mathbf{0}$. Here $L^\vee = I\omega$ is the angular momentum vector written in body-fixed coordinates and $\omega = \Omega^\vee$. We can go a step further and define the linear operator $\Lambda(L)$ such that

$$(\Lambda(L))^\vee = I^{-1}(L)^\vee.$$

This allows us to write (15.36) as $\dot{L} = [L, \Lambda(L)]$. We can show that

$$\Lambda(L) = \{\Lambda, L\} \doteq \mathcal{I}L + L\mathcal{I}$$

for an appropriate symmetric matrix $\mathcal{I} = \mathcal{I}(I)$. The bracket $\{\cdot, \cdot\}$ is called the anti-commutator, in contrast to $[\cdot, \cdot]$ which has a minus sign instead of a plus. This allows us to write Euler's equations of motion in a "double bracket" form

$$\dot{L} = [L, \{\mathcal{I}, L\}].$$

See [15] and references therein for quantitative results concerning the behavior of double bracket equations.

Other sorts of double-bracket equations have been studied extensively as well by R. Brockett and others as a way for continuous dynamical systems to perform discrete computations. Some of these systems involve double Lie brackets. For example, given a symmetric matrix H_0, the dynamical system [70]

$$\dot{H}(t) = [H(t), [H(t), \Lambda]] \quad H(0) = H_0$$

will diagonalize H_0 with the resulting eigenvalues sorted in an order specified by the ordering of the entries in the diagonal matrix Λ.

And given an arbitrary matrix $A \in GL(n, \mathbb{R})$, the dynamical system [70]

$$\dot{R} = RPA^T R - AP$$
$$\dot{P} = -2P + A^T R + R^T A$$

will converge to the polar decomposition $A = R_\infty P_\infty$ as $t \to \infty$ for almost any initial conditions $R(0) \in O(n)$ and $P(0) = P^T(0) > 0$.

15.8.4 Digital Control on $SO(3)$

In a digital control approach, where thrusters are modeled as instantaneous torques, the angular velocity can be modeled as being piecewise constant. This is possible if a small constant torque of the form $\boldsymbol{\tau} = \boldsymbol{\omega} \times (I\boldsymbol{\omega})$ is applied by the thrusters (so that $I\dot{\boldsymbol{\omega}} = \mathbf{0}$). As a practical matter, a sub-optimal solution can be obtained by three "bang-bang" rotations (one about each of the principle axes of the satellite). When the satellite is starting and ending at rest, this reduces the attitude control problem to three one-dimensional linear equations of the form $I_i \ddot{\theta}_i = \tau_i$. The solution for the torque components τ_i in terms of the desired orientation relative to the initial then reduces to the kinematics problem of finding the Euler angles of the desired relative rotation.

If the time interval $[t_1, t_2]$ is divided into N segments each of duration τ', and the angular velocity in the ith subinterval is the constant $\boldsymbol{\omega}_i = (\Omega_i)^\vee$, then the solution to (15.32) at any $t \in [t_1, t_2]$ will be a product of exponentials of the form

$$R = \exp(\tau' \Omega_1) \cdots \exp(\tau' \Omega_n) \exp((t - n\tau')\Omega_{n+1}) \tag{15.37}$$

when $n\tau' \le t \le (n+1)\tau'$. If over each of these intervals the torque $\boldsymbol{\tau} = \boldsymbol{\omega}_n \times (I\boldsymbol{\omega}_n)$ (which is small when $\|\boldsymbol{\omega}_n\|$ is small) is replaced with $\boldsymbol{\tau} = \mathbf{0}$, then (15.37) is no longer exact.

In this discretized problem, an optimality condition is the minimization of

$$C' = \sum_{n=1}^{N} \boldsymbol{\omega}_n^T I \boldsymbol{\omega}_n = \sum_{n=1}^{N} \mathrm{tr}(\Omega_n^T J \Omega_n)$$

subject to the constraint that

$$R_1^{-1} R_2 = \exp(\tau' \Omega_1) \cdots \exp(\tau' \Omega_N).$$

Since the duration of each of these intervals is small, it makes sense to approximate

$$\exp(\tau' \Omega_n) \approx \mathbb{I} + \tau' \Omega_n.$$

However, in doing so, small errors will make

$$R_1^{-1} R_2 \approx \mathbb{I} + \tau' \sum_{n=1}^{N} \Omega_n$$

a bad approximation unless R_1 and R_2 are close.

In the following section we examine the other half of the satellite attitude problem: estimation. When we cannot assume precise measurements of orientation, methods for attitude estimation become critical if we are to have any hope of developing stochastic versions of the deterministic control approaches presented in this section.

15.8.5 Analogs of LQR for $SO(3)$ and $SE(3)$

One generalization of LQR to the context of matrix Lie groups is reviewed in this subsection. To begin, assume that we have complete control over velocities (angular and/or translational). Then for $g(t) \in G = SO(3)$ or $SE(3)$,

$$\dot{g} = gU \quad \text{or} \quad (g^{-1}\dot{g})^\vee = \mathbf{u}$$

where $\mathbf{u} = (U)^\vee$ is the control input. Variations on this "kinematic control" where we assume the ability to specify velocities (rather than forces or torques) have been studied in [53, 55, 103, 167]. If the desired trajectory is $g_d(t)$, the control is simply $\mathbf{u} = (g_d^{-1}\dot{g}_d)^\vee$. However, if it is desirable to minimize control effort while tracking a trajectory in G, then we seek $\mathbf{u}$ such that the cost functional

$$C(\mathbf{u}) = d^2(g(T), g_d(T)) + \frac{1}{2}\int_0^T \{\alpha \|(g_d^{-1}\dot{g}_d)^\vee - \mathbf{u}\|^2 + \beta \|\mathbf{u}\|^2\} dt$$

is minimized where $d(\cdot, \cdot)$ is a metric on G and $\|\cdot\|$ is a norm on the Lie algebra $\mathcal{G}$. Here α and β weight the relative importance of accurately following the trajectory and minimizing control effort.

Han and Park [67] address a modification of the above problem subject to the constraint that $f(g(t)) = 0$ is satisfied for given $f : G \to \mathbb{R}$ where $G = SE(3)$, $\beta = 0$, and they choose $d(g_1, g_2) = \|\log(g_1^{-1}g_2)\|$. They give a closed-form solution to this problem.

A true application of noncommutative harmonic analysis (as opposed to group theory or differential geometry) is in the estimation problem on Lie groups. This is the subject of the next section. Recursive estimation on Lie groups is a prerequisite for analogs of LQG on groups.

15.9 Dynamic Estimation and Detection of Rotational Processes

In this section we explore some of the techniques used in the estimation of time-varying rotational processes. The estimation of rotational processes has been addressed extensively in the satellite attitude estimation and navigation literature using a number of different sensing modalities and algorithms. See e.g., [42, 60, 80, 111, 117, 127, 128, 133, 141, 145, 146, 160, 166] for recent works in this area. One of the more popular kinds of rotation sensors is the fiber-optic gyroscope. See [30, 150] for a description of the operating principles of that device. See [181] for descriptions of traditional mechanical gyros.

In contrast to the applied works mentioned above, a number of theoretical works on estimation on Lie groups and Riemannian manifolds have been presented over the years. For instance, see the work of Duncan [51, 52], Willsky [177, 178], and Lo and Eshleman [110]. Our discussions in the subsections that follow will review those aspects of estimation that use harmonic analysis on $SO(3)$.

15.9.1 Attitude Estimation Using an Inertial Navigation System

A vast literature exists on the attitude (orientation) estimation problem in the satellite guidance and control literature. In addition to those works mentioned previously, see e.g., [33, 101, 131, 132, 148], or almost any recent issue of the *Journal of Guidance, Control, and Dynamics* for up-to-date approaches to this problem.

In this subsection we consider an estimation problem on $SO(3)$ that was introduced in [177, 178].

Consider a rigid object with orientation that changes in time, e.g., a satellite or airplane. It is a common problem to use on-board sensors to estimate the orientation of the object at each instant in time. We now examine several models.

Let $R(t) \in SO(3)$ denote the orientation of a frame fixed in the body relative to a frame fixed in space at time t. From the perspective of the rotating body, the frame fixed in space appears to have the orientation $R'(t) \doteq R^T(t)$. The angular velocity of the body with respect to the space-fixed frame at time t as seen in the space-fixed frame will be $\boldsymbol{\omega}_L = (\dot{R}R^T)^\vee$, and the same angular velocity as seen in the rotating frame will be $\boldsymbol{\omega}_R = R^T \boldsymbol{\omega}_L = (R^T \dot{R})^\vee$ where $\dot{R} = dR/dt$. These are also easily rewritten in terms of R'.

If $\{X_i\}$ denotes the set of basis elements of the Lie algebra $so(3)$:

$$X_1 = \begin{pmatrix} 0 & 0 & 0 \\ 0 & 0 & -1 \\ 0 & 1 & 0 \end{pmatrix}; \quad X_2 = \begin{pmatrix} 0 & 0 & 1 \\ 0 & 0 & 0 \\ -1 & 0 & 0 \end{pmatrix}; \quad X_3 = \begin{pmatrix} 0 & -1 & 0 \\ 1 & 0 & 0 \\ 0 & 0 & 0 \end{pmatrix},$$

then

$$dR(t) = \left(\sum_{i=1}^{3} (\boldsymbol{\omega}_L \cdot \mathbf{e}_i dt) X_i \right) R(t)$$

and

$$dR(t) = R(t) \left(\sum_{i=1}^{3} (\boldsymbol{\omega}_R \cdot \mathbf{e}_i dt) X_i \right).$$

In practice, the increments $du_i = \boldsymbol{\omega}_R \cdot \mathbf{e}_i dt$ are measured indirectly. For instance, an *inertial platform* (such as a gyroscope) within the rotating body is held fixed (as best as possible) with respect to inertial space. This cannot be done exactly, i.e., there is some drift of the platform with respect to inertial space. Let $A \in SO(3)$ denote the orientation of the space-fixed frame relative to the inertial platform in the rotating body. If there were no drift, A would be the identity $\mathbb{I}$. However, in practice there is always drift. This has been modeled as [177][3]:

$$dA(t) = \left(\sum_{i=1}^{3} X_i dW_i(t) \right) A(t) \tag{15.38}$$

---

[3] We are not using the Itô Calculus, but the basic idea is the same.

where the three dimensional Wiener process $\mathbf{W}(t)$ defines the noise model.

The orientation that is directly observable is the relative orientation of the inertial platform with respect to the frame of reference fixed in the rotating body. We denote this as Q^{-1}. In terms of R and A,

$$A^{-1} = RQ^{-1}.$$

Therefore Q, which is the orientation of the rotating body with respect to the inertial platform, is

$$Q = AR.$$

Taking the inverse of both sides gives

$$Q' = R'A'. \tag{15.39}$$

From the chain rule this means

$$dQ' = dR'A' + R'dA'.$$

Substitution of (15.38) and the corresponding expression for dR gives

$$dQ' = \left(\sum_{i=1}^{3} -du_i X_i \right) R'A' + R'A' \left(\sum_{i=1}^{3} X_i dW_i(t) \right).$$

Using (15.39) and defining $x_i dt = -du_i$, this is written as [177]:

$$dQ' = \left(\sum_{i=1}^{3} x_i dt X_i \right) Q' + Q' \left(\sum_{i=1}^{3} X_i dW_i(t) \right). \tag{15.40}$$

In Chapter 16 we discuss how to derive a partial differential equation (a Fokker-Planck equation) corresponding to the stochastic differential equation (15.40). That Fokker-Planck equation can be reduced to a system of simpler equations in Fourier space using techniques from noncommutative harmonic analysis.

In a series of works, other applications of noncommutative harmonic analysis have been explored in the context of estimation of processes on the rotation group. Then following subsection examines one such technique in detail.

15.9.2 Exponential Fourier Densities

In this subsection we review the work of Lo and Eshleman [110] and discuss extensions of their results.

Consider a random rotation, S_0, and the resulting discrete sequence of rotations

$$S_{k+1} = R_k S_k$$

where $\{R_k\}$ is a sequence of known deterministic rotations. It is assumed that S_k is not known directly, but rather, is measured through some noisy measurement process as

$$M_k = V_k S_k$$

where $\{V_k\}$ is s sequence of rotations describing the noise. The estimation problem is then that of finding each S_k given the set of measurements $\{M_1, ..., M_k\}$. In order to do this in analogy with the case of vector measurements, both an appropriate error function (analog of the square of a matrix norm) and class of PDFs describing the noise process (analog of the Gaussian distribution) are required. In Chapter 5 we already discussed metrics on $SO(3)$, and, in principle, the square of any of these can be used. For the sake of concreteness, the function

$$d(R, Q) = \|R - Q\|^2$$

is used here.

Ideally, the PDF describing the noise process should be closed under the operations of conditioning and convolution. Of course, band-limited PDFs on $SO(3)$ will be closed under convolution. Lo and Eshleman chose instead to focus on a set of PDFs that are closed under conditioning. This is achieved by exponentiating a band-limited function. These *exponential Fourier densities* for the random initial rotation, measurement noise, and conditional probability $p(S_{k-1}|M_{k-1}, ..., M_1)$ are all assumed to be of the exponential Fourier form:

$$\rho(S_0) = \exp\left(\sum_{l=0}^{N} \sum_{m,n=-l}^{l} a_{mn}^{l0} U_{mn}^{l}(S_0)\right),$$

$$\rho(V_k) = \exp\left(\sum_{l=0}^{N} \sum_{m,n=-l}^{l} b_{mn}^{lk} U_{mn}^{l}(V_k)\right)$$

and

$$p(S_{k-1}|M_{k-1}, ..., M_1) = \exp\left(\sum_{l=0}^{N} \sum_{m,n=-l}^{l} a_{mn}^{l,k-1} U_{mn}^{l}(S_{k-1})\right).$$

We now show that PDFs of this form are closed under conditioning, i.e., that they are closed under operations of the form

$$\rho(S_k|M_k, ..., M_1) = c_k \rho(M_k|S_k) \rho(S_{k-1}|M_{k-1}, ..., M_1). \tag{15.41}$$

Substitution of $S_{k-1} = R_{k-1}^{-1} S_k$ into the above expression for $\rho(S_{k-1}|M_{k-1}, ..., M_1)$ yields

$$\rho(S_{k-1}|M_{k-1}, ..., M_1) = \exp\left(\sum_{l=0}^{N} \sum_{m,n=-l}^{l} a_{mn}^{l,k-1} U_{mn}^{l}(R_{k-1}^{-1} S_k)\right)$$

$$= \exp\left(\sum_{l=0}^{N} \sum_{m,n=-l}^{l} \left[\sum_{j=-l}^{l} a_{mn}^{l,k-1} U_{mj}^{l}(R_{k-1}^{-1})\right] U_{jn}^{l}(S_k)\right).$$

Similarly, observing that $\rho(M_k|S_k) = \rho(V_k)$, and making the substitution $V_k = M_k S_k^{-1}$ yields

$$\rho(M_k|S_k) = \exp\left(\sum_{l=0}^{N} \sum_{m,n=-l}^{l} b_{mn}^{l,k} U_{mn}^{l}(M_k S_k^{-1})\right)$$

$$= \exp\left(\sum_{l=0}^{N} \sum_{m,n=-l}^{l} \left[(-1)^{m+n} \sum_{j=-l}^{l} b_{j,-m}^{lk} U_{j,\,n}(M_k)\right] U_{mn}^{l}(S_k)\right).$$

By making the definition $\rho(S_0|M_0) = \rho(S_0)$, it is clear that the coefficients a^{lk}_{mn} defining

$$\rho(S_k|M_k, ..., M_1) = \exp \left(\sum_{l=0}^{N} \sum_{m,n=-l}^{l} a^{l,k}_{mn} U^l_{mn}(S_{k-1}) \right)$$

can be recursively updated as

$$a^{lk}_{mn} = \sum_{j=-l}^{l} \left[a^{l,k-1}_{mn} U_{jm}(R^{-1}_{k-1}) + (-1)^{m+n} \sum_{j=-l}^{l} b^{lk}_{j,-m} U_{j,-n}(M_k) \right]$$

where $l, k > 0$. For $l = 0$, the coefficients a^{0k}_{00} serve to enforce the constraint that ρ is a PDF on $SO(3)$.

We note that one of the main features of the exponential Fourier densities introduced by Lo and Eshleman is closure under conditioning. The folded normal distribution for $SO(3)$ discussed in Chapter 14 effectively has this property when the variance is small (which is what we would expect from good sensors). In addition, the folded normal for $SO(3)$ is closed under convolution.

15.10   Towards Stochastic Control on $SO(3)$

In the previous section we reviewed several estimation problems on $SO(3)$ based on different noise models that have been presented in the literature. Here we review some models of noise in the dynamical system to be controlled under the assumption of perfect observation of the state variables. This separation of estimation without regard to control, and control without regard to estimation, is motivated by the linear-system case in which the two problems can be decoupled. To our knowledge, the problem of combined estimation and optimal stochastic control has not been solved in the Lie-group-theoretic setting. However it intuitively makes sense that a feasible (though suboptimal) stochastic controller could be constructed based on suboptimal estimation (using the $SO(3)$-Gaussian function) and stochastic control based on the expected value of the state variables. A first step toward generating such controllers is to have reasonable stochastic system models.

Perhaps the most natural stochastic control model is one where Euler's equations of motion are the governing equations with the torque vector consisting of a deterministic control part and a noise vector:

$$\boldsymbol{\tau} = \mathbf{u} + \mathbf{w}.$$

The question then becomes how to specify $\mathbf{u}(t)$ such that the rotation $R(t)$ and angular velocity $\boldsymbol{\omega}(t)$ are driven to desired values in an optimal way given known statistical properties of the noise $\mathbf{w}dt = d\mathbf{W}$. This appears to be a difficult problem that has not been addressed in the literature.

Another noise model assumes that the rigid body to be controlled is bombarded with random impacts that make the angular momentum (rather than the torque) the noisy quantity. The model we discuss here is a variant of that presented by Liao [107].

Let X_1, X_2, X_3 be the usual basis elements of $so(3)$ normalized so that $(X_i, X_j) = \delta_{ij}$ where $(\cdot, \cdot)$ is the Ad-invariant inner product discussed in Chapter 7. The kinematic equation $\dot{R} = R\Omega$ can be written as

$$dR = R\Lambda Ad(R^{-1})M dt$$

where $M^\vee = R(L)^\vee$ is the angular velocity vector of the body as it appears in the space-fixed frame. In this model, the angular momentum will consist of three parts: the initial angular momentum; the component due to the random disturbances; and the component due to the control (where it is assumed that control implements any desired angular velocity). Then

$$M = \sum_{i=1}^{3} X_i(u_i + dW_i) + M_0$$

where $(M_0)^\vee$ is the initial angular momentum, u_i for $i = 1, 2, 3$ are the controls, and dW_i is a Wiener process representing noise in the system. Then the full system model is

$$dR = R\Lambda Ad(R^{-1}) \sum_{i=1}^{3} X_i(u_i + dW_i) + R\Lambda Ad(R^{-1})M_0 dt. \qquad (15.42)$$

To our knowledge, it remains to find an optimal stochastic control law $\{u_i\}$ for such a model.

15.11 Kinematic Covariance Propagation

Error propagation on the Euclidean motion group arises in a surprising number of different areas in robotics. For example, consider a robotic manipulator for which each joint angle has some backlash. If we describe this backlash as a distribution of possible angles around the nominal one, how will these joint errors add up to produce pose errors at the end effector? As a second example, consider a nonholonomic mobile robot that executes an open loop trajectory. Uncertainties in pose will add up along the path, and if many trials are performed, what will the distribution of terminal poses be? Many such problems in 'probabilistic robotics' can be imagined in the context of Simultaneous Localization and Mapping (or SLAM).

If the errors are small, Jacobian-based methods or first-order error propagation theories can be used. But what if the errors are very large? Here we address the propagation of large errors in rigid-body poses in a coordinate-free way. In this section we show how errors propagated by convolution on the Euclidean motion group, $SE(3)$, can be approximated to second order using the theory of Lie algebras and Lie groups. We then show how errors of moderate size (but not so small that linearization is valid) can be propagated by a recursive formula derived here. This formula takes into account errors to second-order, whereas prior efforts only considered the first-order case. Our formulation is nonparametric in the sense that it will work for probability density functions of any form (not only Gaussians).

In the remainder of this section we review the literature on error propagation, and review the terminology and notation used throughout the paper. In what follows, bold lower case letters denote vectors. N and n are positive integers. G denotes either the groups $SO(3)$ or $SE(3)$. All upper case letters (Roman or Greek) (except for N and G) denote matrices. Lower case letters denote scalars and group elements. A lower case letter followed by parentheses denotes a scalar-valued function.

In Subsection 15.11.1 a brief review of prior literature is presented. In Subsection 15.11.2, important definitions from the basic theory of Lie groups and probability and statistics are reviewed. In Subsection 15.11.3, several theorems are proved. These three subsections form the core of this section.

15.11.1 Literature Review

The representation and estimation of spatial uncertainty have also received attention in the robotics and vision literature. Two classic works in this area are due to Smith and Cheeseman [151] and Su and Lee [158]. More recent work on error propagation by Smith, Drummond, and Roussopoulos [149] describes the concatenation of Gaussian random variables on groups and applies this formalism to mobile robot navigation. In all three of these works, errors are assumed to be small enough that covariances can be propagated by the formula [169, 170]

$$\Sigma_{1*2} = Ad(g_2^{-1})\Sigma_1 Ad^T(g_2^{-1}) + \Sigma_2, \tag{15.43}$$

where Ad is the adjoint operator for $SE(3)$. This equation essentially says that given two 'noisy' frames of reference $g_1, g_2 \in SE(3)$, each of which is a Gaussian random variable with 6×6 covariance matrices Σ_1 and Σ_2, respectively, the covariance of $g_1 \circ g_2$ will be Σ_{1*2}. (Exactly what is meant by a covariance for a Lie group will be explained shortly.) This approximation is very good when errors are very small. We extend this linearized approximation to the quadratic terms in the expansion of the matrix exponential parameterization of $SE(3)$. The origin of (15.43) will become clear for the special case of small errors in our more general nonparametric derivation.

15.11.2 Properties of Mean and Covariance on $SE(3)$

In this section we provide definitions of the mean and covariance of Lie-group-valued functions and illustrate some of their properties. We note in passing that a PDF that is a symmetric function, $\rho(g) = \rho(g^{-1})$, always satisfies the condition

$$\int_G (\log g)^\vee \rho(g)\, dg = \mathbf{0}, \tag{15.44}$$

for $G = SO(3)$ or $G = SE(3)$. This is easy to see if we let $\rho_0(\mathbf{x}) = \rho(e^X)$. Then $\rho_0(\mathbf{x}) = \rho_0(-\mathbf{x})$. This is an even function in the exponential coordinates, and so the odd function $\mathbf{x}\rho_0(\mathbf{x})$ integrates to zero over a symmetric domain of integration in the space of exponential parameters that maps to G. In our case this domain is the ball of radius π (for $SO(3)$), or the Cartesian product of this ball with $\mathbb{R}^3$, both of which are symmetric. Hence the integral in (15.44) vanishes. More generally, if $\rho(g)$ is a symmetric function on $G = SO(3)$ or $G = SE(3)$ then for $\sum_{i=1}^n n_i$ odd,

$$\int_G \prod_{i=1}^n [(\log g)^\vee \cdot \mathbf{e}_i]^{n_i} \rho(g)\, dg = 0. \tag{15.45}$$

This is because the integrand is an odd function of the components of $\mathbf{x}$. For example,

$$\int_G (\log g)^{2k+1} \rho(g)\, dg = 0_n.$$

Definition 15.1. Any $\mu \in G$ for which

$$\int_G [\log(\mu^{-1} \circ g)]^\vee f(g)\, dg = \mathbf{0} \tag{15.46}$$

will be called a *mean* of the PDF $f(g)$, which is a straightforward extension of the Euclidean mean, which satisfies the equation

$$\int_{\mathbb{R}^n} (-\boldsymbol{\mu} + \mathbf{x}) \, f(\mathbf{x}) \, d\mathbf{x} = \mathbf{0}.$$

A difference in the non-Euclidean case is that there may be many means, rather than a unique one. From amongst all of the possible means, we will often choose one for which $d(e, \mu)$ is minimized for some given metric $d : G \times G \to \mathbb{R}_{\geq 0}$.

Furthermore, the *covariance* will be defined as

$$\Sigma \doteq \int_G \log(\mu^{-1} \circ g)^{\vee} [\log(\mu^{-1} \circ g)^{\vee}]^T f(g) \, dg. \tag{15.47}$$

The equality (15.44) can be thought of as a statement of when the mean is at the identity. If $\rho(g)$ has mean at the identity, then $f(g) = \rho(a^{-1} \circ g)$ has mean at a. We will use $\rho(g)$ to denote PDFs with mean at the identity, and $f(g)$ to denote PDFs that can have the mean at some other group element.

Theorem 15.2. *If $f(g)$ has mean μ and covariance Σ, then to second order*

$$\int_G (\log(g))^{\vee} f(g) \, dg = [I + F_1(\Sigma)](\log(\mu))^{\vee}$$

where the matrix-valued function $F_1(\Sigma)$ is defined as

$$F_1(\Sigma) \doteq \frac{1}{12} \sum_{i,j=1}^{6} \sigma_{ij} ad(\tilde{E}_i) ad(\tilde{E}_j)$$

and

$$\int_G (\log(g))^{\vee} ((\log(g))^{\vee})^T f(g) \, dg =$$

$$\Sigma + (\log \mu)^{\vee} ((\log \mu)^{\vee})^T + \frac{1}{2} \left(\Sigma \, ad^T (\log \mu) + ad(\log \mu) \, \Sigma \right).$$

Proof. Let $f(g) = \rho(\mu^{-1} \circ g)$ where $\rho(g)$ has mean at the identity. Then

$$\int_G (\log g)^{\vee} f(g) \, dg = \int_G (\log g)^{\vee} \rho(\mu^{-1} \circ g) \, dg = \int_G (\log(\mu \circ g))^{\vee} \rho(g) \, dg.$$

Using the BCH formula (7.37) with $\mu = \exp X$ and $g = \exp Y$, and using the linearity of the Lie bracket, we find that since $\rho(g)$ is a PDF with mean at the identity,

$$\int_G [\log(\mu \circ g))]^{\vee} \rho(g) \, dg =$$

$$\mathbf{x} + \sum_{i,j=1}^{6} \sigma_{ij} \left\{ \frac{1}{12} [\tilde{E}_i, [\tilde{E}_j, X]] + \frac{1}{48} ([\tilde{E}_i, [X, [\tilde{E}_j, X]]] + [X, [\tilde{E}_i, [\tilde{E}_j, X]]]) \right\}^{\vee} + \cdots.$$

The first expression in the statement of the theorem results from the definition of the adjoint. Likewise,

$$\int_G [\log(\mu \circ g))]^{\vee} ((\log(\mu \circ g))^{\vee})^T \rho(g) \, dg =$$

$$\int_G [\mathbf{x} + \mathbf{y} + \frac{1}{2} ad(X)\mathbf{y} + \cdots][\mathbf{x} + \mathbf{y} + \frac{1}{2} ad(X)\mathbf{y} + \cdots]^T \rho(g) \, dg.$$

Expanding out the product and eliminating terms linear in $\mathbf{y}$ results in the second statement of the theorem.

15.11.3   Propagation of the Mean and Covariance of PDFs on $SE(3)$

Let $\mu_1, \mu_2 \in SE(3)$ be two precise reference frames. Then $\mu_1 \circ \mu_2$ is the frame resulting from stacking one relative to the other. Now suppose that each has some uncertainty. Let $\{h_i\}$ and $\{k_j\}$ be two sets of frames of reference that are distributed around the identity. Let the first have N_1 elements, and the second have N_2. What will the covariance of the set of $N_1 \cdot N_2$ frames $\{(\mu_1 \circ \mu_2)^{-1} \circ \mu_1 \circ h_i \circ \mu_2 \circ k_j\}$ (which are also distributed around the identity) look like?

Let $\rho_i(g)$ be a unimodal PDF with mean at the identity and with a preponderance of its mass concentrated in a unit ball around the identity (where distance from the identity is measured as $\|(\log g)^\vee\|$). Then $\rho_i(\mu_i^{-1} \circ g)$ will be a distribution with the same shape centered at μ_i. In general, the convolution of two PDFs is defined as

$$(f_1 * f_2)(g) \doteq \int_G f_1(h) \, f_2(h^{-1} \circ g) \, dh,$$

and in particular if we make the change of variables $k = \mu_1^{-1} \circ h$, then

$$\rho_1(\mu_1^{-1} \circ g) * \rho_2(\mu_2^{-1} \circ g) = \int_G \rho_1(k)\rho_2(\mu_2^{-1} \circ k^{-1} \circ \mu_1^{-1} \circ g) \, dk.$$

Making the change of variables $g = \mu_1 \circ \mu_2 \circ q$, where q is a relatively small displacement measured from the identity, the above can be written as

$$\rho_{1*2}(\mu_1 \circ \mu_2 \circ q) = \int_G \rho_1(k)\rho_2(\mu_2^{-1} \circ k^{-1} \circ \mu_2 \circ q) \, dk. \tag{15.48}$$

The essence of this section is the efficient approximation of covariances associated with (15.48) when those of ρ_1 and ρ_2 are known. This problem reduces to the efficient approximation of

$$\Sigma_{1*2} = \int_G \int_G \log(q)^\vee [\log(q)^\vee]^T \rho_1(k)\rho_2(\mu_2^{-1} \circ k^{-1} \circ \mu_2 \circ q) \, dk dq \tag{15.49}$$

$$= \int_G \int_G \log(\mu_2^{-1} \circ k \circ \mu_2 \circ q')^\vee [\log(\mu_2^{-1} \circ k \circ \mu_2 \circ q')^\vee]^T \rho_1(k)\rho_2(q') \, dk dq'.$$

Lemma 15.3. *The convolution of PDFs with mean at the identity results (to second order) in a PDF with mean at the identity. Furthermore, if $\rho_1 * \rho_2 = \rho_2 * \rho_1$ and $\rho_i(g) = \rho_i(g^{-1})$, then this result becomes exact.*

Proof.

$$\int_G (\log g)(\rho_1 * \rho_2)(g) \, dg = \int_G \int_G (\log g)\rho_1(h)\rho_2(h^{-1} \circ g) \, dh dg$$

$$= \int_G \int_G \log(h \circ k)\rho_1(h)\rho_2(k) \, dh dk.$$

To second order, all terms in the BCH expansion of $\log(h \circ k)$ are linear in either $\log h$ or $\log k$ (or both), and therefore at least one of the above integrals integrates to zero.

If $\rho_1 * \rho_2 = \rho_2 * \rho_1$ and $\rho_i(g) = \rho_i(g^{-1})$ then it is easy to show that $(\rho_1 * \rho_2)(g) = (\rho_1 * \rho_2)(g^{-1})$, which automatically means that the function $(\rho_1 * \rho_2)(g)$ has mean at the identity.

Theorem 15.4. *If $f_i(g)$ has mean μ_i and covariance Σ_i for $i = 1, 2$, then to second order, the mean and covariance of $(f_1 * f_2)(g)$ are respectively*

$$\mu_{1*2} = \mu_1 \circ \mu_2$$

and

$$\Sigma_{1*2} = A + B + F(A, B) \tag{15.50}$$

where

$$F(A, B) = \frac{1}{4} \sum_{i,j=1}^{d} ad(E_i) B ad(E_j)^T A_{ij}$$

$$+ \frac{1}{12} \left\{ [\sum_{i,j=1}^{d} A'_{ij}] B + B^T [\sum_{i,j=1}^{d} A'_{ij}]^T \right\} \tag{15.51}$$

$$+ \frac{1}{12} \left\{ [\sum_{i,j=1}^{d} B'_{ij}] A + A^T [\sum_{i,j=1}^{d} B'_{ij}]^T \right\}$$

with $A = Ad(\mu_2^{-1}) \Sigma_1 Ad^T(\mu_2^{-1})$, $B = \Sigma_2$, and

$$A'_{ij} = ad(E_i) ad(E_j) A_{ij} \tag{15.52}$$
$$B'_{ij} = ad(E_i) ad(E_j) B_{ij}. \tag{15.53}$$

In the special case when $d = 3$ this is written explicitly as [169, 170]:[4]

$$F(A, B) = \frac{1}{4} C(A, B) + \frac{1}{12} \left[A'' B + (A'' B)^T + B'' A + (B'' A)^T \right],$$

where $C(A, B)$ and A'' are computed as follows:

$$A'' = \begin{pmatrix} A_{11} - tr(A_{11}) \mathbb{I}_3 & \mathbb{O}_3 \\ A_{12} + A_{12}^T - 2tr(A_{12}) \mathbb{I}_3 & A_{11} - tr(A_{11}) \mathbb{I}_3 \end{pmatrix}.$$

B'' is defined in the same way with B replacing A everywhere in the expression. The blocks of C are computed as

$$C_{11} = -D_{11,11}$$
$$C_{12} = -(D_{12,11})^T - D_{11,12} = C_{21}^T$$
$$C_{22} = -D_{22,11} - D_{12,12} - (D_{12,12})^T - D_{11,22}$$

where $D_{ij,kl} = D(A_{ij}, B_{kl})$, and the matrix-valued function $D(\mathcal{A}, \mathcal{B})$ is defined relative to the entries in the 3×3 blocks $\mathcal{A}$ and $\mathcal{B}$ as

$$d_{11} = -\alpha_{33}\beta_{22} + \alpha_{32}\beta_{32} + \alpha_{23}\beta_{23} - \alpha_{22}\beta_{33}$$
$$d_{12} = \alpha_{33}\beta_{21} - \alpha_{32}\beta_{31} - \alpha_{13}\beta_{23} + \alpha_{12}\beta_{33}$$
$$d_{13} = -\alpha_{23}\beta_{21} + \alpha_{22}\beta_{31} + \alpha_{13}\beta_{22} - \alpha_{12}\beta_{32}$$
$$d_{21} = \alpha_{33}\beta_{12} - \alpha_{31}\beta_{32} - \alpha_{23}\beta_{13} + \alpha_{21}\beta_{33}$$

---

[4]Note: Some typographical errors have been corrected in this presentation of that material.

$$d_{22} = -\alpha_{33}\beta_{11} + \alpha_{31}\beta_{31} + \alpha_{13}\beta_{13} - \alpha_{11}\beta_{33}$$
$$d_{23} = \alpha_{23}\beta_{11} - \alpha_{21}\beta_{31} - \alpha_{13}\beta_{12} + \alpha_{11}\beta_{32}$$
$$d_{31} = -\alpha_{32}\beta_{12} + \alpha_{31}\beta_{22} + \alpha_{22}\beta_{13} - \alpha_{21}\beta_{23}$$
$$d_{32} = \alpha_{32}\beta_{11} - \alpha_{31}\beta_{21} - \alpha_{12}\beta_{13} + \alpha_{11}\beta_{23}$$
$$d_{33} = -\alpha_{22}\beta_{11} + \alpha_{21}\beta_{21} + \alpha_{12}\beta_{12} - \alpha_{11}\beta_{22}.$$

Proof. Let $X = \log(\mu_2^{-1}K\mu_2)$ where $k = \exp K$, and let $Y = \log q'$. Using the Baker-Campbell-Hausdorff formula (7.37) to evaluate the log terms in the definition of covariance, and retaining all even terms to second order (since first order terms will integrate to zero), we get

$$\left\{(Z(X,Y))^\vee[(Z(X,Y))^\vee]^T\right\}_{even} = \mathbf{xx}^T + \mathbf{yy}^T + \frac{1}{4}[X,Y]^\vee\left([X,Y]^\vee\right)^T$$
$$+ \frac{1}{12}\mathbf{y}\left([X,[X,Y]]^\vee\right)^T + \frac{1}{12}\mathbf{x}\left([Y,[Y,X]]^\vee\right)^T$$
$$+ \frac{1}{12}[X,[X,Y]]^\vee\mathbf{y}^T + \frac{1}{12}[Y,[Y,X]]^\vee\mathbf{x}^T \quad (15.54)$$

Each of these terms can be expanded using the adjoint concept. For example,

$$[X,Y]^\vee\left([X,Y]^\vee\right)^T = ad(X)\mathbf{yy}^T ad^T(X) \quad \text{and} \quad [X,[X,Y]]^\vee\mathbf{y}^T = ad(X)ad(X)\mathbf{yy}^T.$$
$$(15.55)$$

In our formulation, $X = \mu_2^{-1}K\mu_2$ (where $k = e^K$ and so $\mu_2^{-1} \circ k \circ \mu_2 = \exp(\mu_2^{-1}K\mu_2) = e^X$). Defining the vector $\mathbf{x} = Ad(\mu_2^{-1})\mathbf{k}$, then

$$A = \int_G \mathbf{xx}^T \rho_1(k)\,dk$$
$$= Ad(\mu_2^{-1})\left[\int_G (\log k)^\vee[(\log k)^\vee]^T \rho_1(k)\,dk\right]Ad^T(\mu_2^{-1})$$
$$= Ad(\mu_2^{-1})\Sigma_1 Ad^T(\mu_2^{-1})$$

and since $q' = e^Y$,

$$B = \int_G \mathbf{yy}^T \rho_2(q')\,dq' = \int_G (\log q')^\vee[(\log q')^\vee]^T \rho_2(q')\,dq' = \Sigma_2.$$

The following complicated looking integral (which is nothing more than (15.49) written in exponential coordinates)

$$\Sigma_{1*2} = \int_{q'\in G}\int_{k\in G}\left\{(Z(X,Y))^\vee[(Z(X,Y))^\vee]^T\right\}\rho_1(k)\rho_2(q')\,dkdq'$$
$$= \int_{q'\in G}\int_{k\in G}\left\{(Z(X,Y))^\vee[(Z(X,Y))^\vee]^T\right\}_{even}\rho_1(k)\rho_2(q')\,dkdq'$$

can be simplified. This is because

$$x_i x_j = \mathbf{e}_i^T Ad(\mu_2^{-1})(\log k)^\vee[(\log k)^\vee]^T Ad^T(\mu_2^{-1})\mathbf{e}_j$$

and

$$y_k y_l = \mathbf{e}_k^T(\log q')^\vee[(\log q')^\vee]^T\mathbf{e}_l,$$

and since all terms in $\left\{(Z(X,Y))^{\vee}[(Z(X,Y))^{\vee}]^{T}\right\}_{even}$ can be expressed as weighted sums of such products, it follows that after integration we get

$$\Sigma_{1*2} = A + B + F(A,B). \tag{15.56}$$

For the $SE(3)$ case

$$Ad(g) = \begin{pmatrix} R & \mathbb{O}_3 \\ TR & R \end{pmatrix} \in \mathbb{R}^{6\times6} \quad \text{and} \quad ad(X) = \begin{pmatrix} \Omega & \mathbb{O}_3 \\ V & \Omega \end{pmatrix} \in \mathbb{R}^{6\times6}$$

where $T^{\vee} = \mathbf{t}$, $V^{\vee} = \mathbf{v}$ and $\Omega^{\vee} = \boldsymbol{\omega}$. Then (15.55) becomes:

$$[X,Y]^{\vee}\left([X,Y]^{\vee}\right)^{T} = \begin{pmatrix} \Omega_x & \mathbb{O}_3 \\ V_x & \Omega \end{pmatrix}\begin{pmatrix} \boldsymbol{\omega}_y \\ \mathbf{v}_y \end{pmatrix}[\boldsymbol{\omega}_y^{T}, \mathbf{v}_y^{T}]\begin{pmatrix} -\Omega_x & -V_x \\ \mathbb{O}_3 & -\Omega \end{pmatrix}$$

$$= \begin{pmatrix} \Omega_x & \mathbb{O}_3 \\ V_x & \Omega \end{pmatrix}\begin{pmatrix} \boldsymbol{\omega}_y\boldsymbol{\omega}_y^{T} & \boldsymbol{\omega}_y\mathbf{v}_y^{T} \\ \mathbf{v}_y\boldsymbol{\omega}_y^{T} & \mathbf{v}_y\mathbf{v}_y^{T} \end{pmatrix}\begin{pmatrix} -\Omega_x & -V_x \\ \mathbb{O}_3 & -\Omega \end{pmatrix}$$

and

$$[X,[X,Y]]^{\vee}\mathbf{y}^{T} = \begin{pmatrix} \Omega_x & \mathbb{O}_3 \\ V_x & \Omega \end{pmatrix}\begin{pmatrix} \Omega_x & \mathbb{O}_3 \\ V_x & \Omega \end{pmatrix}\begin{pmatrix} \boldsymbol{\omega}_y \\ \mathbf{v}_y \end{pmatrix}[\boldsymbol{\omega}_y^{T}, \mathbf{v}_y^{T}]$$

$$= \begin{pmatrix} \Omega_x & \mathbb{O}_3 \\ V_x & \Omega \end{pmatrix}\begin{pmatrix} \Omega_x & \mathbb{O}_3 \\ V_x & \Omega \end{pmatrix}\begin{pmatrix} \boldsymbol{\omega}_y\boldsymbol{\omega}_y^{T} & \boldsymbol{\omega}_y\mathbf{v}_y^{T} \\ \mathbf{v}_y\boldsymbol{\omega}_y^{T} & \mathbf{v}_y\mathbf{v}_y^{T} \end{pmatrix}.$$

If we divide the 6×6 symmetric matrices $A = Ad(\mu_2^{-1})\Sigma_1 Ad^{T}(\mu_2^{-1})$ and $B = \Sigma_2$ into 3×3 blocks as

$$A = \begin{pmatrix} A_{11} & A_{12} \\ A_{12}^{T} & A_{22} \end{pmatrix} \quad \text{and} \quad B = \begin{pmatrix} B_{11} & B_{12} \\ B_{12}^{T} & B_{22} \end{pmatrix},$$

then using the specific form of $ad(X)$ and integrating over q' we get

$$\int_{G}[X,[X,Y]]^{\vee}\mathbf{y}^{T}\rho_2(q')\,dq' = \begin{pmatrix} \Omega_x^2 & \mathbb{O}_3 \\ V_x\Omega_x + \Omega_x V_x & \Omega_x^2 \end{pmatrix}B$$

and

$$\int_{G}[X,Y]^{\vee}\left([X,Y]^{\vee}\right)^{T}\rho_2(q')\,dq' =$$

$$\begin{pmatrix} -\Omega_x B_{11}\Omega_x & -(V_x B_{11}\Omega_x)^{T} - \Omega_x B_{12}\Omega_x \\ -V_x B_{11}\Omega_x - (\Omega_x B_{12}\Omega_x)^{T} & -V_x B_{11}V_x - V_x B_{12}\Omega_x - (V_x B_{12}\Omega_x)^{T} - \Omega_x B_{22}\Omega_x \end{pmatrix}.$$

Then integrating over $k \in G$ gives

$$\int_{G}\int_{G}[X,[X,Y]]^{\vee}\mathbf{y}^{T}\rho_1(k)\rho_2(q')\,dk\,dq' = A''B$$

$$\int_{G}\int_{G}[X,Y]^{\vee}\left([X,Y]^{\vee}\right)^{T}\rho_1(k)\rho_2(q')\,dk\,dq' = C(A,B) = \begin{pmatrix} C_{11} & C_{12} \\ C_{12}^{T} & C_{22} \end{pmatrix}.$$

The BCH formula yields several such terms, each of which can be obtained by either transposing those given above or switching the roles of B and A.

The basic ideas in this section have been modified and applied in several different application areas over the years in papers including [10, 34, 37, 92, 112, 126, 134, 180].

15.12   Visual Estimation and Perception

This section reviews mathematical problems in visual perception and the structure of the visual cortex in mammals, and algorithms in computer vision that use the same mathematics to address the problem of occlusions. Subproblems in all of these areas can be formulated as diffusions on $SE(2)$ similar to problems in the pose estimation of stochastically forced rigid bodies.

Tools from geometry and topology have been used by researchers in the field of mathematical psychology for almost as long as these tools have been used by physicists [25, 105, 130]. For other interesting perspectives from the field of psychology on sensory perception see [62].

The work of D.H. Hubel and T.N. Wiesel on the mammalian visual cortex [76, 77, 78] started half a century ago, and was recognized with the 1981 Nobel Prize in Physiology or Medicine (shared with R. W. Sperry). This work has led to mathematical modeling work in the field of mammalian vision that is related to the topic of this book. In particular, the topic of visual perception has been studied from a Lie-group/fiber-bundle perspective for decades [71, 72, 73, 74].

Occlusion is a big problem in vision. Namely, based on a 2D image of the 3D world, how can one make reasonable guesses as to what objects are in front of others (i.e., closer to the eye or camera taking the image) in the 3D world? In order to address this problem, work on stochastic completion fields propagates probabilities on the special Euclidean group $SE(2)$ to make an informed guesses as to where hidden edges of a partially occluded object are likely to intersect. This has been an active area of work spanning a number of years [129, 175, 176, 184].

The mathematics of visual hallucination and its connection to the structure of the visual cortex in the mammalian brain have been studied using differential geometric tools in [21, 136, 137, 41]. Diffusions on $SE(2)$ akin to the stochastic kinematic cart discussed in Chapter 12 have been used in the context of imaging and visual perception using the $SE(2)$ Fourier transform [50, 183] and the semi-discrete motion group Fourier transform [17, 18].

15.13   Summary

In this chapter we have reviewed a number of basic results from the fields of linear systems theory, control, and stochastic processes. Many of these results will be used in subsequent chapters in a variety of contexts. A second goal of this chapter was to compile a number of results from the estimation theory literature in which ideas from noncommutative harmonic analysis are used. In particular, the statistical ideas presented in Chapter 14 have a place in estimation problems on the rotation group.

Of all the topics covered in the original version of this book, perhaps estimation and filtering on $SO(3)$ and $SE(3)$ has seen the largest growth in recent years. This is due to interest in aircraft and spacecraft attitude estimation [11, 16, 38, 46, 56, 66, 98, 99, 100, 120, 121, 143, 144, 152, 153, 159], robot navigation and localization [6, 86, 95, 119, 161], calibration of sensors in underwater vehicles [90, 91], vision and machine learning [4, 5, 39, 43, 64, 68, 116, 135, 164], and sensor networks [48, 162]. See [1, 19, 20, 35, 36, 59, 115, 118, 122] for various numerical and mathematical approaches that are applicable to all of these problems, and [29, 69, 75] for general introductions to geometric mechanics.

An alternative to the approach taken in most of the above references would be to build on the discussion of the unit quaternions in Chapter 5 instead of using rotation matrices. Analogously, the dual quaternions discussed in Chapter 6 could be used to encode full rigid-body motion in 3D without homogeneous transformation matrices. These descriptions have some interesting properties when used in attitude estimation, as described in [57, 58, 65, 182].

References

1. Absil, P.-A., Mahony, R., Sepulchre, R., *Optimization Algorithms on Matrix Manifolds*, Princeton University Press, 2008.
2. Aramanovitch, L.I., "Quaternion Nonlinear Filter for Estimation of Rotating Body Attitude," *Mathematical Methods in the Applied Sciences*, 18(15): 1239–1255, Dec. 1995.
3. Arnol'd, V.I., *Mathematical Methods of Classical Mechanics*, 2nd ed., Springer-Verlag, New York, 2009.
4. Arora, R., Parthasarathy, H., "Optimal Estimation and Detection in Homogeneous Spaces," *IEEE Transactions on Signal Processing*, 58(5): 2623–2635, May 2010.
5. Arora, R., "On Learning Rotations," in *Advances in Neural Information Processing Systems 22*, (NIPS 2009), pp. 55-63, 2009.
6. Baldwin, G., Mahony, R., Trumpf, J., "A Nonlinear Observer for 6 DOF Pose Estimation from Inertial and Bearing Measurements," *Proceedings of the 2009 IEEE International Conference on Robotics and Automation*, (pp. 2237–2242), Kobe, Japan, May, 2009.
7. Bar-Itzhack, I., "Iterative Optimal Orthogonalization of the Strapdown Matrix," *IEEE Transactions on Aerospace and Electronic Systems*, 11(1): 30–37, Jan. 1975.
8. Bar-Itzhack, I.Y., Reiner, J., "Recursive Attitude Determination from Vector Observations: Direction Cosine Matrix Identification," *Journal of Guidance, Control, and Dynamics*, 7(1): 51–56, Jan.–Feb. 1984.
9. Bar-Itzhack, I., Montgomery, P.Y., Gerrick, J.C., "Algorithms for Attitude Determination Using the Global Positioning System," *Journal of Guidance, Control, and Dynamics*, 22(2): 193–201, March–April 1999.
10. Barfoot, T.D., Furgale, P.T., "Associating Uncertainty With Three-Dimensional Poses for Use in Estimation Problems," *IEEE Transactions on Robotics*, 30(3): 679–693, 2014.
11. Barrau, A., Bonnabel, S., "Intrinsic Filtering on $SO(3)$ with Discrete-Time Observations," *Proceedings of the 52nd Annual IEEE Conference on Decision and Control (CDC'13)*, (pp. 3255-3260), December, 2013.
12. Bharadwaj, S., Osipchuk, M., Mease, K.D., Park, F.C., "Geometry and Inverse Optimality in Global Attitude Stabilization," *Journal of Guidance, Control, and Dynamics*, 21(6): 930–939, Nov–Dec. 1998.
13. Björck, A, Bowie, C., "An Iterative Algorithm for Computing the Best Estimate of an Orthogonal Matrix," *SIAM Journal on Numerical Analysis*, 8(2): 358–364, June 1971.
14. Bloch, A.M., Krishnaprasad, P.S., Marsden, J.E., de Alverez, G.S., "Stabilization of Rigid Body Dynamics by Internal and External Torques," *Automatica*, 28(4): 745–756, July 1992.
15. Bloch, A.M., Brockett, R.W., Crouch, P.E., "Double Bracket Equations and Geodesic Flows on Symmetric Spaces," *Communications in Mathematical Physics*, 187(2): 357–373, August 1997.
16. Bonnabel, S., Martin, P., Salaun, E., "Invariant Extended Kalman Filter: Theory and Application to a Velocity-Aided Attitude Estimation Problem," *Proceedings of the 48th IEEE Conference on Decision and Control, held jointly with the 28th Chinese Control Conference. CDC/CCC 2009*, pp. 1297–1304, Dec. 2009.

17. Boscain, U., Duplaix, J., Gauthier, J. P., Rossi, F., "Anthropomorphic Image Reconstruction via Hypoelliptic Diffusion," *SIAM Journal on Control and Optimization*, 50(3): 1309 – 1336, 2012.

18. Boscain, U., Chertovskih, R.A., Gauthier, J.P., Remizov, A.O., "Hypoelliptic Diffusion and Human Vision: A Semidiscrete New Twist," *SIAM Journal on Imaging Sciences*, 7(2): 669 – 695, 2014.

19. Boumal, N., Singer, A., Absil, P.A., "Robust Estimation of Rotations from Relative Measurements by Maximum Likelihood," *Proceedings of the 52nd Annual IEEE Conference on Decision and Control (CDC 2013)*, pp. 1156 – 1161, Firenzi, Italy, Nov. 2013.

20. Bourmaud, G., Mégret, R., Arnaudon, M., Giremus, A., "Continuous-Discrete Extended Kalman Filter on Matrix Lie Groups Using Concentrated Gaussian Distributions," *Journal of Mathematical Imaging and Vision*, 51(1): 209 – 228, 2015.

21. Bressloff, P.C., Cowan, J.D., Golubitsky, M., Thomas, P., Wiener, M., "Geometric Visual Hallucinations, Euclidean Symmetry, and the Functional Architecture of Striate Cortex," *Philosophical Transactions of the Royal Society B: Biological Sciences*, 356(1407): 299 – 330, 2001.

22. Brockett, R.W., "Lie Theory and Control Systems on Spheres," *SIAM Journal on Applied Mathematics*, 25(2): 213 – 225, Sept. 1973.

23. Brockett, R.W., "System Theory on Group Manifolds and Coset Spaces," *SIAM Journal on Control*, 10(2): 265 – 284, May 1972.

24. Brockett, R.W., "Lie Algebras and Lie Groups in Control Theory," in *Geometric Methods in System Theory*, (D.Q. Mayne and R.W. Brockett, eds.), Reidel Publishing Company, Dordrecht-Holland, 1973.

25. Brown, J.F., Voth, A.C., "The Path of Seen Movement as a Function of the Vector Field," *American Journal of Psychology*, 49: 543 – 563, 1937.

26. Brown, R.G., Hwang, P.Y.C., *Introduction to Random Signals and Applied Kalman Filtering* 3rd ed., John Wiley and Sons, New York, 1996.

27. Bryson, A.E., Jr., Ho, Y.-C., *Applied Optimal Control: Optimization, Estimation, and Control*, Hemisphere Publishing Corp., 1975.

28. Bullo, F., Murray, R.M., "Proportional Derivative (PD) control on the Euclidean group," *Proceedings of the 3rd European Control Conference*, pp. 1091-1097, Rome, June 1995. (also see *Caltech CDS Technical Report 95-010*).

29. Bullo, F., Lewis, A.D., *Geometric Control of Mechanical Systems*, Springer, 2004.

30. Burns, W.K., ed., *Optical Fiber Rotation Sensing*, Academic Press, Boston, 1994.

31. Byrnes, C.I., Isidori, A., "On the Attitude Stabilization of Rigid Spacecraft," *Automatica*, 27(1): 87 – 95, 1991.

32. Carta, D.G., Lackowski, D.H., "Estimation of Orthogonal Transformations in Strapdown Inertial Systems," *IEEE Transactions on Automatic Control*, 17(1): 97 – 100, Feb. 1972.

33. Chaudhuri, S., Karandikar, S.S., "Recursive Methods for the Estimation of Rotation Quaternions," *IEEE Transactions on Aerospace and Electronic Systems*, 32(2): 845 – 854, April 1996.

34. Chirikjian, G.S., "Modeling Loop Entropy," *Methods in Enzymology, Part C*, 487: 101 – 130, 2011.

35. Chirikjian, G., Kobilarov, M., "Gaussian Approximation of Non-linear Measurement Models on Lie Groups," *Proceedings of the 53rd Annual IEEE Conference on Decision and Control (CDC'14)*, Dec. 2014.

36. Chiuso, A., Picci, G., Soatto, S., "Wide-Sense Estimation on the Special Orthogonal Group," *Communications in Information and Systems*, 8(3): 185 – 200, 2008.

37. Choi, C., Christensen, H.I., "Robust 3D Visual Tracking Using Particle Filtering on the Special Euclidean Group: A Combined Approach of Keypoint and Edge Features," *The International Journal of Robotics Research*, 31(4): 498 – 519, 2012.

38. Choukroun, D., Bar-Itzhack, I.Y., Osham, Y., "Novel Quaternion Kalman Filter," *IEEE Transactions on Aerospace and Electronic Systems*, 42(1): 174 – 190, 2006.

39. Christensen, H.I., Hager, G.D., "Sensing and Estimation," in *Springer Handbook of Robotics*, pp. 87 – 107, 2008.

40. Chui, C.K., Chen, G., *Kalman Filtering with Real-Time Applications*, 4th ed., Springer, Berlin, 2009.

41. Citti, G., Sarti, A., "A Cortical Based Model of Perceptual Completion in the Roto-Translation Space," *Journal of Mathematical Imaging and Vision*, 24(3): 307 - 326, 2006.

42. Comisel, H., Forster, M., Georgescu, E., Ciobanu, M., Truhlik, V., Vojta, J., "Attitude Estimation of a Near-Earth Satellite Using Magnetometer Data," *Acta Astronautica*, 40(11): 781 - 788, 1997.

43. Cowan, N.J., Chang, D.E., "Geometric Visual Servoing," *IEEE Transactions on Robotics*, 21(6): 1128 - 1138, Dec. 2005.

44. Crassidis, J.L., Lightsey, E.G., Markley, F.L., "Efficient and Optimal Attitude Determination Using Recursive Global Positioning System Signal Operations," *Journal of Guidance, Control, and Dynamics*, 22(2): 193 - 201, March - April 1999.

45. Crassidis, J.L., Markley, F.L., "Predictive Filtering for Attitude Estimation Without Rate Sensors," *Journal of Guidance, Control, and Dynamics*, 20(3): 522 - 527, May-June 1997.

46. Crassidis, J.L., Markley, F.L., Cheng, Y., "Survey of Nonlinear Attitude Estimation Methods," *Journal of Guidance, Control, and Dynamics*, 30(1): 12 - 28, 2007.

47. Crouch, P.E., "Spacecraft Attitude Control and Stabilization: Applications of Geometric Control Theory to Rigid Body Models," *IEEE Transactions on Automatic Control*, 29(4): 321 - 331, April 1984.

48. Cucuringu, M., Lipman, Y., Singer, A., "Sensor Network Localization by Eigenvector Synchronization over the Euclidean Group," *ACM Transactions on Sensor Networks*, 8(3): 19, 2012.

49. Davis, M.H.A., *Linear Estimation and Stochastic Control*, Chapman and Hall, London, 1977.

50. Duits, R., Haije, T.D., Creusen, E., Ghosh, A., "Morphological and Linear Scale Spaces for Fiber Enhancement in DW-MRI," *Journal of Mathematical Imaging and Vision*, 46(3): 326 - 368, 2013.

51. Duncan, T.E., "Stochastic Systems in Riemannian Manifolds," *Journal of Optimization Theory and Applications*, 27(3): 399 - 426, March 1979.

52. Duncan, T.E., "An Estimation Problem in Compact Lie Groups," *Systems & Control Letters*, 10(4): 257 - 263, 1998.

53. Egeland, O., Dalsmo, M., Sordalen, O.J., "Feedback Control of a Nonholonomic Underwater Vehicle with a Constant Desired Configuration," *International Journal of Robotics Research*, 15(1): 24 - 35, Feb. 1996.

54. Elbert, T.F., *Estimation and Control of Systems*, Van Nostrand Reinhold Co., New York, 1984.

55. Enos, M.J., "Optimal Angular Velocity Tracking with Fixed-Endpoint Rigid Body Motions," *SIAM Journal on Control and Optimization*, 32(4): 1186 - 1193, 1994.

56. Farrell, J., Stuelpnagel, J., Wessner, R., Velman, J., Brook, J., "A Least Squares Estimate of Satellite Attitude (Grace Wahba)," *SIAM Review*, 8(3): 384 - 386, 1966.

57. Filipe, N., Tsiotras, P., "Rigid Body Motion Tracking Without Linear and Angular Velocity Feedback Using Dual Quaternions," *Proceedings of the 12th European Control Conference*, pp. 329 - 334, Zurich, Switzerland, July 17 - 19, 2013.

58. Filipe, N., Tsiotras, P., "Simultaneous Position and Attitude Control Without Linear and Angular Velocity Feedback Using Dual Quaternions," *American Control Conference*, pp. 48154820, Washington, DC, June 17 - 19, 2013.

59. Fiori, S., "Auto-Regressive Moving Average Models on Complex-Valued Matrix Lie Groups," *Circuits, Systems, and Signal Processing*, 33(8): 2449 - 2473, 2014.

60. Fujikawa, S.J., Zimbelman, D.F., "Spacecraft Attitude Determination by Kalman Filtering of Global Positioning System Signals," *Journal of Guidance, Control, and Dynamics*, 18(6): 1365 - 1371, 1995.

61. Gardiner, C.W., *Stochastic Methods: A Handbook for the Natural and Social Sciences* 4th ed., Springer-Verlag, Berlin, 2009.

62. Gibson, J.J., *The Senses Considered as Perceptual Systems*, Houghton-Mifflin, Boston, 1966.

63. Gillespie, D.T., *Markov Processes: An Introduction for Physical Scientists*, Academic Press, 1992.

64. Glover, J., Kaelbling, L.P., "Tracking 3-D Rotations with the Quaternion Bingham Filter," http://dspace.mit.edu/handle/1721.1/78248, 2013.

65. Goddard, J.S., "Pose and Motion Estimation from Vision Using Dual Quaternion-Based Extended Kalman Filtering," Ph.D. Dissertation, The University of Tennessee, Knoxville, 1997.

66. Hamel, T., Mahony, R., "Attitude Estimation on $SO(3)$ Based on Direct Inertial Measurements," *Proceedings of the IEEE International Conference on Robotics and Automation (ICRA'06)*, pp. 2170-2175, Orlando, FL., 2006.

67. Han, Y., Park, F.C., "Least Squares Tracking on the Euclidean Group," *IEEE Transactions on Automatic Control*, 46(7): 1127–1132, 2001.

68. Handzel, A.A., Flash, T., "The Geometry of Eye Rotations and Listing's Law," *Advances in Neural Information Processing Systems*, 8: 117–123, 1996.

69. Hatton, R.L., Choset, H., "Geometric Motion Planning: The Local Connection, Stokes Theorem, and the Importance of Coordinate Choice," *International Journal of Robotics Research*, 30(8): 988–1014, 2011.

70. Helmke, U., Moore, J.B., *Optimization and Dynamiical Systems*, Springer-Verlag, New York, 1994.

71. Hoffman, W.C., "The Lie Algebra of Visual Perception," *Journal of Mathematical Psychology*, 3: 65–98, 1966.

72. Hoffman, W.C., "Higher Visual Perception as Prolongation of the Basic Lie Transformation Group," *Mathematical Biosciences*, 6:437–471, 1970.

73. Hoffman, W.C., "Some Reasons Why Algebraic Topology is Important in Neuropsychology: Perceptual and Cognitive Systems as Fibrations," *International Journal of Man-Machine Studies*, 22(6): 613–650, 1985.

74. Hoffman, W.C., "The Visual Cortex is a Contact Bundle," *Applied Mathematics and Computation*, 32(2): 137–167, 1989.

75. Holm, D.D., *Geometric Mechanics, Vol I + II* Imperial College Press, 2008.

76. Hubel, D.H., Wiesel, T.N., "Receptive Fields of Single Neurones in the Cat's Striate Cortex," *Journal of Physiology*, 148(3): 574–591, 1959.

77. Hubel, D.H., Wiesel, T.N., "Receptive Fields, Binocular Interaction and Functional Architecture in the Cat's Visual Cortex," *Journal of Physiology*, 160(1): 106-154, 1962.

78. Hubel, D.H., Wiesel, T.N., "Ferrier Lecture: Functional Architecture of Macaque Monkey Visual Cortex," *Proceedings of the Royal Society of London B: Biological Sciences*, 198(1130): 1–59, 1977.

79. Hughes, P.C., *Spacecraft Attitude Dynamics*, John Wiley and Sons, New York, 1986.

80. Idan, M., "Estimation of Rodrigues Parameters from Vector Observations," *IEEE Transactions on Aerospace and Electronic Systems*, 32(2): 578–586, April 1996.

81. Itô, K., McKean, H.P., *Diffusion Processes and Their Sample Paths*, Springer-Verlag, 1974.

82. Joshi, S.M., Kelkar, A.G., Wen, J.T.-Y., "Robust Attitude Stabilization of Spacecraft Using Nonlinear Quaternion Feedback," *IEEE Transactions on Automatic Control*, 40(10): 1800–1803, Oct. 1995.

83. Junkins, J.L., Turner, J.D., *Optimal Spacecraft Rotational Maneuvers*, Elsevier, New York, 1986.

84. Jurdjevic, V., Sussmann, H.J., "Control Systems on Lie Groups," *Journal of Differential Equations*, 12(2): 313–329, 1972.

85. Kailath, T., Sayed, A.H., Hassibi, B., *Linear Estimation*, Prentice-Hall, Englewood Cliffs, N.J., 2000.

86. Kallem, V., Chang, D.E., Cowan, N.J., "Task-Induced Symmetry and Reduction with Application to Needle Steering," *IEEE Transactions on Automatic Control*, 55(3): 664–673, March 2010.

87. Kalman, R.E., "A New Approach to Linear Filtering and Prediction Problems," *Transactions of the ASME, Journal of Basic Engineering*, 82(1): 35–45, March 1960.

88. Kalman, R. E., Bucy, R.O., "New Results in Linear Filtering and Prediction Theory," *Transactions of the ASME, Journal of Basic Engineering*, 83(1): 95–108, March 1961.

89. Kane, T.R., Likins, P.W., Levinson, D.A., *Spacecraft Dynamics*, McGraw-Hll, New York, 1983.

90. Kinsey, J.C., Whitcomb, L.L., "Adaptive Identification on the Group of Rigid Body Rotations," In *Proceedings of the 2005 IEEE International Conference on Robotics and Automation, ICRA 2005.* (pp. 3256–3261), April, 2005.

91. Kinsey, J.C., Whitcomb, L.L., "Adaptive Identification on the Group of Rigid-Body Rotations and Its Application to Underwater Vehicle Navigation," *IEEE Transactions on Robotics*, 23(1): 124–136, 2007.

92. Knuth, J., Barooah, P., "Error Growth in Position Estimation from Noisy Relative Pose Measurements," *Robotics and Autonomous Systems*, 61(3): 229–244, 2013.

93. Koditschek, D.E., "The Application of Total Energy as a Lyapunov Function for Mechanical Control Systems," in *Dynamics and Control of Multibody Systems* (P.S. Krishnaprasad, J.E. Marsden, J.C. Simo eds.), Vol. 97, pp. 131–157, AMS, 1989.

94. Kolmogorov, A., "Über die analytischen Methoden in der Wahrscheinlichkeitsrechnung," *Mathematische Annalen*, 104: 415–458, 1931.

95. Kwon, J., Choi, M., Park, F.C., Chu, C., "Particle Filtering on the Euclidean Group: Framework and Applications, " *Robotica*, 25(6): 725-737, 2007.

96. Lee, T., Leok, M., McClamroch, N.H., "Optimal Attitude Control of a Rigid Body Using Geometrically Exact Computations on $SO(3)$," *Journal of Dynamical and Control Systems*, 14(4): 465–487, 2008.

97. Lee, T., Leok, M., McClamroch, N.H., "Optimal Control of a Rigid Body Using Geometrically Exact Computations on $SE(3)$," *Proceedings of the 45th IEEE Conference on Decision and Control*, pp. 2710–2715, Dec. 2006.

98. Lee, T., "Stochastic Optimal Motion Planning and Estimation for the Attitude Kinematics on $SO(3)$," *Proceedings of the 52nd Annual IEEE Conference on Decision and Control (CDC 2013)*,(pp. 588–593), December 2013.

99. Lee, T., Leok, M., McClamroch, N.H., "Global Symplectic Uncertainty Propagation on $SO(3)$," *Proceedings of the 47th IEEE Conference on Decision and Control, CDC 2008*, pp. 61–66, December 2008.

100. Lee, T., Leok, M., McClamroch, N.H., "Nonlinear Robust Tracking Control of a Quadrotor UAV on $SE(3)$," *Asian Journal of Control* 15(2): 391–408, 2013.

101. Lefferts, E.J., Markley, F.L., Shuster, M.D., "Kalman Filtering for Spacecraft Attitude Estimation," *Journal of Guidance, Control, and Dynamics*, 5(5): 417–429, 1982.

102. Lemons, D.S., *An Introduction to Stochastic Processes in Physics*, Johns Hopkins University Press, 2002

103. Leonard, N.E., Krishnaprasad, P.S., "Motion Control on Drift-Free, Left-Invariant Systems on Lie Groups," *IEEE Transactions on Automatic Control*, 40(9): 1539–1554, Sept. 1995.

104. Lévy, P., *Processus Stochastiques et Mouvement Brownien*, Gauthier-Villars, Paris, 1948.

105. Lewin, K., *Principles of Topological Psychology*, McGraw-Hill, New York, 1936.

106. Lewis, F.L., *Optimal Control*, Wiley-Interscience Publications, New York, 1986.

107. Liao, M., "Random Motion of a Rigid Body," *Journal of Theoretical Probability*, 10(1): 201–211, 1997.

108. Lin, C.-F., *Modern Navigation, Guidance, and Control Processing*, Prentice Hall, Englewood Cliffs, New Jersey, 1991.

109. Lizarralde, F., Wen, J.T., "Attitude Control without Angular Velocity Measurement: A Passivity Approach, " *IEEE Transactions on Automatic Control*, 41(3): 468–472, March 1996.

110. Lo, J. T.-H., Eshleman, L.R., "Exponential Fourier Densities on $SO(3)$ and Optimal Estimation and Detection for Rotational Processes," *SIAM Journal on Applied Mathematics*, 36(1): 73–82, Feb. 1979.

111. Lo, J.T.-H., "Optimal Estimation for the Satellite Attitude Using Star Tracker Measurements," *Automatica*, 22(4): 477–482, 1986.

112. Long, A.W., Wolfe, K.C., Chirikjian, G.S., "Planar Uncertainty Propagation and a Probabilistic Algorithm for Interception," in *Algorithmic Foundations of Robotics X* (pp. 459-474). Springer Berlin Heidelberg, 2013.

113. Mahony, R., Kumar, V., Corke, P., "Multirotor Aerial Vehicles: Modeling, Estimation, and Control of Quadrotor," *IEEE Robotics and Automation Magazine*, 19(3): 20 – 32, 2012.

114. Mahony, R., Hamel, T., Pflimlin, J.-M., "Nonlinear Complementary Filters on the Special Orthogonal Group," *IEEE Transactions on Automatic Control*, 53(5): 1203 – 1218, June 2008.

115. Maithripala, D.S., Berg, J.M., Dayawansa, W.P., "Almost-Global Tracking of Simple Mechanical Systems on a General Class of Lie Groups" *IEEE Transactions on Automatic Control*, 51(2): 216 – 225, 2006.

116. Malis, E., Hamel, T., Mahony, R., Morin, P., "Dynamic Estimation of Homography Transformations on the Special Linear Group for Visual Servo Control," *IEEE International Conference on Robotics and Automation*, pp. 1498-1503, Kobe, Japan, May 12-17, 2009.

117. Malyshev, V.V., Krasilshikov, M.N., Karlov, V.I., *Optimization of Observation and Control Processes*, AIAA Education Series, J.S. Przemieniecki, series editor. Washington, D.C., 1992.

118. Manton, J.H., "Optimization Algorithms Exploiting Unitary Constraints," *IEEE Transactions on Signal Processing*, 50(3): 635 – 650, 2002.

119. Manyika, J., Durrant-Whyte, H., *Data Fusion and Sensor Management: A Decentralized Information-Theoretic Approach*, Ellis Horwood, New York, 1994.

120. Markley, F.L., "Attitude Filtering on $SO(3)$," *The Journal of the Astronautical Sciences*, 54(3 – 4): 391 – 413, 2006.

121. Markley, F.L., Crassidis, J.L., *Fundamentals of Spacecraft Attitude Determination and Control*, Springer, New York, 2014.

122. Maybeck, P.S., *Stochastic Models, Estimation, and Control*, Vols 1 – 3, Academic Press, 1982.

123. Mayne, D.Q., Brockett, R.W., eds., *Geometric Methods in System Theory*, Reidel Pub. Co., The Netherlands, 1973.

124. McConnell, J., *Rotational Brownian Motion and Dielectric Theory*, Academic Press, New York, 1980.

125. McKean, H.P. Jr., *Stochastic Integrals*, Academic Press, New York, 1969.

126. Miller, L.M., Murphey, T.D., "Trajectory Optimization for Continuous Ergodic Exploration on the Motion Group SE(2)," in 52nd *Annual IEEE Conference on Decision and Control (CDC 2013)*, pp. 4517-4522, Dec. 2013.

127. Mortari, D., "Moon-Sun Attitude Sensor," *Journal of Guidance, Control, and Dynamics*, 34(3): 360 – 364, May – June 1997.

128. Mortari, D., "Euler-q Algorithm for Attitude Determination from Vector Observations," *Journal of Guidance, Control, and Dynamics*, 34(3): 328 – 334, March – April 1998.

129. Mumford, D., "Elastica and Computer Vision," in *Algebraic Geometry and Its Applications*, Chandrajit Bajaj (Ed.), Springer-Verlag: New York, 1994.

130. Orbison, W.D., "Shape as a Function of the Vector-Field," *American Journal of Psychology*, 52:31 – 45, 1939.

131. Oshman, Y., Markley, F.L., "Minimal-Parameter Attitude Matrix Estimation from Vector Observations," *Journal of Guidance, Control, and Dynamics*, 21(4): 595 – 602, July/August 1998.

132. Oshman, Y., Markley, F.L., "Sequential Attitude and Attitude-Rate Estimation Using Integrated-Rate Parameters," *Journal of Guidance, Control, and Dynamics*, 22(3): 385 – 394, May/June 1999.

133. Park, F.C., Kim, J., Kee, C., "Geometric Descent Algorithms for Attitude Determination Using the Global Positioning System," *Journal of Guidance Control and Dynamics*, 23(1): 26 – 33, Jan.–Feb. 2000.

134. Park, W., Liu, Y., Zhou, Y., Moses, M., Chirikjian, G.S., "Kinematic State Estimation and Motion Planning for Stochastic Nonholonomic Systems Using the Exponential Map," *Robotica*, 26(4): 419 – 434, July–Aug. 2008.

105. Pennec, X., "Intrinsic Statistics on Riemannian Manifolds: Basic Tools for Geometric Measurements," *Journal of Mathematical Imaging and Vision*, 25(1): 127 – 154, July 2006.

136. Petitot, J., "The Neurogeometry of Pinwheels as a Sub-Riemannian Contact Structure," *Journal of Physiology - Paris*, 97: 265 - 309, 2003.

137. Petitot, J., *Neurogéométrie de la vision – Modèles mathématiques et physiques des architectures fonctionnelles*, Les Éditions de l'École Polytechnique, 2008.

138. Pisacane, V.L., Moore, R.C., eds., *Fundamentals of Space Systems*, Oxford University Press, New York, 1994.

139. Psiaki, M.L., Martel, F., Pail, P.K., "Three-Axis Attitude Determination via Kalman Filtering of Magnetometer Data," *J. Guidance, Control, and Dynamics*, 13(3): 506 – 514, May–June, 1990.

140. Resnikoff, H.L., "Differential Geometry and Color Perception," *Journal of Mathematical Biology*, 1: 97 – 131, 1974.

141. Reynolds, R.G., "Quaternion Parameterization and a Simple Algorithm for Global Attitude Estimation," *Journal of Guidance, Control, and Dynamics*, 21(4): 669 – 671, July–August 1998.

142. Rugh, W.J., *Linear System Theory*, 2nd ed., Prentice Hall, Englewood Cliffs, N.J., 1996.

143. Sanyal, A.K., Lee, T., Leok, M., McClamroch, N.H., "Global Optimal Attitude Estimation Using Uncertainty Ellipsoids," *Systems and Control Letters*, 57(3): 236 - 245, 2008.

144. Sarlette, A., Sepulchre, R., Leonard, N.E., "Autonomous Rigid Body Attitude Synchronization," *Automatica*, 45(2): 572 – 577, 2009.

145. Savage, P.G., "Strapdown Inertial Navigation Integration Algorithm Design Part 1: Attitude Algorithms," *Journal of Guidance, Control, and Dynamics*, 21(1): 19 – 28, Jan/Feb. 1998.

146. Savage, P.G., "Strapdown Inertial Navigation Integration Algorithm Design Part 2: Velocity and Position Algorithms," *Journal of Guidance, Control, and Dynamics*, 21(2): 208 – 221, March/April 1998.

147. Shrivastava, S.K., Modi, V.J., "Satellite Attitude Dynamics and Control in the Presence of Environmental Torques - A Brief Survey," *Journal of Guidance, Control, and Dynamics*, 6(6): 461 – 471, 1983.

148. Shuster, M.D., Oh, S.D., "Three-Axis Attitude Determination form Vector Observations," *Journal of Guidance, Control, and Dynamics*, 4(1): 70 – 77, 1981.

149. Smith,P., Drummond, T., and Roussopoulos, K., "Computing MAP Trajectories by Representing, Propagating and Combining PDFs Over Groups," *Proceedings of the 9th IEEE International Conference on Computer Vision*, volume II, pages 1275 – 1282, Nice, 2003.

150. Smith, R.B., ed., *Fiber Optic Gyroscopes*, SPIE Milestone Series, Vol. MS 8, SPIE Optical Engineering Press, Washington, D.C., 1989.

151. Smith, R.C., Cheeseman, P., "On the Representation and Estimation of Spatial Uncertainty," *The International Journal of Robotics Research*, 5(4): 56 – 68, 1986.

152. Solo, V., "Attitude Estimation and Brownian Motion on $SO(3)$," *Proceedings of the 49th IEEE Conference on Decision and Control (CDC'10)*, (pp. 4857 – 4862), Dec. 2010.

153. Solo, V., "Stochastic Stability of an Attitude Estimation Algorithm on $SO(3)$," *Proceedings of the 51st IEEE Conference on Decision and Control, CDC'12*, (pp. 1265 – 1270), Dec. 2012.

154. Spindler, K., "Optimal Attitude Control of a Rigid Body," *Applied Mathematics and Optimization*, 34(1): 79 – 90, 1996.

155. Spindler, K., "Optimal Control on Lie Groups with Applications to Attitude Control," *Mathematics of Control, Signals, and Systems*, 11(3): 197 – 219, 1998.

156. Stengel, R.F., *Optimal Control and Estimation*, Dover, New York, 1994.

157. Stratonovich, R.L., *Topics in the Theory of Random Noise: Vols. I and II*, Gordon and Breach Science Publishers, Inc., New York, 1963.

158. Su, S., Lee, C.S.G., "Manipulation and Propagation of Uncertainty and Verification of Applicability of Actions in assembly Tasks," *IEEE Transactions on Systems, Man, and Cybernetics*, 22(6): 1376 – 1389, 1992.

159. Swensen, J.P., Cowan, N.J., "An Almost Global Estimator on $SO(3)$ with Measurement on S^2," *2012 American Control Conference (ACC'12)*, pp. 1780–1786, Montreal, Canada, June 2012.

160. Tekawy, J.A., "Precision Spacecraft Attitude Estimators Using an Optical Payload Pointing System," *Journal of Spacecraft and Rockets*, 35(4): 480–486, July/August 1998.

161. Thrun, S., Burgard, W., Fox, D., *Probabilistic Robotics*, MIT Press, Cambridge, MA, 2005.

162. Tron, R., Vidal, R., Terzis, A., "Distributed Pose Averaging in Camera Networks via Consensus on SE(3)," in *Proceedings of the Second ACM/IEEE International Conference on Distributed Smart Cameras, ICDSC 2008*, pp. 1–10, Stanford, CA, Sept. 7–11, 2008.

163. Tsiotras, P., "Stabilization and Optimality Results for the Attitude Control Problem," *Journal of Guidance, Control, and Dynamics*, 19(4): 772–779, July-August 1996.

164. Tuzel, O., Subbarao, R., Meer, P., "Simultaneous Multiple 3D Motion Estimation via Mode Finding on Lie Groups," *Tenth IEEE International Conference on Computer Vision, ICCV 2005*, (Vol. 1, pp. 18–25), October, 2005.

165. Vadali, S.R., Junkins, J.L., "Optimal Open-Loop and Stable Feedback Control of Rigid Spacecraft Attitude Maneuvers," *Journal of the Astronautical Sciences*, 32(2): 105–122, 1984.

166. Vedder, J.D., "Star Trackers, Star Catalogs, and Attitude Determination: Probabilistic Aspects of System Design," *Journal of Guidance, Control, and Dynamics*, 16(3): 498–504, May-June 1993.

167. Walsh, G.C., Montgomery, R., Sastry, S.S., "Optimal Path Planning on Matrix Lie Groups," *Proc. 33rd Conference on Decision and Control*, pp. 1258–1263, Dec. 1994.

168. Wang, M.C., Uhlenbeck, G.E., "On the Theory of Brownian Motion II," *Rev. Modern Physics.*, 7: 323–342, 1945.

169. Wang, Y., Chirikjian, G.S., "Error Propagation on the Euclidean Group with Applications to Manipulator Kinematics," *IEEE Transactions on Robotics*, 22(4): 591–602, 2006.

170. Wang, Y., Chirikjian, G.S., "Nonparametric Second-Order Theory of Error Propagation on the Euclidean Group," *International Journal of Robotics Research*, 27(1112): 1258-1273, Nov.–Dec. 2008. (Corrigendum: 33(10): 1413, Sept. 2014).

171. Wen, J. T.-Y., Kreutz-Delgado, K., "The Attitude Control Problem," *IEEE Transactions on Automatic Control*, 36(10): 1148–1162, 1991.

172. Wertz, J.R., ed., *Spacecraft Attitude Determination and Control*, Kluwer Academic Publishers, Boston, 1978.

173. Wie, B., Barba, P.M., "Quaternion Feedback for Spacecraft Large Angle Maneuvers," *Journal of Guidance, Control, and Dynamics*, 8(3): 360–365, 1985.

174. Wie, B., Weiss, H., Arapostathis, A., "Quaternion Feedback Regulator for Spacecraft Eigenaxis Rotations," *Journal of Guidance, Control, and Dynamics*, 12(3): 375–380, 1989.

175. Williams, L.R., Jacobs, D.W., "Stochastic Completion Fields: A Neural Model of Illusory Contour Shape and Salience," *Neural Computation*, 9(4): 837-858, 1997.

176. Williams, L.R., Jacobs, D.W., "Local Parallel Computation of Stochastic Completion Fields," *Neural Computation*, 9(4): 859-881, 1997.

177. Willsky, A.S., "Some Estimation Problems on Lie Groups" in *Geometric Methods in System Theory* (D.Q. Mayne and R.W. Brockett, eds.), Reidel Publishing Company, Dordrecht-Holland, 1973.

178. Willsky, A.S., *Dynamical Systems Defined on Groups: Structural Properties and Estimation*, PhD Dissertation, Dept. of Aeronautics and Astronautics, MIT, 1973.

179. Wisniewski, R., *Magnetic Attitude Control for Low Earth Orbit Satellites*, Springer Verlag, New York, 1999.

180. Wolfe, K.C., Chirikjian, G.S., "Signal Detection on Euclidean Groups: Applications to DNA Bends, Robot Localization, and Optical Communication," *IEEE Journal of Selected Topics in Signal Processing* 7(4): 708–719, 2013.

181. Wrigley, W., Hollister, W., Denhard, W., *Gyroscopic Theory, Design, and Instrumentation*, MIT Press, Cambridge, MA 1969.

182. Wu, Y., Hu, X., Hu, D., Li, T., Lian, J., "Strapdown Inertial Navigation System Algorithms Based on Dual Quaternions," *IEEE Transactions on Aerospace and Electronic Systems*, 41(1): 110 - 132, Jan. 2005.

183. Zhang, J., Duits, R., Romeny, B.M., "On Numerical Approaches for Linear Left-invariant Diffusions on $SE(2)$, Their Comparison to Exact Solutions, and Their Applications in Retinal Imaging," arXiv preprint arXiv:1403.3320, 2014.

184. Zweck, J., Williams, L.R., "Euclidean Group Invariant Computation of Stochastic Completion Fields Using Shiftable-Twistable Functions," *Journal of Mathematical Imaging and Vision*, 21: 135 - 154, 2004

Rotational Brownian Motion and Diffusion

In this chapter we examine Brownian motion of a rigid body. Such problems have been studied extensively in the context of molecular motion in liquids (see, e.g., [30, 45, 48, 59], [105] – [107], and [91, 139]). In more abstract settings, non-inertial rotational Brownian motion can be formulated as a diffusion process on the rotation group. This has been studied extensively in the mathematics literature in the context of diffusions on Lie groups and Riemannian manifolds [41, 51, 61, 62, 63, 92, 102, 103, 108, 111, 113, 130, 140, 141, 142]). In the context of molecular motion in liquids, it is usually assumed that viscosity effects and random disturbances are the dominating forces. In contrast, random motion of a rigid body at the macro scale in the absence of a fluid medium is a problem in which inertial terms dominate, and viscous ones are often negligible. Potential applications in this area include the motion of a satellite subjected to random environmental torques. In addition some problems contain viscous and inertial effects which are on the same order of importance. The motion of an underwater robot or high performance airplanes subjected to turbulent currents are examples of this. In these different scenarios the translational and rotational motion often decouples, and so these problems may be solved separately.

In Section 16.1 the classical derivations of translational Brownian motion are reviewed in the context of physical problems. The modern mathematical treatments of Brownian motion (i.e. the Itô and Stratonovich calculus) are not treated here. The interested reader can see, e.g., [23, 60, 64, 70, 77, 101, 122]. In Section 16.2 we review the derivation of the classical Fokker-Planck equation and show how this derivation extends in a straightforward way to processes on manifolds. In Section 16.3 we review rotational Brownian motion from a historical perspective. When inertial effects are important, the diffusion process evolves in both orientation and angular velocity. This is addressed in Section 16.4. Section 16.5 reviews how Fourier expansions on $SO(3)$ are useful in solving PDEs governing the motion of molecules in the non-inertial theory of liquid crystals. Section 16.6 examines a diffusion equation from polymer science that evolves on $SO(3)$.

It is well known that the Abelian Fourier transform is a useful tool for solving problems in translational Brownian motion and heat conduction on the line (see Chapter 2). We review this fact, and show also that the Fourier transform on $SO(3)$ presented in Chapter 9 is a useful tool for rotational Brownian motion when viscous terms dominate. The Fourier techniques presented in Chapter 11 for $\mathbb{R}^3 \times SO(3)$ are relevant in the context of simplifying the form of some PDEs in Section 16.4 where inertial terms cannot be neglected, and rotational displacement and angular velocity are considered to be independent state variables.

16.1 Translational Diffusion

In this section we review the classical physical theories of translational Brownian motion that were developed at the beginning of the twentieth century. In Subsection 16.1.1 we review Albert Einstein's theory of translational Brownian motion. In Subsection 16.1.2 the non-inertial Langevin and Smoluchowski equations are reviewed. Subsection 16.1.3 reviews the classical translational Fokker-Planck equation, which describes how a PDF on position and velocity evolves over time.

16.1.1 Einstein's Theory

In 1828, the botanist Robert Brown observed the random movement of pollen suspended in water (hence the name *Brownian motion*). At the beginning of the twentieth century Albert Einstein, in addition to developing the theory of special relativity, was instrumental in the development of the mathematical theory of Brownian motion. See [38] for the full development of Einstein's treatment of Brownian motion.

In this subsection we review Einstein's approach to the problem. The following subsection reviews the other traditional approaches to translational Brownian motion.

Consider the case of one-dimensional translation. Let $f(x,t)\,dx$ denote the number of particles on the real line in the interval $[x, x+dx]$ at time t. That is, $f(x,t)$ is a number density. Einstein reasoned that the number density of particles per unit x-distance at time t and at time $t + \tau$ for small τ must obey

$$f(x, t + \tau)\,dx = dx \int_{-\infty}^{\infty} f(x + \Delta, t)\phi(\Delta)\,d\Delta \qquad (16.1)$$

where $\phi(\Delta)$ is a symmetric probability density function:

$$\int_{-\infty}^{\infty} \phi(\Delta)\,d\Delta = 1 \quad \text{and} \quad \phi(\Delta) = \phi(-\Delta).$$

Einstein assumed the PDF $\phi(\Delta)$ to be supported only in a small region around the origin. The PDF $\phi(\Delta)$ describes the tendency for a particle to be pushed from side to side.

Since τ is assumed to be small, the first-order Taylor series expansion

$$f(x, t + \tau) \approx f(x, t) + \tau \frac{\partial f}{\partial t}$$

is valid. Likewise, the second-order Taylor series expansion

$$f(x + \Delta, t) \approx f(x, t) + \Delta \frac{\partial f}{\partial x} + \frac{\Delta^2}{2!} \frac{\partial^2 f}{\partial x^2}$$

is taken. Substitution of these expansions in (16.1) yields the heat equation

$$\frac{\partial f}{\partial t} = D \frac{\partial^2 f}{\partial x^2}$$

where

$$D = \frac{1}{\tau} \int_{-\infty}^{\infty} \frac{\Delta^2}{2} \phi(\Delta)\,d\Delta.$$

The Fourier transform for $\mathbb{R}$ is used to generate the solution, as in the case of the heat equation. (See Chapter 2.)

Approaches that came after Einstein's formulation attempted to incorporate physical parameters such as viscosity and inertia. These are described in the next subsection.

16.1.2 The Langevin, Smoluchowski, and Fokker-Planck Equations

In this subsection the relationship between stochastic differential equations governing translational Brownian motion of particles in a viscous fluid and the partial differential equation governing the probability density function of the particles is examined. Our treatment closely follows that of McConnell [91].

The Langevin Equation

In 1908 Langevin [78] considered translational Brownian motion of a spherical particle of mass m and radius R in a fluid with viscosity η with an equation of the form

$$m\frac{d^2x}{dt^2} = -6\pi R\eta\frac{dx}{dt} + X(t). \tag{16.2}$$

Here $X(t)$ is a random force function. It was not until 1930 when Uhlenbeck and Ornstein [131, 135] made assumptions about the behavior of $X(t)$ (in particular that $X(t)\,dt = dW$ where $W(t)$ is a Wiener process) that a PDE governing the behavior of the probability density function, $f(x,t)$, governing the behavior of the particle could be derived directly from the Langevin equation in 16.2.

In particular, if we neglect the inertial term on the left side of (16.2), the corresponding PDE is

$$\frac{\partial f}{\partial t} = \frac{k_B T}{6\pi R\eta}\frac{\partial^2 f}{\partial x^2},$$

where the coefficient results from the assumption that the molecules are Boltzmann distributed in the steady state.

This is the same result as Einstein's, derived by a different means.

The Smoluchowski Equation

Extending this formulation to the three-dimensional case results in the stochastic differential equation

$$m\frac{d^2\mathbf{x}}{dt^2}dt = -6\pi R\eta d\mathbf{x} + \mathbf{F}dt + md\mathbf{W}. \tag{16.3}$$

The Smoluchowski equation [119] is the PDE governing the PDF corresponding to the above SDE in the case when the inertial term on the left is negligible. It is derived using the apparatus of Section 16.2 as

$$\frac{\partial f}{\partial t} = \frac{m^2\sigma^2}{2(6\pi R\eta)^2}\nabla^2 f - \frac{1}{6\pi R\eta}\nabla\cdot(\mathbf{F}f).$$

Assuming the molecules are Boltzmann-distributed in the steady state, it can be shown that $\frac{m^2\sigma^2}{2(6\pi R\eta)^2} = k_B T/(6\pi R\eta)$ and hence in the case when $\mathbf{F} = \mathbf{0}$, the heat equation:

$$\frac{\partial f}{\partial t} = \frac{k_B T}{6\pi R\eta}\nabla^2 f$$

results.

Fourier methods are applicable here for solving this diffusion equation in direct analogy with the one-dimensional case.

16.1.3 The Fokker-Planck Equation

Consider classical Brownian motion as described by the Langevin equation

$$m\ddot{\mathbf{x}}dt = -\xi\dot{\mathbf{x}}dt + d\mathbf{W} \qquad (16.4)$$

where $\xi = 6\pi\nu a$ is the viscosity coefficient. When the inertial effects are negligible in comparison to the viscous ones, we can write this as

$$d\mathbf{x} = Fd\mathbf{W}$$

where $F = \frac{1}{\xi}\mathbb{I}$. If inertial terms are not negligible, and, in addition, a deterministic force is applied to the particle, then the Smoluchowski equation

$$m\ddot{\mathbf{x}}dt = -\xi\dot{\mathbf{x}}dt + \mathbf{h}dt + d\mathbf{W} \qquad (16.5)$$

results. By defining the vector

$$\mathbf{X} = \begin{pmatrix} \mathbf{x} \\ \mathbf{v} \end{pmatrix}$$

where $\mathbf{v} = \dot{\mathbf{x}}$, this equation is then put in the form of (16.8) as

$$\frac{d}{dt}\begin{pmatrix} \mathbf{x} \\ \mathbf{v} \end{pmatrix}dt = \begin{pmatrix} \dot{\mathbf{x}} \\ -\frac{\xi}{m}\mathbf{v} + \frac{1}{m}\mathbf{h} \end{pmatrix}dt + \begin{pmatrix} \mathbb{O}_{3\times3} \\ \mathbb{I}_{3\times3} \end{pmatrix}d\mathbf{W}. \qquad (16.6)$$

A PDE governing the PDF, $F(\mathbf{x},\mathbf{v},t)$, of the particle in phase space may be generated. In the one dimensional case this is

$$\frac{\partial f}{\partial t} = -\frac{\partial}{\partial x}(vf) + \gamma\frac{\partial}{\partial v}(vf) + \frac{\gamma k_B T}{m}\frac{\partial^2 f}{\partial v^2} \qquad (16.7)$$

where $\gamma = 6\pi R\eta$ is the damping coefficient and the random forcing (Weiner process) has strength $2\gamma k_B T/m$ where T is the temperature and k_B is the Boltzmann constant. See [91, 139] for details.

A general methodology for generating partial differential equations for PDFs of random variables described by stochastic differential equations is described in the next section. Equation (16.7) can be derived from (16.6) using this technique. This methodology was first employed by Fokker [47] and Planck [109].

16.2 The General Fokker-Planck Equation

In this section we review Fokker-Planck equations, which are partial differential equations that describe the evolution of probability density functions associated with random variables governed by certain kinds of stochastic differential equations.

That is, we begin with stochastic differential equations of the form

$$d\mathbf{X}(t) = \mathbf{h}(\mathbf{X}(t),t)\,dt + H(\mathbf{X}(t),t)\,d\mathbf{W}(t) \qquad (16.8)$$

result where $\mathbf{X} \in \mathbb{R}^d$ and $\mathbf{W} \in \mathbb{R}^m$. Here the deterministic dynamical system $\dot{\mathbf{X}} = \mathbf{h}(\mathbf{X},t)$ is perturbed at every value of time by noise, or random forcing described by $\mathbf{W}(t)$ where $\mathbf{W}(t)$ is a *Wiener process* and $H \in \mathbb{R}^{m\times m}$. The Fokker-Planck equation is a partial differential equation that governs the time evolution of a probability density function in

X. That is, in place of an infinite ensemble of trajectories $\{\mathbf{X}(t)\}$, the essential properties of the stochastic system can be captured by a time-evolving probability density function, $p(\mathbf{x}, t)$.

In the first subsection we examine the classical Fokker-Planck equation, and in the second subsection we show how the Fokker-Planck equation is derived for processes on certain manifolds. Our presentation is in a style appropriate for physical/engineering scientists, and follows [25, 27, 49, 84, 91, 112, 114, 120, 133].

16.2.1 Derivation of the Classical Fokker-Planck Equation

The goal of this section is to review the derivation of the Fokker-Planck equation, which governs the evolution of the PDF $p(\mathbf{x}, t)$ for a given stationary Markov process, e.g., for a system of the form in (15.21) which is forced by a Wiener process. The derivation reviewed here has a similar flavor to the arguments used in classical variational calculus in the sense that functionals of $p(\mathbf{x})$ and its derivatives, $m(f(\mathbf{x}), f'(\mathbf{x}), ..., \mathbf{x})$, are projected against an "arbitrary" function $\epsilon(\mathbf{x})$, and hence integrals of the form

$$\int_{\mathbb{R}^d} m_i(f(\mathbf{x}), f'(\mathbf{x}), ..., \mathbf{x})\epsilon(\mathbf{x})\, d\mathbf{x} = 0$$

are localized to

$$m(f(\mathbf{x}), f'(\mathbf{x}), ..., \mathbf{x}) = 0$$

using the "arbitrariness" of the function $\epsilon(\mathbf{x})$. The details of this procedure are now examined.

To begin, let $\mathbf{X} = \mathbf{X}(t) \in \mathbb{R}^d$ and $\boldsymbol{\Xi} = \mathbf{X}(t + dt)$ where dt is an infinitesimal time increment. These random variables respectively move in copies of Euclidean space parameterized by $\mathbf{x}$ and $\boldsymbol{\xi}$. Using the properties of $p(\boldsymbol{\xi}, t + dt|\mathbf{x}, t)$ from Chapter 15, and in particular (15.25) and (15.26), we observe that

$$\langle \Xi_i - X_i \rangle \doteq \int_{\mathbb{R}^d} (\xi_i - x_i)p(\boldsymbol{\xi}, t + dt|\mathbf{x}, t)\, d\boldsymbol{\xi} = h_i(\mathbf{x}, t)\, dt \qquad (16.9)$$

and

$$\langle (\Xi_i - X_i)(\Xi_j - X_j) \rangle \doteq \int_{\mathbb{R}^d} (\xi_i - x_i)(\xi_j - x_j)p(\boldsymbol{\xi}, t + dt|\mathbf{x}, t)\, d\boldsymbol{\xi} = \sum_{k=1}^{m} \sigma_k^2 H_{ik}(\mathbf{x}, t)H_{kj}^T(\mathbf{x}, t)\, dt,$$

$$(16.10)$$

where σ_k^2 for $k = 1, ..., m$ are the strengths of the components of the vector Wiener process $\mathbf{W}(t)$.

Now let $\mathbf{Y} = \mathbf{X}(t')$ for some fixed $t' < t$ and let the space of all such starting points be parameterized by the Cartesian coordinates $\mathbf{y}$. We seek a governing equation for the transition probability $p(\mathbf{x}, t|\mathbf{y}, t')$.[1] Using the Chapman-Kolmogorov equation, (15.19), together with the definition of partial derivative, we write

$$\frac{\partial p(\mathbf{x}, t|\mathbf{y}, t')}{\partial t} \cdot dt = p(\mathbf{x}, t + dt|\mathbf{y}, t') - p(\mathbf{x}, t|\mathbf{y}, t')$$

$$= \int_{\mathbb{R}^d} p(\mathbf{x}, t + dt|\mathbf{z}, t)p(\mathbf{z}, t|\mathbf{y}, t')\, d\mathbf{z} - p(\mathbf{x}, t|\mathbf{y}, t').$$

---

[1]In the original version of this book, we used the nonstandard notation $\Pi(\mathbf{y}|\mathbf{x}, t) \doteq p(\mathbf{x}, t|\mathbf{y}, 0)$.

Let $\epsilon(\mathbf{x})$ be an arbitrary compactly supported function for which $\partial\epsilon/\partial x_i$ and $\partial^2\epsilon_i/\partial x_j\partial x_k$ are continuous for all $i,j,k = 1,...,d$. Then the projection of $dt\cdot\partial p/\partial t$ against $\epsilon(\mathbf{x})$ can be expanded as

$$dt\cdot\int_{\mathbb{R}^d}\frac{\partial p(\mathbf{x},t|\mathbf{y},t')}{\partial t}\epsilon(\mathbf{x})\,d\mathbf{x} = \int_{\mathbb{R}^d}\epsilon(\mathbf{x})\,d\mathbf{x}\int_{\mathbb{R}^d}p(\mathbf{x},t+dt|\mathbf{z},t)p(\mathbf{z},t|\mathbf{y},t')\,d\mathbf{z}-\int_{\mathbb{R}^d}p(\mathbf{x},t|\mathbf{y},t')\epsilon(\mathbf{x})\,d\mathbf{x}.$$

Inverting the order of integration in the first term on the right side, and changing the variables of integration in this term as $(\mathbf{x},\mathbf{z})\to(\boldsymbol{\xi},\mathbf{x})$,

$$dt\cdot\int_{\mathbb{R}^d}\frac{\partial p(\mathbf{x},t|\mathbf{y},t')}{\partial t}\epsilon(\mathbf{x})\,d\mathbf{x} = \int_{\mathbb{R}^d}p(\mathbf{x},t|\mathbf{y},t')\left[\int_{\mathbb{R}^d}p(\boldsymbol{\xi},t+dt|\mathbf{x},t)\epsilon(\boldsymbol{\xi})\,d\boldsymbol{\xi} - \epsilon(\mathbf{x})\right]d\mathbf{x}.$$

Expanding the function $\epsilon(\boldsymbol{\xi})$ in its Taylor series about $\mathbf{x}$,

$$\epsilon(\boldsymbol{\xi}) = \epsilon(\mathbf{x}) + \sum_{i=1}^{d}(\xi_i - x_i)\frac{\partial\epsilon}{\partial x_i} + \frac{1}{2}\sum_{i,j=1}^{d}(\xi_i - x_i)(\xi_j - x_j)\frac{\partial^2\epsilon}{\partial x_i\partial x_j} + \cdots$$

and substituting this series into the previous equation results in

$$dt\cdot\int_{\mathbb{R}^d}\frac{\partial p}{\partial t}\epsilon\,d\mathbf{x} = dt\cdot\int_{\mathbb{R}^d}\left[\sum_{i=1}^{d}\frac{\partial\epsilon}{\partial x_i}h_i + \frac{1}{2}\sum_{i,j=1}^{d}\frac{\partial^2\epsilon}{\partial x_i\partial x_j}\sum_{k=1}^{m}\sigma_k^2 H_{ik}H_{kj}^T\right]p\,d\mathbf{x}$$

when (16.9) and (16.10) are observed. Then dt can be removed from both sides.

The final step is to integrate the two terms on the right-hand-side of the above equation by parts and bring everything to the same side of the equation as

$$\int_{\mathbb{R}^d}\left\{\frac{\partial p(\mathbf{x},t|\mathbf{y},t')}{\partial t} + \sum_{i=1}^{d}\frac{\partial}{\partial x_i}(h_i(\mathbf{x},t)p(\mathbf{x},t|\mathbf{y},t'))\right.$$

$$\left. -\frac{1}{2}\sum_{k=1}^{m}\sigma_k^2\sum_{i,j=1}^{d}\frac{\partial^2}{\partial x_i\partial x_j}(H_{ik}(\mathbf{x},t)H_{kj}^T(\mathbf{x},t)p(\mathbf{x},t|\mathbf{y},t'))\right\}\epsilon(\mathbf{x})\,d\mathbf{x} = 0 \qquad (16.11)$$

Then the standard localization argument gives that the term in brackets must be zero. That is,

$$\boxed{\frac{\partial p}{\partial t} + \sum_{i=1}^{d}\frac{\partial}{\partial x_i}(h_i p) - \frac{1}{2}\sum_{k=1}^{m}\sigma_k^2\sum_{i,j=1}^{d}\frac{\partial^2}{\partial x_i\partial x_j}(H_{ik}H_{kj}^T p) = 0\,.} \qquad (16.12)$$

Here $p = p(\mathbf{x},t|\mathbf{y},t')$ for any t'. Note also that regardless of the initial conditions, due to the linearity of the Fokker-Planck equation we have that

$$p(\mathbf{x},t) = \int_{\mathbb{R}^d}p(\mathbf{x},t|\mathbf{y},t')p(\mathbf{y},t')\,d\mathbf{y}$$

also satisfies it.[2]

---

[2]Throughout this book $p(\mathbf{x},t)$ is often written as $f(\mathbf{x},t)$.

16.2.2 The Fokker-Planck Equation on Manifolds

The derivation of the Fokker-Planck equation governing the time-evolution of PDFs on a Riemannian manifold proceeds in an analogous way to the derivation of the previous section. This subject has been studied extensively in the mathematics literature. See for example, the works of Yosida [140], Itô [61, 62, 63], McKean [92], and Emery [40]. Aspects of diffusion processes on manifolds and Lie groups remain of interest today (see e.g., [2, 39, 40, 103, 111, 130]), as do generalizations where the stochastic forcing is a Lévy process [80][3] rather than a Weiner/Brownian-motion process [3, 82].

Unlike many derivations in the modern mathematics literature, our derivation of the Fokker-Planck equation for the case of a Riemannian manifold is strictly coordinate-dependent.

We start by taking (16.8) to describe a stochastic process in an n-dimensional Riemannian manifold, M, with metric tensor $g_{ij}(x_1, ..., x_n)$. Here x_i are coordinates which do not necessarily form a vector in $\mathbb{R}^n$ (e.g., spherical coordinates). The array of coordinates is written as $x = (x_1, ..., x_n)$. In this context $X_i(t)$ denotes a stochastic process corresponding to the coordinate x_i. The definition of the Wiener process and everything proceeds as before. It appears that the only difference now is that the integration-by-parts required to isolate the function $\epsilon(x)$ from the rest of the integrand in the final steps in the derivation in the previous section will be weighted by $\sqrt{\det G}$ due to the fact that the volume element in M is $dV(x) = \sqrt{\det G}\, dx$ where $dx = dx_1 \cdots dx_n$. In particular,

$$\int_M \frac{\partial \epsilon}{\partial x_i} h_i(x,t) p(x,t|y,t')\, dV(x) = \int_{\mathbb{R}^d} \frac{\partial \epsilon}{\partial x_i} h_i(x,t) p(x,t|y,t') \sqrt{\det G(x)}\, dx =$$

$$-\int_{\mathbb{R}^d} \epsilon(x) \frac{\partial}{\partial x_i} \left(h_i(x,t) p(x,t|y,t') \sqrt{\det G(x)} \right) dx.$$

The integration over all $\mathbb{R}^d$ is valid since $\epsilon(x)$ and its derivatives vanish outside of a compact subset of $\mathbb{R}^d$ which can be assumed to be contained in the range of a single coordinate chart of M. Then integrating by parts twice yields

$$\int_M \frac{\partial^2 \epsilon}{\partial x_i \partial x_j} h_i(x,t) p(x,t|y,t')\, dV(x) =$$

$$\int_{\mathbb{R}^d} \epsilon(x) \frac{\partial^2}{\partial x_i \partial x_j} \left(h_i(x,t) p(x,t|y,t') \sqrt{\det G(x)} \right) dx.$$

Using the standard localization argument as before, extracting the functional which multiplies $\epsilon(x)$ and setting it equal to zero yields, after division by $\sqrt{\det G}$ and setting $p(x,t) = p(x,t|x_0,0)$,

$$\frac{\partial p(x,t)}{\partial t} + \frac{1}{\sqrt{\det G}} \sum_{i=1}^d \frac{\partial}{\partial x_i} \left(\sqrt{\det G}\, h_i(x,t) p(x,t) \right) = \qquad (16.13)$$

$$\frac{1}{2} \sum_{k=1}^m \sigma_k^2 \sum_{i,j=1}^d \frac{1}{\sqrt{\det G}} \frac{\partial^2}{\partial x_i \partial x_j} \left(\sqrt{\det G}\, H_{ik} H_{kj}^T p(x,t) \right).$$

---

[3]Roughly speaking, this is a Weiner process with additional jumps that occur at random intervals of time and have random magnitude.

The second term on the left-hand side of the equation above can be written as $\mathrm{div}(fw)$, and this leads us to inquire about the differential-geometric interpretation of the right hand side. In many cases, of interest, the matrices H_{ik} will be the inverse of the Jacobian matrix, and hence in these cases $\sum_k H_{ik}H_{kj}^T = \sum_k ((J_{ik})^{-1}((J_{kj})^{-1})^T) = (g_{ij}(x))^{-1} = (g^{ij}(x))$. Therefore, when $\sigma_i = 2K$ for all i, we find in such cases that the Fokker-Planck equation on M becomes

$$\frac{\partial p(x,t)}{\partial t} + \frac{1}{\sqrt{\det G}}\sum_{i=1}^{d}\frac{\partial}{\partial x_i}\left(\sqrt{\det G}h_i(x,t)p(x,t)\right) = \qquad (16.14)$$

$$\frac{K}{\sqrt{\det G}}\sum_{i,j=1}^{d}\frac{\partial^2}{\partial x_i \partial x_j}\left(\sqrt{\det G}(g^{ij}(x))p(x,t)\right).$$

This equation is similar to, though not exactly the same as, the heat equation on M written in coordinate-dependent form. In fact, a straightforward calculation explained by Brockett [20] equates the Fokker-Planck and heat equation

$$\frac{\partial p(x,t)}{\partial t} = \frac{K}{\sqrt{\det G}}\sum_{i,j=1}^{d}\frac{\partial}{\partial x_i}\left(\sqrt{\det G}(g^{ij}(x))\frac{\partial}{\partial x_j}p(x,t)\right) = K\nabla^2 p. \qquad (16.15)$$

in the special case when

$$h_i(x) = K\sum_{j=1}^{n}\left(\frac{1}{\sqrt{\det G}}(g^{ij}(x))\frac{\partial\sqrt{\det G}}{\partial x_j} + \frac{\partial g^{ij}}{\partial x_j}\right).$$

This is clear by expanding the term on the right-hand-side of (16.14) as

$$\frac{K}{\sqrt{\det G}}\sum_{i,j=1}^{d}\frac{\partial}{\partial x_i}\left[\frac{\partial}{\partial x_j}\left(\sqrt{\det G}(g^{ij}(x))p(x,t)\right)\right],$$

and observing the chain rule:

$$\frac{\partial}{\partial x_j}\left(\sqrt{\det G}(g^{ij}(x))p(x,t)\right) = \left((g^{ij}(x))\frac{\partial\sqrt{\det G}}{\partial x_j} + \sqrt{\det G}\frac{\partial g^{ij}}{\partial x_j}\right)p + (g^{ij}(x))\sqrt{\det G}\frac{\partial p}{\partial x_j}.$$

If $1/K\sqrt{\det G}h_i$ is equated to the term in parentheses above, then the heat equation results. This means that when using the Itô interpretation of an SDE evolving in a curved space, a "correction term" is required to obtain the anticipated behavior.

16.3 A Historical Introduction to Rotational Brownian Motion

In this section stochastic differential equations and associate partial differential equations describing rotational Brownian motion are developed. First, in Subsection 16.3.1, theories which do not take into account inertial effects are considered. Then in Subsection 16.3.3 the equations governing the case when inertia cannot be ignored and no viscous forces are present is examined. In Subsection 16.4, the most general case is examined, together with simplifications based on axial and spherical symmetry.

16.3.1 Theories of Debye and Perrin: Exclusion of Inertial Terms

In 1913, P. Debye developed a model describing the motion of polar molecules in a solution subjected to an electric field. See [30] for a complete treatment of Debye's work. Our treatment follows [91].

Debye assumed the molecules to be spheres of radius R, the dipole moment to be μ, and the viscosity of the solution to be η. The deterministic equation of motion governing a single particle is then found by an angular balance which reduces in this case to:

$$I\ddot{\theta} = -\mu F \sin\theta - 8\pi R^3 \eta \dot{\theta}$$

where I is the scalar moment of inertia of the spherical molecule, F is the magnitude of the applied electric field, and θ is the angle between the axis of the dipole and the field.

In Debye's approximation, the inertial terms are assumed to be negligible, and so the equation of motion

$$\dot{\theta} = -\frac{\mu F \sin\theta}{8\pi R^3 \eta} \qquad (16.16)$$

results.

He then proceeded to derive a partial differential equation describing the concentration of these polar molecules at different orientations. We use $f(\theta, t)$ to denote the orientational density, or concentration, of molecules at time t with axis making angle θ with the direction of the applied electric field. Debye reasoned that the flux of $f(\theta, t)$ consisted of two components - one of which is due to the electric field of the form $f(\theta, t)\dot{\theta}$, and the other of which is of the form $-K \partial f / \partial \theta$ due to Fick's law. Combining these, the flux in orientational density is

$$J(\theta, t) = -\frac{\mu F f \sin\theta}{8\pi R^3 \eta} - K\frac{\partial f}{\partial \theta}$$

where K is a diffusion constant that will be determined shortly.

The number of polar molecules changing orientation from θ to $\theta + d\theta$ in time interval dt is $(\partial f / \partial t)\, d\theta dt$. Equating this to the net orientational flux into the sector $(\theta, \theta + d\theta)$ during the interval dt, which is given by $(J(\theta, t) - J(\theta + d\theta, t))\, dt = -(\partial J / \partial \theta)\, dt$, we see that

$$\frac{\partial f}{\partial t} = \frac{\partial}{\partial \theta}\left(\frac{\mu F \sin\theta}{8\pi R^3 \eta} + K\frac{\partial f}{\partial \theta}\right). \qquad (16.17)$$

The diffusion constant K is determined, in analogy with the case of translational diffusion, by observing that in the special case when the electric field is static $F(t) = F_0$, and the orientational distribution f is governed by the Maxwell-Boltzmann distribution $f = C \exp(-E/k_B T)$ for some constant C where $E = T + V = \frac{1}{2}I\dot{\theta}^2 - \mu F \cos\theta$ is the total energy of one molecule, k_B is the Boltzmann constant, and T is temperature. Substitution of this steady state solution back into (16.17) dictates that

$$K = \frac{k_B T}{8\pi R^3 \eta}.$$

See [91] for details.

Note that the same result can be obtained as the Fokker-Planck equation corresponding to the stochastic differential equation

$$8\pi R^2 \eta d\theta = -\frac{\mu F \sin\theta}{8\pi R^3 \eta} dt + I dW.$$

When no electric field is applied, $F(t) = 0$, the result is the heat equation on the circle or, equivalently, $SO(2)$. The solution of the heat equation on the circle is calculated using the Fourier series on $SO(2)$ subject to the normality condition

$$\int_0^{2\pi} f(\theta, t) \, d\theta = 1.$$

Clearly, this is a low degree-of-freedom model to describe a phenomenon which is, by nature, higher dimensional. In 1929, Debye reported an extension of this one-dof model to allow for random pointing (though not rotation about the axis) of the polar molecule. Hence, the model becomes one of symmetric diffusion on $SO(3)/SO(2) \cong \mathbb{S}^2$. Since (α, β) specify points on the unit sphere, it is natural to assume the heat equation on the sphere would arise. However, because of the inherent axial symmetry imposed on the problem by the direction of the electric field, $f(\alpha, \beta, t) = f(\beta, t)$ for this problem as well. Hence, while the volume element for the sphere is $d\mathbf{u}(\alpha, \beta) = \sin \beta \, d\beta \, d\alpha$, we can consider the hoop formed by all volume elements for each value of β. This has volume $\int_0^{2\pi} \sin \beta \, d\beta \, d\alpha = 2\pi \sin \beta \, d\beta$. Again performing a balance on the number of particles pointing with orientations in the hoop $(\beta, \beta + d\beta)$, and dividing both sides by $\sin \beta$ and canceling common factors, the result is

$$8\pi R^2 \eta \frac{\partial f}{\partial t} = \frac{1}{\sin \beta} \frac{\partial}{\partial \beta} \left[\sin \beta \left(k_B T \frac{\partial f}{\partial \beta} + \mu F f \sin \beta \right) \right].$$

The normalization condition in this case is

$$2\pi \int_0^\pi f(\beta, t) \sin \beta \, d\beta = 1.$$

Implicitly, Debye considered all dipole molecules with polar axis pointing in the same cone as being in the same state. Hence, without explicitly stating this, essentially he defined an equivalence relation on $SO(3)$ for all orientations with the same second Euler angle, i.e., the model has as its variable elements of the double coset $SO(2)\backslash SO(3)/SO(2)$. In the case when no electric field is applied, the axial symmetry in the problem disappears, and the heat equation on $\mathbb{S}^2$ results. When the molecule itself lacks axial symmetry, then a diffusion equation on $SO(3)$ must be considered.

Perrin [105, 105, 107] extended the same kind of model to include full rotational freedom of the molecules, including elliptic shape. Here the anisotropic heat equation on $SO(3)$ results, and we seek $f(A(\alpha, \beta, \gamma), t)$ such that

$$\frac{\partial f}{\partial t} = K \nabla_A^2 f.$$

Solutions to this $SO(3)$-heat equation are given in Chapter 14. Note that Perrin's formulation also ignored the inertial terms in the equation of motion. This is the most elementary case of a non-inertial theory of rotational Brownian motion. A general treatment of non-inertial rotational Brownian motion is given in the next subsection.

16.3.2 Non-Inertial Theory of Diffusion on $SO(3)$

Consider the Euler Langevin equation (i.e., the rotational analog of the Langevin equation):

$$\omega_R dt = \frac{1}{c}d\mathbf{W}.$$

This holds when viscous terms far outweigh inertial ones (such as in the case of a polar molecule) and the object is spherical, and there is no external field applied.

If our goal is to find out how the process develops on the rotation group, then it makes sense to substitute this equation with the system

$$(R^T \dot{R})^\vee dt = \frac{1}{c}d\mathbf{W}.$$

In the case when ZXZ Euler angles α, β, γ are used, $(R^T \dot{R})^\vee = J_R(\alpha, \beta, \gamma)(\dot{\alpha}, \dot{\beta}, \dot{\gamma})^T$ and so we write the system of equations

$$\begin{pmatrix} d\alpha \\ d\beta \\ d\gamma \end{pmatrix} = \frac{1}{c}\begin{pmatrix} \sin\beta\sin\gamma & \cos\gamma & 0 \\ \sin\beta\cos\gamma & -\sin\gamma & 0 \\ \cos\beta & 0 & 1 \end{pmatrix}^{-1}\begin{pmatrix} dW_1 \\ dW_2 \\ dW_3 \end{pmatrix}.$$

This is a stochastic differential equation that evolves on $SO(3)$.

The Fokker-Planck equation for the PDF $f(A(\alpha, \beta, \gamma), t)$ corresponding to the above SDE is

$$\frac{\partial f}{\partial t} = \frac{1}{2}\frac{\sigma^2}{c^2}\left[\nabla_A^2 f + \cot\beta\frac{\partial f}{\partial\beta} - f\right] \tag{16.18}$$

where σ^2 is the variance corresponding to $d\mathbf{W}$.

The terms $\cot\beta\frac{\partial f}{\partial\beta} - f$ in the above equation result from the Itô interpretation of the original SDE. In physical problems, an alternative interpretation of the SDE that yields the expected Fokker-Planck equation (i.e., the heat equation) without having to incorporate any correction terms exists. This is the Stratonovich interpretation. For a detailed discussion of the relationship between these two interpretations, as well as derivations of both forms of Fokker-Planck equations (both coordinate-dependent and coordinate-free) see [23].

16.3.3 Diffusion of Angular Velocity

In the case when viscous effects are not dominant, the PDF $f(\omega, t)$ for the evolution of angular velocity distribution corresponding to the equation

$$d\omega = -B\omega\, dt - I^{-1}(\omega \times I\omega)\, dt + d\mathbf{W}$$

as given by Steele [121], Hubbard [55, 56, 57, 58, 59], and Gordon [50] is

$$\frac{\partial f}{\partial t} = -\nabla_\omega \cdot [B\omega f + I^{-1}(\omega \times I\omega)f] + \sum_{i=1}^{3}\sigma_i^2\frac{\partial^2 f}{\partial\omega_i^2}. \tag{16.19}$$

Here $\{\sigma_i\}$ are the strengths of the random forcing in each coordinate direction and I and B are assumed to be diagonal matrices.

If we assume that initially the angular velocity of all the bodies is ω_0, then $f(\omega, 0) = \delta(\omega - \omega_0)$. Likewise, it can be assumed that as $t \to \infty$ the distribution of angular velocities will approach the Maxwell-Boltzmann distribution [55, 121]:

$$\lim_{t\to\infty} f(\omega, t) = \frac{(I_1 I_2 I_3)^{\frac{1}{2}}}{(2\pi k_B T)^{\frac{3}{2}}}\exp[-\omega^T I\omega/2k_B T].$$

In order for this to hold, the strengths σ_i^2 must be of the form

$$\sigma_i^2 = B_i k_B T / I_i.$$

In general (16.19) does not lend itself to simple closed-form analytical solutions. However, for the spherically symmetric case when $I_1 = I_2 = I_3 = I$, this reduces to

$$\frac{\partial f}{\partial t} = -\nabla_{\boldsymbol{\omega}} \cdot [B\boldsymbol{\omega} f] + \sum_{i=1}^{3} \sigma_i^2 \frac{\partial^2 f}{\partial \omega_i^2}. \tag{16.20}$$

Treating $\boldsymbol{\omega}$ as a vector in $\mathbb{R}^3$, application of the Fourier transform for $\mathbb{R}^3$ reduces (16.20) to the following first-order PDE in Fourier space [55]:

$$\left[\frac{\partial}{\partial t} + \sum_{j=1}^{3} \left(B_j p_j \frac{\partial}{\partial p_j} + \sigma_j^2 p_j^2 \right) \right] \hat{f} = 0$$

where

$$\hat{f}(\mathbf{p}, t) = \frac{1}{8\pi^3} \int_{\mathbb{R}^3} f(\boldsymbol{\omega}, t) \, e^{-i\mathbf{p} \cdot \boldsymbol{\omega}} \, d\boldsymbol{\omega}.$$

The above linear first-order PDE can be solved as [121, 55]:

$$\hat{f}(\mathbf{p}, t) = \prod_{j=1}^{3} \exp - \left[\left(\frac{k_B T}{2I} \right) p_j^2 (1 - e^{-2B_j t}) + i(\boldsymbol{\omega}_0 \cdot \mathbf{e}_j) p_j e^{-B_j t} \right].$$

Fourier inversion yields

$$f(\boldsymbol{\omega}, t) = \prod_{j=1}^{3} \left(\frac{I}{2\pi k_B T (1 - e^{-2B_j t})} \right)^{\frac{1}{2}} \exp\left(-\frac{I(\omega_j - (\boldsymbol{\omega}_0 \cdot \mathbf{e}_j) \, e^{-B_j t})^2}{2k_B T (1 - e^{-2B_j t})} \right).$$

16.4 Brownian Motion of a Rigid Body: Diffusion Equations on $SO(3) \times \mathbb{R}^3$

Consider a rigid body immersed in a viscous fluid and subjected to random moments superimposed on a steady state moment due to an external force field. This model could represent a polar molecule in solution as studied in [30, 91, 139], an underwater vehicle performing station keeping maneuvers while being subjected to random currents, or, in the case of zero viscosity, a satellite in orbit subjected to random bombardments of micrometeorites.

The stochastic differential equation of motion describing the most general situation is

$$I d\boldsymbol{\omega} + \boldsymbol{\omega} \times I\boldsymbol{\omega} \, dt = -C\boldsymbol{\omega} \, dt + \mathbf{N} dt + d\mathbf{W}. \tag{16.21}$$

$I = \mathrm{diag}[I_1, I_2, I_3]$ is the moment of inertia matrix in a frame attached to the body at its center of mass and oriented with principal axes. $C = IB = \mathrm{diag}[c_1, c_2, c_3]$ which we assume to be diagonal in this body-fixed frame, but may be otherwise if the geometry of the object differs from its inertia due to inhomogeneous mass density or variations in surface characteristics. $\mathbf{N}$ is the steady external field, and $d\mathbf{W}$ is the white noise forcing.

The corresponding Fokker-Planck equation is [55]:

$$\frac{\partial F}{\partial t} = -\nabla_{\boldsymbol{\omega}} \cdot [B\boldsymbol{\omega}F + I^{-1}(\boldsymbol{\omega} \times I\boldsymbol{\omega})F] + \sum_{i=1}^{3} \left(\sigma_i^2 \frac{\partial^2}{\partial \omega_i^2} - \omega_i X_i^R \right) F. \tag{16.22}$$

Here $F(A, \boldsymbol{\omega}, t)$ for $A \in SO(3)$ is related to $f(\boldsymbol{\omega}, t)$ of the previous section as

$$f(\boldsymbol{\omega}, t) = \int_{SO(3)} F(A, \boldsymbol{\omega}, t)\, dA.$$

Likewise, when $F(A, \boldsymbol{\omega}, t)$ is known, we can find

$$\mathcal{F}(A, t) = \int_{\mathbb{R}^3} F(A, \boldsymbol{\omega}, t)\, d\boldsymbol{\omega}.$$

When the body is spherical, we can calculate the PDF $F(A, \omega, t)$ from the simplified equation

$$\frac{\partial F}{\partial t} = -\nabla_{\boldsymbol{\omega}} \cdot [B\boldsymbol{\omega}F] + \sum_{i=1}^{3} \left(\sigma_i^2 \frac{\partial^2}{\partial \omega_i^2} - \omega_i X_i^R \right) F. \tag{16.23}$$

In this context, the angular momentum space is identified with $\mathbb{R}^3$ and the Fourier transform on $SO(3) \times \mathbb{R}^3$ can be used to simplify this equation.

Steele [121] reasoned that the average of F over all values of angular momentum must satisfy the equation

$$\frac{\partial \mathcal{F}}{\partial t} = R_1(t)\left[-\frac{\partial^2 \mathcal{F}}{\partial \gamma^2} + \nabla_A^2 \mathcal{F} \right] + R_3(t)\frac{\partial^2 \mathcal{F}}{\partial \gamma^2} \tag{16.24}$$

where

$$R_i(t) = (k_B T / B_i I)[1 - \exp(-B_i t)].$$

The operational properties of the $SO(3)$ Fourier transform converts (16.24) to a system of linear time-varying ODEs that can be solved.

While our goal here has been primarily to illustrate how diffusion equations on $SO(3)$ and $SO(3) \times \mathbb{R}^3$ arise in chemical physics, we note that there are many other aspects to rotational Brownian motion that have been addressed in the literature. Classic papers include [1, 21, 28, 35, 43, 45, 46, 48, 74, 75, 90, 93, 94], and references therein. Books that address this topic include [25, 91, 120]. Recent works include [4, 16, 17, 66, 67, 68, 97, 123, 144].

16.5 Liquid Crystals

Liquid crystals are often considered to be a phase of matter that lies somewhere in between solid crystalline structure and an isotropic liquid. The special properties of liquid crystals result from the long-range orientational order induced by the local interactions of molecules. Liquid crystalline phases are usually categorized according to whether the constituent molecules are rod-like or disc-like.[4] In both cases, there is usually an "average direction" along which the molecules tend to line up (at least locally), called the *director*, and denoted as $\hat{\mathbf{n}}$. In the most general case, the director will be a function of position, $\hat{\mathbf{n}}(\mathbf{a})$, where $\mathbf{a} \in \mathbb{R}^3$. Without loss of generality we can take the director at an inertial frame to be $\hat{\mathbf{n}}(\mathbf{0}) = \mathbf{e}_3$.

---

[4] These terms refer to aspect ratios and do not necessarily imply symmetry.

If we consider $F(\mathbf{a}, A)$ to be a probability density function on $SE(3)$ describing the position and orientation of all molecules in a liquid crystal phase, then one formal definition of the director is $\hat{\mathbf{n}} = \mathbf{n}/|\mathbf{n}|$ where

$$\mathbf{n}(\mathbf{a}) = \int_{SO(3)} A\mathbf{e}_3 F(\mathbf{a}, A) \, dA.$$

Another definition, which for engineering purposes is equivalent to this, is that for each $\mathbf{a} \in \mathbb{R}^3$ it is the mode (or "mean" in the sense defined in Chapter 14) of the following function on the sphere:

$$k(\mathbf{u}(\alpha, \beta); \mathbf{a}) = \frac{1}{2\pi} \int_{\gamma \in SO(2)} A\mathbf{e}_3 F(\mathbf{a}, A) \, d\gamma,$$

where α, β, and γ are Euler angles, and by themselves, α and β can be interpreted as the azimuthal and polar angles forming spherical coordinates on $\mathbb{S}^2 \cong SO(3)/SO(2)$. Often, the director does not depend on spatial position, and so $\hat{n}(\mathbf{a}) = \mathbf{e}_3$ is the same as what results from averaging $\hat{n}(\mathbf{a})$ over a finite cube of material. Sometimes, as discussed below, there is a different director in different planar layers of the liquid crystal. In such cases, the director is a function of only one component of translation, which is the component in the direction which the director rotates. We take this to be the component a_1.

Within the rod-like category are the *nematic* and several *smectic* phases. The nematic phase is characterized by the lack of translational order in the molecules. In this case, all the order is most simply described using an *orientational distribution function* (ODF), which is a probability density function on the rotation group. In the case of the various smectic phases, there is translational order, and the combination of orientational and translational order may be described as a function on the motion group.

In the following subsections, we closely follow the analytical descriptions of the evolution of liquid crystal PDFs presented in [26, 33, 69, 85, 86], and recast these descriptions in the terminology of earlier chapters of this book. Other classic works on the evolution of liquid crystal PDFs include [42, 53]. Recent works include [8, 9, 13, 14, 37, 96, 125]. Books on the mechanics and applications of liquid crystals include [15, 22, 31, 32].

16.5.1 The Nematic Phase

In nematogens, the PDF describing the order in the system is only a function of rotation:

$$F(\mathbf{a}, A) = f(A).$$

Given an orientation-dependant physical property of a single molecule, $X(A)$, the ensemble average of this property over all molecules in the nematic phase is calculated as

$$\langle X \rangle = \int_{SO(3)} X(A) f(A) \, dA$$

where it is usually assumed that the ODF $f(A)$ is related to an orientational potential $V(A)$ as

$$f(A) = \frac{\exp[-V(A)/k_B T]}{\int_{SO(3)} \exp[-V(A)/k_B T] dA}. \tag{16.25}$$

Hence, it is clear that Fourier expansions on the rotation group play will play a role in determining ensemble properties.

In the case when the molecules have cylindrical symmetry and no translational order, the functions $X(A)$ and $f(A)$ are constant over $SO(2)$ rotations along the axis of cylindrical symmetry, and all functions on $SO(3)$ (and integrations) are reduced to those on $SO(3)/SO(2) \cong \mathbb{S}^2$. Most nematic phases are constituted of apolar molecules that posses C_2 symmetry about an axis normal to the axis of cylindrical symmetry, and so only the upper hemisphere $\mathbb{S}^2/C_2$ need be considered.

If the medium has orientational order along the inertial $\mathbf{e}_3$ direction, the ODF will be a function on $SO(2)\backslash SO(3)/SO(2)$. In this case the phase is called *uniaxial*.

In the case of uniaxial nematics, the Maier-Saupe potential [87, 88, 89]

$$V(\beta) = -\epsilon S P_2(\cos\beta)$$

is a widely used approximation for the potential function $V(A) = V(A(0,\beta,0))$. Here ϵ scales the intermolecular interaction, and

$$S = \langle P_2 \rangle = \int_0^1 P_2(\cos\beta)\, f(A(0,\beta,0))\, d(\cos\beta)$$

can be determined experimentally from NMR.

Other variants on the nematic phase are *biaxial* nematics and *chiral* nematics. In chiral nematics, the molecules have a handedness which imparts a helical twist to the orientation axis along a direction orthogonal to the major axis).

In the case of chiral nematics, one would expect the orientational PDF to have a spatial dependence of the form

$$F(\mathbf{a}, A) = f(\mathrm{rot}[\mathbf{e}_1, -2\pi a_1/p] \cdot A),$$

where $a_1 = \mathbf{e}_1 \cdot \mathbf{a}$ is distance along the axis of the helical motion, which is normal to the plane in which the director (which at $a_1 = 0$ points in the $\mathbf{e}_3$ direction) rotates. p is the pitch, which is the distance along the a_1 direction required for the director to make one full counter-clockwise rotation.

16.5.2 The Smectic Phases

In the smectic phases (labeled A through I), varying degrees of translational order persists in addition to orientational order. For instance, in the smectic A phase, the molecules posses spatial order in the direction of the director, but no order in the planes normal to the director. In the smectic C phase, there is order in planes whose normals are a fixed angle away from the director. We shall not address the other smectic phases, but rather show how the PDF for the smectic A phase (and by extension the other smectic phases) is viewed as a function on the motion group.

The fact that smectic A phase liquid crystals have a director, and have centers of mass that lie in planes normal to the director has led researchers in the field of liquid crystal research to describe this order with several different PDFs. One is of the form [33]

$$F_D(A(\alpha,\beta,\gamma)), \mathbf{a}) = \sum_{l=0}^{\infty}\sum_{n=0}^{\infty} A_{ln} P_l(\cos\beta)\cos(2\pi n a_3/d) \qquad (16.26)$$

where $a_3 = \mathbf{a} \cdot \mathbf{e}_3$ and d is a characteristic distance between the parallel planes that is on the order of the length of the long axis of the rod-like molecules composing the smectic A phase. In this context, the normalization

$$\int_0^d \int_0^\pi F_D \sin\beta d\beta da_3 = 1$$

is imposed, and the ensemble average of a quantity, X, is calculated as

$$\langle X \rangle = \int_0^d \int_0^\pi X F_D \sin\beta d\beta da_3.$$

Usually the first three terms in the series in (16.26) are kept, and so this model is defined by the parameters

$$\eta = \langle P_2(\cos\beta) \rangle \qquad \tau = \langle \cos(2\pi a_3/d) \rangle \qquad \sigma = \langle P_2(\cos\beta) \cos(2\pi a_3/d).$$

Another model for the PDF of smectic A phase is [93, 94]:

$$F_M(A(\alpha,\beta,\gamma)), \mathbf{a}) = \frac{\exp[-V_M(\beta,a_3)/k_B T]}{\int_0^d \int_0^\pi \exp[-V_M(\beta,a_3)/k_B T] \sin\beta d\beta da_3} \qquad (16.27)$$

where

$$V_M(\beta,a_3) = -v[\delta\alpha_0\tau\cos(2\pi a_3/d) + (\eta + \alpha_0\sigma\cos(2\pi a_3/d))P_2(\cos\beta)]$$

and v, δ, and α_0 are physical constants.

Clearly, both of these PDFs are functions on the motion group, $SE(3)$.

16.5.3 Evolution of Liquid Crystal Orientation

Experimentally, many properties of liquid crystals are determined when the time evolution of an orientational correlation function of the form [8, 9]

$$\phi_{mm',nn'}^{ll'} = \int_{SO(3) \times SO(3)} f(A_0) f(A_0, A, t) \overline{U_{nm}^l(A_0)} U_{m'n'}^{l'}(A) \, dA dA_0$$

is known. Here $U_{mn}^l(A)$ are matrix elements of $SO(3)$ IURs (or equivalently, Wigner D-functions), $f(A_0, A, t)$ is the probability that a molecule will be at orientation A at time t given that it was at A_0 at time $t_0 = 0$, and $f(A)$ is the equilibrium orientional probability density given by (16.25).

In order to calculate the correlation function $\phi_{mm',nn'}^{ll'}$, it is first necessary to determine $f(A_0, A, t)$ for a given model of the orientational potential, $V(A)$. In the most general case, this is written in terms of a band-limited Fourier series on $SO(3)$ as

$$\frac{V(A)}{k_B T} = \sum_{l=0}^N (2l + 1)\text{tr}[\hat{C}^l U^l(A)],$$

where the Fourier coefficients $\hat{C}_{mn}^l$ are given.

The evolution of $f(A_0, A, t)$ is governed by the diffusion equation [33]:

$$\frac{\partial}{\partial t} f(A_0, A, t) = \Gamma f(A_0, A, t), \qquad (16.28)$$

where the operator Γ is defined as

$$\Gamma \doteq \sum_{i,j=1}^{3} D_{ij} X_i^R \left[X_j^R + X_j^R \left(\frac{V(A)}{k_B T} \right) \right],$$

X_i^R are the differential operators defined in Chapter 9, and $D_{ij} = D_i \delta_{ij}$ are components of the diffusion tensor in the body-fixed molecular frame.

It is common to assume a band-limited solution of the form

$$f(A_0, A, t) = \sum_{l=0}^{M} (2l+1) \operatorname{tr}[\hat{f}^l(A_0, t) U^l(A)] \tag{16.29}$$

and the initial conditions $f(A_0, A, 0) = \delta(A_0^T A)$, in which case (16.28) is transformed into the system of ODEs

$$\frac{d\hat{f}}{dt} = \mathcal{R}\hat{f}$$

with initial conditions $\hat{f}(A_0, 0) = U(A_0^{-1})$. The components of $\mathcal{R}$ are calculated as

$$\mathcal{R}_{m',n';m,n}^{l,l'}(A_0) = \int_{SO(3)} \overline{U_{n'm'}^{l'}(A)} \Gamma U_{mn}^{l}(A) \, dA. \tag{16.30}$$

The solution for the Fourier coefficients, in principle, can be thought of as the matrix exponential

$$\hat{f}(A_0, t) = \exp[\mathcal{R}t] U(A_0^{-1}). \tag{16.31}$$

However, in practice the evaluation of this solution is greatly simplified when $\mathcal{R}$ is a normal matrix. This cannot always be guaranteed, and so a symmetrized version of the problem is solved, in which Γ is replaced by the self-adjoint operator

$$\tilde{\Gamma} = f^{-\frac{1}{2}}(A) \Gamma f^{\frac{1}{2}}(A)$$

and we solve the equation

$$\frac{\partial}{\partial t} \tilde{f}(A_0, A, t) = \tilde{\Gamma} \tilde{f}(A_0, A, t)$$

where

$$\tilde{f}(A_0, A, t) = f^{-\frac{1}{2}}(A) f(A_0, A, t) f^{\frac{1}{2}}(A_0). \tag{16.32}$$

In this case $\tilde{\mathcal{R}}$ (the matrix with elements defined as in (16.30) with $\tilde{\Gamma}$ in place of Γ) is unitarily diagonalizable, and the matrix exponential solution ((16.31) with $\hat{\tilde{f}}$ and $\tilde{\mathcal{R}}$ in place of $\hat{f}$ and $\mathcal{R}$) is efficiently calculated. Then, using (16.32), $f(A_0, A, t)$ is recovered from $\tilde{f}(A_0, A, t)$ (which is calculated as in (16.29) with $\tilde{f}$ and $\hat{\tilde{f}}$ replacing f and $\hat{f}$).

16.5.4 Explicit Solution of a Diffusion Equation for Liquid Crystals

The PDF for the orientation distribution of liquid crystals is governed by a viscosity-dominated formulation (zero inertia) resulting in the equation [98, 99, 100, 26]:

$$\frac{\partial f}{\partial t} = [D_1((X_1^R)^2 + (X_2^R)^2) + D_2(X_3^R)^2] f - K \operatorname{div}(f \operatorname{grad} V) \tag{16.33}$$

where V is a mean potential field. This equation is based on the *Maier-Saupe model* of nematic liquid crystals [87, 88, 89].

D_1 and D_2 are viscosity constants (which are different from each other because of the axial symmetry of the molecules), and K is a constant depending on external field strength and temperature. $u = \cos^2 \beta$ is the normalized field potential for their model, and div and grad are defined in the differential-geometric context described in Chapter 4.

After using properties such as (9.74), Kalmykov and Coffey [69] write the $SO(3)$-Fourier transform of (16.33) as a system of the form

$$\frac{d\mathbf{f}}{dt} = A\mathbf{f}$$

where $f_l = \hat{f}^l_{mn}$ and A is a banded infinite-dimensional matrix. If however, the assumption of a band-limited approximation to the solution is sought, then A becomes finite, and the explicit solution

$$\mathbf{f}(t) = \exp(At)\mathbf{f}(0)$$

can be calculated for band-limit L.

16.6 A Rotational Diffusion Equation from Polymer Science

In polymer science the following diffusion-type PDE arises (see Chapter 17 for derivation):

$$\frac{\partial f}{\partial t} = \mathcal{K}f \tag{16.34}$$

where the operator $\mathcal{K}$ is defined as

$$\mathcal{K} \doteq \frac{1}{2} \sum_{i,j=1}^{3} D_{ij} X^R_i X^R_j + \sum_{k=1}^{3} d_k X^R_k$$

and X^R_i are the differential operators defined in Chapter 9, and d_k and D_{ij} are constants.

Using the operational properties of the $SO(3)$ Fourier transform can be used to convert this kind of PDE into a system of linear ODEs. In this section we take the IUR matrix elements of $SO(3)$ to be of the form

$$U^l_{mn}(g(\alpha, \beta, \gamma)) = (-1)^{n-m} e^{-i(m\alpha + n\gamma)} P^l_{mn}(\cos \beta). \tag{16.35}$$

This differs by a factor of i^{n-m} from those in Chapter 9.

By expanding the PDF in the PDE in (16.34) into a Fourier series on $SO(3)$, the solution can be obtained once we know how the differential operators X^R_i transform the matrix elements $U^l_{m,n}(A)$. Adjusted for the definition of U^l_{mn} in (16.35), these operational properties are:

$$X^R_1 U^l_{mn} = \frac{1}{2}c^l_{-n}U^l_{m,n-1} - \frac{1}{2}c^l_n U^l_{m,n+1}; \tag{16.36}$$

$$X^R_2 U^l_{mn} = \frac{1}{2}ic^l_{-n}U^l_{m,n-1} + \frac{1}{2}ic^l_n U^l_{m,n+1}; \tag{16.37}$$

$$X^R_3 U^l_{mn} = -inU^l_{mn}; \tag{16.38}$$

where $c_n^l = \sqrt{(l-n)(l+n+1)}$ for $l \geq |n|$ and $c_n^l = 0$ otherwise. From this definition it is clear that $c_k^k = 0$, $c_{-(n+1)}^l = c_n^l$, $c_{n-1}^l = c_{-n}^l$, and $c_{n-2}^l = c_{-n+1}^l$). Equations (16.36)−(16.38) follow from (16.35) and the fact that

$$\frac{d}{dt}U_{mn}^l(\text{rot}(\mathbf{e}_1,t))|_{t=0} = \frac{1}{2}c_{-n}^l\delta_{m+1,n} - \frac{1}{2}c_n^l\delta_{m-1,n} \tag{16.39}$$

$$\frac{d}{dt}U_{mn}^l(\text{rot}(\mathbf{e}_2,t))|_{t=0} = \frac{i}{2}c_{-n}^l\delta_{m+1,n} + \frac{i}{2}c_n^l\delta_{m-1,n} \tag{16.40}$$

$$\frac{d}{dt}U_{mn}^l(\text{rot}(\mathbf{e}_3,t))|_{t=0} = -in\delta_{m,n} \tag{16.41}$$

By repeated application of these rules we find

$$(X_1^R)^2 U_{mn}^l = \frac{1}{4}c_{-n}^l c_{-n+1}^l U_{m,n-2}^l - \frac{1}{4}(c_{-n}^l c_{n-1}^l + c_n^l c_{-n-1}^l)U_{mn}^l + \frac{1}{4}c_n^l c_{n+1}^l U_{m,n+2}^l;$$

$$(X_2^R)^2 U_{mn}^l = -\frac{1}{4}c_{-n}^l c_{-n+1}^l U_{m,n-2}^l - \frac{1}{4}(c_{-n}^l c_{n-1}^l + c_n^l c_{-n-1}^l)U_{mn}^l - \frac{1}{4}c_n^l c_{n+1}^l U_{m,n+2}^l;$$

$$(X_3^R)^2 U_{mn}^l = -n^2 U_{mn}^l;$$

$$X_1^R X_2^R U_{m,n}^l = \frac{i}{4}c_{-n}^l c_{-n+1}^l U_{m,n-2}^l + \frac{i}{4}(-c_{-n}^l c_{n-1}^l + c_n^l c_{-n-1}^l)U_{m,n}^l - \frac{i}{4}c_n^l c_{n+1}^l U_{m,n+2}^l;$$

$$X_2^R X_1^R U_{m,n}^l = \frac{i}{4}c_{-n}^l c_{-n+1}^l U_{m,n-2}^l + \frac{i}{4}(c_{-n}^l c_{n-1}^l - c_n^l c_{-n-1}^l)U_{m,n}^l - \frac{i}{4}c_n^l c_{n+1}^l U_{m,n+2}^l.$$

$$X_1^R X_3^R U_{m,n}^l = i\frac{n}{2}(-c_{-n}^l U_{m,n-1}^l + c_n^l U_{m,n+1}^l);$$

$$X_3^R X_1^R U_{m,n}^l = -i\frac{n-1}{2}c_{-n}^l U_{m,n-1}^l + i\frac{n+1}{2}c_n^l U_{m,n+1}^l;$$

$$X_3^R X_2^R U_{m,n}^l = \frac{(n-1)}{2}c_{-n}^l U_{m,n-1}^l + \frac{(n+1)}{2}c_n^l U_{m,n+1}^l;$$

$$X_2^R X_3^R U_{m,n}^l = \frac{n}{2}(c_{-n}^l U_{m,n-1}^l + c_n^l U_{m,n+1}^l);$$

As a direct result of the definition of the $SO(3)$-Fourier inversion formula, we observe that if a differential operator X transforms U_{mn}^l as

$$XU_{m,n}^l = x(n)U_{m,n+p}^l,$$

then there is a corresponding operational property of the Fourier transform:

$$\mathcal{F}(Xf)_{m,n}^l = x(m-p)\hat{f}_{m-p,n}^l. \tag{16.42}$$

We use this to write

$$\mathcal{F}(X_1^R f)_{mn}^l = \frac{1}{2}c_{-m-1}^l \hat{f}_{m+1,n}^l - \frac{1}{2}c_{m-1}^l \hat{f}_{m-1,n}^l;$$

$$\mathcal{F}(X_2^R f)_{mn}^l = \frac{1}{2}ic_{-m-1}^l \hat{f}_{m+1,n}^l + \frac{1}{2}ic_{m-1}^l \hat{f}_{m-1,n}^l;$$

$$\mathcal{F}(X_3^R f)_{mn}^l = -im\hat{f}_{mn}^l;$$

$$\mathcal{F}((X_1^R)^2 f)_{mn}^l = \frac{1}{4}c_{m+1}^l c_{-m-1}^l \hat{f}_{m+2,n}^l - \frac{1}{4}(c_{-m}^l c_{m-1}^l + c_m^l c_{-m-1}^l)\hat{f}_{mn}^l + \frac{1}{4}c_{-m+1}^l c_{m-1}^l \hat{f}_{m-2,n}^l;$$

$$\mathcal{F}((X_2^R)^2 f)_{mn}^l = -\frac{1}{4}c_{m+1}^l c_{-m-1}^l \hat{f}_{m+2,n}^l - \frac{1}{4}(c_{-m}^l c_{m-1}^l + c_m^l c_{-m-1}^l)\hat{f}_{mn}^l - \frac{1}{4}c_{-m+1}^l c_{m-1}^l \hat{f}_{m-2,n}^l;$$

$$\mathcal{F}((X_3^R)^2 f)_{mn}^l = -m^2 \hat{f}_{mn}^l;$$

$$\mathcal{F}((X_1^R X_2^R + X_2^R X_1^R)f)_{m,n}^l = \frac{i}{2}c_{m+1}^l c_{-m-1}^l \hat{f}_{m+2,n}^l - \frac{i}{2}c_{-m+1}^l c_{m-1}^l \hat{f}_{m-2,n}^l;$$

$$\mathcal{F}((X_1^R X_3^R + X_1^R X_3^R)f)_{m,n}^l = -i\frac{2m+1}{2}c_{-m-1}^l \hat{f}_{m+1,n}^l + i\frac{2m-1}{2}c_{m-1}^l \hat{f}_{m-1,n}^l;$$

$$\mathcal{F}((X_3^R X_2^R + X_2^R X_3^R)f)_{m,n}^l = \frac{(2m+1)}{2}c_{-m-1}^l \hat{f}_{m+1,n}^l + \frac{(2m-1)}{2}c_{m-1}^l \hat{f}_{m-1,n}^l;$$

Collecting everything together we have

$$\mathcal{F}\left(\frac{1}{2}\sum_{i,j=1}^3 D_{ij}X_i^R X_j^R f + \sum_{i=1}^3 d_i X_i^R f\right)_{mn}^l = \sum_{k=\max(-l,m-2)}^{\min(l,m+2)} \mathcal{A}_{m,k}^l \hat{f}_{k,n}^l,$$

where

$$\mathcal{A}_{m,m+2}^l = \left[\frac{(D_{11}-D_{22})}{8} + \frac{i}{4}D_{12}\right]c_{m+1}^l c_{-m-1}^l;$$

$$\mathcal{A}_{m,m+1}^l = \left[\frac{(2m+1)}{4}(D_{23} - iD_{13}) + \frac{1}{2}(d_1 + id_2)\right]c_{-m-1}^l;$$

$$\mathcal{A}_{m,m}^l = \left[-\frac{(D_{11}+D_{22})}{8}(c_{-m}^l c_{m-1}^l + c_m^l c_{-m-1}^l) - \frac{D_{33}m^2}{2} - id_3 m\right];$$

$$\mathcal{A}_{m,m-1}^l = \left[\frac{(2m-1)}{4}(D_{23} + iD_{13}) + \frac{1}{2}(-d_1 + id_2)\right]c_{m-1}^l;$$

$$\mathcal{A}_{m,m-2}^l = \left[\frac{(D_{11}-D_{22})}{8} - \frac{i}{4}D_{12}\right]c_{-m+1}^l c_{m-1}^l;$$

Hence, application of the $SO(3)$-Fourier transform to (16.34) and corresponding initial conditions reduces (16.34) to a set of linear time-invariant ODEs of the form

$$\frac{d\hat{f}^l}{dL} = \mathcal{A}^l \hat{f}^l \quad \text{with} \quad \hat{f}^l(0) = \mathbb{I}_{2l+1}. \tag{16.43}$$

Here $\mathbb{I}_{2l+1}$ is the $(2l+1) \times (2l+1)$ identity matrix and the banded matrix $\mathcal{A}^l$ are of the following form for $l = 0, 1, 2, 3$:

$$\mathcal{A}^0 = \mathcal{A}_{0,0}^0 = 0; \quad \mathcal{A}^1 = \begin{pmatrix} \mathcal{A}_{-1,-1}^1 & \mathcal{A}_{-1,0}^1 & \mathcal{A}_{-1,1}^1 \\ \mathcal{A}_{0,-1}^l & \mathcal{A}_{0,0}^l & \mathcal{A}_{0,1}^l \\ \mathcal{A}_{1,-1}^1 & \mathcal{A}_{1,0}^1 & \mathcal{A}_{1,1}^1 \end{pmatrix};$$

$$\mathcal{A}^2 - \begin{pmatrix} \mathcal{A}_{-2,-2}^2 & \mathcal{A}_{-2,-1}^2 & \mathcal{A}_{-2,0}^2 & 0 & 0 \\ \mathcal{A}_{-1,-2}^2 & \mathcal{A}_{-1,-1}^2 & \mathcal{A}_{-1,0}^2 & \mathcal{A}_{-1,1}^2 & 0 \\ \mathcal{A}_{0,-2}^2 & \mathcal{A}_{0,-1}^2 & \mathcal{A}_{0,0}^2 & \mathcal{A}_{0,1}^2 & \mathcal{A}_{0,2}^2 \\ 0 & \mathcal{A}_{1,-1}^2 & \mathcal{A}_{1,0}^2 & \mathcal{A}_{1,1}^2 & \mathcal{A}_{1,2}^2 \\ 0 & 0 & \mathcal{A}_{2,0}^2 & \mathcal{A}_{2,1}^2 & \mathcal{A}_{2,2}^2 \end{pmatrix};$$

$$
\mathcal{A}^3 =
\begin{pmatrix}
\mathcal{A}^3_{-3,-3} & \mathcal{A}^3_{-3,-2} & \mathcal{A}^3_{-3,-1} & 0 & 0 & 0 & 0 \\
\mathcal{A}^3_{-2,-3} & \mathcal{A}^3_{-2,-2} & \mathcal{A}^3_{-2,-1} & \mathcal{A}^3_{-2,0} & 0 & 0 & 0 \\
\mathcal{A}^3_{-1,-3} & \mathcal{A}^3_{-1,-2} & \mathcal{A}^3_{-1,-1} & \mathcal{A}^3_{-1,0} & \mathcal{A}^3_{-1,1} & 0 & 0 \\
0 & \mathcal{A}^3_{0,-2} & \mathcal{A}^3_{0,-1} & \mathcal{A}^3_{0,0} & \mathcal{A}^3_{0,1} & \mathcal{A}^3_{0,2} & 0 \\
0 & 0 & \mathcal{A}^3_{1,-1} & \mathcal{A}^3_{1,0} & \mathcal{A}^3_{1,1} & \mathcal{A}^3_{1,2} & \mathcal{A}^3_{1,3} \\
0 & 0 & 0 & \mathcal{A}^3_{2,0} & \mathcal{A}^3_{2,1} & \mathcal{A}^3_{2,2} & \mathcal{A}^3_{2,3} \\
0 & 0 & 0 & 0 & \mathcal{A}^3_{3,1} & \mathcal{A}^3_{3,2} & \mathcal{A}^3_{3,3}
\end{pmatrix}.
$$

As is well known in systems theory, the solution to (16.43) is of the form of a matrix exponential:

$$
\hat{f}^l(L) = e^{L\mathcal{A}^l}. \tag{16.44}
$$

Since $\mathcal{A}^l$ is a band-diagonal matrix for $l > 1$, the matrix exponential can be calculated much more efficiently (either numerically or symbolically) for large values of l than for general matrices of dimension $(2l+1) \times (2l+1)$. We also gain efficiencies in computing the matrix exponential of $L\mathcal{A}^l$ by observing the symmetry

$$
\mathcal{A}^l_{m,n} = (-1)^{m-n} \overline{\mathcal{A}^l_{-m,-n}}.
$$

Matrices with this kind of symmetry have eigenvalues which occur in conjugate pairs, and if x_m are the components of the eigenvector corresponding to the complex eigenvalue λ, then $(-1)^m \overline{x_{-m}}$ will be the components of the eigenvector corresponding to $\overline{\lambda}$ (see Theorem 10.6).

In general, the numerically calculated values of $\hat{f}^l(L)$ may be substituted back into the $SO(3)$-Fourier inversion formula to yield the solution for $f(A; L)$ to any desired accuracy. When $D_{11} = D_{22} = 1/\alpha_0$ and $D_{33} \to \infty$, and every other parameter in D and $\mathbf{d}$ is zero, the matrices $\mathcal{A}^l$ are all diagonal. This implies that the nonzero Fourier coefficients are of the form $\hat{f}^l_{m,m}(L) = \exp(L\mathcal{A}^l_{m,m})$. However, for $m \neq 0$ the value of D_{33} causes $\hat{f}^l_{m,m}(L)$ to be zero and what remains is a series in l with $m = 0$:

$$
f(A; L) = \sum_{l=0}^{\infty} (2l+1) e^{-l(l+1)L/2\alpha_0} U^l_{0,0}(A) = \sum_{l=0}^{\infty} (2l+1) e^{-l(l+1)L/2\alpha_0} P_l(\cos\beta).
$$

This special case corresponds to the "Kratky-Porod" model discussed in Chapter 17.

All of what is described in this section is well known in the literature, though it is rarely expressed using group-theoretical notation and terminology.

16.7 Other Models for Rotational Brownian Motion

While problems in chemical physics are a rich source of rotational Brownian motion problems, there are other fields in which such problems arise. In this section we examine two non-physics-based models in which diffusions on $SO(3)$ occur.

16.7.1 The Evolution of Estimation Error PDFs

In Chapter 15, we encountered the estimation equation (15.40). This may be written in the form

$$
(Q')^T dQ' = (Q')^T \left(\sum_{i=1}^{3} x_i dt X_i \right) Q' + \left(\sum_{i=1}^{3} X_i dW_i(t) \right). \tag{16.45}
$$

Using the tools of this chapter, it is possible to write a corresponding Fokker-Planck equation. First, we take the $\vee$ of both sides, resulting in

$$J_R(\phi)\,d\phi = Ad(Q')\mathbf{x}dt + d\mathbf{W}$$

where ϕ is the array of ZXZ Euler angles corresponding to Q'. If $\mathbf{x} = \mathbf{0}$, the Fokker-Planck equation for the PDF of Q' is the same as given in (16.18) with $c = 1$. Otherwise, the drift term is different.

16.7.2 Rotational Brownian Motion Based on Randomized Angular Momentum

A different model of the random motion of a rigid body is formulated by Liao [81]. In this formulation, the random forces acting on the body are defined such that they instantaneously change the body's angular momentum in random ways. The corresponding stochastic differential equation is [81]:

$$dR = R\mathcal{I}Ad(R^{-1})\sum_{i=1}^{3} X_i dW_i + R\mathcal{I}Ad(R^{-1})M_0 dt$$

where $M_0 \in so(3)$ is the initial angular momentum of the body in the space-fixed frame and Λ is the inverse inertia operator described in Chapter 15. Multiplying both sides of the above equation by R^{-1} on the left, taking the $\vee$ of both sides, and using the properties of Λ, we find

$$\boldsymbol{\omega}_R dt = I^{-1}\left(Ad(R^{-1})\sum_{i=1}^{3} X_i dW_i\right)^{\vee} + I^{-1}\left(Ad(R^{-1})M_0 dt\right)^{\vee}.$$

Using the fact that $\boldsymbol{\omega}_R = J_R\dot{\phi}$ where $\phi = [\alpha, \beta, \gamma]^T$ is the array of Euler angles, and for $SO(3)$ we have from Chapter 5 that $[Ad(R)] = R$, we find

$$d\phi = J_R^{-1}I^{-1}R^{-1}d\mathbf{W} + J_R^{-1}I^{-1}R^{-1}\mathbf{M}_0$$

where $\mathbf{M}_0 = (M_0)^{\vee}$ and $\mathbf{W} = W^{\vee}$.

The Fokker-Planck equation for this case may be generated using the formulation presented in Subsection 16.2.2. In the special case when $I = \mathbb{I}$ and $\mathbf{M}_0 = \mathbf{0}$ the fact that $RJ_R = J_L$ is used to simplify matters, and we get a "left version" of (16.18) in the sense that now $d\phi = J_L^{-1}d\mathbf{W}$ instead of $d\phi = J_R^{-1}d\mathbf{W}$.

16.7.3 The Effects of Time Lag and Memory

It is not difficult to imagine versions of (16.34) in which there are either time lags or memory effects. In the first case we would have

$$\frac{\partial}{\partial t}f(A,t) = \mathcal{K}f(A, t - \tau)$$

and in the second case,

$$\frac{\partial}{\partial t}f(A,t) = \int_0^t h(t - \tau)\mathcal{K}f(A,\tau)\,d\tau.$$

In both of these cases the *Laplace transform* can be used to convert the time dependence of these equations into an algebraic dependence on the Laplace transform parameter. Recall that the Laplace transform of a bounded function $x(t)$ where $t \in \mathbb{R}_{>0}$ is defined as

$$X(s) = L[x] \doteq \int_0^\infty x(t) e^{-st} dt$$

where $s \in \mathbb{C}$ has positive real part.

Much like the Fourier transform of functions on the line, the Laplace transform has some useful operational properties. Using integration by parts,

$$L\left[\frac{dx}{dt}\right] = sL[x] - x(0).$$

Hence, given a system of linear equations with constant coefficients of the form

$$\frac{d\mathbf{x}}{dt} = A\mathbf{x}, \tag{16.46}$$

we can write

$$sL[\mathbf{x}] - \mathbf{x}(0) = AL[\mathbf{x}].$$

With initial conditions $\mathbf{x}(0)$ given, we can write

$$(s\mathbb{I} - A)L[\mathbf{x}] = \mathbf{x}(0).$$

When $\det(s\mathbb{I} - A) \neq 0$ we can symbolically write

$$\mathbf{x}(t) = L^{-1}[(s\mathbb{I} - A)^{-1}\mathbf{x}(0)]$$

where L^{-1} is the inverse Laplace transform. See [7] for a description of the inversion process. From a practical perspective, it is convenient to write the solution to (16.46) as

$$\mathbf{x}(t) = \exp(tA)\mathbf{x}(0).$$

However, for equations with time lags or memory, the Laplace transform technique is useful. This is because of the two operational properties discussed below.

The Laplace transform also converts convolutions of the form

$$(x * y)(t) = \int_0^t x(\tau)y(t - \tau)\, d\tau$$

to products as

$$L[x * y] = L[x]L[y].$$

Note that the bounds of integration in this convolution are different than in the rest of the book.

An operational property for shifts is [7]:

$$L[x(t - t_0)u(t - t_0)] = e^{-t_0 s} \mathcal{L}[x(t)],$$

where $u(t - t_0)$ is the unit step function equal to zero for $t < t_0$ and equal to unity for $t \geq 0$.

Hence, the $SO(3)$ Fourier transform can be used in combination with the Laplace transform operational properties to handle the time-dependent part of the equations at the beginning of this subsection.

16.8 Closed-Form Solution to Diffusion Equations in the Small-Time Limit

A Fourier expansion can capture the solution of a diffusion equation quite well as time increases because the PDF becomes more and more spread out. But in the small-time limit, Fourier methods become inefficient. Fortunately, an alternative exists. For small values of time, a diffusion process on a unimodular Lie group behaves much like one evolving in Euclidean space. And so a Gaussian approximation in exponential coordinates becomes natural.

The concepts of mean and covariance on matrix Lie groups are not as straightforward. However, for PDFs that are relatively concentrated, definitions used for mean and covariance have been extended from those commonly used for PDFs on $\mathbb{R}^n$. The mean $\mu \in G$ of a PDF $f(g)$ can be defined such that

$$\int_G \log^\vee(\mu^{-1}g)\, f(g)\, dg = \mathbf{0}. \tag{16.47}$$

This concept of mean, which should not be confused with other concepts such as those presented recently in [115], has some particularly useful properties for our application, as described below.

The covariance matrix Σ for these concentrated PDFs can be defined using

$$\Sigma = \int_G \log^\vee(\mu^{-1}g)\left(\log^\vee(\mu^{-1}g)\right)^T f(g)\, dg. \tag{16.48}$$

This too should not be confused with other concepts of covariance that are reviewed in [23, 52, 54].

Along these lines, a PDF inspired by the normal distribution on $\mathbb{R}^n$ has been extended independently by several researchers. (See the discussion and references cited in the context of kinematic covariance propagation in Chapter 15.) This distribution has been called a Gaussian on exponential coordinates and is given by

$$f(g; \mu, \Sigma) = \frac{1}{c(\Sigma)} \exp\left(-\frac{1}{2}\left[\log^\vee(\mu^{-1}g)\right]^T \Sigma^{-1} \log^\vee(\mu^{-1}g)\right). \tag{16.49}$$

Here $c(\Sigma)$ is a normalizing factor that ensures that $f(g; \mu, \Sigma)$ is a PDF. When the covariance is small this normalizing factor can be approximated as

$$c(\Sigma) \approx (2\pi)^{n/2}|\det(\Sigma)|^{1/2}$$

where n is the dimension of $\log^\vee(g)$. For $SE(2)$, $n = 3$. In related work, this approach has also been taken for $SE(3)$ in which case $n = 6$.

Consider the stochastic nonholonomic system described by the SDE (12.24) and Fokker-Planck equation (12.25). In the small-time (or small noise) limit, and starting with the initial conditions $\mu(0) = e$ and $\Sigma(0) = \mathbb{O}$, repeated application of the propagation formulas for the mean and covariance can be iterated to give integrals for these quantities:

$$\mu(t) = \bigcap_{0 \le \tau \le t} \exp\left(\sum_i h_i(\tau)E_i\right) \tag{16.50}$$

and

$$\Sigma(t) = \int_0^t Ad^{-1}\left((\mu(\tau))^{-1}\mu(t)\right) D Ad^{-T}\left((\mu(\tau))^{-1}\mu(t)\right) d\tau. \qquad (16.51)$$

Here $\bigcap_{0 \le \tau \le t}$ denotes a product integral. In the particular case when the motion is in one direction, e.g., E_1,

$$\bigcap_{0 \le \tau \le t} \exp(h(t)E_1) = \exp\left(\int_0^t h(\tau)\,d\tau\, E_1\right)$$

and $(\mu(\tau))^{-1}\mu(t) = \mu(t - \tau)$. Moreover, for any continuous function $f : [0, t] \to \mathbb{R}$ the general property

$$\int_0^t f(t - \tau)\,d\tau = \int_0^t f(\tau)\,d\tau$$

holds. This can be used to simplify (16.51) as

$$\Sigma(t) = \int_0^t Ad^{-1}\left(\mu(\tau)\right) D Ad^{-T}\left(\mu(\tau)\right) d\tau$$

in the special case when $(\mu(\tau))^{-1}\mu(t) = \mu(t - \tau)$. And when $\Sigma(t)$ is small enough, the resulting Gaussian distribution solves the diffusion equation because if $f_t(g) = f(g; t)$, then $f_{t_1+t_2}(g) = (f_{t_1} * f_{t_2})(g)$.

16.9 Closed-Form Folded Solutions

16.9.1 Fourier Series Representation of the Gaussian for $SO(3)$

A natural way to define a Gaussian function for $SO(3)$ is the solution of the heat equation. That is, we seek the solution of the equation

$$\frac{\partial f}{\partial t} = K\nabla^2_{SO(3)}f \qquad (16.52)$$

with an initial condition $f(R, 0) = \delta(R)$. The Laplacian operator for $SO(3)$ in the axis-angle parameterization is given in (9.50).

Since we seek a solution that is a class function, and since every class function for $SO(3)$ is a function only of the angle of rotation θ, (16.52) is simplified into

$$\frac{\partial f}{\partial t} = K\left(\frac{\partial^2 f}{\partial \theta^2} + \cot \theta/2 \frac{\partial f}{\partial \theta}\right). \qquad (16.53)$$

The initial condition $f(R, 0) = \delta(R)$ is expressed as

$$f(\theta, 0) = \sum_{l=0}^{\infty}(2l + 1)\frac{\sin(l + \frac{1}{2})\theta}{\sin \frac{\theta}{2}}.$$

It can be shown that the IURs of $SO(3)$ are matrix-valued eigenfucntions of the Laplacian:

$$\nabla^2 U^l(R) = -l(l + 1)U^l(R),$$

and that the character functions inherit this property:

$$\nabla^2 \chi_l(R) = -l(l+1)\chi_l(R). \tag{16.54}$$

Using (16.54), the solution of (16.53) can be written as

$$f(\theta, t) = \sum_{l=0}^{\infty} (2l+1) \frac{\sin(l+\frac{1}{2})\theta}{\sin \frac{\theta}{2}} e^{-l(l+1)Kt}. \tag{16.55}$$

While (16.55) is a valid solution of the heat equation on $SO(3)$, it has the same drawbacks as the Fourier series solution of the heat equation on the circle. Namely, for small values of Kt, a very large number of terms in the series must be calculated. Even then, the series solution can have significant oscillations. For these reasons, we investigate the Gaussian functions in the following subsection.

16.9.2 Folded Normal Density Solution for $SO(3)$

Chétalet and Chirikjian (see reference in Chapter 14) proposed one possible generalization of the concept of a Gaussian function for the group $SO(3)$. Unlike the Fourier series solution presented in the previous subsection, their solution is analogous to the folded normal density solution on the circle. This candidate for the definition of a Gaussian distribution on $SO(3)$ folds the function

$$f(\theta, t) = \beta^3(t)\theta'(\sin \theta')^{-1} e^{-\pi\beta^2(t)\theta'^2} \tag{16.56}$$

around the circle defined by $-\pi \le \theta \le \pi$, where θ is the angle from the axis-angle parameterization of $SO(3)$, and $\theta' = \theta/2$.

In this subsection we examine the properties of this candidate Gaussian. We also suggest how $\beta(t)$ can be chosen so that the result solves the heat equation on $SO(3)$, and hence can be properly used for the analysis, smoothing, and visualization of orientational data.

Examining the Candidate Function

In this subsection, we perform the calculations required to evaluate the fitness of the trial function in (16.56). We begin by rewriting (16.53) using $\theta' = \theta/2$ as the spatial variable rather than θ. By the chain rule,

$$\frac{\partial f}{\partial \theta} = \frac{\partial f}{\partial \theta'} \frac{\partial \theta'}{\partial \theta} = \frac{1}{2} \frac{\partial f}{\partial \theta'}.$$

(16.53) is rewritten as

$$\frac{\partial f}{\partial t} = \frac{K}{4} \left(\frac{\partial^2 f}{\partial \theta'^2} + 2\cot\theta' \frac{\partial f}{\partial \theta'} \right) \tag{16.57}$$

In order to evaluate if (16.56) satisfies (16.53) for some function $\beta(t)$, we explicitly take derivatives and match both sides of the equation.

$$\frac{\partial f}{\partial t} = 3\beta^2 \dot{\beta}\theta'(\sin\theta')^{-1} e^{-\pi\beta^2\theta'^2} + \beta^3\theta'(\sin\theta')^{-1} e^{-\pi\beta^2\theta'^2}(-2\pi\beta\dot{\beta}\theta'^2)$$

$$= \beta^2\dot{\beta}\theta'(\sin\theta')^{-1} e^{-\pi\beta^2\theta'^2}(3 - 2\pi\beta^2\theta'^2).$$

On the other hand,

$$\frac{\partial f}{\partial \theta'} = \beta^3 (\sin \theta')^{-1} e^{-\pi \beta^2 \theta'^2} - \beta^3 \theta' (\sin \theta')^{-2} \cos \theta' e^{-\pi \beta^2 \theta'^2} - 2\pi \beta^5 \theta'^2 (\sin \theta')^{-1} e^{-\pi \beta^2 \theta'^2}$$

$$= (1 - \theta' (\sin \theta')^{-1} \cos \theta' - 2\pi \theta'^2 \beta^2) \beta^3 (\sin \theta')^{-1} e^{-\pi \beta^2 \theta'^2}.$$

Taking a second derivative with respect to θ', we get

$$\frac{\partial^2 f}{\partial \theta'^2} = -(1 - \theta' (\sin \theta')^{-1} \cos \theta' - 2\pi \theta'^2 \beta^2) \beta^3 (\sin \theta')^{-2} \cos \theta' e^{-\pi \beta^2 \theta'^2}$$

$$-2\pi \beta^5 \theta' (\sin \theta')^{-1} e^{-\pi \beta^2 \theta'^2} (1 - \theta' (\sin \theta')^{-1} \cos \theta' - 2\pi \theta'^2 \beta^2)$$

$$+\beta^3 (\sin \theta')^{-1} e^{-\pi \beta^2 \theta'^2} (-(\sin \theta')^{-1} \cos \theta' + \theta' (\sin \theta')^{-2} \cos^2 \theta' + \theta' - 4\pi \theta' \beta^2).$$

Substituting these expressions into (16.53), we see that the term $(\sin \theta')^{-1} e^{-\pi \beta^2 \theta'^2}$ can be cancelled. In addition many other terms are also cancelled, and the remaining terms are

$$3\beta^2 \dot{\beta} \theta' - 2\pi \beta^4 \dot{\beta} \theta'^3 = \frac{K}{4} [4\pi^2 \beta^7 \theta'^3 - 2\pi \beta^5 \theta' - 4\pi \theta' \beta^5 + \beta^3 \theta']. \tag{16.58}$$

It appears difficult to determine $\beta(t)$. However, it is clear that if the equation

$$\dot{\beta} = -2\pi \beta^3 \cdot \frac{K}{4}$$

holds, many of the terms will be cancelled. The solution of this differential equation under the condition that the limit of $\beta(t)$ as $t \to 0$ is infinity (as must be the case for a delta function), we get

$$\beta(t) = \frac{1}{\sqrt{\pi K t}}.$$

With this function, the only remaining term is the $\beta^3 \theta'$ inside of the brackets in (16.58). This means that when the above differential equation for $\beta(t)$ is solved, then $f(\theta', t)$ will solve the equation

$$\frac{\partial f}{\partial t} = \frac{K}{4} \left(\frac{\partial^2 f}{\partial \theta'^2} + 2 \cot \theta' \frac{\partial f}{\partial \theta'} - f \right), \tag{16.59}$$

which is very similar to (16.57).

Hence, a solution of the above "heat-like" equation (16.59) written in terms of θ (instead of θ') is

$$f(\theta, t) = \frac{1}{2(\pi K t)^{3/2}} \frac{\theta}{\sin \theta/2} e^{-\theta^2/4Kt}. \tag{16.60}$$

Generating a Gaussian from this Candidate

Unfortunately, the function (16.60) does not exactly satisfy the heat equation. In this subsection, we generate the solution of the heat equation by modifying slightly the solution of this heat-like equation.

We are given the heat-like equation

$$\frac{\partial f}{\partial t} = K(\nabla^2_{SO(3)} f - f)$$

with initial conditions $f(R, 0) = \delta(R)$. Let us define an operator $D(\cdot)$ as

$$D(f) = \nabla^2_{SO(3)} f - f.$$

Then our goal is to find the solution of

$$\frac{\partial g}{\partial t} = K(D(g) + g), \qquad (16.61)$$

by using the solution of

$$\frac{\partial f}{\partial t} = K \cdot D(f) \qquad (16.62)$$

under the same initial condition. Here $g(R, t)$ should be related in some way to $f(R, t)$. One of the simplest kinds of transformations between f and g is of the form

$$g(R, t) = \alpha(t) \, f(R, t).$$

Substituting this into (16.61) results in

$$\dot\alpha(t) \, f(R, t) + \alpha(t) \frac{\partial f}{\partial t} = K(D(\alpha(t) \, f(R, t)) + \alpha(t) \, f(R, t)). \qquad (16.63)$$

Since $D(\cdot)$ is a differential operator that acts on the parameters of $R = R(\theta, \lambda, \nu)$, and not t, we can write $D(\alpha(t) \, f(R, t)) = \alpha(t) D(f(R, t))$. Then we can use (16.59) to simplify (16.63) into

$$\dot\alpha f = K \alpha f.$$

Dividing both sides by f, we get a simple ordinary differential equation for $\alpha(t)$, which can be solved as

$$\alpha(t) = \alpha(0) \, e^{Kt}.$$

Since the initial conditions for $f(R, 0)$ and $g(R, 0)$ are identical, we get $\alpha(0) = 1$. Hence, while our initial candidate did not exactly satisfy the heat equation, it is the case that

$$\frac{\partial g}{\partial t} = K \cdot \nabla^2_{SO(3)} g$$

when

$$g(R, t) = f(\theta, t) \, e^{Kt}. \qquad (16.64)$$

with $f(\theta, t)$ defined in (16.60).

16.10   Interpreting Experimental Rotational Anisotropy Spectroscopy Data

This section illustrates how Wigner D-functions (i.e., IURs of $SO(3)$) arise in the context of *Rotational Anisotropy Spectroscopy*, which is an experimental methodology that can be used to relate the shape of protein molecules to macroscopic experimental observations of fluorescence.[5] In typical fluorescence anisotropy experiments on proteins, a single probe (such as a naturally occurring or engineered tryptophan residue) is embedded in the protein molecule. Observations and analysis of the anisotropy decay of a

---

[5] We thank Dr. Dmitri Toptygin for contributing text to this section.

photo-selected sub ensemble of these protein molecules in solution to the uniform orientational distribution are usually interpreted under the assumption of isotropic rotational
Brownian motion. From this, the size and flexibility characteristics of an assumed spherical molecule are assessed. However, when there are two versions of the protein, each
with a probe oriented differently relative to the body-fixed frame, no assumption of
sphericity needs to be made, and information about the aspect ratio of the protein can
be obtained.

16.10.1 Fluorescence Anisotropy for a Single-Rigid-Domain Protein Undergoing Brownian Motion

Fluorescence Anisotropy Spectroscopy is a well-established method for the experimental
observation of molecular orientational dynamics. See, for example, the following references, some of which date back to the 1960s [11, 29, 71, 73, 79, 83, 99, 117, 126, 127,
128, 143]. A fluorescence anisotropy decay experiment produced an anisotropy decay
curve [5, 79]

$$r(t) = \frac{I_{zz} - I_{zx}}{I_{zz} + 2I_{zx}} = \sum_{i=1}^{n} \beta_i \exp(-t/\tau_i). \tag{16.65}$$

Here $n = 3$ for a triaxial rigid macromolecule, but for flexible systems n can be greater,
and for rigid bodies with symmetry n can be less than 3. τ_i is a time constant for each
exponential decay.

The symbol I_{ab} refers to the intensity measured when excitation polarization is
parallel to axis a and emission polarization is parallel to b. I_{zz} and I_{zx} are related
to statistical models of molecular motion described below. The reason why $r(t)$ can
be resolved into a weighted sum of decaying exponentials also is related to molecular
models of rotational Brownian motion. The explanation is somewhat involved, and is
summarized below.

From Chapter 4, we know that any function of direction can be expanded in a series
of spherical harmonics as

$$f(\mathbf{u}(\phi,\theta)) = \sum_{l=0}^{\infty} \sum_{n=-l}^{l} \tilde{f}_n^l Y_l^n(\mathbf{u}(\phi,\theta)). \tag{16.66}$$

Often $Y_l^n(\mathbf{u}(\phi,\theta))$ is written as $Y_l^n(\phi,\theta)$ but it will be convenient to think of them as
functions of unit direction vectors $Y_l^n(\mathbf{u})$ that can be expressed in terms of angles rather
than as functions of angles directly. The spherical harmonics are a complete orthonormal
basis for the set of all well-behaved functions on the sphere. This means that if we rotate
a function of direction as $f(R^T\mathbf{u})$, then it too should have an expansion of a similar
form; only $\tilde{f}_n^l$ will be different and depend on the rotation R. Or stated in a different
way, if $Y_l^n(\mathbf{u})$ is rotated, then the result must be expressible in terms of a weighted sum
of other spherical harmonics. Explicitly, this is written as

$$Y_l^m(R^T\mathbf{u}) = \sum_{n=-l}^{l} D_{mn}^l(R) Y_l^n(\mathbf{u}). \tag{16.67}$$

Here $D_{mn}^l(R)$ for $-l \le m, n \le l$ are the elements of a $(2l+1) \times (2l+1)$ matrix $D^l(R)$
called the lth-order *Wigner matrix* or *Wigner D-function*, which is a matrix-valued
function of rotation.

We are interested in modeling $f(R;t)$ where f is a probability density in orientation (in Euler angles $R = R_z(\alpha)R_x(\beta)R_z(\gamma)$) and t is time. Here $f(R;t) \geq 0$ and

$$\int_{SO(3)} f(R;t)\, dR = \frac{1}{8\pi^2} \int_0^{2\pi} \int_0^{\pi} \int_0^{2\pi} f(R_z(\alpha)R_x(\beta)R_z(\gamma);t) \sin\beta\, d\alpha\, d\beta\, d\gamma = 1$$

and the normalization $1/8\pi^2$ is so that $f(R;t \to \infty) \to 1$ is the uniform orientational density.

The use of Wigner D-functions in the modeling of rotational Brownian motion and fluorescence anisotropy decay is by no means new. See, for example, [36, 116, 124, 136]. Here a short overview of how this modeling tool is connected to the observables in our new formulation is provided.

Suppose that an isolated tryptophan is embedded in a rigid protein. In a reference frame attached to this protein, let the the direction of the transition moment of the tryptophan be denoted as the unit vector $\mathbf{u}$. Since R describes the orientation of the protein relative to the lab frame, $R\mathbf{u}$ is the direction of the transition moment of the tryptophan as seen in the lab frame. Now, if in the lab frame polarized light is shined from the x direction, and if this light has, by design, an electric vector given by unit direction vector $\mathbf{n} = \mathbf{e}_3$, then each protein in the total ensemble will be promoted to the excited state with the probability proportional to $(\mathbf{n}^T R\mathbf{u})^2$, which is $\cos^2\theta$ where θ is the angle between the transition moment of the tryptophan residue and the electric vector of the exciting light.

That is, the effect is to "photo-select," or measure an orientationally biased subset, of the ensemble described by the equilibrium distribution $f_\infty(R)$ (which would normally be the isotropic orientational distribution, but could be anisotropic if the surrounding environment is liquid crystalline or if the proteins under consideration are strongly polar and an external electrical field is applied). In any case, the distribution describing the photo-selected ensemble will certainly not be isotropic. It can be modeled as

$$f_0(R) = \frac{(\mathbf{n}^T R\mathbf{u})^2 f_\infty(R)}{\int_{SO(3)} (\mathbf{n}^T A\mathbf{u})^2 f_\infty(A)\, dA} \tag{16.68}$$

where the denominator is simply the normalization to make $f_0(R)$ a probability density and $A \in SO(3)$ is a dummy variable of integration. The photo-selection process establishes initial conditions for a diffusion equation (to be discussed later in the paper) that defines $f(R;t)$. The solution $f(R;t)$ is then expressed in an $SO(3)$-Fourier series. In other words, $f(R;t)$ that connects $f_0(R)$ and $f_\infty(R)$ *is known for all values of time if the properties of the rotational diffusion process are known.* But of course, this is the issue. If these rotational diffusion coefficients are not known in advance, then they can be measured by performing experiments. And from these, the size characteristics of the molecule can be obtained.

It is important to distinguish between what the photo-selected ensemble of molecules is *doing* (i.e., how the orientational distribution describing its population is evolving), and what is actually *observed* by a fluorescence anisotropy experiment. So far we have addressed the "doing" part. Below we link this to the "observables." The observation of the photo-selected population evolves in two ways. First of all, the total population decays with time because of the radiative decay and non-radiative decay. The total number of molecules in the excited state at time t is $N(t)$, and this number decreases with time. The observation of the photo-selected population also evolves because the molecules change their orientation, and this is what we are interested in here. To separate the effect of rotation from the effect of population decay, we make two measurements. In both measurements the emitted light is observed along the lab frame's y axis (perpendicular to the direction of incoming light). In the first measurement the polarizer is set to have a direction $\mathbf{w} = \mathbf{e}_3$ that is tuned to select photons coming from the sample that have electric vector oriented parallel to the lab frame's z axis.

The probability of emitting such a photon by an excited fluorescent molecule with the transition dipole moment direction $\mathbf{u}$ is proportional to $(\mathbf{e}_3^T R \mathbf{u})^2$; therefore the measured signal is also proportional to $(\mathbf{e}_3^T R \mathbf{u})^2$. In the second measurement the polarizer is oriented to be parallel to the lab frame's x axis ($\mathbf{w} = \mathbf{e}_1$), and the measured signal is proportional to $(\mathbf{e}_1^T R \mathbf{u})^2$.

It is not difficult to see that

$$I_{zz}(t) = [1 + 2r(t)] \cdot N(t) \quad \text{and} \quad I_{zx}(t) = [1 - r(t)] \cdot N(t) \tag{16.69}$$

and therefore

$$I_{zz}(t) + 2I_{zx}(t) = N(t) \quad \text{and} \quad I_{zz}(t) - I_{zx}(t) = r(t) \cdot N(t).$$

Here $r(t)$ depends on rotation only and $N(t)$ dependents on population decay only; $N(t)$ is eliminated when we calculate $r(t)$ from $I_{zz}(t)$ and $I_{zx}(t)$. This explains why we need two measurements, and not just one. In the I_{zz} measurement, the vector $\mathbf{w}$ is parallel to the z axis (and to the vector $\mathbf{n}$). In the I_{zx} measurement, the vector $\mathbf{w}$ is parallel to the x axis (and normal to the vector $\mathbf{n}$). We cannot have a measurement with vector $\mathbf{w}$ parallel to the y axis because the emitted light is observed along the y axis, and electromagnetic waves are transverse waves (the direction of the electric vector, which interacts with the transition moment of the fluorescent probe, is always normal to the direction of propagation).

The experimental setup is described in [129] and the description of experimental observables above and in (16.65) connect to the time-evolving orientational distribution of the photo-selected sub-ensemble because any experimental measurable that depends on the time-evolving ensemble of the photo-selected sub-ensemble of rotors must be of the form

$$m(t) = \langle M \rangle = \int_{SO(3)} M(R) \, f(R; t) \, dR.$$

In particular, the measurables in the current lab setup are of the form

$$m(t, \mathbf{w}) = N(t) \cdot \int_{SO(3)} (\mathbf{w}^T R \mathbf{u})^2 f(R; t) \, dR. \tag{16.70}$$

with $\mathbf{w} = \mathbf{e}_3$ and $\mathbf{w} = \mathbf{e}_1$, as discussed before, where again $N(t)$ is the number of photo-selected molecules. The connection between the orientational distribution $f(R; t)$ and the measurables discussed previously is

$$I_{zz} = m(t, \mathbf{e}_3) \quad \text{and} \quad I_{zx} = m(t, \mathbf{e}_1). \tag{16.71}$$

In completely isotropic systems, where $f_\infty(R) = 1$, and also in uniaxial anisotropic systems, where the director is parallel to the lab frame's z axis (this includes nematic liquid crystals and systems with an alignment potential parallel to the lab frame's z axis), $f_\infty(R)$ is independent of the Euler angle β. In this case

$$I_{xz} = \int_{SO(3)} (\mathbf{e}_1^T R \mathbf{u})^2 f(R; t) \, dR = \int_{SO(3)} (\mathbf{e}_2^T R \mathbf{u})^2 f(R; t) \, dR = I_{yz}.$$

Furthermore, in this case

$$\int_{SO(3)} \left[(\mathbf{e}_1^T R\mathbf{u})^2 + (\mathbf{e}_2^T R\mathbf{u})^2 + (\mathbf{e}_3^T R\mathbf{u})^2 \right] f(R;t) \, dR = 1.$$

Therefore, in these systems, it follows from (16.65) that

$$r_{i,n}(t, \mathbf{u}) = \int_{SO(3)} P_2(\mathbf{e}_3^T R\mathbf{u}) f(R;t) \, dR \qquad (16.72)$$

where $P_2(x)$ is the second-order Legendre polynomial and the subscripts i, n indicate that this expression holds for an isotropic or nematic liquid crystalline solvent. And the dependence of the function $r_{i,n}(t, \mathbf{u})$ on $\mathbf{u}$ indicates that the observed decay curves can depend on the direction of the dipole moment of the the probe in the body-fixed frame of the molecule. Our hypothesis is that if multiple probes are used and multiple curves $r_{i,n}(t, \mathbf{u}_1)$, $r_{i,n}(t, \mathbf{u}_1)$, ..., $r_{i,n}(t, \mathbf{u}_k)$ are observed then for a rigid molecule: (a) more information can be obtained than the case of a single probe; and (b) if $k > 3$, the information becomes redundant, and this redundancy can be used to reject noise.

During the 1970's a number of theories that model the time-evolution of the orientational density of a photo-selected sub-ensemble of rigid molecules were developed [6, 24, 36, 118, 138]. These theories model the anisotropy decay *to the uniform orientational distribution* $f_\infty(R) = 1$, and develop the relationship between $f(R;t)$ and the weighted sum of decaying exponentials in (16.65). Below, methods for linking $f(R;t)$ and mechanical models of rotational Brownian motion are articulated.

16.10.2 Rotational Brownian Motion of Rigid-Body Molecules

Let R denote the orientation of a protein molecule relative to the lab frame. Euler's equations of motion for a rigid molecule subjected to Brownian motion are

$$I\dot{\boldsymbol{\omega}} + \boldsymbol{\omega} \times (I\boldsymbol{\omega}) = -\tilde{\mathbf{X}}^r V - C\boldsymbol{\omega} + S\mathbf{w}. \qquad (16.73)$$

Here $\tilde{\mathbf{X}}^r = [X_1^r, X_2^r, X_3^r]^T$ is the rotational analog of the translational gradient, where

$$(X_i^r f)(R) = \left. \frac{d}{d\epsilon} f(R \cdot \exp(\epsilon E_i)) \right|_{\epsilon=0}. \qquad (16.74)$$

Explicitly in terms of ZXZ Euler angles, if $R = R(\alpha, \beta, \gamma)$, and $f(R(\alpha, \beta, \gamma); t)$ is viewed as a function of these angles directly, $f(\alpha, \beta, \gamma; t)$, then

$$X_1^r f = -\cot\beta \sin\gamma \frac{\partial f}{\partial \gamma} + \frac{\sin\gamma}{\sin\beta} \frac{\partial f}{\partial \alpha} + \cos\gamma \frac{\partial f}{\partial \beta};$$

$$X_2^r f = -\cot\beta \cos\gamma \frac{\partial f}{\partial \gamma} + \frac{\cos\gamma}{\sin\beta} \frac{\partial f}{\partial \alpha} - \sin\gamma \frac{\partial f}{\partial \beta};$$

$$X_3^r f = \frac{\partial f}{\partial \gamma}.$$

I is the moment of inertia of the molecule and C is the anisotropic rotational drag matrix. Here $\mathbf{w} = d\mathbf{W}/dt$ is the time derivative of a Wiener process, and multiplication through by dt gives a system of stochastic differential equations.[6] The matrices C and S are related. Indeed, from the fluctuation-dissipation theorem

$$(2k_BT)C = SS^T, \tag{16.75}$$

where k_B is the Boltzmann constant and T is temperature measured in degrees Kelvin.

It is possible to write the Fokker-Planck/Schmolokowski equations corresponding to (16.73). Indeed, this was done half a century ago and is surveyed in [21, 45, 48, 55, 58, 59, 91]. The result is an *inertial theory* of Brownian motion, pioneered by Smoluchowski in the translational case at the beginning of the twentieth century, and in the rotational case by [121]. Applications include fluorescence anisotropy microscopy [124, 136]. Note that due to the structure of these equations, when C, and S are all properly balanced according to (16.75), the solution to the Fokker-Planck equation corresponding to (16.73) converges to the Boltzmann distribution as $t \to \infty$.

In the *noninertial theory* of rotational Brownian motion (which historically preceded the inertial theory), I is assumed to be negligible and if rotations are parameterized by $\phi = [\alpha, \beta, \gamma]^T$ and $V(R) = V(R(\phi))$, the result is

$$CJ(\phi)\dot{\phi} = -\tilde{\mathbf{X}}^r V(R(\phi)) + S\mathbf{w}. \tag{16.76}$$

Note that (16.76) must be interpreted as a Stratonovich equation because usual Calculus was used to write $\boldsymbol{\omega} = J(\phi)\dot{\phi}$, where $J(\phi)$ is the well-know Jacobian matrix that relates angular velocity expressed in the body-fixed frame to rates of change of Euler angles. (16.76) is equivalent to the coordinate-dependant expression

$$\dot{\phi} = -J^{-1}(\phi)C^{-1}\tilde{\mathbf{X}}^r V(R(\phi)) + J^{-1}(\phi)C^{-1}S\mathbf{w}. \tag{16.77}$$

The noninertial theory of rotational Brownian motion described above becomes equivalent to that developed by Perrin [105, 106, 107] when C is chosen appropriately.

A well-developed theory relates stochastic differential equations such as (16.77) to evolution equations that describe the behavior of time-dependent probability densities of orientation. Such equations are described below for the case when $V = 0$.

16.11   Solving the Equation for Non-Inertial Rotational Diffusion

In the case when there is no external potential, the probability density $f(R, t)$ corresponding to (16.76) (or equivalently (16.77)) is of the form

$$\frac{\partial f}{\partial t} = \mathcal{K}f \tag{16.78}$$

where the operator $\mathcal{K}$ is defined as

$$\mathcal{K} \doteq \frac{1}{2} \sum_{i,j=1}^{3} D_{ij} X_i^r X_j^r$$

and X_i^r are the differential operators defined in (16.74).

---

[6]Though a Wiener process is not differentiable, $d\mathbf{W}_i/dt$ can be interpreted in the sense of finite differences in computer simulations.

The constants D_{ij} in the definition of the diffusion operator $\mathcal{K}$ denote the components of the diffusion tensor,

$$D = C^{-1}SS^{T}C^{-1}.$$

However, from (16.75), it follows that

$$D = (2k_B T)C^{-1}. \tag{16.79}$$

16.12 Extracting Information from Decay Curves for Multiple Independent Probes

Here the mathematical formulation of the previous section is used together with (16.72) to match surface shape characteristics of a protein to experimental measurables.

First, $P_2(e_3^T R\mathbf{u}_k)$ can be expanded in a spherical harmonic expansion of the form

$$P_2(e_3^T R\mathbf{u}_k) = \sum_{m,l} c_{ml} Y_l^m(R\mathbf{u}_k).$$

But by the "addition theorem" for two unit vectors,

$$P_l(\mathbf{a} \cdot \mathbf{b}) = \frac{4\pi}{2l+1} \sum_{m=-l}^{l} Y_l^m(\mathbf{a}) Y_l^{-m}(\mathbf{b}).$$

In our case $l = 2$, $\mathbf{a} = e_3$, and $\mathbf{b} = R\mathbf{u}_k$. Therefore,

$$P_2(e_3^T R\mathbf{u}_k) = \frac{4\pi}{2l+1} \sum_{m=-2}^{2} Y_2^m(e_3) Y_2^{-m}(R\mathbf{u}_k).$$

Furthermore,

$$Y_2^m(e_3) = \begin{cases} \sqrt{5/(4\pi)} & \text{if } m = 0 \\ 0 & \text{if } m \neq 0 \end{cases}.$$

Therefore,

$$P_2(e_3^T R\mathbf{u}_k) = \sqrt{\frac{4\pi}{5}} Y_2^0(R\mathbf{u}_k). \tag{16.80}$$

Then, using (16.67),

$$Y_l^m(R\mathbf{u}_k) = \sum_{n=-l}^{l} (D_{mn}^l(R))^* Y_l^n(\mathbf{u}_k).$$

Combining (16.72) and (16.44) gives a relationship between the parameters D_{ij} and the observables $r_{i,n}(t, \mathbf{u}_k)$. The relationship between the shape of the semi-axis lengths of an approximately ellipsoidal protein can then be extracted using the closed-form relationship between the viscosity tensor and the shape parameters in an ellipsoid developed a century ago by Jeffrey [65].

16.10 Summary

In this chapter we reviewed the classical theories of rotational Brownian motion and their generalizations. Diffusion processes on the rotation group are associated with a diverse array of scenarios ranging from chemical physics and the analysis of liquid crystals to estimation problems. Fokker-Planck equations describing the time evolution of probability density functions in rotational Brownian motions are derived. In the non-inertial theory of rotational Brownian motion, a PDF governed by the Fokker-Planck equation is a function of rotation. In the inertial theory of rotational Brownian motion these PDFs are functions of both rotation and angular velocity.

In addition to the references provided throughout this chapter, topics related to rotational Brownian motion can be found in [19, 34, 44, 104, 108, 110]. In recent years, dipolar interactions [10] have been used with great success in extracting additional information from Nuclear Magnetic Resonance (NMR) experiments.

References

1. Adelman, S.A., "Fokker-Planck Equations for Simple Non-Markovian Systems," *Journal of Chemical Physics*, 64(1): 124–130, 1976.
2. Albeverio, S., Arede, T., Haba, Z., "On Left Invariant Brownian Motions and Heat Kernels on Nilpotent Lie Groups," *Journal of Mathematical Physics*, 31(2): 278–286, Feb. 1990.
3. Applebaum, D., Kunita, H., "Lévy Flows on Manifolds and Lévy Processes on Lie Groups," *Journal of Mathematics of Kyoto University* 33/34:1103–1123, 1993.
4. Balabai, N., Sukharevsky, A., Read, I., Strazisar, B., Kurnikova, M., Hartman, R. S., Coalson, R. D., Waldeck, D. H., "Rotational Diffusion of Organic Solutes: The Role of Dielectric Friction in Polar Solvents and Electrolyte Solutions," *Journal of Molecular Liquids*, 77: 37–60, 1998.
5. Barkley, M.D., Kowalczyk, A.A., Brand, L., "Fluorescence decay studies of anisotropic rotations of small molecules," *Journal of Chemical Physics*, 75(7): 3581–3593, 1981.
6. Belford, G.G., Belford, R.L., Weber, G., "Dynamics of Fluorescence Polarization in Macromolecules," *Proceedings of the National Academy of Sciences of the United States of America*, 69(6): 1392–1393, June 1972
7. Bellman, R.E., Roth, R.S., *The Laplace Transform*, World Scientific, Singapore, 1984.
8. Berggren, E., Tarroni, R., Zannoni, C., "Rotational Diffusion of Uniaxial Probes in Biaxial Liquid Crystal Phases," *Journal of Chemical Physics*, 99(8): 6180–6200, Oct. 1993.
9. Berggren, E., Zannoni, C., "Rotational Diffusion of Biaxial Probes in Biaxial Liquid Crystal Phases," *Molecular Physics*, 85(2): 299–333, 1995.
10. Bernassau, J.M., Black, E.P., Grant, D.M., "Molecular Motion in Anisotropic Medium. I. The Effect of the Dipolar Interaction on Nuclear Spin Relaxation," *Journal of Chemical Physics*, 76(1): 253–256, Jan. 1982.
11. Berne, B.J., Pechukas, P., Harp, G.D., "Molecular Reorientation in Liquids and Gases," *Journal of Chemical Physics*, 49: 3125–3129, 1968.
12. Bialik, C.N., Wolf, B., Rachofsky, E.L., Ross, J.B., Laws, W.R., "Dynamics of Biomolecules: Assignment of Local Motions by Fluorescence Anisotropy Decay," *Biophysical Journal* 75(5): 2564-2573, Nov. 1998.
13. Blenk, S., Ehrentraut, H., Muschik, W. "Statistical Foundation of Macroscopic Balances for Liquid-Crystals in Alignment Tensor Formulation," *Physica A*, 174(1): 119–138, 1991.
14. Blenk, S., Muschik, W. "Orientational Balances for Nematic Liquid-Crystals," *Journal of Non-Equilibrium Thermodynamics*, 16(1): 67–87, 1991.
15. Blinov, L.M., *Electro-Optical and Magneto-Optical Properties of Liquid Crystals*, Wiley-Interscience, 1984.
16. Blokhin, A.P., Gelin, M.F., "Rotation of Nonspherical Molecules in Dense Fluids: A Simple Model Description," *Journal of Physical Chemistry B*, 101(2): 236–243, 1997.
17. Blokhin, A.P., Gelin, M.F., "Rotational Brownian Motion of Spherical Molecules: The Fokker-Planck Equation with Memory," *Physica A*, 229(3): 501–514, 1996.

18. Brenner, H., "The Stokes Resistance of an Arbitrary Particle," *Chemical Engineering Science*, 18:1–25, 1963.

19. Brilliantov, N.V., Vostrikova, N.G., Denisov, V.P., Petrusievich, Yu. M., Revokatov, O.P., "Influence of Dielectric Friction and Near-Surface Increase of Viscosity on Rotational Brownian Motion of Charged Biopolymers in Solution," *Biophysical Chemistry*, 46(3): 227–236, 1993.

20. Brockett, R.W., "Notes on Stochastic Processes on Manifolds," in *Systems and Control in the Twenty-First Century* (C.I. Byrnes et al. eds.), Birkhäuser, Boston, 1997.

21. Budó, A., Fischer, E., Miyamoto, S., "Einfluß der Molekülform auf die dielektrische Relaxation," *Physikalische Zeitschrift*, 40: 337–345, 1939.

22. Chigrinov, V.G., *Liquid Crystal Devices: Physics and Applications*, Artech House, Boston, 1999.

23. Chirikjian, G.S., *Stochastic Models, Information Theory, and Lie Groups, Vols. I+II*, Birkäuser, Boston, 2009/2012.

24. Chuang, T.J., Eisenthal, K.B., "Theory of Fluorescence Depolarization by Anisotropic Rotational Diffusion," *Journal of Chemical Physics* 57(12): 5094–5097, Dec. 1972.

25. Coffey, W., Evans, M., Grigolini, P., *Molecular Diffusion and Spectra*, John Wiley and Sons, New York 1984.

26. Coffey, W.T., Kalmykov, Yu. P., "On the Calculation of the Dielectric Relaxation Times of a Nematic Liquid Crystal from the Non-Inertial Langevin Equation," *Liquid Crystals*, 14(4): 1227–1236, 1993.

27. Coffey, W.T., Kalmykov, Yu. P., *The Langevin Equation: With Applications to Stochastic Problems in Physics, Chemistry and Electrical Engineering*, 3rd ed., World Scientific, Singapore, 2012.

28. Cukier, R.I., "Rotational Relaxation of Molecules in Isotropic and Anisotropic Fluids," *Journal of Chemical Physics*, 60(3): 734–743, Feb. 1974.

29. Dale, R.E., Chen, L.A., Brand, L., "Rotational Relaxation of the "Microviscosity" Probe Diphenylhexatriene in Paraffin Oil and Egg Lecithin Vesicles," *Journal of Biological Chemistry*, 252(21): 7500–7510, 1977.

30. Debye, P., *Polar Molecules*, Dover Publishers, New York, 1929. Also see *The Collected Papers of Peter J.W. Debye*, Interscience Publishers, New York, 1954.

31. de Gennes, P.G., Prost, J., *The Physics of Liquid Crystals*, 2nd ed., Clarendon Press, Oxford, 1998.

32. de Jeu, W.H., *Physical Properties of Liquid Crystalline Materials*, Gordon and Breach, New York, 1980.

33. Dong, R.Y., *Nuclear Magnetic Resonance of Liquid Crystals*, 2nd ed., Springer, New York, 1997. (softcover edition, 2013).

34. Edwards, D., "Steady Motion of a Viscous Liquid in Which an Ellipsoid Is Constrained to Rotate About a Principal Axis," *Quarterly Journal of Pure and Applied Mathematics*, 26: 70–78, 1893.

35. Egelstaff, P.A., "Cooperative Rotation of Spherical Molecules," *The Journal of Chemical Physics*, 53(7): 2590–2598, Oct. 1970.

36. Ehrenberg, M., Rigler, R., "Polarized Fluorescence and Rotational Brownian Motion," *Chemical Physics Letters*, 14(5): 539–544, July 1972.

37. Ehrentraut, H., Muschik, W., Papenfuss, C. "Mesoscopically Derived Orientation Dynamics of Liquid Crystals," *Journal of Non-Equilibrium Thermodynamics*, 22(3): 285–298, 1997.

38. Einstein, A., *Investigations on the Theory of the Brownian Movement*, Dover, New York, 1956.

39. Elworthy, K.D., *Stochastic Differential Equations on Manifolds*, Cambridge University Press, 1982.

40. Emery, M., *Stochastic Calculus in Manifolds*, Springer-Verlag, Berlin, 1989.

41. Epperson, J.B., Lohrenz, T., "Brownian Motion and the Heat Semigroup on the Path Space of a Compact Lie Group," *Pacific Journal of Mathematics*, 161(2): 233–253, 1993.

42. Ericksen, J.L., Kinderlehrer, D., eds., *Theory and Applications of Liquid Crystals*, IMA Volumes in Mathematics and Its Applications, Vol. 5, Springer Verlag, New York, 1987.

43. Evans, G.T., "Orientational Relaxation of a Fokker-Planck Fluid of Symmetric Top Molecules," *The Journal of Chemical Physics*, 67(6): 2911–2915, Sept. 1977.

44. Evans, M.W., Ferrario, M., Grigolini, P., "The Mutual Interaction of Molecular Rotation and Translation," *Molecular Physics*, 39(6): 1369–1389, 1980.

45. Favro, L.D., "Theory of the Rotational Brownian Motion of a Free Rigid Body," *Physical Review*, 119(1): 53–62, July 1960.

46. Fixman, M., Rider, K., "Angular Relaxation of the symmetrical Top," *The Journal of Chemical Physics*, 51(6): 2425–2438, Sept. 1969.

47. Fokker, A.D., "Die mittlere Energie rotierender elektrischer Dipole im Strahlungsfeld," *Annalen der Physik*, 43:810–820, 1914.

48. Furry, W.H., "Isotropic Rotational Brownian Motion," *Physical Review*, 107(1): 7–13, July 1957.

49. Gardiner, C.W., *Handbook of Stochastic Methods* 4th ed., Springer-Verlag, Berlin, 2009.

50. Gordon, R.G., "On the Rotational Diffusion of Molecules," *Journal of Chemical Physics*, 44(5): 1830–1836, 1966.

51. Gorman, C.D., "Brownian Motion of Rotation," *Transactions of the American Mathematical Society*, 94:103–117, 1960.

52. Grenander, U., *Probabilities on Algebraic Structures*, Dover Edition, 2008. (originally, Wiley, 1963).

53. Hess, S. "Fokker-Planck Equation Approach to Flow Alignment in Liquid-Crystals," *Zeitschrift Fur Naturforschung Section A - A Journal of Physical Sciences*, 31(9): 1034–1037, 1976.

54. Heyer, H., *Probability Measures on Locally Compact Groups*, Springer-Verlag, New York, 1977.

55. Hubbard, P.S., "Rotational Brownian Motion," *Physical Review A*, 6(6): 2421–2433, Dec. 1972.

56. Hubbard, P.S., "Rotational Brownian Motion. II. Fourier Transforms for a Spherical Body with Strong Interactions," *Physical Review A*, 8(3): 1429–1436, Sept. 1973.

57. Hubbard, P.S., "Theory of Nuclear Magnetic Relaxation by Spin-Rotational Interactions in Liquids," *Physical Review*, 131(3): 1155–1165, Aug. 1963.

58. Hubbard, P.S., "Nuclear Magnetic Relaxation in Spherical-Top Molecules Undergoing Rotational Brownian Motion," *Physical Review A*, 9(1): 481–494, Jan. 1974.

59. Hubbard, P.S., "Angular Velocity of a Nonspherical Body Undergoing Rotational Brownian Motion," *Physical Review A*, 15(1): 329–336, Jan. 1977.

60. Ikeda, N., Watanabe, S., *Stochastic Differential Equations and Diffusion Processes*, 2nd ed., North-Holland, Amsterdam, 1989.

61. Itô, K., "Brownian Motions in a Lie Group," *Proceedings of the Japan Academy*, 26(8): 4–10, 1950.

62. Itô, K., "Stochastic Differential Equations in a Differentiable Manifold," *Nagoya Mathematical Journal* 1: 35–47, 1950.

63. Itô, K., "Stochastic Differential Equations in a Differentiable Manifold (2)," *Memoirs of the College of Science, University of Kyoto. Series A: Mathematics*, 28(1): 81–85, 1953.

64. Itô, K., McKean, H.P. Jr., *Diffusion Processes and Their Sample Paths*, Springer, 1996.

65. Jeffery, G.B., "The Motion of Ellipsoidal Particles Immersed in a Viscous Fluid," *Proceedings of the Royal Society of London. Series A*, 102(715): 161–179, Nov. 1922.

66. Kalmykov, Y.P., Quinn, K.P., "The Rotational Brownian Motion of a Linear Molecule and Its Application to the Theory of Kerr Effect Relaxation," *Journal of Chemical Physics*, 95(12): 9142–9147, Dec. 1991.

67. Kalmykov, Yu. P., "Rotational Brownian Motion in an External Potential Field: A Method Based on the Langevin Equation," *Chemical Physics Reports*, 16(3): 535–548, 1997.

68. Kalmykov, Y. P., "Rotational Brownian Motion in an External Potential: The Langevin Equation Approach," *Journal of Molecular Liquids*, 69: 117–131, 1996.

69. Kalmykov, Y. P., Coffey, W. T., "Analytical Solutions for Rotational Diffusion in the Mean Field Potential: Application to the Theory of Dielectric Relaxation in Nematic Liquid Crystals," *Liquid Crystals*, 25(3): 329–339, 1998.

70. Karatzas, I., Shreve, S.E., *Brownian Motion and Stochastic Calculus*, 2nd ed., Springer, 1991.

71. Kinosita, K., Jr., Kawato, S., Ikegami, A., "A Theory of Fluorescence Depolarization Decay in Membranes," *Biophysical Journal*, 20: 289–305, 1977.

72. Kinosita, K. Jr, Ikegami, A., Kawato, S., "On the Wobbling-in-Cone Analysis of Fluorescence Anisotropy Decay," *Biophysical Journal*, 37(2): 461–464, 1982.

73. Kizel, V.A., Skylaruk, V.A., Toptygin, D.D., Khanmogometova, S.D., "Kinetics of the Rotational Depolarization of Impurity-Molecule Luminescence in a Liquid-Crystal Matrix," *JETP Letters* 44(1): 34–38, 1986

74. Kobayashi, K.K., "Theory of Translational and Orientational Melting with Application to Liquid Crystals. I," *Journal of the Physical Society of Japan*, 29(1): 101–105, July 1970.

75. Kobayashi, K.K., "Theory of Translational and Orientational Melting with Application to Liquid Crystals," *Molecular Crystals and Liquid Crystals*, 13(2): 137–148, 1971.

76. Kolmogorov, A., "Über die analytischen Methoden in der Wahrscheinlichkeitsrechnung," *Mathematische Annalen*, 104: 415–458, 1931.

77. Kunita, H., *Stochastic Flows and Stochastic Differential Equations*, Cambridge University Press, 1997.

78. Langevin, P., "Sur la théorie du mouvement Brownien," *Comptes. Rendues Acad. Sci. Paris*, 146: 530–533, 1908.

79. Lakowicz, J.R., *Principles of Fluorescence Spectroscopy*, 3rd ed., Springer, New York, 2011.

80. Lévy, P., *Processus Stochastiques et Mouvement Brownien*, Gauthier-Villars, Paris, 1948.

81. Liao, M., "Random Motion of a Rigid Body," *Journal of Theoretical Probability*, 10(1): 201–211, 1997.

82. Liao, M., *Lévy Processes in Lie Groups*, Cambridge University Press, Cambridge, England 2004.

83. Lipari, G., Szabo, A., "Pade Approximants to Correlation Functions for Restricted Rotational Diffusion," *Journal of Chemical Physics*, 75: 2971–2976, 1981.

84. Luckhurst, G.R., Gray, G.W., eds., *The Molecular Physics of Liquid Crystals*, Academic Press, New York, 1979.

85. Luckhurst, G.R., Sanson, A., "Angular Dependent Linewidths for a Spin Probe Dissolved in a Liquid Crystal," *Molecular Physics*, 24(6): 1297–1311, 1972.

86. Luckhurst, G.R., Zannoni, C., Nordio, P.L., Segre, U., "A Molecular Field Theory for Uniaxial Nematic Liquid Crystals Formed by Non-Cylindrically Symmetric Molecules," *Molecular Physics*, 30(5): 1345–1358, 1975.

87. Maier, W., Saupe, A., "Eine einfache molekulare Theorie des nematischen kristallinflüssigen Zustandes," *Zeitschrift für Naturforschung*, 13a: 564–566, 1958.

88. Maier, W., Saupe, A., "Eine einfache molekular-statistische Theorie der nematischen kristallinflüssigen Phase. Teil I," *Zeitschrift für Naturforschung*, 14a: 882–889, 1959.

89. Maier, W., Saupe, A., "Eine einfache molekular-statistische Theorie der nematischen kristallinflüssigen Phase. Teil II," *Zeitschrift für Naturforschung*, 15a: 287–292, 1960.

90. McClung, R.E.D., "The Fokker-Planck-Langevin Model for Rotational Brownian Motion. I. General Theory," *Journal of Chemical Physics*, 73(5): 2435–2442, Sept. 1980.

91. McConnell, J., *Rotational Brownian Motion and Dielectric Theory*, Academic Press, New York, 1980.

92. McKean, H.P., Jr., "Brownian Motions on the 3-Dimensional Rotation Group," *Memoirs of the College of Science, University of Kyoto, Series A*, 33(1): 25–38, 1960.

93. McMillan, W.L., "Simple Molecular Model for the Smectic A Phase of Liquid Crystals," *Physical Review A*, 4(3): 1238–1246, Sept. 1971.

94. McMillan, W.L., "X-Ray Scattering from Liquid Crystals. I. Cholesteryl Nonanoate and Myristate," *Physical Review A*, 6(3): 936–947, Sept. 1972.

95. Millar, D.P., Robbins, R.J., Zewail, A.H., "Torsion and Bending of Nucleic Acids Studied by Subnanosecond Time-Resolved Fluorescence Depolarization of Intercalated Dyes," *Journal of Chemical Physics* 76(4): 2080–2094, Feb. 1982.

96. Muschik, W., Su, B. "Mesoscopic Interpretation of Fokker-Planck Equation Describing Time Behavior of Liquid Crystal Orientation," *Journal of Chemical Physics*, 107(2): 580–584, 1997.

97. Navez, P., Hounkonnou, M.N., "Theory of the Rotational Brownian Motion of a Linear Molecule in 3D: A Statistical Mechanics Study," *Journal of Molecular Liquids*, 70(1): 71–103, 1996.

98. Nordio, P.L., Busolin, P., "Electron Spin Resonance Line Shapes in Partially Oriented Systems," *Journal of Chemical Physics*, 55(12): 5485–5490, Dec. 1971.

99. Nordio, P.L., Rigatti, G., Segre, U., "Spin Relaxation in Nematic Solvents," *Journal of Chemical Physics*, 56(5): 2117–2123, March 1972.

100. Nordio, P.L., Rigatti, G., Segre, U., "Dielectric Relaxation Theory in Nematic Liquids," *Molecular Physics*, 25(1): 129–136, 1973.

101. Øksendal, B., *Stochastic Differential Equations, An Introduction with Applications*, 6th ed., Springer, Berlin, 2010.

102. Orsingher, E., "Stochastic Motions on the 3-Sphere Governed by Wave and Heat Equations," *Journal of Applied Probability*, 24(2): 315–327, 1987.

103. Pap, G., "Construction of Processes with Stationary Independent Increments in Lie Groups," *Archiv der Mathematik*, 69(2): 146–155, 1997.

104. Perico, A., Guenza, M., Mormino, M., "Protein Dynamics: Rotational Diffusion of Rigid and Fluctuating Three Dimensional Structures," *Biopolymers*, 35(1): 47–54, 1995.

105. Perrin, F., "Mouvement Brownien d'un Ellipsoide (I). Dispersion Diélectrique Pour des Molécules Ellipsoidales," *Le Journal de Physique et le Radium*, 7(10): 497–511, Oct. 1934.

106. Perrin, F., "Mouvement Brownien d'un Ellipsoide (II). Rotation Libre et Dépolarisation des Fluorescences. Translation et Diffusion de Molécules Ellipsoidales," *Le Journal de Physique et le Radium*, 7(1): 1–11, Jan. 1936.

107. Perrin, F., "Étude Mathématique du Mouvement Brownien de Rotation," *Annales Scientifiques de l' Ecole Normale Supérieure*, 45: 1–51, 1928.

108. Piccioni, M., Scarlatti, S., "An Iterative Monte Carlo Scheme for Generating Lie Group-Valued Random Variables," *Advances in Applied Probability*, 26(3): 616–628, 1994.

109. Planck, M., "Über einen Satz der statistischen Dynamik und seine Erweiterung in der Quantentheorie," *Sitzungsbericht der Preussischen Akademie der Wissenshaften*, 1917, pp. 324–341, 1917.

110. Polnaszek, C.F., Bruno, G.V., Freed, J.H., "ESR Line Shapes in the Slow-Motional Region: Anisotropic Liquids," *J. Chem. Phys.*, 58(8): 3185–3199, April 1973.

111. Rains, E.M., "Combinatorial Properties of Brownian Motion on the Compact Classical Groups," *Journal of Theoretical Probability*, 10(3): 659–679, 1997.

112. Risken, H., *The Fokker-Planck Equation, Methods of Solution and Applications*, 2nd ed., Springer-Verlag, Berlin, 1989. (3rd printing, 1996).

113. Roberts, P.H., Ursell, H.D., "Random Walk on a Sphere and on a Riemannian Manifold," *Philosophical Transactions of the Royal Society of London*, A252:317–356, 1960.

114. Rothschild, W.G., *Dynamics of Molecular Liquids*, John Wiley and Sons, New York, 1984.

115. Said, S., Manton, J.H., "Extrinsic Mean of Brownian Distributions on Compact Lie groups," *IEEE Transactions on Information Theory*, 58(6): 3521–3535, 2012.

116. Schurr, J.M., Fujimoto, B.S., Wu, P.G., Song, L., "Fluorescence Studies of Nucleic Acids: Dynamics, Rigidities, and Structures," in *Topics in Fluorescence* Vol. 3: Biochemical Applications, pp. 137–229, J.R. Lakowicz (ed.), Plenum Press, New York, 1992.

117. Sklyaruk, V.A., Toptygin, D.D., Khanmagometova, S.D., "Effect of Angular-Distribution of Impurity Molecules in a Liquid-Crystal Matrix on Dipole-Dipole Energy-Transfer," *Optika I Spektroskopiya*, 62(6): 1272–1279, 1987

118. Small, E.W., Isenberg, I., "Hydrodynamic Properties of a Rigid Molecule: Rotational and Linear Diffusion and Fluorescence Anisotropy," *Biopolymers*, 16: 1907–1928, 1977.

119. Smoluchowski, M. v., "Über Brownsche Molekularbewegung unter Einwirkung äußerer Kräfte und deren Zusammenhang mit der verallgemeinerten Diffusionsgleichung," *Annalen der Physik*, 48: 1103–1112, 1915.

120. Steele, D., Yarwood, J., eds., *Spectroscopy and Relaxation of Molecular Liquids*, Elsevier, 1991.

121. Steele, W.A., "Molecular Reorientation in Liquids. I. Distribution Functions and Friction Constants. II. Angular Autocorrelation Functions," *Journal of Chemical Physics*, 38(10): 2404–2410, 2411–2418, 1963.

122. Stratonovich, R.L., *Topics in the Theory of Random Noise: Vols. I and II*, Gordon and Breach Science Publishers,Inc., New York, 1963.

123. Tang, S., Evans, G.T., "Free and Pendular-like Rotation: Orientational Dynamics in Hard Ellipsoid Fluids," *Journal of Chemical Physics*, 103(4): 1553–1560, July 1995.

124. Tao, T., "Time-Dependent Fluorescence Depolarization and Brownian Rotational Diffusion Coefficients of Macromolecules," *Biopolymers* 8(5): 609–632, 1969.

125. Tarroni, R., Zannoni, C., "On the Rotational Diffusion of Asymmetric Molecules in Liquid Crystals," *Journal of Chemical Physics*, 95(6): 4550–4564, Sept. 1991.

126. Toptygin, D., Brand, L., "Fluorescence Decay of DPH in Lipid Membranes: Influence of the External Refractive Index," *Biophysical Chemistry*, 48(2): 205–220, 1993.

127. Toptygin, D., Brand, L., "Determination of DPH Order Parameters in Unoriented Vesicles," *Journal of Fluorescence*, 5(1): 39–50, 1995.

128. Toptygin, D., Savtchenko, R.S., Meadow, N.D., Brand, L., "Homogeneous Spectrally- and Time-Resolved Fluorescence Emission from Single-Tryptophan Mutants of IIA^{Glc} Protein," *Journal of Physical Chemistry B* 105: 2043–2055, 2001.

129. Toptygin, D., "Analysis of Time-Dependent Red Shifts in Fluorescence Emission from Tryptophan Residues in Proteins," as Chapter 9 in *Fluorescence Spectroscopy and Microscopy: Methods and Protocols, Methods in Molecular Biology*, vol. 1076, Yves Engelborghs and Antonie J.W.G. Visser (eds.), pp. 215–256, Springer Science + Business Media, LLC, 2014

130. Tsoi, A.H., "Integration by Parts for a Lie Group Valued Brownian Motion," *Journal of Theoretical Probability*, 6(4): 693–698, 1993.

131. Uhlenbeck, G.E., Ornstein, L.S., "On the Theory of Brownian Motion," *Physical Review*, 36(5): 823–841, 1930.

132. Van Der Meer, W., Pottel, H., Herreman, W., Ameloot, M.,Hendrickx, H., Schröder, H., "Effect of Orientational Order on the Decay of the Fluorescence Anisotropy in Membrane Suspensions: A New Approximate Solution of the Rotational Diffusion Equation," *Biophysical Journal*, 46(4): 515–523, Oct. 1984.

133. van Kampen, N.G., *Stochastic Processes in Physics and Chemistry*, 3rd ed., North Holland, Amsterdam, 2007.

134. Vos, K., van Hoek, A., Visser, A.J., "Application of a Reference Convolution Method to Tryptophan Fluorescence in Proteins. A Refined Description of Rotational Dynamics," *European Journal of Biochemistry*, 165(1): 55–63, May 1987.

135. Wang, M.C., Uhlenbeck, G.E., "On the Theory of Brownian Motion II," *Reviews of Modern Physics*, 7(2-3): 323–342, 1945.

136. Weber, G., "Rotational Brownian Motion and Polarization of the Fluorescence of Solutions," *Advances in Protein Chemistry*, 8: 415–459, 1953.

137. Weber, G., "Theory of Fluorescence Depolarization by Anisotropic Brownian Rotations. Discontinuous Distribution Approach," *Journal of Chemical Physics*, 55(5): 2399–2407, 1971.

138. Weber, G., "Theory of Differential Phase Fluorometry: Detection of Anisotropic Molecular Rotations," *Journal of Chemical Physics*, 66(9): 4081–4091, May 1977.

139. Wyllie, G., "Random Motion and Brownian Rotation," *Physics Reports*, 61(6): 327–376, 1980.

140. Yosida, K., "Integration of Fokker-Planck's Equation in a Compact Riemannian Space," *Arkiv för Matematik*, Band 1, Nr. 9, pp. 71–75, 1949.

141. Yosida, K., "Brownian Motion on the Surface of the 3-Sphere," *Annals of Mathematical Statistics*, 20: 292–296, 1949.

142. Yosida, K., "Brownian Motion in a Homogeneous Riemannian Space," *Pacific Journal of Mathematics*, 2: 263–296, 1952.

143. Zannoni, C., Arcioni, A., Cavatorta, P., "Fluorescence Depolarization in Liquid Crystals and Membrane Bilayers," *Chemistry and Physics of Lipids*, 32(3): 179–250, 1983.

144. Zhang, Y., Bull, T.E., "Nonlinearly Coupled Generalized Fokker-Planck Equation for Rotational Relaxation," *Physical Review E*, 49(6): 4886–4902, June 1994.

Statistical Mechanics of Polymers

In this chapter we apply the concepts of noncommutative convolution and harmonic analysis to study the statistical mechanics of long polymer chains consisting of simpler units. These units can be identical (in which case a homopolymer results) or they may be different (in which case a heteropolymer results). Polymers exist in nature (including DNA and proteins), but usually when referring to polymer theory, it is in the context of the man-made polymers.

Our presentation of the statistical mechanics of polymers is from a very kinematic perspective. We show how certain quantities of interest in polymer physics can be generated numerically using Euclidean-group convolutions. We also show how for wormlike polymer chains, a partial differential equation governs a process that evolves on the motion group and describes the diffusion of end-to-end position and orientation. This equation can be solved using the $SE(3)$-Fourier transform.

Before proceeding to the group-theoretic issues, we present background material on polymer theory in the first few sections that follow. This review material is consolidated from the following books devoted to polymer theory [4, 6, 17, 15, 16, 18, 19, 21, 27, 28, 45, 60, 72, 76, 82].

17.1 General Concepts in Systems of Particles and Serial Chains

Given a collection of n identical particles which are connected by links, the vector from the initial particle to the final one as seen in a frame of reference $\mathcal{F}$ fixed to the initial particle is[1]

$$\mathbf{x}_n = \sum_{i=1}^{n-1} \mathbf{l}_i$$

where $\mathbf{l}_i$ is the vector from particle i to particle $i+1$. The square of the end-to-end distance is then $r^2 = |\mathbf{x}_n|^2$ which can be written as

$$r^2 = \sum_{i=1}^{n-1}\sum_{j=1}^{n-1} \mathbf{l}_i \cdot \mathbf{l}_j = \sum_{i=1}^{n-1} |\mathbf{l}_i|^2 + 2\sum_{i=1}^{n-1}\sum_{j=i+1}^{n-1} \mathbf{l}_i \cdot \mathbf{l}_j. \qquad (17.1)$$

---

[1] In polymer science n usually denotes the number of bonds. Therefore replacing n with $n + 1$ for the number of particles would make our presentation consistent with others in the field.

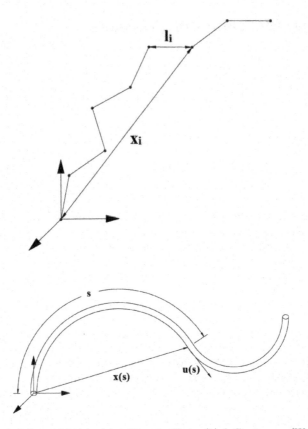

Fig. 17.1. Simple Polymer Models: (a) A Discrete Chain; (b) A Continuous (Wormlike) Chain

In the frame of reference $\mathcal{F}$, the center of mass of the collection of n particles is denoted $\mathbf{x}_{cm}$. The position of the ith particle is $\mathbf{x}_i$ as seen in $\mathcal{F}$ and its position relative to a frame parallel to $\mathcal{F}$ with origin at the center of mass of the system is denoted $\mathbf{r}_i$. Hence

$$\mathbf{x}_i = \mathbf{x}_{cm} + \mathbf{r}_i. \tag{17.2}$$

The position of particle i relative to particle j is

$$\mathbf{x}_{ij} = \mathbf{x}_i - \mathbf{x}_j = \mathbf{r}_i - \mathbf{r}_j.$$

Hence $\mathbf{x}_{i,i+1} = \mathbf{l}_i$ and $\mathbf{x}_{i1} = \mathbf{x}_i$. Furthermore, by definition of the center of mass we have

$$\mathbf{x}_{cm} = \frac{1}{n} \sum_{i=1}^{n} \mathbf{x}_i$$

and summing both sides of (17.2) over all values of $i \in [1, n]$ therefore yields

$$\sum_{i=1}^{n} \mathbf{r}_i = \mathbf{0}. \tag{17.3}$$

The radius of gyration of the system of particles is defined as

$$S \doteq \sqrt{\frac{1}{n} \sum_{i=1}^{n} \mathbf{r}_i \cdot \mathbf{r}_i}. \tag{17.4}$$

It is desirable to have a general statement relating the radius of gyration of a collection of particles to the relative positions of the particles with respect to each other. We now derive such an expression following [21, 39].

Substituting $\mathbf{r}_i = -\mathbf{x}_{cm} + \mathbf{x}_i$ into this equation and observing (17.3) and the fact that

$$\mathbf{x}_{cm} \cdot \mathbf{x}_{cm} = \frac{1}{n^2} \sum_{i=1}^{n} \sum_{j=1}^{n} \mathbf{x}_i \cdot \mathbf{x}_j$$

allows us to write

$$S^2 = \frac{1}{n} \sum_{i=1}^{n} \mathbf{x}_i \cdot \mathbf{x}_i - \frac{1}{n^2} \sum_{i=1}^{n} \sum_{j=1}^{n} \mathbf{x}_i \cdot \mathbf{x}_j$$

In vector form, the law of cosines reads

$$\mathbf{x}_{ij} \cdot \mathbf{x}_{ij} = \mathbf{x}_i \cdot \mathbf{x}_i + \mathbf{x}_j \cdot \mathbf{x}_j - 2\mathbf{x}_i \cdot \mathbf{x}_j$$

or equivalently,

$$\mathbf{x}_i \cdot \mathbf{x}_j = (\mathbf{x}_i \cdot \mathbf{x}_i + \mathbf{x}_j \cdot \mathbf{x}_j - \mathbf{x}_{ij} \cdot \mathbf{x}_{ij})/2.$$

Using this to substitute in for $\mathbf{x}_i \cdot \mathbf{x}_j$ in the above expression for S^2, yields

$$S^2 = \frac{1}{n^2} \left(\sum_{i=1}^{n} \mathbf{x}_i \cdot \mathbf{x}_i + \frac{1}{2} \sum_{i=1}^{n} \sum_{j=1}^{n} \mathbf{x}_{ij} \cdot \mathbf{x}_{ij} \right).$$

Recalling that $\mathbf{x}_i = \mathbf{x}_{1i}$, and using the symmetry of the dot product operation allows us to finally write

$$S = \frac{1}{n} \sqrt{\sum_{1 \leq i < j \leq n} \mathbf{x}_{ij} \cdot \mathbf{x}_{ij}}. \tag{17.5}$$

The above is *Lagrange's theorem*, and the derivation we presented is essentially the same as those in [39, 21].

17.2 Statistical Ensembles and Kinematic Modeling of Polymers

Many of the quantitative results in polymer science depend on the averaging of physical properties over statistical mechanical ensembles. Hence, instead of considering a single polymer chain, one considers the collection of all possible shapes attainable by the chain, each weighted by its likelihood of occurrence. Those with high energy are less likely than those with low energy. In cases when it can be reasoned that all conformations have the same energy, the problem of generating an ensemble is purely geometrical. This purely geometrical, or kinematical, analysis is the subject of this section. The next section reviews classical techniques for incorporating conformational energy in special cases.

Two of the most important quantities that describes the ensemble properties of a polymer chain is the distribution of end-to-end distances and the radius of gyration of the chain over all energy-weighted conformations. In fact, a simple computation shows that the distribution of radii of gyration can be found if the distribution of distances between all points in the chain are known. Hence the problem is really one of finding the distribution of distances between all points in the chain over all energy-weighted conformations. In the simplest models such as the Gaussian chain (or random walk), the freely-jointed chain, and the Kratky-Porod continuous chain where all conformations are assumed to have the same energy and the effects of excluded volume are neglected, closed-form expressions for this distribution (or at least the moments of the distribution) are known. In applications such as relating polymer structure to macroscopic quantities such as viscosity or elasticity, it is usually the statistical mechanical average of the square of the end-to-end distance or radius of gyration which are of importance [5, 43, 74]. In their most general form, these follow from (17.1) and (17.5) as

$$\langle r^2 \rangle = (n-1)l^2 + 2 \sum_{i=1}^{n-1} \sum_{j=i+1}^{n-1} \langle \mathbf{l}_i \cdot \mathbf{l}_j \rangle. \tag{17.6}$$

and

$$\langle S^2 \rangle = \frac{1}{n^2} \sum_{1 \le i < j \le n} \langle \mathbf{x}_{ij} \cdot \mathbf{x}_{ij} \rangle. \tag{17.7}$$

In (17.6) it is assumed that $\langle |\mathbf{l}_i|^2 \rangle = l^2$ is the same for each bond.

The behavior of $\langle \mathbf{l}_i \cdot \mathbf{l}_j \rangle$ and $\langle \mathbf{x}_{ij} \cdot \mathbf{x}_{ij} \rangle$ depends heavily on the particular polymer model

17.2.1 The Gaussian Chain (Random Walk)

Perhaps the most common model for the distribution of end-vector distribution is the Gaussian distribution:

$$W_G(\mathbf{r}) = \left(\frac{3}{2\pi \langle r^2 \rangle} \right)^{\frac{3}{2}} \exp \left[-\frac{3r^2}{2 \langle r^2 \rangle} \right]. \tag{17.8}$$

This distribution is spherically symmetric (and hence depends only on $r = \|\mathbf{r}\|$ where $\mathbf{r}$ is spatial position as measured relative to the proximal end of the chain). It is normalized so that it is a probability density function,

$$\int_{\mathbb{R}^3} W(\mathbf{r}) \, dV = 4\pi \int_0^\infty W(\mathbf{r}) r^2 dr = 1,$$

satisfying

$$\int_{\mathbb{R}^3} W(\mathbf{r}) |\mathbf{r}|^2 dV = 4\pi \int_0^\infty W(\mathbf{r}) r^4 dr = \langle r^2 \rangle.$$

17.2.2 The Freely-Jointed Chain

The freely-jointed chain model assumes that each link is free to move relative to the others with no constraint on the motion and no correlation between the motion of adjacent links. Hence, in this model

$$\langle \mathbf{l}_i \cdot \mathbf{l}_j \rangle = 0 \ \forall \ i \ne j \ \Rightarrow \ \langle r^2 \rangle = (n-1)l^2.$$

Assuming analogous behavior for each subchain in a freely-jointed chain means

$$\langle \mathbf{x}_{ij} \cdot \mathbf{x}_{ij} \rangle = (j-i)l^2 \ \Rightarrow \ \langle S^2 \rangle = \frac{l^2}{2n^2} \sum_{j=1}^{n-1} j(j+1).$$

Using elementary summation formulae, this is written as

$$\langle S^2 \rangle = \frac{(n+1)(n-1)l^2}{6n}. \tag{17.9}$$

For large n, we then have

$$\langle S^2 \rangle \approx \langle r^2 \rangle / 6.$$

Flory [21] derives the statistical distribution of end positions of a freely-jointed chain as a special case of his more general theory, and gives its form as

$$W_F(\mathbf{r}) = \frac{1}{2\pi^2 r} \int_0^\infty \sin(qr)[\sin(ql)/ql]^n q\, dq \tag{17.10}$$

where again $r = |\mathbf{r}|$. Other classical derivations can be found in [7, 61].

Application of the usual Abelian Fourier transform yields

$$\mathcal{F}(W_F(\mathbf{r})) = G(\mathbf{q}) = [\sin(ql)/ql]^n$$

where $q = |\mathbf{q}|$ and $\mathbf{q}$ is the vector of Fourier parameters. From this fact it is clear by the convolution theorem that

$$W_F^{(n_1)} * W_F^{(n_2)} = W_F^{(n_1+n_2)}$$

where the superscript (n) is the number of links in the chain.

Flory [21] derived the following approximate expression for the distribution of end positions:

$$W(\mathbf{r}) = (A/rl^2)\mathcal{L}^{-1}(r/nl)\exp\left[-\frac{1}{l}\int_0^r \mathcal{L}^{-1}(\sigma/nl)\,d\sigma\right] \tag{17.11}$$

where

$$\mathcal{L}(x) = \coth x - 1/x$$

is the *Langevin* function and $\mathcal{L}^{-1}$ is its inverse, which has the series expansion

$$\mathcal{L}^{-1}(a) \approx 3a + \frac{9}{5}a^3 + \frac{297}{175}a^5.$$

Substitution of this into (17.11) provides a quasi-closed-form solution.

Treloar [74] derives that

$$W(\mathbf{r}) = (Ab(r)/rl^2)\left(\frac{\sinh b(r)}{b(r)}\right)^n e^{-b(r)r/l}$$

where

$$b(r) = \mathcal{L}^{-1}(r/nl),$$

and A is the normalization ensuring that $W(\mathbf{r})$ is a probability density function.

17.2.3 The Freely Rotating Chain

In the freely rotating chain model of a polymer, each bond in the chain has the same fixed length, $l = |l_i|$, and each bond vector l_{i+1} makes a bond angle θ with respect to l_i. Hence $l_i \cdot l_{i+1} = l^2 \cos \theta$. Since this is true for each bond in any conformation, it is also true when averaging over all conformations: $\langle l_i \cdot l_{i+1} \rangle = l^2 \cos \theta$.

We can resolve the bond vector l_{i+1} in a direction along bond i and orthogonal to it. Then

$$l_{i+1} = \cos l_i + v_i \tag{17.12}$$

where $v_i \cdot u_i = 0$ by definition. If the unhindered rotation around bond i is parameterized as $\phi_i \in [0, 2\pi]$, then v_i is a function of all the angles $\{\phi_j\}$ for all $j \leq i$, while l_i is a function of all angles $\{\phi_j\}$ for all $j < i$. Hence all conformational averages of the form

$$\langle l_i \cdot v_{i+n} \rangle = 0 \tag{17.13}$$

for all $n \geq 0$ because in this average is an integration over ϕ_{i+n}. That is, for unhindered rotation

$$\int_{\mathbb{S}^1} v_{i+n} d\phi_{i+n} = 0 \;\Rightarrow\; \int_{\mathbb{S}^1} l_i \cdot v_{i+n} d\phi_{i+n} = 0$$

where the second equation above is true because l_i is not a function of ϕ_{i+n} and can be taken out of the integral. Using (17.12) recursively, we can write l_{i+n} as a the sum of $\cos^n \theta l_i$ and a series of terms of the form $a_{i+j}(\theta) v_{i+j}$ for $j = 0, ..., n-1$. Calculating the conformational averages $\langle l_i \cdot l_{i+j} \rangle$ and observing (17.13) we find that

$$\langle l_i \cdot l_{i+j} \rangle = l^2 (\cos \theta)^j. \tag{17.14}$$

It then follows from (17.6) that

$$\langle r^2 \rangle = (n-1)l^2 + 2l^2 \sum_{i=1}^{n-1} \sum_{j=i}^{n-1} (\cos \theta)^{j-i}.$$

Manipulating this expression as in [21, 59], we find that

$$\langle r^2 \rangle = (n-1)l^2 \frac{1 + \cos \theta}{1 - \cos \theta} - 2l^2 \frac{\cos \theta (1 - \cos^{n-1} \theta)}{(1 - \cos \theta)^2}.$$

Similarly, for this model the conformational average of the radius of gyration is

$$\langle S^2 \rangle = \frac{l^2 (n-1)(n+1)(1 + \cos \theta)}{6n(1 - \cos \theta)} - \frac{2l^2 \cos \theta}{n^2 (1 - \cos \theta)^2} \sum_{j=1}^{n-1} \left[j - \left(\frac{1 - \cos^j \theta}{1 - \cos \theta} \right) \cos \theta \right].$$

In the limit as $n \to \infty$, these become

$$(\langle r^2 \rangle / (n-1)l^2)_\infty = \frac{1 + \cos \theta}{1 - \cos \theta} \quad \text{and} \quad (\langle S^2 \rangle / (n-1)l^2)_\infty = \frac{1}{6} (\langle r^2 \rangle / (n-1)l^2)_\infty.$$

17.2.4 Kratky-Porod Chain

The Kratky-Porod (or KP) model [37, 59] is for semi-flexible polymer chains. By imposing the condition that the polymer act like a continuous curve with random curvature,
the evolution of distal-end position and orientation is more gradual than the other models considered previously in this chapter. This model intrinsically takes into account the
properties of chains which are stiff over short segments, but are nonetheless flexible on
the length scale of the whole chain due to their long and slender nature (e.g., DNA).

In this model, the position of a point on an individual polymer chain at an arclength
s from the proximal end is

$$\mathbf{x}(s) = \int_0^s \mathbf{u}(\sigma)\, d\sigma,$$

and the relative position of two points on the chain is given by

$$\mathbf{x}(s_1, s_2) = \mathbf{x}(s_2) - \mathbf{x}(s_1).$$

$\mathbf{u}(s) \in \mathbb{S}^2$ is the unit tangent vector to the curve at s. It is assumed that the proximal end
is attached to the origin of our frame of reference with $\mathbf{u}(0) = \mathbf{e}_3$, and the polymer has
total length L. Denoting the position of the distal end as $\mathbf{r} = \mathbf{x}(L)$, then the expressions
for $\langle |\mathbf{r}|^2 \rangle$ and $\langle S^2 \rangle$ derived for chains with discrete links become

$$\langle |\mathbf{r}|^2 \rangle = \langle \mathbf{r} \cdot \mathbf{r} \rangle = \int_0^L \int_0^L \langle \mathbf{u}(s_1) \cdot \mathbf{u}(s_2) \rangle ds_1 ds_2 \tag{17.15}$$

and

$$\langle S^2 \rangle = \frac{1}{L^2} \int_0^L \left(\int_{s_1}^L \langle \|\mathbf{x}(s_1, s_2)\|^2 \rangle ds_2 \right) ds_1.$$

Yamakawa [82] reasons that $\langle |\mathbf{x}(s_1, s_2)|^2 \rangle = \langle \|\mathbf{x}(s_2 - s_1)\|^2 \rangle$ and hence

$$\langle S^2 \rangle = \frac{1}{L^2} \int_0^L (L - s) \langle \|\mathbf{x}(s)\|^2 \rangle ds. \tag{17.16}$$

Ideally one would like to solve for the probability density function $W_C(\mathbf{u}, \mathbf{x}, s)$ for
the orientation (modulo roll) and position of the frame of reference attached to the
curve for each value of s (and $s = L$ in particular). In Subsection 17.2.4 we examine
the moments $\langle |\mathbf{r}|^2 \rangle$ and $\langle S^2 \rangle$ for this model. In Subsection 17.2.4 we review methods for
approximating the probability density function $W_C(\mathbf{u}, \mathbf{x}, s)$.

Moments of the Continuous Wormlike Chains

Let $\theta(s_1, s_2)$ denote the angle between the tangents to the curve at two different values
of arclength, s_1 and s_2:

$$\cos \theta(s_1, s_2) = \mathbf{u}(s_1) \cdot \mathbf{u}(s_2).$$

If we consider all possible conformations of a continuous curve, then it can be reasoned
based on uniformity of behavior everywhere on the chain that the average $\langle \cos \theta(s_1, s_2) \rangle$
over all conformations is of the form $\langle \cos \theta(s_1 - s_2) \rangle$. Furthermore, as stated in [27], it
is often a realistic assumption that this function has the property:

$$\langle \cos \theta(s_1 - s_2) \rangle = \langle \cos \theta(s_1) \rangle \langle \cos \theta(-s_2) \rangle.$$

The solution of this functional equation is of the form

$$\langle \cos \theta(s) \rangle = e^{-s/a} \tag{17.17}$$

where a is a quantity called the *persistence length* of the polymer. Essentially, it is a parameter which indicates how far one must travel along the polymer before the tangent directions are no longer correlated. Physically this describes how stiff the molecule is. For instance, if $a \to 0$ then $\langle \cos \theta(s) \rangle \approx 0$ for all $s \in [0, L]$, whereas if $a \to \infty$ then $\langle \cos \theta(s) \rangle \approx 1$ for all $s \in [0, L]$. These two situations correspond to completely flexible and completely rigid chains, respectively.

We note that there is a different way to view this same result. The orientation distribution, $w(\mathbf{u}; s)$ corresponding to the ensemble of tangent vectors $\{\mathbf{u}(s)\}$ (each $\mathbf{u}(s)$ measured at arclength s along its respective chain) can be viewed as a diffusion process on the sphere with arclength replacing time. Following this reasoning, we would need to solve the heat equation on the sphere:

$$\frac{\partial w}{\partial t} = K \nabla_u^2 w,$$

where ∇_u^2 is the spherical Laplacian, subject to the initial conditions $w(\mathbf{u}; 0) = \delta(\mathbf{u})$. When $\mathbf{u} = \mathbf{u}(\theta, \phi)$ is parameterized using spherical coordinates,

$$\delta(\mathbf{u}) = \sum_{l=0}^{\infty} \left(\frac{2l+1}{4\pi} \right) P_l(\cos \theta).$$

The solution to the heat equation is therefore independent of ϕ due to the axial symmetry of the initial conditions around the $\mathbf{e}_3$ axis. This solution is written as

$$w(\mathbf{u}; s) = w(\theta; s) = \sum_{l=0}^{\infty} \left(\frac{2l+1}{4\pi} \right) e^{-l(l+1)s/K} P_l(\cos \theta).$$

From this distribution we can calculate

$$\langle \mathbf{u}(s) \cdot \mathbf{u}(0) \rangle = \int_{\mathbb{S}^2} \mathbf{u} \cdot \mathbf{e}_3 w(\mathbf{u}; s) \, d\mathbf{u} = \int_0^\pi \int_0^{2\pi} \cos \theta w(\theta, \phi; s) \sin \theta d\phi d\theta = e^{-s/K}.$$

The denominator of the exponential in this description is called the *persistence length*, a, and so we write the diffusion constant as $K = a$.

Substituting the function in (17.17) into (17.15) we find that

$$\langle r^2 \rangle = 2aL[1 - (a/L)(1 - e^{-L/a})].$$

Substituting this into (17.16) we find that

$$\langle S^2 \rangle = (aL/3) \left(1 - (3a/L)[1 - 2(a/L) + 2(a/L)^2(1 - e^{-L/a})] \right).$$

In the limit as $L \to \infty$ we find that as with the Gaussian chain

$$\langle S^2 \rangle \approx \langle r^2 \rangle / 6.$$

The PDF for the Continuous Chain

As with the Gaussian and freely-jointed chains, it is desirable to have the full distribution of end positions and orientations associated with the model. As has been shown [82, 32, 14], the Fokker-Planck equation for the evolution of the probability density function $W_C(\mathbf{u}, \mathbf{x}; L)$ is

$$\frac{\partial W_C}{\partial L} = K\nabla_u^2 W_C - \mathbf{u} \cdot \nabla_x W_C. \qquad (17.18)$$

The initial conditions corresponding to the PDF at the proximal end are:

$$W_C(\mathbf{u}, \mathbf{x}; 0) = \delta_{\mathbb{S}^2}(\mathbf{u}, \mathbf{e}_3)\,\delta(\mathbf{x}).$$

A series solution to this partial differential equation is given in [82, 26]. In the context of the current work, it suffices to say that while the distribution is not spherically symmetric for nonzero finite values of L and K, we have nonetheless

$$\lim_{L\to\infty} \int_{\mathbb{S}^2} W_C(\mathbf{u}, \mathbf{r}; L)\,d\mathbf{u} = W_G(\|\mathbf{r}\|).$$

While we know of no closed-form solution for the distribution of end-vector position in the continuous curve model, certain approximations can be found in the literature. These include expansions in the ratio a/L (see e.g., [79, 82]).

In very recent work, Thirumalai and Ha [73] relaxed the local constraint $\mathbf{u}(s)\cdot\mathbf{u}(s) = 1$ in the Kratky-Porod model and instead enforced the global constraint $\langle\mathbf{u}(s)\cdot\mathbf{u}(s)\rangle = 1$. They then derived the distribution

$$W_T(r) = \frac{A}{(1 - r^2/L^2)^{\frac{9}{2}}} \exp\left[-\frac{\alpha_0}{(1 - r^2/L^2)^{\frac{9}{2}}}\right] \qquad (17.19)$$

where $\alpha_0 = -9L/8a$ (a is the persistence length) and

$$A = \frac{4}{\pi^{1/2}e^{-\alpha_0}\alpha_0^{-3/2}(1 + 3/\alpha_0 + 15/4\alpha_0^2)}$$

is the constant required to normalize $W_T(\cdot)$ so that it is a probability distribution.

17.3 Theories Including the Effects of Conformational Energy

The models presented in the previous sections are more or less kinematic in nature. Wormlike chain models do somewhat reflect energetics in that the persistence length that defines them can be related to their local bending stiffness. However, the structural and energetic details at the local level are not modeled explicitly. In contrast, the models reviewed in this section take into account these local properties.

17.3.1 Chain with Hindered Rotations

The hindered rotation model is the simplest of several classical models of polymer chains which explicitly include the effects of conformational energy. It assumes that the *bond angles*, $\{\theta_i\}$, are fixed, and each *torsion angle*, ϕ_i, has preferred values dictated by the energy function $E_i(\phi_i)$. The total conformational energy is

$$E = \sum_i E_i(\phi_i).$$

In this model, the probability density function $p(\phi_i)$ describing the distribution of attainable torsion angles is given by[2]

$$p(\phi_i) = \frac{\exp[-E_i(\phi_i)/k_B T]}{\int_0^{2\pi} \exp[-E_i(\phi)/k_B T]d\phi} \tag{17.20}$$

where k_B is the Boltzmann constant and T is the (absolute) temperature measured in degrees Kelvin. Hence, for high temperatures this model reduces to the freely rotating chain model. In any case the conformational average of any function scalar, vector, or matrix-valued function of ϕ_i will be calculated as

$$\langle f \rangle = \int_0^{2\pi} f(\phi_i)p(\phi_i)\,d\phi_i.$$

The probability density function on the whole conformation space of the chain is the product of those for each torsion angle ϕ_i. This means that the conformational average of the product of functions of individual torsion angles will be the product of conformational averages of these functions. In particular, consider the important example given below.

Following the notation in [21], the rotation matrix describing the orientation of a frame attached to bond vector $i+1$ relative to the frame attached to bond i is written as

$$T_i = \begin{pmatrix} \cos\theta_i & \sin\theta_i & 0 \\ \sin\theta_i\cos\phi_i & -\cos\theta_i\cos\phi_i & \sin\phi_i \\ \sin\theta_i\sin\phi_i & -\cos\theta_i\sin\phi_i & -\cos\phi_i \end{pmatrix}. \tag{17.21}$$

In the current context $\theta_i = \theta = const$, and so $T_i = \mathbf{T}_i(\phi_i)$. Assuming that the energy function is symmetric, $E_i(\phi_i) = E_i(-\phi_i)$, then $p(\phi_i)$ is also symmetric and so $\langle \sin\phi_i \rangle = 0$. Hence

$$\langle T_i \rangle = \begin{pmatrix} \cos\theta_i & \sin\theta_i & 0 \\ \sin\theta_i\langle\cos\phi_i\rangle & -\cos\theta_i\langle\cos\phi_i\rangle & 0 \\ 0 & 0 & -\langle\cos\phi_i\rangle \end{pmatrix}.$$

Since the bond vector $\mathbf{l}_i = l\mathbf{e}_1$ in its own coordinate system, and in this same coordinate system $\mathbf{l}_j$ for $j > i$ appears as $lT_jT_{j-1}\cdots T_i\mathbf{e}_1$, it follows that

$$\langle \mathbf{l}_i \cdot \mathbf{l}_j \rangle = l^2\mathbf{e}_1^T\langle T_jT_{j-1}\cdots T_i\rangle\mathbf{e}_1 = l^2\mathbf{e}_1^T\langle T\rangle^{j-i}\mathbf{e}_1, \tag{17.22}$$

where the last product results from the assumption that all of the bond angles and torsional energies are independent of i (and so $\langle T_i \rangle = \langle T \rangle$), and the conformational average of a product of functions of single torsion angles is the product of their conformational averages.

Substitution of (17.22) into (17.1) results in the average of the square of the end-to-end distance given in [21]:

$$\langle r^2 \rangle = (n-1)l^2\mathbf{e}_1^T[(\mathbb{I} + \langle T\rangle)(\mathbb{I} - \langle T\rangle)^{-1} - (2\langle T\rangle/(n-1))(\mathbb{I} - \langle T\rangle^n)(\mathbb{I} - \langle T\rangle)^{-2}]\mathbf{e}_1.$$

[2]Here and everywhere that k_B and T appear next to each other, they are multiplied together, i.e., they are both in the denominator in these expressions.

The average of the square of the radius of gyration follows in a similar way.

In the limit as $n \to \infty$ we find the result from Oka's 1942 paper [56]:

$$\langle\langle r^2 \rangle/(n-1)l^2\rangle_{n\to\infty} = \left(\frac{1+\cos\theta}{1-\cos\theta}\right)\left(\frac{1+\langle\cos\phi\rangle}{1-\langle\cos\phi\rangle}\right) = 6(\langle S^2\rangle/(n-1)l^2)_{n\to\infty}.$$

This reduces to the case of the freely rotating chain when $\langle\cos\phi\rangle = 0$.

17.3.2 Rotational Isomeric State Model

Let the homogeneous transform (element of $SE(3)$) which describes the position and orientation of the frame of reference attached to a serial chain molecule at the intersection of the ith and $(i+1)^{st}$ bond vectors relative to the frame attached at the intersection of the $(i-1)^{st}$ and i^{th} bond vectors be given by

$$g_i = (R_i, \mathbf{l}_i).$$

$\mathbf{l}_i = [l_i, 0, 0]^T$ is the ith bond vector as seen in this frame i (which is attached at the intersection of the $i-1^{st}$ and i^{th} bond vectors)[3]

The vector pointing to the tip of this bond vector from the origin of frame j is $\mathbf{r}_j^i$.

The end-to-end vector for the chain, $\mathbf{r} = \mathbf{r}_0^n$, is the concatenation of these homogeneous transforms applied to $\mathbf{l}_n$:

$$\mathbf{r} = g_1 \circ g_2 \circ \cdots \circ g_n \circ \mathbf{l}_n.$$

Since

$$g_1 \circ g_2 \circ \cdots \circ g_n = \left(\prod_{i=1}^{n} R_i, \sum_{i=1}^{n-1}\left(\prod_{j=0}^{i-1} R_j\right)\mathbf{l}_i\right)$$

(where $R_0 = \mathbb{I}$) we find that

$$\mathbf{r} = \sum_{i=1}^{n}\left(\prod_{j=0}^{i-1} R_j\right)\mathbf{l}_i.$$

From this, the square of the end-to-end distance, $r^2 = \mathbf{r}\cdot\mathbf{r}$, is easily calculated. It is relatively straight forward to show that this can be put in the form [45]:

$$r^2 = [\, 1 \;\; 2\mathbf{l}_1^T R_1 \;\; l_1^2\,] G_2 G_3 \cdots G_{n-1} \begin{bmatrix} l_n^2 \\ \mathbf{l}_n \\ 1 \end{bmatrix} \qquad (17.23)$$

where

$$G_i = \begin{pmatrix} 1 & 2\mathbf{l}_i^T R_i & l_i^2 \\ \mathbf{0} & R_i & \mathbf{l}_i \\ 0 & \mathbf{0}^T & 1 \end{pmatrix}.$$

Similarly, the square of the radius of gyration for a chain of $n+1$ identical atoms is

$$S^2 = \frac{1}{(n+1)^2}\sum_{k=0}^{n-1}\sum_{j=k+1}^{n}\|\mathbf{r}_k^j\|^2.$$

---

[3]In contrast to Section 17.1, here n is the number of bonds, and $n+1$ is the number of backbone atoms.

This can be written as the matrix product [45]:

$$S^2 = [\,1\ 1\ 2\mathbf{l}_1^T R_1\ l_1^2\ l_1^2\,] H_2 H_3 \cdots H_{n-1} \begin{bmatrix} l_n^2 \\ l_n^2 \\ \mathbf{l}_n \\ 1 \\ 1 \end{bmatrix}, \tag{17.24}$$

where

$$H_i = \begin{pmatrix} 1 & 1 & 2\mathbf{l}_i^T R_i & l_i^2 & l_i^2 \\ 0 & 1 & 2\mathbf{l}_i^T R_i & l_i^2 & l_i^2 \\ 0 & 0 & R_i & \mathbf{l}_i & \mathbf{l}_i \\ 0 & 0 & \mathbf{0}^T & 1 & 1 \\ 0 & 0 & \mathbf{0}^T & 1 & 1 \end{pmatrix}.$$

The forms for r^2 and S^2 in (17.23) and (17.24) make it easy to perform conformational averages. These averages are weighted by conformation partition functions which reflect the relative likelihood of one conformation as compared to another based on the energy of each conformation. The *rotational isomeric state model* [21, 63][4] is perhaps the most widely known method to generate the statistical information needed to weight the relative occurrence of conformations in a statistical mechanical ensemble. There are two basic assumptions to the RIS model developed by Flory [21, 22]:

• The conformational energy function for a chain molecule is dominated by interactions between each set of three groups of atoms at the ends of two adjacent bond vectors. Hence the conformational energy function can be written in the form

$$E(\boldsymbol{\phi}) = \sum_{i=2}^{n-1} E_i(\phi_{i-1}, \phi_i)$$

where $\boldsymbol{\phi} = (\phi_1, \phi_2, ..., \phi_{n-1})$ is the set of all rotation angles around bond vectors.

• The value of the conformational partition function $e^{-E(\boldsymbol{\phi})/k_B T}$ is negligible except at the finite number of points where $E(\boldsymbol{\phi})$ is minimized, and hence averages of conformation-dependent functions can be calculated in the following way:

$$\langle f \rangle = \frac{\int_{T^{n-1}} f(\boldsymbol{\phi})\, e^{-E(\boldsymbol{\phi})/k_B T} d\phi_1 \cdots d\phi_{n-1}}{\int_{T^{n-1}} e^{-E(\boldsymbol{\phi})/k_B T} d\phi_1 \cdots d\phi_{n-1}} \approx \frac{\sum_{\boldsymbol{\phi}_\eta \in T^{n-1}} f(\boldsymbol{\phi}_\eta)\, e^{-E(\boldsymbol{\phi}_\eta)/k_B T}}{\sum_{\boldsymbol{\phi}_\eta \in T^{n-1}} e^{-E(\boldsymbol{\phi}_\eta)/k_B T}}.$$

Here the finite set conformations $\{\boldsymbol{\phi}_\eta\}$ are local minima of the energy function $E(\boldsymbol{\phi})$.

As a consequence of the above two assumptions, conformational statistics are generated in the context of the RIS model using statistical weight matrices. A detailed explanation of how this technique is used to generate $\langle r^2 \rangle$ and $\langle S^2 \rangle$ can be found in Flory's classic text [21] (in particular, see Chapter 4 of that book).

One drawback of this technique, as pointed out in [17, 45] is that the actual distribution of end-to-end distance is not calculated using the RIS method. The traditional technique for generating statistical distributions is to use Monte Carlo simulations.

---

[4]Abbreviated as 'RISM' or the 'RIS model'.

While this is a quite effective technique, it has the drawback that it will, by definition, not pick up the tails of a distribution. In contrast, the Euclidean group convolution technique discussed in the sequel guarantees that the tails (which give the border of the distribution) will be captured.

17.3.3 Helical Wormlike Models

The helical wormlike (HW) chain model developed by Yamakawa [82] is a generalization of the Kratky-Porod model, in which the effects of torsional stiffness are included in addition to those due to bending. Hence, in this model, if $\langle r^2 \rangle$ is desired, it is not sufficient to consider only the diffusion of the unit tangent vector $\mathbf{u}(s) \in \mathbb{S}^2$ as s progresses from 0 to L, but also how a frame attached to the curve twists about this tangent. This is a diffusion process on $SO(3)$, and the techniques of noncommutative harmonic analysis play an important role in solving, or at least simplifying, the governing equations.

Let $R(\alpha(s), \beta(s), \gamma(s))$ be the rotation matrix describing the orientation of the frame of reference attached to the curve at s relative to the base frame $R(\alpha(0), \beta(0), \gamma(0)) = \mathbb{I}$ where α, β, γ are ZYZ Euler angles. The unit tangent to the curve is $\mathbf{u}(s) = R(\alpha(s), \beta(s), \gamma(s))\mathbf{e}_3$. The angular velocity matrix

$$\dot{R}R^T = \begin{pmatrix} 0 & -\omega_3 & \omega_2 \\ \omega_3 & 0 & -\omega_1 \\ -\omega_2 & \omega_1 & 0 \end{pmatrix},$$

where

$$\omega_1 = \dot{\beta} \sin\gamma - \dot{\alpha} \sin\beta \cos\gamma \tag{17.25}$$
$$\omega_2 = \dot{\beta} \cos\gamma + \dot{\alpha} \sin\beta \sin\gamma \tag{17.26}$$
$$\omega_3 = \dot{\alpha} \cos\beta + \dot{\gamma}, \tag{17.27}$$

describes the instantaneous evolution in orientation for a small change in s. A dot represents differentiation with respect to s.

In the HW model, Yamakawa assumes the prefered state of the curve (which is assumed to be a solid tube of uniform material) is a helix, which is the result when the following potential energy function is minimized at each point on the curve:

$$V = \frac{1}{2}\alpha_0[\omega_1^2 + (\omega_2 - \kappa_0)^2] + \frac{1}{2}\beta_0(\omega_3 - \tau_0)^2.$$

The equality $\beta_0 = \alpha_0/(1 + \sigma)$ holds in the case when the cross section of the tube is a circular disc and σ is the Poisson's ratio of the material representing the curve $(0 \leq \sigma \leq 1/2)$.

Yamakawa generated equations for the probability density function, $P(\mathbf{R}, s)$, governing the evolution of orientations of the frame attached to the curve at each value of s:

$$\frac{\partial P(R; s)}{\partial s} = \left(\lambda \nabla_R^2 + \lambda\sigma L_3^2 - \kappa_0 L_2 + \tau_0 L_3\right) P(R, s) \qquad P(\mathbf{r}, R; 0) = \delta(R)\,\delta(\mathbf{x}). \tag{17.28}$$

He derived a similar equation for $\mathcal{P}(\mathbf{r}, R; s)$ as

$$\frac{\partial \mathcal{P}(\mathbf{r}, R; s)}{\partial s} = \left(\lambda \nabla_R^2 + \lambda\sigma L_3^2 - \kappa_0 L_2 + \tau_0 L_3 - \mathbf{e}_3^T R^T \nabla_{\mathbf{r}}\right) \mathcal{P}(\mathbf{r}, R; s)$$

$$\tag{17.29}$$

$$\mathcal{P}(\mathbf{r}, R; 0) = \delta(R)\,\delta(\mathbf{r}).$$

The operators L_i in the above equations are the derivatives

$$L_1 = \sin\gamma\frac{\partial}{\partial\beta} - \frac{\cos\gamma}{\sin\beta}\frac{\partial}{\partial\alpha} + \cot\beta\cos\gamma\frac{\partial}{\partial\gamma} \qquad (17.30)$$

$$L_2 = \cos\gamma\frac{\partial}{\partial\beta} + \frac{\sin\gamma}{\sin\beta}\frac{\partial}{\partial\alpha} - \cot\beta\sin\gamma\frac{\partial}{\partial\gamma} \qquad (17.31)$$

$$L_3 = \frac{\partial}{\partial\gamma} \qquad (17.32)$$

(which are the same as X_i^R in ZYZ Euler angles) and $\nabla_R^2 = L_1^2 + L_2^2 + L_3^2$ is the Laplacian for $SO(3)$. The solutions of (17.28) and (17.30) are related as

$$\int_{\mathbb{R}^3} \mathcal{P}(\mathbf{r}, R; s)\, d\mathbf{r} = P(R; s).$$

In fact, (17.28) is derived from (17.30) by integrating over $\mathbb{R}^3$ and observing that

$$\int_{\mathbb{R}^3} \nabla_{\mathbf{r}}\mathcal{P}(\mathbf{r}, R; s)\, d\mathbf{r} = \mathbf{0}.$$

Knowing $P(R, s)$ allows us to find $\langle \mathbf{u}(s) \cdot \mathbf{e}_3\rangle$ as in the Kratky-Porod model, but in this context it is calculated as

$$\langle \mathbf{u}(s) \cdot \mathbf{e}_3\rangle = \int_{SO(3)} \mathbf{e}_3^T Re_3 P(R; s)\, d\mu(R).$$

This then can be substituted into (17.15) and (17.16) to obtain $\langle r^2\rangle$ and $\langle S^2\rangle$. For the case when $\sigma = 0$, Yamakawa [82] has calculated these quantities. In particular,

$$\langle r^2\rangle = \frac{(4+\tau_0^2)L}{4+\nu^2} - \frac{\tau_0^2}{2\nu^2} - \frac{2\kappa_0^2(4-\nu^2)}{\nu^2(4+\nu^2)^2}$$
$$+ \frac{e^{-2L}}{\nu^2}\left(\frac{\tau_0^2}{2} + \frac{2\kappa_0^2}{(4+\nu^2)^2}[(4-\nu^2)\cos\nu L - 4\nu\sin\nu L]\right),$$

where

$$\nu = (\kappa_0^2 + \tau_0^2)^{\frac{1}{2}}.$$

When $\sigma = \tau_0 = \kappa_0 = 0$, we see that (17.28) reduces to the heat equation on $SO(3)$ and has the closed-form series solution given for the KP-model. When $\sigma \neq 0$ and $\tau_0 = \kappa_0 = 0$ the diffusion equation for the axially symmetric rotor results.

It is also possible to integrate (17.30) over $SO(3)$ to solve for $W_{HW}(\mathbf{r}; s)$. In particular, the solution can be expanded in a series of Hermite polynomials as [21, 82]:

$$W_{HW}(\mathbf{r}; L) = (a/\pi)^3 e^{-(ar)^2} \sum_{k=0}^{\infty} c_{2k} H_{2k+1}(ar)/(ar) \qquad (17.33)$$

where,

$$a = \left(\frac{3}{2\langle R^2\rangle}\right)^{\frac{1}{2}},$$

$H_j(x)$ is the jth Hermite polynomial, and

$$c_{2k} = \frac{\langle H_{2k+1}(ar)/(ar)\rangle}{2^{2(k+1)}(2k+1)!}.$$

17.0.4 Other Computational and Analytical Models of Polymers

One of the most widely used traditional computational techniques for determining statistical properties of macromolecular conformations is the Monte Carlo method [65]. The Monte Carlo method is very general, but it has the drawback that it may not capture the "tails" of slowly decreasing probability density functions.

The long range effects of excluded volume are often modeled using self-avoiding walks [1, 20, 25, 33, 42]. The self-avoiding walk is more sophisticated than the traditional random walk (see, e.g., for a discussion of random walks [69]), since self-interpenetration is not allowed. Polymer models based on self-avoiding walks are usually limited to rather short chains ($N \ll 100$) and the polymer is assumed to have backbone atoms that sit on a regular lattice. Other techniques for modeling long-range interactions can be found in [2, 3, 12, 54]. This requires models of the forces of interaction, such as those presented in [40, 71] (and references therein).

In contrast, renormalization group methods [24, 62, 66] are usually limited to very long ones ($N \to \infty$).

Recently, statistics of polymer intertwining has been studied using the concept of a braid group from knot theory [8, 13, 49, 50, 51].

An altogether different approach is to do direct molecular dynamic simulations (see [31, 57] for recent examples). The primary drawback of molecular simulations is the intensive computational requirements.

In the following section, we review an approach introduced in [9] which, in a sense, combines ideas of the self-avoiding walk and renormalization group. Essentially, the statistics of a polymer chain are generated using a weighted convolution on $SE(3)$, where the weighting reflects a bias against interpenetration of the molecule with itself.

17.4 Mass Density, Frame Density, and Euclidean Group Convolutions

In this section we define three statistical properties of macromolecular ensembles. Later it will be shown how these quantities are used to model the interactions of segments which may be far apart as measured along the chain, but are proximal in space. Before introducing these definitions, some notations are required.

Let $\mathbb{R}^3$ denote 3-dimensional Euclidean space, and $SE(3)$ denote the group of rigid-body motions. $\mathbf{x} \in \mathbb{R}^3$ is a position vector, and $g = (R, \mathbf{r}) \in SE(3)$ is a rigid body transformation (or frame of reference). The terminology $SE(3)$ stands for "special Euclidean" group of 3-dimensional space. This group is the semi-direct product of the rotation group $SO(3)$ and the translation group $(\mathbb{R}^3, +)$. The group law is $g_1 \circ g_2 = (R_1 R_2, R_1 \mathbf{r}_2 + \mathbf{r}_1)$, and geometrically represents the concatenation of rigid-body motions, or equivalently, a sequential change of reference frames. The inverse of any element $g \in SE(3)$ and the group identity element are given respectively as $g^{-1} = (R^T, -R^T \mathbf{r})$ and $e = (\mathbb{I}, \mathbf{0})$ where $\mathbb{I}$ is the 3×3 identity. The action of elements of $SE(3)$ on elements of $\mathbb{R}^3$ is defined as $g \circ \mathbf{x} = R\mathbf{x} + \mathbf{r}$. A more detailed review of the group $SE(3)$, including invariant integration and convolution of functions is presented in the appendix.

The three statistical quantities of importance in the present formulation are: (1) The ensemble mass density for the whole chain $\rho(\mathbf{x})$; (2) The ensemble tip frame density $f(g)$ (where g is the frame of reference of the distal end of the chain relative to the proximal); and (3) The function $\mu(g, \mathbf{x})$, which is the ensemble mass density of all configurations which grow from the identity frame fixed to one end of the chain and terminate at the relative frame g at the other end. These quantities are illustrated in Figure 17.2.

The functions ρ, f, and μ are related to each other. Given $\mu(g, \mathbf{x})$, the ensemble mass density is calculated by adding the contribution of each μ for each different end position and orientation:

$$\rho(\mathbf{x}) = \int_G \mu(g, \mathbf{x}) \, dg. \tag{17.34}$$

Here G is shorthand for $SE(3)$. This integration is written as being over all motions of the end of the chain, but only frames g in the support of μ contribute to the integral. dg denotes the invariant integration measure for $SE(3)$ reviewed in the appendix.

In an analogous way, it is not difficult to see that integrating the $\mathbf{x}$-dependence out of μ provides the total mass of configurations of the chain starting at frame e and terminating at frame g. Since each chain has mass M, this means that the frame density $f(g)$ is related to $\mu(g, \mathbf{x})$ as:

$$f(g) = \frac{1}{M} \int_{\mathbb{R}^3} \mu(g, \mathbf{x}) \, d\mathbf{x}. \tag{17.35}$$

We note the total number of frames attained by one end of the chain relative to the other is

$$F = K^N = \int_G f(g) \, dg$$

when each of the chains N degrees of freedom has K preferred states. It then follows that

$$\int_{\mathbb{R}^3} \rho(\mathbf{x}) \, d\mathbf{x} = F \cdot M.$$

If the functions $\rho(\mathbf{x})$ and $f(g)$ are known for the whole chain then a number of important thermodynamic and mechanical properties of the polymer can be determined. For instance, the moments of any positive integer power of the end-to-end distance, $\langle \|\mathbf{r}\|^m \rangle$, can be calculated from $f(g) = f(R, \mathbf{r})$ by first integrating out the orientational dependence:

$$\tilde{\rho}(\mathbf{r}) = \int_{SO(3)} f(R, \mathbf{r}) \, dR,$$

where dR is the normalized invariant integration measure for $SO(3)$. Then

$$\langle \|\mathbf{r}\|^m \rangle = \int_{\mathbb{R}^3} \|\mathbf{r}\|^m \tilde{\rho}'(\mathbf{r}) \, d\mathbf{r}$$

where $d\mathbf{r} = dr_1 dr_2 dr_3$ is the usual integration measure on $\mathbb{R}^3$ and $\tilde{\rho}' = \tilde{\rho}/K^n$ is the normalized version of $\tilde{\rho}$. The PDF of end-to-end distances is given as

$$d(r) = r^2 \int_{\mathbb{S}^2} \tilde{\rho}'(\mathbf{r}) \, d\mathbf{u} \tag{17.36}$$

where in spherical coordinates $\mathbf{r} = r\mathbf{u}$. Here $\mathbf{u} \in \mathbb{S}^2$ is a point on the unit sphere and $d\mathbf{u}$ is the area element for the unit sphere. In the special case when $\tilde{\rho}'(\mathbf{r})$ is spherically symmetric, $d(r) = 4\pi r^2 \tilde{\rho}'(r)$. More generally, it follows that

$$\int_{\mathbb{R}^3} \tilde{\rho}'(\mathbf{r}) \, d\mathbf{r} = \int_0^\infty d(r) \, dr.$$

Fig. 17.2. The Functions $\rho(\mathbf{x})$, $f(g)$ and $\mu(g,\mathbf{x})$ Shown at Top, Middle, and Bottom, Respectively

The functions $\rho(\mathbf{x})$ and $\tilde{\rho}(\mathbf{r})$ are also related in the following way. Imagine the chain is made up of n units,[5] each of which has K preferred energy states, and the ith of which has mass m_i. Then the function $\tilde{\rho}_i(\mathbf{r})$ relating the distribution of points reachable by the distal end of the ith unit relative to the proximal end of the first unit is related to the mass density of the collection of all units as

$$\rho(\mathbf{x}) \approx \frac{1}{K^n} \sum_{i=0}^{n-1} m_i K^{n-i} \tilde{\rho}_i(\mathbf{x}).$$

The weights K^{n-i} reflect the fact that units at the base of the chain will be counted more times than those at the top since they contribute to all subsequent units.

From an analytical viewpoint, all the information above can be calculated if $\mu(g, \mathbf{x})$ is known for the whole chain. If, however, we view the problem from a computational perspective, it is difficult to rationalize storing numerical values of the function $\mu(g, \mathbf{x})$ because $SE(3)$ is a six-dimensional Lie group, $\mathbb{R}^3$ is three-dimensional, and the computational storage required for accurate approximation of a function on the nine-dimensional space $SE(3) \times \mathbb{R}^3$ is prohibitive. For example, a hundred sample points in each coordinate direction would require 10^{18} memory locations. However, calculating $\rho(\mathbf{x})$ and $f(g)$ separately with the same sampling makes the problem literally a million times easier to handle.

Moreover, there are two major problems that must be addressed in order to accurately and efficiently calculate the functions $\rho(\mathbf{x})$ and $f(g)$: (1) For a chain with $N \in [100, 10000]$ units, each neighboring pair of which has K potential wells, it is impossible with current technology to enumerate the K^N conformations of the chain corresponding to all possible combinations of potential wells; (2) The long-range interactions of units that are distant in the chain but proximal in space cannot be modeled using only the functions $\rho(\mathbf{x})$ and $f(g)$. Both of these problems are addressed using the recursive procedure outlined in the following sections.

17.5 Generating Ensemble Properties for Purely Kinematical Models

This section explores mathematical and computational methods for generating statistical ensembles of macroscopically-serial chain macromolecules. In particular, given ensemble properties of subchains, we explore how the properties of the whole chain can be generated if the effects of conformational energy between segments is ignored. While this can be a valid approximation in situations when the energy of interaction is considered negligible, our goal in neglecting energy effects in in this section is purely a matter of pedagogy. For example, introducing useful syntax in the context of a purely kinematical model makes the definitions and concepts used in subsequent sections (where energy effects are incorporated) easier to understand.

Because of the exponential growth in the number of conformations as a function of the number of backbone atoms, it is not possible to generate the functions $\rho(\mathbf{x})$ and $f(g)$ for the whole chain by explicitly enumerating all configurations. This is well known in the polymer science literature. As reviewed in the introduction, the standard approaches to avoiding this exponential growth are: (1) asymptotic approximations in which the ensemble properties are assumed to be Gaussian as $N \to \infty$; and (2) Monte

---

[5]These could be either segments of the chain or single monomer units.

Carlo simulations which choose a small portion of the conformations and approximate overall behavior based on appropriate sampling.

In this section a completely different approach is taken. At the core of this approach remains the philosophy of considering ensemble properties of subsegments of the chain. This idea is not new, its roots go back more than a half century [23, 38]. However, the way in which properties of the whole chain are calculated based on the properties of segments of the chain are quite different than traditional approaches.

The basic idea is that we imagine dividing the chain up into P statistically significant segments. P is chosen large enough so that the continuum approximations to the ensembles $\rho(\mathbf{x})$ and $f(g)$ meet acceptable measures of accuracy. For example, if $N \approx 1000$ and $K = 3$ (as might be the case for torsion angles in a polyethylene molecule) we might choose $P \approx 40$. In this way $K^{N/P}$ is a number which can be managed with a personal computer.

17.5.1 The Mathematical Formulation

For each of the P statistical segments in the chain we can calculate $\rho_i(\mathbf{x})$ and $f_i(g)$ where g is the *relative* frame of reference of the distal end of the segment with respect to the proximal one. For a homogeneous chain, such as polyethylene, these functions are the same for each value of $i = 1, ..., P$.

In the general case of a heterogenous chain, we can calculate the functions $\rho_{i,i+1}(\mathbf{x})$ and $f_{i,i+1}(g)$ for the concatenation of segments i and $i+1$ from those of segments i and $i+1$ separately in the following way:

$$\rho_{i,i+1}(\mathbf{x}) = F_{i+1}\rho_i(\mathbf{x}) + \int_G f_i(h)\rho_{i+1}(h^{-1} \circ \mathbf{x})\,dh \qquad (17.37)$$

and

$$f_{i,i+1}(g) = (f_i * f_{i+1})(g) = \int_G f_i(h)\,f_{i+1}(h^{-1} \circ g)\,dh. \qquad (17.38)$$

In these expressions $h \in G = SE(3)$ is a dummy variable of integration. The meaning of (17.37) is that the mass density of the ensemble of all conformations of two concatenated chain segments results from two contributions. The first is the mass density of all the conformations of the lower segment (weighted by the number of different upper segments it can carry, which is $F_{i+1} = \int_G f_{i+1}dg$). The second contribution results from rotating and translating the mass density of the ensemble of the upper segment, and adding the contribution at each of these poses (positions and orientations). This contribution is weighted by the number of frames that the distal end of the lower segment can attain relative to its base. Mathematically, $L(h)\rho_{i+1}(\mathbf{x}) = \rho_{i+1}(h^{-1} \circ \mathbf{x})$ is a left-shift operation which geometrically has the significance of rigidly translating and rotating the function $\rho_{i+1}(\mathbf{x})$ by the transformation h. The weight $f_i(h)\,dh$ is the number of configurations of the ith segment terminating at frame of reference h.

The meaning of (17.38) is that the distribution of frames of reference at the terminal end of the concatenation of segments i and $i+1$ is the group-theoretical *convolution* of the frame densities of the terminal ends of each of the two segments relative to their respective bases.

Equations (17.37) and (17.38) can be iterated with $\rho_{i,i+1}$ and $f_{i,i+1}$ taking the places of ρ_i and f_i, and $\rho_{i,i+2}$ and $f_{i,i+2}$ taking the places of $\rho_{i,i+1}$ and $f_{i,i+1}$. This is described in detail shortly.

17.5.2 Generating $\mu_{i,i+1}$ from μ_i and μ_{i+1}

Later in the chapter, the effects of interaction between distal segments of a macro-molecule are approximated in an average sense[6] by considering how the functions $\mu_{i,i+j}(g, \mathbf{x})$ and $\mu_{i+j+1,k}(g, \mathbf{x})$ overlap for $k \geq i + j + 1$. In order to calculate the inter-action of these functions, we must first calculate the functions $\mu_{i,i+j}(g, \mathbf{x})$ from the set of functions $(\mu_{i,i+1}(g, \mathbf{x}), \ldots, \mu_{i+j-1,i+j}(g, \mathbf{x}))$. In analogy with the way $\rho_{i,i+j}(\mathbf{x})$ and $f_{i,i+j}(g)$ are calculated by repeating the computations in (17.37) and (17.38), $\mu_{i,i+j}(g, \mathbf{x})$ is constructed recursively by first calculating $\mu_{i,i+1}(g, \mathbf{x})$ from $\mu_i(g, \mathbf{x})$ and $\mu_{i+1}(g, \mathbf{x})$.

This is achieved by observing that

$$\mu_{i,i+1}(g, \mathbf{x}) = \int_G \left(\mu_i(h, \mathbf{x}) f_{i+1}(h^{-1} \circ g) + f_i(h) \mu_{i+1}(h^{-1} \circ g, h^{-1} \circ \mathbf{x}) \right) dh. \quad (17.39)$$

This equation says that there are two contributions to $\mu_{i,i+1}(g, \mathbf{x})$. The first comes from adding up all the contributions due to each $\mu_i(h, \mathbf{x})$. This is weighted by the number of upper segment conformations with distal ends that reach the frame g given that their base is at frame h. The second comes from adding up all shifted (translated and rotated) copies of $\mu_{i+1}(g, \mathbf{x})$, where the shifting is performed by the lower distribution, and the sum is weighted by the number of distinct configurations of the lower segment that terminate at h. This number is $f_1(h) \, dh$.

The veracity of this derivation can be confirmed by integrating the resulting function $\mu_{i,i+1}(g, \mathbf{x})$ over $\mathbb{R}^N$ and $SE(3)$ and comparing with (17.37) and (17.38). Note that segment i has mass M_i, segment $i + 1$ has mass M_{i+1}, and so $M_{i,i+1} = M_i + M_{i+1}$ is the value of M in (17.35) in the context of the present discussion.

17.5.3 Efficient Strategies for Calculating $f(g)$ and $\rho(\mathbf{x})$

Given the above means for calculating the functions $\rho_{i,i+1}(\mathbf{x})$ and $f_{i,i+1}(g)$ from the functions $f_i(g)$, $f_{i+1}(g)$, $\rho_i(\mathbf{x})$, and $\rho_{i+1}(\mathbf{x})$, it is now possible to formulate algorithms for generating these functions for the whole chain: $\rho(\mathbf{x}) = \rho_{1,P}(\mathbf{x})$ and $f(g) = f_{1,P}(g)$.

On a serial processor the most straightforward way to do this is to sequentially start at one end of the chain and repeatedly perform the required integrations. It does not matter at which end we begin. Starting at the base and working toward the distal end, we would calculate the sequence of functions $(\rho_{1,2}(\mathbf{x}), f_{1,2}(g))$, ..., $(\rho_{1,i}(\mathbf{x}), f_{1,i}(g))$, ..., $(\rho_{1,P}(\mathbf{x}), f_{1,P}(g))$. Starting from the other end we would calcu-late $(\rho_{P-1,P}(\mathbf{x}), f_{P-1,P}(g))$, ..., $(\rho_{P-i,P}(\mathbf{x}), f_{P-i,P}(g))$, ..., $(\rho_{1,P}(\mathbf{x}), f_{1,P}(g))$. In either case, viewing the number of computations required to calculate (17.37) and (17.38) as a constant, the recursive computation of these convolution-like integrals requires $\mathcal{O}(P)$ calculations. This is after each of the functions $\rho_i(\mathbf{x})$ and $f_i(g)$ have been calculated, which can be achieved in $\mathcal{O}(P \cdot K^{(N/P)})$ calculations, i.e., $\mathcal{O}(K^{(N/P)})$ calculations to explicitly enumerate configurations of each of the P segments.

The computational speed of the above approach on a parallel computer with P processors is much faster than on a single processor. Clearly, in this case the enumeration of segment conformations is reduced to an $\mathcal{O}(K^{(N/P)})$ time calculation since each of the P ensembles can be calculated separately. Furthermore, instead of explicitly computing a P-fold convolution requiring $\mathcal{O}(P)$ time, convolutions of adjacent functions can be calculated in a pairwise fashion on different processors. This reduces the running time to $\mathcal{O}(\log_2 P)$. For example, if $P = 8$, then the convolutions $f_{1,2} = f_1 * f_2$, $f_{3,4} = f_3 * f_4$,

---

[6]The meaning of this "average sense" is made precise later in the chapter.

$f_{5,6} = f_5 * f_6$, and $f_{7,8} = f_7 * f_8$ are all performed at the same time using $4 = P/2$ processors. Then $f_{1,4} = f_{1,2} * f_{3,4}$ and $f_{5,8} = f_{5,6} * f_{7,8}$ are calculated at the next level using two processors. Finally, $f_{1,8} = f_{1,4} * f_{5,8}$ is performed using a single processor. Thus, we have in this example an 8-fold convolution calculated in the same time as $3 = \log_2(8)$ convolutions on a serial processor.

It is worth noting that the same speed-up achieved for a heterogeneous chain calculated on a parallel processor is valid in the case of a homogeneous chain calculated on a single processor. In this special case $\mathcal{O}(K^{(N/P)})$ time is required for brute force enumeration of one segments of length N/P. Similarly, a P-fold convolution of the same function with itself only requires $\mathcal{O}(\log_2 P)$ distinct convolutions, and thus this order of time. E.g., the three convolutions $f_1 = f * f$, $f_2 = f_1 * f_1$, and $f_3 = f_2 * f_2$ generate the same result as an eight-fold convolution of f with itself.

17.6 Incorporating Conformational Energy Effects

Including the effects of conformational energy, the density function describing the distribution of tip-to-base positions and orientations of a macromolecule can be written in the form

$$f(g) \, \text{Vol}(\Delta(g)) = \int_{\phi \in Im(\Delta(g))} e^{-E(\phi)/k_B T} d\phi \tag{17.40}$$

where in the case of a serial chain $\phi = (\phi_1, \phi_2, ..., \phi_{n-1})$ is the set of all torsion angles, and $d\phi = d\phi_1 \cdots d\phi_{n-1}$.

$\Delta(g)$ is a small 6-dimensional voxel (box) in $SE(3)$ containing the group element g, and $\text{Vol}(\Delta(g))$ is the volume of this voxel. Since the support of f is finite, it can be divided into a finite number of voxels, $\mathcal{M}$. $g(\phi)$ is the end frame of reference of the chain relative to its base for given torsion angles ϕ, and $Im(\Delta(g))$ is the set of all torsion angles such that $g(\phi) \in \Delta(g)$.

Each torsion angle takes its values from the unit circle, T^1, and so the whole collection of angles takes its values from the $n - 1$-dimensional torus $T^{n-1} = T^1 \times \cdots \times T^1$.

We can normalize $f(g)$ in (17.40) by observing that the sum

$$\sum_{i=1}^{\mathcal{M}} f(g_i) \text{Vol}(\Delta(g_i)) = \sum_{i=1}^{\mathcal{M}} \int_{\phi \in Im(\Delta(g_i))} e^{-E(\phi)/k_B T} d\phi$$

becomes

$$\int_G f(g) \, dg = \int_{\phi \in T^{n-1}} e^{-E(\phi)/k_B T} d\phi$$

for sufficiently small voxels.

We note that as $k_B T \to \infty$, $f(g)$ reduces to the purely kinematic model discussed earlier. In the following subsections we examine how the density function in (17.40) is related to the Euclidean-group convolution model under a wide variety of conditions.

17.6.1 Two-State and Nearest-Neighbor Energy Functions

One of the simplest kinds of conformational energy functions is one of the form

$$E(\phi) = \sum_{i=1}^{n-1} E_i(\phi_i). \tag{17.41}$$

This kind of conformational energy function models nearest-neighbor interactions and leads to a separable partition function.

The frame density function for the concatenation of two chain molecules with this kind of energy function is again given by a Euclidean-group convolution. That is, since the energy function is additive, and the partition function is separable, the PDFs of two concatenated segments are multiplied and integrated as

$$(f_1 * f_2)(g) = f(g).$$

This is derived by substituting the energy function in (17.41) into (17.40) for a chain with $n = n_1 + n_2$ torsion angles. The functions ρ and μ are calculated analogously.

17.6.2 Interdependent Potential Functions

When interdependent potential functions are used, it is not possible to completely separate the conformational partition function and perform straight Euclidean-group convolutions. Instead, we can write the conformational energy function as

$$E(\phi) = E_1(\phi_1, ..., \phi_i) + E_{i+1}(\phi_i, \phi_{i+1}) + E_2(\phi_{i+1}, ..., \phi_{n-1}). \qquad (17.42)$$

Then frame densities for the lower and upper segments are generated as before. However, unlike the previously discussed cases, these frame densities are not only functions on $SE(3)$, but also depend on the bond angles contributing to the energy of interaction between the two chain segments. Hence we define f_1 and f_2 by the equalities

$$f_1(g', \phi_i)\, \mathrm{Vol}(\Delta(g')) = \int_{\{T^{i-1}|g(\phi_1,...,\phi_i) \in Im(\Delta(g'))\}} e^{-E_1(\phi_1,...,\phi_i)/k_B T}\, d\phi_1 ... d\phi_{i-1}$$

and

$$f_2(g', \phi_{i+1})\, \mathrm{Vol}(\Delta(g')) = \int_{\{T^{n-i-2}|g(\phi_{i+1},...,\phi_{n-1}) \in Im(\Delta(g'))\}} e^{-E_2(\phi_{i+1},...,\phi_{n-1})/k_B T}\, d\phi_{i+2} ... d\phi_{n-1}.$$

The functions f_1 and f_2 are written more cleanly in the limit of very small voxels using the Dirac delta as

$$f_1(g', \phi_i) = \int_{T^{i-1}} e^{-E_1(\phi_1,...,\phi_i)/k_B T}\, \delta(g^{-1}(\phi_1, ..., \phi_i) \circ g')\, d\phi_1 ... d\phi_{i-1}$$

and

$$f_2(g', \phi_{i+1}) = \int_{T^{n-i-2}} e^{-E_2(\phi_{i+1},...,\phi_{n-1})/k_B T}\, \delta(g^{-1}(\phi_{i+1}, ..., \phi_{n-1}) \circ g')\, d\phi_{i+2} ... d\phi_{n-1}$$

where

$$\delta(g) = \begin{cases} 0 & g \notin \Delta(e) \\ \frac{1}{\mathrm{Vol}(\Delta(e))} & g \in \Delta(e) \end{cases}$$

where e is the identity of G.

These two contributions add to give the composite frame density function using a combination of convolution and weighted integration over the last torsion angle of the first chain segment and first torsion angle of the second chain segment:

$$f(g) = \int_{T^2} \int_{SE(3)} f_1(h, \phi_i)\, f_2(h^{-1} \circ g, \phi_{i+1})\, e^{-E_{i+1}(\phi_i, \phi_{i+1})/k_B T}\, dh\, d\phi_i\, d\phi_{i+1}.$$

This follows from the evaluation of (17.40) with (17.42). From the forms given above for f_1 and f_2, and the properties of the Dirac delta, it is clear that the normalization

$$\int_G f(g)\,dg = \int_{T^{n-1}} e^{-E(\boldsymbol{\phi})/k_B T}\,d\boldsymbol{\phi}$$

holds.

The assumption of discrete sampling of most probable values of torsion angles, as in the RIS model, reduces the integrations over T^2 to summations. Hence, from a computational perspective, interdependent energy functions require K^2 convolutions instead of one, where K is the number of sample points for each torsion angle. Typically for an organic chain molecule $K = 3$.

This procedure is iterated in analogy with the algorithm developed in the previous section for the purely kinematic model. Namely, instead of breaking a chain molecule into two imaginary pieces, it can be broken into an arbitrary number of statistically significant segments, and an energy-weighted convolution of each pair of adjacent segments can be performed.

Hence convolution-like integrals of the form

$$f_{i,l}(g,\phi_i,\phi_l) = \int_{T^2}\int_{SE(3)} f_1(h,\phi_i,\phi_k)\,f_{k+1,l}(h^{-1}\circ g,\phi_{k+1},\phi_l)\,e^{-E_{i+1}(\phi_k,\phi_{k+1})/k_B T}\,dh\,d\phi_k\,d\phi_{k+1}$$

$$(17.43)$$

are performed, where now both the values of base and distal torsion angles must be recorded so that the process can be iterated. The drawback of this is that if the T^2 integral is approximated as a sum over K^2 values, then the K^2 convolutions on $SE(3)$ must be performed for each of the K^2 values of the pairs (ϕ_i,ϕ_l). Hence, K^4 convolutions on $SE(3)$ are required at each step. While this is troublesome, we note that these calculations are independent and hence can be distributed over a parallel computer.

17.6.3 Ensemble Properties Including Long-Range Conformational Energy

In this section we model the longe-range interactions in a macroscopically serial chain using an averaging approach which builds on the formulation of the previous sections. The model uses the functions $\mu(g,\mathbf{x})$ which were not explicitly needed in the case when energy was not important.

Accurately modeling interactions of atoms which are distal in the chain but proximal in space (due to bending of the chain) is one of the most difficult problems in the study of macromolecules, independent of whether they are man-made polymers, proteins, or DNA. Explicitly accounting for all such interactions for all possible configurations by brute force enumeration requires a mind-boggling amount of computational time. At the other extreme, the simplified closed-form analytical models such as the Gaussian random walk do not explicitly account for these interactions. Furthermore, models based on self-avoiding walks and renormalization group methods are respectively limited to rather short chains ($N \ll 100$) or very long ones ($N \to \infty$).

A number of different approaches are considered here to incorporate the effects of energy. Perhaps the most straightforward is to penalize contributions in (17.39) so that when the support of appropriately shifted functions μ_i and μ_{i+1} intersect, these functions would be disallowed from contributing to the computation of $\mu_{i,i+1}$. Clearly this would generate ensemble statistical distributions which would be lower estimates of those generated from the self-avoiding walk model. A slightly more sophisticated model would calculate the energy of interaction of the distributions μ_i and μ_{i+1} and use this information in an appropriate conformational partition function. We now quantify this discussion. The interaction of segment i and $i+1$ is approximated by considering the interaction of the corresponding functions μ_i and μ_{i+1} as:

$$E_{i,i+1}(h,g) = \int_{\mathbb{R}^3} \int_{\mathbb{R}^3} \mu_i(h,\mathbf{x})\mu_{i+1}(h^{-1}\circ g, h^{-1}\circ \mathbf{y})V(\mathbf{x}-\mathbf{y})\,d\mathbf{x}d\mathbf{y}. \qquad (17.44)$$

We assume here that the potential between any two atoms located at positions $\mathbf{x}$ and $\mathbf{y}$ is $V(\mathbf{x}-\mathbf{y})$. Then (17.44) is an approximation of the interaction of all configurations of segment i which terminate at frame h and all configurations of segment $i+1$ with distal end at g and proximal end at h (hence the relative displacement $h^{-1}\circ g$). In the case when the potential function is used to represent pure hard-sphere repulsion and no attraction, then a reasonable model for $V(\cdot)$ is

$$V(\mathbf{x}) = E_0\delta(\mathbf{x}).$$

In this case we write

$$E_{i,i+1}(h,g) = E_0 \int_{\mathbb{R}^3} \mu_i(h,\mathbf{x})\mu_{i+1}(h^{-1}\circ g, h^{-1}\circ \mathbf{x})\,d\mathbf{x}. \qquad (17.45)$$

This then can be used to approximate $\mu_{i,i+1}$ as

$$\mu_{i,i+1}(g,\mathbf{x}) = \int_G \left(\mu_i(h,\mathbf{x})\, f_{i+1}(h^{-1}\circ g) + f_i(h)\mu_{i+1}(h^{-1}\circ g, h^{-1}\circ \mathbf{x})\right) e^{-E_{i,i+1}(h,g)/k_BT}dh. \qquad (17.46)$$

By definition, it follows that

$$f_{i,i+1}(g) = \frac{1}{M_i + M_{i+1}} \int_{\mathbb{R}^3} \mu_{i,i+1}(g,\mathbf{x})\,d\mathbf{x} = \int_G f_i(h)\, f_{i+1}(h^{-1}\circ g)\, e^{-E_{i,i+1}(h,g)/k_BT}dh. \qquad (17.47)$$

In this way, the purely kinematical model is modified so as to take into account the energy of interaction of two adjacent segments. As E_0/k_BT becomes large, the only contributions to $\mu_{i,i+1}$ are from the shifted versions of the functions μ_i and μ_i which do not overlap at all. This extreme case is a lower bound on the $\mu_{i,i+1}$ that the self avoiding walk would generate. This follows for large E_0/k_BT because the model in (17.46) not only disallows the intersection of two adjacent segments, but also disallows any contribution if (17.45) is nonzero, i.e., if μ_i and μ_{i+1} overlap. On the other hand, for smaller values of E_0/k_BT the present model may provide more realistic statistics than lattice self-avoiding walks since there is no artificial restriction on bond and torsion angles that limit conformations to conform to a lattice in the present model. In the next section the group-theoretical formulation is compared with a number of traditional models.

17.7 Statistics of Stiff Molecules as Solutions to PDEs on $SO(3)$ and $SE(3)$

Experimental measurements of the stiffness constants of DNA and other stiff (or semi-flexible) macromolecules have been reported in a number of papers, as well as the statistical mechanics of such molecules. See, for example, [29, 34, 35, 41, 44, 46, 47, 48, 53, 52, 55, 58, 67, 68, 70, 75, 77, 78, 79].

As is often the case in theoretical polymer science, analogies between the motion of a particle along a path and the motion of an observer traversing a polymer chain allow for tools from classical and quantum mechanics to be applied.

17.7.1 Model Formulation

In particular, a number of authors have derived potential energies of bending and/or twisting of a stiff chain that are of the form

$$E = \int_0^L U(\boldsymbol{\omega}(s))\, ds$$

where L is the length of the macromolecule and

$$U = \frac{1}{2}\boldsymbol{\omega}^T B \boldsymbol{\omega} - \mathbf{b}^T \boldsymbol{\omega} + \beta'. \tag{17.48}$$

Here $B = B^T \in \mathbb{R}^{3\times3}$ is a positive semi-definite matrix, $\mathbf{b} \in \mathbb{R}^3$, and $\beta' \in \mathbb{R}$. $\boldsymbol{\omega}$ is the "angular velocity" of a frame of reference which traverses the macromolecule, coinciding with each frame $(\mathbf{a}(s), A(s))$ affixed to the backbone of the molecule for each value of arclength s. This "angular velocity" is the dual vector of the skew symmetric matrix $A^T \dot{A}$, where $(\dot{}) = d/ds$. That is, $\boldsymbol{\omega} \times \mathbf{x} = A^T \dot{A} \mathbf{x}$ for all $\mathbf{x} \in \mathbb{R}^3$. This is completely analogous to the definition of angular velocity of a rigid body as seen in the body fixed frame with s taking the place of time. Henceforth, we will use the notation $U = U(\boldsymbol{\omega}) = U(A, \dot{A})$ to denote the fact that the bending energy is a function of the rotation matrix and its derivative through of the definition of $\boldsymbol{\omega}$.

As well-known examples of (17.48) from the polymer science literature, consider:

The Kratky-Porod Model [17]

$$B = \begin{pmatrix} \alpha_0 & 0 & 0 \\ 0 & \alpha_0 & 0 \\ 0 & 0 & 0 \end{pmatrix}; \quad \mathbf{b} = \begin{pmatrix} 0 \\ 0 \\ 0 \end{pmatrix}; \quad \beta' = 0.$$

The Yamakawa (HW) Model [82]

$$B = \begin{pmatrix} \alpha_0 & 0 & 0 \\ 0 & \alpha_0 & 0 \\ 0 & 0 & \beta_0 \end{pmatrix}; \quad \mathbf{b} = \begin{pmatrix} 0 \\ \alpha_0\kappa_0 \\ \beta_0\tau_0 \end{pmatrix}; \quad \beta' = \frac{1}{2}(\beta_0\tau_0^2 + \alpha_0\kappa_0^2).$$

The Marko-Siggia DNA Model [44]

$$B = \begin{pmatrix} A' & 0 & B \\ 0 & A & 0 \\ B & 0 & C \end{pmatrix}; \quad \mathbf{b} = \begin{pmatrix} B\omega_0 \\ 0 \\ C\omega_0 \end{pmatrix}; \quad \beta' = \frac{1}{2}C\omega_0^2.$$

The Modified Marko-Siggia Model [30]

$$B = \begin{pmatrix} A + B^2/C & 0 & B \\ 0 & A & 0 \\ B & 0 & C \end{pmatrix}; \quad \mathbf{b} = \begin{pmatrix} B\omega_0 \\ 0 \\ C\omega_0 \end{pmatrix}; \quad \beta' = \frac{1}{2}C\omega_0^2.$$

Other modifications of these models may be made to include stretching effects, though the current presentation is restricted to the inextensible case.

We now generate the diffusion equation that governs the evolution of the positional and orientation probability density function $F(\mathbf{a}, A; s)$ for all values of $0 \le s \le L$. Assuming that the proximal end is fixed at the frame $(\mathbf{0}, \mathbb{I})$, then $F(\mathbf{a}, A; 0) = \delta(\mathbf{a}) \delta(A)$. Here the Dirac delta function on the motion group is written as the product of those for $\mathbb{R}^3$ and $SO(3)$.

Under the constraint that the molecule is inextensible, and all the frames of reference are attached to the backbone with their local z-axis pointing in the direction of the next frame, we observe that

$$\mathbf{a}(L) = \int_0^L \mathbf{u}(s)\, ds \quad \text{and} \quad \mathbf{u}(s) = A(s)\mathbf{e}_3. \tag{17.49}$$

Hence, the PDF of interest can be formulated as the following path integral over the rotation group:

$$F(\mathbf{a}, A; L) = \int_{A(0)=I}^{A(L)=A} \delta\left(\mathbf{a}(L) - \int_0^L \mathbf{u}(s)\, ds\right) \exp\left[-\int_0^L U(A, \dot{A})\, ds\right] \mathcal{D}[A(s)], \tag{17.50}$$

where it is assumed that the bending energy U is measured in units of $k_B T$. Path integration over the rotation group has been studied extensively in the literature in the context of quantum mechanics (see Appendix E). Our notation and formulation follows that in [82].

Using the classical Fourier transform pair

$$\hat{f}(\mathbf{k}) = \int_{\mathbb{R}^3} f(\mathbf{a})\, e^{-i\mathbf{k}\cdot\mathbf{a}} d^3 a \quad \leftrightarrow \quad f(\mathbf{a}) = \frac{1}{(2\pi)^3} \int_{\mathbb{R}^3} \hat{f}(\mathbf{k})\, e^{i\mathbf{k}\cdot\mathbf{a}} d^3 k, \tag{17.51}$$

we write

$$\hat{F}(\mathbf{k}, A; L) = \int_{A(0)=I}^{A(L)=A} \exp\left[-\int_0^L (i\mathbf{k}\cdot\mathbf{u} + U)\, ds\right] \mathcal{D}[A(s)].$$

Treating the inner most integrand as i times a Lagrangian with kinetic and potential energies,

$$T = \frac{1}{2} i\omega^T B\omega$$

and

$$V = i[\mathbf{b}\cdot\omega - \beta'] + \mathbf{k}\cdot\mathbf{u},$$

(the constant β' can be ignored) we calculate the momenta and Hamiltonian in the usual way, which for this case means

$$p_k = \frac{\partial L}{\partial \omega_k} \rightarrow H = -i[\frac{1}{2}\mathbf{p}^T B^{-1}\mathbf{p}] + \mathbf{b}^T B^{-1}\mathbf{p} + \mathbf{k}\cdot\mathbf{u}. \tag{17.52}$$

Here and henceforth B is assumed to be positive definite (and hence invertible).

The quantization

$$p_i = -iX_i^R \tag{17.53}$$

is used, where the differential operators X_i^R acting on functions on the rotation group are defined as

$$(X_i^R f)(A) \doteq \left.\frac{df(A \cdot \mathrm{rot}[\mathbf{e}_i, t])}{dt}\right|_{t=0} = \left.\frac{df(A(I + tX_i))}{dt}\right|_{t=0},$$ (17.54)

where

$$X_1 = \begin{pmatrix} 0 & 0 & 0 \\ 0 & 0 & -1 \\ 0 & 1 & 0 \end{pmatrix}; \quad X_2 = \begin{pmatrix} 0 & 0 & 1 \\ 0 & 0 & 0 \\ -1 & 0 & 0 \end{pmatrix}; \quad X_3 = \begin{pmatrix} 0 & -1 & 0 \\ 1 & 0 & 0 \\ 0 & 0 & 0 \end{pmatrix}.$$

The superscript R in X_i^R denotes the fact that the infinitesimal rotation $I + tX_i$ is applied on the right of the argument of the function. This corresponds to an infinitesimal motion relative to the body-fixed frame in a rigid body.

Using the ZXZ Euler angles (α, β, γ) these operators have the explicit form

$$X_1^R = -\cot\beta\sin\gamma\frac{\partial}{\partial\gamma} + \frac{\sin\gamma}{\sin\beta}\frac{\partial}{\partial\alpha} + \cos\gamma\frac{\partial}{\partial\beta};$$

$$X_2^R = -\cot\beta\cos\gamma\frac{\partial}{\partial\gamma} + \frac{\cos\gamma}{\sin\beta}\frac{\partial}{\partial\alpha} - \sin\gamma\frac{\partial}{\partial\beta};$$ (17.55)

$$X_3^R = \frac{\partial}{\partial\gamma}.$$

The Schrödinger-like equation corresponding to the Hamiltonian (17.52) and quantization (17.53) is

$$i\frac{\partial\hat{F}}{\partial L} = H\hat{F}.$$

This takes the explicit form

$$\left(\frac{\partial}{\partial L} - \frac{1}{2}\sum_{k,l=1}^{3}(B_{lk}^{-1}X_l^R X_k^R - 2B_{lk}^{-1}b_k X_l^R) + i\mathbf{k}\cdot\mathbf{u}\right)\hat{F} = 0.$$ (17.56)

Henceforth we will use the quantities $D = B^{-1}$ and $\mathbf{d} = -B^{-1}\mathbf{b}$.

The classical Fourier inversion formula (17.51) then converts (17.56) to

$$\left(\frac{\partial}{\partial L} - \frac{1}{2}\sum_{k,l=1}^{3} D_{lk}X_l^R X_k^R - \sum_{l=1}^{3} d_l X_l^R + \mathbf{u}\cdot\nabla_{\mathbf{a}}\right)F = 0,$$ (17.57)

or equivalently

$$\left(\frac{\partial}{\partial L} - \frac{1}{2}\sum_{k,l=1}^{3} D_{lk}\tilde{X}_l^R \tilde{X}_k^R - \sum_{l=1}^{3} d_l \tilde{X}_l^R + \tilde{X}_6^R\right)F = 0,$$ (17.58)

which is a PDE on the motion group, $SE(3)$. The initial conditions are $F(\mathbf{a}, A; 0) = \delta(\mathbf{a})\,\delta(A)$.

Integrating F over all positions, $\mathbf{a} \in \mathbb{R}^3$, results in a purely orientational density function:

$$f(A; s) = \int_{\mathbb{R}^3} F(\mathbf{a}, A; s)\, d^3 a.$$

Performing this integration over the initial conditions and (17.57) results in the $SO(3)$-diffusion equation

$$\left(\frac{\partial}{\partial L} - \frac{1}{2} \sum_{k,l=1}^{3} D_{lk} X_l^R X_k^R - \sum_{l=1}^{3} d_l X_l^R \right) f = 0 \qquad (17.59)$$

with initial conditions $f(A; 0) = \delta(A)$.

Equation (17.59) is a partial differential equation which governs the evolution of the function f on the rotation group, $SO(3)$. It is solved in series form in the next section using techniques from noncommutative harmonic analysis.

We note that as a direct result of the structure of IUR matrix elements for $SE(3)$ and the Fourier inversion formula (Chapter 10) that

$$\int_{SO(3)} F(\mathbf{a}, A) \, dA = \frac{1}{2\pi^2} \sum_{l'=0}^{\infty} \sum_{m'=-l'}^{l'} \int_0^{\infty} p^2 \, dp \, \hat{F}^0_{0,0;l',m'}(p)[l', m' \,|\, p, s \,|\, 0, 0](\mathbf{a}).$$

If this distribution of end positions is then integrated over the surface of a sphere with radius $a = |\mathbf{a}|$, the result is the end-to-end distance distribution:

$$\frac{a^2}{2\pi^2} \int_{\mathbb{S}^2} \int_{SO(3)} F(a\mathbf{u}, A) \, d\mathbf{u} \, dA = \frac{2}{\pi} a^2 \int_0^{\infty} p^2 \, dp \, \hat{F}^0_{0,0;0,0}(p)[0, 0 \,|\, p, s \,|\, 0, 0](\mathbf{a}).$$

It is easy to verify that $[0, 0 \,|\, p, s \,|\, 0, 0](\mathbf{a}) = j_0(pa) = \sin(pa)/pa$ (zeroth-order spherical Bessel function).

These expressions provide a means of addressing PDFs of end-to-end relative position and end-to-end distance when knowledge of orientation is not critical.

17.7.2 Operational Properties and Solutions of PDEs

By the definition of the $SE(3)$-Fourier transform $\mathcal{F}[\cdot]$ and operators $\tilde{X}_i^R$ reviewed in earlier subsections of this section, we observe that

$$(\widetilde{X_i^R F})(p) = \int_{SE(3)} \frac{d}{dt} \left(F(g \circ \exp(t\tilde{X}_i)) \right) \bigg|_{t=0} U^s(g^{-1}, p) \, dg. \qquad (17.60)$$

Here g can be thought of as $H(g)$ and $\exp(t\tilde{X}_i)$ is an element of the subgroup of $SE(3)$ generated by $\tilde{X}_i$, which for small values of t is approximated as $I + t\tilde{X}_i$. By performing the change of variables $h = g \circ \exp(t\tilde{X}_i)$ and using the homomorphism property of the representations $U^s(\cdot)$, we find that

$$(\widetilde{X_i^R F})(p) = \int_{SE(3)} F(h) \frac{d}{dt} \left(U^s(\exp(t\tilde{X}_i) \circ h^{-1}, p) \right) \bigg|_{t=0} dh \qquad (17.61)$$

$$= \frac{d}{dt} \left(U^s(\exp(t\tilde{X}_i), p) \right) \bigg|_{t=0} \int_{SE(3)} F(h) U^s(h^{-1}, p) \, dh. \qquad (17.62)$$

By defining

$$u^s(\tilde{X}_i, p) = \frac{d}{dt} \left(U^s(\exp(t\tilde{X}_i), p) \right) \bigg|_{t=0},$$

we write

$$(\widetilde{X_i^R F})(p) = u^s(\tilde{X}_i, p) \hat{F}^s(p).$$

Hence, (17.58) can be transformed to the infinite system of linear differential equations:

$$\frac{dF^s}{dL} = \mathcal{B}^s \hat{F}^s, \tag{17.63}$$

where

$$\mathcal{B}^s = \frac{1}{2} \sum_{k,l=1}^{3} D_{lk} u^s(\tilde{X}_l, p) u^s(\tilde{X}_k, p) + \sum_{l=1}^{3} d_l u^s(\tilde{X}_l, p) - u^s(\tilde{X}_6, p).$$

In principle, $F(\mathbf{a}, A; L)$ is then found by simply substituting $\hat{F}^s(p; L) = \exp(L\mathcal{B}^s)$ into the $SE(3)$-Fourier inversion formula. In practice, however, exponentiation of a non-diagonal infinite-dimensional matrix poses some difficulties which need to be addressed in subsequent work.

Explicitly, for $i = 1, 2, 3$ we have

$$u^s(\tilde{X}_i, p) = \frac{d}{dt} U^s_{l',m';l,m}(\mathbf{0}, \exp[tX_i]; p)\Big|_{t=0} = \delta_{l,l'} \frac{d}{dt} U^l_{m',m}(\exp[tX_i])\Big|_{t=0}.$$

The second equality above follows easily from the structure of the matrix elements $U^s_{l',m';l,m}$, and $\frac{d}{dt} U^l_{m,n}(\exp[tX_i])\big|_{t=0}$ are given explicitly in Chapter 9. This, together with the operational property (Chapter 10):

$$u^s_{l',m';l,m}(\tilde{X}_6, p) = \frac{d}{dt} U^s_{l',m';l,m}(t\mathbf{e}_3, I; p)|_{t=0} =$$

$$ip\kappa^s_{l',m'}\delta_{l'-1,l}\delta_{m',m} + ip\frac{m's}{l'(l'+1)} \delta_{l'l}\delta_{m',m} + ip\kappa^s_{l,m}\delta_{l',l-1}\delta_{m',m}$$

where

$$\kappa^s_{l',m'} = \left(\frac{(l'^2 - m'^2)(l'^2 - s^2)}{(2l'+1)(2l'-1)l'^2} \right)^{1/2}$$

allows us to write the elements of $\mathcal{B}^s(p)$ as

$$\mathcal{B}^s_{l',m';l,m} = \mathcal{A}^l_{m',m}\delta_{l',l} - ip\kappa^s_{l',m'}\delta_{l'-1,l}\delta_{m',m} - ip\frac{m's}{l'(l'+1)} \delta_{l'l}\delta_{m',m} - ip\kappa^s_{l,m}\delta_{l',l-1}\delta_{m',m}.$$

This explicit form of the elements of the matrix $\mathcal{B}^s$, together with the Fourier inversion formula for $SE(3)$ allows us to numerically solve for the probability density function $F(\mathbf{a}, A; L)$ for each L, and more specifically the probability density of end-to-end distances. From this, any desired moments can be computed. Numerical implementation by truncating and exponentiating the infinite-dimensional matrices $\mathcal{B}^s$ is described in [10].

17.8 Summary

In this chapter, we presented an introduction to the statistical mechanics of polymer chains. We reviewed classical analytical techniques used to model the statistics of both discrete and continuous chains. We then illustrated how noncommutative harmonic analysis can be used in theoretical polymer science. In particular, we showed how Euclidean-group convolutions can be used as a numerical tool for generating conformational statistics. In the special case of stiff macromolecules, we showed how the Fourier transform for $SE(3)$ converts a diffusion-type partial differential equation evolving on $SE(3)$ into a system of linear ordinary differential equations with constant coefficients.

Whereas classical techniques generate moments of probability density functions describing the relative occurrence of end-to-end positions and orientations, the group-theoretic approach allows us to calculate the whole PDF. The $SE(3)$-diffusion models for DNA conformational statistics presented here have been extended to include the case of bends and internal joints in [83, 84]. For more on the numerical implementation of these ideas see [9, 10, 81]. For recent works that model DNA conformational distributions as $SE(3)$-Gaussians in exponential coordinates, and make use of the covariance propagation formulas presented in Chapter 15, see [11, 80]. For alternative approaches to solving the same sorts of left-invariant diffusion equations on $SE(3)$ that were described in this chapter, see [64].

References

1. Alexandrowicz, Z., "Monte Carlo of Chains with Excluded Volume: A Way to Evade Sample Attrition," *Journal of Chemical Physics*, 51(2): 561–565, July 1969.
2. Allegra G., Ganazzoli F., Bontempelli S., "Good- and Bad-Solvent Effect on the Rotational Statistics of a Long Chain Molecule," *Computational and Theoretical Polymer Science*, 8(1–2): 209–218, 1998
3. Batoulis, J., Kremer, K., "Statistical Properties of Biased Sampling Methods for Long Polymer Chains," *Journal of Physics A: Mathematics and General*, 21(1): 127–146, 1988.
4. Birshtein, T.M., Ptitsyn, O.B., *Conformations of Macromolecules*, Interscience, New York, 1966.
5. Bohdanecký, M., Kovár, J., *Viscosity of Polymer Solutions*, Elsevier, New York, 1982.
6. Boyd, R.H., Phillips, P.J., *The Science of Polymer Molecules*, Cambridge Solid State Science Series, Cambridge University Press, Cambridge, 1993.
7. Chandrasekhar, S., "Stochastic Problems in Physics and Astronomy," *Reviews of Modern Physics*, 15(1): 1–89, Jan. 1943.
8. Chen, S.-J., Dill, K.A., "Symmetries in Proteins: A Knot Theory Approach," *Journal of Chemical Physics* 104(15): 5964–5973, April 1996.
9. Chirikjian, G.S., "Conformational Statistics of Macromolecules Using Generalized Convolution," *Computational and Theoretical Polymer Science*, 11(2): 143–153, 2001.
10. Chirikjian, G.S., Wang, Y., "Conformational Statistics of Stiff Macromolecules as Solutions to Partial Differential Equations on the Rotation and Motion Groups," *Physical Review E*, 62(1): 880–892, 2000.
11. Chirikjian, G.S., "Modeling Loop Entropy," *Methods in Enzymology*, Part C, 487: 101–130, 2011.
12. Collet, O., Premilat, S., "Calculation of Conformational Free Energy and Entropy of Chain Molecules with Long-Range Interactions," *Macromolecules*, 26: 6076–6080, 1993.
13. Comtet, A., Nechaev, S., "Random Operator Approach for Word Enumeration in Braid Groups," *Journal of Physics A: Mathematical and General*, 31(26): 5609–5630, 1998.
14. Daniels, H.E., "The Statistical Theory of Stiff Chains," *Proceedings of the Royal Society* (Edinburgh) Vol. A63: 290–311, 1952.
15. de Gennes, P.G., *Introduction to Polymer Dynamics*, Cambridge University Press, 1990.
16. de Gennes, P.G., *Scaling Concepts in Polymer Physics*, Cornell University Press, 1979.
17. des Cloizeaux, J., Jannink, G., *Polymers in Solution: Their Modelling and Structure*, Clarendon Press, Oxford, 1990.
18. Doi, M., Edwards, S.F., *The Theory of Polymer Dynamics*, Clarendon Press, Oxford, 1986.
19. Doi, M., *Introduction to Polymer Physics*, Oxford University Press, 1996.
20. Fisher, M.E., "Shape of a Self-Avoiding Walk or Polymer Chain," *Journal of Chemical Physics*, 44(2): 616–622, Jan. 1966.
21. Flory, P.J., *Statistical Mechanics of Chain Molecules*, John Wiley & Sons, 1969 (reprinted Hanser Publishers, Munich 1989).

22. Flory, P. J., "Foundations of Rotational Isomeric State Theory and General Methods for Generating Configurational Averages," *Rotational Isomeric State Theory and Methods*, 7(3): 381–392, May/June 1974.

23. Flory, P.J., "The Configuration of Real Polymer Chains," *Journal of Chemical Physics*, 17(3): 303–310, March 1949.

24. Freed, K.F., *Renormalization Group Theory of Macromolecules*, Wiley, New York, 1987.

25. Freed, K.F., "Polymers as Self-Avoiding Walks," *Annals of Probability*, 9(4): 537–556, 1981.

26. Gobush, W., Yamakawa, H., Stockmayer, W.H., Magee, W.S., "Statistical Mechanics of Wormlike Chains. I. Asymptotic Behavior," *Journal of Chemical Physics*, 57(7): 2839–2843, Oct. 1972.

27. Grosberg, A. Yu., Khokhlov, A.R., *Statistical Physics of Macromolecules*, American Institute of Physics, New York, 1994.

28. Grosberg, A. (ed.), *Theoretical and Mathematical Models in Polymer Research*, Academic Press, Boston, 1998.

29. Hagerman, P.J., "Analysis of the Ring-Closure Probabilities of Isotropic Wormlike Chains: Application to Duplex DNA," *Biopolymers*, 24(10): 1881–1897, 1985.

30. Haijun, Zhou, and Ou-Yang Zhong-Can, "Bending and Twisting Elasticity: A Revised Marko-Siggia Model on DNA Chirality," *Physical Review E*, 58(4): 4816–4819, Oct. 1998.

31. He, S., Scheraga, H.A., "Macromolecular Conformational Dynamics in Torsional Angle Space," *Journal of Chemical Physics*, 108(1): 271–286, 1998.

32. Hermans, J.J., Ullman, R., "The Statistics of Stiff Chains, with Applications to Light Scattering," *Physica*, 18(11): 951–971, 1952.

33. Honeycutt, J.D., "A General Simulation Method for Computing Conformational Properties of Single Polymer Chains," *Computational and Theoretical Polymer Science*, 8(1–2): 1–8, 1998.

34. Horowitz, D.S., Wang, J.C., "Torsional Rigidity of DNA and Length Dependence of the Free Energy of DNA Supercoiling," *Journal of Molecular Biology*, 173(1): 75–91, 1984.

35. Klenin, K., Merlitz, H., Langowski, J., "A Brownian Dynamics Program for the Simulation of Linear and Circular DNA and Other Wormlike Chain Polyelectrolytes," *Biophysical Journal*, 74(2): 780–788, Feb. 1998.

36. Kloczkowski, A., Jernigan, R.L., "Computer Generation and Enumeration of Compact Self-Avoiding Walks within Simple Geometries on Lattices," *Computational and Theoretical Polymer Science*, (3–4): 163–173, 1997.

37. Kratky, O., Porod, G., "Röntgenuntersuchung gelöster Fadenmoleküle," *Recueil des Travaux Chimiques des Pays-Bas*, 68(12): 1106–1122, 1949.

38. Kuhn, W., "Über die Gestalt fadenförmiger Moleküle in Lösungen," *Kolloid-Zeitschrift*, 68(1): 2–15, 1934.

39. Lagrange, J.L., "Sur Une Nouvelle Propriété du Centre de Gravité," *Oeuvres de Lagrange*, Vol. 5, pp. 535–540, 1870.

40. Lennard-Jones, J.E., "The Equation of State of Gases and Critical Phenomena," *Physica*, 4(10): 941–956, 1937.

41. Levene, S.D., Crothers, D.M., "Ring Closure Probabilities for DNA Fragments by Monte Carlo Simulation," *Journal of Molecular Biology*, 189(1): 61–72, 1986.

42. Madras, N., Slade, G., *The Self-Avoiding Walk*, Springer, 2012.

43. Mark, J.E., Erman, B., *Rubberlike Elasticity: A Molecular Primer*, Cambridge University Press, 2007.

44. Marko, J. F., Siggia, E.D., "Bending and Twisting Elasticity of DNA," *Macromolecules*, 27(4): 981–988, 1994.

45. Mattice, W.L., Suter, U.W., *Conformational Theory of Large Molecules, The Rotational Isomeric State Model in Macromolecular Systems*, Wiley, New York, 1994.

46. Mondescu, R. P., Muthukumar, M., "Brownian Motion and Polymer Statistics on Certain Curved Manifolds," *Physical Review E*, 57(4): 4411–4419, April 1998.

47. Moroz, J.D., Nelson, P., "Torsional Directed Walks, Entropic Elasticity, and DNA Twist Stiffness," *Proceedings of the National Academy of Sciences of the United States of America*, 94(26): 14,418 – 14,422, 1997.
48. Moroz, J.D., Nelson, P., "Entropic Elasticity of Twist-Storing Polymers," *Macromolecules*, 31(18): 6333 – 6347, 1998.
49. Nechaev, S.K., Grosberg, A.Yu., Vershik, A.M., "Random Walks on Braid Groups: Brownian Bridges, Complexity and Statisics," *Journal of Physics A: Mathematical and General*, 29(10): 2411 – 2433, 1996.
50. Nechaev, S., Desbois, J., "Statistical Mechanics of Braided Markov Chains: I. Analytic Methds and Numerical Simulations," *Journal of Statistical Physics*, 88(1 – 2): 201 – 229, 1997.
51. Nechaev, S.K., *Statistics of Knots and Entangled Random Walks*, World Scientific, Singapore, 1996.
52. Nelson, P., "Sequence-Disorder Effects on DNA Entropic Elasticity," *Physical Review Letters*, 80(26): 5810 – 5812, 1998.
53. Nelson, P., "New Measurements of DNA Twist Elasticity," *Biophysical Journal*, 74(5): 2501 – 2503, 1998.
54. Norisuye, T., Tsuboi, A., Teramoto, A., "Remarks on Excluded-Volume Effects in Semi-Flexible Polymer Solutions," *Polymer Journal*, 28(4): 357 – 361, 1996.
55. Odijk, T., "Physics of Tightly Curved Semiflexible Polymer Chains," *Macromolecules*, 26(25): 6897 – 6902, 1993.
56. Oka, S., "Zur Theorie der statistischen Molekülgestalt hochpolymerer Kettenmoleküle unter Berücksichtigung der Behinderung der freien Drehbarkeit," *Proceedings of the Physico-Mathematical Society of Japan*, 24:657 – 672, 1942.
57. Plimpton, S., Hendrickson, B., "A New Parallel Method for Molecular Dynamics Simulation of Macromolecular Systems," *Journal of Computational Chemistry*, 17(3): 326 – 337, 1996.
58. Podtelezhnikov, A.A., Cozzarelli, N.R., Vologodskii, A.V., "Equilibrium Distributions of Topological States in Circular DNA: Interplay of Supercoiling and Knotting," *Proceedings of the National Academy of Sciences of the United States of America*, 96(23): 12,974 – 12,979, 1999.
59. Porod, G., "X-Ray and Light Scattering by Chain Molecules in Solution," *Journal of Polymer Science*, 10(2): 157 – 166, 1953.
60. Prohofsky, E., *Statistical Mechanics and Stability of Macromolecules*, Cambridge University Press, 1995.
61. Lord Rayleigh, "On the Problem of Random Vibrations, and of Random Flights in One, Two, or Three Dimensions," *Philosophical Magazine*, 37(6): 321 – 347, 1919.
62. Redner, S., Reynolds, P.J., "Position-Space Renormalisation Group for Isolated Polymer Chains," *Journal of Physics A: Mathematical and General*, 14(10): 2679 – 2703, 1981.
63. Rehahn, M., Mattice, W.L., Suter, U.W., "Rotational Isomeric State Models in Macromolecular Systems," *Advances in Polymer Science*, 131/132: 1 – 6, 1997.
64. Reisert, M., Skibbe, H., "Left-Invariant Diffusion on the Motion Group in Terms of the Irreducible Representations of $SO(3)$," arXiv:1202.5414v1 [math.AP] 24 Feb. 2012.
65. Rosenbluth, M.N., Rosenbluth, A.W., "Monte Carlo Calculation of the Average Extension of Molecular Chains," *Journal of Chemical Physics*, 23(2): 356 – 359, Feb. 1955.
66. Shalloway, D., "Application of the Renormalization Group to Deterministic Global Minimization of Molecular Conformation Energy Functions," *Journal of Global Optimization*, 2(3): 281 – 311, 1993.
67. Shi, Y., He, S., Hearst, J.E., "Statistical Mechanics of the Extensible and Shearable Elastic Rod and of DNA," *Journal of Chemical Physics*, 105(2): 714 – 731, July 1996.
68. Shimada, J., Yamakawa, H., "Statistical Mechanics of DNA Topoisomers," *Journal of Molecular Biology*, 184(2): 319 – 329, 1985.
69. Solc, K., Stockmayer, W.H., "Shape of a Random-Flight Chain," *The Journal of Chemical Physics*, 54(6): 2756 – 2757, 1971.
70. Shore, D., Baldwin, R.L., "Energetics of DNA Twisting," *Journal of Molecular Biology*, 170(4): 957 – 981, 1983.

71. Stellman, S.D., Gans, P.J., "Efficient Computer Simulation of Polymer Conformation. I. Geometric Properties of the Hard-Sphere Model," *Macromolecules*, 5(4): 516 – 526, July – Aug. 1972.

72. Sun, S.F., *Physical Chemistry of Macromolecules: Basic Principles and Issues*, John Wiley & Sons, Inc., New York, 1994.

73. Thirumalai, D., Ha, B.-Y., "Statistical Mechanics of Semiflexible Chains: A Mean Field Variational Approach," pp. 1 – 35 in *Theoretical and Mathematical Models in Polymer Research*, A. Grosberg, ed., Academic Press, 1998.

74. Treloar, L.R.G., *The Physics of Rubber Elasticity*, Oxford University Press, 1958.

75. Vinograd, J., Lebowitz, J., Radloff, R., Watson, R., Laipis, P., "The Twisted Circular Form of Polyoma Viral DNA," *Proceedings of the National Academy of Sciences of the USA*, 53(5): 1104 – 1111, 1965.

76. Volkenstein, M.V., *Conformational Statistics of Polymeric Chains*, Interscience, New York 1963 (originally published in Russian by: IzdRtel'stvo Akad. Nauk SSSR, Moscow 1959)

77. Vologodskii, A.V., Anshelevich, V.V., Lukashin, A.V., Frank-Kamenetskii, M.D., "Statistical Mechanics of Supercoils and the Torsional Stiffness of the DNA Double Helix," *Nature*, 280(5720): 294 – 298, July 1979.

78. White, J.H., Bauer, W.R., "Calculation of the Twist and the Writhe for Representative Models of DNA," *Journal of Molecular Biology*, 189(2): 329 – 341, 1986.

79. Wilhelm, J., Frey, E., "Radial Distribution Function of Semiflexible Polymers," *Physical Review Letters*, 77(12): 2581 – 2584, Sept. 1996.

80. Wolfe, K.C., Hastings, W.A., Dutta, S., Long, A., Shapiro, B.A., Woolf, T.B., Guthold, M., Chirikjian, G.S., "Multiscale Modeling of Double-Helical DNA and RNA: A Unification through Lie Groups," *Journal of Physical Chemistry B*, 116(29): 8556 - 8572, 2012.

81. Wolfe, K.C., Chirikjian, G.S., "Signal Detection on Euclidean Groups: Applications to DNA Bends, Robot Localization, and Optical Communication" IEEE Journal of Selected Topics in Signal Processing, 7(4): 708 – 719, Aug. 2013.

82. Yamakawa, H., *Helical Wormlike Chains in Polymer Solutions*, Springer-Verlag, Berlin, 1997.

83. Zhou, Y., Chirikjian, G.S., "Conformational Statistics of Semi-Flexible Macromolecular Chains with Internal Joints," *Macromolecules*, 39(5): 1950 – 1960, 2006.

84. Zhou, Y., Chirikjian, G.S., "Conformational Statistics of Bent Semiflexible Polymers," *Journal of Chemical Physics* 119(9): 4962 – 4970, 2003.

18

Mechanics and Texture Analysis

Classical mechanics is the study of the interplay between forces acting on matter and the resulting motion of matter. Quantitative results are based on four essential balances of quantities in any system composed of matter: (1) balancing the mass entering and leaving the system; (2) balancing the change in momentum of the system with applied force; (3) balancing the change in angular momentum with applied moments of force; and (4) balancing the total energy in the system with the net contributions from external sources. The laws of conservation of mass, momentum, angular momentum, and energy follow from these balances in the absence of external influences.

Traditionally, matter is assumed to exist in the solid, liquid, or gaseous states. Recently the techniques of classical mechanics have also been applied to what are considered the new states of matter such as plasmas, liquid crystals, and gels. At the core of mechanics is the approximation of fluids (including liquids and gases) and solids as continuous distributions of matter, hence the term *continuum mechanics*. In the continuum model, properties of matter are averaged, or smoothed, over space instead of considering the huge collection of molecules that constitute the system. Recent research in mechanics has attempted to bridge the gap between the description of matter at the various length scales of engineering interest ranging from atomic to microscopic to mesoscopic to macroscopic scales.

The material in this chapter illustrates why group theory, functions on the group of rotations, averaging over rotations, and diffusion on the rotation group have found applications in mechanics. We begin in Section 18.1 with a review of continuum mechanics applicable to both fluids and solids. In Section 18.2 we illustrate how the strength of polycrystalline solids and composites can be described as an orientational average of various constituents in a material. In Section 18.3 we show how material properties of a polycrystalline material can be found by averaging single crystal properties over orientations. In Section 18.4 we illustrate how the concept of convolution on groups enters in materials science. In Section 18.5 we show how symmetries in solids, such as the crystallographic point groups, can be used to reduce the complexity of the search for analytical relationships between stress and strain in solid materials. Section 18.6 points to literature on the mechanics of liquid crystals, and analogous descriptions of the dynamics of short fiber strands and ellipsoidal particles suspended in a flowing viscous liquid.

18.1 Review of Basic Continuum Mechanics

In this section we review the basic mechanics of solids and fluids. At the core of both discussions is the description of the relationship between external and internal forces (in the form of surface tractions and internal stresses) and the resulting deformation of a continuum. In the following subsections we review the mathematical tools required to describe the interaction between forces and motions of solids and fluids.

18.1.1 Tensors and How They Transform

Continuum mechanics is based on the relationship between various Cartesian *tensors* describing deformation and force. A tensorial quantity is one which transforms in a prescribed way under rotations. A zeroth-order tensor is simply a scalar quantity and has the same value under all changes in orientation. A first-order tensor is one which has components that transform as

$$x_i = \sum_{j=1}^{3} Q_{ij} x_j' \quad i,j = 1,2,3. \tag{18.1}$$

where Q_{ij} are the elements of a rotation matrix that transforms components defined in a rotated (body-fixed) frame to components defined in a space-fixed frame. We note that the convention in mechanics for the use of primes and how the rotation matrix is defined is different from that in rigid-body kinematics. In Chapter 5 we would write

$$\mathbf{x}_L = Q \mathbf{x}_R \tag{18.2}$$

where the meaning of Q is the rotation matrix describing motion from space-fixed to moving coordinates as seen from the perspective of the space-fixed coordinates, $\mathbf{x}_R$ is the position of a point before motion, and $\mathbf{x}_L$ is the position of the same point after motion (both described in space-fixed coordinates). In contrast, in continuum mechanics, a vector $\mathbf{x}$ (viewed as a first order tensor) is independent of the reference frame. However its components are x_i' when viewed in the body-fixed frame and x_i are components of absolute position. That is, these components are how $\mathbf{x}$ *appears* if we change our frame of reference relative to a space-fixed frame. This means that *in continuum mechanics one would never write* (18.2) (since in (18.2) $\mathbf{x}_L = [x_i]$ and $\mathbf{x}_R = [x_i']$ are 3×1 arrays of numbers that represent the same vector quantity). Rather (18.1) would be used.

In continuum mechanics we would not put a prime over the vector $\mathbf{x}$, but rather only over its components:

$$\mathbf{x} = x_1\mathbf{e}_1 + x_2\mathbf{e}_2 + x_3\mathbf{e}_3 = x_1'\mathbf{e}_1' + x_2'\mathbf{e}_2' + x_3'\mathbf{e}_3' \tag{18.3}$$

where $\mathbf{e}_i' \doteq Q^T\mathbf{e}_i$ and $(\mathbf{e}_i)_j = \delta_{ij}$. Then,

$$x_i' = \mathbf{e}_i' \cdot \mathbf{x} = (Q^T\mathbf{e}_i)^T \left(\sum_{j=1}^{3} x_j\mathbf{e}_j \right) = \sum_{j=1}^{3} x_j(\mathbf{e}_i^T Q\mathbf{e}_j).$$

Since $\mathbf{e}_i' \cdot \mathbf{e}_j = \mathbf{e}_i^T Q\mathbf{e}_j = Q_{ij}$, this is the same as (18.1).

Hence, a first-order tensor is a vector, and the transformation law is the description of the elements of the vector in coordinate systems with different orientations and the same origin. Usually indicial notation, where summation is assumed over repeated indices, is used to write this more compactly as $x_i = Q_{ij}x'_j$. A second-order tensor is one which transforms as

$$X_{ij} = Q_{im}Q_{jn}X'_{mn} \quad i,j,m,n \in \{1,2,3\}. \tag{18.4}$$

The moment-of-inertia tensor for a rigid body is an example of a second-order tensor, i.e., the form of the moment-of-inertia matrix depends on the frame in which it is defined. In Chapter 5 we wrote the relationship between moment of inertia matrices $I = QI_0Q^T$ where I_0 was for the body-fixed frame. In the current notation, we would write $I = [I_{ij}]$ and $I_0 = [I'_{ij}]$. Angular velocity, when viewed as a vector, transforms like a vector, and when viewed as a skew-symmetric matrix in a given coordinate system, transforms between different coordinate systems like a second-order tensor. In general an nth-order tensor is one which transforms as

$$X_{i_1,i_2,...,i_n} = Q_{i_1j_1}Q_{i_2j_2}\cdots Q_{i_nj_n}X'_{j_1,j_2,...,j_n} \quad i_1,...,i_n,j_1,...,j_n \in \{1,2,3\}. \tag{18.5}$$

A tensor is completely defined by its components given in a frame of reference at a particular orientation. That is, the 3^n numbers $X_{j_1,j_2,...,j_n}$ and the frame of reference $\{\mathbf{e}_1,\mathbf{e}_2,\mathbf{e}_3\}$ completely define an nth-order tensor. The tensor is written in terms of its components and the basis vectors as

$$X = X_{j_1,j_2,...,j_n}\mathbf{e}_{j_1} \otimes \mathbf{e}_{j_2} \otimes \cdots \otimes \mathbf{e}_{j_n}$$

for $j_1,...,j_n \in \{1,2,3\}$. Here $\otimes$ denotes the *tensor product* of two vectors. For a second-order tensor, we can think of $\mathbf{e}_{j_1} \otimes \mathbf{e}_{j_2}$ as a matrix with $(i,j)^{th}$ element $\delta_{j_1,i}\delta_{j,j_2}$ when $(\mathbf{e}_i)_j = \delta_{ij}$.

We note that (18.5) is also written as

$$X_{j_1,j_2,...,j_n}\mathbf{e}_{j_1} \otimes \mathbf{e}_{j_2} \otimes \cdots \otimes \mathbf{e}_{j_n} = X'_{j_1,j_2,...,j_n}\mathbf{e}'_{j_1} \otimes \mathbf{e}'_{j_2} \otimes \cdots \otimes \mathbf{e}'_{j_n}$$

where $\mathbf{e}'_j = Q^T\mathbf{e}_j$.

Another notation that says the same thing is

$$X = X_{j_1,j_2,...,j_n}(Q)\,\mathbf{e}_{j_1}(Q) \otimes \mathbf{e}_{j_2}(Q) \otimes \cdots \otimes \mathbf{e}_{j_n}(Q) \tag{18.6}$$

where $\mathbf{e}_j(Q) = Q^T\mathbf{e}_j$ and

$$X_{i_1,i_2,...,i_n}(Q) = Q_{i_1j_1}Q_{i_2j_2}\cdots Q_{i_nj_n}X'_{j_1,j_2,...,j_n} \tag{18.7}$$

for any $Q \in SO(3)$ and $X_{j_1,j_2,...,j_n}(\mathbb{I}) = X'_{j_1,j_2,...,j_n}$. That is, when $Q = \mathbb{I}$, the components of a tensor in the space-fixed and body-fixed frames coincide.

The constraint imposed by (18.7) implies that a concatenation of changes of relative reference orientation by R then A results in the tensor elements

$$(R(A)X_{j_1,j_2,...,j_n})(Q) = X_{j_1,j_2,...,j_n}(QA),$$

where $R(\cdot)$ is the right regular representation. When

$$X_{j_1,j_2,...,j_n}(QA) = X_{j_1,j_2,...,j_n}(Q), \tag{18.8}$$

for all A in a subgroup of $SO(3)$ and for all $j_1,j_2,...,j_n \in \{1,2,3\}$, this indicates a symmetry in the physical phenomenon under investigation. For instance, for the mechanics of a crystalline material (18.8) should hold for all A in one of the 32 crystallographic point groups (11 of which are subgroups of $SO(3)$, and all of which are finite subgroups of $O(3)$).

The form of the transformation law for the nth-order tensor in (18.6) makes it clear that tensor components $X_{j_1,j_2,...,j_n}(Q)$ constitute a set of 3^n scalar functions on $SO(3)$.

In the discussion above, we have considered tensors that are constant. That is, they are not functions of position in $\mathbb{R}^3$. In fact, they are not functions of anything (even though the components of a tensor in a given basis can be considered as functions of orientation). We now consider some issues regarding tensor functions of position, or *tensor fields*. In the same way that a scalar function $f(\mathbf{x})$ can be used to describe the relative density or temperature at a point $\mathbf{x}$, tensor quantities can vary over $\mathbb{R}^3$ as well. The components of a tensor field are then of the form $F_{j_1,j_2,...,j_n}(\mathbf{x}, Q)$. Each of these components can be viewed as a function on $SE(3)$.

Something that will be useful later is to know how a tensor field transforms under rigid-body motion. That is, if we have a tensor field $F(\mathbf{x})$, and we rigidly move it by $g = (\mathbf{a}, A) \in SE(3)$, we want to know what the resulting tensor field looks like in the original frame of reference. For the scalar case we have seen in previous chapters that $f(\mathbf{x})$ transforms to $f(g^{-1} \cdot \mathbf{x}) = f(A^{-1}(\mathbf{x} - \mathbf{a}))$ under the motion $g \in SE(3)$. In the case of an nth-order tensor field, a motion by $g = (\mathbf{a}, A)$ results in the transformation of tensor components:

$$F_{i_1,i_2,...,i_n}(\mathbf{x}) \to A_{i_1 j_1} A_{i_2 j_2} \cdots A_{i_n j_n} F_{j_1,j_2,...,j_n}(A^{-1}(\mathbf{x} - \mathbf{a})), \qquad (18.9)$$

where $F_{i_1,i_2,...,i_n}(\mathbf{x})$ are the components of $F(\mathbf{x})$ at $Q = \mathbb{I}$.

In the particular case of a vector field $\mathbf{v}(\mathbf{x})$ expressed in the frame $(\mathbf{0}, \mathbb{I})$ as the 3×1 array with components $v_i(\mathbf{x})$ we have:[1]

$$v_i(\mathbf{x}) \to A_{ij} v_j(A^T(\mathbf{x} - \mathbf{a})), \qquad (18.10)$$

and for a 2-tensor field $F(\mathbf{x})$ expressed in the frame $(\mathbf{0}, \mathbb{I})$ as the 3×3 matrix with components $F_{ij}(\mathbf{x})$ we have

$$F_{ij}(\mathbf{x}) \to A_{im} A_{jn} F_{mn}(A^T(\mathbf{x} - \mathbf{a})). \qquad (18.11)$$

The following subsections examine how tensors arise in various contexts in mechanics. Subsection 18.1.2 addresses solid mechanics, while Subsection 18.1.3 addresses fluids. These subsections summarize results that can be found in greater detail in classic texts on mechanics such as [78, 86, 89].

18.1.2 Mechanics of Elastic Solids

The discussion here is limited to the statics of a continuous solid which responds elastically to applied load. Within this context the only important factors are the load applied to the solid, the resulting internal stresses, the material elastic properties, and how the material deforms as a result of the state of stress in the solid.

In Chapter 5 we already introduced aspects of the kinematics of continuous solids. Any point initially at coordinates $\mathbf{X} \in \mathbb{R}^3$ are moved to the points $\mathbf{x}(\mathbf{X}, t)$ at time t. This is called a referential, material, or Lagrangian description of the solid. The time t_0

---

[1] We often use the word vector to denote arrays such as $[v_i]$, but in this context it is important to make the distinction between the (physical) vector $\mathbf{v}$ and the array of numbers that describes this vector in a particular frame of reference.

when $\mathbf{x}(\mathbf{X}, t_0) = \mathbf{X}$ designates the referential state of the material. Usually this is when the material is not subjected to loads.

The mass density per unit volume at a point $\mathbf{X}$ in the referential configuration is $\rho(\mathbf{X})$. This is a scalar field, i.e., the value of density is independent of the orientation with which it is observed. The conservation of mass is automatically satisfied as $d\rho/dt = 0$ if the deformation $\mathbf{x}(\mathbf{X}, t)$ is *admissible*. Namely, if the Jacobian matrix with elements $\partial x_i/\partial X_j$ is not singular, and interpenetration of material is disallowed.

The velocity and acceleration of a material particle initially at referential position $\mathbf{X}$ are

$$\mathbf{v} = \frac{\partial \mathbf{x}}{\partial t} \quad \text{and} \quad \mathbf{a} = \frac{\partial \mathbf{v}}{\partial t} = \frac{\partial^2 \mathbf{x}}{\partial t^2}.$$

These definitions are valid for any value of t.

The relationship between the position of a particle and its near neighbors, and how this relationship changes when the material deforms, is quantified by the concept of strain. The *Lagrangian strain tensor* is essentially a constant multiple of the metric tensor for a deformed volume in $\mathbb{R}^3$ which is parameterized with Lagrangian coordinates:

$$E^L(\mathbf{X}, t) = \frac{1}{2} J^T(\mathbf{X}, t) J(\mathbf{X}, t)$$

where $J = \nabla \mathbf{x}$ is the Jacobian matrix with elements $J_{ij} = \partial x_i/\partial X_j$. Hence, E^L contains all the information regarding how each infinitesimal volume is distorted under the deformation.

The *displacement*, $\mathbf{u}(\mathbf{X}, t)$, corresponding to a deformation $\mathbf{x}(\mathbf{X}, t)$ indicates how much each point moves from its original position: $\mathbf{u}(\mathbf{X}, t) = \mathbf{x}(\mathbf{X}, t) - \mathbf{X}$. The *displacement gradient* is the matrix with elements $(\nabla \mathbf{u})_{ij} = \partial u_i/\partial X_j$. This is a useful definition in the context of small deformations, because in this case the Lagrangian strain tensor is replaced by the *infinitesimal strain tensor*:

$$E = \frac{1}{2}((\nabla \mathbf{u}) + (\nabla \mathbf{u})^T).$$

That is, E is a first-order approximation to E^L which becomes exact for infinitesimally small deformations.

Given a force, $\mathbf{t}$, acting on the surface of any infinitesimal volume within a solid body, the *stress tensor*, $\mathbf{T}$, within the volume relates the applied surface force and the normal to the surface as

$$\mathbf{t}(\mathbf{x}, \mathbf{n}, t) = T(\mathbf{x}, t)\mathbf{n}.$$

The vector $\mathbf{t}$ is called a surface traction acting on the infinitesimal volume. The linear relationship between $\mathbf{t}$ and $\mathbf{n}$ via the stress tensor is a result of *Cauchy's stress principle*.

Performing a balance of moments on any infinitesimal volume within a solid indicates that [78, 89]:

$$T = T^T$$

as the volume shrinks to zero regardless of whether or not the material is in mechanical equilibrium since inertial terms (which depend on the volume of the material element) become smaller much faster than area-dependent quantities such as T.

By performing a momentum balance on the infinitesimal volume, we arrive at the equilibrium equations

$$\nabla_{\mathbf{x}} \cdot T + \rho \mathbf{b} = \rho \mathbf{a}, \qquad (18.12)$$

where $\rho(\mathbf{x})$ is the mass density per unit volume in the material at point $\mathbf{x}$, $\mathbf{a}$ is the acceleration, and $\mathbf{b}$ is body force per unit mass. In the case of static equilibrium $\mathbf{a} = \mathbf{0}$.

For a solid undergoing small deformations,

$$\frac{\partial T_{ij}}{\partial x_j} \approx \frac{\partial T_{ij}}{\partial X_j} \quad \text{and} \quad \rho(\mathbf{x}) \approx \rho(\mathbf{X}). \tag{18.13}$$

Hence, if the stress tensor T can be related to the strain tensor E, then a partial differential equation in the displacements $u_i(\mathbf{X}, t)$ can be written with independent variables X_i and t.

In solid mechanics, the problem of finding the relationship $T = T(E)$ for particular materials (which is called a *constitutive equation*) is a fundamental and significant problem because knowing this allows us to simulate the static and dynamic behavior of any solid material. The most common kind of constitutive law in solid mechanics is of the form

$$T_{ij} = C_{ijkl} E_{kl},$$

or equivalently,

$$E_{ij} = S_{ijkl} T_{kl},$$

where C is called the stiffness (or elasticity) tensor, and $S = C^{-1}$ is called the compliance tensor. Due to the symmetries $T_{ij} = T_{ji}$ and $E_{ij} = E_{ji}$, it follows that in general

$$C_{ijkl} = C_{jikl} = C_{ijlk}.$$

Using group theory to help make the search for constitutive laws (i.e. determining additional structure in the coefficients C_{ijkl}) more tractable is the subject of Section 18.5.

18.1.3 Fluid Mechanics

Whereas in the mechanics of solids the deformations are generally small and it is convenient to use referential descriptions, this is usually not the case in fluid mechanics. Instead, it is convenient to work in *spatial* or *Eulerian* coordinates. In this description a fluid particle is usually not tracked over time, but rather the velocity of fluid passing through the point $\mathbf{x}$ at time t is observed. The result is a velocity field $\mathbf{v}(\mathbf{x}, t)$ defined at each spatial coordinate and time. Since $\mathbf{x} = \mathbf{x}(\mathbf{X}, t)$, the observation of conservation of mass, $d\rho/dt = 0$ is no longer automatically guaranteed as it is in solid mechanics by specifying it in the referential coordinates. Instead, density at a particular instance in time must satisfy the equation

$$\frac{d\rho}{dt} = \frac{\partial \rho}{\partial t} + \mathbf{v} \cdot \nabla \rho = 0 \tag{18.14}$$

where $\mathbf{v}$ is the velocity of fluid passing through the spatial point $\mathbf{x}$ at time t, and $\nabla \rho$ is the gradient of ρ with respect to $\mathbf{x}$. Equation 18.14, which is simply an application of the chain rule, can be viewed as a particular example of a general scalar conservation law in spatial coordinates.

The acceleration of a particle of fluid at $\mathbf{x}$ at time t is also computed using the chain rule:

$$\mathbf{a}(\mathbf{x}, t) = \frac{d\mathbf{v}}{dt} = \frac{\partial \mathbf{v}}{\partial t} + (\nabla \mathbf{v})\mathbf{v}, \tag{18.15}$$

where the components of $\nabla \mathbf{v}$ are $\partial v_i / \partial x_j$.

The force balance in (18.12) is equally valid for fluid mechanics as it was in the case of solid mechanics. However, the assumptions in (18.13) are no longer valid since deformations are large. Furthermore, since we are no longer using a referential description, the acceleration must now be calculated as in (18.15).

Again in order to write a partial differential equation which describes the fluid motion, a constitutive law is needed. In the context of fluid mechanics, laws of the form $T = T(\mathbf{v}, \nabla\mathbf{v})$ are common, the most common being the *Newtonian fluids* with:

$$T = -p\,\mathbb{I} + \lambda\,\mathrm{tr}(\nabla\mathbf{v})\,\mathbb{I} + \mu(\nabla\mathbf{v} + (\nabla\mathbf{v})^T),$$

where $p(\mathbf{x}, t)$ is the pressure, and λ, μ are constants specific to the fluid.

In the case of incompressible Newtonian fluids, the conservation of mass expression reduces to

$$\nabla \cdot \mathbf{v} = 0, \tag{18.16}$$

and (18.12) becomes

$$\rho\left(\frac{\partial\mathbf{v}}{\partial t} + (\nabla\mathbf{v})\mathbf{v}\right) = \rho\mathbf{b} - \nabla p + \mu\nabla^2\mathbf{v}. \tag{18.17}$$

Hence, (18.16) and (18.17) constitute four PDEs in the four variables v_1, v_2, v_3, and p.

In the special case when the density is constant throughout the fluid: $\rho(\mathbf{x}, t) = \rho_0$ and the body force $\mathbf{b}$ is conservative ($\mathbf{b} = -\nabla V$) the Navier-Stokes equations are transformed by taking the curl of both sides resulting in the form

$$\frac{\partial\boldsymbol{\xi}}{\partial t} + (\nabla\boldsymbol{\xi})\mathbf{v} = (\nabla\mathbf{v})\boldsymbol{\xi} + \nu\nabla^2\boldsymbol{\xi} \tag{18.18}$$

where

$$\boldsymbol{\xi} = \nabla \times \mathbf{v} = ((\nabla\mathbf{v}) - (\nabla\mathbf{v})^T)^\vee \tag{18.19}$$

is called the *vorticity* and $\nu = \mu/\rho$.

In section 18.6.1 we discuss how these equations are generalized in the literature to include the dynamics of fibers and elliptic particles embedded in a flow.

18.2 Orientational and Motion-Group Averaging in Solid Mechanics

A topic that has become popular in the fields of mechanics and materials in recent years is how to bridge well-understood properties at very small scales with the observed bulk behavior. See, for example, [109]. We review mathematical approaches that attempt to bridge these properties by averaging the properties of individual crystals that have a known orientation distribution in the bulk material.

In the first two subsections of this section we describe techniques for averaging tensor properties over orientations. For these sections we follow the formulation in [77]. In the third subsection, we extend this formulation to include averages over rigid-body motion.

18.2.1 Tensor-Valued Functions of Orientation

Recall from (18.6) that every n^{th}-order tensor quantity is defined by 3^n scalar functions on $SO(3)$ of the form (18.7). However, a tensor quantity itself is, by definition, independent of the orientation from which it is viewed (although its components vary with changes in orientation). However, it is possible to define a tensor-valued function of orientation as

$$X(R) = X_{j_1, j_2, \ldots, j_n}(R, Q) \mathbf{e}_{j_1}(Q) \otimes \mathbf{e}_{j_2}(Q) \otimes \cdots \otimes \mathbf{e}_{j_n}(Q), \tag{18.20}$$

where for each fixed R, the Q-dependence of $X_{j_1, j_2, \ldots, j_n}(R, Q)$ satisfies (18.7).

For instance, $X(R)$ may be a tensor where R denotes the orientation of the physical relative to fixed space, and Q denotes the orientation relative to fixed space from which we observe the tensor quantity. If there is only one tensor quantity of interest, there is no reason to use both R and Q (a relative orientation of one with respect to the other will suffice). However, if many tensor quantities of the same kind occupy a localized region of space, then the average

$$\langle X \rangle \doteq \int_{SO(3)} X(R) \rho(R) \, dR$$

can be a useful approximation of the average tensorial properties, where $\rho(R)$ is a PDF describing the distribution of physical quantities.

In component form we observe that

$$\langle X \rangle_{j_1, j_2, \ldots, j_n}(Q) = \int_{SO(3)} X_{j_1, j_2, \ldots, j_n}(R, Q) \rho(R) \, dR$$

defines a tensor whenever $X_{j_1, j_2, \ldots, j_n}(R, Q)$ does for each fixed R.

In [77] the case

$$X_{i_1, i_2, \ldots, i_n}(R, Q) = (QR^T)_{i_1 j_1} (QR^T)_{i_2 j_2} \cdots (QR^T)_{i_n, j_n} X_{j_1, j_2, \ldots, j_n}(R)$$

is considered. Using the notation

$$l_{ij}(Q) = Q_{ij} = \mathbf{e}_i^T Q \mathbf{e}_j = l_{ji}(Q^T)$$

to emphasize that the $(i, j)^{th}$ component of Q is a function of Q, and using the bi-invariance of integration on $SO(3)$, we see that [77]:

$$\langle X \rangle_{i_1, i_2, \ldots, i_n}(Q) = \int_{SO(3)} l_{i_1 j_1}(QR^T) l_{i_2 j_2}(QR^T) \cdots l_{i_n, j_n}(QR^T) X_{j_1, j_2, \ldots, j_n}(R) \, dR$$

$$= \int_{SO(3)} l_{j_1 i_1}(R) l_{j_2 i_2}(R) \cdots l_{j_n, i_n}(R) X_{j_1, j_2, \ldots, j_n}(RQ) \, dR.$$

Also using the invariance of integration, and the fact that $l_{ij}(QR) = l_{ik}(Q) l_{kj}(R)$ (using summation notation), it can be shown that [77]:

$$\langle X \rangle_{i_1, i_2, \ldots, i_n}(AQ) = A_{i_1 j_1} A_{i_2 j_2} \cdots A_{i_n, j_n} \langle X \rangle_{j_1, j_2, \ldots, j_n}(Q).$$

Hence $\langle X \rangle$ is a tensor.

18.2.2 Orientational Averaging of Tensor Components

When averaging components of an nth-order tensor over all orientations with $\rho(A) = 1$, it is natural to seek closed-form expressions for integrals of the form

$$I_{i_1,\ldots,i_n;j_1,\ldots,j_n} = \int_{SO(3)} l_{i_1,j_1}(A) l_{i_2,j_2}(A) \cdots l_{i_n,j_n}(A) \, dA. \tag{18.21}$$

One way to approach this systematically is to rewrite (9.6) as

$$A = T_1 U^1(A) T_1^{-1},$$

and to sequentially rewrite products of rotation matrix elements in terms of sums of IUR matrix elements and Clebsch-Gordan coefficients. That is,

$$l_{i_1,j_1}(A) l_{i_2,j_2}(A) \cdots l_{i_n,j_n}(A) = \sum_{mnl} c^{mnl}_{i_1,j_1,\ldots,i_n,j_n} U^l_{mn},$$

where c_{mnl} is some combination of Clebsch-Gordan coefficients. Then the fact that

$$\int_{SO(3)} U^l_{mn}(A) \, dA = \delta_{l,0} \delta_{m,0} \delta_{n,0}$$

can be used to eliminate all contributions to the integral (18.21) other than the coefficients of $U^0_{00}(A) = 1$.

While this is a systematic approach to the problem for a tensor of any order, the tensors in mechanics tend to be either of second order or fourth order. It is sufficient to evaluate

$$I_{i_1,i_2;j_1,j_2} = \int_{SO(3)} l_{i_1,j_1}(A) l_{i_2,j_2}(A) \, dA. \tag{18.22}$$

at

$$\sum_{n=1}^{6} \sum_{m=1}^{n} 1 = \sum_{n=1}^{6} n = \frac{6 \cdot 7}{2} = 21 \tag{18.23}$$

independent values of the arguments of $I_{i_1,i_2;j_1,j_2}$.

Equation (18.23) comes from the symmetries in $I_{i_1,i_2;j_1,j_2}$. In particular, from the invariance of integration under inversion of the argument,

$$I_{i_1,i_2;j_1,j_2} = \int_{SO(3)} l_{i_1,j_1}(A^T) l_{i_2,j_2}(A^T) \, dA = \int_{SO(3)} l_{j_1,i_1}(A) l_{j_2,i_2}(A) \, dA = I_{i_2,i_1;j_2,j_1}. \tag{18.24}$$

This is why the first sum on the left-hand side of (18.23) is to 6 instead of 9. The second sum on the left hand side of (18.23) is to n instead of 6 because

$$I_{i_1,i_2;j_1,j_2} = I_{j_1,j_2;i_1,i_2}.$$

Similarly for the fourth-order case, due to symmetries, there are not 9^4, but rather at most

$$\sum_{n=1}^{6} \sum_{m=1}^{n} \sum_{l=1}^{m} \sum_{k=1}^{l} 1 = \sum_{n=1}^{6} \sum_{m=1}^{n} \frac{m(m+1)}{2} = 126.$$

integrals that need to be calculated.

We now enumerate some of the explicit results. Recall that ZXZ Euler angles represent a double-coset decomposition of $SO(3)$, both $SO(3)/SO(2)$ and $SO(2)\backslash SO(3)$ can be identified with the unit sphere $\mathbb{S}^2$, and the subgroups parameterized by $(\alpha, \beta, \gamma) = (\alpha, 0, 0)$ and $(0, 0, \gamma)$ are both isomorphic to $SO(2)$. With $G = SO(3)$ and $H = SO(2)$ we write

$$\int_{G/H} f(gH)\,d(gH) \doteq \int_{\mathbb{S}^2} f(\mathbf{u}'(\beta,\alpha))\,d\mathbf{u}'(\beta,\alpha),$$

and

$$\int_{H\backslash G} f(Hg)\,d(Hg) \doteq \int_{\mathbb{S}^2} f(\mathbf{u}''(\beta,\gamma))\,d\mathbf{u}''(\beta,\gamma).$$

For $SO(3)/SO(2)$ we take

$$\mathbf{u}'(\beta,\alpha) \doteq A\mathbf{e}_3$$

and for $SO(2)\backslash SO(3)$ and we take

$$\mathbf{u}''(\beta,\gamma) \doteq A^T\mathbf{e}_3.$$

We note that even though

$$\mathbf{u}(\theta,\phi) \neq \mathbf{u}'(\theta,\phi) \neq \mathbf{u}''(\theta,\phi),$$

it is nonetheless true that

$$d\mathbf{u}(\theta,\phi) = d\mathbf{u}'(\theta,\phi) = d\mathbf{u}''(\theta,\phi).$$

where

$$\mathbf{u}(\theta,\phi) = [\sin\theta\cos\phi, \sin\theta\sin\phi, \cos\theta]^T$$

is the standard parameterization of $\mathbb{S}^2$.

A useful tool in evaluating integrals of products of rotation matrix elements over $SO(3)$ is to first determine integrals over $SO(2)$ and $\mathbb{S}^2$. For instance[2] [77]:

$$\frac{1}{2\pi}\int_0^{2\pi} l_{ki}(A)l_{kj}(A)\,d\alpha = \frac{1}{2}\mu_{ij}$$

for $k = 1, 2$ where

$$\mu_{ij} \doteq \delta_{ij} - l_{3i}(A)l_{3j}(A)$$

and

$$\int_{\mathbb{S}^2} l_{3i}(A)l_{3j}(A)\,d\mathbf{u}''(\beta,\gamma) = \frac{1}{3}\delta_{ij}$$

combine to yield

$$\int_{SO(3)} l_{ki}(A)l_{kj}(A)\,dA = \frac{1}{3}\delta_{ij}.$$

Similarly,

$$\frac{1}{2\pi}\int_0^{2\pi} l_{1,i}(A)l_{2,j}(A)\,d\alpha = -\frac{1}{2\pi}\int_0^{2\pi} l_{2,i}(A)l_{1,j}(A)\,d\alpha$$

$$= \frac{1}{2}(\epsilon_{1,i,j}l_{3,1}(A) + \epsilon_{2,i,j}l_{3,2}(A) + \epsilon_{3,i,j}l_{3,3}(A))$$

(where ϵ_{ijk} is the alternating tensor) and the fact that

$$\int_{\mathbb{S}^2} l_{3,i}(A)\,d\mathbf{u}''(\beta,\gamma) = 0$$

---

[2]In this section repeated indices do not indicate summation.

implies that

$$\int_{SO(3)} l_{1,i}(A)l_{2,j}(A)\,dA = \int_{SO(3)} l_{2,i}(A)l_{1,j}(A)\,dA = 0\,.$$

The decomposition of the $SO(3)$ integral into an integral over $SO(2)$ and $SO(3)/SO(2)$ yields a similar result. However, we only need to work with one decomposition because of the symmetry (18.24).

For averaging over fourth-order tensor components, the following are useful [77]:

$$\frac{1}{2\pi}\int_0^{2\pi} l_{mi}(A)l_{mj}(A)l_{mk}(A)l_{ml}(A)\,d\alpha = \frac{1}{8}(\mu_{ij}\mu_{kl} + \mu_{ik}\mu_{jl} + \mu_{il}\mu_{jk})$$

for $m = 1, 2$;

$$\frac{1}{2\pi}\int_0^{2\pi} l_{1,i}(A)l_{2,j}(A)l_{2,k}(A)l_{2,l}(A)\,d\alpha = -\frac{1}{2\pi}\int_0^{2\pi} l_{2,i}(A)l_{1,j}(A)l_{1,k}(A)l_{1,l}(A)\,d\alpha$$

$$\int_{\mathbb{S}^2} l_{3,i}(A)l_{3,j}(A)l_{3,k}(A)l_{3,l}(A)\,d\mathbf{u}(\beta,\gamma) = \frac{1}{15}(\delta_{ij}\delta_{kl} + \delta_{ik}\delta_{jl} + \delta_{il}\delta_{jk});$$

$$\int_{\mathbb{S}^2} (l_{3,1}(A))^{m_1}(l_{3,2}(A))^{m_2}(l_{3,3}(A))^{m_3}\,d\mathbf{u}''(\beta,\gamma) = 0$$

for $m_1 + m_2 + m_3 = 4$ and $m_i > 0$. Using the fact that

$$\int_{\mathbb{S}^2} \mu_{ij}(\beta,\gamma)\,d\mathbf{u}''(\beta,\gamma) = \frac{2}{3}\delta_{ij},$$

the above equations combine to yield

$$\int_{SO(3)} l_{1,i}(A)l_{2,j}(A)l_{2,k}(A)l_{2,l}(A)\,dA = -\int_{SO(3)} l_{2,i}(A)l_{1,j}(A)l_{1,k}(A)l_{1,l}(A)\,dA$$

and

$$\int_{SO(3)} (l_{3,1}(A))^{m_1}(l_{3,2}(A))^{m_2}(l_{3,3}(A))^{m_3}\,dA = 0$$

when $m_1 + m_2 + m_3 = 4$ and $m_i > 0$.

18.2.3 Motion-Group Averaging: Modeling Distributed Small Cracks and Dilute Composites

Two problems for which ideas of noncommutative harmonic analysis such as integration and convolution on groups can be used in mechanics are: (1) modeling the effective properties of a solid body with a distribution of cracks with the same shape and size; and (2) modeling the effective properties of a solid body containing a dilute collection of identical inclusions.

A large body of literature in mechanics has considered the stress/strain tensor fields around cracks under various loading conditions, and the effective strength of a homogenized model that replaces the region containing the crack or inclusion with a continuum model that behaves similarly. If $F(\mathbf{x})$ is a tensor field quantity (such as effective stiffness, compliance, or elasticity) that replaces the crack or inclusion, then an approximation for the tensor field for the whole body can be written as

$$\langle F \rangle(\mathbf{x}) = \int_{SE(3)} \rho(g) F(g^{-1} \cdot \mathbf{x}) \, d(g) \qquad (18.25)$$

where $\rho(g)$ is the positional and orientational density of the cracks/inclusions at $g = (\mathbf{a}, A) \in SE(3)$. For instance, if $F(\mathbf{x})$ is a 2-tensor field with elements $F_{ij}(\mathbf{x})$ in the frame $(\mathbf{0}, \mathbb{I})$, then (18.25) is written as

$$\langle F \rangle_{ij}(\mathbf{x}) = \int_{SE(3)} \rho(g) A_{ii_1} A_{jj_1} F_{i_1 j_1} (g^{-1} \cdot \mathbf{x}) \, d(g).$$

While in our discussion we have not assumed any particular form (or order) for the tensor fields, numerous references have considered the tensor fields around inclusions and cracks. For works that address the tensor fields in cracked media and those with rigid inclusions and techniques for homogenization (averaging) see [28, 29, 38, 42, 56, 60, 64, 80, 81, 82, 112, 137, 149].

We note that this formulation also holds for non-mechanical tensor fields such as in the optical properties of liquid crystals.

18.3 The Orientational Distribution of Polycrystals

Polycrystalline materials are composed of many identical crystals oriented in different ways. When it is the case that the crystals in any small volume element in $\mathbb{R}^3$ are evenly distributed in their orientation, then the orientational distribution function (ODF) is assumed to be constant: $\rho(A) = 1$, and the overall material is treated as isotropic.

However, when $\rho(A) \neq 1$, the question arises as to how the mechanical properties of each crystal contribute to those properties of the bulk material. If each crystal in the aggregate has a rotational symmetry group $G_C < SO(3)$, then the ODF will satisfy

$$\rho(A) = \rho(AB)$$

for all $B \in G_C$. In contrast, it is also possible for the material sample to have symmetry, in which case

$$\rho(A) = \rho(B'A)$$

for all $B' \in G_S < SO(3)$. The symmetry groups G_C and G_S are, respectively, said to reflect crystal and sample symmetry [20]. For a deeper discussion of ODF symmetries see [7, 22, 47]. We note that our notation of using relative (as opposed to absolute) rotations gives the above expressions a different form than those found in [20].

The analysis of texture (i.e., the form of the function $\rho(A)$) is used in the context of geology (see, e.g., [32, 74, 107, 118, 139]) as well as in quantitative metallurgy (see. e.g., [9, 23, 75, 91, 126, 141, 147]).

In addition to the orientation distribution function $\rho(A)$, it has been speculated that a second function, called the *misorientation distribution function* (MODF) plays a role in the material properties of crystalline aggregates [26, 50, 110, 111, 159]. The use of ODFs to describe polycrystalline materials can also be found in [1, 3, 27].

Whereas the ODF describes how the orientations of crystals in the aggregate, the MDOF describes the distribution of differences in orientation between adjacent crystals. That is, if one crystal has orientation A_1 and one of its neighbors has orientation A_2, then instead of summing the ODFs $\delta(A_i^{-1}A)$ over all crystals enumerated by i, the MODF sums $\delta(B_j^{-1}A)$ over all shared planes between crystals. B_j is a relative orientation such as $A_1^{-1}A_2$.

In Subsection 18.3.1 we examine the forward problem of approximating bulk properties from given single-crystal properties. In Subsection 18.3.2 we examine the inverse problem of approximating single-crystal properties from experimentally measured bulk material properties and a known orientation distribution.

18.3.1 Averaging Single-Crystal Properties over Orientations

A natural assumption is that the tensor components of the bulk material correspond to those of the crystal averaged over all orientations. However, the question immediately arises as to what tensor properties are to be averaged. That is, we must determine if it makes sense to average stiffness:

$$\langle C \rangle_{ijkl} = C_{i_1,j_1,k_1,l_1} \int_{SO(3)} \rho(A) l_{i,i_1}(A) l_{j,j_1}(A) l_{k,k_1}(A) l_{l,l_1}(A) \, dA, \tag{18.26}$$

or to average compliance:

$$\langle S \rangle_{ijkl} = S_{i_1,j_1,k_1,l_1} \int_{SO(3)} \rho(A) l_{i,i_1}(A) l_{j,j_1}(A) l_{k,k_1}(A) l_{l,l_1}(A) dA. \tag{18.27}$$

We note that while for a single crystal $S = C^{-1}$ and for the bulk material $S_{bulk} = C_{bulk}^{-1}$, in general

$$\langle S \rangle \neq \langle C \rangle^{-1}.$$

Hence, the computations in (18.26) or (18.27) constitute choice. If we choose to average stiffness as in (18.26) this is called a *Voigt* model [145], and if we choose to average compliance as in (18.27) this is called a *Reuss* model [116]).

Another approximation due to Hill [52] is to set the bulk properties as

$$C_{Hill} = \frac{1}{2}(\langle C \rangle + \langle S \rangle^{-1}); \qquad S_{Hill} = \frac{1}{2}(\langle S \rangle + \langle C \rangle^{-1}).$$

Apparently $C_{bulk} \approx C_{Hill}$ and $S_{bulk} \approx S_{Hill}$ can be useful approximations, even though mathematically $(C_{Hill})^{-1} \neq S_{Hill}$ and $(S_{Hill})^{-1} \neq C_{Hill}$.

In the remainder of this section we review another technique that others have proposed to circumvent this problem. In particular, we follow [5, 79, 96, 97, 99] and Chapter 7 of [69].

In general, C_{ijkl} can be decomposed as [79]:

$$C_{ijkl} = C_{I,J} B_{ij}^{(I)} B_{kl}^{(J)}$$

where

$$B^{(1)} = \begin{pmatrix} 1 & 0 & 0 \\ 0 & 0 & 0 \\ 0 & 0 & 0 \end{pmatrix}; \quad B^{(2)} = \begin{pmatrix} 0 & 0 & 0 \\ 0 & 1 & 0 \\ 0 & 0 & 0 \end{pmatrix}; \quad B^{(3)} = \begin{pmatrix} 0 & 0 & 0 \\ 0 & 0 & 0 \\ 0 & 0 & 1 \end{pmatrix};$$

$$B^{(4)} = \frac{1}{\sqrt{2}} \begin{pmatrix} 0 & 1 & 0 \\ 1 & 0 & 0 \\ 0 & 0 & 0 \end{pmatrix}; \quad B^{(5)} = \frac{1}{\sqrt{2}} \begin{pmatrix} 0 & 0 & 1 \\ 0 & 0 & 0 \\ 1 & 0 & 0 \end{pmatrix}; \quad B^{(6)} = \frac{1}{\sqrt{2}} \begin{pmatrix} 0 & 0 & 0 \\ 0 & 0 & 1 \\ 0 & 1 & 0 \end{pmatrix};$$

The matrices $B^{(I)}$ form an orthogonal basis for the set of all 3×3 symmetric matrices, in the sense that

$$\mathrm{tr}(B^{(I)} B^{(J)}) = \delta_{IJ}$$

and so we can define the Ith element of a 6×1 array $\mathbf{X}$ as $\mathrm{tr}(XB^I)$ for any 2-tensor X. Likewise, a 6×6 stiffness matrix is defined as

$$C_{IJ} \doteq B_{ij}^{(I)} C_{ijkl} B_{kl}^{(J)}.$$

This means that the relationship $T_{ij} = C_{ijkl} E_{kl}$ can be rewritten as the matrix expression

$$T_I = C_{IJ} E_J.$$

Assume, as in [79], that in addition to those symmetries which are induced by the symmetries of T and E, that $C_{ijkl} = C_{klij}$ (so that the matrix C_{IJ} is symmetric). Then it is possible to diagonalize C_{IJ} by a proper change of basis. Let $\{X^{(I)}\}$ for $I = 1, 2, ..., 6$ denote the basis for which $C_{IJ} = C^{(I)} \delta_{IJ}$ (no sum). That is,

$$C_{ijkl} X_{kl}^{(I)} = C^{(I)} X_{ij}^{(I)} \quad (\text{no sum on } I).$$

Then it is possible to write [69]:

$$C_{ijkl} = \sum_{I=1}^{6} C^{(I)} X_{ij}^{(I)} X_{kl}^{(I)}. \tag{18.28}$$

Since $C^{(I)} > 0$ for all $I = 1, 2, ..., 6$, it makes sense to define the logarithm of C component-by-component as

$$(\log C)_{ijkl} = \sum_{I=1}^{6} \log(C^{(I)}) X_{ij}^{(I)} X_{kl}^{(I)}.$$

In this way,

$$C = \exp(\log C).$$

The components of the average of $\log C$ over all orientations are then

$$\langle \log C \rangle_{ijkl} = \sum_{i=1}^{6} \log(C^{(I)}) X_{i_1, j_1}^{(I)} X_{k_1, l_1}^{(I)} \int_{SO(3)} \rho(A) l_{i,i_1}(A) l_{j,j_1}(A) l_{k,k_1}(A) l_{l,l_1}(A) \, dA.$$

An assumption that has been made in the literature for mathematical convenience is that the logarithm of the bulk stiffness tensor is approximately equal to the average of the logarithms of each crystal stiffness tensor [5, 69]:

$$\tilde{C} \doteq \exp(\langle \log C \rangle).$$

Here $\tilde{C}$ is an approximation to C_{bulk}. In this way,

$$(\tilde{C})^{-1} = \exp(-\langle \log C \rangle) = \exp(\langle \log C^{-1} \rangle) = \exp(\langle \log S \rangle) \doteq \tilde{S}.$$

While this is a mathematical assumption that allows us to average single-crystal properties over all orientations to yield the approximations $C_{bulk} \approx \tilde{C}$ and $S_{bulk} \approx \tilde{S}$ for a bulk polycrystalline material, and it has the desired property that $(\tilde{C})^{-1} = \tilde{S}$ (which is consistent with the requirement that $C_{bulk}^{-1} = S_{bulk}$), this is a purely mathematical construction. The only way to determine if this construction (or the whole hypothesis of orientational averaging) is valid is through experimentation. See [69] for discussions.

For other approaches to the calculation of mean (bulk) material properties (and bounds on these properties) using the geometric mean and other averaging techniques see [12, 14, 43, 65, 66, 72, 73, 90, 92, 96, 100, 108, 153].

18.3.2 Determining Single Crystal Strength Properties from Bulk Measurements of Polycrystalline Materials

In this subsection we examine the problem of finding the elasticity and compliance tensor components C_{ijkl} and S_{ijkl} when the ODF is known and either $\langle C \rangle_{ijkl}$ or $\langle S \rangle_{ijkl}$ are measured from experiment.

Let

$$M_{ii'jj'kk'll'} = \int_{SO(3)} \rho(A) l_{i,i'}(A) l_{j,j'}(A) l_{k,k'}(A) l_{l,l'}(A) \, dA$$

be the eighth-order tensor that is known when the ODF $\rho(A)$ is known. In principle, if the Voigt model is used, the stiffness tensor is found from the corresponding bulk property by the simple inversion

$$C_{ijkl} = \langle C \rangle_{i'j'k'l'} M^{-1}_{ii'jj'kk'll'}.$$

If the Reuss model is used, the compliance tensor is found as

$$S_{ijkl} = \langle S \rangle_{i'j'k'l'} M^{-1}_{ii'jj'kk'll'}.$$

More complicated calculations are involved in inverting the Hill and geometric mean approaches. For approaches to solving for the single crystal properties from bulk measurements see [83, 155].

18.4 Convolution Equations in Texture Analysis

In Sections 18.2 and 18.3 it is assumed that the orientational distribution is known, and is a purely kinematic quantity, i.e., it does not explicitly depend on the magnitude of stress or strain in the material. In this section we examine two equations from the field of *texture analysis*.[3] For developments in the field of texture analysis see [20, 39, 50, 62, 67, 69, 88, 95, 117, 121, 122, 133, 136, 144, 150, 151, 152, 157].

The two equations we examine respectively address the problem of determining the orientation distribution function from experimental measurements, and modeling how the ODF changes under applied loads. Both can be formulated as convolutions on $SO(3)$.

Determination of the ODF from Experimental Measurements

In practice, the orientation distribution function in a material sample is not known prior to performing crystallographic experiments. These experiments use techniques in which the full orientations of crystals are not determined directly. Rather, a distribution of crystal axes is determined. This is called a *pole figure*, and can be represented as a function on the sphere $\mathbb{S}^2$ for each orientation of the experiment. The reconstruction of structure from pole figures is part of a field called *stereology* (see [120, 140] for more on this topic).

By rotating the sample of material, and taking an infinite number of projections, the ODF can be reconstructed from the distribution of crystal axes. In a sense, this is like a Radon transform for the rotation group. Mathematically, we can write the equation relating the orientational distribution $\rho(A)$ to experimental measurements as

---

[3]This refers to the area of materials science concerned with modeling the bulk properties of crystal aggregates based on individual crystal properties.

$$\int_{SO(3)} \rho(A)\, \delta(\mathbf{y}^T A\mathbf{h} - 1)\, dA = p(\mathbf{h}, \mathbf{y}). \tag{18.29}$$

where $\mathbf{h} \in \mathbb{S}^2$ represents an axis in the crystal frame and $\mathbf{y} \in \mathbb{S}^2$ represents an axis in the sample frame, and these two frames are related by the rotation A. We take the sample frame to be fixed and A to denote the orientation of the crystal frame relative to it. If the opposite convention is used, then (18.29) is changed by replacing $\rho(A)$ with $\tilde{\rho}(A) = \rho(A^{-1})$.

When a projection is taken with $\mathbf{h}$ fixed, the resulting function $p_{\mathbf{h}}(\mathbf{y}) \doteq p(\mathbf{h}, \mathbf{y})$ is a pole figure. When a projection is taken with $\mathbf{y}$ fixed, $p_{\mathbf{h}}(\mathbf{y}) \doteq p(\mathbf{h}, \mathbf{y})$ is called an *inverse pole figure*.

Usually in materials science (18.29) is not written in the form we have given, though it is equivalent (see, e.g., [20], [69]). The issue of approximating the orientation distribution function from a finite number of pole figures has been addressed extensively in the literature [57, 58, 93, 94, 102, 128]. For other techniques to do pole figure inversion and ODF determination see [17, 40, 71].

We mention this problem here because by defining

$$\Delta_{\mathbf{h}}(\mathbf{u}) \doteq \delta(\mathbf{h}^T\mathbf{u} - 1)$$

for all $\mathbf{u} \in \mathbb{S}^2$, we can write (18.29) as the convolution equation

$$\int_{SO(3)} \rho(A)\Delta_{\mathbf{h}}(A^T\mathbf{y})\, dA = \rho * \Delta_{\mathbf{h}} = p_{\mathbf{h}}(\mathbf{y}).$$

The determination of $\rho(A)$ from a single equation like the one above is not possible (as we would expect from the fact that the $SO(3)$-Fourier transform matrix of the function $\Delta_{\mathbf{h}}(\mathbf{u})$ is singular).

We note that (18.29) can be written in the equivalent form [20]:

$$p(\mathbf{h}, \mathbf{y}) = \frac{1}{2\pi} \int_0^{2\pi} \rho(R(\mathbf{e}_3, \mathbf{y})\mathrm{rot}[\mathbf{e}_3, \theta]R(\mathbf{h}, \mathbf{e}_3))\, d\theta$$

where $R(\mathbf{a}, \mathbf{b})$ is the rotation matrix (5.27) that takes the vector $\mathbf{a}$ into $\mathbf{b}$. Expansion of $\rho(A)$ in a Fourier series on $SO(3)$ and using the property

$$U^l(R(\mathbf{e}_3, \mathbf{y})\mathrm{rot}[\mathbf{e}_3, \theta]R(\mathbf{h}, \mathbf{e}_3)) = U^l(R(\mathbf{e}_3, \mathbf{y}))U^l(\mathrm{rot}[\mathbf{e}_3, \theta])U^l(R(\mathbf{h}, \mathbf{e}_3))$$

allows us to integrate out the θ in closed form and reduces the pole inversion problem to the solution of a set of linear algebraic equations. Sample and/or crystal symmetries are useful in reducing the dimension of the system of equations. A problem that has been observed is that the solution obtained by inverting pole figures can have negative values. This is called the *ghost* problem [87, 142, 143].

Modern experimental techniques exist that can more directly determine the orientation of single crystals (grains) in an aggregate polycrystalline material. These methods of imaging and orientation mapping are based on the measurement of Kikuchi patterns. See [4, 11, 33, 63, 85, 98, 123, 154] for descriptions of experimental methods.

Texture Transformation/Evolution

The ODF of a polycrystalline material, $f_1(A)$, can be changed by various mechanical processes to result in an ODF $f_2(A)$ (see [20], Chapter 8). The change $f_1(A) \rightarrow f_2(A)$

is called texture transformation (or texture evolution). It has been postulated [20] that the process of texture transformation can be posed as the convolution

$$(w * f_1)(A) = f_2(A) \tag{18.30}$$

where here convolution is on the group $SO(3)$, and the function $w(A)$ is characteristic of the particular transformation process (e.g., phase transformation, recrystallization, etc.).

In this context, the three problems of interest that center around (18.30) are [20]: (1) finding $f_2(A)$ for given $f_1(A)$ and $w(A)$; (2) finding the original $f_1(A)$ for given $f_2(A)$ and $w(A)$; finding $w(A)$ for given $f_2(A)$ and $f_1(A)$. The first of these problems is a convolution equation (which can be calculated efficiently numerically when $w(A)$ and $f_1(A)$ are band-limited functions). The second and third problem are inverse problems which generally require regularization techniques to solve. The issues in the regularization of convolution equations on $SO(3)$ are similar to those for $SE(3)$ (as described in Chapters 12 and 13).

For more on the idea of texture transformation, see [24, 25, 61, 84, 119].

18.5 Constitutive Laws in Solid Mechanics

In this section we review a few of the simplest models for the relationship between stress and strain in solids. When the material is isotropic (has the same properties when a given point is viewed from all different orientations) this imposes severe restrictions on how stress and strain are related. Also, when the material is anisotropic, but has certain symmetries, this also imposes constraints that must be taken into account in the constitutive equations. The material we review in this section is a summary of material that can be found in [8, 78, 127]

A constitutive law is a relationship between stress and strain. In solid mechanics, this is essentially a generalization of Hooke's law:

$$F = \varphi(x) = kx \tag{18.31}$$

where F is the force applied to a spring, and x is the amount of displacement or elongation of the spring. Hooke's law is a linear relationship, with the stiffness k serving as the constant of proportionality. However, for large displacements this model breaks down, and taking $\varphi(x) = kx + \epsilon x^3$ extends the range of applicability of the model for larger deformations. When $\epsilon > 0$ the spring is called *stiffening*, and for $\epsilon < 0$ it is called *softening*.

18.5.1 Isotropic Elastic Materials

For a three-dimensional element of material, a constitutive law relates stress and strain. The simplest kind of deformation is an elastic one:

$$T = \varphi(E). \tag{18.32}$$

An elastic deformation is necessarily reversible, and does not depend on the history of strain states prior to the current time. The simplest of the elastic deformations is the linearly elastic model (which is a 3D analog of the 1D version of (18.31)). This is written in indicial notation as

$$T_{ij} = C_{ijkl}E_{kl}, \tag{18.33}$$

where C_{ijkl} are the components of a fourth-order tensor called the *stiffness (or elasticity) tensor*. We can also write

$$E_{ij} = S_{ijkl}T_{kl},$$

where S_{ijkl} are components of the *compliance tensor* $S = C^{-1}$. In fact, we have seen these expressions earlier in this chapter. The tensorial nature of C with components C_{ijkl} follows from the fact that T and E are both second-order tensors, i.e., $T_{ij}(Q) = C_{ijkl}(Q)E_{kl}(Q)$ implies

$$Q_{i,i_1}Q_{j,j_1}T'_{i_1,j_1} = C_{ijkl}(Q)Q_{k,k_1}Q_{l,l_1}E'_{k_1,l_1}.$$

Then for (18.33) to hold in the primed coordinate system, $T'_{ij} = C'_{ijkl}E'_{kl}$, it must be that

$$Q_{i,i_1}Q_{j,j_1}C'_{i_1,j_1,k,l}E'_{kl} = C_{ijkl}(Q)Q_{k,k_1}Q_{l,l_1}E'_{k_1,l_1}.$$

Multiplying both sides by $Q_{i,m}Q_{j,n}$ and summing over i and j yields

$$\delta_{i_1,m}\delta_{j_1,n}C'_{i_1,j_1,k,l}E'_{kl} = Q_{i,m}Q_{j,n}C_{ijkl}(Q)Q_{k,k_1}Q_{l,l_1}E'_{k_1,l_1}.$$

Using the localization argument that this must hold for all possible values of E'_{ij} (see, e.g., [78]), and changing the indices of summation on the left from k,l to k_1,l_1, we find

$$C'_{m,n,k_1,l_1} = Q_{i,m}Q_{j,n}Q_{k,k_1}Q_{l,l_1}C_{ijkl}(Q).$$

This can also be written in the form

$$C_{ijkl}(Q) = Q_{i,i_1}Q_{j,j_1}Q_{k,k_1}Q_{l,l_1}C'_{i_1,j_1,k_1,l_1}.$$

This verifies that $C_{ijkl}(Q)$ transform as the components of a fourth-order tensor. A similar calculation shows that $S_{ijkl}(Q)$ transforms as tensor components.

At first glance, it may appear that C_{ijkl} for $i,j,k,l \in \{1,2,3\}$ constitute a set of $3^4 = 81$ independent parameters. However, C_{ijkl} inherits symmetries from those of T and E. That is, $T_{ij} = T_{ji}$ and $E_{kl} = E_{lk}$ imply that

$$C_{ijkl} = C_{jikl} = C_{ijlk}. \tag{18.34}$$

Hence there are only 36 free parameters (6 for all i,j and each fixed k,l, and vice versa).

Further symmetry relations exist when the material has symmetry group $G \le SO(3)$ because then for all $A \in G$

$$C_{ijkl}(A) = C_{ijkl}.$$

The most restrictive case is $G = SO(3)$. In this case, it can be shown (see, e.g., [78, 127]) that

$$C_{ijkl} = \lambda\delta_{ij}\delta_{kl} + 2\mu\delta_{ik}\delta_{jl} \tag{18.35}$$

is the most general isotropic fourth-order tensor which also observes the symmetries (18.34). The two free constants λ and μ are called the Lamé constants. The corresponding elements of the compliance tensor are written as

$$S_{ijkl} = -\frac{\lambda}{2\mu}\delta_{ij}\delta_{kl} + \frac{1}{2\mu}\delta_{ik}\delta_{jl}. \tag{18.36}$$

For an isotropic linearly elastic solid, it is common to write the constitutive law with stiffness (18.35) as [78, 86, 89]:

$$
\begin{pmatrix} T_{11} \\ T_{22} \\ T_{33} \\ T_{12} \\ T_{13} \\ T_{23} \end{pmatrix} = \begin{pmatrix} \lambda+2\mu & \lambda & \lambda & 0 & 0 & 0 \\ \lambda & \lambda+2\mu & \lambda & 0 & 0 & 0 \\ \lambda & \lambda & \lambda+2\mu & 0 & 0 & 0 \\ 0 & 0 & 0 & \mu & 0 & 0 \\ 0 & 0 & 0 & 0 & \mu & 0 \\ 0 & 0 & 0 & 0 & 0 & \mu \end{pmatrix} \begin{pmatrix} E_{11} \\ E_{22} \\ E_{33} \\ 2E_{12} \\ 2E_{13} \\ 2E_{23} \end{pmatrix}. \tag{18.37}
$$

The reason for the factor of 2 in front of the E_{ij} terms (as opposed to multiplying μ in the $(4,4)$, $(5,5)$, and $(6,6)$ elements in the above matrix) is explained in the next subsection.

More generally, a constitutive law (which is elastic but not necessarily linear) must also preserve the fact that T and E are both tensors. This imposes the condition on (18.32):

$$
Q_{i,i_1} Q_{j,j_1} \varphi_{i_1,j_1}(E_{kl}) = \varphi_{ij}(Q_{k,k_1} Q_{l,l_1} E_{k_1,l_1}).
$$

18.5.2 Materials with Crystalline Symmetry

In the case when the symmetry group of a material is a finite subgroup $G < SO(3)$, the 36 independent constants that define the tensor elements C_{ijkl} do not reduce to only the two Lamé constants. The construction of C_{ijkl} for materials with various crystallographic symmetries can be found in [8, 127], and references therein.

Using the symmetries of T and E, and writing the equation $T_{ij} = C_{ijkl} E_{kl}$ in the matrix form

$$
\begin{pmatrix} T_{11} \\ T_{22} \\ T_{33} \\ T_{12} \\ T_{13} \\ T_{23} \end{pmatrix} = \begin{pmatrix} C_{1111} & C_{1122} & C_{1133} & C_{1112} & C_{1113} & C_{1123} \\ C_{2211} & C_{2222} & C_{2233} & C_{2212} & C_{2213} & C_{2223} \\ C_{3311} & C_{3322} & C_{3333} & C_{3312} & C_{3313} & C_{3323} \\ C_{1211} & C_{1222} & C_{1233} & C_{1212} & C_{1213} & C_{1223} \\ C_{1311} & C_{1322} & C_{1333} & C_{1312} & C_{1313} & C_{1323} \\ C_{2311} & C_{2322} & C_{2333} & C_{2312} & C_{2313} & C_{2323} \end{pmatrix} \begin{pmatrix} E_{11} \\ E_{22} \\ E_{33} \\ 2E_{12} \\ 2E_{13} \\ 2E_{23} \end{pmatrix}, \tag{18.38}
$$

the problem of characterizing material properties when symmetries exist is the same as determining the structure of the 6×6 stiffness matrix above.[4] We will write (18.38) as

$$
\mathbf{T} = \hat{C}\mathbf{E}
$$

to distinguish it from (18.33).

Perhaps the most straightforward way to construct elasticity tensors for materials with symmetry is to enforce the equality

$$
C_{ijkl} = A_{i,i_1} A_{j,j_1} A_{k,k_1} A_{l,l_1} C_{i_1,j_1,k_1,l_1}
$$

for all $A \in G$. This results in a degenerate set of $36 \cdot |G|$ simultaneous equations which places a restriction on the form of the coefficients C_{ijkl}.

While this works for a finite group G, we cannot enumerate all the elements of a Lie group such as $SO(3)$.

Another way is to start with an arbitrary set of 36 elasticity constants C_{ijkl}, average over all elements of G by defining

---

[4]Note the 2s in (18.38) are required to make this matrix equation the same as (18.33) since E_{21}, E_{31}, E_{32} do not explicitly appear.

$$\langle C \rangle_{ijkl} = \frac{1}{|G|} \sum_{A \in G} A_{i,i_1} A_{j,j_1} A_{k,k_1} A_{l,l_1} C_{i_1,j_1,k_1,l_1},$$

and determine the form of each $\langle C \rangle_{ijkl}$ as a function of all the C_{i_1,j_1,k_1,l_1}.

Clearly, for any $Q \in G$

$$Q_{i_2,i} Q_{j_2,j} Q_{k_2,k} Q_{l_2,l} \langle C \rangle_{ijkl} = \frac{1}{|G|} \sum_{A \in G} Q_{i_2,i} Q_{j_2,j} Q_{k_2,k} Q_{l_2,l} A_{i,i_1} A_{j,j_1} A_{k,k_1} A_{l,l_1} C_{i_1,j_1,k_1,l_1}$$

$$= \frac{1}{|G|} \sum_{A \in G} (QA)_{i_2,i_1} (QA)_{j_2,j_1} (QA)_{k_2,k_1} (QA)_{l_2,l_1} C_{i_1,j_1,k_1,l_1}$$

$$= \langle C \rangle_{i_2,j_2,k_2,l_2}.$$

This is because the sum of any function on a finite group over all group elements is invariant under shifts. (See Chapters 7 and 8.)

The same approach can be used for continuous symmetries, such as the isotropic case, with the summation being replaced with integration over the group. That is, we can calculate

$$\langle C \rangle_{ijkl} = \int_{SO(3)} A_{i,i_1} A_{j,j_1} A_{k,k_1} A_{l,l_1} C_{i_1,j_1,k_1,l_1} dA$$

and identify how many repeated sets of parameters from C_{i_1,j_1,k_1,l_1} are present in $\langle C \rangle_{ijkl}$.

While this method can be used in principle, it too can be quite tedious when performing calculations by hand.

Some special cases of importance are [89]:

$$\hat{C} = \hat{C}^T,$$

in which case there are 21 rather than 36 free parameters;

$$\hat{C} = \begin{pmatrix} C_{1111} & C_{1122} & C_{1133} & 0 & 0 & C_{1123} \\ C_{2211} & C_{2222} & C_{2233} & 0 & 0 & C_{2223} \\ C_{3311} & C_{3322} & C_{3333} & 0 & 0 & C_{3323} \\ 0 & 0 & 0 & C_{1212} & C_{1213} & 0 \\ 0 & 0 & 0 & C_{1312} & C_{1313} & 0 \\ C_{2311} & C_{2322} & C_{2333} & 0 & 0 & C_{2323} \end{pmatrix},$$

which indicates a plane of elastic symmetry;

$$\hat{C} = \begin{pmatrix} C_{1111} & C_{1122} & C_{1133} & 0 & 0 & 0 \\ C_{2211} & C_{2222} & C_{2233} & 0 & 0 & 0 \\ C_{3311} & C_{3322} & C_{3333} & 0 & 0 & 0 \\ 0 & 0 & 0 & C_{1212} & 0 & 0 \\ 0 & 0 & 0 & 0 & C_{1313} & 0 \\ 0 & 0 & 0 & 0 & 0 & C_{2323} \end{pmatrix},$$

which indicates orthotropic (three orthogonal planes of) symmetry.

18.6 Orientational Distribution Functions for Non-Solid Media

Our discussion in this chapter primarily addresses the application of techniques from noncommutative harmonic analysis in the mechanics of solids and quantitative materials science. In this short section we provide pointers to the literature for the reader with an interest in fluid mechanics and rheology.

18.6.1 Orientational Dynamics of Fibers in Fluid Suspensions

The problem of the evolution of orientations of fibers and rigid particles in fluid has been addressed extensively. See, for example, [10, 19, 30, 41, 45, 46, 53, 59, 68, 113, 114, 115, 124, 125, 131, 132]. Whereas problems in solid mechanics and quantitative materials science can be described well using the ODF, in fluid mechanics the ODF is not sufficient. Rather, a model of the interaction of the ODF (which becomes a function of time as well as in the context of fluids) and the fluid surrounding the collection of particles/fibers becomes critical. Therefore, the importance of harmonic analysis of the ODF alone would seem to be less relevant in the context of fluid mechanics under low viscosity conditions.

18.6.2 Applications to Liquid Crystals

In the analysis of optical and mechanical properties of liquid crystals a number of tensor quantities arise. See, e.g., [34, 35, 55]. The ODF for liquid crystals is addressed in Chapter 16. Given this ODF, orientational averaging can be performed as in solid mechanics to determine bulk properties. The assumption of high viscosity makes a non-inertial model for liquid crystals appropriate, and the limitations in the application of harmonic analysis in fluid mechanics are not as severe.

18.7 Summary

In this chapter we have examined a variety of orientation-dependent issues in the mechanics of anisotropic solids and liquids containing rigid inclusions. The idea of averaging over orientations, as well as modeling the time evolution of certain processes using convolution on $SO(3)$ both naturally arose.

In our presentation, most of the phenomena were anisotropic but spatially uniform, and so ideas from harmonic analysis on $SO(3)$ entered. For properties that vary over both orientation and spatial position, a motion-group distribution function $f(\mathbf{a}, A)$ is a natural replacement for the ODF $\rho(A)$. Integration of such a function over orientation provides the spatial variability of properties, whereas integration over the translation (position) provides an average ODF for the whole sample.

While the ODF we have considered is a function on $SO(3)$, symmetries in the physical phenomena are represented in the form of symmetries of the ODF. In the case of fibers with axial symmetry, the ODF can be viewed as a function on $\mathbb{S}^2$. For example, the problem of determining effective properties of fiber-reinforced composites using orientational averaging has been considered in a number of works including [6, 16, 44, 104, 129, 134, 135, 148]. The fact that such averages are usually taken over $\mathbb{S}^2$ instead of $SO(3)$ reduces the need for knowledge of noncommutative harmonic analysis in that particular context.

Since the original publication of our book, additional relevant references have come to our attention in the area of texture and orientation in mechanics. These include [2, 36, 51, 101, 130, 158].

Finally, we mention that, though the emphasis of this chapter has been on orientational averaging, many other mathematical methods have been employed in mechanics over the past century. Pioneering works include the group-theoretic formulations in [13, 31, 37, 49, 138], and differential-geometric formulations in [15, 106, 146]. For a modern perspective on these topics see [48].

References

1. Adams, B.L., "Description of the Intercrystalline Structure Distribution in Polycrystalline Metals," *Metallurgical Transactions A*, 17(12): 2199–2207, 1986.
2. Adams, B.L., Boehler, J.P., Guidi, M., Onat, E.T., "Group Theory and Representation of Microstructure and Mechanical Behavior of Polycrystals," *Journal of the Mechanics and Physics of Solids*, 40(4): 723–737, 1992.
3. Adams, B.L., Field, D.P., "A Statistical Theory of Creep in Polycrystalline Materials," *Acta Metallurgica et Materialia*, 39(10): 2405–2417, 1991.
4. Adams, B.L., Wright, S.I., Kunze, K., "Orientation Imaging: The Emergence of a New Microscopy," *Metallurgical Transactions A-Physical Metallurgy and Materials Science* 24(4): 819–831, April 1993.
5. Aleksandrov, K.S., Aisenberg, L.A., "Method of Calculating the Physical Constants of Polycrystalline Materials," *Soviet Physics-Doklady*, 11: 323–325, 1966.
6. Alwan, J.M., Naaman, A.E., "New Formulation for Elastic Modulus of Fiber-Reinforced, Quasibrittle Matrices," *Journal of Engineering Mechanics*, 120(11): 2443–2461, Nov. 1994.
7. Baker, D.W., "On the Symmetry of Orientation Distribution in Crystal Aggregates," in *Advances in X-ray Analysis*, Vol. 13, pp. 435–454, Springer, 1970.
8. Bao, G., *Application of Group and Invariant-Theoretic Methods to the Generation of Constitutive Equations*, Ph.D. Dissertation, Lehigh University, 1987.
9. Barrett, C.S., Massalski, T.B., *Structure of Metals*, 3rd ed., McGraw-Hill, New York, 1980.
10. Batchelor, G.K., "The Stress Generated in a Non-Dilute Suspension of Elongated Particles by Pure Straining Motion," *Journal of Fluid Mechanics*, 46(4): 813–829, 1971.
11. Baudin, T., Penelle, R., "Determination of the Total Texture Function from Individual Orientation Measurements by Electron Backscattering Pattern," *Metallurgical Transactions A- Physical Metallurgy and Materials Science*, 24(10): 2299–2311, Oct. 1993.
12. Becker, R., Panchanadeeswaran, S., "Crystal Rotations Represented as Rodrigues Vectors," *Textures and Microstructures*, 10(3): 167–194, 1989.
13. Belinfante, J.G., "Lie Algebras and Inhomogeneous Simple Materials," *SIAM Journal on Applied Mathematics*, 25(2): 260–268, 1973.
14. Beran, M.J., Mason, T.A., Adams, B.L., Olsen, T., "Bounding Elastic Constants of an Orthotropic Polycrystal Using Measurements of the Microstructure," *Journal of the Mechanics and Physics of Solids*, 44(9): 1543–1563, Sept. 1996.
15. Bloom, F., "Modern Differential Geometric Techniques in the Theory of Continuous Distributions of Dislocations," Lecture Notes in Mathematics 733, Springer-Verlag, Berlin Heidelberg, 1979.
16. Boutin, C., "Microstructural Effects in Elastic Composites," *International Journal of Solids and Structures*, 33(7): 1023–1051, 1996.
17. Bowman, K.J., "Materials Concepts Using MathCAD .1. Euler Angle Rotations and Stereographic Projection," *JOM-Journal of the Minerals Metals and Materials Society*, 47(3): 66–68, March 1995.

18. Brady, J.F., Phillips, R.J., Lester, J.C., Bossis, G. "Dynamic Simulation of Hydrodynam-ically Interacting Suspensions," *Journal of Fluid Mechanics*, 195: 257–280, 1988.

19. Bretherton, F.P., "The Motion of Rigid Particles in a Shear Flow at Low Reynolds Num-ber," *Journal of Fluid Mechanics*, 14: 284–304, 1962.

20. Bunge, H.-J., *Texture Analysis in Materials Science*, Butterworths, London, 1982; Also *Mathematische Methoden der Texturanalyse*, Akademie-Verlag, Berlin, 1969.

21. Bunge, H.-J., Esling, C., Dahlem, E., Klein, H., "The Development of Deformation Textures described by an Orientation Flow Field," *Textures and Microstructures*, 6(3): 181–200, 1986.

22. Bunge, H.-J., "Zur Darstellung allgemeiner Texturen," *Zeitschrift für Metallkunde*, 56: 872–874, 1965; also see *Mathematische Methoden der Texturanalyse*, Akademie-Verlag, Berlin, 1969.

23. Bunge, H.-J., Esling, C., *Quantitative Texture Analysis*, Dtsch. Gesell. Metallkde., Oberursel, 1982.

24. Bunge, H.J., Humbert, M., Welch, P.I., "Texture Transformation," *Textures and Mi-crostructures*, 6(2): 81–95, 1984.

25. Bunge, H.-J., "3-Dimensional Texture Analysis," *International Materials Reviews* 32(6): 265–291, 1987.

26. Bunge, H.-J., Weiland, H., "Orientation Correlation in Grain and Phase Boundaries," *Textures and Microstructures* 7(4): 231–263, 1988.

27. Bunge, H.-J., Kiewel, R., Reinert, T., Fritsche, L., "Elastic Properties of Polycrystals - Influence of Texture and Stereology," *Journal of the Mechanics and Physics of Solids* 48(1): 29–66, Jan. 2000.

28. Castaneda, P.P., "The Effective Mechanical Properties of Nonlinear Isotropic Composites," *Journal of Mechanics and Physics of Solids*, 39(1): 45–71, 1991.

29. Castaneda, P. P., Willis, J. R., "The Effect of Spatial Distribution on the Effective Behavior of Composite Materials and Cracked Media," *Journal of Mechanics and Physics of Solids*, 43(12): 1919–1951, 1995.

30. Claeys, I.L., Brady, J.F., "Suspensions of Prolate Spheroids in Stokes Flow. Part 2. Sta-tistically Homogeneous Dispersions," *Journal of Fluid Mechanics*, 251: 443–477, 1993.

31. Coleman, B.D., Noll, W., "Material Symmetry and Thermostatic Inequalities in Finite Elastic Deformations," *Archive for Rational Mechanics and Analysis*, 15: 87–111, 1964.

32. Coulomb, P., *Les Textures dans les Metaux de Reseau Cubique*, Dunood, Paris, 1982.

33. Davies, R.K., Randle, V., "Application of Crystal Orientation Mapping to Local Orienta-tion Perturbations," *European Physical Journal-Applied Physics*, 7(1): 25–32, July 1999.

34. de Gennes, P.G., Prost, J., *The Physics of Liquid Crystals*, 2nd ed., Oxford University Press, New York, 1998.

35. de Jeu, W.H., *Physical Properties of Liquid Crystalline Materials*, Gordon and Breach, New York, 1980.

36. Engler, O., Randle, V., *Introduction to Texture Analysis: Macrotexture, Microtexture, and Orientation Mapping*, 2nd ed., CRC Press, Boca Raton, FL, 2010.

37. Ericksen, J.L., "On the Symmetry and Stability of Thermoelastic Solids," *Journal of Ap-plied Mechanics*, 45(4): 740–744, 1978.

38. Eshelby, J., "The Determination of the Elastic Field of an Ellipsoidal Inclusion, and Related Problems," *Proceedings of the Royal Society of London*, Series A, 241: 376–396, 1957.

39. Esling, C., Humbert, M., Philippe, M.J., Wagner, F., "H.J. Bunge's Cooperation with Metz: Initial Work and Prospective Development," *Texture and Anisotropy of Polycrystals*, 273(2): 15–28, 1998.

40. Ewald, P.P., "The 'Poststift' - A Model for the Theory of Pole Figures," *Journal of the Less Common Metals*, 28(1): 1–5, 1972.

41. Folgar, F., Tucker, C.L., "Orientation Behavior of Fibers in Concentrated Suspensions," *Journal of Reinforced Plastics and Composites*, 3(2): 98–119, 1984.

42. Forest, S., "Mechanics of Generalized Continua: Construction by Homogenization," *Le Journal de Physique IV*, 8(4): 39–48, 1998.

43. Frank, F.C., "The Conformal Neo-Eulerian Orientation Map," *Philosophical Magazine A*, 65(5): 1141–1149, 1992.

44. Fu, S., Lauke, B., "The Elastic Modulus of Misaligned Short-Fiber-Reinforced Polymers," *Composites Science and Technology*, 58(3): 389–400, 1998.

45. Ganani, E., Powell, R.L., "Rheological Properties of Rodlike Particles in a Newtonian and a Non-Newtonian Fluid," *Journal of Rheology*, 30(5): 995–1013, 1986.

46. Graham, A.L., Mondy, L.A., Gottlieb, M., Powell, R.L., "Rheological Behavior of a Suspension of Randomly Oriented Rods," *Applied Physics Letters*, 50(3): 127–129, Jan. 1987.

47. Guidi, M., Adams, B.L., Onat, E.T., "Tensorial Representations of the Orientation Distribution Function in Cubic Polycrystals," *Textures and Microstructures*, 19: 147–167, 1992.

48. Gurtin, M.E., Fried, E., Anand, L., *The Mechanics and Thermodynamics of Continua*, Cambridge University Press, Cambridge, England, 2010.

49. Gurtin, M.E., Williams, W.O., "On the Inclusion of the Complete Symmetry Group in the Unimodular Group," *Archive for Rational Mechanics and Analysis*, 23(3): 163–172, 1966.

50. Haessner, F., Pospiech, J., Sztwiertnia, K., "Spatial Arrangement of Orientations in Rolled Copper," *Materials Science and Engineering*, 57(1): 1–14, 1983.

51. He, Y., Godet, S., Jonas, J.J., "Representation of Misorientations in RodriguesFrank Space: Application to the Bain, KurdjumovSachs, NishiyamaWassermann and Pitsch Orientation Relationships in the Gibeon Meteorite," *Acta Materialia*, 53(4): 1179–1190, 2005.

52. Hill, R., "The Elastic Behaviour of a Crystalline Aggregate," *Proceedings of the Physical Society*, Section A, 65(5): 349–354, 1952.

53. Hinch, E.J., Leal, L.G., "The Effect of Brownian Motion on the Rheological Properties of a Suspension of Non-Spherical Particles," *Journal of Fluid Mechanics*, 52(4): 683–712, 1972.

54. Hinch, E.J. "Averaged-Equation Approach to Particle Interactions in a Fluid Suspension," *Journal of Fluid Mechanics*, 83(4): 695–720, 1977.

55. Hsu, P., Poulin, P., Weitz, D.A., "Rotational Diffusion of Monodisperse Liquid Crystal Droplets," *Journal of Colloid and Interface Science*, 200(1): 182–184, 1998.

56. Hutchinson, J.W., "Elastic-plastic Behaviour of Polycrystalline Metals and Composites," *Proceedings of the Royal Society of London A: Mathematical, Physical and Engineering Sciences*, 319(1537): 247–272, 1970.

57. Imhof, J., "Die Bestimmung einer Nährung für die Funktion der Orientierungsverteilung aus einer Polfigur," *Zeitschrift für Metallkunde*, 68: 38–43, 1977.

58. Imhof, J. "Texture Analysis by Iteration I. General Solution of the Fundamental Problem," *Physica Status Solidi B - Basic Reserch*, 119(2): 693–701, 1983.

59. Jeffery, G.B., "The Motion of Ellipsoidal Particles Immersed in a Viscous Fluid," *Proceedings of the Royal Society of London A: Mathematical, Physical and Engineering Sciences*, 102(715): 161–179, 1922.

60. Kailasam, M., Ponte Castaneda, P., "A General Constitutive Theory for Linear and Nonlinear Particulate Media with Microstructure Evolution," *Journal of Mechanics and Physics of Solids*, 46(3): 427–465, 1998.

61. Kallend, J.S., Morris, P.P., Davies, G.J., "Texture Transformations: The Misorientation Distribution Function," *Acta Metallurgica*, 24(4): 361–370, 1976.

62. Kallend, J.S., Kocks, U.F., Rollett, A.D., Wenk, H.R., "Operational Texture Analysis," *Materials Science and Engineering A-Structural Materials Properties Microstructure and Processing*, 132: 1–11, Feb. 1991.

63. Katrakova, D., Maas, C., Hohnerlein, D., Mucklich, F., "Experiences on Contrasting Microstructure Using Orientation Imaging Microscopy," *Praktische Metallographie-Practical Metallography*, 35(1): 4–20, Feb. 1998.

64. Kassir, M.K., Sih, G.C., "Three-Dimensional Stress Distribution Around an Elliptical Crack Under Arbitrary Loadings," *Journal of Applied Mechanics*, 33(3): 601–611, 1966.

65. Kiewel, H., Fritsche, L., "Calculation of Effective Elastic Moduli of Polycrystalline Materials Including Nontextured Samples and Fiber Textures," *Physical Review B*, 50(1): 5–16, 1994.

66. Kneer, G., "Über die Berechnung der Elastizitätsmoduln vielkristalliner Aggregate mit Textur," *Physica Status Solidi*, Series B, 9(3): 825 – 838, 1965.
67. Knorr, D.B., Weiland, H., Szpunar, J.A., "Applying Texture Analysis to Materials Engineering Problems," *JOM-Journal of the Minerals Metals and Materials Society*, 46(9): 32 – 36, Sept. 1994.
68. Koch, D.L., "A Model for Orientational Diffusion in Fiber Suspensions," *Physics of Fluids*, 7(8): 2086 – 2088, August 1995.
69. Kocks, U.F., Tomé, C.N., Wenk, H.-R., *Texture and Anisotropy: Prefered Orientations in Polycrystals and Their Effect on Materials Properties*, Cambridge University Press, Cambridge, 1998.
70. Kondo, K., "Non-Riemannian Geometry of Imperfect Crystals from a Macroscopic Viewpoint," *Memoirs of the Unifying Study of the Basic Problems in Engineering Science by Means of Geometry*, 1: 6 – 17, 1955.
71. Krigbaum, W.R., "A Refinement Procedure for Determining the Crystallite Orientation Distribution Function," *Journal of Physical Chemistry*, 74(5): 1108 – 1113, 1970.
72. Kröner, E., "Berechnung der elastischen Konstanten des Vielkristalls aus den Konstanten der Einkristalle," *Zeitschrift für Physik*, 151(1): 504 – 518, 1958.
73. Kröner, E., "Bounds for Effective Elastic Moduli of Disordered Materials," *Journal of the Mechanics and Physics of Solids*, 25(2): 137 – 155, 1977.
74. Krumbein, W.C., "Preferred Orientation of Pebbles in Sedimentary Deposits," *Journal of Geology*, 47(7): 673 – 706, 1939.
75. Kudriawzew, I.P., *Textures in Metals and Alloys* (in Russian), Isdatel'stwo Metallurgija, Moscow, 1965.
76. Lagzdiņš, A., Tamužs, V., "Tensorial Representation of the Orientation Distribution Function of Internal Structure for Heterogeneous Solids," *Mathematics and Mechanics of Solids*, 1: 193 – 205, 1996,.
77. Lagzdiņš, A., Tamužs, V., Teters, G., Krēgers, A., *Orientational Averaging in Solid Mechanics*, Pitman Research Notes in Mathematics Series, 265, Longman Scientific & Technical, London, 1992.
78. Lai, W.M., Rubin, D., Krempl, E., *Introduction to Continuum Mechanics*, Pergamon Press, New York, 1978.
79. Leibfried, G., Breuer, N., *Point Defects in Metals. I: Introduction to the Theory*, Springer-Verlag, Berlin, 1978.
80. Li, B., "Effective Constitutive Behavior of Nonlinear Solids Containing Penny-Shaped Cracks," *International Journal of Plasticity*, 10(4): 405 – 429, 1994.
81. Li, G., Castaneda, P.P., "The Effect on Particle Shape and Stiffness on the Constitutive Behavior of Metal-Matrix Composites," *International Journal of Solids and Structures*, 30(23): 3189 – 3209, 1993.
82. Li, G., Castaneda, P.P., Douglas, A.S., "Constitutive Models for Ductile Solids Reinforced by Rigid Spheroidal Inclusions," *Mechanics of Materials*, 15(4): 279 – 300, 1993.
83. Li, D.Y., Szpunar, J.A., "Determination of Single-Crystals Elastic-Constants from the Measurement of Ultrasonic Velocity in the Polycrystalline Material," *Acta Metallurgica et Materialia*, 40(12): 3277 – 3283, Dec. 1992.
84. Lindsay, R., Chapman, J.N., Craven, A.J., McBain, D., "A Quantitative Determination of the Development of Texture in Thin Films," *Ultramicroscopy*, 80(1): 41 – 50, August 1999.
85. Liu, Q., "A Simple and Rapid Method for Determining Orientations and Misorientations of Crystalline Specimens in TEM," *Ultramicroscopy*, 60(1): 81 – 89, August 1995.
86. Love, A.E.H., *A Treatise on the Mathematical Theory of Elasticity*, 4th ed., Dover Publications, New York, 1927.
87. Lücke‘ K., Pospiech, J., Jura, J., Hirsch, J., "On the Presentation of Orientation Distribution Functions by Model Functions," *Zeitschrift für Metallkunde*, 77(5): 312 – 321, 1986.
88. Luzin, V., "Optimization of Texture Measurements. III. Statistical Relevance of ODF Represented by Individual Orientations," *Materials Science Forum*, 273 – 275: 107 – 112, 1998.
89. Malvern, L.E., *Introduction to the Mechanics of a Continuous Medium*, Prentice-Hall, Inc., Englewood Cliffs, NJ, 1969.

90. Mason, T.A., "Simulation of the Variation of Material Tensor Properties of Polycrystals Achieved Through Modification of the Crystallographic Texture," *Scripta Materialia*, 39(11): 1537–1543, Nov. 1998.

91. Matthies, S., *Aktuelle Probleme der Texturanalyse*, Akademie der Wissenschaften, D.D.R., Zentralinstitut fur Kernforschung, Rossendorf-Dresden, 1982.

92. Matthies, S., Vinel, G.W., Helming, K., *Standard Distributions in Texture Analysis: Maps for the Case of Cubic-Orthorhombic Symmetry*, Akademie Verlag, Berlin, 1987.

93. Matthies, S., "Reproducibility of the Orientation Distribution Function of Texture Samples from Pole Figures (Ghost Phenomena)," *Physica Status Solidi B-Basic Research*, 92(2): K135–K138, 1979.

94. Matthies, S., "On the Reproducibility of the Orientation Distribution Function of Texture Samples from Pole Figures (VII): Reliability of Reproduced ODF's; Standard Functions," *Kristall und Technik*, 16(9): 1061–1071, 1981.

95. Matthies, S., Helming, K., Kunze, K., "On the Representation of Orientation Distributions by σ-Sections -I. General Properties of σ-Sections; II. Consideration of Crystal and Sample Symmetry, Examples," *Physica Status Solidi B*, 157: 71–83, 489–507, 1990.

96. Matthies, S., Humbert, M., "The Realization of the Concept of the Geometric Mean for Calculating Physical Constants of Polycrystalline Materials," *Physica Status Solidi B*, 177(2): K47–K50, 1993.

97. Matthies, S., Humbert, M., "On the Principle of a Geometric Mean of Even-Rank Symmetrical Tensors for Textured Polycrystals," *Journal of Applied Crystallography*, 28(3): 254–266, June 1995.

98. Morawiec, A., "Automatic Orientation Determination from Kikuchi Patterns," *Journal of Applied Crystallography*, 32(4): 788–798, August 1999.

99. Morawiec, A., "Calculation of Polycrystal Elastic Constants from Single Crystal Data," *Physica Status Solidi B*, 154(2): 535–541, 1989.

100. Morawiec, A., "Review of Deterministic Methods of Calculation of Polycrystal Elastic Constants," *Textures and Microstructures*, 22: 139-167, 1994.

101. Morawiec, A., *Orientations and Rotations: Computations in Crystallographic Textures*, Springer, 2004 (softcover ed., 2013).

102. Muller, J., Esling, C., Bunge, H.J., "An Inversion Formula Expressing the Texture Function in Terms of Angular Distribution Functions," *Journal of Physics*, 42(2): 161–165, 1981.

103. Nikolayev, D.I., Savyolova, T.I., Feldman, K., "Approximation of the Orientation Distribution of Grains in Polycrystalline Samples by Means of Gaussians," *Textures and Microstructures*, 19: 9–27, 1992.

104. Ngollé, A., Péra, J., "Microstructural Based Modelling of the Elastic Modulus of Fiber Reinforced Cement Composites," *Advanced Cement Based Materials*, 6(3): 130–137, Oct./Nov. 1997.

105. Noll, W., "A Mathematical Theory of the Mechanical Behavior of Continuous Media," *Archive for Rational Mechanics and Analysis*, 2(1): 197–226, Jan. 1958.

106. Nôno T., "On the Symmetry Groups of Simple Materials: An Application of the Theory of Lie Groups," *Journal of Mathematical Analysis and Applications*, 24(1): 110–35, Oct. 1968.

107. Owens, W.H., "Strain Modification of Angular Density Distributions," *Tectonophysics*, 16(3): 249–261, 1973.

108. Park, N.J., Bunge, H.J., Kiewel, H., Fritsche, L., "Calculation of Effective Elastic Moduli of Textured Materials," *Textures and Microstructures*, 23: 43–59, 1995.

109. Phillips, R., *Crystals, Defects and Microstructures: Modeling Across Scales*, Cambridge University Press, 2001.

110. Pospiech, J., Lücke, K., Sztwiertnia, J., "Orientation Distributions and Orientation Correlation Functions for Description of Microstructures," *Acta Metallurgica et Materialia*, 41(1): 305–321, 1993.

111. Pospiech, J., Sztwiertnia, J., Haessner, F., "The Misorientation Distribution Function," *Textures and Microstructures*, 6: 201–215, 1986.

112. Pursey, H., Cox, H.L., "The Correction of Elasticity Measurements on Slightly Anisotropic Materials," *Philosophical Magazine*, 45(362): 295–302, 1954.
113. Rahnama, M., Koch, D.L., "The Effect of Hydrodynamic Interactions on the Orientation Distribution in a Fiber Suspension Subject to Simple Shear Flow," *Physics of Fluids*, 7(3): 487–506, March 1995.
114. Ralambotiana, T., Blanc, R., Chaouche, M., "Viscosity Scaling in Suspensions of Non-Brownian Rodlike Particles," *Physics of Fluids*, 9(12): 3588–3594, Dec. 1997.
115. Rao, B.N., Tang, L., Altan, M.C., "Rheological Properties of Non-Brownian Spheroidal Particle Suspensions," *Journal of Rheology*, 38(5): 1335–1351, 1994.
116. Reuss, A., "Berechnung der Fliessgrenze von Mischkristallen auf Grund der Plastizitätsbedingung für Einkristalle," *Zeitschrift für Angewandte Mathematik und Mechanik*, 9: 49–58, 1929.
117. Roe, R.-J., "Description of Crystallite Orientation in Polycrystalline Materials III, General Solution to Pole Figure Inversion," *Journal of Applied Physics*, 36(6): 2024–2031, 1965.
118. Sander, B., *An Introduction to the Study of Fabrics of Geological Bodies*, Pergamon, New York, 1970.
119. Sargent, C.M., "Texture Transformation," *Scripta Metallica*, 8(7): 821–824, 1974.
120. Saxl, I., *Stereology of Objects with Internal Structure*, Materials Science Monographs 50, Elsevier, Amsterdam, 1989.
121. Schaeben, H., "Parameterizations and Probability Distributions of Orientations," *Textures and Microstructures*, 13: 51–54, 1990.
122. Schaeben, H., "A Note on a Generalized Standard Orientation Distribution in PDF-Component Fit Methods," *Textures and Microstructures*, 23: 1–5, 1995.
123. Schwarzer, R.A., "Automated Crystal Lattice Orientation Mapping Using a Computer-Controlled SEM," *Micron*, 28(3): 249–265, June 1997.
124. Shaqfeh, E.S.G., Fredrickson, G.H., "The Hydrodynamic Stress in a Suspension of Rods," *Physics of Fluids*, 2(1): 7–24, Jan. 1990.
125. Shaqfeh, E.S.G., Koch, D.L., "Orientational Dispersion of Fibers in Extensional Flows," *Physics of Fluids*, 2(7): 1077–1093, July 1990.
126. Smallman, R.E., *Physical Metallurgy*, Pergamon, Oxford, 1970.
127. Smith, G.F., *Constitutive Equations for Anisotropic and Isotropic Materials*, North-Holland, New York, 1994.
128. Starkey, J., "A Crystallographic Approach to the Calculation of Orientation Diagrams," *Canadian Journal of Earth Sciences*, 20(6): 932–952, 1983.
129. Suquet, P., Garajeu, M., "Effective Properties of Porous Ideally Plastic or Viscoplastic Materials Containing Rigid Particles," *Journal of Mechanics and Physics of Solids*, 45(6): 873–902, 1997.
130. Suwas, S., Ray, R.K., *Crystallographic Texture of Materials*, Springer-Verlag, London, 2014.
131. Szeri, A.J., Lin, J.D. "A Deformation Tensor Model of Brownian Suspensions of Orientable Particles – The Nonlinear Dynamics of Closure Models," *Journal of Non-Newtonian Fluid Mechanics*, 64(1): 43–69, 1996.
132. Szeri, A.J., Milliken, W.J., Leal, L.G., "Rigid Particles Suspended in Time-Dependent Flows: Irregular Versus Regular Motion, Disorder Versus Order," *Journal of Fluid Mechanics*, 237: 33–56, 1992.
133. Tavard, C., Royer, F., "Indices de Texture partiels des Solides Polycristallines et Interprétation de l'Approximation de Williams," *Comptes rendus de l'Académie des Sciences Paris*, 284: 247, 1977.
134. Theocaris, P.S., Stavroulakis, G.E., Panagiotopoulos, P.D., "Calculation of Effective Transverse Elastic Moduli of Fiber-Reinforced Composites by Numerical Homogenization," *Composites Science and Technology*, 57: 573–586, 1997.
135. Theocaris, P. S., Stavroulakis G. E., "The Homogenization Method for the Study of Variation of Poisson's Ratio in Fiber Composites," *Archive of Applied Mechanics*, 68: 281–295, 1998.

136. Tóth, L.S., Van Houtte, P., "Discretization Techniques for Orientation Distribution Functions," *Textures and Microstructures*, 19: 229–244, 1992.

137. Tresca, H., "Mémoire sur l'écoulement des corps solides soumis à des fortes pressions," *Comptes Rendus Hébdom. Acad. Sci. Paris*, 59: 754-758, 1864.

138. Truesdell, C., "A Theorem on the Isotropy Groups of a Hyperelastic Material," *Proceedings of the National Academy of Sciences of the United States of America*, 52(4): 1081–1083, 1964.

139. Turner, F.J., Weiss, L.E., *Structural Analysis of Metamorphic Tectonics*, McGraw-Hill, New York, 1963.

140. Underwood, E.E., *Quantitative Stereology*, Addison-Wesley, Reading, Mass., 1970.

141. Underwood, F.A., *Textures in Metal Sheets*, MacDonald, London, 1961.

142. Van Houtte, P., "The Use of a Quadratic Form for the Determination of Non-Negative Texture Functions," *Textures and Microstructure*, 6: 1–20, 1983.

143. Van Houtte, P., "A Method for the Generation of Various Ghost Correction Algorithms - The Example of the Positivity Method and the Exponential Method," *Textures and Microstructure*, 13: 199–212, 1991.

144. Viglin, A.S., "A Quantitative Measure of the Texture of a Polycrystalline Material - Texture Function," *Fiz. Tverd. Tela*, 2: 2463–2476, 1960.

145. Voigt, W., *Lehrbuch der Kristallphysik*, Teubner, Leipzig, 1928.

146. Wang, C.-C., "On the Geometric Structures of Simple Bodies, A Mathematical Foundation for the Theory of Continuous Distributions of Dislocations," *Archive for Rational Mechanics and Analysis*, 27(1): 33–94, 1967.

147. Wassermann, G., Grewen, J., *Texturen Metallischer Werkstoffe*, Springer, Berlin, 1962.

148. Watt, J.P., Davies, G.F., O'Connell, R.J., " Elastic Properties of Composite Materials," *Reviews of Geophysics*, 14(4): 541–563, 1976.

149. Weibull, W.A., *A Statistical Theory of the Strength of Materials*, Proceedings of the Royal Swedish Institite for Engineering Research, vol. 151, 1939.

150. Weiland, H., "Microtexture Determination and Its Application to Materials Science," *JOM-Journal of the Minerals Metals and Materials Society*, 46(9): 37–41, Sept. 1994.

151. Weissenberg, K., "Statistische Anisotropie in kristallinen Medien und ihre röntgenographische Bestimmung," *Annalen der Physik* 69: 409–435, 1922.

152. Wenk, H.-R., ed., *Preferred Orientation in Deformed Metals and Rocks: An Introduction to Modern Texture Analysis*, Academic Press, New York, 1985.

153. Wenk, H.R., Matthies, S., Donovan, J., Chateigner, D., "BEARTEX: a Windows-Based Program System for Quantitative Texture Analysis," *Journal of Applied Crystallography*, 31(2): 262–269, April 1998.

154. Wright, S.I., Zhao, J.W., Adams, B.L., "Automated-Determination of Lattice Orientation from Electron Backscattered Kikuchi Diffraction Patterns," *Textures and Microstructures*, 13(2–3): 123–131, 1991.

155. Wright, S.I., "Estimation of Single-Crystal Elastic-Constants from Textured Polycrystal Measurements," *Journal of Applied Crystallography*, 27(5): 794–801, Oct. 1994.

156. Yamane, Y., Kaneda, Y., Dio, M. "Numerical-Simulation of Semidilute Suspensions of Rodlike Particles in Shear-Flow," *Journal of Non-Newtonian Fluid Mechanics*, 54: 405–421, 1994.

157. Yashnikov, V.P., Bunge, H.-J., "Group-Theoretical Approach to Reduced Orientation Spaces for Crystallographic Textures," *Textures and Microstructures*, 23: 201–219, 1995.

158. Zheng, Q.S., "Theory of Representations for Tensor Functions - A Unified Invariant Approach to Constitutive Equations," *Applied Mechanics Reviews*, 47(11): 545–587, 1994.

159. Zhuo, L., Watanabe, T., Esling, C., "A Theoretical Approach to Grain-Boundary-Character-Distribution (GBCD) in Textured Polycrystalline Materials," *Zeitscrift für Metallkunde*, 85(8): 554–558, Aug. 1994.

Protein Kinematics

This chapter covers applications of harmonic analysis related to the three-dimensional structure and motions of protein molecules and the complexes that they form. The topics which collectively constitute "protein kinematics" include: (1) protein-protein docking; (2) the molecular theory of solvation; (3) structural bioinformatics; (4) experimental structure determination (including x-ray crystallography and cryo-electron microscopy); and (5) modeling mobility of loops. Each of these topics could individually be a whole book if treated in detail. We only touch on those aspects that relate to new parameterizations and sampling methods for rigid-body motion, convolution, covariance propagation, and harmonic analysis.

We begin with mathematical models of the interactions between pairs of protein molecules. Following this we illustrate how methods of Lie theory and harmonic analysis can be used to interpret statistics of geometric patterns observed in large databases of protein structures. Then we discuss mathematical models of the experimental methods used to obtain information about biomolecular structure, including protein crystallography and cryo-electron microscopy. Finally, we model "loop entropy" in biomolecular chains with end constraints. The methods used in all of these areas are similar in many ways with those used in Chapters 5, 6, 12, 13, and 14. In fact, several of the problems discussed in earlier chapters were first explored while studying biomolecular structure and function. For example, generating robot C-space obstacles by convolution (Chapter 12) is essentially the same as exploring candidate positions and orientations in protein-protein docking, as studied in [79]. The computation of FFTs on the rotation group $SO(3)$ using Euler angles (Chapter 9) was studied in the 1970s in the context of x-ray crystallography of proteins [47] (though it was fast in α and γ, but not in β). Also, the Procrustes problem discussed in Chapter 14 was addressed in the context of aligning protein structures starting in the 1970s [9, 42, 76, 77, 88].

The remainder of this chapter is organized as follows. Section 19.1 discusses the protein-protein docking problem and rigid-body parameterizations that are natural in this context. Section 19.2 examines the molecular theory of liquids, such as dilute protein solutions. Section 19.3 examines the relationship between potential energies of interacting substructures within individual molecules and the frequency of occurrence of such interactions observed in large databases. This is a version of the "structural bioinformatics" problem. Sections 19.4 focuses on mathematical problems that arise in x-ray crystallography of proteins. Closely related to this is the issue of how to deterministically sample rotations in "almost uniform" ways for efficient global optimization. This problem, which is also relevant to other topics in this chapter (and in other fields, such as robotics) is examined in Section 19.5. Section 19.6 shows how the class averaging problem in cryo-electron microscopy can be formulated as an $SE(2)$ convolution. And finally, Section 19.7 models the distribution of conformational states of end-constrained loop regions in biomolecular structures and explains how this problem can be formulated as an $SE(3)$ convolution.

19.1 The Mathematics of Protein-Protein Docking

This section discusses new parameterizations of rotations and full rigid-body motions in three-dimensional space that are specifically well suited to the protein-protein docking problem, and it follows [29]. These new parameterizations have the special property that their form is the same when the moving frame is viewed from the world frame, and vice versa, and the values of the parameters in the forward and reverse transformations have a simple algebraic relationship to each other. This type of parameterization may be particularly useful in biomolecular applications such as protein-protein interactions, and analogous macroscopic engineering problems, such as spacecraft docking. In these problems, unlike in mechanisms and robotics, there is no clear distinction between a ground link and a distal link. Therefore parameterizing motions such that the original motion and its inverse have the same form is valuable for ease in switching back and forth between different perspectives.

19.1.1 Review of the Literature on Protein-Protein Interactions

The protein-protein docking problem can be modeled in a very similar way to the C-space obstacle generation problem in robot motion planning (Chapter 12) and the template matching problem in image processing and pattern recognition (Chapter 13). The goal in protein-protein docking is to find rigid-body motions to optimize shape and chemical complementarily for two proteins to adhere to each other. By treating one protein as fixed and the other as moving, and defining functions that take a constant positive value on the interior of each body, an $SE(3)$ correlation of these functions can be used to describe the region in $SE(3)$ in which the bodies would have non-physical overlaps. The five-dimensional boundary of this region constitutes all possible candidates for rigid docking.

A number of works have sought to merge ideas similar to those in Chapters 10–13 with the seminal work in [79] for fast prediction of protein-protein docking [5, 6, 137, 138]. This is an interesting application of harmonic analysis, which uses the same basic idea as in [79], but casts the problem in the form of Euclidean-group correlations as in Chapters 12 and 13. Some of the recent works in this area appear to be unaware of many of the references in those chapters, despite the obvious overlap in methodologies. Often methods developed in one field get reinvented in another when researchers fail to explore outside of their narrow niches. By taking a wider view, protein-protein docking and molecular recognition can benefit from methods developed in other fields such as spacecraft docking, advances in computer graphics, and robotics, etc. Then problems that are essentially the same across these fields can benefit from the efforts of a larger pool of talent.

Recent work in this area breaks down roughly into categories the following according to use of the following methodologies: spherical-harmonics/spherical-Bessel-functions [59, 137, 138, 155]; Zernike-polynomials [168, 169]; fast Wigner D-function evaluation and rotation-invariant features for fast screening [32, 82, 83, 90]; ultra-fast Fourier methods [139]; motion-group Fourier methods [5, 6, 91] (which are very similar to those expressed in the original version of our book and the references therein); and non-Fourier approaches [62, 108, 147].

Since the motion-group Fourier formulation of the problem is a correlation that is indistinguishable from the robotics and imaging applications discussed in earlier chapters, the main topic in our discussion of protein-protein docking will be a special kind of parameterization of rotations and rigid-body motions that is "symmetrical" in the sense that its form is very similar from the perspective of both molecules.

Stated in a technical way, the main goal of this section is to introduce parameterizations of 3×3 rotation matrices, $R(\mathbf{q})$, and 4×4 homogeneous transformations, $H(\mathbf{q}')$, such that: (1) The singularities of these parameterizations are outside of the range of parameter values that are normally encountered in applications; and (2) It is possible to write $[R(\mathbf{q}_1)]^T = R(\mathbf{q}_2)$ and $[H(\mathbf{q}'_1)]^{-1} = H(\mathbf{q}'_2)$ where the relationship between $\mathbf{q}_1$ and $\mathbf{q}_2$ and between $\mathbf{q}'_1$ and $\mathbf{q}'_2$ involve simple symmetry operations on their arguments that do not involve any transcendental calculations.

19.1.2 New Symmetrical Rotation Parameterizations for Protein-Protein Interactions

We begin with the case of pure rotations and review some properties of the rotations $R(\mathbf{a}, \mathbf{b})$ (introduced in Chapter 5) that transfer the vector $\mathbf{a}$ to the vector $\mathbf{b}$.

Rotating Unit Vector Into Another

Let $R(\mathbf{a}, \mathbf{b})$ denote the rotation along the fixed axis defined by the direction $\mathbf{a} \times \mathbf{b}$ such that $R(\mathbf{a}, \mathbf{b})\mathbf{a} = \mathbf{b}$. The form of this matrix was given explicitly in (5.27) as

$$R(\mathbf{a}, \mathbf{b}) \doteq \exp\left(\theta_{ab} \cdot \widehat{\frac{\mathbf{a} \times \mathbf{b}}{\|\mathbf{a} \times \mathbf{b}\|}}\right) \tag{19.1}$$

$$= \mathbb{I} + \widehat{\mathbf{a} \times \mathbf{b}} + \frac{(1 - \mathbf{a} \cdot \mathbf{b})}{\|\mathbf{a} \times \mathbf{b}\|^2}\left(\widehat{\mathbf{a} \times \mathbf{b}}\right)^2 \tag{19.2}$$

where θ_{ab} is the unique angle in the range $[0, \pi]$ such that

$$\sin\theta_{ab} = \|\mathbf{a} \times \mathbf{b}\| \quad \text{and} \quad \cos\theta_{ab} = \mathbf{a} \cdot \mathbf{b}.$$

The above expression defines a finite rotation of $\mathbf{a}$ to $\mathbf{b}$. It is by no means the only rotation matrix that performs this operation. For example pre- and post- multiplying by rotations around axes parallel to $\mathbf{b}$ and $\mathbf{a}$, respectively, will have the same net effect. But among all rotations that convert $\mathbf{a}$ to $\mathbf{b}$, $R(\mathbf{a}, \mathbf{b})$ is special in that it is the 'most direct'. For example, this can be seen when θ_{ab} is small. In this case if the angle of rotation of the product $\exp(\epsilon\hat{\mathbf{b}}/\|\mathbf{b}\|) R(\mathbf{a}, \mathbf{b})$ is calculated, then the Baker-Campbell-Hausdorf formula gives[1]

$$\log^\vee(\exp(\epsilon\hat{\mathbf{b}}/\|\mathbf{b}\|) R(\mathbf{a}, \mathbf{b})) \approx \theta_{ab} \cdot \frac{\mathbf{a} \times \mathbf{b}}{\|\mathbf{a} \times \mathbf{b}\|} + \epsilon\frac{\mathbf{b}}{\|\mathbf{b}\|} + O(\theta_{ab} \cdot \epsilon).$$

---

[1] Here $\log^\vee(\cdot)$ is shorthand for first taking the log, and then converting the resulting skew-symmetric matrix to the corresponding dual vector.

Because $\mathbf{a} \times \mathbf{b}$ and $\mathbf{b}$ are orthogonal, taking the norm we get

$$\theta_{new} = \| \log(\exp(\epsilon \hat{\mathbf{b}}/\|\mathbf{b}\|) R(\mathbf{a}, \mathbf{b})) \| \approx \sqrt{\theta_{ab}^2 + \epsilon^2} + O(\theta_{ab} \cdot \epsilon).$$

The same sort of thing would happen if we had post-multiplied by a rotation around $\mathbf{a}$. And so, it is not difficult to see that perturbations of this rotation that perform the the same transition from $\mathbf{b}$ to $\mathbf{a}$ will have a larger rotation angle, and hence describe a less direct transformation.

From (19.2) it is easy to see that

$$[R(\mathbf{a}, \mathbf{b})]^{-1} = R(\mathbf{b}, \mathbf{a}). \tag{19.3}$$

Moreover, from (19.2)

$$R(-\mathbf{a}, -\mathbf{b}) = R(\mathbf{a}, \mathbf{b})$$

and

$$R(\mathbf{b}, \mathbf{a}) = R(-\mathbf{a}, \mathbf{b}) = R(\mathbf{a}, -\mathbf{b}). \tag{19.4}$$

Also, since $\text{rot}(\mathbf{a}, \alpha)\mathbf{a} = \mathbf{a}$, for any rotation angle α, it follows that

$$\text{rot}(\mathbf{b}, -\beta)R(\mathbf{a}, \mathbf{b})\text{rot}(\mathbf{a}, \alpha)\mathbf{a} = \mathbf{b} \tag{19.5}$$

for any angles α and β.

One thing to observe is that even though

$$R(\mathbf{b}, \mathbf{c})R(\mathbf{a}, \mathbf{b})\mathbf{a} = R(\mathbf{a}, \mathbf{c})\mathbf{a} = \mathbf{c},$$

generally

$$R(\mathbf{b}, \mathbf{c})R(\mathbf{a}, \mathbf{b}) \neq R(\mathbf{a}, \mathbf{c}).$$

This is easy to see with a simple example. Suppose $\mathbf{a} = \mathbf{e}_1$, $\mathbf{b} = \mathbf{e}_2$, $\mathbf{c} = \mathbf{e}_1$, Then

$$R(\mathbf{e}_2, \mathbf{e}_3)R(\mathbf{e}_1, \mathbf{e}_2) = R_1(\pi/2)R_3(\pi/2)$$

but

$$R(\mathbf{e}_1, \mathbf{e}_3) = R_2(-\pi/2).$$

These matrices, while having the same effect of transforming $\mathbf{e}_1$ to $\mathbf{e}_3$, are not the same. Note that since for any $Q \in SO(3)$

$$Q \left(\widehat{\mathbf{a} \times \mathbf{b}} \right) Q^T = Q\widehat{[\mathbf{a} \times \mathbf{b}]} = \widehat{(Q\mathbf{a}) \times (Q\mathbf{b})},$$

and from this it follows that

$$R(Q\mathbf{a}, Q\mathbf{b}) = QR(\mathbf{a}, \mathbf{b})Q^T. \tag{19.6}$$

A Symmetrical Parameterization for Pure Rotations

Here the following "symmetrical" parameterization for rotations is defined as

$$\boxed{R_{12}(\mathbf{u}_1, \mathbf{u}_2, \nu) \doteq \text{rot}(\mathbf{u}_1, -\nu/2)R(\mathbf{u}_1, \mathbf{u}_2)\text{rot}(\mathbf{u}_2, \nu/2).} \tag{19.7}$$

This is one of the most critical equations in this section (which is why it is boxed).

While the above parameterization could have been defined without the negative sign, including it makes the expression for the inverse rotation symmetric in the parameters:

$$R_{12}^{-1} = [\text{rot}(\mathbf{u}_2, \nu/2)]^{-1} [R(\mathbf{u}_1, \mathbf{u}_2)]^{-1} [\text{rot}(\mathbf{u}_1, -\nu/2)]^{-1}$$

$$= \text{rot}(\mathbf{u}_2, -\nu/2) R(\mathbf{u}_2, \mathbf{u}_1) \text{rot}(\mathbf{u}_1, \nu/2) = R_{21}. \qquad (19.8)$$

The above manipulations result in the very pleasant property that under inversion

$$(\mathbf{u}_1, \mathbf{u}_2, \nu) \rightarrow (\mathbf{u}_2, \mathbf{u}_1, \nu).$$

This parameterization shares some features of the Euler angles and some of the axis-angle parameterization. Namely, it is a product of exponentials formula (like Euler angles), but has a single rotation angle, ν, and a single unit vector like the axis-angle parameters. But this description is symmetrical, and the singularities can be placed as desired by fixing $\mathbf{u}_1$ appropriately. If $\mathbf{u}_1$ is not fixed, then $R_{12}(\mathbf{u}_1, \mathbf{u}_2, \nu)$ has five degrees of freedom (two for each unit vector), and is hence redundant. This redundancy will be important in Section 19.1.3 for defining symmetrical parameterizations of full rigid-body motion that will use the extra freedom built into $R_{12}(\mathbf{u}_1, \mathbf{u}_2, \nu)$.

Why Not Use Exponential Coordinates?

A natural question to ask is: "Why use the parameterization in (19.7) instead of exponential coordinates or Euler angles?" Our answer to this is that while exponential coordinates have the nice property that singularities are pushed far away from the identity, and hence are ideal for many applications (including the loop entropy problem discussed later in this chapter), the fact that they have singularities when bodies are rotated by π relative to each other is problematic in docking since this is a common rotation that occurs when proteins form dimers. Euler angles are very convenient for harmonic analysis because they are a double-coset decomposition, but they have all of the well-known problems of singularities and do not transform nicely under inversion of the rotation matrix. These undesirable properties of the exponential and Euler-angle parameterizations are inherited when going from pure rotation to the case of full rigid-body motions.

19.1.3 Symmetrical Parameterizations of Full Rigid-Body Motions

Building on the symmetrical parameterization of rotations given above, we now discuss the case of full rigid-body motions including both rotations and translations. We begin with a candidate based on screw-theory before settling on our final answer to the problem.

Screw-Screw-Screw Parameterizations

Any element of $SE(3)$ can be parameterized in a way that is analogous to Euler angles, as a product of three independent screw motions [98, 100]:

$$g(a, \alpha; b, \beta; c, \gamma) = \text{screw}(\mathbf{e}_3, a, \alpha)\, \text{screw}(\mathbf{e}_1, b, \beta)\, \text{screw}(\mathbf{e}_3, c, \gamma) \qquad (19.9)$$

where

$$\text{screw}(\mathbf{e}_3, a, \alpha) = \exp(\alpha \tilde{X}_3 + a \tilde{X}_6) = \begin{pmatrix} R_3(\alpha) & a\mathbf{e}_3 \\ \mathbf{0}^T & 1 \end{pmatrix}$$

and

$$\text{screw}(\mathbf{e}_1, a, \alpha) = \exp(\alpha \tilde{X}_1 + a \tilde{X}_4) = \begin{pmatrix} R_1(\alpha) & a\mathbf{e}_1 \\ \mathbf{0}^T & 1 \end{pmatrix}.$$

If only two such screws are concatenated, then the DH parameters result. The difference here is that three screws are used.

The inverse of the product of screw motions is the product of the inverses in reverse order, and the inverse of each screw motion is simply of the form

$$[\text{screw}(\mathbf{e}_i, a, \alpha)]^{-1} = \text{screw}(\mathbf{e}_i, -a, -\alpha).$$

This means that that, for example, when using 313 screws, the inverse has a pleasantly symmetric form and the parameters simply interchange as

$$(a, \alpha) \to (-c, -\gamma); (b, \beta) \to (-b, -\beta); (c, \gamma) \to (-a, -\alpha).$$

This is promising, but the questions of appropriate range of parameters and singularities arise. For example, if β is restricted to the same range as in the Euler angles, then $\beta \to -\beta$ is not a valid transformation. Moreover, if singularities are present in undesirable locations, this could be a problem. Therefore, singularities are analyzed now. After performing the matrix multiplications,

$$g(a, \alpha; b, \beta; c, \gamma) = \begin{pmatrix} R_3(\alpha) R_1(\beta) R_3(\gamma) & c \cdot R_3(\alpha) R_1(\beta) \mathbf{e}_3 + b \cdot R_3(\alpha) \mathbf{e}_1 + a \cdot \mathbf{e}_3 \\ \mathbf{0}^T & 1 \end{pmatrix}.$$

Observe that this is of the form

$$g(a, \alpha; b, \beta; c, \gamma) = \begin{pmatrix} R(\boldsymbol{\alpha}) & J_L(\boldsymbol{\alpha})\mathbf{a} \\ \mathbf{0}^T & 1 \end{pmatrix}$$

where $\boldsymbol{\alpha} = [\alpha, \beta, \gamma]^T$ and $\mathbf{a} = [a, b, c]^T$. $R(\boldsymbol{\alpha})$ is the ZXZ Euler angle parameterization and $J_L(\boldsymbol{\alpha})$ is the left Jacobian for $SO(3)$ in that parameterization given in Chapter 5.

Due to the structure of this matrix, the columns of the right Jacobian will be of the form

$$J_R \mathbf{e}_i = \begin{pmatrix} \left(R^T \frac{\partial R}{\partial \alpha_i} \right)^\vee \\ R^T \frac{\partial J_L}{\partial \alpha_i} \mathbf{a} \end{pmatrix} \quad \text{for} \quad i = 1, 2, 3$$

and

$$J_R \mathbf{e}_i = \begin{pmatrix} \mathbf{0} \\ R^T J_L \mathbf{e}_{i-3} \end{pmatrix} \quad \text{for} \quad i = 4, 5, 6.$$

This means that

$$\det J_R = \det J_R \cdot \det R^T J_L = (\det J_R)^2.$$

Since the Jacobian determinant is related in this way to that of the Euler angles, it suffers from the same singularities.

Symmetrical Rigid-Body Parameters For Molecular Docking Applications

We now combine the results presented previously in this section to obtain symmetrical parameterizations of full rigid-body motion with singularities located out of range of where they can cause trouble in molecular docking applications.

Here a symmetrical parameterization of 3D rigid-body motion is build from the symmetrical rotation rotational parameterization defined in (19.7). Only now, the freedom to choose $\mathbf{u}_1$ will be used up, and this unit vector will become part of the parameterization.

A symmetrical parameterization of $SE(3)$ results when using (19.7) and writing

$$H(g_{12}) = H(R_{12}, r\mathbf{u}_1) = \begin{pmatrix} R_{12} & r\mathbf{u}_1 \\ \mathbf{0}^T & 1 \end{pmatrix}.$$

Or more explicitly,

$$H(\theta_1, \phi_1, \theta_2, \phi_2, \nu, r) = \begin{pmatrix} R(\mathbf{u}(\theta_1, \phi_1), \mathbf{u}(\theta_2, \phi_2), \nu) & r \cdot \mathbf{u}(\theta_1, \phi_1) \\ \mathbf{0}^T & 1 \end{pmatrix}.$$

Consider two bodies, each with an attached reference frame. The position and orientation of the second body appears to be $g_{12} = (R_{12}, r\mathbf{u}_1)$ as seen in the reference frame of body 1. Likewise, the position and orientation of body 1 appears to be $g_{21} = (R_{21}, r\mathbf{u}_2)$ in the frame attached to body 2. It then follows that $g_{12} \circ g_{21} = g_{21} \circ g_{12} = e$ and that

$$g_{21} = g_{12}^{-1} = (R_{12}^T, -rR_{12}^T\mathbf{u}_1).$$

The vectors $\mathbf{u}_i$ lie on the same line in space, i.e., the line connecting the origins of the two reference frames, though they are defined in their respective reference frames and define opposite directions. From (19.3) and (19.8), it follows that

$$\begin{aligned} -rR_{12}^T\mathbf{u}_1 &= -r \cdot \text{rot}(\mathbf{u}_2, -\nu/2)R(\mathbf{u}_2, \mathbf{u}_1)\text{rot}(\mathbf{u}_1, \nu/2)\mathbf{u}_1 \\ &= -r \cdot \text{rot}(\mathbf{u}_2, -\nu/2)R(\mathbf{u}_1, -\mathbf{u}_2)\text{rot}(\mathbf{u}_1, \nu/2)\mathbf{u}_1 \\ &= -r \cdot \text{rot}(\mathbf{u}_2, -\nu/2)R(\mathbf{u}_1, -\mathbf{u}_2)\mathbf{u}_1 \\ &= -r \cdot \text{rot}(\mathbf{u}_2, -\nu/2)(-\mathbf{u}_2) \\ &= r \cdot \mathbf{u}_2. \end{aligned}$$

By choosing the unit vector $\mathbf{u}_1$ as the direction of the translational motion for body 1 (i.e., as the unit vector pointing to the origin of reference frame 2 as seen in reference frame 1), it is no longer an arbitrary fixed vector.

If $\mathbf{u}_i = \mathbf{u}(\theta_i, \phi_i)$ is a spherical-coordinate parameterization, then the six dimensions of rigid-body motion can be described with the parameters $\theta_1, \phi_1, \theta_2, \phi_2, r$ together with a rotation by an angle ν around the line shared by the vectors $\mathbf{u}_i$.

Following from the discussion of the pure rotation case, under inversion these parameters behave as:

$$\boxed{(\mathbf{u}_1, \mathbf{u}_2, r, \nu) \to (\mathbf{u}_2, \mathbf{u}_1, r, \nu),}$$

or, in terms of components,

$$(\theta_1, \phi_1, \theta_2, \phi_2, r, \nu) \to (\theta_2, \phi_2, \theta_1, \phi_1, r, \nu).$$

This is symmetrical, and in analogy with the planar case, has singularities that can be easily avoided by appropriately attaching the reference frames inside the molecules.

19.2 Molecular Theory of Liquids as $SE(3)$ Convolution

Here we consider a dilute solution confined to a region of space containing N copies of a large rigid molecule dissolved in a solvent of many more small molecules.[2] The behavior of each large molecule is expected to act independently of the others. The configurational Boltzmann distribution for all N copies would be the product of those for each individual one. However, as the number density per unit volume of these molecules becomes greater, and in the extreme case when a pure liquid consisting of these large molecules is considered, then the behavior of each copy is no longer independent.

Let $C \subset \mathbb{R}^3$ denote the region in which the N molecules are constrained to move, and let $D = C \times SO(3)$ be a finite-volume domain within $SE(3)$. Then the configurational Boltzmann distribution will be of the form

$$f(g_1, g_2, ..., g_N; \beta, D) = \frac{1}{Z_c(\beta, N, D)} e^{-\beta V(g_1, g_2, ..., g_N)}$$

where each $g_i \in D$ and

$$Z_c(\beta, N, D) = \int_{D^N \subset SE(3)^N} e^{-\beta V(g_1, g_2, ..., g_N)} \, dg_1 dg_2 \cdots dg_N.$$

Note that $SE(3)^N$ is shorthand for the N-fold product $SE(3) \times SE(3) \times \cdots \times SE(3)$ and similarly for D^N. The integration takes place over this finite-volume domain. Alternatively, the shape of the region could be absorbed into the definitions of $f(\cdot)$ by using window functions enforced by an appropriate potential, and then the integral can be extended over all of $SE(3)^N$.

19.2.1 Probability Density and Correlation Functions

Since each molecule can exist in each location in this $6N$-dimensional configuration space, it is common to define the *generic distribution function* [61, 66, 68]

$$f'(g_1, g_2, ..., g_N; \beta, D) \doteq N! \, f(g_1, g_2, ..., g_N; \beta),$$

which reflects that the position and orientation of any molecule can be swapped with any of the others.

If we are only interested in the generic distribution function for the first m of these molecules without regard to the others, then

$$f'(g_1, g_2, ..., g_m; \beta, D) = \frac{1}{(N-m)!} \int_{D^{N-m}} f'(g_1, g_2, ..., g_N; \beta, D) \, dg_{m+1} \, dg_{m+2} \cdots dg_N.$$

Integrating $f(g_1, g_2, ..., g_m; \beta, D)$ over all copies except for the ith gives $f(g_i; \beta, D)$, and $f'(g_i; \beta, D)$ is defined in an analogous way. A quantitative indicator of how independently the N copies of the rigid molecules behave is the value of the so-called (generic) *correlation function*

$$\gamma(g_1, g_2, ..., g_m; \beta, D) \doteq \frac{f'(g_1, g_2, ..., g_m; \beta, D)}{\prod_{i=1}^{m} f'(g_i; \beta, D)} = K(N, m) \frac{f(g_1, g_2, ..., g_m; \beta, D)}{\prod_{i=1}^{m} f(g_i; \beta, D)}.$$

Here for $m \ll N$,

---

[2]The material in this section is a truncated version of the longer discussion given in [27].

$$K(N, m) \doteq N^{-m} \frac{N!}{(N-m)!} \approx 1 - \frac{m(m-1)}{2N},$$

which approaches unity as $N \to \infty$ and m is held fixed. Note that the closer γ is to 1 for all values of the argument, the more independent the behaviors are.

If each molecule has unit mass, then the mass density per unit volume at a given temperature can be computed as [61, 66, 68][3]

$$\rho(\mathbf{r}; \beta, D) = \frac{N}{Z_c(\beta, N, D)} \int_{D^{N-1}} \int_{SO(3)} e^{-\beta V(g, g_2, g_3, \dots, g_N)} \, dR \, dg_2 \cdots dg_N$$

$$= \left\langle \sum_{i=1}^{N} \delta(g_i^{-1} \circ g) \right\rangle \tag{19.10}$$

where $g = (\mathbf{r}, R) \in SE(3)$, dR is the normalized Haar measure for $SO(3)$, and $\langle \cdot \rangle$ denotes the average over the Boltzmann-weighted ensemble.

Now consider the case of an isotropic homogeneous fluid where $f'(g_i; \beta, D) = \rho(\beta, D)$, i.e., the mass density per unit volume is constant at a particular temperature within the region containing the molecules. Then we can write the (pair) correlation function as [61, 66, 68]

$$\gamma(g, g'; \beta, D) = \frac{N(N-1)}{\rho^2(\beta) \cdot Z_c(\beta, N, D)} \int_{D^{N-2}} e^{-\beta V(g, g', g_3, \dots, g_N)} \, dg_3 \cdots dg_N$$

$$= \frac{1}{\rho^2(\beta)} \left\langle \sum_{i \neq j} \delta(g_i^{-1} \circ g) \, \delta(g_j^{-1} \circ g') \right\rangle. \tag{19.11}$$

19.2.2 Relationship to Convolutions on Groups

The *total correlation function* between two molecules is defined as

$$\tilde{h}(g_1, g_2; \beta, D) \doteq \gamma(g_1, g_2; \beta, D) - 1.$$

The *direct correlation function* between two molecules, $\tilde{c}(g_1, g_2; \beta, D)$, is defined to satisfy the equation

$$\tilde{h}(g_1, g_2; \beta, D) = \tilde{c}(g_1, g_2; \beta, D) + \rho \int_{D \subset SE(3)} \tilde{c}(g_1, g_3; \beta, D) \tilde{h}(g_3, g_2; \beta, D) \, dg_3.$$

Each g_i is an absolute motion relative to the lab frame, and all interactions between molecules are relative. We can then write the above integral as $\tilde{h}(g_1, g_2; \beta, D) = h(g_1^{-1} \circ g_2)$ and $\tilde{c}(g_1, g_2; \beta, D) = c(g_1^{-1} \circ g_2)$, were $h(\cdot)$ and $c(\cdot)$ depend only on relative pose, and the dependence on β, D has been absorbed into how these functions are defined. Moreover, if these functions decay rapidly, and if the molecules are not located at the boundary of the containing region, then integrals over D and $SE(3)$ are indistinguishable. Therefore,

$$h(g_1^{-1} \circ g_2) = c(g_1^{-1} \circ g_2) + \rho \int_{SE(3)} c(g_1^{-1} \circ g_3) h(g_3^{-1} \circ g_2) \, dg_3.$$

---

[3]In this field, orientational integrals are usually not normalized, and so an additional multiplicative factor of $\Omega \doteq 8\pi^2$ appears in classical works for molecules without any symmetry.

Letting $k = g_1^{-1} \circ g_3$, observing that $dk = dg_3$, and $g_3^{-1} \circ g_2 = k^{-1} \circ g_1^{-1} \circ g_2$ means that the above equation can be written using the concept of convolution on $SE(3)$ as

$$h(g_{12}) = c(g_{12}) + (c * h)(g_{12}) \quad \text{where} \quad g_{12} \doteq g_1^{-1} \circ g_2. \tag{19.12}$$

This form of the 6D version of the *Ornstein-Zernike* equation [123] and its extensions in the theories of *Percus-Yevick* [130, 131] and *Chandler-Andersen* [23] is usually written differently in the literature [3, 61, 66, 68]. The group-theoretic notation in (19.12) allows it to be more succinctly stated than the way it is in the literature. Furthermore, the $SE(3)$ Fourier transform can be used to solve for h in terms of c (or vice versa) because the convolution theorem gives

$$\hat{h}(p) = [\mathbb{I} + \hat{h}(p)]\hat{c}(p) \implies \hat{c}(p) = [\mathbb{I} + \hat{h}(p)]^{-1}\hat{h}(p) \tag{19.13}$$

or, equivalently,

$$\hat{c}(p) = \hat{h}(p)[\mathbb{I} - \hat{c}(p)] \implies \hat{h}(p) = \hat{c}(p)[\mathbb{I} - \hat{c}(p)]^{-1} \tag{19.14}$$

if the inverses exist.

This formulation by itself does not solve the problem, but it provides a constraint which, together with a "closure relation," characterizes the situation. Two popular closure relations are the Percus-Yevick and hyper-netted-chain approximations.

For other approaches to modeling molecular liquids and solvation effects see [13, 53, 56]

19.3 Structural Bioinformatics

This section reviews one kind of statistical problem concerned with geometric relationships between substructures in proteins, following the approach first presented in [99]. In particular, we are interested in relating the observed frequency of occurrence of pairs of interacting helices in globular proteins to the underlying potential energy that causes these interactions.

Proteins are comprised of substructures. These so-called *secondary structures* fall into broad categories including α-helices, β-sheets (of which there are two kinds), and loop regions. In this section we focus on the problem of how α-helices prefer to pack against each other within globular proteins (which are the water-soluble proteins for which the most information is available).

An implicit assumption in the study of helix-helix interactions is that α-helices are essentially rigid objects. Though this is not exactly true, we focus the present discussion on those helix pairs that in which the constituent helices are close to the ideal α-helical shape, and hence have not exhibited significant deformation.

Recall that the relative position and orientation between rigid bodies can be expressed with a pair $(A, \mathbf{a})$ where A is a 3×3 rotation matrix and $\mathbf{a}$ is a 3D translation vector, and the corresponding 4×4 homogeneous transformation matrices are of the form

$$H(A, \mathbf{a}) = \begin{pmatrix} A & \mathbf{a} \\ \mathbf{0}^T & 1 \end{pmatrix},$$

and matrix multiplication corresponds to the $SE(3)$ group operation:

$$H(A_1, \mathbf{a}_1)H(A_2, \mathbf{a}_2) = H((A_1, \mathbf{a}_1) \circ (A_2, \mathbf{a}_2)) = H(A_1 A_2, A_1 \mathbf{a}_2 + \mathbf{a}_1).$$

Two special kinds of homogeneous transformations are pure rotations and pure translations along the axes of local coordinate systems:

$$\mathrm{rot}(\mathbf{e}_i, \theta) \doteq \begin{pmatrix} R_i(\theta) & \mathbf{0} \\ \mathbf{0}^T & 1 \end{pmatrix}$$

and

$$\mathrm{trans}(\mathbf{e}_i, x) \doteq \begin{pmatrix} \mathbb{I} & x\mathbf{e}_i \\ \mathbf{0}^T & 1 \end{pmatrix}.$$

As we have seen earlier, the Lie group of rigid-body motions in three-dimensional space possesses a unique bi-invariant integration measure. That is, there is only one correct way to integrate over rigid-body motions. In particular, suppose we are given a function $f(A, \mathbf{a})$ describing the relative pose (position and orientation) of the frame of reference attached to body 2 relative to the frame of reference attached to body 1. If the parameters defining A and those defining $\mathbf{a}$ are independent, then there is only one correct way to integrate it as

$$I = \int_{A \in SO(3)} \int_{\mathbf{a} \in \mathbb{R}^3} f(A, \mathbf{a}) \, d\mathbf{a} \, dA.$$

Here $SO(3)$ is the group of rotations in three-dimensional space and dA is its bi-invariant integration measure. If $A = A(\alpha, \beta, \gamma)$ is the common ZXZ Euler-angle parameterization then, as discussed in Chapter 5,

$$dA = \frac{1}{8\pi^2} \sin \beta \, d\alpha \, d\beta \, d\gamma.$$

In contrast, if the position of the origin of frame 2 is described in Cartesian coordinates relative to frame 1, then

$$d\mathbf{a} = dx \, dy \, dz.$$

For the three cases shown in Figure 19.1, spatial rigid-body motions of helix 2 relative to helix 1 are parameterized in three different ways. It is not obvious a priori how the correct integration measure should be expressed in terms of the parameters describing each of those models. Determining this is essential in order to correctly account for the statistical biases inherent in the three data sets. For this reason, the general method for determining the volume element for integrating over rigid-body motions is derived here. The results for all three cases in Figure 19.1 are then given. The explicit calculations are contained in the following subsections.

For "small" rigid-body motions,

$$H \approx \mathbb{I} + \begin{pmatrix} \Omega & \mathbf{v} \\ \mathbf{0}^T & 0 \end{pmatrix} \Delta t \tag{19.15}$$

where $\Omega = -\Omega^T$. The angular velocity vector can be extracted from the skew-symmetric matrix Ω as $\Omega^\vee = \boldsymbol{\omega}$ to describe the rotational part of the displacement. Since the second term in (19.15) consists mostly of zeros, it is common to extract the information necessary to describe the motion as

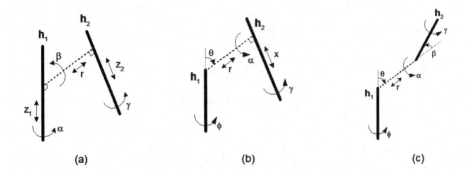

Fig. 19.1. Different Ways That Finite Helical Axes Can Interact. (Reprinted from Lee, S., Chirikjian, G.S., "Inter–Helical Angle and Distance Preferences in Globular Proteins," *Biophysical Journal*, 86(2): 1105–1117, 2004, with permission from Elsevier)

$$\begin{pmatrix} \Omega & \mathbf{v} \\ \mathbf{0}^T & 0 \end{pmatrix}^{\vee} = \begin{pmatrix} \boldsymbol{\omega} \\ \mathbf{v} \end{pmatrix}.$$

This six-dimensional vector is called an *infinitesimal* screw motion or *infinitesimal twist*.

Given a homogeneous transform consisting of motions that are not necessarily small,

$$H(\mathbf{q}) = \begin{pmatrix} R(\mathbf{q}) & \mathbf{b}(\mathbf{q}) \\ \mathbf{0}^T & 0 \end{pmatrix}$$

parametrized with coordinates $(q_1, ..., q_6)$, which we write as a vector $\mathbf{q} \in \mathbb{R}^6$, we can express the homogeneous transform corresponding to a slightly changed set of parameters as the truncated Taylor series

$$H(\mathbf{q} + \delta\mathbf{q}) = H(\mathbf{q}) + \sum_{i=1}^{6} \Delta q_i \frac{\partial H}{\partial q_i}(\mathbf{q}).$$

This result can be shifted to the identity transformation by multiplying on the left by H^{-1} to define an equivalent relative infinitesimal motion. In this case we write

$$\begin{pmatrix} \boldsymbol{\omega}_R \\ \mathbf{v}_R \end{pmatrix} = \mathcal{J}_R(\mathbf{q})\dot{\mathbf{q}} \quad \text{where} \quad \mathcal{J}_R(\mathbf{q}) = \left[\left(H^{-1}\frac{\partial H}{\partial q_1} \right)^{\vee}, \cdots, \left(H^{-1}\frac{\partial H}{\partial q_6} \right)^{\vee} \right]. \quad (19.16)$$

Here

$$\mathbf{v}_R = A^T \dot{\mathbf{a}}.$$

The volume element for integrating over rigid-body motions in the coordinates $q_1, ..., q_6$ is:

$$dH = |\det \mathcal{J}_R| dq_1 \cdots dq_6.$$

In cases where the two rigid bodies have symmetries, and hence the function $f(A, \mathbf{a})$ is constant over certain coordinates, it makes sense to use a parameterization which captures this fact and then integrate out all such coordinates. In this way marginal probability densities on a space of reduced-dimension can be examined. Below, the form of the volume elements is given, and the proper reductions are performed for the three cases shown in Figure 19.1.

19.3.1 Case 1: Line-to-Line Interaction

In Figure 19.1(a), the series of rigid-body motions that result in the frame attached at the base of helix 1 being moved to the base of helix 2 parameterize the homogeneous transformation

$$H(\alpha, z_1, \beta, r, z_2, \gamma) = \text{rot}(\mathbf{e}_3, \alpha)\text{trans}(\mathbf{e}_3, z_1)\text{rot}(\mathbf{e}_1, \beta)\text{trans}(\mathbf{e}_1, r)\text{rot}(\mathbf{e}_3, \gamma)\text{trans}(\mathbf{e}_3, z_2).$$
$$(19.17)$$

Substitution into (19.16), and following the detailed calculations at the end of this subsection results, to within an arbitrary multiplicative constant, in

$$\boxed{|\det \mathcal{J}_R| = \sin^2 \beta.}$$
$$(19.18)$$

Integrating z_1 over the range $[-L_2 \sin \beta, L_1 - L_2 \sin \beta]$ and z_2 over the range $[-L_2, 0]$ (which are the allowable range of variables that defines the line-on-line contact model), and integrating over the angles α and γ, one obtains a volume element for inter-helical angle and distance which, to within an arbitrary multiplicative constant, is of the form:

$$dV(\beta, r) = L_1 L_2 \sin^2 \beta \, \beta dr.$$

The detailed calculations resulting in (19.18) are as follows. Performing the multiplications in Equation (19.17),

$$H = \begin{pmatrix} R_3(\alpha)R_1(\beta)R_3(\gamma) & z_2 \cdot R_3(\alpha)R_1(\beta)\mathbf{e}_3 + r \cdot R_3(\alpha)\mathbf{e}_1 + z_1\mathbf{e}_3 \\ \mathbf{0}^T & 1 \end{pmatrix}.$$
$$(19.19)$$

Unlike the other two cases, three variables which appear in the translation part of the homogeneous transformation matrix do not appear in the rotation part. This gives a block structure to the Jacobian matrix, and makes the determinant easy to compute.

In particular, if we group the variables as $\mathbf{q}_1 = (\alpha, \beta, \gamma)$ and $\mathbf{q}_2 = (r, z_1, z_2)$, then the Jacobian will have the form

$$\mathcal{J}_R = \begin{pmatrix} J_R & \mathbb{O}_{3 \times 3} \\ A^T \frac{\partial \mathbf{a}}{\partial \mathbf{q}_1} & A^T \frac{\partial \mathbf{a}}{\partial \mathbf{q}_2} \end{pmatrix}.$$

Here the matrix J_R is

$$J_R(A) = \left[\left(A^T \frac{\partial A}{\partial \alpha}\right)^\vee, \left(A^T \frac{\partial A}{\partial \beta}\right)^\vee, \left(A^T \frac{\partial A}{\partial \gamma}\right)^\vee \right] = \begin{pmatrix} \sin \beta \sin \gamma & \cos \gamma & 0 \\ \sin \beta \cos \gamma & -\sin \gamma & 0 \\ \cos \beta & 0 & 1 \end{pmatrix}.$$

Due to the block lower diagonal form of this matrix, and the fact that A is a rotation matrix and therefore $\det A = +1$, it is clear that

$$|\det \mathcal{J}_R| = |\det J_R| \cdot \left| \det \frac{\partial \mathbf{a}}{\partial \mathbf{q}_2} \right|.$$

From (19.19),

$$\frac{\partial \mathbf{a}}{\partial z_1} = \mathbf{e}_3; \quad \frac{\partial \mathbf{a}}{\partial r} = R_3(\alpha)\mathbf{e}_1; \quad \frac{\partial \mathbf{a}}{\partial z_2} = R_3(\alpha)R_1(\beta)\mathbf{e}_3.$$

Therefore, a small computation shows that $\frac{\partial \mathbf{a}}{\partial \mathbf{q}_2} = J_R$. Since $|\det J_R| = \sin \beta$, it follows that

$$|\det \mathcal{J}_R| = \sin^2 \beta.$$

19.3.2 Case 2: End-to-Line Interaction

Observing Figure 19.1(b), it is clear that the rigid-body motion taking frame 1 into frame 2 is of the form:

$$H(\phi, \theta, r, \alpha, x, \gamma) = \text{rot}(\mathbf{e}_3, \phi)\text{rot}(\mathbf{e}_1, \theta)\text{trans}(\mathbf{e}_3, r)\text{rot}(\mathbf{e}_3, \alpha)\text{trans}(\mathbf{e}_1, x)\text{rot}(\mathbf{e}_1, \gamma).$$
(19.20)

Following the calculations at the end of this section,

$$\boxed{|\det \mathcal{J}_R| = r\sin\theta.}$$
(19.21)

In this case it clearly makes sense to integrate over ϕ, γ and x. The resulting volume element depends on the remaining parameters as

$$dV(\theta, r, \alpha) = r\sin\theta \, dr d\theta d\alpha.$$

The detailed calculations resulting in (19.21) are as follows. Performing the multiplications in Equation (19.20),[4]

$$\left(H^{-1}\frac{\partial H}{\partial \phi}\right)^{\vee} = \begin{pmatrix} R_1^T(\gamma)R_3^T(\alpha)R_1^T(\theta)\mathbf{e}_3 \\ rR_1^T(\gamma)R_3^T(\alpha)R_1^T(\theta)E_3R_1(\theta)\mathbf{e}_3 + xR_1^T(\gamma)R_3^T(\alpha)R_1^T(\theta)E_3R_1(\theta)R_3(\alpha)\mathbf{e}_1 \end{pmatrix}$$

$$= \begin{pmatrix} \sin\theta\sin\alpha \\ \sin\theta\cos\alpha\cos\gamma + \cos\theta\sin\gamma \\ \sin\theta\cos\alpha\sin\gamma - \cos\theta\cos\gamma \\ r\sin\theta\cos\alpha \\ -\sin\theta(r\sin\alpha\cos\gamma + x\cos\alpha\sin\gamma) + x\cos\theta\cos\gamma \\ \sin\theta(r\sin\alpha\sin\gamma - x\cos\alpha\cos\gamma) - x\cos\theta\sin\gamma \end{pmatrix}$$

$$\left(H^{-1}\frac{\partial H}{\partial \theta}\right)^{\vee} = \begin{pmatrix} R_1^T(\gamma)R_3^T(\alpha)\mathbf{e}_1 \\ -rR_1^T(\gamma)R_3^T(\alpha)\mathbf{e}_2 + xR_1^T(\gamma)R_3^T(\alpha)E_1R_3(\alpha)\mathbf{e}_1 \end{pmatrix}$$

$$= \begin{pmatrix} \cos\alpha \\ \sin\alpha\cos\gamma \\ \sin\alpha\sin\gamma \\ -r\sin\alpha \\ -r\cos\alpha\cos\gamma + x\sin\alpha\sin\gamma \\ r\cos\alpha\sin\gamma + x\sin\alpha\cos\gamma \end{pmatrix}$$

$$\left(H^{-1}\frac{\partial H}{\partial r}\right)^{\vee} = \begin{pmatrix} \mathbf{0} \\ R_1^T(\gamma)\mathbf{e}_3 \end{pmatrix} = \begin{pmatrix} 0 \\ 0 \\ 0 \\ 0 \\ \sin\gamma \\ \cos\gamma \end{pmatrix}$$

---

[4] In this section, E_i denotes the 3×3 skew-symmetric matrix such that $E_i \mathbf{x} = \mathbf{e}_i \times \mathbf{x}$.

$$\left(H^{-1}\frac{\partial H}{\partial \alpha}\right)^{\vee} = \begin{pmatrix} R_1^T(\gamma)\mathbf{e}_3 \\ xR_1^T(\gamma)\mathbf{e}_2 \end{pmatrix} = \begin{pmatrix} 0 \\ \sin\gamma \\ \cos\gamma \\ 0 \\ x\cos\gamma \\ -x\sin\gamma \end{pmatrix}$$

$$\left(H^{-1}\frac{\partial H}{\partial x}\right)^{\vee} = \begin{pmatrix} \mathbf{0} \\ \mathbf{e}_1 \end{pmatrix} = \begin{pmatrix} 0 \\ 0 \\ 0 \\ 1 \\ 0 \\ 0 \end{pmatrix}$$

$$\left(H^{-1}\frac{\partial H}{\partial \gamma}\right)^{\vee} = \begin{pmatrix} \mathbf{e}_1 \\ \mathbf{0} \end{pmatrix} = \begin{pmatrix} 1 \\ 0 \\ 0 \\ 0 \\ 0 \\ 0 \end{pmatrix}$$

Taking the determinant results in (19.21).

19.3.3 Case 3: End-to-End Interaction

Looking at Figure 19.1(c), the sequence of concatenated rigid-body motions that takes frame 1 to frame 2 is:

$$H(\phi,\theta,r,\alpha,\beta,\gamma) = \mathrm{rot}(\mathbf{e}_3,\phi)\mathrm{rot}(\mathbf{e}_1,\theta)\mathrm{trans}(\mathbf{e}_3,r)\mathrm{rot}(\mathbf{e}_3,\theta)\mathrm{rot}(\mathbf{e}_1,\beta)\mathrm{rot}(\mathbf{e}_3,\gamma).$$
(19.22)

Following the calculations at the end of this subsection,

$$\boxed{|\det \mathcal{J}_R| = r^2 \sin\beta \sin\theta.}$$
(19.23)

In this case it clearly makes sense to integrate over ϕ and γ. The resulting volume element depends on the remaining parameters as:

$$dV(r,\theta,\alpha,\beta) = r^2 \sin\beta \sin\theta dr d\theta d\beta d\alpha$$

The detailed calculations are as follows. Performing the multiplications in Equation (19.22),

$$H = \begin{pmatrix} R_3(\phi)R_1(\theta)R_3(\alpha)R_1(\beta)R_3(\gamma) & r\cdot R_3(\phi)R_1(\theta)\mathbf{e}_3 \\ \mathbf{0}^T & 1 \end{pmatrix},$$

$$H^{-1} = \begin{pmatrix} R_3^T(\gamma)R_1^T(\beta)R_3^T(\alpha)R_1^T(\theta)R_3^T(\phi) & -r\cdot R_3^T(\gamma)R_1^T(\beta)\mathbf{e}_3 \\ \mathbf{0}^T & 1 \end{pmatrix},$$

$$
\left(H^{-1}\frac{\partial H}{\partial r}\right)^{\vee} = \begin{pmatrix} \mathbf{0} \\ R_3^T(\gamma)R_1^T(\beta)\mathbf{e}_3 \end{pmatrix} = \begin{pmatrix} 0 \\ 0 \\ 0 \\ \sin\beta\sin\gamma \\ \sin\beta\cos\gamma \\ \cos\beta \end{pmatrix}
$$

$$
\left(H^{-1}\frac{\partial H}{\partial \alpha}\right)^{\vee} = \begin{pmatrix} R_3^T(\gamma)R_1^T(\beta)\mathbf{e}_3 \\ \mathbf{0} \end{pmatrix} = \begin{pmatrix} \sin\beta\sin\gamma \\ \sin\beta\cos\gamma \\ \cos\beta \\ 0 \\ 0 \\ 0 \end{pmatrix}
$$

$$
\left(H^{-1}\frac{\partial H}{\partial \beta}\right)^{\vee} = \begin{pmatrix} R_3^T(\gamma)\mathbf{e}_1 \\ \mathbf{0} \end{pmatrix} = \begin{pmatrix} \cos\gamma \\ -\sin\gamma \\ 0 \\ 0 \\ 0 \\ 0 \end{pmatrix}
$$

$$
\left(H^{-1}\frac{\partial H}{\partial \gamma}\right)^{\vee} = \begin{pmatrix} \mathbf{e}_3 \\ \mathbf{0} \end{pmatrix} = \begin{pmatrix} 0 \\ 0 \\ 1 \\ 0 \\ 0 \\ 0 \end{pmatrix}
$$

$$
\left(H^{-1}\frac{\partial H}{\partial \phi}\right)^{\vee} = \begin{pmatrix} R_3^T(\gamma)R_1^T(\beta)R_3^T(\alpha)R_1^T(\theta)\mathbf{e}_3 \\ r\cdot R_3^T(\gamma)R_1^T(\beta)R_3^T(\alpha)R_1^T(\theta)E_3 R_1(\theta)\mathbf{e}_3 \end{pmatrix}
$$

$$
= \begin{pmatrix} \sin\theta(\sin\alpha\cos\gamma + \cos\alpha\cos\beta\sin\gamma) + \cos\theta\sin\beta\sin\gamma \\ -\sin\theta(\sin\alpha\sin\gamma - \cos\alpha\cos\beta\cos\gamma) + \cos\theta\sin\beta\cos\gamma \\ \sin\theta\cos\alpha\sin\beta - \cos\theta\cos\beta \\ r\sin\theta(\cos\alpha\cos\gamma - \sin\alpha\cos\beta\sin\gamma) \\ -r\sin\theta(\cos\alpha\sin\gamma + \sin\alpha\cos\beta\cos\gamma) \\ r\sin\alpha\sin\beta\sin\theta \end{pmatrix}
$$

$$\left(H^{-1}\frac{\partial H}{\partial \theta}\right)^{\vee} = \begin{pmatrix} R_3^T(\gamma)R_1^T(\beta)R_3^T(\alpha)\mathbf{e}_1 \\ \\ -r\cdot R_3^T(\gamma)R_1^T(\beta)R_3^T(\alpha)\mathbf{e}_2 \end{pmatrix}$$

$$= \begin{pmatrix} \cos\alpha\cos\gamma - \sin\alpha\cos\beta\sin\gamma \\ -\cos\alpha\sin\gamma - \sin\alpha\cos\beta\cos\gamma \\ \sin\alpha\sin\beta \\ -r(\sin\alpha\cos\gamma + \cos\alpha\cos\beta\sin\gamma) \\ r(\sin\alpha\sin\gamma - \cos\alpha\cos\beta\cos\gamma) \\ r\cos\alpha\sin\beta \end{pmatrix}$$

Taking the determinant of $|\det \mathcal{J}_R|$ results in (19.23).

19.3.4 Implications for Extracting Statistical Potentials

The problem of determining helix-helix packing preferences has a long history [38, 39, 44, 101, 165]. Also of interest is to know how individual amino acid residues prefer to position, orient, and change conformation relative to each other within proteins [22, 99, 102]. Once probabilities are obtained, statistical potentials can be extracted under the assumption that the observed conformational states are Boltzmann distributed [116, 151, 170]. But in this procedure the correct integration measure is important, because without it, an effective artificial biasing potential will result. (A similar problem arises in the simulation of polymers using rigid-bond models, which is known as the "Fixman metric tensor effect" as described in [127] and references therein.) This is why it was important to compute the Jacobian matrices presented in this section. Other works which derive similar results to those presented here (for the line-on-line case only) are [171, 172], which in turn build on [16], which did not use the correct normalization.

Essentially, if $g_{12} \in SE(3)$ denotes the relative position and orientation between bodies 1 and 2, then the potential energy

$$V_{12}(g_{12}) = V_{21}(g_{21}) \text{ where } g_{21} = g_{12}^{-1}$$

and if everything is Boltzmann distributed, then the pose probability density will be of the form

$$p_{12}(g_{12}) = \frac{1}{Z}\exp(-V_{12}(g_{12})/(k_B T)) \text{ where } Z \doteq \int_{SE(3)} \exp(-V_{12}(g_{12})/(k_B T))\,dg_{12}.$$

Then going the other way and extracting potentials from probabilities means that

$$V_{12}(g_{12}) = -k_B T(\log_e p_{12}(g_{12}) + \log_e Z)$$

where $\exp(\cdot)$ is the classical (scalar) exponential function, and $\log_e(\cdot)$ is the logarithm function in base e (i.e., the natural logarithm). Having the correct Jacobian is important both to compute Z correctly and to ensure that when using coordinates $p_{12}(g_{12}(\mathbf{q}))$ is not viewed as a probability with respect to the measure $d\mathbf{q}$, but rather with respect to $|\mathcal{J}(\mathbf{q})|d\mathbf{q}$.

19.4 Mathematical Aspects of Protein Crystallography

In this section, we show how rigid-body kinematics can be used to assist in determining the atomic structure of proteins when using x-ray crystallography, which is a powerful method for structure determination. This section builds on the presentations in [28, 33, 34, 35, 36]

The importance of determining molecular structures for understanding biological processes and for the design of new drugs is well known. Phasing is a necessary step in determining the three-dimensional structure of molecules from x-ray diffraction patterns. A computational approach called molecular replacement (MR) is a well-established method for phasing x-ray diffraction patterns for crystals composed of biological macromolecules. Popular and effective software packages for doing molecular replacement include [1, 19, 40, 71, 73, 87, 118, 149, 154, 167]. In MR, a search is performed over positions and orientations of a known biomolecular structure within a model of the crystallographic asymmetric unit, or, equivalently, multiple symmetry-related molecules in the crystallographic unit cell. Unlike the discrete space groups known to crystallographers and the continuous rigid-body motions known to kinematicians, the set of motions over which molecular replacement searches are performed does not form a group. Rather, it is a coset space of the group of continuous rigid-body motions, $SE(3)$, with respect to the crystallographic space group of the crystal, which is a discrete subgroup of $SE(3)$. Properties of these 'motion spaces' (which are compact manifolds) are investigated here. The following notation is used throughout this section:

- $\mathbb{R}^n$: n-dimensional Euclidean space
- X : shorthand for $\mathbb{R}^n$ (parameterized in Cartesian coordinates $\{x_i\}$)
- $SE(n)$: the special Euclidean group of n-dimensional space
- G : shorthand for $SE(3)$, a six-dimensional Lie group
- Γ : a Sohncke crystallographic space group (i.e., one which preserves chirality)
- $\mathbb{L}$: a lattice in Euclidean space
- T : the discrete group of translational symmetries of a lattice
- $\mathbb{F}$: the factor group $\Gamma/T = T\backslash\Gamma$
- $\mathbb{T}^n$: the n-dimensional torus
- $F_{\Gamma\backslash G}$: the fundamental domain in G corresponding to $\Gamma\backslash G$
- $\mathbb{Z}$: the integers.

In the field of biomolecular crystallography, the motions of interest are those that describe the discrete symmetries of a crystal lattice [17, 18, 72, 75, 93]. At the beginning of the twentieth century, all possible lattices and their symmetries were characterized independently by [11, 55, 145] as summarized in [67]. For in-depth introductions to the field see [54, 64, 65, 86, 104, 113, 175] For more modern mathematical treatments see [41, 52, 70, 74, 158]. Moreover, crystallography continues to play an important role in the modern world [121], and excellent internet-based tools exist for crystallographic computations [4, 92].

The result of those studies concluded that that there are 230 distinct classes of crystallographic space groups. Of these, the vast majority have mirror reflections or glide planes, and only 65 are "Sohncke" (i.e., are 'special' in the sense that they preserve handedness of chiral structures). In the most up-to-date crystallographic terminology, such space groups are called "Sohncke" groups, after the German physicist, Leonhard Sohncke, who characterized them in [152], though the word "chiral" can also be found in reference to these groups. The Sohncke space groups are particularly important in macromolecular crystallography, which has been responsible for determining the shape of 80 percent of the approximately 100,000 protein structures deposited in the protein data bank (PDB) [14].

Interestingly, the overlap between kinematics and crystallography over the past 100 years has been minimal despite the fact that the Sohncke crystallographic space groups are discrete subgroups of $SE(3)$. The purpose of this section is to examine hitherto unrecognized relationships between these fields. In particular, it is shown that if $G = SE(3)$ and Γ is a crystallographic subgroup of G, then the coset space $\Gamma \backslash G$ is a compact manifold that describes the configuration space of all allowable poses of a protein in a crystal. This is the 6D-search space of interest in the field of molecular replacement [141, 142]. The relationship between molecular replacement searches and fast computation of Wigner-D-function expansions is well known [47, 119, 120, 164]. In fact, as mentioned earlier, Crowther's "Fast Rotation Function" was, to the best of our knowledge, the first FFT on $SO(3)$.

The remainder of this section is organized as follows. First, a review of concepts from theoretical kinematics and crystallography is presented. Then the concept is illustrated in the 2D case in a concrete way. This is followed by discussion of the molecular replacement problem and the role of motion spaces in the 3D case.

19.4.1 Structure of Sohncke Space Groups

We now briefly review the mathematical structure of crystallographic space groups. Recall that altogether there are 230 types of crystallographic groups in total in the three-dimensional case, 65 of which are proper/Sohncke [112, 136, 143]. When referring to "space groups" one is really referring to "equivalence classes of space groups." That is, if Γ is a space group then when $\alpha \Gamma \alpha^{-1} \doteq \{\alpha \circ \gamma \circ \alpha^{-1} \,|\, \gamma \in \Gamma\}$ is also a space group for some fixed affine transformation α, Γ and $\alpha \Gamma \alpha^{-1}$ are considered to be equivalent. Here α is a special affine transformation (i.e., $\alpha = (A, \mathbf{a})$ with $\det A > 0$) that converts one space group to another in the same class, and in general this imposes severe restrictions on the allowable α. The enumeration of equivalence classes of space groups is discrete, but since the affine group is continuous, an uncountably infinite number of equivalent space groups can be generated. In the planar case there are 17 wallpaper groups, five of which preserve chirality $(p_1, p_2, p_3, p_4, p_6)$. Space groups can be divided into two broad classes: the symmorphic groups, which can be written as a semi-direct product of the translation and rotation subgroups, and the nonsymmorphic groups, which cannot. (Though all crystallographic groups have a subgroup of primitive translations which is the same as that for the Bravais lattice, the symmorphic nature of the Bravais-lattice-space-groups is not inherited by the majority of the crystallographic space groups.) In the planar case, all chirality-preserving wallpaper groups are symmorphic. In the three-dimensional case there are 73 symmorphic and 157 nonsymmorphic space groups. Of the 65 Sohncke space groups, 24 are symmorphic and 41 are nonsymmorphic. The most commonly adopted space group for crystals of biological macromolecules is called $P2_1 2_1 2_1$ and is group number 19 of the 230 as listed in the *International Tables of Crystallography* and the Bilbao Server (http://www.cryst.ehu.es). This is a nonsymmorphic symmetry group.

In both the symmorphic and nonsymmorphic cases it is possible to write

$$\Gamma = \bigcup_{i=1}^{|\mathbb{F}|} T(R_i, \mathbf{v}_{R_i}) = \bigcup_{i=1}^{|\mathbb{F}|} (R_i, \mathbf{v}_{R_i}) T \tag{19.24}$$

where $\mathbb{F} \doteq T\backslash\Gamma$ is the "factor group" and a finite set of symmetry operations with elements $g_i = (R_i, \mathbf{v}_{R_i})$ can be formed where the translations $\mathbf{v}_{R_i}$ are all contained inside a unit cell. T is the translation group for the lattice $\mathbb{L}$ (which is common to each Bravais space group and the crystallographic groups that result from filling the primitive Bravais unit cells with crystal motifs). For symmorphic crystals it is always possible to choose an origin of the coordinate system such that $\mathbf{v}_{R_i} = \mathbf{0}$ for all $i \in \{1, ..., |\mathbb{F}|\}$. In this case, the set of symmetry operations $\{g_i | i = 1, ..., |\mathbb{F}|\}$ forms a finite subgroup of Γ. In the nonsymmorphic case, at least some $\mathbf{v}_{R_i} \neq \mathbf{0}$ regardless of how the origin of the coordinate system is chosen, and the resulting set of coset representatives (symmetry operators) does not form a subgroup of Γ. When Γ is a nonsymmorphic Sohncke group and $\mathbf{v}_{R_i}$ are not zero, $(R_i, \mathbf{v}_{R_i})$ describes a screw transformation.

In analogy to the way that a group is divided into coset spaces, a space, X, on which a group, G, acts is also be divided into orbits. For example, $\mathcal{R} \doteq SO(3)$ acting on X divides it into spheres. The set of all of these orbits is denoted as $G\backslash X$. Some books denote this as X/G, but to be consistent with the definition of action, in which g acts on the left of $\mathbf{x}$, it makes more sense to write $G\backslash X$ in analogy with the way that $H\backslash G$ preserves the order of $h \circ g$ in the definition of $Hg \in H\backslash G$.

The fundamental domains $F_{\Gamma\backslash X}$ have been calculated in [63], and these are "exact" up to co-dimension 2, but are only exact in a rigourous mathematical sense for the the small subset of space groups called "Bieberbach" groups. In this case $\Gamma_B\backslash X$ is a flat manifold [24, 174]. More generally, the quotient space $\Gamma\backslash X$ is not a manifold. This type of space has been studied extensively by mathematicians and is called a "Euclidean orbifold" [15, 51, 117, 122, 144, 161, 173].

An immediate crystallographic consequence of these definitions is that if Γ is the full Sohncke symmetry group of a macromolecular crystal, then $F_{\Gamma\backslash X}$ can be identified with the asymmetric unit, where the quotient notation here is due to the fact that Γ acts on $X = \mathbb{R}^3$. $T < \Gamma$ is the largest discrete translation group of the crystal (and so $T < \mathcal{T} \cong (\mathbb{R}^3, +)$), and $F_{T\backslash X}$ and $F_{T\backslash\mathcal{T}}$ can be identified with the primitive unit cell. The distinction between $T\backslash X$ and $T\backslash\mathcal{T}$ is subtle. The latter is a coset space, and the former is simply a space of orbits. Since T is a normal subgroup of $\mathcal{T}$, the unit cell can be endowed with a group structure, namely periodic addition. For this reason, a unit cell in n-dimensional space is equivalent to an n-dimensional torus, $T\backslash\mathcal{T} \cong \mathbb{T}^n$. This fact is implicitly and extensively used in crystallography to expand the density in a unit cell in terms of Fourier series. Furthermore, the translational motion of the contents of a unit cell is easy to handle within the framework of classical mathematics. However, if one wishes to focus attention in MR searches on the asymmetric unit $F_{\Gamma\backslash X} \subset F_{T\backslash X}$, which is smaller, and therefore advantageous from the perspective of the number of grid points required to describe it, then there is no associated group operation. Furthermore, even in the case when the whole unit cell is considered, though periodic translations are handled in an effortless way within the context of classical Fourier analysis, rotations of the rigid contents within a unit cell of a crystal are somewhat problematic within the classical framework, which provides the motivation for the current work.

The coset space $\Gamma\backslash G$ again can be visualized as a "fundamental domain" $F_{\Gamma\backslash G} \subset G$ in which a single point in G from each coset is recorded. To demonstrate this concept, we consider $G = SE(2)$ and $\Gamma = p4$, the planar crystallographic (or "wallpaper") group corresponding to the translational and rotational symmetries of the square lattice depicted in Figure 7.2. In this example, we can take the closure of the fundamental domain as

$$\overline{F_{p4\backslash SE(2)}} \cong \{(x, y, \theta) \in [0, 1] \times [0, 1] \times [0, \pi/2]\}. \tag{19.25}$$

The coset space itself, $p_4\backslash SE(2)$, can be viewed as this fundamental domain with opposing faces appropriately glued. However, unlike the case of $p1$ where the opposing faces are glued in the obvious way, this time the gluing is more complicated. When done properly, this produces a three-manifold with the property that when Γ acts on the left, it leaves all points fixed, and similarly when G acts on the right.

The reason $\cong$ is used in the above expression, rather than $=$, is that there are many equivalent (but different) ways to define fundamental domains. For example it is also possible to say

$$\overline{F_{p4\backslash SE(2)}} \cong \{(x,y,\theta) \in [0,1/2] \times [0,1/2] \times [0,2\pi]\}. \tag{19.26}$$

In both cases, left action by $p4$ on these closures of fundamental domains will "tile" $SE(2)$ "with a little grout left over" corresponding to overlapping sets of measure zero where faces meet. In short, in this context $S_1 \cong S_2$ means

$$\bigcup_{\gamma \in \Gamma} \gamma \cdot S_1 = \bigcup_{\gamma \in \Gamma} \gamma \cdot S_2 = G$$

and

$$Vol((\gamma \cdot S_i) \cap (\gamma' \cdot S_i)) = 0$$

for all $\gamma, \gamma' \in \Gamma$ and $i = 1, 2$. It is not a statement of topological equivalence or isomorphism as groups (in fact, the objects of study here are not groups, since Γ is not a normal subgroup of G).

The fundamental domain itself is a tricky thing to define because it is a half-open half-closed object. But its closure, as defined above, is easy to define, as is its interior $F^{\circ}_{p4\backslash SE(2)}$ (which is defined in the same way as the closure, but with open intervals $(0,1)$ and $(0,\pi/2)$). The true fundamental domain is related to these as

$$F^{\circ}_{p4\backslash SE(2)} \subset F_{p4\backslash SE(2)} \subset \overline{F_{p4\backslash SE(2)}}$$

and all three differ by only sets of measure zero (i.e., points on the bounding faces).

In the case of planar motion spaces, the result after gluing the first instance of a fundamental domain is more complicated than a three-torus because in identify g and $\gamma \circ g$, the faces of $\overline{F_{p4\backslash SE(2)}}$ must be twisted before they are glued together. Examining the action of $p4$ on $SE(2)$, the following can happen:

$$(x,y,\theta) \to (x + z_1, y + z_2, \theta),$$
$$(x,y,\theta) \to (-y + z_1, x + z_2, \theta + \pi/2),$$
$$(x,y,\theta) \to (-x + z_1, -y + z_2, \theta + \pi),$$
$$(x,y,\theta) \to (y + z_1, -x + z_2, \theta + 3\pi/2).$$

(The first of these is the $p1$ action, which will be present in all planar space groups.)

19.4.2 A Mathematical Formulation of Molecular Replacement

This section addresses why coset-space manifolds of the form $\Gamma \backslash G$ are important for biomolecular crystallography in the case of $G = SE(3)$ and Γ being one of the 65 Sohncke space groups. The simplest of these space groups is called $P1$ (capital P rather than lower case p). It is the group of discrete translations of a lattice. This can be thought of as the set of homogeneous transformations of the form $P1 \doteq \{\text{trans}([z_1, z_2, z_3]^T) \mid z_i \in$

$\mathbb{Z}$}. As in the planar case, every space group contains this group of lattice translations as a normal subgroup. This condition is written as $T \triangleleft \Gamma$. The space group Γ itself will generally contain these translations, pure discrete rotations, and discrete screw displacements.

Suppose that the macromolecular structure of interest has an electron density $\rho(\mathbf{x})$. That is, there exists a function $\rho : X \to \mathbb{R}$. This function may be constructed by adding densities of individual domains within the structure.

This means that the total electron density of the non-solvent part of the crystal will be

$$\rho_{\Gamma \backslash X}(\mathbf{x}) \doteq \sum_{\gamma \in \Gamma} \rho(\gamma^{-1} \cdot \mathbf{x}).$$

The symmetry group, Γ, and number of copies of the molecule in a given unit cell can both be estimated directly from the experimental data. The inverse of γ is applied under the function to move $\rho(\mathbf{x})$ by γ in analogy with the way a function on the real line, $f(x)$, is translated one unit in the *positive* direction along the x axis by evaluating $f(x-1)$. Note that such a function $\rho_{\Gamma \backslash X}(\mathbf{x})$ is "Γ-periodic" in the sense that for any $\gamma_0 \in \Gamma$,

$$\rho_{\Gamma \backslash X}(\gamma_0^{-1} \cdot \mathbf{x}) = \rho_{\Gamma \backslash X}(\mathbf{x}). \tag{19.27}$$

Now suppose that before constructing symmetry-related copies of the density $\rho(\mathbf{x})$, we first move it by an arbitrary $g \in G$. The result will be $\rho(\mathbf{x}; g) \doteq \rho(g^{-1} \cdot \mathbf{x}) = \rho(g^{-1} \cdot \mathbf{x}; e)$. It is easy to see that

$$\rho_{\Gamma \backslash X}(\mathbf{x}; g) \doteq \sum_{\gamma \in \Gamma} \rho(\gamma^{-1} \cdot \mathbf{x}; g) = \sum_{\gamma \in \Gamma} \rho((\gamma \circ g)^{-1} \cdot \mathbf{x}).$$

In an x-ray diffraction experiment for a single-domain protein, $\rho(\mathbf{x})$ is not obtained directly. Rather, the magnitude of the classical Fourier transform of $\rho_{\Gamma \backslash X}(\mathbf{x}; g)$ is obtained. In general, if $\{\mathbf{a}_i \,|\, i = 1, ..., n\}$ are the vectors describing lattice directions, so that each element of the group T consists of translations of the form $\mathbf{t}(k_1, k_2, ..., k_n) = \sum_{j=1}^{n} k_j \mathbf{a}_j \in T$, then the classical Fourier series coefficients for $\rho_{\Gamma \backslash X}(\mathbf{x}; g)$ (which for each fixed $g \in G$ is a function on $T \backslash \mathcal{T}$) are denoted as $\hat{\rho}_{\Gamma \backslash X}(\mathbf{k}; g)$. There is duality between the Fourier expansions for T and $T \backslash \mathcal{T}$, and, likewise, $\hat{U} \cong \mathbb{Z}^n$ is the unitary dual of U.

Now, the g in each of these expressions can be taken to be in G, but this is wasteful because G extends to infinity, and the same result appears whether g or $\gamma \circ g$ is used for any $\gamma \in \Gamma$. Therefore, the rigid-body motions of interest are those that can be taken one from each coset $\Gamma g \in \Gamma \backslash G$. If all such representatives of cosets $\Gamma g \in \Gamma \backslash G$ are collected, the result will be the fundamental region $F_{\Gamma \backslash G} \subset G$. A goal of molecular replacement is then to find $g \in F_{\Gamma \backslash G}$ such that $\left| \hat{\rho}_{\Gamma \backslash X}(\mathbf{k}; g) \right|^2$ best matches with the Patterson function, $\hat{P}(\mathbf{k})$, which contains equivalent information to that in the diffraction pattern. In other words, a fundamental goal of molecular replacement is to minimize a cost function of the form

$$C(g) = \sum_{\mathbf{k} \in \hat{U}} d\left(\left| \hat{\rho}_{\Gamma \backslash X}(\mathbf{k}; g) \right|^2, \hat{P}(\mathbf{k}) \right) \tag{19.28}$$

where $d(\cdot, \cdot)$ is some measure of distance, discrepancy, or distortion between densities or intensities. No matter the choice of $d(\cdot, \cdot)$, the cost functions $C(g)$ in (19.28) inherit the symmetry of $\rho_{\Gamma \backslash X}(\mathbf{x})$ in (19.27) in the sense that

$$C(g) = C(\gamma \circ g) \ \forall \gamma \in \Gamma. \tag{19.29}$$

This makes $C(g)$ a function on $\Gamma \backslash G$ (or, equivalently, $F_{\Gamma \backslash G}$), in analogy with the way that a periodic function on the real line can be viewed as a function on the circle. We note that spaces such as the quotient of the the normalizer of Γ in G by the centralizer of Γ in G have been studied in the crystallography literature [69], but very little has been said about $\Gamma \backslash G$ and $F_{\Gamma \backslash G}$.

Finding maximal values of $C(g)$ is essentially what molecular replacement is about. With this, diffraction patterns can be phased, and the inverse Fourier transform can be used to recover the electron density of the macromolecule of interest. Phasing is a necessary step in determining their three-dimensional structure. The importance of these structures in understanding biological processes and in designing new drugs is well known [143].

19.4.3 Details Regarding the Structure of Sohncke Space Groups in Three-Dimensional Space

In order to understand more about the "motion space" $F_{\Gamma \backslash G}$ and the "motion manifold" $\Gamma \backslash G$, it is important to first examine the structure of the Sohncke space group Γ, including its translational, rotational, and screw symmetry operations.

Every Bravais lattice has a space group that can be written as the semi-direct product of the lattice translation group, T, and a point group of a Bravais lattice, P. That is, $\Gamma_L = T \rtimes P$. Moreover, every Sohncke crystallographic space group, Γ, has the group of lattice translations as a subgroup, $T \leq \Gamma$ such that $T \backslash \Gamma = \mathbb{F}$ is isomorphic to a point group formed by stripping away the translational parts of any screw transformations in $\mathbb{F}$ and keeping only the pure rotations that remain from these and from any pure rotations in $\mathbb{F}$.

Three-dimensional space groups are divided into two categories [17, 18, 72, 146]. The first, called *symmorphic* (or simple), consists of semi-direct products of a group of discrete lattice translations and discrete lattice rotations. These can be thought of as an analogue of $SE(3)$ which is a semidirect product of continuous translations and rotations. The second category of space groups is called *nonsymmorphic*, and surprisingly, these groups cannot be written simply as a semi-direct product of pure lattice rotations and pure lattice translations. That is, they always contain residual lattice screw motions. The easiest nonsymmorphic space group is $P2_1$. It is generated by concatenating the following transformations: $(x_1, x_2, x_3) \rightarrow (x_1 + z_1, x_2 + z_2, x_3 + z_3)$ (the lattice translations where $z_i \in \mathbb{Z}$, as with any space group) and $(x_1, x_2, x_3) \rightarrow (-x_1, -x_2, x_3 + 1/2)$. That is all! The second transformation is a 2_1 screw displacement along the x_3 direction. Squaring it results in a lattice translation by one unit in that direction. The group generated by all such translations and screw displacements cannot be written as a semi-direct product of lattice rotations and lattice translations, since the translation by $1/2$ is not a lattice translation and cannot be decoupled from the rotation by π that is part of the screw motion.

Let Γ_s and Γ_{ns}, respectively, denote symmorphic and nonsymmorphic Sohncke crystallographic space groups. Then $\Gamma_s = T \rtimes P_s$ and it is possible to select elements of Γ_s to form a fundamental domain with a finite number of elements, $F_{T \backslash \Gamma_s} \cong \mathbb{F}_s$, where

$$F_{T \backslash \Gamma_s} = \{\mathbf{0}\} \rtimes P_s \leq \{\mathbf{0}\} \rtimes P. \tag{19.30}$$

Note here the equality (as groups) and inequality (denoting a subgroup-group relationship) rather than the weaker statements that would be implied if these were, respectively, congruence and set inclusion signs. Moreover, it follows from $P_s \leq P$ that

$$F_{T\backslash\Gamma_s} \leq \Gamma_s \leq \Gamma_L. \tag{19.31}$$

In the nonsymmorphic case it is still possible to write

$$F_{T\backslash\Gamma_{ns}} \cong \mathbb{F}_{ns} = T\backslash\Gamma_{ns} \cong P_{ns} \leq P, \tag{19.32}$$

and $F_{T\backslash\Gamma_{ns}} \subseteq \Gamma_{ns}$, but it is no longer possible to write (19.30), and in general it is not the case that $F_{T\backslash\Gamma_{ns}} \subseteq \Gamma_L$ or $\{\mathbf{0}\} \rtimes P_{ns} \leq \Gamma_{ns}$. Therefore, statements analogous to (19.31) do not hold in the nonsymmorphic case.

The fact that $\mathbb{F}$ (which can be either P_s or $\mathbb{F}_{ns}$) is a group follows from the fact that T is not only a subgroup, but is also normal in all cases. This means that $T\backslash\Gamma = \Gamma/T = \mathbb{F}$ is a group in its own right.

Let $X = \mathbb{R}^n$ and $G = SE(n)$. The structure of space groups is important in the context of characterizing allowable motions within asymmetric units because the space $\Gamma\backslash G$ can be decomposed in different ways. For example, it is always possible to write

$$F_{\Gamma\backslash G} \cong \left(F_{\Gamma\backslash X}\right) \times SO(3).$$

In other words, the fundamental region of the quasi-group can be taken as the product of translations within the asymmetric unit and the whole of the rotation group. Moreover, if Γ is symmorphic it is possible to write

$$F_{\Gamma_s\backslash G} = \left(F_{T\backslash X}\right) \times \left(F_{P_s\backslash SO(3)}\right). \tag{19.33}$$

This is convenient because when the fundamental region for $T\backslash X$ is taken to be the Wigner-Seitz cell, which has the point symmetry group P_s, then the fundamental region for $\Gamma_s\backslash G$ is invariant under conjugation with respect to elements of P_s.

In the nonsymmorphic case, the Wigner-Seitz unit cell can have some rotational symmetries (corresponding to the subset of symmetry operations with $\mathbf{v}_{R_i} = \mathbf{0}$). If there exists a nonempty subset of these operators that is closed under multiplication, then the resulting point group formed by these elements, P_u, will be a subgroup of both $\mathbb{F}_{ns}$.[5] Moreover, $T \rtimes P_u$ will be a symmorphic subgroup of the nonsymmorphic group Γ_{ns} and P_u will be a subgroup of the point group P of the Wigner-Seitz unit cell corresponding to $T\backslash X$. This means that

$$F_{(T\rtimes P_u)\backslash X} = F_{P_u\backslash(T\backslash X)}.$$

This is relevant to the main topic of this section because most macromolecular crystals are nonsymmorphic, and so it cannot be assumed that (19.33) holds, and we seek alternative admissible ways to decompose $\Gamma\backslash G$. In prior work, sampling schemes for these spaces have been investigated [177, 178, 179].

In all cases

$$\mathrm{Vol}(T\backslash X) = |\mathbb{F}| \cdot \mathrm{Vol}(\Gamma\backslash X).$$

Therefore,

$$\mathrm{Vol}(T\backslash X) = |P_u| \cdot \mathrm{Vol}((T \rtimes P_u)\backslash X)$$

and

$$\mathrm{Vol}(T\backslash X) = |\mathbb{F}_{ns}| \cdot \mathrm{Vol}(\Gamma_{ns}\backslash X).$$

Equating these two expressions gives

––––––––––––––––––––
[5] In the symmorphic case $P_u = \mathbb{F}_{ns} = P_{ns}$.

$$\mathrm{Vol}((T \rtimes P_u)\backslash X) = \frac{|\mathbb{F}_{ns}|}{|P_u|} \cdot \mathrm{Vol}(\Gamma_{ns}\backslash X). \qquad (19.34)$$

If $|\mathbb{F}_{ns}| = |P_u| + |S|$ (where $|S|$ is the number of screw symmetry operations) and since $P_u < \mathbb{F}_{ns}$, then, from Lagrange's theorem, $|P_u|\backslash|\mathbb{F}_{ns}| = |P_u\backslash\mathbb{F}_{ns}|$ must be a positive integer. Furthermore, $|P_u|\backslash|\mathbb{F}_{ns}| = 1 + |P_u|\backslash|S|$, and so $|P_u|\backslash|S|$ must be a nonnegative integer. So the number of screw transformations in the set of coset representatives of $\mathbb{F}_{ns} = T\backslash\Gamma_{ns}$ must be an integer multiple of the number of coset representatives that are rotational transformations.

19.4.4 Coset Representative for $T\backslash\Gamma$ for the Most Common Macromolecular Space Groups

The most commonly encountered space groups, Γ, in the Protein Data Bank (PDB) are listed below together with coset representatives of $P1\backslash\Gamma$ computed from the Bilbao Crystallographic Server (http://www.cryst.ehu.es). When Γ has a name starting with 'P' then $P1 = T$, whereas for other space groups, such as those that start with 'C', $P1 < T$. The coset reps, γ, are listed as $\gamma \cdot \mathbf{x}$ where $\mathbf{x} = [x, y, z]^T$ is an arbitrary position in the asymmetric unit. This is done instead of using the 4×4 matrix $H(g)$ to save space. A few of the most common space groups (defined by their action on arbitrary $\mathbf{x} = (x, y, z)^T \in \mathbb{R}^3$) are listed below.

$P2_12_12_1 : (x, y, z); (-x + 1/2, -y, z + 1/2); (-x, y + 1/2, -z + 1/2); (x + 1/2, -y + 1/2, -z);$

$P2_1 \quad : (x, y, z); (-x, y + 1/2, -z);$

$C2 \quad : (x, y, z); (-x, y, -z); (x + 1/2, y + 1/2, z); (-x + 1/2, y + 1/2, -z);$

$P2_12_12 : (x, y, z); (-x, -y, z); (-x + 1/2, y + 1/2, -z); (x + 1/2, -y + 1/2, -z);$

$C222_1 \quad : (x, y, z); (-x, -y, z + 1/2); (-x, y, -z + 1/2); (x, -y, -z); (x + 1/2, y + 1/2, z);$
$\qquad\quad (-x + 1/2, -y + 1/2, z + 1/2); (-x + 1/2, y + 1/2, -z + 1/2); (x + 1/2, -y + 1/2, -z);$

$P4_32_12 : (x, y, z); (-x, -y, z + 1/2); (-y + 1/2, x + 1/2, z + 3/4); (y + 1/2, -x + 1/2, z + 1/4);$
$\qquad\quad (-x + 1/2, y + 1/2, -z + 3/4); (x + 1/2, -y + 1/2, -z + 1/4); (y, x, -z); (-y, -x, -z + 1/2).$

The value of $|P_u|\backslash|S|$ in these six cases are respectively 3, 1,1, 1, 3, and 3.

For a more detailed description of how Sohncke space groups decompose into symmorphic and nonsymmorphic parts, see [37].

19.5 Methods for "Optimal" Deterministic Sampling of Rotations

Motivated by problems in molecular replacement, protein-protein docking, and robotics, a number of authors have proposed "optimal" rotation sampling methods. Different criteria have been used to define optimality, and generally the methods proposed are not actually exactly optimal, but perform well in practice. In this section we review several of these methods.

19.5.1 Lattman Angles

The Lattman angles [94, 95, 96] are widely used in orientation searches in molecular replacement as a variant of the Euler angles. Here it is shown using variational principles that the "optimal" sampling properties of the Lattman variant of the Euler angles can be further improved.

Let $\mathbf{q} = (\alpha, \beta, \gamma)$. Then a rotation matrix parameterized by ZXZ Euler angles is

$$R(\mathbf{q}) = R_3(\alpha)R_1(\beta)R_3(\gamma)$$

where $R_i(\theta)$ denotes a counter-clockwise rotation around the ith coordinate axis in three-dimensional space. The associated Jacobian matrix, which relates the body-fixed description of angular velocity to the time rate of change of $\mathbf{q}$, $\boldsymbol{\omega} = J(\mathbf{q})\dot{\mathbf{q}}$, is

$$J(\mathbf{q}) = \begin{pmatrix} \sin\beta\sin\gamma & \cos\gamma & 0 \\ \sin\beta\cos\gamma & -\sin\gamma & 0 \\ \cos\beta & 0 & 1 \end{pmatrix}.$$

The associated metric tensor is then

$$G(\mathbf{q}) = J^T(\mathbf{q})J(\mathbf{q}) = \begin{pmatrix} 1 & 0 & \cos\beta \\ 0 & 1 & 0 \\ \cos\beta & 0 & 1 \end{pmatrix}.$$

The Lattman angles $\mathbf{q}' = (\alpha', \beta', \gamma')$ are defined relative to the Euler angles as the following alternative parameterization of rotations:

$$\alpha' = \frac{1}{\sqrt{2}}(\alpha - \gamma); \quad \beta' = \beta; \quad \gamma' = \frac{1}{\sqrt{2}}(\alpha + \gamma).$$

Sometimes the Lattman angles are defined with $\sqrt{2}$ replaced by 1 or by 2. We use the convention given above because then

$$\mathbf{q}' = \mathbf{f}_1(\mathbf{q}) = R_2(-\pi/4)\mathbf{q}$$

and

$$\mathbf{q} = \mathbf{f}_1^{-1}(\mathbf{q}') = R_2(\pi/4)\mathbf{q}', \implies \frac{\partial\mathbf{q}}{\partial\mathbf{q}'} = R_2(\pi/4).$$

The Jacobian and metric tensor for the Lattman angles are then

$$J(\mathbf{q}') = J(\mathbf{q})\frac{\partial\mathbf{q}}{\partial\mathbf{q}'} = J(\mathbf{q})R_2(\pi/4)$$

and

$$G'(\mathbf{q}') = R_2^T(\pi/4)G(\mathbf{q})R_2(\pi/4) = \begin{pmatrix} \sin^2\frac{\beta'}{2} & 0 & 0 \\ 0 & 1 & 0 \\ 0 & 0 & \cos^2\frac{\beta'}{2} \end{pmatrix}.$$

The intuition behind this choice is that the metric tensor then looks more like the 3×3 identity matrix.

However, Lattman did not stop there. He defined a sampling scheme that is essentially a second change of variables of the form

$$\alpha'' = \alpha' \sin \frac{\beta'}{2}; \quad \beta'' = \beta'; \quad \gamma'' = \gamma' \cos \frac{\beta'}{2}.$$

This represents a nonlinear transformation of coordinates of the form

$$\mathbf{q}'' = \mathbf{f}_2(\mathbf{q}')$$

with associated Jacobian

$$\frac{\partial \mathbf{q}''}{\partial \mathbf{q}'} = \begin{pmatrix} \sin \frac{\beta'}{2} & \frac{1}{2}\alpha' \cos \frac{\beta'}{2} & 0 \\ 0 & 1 & 0 \\ 0 & -\frac{1}{2}\gamma' \sin \frac{\beta'}{2} & \cos \frac{\beta'}{2} \end{pmatrix}. \tag{19.35}$$

From basic differential geometry, this is related to the Jacobian of the inverse transformation as

$$\frac{\partial \mathbf{q}'}{\partial \mathbf{q}''} = \left[\frac{\partial \mathbf{q}''}{\partial \mathbf{q}'}\right]^{-1}\bigg|_{\mathbf{q}'=\mathbf{f}_2^{-1}(\mathbf{q}'')}.$$

The Jacobian in (19.35) has the same kind of $\sin\beta$ singularity as the Euler angles, and so

$$G''(\mathbf{q}'') = \left(\frac{\partial \mathbf{q}'}{\partial \mathbf{q}''}\right)^T G'(\mathbf{q}')\bigg|_{\mathbf{q}'=\mathbf{f}_2^{-1}(\mathbf{q}'')} \frac{\partial \mathbf{q}'}{\partial \mathbf{q}''}$$

will have a unit determinant.

Explicitly,

$$G''(\mathbf{q}'') = \begin{pmatrix} 1 & -\frac{1}{2}\alpha' \cos\frac{\beta}{2} & 0 \\ -\frac{1}{2}\alpha' \cos\frac{\beta}{2} & 1 + \frac{1}{4}\left[(\alpha')^2 \cos^2\frac{\beta}{2} + (\gamma')^2 \sin^2\frac{\beta}{2}\right] & \frac{1}{2}\gamma' \sin\frac{\beta}{2} \\ 0 & \frac{1}{2}\gamma' \sin\frac{\beta}{2} & 1 \end{pmatrix}\bigg|_{\mathbf{q}'=\mathbf{f}_2^{-1}(\mathbf{q}'')}.$$

As a result, when the Lattman angles are close to zero, the resulting metric tensor looks even more like the 3×3 identity matrix.

19.5.2 Almost Optimal Deterministic Sampling of $SO(3)$ Based on Fundamental for Finite Subgroups

Here we review a completely different approach to almost-uniform sampling using the exponential map and the division of a Lie group into fundamental domains, as was done in [34, 177].

If G denotes $SO(3)$ or $SE(2)$, or any Lie group, and $\Gamma, \Gamma' < G$ denote discrete subgroups, then right- and left-coset-spaces are defined as

$$\Gamma \backslash G \doteq \{\Gamma g \,|\, g \in G\}$$

and

$$G/\Gamma' \doteq \{g\Gamma' \,|\, g \in G\}.$$

A double coset space is defined as

$$\Gamma\backslash G/\Gamma' \doteq \{\Gamma g \Gamma' \,|\, g \in G\}.$$

Associated with any coset (or double-coset), it is possible to define a set of distinguished coset (or double-coset) representatives, exactly one per coset. Such a set defines a fundamental domain in G that has the same dimension as G, but lesser volume. Under the left action by Γ, the fundamental domain $F_{\Gamma\backslash G}$ is translated and the closure of the union of all translates covers G without gaps[6] or duplicates. Similarly, right action by Γ' on the fundamental domain $F_{G/\Gamma'}$ and the double-sided-action of $\Gamma \times \Gamma'$ on $F_{\Gamma\backslash G/\Gamma'}$ produces translates the closure of which cover G.

One way to construct fundamental domains is as Voronoi cells within G. Since G is a Riemannian manifold, a proper distance function $d : G \times G \longrightarrow \mathbb{R}_{\geq 0}$ can be constructed, and we can define

$$F_{\Gamma\backslash G} \doteq \{g \in G \,|\, d(e, g) < d(e, \gamma \circ g) \; \forall \gamma \in \Gamma\},$$

$$F_{G/\Gamma'} \doteq \{g \in G \,|\, d(e, g) < d(e, \Gamma' \circ g) \; \forall \Gamma' \in \Gamma'\},$$

and

$$F_{\Gamma\backslash G/\Gamma'} \doteq \{g \in G \,|\, d(e, g) < d(e, \gamma \circ g \circ \Gamma') \; \forall (\gamma, \Gamma') \in \Gamma \times \Gamma'\}.$$

Explicitly, distance functions for $SO(3)$ and $SE(2)$ can be defined as

$$d_{SO(3)}^{(1)}(R_1, R_2) = \|R_1 - R_2\|$$

where $\|A\| = \sqrt{\text{tr}(AA^T)}$ is the Frobenius norm, and

$$d_{SE(2)}^{(1)}(g_1, g_2) = \|g_1 - g_2\|_W$$

where an arbitrary element of $SE(2)$ is of the form

$$g = \begin{pmatrix} \cos\theta & -\sin\theta & x \\ \sin\theta & \cos\theta & y \\ 0 & 0 & 1 \end{pmatrix},$$

$W = W^T$ as a 3×3 weighting matrix, and $\|A\|_W = \sqrt{\text{tr}(AWA^T)}$ The above distance measures are 'extrinsic' in the sense that they rely on how these matrix Lie groups are embedded in $\mathbb{R}^{3\times3}$.

It is also possible to define 'intrinsic' measures of distance using the logarithm function. Since both $SO(3)$ and $SE(2)$ are matrix-Lie-groups, their exponential maps are the matrix exponentials. Explicitly for $SO(3)$, elements of the associated Lie algebra, $so(3)$ are skew-symmetric matrices

$$X = \begin{pmatrix} 0 & -x_3 & x_2 \\ x_3 & 0 & -x_1 \\ -x_2 & x_1 & 0 \end{pmatrix}$$

and the exponential gives

---

[6] In practice, fundamental domains often defined to be open sets, and so the union of translates themselves does not completely cover G, as it has gaps of measure zero. On the other hand, the union of the closure of fundamental domains will cover, but with a set of measure zero of duplicates. This distinction between a fundamental domain, its interior, and its closure are inconsequential for our purposes.

$$R = \exp X = \mathbb{I} + \frac{\sin \|\mathbf{x}\|}{\|\mathbf{x}\|} X + \frac{1 - \cos \|\mathbf{x}\|}{\|\mathbf{x}\|^2} X^2$$

where $\mathbf{x} = [x_1, x_2, x_3]^T = X^\vee$ is the dual vector corresponding to X. The opposite operation gives $\hat{\mathbf{r}} - X$. And for $SE(2)$ elements of the corresponding Lie algebra, $se(2)$, are of the form

$$X = \begin{pmatrix} 0 & -\theta & v_1 \\ \theta & 0 & v_2 \\ 0 & 0 & 0 \end{pmatrix},$$

and the exponential is

$$g = \exp(X) = \begin{pmatrix} \cos\theta & -\sin\theta & x(\theta, v_1, v_2) \\ \sin\theta & \cos\theta & y(\theta, v_1, v_2) \\ 0 & 0 & 1 \end{pmatrix},$$

where the functions $x(\theta, v_1, v_2)$ and $y(\theta, v_1, v_2)$ have been computed in closed form in Chapter 6. The inverse map for each is the matrix logarithm. This degenerates when $\|\mathbf{x}\|$ or θ is π. By restricting the discussion to the case when $\|\mathbf{x}\|, \theta < \pi$, log is uniquely defined on a subset of G depleted by a set of measure zero. This depletion will have no effect on our formulation. For example, it becomes possible to define

$$d_{SO(3)}^{(2)}(R_1, R_2) = \| \log(R_1^T R_2) \|$$

when $R_1^T R_2$ is not a rotation by π. Otherwise $d_{SO(3)}^{(2)}(R_1, R_2) = \pi$ and, similarly,

$$d_{SE(2)}^{(2)}(g_1, g_2) = \| \log(g_1^{-1} \circ g_2) \|_{W'}$$

where W' could be different than W. As in the $SO(3)$ case, a map from $se(2)$ to $\mathbb{R}^3$ can be defined as $X^\vee = [v_1, v_2, \theta]^T$.

It is interesting to note that regardless of whether the intrinsic or extrinsic measures are used, the above distance functions for $G = SO(3)$ are bi-invariant:

$$d_{SO(3)}(R_1, R_2) = d_{SO(3)}(QR_1, QR_2) = d_{SO(3)}(R_1 Q, R_2 Q)$$

for arbitrary $Q \in G$, whereas no such bi-invariant metric for $SE(2)$ is possible.

By dividing a Lie group such as $SO(3)$, $SE(2)$, or $SE(3)$ into small fundamental domains corresponding to a right-coset space relative to a discrete subgroup, the exponential map can be used to parameterize the contents of the cell centered at the identity element. This cell will be relatively small, and so any sampling method used in Euclidean space can be used in the Lie algebra to produce samples with any desired property in the cell. This is due to the simple fact that when X is small $\exp X \approx \mathbb{I} + X$. That is, within small cells centered at the identity, the exponential map is almost linear. Then samples throughout the Lie group can be obtained by group action on the samples in the cells as discussed in [177, 178, 179].

Applications to Optimization

In many problems (including the structural biology one), but also in attitude estimation, medical image registration, and in maximum likelihood computations in robot localization, one seeks to minimize a function on a Lie group (typically either $SO(n)$ or

$SE(n)$ or their products). Such minimizations can be performed either using gradient descent, or using a grid-based search strategy. Grid-based strategies are often favored when the terrain of the function is rugged with many local minima. Here we establish a group-theory based method for establishing grids with points that are spaced well, in contrast to a uniform grid in Euler angles. In so doing, computational resources are not wasted on poorly constructed grid searches.

Given two discrete subgroups $H, K < G$, where G is compact (e.g., $SO(3)$), the region defined as

$$\tilde{F}_{H\backslash G/K} \doteq \{g \in G \,|\, d(e,g) < d(e,g) < d(e, h \circ g \circ k)$$

for all $(h, k) \in H \times K$ will occupy a volume with fraction $1/N$ that of $Vol(SO(3)) = 8\pi^2$ where

$$N = \frac{|H| \cdot |K|}{|H \cap K|}.$$

In the case when $H = \{e\}$ or $K = \{e\}$ this reduces to a factor of $|K|$ or $|H|$, corresponding to the fact that the double coset in this case reduces to the (single) left- and right-coset cases. When $H \leq K$ then the factor is again $|K|$. This means that $\tilde{F}_{H\backslash G/K}$ as defined above cannot be the fundamental domain $F_{H\backslash G/K}$ for the double-coset space $H\backslash G/K$ because it is of the same volume as the fundamental domain for a single coset. However, we can say that

$$F_{H\backslash G/K} \subseteq \tilde{F}_{H\backslash G/K} \subseteq F_{G/K}, F_{H\backslash G}$$

When $H \cap K = \{e\}$, then we will in fact get $|H| \cdot |K|$ Voronoi cells that tile G with gaps of measure zero.

In the other extreme case when $K = H$ is a point group, the bi-invariance of the metric gives Vornoi cells $F_{H\backslash G} = F_{G/H} = \tilde{F}_{H\backslash G/H}$. In this scenario, the way to interpret $F_{H\backslash G/H}$ is as a barycentric subdivision of $F_{H\backslash G}$ such that the original coset space can be reconstructed (up to a missing set of measure zero) by the following combination of adjoint action and union

$$F_{H\backslash G} = \bigcup_{h \in H} h \, F_{H\backslash G/H} \, h^{-1}.$$

For example, if H is the group of rotational symmetry operations of the icosahedron, and $G = SO(3)$, then $|H| = 60$ and $\log(F_{H\backslash G})$ can be viewed as a dodecahedral cell centered at the origin of the Lie algebra, $\mathcal{G} = so(3)$, and each $F_{H\backslash G/H}$ can be viewed as the tetrahedron in Figure 19.2

Similarly, if $K < H$, then a $|K|$-fold division of $F_{H\backslash G}$ can be computed to represent $F_{H\backslash G/K}$ and these pieces can be reconstructed by adjoint action.

This means that we can divide $SO(3)$ into 3600 tetrahedral pieces of equal size and shape. A reasonably uniform and fine sampling can be obtained by choosing the point at the barycenter of each of these tetrahedra (and at their vertices if desired). If even finer samples are required, then sampling from an arbitrarily fine Cartesian grid in the Lie algebra can be imposed within each tetrahedron.

For alternative methods for sampling rotations, see the discussion in Chapter 5 and in [27, 114, 115, 180]

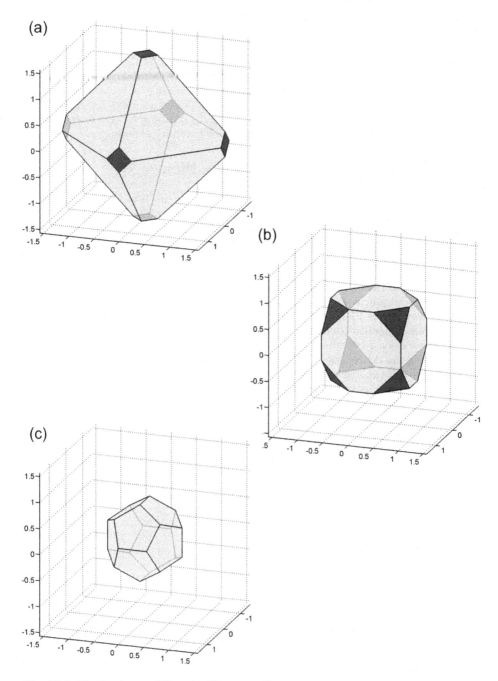

Fig. 19.2. The Fundamental Domains $F_{Tet\backslash SO(3)}$, $F_{Cub\backslash SO(3)}$, $F_{Icos\backslash SO(3)}$ Viewed in Exponential Coordinates. (Reprinted from Chirikjian, G.S., Yan, Y., "Mathematical Aspects of Molecular Replacement: II. Geometry of Motion Spaces," *Acta Crystallographica A*, 68(2): 208–221, 2012, with permission of the International Union of Crystallography)

FFTs and Fast Convolutions Based on Coset Decompositions

In problems ranging from robot arms to biopolymers such as DNA, the problem of characterizing the distribution of positions and orientations of the distal end relative to its proximal end arises, as discussed in Chapter 17. The probability density function $f : SE(n) \to \mathbb{R}_{\geq 0}$ contains useful information in a wide range of scenarios. If around each of m joints, or chemical bonds, in a serial chain structure there are K preferred states, then $f_m(g)$ will describe the distribution of the resulting K^m reference frames.

In Chapter 12 it was shown that such functions have the property

$$f_{m_1+m_2}(g) = (f_{m_1} * f_{m_2})(g) = \int_G f_{m_1}(h) f_{m_2}(h^{-1} \circ g)\, dh \qquad (19.36)$$

where dh is the natural integration measure for $G = SE(n)$. The first equality above is an observation about the relationship between densities in concatenated chains. The second is a definition of convolution of two functions on a (unimodular) Lie group. Since the integral over $SE(3)$ has an integral over $SO(3)$ buried inside, the convolution over $SE(3)$ has many convolutions over $SO(3)$ (one for each pair of translations $(\mathbf{t}_h, \mathbf{t}_g)$) buried inside.

Efficient algorithms for computing convolutions on rotation and motion groups have been developed previously, as discussed in Chapter 9 using Euler angles α, β, γ. These are all based on the FFT for translation and in the α and γ Euler angles, and various fast Wigner-d-function transforms in the β variable.

Here we discuss two alternatives to this approach: (1) computing convolutions directly using the double-coset decompositions described earlier, without converting to the Fourier domain; and (2) by rapidly computing Fourier transforms on groups using double-coset decompositions of the sort derived earlier.

Fast Convolutions by Direct Evaluation

Any integral over G can be decomposed as

$$\int_G f(g)\, dg = \sum_{(h,k) \in H \times K} \int_{F_{H \backslash G / K}} f(h \circ g' \circ k)\, dg' \qquad (19.37)$$

where dg' is the same volume volume element as for G, but restricted to $F_{H \backslash G / K} < G$.

In analogy with the FFT, which predicates speed only under the restriction that $f(g)$ is a band-limited function, we can assume that $f(g)$ can be approximated to any desired precision on $F_{H \backslash G / K}$ as a polynomial in exponential coordinates. Then, if we seek to compute (19.37) we can compute each $\int_{F_{H \backslash G / K}} f(h \circ g' \circ k)\, dg'$ rapidly via a combination of the divergence theorem and closed-form quadrature rules. The effect of this is that an integral such as (19.37), instead of taking $\mathcal{O}(N^{\dim(G)})$ computations, will take $\mathcal{O}(N^{\dim(G)-1})$ (and some constant factor that depends on the quadrature). Since a convolution involves an integral for each value of the argument, when using brute force, it requires $\mathcal{O}(N^{2 \cdot \dim(G)})$. However, since the values of a polynomial function can be reconstructed on the interior of the cells $F_{H \backslash G / K}$ from values on their faces, this means that effectively convolutions can be computed in $\mathcal{O}(N^{2 \cdot \dim(G)-2})$. For example, for $SO(3)$ and $SE(2)$, this is $\mathcal{O}(N^4)$, which is much more manageable than $\mathcal{O}(N^6)$.

Double-Coset Decompositions and Fast Approximate FFTs on $SO(3)$ and $SE(2)$

Euler angles are by no means the only way to parameterize $SO(3)$. We can imagine different FFT algorithms based on different parameterizations and coset decompositions with respect to subgroups other than $SO(2)$.

The specific property of IURs that will be used in this section is

$$U(\exp X, l) = \exp\left(W(X, l)\right) \tag{19.38}$$

where $W(X, l) = \sum_{i=1}^{3} x_i W_i(l)$ with

$$(W_1(l))mn = -\frac{i}{2}c^l_{-n}\delta_{m+1,n} - \frac{i}{2}c^l_n\delta_{m-1,n}$$

$$(W_2(l))mn = +\frac{i}{2}c^l_{-n}\delta_{m+1,n} - \frac{i}{2}c^l_n\delta_{m-1,n}$$

$$(W_3(l))mn = -in\delta_{m,n}$$

where $\delta_{m,n}$ is the Kronecker delta, and $c^l_n = \sqrt{(l-n)(l+n+1)}$ for $|n| \le l$. We use the property in (19.38) together with the double-coset decomposition. Explicitly,

$$\hat{f}(l) = \sum_{(P,Q)\in \mathbb{P}\times\mathbb{Q}} \int_{\mathbb{P}\backslash SO(3)/\mathbb{Q}} f(PRQ)U((PRQ)^T, l)\, dR,$$

which can be rewritten as

$$\sum_{(P,Q)\in \mathbb{P}\times\mathbb{Q}} U(Q^T, l) \left[\int_{\mathbb{P}\backslash SO(3)/\mathbb{Q}} f(PRQ)U(R^T, l)\, dR \right] U(P^T, l).$$

We can think of $f(PRQ)$ as the function $f(R)$ being moved under left and right action so that the point initially at R is now in the vicinity of $\mathbb{I}$. The fact that $U(\exp X, l)$ can be expressed as a truncated Taylor series in the variable X within the fundamental domain centered on the identity is then very useful because $W(X, l)$ will have polynomial entries, each of which can be computed by evaluation on their boundary. And hence the computation of the integral over $\mathbb{P}\backslash SO(3)/\mathbb{Q}$ is efficient.

The number of terms needed in the truncation of $U(\exp X, l)$ is related to how fine the subdivision of $SO(3)$ into cells is – the finer the subdivision, the fewer terms needed in the Taylor series.

Regarding the Fourier inversion formula, the double-coset decompositions of the sort used throughout this section have the benefit that they can use the tri-diagonal nature of the W_i matrices (when each PRQ is close to the identity, which is the case by our choice of fundamental domain for double cosets). This is in place of the recurrence relations that are usually used for matrix elements of $D(R, l)$, which can be unstable.

19.6 Cryo-Electron Microscopy

In *cryo-electron microscopy* (cryo-EM) a purified solution consisting of many copies of large biomolecular complexes is flash frozen as a thin layer. The biomolecular complexes are captured at many randomized orientations embedded in the thin layer of surrounding vitreous ice. This sample, which may have tens of thousands of copies of the complex in it, is then taken to an electron microscopy which takes projections along a fixed direction relative to a space-fixed reference frame. This corresponds to projections relative to random and unknown directions relative to the body-fixed reference frame of the molecule. The goal is to reconstruct the 3D shape of the biomolecular complex from these projections [57]. Our discussion of the problem breaks it up into two parts that are addressed separately in the following two subsections.

19.6.1 The Single-Particle Reconstruction Problem

In cryo-electron-microscopy the goal is to reconstruct as accurately as possible the 3D shape of biomolecular complexes composed of multiple proteins from noisy projection data [7, 25, 45, 46, 58, 106, 129, 176]. That is, the goal is to obtain from these projections, the 3D electron density function, $\rho(\mathbf{x})$, that describes the shape of the biomolecular complex. Since the complexes appear at many different orientations relative to the electron beam in the microscope, this is equivalent to taking projections through many different directions in the body fixed frame of a single complex when it can be considered to be rigid. This is related to the computed tomography problem discussed in Chapter 13, but there are several added difficulties associated with the cryo-EM reconstruction problem. First, unlike the computed tomography problem, the projection directions are not known a priori and are not confined to a known plane. Second, the distribution of molecular rotations may not be uniformly random and may not fully sample the space of all projection directions. Third, the projections are noisy due in part to the interaction of the electron beam and imperfections in the ice that surrounds the biomolecular structures. In other words, if $\mathbf{x} = [x, y, z]^T$, a typical projection will be of the form

$$\rho_i'(x, y) = \int_{z \in \mathbb{R}} \rho(R_i^T \mathbf{x}) \, dx + n_i(x, y) \tag{19.39}$$

where $n_i(x, y)$ is a Gaussian noise field consisting of white noises that may be correlated in some way between different pixels in the projection images, and R_i is unknown. From a large number of these measured planar projections, $\rho_i'(x, y)$ for $i = 1, ..., N$, the goal is to recover $\rho(\mathbf{x})$. Typically N is on the order of 10,000 or more.

Cryo-EM has been used extensively in structural biology to understand how individual proteins assemble into large complexes such as viral capsids, ribosomes, and ion channels. Since these complexes are a linchpin between molecular and cellular biophysics, they will remain of interest in the decades to come.

The most effective way to do this in practice has been to use clustering algorithms that divide the N images into M classes that share common characteristics, where M is on the order of 100. The reasoning is that images in the same class correspond to projections from similar directions. Images within the same class are then aligned as best as possible, so that the signal in the projection is reinforced and the noise terms cancel each other. A popular method to implement this "class averageing" is reference-free alignment [111, 128].

19.6.2 Class Averaging

Within each "class" of projections that share a sufficient number of similar attributes to assume that they are coming from the same projection direction, one way to increase the signal-to-noise ration is by "class averaging." In this procedure noisy images are aligned to reinforce the signal and to cause background noise to cancel out [150, 181]. But in the process of doing this superposition, the net effect is to blur the underlying ideal projection of interest.

Since alignment cannot be perfect, the result of any class averaging process will be a blurred version of the true projection that is sought, where the amount of blurring depends on the characteristics of the background noise. In a series of papers [124, 126] it has been shown that this blurring in the image plane of each class is a convolution on $SE(2)$ of the form

$$\rho'(\mathbf{x}') = \int_{SE(2)} f(g)\rho'_0(g^{-1}\mathbf{x}') \, dg \tag{19.40}$$

where $\mathbf{x}' = [x, y]^T$ and $\rho'_0(\mathbf{x}')$ is the pristine image that is desired.

As derived in [125], the probability density function for the misalignments after the matching the centers of mass and principal axes of inertia of the images is given as

$$f(g(x,y,\theta); \xi_\sigma, \xi_\theta) = \frac{1}{2\pi\xi_\sigma^2} e^{-\left(\frac{x^2+y^2}{2\xi_\sigma^2}\right)} \frac{1}{2\sqrt{2\pi}\xi_\theta} \sum_{k=-\infty}^{\infty} \left(e^{\frac{-(\theta-2\pi k)^2}{2\xi_\theta^2}} + e^{\frac{-(\theta-(2k+1)\pi)^2}{2\xi_\theta^2}} \right).$$
$$\tag{19.41}$$

Here $g(x,y,\theta) \in SE(2)$.

While the misalignments of translation forms a unimodal Gaussian distribution, the misalignments of rotation forms a bimodal distribution. This is because an image has two equivalent principal axes whose directions are opposite to each other. Though this ambiguity makes it difficult to determine the rotational alignment, it is easy to have the resulting distribution for the rotational misalignment. It is essentially the average of two Gaussian functions wrapped around the circle with the same standard deviation ξ_θ and two different means, 0 and π. For $N \times N$ images, the parameters in (19.41) are computed directly from the background noise properties as in [125].

Various methods have been applied to solve the inverse problem of recovering ρ'_0 from ρ'. One of the most popular is the maximum likelihood method [150]. In contrast, newer approaches to the problem approach it as a deconvolution over $SE(2)$ [124, 126].

Once this problem of obtaining the best possible class average is solved, the next problem is to recover $\rho(\mathbf{x})$ from the projections $\rho'_m(\mathbf{x}')$ for $m = 1, ..., N$. A classical approach to this is the "method of common lines" [45, 46]. Other new and interesting approaches have been put forth in [181].

19.7 Conformational Mobility of End-Constrained Biopolymers: Modeling Loop Entropy

In Chapter 17, probability densities describing the position and orientation of the distal end of a polymer chain relative to its own proximal end were derived using convolution and harmonic analysis on $SE(3)$. Here we revisit the problem for the case when both the proximal and distal ends are constrained and we examine how much freedom each intermediate residue in the chain has to move.

Modeling the conformational flexibility of loops, and the associated entropy, can be viewed as a more limited version of the protein folding problem. In protein folding, the ultimate goal is to predict the folded structure of a protein from nothing more than a priori knowledge of the amino acid sequence [10, 49, 78]. Very successful algorithms currently rely on additional information from databases of existing structures [140]. Methods from robotics have been brought to bear on both aspects of protein conformation [97, 105] and RNA folding [159, 160]. In the more restricted problem of modeling the conformational variability of loops, the end positions and orientations are given, and each loop represents only a small fraction of the length of a polypeptide chain. Methods for analyzing conformational variability typically involve using methods from robot-arm kinematics, and sampling different discrete solutions, such as in [21, 81, 110, 148]. For more recent work in this area see [2, 43, 107, 133, 156, 166]. Others have studied DNA loops [8, 60, 109, 157], but usually not from the perspective of conformational ensembles, though the methods described here apply equally well to DNA, RNA, and proteins. A few other works address this problem, including [153, 182, 183].

In contrast, the presentation given here views the full ensemble of conformations that satisfy end constraints as a probabilistic problem, the solution to which is a probability density on multiple copies of the motion group $SE(3)$. This follows the presentations in [30, 31]. For example, if probabilites in the Ramachandran plot [135] can be used to obtain polymer-like end-pose distributions for proteins [84] for the whole chain using the methods from Chapter 17 or [26]. However, in this section we use an $SE(3)$-Gaussian approach rather than harmonic analysis.

When considering models of polypeptide chains, it will often be convenient to treat parts of the chain as rigid. For example, the plane of the peptide bond can be considered rigid, as can a cluster of sidechain atoms such a methyl group. At a courser level, one might consider an alpha helix to be a rigid object. At a courser level still, a whole domain might be approximated as a rigid body.

Therefore, it is clear that at various levels of detail, when characterizing the conformational entropy of a protein, it is conceivable that attaching reference frames to the rigid elements and recording the set of all possible rigid-body motions between these elements is a way to describe the conformational part of the Boltzmann distribution, and therefore get at the conformational entropy via Gibbs' formula.

19.7.1 Mathematics of Rigid-Body Motion

The group of rigid-body motions, which is also called the Special Euclidean group and is denoted $SE(3)$, is the semi direct product of $\mathbb{R}^3$ with the special orthogonal group, $SO(3)$. In both instances, the word 'special' means that reflections are excluded and only physically allowable isometries of three-dimensional space are allowed.

We denote elements of $SE(3)$ as $g = (\mathbf{a}, A) \in SE(3)$ where $A \in SO(3)$ and $\mathbf{a} \in \mathbb{R}^3$. For any $g = (\mathbf{a}, A)$ and $h = (\mathbf{r}, R) \in SE(3)$, the group law is written as $g \circ h = (\mathbf{a} + A\mathbf{r}, AR)$, and $g^{-1} = (-A^T\mathbf{a}, A^T)$. Alternately, one may represent any element of $SE(3)$ as a 4×4 homogeneous transformation matrix of the form

$$g = \begin{pmatrix} A & \mathbf{a} \\ \mathbf{0}^T & 1 \end{pmatrix},$$

in which case the group law is matrix multiplication.

For small translational (rotational) displacements from the identity along (about) the ith coordinate axis, the homogeneous transforms representing infinitesimal motions look like[7]

---

[7] These E_i are exactly the $\tilde{X}_i$ discussed in Chapter 10.

$$g_i(\epsilon) \doteq \exp(\epsilon E_i) \approx I + \epsilon E_i$$

where

$$E_1 = \begin{pmatrix} 0 & 0 & 0 & 0 \\ 0 & 0 & -1 & 0 \\ 0 & 1 & 0 & 0 \\ 0 & 0 & 0 & 0 \end{pmatrix} ; \quad E_2 = \begin{pmatrix} 0 & 0 & 1 & 0 \\ 0 & 0 & 0 & 0 \\ -1 & 0 & 0 & 0 \\ 0 & 0 & 0 & 0 \end{pmatrix},$$

$$E_3 = \begin{pmatrix} 0 & -1 & 0 & 0 \\ 1 & 0 & 0 & 0 \\ 0 & 0 & 0 & 0 \\ 0 & 0 & 0 & 0 \end{pmatrix} ; \quad E_4 = \begin{pmatrix} 0 & 0 & 0 & 1 \\ 0 & 0 & 0 & 0 \\ 0 & 0 & 0 & 0 \\ 0 & 0 & 0 & 0 \end{pmatrix} ;$$

$$E_5 = \begin{pmatrix} 0 & 0 & 0 & 0 \\ 0 & 0 & 0 & 1 \\ 0 & 0 & 0 & 0 \\ 0 & 0 & 0 & 0 \end{pmatrix} ; \quad E_6 = \begin{pmatrix} 0 & 0 & 0 & 0 \\ 0 & 0 & 0 & 0 \\ 0 & 0 & 0 & 1 \\ 0 & 0 & 0 & 0 \end{pmatrix},$$

I is the 4×4 identity matrix, ϵ is a real number with small absolute value, and $\exp(X) = I + X + X^2/2 + \cdots$ is the matrix exponential defined by the Taylor series of the usual exponential function evaluated with a matrix rather than a scalar.

The 'exponential parametrization'

$$g = g(\chi_1, \chi_2, ..., \chi_6) = \exp\left(\sum_{i=1}^{6} \chi_i E_i\right) \tag{19.42}$$

is a useful way to describe relatively small rigid-body motions because, unlike the Euler angles, it does not have singularities near the identity.

One defines the 'vee' operator, $\vee$, such that for any

$$X = \sum_{i=1}^{6} \chi_i E_i,$$

$$X^\vee = \begin{pmatrix} \chi_1 \\ \chi_2 \\ \chi_3 \\ \vdots \\ \chi_6 \end{pmatrix}.$$

The 6×6 adjoint matrix, Ad_g, is defined by the expression

$$Ad_g(X^\vee) = (gXg^{-1})^\vee,$$

and explicitly if $g = (\mathbf{a}, A)$ then

$$Ad_g = \begin{pmatrix} A & 0 \\ \hat{\mathbf{a}}A & A \end{pmatrix},$$

where $\hat{\mathbf{a}}A$ denotes the matrix resulting from the cross product of $\mathbf{a}$ with each column of A.

The vector of exponential parameters, $\chi \in \mathbb{R}^6$, can be obtained from $g \in G$ with the formula

$$\chi = (\log g)^\vee. \tag{19.43}$$

The action of an element of the motion group, $g = (A, \mathbf{a})$, on a vector $\mathbf{x}$ in three-dimensional space is defined as $g \cdot \mathbf{x} = A\mathbf{x} + \mathbf{a}$. In contrast, given a function $f(\mathbf{x})$, we can translate and rotate the function by g as $f(g^{-1} \cdot \mathbf{x})$. The fact that the inverse of the transformation applies under the function (rather than the transformation itself) in order to implement the desired motion is directly analogous to the case of translation on the real line. For example, given a function $f(x)$ with its mode at $x = 0$, if we want to translate the whole function in the positive x direction by amount ξ so that the mode is at $x = \xi$, we compute $f(x - \xi)$ (not $f(x + \xi)$). This is a very important point to understand in order for the rest of the chapter to make sense.

19.7.2 Manipulations of Functions of Rigid-Body Motion

Suppose that three rigid bodies labled 0, 1 and 2 are given, with reference frames attached to each, and assume that only sequentially adjacent bodies interact. Suppose also that body 0 is fixed in space and the ensemble of all possible motions of body 1 with respect to 0 are recorded, and motions of 2 with respect to 1 are also recorded. Then we have two functions of motion, $f_{0,1}(g)$ and $f_{1,2}(g)$ which together describe the conformational variability of this simple system. If we are interested in knowing the probability distribution describing the ensemble of all possible ways that body 2 can move relative to body 0, how is this obtained? In fact, it is computed via the convolution on $SE(3)$:

$$f_{0,2}(g) = (f_{0,1} * f_{1,2})(g) = \int_G f_{0,1}(h) \, f_{1,2}(h^{-1} \circ g) \, dh.$$

What this says, is that the distribution $f_{1,2}(g)$ is shifted through all possible rigid-body motions, h, weighted by the frequency of occurrence of these motions, $f_{0,1}(h)$, and integrated over all values of $h \in G$ (G is just short for 'Group', which, throughout this section, is the group of rigid-body motions, $SE(3)$).

Explicitly, what is meant by this integral? Let us assume for the moment that rotations are parameterized using Euler angles. The range of the Euler angles is $0 \leq \alpha, \gamma \leq 2\pi$ and $0 \leq \beta \leq \pi$. In this parameterization the volume element for G is given by

$$dg = \frac{1}{8\pi^2} \sin \beta d\alpha d\beta d\gamma dr_1 dr_2 dr_3,$$

which is the product of the volume elements for $\mathbb{R}^3$ ($d\mathbf{r} = dr_1 dr_2 dr_3$), and for $SO(3)$ ($dR = \frac{1}{8\pi^2} \sin \beta d\alpha d\beta d\gamma$). The normalization factor in the definition of dR is so that $\int_{SO(3)} dR = 1$. The volume element for $SE(3)$ can also be expressed in the exponential coordinates described in the previous subsection, in which case

$$dg = |J(\chi)| d\chi_1 \cdots d\chi_6$$

where $|J(\chi)|$ is a Jacobian determinant for this parameterization. The Jacobian can be computed using the formula

$$J(\chi) = \left[\left(g^{-1} \frac{\partial g}{\partial \chi_1} \right)^\vee, \cdots, \left(g^{-1} \frac{\partial g}{\partial \chi_6} \right)^\vee \right]$$

and it can be shown that $|J(\mathbf{0})| = 1$ and so close to the identity the Jacobian factor in this parameterization can be ignored (which is not true for many other parameterizations, including the Euler angles).

The fact that the volume element is invariant to right and left translations, i.e,

$$dg = d(h \cap g) - d(g \cap h),$$

was discussed in detail in Chapter 6.

A convolution integral of the form

$$(f_{0,1} * f_{1,2})(g) = \int_G f_{0,1}(h) \, f_{1,2}(h^{-1} \circ g) \, dh$$

can be written in the following equivalent ways:

$$(f_{0,1} * f_{1,2})(g) = \int_G f_{0,1}(z^{-1}) \, f_{1,2}(z \circ g) \, dz = \int_G f_{0,1}(g \circ k^{-1}) \, f_{1,2}(k) \, dk \qquad (19.44)$$

where the substitutions $z = h^{-1}$ and $k = h^{-1} \circ g$ have been made, and the invariance of integration under shifts and inversions is used.

The concept of convolution on $SE(3)$ will be central in the formulation that follows.

One can define a Gaussian distribution on the six-dimensional Lie group $SE(3)$ much in the same way as is done on $\mathbb{R}^6$ provided that: (1) the covariances are small; and (2) the mean is located at the identity. The reason for these conditions is that near the identity, $SE(3)$ resembles $\mathbb{R}^6$ which means that $dg \approx d\chi_1 \cdots d\chi_6$. Therefore, we can define the Gaussian in the exponential parameters as

$$f(g(\boldsymbol{\chi})) = \frac{1}{(2\pi)^3 |\Sigma|^{\frac{1}{2}}} \exp(-\frac{1}{2}\boldsymbol{\chi}^T \Sigma^{-1} \boldsymbol{\chi}). \qquad (19.45)$$

Consider two such distributions enumerated by consecutive values of the integer i: $f_{i,i+1}(g)$ and $f_{i+1,i+2}(g)$. Each can be shifted to its mean, e.g., $f_{i,i+1}(g_{i,i+1}^{-1} \circ g)$. Moreover, each will have a 6×6 covariance, e.g., $\Sigma_{i,i+1}$. We seek the mean and covariance of their convolution. Without loss of generality, take $i = 0$. It can be shown from the discussion of covariance propagation in Chapter 15 that when the covariances of the two functions being convolved are small, the mean of the convolution $f_{0,1}(g_{0,1}^{-1} \circ g) * f_{1,2}(g_{1,2}^{-1} \circ g)$ will be $g_{0,2} \approx g_{0,1} \circ g_{1,2}$ and the covariance of the convolution will be

$$\Sigma_{0,2} \approx Ad_{g_{1,2}}^{-1} \Sigma_{0,1} Ad_{g_{1,2}}^{-T} + \Sigma_{1,2}. \qquad (19.46)$$

This provides a method for computing covariances of two concatenated segments, and this formula can be iterated to compute covariances of chains without having to compute convolutions directly.

19.7.3 Approximating Entropy of the Loops in the Folded Ensemble

The native ensemble of a protein is characterized by a relatively high degree of order. However, the native form is not completely rigid. In particular, loop/coil regions connecting secondary structures can exhibit large motions. Bounds on the contribution of loop motions to overall entropy are discussed here.

If $f(g_1, ..., g_n)$ is the conformational distribution function describing the positions and orientations of all bodies in a chain with respect to its proximal end, and if we fix the distal end at a specific pose, g_{end}, then the resulting distribution will the the conditional density

$$f^{fix}(g_1, g_2, ..., g_{n-1}; g_{end}) = f(g_1, ..., g_{n-1}|g_n = g_{end}) = f(g_1, ..., g_{n-1}, g_{end})/f_{0,n}(g_{end}).$$
(19.47)

The entropy of this distribution in some cases can either be computed directly, or each of the marginals can be computed as

$$f^{fix}_{0,i}(g_i; g_{end}) = f_{0,i}(g_i)\, f_{i,n}(g_i^{-1} \circ g_{end})/f_{0,n}(g_{end}).$$
(19.48)

The reason is that in the definition of $f(g_1, ..., g_n)$, the variable g_i appears in only the two multiplied terms: $f_{i-1,i}(g_{i-1}^{-1} \circ g_i) \cdot f_{i,i+1}(g_i^{-1} \circ g_{i+1})$. Marginalizing over g_1 through g_{i-1} results in $f_{0,i}(g_i)$. If g_i was the identity element, then marginalizing over g_{i+1} through g_{n-1} would yield $f_{i,n}(g_{end})$. However, because $g_i \neq e$, this result is shifted by g_i to yield $f_{i,n}(g_i^{-1} \circ g_{end})$. Division by $f_{0,n}(g_{end})$ is the nornalization required to make the result a PDF (since integration of the numerator in (19.48) over g_i is a convolution). This denominator is carried along from (19.47).

Intuitively, the entropy of a chain with fixed ends must be smaller than that of a chain with freely moving ends. This an be quantified when using (19.47) and (19.48), as will be shown in the examples in the next subsection.

19.7.4 Short Loops Modeled as Semiflexible Polymers

Suppose that we have a semiflexible loop. Then each g_i will deviate only a relatively small amount from a constant reference pose, h_i, and so we write $g_i = h_i \circ \exp \chi_i$ where $\|\chi_i\| < 1$. The relative motion between adjacent reference frames will be

$$g_i^{-1} \circ g_{i+1} = \exp -\chi_i \circ h_i^{-1} \circ h_{i+1} \circ \exp \chi_{i+1}.$$

If the probability density $f_{i,i+1}$ is a Gaussian with mean at $h_i^{-1} \circ h_{i+1}$, then it will be of the form

$$f_{i,i+1}(g) = F_{\Sigma_{i,i+1}}((h_i^{-1} \circ h_{i+1})^{-1} \circ g)$$

where

$$F_{\Sigma_{i,i+1}}(\exp \chi) = \frac{1}{(2\pi)^3 |\Sigma_{i,i+1}|^{\frac{1}{2}}} \exp(-\frac{1}{2}\chi^T \Sigma_{i,i+1}^{-1} \chi).$$

Therefore,

$$f_{i,i+1}(g_i^{-1} \circ g_{i+1}) = F_{\Sigma_{i,i+1}}((h_i^{-1} \circ h_{i+1})^{-1} \circ \exp -\chi_i \circ h_i^{-1} \circ h_{i+1} \circ \exp \chi_{i+1}).$$

For small motions between adjacent bodies, the approximation

$$[\log((h_i^{-1} \circ h_{i+1})^{-1} \circ \exp -\chi_i \circ h_i^{-1} \circ h_{i+1} \circ \exp \chi_{i+1})]^\vee \approx \chi_{i+1} - Ad_{h_i^{-1} \circ h_{i+1}}^{-1} \chi_i$$

is accurate. If we use the shorthand $A_{i,i+1} \doteq Ad_{h_i^{-1} \circ h_{i+1}}$, then

$$f_{i,i+1}(g_i^{-1} \circ g_{i+1}) =$$

$$\frac{1}{(2\pi)^3 |\Sigma_{i,i+1}|^{\frac{1}{2}}} \exp -\frac{1}{2}[\chi_i^T, \chi_{i+1}^T] \begin{pmatrix} \Sigma_{i,i+1}^{-1} & -A_{i,i+1}^{-T}\Sigma_{i,i+1}^{-1} \\ -\Sigma_{i,i+1}^{-1}A_{i,i+1}^{-1} & A_{i,i+1}^{-T}\Sigma_{i,i+1}^{-1}A_{i,i+1}^{-1} \end{pmatrix} \begin{bmatrix} \chi_i \\ \chi_{i+1} \end{bmatrix}.$$
(19.49)

This, together with the assumption of pairwise independence leads $f(g_1, g_2, ..., g_n)$ to be a Gaussian distribution on $SE(3)^n$ of the form

$$f(g_1, g_2, ..., g_n) = \prod_{i=0}^{n-1} f_{i,i+1}(g_i^{-1} \circ g_{i+1})$$

in the variable $\chi = [\chi_1^T, ..., \chi_n^T]^T$ with an inverse covariance of the form

$$\Sigma^{-1} =$$

$$
\begin{pmatrix}
\Sigma_{0,1}^{-1} + \Sigma_{1,2}'^{-1} & -A_{1,2}^{-T}\Sigma_{1,2}^{-1} & 0 & 0 & \cdots & 0 & 0 \\
-\Sigma_{1,2}^{-1}A_{1,2}^{-1} & \Sigma_{1,2}^{-1} + \Sigma_{2,3}'^{-1} & -A_{2,3}^{-T}\Sigma_{2,3}^{-1} & 0 & \cdots & 0 & 0 \\
0 & -\Sigma_{2,3}^{-1}A_{2,3}^{-1} & \ddots & \ddots & \ddots & \vdots & \vdots \\
0 & 0 & \ddots & \ddots & \ddots & 0 & 0 \\
\vdots & \vdots & \ddots & -\Sigma_{n-3,n-2}^{-1}A_{n-3,n-2}^{-1} & \Sigma_{n-2,n-1}^{-1} + \Sigma_{n-3,n-2}'^{-1} & -A_{n-2,n-1}^{-T}\Sigma_{n-2,n-1}^{-1} & 0 \\
0 & \cdots & 0 & 0 & -\Sigma_{n-2,n-1}^{-1}A_{n-2,n-1}^{-1} & \Sigma_{n-2,n-1}^{-1} + \Sigma_{n-1,n}'^{-1} & -A_{n-1,m}^{-T}\Sigma_{n-1,n}^{-1} \\
0 & \cdots & 0 & 0 & 0 & 0 & \Sigma_{n-1,n}^{-1}
\end{pmatrix}
$$

$$(19.50)$$

where $\Sigma_{i,i+1}' = A_{i,i+1}\Sigma_{i,i+1}A_{i,i+1}^T$ and $g_0 \doteq e$.

The entropy for the case with free ends is then given by a closed-form formula in [31], which is relatively efficient to compute due to the block tri-diagonal form of Σ^{-1}.

Obtaining the entropy for the case when both ends are fixed in position and orientation is also possible withing this model. Using the fact that the adjoint is a homomorphism, i.e., $Ad(g_1 \circ g_2) = Ad(g_1)Ad(g_2)$ and $Ad(g^{-1}) = Ad^{-1}(g)$, this generalizes to the concatenation of n reference frames that vary around values h_i as

$$\Sigma_{0*1*\cdots*n} = \sum_{k=0}^{n} Ad_{h_{k,n}}^{-1} \Sigma_k \, Ad_{h_{k,n}}^{-T} \,. \tag{19.51}$$

In order to compute the entropy for the case when the distal end of the chain is fixed at $g_n = g_{end}$, we would use (19.47) with the covariance of $f_{0,n}(g)$ being given by (19.51). Conditioning of Gaussians by Gaussians yields Gaussians, and the entropy of which can be computed in closed form as discussed in [31]. Modeling loop entropy has a number of potential applications, including assisting in understanding the change in entropy in going from unfolded to folded protein structures, as well as computing entropy changes in going from free to packed nucleic acids, as studied in [134].

19.8 Summary

Problems that arise in studying biomolecular structure, motion, and interactions involve rigid-body motions and functions of motion. Moreover, the analysis of geometric patterns in large databases of molecules, and experimental methodologies for obtaining information about the shape of molecules both involve group representation theory and harmonic analysis at various levels. This chapter has provided has addressed selected topics in these areas. Other experimental modalities such as small-angle x-ray scattering (SAXS) [89, 132] and residual dipolar couplings (RDCs)[12, 162, 163] in the context of nuclear magnetic resonance experiments are other extremely promising methods that may benefit from harmonic analysis, but are not covered here. For further reading on how motion and experimental biophysical measurements are related see [20, 50, 103]. For other computational aspects of "protein kinematics" not covered here such as analysis of conformational mobility of proteins see [48, 85].

References

1. Adams, P.D., Afonine, P.V., Bunkóczi, G., Chen, V.B., Davis, I.W., Echols, N., Headd, J.J., Hung, L.-W., Kapral, G.J., Grosse-Kunstleve, R.W., McCoy, A.J., Moriarty, N.W., Oeffner, R., Read, R.J., Richardson, D.C., Richardson, J.S., Terwilliger, T.C., Zwart, P.H., "PHENIX: A Comprehensive Python-Based System for Macromolecular Structure Solution," *Acta Crystallographica Section D: Biological Crystallography*, 66(2): 213 – 221, 2010.
2. Al-Nasr, K., He, J., "An Effective Convergence Independent Loop Closure Method Using Forward-Backward Cyclic Coordinate Descent," *International Journal of Data Mining and Bioinformatics*, 3(3): 346 – 361, 2009.
3. Allen, M.P., Tildesley, D.J., *Computer Simulation of Liquids*, Oxford University Press, Oxford, England, 1987.
4. Aroyo, M.I., Perez-Mato J.M., Capillas, C., Kroumova, E., Ivantchev, S., Madariaga, G., Kirov, A., Wondratschek, H., "Bilbao crystallographic server: I. Databases and Crystallographic Computing Programs," *Zeitschruft für Kristallographie*, 221(1): 15 – 27, 2006.
5. Bajaj, C. L., Chowdhury, R., Siddahanavalli, V., "F^2 Dock: Fast Fourier Protein – Protein Docking," *IEEE/ACM Transactions on Computational Biology and Bioinformatics (TCBB)*, 8(1): 45 – 58, 2011.
6. Bajaj, C., Bauer, B., Bettadapura, R., Vollrath, A., "Nonuniform Fourier Transforms for Rigid – Body and Multidimensional Rotational Correlations," *SIAM Journal on Scientific Computing*, 35(4): B821 – B845, 2013.
7. Baker, T.S., Cheng, R.H., "A Model-Based Approach for Determining Orientations of Biological Macromolecules Imaged by Cryo-Electron Microscopy," *Journal of Structural Biology*, 116(1): 120 – 130, 1996.
8. Balaeff, A., Mahadevan, L., Schulten, K., "Modeling DNA Loops Using the Theory Of Elasticity," *Physical Review E*, 73(3): 031919, 2006.
9. Baldwin, P.R., Penczek, P.A., "Estimating Alignment Errors in Sets of 2 – D Images," *Journal of Structural Biology*, 150(2): 211 – 225, 2005.
10. Baldwin, R.L., Rose, G.D., "Is Protein Folding Hierarchic? I. Local Structure and Peptide Folding," *Trends in Biochemical Sciences*, 24(1): 26 – 33, 1999.
11. Barlow, W., "Über die Geometrischen Eigenschaften homogener starrer Strukturen und ihre Anwendung auf Krystalle," Zeitschrift für Krystallographie und Mineralogie, 23: 1-63, 1894.
12. Bax, A., Grishaev, A., "Weak Alignment NMR: A Hawk-Eyed View of Biomolecular Structure," *Current Opinion in Structural Biology*, 15(5): 563 – 570, 2005.
13. Beck, T.L., Paulaitis, M.E., Pratt, L.R., *The Potential Distribution Theorem and Models of Molecular Solutions*, Cambridge University Press, 2006.
14. Berman, H.M., Battistuz, T., Bhat, T.N., Bluhm, W.F., Bourne, P.E., Burkhardt, K., Feng, Z., Gilliland, G.L., Iype, L., Jain, S., Fagan, P., Marvin, J., Padilla, D., Ravichandran, V., Schneider, B., Thanki, N., Weissig, H., Westbrook, J.D., Zardecki, C., "The Protein Data Bank," *Acta Crystallographica D*, 58: 899 – 907, 2002.

15. Bonahon, F., Siebenmann, L., "The Classification of Seifert Fibred 3-Orbifolds," in *Low Dimensional Topology*, R. Fenn, ed., London Mathematical Society Lecture Notes, 95, Cambridge University Press, 1985.

16. Bowie, J.U., "Helix Packing Angle Preferences," *Nature Structural Biology*, 4(11): 915–917, 1997.

17. Bradley, C., Cracknell, A., *The Mathematical Theory of Symmetry in Solids*, Oxford University Press, 1972/2010.

18. Burns, G., Glazer, A.M., *Space Groups for Solid State Scientists*, 2nd ed., Academic Press, Boston, 1990.

19. Caliandro, R., Carrozzini, B., Cascarano, G.L., Giacovazzo, C., Mazzone, A., Siliqi, D., "Molecular Replacement: The Probabilistic Approach of the Program REMO09 and Its Applications," *Acta Crystallographica A*, 65: 512-527, 2009.

20. Cantor, C.R., Schimmel, P.R., *Biophysical Chemistry, Parts I-III*, W.H. Freeman and Company, New York, 1980.

21. Canutescu, A.A., Dunbrack, R.L., "Cyclic coordinate descent: A robotics algorithm for protein loop closure," *Protein Science*, 12(5): 963–972, May 2003.

22. Chakrabarti, P., Debnath, P., "The Interrelationships of Side-chain and Main-chain Conformations in Proteins," *Progress in Biophysics & Molecular Biology*, 76: 1–102, 2001.

23. Chandler, D., Andersen, H.C., "Optimized Cluster Expansions for Classical Fluids. II Theory of Molecular Liquids," *Journal of Chemical Physics*, 57(5): 1930–1937, 1972.

24. Charlap, L.S., *Bieberbach Groups and Flat Manifolds*, Springer-Verlag, New York, 1986.

25. Cheng, Y., Walz, T., "The Advent of Near-Atomic Resolution in Single-Particle Electron Microscopy," *Annual Review of Biochemistry*, 78: 723–742, 2009.

26. Chirikjian, G.S., Kyatkin, A.B., *Engineering Applications of Noncommutative Harmonic Analysis*, CRC Press, Boca Raton, FL, 2001.

27. Chirikjian, G.S., *Stochastic Models, Information Theory, and Lie Groups: Volumes I + II*, Birkhäuser, Boston, 2009/2012.

28. Chirikjian, G.S., "Kinematics Meets Crystallography: The Concept of a Motion Space," *Journal of Computing and Information Science in Engineering* 15(1): 011012, 2015.

29. Chirikjian, G.S., "Rigid–Body Parameters for Molecular Docking Applications," Proc. ASME Design Technical Meetings, DETC2014–34246, Buffalo, NY August 2014.

30. Chirikjian, G.S., "Group Theory and Biomolecular Conformation, I.: Mathematical and Computational Models, *Journal of Physics: Condensed Matter*, 22: 323103, 2010.

31. Chirikjian, G.S., "Modeling Loop Entropy," *Methods in Enzymology*, Part C, 487: 101–130, 2011.

32. Chirikjian, G.S., Kim, P.T., Koo, J.Y., Lee, C.H., "Rotational Matching Problems," *International Journal of Computational Intelligence and Applications*, 4(4): 401–416, Dec. 2004.

33. Chirikjian, G.S., "Mathematical Aspects of Molecular Replacement: I. Algebraic Properties of Motion Spaces," *Acta Crystallographica A*, 67(5): 435-446, 2011.

34. Chirikjian, G.S., Yan, Y., "Mathematical Aspects of Molecular Replacement: II. Geometry of Motion Spaces," *Acta Crystallographica A*, 68(2): 208–221, 2012.

35. Chirikjian, G.S., Sajjadi, S., Toptygin, D., "Mathematical Aspects of Molecular Replacement. III. Properties of Space Groups Preferred by Proteins in the Protein Data Bank," *Acta Crystallographica A*, 71(2): 186-194, March 2015.

36. Chirikjian, G.S., Shiffman, B., "Collision-Free Configuration-Spaces in Macromolecular Crystals," Workshop on Robotics Methods for Structural and Dynamic Modeling of Molecular Systems, Robotics Science and Systems Conference, Berkeley, California, July 12, 2014 https://cs.unm.edu/amprg/rss14workshop/PAPERS/Chirikjian.pdf

37. Chirikjian, G.S., Ratnayake, K., Sajjadi, S., "Decomposition of Sohncke Space Groups into Products of Bieberbach and Symmorphic Parts," *Zeitschrift für Kristallographie*, 230(12): 719–741, 2015.

38. Chothia, C., Levitt, M., Richardson, D., "Structure of Proteins: Packing of α-Helices and Pleated Sheet," *Proceedings of the National Academy of Sciences of the USA*, 74(10): 4130–4134, 1977.

39. Chothia, C., Levitt, M., Richardson, D., "Helix to Helix Packing in Proteins," *Journal of Molecular Biology*, 145(1): 215–250, 1981.

40. Collaborative Computational Project Number 4, "The CCP4 Suite: Programs for Protein Crystallography," *Acta Crystallographica D*, 50: 760–766, 1994. http://www.ccp4.ac.uk/

41. Conway, J.H., Delgado Friedrichs, O., Huson, D.H., Thurston, W.P., "On Three-Dimensional Space Groups," *Beiträge zur Algebra und Geometrie*, 42(2): 475–507, 2001.

42. Coutsias, E.A., Seok, C., Dill, K.A., "Using Quaternions to Calculate RMSD," *Journal of Computational Chemistry*, 25(15): 1849-1857, 2004.

43. Coutsias, E.A., Seok, C., Jacobson, M.P., Dill, K.A., "A Kinematic View of Loop Closure.," *Journal of Computational Chemistry*, 25(4): 510–528, 2004.

44. Crick, F., "The Packing of α–Helices: Simple Coiled-Coils," *Acta Crystallographica*, 6(8–9): 689–697, 1953.

45. Crowther, R.A., Amos, L.A., Klug, A., "Three Dimensional Reconstruction of Spherical Viruses by Fourier Synthesis from Electron Micrographs," *Nature*, 226: 421–425, 1970.

46. Crowther, R.A., DeRosier, D.J., Klug, A., "The Reconstruction of a Three-Dimensional Structure from Projections and Its Application to Electron Microscopy," *Proceedings of the Royal Society of London. A. Mathematical and Physical Sciences*, 317(1530): 319–340, 1970.

47. Crowther, R.A., "The Fast Rotation Function," in *The Molecular Replacement Method*, M.G. Rossmann, ed., pp. 173–178. Gordon & Breach, 1972.

48. Das, P., Moll, M., Stamati, H., Kavraki, L.E., Clementi, C., "Low-Dimensional, Free-Energy Landscapes of Protein-Folding Reactions by Nonlinear Dimensionality Reduction," *Proceedings of the National Academy of Sciences*, 103(26): 9885–9890, 2006.

49. Dill, K.A., Bromberg, S., *Molecular Driving Forces: Statistical Thermodynamics in Chemistry and Biology*, Garland Science/Taylor and Francis, New York, 2003.

50. Donald, B.R., *Algorithms in Structural Molecular Biology*. The MIT Press, 2011.

51. Dunbar, W.D., *Fibered Orbifolds and Crystallographic Groups*, PhD Dissertation, Department of Mathematics, Priceton University, June, 1981.

52. Engel, P., *Geometric Crystallography: An Axiomatic Introduction to Crystallography*, D. Reidel Publishing Co., Kluwer, Boston, 1986.

53. Evans, D.J., Morriss, G., *Statistical Mechanics of Nonequilibrium Liquids*, 2nd ed, Cambridge University Press, Cambridge, 2008.

54. Evarestov, R.A., Smirnov, V.P., *Site Symmetry in Crystals: Theory and Applications*, 2nd ed., Springer, New York, 1993.

55. Fedorov, E.S., *Symmetry of Crystals*, ACA Monograph Number 7, 1971 (English translation of 1885 book in Russian).

56. Feig, M., ed., *Modeling Solvent Environments: Applications to Simulations and Biomolecules*, Wiley–VCH, Weinheim, 2010.

57. Frank, J., *Three Dimensional Electron Microscopy of Macromolecular Assemblies*, 2nd ed., Academic Press, 2006.

58. Frank, J., Radermacher, M., Penczek, P., Zhu, J., Li, Y., Ladjadj, M., Leith, A., "SPIDER and WEB: Processing and Visualization of Images in 3D Electron Microscopy and Related Fields," *Journal of Structural Biology*, 116(1): 190–199, 1996.

59. Garzon, J. I., Lopéz–Blanco, J. R., Pons, C., Kovacs, J., Abagyan, R., Fernandez–Recio, J., Chacon, P., "FRODOCK: A New Approach for Fast Rotational Protein-Protein Docking," *Bioinformatics*, 25(19): 2544–2551, 2009.

60. Goyal, S., Lillian, T., Blumberg, S., Meiners, J.C., Meyhöfer, E., Perkins, N.C., "Intrinsic Curvature of DNA Influences LacR-Mediated Looping," *Biophysical Journal*, 93(12): 4342–4359, 2007.

61. Gray, C.G., Gubbins, K.E., *Theory of Molecular Fluids, Vol. 1: Fundamentals*, Clarendon Press, Oxford, England, 1984.

62. Gray, J.J., Moughon, S., Wang, C., Schueler-Furman, O., Kuhlman, B., Rohl, C.A., Baker, D., "Protein-Protein Docking with Simultaneous Optimization of Rigid–Body Displacement and Side–Chain Conformations," *Journal of Molecular Biology*, 331(1): 281–299, 2003.

63. Grosse-Kunstleve, R.W., Wong, B., Mustyakimov, M., Adams, P.D., "Exact Direct-Space Asymmetric Units for the 230 Crystallographic Space Groups," *Acta Crystallographica A*, 67(3): 269–275, 2011.
64. Hahn, T., ed., *International Tables for Crystallography, Space Group Symmetry: Brief Teaching Edition*, 2002.
65. Hammond, C., *The Basics of Crystallography and Diffraction*, Oxford University Press, Oxford, England, 1997.
66. Hansen, J.-P., McDonald, I.R., *Theory of Simple Liquids*, 3rd ed., Academic Press, 2006.
67. Hilton, H., *Mathematical Crystallography and the Theory of Groups of Movements*, 1903, (Dover Publications Inc. 1963).
68. Hirata, F. (ed.), *Molecular Theory of Solvation*, Kluwer Academic Publishers, Dordrecht, The Netherlands, 2003.
69. Hirshfeld, F.L., "Symmetry in the Generation of Trial Structures," *Acta Crystallographica A* 24: 301–311, 1968.
70. Iversen, B., *Lectures on Crystallographic Groups*, Aarhus Universitet Matematisk Institut Lecture Series 1990/91 No. 60, Fall 1990.
71. Jamrog, D.C., Zhang, Y., Phillips Jr., G.N., "SOMoRe: A Multi-Dimensional Search and Optimization Approach to Molecular Replacement," *Acta Crystallographica D*, 59:304–314, 2003.
72. Janssen, T., *Crystallographic Groups*, North Holland/Elsevier, New York, 1973.
73. Jeong, J., Lattman, E., Chirikjian, G.S., "A Method for Finding Candidate Conformations for Molecular Replacement Using Relative Rotation Between Domains of a Known Structure," *Acta Crstallographica D*, D62: 398–409, 2006.
74. Johnson, C.K., Burnett, M.N., Dunbar, W.D., "Crystallographic Topology and Its Applications," in *Crystallographic Computing 7 Macromolecular Cystallographic Data*, edited by P. E. Bourne and K. D. Watenpaugh, Oxford University Press, 1997. http://www.ornl.gov/sci/ortep/topology/preprint.html
75. Julian, M.M., *Foundations of Crystallography with Computer Applications*, CRC Press, Taylor and Francis Group, 2008.
76. Kabsch, W., "A Solution for the Best Rotation to Relate Two Sets of Vectors," *Acta Crystallographica A*, 32(5): 922–923, 1976.
77. Kabsch, W., "A Discussion of the Solution for the Best Rotation to Relate Two Sets of Vectors," *Acta Crystallographica A*, 34(5): 827–828, 1978.
78. Karplus, M., Weaver, D.L., "Protein-Folding Dynamics," *Nature*, 260(5550): 404–406, 1976.
79. Katchalski-Katzir, E., Shariv, I., Eisenstein, M., Friesem, A., Aflalo, C., Vakser, I., "Molecular Surface Recognition: Determination of Geometric Fit Between Proteins and Their Ligands by Correlation Techniques," *Proceedings of the National Academy of Science of the USA*, 89: 2195-2199, 1992.
80. Kavraki, L.E., Svestka, P., Latombe, J.C., Overmars, M.H., " Probabilistic Roadmaps for Path Planning in High-Dimensional Configuration Spaces," *IEEE Transactions on Robotics and Automation*, 12(4): 566–580, Aug. 1996.
81. Kazerounian, K., Latif, K., Rodriguez, K., Alvarado, C., "Nano-Kinematics for Analysis of Protein Molecules," *Journal of Mechanical Design*, 127(4): 699–711, July 2005.
82. Kazhdan, M., "An Approximate and Efficient Method for Optimal Rotation Alignment of 3d Models," *IEEE Transactions on Pattern Analysis and Machine Intelligence*, 29(7): 1221–1229, 2007.
83. Kazhdan, M., Funkhouser, T., Rusinkiewicz, S., "Rotation Invariant Spherical Harmonic Representation of 3D Shape Descriptors," in *Proceedings of the 2003 Eurographics/ACM SIGGRAPH symposium on Geometry processing. SGP 03*, pp. 156-164, Aire–la–Ville, Switzerland, Switzerland: Eurographics Association, 2003.
84. Kim, J.S., Chirikjian, G.S., "A Unified Approach to Conformational Statistics of Classical Polymer and Polypeptide Models," *Polymer*, 46(25): 11,904–11,917, Nov. 28 2005.

85. Kim, M.K., Jernigan, R.L., Chirikjian, G.S., "Rigid–Cluster Models of Conformational Transitions in Macromolecular Machines and Assemblies," *Biophysical Journal*, 89(1): 43–55, July 2005.

86. Kim, S.K., *Group Theoretical Methods and Applications to Molecules and Crystals*, Cambridge University Press, 1999.

87. Kissinger, C.R., Gelhaar, D.K., Fogel, D.B., "Rapid Automated Molecular Replacement by Evolutionary Search," *Acta Crystallographica D*, 55(2): 484–491, 1999.

88. Kneller, G.R., "Comment on 'Using Quaternions to Calculate RMSD' [J. Comp. Chem. 25, 1849 (2004)]," *Journal of Computational Chemistry*, 26(15): 1660–1662, 2005.

89. Konarev, P.V., Volkov, V.V., Sokolova, A.V., Koch, M.H., Svergun, D.I., "PRIMUS: A Windows PC-Based System for Small-Angle Scattering Data Analysis," *Journal of Applied Crystallography*, 36(5): 1277–1282, 2003.

90. Kovacs, J. A., Wriggers, W., "Fast Rotational Matching," *Acta Crystallographica Section D: Biological Crystallography*, 58(8): 1282–1286, 2002.

91. Kovacs, J.A., Chacón, P., Cong, Y., Metwally, E., Wriggers, W., "Fast Rotational Matching of Rigid Bodies by Fast Fourier Transform Acceleration of Five Degrees of Freedom," *Acta Crystallographica Section D: Biological Crystallography*, D59(8): 1371–1376, 2003.

92. Kroumova, E., Capillas, C., Aroyo, M.I., Perez-Mato, J.M., Ivantchev, S., Madariaga, G., Kirov, A., Wondratschek, H., Stokes, H., Hatch, D., "The Bilbao Crystallographic Server," IOP Conference Series 173:383-386, 2003.

93. Ladd, M.F.C., *Symmetry in Molecules and Crystals*, Ellis Horwood Limited/John Wiley and Sons, New York, 1989.

94. Lattman, E.E., Love, W.E., "A Rotational Search Procedure for Detecting a Known Molecule in a Crystal," *Acta Crystallographica B*26: 1854–1857, 1970.

95. Lattman, E.E., Loll, P.J., *Protein Crystallography: A Concise Guide*, The Johns Hopkins University Press, Baltimore, Maryland, 2008.

96. Lattman, E.E., "Optimal Sampling of the Rotation Function," *Acta Crystallographica*, B28: 1065–1068,1972.

97. Lavalle, S.M., Finn, P.W., Kavraki, L.E., Latombe, J.C., "A Randomized Kinematics-Based Approach to Pharmacophore-Constrained Conformational Search and Database Screening," *Journal of Computational Chemistry*, 21(9): 731–747, July 2000.

98. Lee, S., Chirikjian, G.S., "Inter–Helical Angle and Distance Preferences in Globular Proteins," *Biophysical Journal*, 86(2): 1105–1117, Feb. 2004.

99. Lee S., Chirikjian G.S., "Pose Analysis of Alpha–Carbons in Proteins," *International Journal of Robotics Research*, 24(2–3): 183–210, Feb.–March 2005

100. Lee, K., Wang, Y., Chirikjian, G.S.. "$O(n)$ Mass Matrix Inversion for Serial Manipulators and Polypeptide Chains Using Lie Derivatives," *Robotica*, 25(6): 739–750, 2007.

101. Levitt, M., Chothia, C., "Structural Patterns in Globular Proteins," *Nature*, 261(5561): 552–558, 1976.

102. Levitt, M., "Conformational Preferences of Amino Acids in Globular Proteins," *Biochemistry*, 17(20): 4277–4285, 1978.

103. Levitt, M.H., *Spin Dynamics: Basics of Nuclear Magnetic Resonance*, 2nd ed., John Wiley and Sons, New York, 2008.

104. Lockwood, E.H., MacMillan, R.H., *Geometric Symmetry*, Cambridge University Press, Cambridge England, 1978.

105. Lotan, I., Schwarzer, F., Halperin, D., Latombe, J.C., "Algorithm and Data Structures for Efficient Energy Maintenance During Monte Carlo Simulation of Proteins," *Journal of Computational Biology*, 11(5): 902–932, Oct. 2004.

106. Ludtke, S.J., Baldwin, P.R., Chiu, W., "EMAN: Semiautomated Software for High-Resolution Single-Particle Reconstructions," *Journal of Structural Biology*, 128(1): 82–97, 1999.

107. Mandell, D.J., Coutsias, E.A., Kortemme, T., "Sub-Angstrom Accuracy in Protein Loop Reconstruction by Robotics-Inspired Conformational Sampling," *Nature Methods*, 6: 551-552, 2009.

108. Mandell, J.G., Roberts, V.A., Pique, M.E., Kotlovyi, V., Mitchell, J.C., Nelson, E., Tsigelny, I., Eyck, L.F.T., "Protein Docking Using Continuum Electrostatics and Geometric Fit," *Protein Engineering Design and Selection*, 14(2): 105-113, 2000.
109. Manning, R.S., Maddocks, J.H., Kahn, J.D., "A Continuum Rod Model of Sequence–Dependent DNA Structure," *Journal of Chemical Physics*, 105(13): 5626–5646, 1996.
110. Manocha, D., Zhu, Y.S., Wright, W., "Conformational-Analysis of Molecular Chains Using Nano-Kinematics," *Computer Applications in the Biosciences*, 11(1): 71–86, Feb. 1995.
111. Marco, S., Chagoyen, M., de la Fraga, L.G., Carazo, J.M., Carrascosa, J.L., "A Variant to the 'Random Approximation' of the Reference–Free Alignment Algorithm," *Ultramicroscopy*, 66(1–2): 5–10, 1996.
112. McPherson, A., *Introduction to Macromolecular Crystallography*, John Wiley and Sons, Hoboken, NJ, 2003.
113. Miller, W. Jr., *Symmetry Groups and Their Applications*, Academic Press, New York, 1972.
114. Mitchell, J.C., "Discrete Uniform Sampling of Rotation Groups Using Orthogonal Images," *SIAM Journal of Scientific Computing*, 30(1): 525-547, 2007.
115. Mitchell, J. C., "Sampling Rotation Groups by Successive Orthogonal Images," *SIAM Journal on Scientific Computing*, 30(1): 525–547, 2008.
116. Miyazawa, S., Jernigan, R.L., "Estimation of Effective Interresidue Contact Energies from Protein Crystal-Structures– Quasi-Chemical Approximation," *Macromolecules*, 18(3): 534–552, 1985.
117. Montesinos, J.M., *Classical Tessellations and Three-Manifolds*, Springer-Verlag, Berlin, 1987.
118. Navaza, J., "AMoRe: An Automated Package for Molecular Replacement," *Acta Crystallographica A*, 50(2): 157–163, 1994.
119. Navaza, J., "On the Fast Rotation Function," *Acta Crystallographica A*, 43(5): 645–653, 1987.
120. Navaza, J., "On the Computation of the Fast Rotation Function," *Acta Crystallographica D*, 49(6): 588–591, 1993.
121. Nespolo, M., "Does Mathematical Crystallography Still Have a Role in the XXI Century?" *Acta Crystallographica A*, 64: 96-111, 2008.
122. Nikulin, V.V., Shafarevich, I.R., *Geometries and Groups*, (M. Reid, translator), Springer, New York, 1987.
123. Ornstein, L.S., Zernike, F., "Accidental Deviations of Density and Opalescence at the Critical Point of aSingle Substance," *Proc. Akad. Sci. (Amst.)* 17:793, 1914.
124. Park, W., Chirikjian, G.S., "An Assembly Automation Approach to Alignment of Non–Circular Projections in Electron Microscopy," *IEEE Transactions on Automation Science and Engineering*, 11(3): 668–679, 2014.
125. Park, W., Midgett, C.R., Madden, D.R., Chirikjian, G.S., "A Stochastic Kinematic Model of Class Averaging in Single-Particle Electron Microscopy," *International Journal of Robotics Research*, 30(6): 730–754, May 2011.
126. Park, W., Madden, D.R., Rockmore, D.N., Chirikjian, G.S., "Deblurring of Class-Averaged Projections in Electron Tomography," *Inverse Problems*, 26(3): 3500502, March 2010.
127. Patriciu, A., Chirikjian, G.S., Pappu, R.V., "Analysis of the Conformational Dependence of Mass-Metric Tensor Determinants in Serial Polymers with Constraints," *Journal of Chemical Physics*, 121(24): 12708–12720, Dec. 2004.
128. Penczek, P., Radermacher, M., Frank, J., *Three-Dimensional Reconstruction of Single Particles Embedded in Ice*, Ultramicroscopy, 40(1): 33–53, 1992.
129. Penczek, P.A., Zhu, J., Frank, J., "A Common-Lines Based Method for Determining Orientations for $N > 3$ Particle Projections Simultaneously," *Ultramicroscopy*, 63(3): 205–218, 1996.
130. Percus, J.K., Yevick, G.J., "Analysis of Classical Statistical Mechanics by Means of Collective Coordinates," *Physical Review*, 110: 1, 1958.

131. Percus, J.K., "Approximation Methods in Classical Statistical Mechanics," *Physical Review Letters*, 8(11): 462, 1962.

132. Petoukhov, M.V., Svergun, D.I., "Global Rigid Body Modeling of Macromolecular Complexes Against Small-Angle Scattering Data," *Biophysical Journal*, 89(2): 1237–1250, 2005.

133. Porta, J.M., Ros, L., Thomas, F., Corcho, F., Canto, J., Perez, J.J., "Complete Maps of Molecular Loop Conformational Spaces," *Journal of Computational Chemistry*, 28(13): 2170-2189, Oct. 2007.

134. Purohit, P.K., Kondev, J., Phillips, R., "Mechanics of DNA Packaging in Viruses," *Proceedings of the National Academy of Sciences of the USA*, 100(6): 3173–3178, 2003.

135. Ramachandran, G.N., Sasisekharan, V., "Conformation of Polypeptides and Proteins," *Advances in Protein Chemistry* 28: 283–437, 1968.

136. Rhodes, G., *Crystallography Made Crystal Clear*, 2nd ed., Academic Press, San Diego, CA, 2000.

137. Ritchie, D.W., "High-Order Analytic Translation Matrix Elements for Real-Space Six-Dimensional Polar Fourier Correlations," *Journal of Applied Crystallography*, 38(5): 808–818, 2005.

138. Ritchie, D. W., Kozakov, D., Vajda, S., "Accelerating and Focusing ProteinProtein Docking Correlations Using Multi-Dimensional Rotational FFT Generating Functions," *Bioinformatics*, 24(17): 1865–1873, 2008.

139. Ritchie, D. W., Venkatraman, V., "Ultra-fast FFT Protein Docking on Graphics Processors," *Bioinformatics*, 26(19): 2398–2405, 2010.

140. Rohl, C.A., Strauss, C.E., Misura, K.M., Baker, D., "Protein Structure Prediction Using Rosetta," *Methods in Enzymology*, 383: 66–93, 2004.

141. Rossmann, M.G., Blow, D.M., "The Detection of Sub-Units within the Crystallographic Asymmetric Unit," *Acta Crystallographica*, 15: 24–31, 1962.

142. Rossmann, M.G., "Molecular Replacement – Historical Background," *Acta Crystallographica D*, 57: 1360–1366, 2001.

143. Rupp, B., *Biomolecular Crystallography: Principles, Practice, and Application to Structural Biology*, Garland Science, Taylor and Francis Group, New York, 2010.

144. Satake, I., "On a Generalization of the Notion of a Manifold," *Proceedings of the National Academy of Science of the USA*, 42: 359–363, 1956.

145. Schoenflies, A., *Theory of Crystal Structure*, Teubner, Leipzig (in German), 1891.

146. Senechal, M., "A Simple Characterization of the Subgroups of Space Groups," *Acta Crystallographica A*, 36: 845–850, 1980.

147. Schneidman-Duhovny, D., Inbar, Y., Nussinov, R., Wolfson, H.J., "Geometry-Based Flexible and Symmetric Protein Docking," *Proteins: Structure, Function, and Bioinformatics*, 60(2): 224-231, 2005.

148. Shehu, A., Clementi, C., Kavraki, L.E., "Modeling Protein Conformational Ensembles: From Missing Loops to Equilibrium Fluctuations," *Proteins: Structure, Function, and Bioinformatics*, 65(16): 164–179, 2006.

149. Sheriff, S., Klei, H.E., Davis, M.E., "Implementation of a Six-Dimensional Search Using the AMoRe Translation Function for Difficult Molecular-Replacement Problems," *Journal of Applied Crystallography*, 32: 98–101, 1999.

150. Sigworth, F.J., "A Maximum-Likelihood Approach to Single-Particle Image Refinement," *Journal of Structural Biology*, 122(3): 328–339, 1998.

151. Sippl, M.J., "Knowledge-Based Potentials for Proteins," *Current Opinion in Structural Biology*, 5(2): 229–235, 1995.

152. Sohncke, L., *Entwickelung einer Theorie der Krystallstruktur*, B.G. Teubner, Leipzig 1879.

153. Spakowitz, A.J., "Wormlike Chain Statistics with Twist and Fixed Ends," *Europhysics Letters*, 73(5): 684–690, 2006

154. Storoni, L.C., McCoy, A.J., Read, R.J., "Likelihood – Enhanced Fast Rotation Functions," *Acta Crystallographica Section D: Biological Crystallography*, 60(3): 432–438, 2004.

155. Sumikoshi, K., Terada, T., Nakamura, S., Shimizu, K., "A Fast Protein-Protein Docking Algorithm Using Series Expansions in Terms of Spherical Basis Functions," *Genome Informatics*, 16: 161–193, 2005.

156. Stein, A., Kortemme, T., "Improvements to Robotics-Inspired Conformational Sampling in Rosetta," *PloS One*, 8(5), e63090, 2013.

157. Swigon, D., Coleman, B.D., Olson, W.K., "Modeling the Lac Repressor-Operator Assembly: The Influence of DNA Looping on Lac Repressor Conformation," *Proceedings of the National Academy of Sciences of the USA*, 103(26): 9879–9884, 2006.

158. Szczepański, A., *Geometry of Crystallographic Groups*, World Scientific Publishing Company, Singapore, 2012.

159. Tang, X., Kirkpatrick, B., Thomas, S., Song, G., Amato, N.M., "Using Motion Planning to Study RNA Folding Kinetics," *Journal of Computational Biology*, 12(6): 862–881, 2005.

160. Thomas, S., Song, G., Amato, N.M., "Protein Folding by Motion Planning," *Physical Biology*, 2(4): S148–S155, Dec. 2005

161. Thurston, W.P., *Three-Dimensional Geometry and Topology*, (edited by Silvio Levy), Princeton University Press, Princeton, New Jersey, 1997.

162. Tjandra, N., Bax, A., "Direct Measurement of Distances and Angles in Biomolecules by NMR in a Dilute Liquid Crystalline Medium," *Science*, 278(5340): 1111–1114, 1997.

163. Tolman, J.R., Al-Hashimi, H.M., Kay, L.E., Prestegard, J.H., "Structural and Dynamic Analysis of Residual Dipolar Coupling Data for Proteins," *Journal of the American Chemical Society*, 123(7): 1416–1424, 2001.

164. Trapani, S., Navaza, J., "Calculation of Spherical Harmonics and Wigner D–Functions by FFT: Applications to Fast Rotational Matching in Molecular Replacement and Implementation into AMoRe," *Acta Crystallographica Section A: Foundations of Crystallography*, 62(4): 262–269, 2006.

165. Trovato, A., Seno, F., "A New Perspective on Analysis of HelixHelix Packing Preferences in Globular Proteins," *PROTEINS: Structure, Function, and Bioinformatics*, 55(4): 1014–1022, 2004.

166. Tyka, M.D., Jung, K., Baker, D., "Efficient Sampling of Protein Conformational Space Using Fast Loop Building and Batch Minimization on Highly Parallel Computers," *Journal of Computational Chemistry*, 33(31): 2483–2491, 2012.

167. Vagin, A., Teplyakov, A., "Molecular Replacement with MOLREP," *Acta Crystallographica D*, 66: 22-25, 2010.

168. Venkatraman, V., Sael, L., Kihara, D., "Potential for Protein Surface Shape Analysis Using Spherical Harmonics and 3D Zernike Descriptors," *Cell Biochemistry and Biophysics*, 54(1–3): 23–32, 2009.

169. Venkatraman, V., Yang, Y.D., Sael, L., Kihara, D., "Protein–Protein Docking Using Region–Based 3D Zernike Descriptors," *BMC Bioinformatics*, 10(1): 407, 2009.

170. Vajda, S., Sippl, M., Novotny, J., "Empirical Potentials and Functions for Protein Folding and Binding," *Current Opinion in Structural Biology*, 7(2): 222–228, 1997.

171. Walther, D., Eisenhaber, F., Argos, P., "Principles of Helix-Helix Packing in Proteins: The Helical Lattice Superposition Model," *Journal of Molecular Biology*, 255: 536–553, 1996.

172. Walther, D., Springer, C., Cohen, F.E., "Helix-Helix Packing Angle Preference for Finite Helix Axes," *PROTEINS: Structure, Function, and Genetics*, 33: 457–459, 1998.

173. Weeks, J.R., *The Shape of Space*, Marcel Dekker, Inc., New York, 1985.

174. Wolf, J.A., *Spaces of Constant Curvature*, AMS Chelsea Publishing), 6th ed., 2010.

175. Wondratschek, H., Müller, U., eds., *International Tables for Crystallography, Volume A1: Symmetry Relations Between Space Groups*, Springer, 2nd ed., 2008.

176. Wriggers, W., Milligan, R.A., McCammon, J.A., "Situs: A Package for Docking Crystal Structures into Low–Resolution Maps from Electron Microscopy," *Journal of Structural Biology*, 125(2): 185–195, 1999.

177. Yan, Y., Chirikjian, G.S., "Almost-Uniform Sampling of Rotations for Conformational Searches in Robotics and Structural Biology," *Proceedings of the 2012 IEEE International Conference on Robotics and Automation*, ICRA 2012, pp. 4254–4259, Minneapolis, MN, 14–18 May 2012.

178. Yan, Y., Chirikjian, G.S., "Molecular Replacement for Multi-Domain Structures Using Packing Models," *Proceedings of the ASME 2011 International Design Engineering Technical Conferences & Computers and Information in Engineering Conference* IDETC/CIE 2011, paper DETC2011–48583, August 28–31, 2011, Washington, DC, USA.

179. Yan, Y., Chirikjian, G.S., "Voronoi Cells in Lie Groups and Coset Decompositions: Implications for Optimization, Integration, and Fourier Analysis," *Proceedings of the 52nd IEEE Conference on Decision and Control*, pp. 1137-1143, Firenze, Italy, December 10–13, 2013.

180. Yershova, A., Jain, S., LaValle, S., Mitchell, J.C., "Generating Uniform Incremental Grids on SO(3) Using the Hopf Fibration," *The International Journal of Robotics Research*, 29(7): 801-812, 2010.

181. Zhao, Z., Singer, A., "Rotationally Invariant Image Representation for Viewing Direction Classification in Cryo–EM," *Journal of Structural Biology*, 186(1): 153–166, 2014.

182. Zhang, J., Lin, M., Chen, R., Wang, W., Liang, J., "Discrete State Model and Accurate Estimation of Loop Entropy of RNA Secondary Structures," *Journal of Chemical Physics*, 128(12): 125107, 2008.

183. Zhou, H.-X., "Loops in Proteins Can Be Modeled as Worm-Like Chains," *Journal of Physical Chemistry B*, 105(29): 6763–6766, 2001.

A

Computational Complexity and Polynomials

In this appendix we review notation such as the "Big $\mathcal{O}$" symbol, and examine the computational complexity of operations on polynomials such as multiplication, division, and evaluation. We also review the complexity of matrix multiplication. All of these operations are important in assessing the computational complexity of numerically computing Fourier transforms and Fourier inversion formulas on groups.

A.1 The Sum, Product, and Big-$\mathcal{O}$ Symbols

The symbols $\sum$ and $\prod$ are generally used to denote sums and products, respectively. Usually these symbols will have integer subscripts and superscripts:

$$\sum_{i=m}^{n} a_i = a_m + a_{m+1} + \cdots + a_n$$

and

$$\prod_{i=m}^{n} a_i = a_m \cdot a_{m+1} \cdot \cdots \cdot a_n$$

where m and n are integers and $m < n$.

As a special case, x raised to the power k is

$$x^k = \prod_{j=1}^{k} x_j \quad \text{for} \quad x_j = x. \tag{A.1}$$

In order to illustrate summation notation, consider the following example. Given positive integers $m, n \in \mathbb{Z}_{>0}$, an $m \times n$ *matrix*, A, is an array of numbers of the form

$$A = \begin{pmatrix} a_{11} & a_{12} & \cdots & a_{1n} \\ a_{21} & a_{22} & \cdots & \vdots \\ \vdots & \vdots & \ddots & a_{m-1,n} \\ a_{m1} & \cdots & a_{m,n-1} & a_{mn} \end{pmatrix}.$$

The element (or entry) in the ith row and j^{th} column of A is denoted a_{ij} (and likewise, the elements of any matrix denoted with an upper-case letter are generally written as subscripted lower case letters). Given an $n \times p$ matrix B the $(i,j)^{th}$ element of the product AB is defined as:

$$(AB)_{ij} \doteq \sum_{k=1}^{n} a_{ik} b_{kj}.$$

When $m = n$, the *trace* of A is defined as

$$\text{tr}(A) \doteq \sum_{i=1}^{n} a_{ii}.$$

When a computational procedure (such as summing or multiplying n numbers, or multiplying two matrices) can be performed in $g(n)$ arithmetic operations, we say it has $\mathcal{O}(f(n))$ complexity if there exists a positive number N such that

$$|g(n)| \leq c f(n) \tag{A.2}$$

for some finite and positive constant c and all $n \geq N$.

The "$\mathcal{O}$" notation hides constants and lower order terms. For instance,

$$10^{-10} n^2 + 10^6 n + 10^{10} = \mathcal{O}(n^2).$$

Similarly, when evaluating computational complexity, one often finds expressions like $\mathcal{O}(n^p (\log n)^r)$. The base of the logarithm need not be specified when only the order of computation is concerned since if $n = b^K$, then

$$n = (10^{\log_{10} b})^K = 10^{K \log_{10} b}, \tag{A.3}$$

and since b is a constant $\mathcal{O}(K) = \mathcal{O}(K \log_{10} b)$, which in combination with (A.3) leads to

$$\mathcal{O}(\log_b n) = \mathcal{O}(K) = \mathcal{O}(K \log_{10} b) = \mathcal{O}(\log_{10} n).$$

As examples of the usefulness of this notation, consider the sums

$$\sum_{k=1}^{n} 1 = n \quad \text{and} \quad \sum_{k=1}^{n} k = \frac{n(n+1)}{2}.$$

Whereas these are easy to verify, it is often more convenient to write

$$\sum_{k=1}^{n} k^p = \mathcal{O}(n^{p+1})$$

than to find the specific formula. (Intuitively the above makes sense because integrating x^p from 1 to n provides an approximation that grows in the same way as the sum for $p = 0, 1, 2, ...$)

Often in computational procedures, loops are used to perform calculations. A common loop is to perform a sum of sums. In this regard we note the following convenient expression:

$$\sum_{i=g}^{h} \left(\sum_{k=g}^{i} a_k \right) b_i = \sum_{i=g}^{h} a_i \left(\sum_{k=i}^{h} b_k \right) \tag{A.4}$$

where g, h, i and k are finite integers, and $g \leq h$.

A.2 The Complexity of Matrix Multiplication

Given $n \times n$ matrices $A = [a_{ij}]$ and $B = [b_{ij}]$, computing the product $C = AB$ by the definition

$$c_{ik} = \sum_{j=1}^{n} a_{ij} b_{jk}$$

uses n multiplications and $n - 1$ additions for each fixed pair of (i, k). Doing this for all $i, k \in \{1, ..., n\}$ then requires $\mathcal{O}(n^3)$ operations.

While this is the most straightforward way to compute matrix multiplications, theoretically, it is not the most efficient computational technique. We now review a method introduced in [61] and discussed in detail in [2, 7]. Other methods for reducing the complexity of matrix multiplication can be found in [50, 51].

Strassen's algorithm [61] is based on the calculation of the product of 2×2 matrices $A = [a_{ij}]$ and $B = [b_{ij}]$ to yield $C = AB$ using the following steps. Let

$$d_1 = (a_{12} - a_{22})(b_{21} - b_{22})$$
$$d_2 = (a_{11} + a_{22})(b_{11} + b_{22})$$
$$d_3 = (a_{11} - a_{21})(b_{11} + b_{12})$$
$$d_4 = (a_{11} + a_{12})b_{22}$$
$$d_5 = a_{11}(b_{12} - b_{22})$$
$$d_6 = a_{22}(b_{21} - b_{11})$$
$$d_7 = (a_{21} + a_{22})b_{11}.$$

From these intermediate calculations the entries in the 2×2 matrix $C = [c_{ij}]$ are computed as

$$c_{11} = d_1 + d_2 - d_4 + d_6$$
$$c_{12} = d_4 + d_5$$
$$c_{21} = d_6 + d_7$$
$$c_{22} = d_2 - d_3 + d_5 - d_7.$$

If the only goal were to multiply 2×2 matrices efficiently this algorithm (which uses 7 multiplications and 18 additions/subtractions) would be far less desirable than the straightforward way using the definition of matrix multiplication. However, the goal is to recursively use this algorithm for matrices that have dimensions $n = 2^K$ (since any matrix of dimensions not of this form can be embedded in a matrix of this form at little cost). This is possible because Strassen's algorithm outlined above holds not only when a_{ij} and b_{ij} are scalars, but also when these elements are themselves matrices.

It is not difficult to see from the 2×2 case that the number $T(n)$ of arithmetic operations required to compute the product of two $n \times n$ matrices using this algorithm recursively satisfies the recurrence [2]

$$T(n) = 7 T(n/2) + 18 (n/2)^2.$$

The complexity of any algorithm for which $T(n)$ satisfies such a recurrence relation will be of $\mathcal{O}(n^{\log_2 7})$ [2]. Hence as n becomes large, this, and other, recursive approaches to matrix multiplication become more efficient than direct evaluation of the definition.

Throughout the text, when we discuss matrix computations we use the complexity $\mathcal{O}(n^\gamma)$ keeping in mind that for direct evaluation of the definition $\gamma = 3$ and for Strassen's algorithm $\gamma \approx 2.81$. For other approaches and bounds developed up to the early 1980s, see [50, 51]. More recent algorithms include the work of Coppersmith and Winograd [18], which has a theoretical performance od $\gamma = 2.38$. The theoretical complexity of matrix multiplication has continued to attract interest [53]. Algorithms and analyses based on group theory and the FFT with the potential to be fast and numerically stable have been developed rather recently [14, 15, 16, 20, 60, 69].

A.3 Polynomials

An nth-degree real(complex) polynomial is a sum of the form

$$p(x) = a_0 x^n + a_1 x^{n-1} + \cdots + a_{n-1}x + a_n = \sum_{k=0}^{n} a_k x^{n-k}$$

where $a_i \in \mathbb{R}$ (or $\mathbb{C}$), $a_0 \neq 0$ and $k, n \in \{0, 1, 2, ...\}$.

A naive evaluation of $p(x)$ at a single value of x would be to calculate $z_k = x^k$ as an k-fold product for each $k \in \{1, ..., n\}$. This would use $k - 1$ multiplications for each value of k, and hence a total of

$$\sum_{k=1}^{n}(k - 1) = \frac{n(n + 1)}{2} - n = \mathcal{O}(n^2)$$

multiplications. Then evaluating $\sum_{i=0}^{n} a_i z_{n-i}$ can be performed with n additions and n multiplications. The resulting cost of polynomial evaluation when taking this naive approach is $n(n + 1)/2 + n = \mathcal{O}(n^2)$ arithmetic operations.

Better computational performance can be achieved by observing that $z_k = x \cdot z_{k-1}$. This is an example of a *recurrence relation*. Using this, the calculation of all x^k for $k = 1, ..., n$ is achieved using n multiplications, and the evaluation of $p(x)$ at a particular value of x is reduced to $3n = \mathcal{O}(n)$ arithmetic operations.

This can be improved further using the *Newton-Horner* recursive scheme:

$$p_0 = a_0 \qquad p_{i+1} = x \cdot p_i + a_{i+1}$$

for $i = 0, ..., n - 1$. This uses one multiplication and one addition for each value of i, and is hence an algorithm requiring $2n = \mathcal{O}(n)$ computations. This algorithm also has the advantage of not having to store the values $z_1, ..., z_n$.

Various algorithms for the efficient evaluation of a polynomial at a fixed value of x can be found in the literature. See, for example, [23, 49]. For algorithms that efficiently evaluate a polynomial at many points, and/or interpolate a polynomial to fit many points, see [2, 6, 7, 17, 37, 43, 45].

Naive approaches to finding the coefficients of products (by direct expansion) and quotients (using long division) of polynomials, as well as the evaluation of $p_n(x)$ at n distinct points $x_1, ..., x_{n-1}$ lead to $\mathcal{O}(n^2)$ algorithms. The subsequent sections review more efficient algorithms to achieve the same results.

A.3.1 Efficient Multiplication of Polynomials

Consider the complexity of multiplying two n^{th}-degree polynomials:

$$\left(\sum_{i=0}^{n} a_i x^i \right) \left(\sum_{j=0}^{n} b_j x^j \right) = \sum_{k=0}^{2n} c_k x^k, \tag{A.5}$$

A naive computation as

$$\sum_{k=0}^{2n} c_k x^k = \sum_{i=0}^{n} \sum_{j=0}^{n} a_i b_j x^{i+j}$$

might lead one to believe that $\mathcal{O}(n^2)$ computations are required to find the set $\{c_k | k = 0, ..., 2n\}$. This, however, would be a wasteful computation because upon closer examination we see that

$$c_k = \sum_{j=0}^{n} a_j b_{k-j}. \tag{A.6}$$

This is essentially the convolution of two sequences, and the set $\{c_k | k = 0, ..., 2n\}$ can be calculated using the FFT in

$$M(n) \doteq \mathcal{O}(n \log n)$$

arithmetic operations.

A.3.2 Efficient Division of Polynomials

Let p and q be polynomials of degree n and m respectively with $n > m$. Then we can always find two polynomials s and r such that

$$p = s \cdot q + r$$

where the degree of r is less than m, and the degree of s is $n - m$. In this case we say

$$p \equiv r \pmod{q}.$$

Finding r and s by long division can be performed in $\mathcal{O}(mn)$ computations. This is because there are $n - m + 1$ levels to the long division, each requiring at most $\mathcal{O}(m)$ arithmetic operations. For example, consider $p(x) = x^4 + 3x^3 + 2x^2 + x + 1$ and $q(x) = x^2 + 2x + 1$. Performing long division we observe that

$$x^4 + 3x^3 + 2x^2 + x + 1 = (x^2 + x - 1)(x^2 + 2x + 1) + 2x + 2.$$

In analogy with the way polynomial multiplication can be improved upon, so too can division. When $n = 2m$, division can be performed in

$$D(n) \doteq \mathcal{O}(n \log n)$$

computations, as opposed to the $\mathcal{O}(n^2)$ required by long division.

A.3.3 Efficient Polynomial Evaluation

Given n distinct values of x labeled $x_0, ..., x_{n-1}$, the polynomial evaluation problem considered here is that of determining the values of $y_k = p(x_k)$ where $p(x)$ is a polynomial of degree $n - 1$. This is written explicitly as

$$y_k = \sum_{j=0}^{n-1} b_{kj} a_j \quad \text{where} \quad b_{kj} = (x_k)^{n-j-1}. \tag{A.7}$$

A naive computation would be to simply perform the $\mathcal{O}(n^2)$ arithmetic operations indicated by the above matrix-vector product. However this does not take advantage of the special structure of the matrix B comprised of the entries b_{kj}. This matrix, called a *Vandermonde* matrix [30, 35, 52], has the special property that its product with a vector of dimension n can be computed in $\mathcal{O}(n \log^2 n)$ arithmetic operations [22]. Below we will show why this result is so using arguments from [2, 6].

It is well known that $p(x_k)$ is equal to the remainder of the division $p(x)/(x - x_k)$,

$$\frac{p(x)}{x - x_k} \equiv p(x_k) \,(\mathrm{mod}\,(x - x_k)), \tag{A.8}$$

where the "$\equiv$ mod" notation is defined in (B.1). For instance let $p(x) = x^3 + 2x + 1$. Then since

$$x^3 + 2x + 1 = (x - 4)(x^2 + 4x + 18) + 73$$

we calculate $p(4) = 73$.

Let $p_0, p_1, ..., p_{n-1}$ be monomials of the form $p_k = x - x_k$ for a distinct set of given values $x_k \in \mathbb{R}$. If p is a polynomial of degree n, then the set of n residues r_i defined by

$$p = s_i \cdot p_i + r_i$$

where s_i are polynomials can be computed in $\mathcal{O}(M(n)\log n) = \mathcal{O}(n \log^2 n)$ operations. This is better than performing the division algorithm n times (which would use $\mathcal{O}(n^2 \log n)$ computations).

This efficiency is achieved by first calculating the products $p_0\,p_1, p_2\,p_3, p_4\,p_5, ..., p_{n-2}\,p_{n-1}$, then $p_0\,p_1\,p_2\,p_3, p_4\,p_5\,p_6\,p_7, ..., p_{n-4}\,p_{n-3}\,p_{n-2}\,p_{n-1}$, and so on as indicated by the upward-pointing arrow in Figure A.1. Labeling the base of the tree as level zero, at the kth level up from the bottom there are $n/2^k$ polynomials, each of degree 2^k. The product of all polynomials as described above constitutes a total of

$$\mathcal{O}\left(\sum_{k=1}^{\log_2 n} \frac{n}{2^k}(2^k \log_2 2^k)\right) = \mathcal{O}\left(n \sum_{k=1}^{\log_2 n} k\right) = \mathcal{O}(n \log^2 n)$$

calculations.

Next, the $\log_2 n$ divisions corresponding to the downward-pointing arrow in Figure A.1 are performed (each requiring $\mathcal{O}(n \log_2 n)$ operations). The residues (remainders) that result at the end of the process are the desired values $p(x_k)$.

A.3.4 Efficient Polynomial Interpolation

We now review an efficient algorithm for polynomial interpolation. Given n values of x labeled $x_0, ..., x_{n-1}$, the polynomial interpolation problem is to find the n coefficients

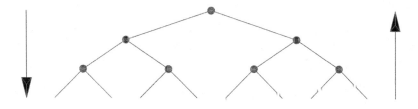

Fig. A.1. Multiplications from Bottom to Top; Divisions from Top to Bottom

$a_0, ..., a_{n-1}$ defining an $n-1$-degree polynomial, $p(x)$, that satisfies the equalities $p(x_k) = y_k$ for the given values $y_0, ..., y_{n-1}$.

In this context the values of x_k are given, and the problem of finding the vector $\mathbf{a} = [a_0, ..., a_{n-1}]^T$ could be written as an inversion of the matrix B in (A.7) followed by a product of B^{-1} with the vector $\mathbf{y}$. Using Gaussian elimination, $\mathcal{O}(n^3)$ arithmetic operations would be required to invert B. In principle faster inversion methods exist for general matrices [61]. Furthermore, the special structure of B again allows us to efficiently find B^{-1} and perform interpolation in $\mathcal{O}(n \log^2 n)$ arithmetic computations.

Another way to achieve the same result is as follows. The *Lagrange interpolation polynomials* (discussed in Subsection 5.2 of Chapter 3) are defined as

$$L_n(x) \doteq \sum_{k=0}^{n} f_k \frac{l^{n+1}(x)}{(x - \xi_k)(l^{n+1})'(\xi_k)} \tag{A.9}$$

where

$$l^{n+1}(x) = \prod_{k=0}^{n} (x - \xi_k).$$

These have the property $L_n(\xi_j) = f_j$ for $j = 0, ..., n$ because for a polynomial $l^{n+1}(x)$ we find from l'Hôpital's rule that

$$\left. \frac{l^{n+1}(x)}{x - \xi_k} \right|_{x=\xi_k} = \left. \frac{dl^{n+1}}{dx} \right|_{x=\xi_k},$$

and when evaluated at $x = \xi_k$ all but one term in the sum in (A.9) vanish.

As a special case of (A.8), we observe that

$$\frac{l^{n+1}(x)}{x - \xi_k} \equiv \left. \frac{dl^{n+1}}{dx} \right|_{x=\xi_k} \mod (x - \xi_k).$$

Following [2], we use these facts to formulate an efficient polynomial interpolation procedure. First define

$$q_{ij} = \prod_{m=i}^{i+2^j-1} (x - \xi_m).$$

Then for $j \neq 0$ define

$$s_{ij} = q_{ij} \sum_{m=i}^{i+2^j-1} \frac{f_m}{(x - \xi_m) \left. \frac{dl^{n+1}}{dx} \right|_{x=\xi_m}},$$

and

$$s_{i0} = \frac{f_i}{\left.\dfrac{dl^{n+1}}{dx}\right|_{x=\xi_i}}.$$

A recursive procedure defined as

$$s_{ij} = s_{i,j-1}q_{i+2^{j-1},j-1} + s_{i+2^{j-1},j-1}q_{i,j-1}$$

is then used to calculate $L_n(x) = s_{0,K}$ when $n = 2^K$. It starts by calculating s_{i0} for $i = 0, ..., n-1$ in unit increments. Then it runs through the values $j = 1, ..., K$ in unit increments and $i = 0, ..., n-1$ in increments of 2^j. This results in an $\mathcal{O}(D(n) \log_2 n) = \mathcal{O}(n(\log_2 n)^2)$ algorithm.

B

Set Theory

In this appendix we review basic definitions and special kinds of sets and associated functions. Set theory is one of the foundations of modern mathematics, and has deep historical roots. We only touch on those topics of direct relevance to computational and applied harmonic analysis.

B.1 Basic Definitions

The concept of a *set* is the foundation on which all of modern mathematics is built. A set is simply a collection of well-defined objects called *elements* or *points*. A set will be denoted here with capital letters. If S is a set, the notation $x, y \in S$ means x and y are elements of S. Sets can contain a finite number of elements or an infinite number. Sets can be discrete in nature or have a continuum of elements. The elements of the set may have an easily recognizable common theme, or the set may be composed of seemingly unrelated elements. Examples of sets include

$$S_1 = \{a, 1, \triangle, \text{Bob}\};$$

$$S_2 = \{0, 1, ..., 11\};$$

$$S_3 = \mathbb{Z} = \{..., -2, -1, 0, 1, 2, ...\}$$

$$S_4 = \mathbb{R};$$

$$S_5 = \{A \in \mathbb{R}^{3 \times 3} \,|\, \det A = 1\};$$

$$S_6 = \{\{0\}, \{0, 1\}, \{0, 1, 2\}\}.$$

S_1 is a finite set containing four elements that are explicitly enumerated. S_2 is a finite set with twelve elements (the numbers zero through eleven) that are not all explicitly listed because the pattern is clear. S_3 is the set of integers. This is a discrete set with an infinite number of elements. S_4 is the set of all real numbers. This is a set that possesses a continuum of values and is therefore infinite. S_5 is a set of 3×3 matrices with real entries that satisfy the condition of unit determinant. Often it is convenient to define sets as parts of other sets with certain conditions imposed. The vertical line in the definition of S_5 above is read as "such that." In other words, the larger set in this case is $\mathbb{R}^{3 \times 3}$, and S_5 is defined as the part of this larger set such that the condition on the right is observed. S_6 illustrates that a set can have elements that are themselves sets.

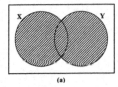 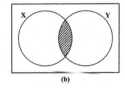 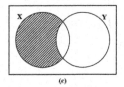

(a)   (b)   (c)

Fig. B.1. Venn Diagrams Depicting: (a) $X \cup Y$; (b) $X \cap Y$; (c) $X - Y$

Unless otherwise specified, the order of elements in a set does not matter. Hence the following sets are all the same

$$\{0,1,2\} = \{1,2,0\} = \{2,1,0\} = \{0,2,1\}.$$

In general, two sets X and Y are equal if for every $x \in X$, it is also true that $x \in Y$, and for every $y \in Y$ it is also true that $y \in X$. In the special case when the order of elements does matter, the set is called an *ordered set*.

Two of the most basic concepts in set theory are those of intersection, $\cap$, and union, $\cup$. The intersection of two sets is the largest set that is shared by both sets:

$$X \cap Y = \{x \mid x \in X \text{ and } x \in Y\}.$$

The union of two sets is the smallest set that contains both sets:

$$X \cup Y = \{x \mid x \in X \text{ or } x \in Y\}.$$

The order in which unions or intersections of two sets are performed does not matter. That is,

$$X \cup Y = Y \cup X \quad \text{and} \quad X \cap Y = Y \cap X.$$

The set-theoretic difference $X - Y$ of two sets X and Y is the set of elements in X but not in Y.

For example,

$$S_1 \cap S_2 = \{1\};$$

$$S_2 \cap S_3 = S_2;$$

$$S_1 \cup S_2 = \{a, 0, 1, 2, 3, 4, 5, 6, 7, 8, 9, 10, 11, \triangle, \text{Bob}\};$$

$$S_1 - S_1 \cap S_2 = \{a, \triangle, \text{Bob}\}.$$

Many concepts from set theory are demonstrated graphically with a *Venn diagram*. Figure B.1 illustrates union, intersection, and difference of sets.

When two sets X and Y have no elements in common, we write

$$X \cap Y = \emptyset.$$

The symbol $\emptyset$ is called the *empty set* and is the set with no elements. This immediately introduces the seemingly paradoxical statement that every set has the empty set as a subset, and therefore the intersection of any two sets contains the empty set. Hence every two sets have at least the empty set in common.[1] If the intersection of two sets

---

[1] Other rules such as a set is not allowed to be an element of itself can be found in books on axiomatic set theory.

is the empty set, the two sets are called *disjoint*. In the examples above, S_1 and S_6 are disjoint since none of the elements in both sets are the same (despite the fact that the element $1 \in S_1$ is contained in two of the elements of S_6).

Given a collection of sets $\{S_1, ..., S_n\}$ where n is a positive integer and $I = \{1, ..., n\}$ is a set of consecutive integers (called an *indexing set*), the union and intersection of the whole collection of sets are denoted respectively as

$$\bigcup_{i \in I} S_i = S_1 \cup S_2 \cup \cdots \cup S_n$$

and

$$\bigcap_{i \in I} S_i = S_1 \cap S_2 \cap \cdots \cap S_n.$$

More generally, the indexing set can be any countably infinite set (e.g. the integers) and need not consist of numbers (consecutive or otherwise).

If a set S_1 is contained in a set S_2 then S_1 is called a *subset* of S_2. The set X is contained in the set Y if all of the elements of X are contained in Y. The notation

$$X \subseteq Y$$

is used to denote that X is contained in Y, and can possibly be equal to Y. When $X \subseteq Y$ and $X \neq Y$ we use the notation $X \subset Y$ and call X a *proper subset* of Y. This notation is not standardized in mathematics. Often in mathematics the symbol $\subset$ is used to denote what we have defined as $\subseteq$. Our notational choice is so that the symbols $\leq$ and $<$ for subgroup and proper subgroup are consistent with the symbols for subset and proper subset.

Clearly it is always true that

$$X \cap Y \subseteq Y \quad \text{and} \quad X \cap Y \subseteq X$$

and also

$$X \subseteq X \cup Y \quad \text{and} \quad Y \subseteq X \cup Y.$$

If $A = \{a_1, ..., a_n\}$ and $B = \{b_1, ..., b_m\}$, then $|A| = n$ and $|B| = m$ are the number of elements in each. The number of elements in a set is called the *order* of the set. The *Cartesian product* of A and B is the set of ordered pairs $A \times B = \{(a, b) | a \in A, b \in B\}$. It follows that $|A \cup B| = |A| + |B| - |A \cap B|$ and $|A \times B| = |A| \cdot |B|$.

Given a set S, the concept of *equivalence* of elements $a, b \in S$ (denoted $a \sim b$) is any relationship between elements that satisfies the properties of reflexivity ($a \sim a$), symmetry (if $a \sim b$ then $b \sim a$), and transitivity (if $a \sim b$ and $b \sim c$, then $a \sim c$). An *equivalence class*, $[a] \subset S$ determined by $a \in S$ is defined as $[a] \doteq \{b \in S : b \sim a\}$. The set S can be decomposed into disjoint equivalence classes since every element of the set is in one and only one equivalence class [47]. The set of equivalence classes is denoted as $S/\sim$. Hence for any $s \in S$, $[s] \in S/\sim$ and

$$[s_1] \cap [s_2] = \begin{cases} \emptyset & \text{for } [s_1] \neq [s_2] \\ [s_1] & \text{for } [s_1] = [s_2] \end{cases}.$$

An important example of equivalence classes is the one established by modular arithmetic. Two numbers a and b are said to be *congruent modulo n* if n divides $a - b$ without a remainder. This is denoted

$$a \equiv b \,(\mathrm{mod}\, n). \tag{B.1}$$

It is easy to check that congruence defined in this way is an equivalence relation.

B.2 Maps and Functions

Another of the basic concepts of set theory is a *map* (or *mapping*) that assigns elements from one set to another. A map $f : X \to Y$ assigns to each element in X one and only one element in Y. Unless otherwise stated, the map $f : X \to Y$ need not "use up" all the elements in Y, and it may be the case that two or more elements in X are assigned the same element in Y. The subset of Y onto which all the elements of X map is called the *image* of f, and is denoted as $f(X)$ or $Im(f)$. The collection of elements in X that map to the same element $y \in Y$ is called the *preimage* or *fiber* of y in X. In the case when all the elements of Y are assigned to elements in X, the map is called *onto* or *surjective* and $f(X) = Y$. Another way to say this is that f is surjective when every point in Y has at least one corresponding point in X. A map that is not necessarily surjective, and satisfies the constraint for any $x_1, x_2 \in X$ that $f(x_1) = f(x_2)$ implies $x_1 = x_2$ is called *one-to-one* or *injective*. Another way to say this is that f maps to each element of $f(X)$ from only one element of X. In the case when f is both surjective and injective it is called *bijective* or *invertible*. In this case there is a unique map

$$f^{-1} : Y \to X$$

that is also bijective, such that $(f \circ f^{-1})(y) = y$ and $(f^{-1} \circ f)(x) = x$ where $\circ$ denotes the composition of maps. That is, if $f : X \to Y$ and $g : Y \to Z$ then $g \circ f : X \to Z$ where $(g \circ f)(x) = g(f(x))$.

A *function* is a map from a set into either the real or complex numbers (or, more generally, any *field* as defined in Section C.3).

Sets with certain kinds of maps or properties associated with them are given special names. For instance, a *metric space* is a set S together with a real-valued function $d : S \times S \to \mathbb{R}$, called a *metric* or *distance function* such that $d(s_1, s_2) > 0$ if $s_1 \neq s_2$; $d(s_1, s_1) = 0$ for all $s_1 \in S$; $d(s_1, s_2) = d(s_2, s_1)$; $d(s_1, s_3) \leq d(s_1, s_2) + d(s_2, s_3)$.

A *topological space* is a set S together with a collection of subsets of S (called a topology of S, and denoted T), such that: (1) the empty set, $\emptyset$, and S are contained in T; (2) the union of any subsets of T are in T; and (3) the intersection of any finite number of subsets of T are also in T.

Examples of topologies of an arbitrary set S include $\{S, \emptyset\}$ (called the indiscrete or trivial topology); the set of all subsets of S (called the discrete topology); and if the set is a metric space, the set of all subsets formed by the infinite union and finite intersection of all "ϵ-balls"

$$B_\epsilon(z) = \{s | d(s, z) < \epsilon\}$$

for $\epsilon \in \mathbb{R}$ and $s, z \in S$ is called a metric topology.

Given a topological space S with topology T, we say that the subset $U \subseteq S$ is *open* if $U \in T$. A subset $V \subseteq S$ is called *closed* if $S - V$ is open. Given a subset $A \subseteq S$, the *interior* of A (denoted $Int(A)$) is the union of all open sets contained in A. The closure of A (denoted $Cl(A)$) is the intersection of all closed sets in S that contain A. $Int(A)$ is an open set and $Cl(A)$ is a closed set, and the following relationship holds:

$$Int(A) \subseteq A \subseteq Cl(A).$$

A topological space X is called a *Hausdorff space* if for each pair of distinct points $x_1, x_2 \in X$ there exist open sets $U_1, U_2 \subset X$ such that $x_1 \in U_1$, $x_2 \in U_2$ and $U_1 \cap U_2 = \emptyset$.

Any map $f : X \to Y$ induces equivalence relations on the two involved sets. For instance, if f is not surjective we can call two elements of Y equivalent if they are both in $f(X)$ or if they are both in $Y - f(X)$. Similarly, if f is not injective, all points in X which map to a given point in Y can be called equivalent. Finally, if f is bijective and an equivalence relation $x_1 \sim x_2$ exists between elements in X, then the elements $f(x_1)$ and $f(x_2)$ can be called equivalent as well.

For example, let $X = \mathbb{R}$, $Y = \{c \in \mathbb{C} \mid c\bar{c} = 1\}$, and $f(x) = e^{ix}$. Then all $x \in X$ that map to the same value $y \in Y$ can be called equivalent. The equivalence defined in this way is the same as congruence mod 2π, and the equivalence class is therefore $X/\sim = \mathbb{R}/2\pi\mathbb{R}$.

In general, since a set is divided into disjoint equivalence classes, we can write

$$\bigcup_{[s] \in S/\sim} [s] = S.$$

Another way to write this is by constructing an indexing set I consisting of one and only one representative from each equivalence class. Then

$$\bigcup_{s \in I} [s] = S.$$

Here we are assuming that $S/\sim$ is a countable set.

For example, consider

$$S = \{0, 1, 2, 3, 4, 5, 6, 7, 8, 9, 10, 11\}.$$

Using the equivalence relation $a \sim b \Leftrightarrow a \equiv b \,(\mathrm{mod}\,3)$ we have the equivalence classes

$$[0] = \{0, 3, 6, 9\}; \quad [1] = \{1, 4, 7, 10\}; \quad [2] = \{2, 5, 8, 11\}.$$

Clearly $S = [0] \cup [1] \cup [2]$.

Given a complex-valued function of set-valued argument $f : S \to \mathbb{C}$, where S is a finite set, we can make the following decomposition:

$$\sum_{s \in S} f(s) = \sum_{[s] \in S/\sim} \left(\sum_{x \in [s]} f(x) \right). \tag{B.2}$$

That is, using the disjointness property, we can first sum the values of the function evaluated for all the elements in each equivalence class, then sum these sums over all equivalence classes. An example of a function on a set is the *characteristic function* (or *indicator function*) of a subset $X \subset S$. This function is defined for all $x \in S$ as

$$\chi_X(x) = \begin{cases} 1 \text{ for } x \in X \\ 0 \text{ for } x \notin X \end{cases}.$$

B.3 Measure and Integration

The area of mathematics called *measure theory* is fundamental in generalizing intuitive concepts of integration. Well-known books in this field include [5, 25, 33, 63]. The object of study is a "measure space" which consists of a set, X, together with a "measure," μ. Let $\mathcal{S}_X$ denote the set of all subsets of X. Then a measure μ is defined as a function

$\mu : \mathcal{S}_X \longrightarrow \mathbb{R}_{\geq 0}$ such that $\mu(\emptyset) = 0$ and, for every $A_1, A_2, ..., A_n \subset X$ satisfying the condition of mutual disjointness,

$$A_i \cap A_j = \emptyset \quad \forall\ 1 \leq i \neq j \leq n,$$

the following equality holds:

$$\mu \left(\bigcup_{i=1}^n A_i \right) = \sum_{i=1}^n \mu(A_i).$$

Moreover, if $A \subseteq X$ is partitioned into disjoint subsets such that $A = \bigcup_{i=1}^n A_i$, and if a function assigns values as $f : \{A_1, ..., A_n\} \to \mathbb{C}$, then we can define

$$\mu(f) \doteq \sum_{i=1}^n f(A_i)\, \mu(A_i).$$

If the space is partitioned into infinitely small disjoint subsets, and if $f : A \to \mathbb{C}$, then we write

$$\mu(f) \doteq \int_{x \in A} f(x)\, d\mu(x). \tag{B.3}$$

Given such a $d\mu(x)$ (which is also referred to as a measure), it becomes possible to talk about function spaces such as $\mathcal{L}^p(A, d\mu)$ – a function space that consists of all functions f for which $\mu(|f|^p)$ is finite.

For example, if X is a Riemannian manifold, and if A is a subset of the same dimension as X which can be parameterized with one coordinate chart, we can take $d\mu(x(q)) = |G(q)|^{\frac{1}{2}} dq$, where $G(q)$ is the metric tensor and q is an array of coordinates that parameterizes the neighborhood of $x \in X$. If f is the set indicator function, which takes a value of unity on A, then the result of (B.3) becomes $\mu(A)$.

The measure of a subset $A \subset X$ which has lower dimension than X is zero, and such sets are called "sets of measure zero" or "sets of non-positive measure."

If $B_1, B_2, ..., B_n \subset X$ and $B_i \cap B_j \neq \emptyset$, then we have the property

$$\mu(B_i \cup B_j) = \mu(B_i) + \mu(B_j) - \mu(B_i \cap B_j).$$

Applying this rule twice gives

$$\begin{aligned}
\mu(B_i \cup B_j \cup B_k) = \ & \mu(B_i) + \mu(B_j) + \mu(B_k) \\
& - \mu(B_i \cap B_j) - \mu(B_i \cap B_k) - \mu(B_k \cap B_j) \\
& + \mu(B_i \cap B_j \cap B_k).
\end{aligned}$$

Iterating gives the general *inclusion-exclusion formula*:

$$\mu \left(\bigcup_{k=1}^n B_k \right) = \sum_{k=1}^n (-1)^{k-1} \sum_{1 \leq i_1 < \cdots < i_k \leq n} \mu(B_{i_1} \cap \cdots \cap B_{i_k}). \tag{B.4}$$

B.4 Invariant Measures and Metrics

In the case when S is not finite, it is useful to have a tool for determining the net value of a function evaluated over all of the elements $s \in S$. This is where measure theory comes into play. Here we extend the discussion from the previous section and follow the definitions given in [33, 59].

A σ-algebra $\mathcal{S}$ is defined as a family of subsets of a set S with the properties that $\emptyset, S \in \mathcal{S}$, and if $S' \in \mathcal{S}$ then $S - (S \cap S') \in \mathcal{S}$, and for any countable set $\{S_1, S_2, ...\}$ with each $S_i \in \mathcal{S}$ it holds that $\cup_{i=1}^{\infty} S_i \in \mathcal{S}$ and $\cap_{i=1}^{\infty} S_i \in \mathcal{S}$. For technical reasons, the use of measures is restricted to σ-algebras rather than arbitrary sets of subsets.

A *Haar measure* is a measure on a locally compact group G (see Chapter 7) for which $\mu(g \circ A) = \mu(A)$ for any $g \in G$ and any compact subset $A \subset G$. If the subset A is divided into disjoint infinitesimal elements, $da \subset A$, then we define $\mu(da) = d\mu(a)$ where $a \in A$. Integration with respect to a given measure is then defined as $\mu(A) \doteq \int_{a \in A} d\mu(a)$. The mathematician A. Haar proved that such a measure always exists on a locally compact group, and all such measures are scalar multiples of each other. Sometimes the term *Hurwitz* measure is used instead of Haar measure for the particular case of a Lie group. When a Haar measure is invariant under right as well as left shifts, it is called a *bi-invariant* measure.

A metric $d : S \times S \to \mathbb{R}_{\geq 0}$ is said to be *G-invariant* if, for some group G that acts on S, $d(A, B) = d(g \cdot A, g \cdot B) \; \forall g \in G$ and all $A, B \in S$. Recall that the closure condition $g \cdot A, g \cdot B \in S$ is a property of group actions.

Two geometrical objects A and B (closed and bounded regions in $\mathbb{R}^n$) are said to be *similar*[2] (or equivalent under the action of a group G) if $d(A, g \cdot B) = 0$ for some $g \in G$ and the G-invariant metric $d(\cdot, \cdot)$ (such as $d(A, B) = Vol(A \cap B)$ when G is locally volume preserving, e.g., rigid-body motion).

Theorem B.1. *The function* $D_G([A], [B]) \doteq \min_{g \in G} d(A, g \cdot B)$ *is a metric on the set of equivalence classes* S/G *with elements of the form*

$$[A] = \{A' \in S : d(A, g \cdot A') = 0 \quad \text{for} \quad \text{some} \quad g \in G\}$$

when $d(A, B)$ *is a G-invariant metric on* S.

Proof. Let

$$[A] = \{A' \in S : d(A, g \cdot A') = 0 \quad \text{for} \quad \text{some} \quad g \in G\}$$

denote the equivalence class containing A. Then $D_G([A], [B])$ is well defined, and is a metric because it satisfies the following three properties:

Positive Definiteness

$$D_G([A], [A]) = 0 \quad \text{(by definition since we can choose } g = e\text{)}.$$

$$D_G([A], [B]) = 0 \implies \min_{g \in G} d(A, g \cdot B) = 0 \implies B \in [A] \quad \forall B \in [B].$$

Likewise, from symmetry, $A \in [B] \quad \forall A \in [A]$, Therefore $[A] = [B]$.

---

[2]Note this definition is a generalization of the standard geometrical concept.

Symmetry

$$D_G([A],[B]) = \min_{g \in G} d(A, g \cdot B)$$
$$= \min_{g \in G} d(g^{-1} \cdot A, B)$$
$$= \min_{g \in G} d(g \cdot A, B)$$
$$= \min_{g \in G} d(B, g \cdot A)$$
$$= D_G([B],[A]).$$

Triangle Inequality

$$D_G([A],[C]) = \min_{g \in G} d(A, g \cdot C)$$
$$\leq \min_{g \in G} \min_{h \in G} (d(A, h \cdot B) + d(h \cdot B, g \cdot C)) \qquad (\text{B.5})$$
$$= \min_{h \in G} d(A, h \cdot B) + \min_{g \in G} \min_{h \in G} d(h \cdot B, g \cdot C)$$
$$= D_G([A],[B]) + \min_{g,h \in G} d(B, (h^{-1} \circ g) \cdot C)$$
$$= D_G([A],[B]) + \min_{k \in G} d(B, k \cdot C)$$
$$= D_G([A],[B]) + D_G([B],[C]).$$

The inequality (B.5) follows from the fact that

$$\min_{g \in G} d(A, g \cdot C) \leq \min_{g \in G} (d(A, h' \cdot B) + d(h' \cdot B, g \cdot C))$$

for *any* $h' \in G$ since $d(\cdot, \cdot)$ is a metric and thus satisfies the triangle inequality.
Therefore all the metric properties are satisfied. $\square$

Such metrics have been used in several application areas including multi-robot systems [10, 48] (with G being a permutation group) and in crystallography [12] (with G being a crystallographic space group). More generally, G can be either a discrete group or a Lie group.

C

Vector Spaces and Algebras

Consider a nonempty set, $\mathcal{X}$, on which the operations of addition of elements $(x, y \in \mathcal{X} \rightarrow x+y \in \mathcal{X})$ and scalar multiplication with elements $(x \in \mathcal{X}$ and $\alpha \in \mathbb{C} \rightarrow \alpha \cdot x \in \mathcal{X})$ are defined. The triplet $(\mathcal{X}, +, \cdot)$ is said to be a *complex vector space*[1] if the following properties are satisfied:[2]

$$x + y = y + x \quad \forall \ x, y \in \mathcal{X} \tag{C.1}$$

$$(x + y) + z = x + (y + z) \quad \forall \ x, y, z \in \mathcal{X} \tag{C.2}$$

$$\exists 0 \in \mathcal{X} \quad \text{s.t.} \quad x + 0 = x \in \mathcal{X} \tag{C.3}$$

$$\exists (-x) \in \mathcal{X} \quad \text{for each} \quad x \in \mathcal{X} \quad \text{s.t.} \quad x + (-x) = 0 \tag{C.4}$$

$$\alpha \cdot (x + y) = \alpha \cdot x + \alpha \cdot y \quad \forall \ \alpha \in \mathbb{C}, x, y \in \mathcal{X} \tag{C.5}$$

$$(\alpha + \beta) \cdot x = \alpha \cdot x + \beta \cdot x \quad \forall \ \alpha, \beta \in \mathbb{C}, x \in \mathcal{X} \tag{C.6}$$

$$(\alpha\beta) \cdot x = \alpha \cdot (\beta \cdot x) \quad \forall \ \alpha, \beta \in \mathbb{C}, x \in \mathcal{X} \tag{C.7}$$

$$1 \cdot x = x \quad \forall \ x \in \mathcal{X}. \tag{C.8}$$

The first four conditions, (C.1)-(C.4), imply that every vector space is an Abelian group under addition.

Vector spaces can be finite-dimensional or infinite-dimensional. For example, $(\mathbb{C}^N, +, \cdot)$ for any finite $N \in \mathbb{Z}_{>0}$ is a finite-dimensional complex vector space, and the set of all complex-valued functions on $\mathbb{R}^N$, equipped with pointwise addition and scalar multiplication, is an infinite-dimensional complex vector space. In contrast, $(\mathbb{R}^N, +, \cdot)$ and the set of all real-valued functions on $\mathbb{R}^N$ are both examples of real vector spaces that are respectively finite- and infinite-dimensional.

Let $V = (\mathcal{X}, +, \cdot)$ be a vector space. The elements of V (elements of $\mathcal{X}$) are called vectors. If a complex vector space V has an *inner* (or scalar) product $(\cdot, \cdot)$ defined between vectors such that $(x, y) \in \mathbb{C}$ for all $x, y \in V$, and satisfies the properties:

---

[1] If $\mathbb{C}$ is replaced with $\mathbb{R}$ in these definitions, the result is a *real vector space*, whereas if $\mathbb{C}$ is replaced by a general field $\mathbb{F}$, the result is called a "vector space over the field $\mathbb{F}$." See Section C.3 for what is meant by a field.

[2] In these statements, the notation $\exists$ is shorthand for "there exists" and *s.t.* is shorthand for "such that."

$$(x,x) \geq 0, \quad (x,x) = 0 \Rightarrow x = 0 \tag{C.9}$$

$$(x, y+z) = (x,y) + (x,z) \tag{C.10}$$

$$(x, \alpha y) = \alpha(x,y) \quad \forall \alpha \in \mathbb{C} \tag{C.11}$$

$$(x,y) = \overline{(y,x)} \tag{C.12}$$

then the pair $(V, (\cdot, \cdot))$ is called an *inner product space*. A real inner product space can be viewed as the special case where $\alpha \in \mathbb{R}$ and (C.12) reduces to $(x,y) = (y,x)$.

Note that there is a difference between the way mathematicians and physicists define these properties. Throughout this book we have used the physicist's convention above whereas a mathematician would use the alternate defining property

$$(\alpha x, y) = \alpha(x,y)$$

in place of (C.11). Using (C.10) and (C.12) the mathematician's convention amounts to linearity in the first argument rather than the second. We will use whichever convention is more convenient for a given application.

The *norm* of a vector is defined as

$$\|x\| \doteq \sqrt{(x,x)}.$$

The usual dot (scalar) product of vectors $\mathbf{x}, \mathbf{y} \in \mathbb{R}^N$ defined by $\mathbf{x} \cdot \mathbf{y} = \sum_{i=1}^N x_i y_i$ is an example of an inner product, and $(\mathbb{R}^N, \cdot)$ is a real inner product space. Given $A, B \in \mathbb{C}^{N \times N}$, $(A, B) = \mathrm{tr}(AB^*)$ is an inner product in the mathematician's convention, where $*$ denotes the complex conjugate transpose. Likewise, for $A, B \in \mathbb{R}^{N \times N}$, $(A, B) = \mathrm{tr}(AB^T)$ is an inner product.

Since points in the plane can be identified with complex numbers, we can define the inner product between two complex numbers $a = a_1 + ia_2$ and $b = b_1 + ib_2$ as

$$(a,b) = \mathrm{Re}(a\bar{b}) = a_1 b_1 + a_2 b_2.$$

This satisfies the definition of an inner product, and even though the vectors in this case are complex numbers, $\mathbb{C}$, in this example, is a *real* vector space, and with the above definition of inner product, $(\mathbb{C}, (\cdot, \cdot))$ is a *real* inner product space.

The product of two real-valued functions on $\mathbb{R}^N$, $\alpha(\mathbf{x})$ and $\beta(\mathbf{x})$, defined by

$$(\alpha, \beta) = \int_{\mathbb{R}^N} \alpha(\mathbf{x})\beta(\mathbf{x}) \, dx_1 ... dx_N < \infty$$

is also an inner product.

A basis for a vector space is a set of linearly independent vectors $\{e_1, ..., e_N\}$ whose linear combination spans the vector space. That is, any $x \in V$ can be uniquely expressed as $x = \sum_{i=1}^N x^{(i)} \cdot e_i$ for appropriate scalars $\{x^{(i)}\}$. In the case when $V = \mathbb{R}^N$, $e_i = \mathbf{e}_i$ where $(\mathbf{e}_i)_j = \delta_{ij}$ is one possible basis. When $N < \infty$ elements are needed to span V, then V is called N-dimensional. If an infinite number of independent basis elements exist, then V is called an infinite-dimensional vector space. Function spaces, such as the set of all real-valued functions on $\mathbb{R}^N$, are usually infinite-dimensional.

C.1 Bounds and Decompositions in Inner Product Spaces

Two general properties of inner-product spaces are that the *Cauchy-Schwarz inequality*

$$(x, y)^2 \leq (x, x)(y, y) \tag{C.13}$$

holds for any two vectors $x, y \in V$, and given an arbitrary basis for V, an orthonormal one can be constructed using the *Gram-Schmidt orthogonalization process*. We now examine these in detail.

C.1.1 The Cauchy-Schwarz Inequality

To see that the Cauchy-Schwarz inequality holds, we only need to observe that

$$f(t) = (x + ty, x + ty) = \|x + ty\|^2 \geq 0.$$

Expanding out, we find the quadratic equation in t of the form

$$f(t) = (x, x) + 2(x, y)t + (y, y)t^2 \geq 0.$$

Since the minimum of $f(t)$ occurs when $f'(t) = 0$ (i.e., when $t = -(x, y)/(y, y)$), the minimal value of $f(t)$ is

$$f(-(x, y)/(y, y)) = (x, x) - (x, y)^2/(y, y)$$

when $y \neq 0$. Since $f(t) \geq 0$ for all values of t, the Cauchy-Schwarz inequality follows. In the $y = 0$ case, (C.13) reduces to the equality $0 = 0$.

When the vector space is the space of square-integrable, real-valued functions on the unit interval with respect to the measure dx, $\mathcal{L}^2([0, 1], dx)$, the Cauchy-Schwartz inequality takes the form

$$\int_0^1 f_1(x) f_2(x) \, dx \leq \left(\int_0^1 f_1^2(x) \, dx \right)^{\frac{1}{2}} \left(\int_0^1 f_2^2(x) \, dx \right)^{\frac{1}{2}}.$$

Choosing $f_1(x) = f(x)$ and $f_2(x) = \text{sign}(f(x))$ and squaring both sides, it follows that

$$\left(\int_0^1 |f(x)| dx \right)^2 \leq \int_0^1 f^2(x) \, dx. \tag{C.14}$$

C.1.2 The Gram-Schmidt Orthogonalization Process

Let V be an n-dimensional inner product space, and let $\{v_1, ..., v_n\}$ be a basis for V. Then an orthonormal basis for V can be constructed as follows. First normalize one of the original basis vectors (e.g., v_1) and define:

$$u_1 \doteq v_1/\|v_1\|.$$

Then define u_2 by taking away the part of v_2 that is parallel to u_1 and normalizing what remains:

$$u_2 \doteq (v_2 - (v_2, u_1)u_1)/\|v_2 - (v_2, u_1)u_1\|.$$

It is easy to see that $(u_1, u_2) = 0$ and they are both unit vectors. The process then is recursively performed by taking away the parts of v_i that are parallel to each of the $\{u_1, ..., u_{i-1}\}$. Then u_i is defined as the unit vector of what remains, i.e.,

$$u_i \doteq \frac{v_i - \sum_{k=1}^{i-1}(v_i, u_k)u_k}{\left\| v_i - \sum_{k=1}^{i-1}(v_i, u_k)u_k \right\|}.$$

This process is repeated until a full set of orthonormal basis vectors $\{u_1, ..., u_n\}$ is constructed.

C.2 Algebras

A real vector space, V, for which an additional operation $\wedge$ is defined such that $x \wedge y \in V$ for every $x, y \in V$, is called a real *algebra* if for every $x, y, z \in \mathcal{X}$ and $\alpha \in \mathbb{R}$,

$$(x + y) \wedge z = x \wedge z + y \wedge z \tag{C.15}$$
$$z \wedge (x + y) = z \wedge x + z \wedge y \tag{C.16}$$
$$(\alpha \cdot x) \wedge y = x \wedge (\alpha \cdot y) = \alpha \cdot (x \wedge y). \tag{C.17}$$

For every N-dimensional algebra, we can expand elements x and y in a basis $\{e_i\}$ so that $x = \sum_{i=1}^{N} x^{(i)} e_i$, $y = \sum_{i=1}^{N} y^{(i)} e_i$, and

$$x \wedge y = \sum_{i=1}^{N} x^{(i)} y^{(j)} (e_i \wedge e_j).$$

An algebra, $(V, \wedge)$, is called an *associative algebra* if for every $x, y, z \in \mathcal{X}$

$$x \wedge (y \wedge z) = (x \wedge y) \wedge z.$$

$(V, \wedge)$ is called a *commutative algebra* if

$$x \wedge y = y \wedge x.$$

An example of an associative algebra is the set of $N \times N$ real matrices $\mathbb{R}^{N \times N}$, with $\wedge$ representing matrix multiplication. The subset of diagonal $N \times N$ real matrices is both an associative and commutative algebra under the operation of matrix multiplication.

A *Lie algebra* is an algebra where the operation $x \wedge y$ (denoted as $[x, y]$ in this special context) satisfies for $x, y, z \in (V, [,])$ the additional properties

$$[x, y] = -[y, x] \qquad [x, [y, z]] + [y, [z, x]] + [z, [x, y]] = 0. \tag{C.18}$$

These properties are, respectively, called anti-symmetry and the Jacobi identity. The operation $[x, y]$ is called the *Lie bracket* of the vectors x and y. The property $[x, x] = 0$ follows automatically. Note that Lie algebras are not generally associative.

Examples of Lie algebras are: (1) $\mathbb{R}^3$ with the cross-product operation between two vectors; and (2) $\mathbb{R}^{N \times N}$ with the matrix commutator operation: $[A, B] = AB - BA$ for $A, B \in \mathbb{R}^{N \times N}$.

C.3 Rings and Fields

In this section we follow the definitions given in Herstein [34].

An *associative ring* (or simply a *ring*) R is a nonempty set with two operations, denoted $+$ and $\cdot$, such that for all $a, b, c \in R$ and the special element $e \in R$ the following properties are satisfied:

$$a + b \in R \tag{C.19}$$
$$a + b = b + a \tag{C.20}$$
$$(a + b) + c = a + (b + c) \tag{C.21}$$
$$\exists\, 0 \in R \quad \text{s.t.} \quad a + 0 = a \tag{C.22}$$
$$\exists\, (-a) \in R \quad \text{s.t.} \quad a + (-a) = 0 \tag{C.23}$$
$$a \cdot b \in R \tag{C.24}$$
$$a \cdot (b \cdot c) = (a \cdot b) \cdot c \tag{C.25}$$
$$a \cdot (b + c) = a \cdot b + a \cdot c \tag{C.26}$$
$$(b + c) \cdot a = b \cdot a + c \cdot a \tag{C.27}$$

When, in addition to the additive identity element, 0, there is a multiplicative identity element $1 \in R$, then R is called a *ring with unit element* (or *ring with unity*). If $a \cdot b = b \cdot a$, then R is called a *commutative ring*.

A standard example of a ring is the set of integers with usual multiplication and addition (this is a commutative ring with unit element). A more exotic example of a ring is the set of all square-integrable functions on a locally compact group, with the operations of addition and convolution of functions.

A *field* is a commutative ring with unit element in which every nonzero element has a multiplicative inverse. A *skew field* is an associative (but not commutative) ring in which every nonzero element has a multiplicative inverse.

The real and complex numbers are both fields. The quaternions are an example of a skew field.

D

Matrix Functions and Decompositions

In this appendix, different ways to decompose matrices into products of simpler matrices are examined. At the core of many of these decompositions is the concept of an eigenvalue/eigenvector pair. Given a matrix $A \in \mathbb{C}^{n \times n}$, an eigenvalue, λ, and unit eigenvector, $\mathbf{v}$, are defined to satisfy the equality

$$A\mathbf{v} = \lambda \mathbf{v}. \tag{D.1}$$

Requiring $\|\mathbf{v}\| = 1$ does not impose severe restrictions, since an eigenvector that does not have unit length can always be generated from one that has unit length by multiplying $\mathbf{v}$ by a nonzero complex number.

In the sections that follow this concept will be used extensively.

D.1 Special Properties of Symmetric Matrices

The following properties of symmetric real matrices, and their generalizations, are very important in all areas of engineering mathematics. The proofs given for real symmetric matrices below follow for the case of Hermitian matrices[1] (and operators defined in Appendix E.2) as well when the appropriate inner product is used. In the real cases below, we take $(\mathbf{u}, \mathbf{v}) \doteq \mathbf{u} \cdot \mathbf{v} = \mathbf{u}^T \mathbf{v}$, whereas in the complex case $(\mathbf{u}, \mathbf{v}) \doteq \mathbf{u}^* \mathbf{v}$.

Theorem D.1. *Eigenvectors of real symmetric matrices corresponding to different eigenvalues are orthogonal.*

Proof. Given a real symmetric matrix A and two of its different eigenvalues λ_i and λ_j with corresponding eigenvectors $\mathbf{u}_i$ and $\mathbf{u}_j$, by definition the following is true: $A\mathbf{u}_i = \lambda_i \mathbf{u}_i$, and $A\mathbf{u}_j = \lambda_j \mathbf{u}_j$. Multiplying the first of these on the left by $\mathbf{u}_j^T$, multiplying the second one on the left by $\mathbf{u}_i^T$, and subtracting, we get

$$(\mathbf{u}_j, A\mathbf{u}_i) - (\mathbf{u}_i, A\mathbf{u}_j) = (\mathbf{u}_j, \lambda_i \mathbf{u}_i) - (\mathbf{u}_i, \lambda_j \mathbf{u}_j) = (\lambda_i - \lambda_j)(\mathbf{u}_i, \mathbf{u}_j).$$

Since A is symmetric, the left-hand side of the equation is zero.[2] Since the eigenvalues are different, we can divide by their difference, and we are left with $\mathbf{u}_i \cdot \mathbf{u}_j = 0$.

Theorem D.2. *Eigenvalues of real symmetric matrices are real.*

---

[1] $H \in \mathbb{C}^{N \times N}$ such that $H = H^*$ where by definition $H^* = \overline{H^T}$.

[2] This follows from the transpose rule $(A\mathbf{u})^T = \mathbf{u}^T A^T$.

Proof. To show that something is real, all we have to do is show that it is equal to its complex conjugate. Recall that given $a, b \in \mathbb{R}$, a complex number $c = a + b\sqrt{-1}$ has a conjugate $\bar{c} = a - b\sqrt{-1}$. If $c = \bar{c}$, then $b = 0$ and $c = a$ is real. The complex conjugate of a vector or matrix is just the complex conjugate of its elements. Furthermore, the complex conjugate of a product is the product of the complex conjugates. Therefore, given $A\mathbf{u}_i = \lambda_i \mathbf{u}_i$, we can take the complex conjugate of both sides to get

$$\overline{(A\mathbf{u}_i)} = \overline{A}\overline{\mathbf{u}}_i = A\overline{\mathbf{u}}_i = \overline{(\lambda_i \mathbf{u}_i)} = \overline{\lambda}_i \overline{\mathbf{u}}_i.$$

In this derivation $\overline{A} = A$ because the elements of A are real.

Then, using an argument similar to the one in Theorem D.1, we compute

$$(\overline{\mathbf{u}}_i, A\mathbf{u}_i) = (\overline{\mathbf{u}}_i, \lambda_i \mathbf{u}_i) \qquad \text{and} \qquad (\mathbf{u}_i, \overline{A}\overline{\mathbf{u}}_i) = (\mathbf{u}_i, \overline{\lambda}_i \overline{\mathbf{u}}_i).$$

Subtracting the second from the first, and using the fact that A is symmetric, we get

$$(\overline{\mathbf{u}}_i, A\mathbf{u}_i) - (\mathbf{u}_i, A\overline{\mathbf{u}}_i) = 0 = (\overline{\mathbf{u}}_i, \lambda_i \mathbf{u}_i) - (\mathbf{u}_i, \overline{\lambda}_i \overline{\mathbf{u}}_i) = (\lambda_i - \overline{\lambda}_i)(\mathbf{u}_i, \overline{\mathbf{u}}_i).$$

Dividing by $(\mathbf{u}_i, \overline{\mathbf{u}}_i)$, which is a positive real number, we get $\lambda_i - \overline{\lambda}_i = 0$ which means that the imaginary part of λ_i is zero, or equivalently $\lambda_i \in \mathbb{R}$.

What these results mean is that given a real symmetric matrix with distinct eigenvalues (i.e., if all eigenvalues are different), we can write

$$A[\mathbf{u}_1, ..., \mathbf{u}_n] = [\lambda_1 \mathbf{u}_1, ..., \lambda_n \mathbf{u}_n] \qquad \text{or} \qquad AQ = Q\Lambda,$$

where

$$Q = [\mathbf{u}_1, ..., \mathbf{u}_n] \qquad \text{and} \qquad \Lambda = \begin{pmatrix} \lambda_1 & 0 & ... & 0 \\ 0 & \lambda_2 & 0 & ... \\ 0 & ... & \lambda_{n-1} & 0 \\ & & & \\ 0 & ... & 0 & \lambda_n \end{pmatrix}.$$

In fact, this would be true even if there are repeated eigenvalues, but we won't prove that here.

A *positive-definite* matrix $A \in \mathbb{R}^{n \times n}$ is one for which

$$\mathbf{x}^T A \mathbf{x} \geq 0$$

with equality holding only when $\mathbf{x} = \mathbf{0}$.

Theorem D.3. *All eigenvalues of a real, positive-definite matrix are positive.*

Proof. Using the information given above, we can rewrite $(\mathbf{x}, A\mathbf{x})$ (whose sign must be positive for all possible choices of $\mathbf{x} \in \mathbb{R}^n$ if it is positive definite) as $(\mathbf{x}, Q\Lambda Q^T \mathbf{x}) = (Q^T \mathbf{x}, \Lambda(Q^T \mathbf{x}))$. Defining $\mathbf{y} = Q^T \mathbf{x}$, we see that

$$(\mathbf{x}, A\mathbf{x}) = (\mathbf{y}, \Lambda \mathbf{y}) = \sum_{i=1}^{n} \lambda_i y_i^2.$$

Since this must be positive for arbitrary vectors $\mathbf{x}$, and therefore for arbitrary vectors $\mathbf{y}$, each λ_i must be greater than zero. If they are not, then it would be possible to find a vector for which the quadratic form is not positive.

As a result of these theorems, we will always be able to write a symmetric positive definite matrix $A = A^T \in \mathbb{R}^{n \times n}$ in the form

$$A = Q \Lambda Q^T$$

where $Q \in SO(n)$ is a special orthogonal matrix, and Λ is diagonal with positive elements on the diagonal.[3]

Matrix Functions

In the same way that it is possible to expand certain functions of a real argument x in a convergent Taylor series about $x = 0$:

$$f(x) = \sum_{n=0}^{\infty} \frac{f^{(n)}(0)}{n!} x^n$$

(where $f^{(n)}(x)$ is the nth derivative of $f(x)$ with respect to x in some open interval about $x = 0$), it is also possible to define certain *matrix functions* using a Taylor series. In the same way that the pth power of a matrix A is

$$A^p = \underbrace{AA \cdots A}_{p \text{ times}},$$

the exponential of a matrix A is

$$\exp(A) = \sum_{n=0}^{\infty} \frac{1}{n!} A^n.$$

Similarly, $\cos(A)$ and $\sin(A)$ are well defined. Fractional powers of matrices cannot generally be represented using a Taylor series (e.g., try writing the Taylor series for $f(x) = \sqrt{x}$ - it cannot be done because the slope of this function is infinite at $x = 0$). In fact, it is not generally true that the square root of a matrix can even be calculated. For example, try to find a matrix A such that

$$A^2 = \begin{pmatrix} 0 & 1 \\ 0 & 0 \end{pmatrix}.$$

However, in the case of symmetric real matrices, roots can always be taken.

As shown in the previous section, a symmetric matrix can always be represented using the form

$$A = Q \Lambda Q^T.$$

It follows naturally that any positive power p of the matrix A is of the form

$$A^p = Q \Lambda^p Q^T.$$

What may be surprising is that for fractional and negative powers this expression also holds. That is one reason why the inverse of a matrix A can be written as A^{-1} where

$$A^{-1} = \left(Q \Lambda Q^T \right)^{-1} = Q \Lambda^{-1} Q^T,$$

---

[3]In the Hermitian case, $H = H^*$, the analog to this is that $H = U \Lambda U^*$ where $U \in U(n)$.

and Λ^{-1} is the diagonal matrix with diagonal elements $(\Lambda^{-1})_{ii} = 1/(\Lambda)_{ii}$. Clearly if any of the elements of Λ are zero, this is going to cause a problem (as we would expect, a matrix with a zero eigenvalue is not invertible).

We can define the square root of a positive semi-definite matrix as:

$$M^{\frac{1}{2}} = Q\Lambda^{\frac{1}{2}}Q^{T}$$

such that

$$M^{\frac{1}{2}}M^{\frac{1}{2}} = (Q\Lambda^{\frac{1}{2}}Q^{T})(Q\Lambda^{\frac{1}{2}}Q^{T}) = Q\Lambda^{\frac{1}{2}}(Q^{T}Q)\Lambda^{\frac{1}{2}}Q^{T} = Q\Lambda^{\frac{1}{2}}\Lambda^{\frac{1}{2}}Q^{T} = Q\Lambda Q^{T} = M.$$

Note that the matrix functions defined as Taylor series or roots are *not* simply the functions or roots applied to matrix entries. In fact, for any square matrix, $A = [a_{ij}]$, of dimension greater than 1×1, it is always the case (even for diagonal matrices) that

$$(e^{A})_{ij} \neq e^{a_{ij}}.$$

The matrix roots defined above are only for symmetric matrices, and it is **not** simply the square root of the elements of the matrix. Also, the square root of a real symmetric matrix is symmetric, but it is only real if the original matrix is positive semi-definite.

D.2 Special Properties of the Matrix Exponential

Given the $n \times n$ matrices X and A where $X = X(t)$ is a function of time and A is constant, the solution to the differential equation

$$\frac{d}{dt}(X) = AX \tag{D.2}$$

subject to the initial conditions $X(0) = \mathbb{I}_{n \times n}$ is

$$X(t) = \exp(tA).$$

We now show that

$$\boxed{\det(\exp A) = e^{\mathrm{tr}(A)}} \tag{D.3}$$

where

$$\mathrm{tr}(A) = \sum_{i=1}^{n} a_{ii} \tag{D.4}$$

and using the permutation notation introduced in Chapter 7,

$$\det(X) = \sum_{\sigma \in S_n} (\mathrm{sign}\,\sigma) x_{\sigma(1),1} \cdots x_{\sigma(n),n} = \sum_{\sigma \in S_n} (\mathrm{sign}\,\sigma) x_{1,\sigma(1)} \cdots x_{n,\sigma(n)}. \tag{D.5}$$

Here $\mathrm{sign}\,\sigma \in \{-1, 1\}$ with the sign determined by how the permutation decomposed into basic *transpositions* (i.e., permutations of only two numbers, leaving the remaining $n-2$ numbers unchanged). An even/odd number of transpositions corresponds to $+/-$, respectively.

Using the notation

$$\det X = \begin{vmatrix} x_{11} & x_{12} & \cdots & x_{1n} \\ x_{21} & x_{22} & \cdots & \vdots \\ \vdots & \vdots & \ddots & \vdots \\ x_{n\,1} & x_{n\,2} & \cdots & x_{n\,n} \end{vmatrix},$$

it follows from the product rule for differentiation and (D.5) that

$$\frac{d}{dt}(\det X) = \begin{vmatrix} \frac{dx_{11}}{dt} & \frac{dx_{12}}{dt} & \cdots & \frac{dx_{1n}}{dt} \\ x_{21} & x_{22} & \cdots & \vdots \\ \vdots & \vdots & \ddots & \vdots \\ x_{n\,1} & x_{n\,2} & \cdots & x_{n\,n} \end{vmatrix} + \begin{vmatrix} x_{11} & x_{12} & \cdots & x_{1n} \\ \frac{dx_{21}}{dt} & \frac{dx_{22}}{dt} & \cdots & \vdots \\ \vdots & \vdots & \ddots & \vdots \\ x_{n\,1} & x_{n\,2} & \cdots & x_{n\,n} \end{vmatrix} + \cdots +$$

$$\begin{vmatrix} x_{11} & x_{12} & \cdots & x_{1n} \\ \vdots & \vdots & \ddots & \vdots \\ \frac{dx_{n-1,\,1}}{dt} & \frac{dx_{n-1,\,2}}{dt} & \cdots & \frac{dx_{n-1,\,n}}{dt} \\ x_{n\,1} & x_{n\,2} & \cdots & x_{n\,n} \end{vmatrix} + \begin{vmatrix} x_{11} & x_{12} & \cdots & x_{1n} \\ x_{21} & x_{22} & \cdots & \vdots \\ \vdots & \vdots & \ddots & \vdots \\ \frac{dx_{n\,1}}{dt} & \frac{dx_{n\,2}}{dt} & \cdots & \frac{dx_{n\,n}}{dt} \end{vmatrix}. \tag{D.6}$$

Equation (D.2) is written in component form as

$$\frac{dx_{ik}}{dt} = \sum_{j=1}^{n} a_{ij} x_{jk}.$$

After making this substitution, one observes that the ith term in (40) becomes

$$\begin{vmatrix} x_{11} & x_{12} & \cdots & x_{1n} \\ \vdots & \vdots & \ddots & \vdots \\ \sum_{j=1}^{n} a_{ij} x_{j\,1} & \sum_{j=1}^{n} a_{ij} x_{j\,2} & \cdots & \vdots \\ \vdots & \vdots & \ddots & \vdots \\ x_{n\,1} & x_{n\,2} & \cdots & x_{n\,n} \end{vmatrix} = \begin{vmatrix} x_{11} & x_{12} & \cdots & x_{1n} \\ \vdots & \vdots & \ddots & \vdots \\ a_{ii} x_{i\,1} & a_{ii} x_{i\,2} & \cdots & \vdots \\ \vdots & \vdots & \ddots & \vdots \\ x_{n\,1} & x_{n\,2} & \cdots & x_{n\,n} \end{vmatrix}.$$

This follows by subtracting a_{ij} times the jth row of X from the ith row of the left side for all $j \neq i$.

The result is then

$$\frac{d}{dt}(\det X) = \mathrm{tr}(A)(\det X).$$

Since $\det X(0) = 1$, this implies

$$\det X = \exp(\mathrm{tr}(A)t).$$

Evaluation of both sides at $t = 1$ yields (D.3).

The equalities

$$\exp(A + B) = \exp A \exp B = \exp B \exp A \tag{D.7}$$

can be easily observed when

$$AB = BA$$

by expanding both sides in a Taylor series and equating term by term. What is perhaps less obvious is that sometimes the first and/or the second equality in (D.7) can be true when A and B do not commute. For example, Fréchet [28] observes that when

$$A = \begin{pmatrix} 0 & 2\pi \\ -2\pi & 0 \end{pmatrix} \quad B = \begin{pmatrix} 1 & 0 \\ 0 & -1 \end{pmatrix}$$

then $AB \neq BA$ but it is nonetheless true that

$$e^A e^B = e^B e^A.$$

Another example of a choice for the matrices A and B can be found in [68] for which $AB \neq BA$ and $e^{A+B} = e^A e^B \neq e^B e^A$.

Having said this, $AB = BA$ is in fact a necessary condition for $\exp t(A + B) = \exp(tA)\exp(tB) = \exp(tB)\exp(tA)$ to hold for all values of $t \in \mathbb{R}_{>0}$.

D.3 Matrix Decompositions

In this section we state without proof some useful definitions and theorems from matrix theory.

D.3.1 Decomposition of Complex Matrices

Theorem D.4. *(QR Decomposition) [36]: For any $n \times n$ matrix, A, with complex entries, it is possible to find a $Q \in U(n)$, the group of $n \times n$ unitary matrices, such that*

$$A = QR$$

where R is upper triangular. In the case when A is real, one can take $Q \in O(n)$, the group of $n \times n$ real orthogonal matrices, and R real.

Theorem D.5. *(Cholesky Decomposition) [36]: Any $n \times n$ complex matrix B that is decomposable as $B = A^*A$ for some $n \times n$ complex matrix A can be decomposed as $B = LL^*$ where L is lower triangular with nonnegative diagonal entries.*

Theorem D.6. *(Schur's Unitary Triangularization Theorem)[36] For any $A \in \mathbb{C}^{n \times n}$, it is possible to find a matrix $U \in U(n)$ such that*

$$U^* A U = T$$

where T is upper (or lower) triangular with the eigenvalues of A on its diagonal.

Note: This does *not* mean that for real A that U will necessarily be real orthogonal. For example, if $A = -A^T$ then $Q^T A Q$ for $Q \in O(n)$ will also be skew symmetric and hence cannot be upper triangular.

A *Jordan block* corresponding to a k-fold repeated eigenvalue is a $k \times k$ matrix with the repeated eigenvalue on its diagonal and the number 1 in the super diagonal. For example,

$$J_2(\lambda) = \begin{pmatrix} \lambda & 1 \\ 0 & \lambda \end{pmatrix},$$

$$J_3(\lambda) = \begin{pmatrix} \lambda & 1 & 0 \\ 0 & \lambda & 1 \\ 0 & 0 & \lambda \end{pmatrix},$$

etc.

The *direct sum* of two square matrices, $A \in \mathbb{C}^{n \times n}$ and $B \in \mathbb{C}^{m \times m}$, is the $(m+n) \times (m+n)$ block-diagonal matrix

$$A \oplus B = \begin{pmatrix} A & \mathbb{O}_{n \times m} \\ \mathbb{O}_{m \times n} & B \end{pmatrix}.$$

Clearly

$$\text{tr}(A \oplus B) = \text{tr}(A) + \text{tr}(B)$$

and

$$\det(A \oplus B) = \det(A) \cdot \det(B).$$

The notation $n_i J_i$ stands for the n_i-fold direct sum of J_i with itself:

$$n_i J_i = J_i \oplus J_i \oplus \cdots \oplus J_i = \begin{pmatrix} J_i & 0 & 0 \\ 0 & \ddots & 0 \\ 0 & 0 & J_i \end{pmatrix}.$$

The notation

$$\sum_{i=1}^{m} \bigoplus A_i = A_1 \oplus A_2 \oplus \cdots \oplus A_m$$

is common, and we use it in Chapter 8, as well as below. Note that $\dim(J_i) = i$ and

$$\dim \left(\sum_{i=1}^{m} \bigoplus A_i \right) = \sum_{i=1}^{m} \dim A_i.$$

Every matrix $A \in \mathbb{C}^{n \times n}$ can be written in the Jordan normal form

$$A = TJT^{-1}$$

where T is an invertible matrix and

$$J = \sum_{j=1}^{q} \bigoplus \left(\sum_{i=1}^{m} \bigoplus n_i^j J_i(\lambda_j) \right)$$

is the direct sum of a direct sum of Jordan blocks with m being the dimension of the largest Jordan block, q being the number of different eigenvalues, and n_i^j being the number of times $J_i(\lambda_j)$ is repeated in the decomposition of A. Note that

$$\sum_{j=1}^{q} \sum_{i=1}^{m} i \cdot n_i^j = \dim(A).$$

For instance, if

$$J = J_1(\lambda_1) \oplus J_1(\lambda_2) \oplus J_2(\lambda_3) \oplus J_2(\lambda_3) \oplus J_3(\lambda_4) \oplus J_5(\lambda_5) \oplus J_6(\lambda_5),$$

then $m = 6$, $q = 5$, and all values of n_i^j are zero accept for

$$n_1^1 = n_1^2 = n_3^4 = n_5^5 = n_6^5 = 1; \quad n_2^3 = 2.$$

In the special case when A is real *and* all of its eigenvalues are real, then T can be taken to be real.

A square matrix A with complex entries is called *normal* if

$$A^*A = AA^*.$$

Examples of normal matrices include Hermitian, skew-Hermitian, unitary, real orthogonal, and real symmetric matrices.

Theorem D.7. *Normal matrices are unitarily diagonalizable [36].*

D.3.2 Numerical Methods for Real Matrices

In this section we review some numerical methods used for real matrices.

Theorem D.8. *(Singular Value Decomposition (SVD)) [30]: For any real $m \times n$ matrix A, there exist orthogonal matrices $U \in O(m)$ and $V \in O(n)$ such that*

$$A = U\Sigma V^T$$

where Σ is an $m \times n$ matrix with entries $\Sigma_{ij} = \sigma_i \delta_{ij}$. The value σ_i is the ith largest singular value of A.

Corollary D.9. *Every real square matrix A can be written as the product of a symmetric matrix $S_1 = U\Lambda U^T$ and the orthogonal matrix*

$$R = UV^T, \tag{D.8}$$

or as the product of R and $S_2 = V\Lambda V^T$. Hence we write

$$A = S_1 R = RS_2. \tag{D.9}$$

This is called the *polar decomposition*.

In the case when $\det A \neq 0$, we calculate

$$R \doteq A(A^T A)^{-\frac{1}{2}} \tag{D.10}$$

(the negative fractional root makes sense for a symmetric positive definite matrix). We note that (D.8) is always a stable numerical technique for finding R, whereas (D.10) becomes unstable as $\det(A)$ becomes small.

Many other matrix decompositions exist. For example, for invertible $n \times n$ real matrices such that $\det(A_i) \neq 0$ for $i = 1, ..., n$, [4] it is possible to write

$$A = LU$$

where L is a unique lower triangular matrix and U is a unique upper triangular matrix. This is called an *LU-decomposition* [30].

See the classic references [29, 67] for other general properties of matrices and decompositions.

---

[4] The notation A_i denotes the $i \times i$ matrix formed by the first i rows and i columns of the matrix A.

E

Techniques from Mathematical Physics

In this appendix we review the Dirac delta function, self-adjoint differential operators, path integration, and quantization. These methods from mathematical physics find their way into several of the chapters.

E.1 The Dirac Delta Function

In several chapters we use properties of the Dirac delta function that are presented here. Our treatment follows those in [40, 56].

The Dirac delta "function" is actually not a function in the most rigorous sense. It is an example of a class of mathematical objects called *generalized functions* or *distributions*. While the Dirac delta function has been used in the physical sciences to represent highly concentrated or localized phenomena for almost a century, the development of a rigorous mathematical theory of distributions came much more recently. It was not until the middle of the twentieth century that the French mathematician Laurent Schwartz articulated a precise mathematical theory of distributions, building on the foundations of the Russian mathematician Sergei Sobolev.

Intuitively, a Dirac delta function is a "spike" that integrates to unity and has the property that it "picks off" values of a continuous square-integrable function when multiplied and integrated:

$$\int_{-\infty}^{\infty} f(x)\,\delta(x)\,dx = f(0).$$

There are many ways to express the Dirac delta function as the limit of a sequence of functions substituted in the above integral. That is, there are many sequences of functions $\{\delta_m^{(k)}(x)\}$ for $m = 1, 2, \ldots$ for which

$$\lim_{m \to \infty} \int_{-\infty}^{\infty} f(x)\,\delta_m^{(k)}(x)\,dx = f(0) \tag{E.1}$$

is satisfied. Here we use k to specify each of a number of classes of such functions, and m enumerates individual functions for fixed k. For instance,

$$\delta_m^{(1)}(x) = \frac{\sin mx}{\pi x} = \frac{1}{2\pi} \int_{-m}^{m} e^{i\omega x}\,d\omega;$$

$$\delta_m^{(2)}(x) = \frac{1}{\pi}\frac{m}{1+m^2 x^2};$$

$$\delta_m^{(3)}(x) = \left(\frac{m}{\pi}\right)^{\frac{1}{2}} e^{-mx^2};$$

$$\delta_m^{(4)}(x) = \begin{cases} \dfrac{\exp\left(\frac{-1}{1-m^2 x^2}\right)}{\int_{-\frac{1}{m}}^{\frac{1}{m}} \exp\left(\frac{-1}{1-m^2 r^2}\right) dr} & \text{for } |x| < 1/m \\[2ex] 0 & \text{for } |x| \geq 1/m \end{cases};$$

$$\delta_m^{(5)}(x) = \begin{cases} 1/(2m) & \text{for } |x| < 1/m \\ 0 & \text{for } |x| \geq 1/m \end{cases};$$

$$\delta_m^{(6)}(x) = \begin{cases} 1/m & \text{for } 0 \leq x < 1/m \\ 0 & \text{elsewhere} \end{cases};$$

$$\delta_m^{(7)}(x) = \begin{cases} -m^2|x| + m & \text{for } |x| < 1/m \\ 0 & \text{elsewhere} \end{cases};$$

Each of the sequences $\{\delta_m^{(k)}(x)\}$ has certain advantages. For instance, for $k = 1, ..., 4$, the functions are continuous for finite values of m. The cases $k = 5, 6, 7$ are intuitive definitions given in engineering texts. In the cases $k = 4-7$, the functions $\delta_m^{(k)}(x)$ all have finite support. The case $k = 1$ is used in the Shannon sampling theorem, and we use $k = 3$ (the Gaussian) in the proof of the Fourier inversion formula in Chapter 2. Others are possible as well, but our discussion is restricted to those mentioned above.

Regardless of which sequence of distributions is used to describe the Dirac delta function, we note that $\delta(x)$ is only rigorously defined when it is used inside of an integral, such as (E.1). Under this restriction, we use the notation

$$\delta(x) =_w \lim_{m \to \infty} \delta_m^{(k)}(x)$$

for any $k \in \{1, ..., 7\}$ to mean that (E.1) holds for any continuous and square integrable $f(x)$. We will call $=_w$ "weak equality". If $=_w$ is replaced with a strict equality sign, the limit does not rigorously exist. However, to avoid the proliferation of unwieldy notation throughout the book, *every equality involving delta functions on the outside of an integral will be assumed to be equality in the sense of* $=_w$. This is not such a terrible abuse, since $\delta(x)$ is only used inside of an integral anyway.

E.1.1 Derivatives of Delta Functions

Considering the "spike-like" nature of the Dirac delta function (which is so singular it is not even really a function), it may appear implausible to consider differentiating $\delta(x)$. However, since $\delta(x)$ is only used under an integral in which it is multiplied by a continuous (and in the present context, an n-times differentiable) function, we can use integration by parts to define

$$\int_{-\infty}^{\infty} f(x) \frac{d^n}{dx^n}\delta(x)\,dx = (-1)^n \int_{-\infty}^{\infty} \frac{d^n f}{dx^n}\delta(x)\,dx = (-1)^n \frac{d^n f}{dx^n}\bigg|_{x=0}.$$

In this context, we do not use the $k = 5, 6, 7$ definitions of the sequence $\delta_m^{(k)}(x)$ since they are not differentiable by definition. For the others, it is often possible to obtain analytical expressions for the integral of $\frac{d^n f}{dx^n} \delta_m^{(k)}(x)$, from which the limit can be calculated explicitly.

Other more intricate formulas involving derivatives of Dirac delta functions can be found in [40].

E.1.2 Dirac Delta Functions with Functional Arguments

In the context of orthogonal expansions and transforms in curvilinear coordinates, it is very useful to decompose Dirac delta functions evaluated at a function of the coordinates into delta functions in the coordinates themselves. For instance, in polar coordinates $x_1 = r \cos \theta$ and $x_2 = r \sin \theta$. Therefore,

$$\delta(\mathbf{x} - \mathbf{x}^0) = \delta(x_1 - x_1^0)\,\delta(x_2 - x_2^0) = \delta(r \cos \theta - x_1^0)\,\delta(r \sin \theta - x_2^0).$$

It is useful to write this in terms of separate delta functions in r and θ. The first step in doing this is to write $\delta(y(x))$ in terms of $\delta(x)$ when $y(\cdot)$ is a differentiable and invertible transformation from the real line to itself with x_0 defined by the condition $y(x_0) = 0$. Using the standard change-of-coordinates formula

$$\int_{y(a)}^{y(b)} F(y)\,dy = \int_a^b F(y(x))|dy/dx|dx$$

in reverse, we write

$$\int_{-\infty}^{\infty} \delta(y(x))\,f(x)\,dx = \int_{-\infty}^{\infty} \delta(y)\,f(x(y))|dx/dy|dy.$$

Since $\delta(y)$ "picks off" the value of the function in the second integral above at $y = 0$, we have

$$\int_{-\infty}^{\infty} \delta(y(x))\,f(x)\,dx = f(x(y))|dx/dy|\big|_{y=0} = f(x)/|dy/dx|\big|_{x=x_0}.$$

Since

$$[f(x)/|dy/dx|]_{x=x_0} = \int_{-\infty}^{\infty} \delta(x - x_0)\,f(x)/|dx/dx|dx,$$

we conclude that

$$\delta(y(x)) = \delta(x - x_0)/|dy/dx|.$$

Similarly, when $y(x)$ is not monotonic (and hence not invertible) with $y(x_i) = 0$ and $(dy/dx)_{x=x_i} \neq 0$ for $i = 0, ..., n - 1$, we have

$$\delta(y(x)) = \sum_{i=0}^{n-1} \frac{\delta(x - x_i)}{|y'(x)|} \tag{E.2}$$

where $y' = dy/dx$. As always, the equality sign in (E.2) is interpreted as equality under the integral, and this leads to multiple possible definitions of $\delta(y(x))$ that act in the same way. For instance, we can write $y'(x_i)$ in place of $y'(x)$ in (E.2) or multiply the right-hand side of (E.2) by any continuous function $\phi(x)$ satisfying $\phi(x_i) = 1$. Hence, instead of considering the discussion above as a derivation of (E.2), it can be considered as a justification for using (E.2) as a definition.

E.1.3 The Delta Function in Curvilinear Coordinates: What to Do at Singularities

Given a multi-dimensional coordinate transformation $\mathbf{y} = \mathbf{y}(\mathbf{x})$ (which is written in components as $y_i = y_i(x_1, ..., x_n)$ for $i = 1, ..., n$), the following well-known integration rule holds:

$$\int_{Im(Box(\mathbf{a},\mathbf{b}))} F(y_1, ..., y_n) \, dy_1 \cdots dy_n =$$

$$\int_{a_1}^{b_1} \cdots \int_{a_n}^{b_n} F(y_1(x_1, ..., x_n), ..., y_n(x_1, ..., x_n)) |\det J| dx_1 \cdots dx_n$$

where

$$J = \left[\frac{\partial \mathbf{y}}{\partial x_1}, ..., \frac{\partial \mathbf{y}}{\partial x_n} \right]$$

is the Jacobian matrix of the transformation and $|\det J|$ gives a measure of local volume change, and $Im(Box(\mathbf{a}, \mathbf{b}))$ is the "image" of the box defined by $a_i \leq x_i \leq b_i$ for $i = 1, ..., n$ in y-coordinates.

Using the same reasoning as in the one-dimensional case and assuming the function $\mathbf{y}$ is differentiable and one-to-one, we write

$$\delta(\mathbf{y}(\mathbf{x})) = \delta(\mathbf{x} - \mathbf{x}_0)/|\det J(\mathbf{x})|$$

where $\mathbf{y}(\mathbf{x}_0) = \mathbf{0}$, and it is assumed that $|\det J(\mathbf{x}_0)| \neq 0$.

In the case when the independent variables are curvilinear coordinates $\mathbf{u} = (u_1, ..., u_n)$ and the dependent variable is the position $\mathbf{x} \in \mathbb{R}^n$, we have the parameterization $\mathbf{x}(\mathbf{u})$ and

$$\delta(\mathbf{x}-\mathbf{p}) = \delta(\mathbf{x}(\mathbf{u})-\mathbf{p}) = \delta(\mathbf{u}-\mathbf{q})/|\det J(\mathbf{u})| = \delta(u_1-q_1) \cdots \delta(u_n-q_n)/|\det J(u_1, ..., u_n)| \tag{E.3}$$

where $\mathbf{p} = \mathbf{x}(\mathbf{q})$, and we assume that $|\det J(\mathbf{q})| \neq 0$.

At a singularity $\mathbf{x}(\boldsymbol{\sigma}) = \mathbf{s}$, we have $|\det J(\boldsymbol{\sigma})| = 0$, and the calculation of the Dirac delta is a bit different. At a singularity, the curvilinear coordinate description breaks down in the sense that a continuum of coordinate values (as opposed to a single discrete value) map to one point. For instance, in spherical coordinates, when $\theta = 0$, all values of ϕ describe the same point on the x_3-axis, and when $r = 0$, all values of θ and ϕ describe the same point $(\mathbf{x} = \mathbf{0})$.

More generally, in curvilinear coordinates $\mathbf{u} = (u_1, ..., u_n)$ such that $\mathbf{x} = \mathbf{x}(\mathbf{u})$, we have as a special case that at a singular point

$$\mathbf{s} = \mathbf{x}(v_1, ..., v_k, \sigma_{k+1}, ..., \sigma_n)$$

where $\sigma_{k+1}, ..., \sigma_n$ are fixed parameter values that define $\mathbf{s}$, and $v_1, ..., v_k$ can take any values in a continuous set of values (denoted below as C_σ) for which they are defined. The definition of the Dirac delta function in curvilinear coordinates (E.3) is then modified at singularities as:

$$\delta(\mathbf{x} - \mathbf{s}) = \frac{\delta(u_{k+1} - \sigma_{k+1}) \cdots \delta(u_n - \sigma_n)}{\int_{C_\sigma} |\det J(v_1, ..., v_k, u_{k+1}, ..., u_n)| dv_1 \cdots dv_k}. \tag{E.4}$$

For example, when considering polar coordinates in the plane, $\mathbf{x} = \mathbf{x}(\mathbf{u})$ where $\mathbf{q} = [r, \phi]^T$, and $\mathbf{p} = [r_0, \phi_0]^T$, the Dirac delta function is

$$\delta(\mathbf{x} - \mathbf{p}) = \begin{cases} \frac{\delta(r-r_0)\,\delta(\phi-\phi_0)}{r} & \text{where} \quad r_0 \neq 0 \\[2ex] \frac{\delta(r-r_0)}{2\pi r} & \text{where} \quad r_0 = 0 \end{cases} \tag{E.5}$$

In the first case we use (E.3), and in the second we use (E.4) with ϕ integrated out.

As a more complicated example, using the spherical coordinates described in Chapter 4, and the usual (unnormalized) integral over the sphere, we write the Dirac delta function at a point $\mathbf{p} = \mathbf{x}(r_0, \theta_0, \phi_0)$ with $\mathbf{q} = [r_0, \theta_0, \phi_0]^T$ as

$$\delta(\mathbf{x} - \mathbf{p}) = \begin{cases} \frac{\delta(r-r_0)\,\delta(\theta-\theta_0)\,\delta(\phi-\phi_0)}{r^2 \sin\theta} & \text{where} \quad r_0 \neq 0,\ \sin\theta_0 \neq 0 \\[2ex] \frac{\delta(r-r_0)\,\delta(\theta-\theta_0)}{2\pi r^2 \sin\theta} & \text{where} \quad r_0 \neq 0,\ \sin\theta_0 = 0 \\[2ex] \frac{\delta(r-r_0)}{4\pi r^2} & \text{where} \quad r_0 = 0 \end{cases} \tag{E.6}$$

In the first case above, there is no singularity and (E.3) is used. In the second case, there is a whole line of singular points, each of which is defined by a value of r_0 and θ_0, but ϕ becomes superfluous and gets integrated out. In the third case, only the value of $r_0 = 0$ defines the singular point and both ϕ and θ get integrated out.

For a continuous function $f(\mathbf{x}) \in \mathcal{L}^2(\mathbb{R})$ for $\mathbf{x} = \mathbf{x}(\mathbf{u})$, this has the desired effect of yielding the same result as the definition

$$\int_{\mathbb{R}^n} f(\mathbf{x})\,\delta(\mathbf{x} - \mathbf{s})\,d\mathbf{x} = f(\mathbf{s})$$

in Cartesian coordinates. However, (E.4) *only* holds at singularities.

The extension of these concepts to m-dimensional surfaces in $\mathbb{R}^n$ and Lie groups follows in exactly the same way, with the appropriate definition of the Jacobian. Examples are provided throughout the book, including (4.23), (9.18), and (9.36).

E.2 Self-Adjoint Differential Operators

Linear second-order differential operators acting on complex-valued functions $u(x)$ defined on an interval $[0, 1]$ are written as

$$Lu = a_0(x)u'' + a_1(x)u' + a_2(x)u. \tag{E.7}$$

We will assume that the complex-valued functions $a_i(x)$ are at least twice differentiable, and $a_0(x) \neq 0$ for $0 < x < 1$.

Let $\{u_i(x)\}$ be a complete orthonormal basis for $\mathcal{L}^2([0, 1], \mathbb{C}, dx)$. The *matrix elements* of L in this basis are defined relative to the inner product

$$(u_i, u_j) = \int_0^1 \overline{u_1} u_2 dx$$

as

$$L_{ij} = (u_i, Lu_j) = \int_0^1 \overline{u_i} Lu_j dx.$$

The operator L is called *Hermitian* if [3]

$$(u_i, Lu_j) = (Lu_i, u_j). \tag{E.8}$$

That is, the operator L is called Hermitian when the matrix with elements $L_{ij} = (u_i, Lu_j)$ is a Hermitian. Since the basis in this case has an infinite number of elements, a Hermitian operator corresponds to an infinite-dimensional Hermitian matrix.

The *adjoint*, L^*, of the operator L in (E.7) is defined by the equality

$$(u_i, L^* u_j) = (Lu_i, u_j). \qquad (\text{F}.9)$$

A self-adjoint operator $(L = L^*)$ is always Hermitian with respect to a basis $\{u_i\}$ where each $u_i(x)$ satisfies certain boundary conditions.

Explicitly, the form of the adjoint of the operator in (E.7) is

$$L^* u = [\overline{a_0(x)}u]'' - [\overline{a_1(x)}u]' + \overline{a_2(x)}u.$$

This is found by integrating $(u, L^* u)$ by parts and assuming $u(x)$ satisfies certain boundary conditions.

The concepts of self-adjoint and Hermitian operators extend to functions on domains other than the unit interval. In the case of the circle and rotation groups, the boundary conditions imposed on the functions u_i are satisfied automatically due to the continuity of these functions. The definitions in (E.8) and (E.9) are exactly the same for these cases with the inner product being defined with the appropriate integration measures.

E.3 Functional Integration

The functional integral has been used extensively in theoretical physics to describe quantum and statistical phenomena [26, 38, 55].

Functional integration can be considered as a generalization of multi-dimensional integration for the case of an infinite-dimensional domain. Consider the n-dimensional integral

$$\int \dots \int \exp\left(-f(x_1, \dots, x_n)\right) dx_1 \dots dx_n. \qquad (\text{E}.10)$$

We assume that integral (E.10) is well defined. The functional integral is the generalization of (E.10) for the case of infinite-dimensional domains (for real- or complex-valued functions). Let us consider $f(x)$ (we assume that $f(x) \in \mathcal{L}^2([x_-, x_+])$) on the one-dimensional interval $[x_-, x_+]$ (which may be considered considered finite or infinite; in physical application it is frequently infinite). We also consider some functional $S(f) = \int L(f(x)) \, dx$. This functional has physical meaning (usually it is an action functional) and may contain differential operators. An expression of the type $\exp(-S)$ has practical importance for physical applications.

We discretize the interval, and write $\int L(f(x)) \, dx$ as the sum

$$\sum_n L(f(x_n))\Delta x_n = \sum_n \tilde{L}(f(x_n))$$

and integrate the product of function values $\exp S(f(x_n))$ for all possible values of $f(x_n)$:

$$Z = \lim_{N \to \infty, \Delta x_n \to 0} \int \dots \int \exp\left[-\sum_{n=0}^{N} \tilde{L}(f(x_n))\right] df_1 \dots df_N. \qquad (\text{E}.11)$$

We assume that expression (E.11) exists. In this case, this expression defines the functional integral

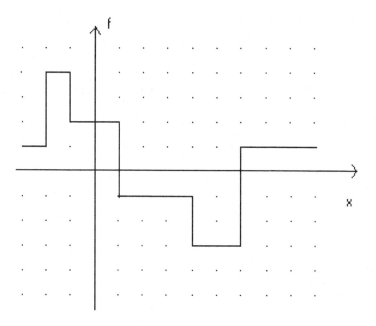

Fig. E.1. Illustration of the Path Integral

$$Z = \int \exp\left(-S(f(x))\right) Df \qquad (E.12)$$

where Df stands for an infinite-dimensional integration measure. It is assumed that the function $f(x)$ has well-defined limits f_-, f_+ at $x \to \pm\infty$ (or for $x = x_\pm$ on finite intervals). Specific boundary conditions may be imposed on $f(x)$ also at finite x.

The expression (E.12) defines the one-dimensional functional integral (because $f(x)$ is a function of the one dimensional parameter x), the functional integral may be also generalized for m-dimensional space x_i, $i = 1, ..., m$.

The term *path integration* is also frequently used for functional integration for the following reason. Let us consider a discrete set of values $f(x_n)$ for all possible values of x_n. The set of all possible values $f(x_n)$ can be depicted as the two-dimensional grid in Figure E.1. Then the functional integral (E.11) may be considered as a sum of the products of weight factors $\exp\left(-\tilde{L}(f(x_n))\right)$ taken along all possible paths on the grid of Figure E.1. For example, for the path in Figure E.1, we have to multiply factors $\exp\left(-\tilde{L}(f^i(x_n))\right)$ at each point x_n (where the function has some value $f^i(x_n)$) to get a contribution of this path to the functional integral. Summation of contributions along all possible paths satisfying boundary conditions $f(x) \to f_\pm$ (for $x \to \pm\infty$) defines functional integral (E.12) in the limit $\Delta x_n \to 0$, $N \to \infty$.

The contribution with the minimal value of product factors $\exp\left(-\tilde{L}(f^i(x_n))\right)$ (which defines the function $f(x)$ with the minimal action $S(f(x))$ for the case of real-valued functional) gives the largest contribution to the functional integral, and other path contributions describe additional quantum or statistical corrections.

For quadratic (Gaussian) functionals $S(f(x))$ the integration may be performed exactly. Let us consider first the multi-dimensional integral

$$\int_{-\infty}^{\infty} ... \int_{-\infty}^{\infty} \exp\left(-1/2 \sum_{j=1}^{n} a_j x_j^2\right) dx_1...dx_n = \frac{(2\pi)^{n/2}}{\Pi_{j=1}^{n} a_j^{1/2}}.$$

We can consider the set of coordinates $(x_1, ..., x_n)$ as a vector in n-dimensional space and write the sum as a scalar product of vectors $\mathbf{x}$ and $A\mathbf{x}$, where A is a diagonal matrix with elements a_n

$$\sum_{j=1}^{n} a_j x_j^2 = (\mathbf{x}, A\mathbf{x}).$$

We assume that a $1/(2\pi)^{1/2}$ factor is included in the definition of each integration measure. Then the result of the integration may be written as

$$\int \exp\left(-1/2\,(\mathbf{x}, A\mathbf{x})\right) d^n x/(2\pi)^{n/2} = (\det A)^{-1/2}.$$

This expression is also valid for the case of real symmetric matrices, and the Gaussian integration also may be performed for Hermitian matrices for the case of complex-valued functions. We also mention the expression

$$\int \exp\left(-[1/2(\mathbf{x}, A\mathbf{x}) + (\mathbf{b}, \mathbf{x}) + c]\right) d^n x/(2\pi)^{n/2} =$$

$$\exp\left(1/2\,(\mathbf{b}, A^{-1}\mathbf{b}) - c\right) (\det(A))^{-1/2}.$$

The functional integral is the integral in the infinite-dimensional space of possible values of $f(x)$. In this case when A is an operator (usually differential) acting in this function space, the quadratic functional $S = \int f(x)Af(x)\,dx$ is a scalar product (for real-valued functions) (f, Af) in the functional space. The value of the integral (where we assumed each factor $(2\pi)^{1/2}$ is absorbed in the definition of integration measure) is

$$\int \exp\left(-\frac{1}{2}\int f(x)Af(x)\,dx\right) Df = (\det A)^{-1/2}$$

where the determinant of operator A is defined as the product of its eigenvalues.

See [41, 57, 62, 64] for various aspects of path integration.

E.4 Quantization of Classical Equations of Motion

The functional integral of the form

$$Z = \int \exp\left(-S(f(s))\right) Df \tag{E.13}$$

where Df stands for an infinite-dimensional integration measure on the one-dimensional line (or segment) parameterized by s, describes a partition function of a statistical system in statistical physics. Similar integrals have been used to describe the amplitude of transitions and Green's functions in quantum field theory. As we have discussed previously, the path with the smallest value of the action functional $S(f) = \int L(f(s), f'(s))\,ds$ (where $L(f(s), f'(s))$ is the Lagrangian of the system) gives the largest contribution to the functional integral. The path with the minimal action gives a classical trajectory in classical mechanics. The variation of the action $S(f)$ with respect to f gives the well-known Euler-Lagrange equation (see Appendix F for derivation):

$$\frac{\partial L}{\partial f} = \frac{d}{ds}\left(\frac{\partial L}{\partial f'}\right).$$

An equivalent equation of motion can be found also in the Hamiltonian formalism. We remind the reader that the Hamiltonian of the system may be found from the Lagrangian L as

$$H = pf' - L$$

where $p = \partial L/\partial f'$ is the classical momentum, and $f' = df/ds$.

The Hamiltonian for the system should be expressed as a function of p and f.[1] The classical equations of motion in the Hamiltonian formalism are written as

$$f' = \frac{\partial H(p, f)}{\partial p}$$

and

$$p' = -\frac{\partial H(p, f)}{\partial f}.$$

The *Poisson bracket*, defined as

$$\{H, A\} \doteq \frac{\partial H}{\partial p}\frac{\partial A}{\partial f} - \frac{\partial H}{\partial f}\frac{\partial A}{\partial p},$$

describes the evolution of a dynamical quantity with the "time"-coordinate, which in the current context is s, as

$$\frac{dA}{ds} = \{H, A\}.$$

We note

$$\{p, f\} = 1. \tag{E.14}$$

In the quantum case, momenta and coordinates become operators. This reflects the fact that they do not commute with each other as in the classical case. For a system with a single coordinate and momentum

$$[\hat{p}, \hat{f}] = \hat{p}\hat{f} - \hat{f}\hat{p} = -i. \tag{E.15}$$

This is a quantum analogy of (E.14).

It is assumed that this equation is valid when it acts on the wave functions describing the quantum system. Then the Hamiltonian $\hat{H}(\hat{p}, \hat{f})$ describing the system becomes an operator and the Scrödinger-like equation

$$i\frac{\partial \psi(f, t)}{\partial t} = \hat{H}\psi(f, t)$$

can be used to describe a non-relativistic quantum system.

---

[1] f is an analogy of the coordinate and s is an analogy of time in this notation.

Thus, the main step in quantization consists of replacing classical momentum by the operator

$$p \to \hat{p} = -i\frac{\partial}{\partial f}.$$

The coordinate operators act as simple multiplication in this representation, i.e.,

$$\hat{f}\psi = f\psi.$$

The physical reason for quantization is the *uncertainty principle*, i.e. the principle stating that coordinates and momenta cannot be measured precisely at the same time. More rigorously, for the one-dimensional case:

$$\delta f \delta p \geq 1/2 \, h/(2\pi)$$

where δf and δp are the uncertainty in measurements of the coordinate and momentum. Here h is Planck's constant (units where $h/(2\pi)$ is set to 1 are used frequently; we assume this in this section). Thus, a quantum particle with precise value of the momentum p should have a completely uncertain coordinate f. The wave function

$$\psi = \exp(i\,p\,f)$$

satisfies this property. We also observe that the operator

$$\hat{p} = -i\frac{\partial}{\partial f}$$

acts on ψ as

$$\hat{p}\psi = p\psi.$$

That is, this operator describes the momentum of a quantum particle. It can be easily checked that for operators $\hat{p}$ and $\hat{f}$ (where $\hat{f}\psi = f\psi$) the equation

$$(\hat{p}\hat{f} - \hat{f}\hat{p})\psi(f) = -i\frac{\partial(f\psi)}{\partial f} - (-i\frac{f\partial\psi}{\partial f}) = -i\psi$$

is valid, which is just the commutation relation (E.15).

Another approach to look at quantization is a group theoretical approach. As we mentioned before, each invariant group of transformations of a system corresponds to a conservation law for the system (an m-parameter group corresponds to m conserving dynamical invariants for the system, which is a statement of *Noether's Theorem*). The momentum of the system may be considered as an infinitesimal translation operator

$$\hat{p}\psi = -i\frac{d\psi(f+\Delta)}{d\Delta}\Big|_{\Delta=0} = -i\frac{\partial\psi}{\partial f}.$$

The eigenfunction of the momentum operator is $\psi = \exp ipf$ and the commutation relation (E.15) is valid. The substitution of classical momentum p and coordinate f in Hamiltonian by $\hat{p}$ and $\hat{f}$ is a quantization of the system. In the case of rotational Brownian motion and the polymer systems studied in Chapters 16 and 17, the infinitesimal rotation operators X_i^R acting on functions on $SO(3)$ are used to quantize the classical angular momentum of the system.

E.5 Integral Equations

Consider a measure space (X, μ) and let $K : X \times X \to \mathbb{C}$. In the applications in this book, X is either a Euclidean space, a measurable subset thereof, or a Riemannian manifold.

"Fredholm integral equations" often arise in applications.[2] These equations are of the form

$$g(x) + \lambda u(x) = \int_X K(x, y)\, u(y)\, d\mu(y)$$

where we are given the following: K (which is called a *kernel*), the function $g : X \to \mathbb{C}$, and the constant $\lambda \in \mathbb{C}$.[3] The function $u : X \to \mathbb{C}$ is not given. If $\lambda \doteq 0$, then the equation is said to be of the first kind, whereas if $\lambda \neq 0$, it is of the second kind. If $g(x) \doteq 0$ and λ is allowed to vary, then an eigenfunction/eignevalue problem results.

Methods for solving Fredholm integral equations are well established, and there is a close relationship between solution methods for such equations and linear algebra. This is because given any pair of functions of the form $f_i : X \to \mathbb{C}$ for $i = 1, 2$ and an inner product,

$$(f_1, f_2) \doteq \int_X \overline{f_1(x)} f_2(x)\, d\mu(x),$$

an orthonormal basis $\{\phi_i\}$ for $\mathcal{L}^2(X, d\mu)$ can be constructed with respect to the inner product such that $(\phi_i, \phi_j) = \delta_{ij}$. This allows us to convert the functions u and g into infinite-dimensional vectors by projecting onto this basis. For example, a Fredholm integral equation of the first kind becomes a matrix equation of the form

$$\mathbf{g} = K\mathbf{u}$$

where $\mathbf{g}$, $\mathbf{u}$, and K have components of the form $g_i = (\phi_i, g)$, $u_i = (\phi_i, u)$, and

$$k_{ij} = (\phi_i, K\phi_j),$$

respectively.

The adjoint of the operator K is the operator K^* such that

$$(\phi_i, K\phi_j) = (K^*\phi_i, \phi_j).$$

An operator for which $K = K^*$ is called *self-adjoint*.

After introducing a basis, it then becomes possible to talk about the eigenvalues, the trace, and the Hilbert-Schmidt norm of K in concrete terms. In particular, the trace of K is defined as

$$\operatorname{tr}(K) \doteq \sum_i k_{ii}$$

and the Hilbert-Schmidt norm is defined as

$$\|K\|_{H.S.} \doteq \sqrt{\operatorname{tr}(KK^*)}.$$

An important result that we will use is that, under some mild conditions, if the trace of operator is computed as a finite value in one orthonormal basis, it will be finite in

---

[2] Named after Swedish mathematician Erik Ivar Fredholm (1866-1927).

[3] Often such equations are written with λ multiplied with the integral on the right-hand side instead of multiplied with $u(x)$, and with $g(x)$ moved to the right-hand side of the equation. These differences are only superficial.

all orthonormal bases, and its value does not depend on the basis chosen. The following theorem gives a way to compute the trace (and hence the Hilbert-Schmidt norm as well) without introducing a basis. Kernels that satisfy the conditions of this theorem are called *Hilbert-Schmidt operators of trace class*.

Theorem E.1. *Let X be a compact subset of $\mathbb{R}^n$ and let $K : X \times X \to \mathbb{C}$. Then the following pairs of conditions are equivalent:*[4]

- $\mathrm{tr}(K)$ *and* $\|K\|_{H.S.}$ *are both finite when computed in any countable basis.*
- *The integrals* $I_1(K)$ *and* $I_2(K)$ *defined below are both finite:*

$$I_1(K) \doteq \int_X K(x,x)\, d\mu(x) \quad \text{and} \quad I_2(K) \doteq \int_X \int_X K(x,y) K^*(y,x)\, d\mu(x)\, d\mu(y).$$

Moreover, if the above conditions hold, then $\mathrm{tr}(K) = I_1(K)$ *and* $\|K\|_{H.S.}^2 = I_2(K)$.

Proof. If $\{\phi_i\}$ is a countable basis for $\mathcal{L}^2(X, d\mu)$, then $\{\phi_i \cdot \phi_j\}$ is a countable basis for $\mathcal{L}^2(X \times X, d\mu \times d\mu)$. By definition, if $K \in \mathcal{L}^2(X \times X, d\mu \times d\mu)$, then $I_2(K)$ must be finite, and we can write

$$K(x,y) = \sum_{i,j} k_{ij}\, \phi_i(x)\overline{\phi_j(y)} \tag{E.16}$$

and

$$K^*(x,y) = \sum_{i,j} k_{ij}^*\, \phi_i(x)\overline{\phi_j(y)} = \overline{K(y,x)} \quad \text{where} \quad k_{ij}^* \doteq \overline{k_{ji}}\,.$$

Then

$$
\begin{aligned}
(\phi_{i'}, K\phi_{j'}) &= \int_X \overline{\phi_{i'}(x)} \left(\int_X \left[\sum_{i,j} k_{ij}\, \phi_i(x)\overline{\phi_j(y)} \right] \phi_{j'}(y)\, d\mu(y) \right) d\mu(x) \\
&= \sum_{i,j} k_{ij} \int_X \int_X \overline{\phi_{i'}(x)}\phi_i(x)\overline{\phi_j(y)}\phi_{j'}(y)\, d\mu(x) d\mu(y) \\
&= \sum_{i,j} k_{ij} \int_X \overline{\phi_{i'}(x)}\phi_i(x)\, d\mu(x) \int_X \overline{\phi_j(y)}\phi_{j'}(y)\, d\mu(y) \\
&= \sum_{i,j} k_{ij}\, \delta_{i,i'}\delta_{j,j'} = k_{i'j'}
\end{aligned}
$$

and

$$\sum_{i'} (\phi_{i'}, K\phi_{i'}) = \sum_{i'} k_{i'i'}\,.$$

On the other hand, evaluating $I_1(K)$ using (E.16) gives

$$
\begin{aligned}
I_1(K) &= \int_X \left(\sum_{i,j} k_{ij}\phi_i(x)\overline{\phi_j(x)} \right) d\mu(x) \\
&= \sum_{i,j} k_{ij} \int_X \phi_i(x)\overline{\phi_j(x)}\, d\mu(x) \\
&= \sum_{i,j} k_{ij}\, \delta_{ij} = \sum_i k_{ii}\,.
\end{aligned}
$$

---

[4]Note that the kernel $K(x,y)$ need not be self-adjoint.

In the previous calculations, the multiple instances of interchanging summations and integrals are only justified if $\sum_{i'} k_{i'i'} < \infty$.

Substituting (E.16) into the definition of $I_2(K)$ and performing similar calculations, again using orthogonality, gives

$$I_2(K) = \sum_{ij} |k_{ij}|^2 = \|K\|_{H.S.}^2. \qquad \square$$

In more casual terms, Theorem E.1 says that if K is well-behaved, it is possible to compute the trace and the Hilbert-Schmidt norm of K without introducing a basis. This result is used in Chapter 10 in proving the completeness of IURs for $SE(2)$ and $SE(3)$ without any knowledge of special functions.

F

Variational Calculus

From elementary calculus we all know that the extremal values (minima, maxima, or points of inflection) of a differentiable function $f(x)$ can be found by taking the derivative of the function and setting it equal to zero, i.e.,

$$\left.\frac{df}{dx}\right|_{x=y} = 0 \quad \Leftrightarrow \quad y \in \mathcal{Y}$$

where $\mathcal{Y}$ is defined to be the set of extrema.

Determining whether or not the values $y \in \mathcal{Y}$ yield local minima, maxima or points of inflection can be achieved by looking at whether the neighboring values of $f(y)$ for $y \in \mathcal{Y}$ are greater or less than these values themselves. This is equivalent to taking the second derivative of $f(y)$, and checking its sign when evaluated at the points $y \in \mathcal{Y}$.

While engineers and scientists generally are familiar with the extension of elementary calculus to the case of several variables (called multivariable calculus), there is another very useful extension that may not be as familiar. Suppose that instead of finding the values of a variable y that extremizes a function $f(y)$, we are interested in extremizing a quantity of the form:

$$J = \int_{x_1}^{x_2} f\left(y, \frac{dy}{dx}, x\right) dx \tag{F.1}$$

with respect to all possible *functions* $y(x)$. How can this be done? Two examples of when such problems arise are now given.

Example 1: Supposed we want to find the curve $y = y(x)$ in the x-y plane which connects the points (x_1, y_1) and (x_2, y_2), and which has minimal length. The length of the curve is

$$J = \int_{x_1}^{x_2} \sqrt{1 + \left(\frac{dy}{dx}\right)^2} \, dx.$$

While we already know the answer to this question (the straight line segment connecting the two points), this example illustrates the kinds of problems to which variational calculus can be applied.

Example 2 : As another example, we restate a problem of an economic/management nature found in [39]. Suppose that a small company produces a single product. The amount of the product in stock is $x(t)$ (it is a function of time because depending on sales and the rate at which it is produced, the product inventory changes). The rate with which the product can be made is $dx/dt = \dot{x}(t)$ (units per day). Suppose the company receives an order to deliver an amount A in T days from the time of the order, i.e., $x(T) = A$. Assume nothing is in inventory when the order is placed, so $x(0) = 0$. Now, the decision as to what production schedule to use (plot of $\dot{x}$ as a function of time) depends on: (1) the cost of keeping inventory; and (2) the cost associated with production. Suppose the cost of inventory is proportional to the amount of product that is currently being held in inventory : $C_I = c_2 x(t)$ (such as would be the case if the company pays rent by the square foot). Suppose the cost of producing one unit of product is linear in the rate at which it is produced: $C_P = c_1 \dot{x}(t)$. This can be a realistic scenario because if the company needs to produce more, it will have to pay overtime and hire part-time workers who would otherwise not be employed by the company. The cost per unit time for production will be the cost per unit produced times the rate of production: $C_P \dot{x}$. The total amount of money being expended at time t is $f(x, \dot{x}) = c_1(\dot{x})^2 + c_2 x$. The total expenditure between now and the time the shipment is due is $J = \int_0^T f(x, \dot{x}) \, dt$. Thus, this problem is also of the general form of a variational calculus problem.

Note that nothing has been solved in the above examples; we have only *formulated* the problem.

One possible approach to solve the problem is to assume the solution has a particular form. For instance, we could assume the function $y(x) = a_0 + a_1 x + a_2 x^2 + \ldots$. Then substituting this into the expression for J in (F.1) and integrating gives $J = J(a_0, \ldots)$. The optimal values for a_i could then be found using standard methods of multivariable calculus. Instead of a power series, we could have used a Fourier series, or any other complete series representation. The solution obtained in this way generally will be suboptimal because the set of coefficients, $\{a_i\}$, will be finite.

There is another way of solving the problem that is extremely useful in fields as different as classical mechanics and economics. It goes by many names including: variational calculus, calculus of variations, optimal control, Hamilton's Principle, Principle of Least Action, and dynamic optimization. We will now explore the basis of this method.

F.1 Derivation of the Euler-Lagrange Equation

Loosely speaking, a variational operator,[1] denoted as δ, finds functions $y(x)$ that yield extremal values of the integral:

$$J = \int_{x_1}^{x_2} f(y(x), y'(x), x) \, dx$$

for a given function $f(\cdot)$ in the same way that the operator d/dy finds the extremal values of a function $f(y)$. This new problem may be subject to boundary conditions $y(x_1) = y_1$ and $y(x_2) = y_2$, or the boundary conditions may be free.

If we were to *assume* that the optimal solution to the problem is $y(x)$, then for any $\alpha \in \mathbb{R}$ other than $\alpha = 0$, the following will *not* be the optimal value of the integral

$$\hat{J}(\alpha) = \int_{x_1}^{x_2} f(Y, Y', x) \, dx,$$

---

[1]The use of δ as the standard notation for variational operator on the one hand, and for the unrelated concept of Dirac delta function on the other hand, is unfortunate. But in this book there are no contexts in which both arise together, and so there should be no confusion.

where

$$Y = Y(x, \alpha) = y(x) + \alpha \epsilon(x)$$

and $\epsilon(x)$ is any continuous function such that

$$\epsilon(x_1) = \epsilon(x_2) = 0. \tag{F.2}$$

The notation $\hat{J}$ is introduced to distinguish between the integral J resulting from the assumed function $y(x)$ and the value of the same integral evaluated with $Y(x, \alpha)$. That is, $J = \hat{J}(0)$, and $Y(x, 0) = y(x)$.

Note that $Y(x)$ satisfies the same boundary conditions as $y(x)$, but by definition it must be the case that $\hat{J}(\alpha) \geq J$. We can introduce the concept of a variational operator as follows:

$$\delta \hat{J} \doteq \left. \frac{\partial \hat{J}}{\partial \alpha} \right|_{\alpha=0} d\alpha. \tag{F.3}$$

Here α is a variable which is introduced into the calculus of variations problem to distinguish all functions 'within the neighborhood' of the desired function and meeting the boundary conditions $Y(x_1, \alpha) = y_1$ and $Y(x_2, \alpha) = y_2$. This is shown in Figure F.1

δ is nothing more than shorthand for the operation in (F.3). It is used like a derivative. There are four properties of δ (which follow naturally from the above equations). These are:

- It commutes with integrals: $\delta \int_{x_1}^{x_2} f(Y, Y', x)\, dx = \int_{x_1}^{x_2} \delta f(Y, Y', x)\, dx$. This follows because δ is basically a derivative, and taking a derivative of an integral when the variables of integration and differentiation are different can be done in any order when the bounds of integration are finite.
- It acts like a derivative on Y and Y' when it is applied to the function $f(Y, Y', x)$ but treats independent variables such, as x, like constants: $\delta f = \frac{\partial f}{\partial Y} \delta Y + \frac{\partial f}{\partial Y'} \delta Y'$. This is because Y depends on α by definition and x does not.
- It commutes with derivatives: $\delta \left(\frac{\partial Y}{\partial x} \right) = \frac{\partial}{\partial x}(\delta Y)$. This follows because δ is basically a derivative, and derivatives with respect to independent variables commute.
- The variation of $Y(x)$ vanishes at the endpoints : $\delta Y(x_1) = \delta Y(x_2) = 0$. This follows from F.2.

We can use these properties to generate conditions which will yield the extremal solution $y(x)$ that we seek. Namely,

$$\delta \hat{J} = \delta \int_{x_1}^{x_2} f(Y, Y', x)\, dx = \int_{x_1}^{x_2} \delta f(Y, Y', x)\, dx = \int_{x_1}^{x_2} \left(\frac{\partial f}{\partial Y} \delta Y + \frac{\partial f}{\partial Y'} \delta Y' \right) dx \tag{F.4}$$

$$= \int_{x_1}^{x_2} \left(\frac{\partial f}{\partial y} \delta Y + \frac{\partial f}{\partial y'} \delta Y' \right) dx.$$

The last step is true, just by the chain rule.[2]

Now we will use the final property to rewrite the second part of this expression as:

$$\frac{\partial f}{\partial y'} \delta Y' = \frac{\partial f}{\partial y'} \frac{d}{dx}(\delta Y).$$

---

[2] $\partial/\partial y(f(g(y), g'(y'), x)) = (\partial f/\partial g)(\partial g/\partial y)$. In our case, $g = Y$ and $\partial Y/\partial y = 1$. Therefore $\partial f/\partial y = \partial f/\partial Y$. The same is true for $\partial f/\partial y' = \partial f/\partial Y'$.

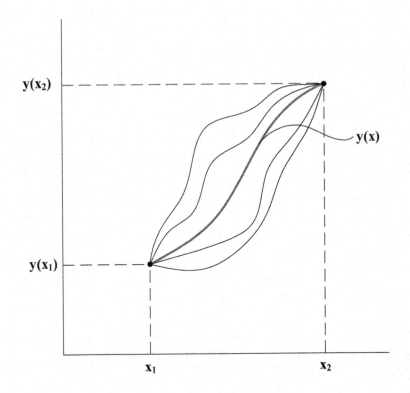

Fig. F.1. The Optimal Path and Surrounding Candidate Paths

Using integration by parts[3] means:

$$\int_{x_1}^{x_2} \frac{\partial f}{\partial y'} \frac{d}{dx}(\delta Y)\, dx = \frac{\partial f}{\partial y'} \delta Y \big|_{x_1}^{x_2} - \int_{x_1}^{x_2} \frac{d}{dx}\left(\frac{\partial f}{\partial y'}\right) \delta Y\, dx.$$

Since $\delta Y(x) = \epsilon(x)$ vanishes at endpoints, the first term is zero, and we get

$$\delta \hat{J} = \int_{x_1}^{x_2} \left(\frac{\partial f}{\partial Y}\delta Y - \frac{d}{dx}\left(\frac{\partial f}{\partial Y'}\right)\delta Y\right)\, dx.$$

This is easily rewritten as

---

[3] $\int_a^b u\, dv = uv\big|_a^b - \int_a^b v\, du.$

$$\delta \hat{J} = \int_{x_1}^{x_2} \left(\frac{\partial f}{\partial y} - \frac{d}{dx} \left(\frac{\partial f}{\partial y'} \right) \right) \epsilon(x)\, dx. \tag{F.5}$$

Using a general theorem that says that if

$$\int_a^b M(x) \epsilon(x)\, dx = 0$$

for all possible differentiable functions $\epsilon(x)$ for which $\epsilon(a) = \epsilon(b) = 0$, then the function $M(x) = 0$, the integrand in (F.5) can be set equal to zero, and so

$$\frac{\partial f}{\partial y} - \frac{d}{dx} \left(\frac{\partial f}{\partial y'} \right) = 0. \tag{F.6}$$

This is called the *Euler-Lagrange* equation Note that if the function f does not depend on y', the second term vanishes, and we get the familiar minimization problem from calculus.

We now have the tools to solve the two example problems stated at the beginning of this appendix.

Solution for Example 1: If $f = (1 + (y')^2)^{1/2}$, there are no y terms so the Euler-Lagrange equation reduces to

$$\frac{d}{dx} \left(\frac{\partial f}{\partial y'} \right) = 0 \quad \Rightarrow \quad \frac{\partial f}{\partial y'} = c.$$

In this particular problem, this gives $(1 + (y')^2)^{-1/2} y' = c$. This can be rewritten as $c^2(1 + (y')^2) = (y')^2$. Solving for y' yields $y' = [c^2/(1 - c^2)]^{1/2} = a$. Since c was arbitrary, we have replaced the fraction with a, another arbitrary constant. Integration yields $y = ax + b$, where b is arbitrary. The boundary conditions $y(x_1) = y_1$ and $y(x_2) = y_2$ specify a and b completely.

Solution for Example 2: Again we use the Euler-Lagrange equation, this time with $f = c_1(\dot{x})^2 + c_2 x$, and we get $\frac{\partial f}{\partial x} - \frac{d}{dt} \left(\frac{\partial f}{\partial \dot{x}} \right) = c_2 - 2c_1 \ddot{x} = 0$. Rewriting as $\ddot{x} = \frac{1}{2} c_2/c_1$ and integrating twice, we get $x(t) = \frac{1}{4} c_2/c_1 t^2 + at + b$. We can solve for a and b given that $x(0) = 0$ and $x(T) = A$.

While we have presented the one-dimensional variational problem, the same methods are used for a functional dependent on many variables where

$$J = \int_a^b f(y_1, ..., y_n, y'_1, ..., y'_n, x)\, dx.$$

In this case, a set of simultaneous Euler-Lagrange equations are generated as

$$\frac{\partial f}{\partial y_i} - \frac{d}{dx} \left(\frac{\partial f}{\partial y'_i} \right) = 0. \tag{F.7}$$

Likewise, the treatment of problems where $f(\cdot)$ depends on higher derivatives of y is straightforward. The derivation above is simply extended by integrating by parts once for each derivative of y. See [8, 24] for further reading on the classical calculus of variations and [31, 39] for the more modern extensions of optimal control and dynamic optimization.

F.2 Sufficient Conditions for Optimality

The Euler-Lagrange equations provide *necessary* conditions for optimality, but there is usually no guarantee that a solution of the Euler-Lagrange equations will be optimal. However, in certain situations, the structure of the function $f(\cdot)$ will guarantee that the solution generated by the Euler-Lagrange equations is a globally optimal solution. We now examine one such case of relevance to the discussion in Chapter 6.

Suppose the integrand in the cost functional is of the form

$$f(y, y', x) = (y')^2 g(y)$$

where $g(y) > 0$ for all values of y. The corresponding Euler-Lagrange equation is

$$2y''g + (y')^2 \frac{\partial g}{\partial y} = 0.$$

Multiplying both sides by y' and integrating yields the exact differential

$$((y')^2 g)' = 0.$$

Integrating both sides with respect to x and isolating y' yields

$$y' = c\, g^{-\frac{1}{2}}(y)$$

where c is the arbitrary constant of integration. With the boundary conditions $y(0) = 0$ and $y(1) = 1$, we can then write

$$F(y) \doteq \frac{1}{c} \int_0^y g^{\frac{1}{2}}(\sigma)\, d\sigma = x,$$

where

$$c = \int_0^1 g^{\frac{1}{2}}(\sigma)\, d\sigma.$$

The expression $F(y) = x$ can be inverted (F is monotonically increasing since $g > 0$) to yield $y = F^{-1}(x)$.

To see that this solution is optimal, one need only substitute

$$y' = g^{-\frac{1}{2}}(y) \int_0^1 g^{\frac{1}{2}}(\sigma)\, d\sigma \qquad (F.8)$$

into the cost functional

$$J(Y) = \int_0^1 g(Y)(Y')^2 dx$$

to see that

$$J(y) = \left(\int_0^1 g^{\frac{1}{2}}(y)\, dy \right)^2 = \left(\int_0^1 g^{\frac{1}{2}}(y(x)) y'\, dx \right)^2.$$

Since in general from (C.14)

$$\int_0^1 [f(x)]^2 dx \geq \left(\int_0^1 f(x)\, dx \right)^2,$$

we see that by letting $f(x) = g^{\frac{1}{2}}(y(x)) y'(x)$ that $J(y) \leq J(Y)$ where y is the solution generated by the Euler-Lagrange equation and $Y(x)$ is any other continuous function satisfying $Y(0) = 0$ and $Y(1) = 1$.

Multi-dimensional examples in which variational calculus provides globally optimal solutions are addressed in [11].

Manifolds and Riemannian Metrics

In this section we present definitions from elementary differential geometry and topology used throughout the text. Many works on differential geometry and its applications can be found. These include the very rigorous classic texts by Kobayashi and Nomizu [42], Abraham and Marsden [1], Guillemin and Pollack [32] and Warner [66]. We have found the classic text of Arnol'd [4] to be particularly readable, and the books by Choquet-Bruhat, et al. [13], Schutz [58], and Burke [9] to be useful in connecting differential geometric methods to physical problems. We also have found the introductory texts of Millman and Parker [46] and Do Carmo [21] to serve as an excellent connection between differential geometry in two and three dimensions and its abstraction to coordinate-free descriptions in higher dimensions. Finally, we have found the books by Lee [44] and Rosenberg [54] to be rigorous introductions written in a style appropriate for mathematically-oriented engineers and scientists.

We note before proceeding that the use of differential forms has been avoided in order to reduce the scope of this brief overview of basic differential geometry. The interested reader is refered to the books by Chirikjian [11], Darling [19], Flanders [27] (and references therein) for information on this subject.

G.1 Open Neighborhoods and Homeomorphisms

We have already seen the definition of a metric, $d(\cdot,\cdot)$. Recall that when a metric is defined on a set X, the pair (X,d) is called a *metric space*. Henceforth it will be clear when the shorthand X is used to refer to the metric space (X,d).

An *open ball of radius* ϵ around $x \in X$ (denoted $B_\epsilon(x)$) is the set of all $y \in X$ such that $B_\epsilon(x) = \{y \in X | d(x,y) < \epsilon\}$. A subset $A \subset X$ is called *open* if for every point $x \in A$ it is possible to find an open ball $B_\epsilon(x) \subset A$, i.e., for every point $x \in A$, there must exist a $B_\epsilon(x)$ that is also in A for some value of $\epsilon \in \mathbb{R}_{>0}$. Any open set containing a point $x \in X$ is called an *open neighborhood* of x.

As a consequence of these definitions (or the more general properties of topological spaces) it follows that the union of two open sets is open, as is their intersection. It also can be shown that given any two distinct points in a metric space, it is possible to find neighborhoods of the points that do not intersect.

The concept of open sets is important in the definition of continuity of functions between metric spaces.

For a continuous map from metric space X to metric space Y, $\phi : X \to Y$, every open set $V \subset Y$ has a preimage $\phi^{-1}(V) \subset X$ that is also open.

The concept of continuity of mappings between metric spaces is of particular importance when one seeks an intrinsic description of higher-dimensional analogs of curves and surfaces. In this case, one of the metric spaces, Y, is taken to be $\mathbb{R}^n$ for some positive integer n, and X is denoted as M (for manifold). Since we have seen how various coordinates can be defined for the whole of $\mathbb{R}^n$ (and hence any of its open subsets), it follows that ϕ^{-1} provides a tool to map coordinates onto an open subset of M when a bijective mapping exists between that open subset and an open subset of $\mathbb{R}^n$, and both ϕ and ϕ^{-1} are continuous. More generally, a bijection f between two toplogical spaces X and Y is called a *homeomorphism* if both $f : X \to Y$ and $f^{-1} : Y \to X$ are continuous.

G.2 Coordinate Charts and Orientation

An *n-dimensional proper coordinate chart about* $x \in M$ is the pair (U, ϕ) where U is a neighborhood of x and $\phi : U \to \mathbb{R}^n$ is a homeomorphism. It then follows that $\phi(U)$ is open in $\mathbb{R}^n$. A collection of coordinate charts $\{U_i, \phi_i\}$ for $i \in I$ (I is a set that indexes the charts) is called an *atlas*.

This allows us to rigorously define what is meant by a space that locally "looks like" $\mathbb{R}^n$ everywhere.

A *manifold*, M, is a metric space for which there is an atlas with the following properties:

- $\{U_i, \phi_i\}$ exists so that for each $x \in M$, $x \in U_i$ for some $i \in I$.
- If (U_i, ϕ_i) and (U_j, ϕ_j) are any two coordinate charts in the atlas for which $(U_i, \phi_i) \cap (U_j, \phi_j) \neq \emptyset$, then the composed map

$$\phi_j \circ \phi_i^{-1} : \phi_i(U_i \cap U_j) \to \phi_j(U_i \cap U_j)$$

is continuous.
- All possible charts with the above two properties are contained in the atlas.

When all the mappings $\phi_j \circ \phi_i^{-1}$ are infinitely differentiable, the manifold is called a *differentiable manifold*. If, in addition, each of these mappings has a convergent Taylor series it is called an *analytic manifold*. Henceforth, when referring to a manifold, we mean an analytic manifold. Complex analytic manifolds are defined in an analogous way (see e.g., [65]). In the context of our treatment it suffices to consider complex manifolds of complex dimension n as real manifolds of dimension $2n$ with additional properties.

In a manifold, a curve $x(t)$ is defined as $x : \mathbb{R} \to M$. One can generally construct a set of curves $\{x_k(t)\}$ for $k = 1, ..., n$ that intersect at $x(0)$ such that $x_k(0) \in U_i$ for all $k = 1, ..., n$. Furthermore, the set of curves $\{x_k(t)\}$ can be constructed so that the tangents to the *images* of the curves $\{\phi(x_k(t))\}$ form a basis for $\mathbb{R}^n$. We call such a set of curves *nondegenerate*. An *orientation* is assigned to $x(0)$ by taking the sign of the determinant of the Jacobian matrix whose i^{th} column is the image of the tangent to $\phi(x_k(t))$ at $t = 0$. A particular set of nondegenerate curves that is of importance is generated for U_i as $\phi_i^{-1}(u_k \mathbf{e}_k)$ for $k = 1, ..., n$ and $u_k \in \mathbb{R}$. We will call this set of curves *coordinate curves*, since they take Cartesian coordinates in $\phi(U_i) \subset \mathbb{R}^n$ and map them to $U_i \subset M$. Clearly the orientation assigned to each set of intersecting coordinate curves is $\det(\mathbf{e}_1, ..., \mathbf{e}_n) = +1$.

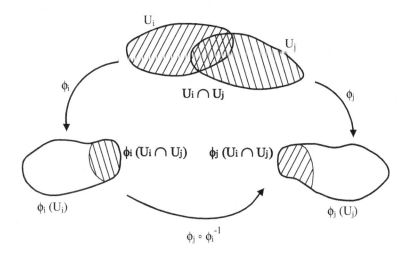

Fig. G.1. Relationships Between Coordinate Charts

G.3 Tangent Vectors and Tangent Bundles

The tangent vectors themselves are harder to visualize. Formally, a tangent vector is defined as an operator on the space of all smooth real-valued functions on M (denoted $C^\infty(M)$). A tangent vector at $x \in M$ is an operator

$$v_x : C^\infty(M) \to \mathbb{R}$$

satisfying linearity

$$v_x(af + bg) = av_x(f) + bv_x(g)$$

and the condition

$$v_x(f \cdot g) = g(x)v_x(f) + f(x)v_x(f)$$

that is reminiscent of the product rule for differentiation. The set of all tangent vectors at $x \in M$ is denoted $T_x M$ and is called the *tangent space* of M at x.

One defines

$$TM = \{(x, v_x) | x \in M, v_x \in T_x M\}.$$

The *natural projection* is the mapping

$$\pi : TM \to M.$$

The *tangent bundle* consists of the *total space* TM, the *base space* M, the *fibers* $T_x M$, and the projection $\pi(x, v_x) = x$.

A manifold is called *orientable* if the sign of the Jacobian determinant of $\phi_j \circ \phi_i^{-1}$ is positive for all i, j for which $U_i \cap U_j \neq \emptyset$. Since $\phi_j \circ \phi_i^{-1}$ is a mapping between two open sets in $\mathbb{R}^n$, the Jacobian is calculated in a similar way as for curvilinear coordinates in Chapter 4. Orientability essentially means that it is possible to define coordinate charts so that a set of coordinate curves of any point in one chart is positively oriented in any other chart containing the same point.

All the Lie groups considered in the text are orientable manifolds, as are all Lie subgroups of $GL(n, \mathbb{R})$ and $GL(n, \mathbb{C})$ [19].

A *Riemannian metric* on a manifold M is a family of smoothly varying positive definite inner products on $T_x M$ for each $x \in M$. For any two vectors $v_1, v_2 \in T_x M$, the value of the metric is denoted $g_x(v_1, v_2)$.

Note that a Riemannian metric might not exist for an arbitrary manifold. Hence the following definition is required.

A *Riemannian manifold* is the pair (M, g) where M is a smooth manifold and $g = g_x$ is a Riemannian metric.

An important result from differential topology that we use implicitly (without proof) in the text is that every N-dimensional differentiable manifold is homeomorphic to a subset of $\mathbb{R}^{2N+1}$. As an example, the manifold of $SO(3)$ has an image that is a subset of $\mathbb{R}^6$ by writing a rotation matrix as $R = [\mathbf{u}, \mathbf{v}, \mathbf{u} \times \mathbf{v}]$ where $\mathbf{u}$ and $\mathbf{v} \in \mathbb{R}^3$ are unit vectors, and the correspondence

$$R \Leftrightarrow [\mathbf{u}^T, \mathbf{v}^T]^T$$

defines a homeomorphism between $SO(3)$ and a subset of $\mathbb{R}^6$. Since a subset of $\mathbb{R}^6$ is also one of $\mathbb{R}^7$, this example may be viewed as a demonstration of the theorem.

References

1. Abraham, R., Marsden, J.E., *Foundations of Mechanics*, 2nd edition, Benjamin/Cummings, San Mateo, CA, 1978.
2. Aho, A.V., Hopcroft, J.E., Ullman, J.D., *The Design and Analysis of Computer Algorithms*, Addison-Wesley Publishing Company, Reading Mass., 1974.
3. Arfken, G.B., Weber, H.J., *Mathematical Methods for Physicists*, 7th ed., Academic Press, San Diego, 2012.
4. Arnol'd, V.I., *Mathematical Methods of Classical Mechanics*, Springer-Verlag, New York, 1978.
5. Berberian, S.K., *Measure and Integration*, Macmillan, New York, 1965. (Reprinted by the American Mathematical Society, Providence, RI, 2011.)
6. Borodin, A., Munro, I., *The Computational Complexity of Algebraic and Numerical Problems*, Elsevier, New York, 1975.
7. Borodin, A.B., Munro, I., "Evaluating Polynomials at Many Points," *Information Processing Letters*, 1(2): 66 – 68, 1971.
8. Brechtken-Manderscheid, U., *Introduction to the Calculus of Variations*, Chapman and Hall, New York, 1991.
9. Burke, W.L., *Applied Differential Geometry*, Cambridge University Press, Cambridge, 1985.
10. Chiang, C.-J., Chirikjian, G.S., "Similarity Metrics with Applications in Modular Robot Motion Planning," *Autonomous Robots*, 10(1): 91 – 106, Jan. 2001.
11. Chirikjian, G.S., *Stochastic Models, Information Theory, and Lie Groups, Vols 1+2.*, Birkhäuser, Boston, 2009,2011.
12. Chirikjian, G.S., Yan, Y., "Mathematical Aspects of Molecular Replacement: II. Geometry of Motion Spaces," *Acta Crystallographica A*, 68(2): 208 – 221, 2012.
13. Choquet-Bruhat, Y., DeWitt-Morette, C., Dillard-Bleick, M., *Analysis, Manifolds and Physics*, North-Holland, Amsterdam, 1982.
14. Cohn, H., Umans, C., "A Group-Theoretic Approach to Fast Matrix Multiplication," in *Proceedings of the 44th Annual IEEE Symposium on Foundations of Computer Science*, pp. 438 – 449, Oct. 2003.
15. Cohn, H., Kleinberg, R., Szegedy, B., Umans, C., "Group-Theoretic Algorithms for Matrix Multiplication," in *Proceedings of the 46th Annual IEEE Symposium on Foundations of Computer Science*, pp. 379 – 388, Oct. 2005.
16. Cohn, H., Umans, C., "Fast Matrix Multiplication Using Coherent Configurations," *Proceedings of the SIAM International Conference on Discrete Algorithms (SODA)*, pp. 1074 – 1086, Jan. 2013.
17. Collins, G.E., "Computer Algebra of Polynomials and Rational Functions," *American Mathematical Monthly*, 80(7): 725 – 754, 1973.
18. Coppersmith, D., Winograd, S., "Matrix Multiplication via Arithmetic Progressions," *Journal of Symbolic Computation*, 9(3): 251 – 280, 1990.

19. Darling, R.W.R., *Differential Forms and Connections*, Cambridge University Press, 1994.
20. Demmel, J., Dumitriu, I., Holtz, O., Kleinberg, R., "Fast Matrix Multiplication is Stable," *Numerische Mathematik*, 106(2): 199 – 224, 2007.
21. Do Carmo, M., *Differential Geometry of Curves and Surfaces*, Prentice-Hall, Englewood Cliffs, NJ, 1976.
22. Driscoll, J.R., Healy, D.M., Jr., Rockmore, D.N., "Fast Discrete Polynomial Transforms with Applications to Data Analysis for Distance Transitive Graphs," *SIAM Journal on Computing* 26: 1066 – 1099, 1997.
23. Eve, J., "The Evaluation of Polynomials," *Numerische Mathematik*, 6: 17 – 21, 1964.
24. Ewing, G.M., *Calculus of Variations With Applications*, Dover, 1985.
25. Federer, H., *Geometric Measure Theory*, Springer, Berlin, 1969 (Reprinted, 2013).
26. Feynman, R.P., Hibbs, A.R., *Quantum Mechanics and Path Integrals* (Emended by D.F.Styer), Dover, 2010.
27. Flanders, H., *Differential Forms with Applications to the Physical Sciences*, Dover Publications, Inc., New York, 1989.
28. Fréchet, M., "Les Solutions Non Commutables De L'Equation Matricielle $e^X \cdot e^Y = e^{X+Y}$," *Rendiconti Del Circulo Matematico Di Palermo*, Series 2, 1: 11 – 21, 1952; and 2: 71 – 72, 1953.
29. Gantmacher, F.R., *The Theory of Matrices*, Chelsea, New York, 1959.
30. Golub, G.H., Van Loan, C.F., *Matrix Computations*, 4th ed., The Johns Hopkins University Press, Balrimore, 2012.
31. Gruver, W.A., Sachs, E., *Algorithmic Methods in Optimal Control*, Pitman Publishing, Ltd., Boston, 1980.
32. Guillemin, V., Pollack, A., *Differential Topology*, Prentice-Hall, Englewood Cliffs, NJ, 1974.
33. Halmos, P.R., *Measure Theory*, Springer-Verlag, New York, 1974.
34. Herstein, I.N., *Topics in Algebra*, 2nd edition, John Wiley and Sons, 1975.
35. Higham, N.J., "Fast solution of Vandermonde-ike systems involving orthogonal polynomials," *IMA Journal on Numerical Analysis*, 8: 473 – 486, 1988
36. Horn, R.A., Johnson, C.R., *Matrix Analysis*, 2nd ed., Cambridge University Press, New York, 2012.
37. Horowitz, E., "A Fast Method for Interpolation Using Preconditioning," *Information Processing Letters*, 1(4): 157 – 163, 1972.
38. Itzykson C., Zuber J.B., *Quantum Field Theory*, Dover, 2006.
39. Kamien, M.I., Schwartz, N.L., *Dynamic Optimization: The Calculus of Variations and Optimal Control in Economics and Management*, 2nd ed., North-Holland, New York, 1991. (Dover edition, 2012).
40. Kanwal, R.P., *Generalized Functions, Theory and Technique*, 2nd Edition, Birkhäuser, Boston, 1998. (paperback, 2012).
41. Kleinert, H., *Path Integrals in Quantum Mechanics, Statistics and Polymer Physics, and Financial Markets*, 4th ed., World Scientific, Singapore, 2009.
42. Kobayashi, S., Nomizu, K., *Foundations of Differential Geometry Vols. I and II*, John Wiley and Sons, 1963 (Wiley Classics Library Edition, 1996).
43. Kung, H.T., "Fast Evaluation and Interpolation," Dept. of Compter Science, CMU, Pittsburgh, 1973.
44. Lee, J.L., *Riemannian Manifolds: An Introduction to Curvature*, Springer-Verlag, New York, 1997.
45. Lipson, J., "Chinese Remaindering and Interpolation Algorithms," *Proceedings of the 2nd Symposium on Symbolic and Algebraic Manipulation*, pp. 372 – 391, 1971.
46. Millman, R.S., Parker, G.D., *Elements of Differential Geometry*, Prentice-Hall Inc., Englewood Cliffs, NJ 1977.
47. Munkres, J.R., *Topology*, 2nd ed., Prentice-Hall, Inc., Englewood Cliffs, NJ, 2000.
48. Pamecha, A., Ebert-Uphoff, I., Chirikjian, G.S., "Useful Metrics for Modular Robot Motion Planning," *IEEE Transactions on Robotics and Automation*, 13(4): 531 – 545, August 1997.
49. Pan, V.Y. "Methods of Computing Values of Polynomials," *Russian Mathematical Surveys*, 21(1): 105 – 136, 1966.

50. Pan, V.Y., "How Can We Speed Up Matrix Multiplication," *SIAM Review* 26: 393 – 416, 1984.

51. Pan, V.Y., *How to Multiply Matrices Fast*, Springer-Verlag, Berlin, Heidelberg 1984.

52. Press, W.H., Teukolsky, S.A., Vetterling, W.T., Flannery, B.P., *Numerical Recipes: The Art of Scientific Computing*, 3rd ed., Cambridge University Press, 2007.

53. Raz, R., "On the Complexity of Matrix Product," in *Proceedings of the 34th Annual ACM Symposium on Theory of Computing*, pp. 144 – 151, May 2002.

54. Rosenberg, S., *The Laplacian on a Riemannian Manifold: An Introduction to Analysis on Manifolds,* (London Mathematical Society Student Texts , No 31), Cambridge University Press, 1997.

55. Ryder L.H., *Quantum Field Theory*, 2nd ed., Cambridge University Press, 1996.

56. Saichev, A.I., Woyczyński, W.A., *Distributions in the Physical and Engineering Sciences*, Applied and Numerical Harmonic Analysis Series (J.J. Benedetto, series editor) Birkhäuser, Boston, 1997.

57. Schulman, L.S., *Techniques and Applications of Path Integration*, Dover, 2005.

58. Schutz, B.F., *Geometrical Methods of Mathematical Physics*, Cambridge University Press, 1980.

59. Stokey, N.L., Lucas, R.E., Jr., Prescott, E.C., *Recursive Methods in Economic Dynamics*, Harvard University Press, Cambridge, Mass., 1989.

60. Stothers, A., *On the Complexity of Matrix Multiplication*, Ph.D. Dissertation, University of Edinburgh, 2010.

61. Strassen, V., "Gaussian Elimination is Not Optimal," *Numerische Mathematik*, 13: 354 – 356, 1969.

62. Swanson, M.S., *Path Integrals and Quantum Processes*, Dover, 2014.

63. Tao, T., *An Introduction to Measure Theory*, American Mathematical Society, Providence, RI, 2011.

64. Tomé, W., *Path Integrals on Group Manifolds*, World Scientific, Singapore, 1998.

65. Varadarajan, V.S., *Lie Groups, Lie Algebras, and Their Representations*, Springer-Verlag, New York 1984.

66. Warner, F.W., *Foundations of Differentiable Manifolds and Lie Groups*, Springer-Verlag, New York, 1983.

67. Wedderburn, J.H.M., *Lectures on Matrices*, Dover, New York, 2004.

68. Wermuth, E.M.E., Brenner, J., Leite, F.S., Queiro, J.F., "Computing Matrix Exponentials," *SIAM Review*, 31(1): 125 – 126, 1989.

69. Williams, V.V., "Multiplying Matrices Faster Than Coppersmith-Winograd," *Proceedings of the 44th ACM Symposium on Theory of Computing*, New York, NY, pp. 887 - 898, May 19 - 22, 2012.

Index